BIOLOGY
Discovering Life

About the Cover: Two influences arising from human activities have had a very serious effect on the bluebird population in America. First, starlings and sparrows were brought to Central Park from England. With their aggressive behavior, these birds were able to outcompete the bluebirds for nesting sites and to proliferate and expand their range throughout the country. Second, in rural areas, when farmers replaced wooden fence posts with metal ones, the bluebirds lost even more nesting sites. Since the 1950s however, conservationists have been building and placing nesting boxes at desirable nesting sites, and the bluebird population is recovering. This example shows how human interference with nature resulted in a decline in the population of a species, and how subsequent intervention on behalf of that species resulted in a recovery of the population.

BIOLOGY

Discovering Life

Joseph S. Levine

Kenneth R. Miller
Brown University

D. C. HEATH AND COMPANY
Lexington, Massachusetts   Toronto

Authors' Dedication
To Carol and Bob, Marion and Ray, our parents, who gave us our lives, the best of our values, our love of learning, and our appreciation for the importance of education in a complex world.

Publisher's Dedication
All of us involved with this project will remember with affection Mary Le Quesne, our editor, mentor, and friend. Her enthusiasm, laughter, and love of publishing will remain an inspiration to all who worked with her.

Acquisitions Editors: Mary Le Quesne, Elizabeth Coolidge-Stolz
Editorial Director: Kent Porter Hamann
Developmental Editor: Barbara Withington Meglis
Production Editors: Cormac Joseph Morrissey, Marret McCorkle, Jill Hobbs
Editorial Assistants: Jennifer Raymond, Joanne M. Williams
Copyeditor: Connie Day
Proofreader: Phyllis Coyne
Indexer: Michael Loo
Designer: Cornelia Boynton
Art Editors: Laurel Smith, Gary Crespo
Design Assistants: Penny Peters, Tama Hochbaum, Laura Fredericks, Marsha Goldberg
Production Coordinator: Michael O'Dea
Photo Researchers: Billie Ingram, Carole Frolich, Linda Finnegan, Mary Lang, Sharon Donahue
Text Permissions Editor: Margaret Roll
Artists: Patrice Rossi, Mickey Senkarik, Marlene DenHouter, Arleen Frasca, Lynette Cook, Marcia Smith, Horvath & Cuthbertson, Lyrl Ahern, Charles Boyter, Andrew Robinson, Michael Woods, Illustrious Inc., Sanderson Associates
Compositor: Graphic Typesetting Service, Inc.
Film Preparation: Colotone Separations
Printer: R.R. Donnelley and Sons
Cover: (*front*) Wildlife Photography/Michael L. Smith, (*back*) Greg Wenzel

Published simultaneously in Canada.

Printed in the United States of America.

International Standard Book Number: 0–669–12008–1 (Student Edition)
International Standard Book Number: 0–669–24487–2 (Instructor's Edition)

Library of Congress Catalog Card Number: 90-82264

10 9 8 7 6 5 4 3 2

Joseph S. Levine

Joseph Levine completed his undergraduate work at Tufts University, obtained a Master's degree at the Boston University Marine Program, and received his Ph.D. from Harvard University. Joe developed innovative laboratory and lecture courses in biology at several schools, including Tufts, Harvard, and Boston College. His research on the physiology, ecology, and evolution of visual systems in aquatic animals has appeared in journals ranging from *Science* to *Scientific American*.

Actively involved in several pursuits that combine education with his research interests, Joe has served as advisor for the PBS series "Living Wild" and "NOVA." During a fellowship at WGBH in Boston, he coproduced the prime-time *Science Gazette* and science features for National Public Radio's "Morning Edition," and "All Things Considered." With the NOVA staff, he assembled a proposal for a series on molecular biology entitled "Life: Cracking the Code." He has also served as senior program designer for state aquarium projects in Texas, New Jersey, and Florida, and conceived an AIDS awareness exhibit for the California Museum of Science and Industry in Los Angeles.

Joe is currently vice president of Boston Science Communications, Inc., specializing in the production of educational films, videos, and multimedia educational products, and he is an associate at the Museum of Comparative Zoology at Harvard University.

Kenneth R. Miller

Professor of Biology at Brown University, Ken Miller is a plant cell biologist. Miller completed his undergraduate work in biology at Brown University and was awarded a Ph.D. in biology from the University of Colorado. His teaching credits include Harvard University, the University of Colorado, and Brown University, where he currently teaches biology to both majors and nonmajors.

Ken is editor-in-chief of *Advances in Cell Biology*, editor of *Journal of Cell Science*, and former editor of *Journal of Cell Biology*. He is also chairman of the Education Committee of the American Society for Cell Biology. Professor Miller has published over 50 articles in professional and lay scientific journals. Among his recent published work are a cover story on photosynthesis in *Nature* and articles in *Scientific American*, *Journal of Cell Biology*, *Proceedings of the National Academy of Science*, *Endeavor*, and *Trends in Biochemical Sciences*.

A nationally recognized spokesman for evolution in the evolution–creation controversy, Ken Miller maintains an active involvement in the public examination of this issue.

PREFACE

A growing number of people in this country have no understanding of science whatsoever; they either believe in science or they don't believe in science, just as they either believe in ghosts or don't believe in ghosts. They therefore treat science as a belief system—not unlike superstition—rather than as a way of interacting with the world in a rational manner. To remedy this situation in biology without making drastic changes in the body of material generally covered in introductory courses, we have employed three techniques. One is to lead students through narratives of observation, inquiry, discovery, controversy, and social relevance, rather than to present them with a list of "established" facts and conclusions. The second approach, wherever the subject matter permits, is to place the student directly in the midst of current research in a way that gives a sense of involvement and immediacy with the material. The third approach is to structure the entire book with a student's world view in mind. We believe strongly that students are most easily brought into a study of biology by biology itself.

Structure of the Text

Our book opens with an examination of biology and the nature of science (Part One: Introduction to Biology), and then immediately moves on to Organisms and Ecology (Part Two). This contrasts sharply with an approach that buries students in several chapters on chemistry before they are allowed to glimpse biology itself.

We do attach great importance to the study of essential chemistry, and have covered it thoroughly. We have, however, placed biological chemistry in its proper historical and scientific context. Chemistry was originally quite separate from biology, and chemical concepts were incorporated into biological studies only gradually. Evolutionary theory and Mendelian genetics, two cornerstones of nineteenth century biology, for example, developed without the aid of chemistry and were largely responsible for the interest of biologists in the chemical nature of genes and living cells. Our sequence of presentation (coverage of biological chemistry after genetics and evolution) thus allows students to follow in researchers' footsteps and to appreciate the need for understanding the chemical nature of life. We feel from our own experience that this approach not only demystifies chemistry, but grounds this most abstract of subjects in a biological reality to which students can relate.

We also take into account the reality that some students entering introductory classes may not have had sufficient exposure to evolutionary thought. Our solution to this problem is to structure our section on Mendelian Genetics and Evolution (Part Three) differently than most other books. We begin with a historical narrative that explains the Western philosophical world view before Darwin, details the intellectual and philosophical leaps Darwin and his contemporaries had to make, and progresses through a long list of evidence that evolution has occurred. We raise Darwin's vital questions about the nature of inheritance as an introduction to a historical and experimental approach to Mendelian and population genetics. Only then do we discuss the more esoteric details of modern evolutionary population genetics.

Throughout this process, we link biology not only to other sciences, but to philosophy, sociology, and other disciplines in the humanities as well. We present the historical sections in a manner that allows students to trace the development of vital ideas in biology. We regularly introduce experimental evidence to explain the thought processes that have expanded and changed our understanding over time.

We then present the Diversity of Life section (Part Four) in the format of an "Evolutionary History of Life." Here, aided by a unique illustration package, we present two parallel tracks of information: the evolutionary history of major groups of organisms, and the salient anatomical details of each group. By so doing, we have tied this often disjointed material together into a narrative that stresses the evolutionary interactions between organisms and their changing environments and between plants and animals. These chapters should thus be particularly welcomed by professors who emphasize evolutionary processes rather than classification and anatomical details. At the same time, we are confident that the special illustrations we have developed for these chapters will be valuable teaching and study aids in those classes where anatomy and life-cycles are accentuated.

As mentioned previously, we have tried to place the importance of chemical principles more in context and have used our treatment of genetics covered in Part Three to develop a sense of interest in biochemistry. We hope that students will begin this section, The Molecules of Life (Part Five), wondering about the chemical nature of the gene, interested in how genotype actually determines phenotype, and curious about the mechanisms by which genes exert their effects.

Each of the chapters in Part Five explores one aspect of the connections between the material world and the living world. At first, these connections are on the most basic level: the nature of chemical reactions, the role of energy in biochemistry, and the structure of macromolecules. Gradually, however, these connections become more profound. We have made a special effort to involve the reader in the process of discovery, which led to the double helix model of DNA structure, and this approach extends to advanced topics in molecular biology as well. In addition to writing these chapters to emphasize how we know what we know, we also make the point that there is still a great deal we do *not* understand.

Throughout the text, we have emphasized the unity of living organisms. Plants and animals are used freely as examples for important concepts in genetics, molecular biology, evolution, and ecology. Despite this evenhanded emphasis on the diversity of life, there are a number of key concepts in plant development and physiology that are best dealt with in a specialized way. The four chapters in our Plant Systems Section (Part Six) ask what it is like to be a plant. In their own way, each chapter challenges the reader to imagine the life of a plant and the ways in which plants respond to their environment. By placing this section immediately after material on biological energy and molecular biology, we are able to examine the central role of bioenergetics and molecular biology in the study of plants, and to connect plants to the rest of the living world. The physiological principles emphasized in this section also pave the way for a discussion of animal physiology in the section to follow.

The final section, Animal Systems (Part Seven), includes 13 chapters that cover the field of organismal physiology from basic concepts of homeostasis through animal behavior. While most of these chapters follow more or less traditional organization, each, we feel, brings a fresh, up-to-date, and vigorous approach to the material. Not only do we include the latest discoveries in such rapidly changing areas as development, cancer, and AIDS, but we relate these developments to breakthroughs in other areas covered earlier in the text. Our coverage of immunology begins with a historical look at biology, medicine, and disease, and proceeds through the latest developments in this dynamic and vitally important area. The chapters on the nervous, sensory, and muscular systems lead logically and progressively up to an integrated treatment of animal behavior that concludes the book.

We believe that this overall structure, which many instructors throughout the country have used successfully, is the best way to present biology to the mix of nonscience *and* science concentrators enrolled in the introductory biology course. Nonetheless, we realize that there is more than one way to teach biology effectively. We have therefore organized our sections in a modular fashion that allows skipping around to suit an individually designed presentation.

We hope that this book will bring to your students some of the excitement and passion scientists bring to their work. If we can help you as instructors to pass this knowledge to a new generation of students, the kindness of the people who have shared their gifts with both of us in the past will be partly repaid. More than that, no teacher or former student can ask.

Comprehensive Teaching and Learning Package

The following supplements accompany *Biology: Discovering Life*:

- Study Guide by George Karleskint, Jr. (12009–X)
- Instructor's Guide by Gail R. Patt (12014–6)
- Investigations in Biology by Richard J. Montgomery and William D. Elliott (12010–3)
- Instructor's Guide for Investigations in Biology by Richard J. Montgomery and William D. Elliott (24447–3)
- Test Item File by Bernard L. Frye and Joyce A. Blinn (12011–1)
- D. C. Heath Exam Computerized Test Bank for IBM PC, Apple, and Macintosh
- Transparencies with Transparency Lecture Guide by George Karleskint, Jr. (12012–X)

Acknowledgments

Each of us once thought that textbook writing was a solitary art. From time to time, as we worked quietly on drafts of this book, it was possible to entertain the notion that this might be true. As *Biology: Discovering Life* neared the final stages, however, the folly of this idea became clear. We are enormously grateful for the kindness and support of our many scientific friends and colleagues who contributed time, advice, photographs, and encouragement to this project. We are also grateful for the patient understanding of the friends, family members, and children who endured our seeming lack of interest in anything outside this project for months at a time. In many ways, they have made the greatest sacrifice associated with this project, and we will always be in their debt.

We soon grew to appreciate the talented work of the editors, artists, photo researchers, and production workers who made this book a reality. There were a few disagreements and (in retrospect) laughable squabbles. But every page of this book is the carefully crafted work of a small army of skilled and dedicated individuals. Throughout this project, we have felt ourselves lucky to know and work with dozens of people who are the very best in their fields. We are fortunate to have made these associations, and we know that the book has become a joint product in which many people take justifiable pride.

Our thanks go out to the entire staff at D. C. Heath, especially to the late Mary Le Quesne, the senior acquisitions editor who played a vital role in the development of this project. Her love of the material, her dedication to her work, her warmth of spirit, and her unflagging enthusiasm in the face of terminal cancer inspired us and all of her co-workers at Heath.

We have also learned that during a writing and publishing project as long as this one, there are invariably significant staff changes. Luckily for us and for the quality of this book, the members of our supervisory group at Heath were not only of the highest professional caliber but also dedicated themselves totally to our project. We gratefully acknowledge the extraordinary level of commitment and the highly skilled and successful labors of Kent Porter Hamann, editorial director; Kathi Prancan and Elizabeth Coolidge-Stolz, acquisitions editors; Barbara Withington Meglis, developmental editor; and Cormac Morrissey, production editor. All of these people were brought into the project after it had already taken shape but early enough for the quality of their work to influence it powerfully. We also want to acknowledge the contribution of Cornelia Boynton, Laurel Smith, and Gary Crespo, who lent their remarkable artistic vision to a staggeringly complex illustration program. We thank them all heartily.

J.S.L.
K.R.M.

TO THE STUDENT

You are living in the midst of a revolution—a revolution in our understanding of life. As this revolution progresses, it will forever change the relationship between biology and human society because it will involve all the practical applications of biological knowledge. These applications span the full spectrum of human concerns, from the desperate battle against AIDS, to debates concerning the uniqueness of the human mind, to actions that affect the future of your life and all life on earth.

Part of this revolution has been launched by breakthroughs in our understanding of the molecules that make up living tissue. On this submicroscopic frontier, our challenge is to understand how genes do the many things they do: create our cells, direct those cells to form bones in our legs and thoughts in our brains, change with age, and evolve from generation to generation. Tied into this research, the fledgling biotechnology industry promises to revolutionize the way we grow our food, the way we are born, and the way we die.

At the other extreme of life's phenomena, on a scale so vast that it encompasses our entire planet, our challenge is to protect the global environments that give us life. Ecologists have discovered that some of our actions may be causing a planet-wide climate change whose long-term effects we can scarcely comprehend. Other human activities are destroying Earth's protective ozone layer. And still others are threatening the supplies of clean air, drinkable water, and food so vital to Earth's constantly growing human population.

The biological revolution has already begun to bestow on us a range of potent techniques and capabilities that just a decade ago existed only in the realm of science fiction. But the promise of these newfound powers carries with it the responsibility to use these powers wisely. Genetic engineering raises serious ecological, legal, and moral questions about the creation of new life forms. Biomedicine, too, confronts sobering dilemmas: Is it worth opening an ethical Pandora's box if we can cure genetic disease by altering human DNA? How can we weigh moral concerns about human embryo experimentation against possible practical benefits?

This book will not presume to answer these or other socially vital questions for you. As authors and teachers we hope to provide you with enough information about the biology behind these issues to allow you to examine these problems critically and to participate in essential public debate about them. To accomplish these goals, we intend to propel your understanding of biology beyond media images of laboratory-created monsters, imminent catastrophes in global ecology, and miracle cancer cures.

And finally, we want to share with you the excitement and vitality that we, as biologists, see in our field. In many other disciplines, where exciting times and events seem out of reach, locked in the past, the present may appear ordinary by comparison. But that isn't true in biology today. To prove it, we invite you to partake in the delight and the excitement, the gifts and the dangers, of a revolution still very much in progress.

J.S.L.
K.R.M.

REVIEWER ACKNOWLEDGMENTS

Joseph Allamong, *Ball State University*
Margaret Anglin, *St. Louis Community College at Meramec*
Lois Bailey, *Southeastern Community College*
Frank Baron, *Duquesne University*
May R. Berenbaum, *University of Illinois at Urbana-Champaign*
Joseph S. Bettencourt, *Marist College*
P. K. Bhattacharya, *Indiana University Northwest*
David L. Bishop, *Moorpark College*
Donald Bissing, *Southern Illinois University at Carbondale*
Robert V. Blystone, *Trinity University*
Richard Boohar, *University of Nebraska–Lincoln*
Maynard C. Bowers, *Northern Michigan University*
Jean A. Bowles, *Metro State College*
J. H. Brown, *University of Houston*
Peter Burn, *Suffolk University*
Thomas E. Byrne, *Roane State Community College*
William Cain, *University of Delaware*
Thomas R. Campbell, *Los Angeles Pierce College*
Joseph Chinnici, *Virginia Commonwealth University*
Norm Christensen, *Duke University*
David Cotter, *Georgia College*
Joe R. Cowles, *University of Houston*
Alfred H. Crawford, *Southern Florida Community College*
Judy Daniels, *Eastern Michigan University*
Manuel E. Daniels, Jr., *Tallahassee Community College*
Deborah S. Dempsey, *Northern Kentucky University*
Raymond Dillon, *University of South Dakota*
Linda Dion, *University of Delaware*
Nathan Dubowsky, *Westchester Community College*
Frank Duroy, *Essex County College*
M. Duvall, *Smithsonian Institution*
David Eberiel, *University of Lowell*
H. W. Elmore, *Marshall University*
Susan Ernst, *Tufts University*
Robert C. Evans, *Rutgers University*
Darrel Falk, *Point Loma Nazarene College*
Carl D. Finstad, *University of Wisconsin–River Falls*
Jim Fowler, *State Fair Community College*
David J. Fox, *University of Tennessee, Knoxville*
Bernard L. Frye, *University of Texas at Arlington*
Ric Garcia, *Clemson University*
Michael S. Gaines, *University of Kansas*
Thomas C. Gray, *University of Kentucky*
Elizabeth C. Hager, *Trenton State College*
Thomas Earl Hanson, *Temple University*
Robert E. Herrington, *Georgia Southwestern College*
Carl Hoegler, *Marymount College*
Alfred J. Hopwood, *Saint Cloud State University*
Daniel Hornbach, *Macalester College*
Patricia J. Humphrey, *Ohio University*
Robert N. Hurst, *Purdue University*
Mary Keim, *Seminole Community College*
Donald L. Kimmel, Jr., *Davidson College*
Valerie M. Kish, *Hobart and William Smith Colleges*
F. M. Knapp, *Stetson University*
Helen G. Koritz, *College of Mount Saint Vincent*
James Lampky, *Central Michigan University*
John H. Langdon, *University of Indianapolis*
Ron W. Leavitt, *Brigham Young University*

Michael Lee, *Joliet Junior College*
James Luken, *Northern Kentucky University*
Joseph Marshall, *West Virginia University*
John Matsui, *University of California at Berkeley*
John Mattox, *Northeast Missouri Community College*
Fred McCorkle, *Central Michigan University*
Dorothy Minkoff, *Trenton State College*
David J. Morafka, *California State University, Dominquez Hills*
Robert L. Neill, *University of Texas at Arlington*
Mary Nossek, *Ohio University*
Thaddeus Osmolski, *University of Lowell*
Joel Ostroff, *Brevard Community College*
Helen Oujesky, *University of Texas at San Antonio*
Gail R. Patt, *Boston University*
Peter Pedersen, *Cuesta College*
Kathryn Stanley Podwall, *Nassau Community College*
David M. Prescott, *University of Colorado*
Jeffrey Pudney, *Harvard Medical School*
Ralph Reiner, *College of the Redwoods*
Donald Reinhardt, *Georgia State University*
Robert Rinehart, *San Diego State University*
Robert Romans, *Bowling Green State University*
David Rose, *Trenton State College*
Peter E. Russel, *Chaffey Community College*
A. G. Scarbrough, *Towson State University*
Allen B. Schlesinger, *Creighton University*
Robert Schodorf, *Lake Michigan College*
Erik Scully, *Towson State University*
Richard W. Search, *Thomas College*
Margaret Simpson, *Sweet Briar College*
Susan Singer, *Carlton College*
Jerry W. Smith, *St. Petersburg Junior College*
Marshall Smith, *Los Angeles Mission College*
Beatrice L. Snow, *Suffolk University*
Gilbert D. Starks, *Central Michigan University*
Howard J. Stein, *Grand Valley State University*
Philip Stein, *State University of New York College at New Paltz*
Robert W. Sterner, *University of Texas at Arlington*
Eric Strauss, *Tufts University*
Stephen Subtelny, *Rice University*
Gerald Summers, *University of Missouri–Columbia*
Marshall Sundberg, *Louisiana State University*
Daryl Sweeney, *University of Illinois at Urbana–Champaign*
Dan Tallman, *Northern State University*
William J. Thieman, *Ventura College*
Kathy S. Thompson, *Louisiana State University*
Bruce Tomlinson, *State University of New York College at Fredonia*
Michael Treshow, *University of Utah*
Marenes Tripp, *University of Delaware*
Kent M. Van De Graaff, *Brigham Young University*
Pauline E. Washington, *Forest Park Community College*
Jean Werth, *William Paterson College*
George J. Wilder, *Cleveland State University*
Garrison Wilkes, *University of Massachusetts at Boston*
Wayne Wofford, *Union University*
Paul Wright, *Western Carolina University*
Tommy Elmer Wynn, *North Carolina State University*
Gerald W. Zimmerman, *University of Indianapolis*

Art Reviewers:
H. W. Elmore, *Marshall University*
Michael C. Kennedy, *Hahnemann University*
Richard E. Morel

BRIEF CONTENTS

PART ONE *Introduction to Biology*

1 Understanding Life: Our Responsibility   2
2 Science and Society   19

PART TWO *Organisms and Ecology*

3 Energy and Nutrients in Ecosystems   34
4 Environments, Organisms and Environmental
   Variables   54
5 Population Growth and Control   74
6 Ecosystems: The Working Units of
   the Biosphere   93
7 Human Ecology   112

PART THREE *Evolution and Mendelian Genetics*

8 Darwin's Dilemma: The Birth of Evolutionary
   Theory   140
9 Continuity of Life: Cellular Reproduction   165
10 Genetics: The Science of Inheritance   180
11 Human Genetics   208
12 Darwinian Theory Evolves   229
13 Evolution of Species   247

PART FOUR *The Diversity of Life*

14 Biological Classification and the Origin of Life   264
15 The Early History of Life: The Invertebrates
   and the Algae   286
16 The Silurian Period: The First Lower Plants and
   Jawed Vertebrates   308
17 Emergence of the Modern World   325
18 Evolution of Mammals and the Ascent of
   Homo sapiens   347

PART FIVE *Molecules of Life*

19 The Molecules of Life   362
20 Macromolecules   377
21 Chemical Reactions and Energy   394
22 Molecules and Genes   408
23 Evolution at the Molecular Level: Natural
   and Artificial   436
24 Cell Organization   457
25 Cells and Energy   485
26 Photosynthesis   503

PART SIX *Plant Systems*

27 Plant Structure and Function   520
28 Reproduction in Flowering Plants   542
29 Plant Growth and Nutrition   561
30 Control of Plant Growth and Development   573

PART SEVEN *Animal Systems*

31 Multicellular Organization and Homeostasis   592
32 Circulation   608
33 Respiration   629
34 Nutrition and Digestion   647
35 Regulation of the Cellular Environment   677
36 Chemical Communication   699
37 Reproduction   723
38 Embryology and Development   744
39 Immunity and Disease   777
40 Nervous Control   807
41 The Sensory System   835
42 The Musculo-Skeletal System   859
43 Animal Behavior   878

CONTENTS

PART ONE Introduction to Biology

CHAPTER 1 Understanding Life: Our Responsibility   2

The Biological Revolution   2
Life: A Global Phenomenon   4
  From DNA to Biosphere: Spaceship Earth?
  Or a Living Plant? 5
The Nature of Science   6
  Science and Religion 6
Current Controversies: *The Gaia Hypothesis*   7
  Science as a Way of Knowing 8
  Science as a Mirror of Society 9
At the Heart of Science: The Scientific Method   10
  Observation 10
  Forming Hypothesis 10
  Experimentation 11
  Reevaluation 12
  Scientific Theories 12
The Scientific Method in the Public Eye:
Case Studies   12
  Medication for Bacterial Infections in
  Aquarium Fishes 13
    Hypothesis • Background • Experimental
    design • Results • Conclusions according to
    company spokesman • Possible alternative
    conclusions • Discussion
  Clinical Trial of Azidothymidine (AZT) in the Treatment
  of AIDS 15
    Hypothesis • Background • Experimental
    design • Results • Conclusions • Discussion
  Science in the Real World: Between a Rock
  and a Hard Place 17

CHAPTER 2 Science and Society   19

The Impact of Science on Society   19
The Effects of Society on Science   20
Society, Disease, and Medicine: In Time of Plague   20
  AIDS: Society's Reaction Makes a Bad
  Situation Worse 22

Ecology and Economics: Tending Our Houses   24
  A Historical Perspective on Ecology and Economics 25
  The Greenhouse Effect: Atmosphere as Incubator 27
  Carbon Dioxide and Its Effect on
  Global Temperature 27
    The geochemical view • The biochemical view
  Human Involvement in Global Warming 29

PART TWO Organisms and Ecology

CHAPTER 3 Energy and Nutrients in Ecosystems   34

Autotrophs and Energy   34
Life's Power Supply   34
Heterotrophs: Obtaining Energy Captured by Plants   36
  Carbohydrates: Energy Carriers in the Biosphere 36
    Sugars and edible carbohydrates • Cellulose
Ecological Strategies of Heterotrophs   37
  Herbivores: The Diverse Plant Eaters 37
    Leaf eaters • Seed, fruit, and berry pickers
  Carnivores: The meat eaters 39
  Decomposers: Organisms of decay 40
  Other Feeding Techniques 40
Energy and Life   41
  Energy and Life in the Sea 41
Food Chains and Food Webs   42
Energy Flow Through Ecosystems   44
  Trophic Levels 45
The Cycling of Nutrients   46
  The Hydrological Cycle 46
  The Carbon Cycle 47
  The Nitrogen Cycle 48
    From primary production to
    consumers • From organisms to the
    environment • Microorganisms in the
    environment • Completing the
    cycle • Earth's nitrogen reserve:
    the atmosphere
Theory in Action: *The Hidden Water Crisis*   49
  Nutrient Limitation 52

CHAPTER 4   *Organisms, Environments, and Environmental Variables*   **54**

The Importance of Understanding Environmental Variables   54
Environments, Habitats, and Niches   54
 Habitat: Where an Organism Lives   54
 Niche: How an Organism Makes Its Living   57
  Physical aspects of the niche • Biological aspects of the niche
Environmental Variables   58
 Sunlight   58
 Temperature   59
 Water and Dissolved Salts   59
 Oxygen   60
 Metabolic Wastes   60
 Nutrients   60
The Impact of Environmental Variables on Organisms   61
 Keeping Internal Conditions Constant   61
 Law of Tolerance   61
Theory in Action: *Where Are the Bass?*   62
Global Climate and Life   63
 Climate Versus Weather   63
 Solar Energy and Temperature   64
 Global Air Movement   64
  Winds and oceanic surface currents • Winds, mountains, and rainfall • Lakes and oceans: climate moderators
 Effects of Vegetation on Climate   67
Long-Term Changes in Environments   68
 Global Climatic Changes   68
  Astronomical causes • Geological causes
Theory in Action: *El Niño*   69
 Biological Causes of Environmental Change: Succession   70
  Primary succession • Secondary succession
 Climax Communities: Stability and Replacement   72

CHAPTER 5   *Population Growth and Control*   **74**

Dynamics of Populations   74
 Functional Units of Ecology   74
 Exponential Growth   75

Logistic Growth   76
 The Concept of Carrying Capacity   76
  Factors determining carrying capacity
 Population Growth in Nature   78
Demography: The Study of Populations   78
 Population Age Structure   78
  Mortality and survivorship • Fertility • Age structure and population growth
Human Population Growth   81
Population-Regulating Factors   81
 Competition   82
  Competitive exclusion • Competition and the niche • Competition in nature
Current Controversies: *Human Population Growth*   84
 Predation   85
  Predator-prey oscillations
 Parasitism   86
Theory in Action: *Competition, Predation, and the Complexity of Natural Ecosystems*   87
 Other Density-Dependent Factors   88
 Positive Interactions Between Organisms   89
  Commensalism • Mutualism
 Density-Independent Population Regulations   89
The Effect of Environmental Predictability   90
 Unpredictable Environments   90
 Predictable Environments   91
Can We Predict (And Control) Our Own Impact?   91

CHAPTER 6   *Ecosystems: The Working Units of the Biosphere*   **93**

The Importance of Biological and Ecological Diversity   93
 The Resilience of Natural Systems   93
 Biomes   94
Life in Terrestrial Ecosystems   94
 The Importance of Soil   94
 Environmental Variables and Elevation   96
Terrestrial Biomes   96
 Tundra   96
 Taiga   97
 Temperate Forest   98
 Desert   99
 Grassland   100
 Tropical Rain Forest   101

Life in Aquatic Ecosystems 102
    The Importance of Light and Temperature 102
    Freshwater Ecosystems 103
    Brackish-Water Environments 103
        Estuaries • Temperate zone salt marshes • Mangroves • Marsh and mangrove ecology • Spawning grounds
Theory in Action: *The Fishes and the Forest* 106
    The Open Sea 107
    Between the Tides: Life on Rocky Shores 108
    Kelp Forests 108
    Coral Reefs 109
Interaction Among Ecosystems 110

CHAPTER 7 *Human Ecology* 112

Ecology and Civilization 112
    Island Earth 112
The State of the World 113
Human Impact on the Biosphere 114
    Water Pollution 114
        Chronic surface-water pollution • Industrial water pollutants • Residential sewage • Contributing factors • Groundwater contamination
    Air Pollution 117
        Smog • Carbon dioxide and global climate • Acid rain • Changes in the ozone layer
    Habitat Destruction and Loss of Biological Diversity 120
        Deforestation • Marine and estuarine habitat destruction • Loss of species • Biodiversity and agriculture
    Sustainability: Living Within Our Means 123
Environmental Problems: Causes and Solutions 124
    Human Population Growth 124
    World Food Production 125
Theory in Action: *The Goddess and the Computer* 126
        The Green Revolution: successes and problems • Monoculture and chemicals • Soil depletion • Irrigation and salinization • World agriculture: trends and prospects • Toward a sustainable agriculture
    Fisheries 130
        Aquaculture
Energy 131
    Nonrenewable Energy 131
        Fossil fuels • Nuclear power
    Renewable Energy 132

Tackling the Dilemma 133
    The Impact of Economics on Ecology 133
    The Impact of Ecology on Economics 133
Current Controversies: *Ecology, Engineering, Banks, and Politics* 134
    Changing perspectives
Environmental Policy: Ethics and Reality 135

PART THREE *Evolution and Mendelian Genetics*

CHAPTER 8 *Darwin's Dilemma: The Birth of Evolutionary Theory* 140

The Western World View Before Darwin 140
    Stability in Biology and Society 141
Darwin's Time: A World in Flux 141
    Fossils and Catastrophism 141
Geologists Challenge the Static Earth 143
Biological Science and Change 144
    Lamarck's Theory of Evolution 144
    Darwin's Contribution 144
Darwin on the *Beagle* 144
    Geological Observations 144
Theory in Action: *Darwin's Delay and Wallace's Insight* 145
    Biological Diversity 146
    Fitness 147
    The Galapagos Islands 147
Darwin's Theory of Evolution by Natural Selection 149
    Variation 149
    Artificial Selection 150
    Natural Selection 151
    Adaptation 151
Summary of Natural Selection as Presented by Darwin 152
    Scientific and Philosophical Significance 152
        Philosophical ramifications • Darwin and politics • Evolution and other sciences • Biological significance
Data Supporting the Fact of Evolutionary Change 153
    Similarities in Anatomy and Development 153
        Homologous structures and adaptive radiation • Analogous structures and convergent evolution • Vestigial structures
    Darwin and Fossils: Possibilities and Problems 157
    Paleontology and Evolution Today 158
    Coevolution: Support from Plant and Animal Relationships 159
        Plant–pollinator coevolution • Plant–herbivore coevolution

Symbiosis: Life Together   160
    Parasitism   •   Commensalism   •   Mutualism
Mimicry   160
Support from Comparative Biochemistry   160
Darwinian Theory Evolves   162
    The Contribution of Genetics   163

CHAPTER 9   Continuity of Life: Cellular Reproduction   165

The Diversity of Cells   165
    Types of cells
Growth and Development: The Problem of Reproducing a Cell   168
    Cellular Growth   168
    The Cellular Life Cycle   169
Cell Division in Prokaryotes   170
Cell Division in Eukaryotes   170
    The Eukaryotic Cell Cycle   170
        Cellular changes during the cycle
    Mitosis: The Beginning of Division   171
        Prophase   •   Metaphase   •   Anaphase   •   Telophase
    Cytokinesis   175
Theory in Action: Where Does the Nuclear Envelope Go?   176
    Control of Cell Division   177
        Contact inhibition   •   External controls on cell division   •   Cancer: When control is lost

CHAPTER 10   Genetics: The Science of Inheritance   180

Inheritance: A Problem   180
Mendel and the Birth of Genetics   181
    Mendel's First Experiments   181
    A Theory of Particulate Inheritance   182
        The Punnett square
    Dominance   183
    Mendel's First Principles   184
    Genotype and Phenotype   184
    Independent Assortment   185
    Problem Solving with Mendelian Genetics   186
    Genetics and the Cell   186
Meiosis: Forming a New Organism   188
    Reduction Division   188
        The first meiotic division   •   The second meiotic division
    Meiosis and the Rise of Mendelian Genetics   191

Theory in Action: Genes That Cheat at Meiosis   192
    Gene linkage and Crossing Over   192
Genetic Maps   195
    Giant Chromosomes in Tiny Organisms   196
The Inheritance of Sex   197
    Inheritance of the Sex Chromosomes   197
    Sex Linkage   198
Theory in Action: Nettie Stevens and the Y Chromosome   200
    Other Patterns of Sex Determination   200
Other Forms of Genetic Variation   201
    Incomplete Dominance   201
    Codominance   201
    Multiple Alleles   202
    Traits Controlled by Gene Interaction   202
    Polygenic Systems   203
Theory in Action: Using Genetics to Solve Problems   204
    Environmental Effects on Gene Expression   206

CHAPTER 11   Human Genetics   208

The Human Genetic System   208
    Genetics with Human Subjects: The Pedigree   210
    Finding Genes: Abnormalities and Disorders   210
Human Genes   210
    Autosomal Recessive Inheritance   211
        Albinism   •   Tay-Sachs disease   •   Cystic fibrosis   •   Other autosomal recessives
    Autosomal Dominant Inheritance   212
        Darwin Tubercle   •   Achondroplasia   •   Huntington's disease   •   Polydactyly
Theory in Action: Genetic Counseling: Knowing the Odds   212
    Multiple Alleles   213
    Polygenic Traits or "You've Got Your Uncle's Nose!"   215
        Size and shape   •   Fingerprints   •   Skin color
Current Controversies: Human Intelligence: Are There Genes for IQ?   217
Sex-Linked Human Inheritance   218
    Sex-Linked Disorders   219
        Hemophilia   •   Colorblindness   •   Lesch–Nyhan syndrome   •   Retinitis pigmentosa   •   Duchenne muscular dystrophy   •   Hypophosphatemia
Chromosomal Inheritance   222
    Disjunction Abnormalities   222
        Turner syndrome   •   Klinefelter syndrome   •   Other sex chromosome abnormalities
    Autosomal Trisomy   223
Theory in Action: Prenatal Genetics   224
    X-Chromosome Inactivation   225

Chromosome Mapping 225
    Mapping Human Genes 225
    Gene Mapping by Cell Fusion and Culture 226
    Chromosome Deletions and Translocations 226

CHAPTER 12   Darwinian Theory Evolves   229

Species and Fitness: Genetic Definitions 229
    The Species 229
    Evolutionary Fitness and Adaptation 230
Evolution and Natural Selection 230
    Observable Variation in Organisms 230
    Natural Selection: Effects on Phenotype 232
        The hypothetical population • Stabilizing
        selection • Directional selection •
        Disruptive selection
Variation and Selection: A Case Study 233
    Variation in Beak Size: Raw Material 235
        Beak size and seed-handling ability •
        The importance of food limitation
    Response to Drought: Directional Selection 236
Other Experimental Studies of Natural Selection 236
    Effects of Stabilizing Selection 238
    Effects of Directional Selection 238
    Effects of Disruptive Selection 238
The Hardy-Weinberg Law 238
    When Evolution Will Occur and When It Will Not 239
Theory in Action: The Hardy-Weinberg Equilibrium 240
Genetic Drift 241
    The Founder Effect 242
Pleiotropy and Heterozygous Advantage 242
Modern Studies of Genetic Variation 243
    Evolution Due to Human Activity: British Moths 244
    Evolution Today: Out of Control 245

CHAPTER 13   Evolution of Species   247

The Origin of Species 247
    Prezygotic Isolating Mechanisms 247
        Mechanical isolation • Behavioral isolation •
        Temporal isolation • Gamete incompatibility
    Postzygotic Isolating Mechanisms 248
        Ecological isolation
Mechanisms of Speciation: How Reproductive
Isolation Develops 249
    Allopatric Speciation 249
        Effects of separation

Theory in Action: Worlds Apart: Island Mountains and Inland Seas 253
    A Case Study of Allopatric Speciation:
    Galapagos Finches 254
        Colonization • Genetic divergence in
        allopatry • Reproductive
        isolation • Competition and further
        divergence • Further speciation
Theory in Action: Why Shouldn't Species Interbreed? 255
    Parapatric Speciation: Life on the Fringe 256
    Sympatric Speciation: Alone in a Crowd 256
Current Debate on Evolutionary Theory 257
    The Darwinian View of Species 257
    Species and Gradual Change 259
    Punctuated Equilibrium 259

PART FOUR   The Diversity of Life

CHAPTER 14   Biological Classification and the Origin of Life   264

Finding Order in Diversity Taxonomy:
The Science of Names 264
    The System of Binomial Nomenclature 264
    Higher Taxonomic Categories 266
Systematics: The Science of Classification 266
    Molecular Sytematics 267
The Five Kingdoms 267
The Origins of Life 268
    Conditions on the Early Earth 268
    The Organic Soup: Life from Non-Life 270
    From Molecules to Life 270
The Earliest Life Forms 271
    The Road to Autotrophy 271
Theory in Action: Submarine Hot Springs: Ancient
Bacteria, Symbiosis, and the Origin of Life 272
    Photosynthesizers Change the World 273
    Evidence of an Aerobic World 273
The Rise of Eukaryotes 274
    The Symbiotic Theory of Eukaryote Origins 275
    Living Prokaryotes and Single-Celled Eukaryotes 276
Bacteria: Kingdom Monera 276
    "Ancient" Bacteria: Division Archaebacteria 277
    The "Blue-Greens": Division Cyanobacteria 278
    "True" Bacteria: Division Eubacteria 279
    Bacteria in Nature and Technology 279
        Bacteria and the human body

Unicellular Eukaryotes: Kingdom Protista 279
    Animal-Like Protists 280
        Phylum Mastigophora • Phylum Sarcodina •
        Phylum Sporozoa • Phylum Cillata
    Fungus-Like Protists 281
    Plant-Like Protists 282
        Phylum Euglenophyta • Phylum
        Chrysophyta • Phylum Pyrrophyta
The Viruses 283
    Viral Diversity 284

CHAPTER 15   **The Early History of Life: The Invertebrates and the Algae**   **286**

Late Proterozoic (or Precambrian) Era 286
    The Dawn of Multicellular Life 286
Paleozoic Era: The Cambrian Period 288
    Ecological Diversity 289
    Modern Phyla First Known From the Cambrian 289
The Invertebrates 289
    The Sponges: Phylum Porifera 289
    Jellyfish and Sea Anemones: Phylum Cnidaria 289
    Worm-Like Phyla 292
        Body cavities
    The Flatworms: Phylum Platyhelminthes 292
    The Roundworms: Phylum Nematoda 294
The Higher Invertebrates 294
    The Snails, Clams, and Squid: Phylum Mollusca 297
        The Chitons: Class Polyplacophora • Snails and
        slugs: Class Gastropoda • Clams, oysters, and
        scallops: Class Bivalvia • Octopi, squid, and
        nautilus: Class Cephalopoda
    The Segmented Worms: Phylum Annelida 299
    The First Arthropods: Phylum Arthropoda 299
        The extinct trilobites: Subphylum
        Trilobita • Chelicerates: Subphylum
        Chelicerata
    Starfish, Sea Lillies, and Sea Cucumbers:
    Phylum Echinodermata 301
Theory in Action: *Peripatus: The Worm That Time Forgot* 303
Paleozoic Era: The Ordovician Period 303
The Algae 303
    The Green Algae: Division Chlorophyta 304
    The Brown Algae: Division Phaeophyta 306
    The Red Algae: Division Rhodophyta 306

CHAPTER 16   **The Silurian Period: The First Lower Plants and Jawed Vertebrates**   **308**

Paleozoic Era: The Silurian Period 308
Origins of Life on Land 308
The First Land Plants 310
    Mosses, Liverworts, Hornworts: The Division
    Bryophyta 310
Early Vascular Plants 312
    Club Mosses and Horsetails: Divisions Lycophyta
    and Sphenophyta 312
    Ferns: Division Pterophyta 312
    Evolutionary Trends in Plant Reproduction 314
Kingdom Fungi 314
    Fungi as Decomposers 314
    Fungi as Symbionts 314
        Lichens • Mycorrhizae
Main Groups of Fungi 316
    Common Molds: Division Zygomycota 316
    Sac Fungi and Yeasts: Division Ascomycota 316
    Mushrooms: Division Basidiomycota 316
    The Water Molds: Division Oomycota 316
    The Imperfect Fungi: Division Deuteromycota 318
The First Land Animals 318
    Arthropods 318
        Scorpions, spiders, and mites: Subphylum
        Chelicerata • Subphyla Crustacea and
        Uniramia • Shrimps, crabs, and lobsters:
        Subphylum Crustacea • Centipedes, millipedes,
        and insects: Subphylum Uniramia •
        Centipedes, and millipedes: Classes Chilopoda
        and Diplopoda
Vertebrate Origins 320
    Chordates: Phylum Chordata 320
        Tunicates and lancelets: Subphyla Urochordata
        and Cephalochordata • Tunicates •
        Lancelets • Subphylum Vertebrata • Jawless
        fishes: The first vertebrates • Placoderms:
        The first jawed fishes
    Evolutionary Innovations and Adaptive Radiations 323

CHAPTER 17   **Emergence of the Modern World**   **325**

The Devonian Period: The Reign of
Vertebrates Begins 325
    The Cartilaginous Fishes: Class Chondrichthyes 325
    The Bony Fishes: Class Osteichthyes 326
        The ray-finned fishes • The fleshy-finned
        fishes

Colonization of the Land Begins 330
   Plants 330
      Early Terrestrial Vertebrates: Class Amphibia 331
The Carboniferous Period 333
   Insects: The Arthropod Class Insecta 334
   The First Reptiles: Class Reptilia 335
The Permian Period 335
   The First Higher Plants: Gymnosperms 337
      Cycads and ginkgoes: Divisions Cycadophyta and
      Ginkgophyta • Conifers: Division Coniferophyta
   The Rise of the Reptiles 337
The Mesozoic Era 338
Current Controversies: *Hot-Blooded Dinosaurs?* 338
   The Ruling Reptiles 339
      Dinosaurs • Birds
   The First Mammals 341
   The Rise of Seed Plants 342
      Flowering plants: Division Anthophyta–Classes
      Monocotyledonae and Dicotyledonae
Current Controversies: *How Does Half a Bird Fly?* 344

CHAPTER 18 *Evolution of Mammals and the Ascent of Homo sapiens* 347

The Cenozoic Era 347
Mammals: Class Mammalia 348
   Egg-Laying Mammals: Subclass Prototheria 349
   Pouched Mammals: Subclass Theria,
   Order Metatheria 349
   Placental Mammals: The Eutherian Orders 350
Primate and Human Evolution 351
   Division of the Primate Line 351
   The Hominoid Lines 352
      Hominid radiation
   The Australopithecines 354
   Evolutionary Trends in Hominids 356
   Genus *Homo* 356
      Neanderthals: The first *Homo sapiens* •
      Cro-Magnons

PART FIVE *The Molecules of Life*

CHAPTER 19 *The Molecules of Life* 362

Life Is Chemical 362
Atoms and Molecules 363
   Chemical and Physical Properties 364
   Atomic Numbers and Atomic Weights 365
   Isotopes 365

Chemical Compounds and Chemical Bonds 366
   Ionic Bonds 367
Theory in Action: *Making Chemical Solutions: The Mole* 367
   Covalent Bonds 368
   The Differences Between Covalent and
   Ionic Bonds 369
Water: Something Special 369
   Acids, Bases, and Buffers 371
   Other Properties of Water 373
The Molecules of Living Things 373
   "Organic" Chemistry 373
   Chemical Groups 374

CHAPTER 20 *Macromolecules* 377

Macromolecules and Polymerization 377
Carbohydrates 377
   Monosaccharides 379
   Disaccharides 379
   Polysaccharides 381
Proteins 381
   Amino Acids 383
   The Peptide Bond 383
Protein Structure 384
   The Possibilities of a Polypeptide 384
   Polypeptides and Proteins 384
   Primary Structure 384
   Secondary Structure 385
      The alpha helix • The beta sheet
   Tertiary Structure 386
   Quaternary Structure 386
   Prosthetic Groups 388
Lipids 388
   Triglycerides and Neutral Lipids 390
   Polar Lipids 390
   Steroids 392
Nucleic Acids 392

CHAPTER 21 *Chemical Reactions and Energy* 394

Energy and Life 394
Energy and Chemical Reactions 395
   Thermodynamics 395
   The Chemical Balance of Energy 397
Starting a Chemical Reaction 398
   Activation Energy 398
   Catalysts 399
Enzymes 399
   Carboxypeptidase—A Model Enzyme 400

The Characteristics of Enzymes 400
 Enzymes Obey the Laws of Chemistry 401
 Enzymes and the Mass-Action Principle 403
 Enzyme Regulation 403
Theory in Action: *The Enzyme That Isn't a Protein* 404
 Allosteric Effects and Feedback Pathways 405
 Cofactors 405
 What Can Enzymes Do? 405
Epilogue: A Link Between Biochemistry
and Genetics? 406
 New Laws of Physics? 406

CHAPTER 22 **Molecules and Genes** **408**

The Chemical Nature of the Gene 408
 Coding Capacity: Proteins Seem to Be the Best Bet 408
 The Transforming Principle 408
 The Transforming Principle is DNA 409
Clues to the Structure of DNA 410
 Base Composition 410
 The X-ray Pattern 410
 Interpreting the X-ray Pattern 411
 The Double Helix Model 412
Theory in Action: *The Prize* 414
DNA Replication 415
 DNA Polymerase 415
The Flow of Information 416
 The Role of DNA as a Coding Molecule 416
 Nucleic Acid Templates 416
 The Role of RNA 416
Protein Synthesis 416
 The Genetic Code 416
 Transfer RNA 417
 The Ribosome 418
Theory in Action: *Seeing Genes at Work* 423
The Viruses: A Lesson in Information Transfer 423
 Bacteria Eaters 424
 Other Viruses 425
Chromosomes and DNA 426
 Prokaryotic Chromosomes 426
 Eukaryotic Chromosomes 427
The Control of Gene Expression in Prokaryotes 428
 The *lac* Operon 429
Gene Regulation in Eukaryotes 430
 Regulation at Many Levels 430
Theory in Action: *Prokaryotes and Eukaryotes: Differences in Style and Substance* 431
 Genes That Control the Expression of Other Genes 432
 Transcription Factors 432
 Intervening Sequences and the Control of
 RNA Processing 433
 Genome Structure 434

CHAPTER 23 **Evolution at the Molecular Level: Natural and Artificial** **436**

Mutations as a Source of Genetic Change 436
 Mutations Are Heritable Changes in DNA 436
 The Effects of Mutations 437
 Mutagens 438
Theory in Action: *A Molecular Disease: The Good and Bad Aspects of One Mutation* 440
Mutations and Evolution 442
 Neutral Mutations in the Human Genome 442
 Introns and Evolution 443
 Tracing the Course of Evolution 443
 Evolution and Molecular Change 444
Genetic Engineering: Artificial Evolution 445
 The Technical Basis of Genetic Engineering 445
  Restriction enzymes • Electrophoresis:
  Separating the fragments • Reading the
  sequence • Plasmids
Theory in Action: *DNA: The Ultimate Fingerprint?* 448
 Making the Chimera 449
 Transformation 449
 Why Clone? 450
 Transforming Other Cells 451
Is Genetic Engineering Safe? 452
The New Human Genetics 453
 RFLPs and Genetic Disease 453
Theory in Action: *Using RFLPs to Detect a Gene Defect* 454
 Cystic Fibrosis: New Light on a Mysterious Killer 454
Current Controversies: *Would You Really Want to Know?* 455

CHAPTER 24 **Cell Organization** **457**

The Microscope: Extending the Senses 457
 The Invention of the Microscope 458
 The Light Microscope 459
 The Electron Microscope 460
The Cell: A Detailed Look 462
Biological Membranes 463
 The Basic Structure of Membranes 463
  The lipid bilayer • Membrane proteins
Theory in Action: *Getting Lucky with Mistakes: How Science Sometimes Works* 464
 The Diversity of Cellular Membranes 466
 The Functions of Biological Membranes 466
  Membranes as barriers
The Movement of Molecules Across Membranes 467
 The Laws of Diffusion 467
 Diffusion Across a Biological Membrane 468
  Simple diffusion • Facilitated diffusion •
  Active transport

Osmosis: Diffusion of Water Across a Selectively
Permeable Membrane 470
Effects of Osmosis on a Living Cell 471
Cell Organelles 472
The Nucleus: The First Organelle 472
The Ribosome 473
The Endoplasmic Reticulum 474
The Golgi Apparatus 476
Secretory Vesicles 476
Theory in Action: *Why Are There So Many Types
of Microscopes?* 477
Vacuoles 478
Lysosomes 478
The Cytoskeleton 478
Microtubules 479
Microfilaments and Intermediate Filaments 481
Energy-Producing Organelles 483
Mitochondria and Chloroplasts 483

CHAPTER 25 *Cells and Energy* 485

Metabolism 485
Glucose: The Energy in One Food Molecule 486
A Logical Molecule to Study 486
Oxidation and Reduction 486
Twelve Electrons Are Removed During Glucose
Oxidation 487
Energy from Glucose is Trapped in the Form
of ATP 487
Glycolysis: The First Pathway 488
Investing Some Energy—It Takes ATP to
Make ATP 489
Making ATP: Substrate Phosphorylation 490
Glycolysis Releases Only a Small Fraction of
Available Energy 490
The Krebs Cycle 491
Releasing Energy in the Presence of Oxygen 491
A Closer Look at the Krebs Cycle 493
Carbon • ATP • Oxidation-reduction
Electron Transport 493
Energy from Oxidation and Reduction 493
Chemiosmosis: Why the Electron Transport Chain
Makes Sense 494
Cellular Respiration 497
Summing Up the Oxidation of Glucose 498
How Other Molecules Enter the
Respiratory Pathways 498
Fermentation 499
Different Types of Fermentation 499
Alcohol fermenters • Lactate fermenters
The By-products of Fermentation 500

CHAPTER 26 *Photosynthesis* 503

Energy from Sunlight 503
Chlorophyll and Sunlight 504
The Light-Dependent Reactions of Photosynthesis 505
The Photosynthetic Reaction Center 507
Photosynthetic Electron Transport:
Oxygen Evolution 508
Photosynthetic Electron Transport: Noncyclic Flow 510
Cyclic Electron Flow 510
Photophosphorylation: Chemiosmosis in
the Chloroplast 510
Summary of the Light-Dependent Reactions 511
The Light-Independent Reactions of Photosynthesis 511
The Calvin Cycle 512
C4 Photosynthesis 514
Mitochondria and Chloroplasts: Remarkably
Similar Organelles 515

PART SIX *Plant Systems*

CHAPTER 27 *Plant Structure and Function* 520

What Is a Plant? 520
The Need for Sunlight 520
The Need for Water 521
The Need to Reproduce 522
Plant Structure: From Water to Land 522
Plants Restricted to Wet Environments 522
Plant Life on Land 524
The Vascular Plants 525
Theory in Action: *Plants and Medicine* 526
Gymnosperms and Angiosperms 527
Plant Organization 527
The Roots: Starting From the Ground Up 528
Root Structure 528
Xylem 529
Phloem 531
Water Movement Into the Roots 532
Theory in Action: *Another Use for Phloem Sap* 533
Growth Patterns in Roots 534
The Stems: Maintaining the Flow 535
Growth Patterns in Stems 535
Growth Patterns in Tree Stems 536
Reaction to Injury 536
The Leaves: Harvesting Sunlight 538
Regulation of Gas Exchange 538
Internal Leaf Tissue 540

CHAPTER 28 *Reproduction in Flowering Plants* 542

Asexual and Sexual Reproduction 542
  Asexual Reproduction 542
    Vegetative reproduction in nature • Vegetative reproduction in horticulture
  Sexual Reproduction 544
Alternation of Generations 544
  The Principle of Alternation 544
  Alternation of Generations in the Angiosperms 545
The Details of Plant Reproduction 546
  The Organs of Reproduction 546
  Formation of Spores and Gametes 546
    The female tissues: megaspores and eggs •
    The male tissues: microspores and pollen
  Pollination and Fertilization 547
    Initial events: the growth of the pollen tube •
    Double fertilization—embryo and endosperm
  Embryonic Development 547
  Fruits 550
    Fruit types
Current Controversies: *Evolutionary Pressure in a Very Small Arena* 552
Plants and Animals: Flowers, Pollen, and Seeds 553
  Flower Specialization 553
  Techniques of Pollination 553
    Wind pollination • Insect pollination • Bird pollination • Pollination by other animals • Pollination and self–incompatibility
Theory in Action: *The Grand Masquerade: Plants as Mimics* 555
  Techniques of Seed Dispersal 557
    Wind dispersal • Water dispersal • Dispersal by animals
Current Controversies: *Is Sex Doomed? Apomixis* 559

CHAPTER 29 *Plant Growth and Nutrition* 561

Germination and Development 561
  Dormancy 561
  Germination 563
  The Beginning of Growth 563
Nutrient Requirements 564
  Plant Chemical Composition 564
  Plant Nutrients 564
  Deficiencies in Mineral Nutrients 565
Fulfilling the Nutrient Needs of Plants 565
  Soil and Its Nutrients 565
  Soil Fertilization and Management 567

Theory in Action: *Solar-Powered Marine Animals* 569
  Symbiosis and Plant Nutrition 569
    Mycorrhizae • Nitrogen fixation
  Meat Eaters 571

CHAPTER 30 *Control of Plant Growth and Development* 573

Control of Plant Growth Patterns 573
  The Nature of Plant Growth 573
  Responses to the Environment 574
    Geotropism • Phototropism • Thigmotropism
Plant Hormones 575
Theory in Action: *Chemical Defense Mechanisms: How Plants Fight Back* 576
  Auxin 576
  Gibberellin 580
  Cytokinins 580
  Abscisic Acid 581
  Ethylene 581
Phytochrome: The Regulation of Plant Growth 582
  The Discovery of Phytochrome 582
  Photoperiodism 584
  Florigen 585
Autumn: A Case Study 585
Tomorrow's Plants 586
  Plant Cell Culture 586
  Plant Genetic Engineering 586

PART SEVEN *Animal Systems*

CHAPTER 31 *Multicellular Organization and Homeostasis* 592

Maintaining the Balance 592
The Range of Cells and Tissues 593
  The Diversity of Cells 593
  Epithelial Tissue 593
    Cell junctions
  Connective Tissue 596
    Supporting tissue • Circulating tissue
  Nervous Tissue 598
  Muscle Tissue 598
Organs and Organ Systems 599
Major Organ Systems 600
Studying the Human Body 600
  The Basic Body Plan 600

Homeostasis 602
    Maintaining the Internal Environment 602
    The Concept of Feedback 603
        Negative feedback • Positive feedback
    Homeostasis in Action: Eat, Drink, and Be Merry 604
        Water balance • Sugar balance

CHAPTER 32 *Circulation*

The Importance of Internal Transport 608
    Open Circulatory Systems 609
    Closed Circulatory Systems 609
The Vertebrate Circulatory System 609
    Vertebrate Blood Vessels 610
    The Vertebrate Heart 611
    The Structure of the Human Heart 613
    Blood Flow Through the Heart 613
        Blood supply to the heart muscle is critical
    Control of Heartbeat 614
    The Cardiac Cycle 616
Circulatory Patterns in the Human Body 616
    The Pulmonary Circulation 616
    The Systemic Circulation 616
    Control of Blood Circulation Patterns 617
    Material Exchange in the Capillaries 617
    Blood Pressure 618
        Arterial blood pressure • Factors affecting
        blood pressure
Disorders of the Circulatory System 619
    Hypertension 619
    Atherosclerosis 620
    Heart Attack 620
        Risk factors in heart disease
Theory in Action: *EKGs* 621
    Stroke 622
The Lymphatic System 622
Blood Composition 623
    Plasma 623
    Blood Cells 623
        Red blood cells • White blood cells •
        Platelets and blood clotting
Theory in Action: *Artificial Hearts* 624

CHAPTER 33 *Respiration* 629

The Cellular Roots of Respiration 629
    Respiration in Water and Air 629
The Variety of Respiratory Systems 630
    Exchange Through Body Surfaces 630
    Exchange Through Gills 631

Theory in Action: *The Ice Fish: Blood Without Hemoglobin* 633
    Exchange Through Tracheae 633
    Lungs 634
        The avian lung—increased efficiency
    Conserving Moisture 635
The Human Respiratory System 636
    Breathing 636
    The Movement of Air 638
Gas Exchange 638
    The Alveoli 638
Theory in Action: *Bhopal* 639
    The Role of Hemoglobin 639
    Gas Exchange Between Blood and Tissues 640
        Partial pressure • Hemoglobin and oxygen
    The Transport of Carbon Dioxide 642
Control of Respiration 642
Damage to the Respiratory System 642
    Subverting the System: Carbon Monoxide 642
Theory in Action: *The Bends* 643
    Smoking and Lung Damage 643
Theory in Action: *Saving a Life: The Abdominal Thrust* 644
    Emphysema 645
    Asthma 645

CHAPTER 34 *Nutrition and Digestion* 647

Nutritional Needs of Heterotrophs 647
    Food Energy and Calories 647
    Food and Essential Nutrients 648
        Water • Carbohydrates • Proteins
        • Fats • Vitamins • Minerals
Theory in Action: *Cholesterol and Heart Disease: Research and Changing Perspectives* 652
    Balancing Nutrient Input and Requirements 656
Theory in Action: *The Hundred-Watt Bulbs* 656
    Obesity • Losing weight • Eating disorders
Obtaining Food: The Universal Preoccupation 658
    Feeding Techniques 658
    Cellular Digestion 659
    Animals with Gastrovascular Cavities 661
    Animals with Alimentary Canals 661
The Human Digestive System 662
    The Oral Cavity 663
Theory in Action: *Intestinal Bacteria, The Essential Dinner Guests* 664
    Swallowing 665
    The Esophagus 666
    The Stomach 666
        Ulcers
    The Duodenum 667

Theory in Action: *Alcohol: The Deadly Nutrient* 668
    Pancreatic secretions • Liver secretions
    The Small Intestine 670
    The Liver and the Hepatic Portal Circulation 672
    The Large Intestine 672
Theory in Action: *Diarrhea: When the Large Intestine Fails* 673
    Timing 673

CHAPTER 35   *Regulation of the Cellular Environment*   677

Water, Wastes, and Temperature 677
    Toxic Ammonia 678
    Dealing with Ammonia: Three Strategies 678
        Ammonia excretion • Uric acid • Urea
Excretory Systems 679
The Human Excretory System 681
    The Kidney 681
    The Nephron 682
        Formation of the filtrate • Resorption •
        Concentration • Secretion • Urination
Theory in Action: *The Artifical Kidney* 684
Theory in Action: *Urine Testing* 687
    Control of Water Balance 688
        ADH and the collecting ducts • Thirst
        • Aldosterone • Acid-base regulation
Theory in Action: *Dealing with Nitrogen: Another Style of Life* 691
Water Balance in Other Animals 691
    Aquatic animals 691
    Land animals 692
Temperature Regulation 693
    Ectotherms 694
    Endotherms 694
        Adaptations for temperature regulation
    Thermoregulation in Humans 696
        Response to cold • Response to heat
    Homeostasis and Thermoregulation 697

CHAPTER 36   *Chemical Communication*   699

The Body's Vital Messengers 699
The Endocrine System 700
    The Principle of Chemical Signaling 700
    Hormonal Control in Insects 701
Components of the Endocrine System 702
Types of Hormones 703
Theory in Action: *The Pineal Gland* 704

The Nature of Hormone Action 705
    The Importance of Receptors 705
    Steroids 706
    Peptides and Their Derivatives 706
    Second Messengers as Amplifiers 706
Control of the Endocrine System 710
    Hormonal Control Systems: Two Techniques 710
        Negative feedback loops • Complementary
        hormone action
    The Pituitary Gland 712
    The Posterior Pituitary and Its Hormones 712
Theory in Action: *The Heart is an Endocrine Gland Too* 713
    The Anterior Pituitary and Its Hormones 713
        Releasing hormones control the anterior
        pituitary • Anterior pituitary hormones
Major Endocrine Glands and Their Hormones 715
    The Thyroid 715
        Feedback control of thyroxine levels
    The Parathyroids 717
    The Pancreas and Control of Blood Glucose 717
        Diabetes
    The Adrenal Glands 719
        The adrenal cortex • The adrenal medulla
    The Gonads 720
    The Thymus 720
    Other Endocrine Glands 720
    Prostaglandins 721

CHAPTER 37   *Reproduction*   723

Forming a New Generation 723
The Male Reproductive System 724
    The Testes 725
    The Male Reproductive Tract 727
    Male Sex Hormones 727
    Puberty in Males 728
Theory in Action: *For Want of a Receptor* 728
The Female Reproductive System 729
    The Ovaries 729
    The Female Reproductive Tract 730
    Female Sex Hormones and the Reproductive Cycle 731
    The Conclusion of the Cycle 731
        Menstruation: When fertilization does not
        occur • Pregnancy: When fertilization
        does occur
Theory in Action: *Other Reproductive Cycles* 733
        Timing of the menstrual cycle
    Hormones, Behavior, and Reproduction in
    Other Mammals 735
Sexual Intercourse 735
Contraception 736
    Infertility 738

Sexually Transmitted Diseases 739
Theory in Action: *In Vitro Fertilization* 739
    The Spread of STDs 740
        Syphilis • Gonorrhea • Chlamydia •
        Herpes • Genital warts • AIDS
    The Prevention of Sexually Transmitted Diseases 742
Current Controversies: *Is Sex the Cornerstone?* 742

Theory in Action: *Metastasis: How Does a Cancer Cell Invade?* 771
    Theories of Oncogene Action 772
    Treatment for Cancer 773
    Prevention and Control of Cancer 773
Theory in Action: *A Specific Gene for a Heritable Cancer* 774
    The Clearest Cause of Cancer 775

CHAPTER 38 **Embryology and Development** 744

The Patterns of Development 744
    The Egg: Protecting and Nurturing the Embryo 744
Early Development 746
    Gametogenesis 746
    Fertilization 746
        Response of the sperm • Response of the egg
    Cleavage in the Embryo: Cell Division Begins 747
        The blastula
    Gastrulation and Germ Layer Formation 748
    Neurulation 751
    Organogenesis 751
    Mechanisms of Development 752
Human Embryology 755
    Gametogenesis and Fertilization 755
    The First Month of Development 755
    The First Trimester 758
    The Second Trimester 758
    The Third Trimester and Birth 759
    Maturation: Development Continues 761
    Aging 761
Coordination of Development 762
    Development in Drosophila 763
    Caenorhabditis: The World's Best-Known Embryo? 763
    Developmental Anomalies as Research Tools 763
    Master Control Genes 764
        Homeotic mutants • The homeobox
        sequence • Sex determination in humans
    The Influence of Egg Cytoplasm 766
    Embryonic induction 766
    Further Control During Maturation 768
Theory in Action: *Hen's Teeth!* 768
Cancer: When Cells Rebel 770
    Failure to Control Cell Growth 770
    Cancer Traced to Altered Genes 770
        Studies of cancer that seemed to be
        inherited • Studies of environmental
        carcinogens • Studies of cancer-causing
        viruses • Studies of gene transfer using
        tumor cells

CHAPTER 39 **Immunity and Disease** 777

Disease Through the Ages 777
    The First Vaccination: A Shot in the Dark 777
    The Agents of Disease 778
Pathogenic Organisms 780
    What Makes a Pathogen? 780
    The Diversity of Pathogens 780
The Immune System: Defense Against Infection 781
    Nonspecific Defenses 781
        The skin • Enzymatic defenses •
        Inflammation
    Specific Defenses 783
Components of the Immune System 784
    Cells and Organs 784
    Antibodies: The Primary Molecules of the
    Immune System 785
The Immune System in Action 786
    Antigen Stimulation 787
    B-Cell Growth and Differentiation 788
Theory in Action: *Millions of Antibodies from a Few Genes* 788
        Somatic mutation
    Effects of the Humoral Response 790
    T-Cells and the Cell-Mediated Immune Response 791
        T-cell actions
    Self and Non-Self 792
        Transplant rejection
When the Immune System Works and When It Fails: Case Studies 794
    Recovering from Disease with Permanent Immunity 794
Theory in Action: *Monoclonal Antibodies* 795
    Recovering from Disease with Temporary Immunity 797
    A Pathogen That Defeats the Immune System 797
Disorders of the Immune System 797
    Allergies 797
    Autoimmune Diseases 798
AIDS 798
    The History of AIDS in the United States 799
    The Biology of HIV Infection 799

The Spread of HIV Infection 801
How AIDS Is *Not* Transmitted 801
How AIDS Can Be Transmitted 802
    Sexual transmission • Sharing of syringes •
    Transmission from mother to fetus •
    Transfusion-borne AIDS
Theory in Action: *Protecting Yourself Against AIDS* 804
    The Outlook for Development of Vaccines
    and Cures 805

CHAPTER 40 *Nervous Control* 807

The Miracle of Mind 807
Why Nervous Systems? 808
    Components of Nervous Systems 808
        Types of neurons • Neural circuits
How Neurons Work 811
    The Difference Between Potential and Current 811
Membrane Pumps, Leaks, and Electric Potentials 811
    Ion Channels and Excitability 812
The Action Potential: Electrically Gated Channels
in Action 814
    Depolarization and Threshold 814
    The Traveling Wave: An All-or-Nothing Event 815
    Summary of Action Potential 816
    The Role of Myelin 816
Synapses: Sites of Neural Control and Integration 817
    Electrical Synapses 817
    Chemical Synapses 817
    A Simple Synapse: The Neuromuscular Junction 818
    Synapses Between Neurons 820
        Summation: the key to information processing
    A Diversity of Transmitters 822
        Endorphins: pain and the brain
Theory in Action: *Drugs, Synapses and the Brain* 823
    Changes at the Synapse: Learning and Memory 824
        Experience and synaptic function • Experience
        and change in neural structures
From Neuron to Brain: The Evolution of
Nervous Systems 824
The Human Nervous System 825
    The Peripheral Nervous System 826
        The somatic nervous system • The autonomic
        nervous system
    The Central Nervous System 827
    The Brain 828
        The forebrain: cerebrum and thalamus •
        The midbrain • The hindbrain
    The Spinal Cord 831
So Elegant an Enigma: The Biology of the Mind 832

CHAPTER 41 *The Sensory System* 835

The World, the Senses, and Reality 835
The Nature of Sensory Stimuli 836
Essentials of Sensory Function 836
    Sensory Transduction 836
    Information Processing in the Brain 838
The Diversity of Sensory Systems 839
    Senses and Environments 839
    Similarities Among the Senses 839
Vision: A Model of Sensory Function 839
    Light and the Visual Environment 840
    Structures and Functions of the Eye 840
    The Retina 842
    The Photoreceptors 844
    Processing of Visual Information 846
        Adaptation • Color vision • Lessons from
        vision applied to the other senses
Audition 848
    What is Sound? 848
    Hearing in Air: The Human Ear 849
    Sound in Water: The Lateral Line Sense 850
    Echolocation 850
Balance and Acceleration 851
The Chemical Senses 853
    Olfaction 853
    Gustation 853
    The Common Chemical Sense 854
The General Senses 854
Special Senses of Animals 856
    Infrared Detection 856
    Electroreception 856

CHAPTER 42 *The Musculo-Skeletal System* 859

The Wonder of Controlled Movement 859
Structure and Function in Muscle Tissue 860
    Skeletal Muscle 860
    Smooth Muscle 861
    Cardiac Muscle 861
    The Structure of Skeletal Muscle 861
    Contraction in Individual Muscle Cells 864
        The stimulus • The force generators: sliding
        filaments • Contractions of whole
        muscles • Motor units • Single twitch,
        summation, and tetanus
    Muscle Physiology: The Power Behind the
    Force 867
        Fatigue

The Skeletal System: Levers and Hinges   869
   Axial Skeleton   869
   Appendicular Skeleton   869
Theory in Action: *Exercise and Skeletal Muscles*   **870**
   Cartilage   871
   Bone   871
     The structure of compact bone • Bone growth
     and remodeling
   Joint Structure and Function   874
Muscles and Bones Together: A Dynamic System   875
   Coordination of Muscular Activity   876
   Walking: A Complex "Simple" Activity   876

CHAPTER 43   *Animal Behavior*   **878**

The Behavioral Sciences   878
   Terminology and Approach in Behavioral Studies   878
Elements of Behavior   879
   Sensory Worlds   879
   Simple Behaviors: Programmed Responses   880
   Complex Behaviors: Inherited and Acquired   881
     Instinctive behaviors • Learning
   Instinctively Guided Learning   883
     Migration and homing
   Genetic Influences on Song Learning in Birds   884
   Behavioral Genetics   885
     Genes and mating behavior • Genes
     and learning
Social Behavior   886
   Communication   887
     Intraspecific communication • Courtship
     behavior • Male-male competition and
     female choice
   Language   590
     The dance language of the bees • Language
     in primates
   The Evolution of Social Behavior   892
   Insect Societies   893
     The insect colony
   Primate Societies   896
Human Behavior   896

APPENDIX

Answers   A-2
Genetics Problem Set: Mitosis, Meiosis, and
Mendelian Genetics   A-4
Solutions to Genetics Problem Set   A-5
The Metric System   A-7
Classification Scheme   A-8
Readings   A-10
Glossary   A-13
Illustration Credits   A-30
Index   A-35

A Note from the Authors

■ Between us, we've taught biology courses for majors and non-majors for more than 25 years. We've discovered that both groups of students share an appreciation for teaching that begins with their own experiences and then leads gradually into the key scientific questions of the day.

We start each chapter by making exactly those sorts of connections. They will engage your students and prepare them for the material that follows.

■ Have you ever wanted to mark a text to tell students that a topic is related to key information in another chapter?

Well, we do that. Throughout the text you'll find arrows that let the student see that another chapter contains information they might find useful.

DNA: The Ultimate Fingerprint?

Late in the nineteenth century, law enforcement officers began to use fingerprints as a means of identifying criminals. Fingerprints are ideal for this purpose, because they differ so widely from one individual to the next. But fingerprints are useful only when a criminal leaves a number of clean, complete prints that can be matched with police records. In most crimes, fingerprint evidence is not good enough to be used for identification. Today, however, molecular biology has developed a new tool that may become the ultimate weapon of criminology: the *DNA fingerprint*.

DNA fingerprinting makes use of the fact that certain DNA regions between genes are extremely variable from one individual to the next. Many of these "hypervariable" regions contain multiple copies of simple base sequences, and the numbers of copies differ from one individual to the next: One person might have 40 repeats between two genes, another 31, and another 15. By constructing a small DNA fragment known as a *probe*, which is complementary to one of these simple sequences, researchers can examine a gel of total DNA and then use the probe to locate fragments of different sizes on that gel. If the probe recognizes a large number of fragments from around the genome, a single gel can produce a "fingerprint" that is unique for one individual. If two probes are used, the resulting gel pattern is so specific that it can be distinguished from the pattern of any other individual in the world.

DNA fingerprints can be prepared from cells found in a drop of blood or semen and even from fragments of skin caught under the fingernails of a crime victim. In 1988 this technique came to the aid of a 27-year-old com-

puter operator at Disney World in Florida who had been attacked, beaten, and raped in her home in Orlando in 1986. Although she caught a brief glimpse of her assailant's face during the rape, she faced the terrifying prospect of having to convince a jury that her brief eyewitness identification of the suspect, 24-year-old Tommie Lee Andrews, was absolutely certain. Andrews claimed that he was at home during the evening the rape occurred, and he even produced a witness to substantiate that claim. It was a classic case of the victim's word against that of the accused. The prosecuting attorneys sought the aid of a molecular biologist

to perform the "DNA fingerprinting" test on a semen sample taken by police the night of the rape. When the DNA fingerprint of this sample was compared with a blood sample taken from Andrews, the results were conclusive: It was a perfect match. The jury returned a verdict of guilty, resulting in a jail sentence of 22 years for Andrews.

DNA fingerprinting is not, however, considered "absolute" evidence because there are no uniform standards and procedures for the fingerprinting process. When uniform standards are developed, DNA fingerprinting should prove to be the ultimate weapon of criminology.

The first use of DNA fingerprinting in the United States is shown in this photograph of DNA patterns matching semen samples from a rape victim with blood samples from an accused suspect. Two different restriction enzymes were used, and in each case the result was a match between semen from the rapist and blood drawn from the accused.

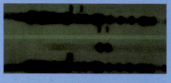

blood

semen

blood

semen

■ No text is complete unless it is up-to-date and immediate. The feature to the left is indicative of what we try to do throughout the text: maximize the personal impact of science on the individual student.

When DNA fingerprinting was first used to convict a criminal in the United States, we brought that case into the classroom.

We wrote a feature that described the crime, explained the evidence, and showed the actual gels presented in court. Ken used this feature in his biology class and discovered that it galvanized students. We bet that it will keep yours on the edge of their seats, too.

...uding Lap-Chee ...s of Michigan ...e. They started ...he CF trait and ...the long arm of ...ng and sequenc- ...sociated with CF. ...understand CF ...gene itself may

...s the amino acid ...mbrane protein, ...a structure for

...otein transports ...may explain the ...t could produce ...ight also explain ...experience. CF ...y is on the verge ...ne of our most ...battling killers.

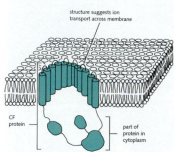

structure suggests ion transport across membrane

CF protein

part of protein in cytoplasm

Figure 23.17 The sequence of the cystic fibrosis gene suggests that it may be a membrane protein involved in ion transport. This drawing shows a structural model for this protein in a biological membrane.

■ We have tried to collect the best of recent and lesser-known stories to share with you and your students. We discuss biomedical advances and their real-life consequences. Would you want to know if you are likely to develop a fatal disease later in life? Stories like this place science in perspective, helping students realize how science can create ethical dilemmas.

Would You Really Want to Know?

The availability of a test for Huntington's disease is a great medical breakthrough. But it confronts the children of Huntington's sufferers with a dilemma: *to test or not to test.* Universal application of the test might make it possible to eliminate the disease. If all potential carriers were to be tested, and if all agreed not to have children if they tested positive, the incidence of the disease might be cut dramatically. Many young adults living in fear and uncertainty would have the relief of learning that they will never develop the disease, and those not so lucky might be grateful for years of

advance warning to plan for the onset of the disease.

But many people who may one day succumb to Huntington's have decided that they do *not* want to take the test. Some of these people have said that they prefer uncertainty to the possibility of discovering that they *do* have the disease. Others have said that they will have children regardless of their status on the Huntington's test, seeing nothing tragic about bringing into the world a child who might develop the disease. After all, Huntington's sufferers lead full lives until the disease develops. In fact,

about 50 percent of the possible Huntington's carriers who were offered the test in the Boston area in 1989 declined to be tested. Molecular biology may produce powerful new tools with which we can search for genetic diseases, but deciding how to use those tools will remain a matter of individual choice.

Ferns: Division Pterophyta

haploid
diploid

zygote
young sporophyte
parent gametophyte
Fertilization
archegonium
egg
sperm
mature sporophyte
antheridium
mature gametophyte
Meiosis
spores
sporangium

Gametophyte
archegonium
egg
antheridium
rhizoids
blade
frond
stipe
rhizome
roots
sori (clusters of sporangia)
Mature Sporophyte

Life cycle: Haploid spores are released from sporangia to be carried on the wind. The spores germinate into small, haploid gametophytes. Sperm produced in antheridia swim to eggs in archegonia; the zygote produces a diploid sporophyte that grows out of and soon dwarfs its parent.

Anatomy: Leafy fronds, composed of *blade* and *stipe*, belong to diploid fern sporophytes, which are larger and more independent than those of mosses. Fronds unfurl from creeping *rhizomes* that grow underground or along the surface. Clusters of sporangia called *sori* form on the undersides of mature fronds. Well-developed roots extend into the soil.

Living ferns range from small species common in moist-temperate environments to giant tropical treeferns.

■ Teaching diversity is tough, which is why we present diversity by relating it to evolutionary history; to create a coherent context that allows us to integrate diversity—past and present—with plate tectonics, climate change, and interactions among plants and animals. To organize these relationships we use two types of illustrations: "period boxes" that summarize information about major stages in earth's history, and "taxon" boxes of taxonomic and anatomical information for each major group.

r in tropical rain
insects have been
wn species range

...les, which first
... their long-lived
... reptiles' adaptive
... long after their
...d, but the group
...alf of the Carbon-
...ized upon insects
...dvantage of much
...em to *bite*, rather
...lso evolved more
...But the key inno-

vation that spurred the reptilian radiation was the **amniotic egg,** a watertight egg produced by internal fertilization and wrapped in three protective membranes. Amniotic eggs serve land vertebrates much the same way seeds serve land plants—by making it easier for them to reproduce without standing water. Amniotic eggs protect developing embryos from desiccation, nourish them as they grow, provide a space for the storage and ultimate disposal of waste products, and enable the egg to exchange respiratory gases with the surrounding air.

THE PERMIAN PERIOD

The Permian period was a time of great environmental stress and innovation for all plants and animals, as geological changes in Pangaea produced cooler, drier climates. A mass extinction claimed more than 50 percent of all terrestrial animal families and more than 95

The Permian Period

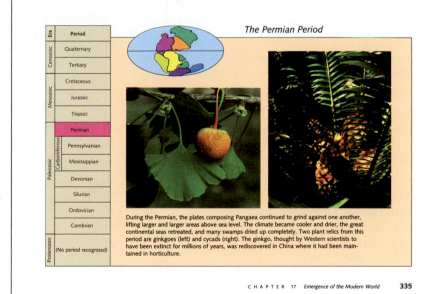

Era	Period
Cenozoic	Quaternary
	Tertiary
Mesozoic	Cretaceous
	Jurassic
	Triassic
Paleozoic	Permian
Carboniferous	Pennsylvanian
	Mississippian
	Devonian
	Silurian
	Ordovician
	Cambrian
Proterozoic	(No period recognized)

During the Permian, the plates composing Pangaea continued to grind against one another, lifting larger and larger areas above sea level. The climate became cooler and drier, the great continental seas retreated, and many swamps dried up completely. Two plant relics from this period are ginkgoes (left) and cycads (right). The ginkgo, thought by Western scientists to have been extinct for millions of years, was rediscovered in China where it had been maintained in horticulture.

A few scientists, however, believed that DNA was such an interesting molecule that X-ray diffraction should be attempted even if perfect crystals couldn't be formed. One such person was Rosalind Franklin, a young scientist working with Maurice Wilkins, a crystallographer in London. Franklin drew a thick suspension of the fiber-like DNA molecules up into a glass capillary tube and used this DNA sample to scatter X rays. In the tube, she hoped, the thick suspension of DNA molecules would be forced to line up so that if the molecules were parallel to the tube. Like spaghetti drawn through a straw, the molecules were all arranged in the same direction—not perfect enough to give a crystal-like pattern, but just good enough to yield a few clues about the structure of the DNA molecule.

Interpreting the X-ray Pattern

One of the X-ray patterns produced by Franklin's DNA samples is shown in Figure 22.3. The pattern contains two critical clues to the structure of the DNA molecule (graphically summarized in the figure).

Clue number 1: The two large dark patches at the top and bottom of the figure showed that some structure in the molecule was arranged at a right angle to the long axis of DNA and repeated at a distance of 3.4 Å (see Appendix: The Metric System). In other words, something in the molecule was arranged like the rungs of a ladder.

Clue number 2: The X-like mark in the center of the pattern showed that something in the molecule was

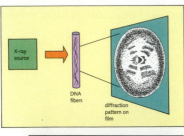

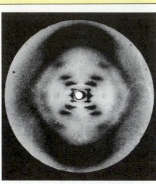

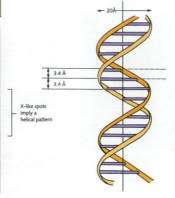

Figure 22.3 (left) DNA fibers were taken up in a thin tube so that most of them were oriented in the same direction. An X-ray diffraction pattern was then recorded on film. (bottom) X-ray diffraction pattern of DNA in the "B" form, as taken by Rosalind Franklin in 1952. Franklin's X-ray pattern contained two important clues to the structure of DNA. The large spots on the top and bottom of the pattern indicate that there is a regular spacing of 3.4 Å along the length of the fiber. The "X"-shaped pattern in the center indicates that there is a zigzag feature in the molecule, which might be consistent with a helix.

CHAPTER 22 *Molecules and Genes* **411**

■ We explain the details of one of the key advances in 20th century biology—the development of the Double Helix model. Rather than skip over Franklin's X-ray photograph, we explain what it says about the structure of DNA.

We present students with clues to the structure of DNA—clues just like those the actual scientists considered in the early 1950s—and challenge students to organize these clues into a model.

Students enjoy being active participants who can understand and interpret experimental results.

■ Instructors use genetics to solve problems and make predictions.

We decided to show examples of *how* a student might solve a genetics problem. We begin with a simple problem, similar to what might be assigned, and work the solution through, step-by-step.

We emphasize both the predictive power of genetics and its limitations. We think you'll find that this gives students an advantage in understanding the key principles of inheritance.

F₂ generation), half produced flies with wild-type eye color, and the other half produced 50 percent flies with wild-type eyes and 50 percent flies with purple eyes. *Determine the genotype of the original parents.*

By reasoning backwards from the last cross, we can solve the problem. Purple is a recessive trait, so each of the purple-eyed flies must have been genotype *ww*. Because half of the F₁ flies that were mated to the purple flies produced offspring with 50 percent purple eyes, these F₁ flies must have had a copy of the *w* gene. Their own eye color was wild type, so they must also have had a copy of the wild-type gene, making them genotype *Ww*. The other half of the F₁ generation did not produce any purple-eyed flies in their F₂ offspring. This means that their genotype must have been *WW*. What would the genotypes of the two original fly parents have to have been to produce an F₁ generation that was half *WW* and half *Ww*? The answer is that they must have been *WW* and *Ww*.

It is also possible to use genetics to predict future events. You will remember that Mendel discovered that tall is dominant over short in garden peas. Two tall peas are crossed and 100 seeds are collected. Two seeds are planted and grown under identical conditions. One produces a tall plant and the other a short plant. On the basis of these results, *determine*

how many of the remaining 98 plants can be expected to grow into tall plants.

Remember that we were not told the genotypes of the original parents—only their phenotypes (they were both tall). However, because they were both tall, we know that each had at least one of the dominant genes for tallness (*T*). The single seed of the first filial generation that grew into a short plant had to be genotype *tt*, because the *t* gene is recessive. The

only way in which a short plant could have been produced was for each of the parents to have had at least one *t* gene. Therefore the two parents were both genotype *Tt*. Simple genetic analysis tells us that 25 percent of the seeds in a cross between two such plants will grow to display the short phenotype. Because 75 percent of 98 = 73.5, 73 or 74 of the remaining seedlings can be expected to be tall.

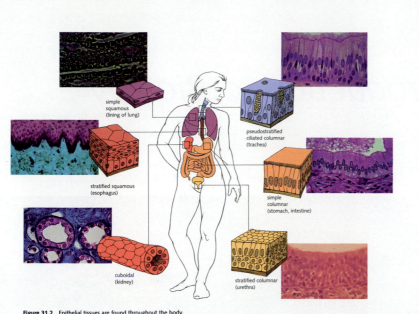

simple
squamous
(lining of lung)

pseudostratified
ciliated columnar
(trachea)

stratified squamous
(esophagus)

simple
columnar
(stomach, intestine)

cuboidal
(kidney)

stratified columnar
(urethra)

Figure 31.2 Epithelial tissues are found throughout the body. They are placed into categories based on the structural organization of the cells of which they are made. Photomicrographs of six major types of epithelial tissue. Clockwise (from the upper left), simple squamous epithelium (from the lung), pseudostratified columnar epithelium (trachea), simple columnar epithelium (stomach), stratified columnar epithelium (urethra), simple cuboidal epithelium (from kidney), and stratified squamous epithelium (esophagus).

594   P A R T 7   *Animal Systems*

found in tissues
mesoderm, are u
ings) of epithelia

Epithelial tissu
of the cells that f
of *squamous, cub
epithelium consi
epithelium consi
**dostratified epit
yet all the cells a

Cell junctions I
compose them m
The cells of man
communication

■ In a single diagram students get three important pieces of information about epithelial tissue: its appearance in the microscope, its cellular structure, and its location in the body. This one-stop approach to illustration makes teaching and learning much easier.

■ This sequence of drawings, at three levels of magnification, gives the reader a feeling of intimacy with the biological system—and that is exactly our goal: to draw students into the process of discovery.

J.S.L.
K.R.M.

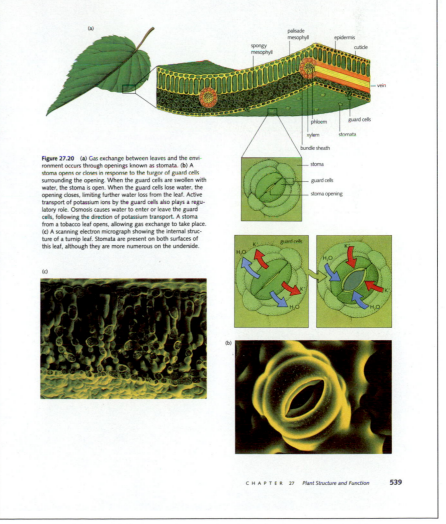

palisade
mesophyll

epidermis

cuticle

spongy
mesophyll

vein

phloem

guard cells

xylem

stomata

bundle sheath

stoma

guard cells

stoma opening

guard cells

K^+ H_2O

Figure 27.20 (a) Gas exchange between leaves and the environment occurs through openings known as stomata. (b) A stoma opens or closes in response to the turgor of guard cells surrounding the opening. When the guard cells are swollen with water, the stoma is open. When the guard cells lose water, the opening closes, limiting further water loss from the leaf. Active transport of potassium ions by the guard cells also plays a regulatory role. Osmosis causes water to enter or leave the guard cells, following the direction of potassium transport. A stoma from a tobacco leaf opens, allowing gas exchange to take place. (c) A scanning electron micrograph showing the internal structure of a turnip leaf. Stomata are present on both surfaces of this leaf, although they are more numerous on the underside.

C H A P T E R 27   *Plant Structure and Function*   **539**

A N O T E F R O M T H E A U T H O R S   **xxxi**

PART ONE

Introduction to Biology

The question of all questions for humanity, the problem which lies beyond all others and is more interesting than any of them is that of the determination of man's place in Nature and his relation to the Cosmos. Whence our race came, what sort of limits are set to our power over Nature and to Nature's power over us, to what goal are we striving, . . . [these] . . . are the problems which present themselves afresh, with undiminished interest, to every human being born on earth.

—T. H. Huxley, 1863

These monumental questions, so eloquently articulated by Huxley, have puzzled and inspired human beings for thousands of years. Today we seek to answer most of them through the disciplines we call the sciences: biology, chemistry, physics, astronomy, psychology, sociology, and others. But what, exactly, is science?

The first part of our text explores the nature of the biological sciences. Chapter 1 presents science as a way of relating to the world around us and introduces the methods of scientific investigation. Chapter 2 uses specific examples to show both contemporary and historical connections between the biological sciences and human society.

◀ Our species is neither the largest on Earth nor the most numerous, but it is the most influential of all multicellular animals. As our numbers grow, our global impact expands; as technology advances, the world seems to shrink around us. This false-color satellite image spans much of the Amazon Basin, an area once considered impossibly remote. Now a network of roads (blue lines) penetrates the former wilderness, dwellings (blue boxes) house new settlers, and plumes of white smoke signal the burning of the rain forest.

CHAPTER 1

Understanding Life: Our Responsibility

You are living in the midst of a revolution in our understanding of life. As this revolution progresses, it is changing both the face of human society and the relationship between our species and the rest of the living world. Never has our understanding of the living world been in such constant ferment. Rarely—if ever—in history have our horizons expanded with such dizzying speed. And never has it been so vitally important for *everyone,* scientist and nonscientist alike, to understand both the power and the responsibility conferred on us by our new knowledge of life's phenomena.

Although these might seem like overstatements, they are not; the practical applications of biology span the spectrum of human concerns from the battle against AIDS, through debates about the uniqueness of the human mind, to actions that affect the future of all life on earth (Figure 1.1). If you doubt the truth of these statements, take a moment to make a list of the most serious problems facing you as an individual and as a global citizen today. Consider the sorts of events and processes that have the potential to either shorten your life and the lives of others.

Chances are that any list you make contains drug and alcohol abuse, smoking and lung disease, AIDS, cancer, heart disease, air pollution, and disposal of toxic wastes. Now think about these problems briefly. All of them fall within the province of either biology or medicine. All of them involve social, ethical, and moral considerations along with scientific ones. And all are laden with emotion and suffused with strongly held, often irrational public attitudes.

THE BIOLOGICAL REVOLUTION

"But seriously," you might ask, "has biology really changed that much? And even if it has, does any of this affect me?"

Biology has indeed changed. And yes, it affects and will continue to affect your daily life and future in many ways.

Even a few decades ago, biology was primarily a descriptive science that aspired mainly to examine, characterize, and name the diverse living things that inhabit the earth. Our planet seemed a vast and inexhaustible treasure trove of new and exotic species, unexplored wilderness, and abundant natural resources. And though researchers worked hard and long, both in the field and in the laboratory, they could describe the things they saw only in fairly superficial terms, for they were unable to fathom the invisible biochemical processes behind life's mysteries.

Furthermore, although it might seem inconceivable to you as a child of the 1970s, medicine had very few real remedies as recently as 50 years ago. In those days, doctors—like biologists in general—were but observers; though trained to diagnose and predict the course of disease, they were usually helpless in the face of serious illness. Treatment was, by default, the least important part of the medical school curriculum, and hospitals—far from being treatment facilities where people went to be cured—were little more than observation wards where poor people went to die.

But today all that has changed, and it continues to change at breakneck speed. Part of this revolution has been launched by breakthroughs in our understanding of life's tiniest particles, the molecules that make up living tissue. On this submicroscopic frontier, our challenge is to understand how genes dictate the creation of our cells,

Figure 1.1 Science affects modern life on a daily basis. (left) In the wake of accidents at the Three Mile Island and Chernobyl nuclear power plants, many Americans are confused and concerned about the safety of nuclear power. (right) This great quilt, each square of which commemorates a casualty of AIDS, is one of the more visible indicators of the toll this disease has exacted. Yet more than a decade into the epidemic, most Americans are still tragically misinformed about AIDS, and vital information is not adequately disseminated—primarily for political (rather than scientific) reasons.

how they direct those cells to form bones in our legs and thoughts in our brains, how they change with age, and how they evolve from generation to generation. Tied in to this research, the fledgling biotechnology industry promises to revolutionize the way we grow our food, the way we are born, the way we fight disease and illness, and the way we die.

At the other extreme of life's phenomena, on a scale so vast that it encompasses our entire planet, our challenge is to protect the global environments that give us life. Human actions ranging from the destruction of global forests to the burning of coal and oil may be causing a planetwide warming trend whose long-term effects we can scarcely comprehend. Other human activities are destroying the earth's protective ozone layer. And still others are threatening the supplies of clean air, water, and food that are vital to the constantly growing human population.

The biological revolution has already begun to bestow on us the kind of power over ourselves and our surroundings that just a decade ago existed only in the realm of science fiction. But the promise of power carries with it the responsibility to use that power wisely. Do we know enough to create new forms of life in the laboratory safely? Is it worth opening an ethical Pandora's Box if by altering human genes we can cure inherited diseases? How can we weigh moral concerns about experimentation on human embryos against the potentially life-saving benefits of those procedures? What can be done to minimize the impact of human society on the global environment? All these issues demand the attention of scientifically informed citizens.

We cannot hope—and, indeed, would not presume—to provide definitive answers to many of these questions in this book. We do hope, however, to provide you with enough basic information about life and the science of life to enable you to make rational judgments on biological matters that concern you. As first steps in that direction, this chapter and the next will introduce you to three basic themes that recur throughout the book: the nature of life, the nature of science, and the relationship of science to society.

Figure 1.2 A NASA photograph showing the spectacle of the entire planet Earth— a scene astronauts have described as both breathtakingly beautiful and spiritually uplifting.

Figure 1.3 Living cells display a wide diversity of structures and adaptations. (left) Macrophage: a key player in the defense against infection and disease as we will see in Chapter 39. (center) Sperm: a starfish sperm on the egg surface. (right) Protozoan: *Actinophrys sol*, one of the many varieties of single-celled organisms.

LIFE: A GLOBAL PHENOMENON

From space, Earth is a dazzling sphere colored blue by oceans, green by plants, and white by ice and clouds (Figure 1.2). Against the stark background of interplanetary space, our planet hovers in striking contrast to the desolate globes that share our sun; warm, moist, and inviting, the earth is very much alive. This awe-inspiring view allows us to appreciate life as a global phenomenon, for it reveals the entire **biosphere,** the majestic planetwide network of Earth's physical environments and living plants and animals.

That biosphere is composed of countless living systems that encompass many levels of organization, span an enormous range of sizes, and operate on time scales ranging from infinitesimally small fractions of a second to many millions of years. Yet all living things share certain traits: They take in and process materials from their surroundings, they grow and reproduce, and they respond to events around them. Over long periods of time, living things often change, or *evolve,* in ways that produce new varieties. A salmon, an orchid plant, a monkey, and a mosquito are all alive by these criteria; as easily recognizable assemblages of active, interdependent living parts, they are called *organisms.*

Organisms, in turn, are composed of *cells;* the basic units of living things. Cells are the smallest units of any organism that can actually be considered *alive* in the fullest sense of the word, for they can grow, reproduce, respond to their surroundings, and adapt to change. Though all cells share many properties, they are extraordinarily diverse; the human body alone is made up of at least 85 completely different cell types. Elsewhere, cells exist in an almost infinite variety of sizes, shapes, and colors (Figure 1.3). A great many organisms, in fact, consist of nothing more than a single living cell.

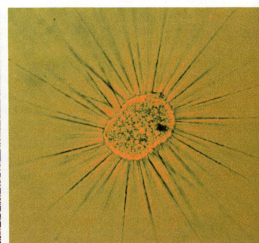

If we restrict our gaze for a moment, however, we can make a stab at describing a typical cell (Figure 1.4). Such a cell would be about 20 microns (2.0×10^{-5} meters) in diameter. It would have a prominent, centrally placed structure about 6 microns in diameter that we call a **nucleus** after a Latin word meaning "kernel," because it looks like a seed in the center of a fruit. It would be surrounded by a *cell membrane* that separates it from its surroundings. And it would contain many tiny structures called *organelles* that function much like miniature organs.

In a sense, every living cell is a storehouse of the very information about the nature of life that biologists have always sought. If we take a handful of cells and dissolve them in detergent, we release a clear, sticky substance that can be twisted like caramel around a glass rod. This is *deoxyribonucleic acid,* or *DNA,* the molecule of heredity (Figure 1.5). Each DNA molecule is a double string of billions of atoms coiled up tightly into a beautiful and elegant spiral chain. Inscribed on that chain is a wondrous code, a linear sequence of four simple molecules that carries the information necessary to build and operate a fly, an oak tree, or a human being. It is a code that was conceived along with life itself nearly three and a half billion years ago, and a code that we are just beginning to decipher.

From DNA to Biosphere: Spaceship Earth? Or a Living Planet?

Making the conceptual leap between the biosphere and DNA is no mean feat. And as we concentrate on various levels of biological organization later in the book, it will be easy to lose track of the connections between the DNA molecules that direct life's processes and the planet that houses us all. So while we still have in mind both these extremes of life's processes, let us look briefly at two different views of the connections between life and our planet.

Years ago, Buckminster Fuller coined the term *spaceship Earth* (Figure 1.6) to suggest that our planet is a giant spaceship on whose life-support systems all organisms depend. Expanding Fuller's concept into a model of planetary function, many scientists believe Earth operates as a sort of giant orbiting geochemical machine whose workings fortuitously sustain life but whose processes are not controlled by life. In this view, living organisms are simply "passengers" on a planet whose environment they may affect but do not control.

Other scientists postulate that physical and chemical conditions on Earth have been, and continue to be, shaped by life's processes. These scientists do not deny that living organisms adapt over time to changing conditions around them. They simply argue that as life on Earth has evolved, it has brought about major changes in global conditions that would not have occurred otherwise. In a sense, these

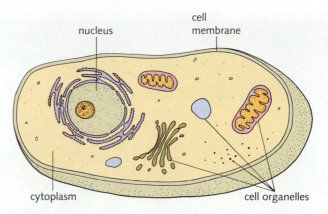

Figure 1.4 The "typical" animal cell exhibits familiar structures common to most multicellular animals.

Figure 1.5 A computer graphic reconstruction shows the beauty and the complexity of the extraordinary molecule called DNA.

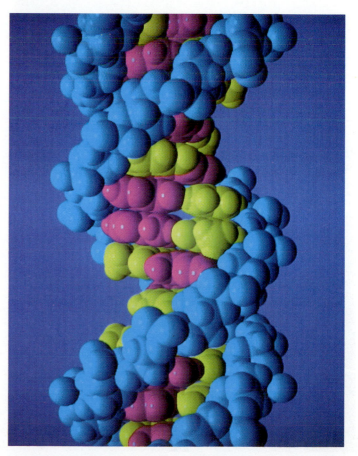

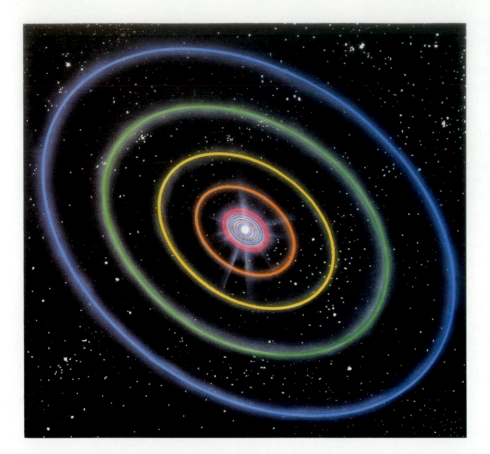

Figure 1.6 A view of Earth within the solar system allows us to see our planet as we know it really is. No longer infinite and inexhaustible, Earth is clearly a finite globe hurtling through space. In that respect it can be compared to a giant spaceship.

researchers believe that those invisible DNA molecules contain information that directs not only the lives of individual organisms but also transformations that encompass our entire planet. As farfetched as this idea might seem to you now, you will learn over the next several chapters that even single-celled organisms (in sufficient numbers) have had profound effects on both the face of the earth and its atmosphere.

Some researchers go even further, maintaining that the biosphere is not just a collection of plants and animals but a single, global "superorganism" whose components have evolved in ways that help maintain the delicate balance necessary for planetary life. This view of life has been termed the **Gaia hypothesis** after the earth goddess of the ancient Greeks (see Current Controversies, opposite page).

Keep these two points of view in mind throughout this course. For if you believe that we are simply passengers on a planet governed by nonliving factors, you can infer that the activities of living organisms, including humans, have little effect on that planet. If you grant, on the other hand, that life's activities have a major influence on global conditions, you are forced to acknowledge that human actions may have planetwide consequences.

THE NATURE OF SCIENCE

Most of this book deals exclusively with science itself, with the study of living systems on many levels. But the scientific world does not exist in a vacuum; it affects, and is affected by, the complex fabric of human society in many ways. Although these interactions are not always obvious, they determine today—just as they have determined throughout history—not only the *kind* of scientific research that takes place but also the way in which that research is *interpreted* and *applied*. To appreciate fully this subtle yet crucial interplay between science and society, we must appreciate both the nature of science as a human endeavor and its relationship to the psychological and social forces that shape our culture.

Science and Religion

Contemplating the living world today, with generations of scholarship behind us, we are still fascinated by natural phenomena. We gawk in amazement at elephants and hummingbirds, stand in awe of devastating hurricanes, and marvel each year at the inexorable changing of the

The Gaia Hypothesis

James Lovelock, a British biochemist, and his American colleague Lynn Margulis do not believe that long-term constancy in Earth's temperature and the life-sustaining levels of oxygen in the atmosphere have arisen incidentally.

Life's influence on planetary conditions struck Lovelock as so powerful and so precisely regulated that he proposed the *Gaia hypothesis*. According to Lovelock, "the physical and chemical condition of the earth's surface, of the atmosphere, and of the oceans has been and is actively made fit and comfortable by the presence of life itself."

His hypothesis rests on evidence that living organisms *interact* with, and powerfully affect, Earth's atmosphere and geochemical cycles. Lovelock goes further, however; he proposes that all life on Earth has evolved into a global superorganism—Gaia—whose parts monitor and manipulate carbon dioxide concentration, oxygen levels, and other environmental parameters. He argues that only this active monitoring and correction keep global conditions within the narrow margins essential for life.

In this view, Gaia's atmosphere and oceans act like a global circulatory system, carrying compounds across the globe and dumping them where nec-

essary. Plants and animals living and dead process and store such critical compounds as oxygen and carbon dioxide, releasing them as necessary to control Earth's temperature and atmospheric composition.

Homeostasis is the term physiologists use to describe an organism's maintenance of stable internal conditions in the face of a changing external environment. Lovelock was bold enough to propose that this global superorganism has actively maintained planetary homeostasis over the three and a half billion years it has been alive.

The Gaia hypothesis is controversial, but it is extremely useful as an alternative point of view. It reminds us that though we tend to think of Earth as stable, we as living organisms have altered it in many ways. As we will see in Chapter 7, concentrations of carbon dioxide and ozone in the atmosphere are changing, undoubtedly as a result of human activities. What will come of those changes no one knows. In the meantime, Lovelock's ideas can help us view the evolution of life and the physical evolution of Earth not as two separate series of events, but as a single, tightly integrated process.

---

seasons. Small wonder, then, that ever since our species became human, we have pondered the origin of living things and the forces of nature.

At first, people explained these phenomena with tales of nature gods and goddesses, of magic, and of witchcraft. Then, over time, various human interpretations of the world and the forces that guide it evolved and converged into two lines of thought—science and religion—

that represent humanity's two main efforts to put human life and the natural world in perspective. Although science and religion are often portrayed as conflicting, they need not contradict each other, for their aspirations and emphasis are as different as night and day.

Typically, formal religions are organized around a series of dogmas, usually based on divine revelation, that are often not open to interpretation and revision. Orthodox

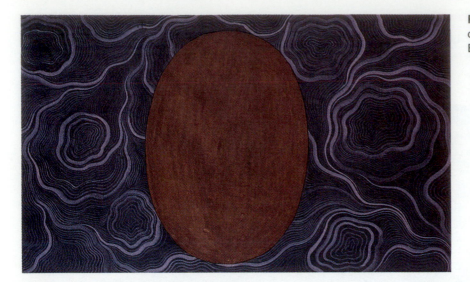

Figure 1.7 A 1730 representation of the Golden Egg from which Brahma appeared.

religion, especially, is often based on trust in inviolable scriptures and depends on unquestioning belief. In most Western societies today, religious leaders see religion's primary purpose as providing the faithful with moral guidance, spiritual meaning for existence, and rituals to help them through high and low points in their lives. Inasmuch as human judgments on personal conduct, morals, and ethics often coincide, the world's great religions can often find common ground, as interfaith councils affirm.

Religious explanations of natural phenomena, in contrast, have always varied extensively from culture to culture. According to the Incas, the sun god placed the first humans on two small islands in mountaintop Lake Titicaca. The Chewa tribe of Malawi credits a lonely chameleon with giving birth to all other forms of animal life. The Hindu Manava Sastra describes how God dispelled the darkness with an egg as brilliant as the sun, from which Brahma, the creator of all thinking beings, emerged (Figure 1.7). And according to James Ussher, archbishop of Armagh and a distinguished Irish scholar of the 1600s, the earth was created at 9 A.M. on Sunday the 23rd of October in the year 4004 B.C., and Adam and Eve were evicted from Eden 18 days later. The accuracy of these stories cannot be tested in any practical sense, but each has been accepted within its own culture, and each serves its purpose within its faith.

Science as a Way of Knowing

Science, on the other hand, deals exclusively with the natural world and utilizes a completely different way of explaining natural events. The main concerns of modern science were eloquently summarized by Francisco Ayala, a leading biologist, as follows:

1. Science tries to collect and organize information about the world in a systematic way. While engaged in this process, scientists constantly look for recurring patterns and relationships among events and processes.
2. Science attempts to explain observed events and processes in terms of natural phenomena that can themselves be observed or demonstrated in a scientific fashion.
3. The explanations of events proposed by scientists—called *hypotheses*—must be *testable*. This means that any explanation may be altered or rejected if it no longer fits all the available evidence.
4. Scientists continually attempt to construct powerful hypotheses that satisfactorily explain a large number of observed events.

Most important, science is neither dogma nor an assemblage of immutable facts. Science is, rather, a "way of knowing"; it is an ongoing process of observing the world around us, forming ideas about how that world operates, conducting tests of those ideas, and continually revising our conclusions. It requires both the ability to place events in perspective across time and distance and the talent to combine the unpredictable power of the human mind with the precision of meticulous measurement. When scientists attempt to explain "why" something happens, that explanation concerns only the actions of natural forces that are themselves understandable in scientific terms. Although biology endeavors to explain what life is, how life began, and how humans evolved, it cannot address issues such as why life exists, what the meaning of life is, or what the purpose of human effort in the world should be.

Scientific concerns are thus completely separate and distinct from (rather than antagonistic to) the goals of most religions. Many leading scientists throughout history have been devoutly religious. Generations of these scientists have had no problem believing that a god who controls the universe with immutable, physical laws is just as awe-inspiring and worthy of worship as one who governs by intervening through the use of supernatural power.

Whenever you evaluate scientific work, remember these differences. Science and scientists should *never* be approached with the faith better reserved for religion. Scientific ideas are *always* open to question. Controversy and constant change lie at the heart of the scientific process.

Science as a Mirror of Society

Science is a very human endeavor, and as such it is subject to both the advantages and the pitfalls of the human condition. On the positive side, science benefits from majestic leaps of intuition that are the hallmark of human thought. From the truly revolutionary ideas of Copernicus, who first understood the sun to be the center of the solar system, to the discovery of the structure of DNA, which illuminated the mechanism of heredity, science has depended on the extraordinary power of human creativity.

But precisely *because* scientific thought is so abstract and creative, it can be colored by the nature of the mind that creates it. Scientists are subject to emotions, stubbornly held opinions, pride, and prejudices, just like everyone else. The numbers gathered by honest scientists in quantitative experiments are seldom open to question. The important thing, however, is not just what *numbers* scientists gather but what *inferences* they draw from those numbers. And while researchers are drawing those inferences, prejudice can cloud clear thinking.

The latter half of the nineteenth century offers many examples of the effects of prejudice on science, for it was a time of great social upheaval in Western civilization. In the midst of this intellectual and philosophical ferment in Europe, several white, male researchers employed the scientific method—consciously or unconsciously—to justify their own prejudices and reinforce their privileged positions in the fragile status quo.

One branch of contemporary "science" called craniometry, for example, claimed to evaluate intelligence and general mental superiority by examining the sizes and shapes of human skulls and brains (Figure 1.8). One group of craniometers led by Paul Broca, a respected professor of surgery at the Faculty of Medicine in Paris, tried to prove that women and the members of all nonwhite races were mentally inferior to Caucasian men. Today we easily dismiss these discredited assertions of Broca and his colleagues as groundless pseudoscience. In their day, however, the "scientific" authority of Broca's school lent these highly prejudiced statements a credibility not merited by

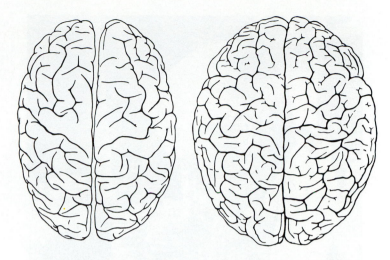

Figure 1.8 The renowned mathematician K. F. Gauss's brain (right) weighed 1,492 grams, only slightly larger than the average brain (left). This dispelled a number of theories regarding brain size and intelligence. Gauss's brain did, however, prove to be more convoluted than the average brain.

the flimsy data on which they were ostensibly based. And it does not take much imagination to extrapolate such "findings" to exclusionary racist policies regarding education, employment, and equal protection under the law.

This is not to say that all (or even many) scientists either have an ax to grind or are sloppy thinkers. But the influence of preexisting ideas on scientific thought is often subtle and unpredictable, and the best researchers take extra care in their experimental design to eliminate any effects of their unavoidable biases.

These precautions are necessary because our view of the world is always based on limited information, as you will learn in detail in Chapter 41. Sometimes, when our senses gather insufficient information for us to make a judgment, the brain "manufactures" additional details on the basis of our expectations. How many times have you been certain that you heard a familiar voice or caught a glimpse of a familiar face, only to find out that you were mistaken? At other times, our brains selectively pay attention to certain bits of information and ignore others on the basis of prior expectations. Have you ever searched endlessly for a book or package that was right in front of you because you thought it was green when it was actually blue? If they are not extremely careful, the same sort of thing can happen to scientists while they are gathering and analyzing data. Depending on what they *expect* to find, researchers may collect different data and may interpret the same sets of data in different ways.

Figure 1.9 Alexander Fleming working in his laboratory.

The important point of this discussion is that prejudices, societal attitudes, and researchers' expectations affect the process of science today just as surely as they did in Broca's time. Because we are part of the society that generates those attitudes, however, we may not see their effects on scientific judgment as readily as we can spot mistakes of the past. For that reason, it is dangerous to place blind faith in the judgment of any single "expert" on scientifically grounded matters of public health or environmental policy. In such matters, science improperly interpreted, accepted with unquestioning faith, or applied incautiously can be useless or even dangerous. That's why, in a democracy such as ours, the public *must* understand science sufficiently to weigh various arguments and make reasoned judgments at election time.

AT THE HEART OF SCIENCE: THE SCIENTIFIC METHOD

The power of science resides in its ability to organize individual observations into a logical system that explains past events and predicts the outcome of current phenomena. In the words of Preston Cloud, "A subject becomes a science when it is brought within the bounds of natural law and a body of ordering hypotheses that makes its hypotheses testable—where evidence takes priority over authority and revelation." To accomplish that goal, science proceeds not with faith but with a series of ordered steps called the **scientific method.** In reality, science depends on much more than such "cookbook" methodology. Yet examining the idealized form of the scientific method provides vital insights into the nature of science.

Observation

Every scientific investigation begins with *observations*. In some cases, these observations are made over a long period; Charles Darwin spent years traveling around the world and taking note of natural phenomena. Other observations may occur in an instant, as did Alexander Fleming's discovery of penicillin. Fleming noticed one day that one of his culture dishes of bacteria was contaminated with a fungus. He also noticed that there were no bacteria growing around the fungal colony. He immediately realized that the fungus might be producing a substance that killed bacteria (Figure 1.9).

Of course, new and exciting discoveries in science are not generated by people who simply note facts, but rather by people with a special gift for combining observation with creative thinking. (Fleming, for example, could have said "Damn! Another contaminated culture!" and thrown the fateful dish away.) A good researcher is able, as the philosopher Schopenhauer pointed out, "to think something that nobody has thought yet, while looking at something that everybody sees."

Forming an Hypothesis

If a series of observations is reproducible, scientists invoke **inductive reasoning** to generate an **hypothesis.** In reasoning inductively, scientists work from observations of *particular* events to formulate *general principles*.

A simple hypothesis may propose a cause-and-effect relationship to explain a series of observations, although most hypotheses make general predictions that extend beyond the original events. A more complex hypothesis

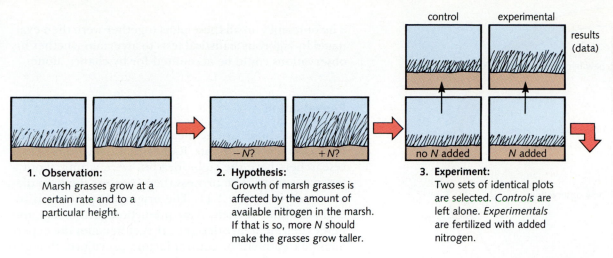

1. **Observation:**
Marsh grasses grow at a certain rate and to a particular height.

2. **Hypothesis:**
Growth of marsh grasses is affected by the amount of available nitrogen in the marsh. If that is so, more *N* should make the grasses grow taller.

3. **Experiment:**
Two sets of identical plots are selected. *Controls* are left alone. *Experimentals* are fertilized with added nitrogen.

may propose a *model* of the way a complex system operates. Darwin's early hypotheses about evolutionary change created such a model to explain the changes he had observed to occur in plants and animals over time.

Experimentation

The next step is to employ **deductive reasoning** to make specific new predictions from the general statements of the hypothesis and from other general principles assumed or believed to be true. These predictions are then tested by performing a series of **controlled experiments.**

In controlled experiments, scientists perform two sets of parallel trials. In one set of trials they vary a single factor, which is called the **experimental variable.** This is known as the *experimental* set. The other set of trials, known as the **control,** is subjected to all the same conditions *except* the changes in the experimental variable. Whenever possible, scientists design such experiments to provide *quantitative data*—results that can be described in numerical form and can thus be subjected to statistical analysis. Designing experiments in this manner helps ensure that any effect observed in the experiment is due to the factor or factors under investigation.

While Ivan Valiela was studying the ecology of a New England salt marsh, for example, his observations led him to hypothesize that the growth of marsh grass was limited by the availability of nitrogen, a nutrient necessary for plant growth (Figure 1.10). A straightforward prediction based on this hypothesis is that marsh grasses will grow faster or larger (or both) when more nitrogen is available.

To test this hypothesis, Valiela chose several marsh plots that were as similar to one another as possible in

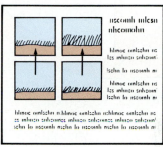

4. **Observation:**
Fertilized plots grow taller and more rapidly

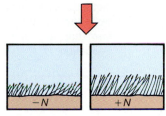

5. **Interpretation and conclusion:**
Growth of marsh grasses is affected by *N*.

Figure 1.10 A diagrammatic representation of a well-designed experiment that examined the effects of nitrogen on the growth of salt marsh grasses.

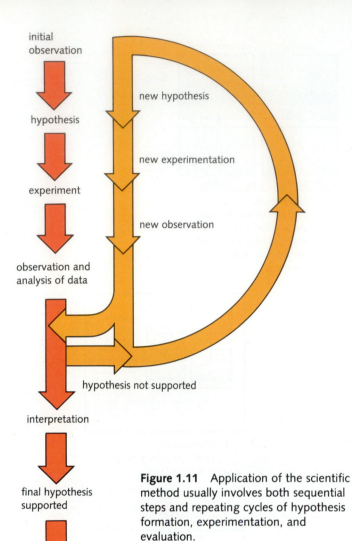

initial
observation

hypothesis

experiment

observation and
analysis of data

new hypothesis

new experimentation

new observation

hypothesis not supported

interpretation

final hypothesis
supported

theory

Figure 1.11 Application of the scientific method usually involves both sequential steps and repeating cycles of hypothesis formation, experimentation, and evaluation.

plant density, soil type, freshwater input, and height above the average tide level. In a series of *experimental* plots, he added nitrogen fertilizer, thus manipulating the variable that was the subject of his hypothesis. Other plots, selected as *controls*, were exposed to all the same environmental conditions as the experimental plots *except* for the addition of nitrogen.

Throughout the growing season, Valiela sampled the plots to measure growth rates, chemical composition of the leaves, decay rates of dead leaves, and so on. By statistically comparing the results of these tests, he determined—among other things—that several important marsh grasses grew taller and larger than controls when they were given additional nitrogen. The experiments were checked and the results replicated. The hypothesis was supported.

Note that a single, isolated experiment is never sufficient; scientists must be able to *reproduce* or *replicate* their observations to take them seriously. Valiela used several control and experimental plots, rather than a single pair.

The outcomes on all these plots together were then evaluated by rigorous statistical tests to ascertain whether his observations could be accounted for by chance alone.

Reevaluation

Often hypotheses are neither supported nor discredited by one set of experiments. Rather, new data indicate that researchers have the right idea but were wrong about a few particulars. The process then reenters the loop diagrammed in Figure 1.11. The original hypothesis is reevaluated and revised. New predictions are made, and new experiments are designed that either alter the experimental treatment or control factors previously thought extraneous. Many circuits around this loop are often necessary before the final hypothesis is supported.

Scientific Theories

Sometimes an hypothesis will continue to expand in complexity and grow in the power and accuracy of its predictions until it becomes worthy of the name **theory.** Charles Darwin's original sets of observations and hypotheses about the processes that shape the evolution of species grew and expanded for several years before he presented it to the world as the *theory of evolution by natural selection.*

Note that the meaning of the word *theory* as it is used in science is very different from its meaning as you might use the word in daily life. When you say, "I have a theory," you mean something like "I have a hunch." But when scientists talk about gravitational theory or evolutionary theory, they are referring to a long-established body of observations and experimental proofs.

THE SCIENTIFIC METHOD IN THE PUBLIC EYE: CASE STUDIES

Applying the scientific method to noncontroversial events in nature can be straightforward. But today, scientists are often called upon to pass judgments on matters of intense emotional or financial concern. People with cancer or AIDS turn desperately to medical researchers demanding to know which treatments will save their lives. Governments, corporations, and individuals inquire with equal intensity about environmental matters: How much waste treatment is enough? Will another coastal housing development destroy local shellfish beds? And advertisements for everything from toothpaste to thermal underwear claim that the quality of their products has been "scientifically proven."

Whenever human lives or large sums of money depend on experimental outcomes, the design and evaluation of controlled experiments are much more difficult. In such cases it may be impossible to devise an experiment that satisfies everyone. Because experimentation stands at the heart of science, we will take the time here to present two very different examples of experiments. For each we will present the hypothesis being tested, background information on the investigation, some details of experimental design, and a final evaluation.

Medication for Bacterial Infections in Aquarium Fishes

Hypothesis "Brand X" antibiotic is effective in treating certain diseases of ornamental aquarium fishes.

Background Fishes in home aquariums often get sick, so hobbyists are always looking for medications to treat fish diseases. Because the term *scientifically proven* carries great weight with consumers, companies scramble to put it on their product labels. There is, however, no equivalent of the Food and Drug Administration (FDA) to set standards for "proving" the medical value of aquarium medications, and *Consumer Reports* has thus far not evaluated these products either. As a result, pet supply manufacturers are left on their own to test, evaluate, and advertise their products. With no set standard for the industry, experiments may or may not be competently designed.

Experimental design The manufacturer of "Brand X" aquarium antibiotic hired two independent academic researchers to conduct experimental tests of drug action on four bacterial diseases of fishes. Thirty-six aquarium fishes were chosen and observed for a time to make certain that they were healthy and vigorous specimens. Then four types of bacteria (labeled here B1, B2, B3, and B4) were inoculated into 9 fishes each. Disease appeared in all injected individuals after 3 days, but no treatment was instituted until the eighth day to ensure that several animals in each group would be seriously ill. Medication was then administered to all individuals at recommended doses, and the results were followed for 5 days.

Results The results were presented in two forms. Survival of animals is shown in Table 1.1. An additional part of the report contains verbal descriptions of visible changes in the degree of infection of survivors. Those verbal reports generally indicated lessening of disease symptoms, although in only two cases did the animals recover completely before the end of the experiment. The experiment is summarized in a different manner in Figure 1.12.

Table 1.1 *Survival of Fishes After 7-Day Infections and 5-Day Treatment*

	Initially	7 Days After Infection	After 5 Days of Treatment
B1	9	3	3
B2	9	3	3
B3	9	5	3
B4	9	6	6
Total	36	17	15

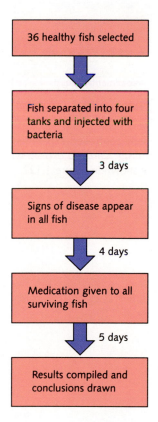

Figure 1.12 A procedural outline of the fish drug experiment, as conducted, summarizes the investigators' reasoning. Do you see any problems with this experimental design?

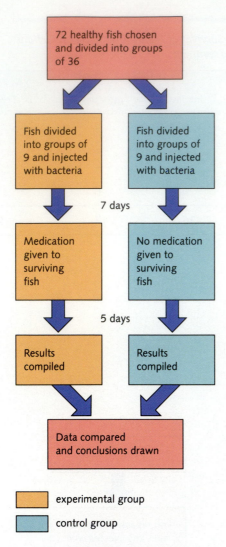

72 healthy fish chosen and divided into groups of 36

Fish divided into groups of 9 and injected with bacteria

Fish divided into groups of 9 and injected with bacteria

7 days

Medication given to surviving fish

No medication given to surviving fish

5 days

Results compiled

Results compiled

Data compared and conclusions drawn

☐ experimental group

☐ control group

Figure 1.13 A properly controlled experiment to test the efficacy of an aquarium drug. Can you explain why this design is preferable to the one shown in Figure 1.12?

Conclusions according to company spokesman Before treatment, fishes were dying at a rate of approximately 2 per day. After treatment began, only 2 died in 5 days. Other surviving fishes were infected but showed significant improvement in condition during treatment. Therefore, Brand X is of significant value in treating these diseases in these fishes.

Possible alternative conclusions Because of the way this experiment was designed, the data it yielded did not exclude the following alternative interpretations of what happened:

1. It is possible that the fishes that survived 7 days of illness were strong enough to have recovered on their own without treatment. Fishes, like humans, have immune systems—bodywide defenses against infection—that may have successfully battled the bacteria. Fish immune systems do, however, usually take about a week to mobilize fully. Thus the infections could have killed physically or genetically weaker individuals before their immune systems sprang into action. Stronger individuals, on the other hand, might have survived without antibiotic treatment.

2. It is possible that other procedures associated with medication (such as water changes) reduced mortality independently of drug action.

Discussion The manufacturer's interpretation of the results follows the logic "The fish had disease; they were treated; they stopped dying of the disease; therefore the medication works." But this sort of "before and after" interpretation is not valid in the absence of properly controlled experiments. In order to prove the efficacy of Brand X, the experiment should have involved twice as many fishes divided into groups, as shown in Figure 1.13. One set (the experimentals) would have been treated just as these were. Another set (the controls) would have been treated exactly the same way until the eighth day. At that point, instead of adding medication to the controls, the researchers would have added what is called a *placebo*: a look-alike substance that contained no antibiotic. Then, by *comparing* survivorship and general health of controls and experimentals, the researchers could have made a judgment about whether the medication had any effect on the course of disease.

Furthermore, the verbal descriptions of surviving fishes should have been made by trained observers *who did not know which animals received medication and which did not*. In this manner, the subjective process of describing the animals' condition could not have been biased by expectations that medicated fish would be better off. Finally, those descriptions should have been evaluated by the head researchers, who (for the same reason) *should not have known which descriptions applied to experimentals and controls until after final evaluation was complete*. This would have constituted a **double-blind study** because it contained two sets of "blind" evaluations that could not have been influenced by anticipated or hoped-for results.

As a result of these two major flaws, had this study been submitted to a scientific journal in which articles are reviewed by other scientists, it would have been laughed off the editors' desks. Unfortunately, the flawed logic illustrated here is accepted by many nonscientists. Experiments not unlike this one are the source of the "scientific evidence" used in marketing a variety of products. At the same time, people who are in the grip of serious disease often accept this sort of reasoning and rush to purchase "miracle" cures such as the widely touted but worthless "cancer drug" laetrile. These unfortunate individuals hear that some people were sick, that they took the medication, and that they later recovered. Although that sequence of events may have occurred, the sequence as described proves nothing about the drug's ability to cure disease.

Clinical Trial of Azidothymidine (AZT) in the Treatment of AIDS

Hypothesis   The drug azidothymidine (AZT) is effective in the treatment of acquired immune deficiency syndrome (AIDS).

Background   AIDS is caused by a virus that cripples the body's ability to fight disease. (We will cover AIDS in detail in Chapter 39.) Most people infected with this virus die of what are called *opportunistic infections,* diseases that the body can normally fight off but that ravage people with AIDS. Once afflicted with full-blown AIDS, patients have little or no hope of survival without treatment. AZT first showed promise by interfering with the reproduction of HIV in test-tube experiments. Preliminary trials in seriously ill patients were not conclusive.

Experimental design   This experiment, sponsored by the manufacturer of AZT, was designed as a double-blind, randomized, placebo-controlled study. Obviously, researchers did not inject healthy individuals with the virus; people already afflicted with AIDS were asked to volunteer for the study. A computer-generated code randomly assigned each subject to receive either AZT or a placebo. Neither the subjects nor the researchers evaluating their condition knew which patients received AZT. The placebo itself was designed to look and taste just like AZT. Subjects were asked not to take any other medications without the permission of the researchers. No other treatment to prevent or control opportunistic infections was permitted to either experimental or control subjects during the course of the study. Data were rigorously analyzed by advanced statistical methods to determine whether differences between experimentals and controls could have been due to chance alone (Figure 1.14).

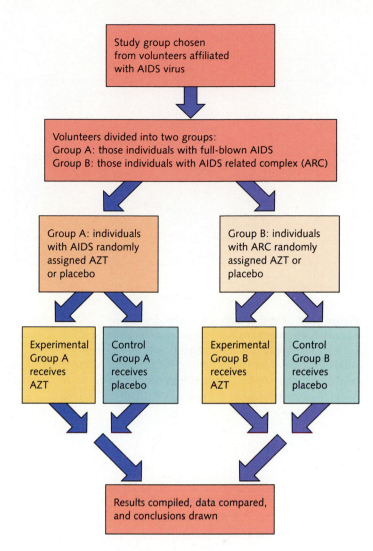

Figure 1.14   Schematic representation of AZT clinical trial design. Because this experiment dealt with human subjects, researchers had no control over the initial selection of individuals infected with HIV. (They could certainly not select healthy individuals and infect them with the virus for the purposes of the study!) Beyond that point, however, patients were diagnosed, divided into two medical categories (AIDS and ARC) and further subdivided into experimentals and controls. Double-blind procedures ensured that neither experimenters' expectations nor those of the patients would affect data gathering or interpretation during the experiment.

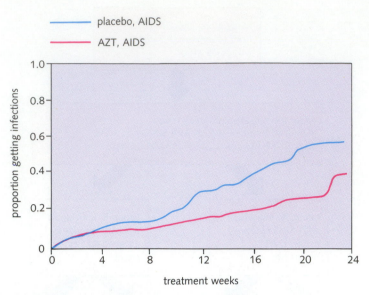

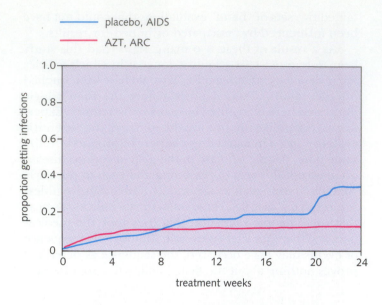

Figure 1.15 Patients with AIDS who developed opportunistic infections during the treatment period. (a) Treatment results among patients with full-blown AIDS. (b) Treatment results among patients with AIDS-related complex (ARC). Adapted from the paper by Fischl *et al.* referenced at the end of the chapter.

Table 1.2 *Projected Probability of 24-Week Survival*

Experimental Group	Survival Rate
AIDS	
AZT	96%
Placebo	76%
AIDS-Related complex (ARC)	
AZT	100%
Placebo	81%

Note: This table was adapted and simplified from Table 1 in the paper by Fischl *et. al.* referenced at the end of the chapter. Those interested in the statistical analyses accompanying these data are encouraged to read the paper in its orignial form.

Ethical considerations weighed heavily in experimental design. The study was approved by committees at each of 12 participating medical centers across the country. An additional independent board was established to review the study data on a regular basis. It was this group's responsibility to protect the subjects if it appeared that the medication was causing unacceptable toxic side effects. This board also kept watch for any sign that the drug was clearly benefiting those who received it. Every subject signed a consent form certifying that he or she understood the risks and benefits of the experimental procedure. (Imagine the courage of these participants; they knew that half of them were voluntarily forgoing treatment that might save or prolong their lives. They did so in order to help prove whether or not the drug worked.)

Results Because many of the data from this experiment deal with complex phenomena that we will not cover until Chapter 39, we will present only the most basic results here. Figure 1.15 (a) and (b) are two graphs showing the number of patients who developed opportunistic infections during the treatment period. Figure 1.15(*a*) shows results among those people with full-blown AIDS; Figure 1.15(*b*) shows similar information for subjects with the less severe AIDS-related complex. Note that in both groups, subjects receiving AZT had a lower rate of opportunistic infections than did those receiving the placebo. Table 1.2 shows that AZT recipients also had significantly higher chances of surviving the 24-week treatment period.

Additional data indicated that AZT caused side effects of varying severity in different subjects. Side effects ranged from muscle aches, insomnia, and nausea to serious anemia and the loss of certain classes of white blood cells.

Figure 1.16 While the short-term effects of a major oil spill—such as the spill from the Exxon Valdez, shown here—are dramatic and easily seen, long term, subtle effects on ecosystems are more difficult to determine. One problem in this regard is that scientists often lack detailed data on the state of the system before the spill, which would function as controls.

Conclusions This study indicated that AZT produced the best results of any medication available for AIDS treatment at the time it was conducted. Researchers therefore recommended the use of AZT in the treatment of AIDS and AIDS-related complex (ARC) in any patients who could tolerate its side effects.

Discussion This study was terminated earlier than planned because emerging data showed clearly that AZT produced positive results. When the study was terminated, all placebo recipients were offered the opportunity to receive AZT. Additional information summarized three months after the study ended credited AZT with a fourfold to sixfold reduction in deaths in the study population over the entire period.

The data from this study, along with other data not discussed here, strongly influenced the continued use of AZT as the drug of choice for treating AIDS through the closing years of the 1980s. These data were also used to support the contention that AZT treatment might be useful for individuals infected with HIV but not yet showing visible symptoms of the disease. Can you see how these data hint at that possibility? What sort of experiment might you design to test that hypothesis specifically? Can you see the problems inherent in actually carrying out such a study?

Science in the Real World: Between a Rock and a Hard Place

As you can see, designing conclusive experiments is far from simple. The first example we examined was poorly designed and, as a result, could neither support nor discredit the hypothesis it was intended to test. The second, though extraordinarily difficult to conduct, was well designed and—despite inevitable shortcomings associated with experiments involving humans—has become accepted as a standard in its field.

There are many other problems in applying the scientific method in the real world. Evaluating the effect of an oil spill, for example, is difficult because in most cases there are insufficient data to provide a "control" (the environment before the spill) for comparison to the "experimental" (the environment after the spill) (Figure 1.16). And returning to the medical field, the personnel and time required for controlled clinical trials have begun to delay the approval of new drugs that might be useful in treating cancer, AIDS, and certain other diseases. Such difficult situations often pose tough questions to those attempting to use science for humanity's benefit. These difficulties are among the reasons why nonscientists need to understand as completely as possible what science is and how it works.

Fischl, Margaret, *et al.* 1987. The efficacy of Azidothymidine in the treatment of patients with AIDS and Aids related complex—A double-blind, placebo-controlled trial. *New England Journal of Medicine,* *317*:185–191.

Richman, Douglas D., *et al.* 1987. The toxicity of Azidothymidine in the treatment of patients with AIDS and Aids related complex—A double-blind, placebo-controlled trial. *New England Journal of Medicine,* *317*:192–197.

SUMMARY

Life encompasses a great range of phenomena, from interactions among invisible molecules to events that affect our entire planet. These phenomena are all interconnected, and all represent the complexity of life. All living things grow, reproduce, take in and process materials from their surroundings, and respond to their environment. At the root of all life is the biological information stored in molecules of DNA.

Throughout history, humans have sought to explain the natural world. The development of science has been one response to that attempt to explain. Scientific knowledge is assembled from direct experience with nature, depends on experimentation, is constrained by the boundaries of our senses and the limits of our imagination, and is never final. The scientific method is an investigation in which a scientist proposes an hypothesis, devises and executes experiments to test that hypothesis, and then interprets the results in light of that hypothesis. Though apparently straightforward, this process is complex and highly subjective and is often colored, therefore, by human limitations such as cultural bias.

STUDY FOCUS

After studying this chapter, you should be able to:

■ Describe some of the ways in which life affects the Earth.
■ Understand the essential attributes of scientific thought.
■ Describe the scientific method.
■ Analyze and, where necessary, criticize experimental results and conclusions.
■ Appreciate some of the effects of cultural values on science.

SELECTED TERMS

biosphere *p. 4*
Gaia hypothesis *p. 6*
scientific method *p. 10*
hypothesis *p. 10*
controlled experiment *p. 11*
experimental variable *p. 11*
control *p. 11*
theory *p. 12*
double-blind study *p. 14*

REVIEW

Discussion Questions

1. Describe the steps in the scientific method. How does the ideal scientific method differ from real scientific experimentation?

2. What are some current issues that involve both biology and questions of ethics or of public policy? Can science offer a definitive solution to any of these issues?

3. Can science ever provide the absolute truth about anything in nature? Why or why not? Give an example of something that was once accepted as a "truth" in any branch of science but has since been proved false.

Objective Questions   (Answers in Appendix)

4. The Gaia Hypothesis states that
   (a) life processes are regulated solely by planetary conditions.
   (b) over millions of years, global conditions have fluctuated greatly.
   (c) global conditions are regulated by life processes.
   (d) living organisms have no influence on global conditions.

5. What does a scientist do with a hypothesis?
   (a) controls it
   (b) changes it to a theory
   (c) refers it to a world authority
   (d) tests it

6. The following steps are part of the scientific method. What is the last step in the scientific method?
   (a) make observations over a long period
   (b) conduct many experiments
   (c) generalize from test results
   (d) collect and organize test results

7. What happens when a hypothesis is neither supported nor discredited by one set of experiments?
   (a) the results are published
   (b) the steps involved in the scientific method are repeated
   (c) the original observations and hypotheses become a theory
   (d) more variables are added to the experiment

CHAPTER 2

Science and Society

When Mark O'Donnell of Holbrook, Massachusetts, was a boy, he and his friends loved to play in the fields they called "the pastures." The pastures seemed to be a children's wilderness paradise, offering several acres to roam around in and all sorts of things to play with. Among the kids' favorite toys were abandoned barrels filled with a green, jelly-like substance they called "moon glob." Mark and his best friend spent many happy hours rolling in empty barrels and tossing handfuls of moon glob at each other.

At age 27, Mark was diagnosed as suffering from an extremely rare cancer of the adrenal glands. He died a year later. Mark's best friend now has Hodgkin's disease, a painless but usually fatal cancer that attacks cells of the immune system. Within a 3-year period, 10 other young people in that same small town—most of whom played with Mark in the pastures—were also diagnosed with cancer. And on one street not too far from Mark's house, a woman died of breast cancer in nearly every other house.

Then, in 1982, an evening news program listed a chemical plant near the pastures as among the most toxic places in the nation. Within minutes, Mark's mother Joanne got a phone call. "That's what got Mark," her friend said. It didn't take long for Joanne O'Donnell to become the town's most outspoken environmentalist, campaigning to close down the plant. In addition to dumping toxic compounds onto the ground, it turned out, the plant had also regularly poured chemicals into a brook that fed the local reservoir.

THE IMPACT OF SCIENCE ON SOCIETY

The point of the Mark O'Donnell story is that although some issues pertaining to science and technology stay neatly tucked away in laboratories or in textbooks, others thrust themselves into our daily lives. In order that we not become pawns in a chess game whose rules we do not understand, all of us must be familiar enough with scientific information to evaluate what is relevant to us. Joanne O'Donnell wasn't trained as an ecologist. She hadn't even given environmental issues much thought before Mark's illness. Nonetheless, because of events around her, she found herself drawn into a tangle of legal and scientific issues surrounding the toxicity and cancer-causing potential of such compounds as dioxin, arsenic, DDT, and chlordane (Figure 2.1).

The O'Donnells' experience, in addition to providing a tragic story about toxic wastes and personal loss, is also a parable for many more interactions between society and scientific matters. Our lives will be far safer and more enjoyable if we can control the way science and technology affect us, rather than turning to science (and to the courts) to seek redress after something has gone awry.

Joanne O'Donnell has been able to win some battles, but only in a limited sense. True, the plant has been closed, and the Environmental Protection Agency has placed the pastures and the plant high on its superfund cleanup list. It is also true that the company that is alleged to have polluted the local reservoir faces several negligence lawsuits stemming from deaths such as Mark's. That corporation, however, has declared itself bankrupt, and its top officials have maneuvered legally to protect their personal assets. But the O'Donnell family's most treasured assets weren't financial; they were each other. And in addition to the loss of Mark, one of Joanne's daughters has lost her spleen, another has had a tumor removed

Figure 2.1 Nearly all industries—from those that make paper to those that produce nuclear power—generate waste products. How dangerous are these wastes? Can they be recycled? If not, can they be disposed of safely? These issues are particularly relevant in the case of low-level radioactive wastes, shown here in a trench near Hanford, Washington.

from her pituitary, and a third seems to be developing a similar tumor. Medical science may be able to help the O'Donnells live more or less normal lives, but not without great emotional hardship and substantial medical bills. Proper attention to the impact of environmental degradation on Holbrook and its people could have avoided a lot of pain and suffering—and could incidentally have saved large amounts of money that must now be spent on doctors, lawyers, and cleaning up the mess.

THE EFFECTS OF SOCIETY ON SCIENCE

Just as science inevitably affects society, the psychological, economic, and philosophical context of society influences science at several steps in the theoretically objective processes of the scientific method. Societal attitudes affect the sorts of questions scientists are interested in, the specific questions they ask as they conduct their investigations, the work recommended for funding by federal granting agencies, the amount of money appropriated by the federal government for those agencies to spend, the funds offered to support research by corporations and foundations, the treatment of experimental results by the media, and the response of the general public to the findings.

If society is somehow "primed" to receive an issue, for example, or if the mass media decide to grease the wheels a bit, even a little scientific data can have an enormous effect on public attitudes and opinions. On the other hand, if society (or a particularly powerful part of

society) doesn't want to hear about an issue, scientists can shout at the top of their lungs with little result.

To give concrete illustrations of typical interactions between science and society over time, we will look at two areas that are very much in the public consciousness today. We will first examine the history of society's reaction to disease as it affected research and public education surrounding the AIDS epidemic. Our goal in presenting this example is to show how societal attitudes can either help or hinder biomedical research, development, and public education. Next we will take a long view of human interactions with local and global environments. Our goal in this case will be to show how the public's perception of ecological "reality"—which has more to do with mass psychology than with fact—has influenced the willingness of national and international politicians to deal with the economic realities of pollution.

SOCIETY, DISEASE, AND MEDICINE: IN TIME OF PLAGUE

Plague. The word has an ominous, biblical ring to it—and with good reason. Diseases have devastated human populations from prehistoric times to the present, and their names are infamous even today: yellow fever, bubonic plague, smallpox, syphilis, leprosy, tuberculosis, AIDS. One plague wiped out nearly a quarter of Europe's population in the fourteenth century. Another, in 1970, killed 10,000 people in the Indian State of Bihar alone. And by 1989, the number of officially diagnosed AIDS cases in the United States surpassed 100,000, with 60,000 of the infected already dead.

In every part of the globe, in every culture, people have lived with disease. As it is human nature to do, they have always tried their best to make that suffering an intelligible part of the world around them. For most of history, and in most places around the world, religion provided the glue that held the world together. In Asia, where smallpox was an all-too-common fact of life, parents knew that if sick children recovered they would be safe from future attacks. Therefore they prayed to a smallpox goddess, such as Jyeshtha (Figure 2.2), to spare their sons and daughters.

In Europe, too, diseases were mysterious scourges for which people blamed unknown poisons, bad air, or divine retribution. Disease was often seen as God's punishment for sin or for the aberrant behaviors of the poor, minority groups, or foreigners. Epidemics were nurtured as much by ignorance and fear as by crowding and poor sanitation. The primary reactions to early outbreaks of bubonic plague in Europe, for example, were panic, fear, and denial, as people sought desperately for someone to blame. Many unfortunate scapegoats—almost invariably outside the mainstream of society—were tortured and executed rather than treated (Figure 2.3). By finding a cause for disease in the actions of individuals viewed as "different," people accomplished two goals: They gave themselves something "positive" to do (persecuting the suffering) in a situation in which they were otherwise helpless, and they offered one another a psychological defense

Figure 2.2 By praying to gods and goddesses such as Jyeshtha, an Indian smallpox goddess depicted here in a stone carving from the 8th or 9th century, many people around the world sought protection from the ravages of disease.

Figure 2.3 In Europe, diseases were often seen, not only as divine actions, but specifically as punishments for wrongdoing against the laws of God and man. As a result, public response to infection was often less than enlightened. This scene shows plague sufferers in the Italian city of Milan being tortured and executed.

against fear by assuring themselves that only "that other sort of person" would get ill.

In and amongst the fear and superstition, however, useful medical knowledge began to accumulate. By the eighteenth century, many physicians suspected that certain diseases were contagious, but the mechanisms of illness were still obscure. Not until the late nineteenth century was there a fundamental change in the way people perceived disease. Finally, it was generally realized that specific microorganisms could make people sick and that proper treatment could cure many ailments. The twentieth century saw the rise of modern medicine and with it the assumption that most diseases could be cured (or at least controlled) by scientific means. In industrialized countries, ignorance, fear, and superstition gave way to confidence in physicians' power to cure and prevent infections. Response to most diseases became fairly straightforward: Identify the problem, determine how the disease is spread, find a way to relieve symptoms, search for a cure, and educate the public on how to avoid infection. Interestingly, according to many health professionals, programs of rational public health education have been as important as, or even more important than, medical advances in curbing the destruction caused by epidemics. In other words, telling people how to avoid disease is invariably more effective (and less costly) than attempting to cure the sick.

One would have thought that by the last quarter of the twentieth century, any new disease would be attacked with the same calm resolve, scientific determination, and public education that had triumphed over so many killers of the past. That comfortable self-assurance was destroyed early in the 1980s by a disease called AIDS.

AIDS: Society's Reaction Makes a Bad Situation Worse

Acquired immune deficiency syndrome, or AIDS (described in detail in Chapter 39), set the American medical community on its ear when it surfaced here in the early 1980s. The first cases were a baffling collection of strange ailments—a previously rare form of cancer and several fatal infections from microorganisms that had almost never been known to cause disease in healthy people. The first thing about the disease that became clear was its mode of action; somehow it crippled the immune systems of these patients, and left their bodies helpless against a wide range of ailments. The disease first appeared among members of the gay community, in which it began to spread rapidly. Soon, cases began appearing among intravenous (IV) drug users and their sexual partners.

Many medical researchers and science journalists note that the challenge of this new and apparently fatal disease brought out both the best and the worst in modern society. Within three years, researchers not only determined that AIDS was caused by an infectious agent, but also isolated and identified the previously unknown virus responsible for the disease. Given the difficulties of locating this particular sort of virus, that speed is little short of astonishing. The history of these findings is a modern-day scientific detective story that is both fascinating and inspiring.

Epidemiologists conducted studies among the growing number of gay and IV-drug-using AIDS patients. They suggested by the fall of 1981 that AIDS was a contagious disease that might be spread through blood (via the sharing of contaminated needles) and through sexual intercourse. By 1982, the first cases of AIDS were reported among infants born to infected mothers and among hemophiliacs and surgical patients who had received transfusions of blood or blood products (Figure 2.4). Those physicians closest to the data became convinced that AIDS was caused by an infectious agent contained in semen and contaminated blood. They soon determined that this agent could be spread from one person to another not by casual contact, but by only three means: through sexual intercourse, through injection (or transfusion) of contaminated blood, and from infected mother to unborn child.

In this respect, an important part of the scientific community responded admirably—even courageously—to the challenge of AIDS. In recent years, new drugs and medications have been developed at an astonishing rate and rushed through testing procedures to help prolong the lives of people afflicted with AIDS. One researcher attempting to develop a vaccine for the disease showed his boldness and dedication by experimenting on his own body; he injected himself with a test vaccine.

Serious problems arose, however, in translating this rapidly increasing body of knowledge about AIDS into the sort of public health education programs that have been so important in controlling epidemics of the past. Large portions of American society—including certain physicians, researchers, journalists, and members of the general public—were unable to face this lethal, and (at least temporarily) incurable disease in a rational manner. Instead, the same sort of fear, ignorance, and denial that had driven public responses to plagues of old fueled a return to prejudice and superstition. Minority groups (gay men and intravenous drug users) and outsiders (Haitians) were singled out for blame. The mass media and many otherwise responsible physicians alternated between denial that AIDS was anything more than a "gay" disease and hysterical headlines warning that "Now, no one is safe."

These reactions hampered our ability to understand and control the disease until thousands of people had died and hundreds of thousands had been infected. Many in the medical community were shockingly slow to react, and a few refused to treat people with AIDS. Certain powerful individuals in government blocked both the release of funds for AIDS research and effective public health actions in the first critical years of the decade. In 1983, nearly a year after the first cases of transfusion-related AIDS were reported and confirmed, those who administered the nation's blood banks still denied that there was enough risk to merit screening donated blood. As a result of this denial, which was based on the opinion that the evidence in hand did not yet merit the additional expense that screening required, an as-yet-unknown number of hemophiliacs and other transfusion recipients who received blood during the first half of the decade have been infected with the AIDS virus. Because the disease may take more than a decade to cause visible problems, these people and their sexual partners and newborns are at high risk for contracting the disease.

As Dickens wrote about another difficult period in human history, "It was the best of times, it was the worst of times." In one small town, a 9-year-old boy with AIDS was a rallying point for support from his community and classmates, who welcomed him to school and raised thousands of dollars for AIDS research. Yet in another small town, two hemophiliac children who had contracted AIDS from blood transfusions were barred from school. When their parents fought to have the children readmitted to school, extremists set fire to the family's house in a desperate effort to drive them out of town.

This sort of hysterical reaction served the same psychological purposes as did public reactions to plagues during the ostensibly less-well-informed Middle Ages: By insisting that the disease is caused by and restricted to "others," people vent their frustrations and make themselves feel safer. But such reactions also hindered—and continue to hinder—the task of fighting the disease. Prejudice and American embarrassment about sexual activity have created major roadblocks to government-sponsored dissemination of information that could prevent thousands of additional AIDS cases.

An editorial that appeared in 1988 in the British scientific magazine *New Scientist* still sums up the despair of many public health officials:

The extent to which AIDS and AIDS sufferers are still stigmatized in the US has meant that many federal public health and education programmes still remain in limbo, . . . victim to . . . sensibilities over the public use of words such as "condom" . . . and . . . contentment that the deaths should continue provided that they decimate only the unpopular, the inarticulate, or the marginal: gay men, drug users, blacks, hispanics, the poor, and far-off populations in Central Africa.

At the time of this writing, an average of 198 AIDS cases are reported to the Centers for Disease Control

Figure 2.4 No one in this family fit the media's image of a "high risk group" in the early 1980's. But by the time this picture was taken, the father—a hemophiliac—had acquired AIDS from a transfusion, and had unwittingly passed it on to his wife. She, in turn, passed it to her newborn son during pregnancy or through breast-feeding. Father and baby have both died. The young girl—though she lived in close contact with three infected family members— was not infected.

every day, including an increasing number of babies infected with the virus while still in the womb. By the end of 1989, more Americans had died from this disease than were killed in the Vietnam war. That number represents not only huge numbers in New York and San Francisco but also scores of shattered lives in other towns across the country: 84 in Allentown, Pennsylvania, 122 in Tulsa, Oklahoma, and 76 in Omaha, Nebraska. And the United States Public Health Service estimates that between 1 and 1.5 million Americans are already infected with the virus. The mounting AIDS caseload threatens to overwhelm the health care industry in several cities.

It would be comforting to report that the initial outbursts of denial, prejudice, and persecution of the victim are over, but that is not yet true. Serious AIDS prevention programs have been started in only a handful of areas,

Figure 2.5 Rubber tappers—who settle in the rainforest and exploit its resources in a potentially sustainable way—are often caught in the center of economic and ecological controversy. On one side are indigenous forest peoples who resent any intruders. On the other side are developers, cattle ranchers, and an increasing number of subsistence farmers whose actions cause wholesale forest destruction.

and these have been handicapped by a lack of reliable statistical information on the sexual behavior of Americans. Experts warn that without those solid data they will not be able either to predict the spread of AIDS or to design effective AIDS-prevention education programs. Remarkably, attempts by the National Institute of Child Health and Human Development to gather that information have been stymied by pressure from certain members of Congress who oppose the research for political reasons. It would be difficult to conceive of similar reactions to efforts aimed at curtailing an epidemic of measles, chickenpox, or flu. But such is the continuing battle that health workers face, even in the United States in the twentieth century—fighting disease on the one hand and fear and ignorance on the other.

In summary, it is important to recognize in this story a specific example of a general lesson. Even the most useful knowledge can benefit society only to the extent that society is willing to accept that knowledge and to act on it in a rational manner. It is also important to realize that the sorts of responses discussed here are by no means confined to American or even Western society. In Africa

(where, as you will see in Chapter 39, AIDS has spread through heterosexual rather than homosexual intercourse), the disease is treated very differently. Yet behind part of that difference lies a similar societal response to the threat; among Africans, one of AIDS's common names is "the European disease."

ECOLOGY AND ECONOMICS: TENDING OUR HOUSES

Francisco (Chico) Mendez wasn't born into circumstances that you'd expect to propel him onto the front pages of international newspapers. He spent most of his life in a jungle town called Xapuri, deep in the Amazon, where he earned a living by tapping rubber trees for their sap. But as the leader of a rubber tappers union, Mendez found himself smack in the middle of a national dispute that made global headlines.

Rubber tappers, you see, are people who harvest products of the rain forest for a living (Figure 2.5). By virtue of both their temperament and their profession, they prefer to see most of the jungle left standing. They correctly surmise that their region will produce more valuable goods for a longer period of time if the rain forest remains intact, allowing them to harvest its rich and valuable store of plants and animals as renewable resources. That philosophy puts rubber tappers squarely at odds with wealthy and powerful ranchers, who prefer to destroy the rain forest to plant forage grasses for cattle.

Campaigning to save the rain forest, Mendez found himself first labeled an environmentalist, then lauded as a self-styled ecologist. His precious rain forests received more and more national attention. Mendez received an award from the United Nations for his efforts. Things seemed to be going well. But in December 1988, at the age of 44, he was shot and killed outside his jungle home by the son of a cattle rancher.

Like Joanne O'Donnell, Chico Mendez was caught up in a firestorm that, although heavily involved with ecological science, was driven by economics and politics. To those who study the history of human interactions with the environment, these complex stories of environment, intrigue, money, and power are not surprising. That's because the disciplines of ecology and economics have affected one another since ancient times. We will not trace the details of human interactions with the global environment here; that presentation must wait for Chapter 7, which follows several chapters on the workings of the biosphere. Here we will concentrate on certain pivotal points in the relationship between human society and its environment, illustrating our current difficulties with a single major issue of global significance.

A Historical Perspective on Ecology and Economics

The words *ecology* and *economics* are both derived from the Greek word *oikos*, which means "house." Over time, economics has become concerned primarily with the art of human household management, whereas ecology deals with the study of nature's "houses" and their inhabitants. For some time the disciplines ran in parallel, but by the eighteenth century their philosophies had diverged.

The split began in 1749, when Swedish botanist Carl von Linné (Linnaeus) wrote an essay entitled "The Oeconomy of Nature," in which he argued that God created "nature's economy" exclusively to serve the human economy. Adam Smith, the founder of modern economics and an avid disciple of Linnaeus, adopted a similar view of the biosphere as a global warehouse of raw materials for human use. Smith also believed that it was humanity's birthright to dominate nature through hard work and technological advancement. Thus the writings of a biologist spawned a very human-centered view of people's interactions with nature.

As Smith's successors, most economists adopted a human-centered view of human interactions with nature and developed economic models that analyzed some aspects of human activity quite accurately. But until recently, economic theory failed to consider the long-term ecological effects of human activities: depletion of nonrenewable resources such as coal and oil, loss or pollution of renewable resources such as air and water, and the disappearance of species and ecosystems.

In contrast, biologists and natural philosophers, as far back as Linnaeus's time, realized that unrestricted growth and development could have serious environmental consequences. The proponents of this view who were best known to the public were not ecological researchers but gentleman–naturalist–authors such as Henry David Thoreau. Thoreau, an early champion of a less human-centered perspective, embraced the natural world as home to *all* life. He and his successors counseled society to recognize, and learn to live within, environmental limitations.

During the mid-twentieth century, it became apparent that human activities were, in fact, disturbing many local ecosystems. Then, in 1962, biologist Rachel Carson published *Silent Spring*, a disturbing documentary of the destructive power of chemical pesticides then widely used in agriculture. *Silent Spring* inaugurated the environmental movement by warning of ecological apocalypse—a spring without the sounds of birds and insects.

The scientific environmental movement, often at odds with local and national governments and economists, grew during the 1960s and early 1970s. Throughout this period, ecological researchers strengthened the scientific underpinnings of their predecessors' philosophical and emo-

Figure 2.6 Public demonstrations and other actions (not always as colorful or as high-profile as this one) have been—and will continue to be—vital in directing both federal and state environmental legislation and the attitudes and practices of major corporations.

tional appeals. The concern over environmental matters shown by the general public was reflected in several important pieces of legislation designed to protect the environment (Figure 2.6).

Unfortunately, the public image of "ecology" fell on hard times in the mid-1970s and early 1980s. The country suffered an economic recession, triggered in part by the sudden shortage of imported oil created by a cartel of oil-producing nations. It became popular during those years to attribute financial hardships to convenient local scapegoats, including environmental activists. One bumper sticker that was popular in New England during those years read "Hungry? Cold? Eat an environmentalist!"

Ecologists kept working, of course, and they continued to amass important data about the local and global effects of human activities on the environment. But that increasing body of data did not—for an unfortunately long time—have any tangible effect on either public perceptions or national environmental policy. All politicians paid lip service to the environment, but given the public mood, few wanted to confront their constituents with the cost of paying for a cleaner environment. Although the environmental agencies set up earlier remained in place, several were decimated by budget cuts and executive-branch decisions not to spend money that Congress had earmarked for environmental action. Offered a choice between investing large sums of money in waste treatment and quietly dumping toxic materials where they could, many corporations similarly chose the easy route to short-term profits.

Then, sometime between the beginning of 1988 and the middle of 1989, the world at large suddenly woke up. During the summer of 1988, much of the country sweltered under a heat wave with temperatures regularly above 100°F. At the same time, agricultural areas from California to Georgia were scorched by a drought that cut grain harvests by 31 percent. Scientists detected dangerous holes in the earth's protective ozone layer. Raw sewage, garbage, and discarded syringes washed up, not only onto public beaches around New York and Boston, but onto exclusive resorts on the islands of Nantucket and Martha's Vineyard. During one weekend, coastal waters in Alaska, Rhode Island, California, and Delaware were all being fouled by oil from separate, major tanker accidents (Figure 2.7). One of those spills, caused by navigational errors aboard the tanker Exxon *Valdez*, befouled more than 1355 miles of formerly pristine Alaska coastline.

None of these events was either new or unexpected from a scientific point of view. Scientists even rushed to point out that the American heat wave and drought could not definitively be cited as evidence that a global warming trend had begun. But somehow, this time, the total of these human-caused disturbances in the planetary environment finally shifted the balance and changed the attitudes of both the mass media and the public. *Time* magazine replaced its "Man or Woman of the Year" cover story with "Endangered Earth: Planet of the Year" (Figure 2.8), and *National Geographic* ran a similar global issue. The public buzzed. Economists and local politicians suddenly sat up and took notice. *Money* magazine ran an article about the *Valdez* spill entitled "Tanker from Hell" that described not the effects of the much-publicized spill on the environment, but its long-term effect on business. "Exxon's oil spill," the article warned, "may usher in an age when companies—and their shareholders—will be penalized for a devil-may-care approach to the environment." *The Wall Street Journal* began regularly running such headlines as "Debate Over Pollution and Global Warming Has Detroit Sweating," and "Car Makers, Fearing a Wave of 'Reregulation,' Brace for New Emissions Rules."

Finally, world political leaders felt obliged to get into the act. During a global economic meeting among leaders of the world's seven largest industrial nations in the summer of 1989, environmental issues took center stage. Their meeting, in fact, came to be called "The Green Summit." "What defense has been to world leaders for the past 40 years," an editorial in Britain's *The Economist* predicted at that time, "the environment will be for the next 40."

To give you a taste of what all the excitement is about and a sample of the ongoing debates among scientists

Figure 2.7 Coastal waters and beaches suffer incalculable harm from oil spills such as this one near Port Angeles, Washington.

over uncertainties in models of the way the world works, we will take a brief look at one of the most talked-about phenomena in ecology today: global warming and the greenhouse effect.

The Greenhouse Effect: Atmosphere as Incubator

Temperatures in our part of the solar system vary enormously, and virtually everywhere except on Earth, conditions are lethal to life as we know it. On nearby Venus, daytime temperatures hover near 480°C (over 900°F), hot enough to melt lead. And on Mars, temperatures vary more than 150°C (270°F) between noon and midnight, plummeting to an icy −128°C on the planet's dark side. On Earth, by comparison, the average surface temperature is a comfortable 15°C (59°F), and the most extreme temperatures, from summer in tropical deserts to midwinter in Antarctica, cover only a fraction of the temperature ranges found on our nearest planetary neighbors.

Furthermore, Earth's average temperature has remained nearly constant for more than 3 billion years; it has never dropped below 5°C (41°F) or risen above 25°C (77°F) since the earliest forms of life evolved. This stability is truly remarkable because the sun has been getting gradually hotter. Astronomers calculate that the sun's heat production is now 30 percent higher than it was when life began on Earth.

We owe our planet's hospitable temperature in large part to levels of atmospheric carbon dioxide, water vapor, and certain other gases that readily admit sunlight to the planet's surface but retard the escape of heat. Because these gases act in much the same way as the glass in a greenhouse, this phenomenon is called the **greenhouse effect,** and the gases involved are often called *greenhouse gases* (Figure 2.9).

Greenhouse gases are critically important in balancing incoming solar radiation and escaping heat. If their concentration were to increase beyond a certain point, more heat would be retained and the planet would get warmer. And if Earth's average temperature were to rise by only a few degrees, global weather patterns could change, heat waves and droughts would plague important agricultural areas, and polar ice caps could melt, flooding coastal cities. If, on the other hand, greenhouse gas concentrations were to fall, Earth's temperature would drop. And a fall of only a few degrees in average temperature could trigger an ice age.

If this much is clear to most scientists, why does so much controversy exist among scientists and politicians regarding "the greenhouse effect?" The answer—as is usually the case in science—is complex.

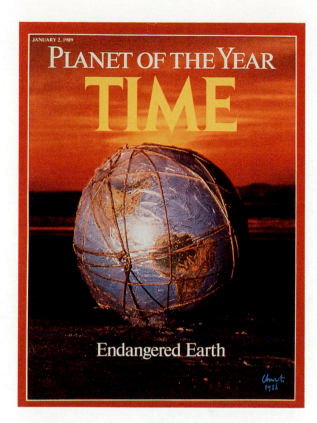

Figure 2.8 This cover illustration is emblematic of the mass media's sudden "rediscovery" of global ecological issues.

Carbon Dioxide and Its Effect on Global Temperature

The key to Earth's constant temperature lies in the changes our atmosphere's carbon dioxide concentration has undergone through time. During Earth's early years, our atmosphere had close to 1000 times as much carbon dioxide as it does today. The greenhouse effect of all that carbon dioxide kept Earth warm, even though the sun at the time was much dimmer. But over millions of years, as the sun grew warmer, much of the planet's early surplus of carbon dioxide dissolved in the oceans, formed carbonate rocks, and thus left the atmosphere. Later, after life evolved, many living organisms took up carbon dioxide and incorporated it either into their bodies or into protective body coverings.

Since that time, atmospheric carbon dioxide concentrations have dropped to a present level of about 0.03 percent, allowing more and more heat to escape from Earth's surface. Somehow—and scientists still do not agree on just how—the decrease in the greenhouse effect paralleled the rise in the sun's output so closely that planetary temperatures stayed nearly constant.

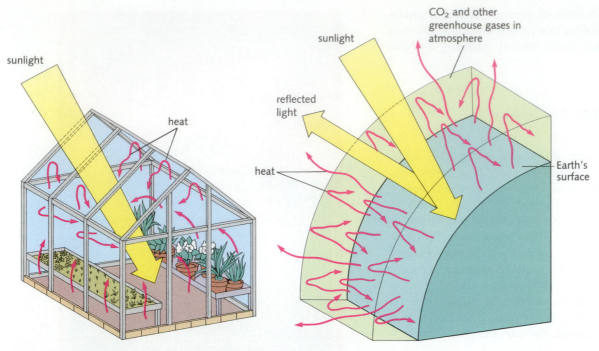

Figure 2.9 (left) When sunlight enters a greenhouse, much of its energy is absorbed and reflected by objects inside. Some of the absorbed energy is reradiated as heat. But, because greenhouse glass doesn't let infrared energy pass through, much of that reradiated heat is trapped inside, warming the greenhouse. (right) Greenhouse gases in the atmosphere allow sunlight to enter and strike Earth's surface. Much of the solar energy is reradiated as heat from the Earth's surface, only to be absorbed by the greenhouse gases and reradiated again in all directions. About half escapes into space, but the other half is directed back at Earth's surface, where it warms the planet.

Was this blind luck, or does Earth have some sort of "thermostat" that regulates its temperature? There are two different sets of explanations. Some researchers suggest that the control mechanisms are the result of geological and chemical processes alone. The other explanation involves living organisms as well.

The geochemical view Those who credit geochemistry alone as a temperature regulator argue that Earth's carbon dioxide revolves in a giant cycle the most important pathways of which involve strictly physical processes (Figure 2.10). Atmospheric carbon dioxide dissolves in rainwater, forming a weak acid that erodes rocks composed of calcium, silicon, and oxygen. The resulting mixture of elements travels through rivers and streams to the oceans, where the carbon dioxide may combine with calcium and magnesium to form insoluble compounds called carbonates. These carbonates precipitate out of solution and accumulate on the ocean bottom, where they harden into the sedimentary rocks called limestone and dolomite. In certain places, geological activity forces those rocks underneath the continents, sometimes so deeply that

intense heat drives the carbon dioxide out in gaseous form. When volcanoes erupt, this underground carbon dioxide is reinjected into the atmosphere.

According to the strictly geochemical view, changes in atmospheric carbon dioxide content result exclusively from changes in this cycle that are caused by variations in planetary temperature. Although those who hold this view do not deny the effect of living organisms on atmospheric carbon dioxide, they believe that geologic processes are far more important.

The biogeochemical view Other scientists believe that living organisms play a major role in regulating carbon dioxide levels. They point out that most carbonate rocks created today are formed not geochemically, but biologically; myriads of single-celled marine organisms incorporate dissolved carbon dioxide into their shells. When these organisms die, their shells fall to the ocean bottom and accumulate into vast deposits such as, for example, the famous white cliffs of Dover. As this process removes carbon dioxide from the sea, more of the gas in the atmosphere goes into solution to replace it. At the same time, terrestrial plants incorporate carbon dioxide from the air into living tissue. Much of this organic carbon is continually recycled, as we will discuss shortly, but a great deal was buried in the form of vast organic deposits that, over time, became coal and oil.

The biogeochemical view argues that changes in the growth rates of plants and other organisms significantly affect the amount of carbon dioxide left in the atmosphere. Such organic processes, which could easily be affected by any change in global temperature, may have acted as a sort of "living thermostat," adjusting the atmosphere's carbon dioxide concentration ever since life became established.

Which theory is correct? We don't know yet. At the present time, insufficient data are available for us to determine whether strictly geochemical or biogeochemical processes are dominant today. Even fewer data are available about what happened 3 billion years ago.

Human Involvement in Global Warming

Although scientific debates such as this one may seem abstract, they are potentially vital to the future health of our species and our planet. As you will see in more detail in Chapter 7 (and have undoubtedly learned about in the media), we know that the burning of coal and oil and the destruction of global forests are releasing enough stored organic carbon to raise atmospheric carbon dioxide levels slowly but steadily every year. Is this steady addition of

Figure 2.10 This schematic shows the important parts of the global carbonate cycle. The processes are separated to show biological, biogeochemical and geochemical processes, as well as the effects of human activity.

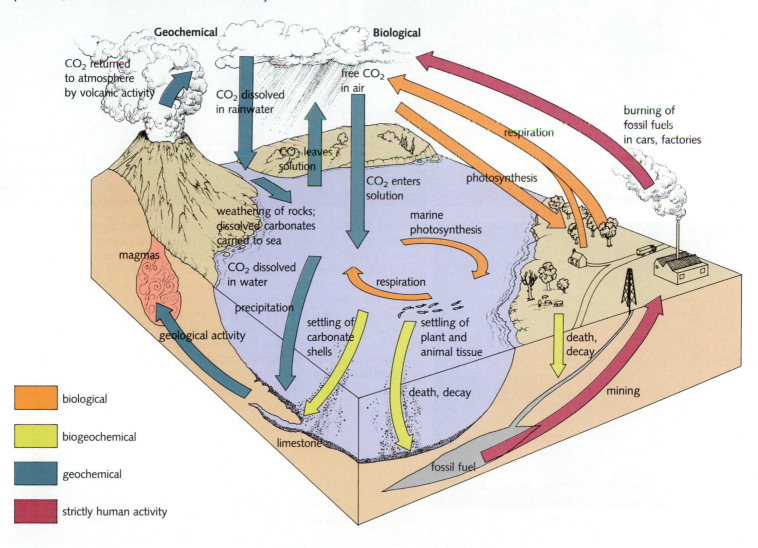

carbon dioxide sufficient to set off a significant global warming trend? And if so, how will global climate respond? No one knows. Although evidence in favor of global warming is beginning to accumulate, it will be at least two decades before we know for sure.

One thing is clear, however: It will be easier to prevent runaway change in climate if we begin to act *now* to control our production of carbon dioxide and other greenhouse gases. It is also clear that any actions taken to slow down the release of carbon dioxide must involve all the world's countries. It has been easy—though both scientifically and politically inaccurate—for many mass media reports on this subject to place the blame only on developing nations such as Brazil, where huge tracts of virgin rain forest are being destroyed every day (Figure 2.11). That exclusive emphasis is hardly fair. In industrialized countries such as ours, most virgin forests are already gone, others are being cut constantly, and the burning of fossil fuels vastly exceeds that of any Third World nation (Figure 2.11).

Ultimately, in this and all other parts of the global ecological arena, humans in all societies must be willing to make significant changes in their various styles of life if we are to inhabit the earth responsibly. In order to understand why those changes are necessary, we must all understand the rules that govern the miraculous, global phenomenon we call life. Furthering that vital understanding is a major goal of this book.

Figure 2.11 (top, left) The burning of fossil fuels in developed nations—both in industry and in private homes and automobiles—contributes significantly to increases in atmospheric carbon dioxide. (top, right) This area of clearcut rainforest near Manaus, Brazil, presents a picture now commonly displayed in media discussions of deforestation. These areas seldom recover from disturbance, and often revert to wasteland in less than a decade. (bottom, left) This clearcut area is closer to home—near Booth's Bay in Washington state. Many American forests are being destroyed with frightening speed, though with little fanfare. While it is true that temperate forests can often be replaced with tree farms, single-species fields of trees are not ecological replacements for old-growth forests.

SUMMARY

Science has both positive and negative effects on society. Modern industries have produced pollutants which threaten other organisms as well as ourselves, and will challenge our skills to clean them up. The attitudes of society, in turn, affect the way in which science is conducted and the uses to which it is put. We might have hoped, for example, that the triumphs of modern medicine had prepared society to react with intelligence and compassion to the outbreak of a serious and fatal illness like AIDS. However, in many ways the fear and ignorance that many elements in society have displayed in the face of the AIDS challenge have served to hinder scientific efforts to fight the disease. In many ways, dealing with such fear and ignorance provides a challenge at least as great as conquering AIDS in medical and biological terms.

On a global scale, we face environmental challenges so severe that ecosystems as massive as the Amazon rain forests are threatened and may require bold action and personal courage to save them. Many of our planet's problems stem from a failure to reconcile the related disciplines of *ecology* and *economics*. One of these emerging problems is a gradual increase in the percentage of carbon dioxide in the atmosphere. This buildup may increase temperatures throughout the year and lead to dramatic changes in world climate. Dealing with such problems will require a greater understanding of the nature of life and of the interactions of living organisms with planet Earth.

STUDY FOCUS

After studying this chapter, you should be able to:

- Describe some of the interactions between science and society.
- Describe some of the obstacles that societal attitudes have placed in the way of fighting the spread of AIDS.
- Relate some of the efforts that environmentalists have made to save the Amazon rain forests.
- Explain how increasing levels of atmospheric carbon dioxide may affect the global climate.

SELECTED TERMS

greenhouse effect *p. 27*   biogeochemical view *p. 28*
geochemical view *p. 28*

REVIEW

Discussion Questions

1. What is the greenhouse effect?

2. Why are the concentrations of greenhouse gases important in determining conditions for life on Earth?

3. Though no concentrations of living beings even remotely approach Earth in size, the actions of all organisms together clearly affect the atmosphere. Describe how processes in the biosphere participate in controlling planetary carbon dioxide levels.

4. The burning of fossil fuels is causing an increase in the amount of carbon dioxide in the atmosphere. What could happen to global climate as a result?

5. Whether or not the Gaia hypothesis discussed in the last chapter is "true" in an absolute sense, its thesis has value in directing studies of the global environment. How?

Objective Questions (Answers in Appendix)

6. A modern publication that inspired public awareness of environmental concerns was
   (a) Adam Smith's *Wealth of Nations.*
   (b) Rachel Carlson's *The Silent Spring.*
   (c) Carl von Linneaeus' *The Oeconomy of Nature.*
   (d) William Fitzgerald's *The Exxon Valdez.*

7. A factor that contributes to the "greenhouse effect" is the
   (a) destruction of tropical rain forests.
   (b) use of carbon dioxide by terrestrial plants.
   (c) depletion of minerals in the soil by overfarming.
   (d) use of glass solar panels to capture radiant energy.

8. The location of *most* of the earth's carbon is
   (a) the atmosphere and ocean.
   (b) oil and shale deposits.
   (c) sedimentary rocks.
   (d) igneous and metamorphic rocks.

9. The acronym "AIDS" stands for
   (a) Acquired Immune Deficiency Symptoms.
   (b) Acquired Immune Deficiency System.
   (c) Acquired Immune Deficiency Syndrome.
   (d) Arresting Intravenous Disease System.

10. AIDS is *not* transmitted by
    (a) contaminated food and water.
    (b) blood.
    (c) sexual intercourse.
    (d) birth.

PART TWO

Organisms and Ecology

The biosphere is in many ways like an extraordinarily complicated jigsaw puzzle in which each living species is a single piece. Unlike normal jigsaw puzzles, however, the biosphere and its pieces are alive; they interact with each other constantly and powerfully. The branch of biology that examines those interactions and attempts to predict the response of living systems to future environmental change is called *ecology*.

Ecology's predictive ability is more important today than ever before in history, because earth's burgeoning human population has begun to tax the planet's physical and biological support systems. The atmosphere's ozone layer, which protects us from damaging ultraviolet radiation, is thinning out. Africa's expanding deserts threaten food production in developing nations. From all over the United States there are reports of ground water unfit to drink, air unsafe to breathe, acid rain that damages forests and streams, and seafood contaminated with cancer-causing chemicals. Each of these changes represents an ecological consequence of human activity.

Today more than ever before, we need to understand how the biosphere functions so that we can evaluate the effects of human activity and plan wisely for the future.

In the past, decisions about industrial and agricultural development were based almost entirely on political and economic considerations, but making decisions in that manner is no longer feasible. There are basic laws of nature that affect *all* forms of life, including humans. If we continue to ignore these laws, we risk upsetting the balance of life that has existed on the earth for millions of years.

Many ecologists are seriously concerned about the state of global ecological conditions today. In order to evaluate those concerns, in Part Two of this text we examine the way organisms interact with each other and with their environments. Chapter 3 traces the flow of energy and the cycling of nutrients through the biosphere. Chapter 4 describes the interactions of global weather and climate patterns with living organisms. Chapter 5 explains the factors that control the growth of plant and animal populations. Chapter 6 looks at the different characteristics of ecosystems (ecological communities together with their physical environments) around the globe. Finally, Chapter 7 relates modern ecological understanding to the current status and future position of humans in the living world.

◄ On the floor of a marsh in the Pacific Northwest, a lifeless snail and the decaying leaf upon which it rests elegantly symbolize the cycle of birth, growth, death and decay that rules the natural world. As these organisms decompose, the nutrients they contain will be returned to their environment, making possible the continued growth of generations to come.

CHAPTER **3**

Energy and Nutrients in Ecosystems

Across Africa's Serengeti plain, grasses as tall as a person bend in the wind that carries seasonal rains. Growing with astonishing speed when water is available, these grasses wait out the dry season in protective dormancy (Figure 3.1). In the lowland forests of Burma and Ceylon, rain forest trees form an emerald canopy with leaves spread to catch sunlight over 100 feet from the ground. Blessed by abundant rainfall and warm temperatures year-round, many of these trees constantly grow new leaves and lose old ones. In the cold, clear water off the California coast, a 30-foot forest of yellow-green kelp sways gently in the current, adding new tissue to its blades at the remarkable rate of 50 cm/day. And in the stormy waters of George's Bank off the New England coast, floating, single-celled algae grow and divide as they drift with the currents.

These very different organisms share an ability vital to life on earth: They can harness the power of solar energy to assemble simple, inorganic compounds into the complex molecules they require. For this reason, they are called **autotrophs** (*auto* means "self"; *troph* means "food or feeding"; *autotroph* means "self-feeding"). Such organisms are also called **primary producers** because they are the first producers of complex compounds (Figure 3.2).

AUTOTROPHS AND ENERGY

The earth's most important primary producers carry on **photosynthesis,** the process by which the green pigment *chlorophyll* harnesses solar energy to convert carbon dioxide and water into energy-rich organic compounds known as **carbohydrates** (*carbo* means "containing carbon"; *hydrate* means "containing hydrogen and oxygen").

The simplest and most biologically important carbohydrates are the *sugars*, such as glucose, sucrose, and fructose. All sugars are composed of individual units known as simple sugars or *monosaccharides*, each of which contains from 3 to 7 carbons. One of the most important simple sugars is an energy-rich molecule called *glucose* (Figure 3.3) which plants make by expending the energy captured from sunlight. The energy that plants pour into glucose molecules can be released by breaking those molecules back down to carbon dioxide and water in the presence of oxygen. That is precisely what happens in living cells; both animals and plants power their essential life processes by burning glucose as fuel in the process known as cellular respiration (see Chapter 25).

Glucose is also important because it provides a convenient way for organisms to store energy. Living cells contain a great deal of water, and sugars are readily soluble in water. This means that sugar can be held in solution anywhere in the living cell and kept available as an energy source.

LIFE'S POWER SUPPLY

In some forms, energy is easy to measure and to understand. The heat energy of a roaring bonfire is easy to appreciate. But how can energy be stored in chemical form?

The answer is the *chemical bond*, the linkage that binds atoms together to form molecules, the basic units of chemical compounds. Energy is required to form a chemical bond between two atoms, and we can think of that energy as being stored in the bond, to be released whenever and however the bond is broken. The fact that some bonds contain more energy than others is very important for biology. We can see why in a simple example. Figure 3.4 shows a photovoltaic cell, which converts energy from sunlight into an electric current. That current is applied to two metal rods (electrodes) in a beaker containing water, which is composed of hydrogen and oxygen. The energy in the current breaks apart the bonds that link hydrogen to oxygen in the water molecules. Oxygen gas (O_2) bubbles up along the positive electrode, and hydrogen gas (H_2) along the negative electrode. What has actually happened is that H–O bonds have been

Figure 3.2 This freshwater alga is an aquatic primary producer.

Figure 3.1 Africa's Serengeti plain.

Figure 3.3 The structure of glucose, an important carbohydrate molecule involved in energy-storing and energy-releasing pathways within living organisms.

replaced by O=O and H–H bonds. This means that bonds were broken and new bonds were formed. In this case, the energy required to make the new set of bonds is *greater* than the energy released by breaking the old bonds, so the reaction requires energy to run. The electric current from the photovoltaic cell provides that energy.

But the energy poured into this reaction by the photovoltaic cell is not lost. The hydrogen and oxygen produced by the apparatus can be gathered into separate vessels, combined as shown, and a spark can be used to ignite the hydrogen–oxygen mixture. Once ignited, the hydrogen burns brightly (combines chemically with the oxygen), giving off a tremendous amount of energy in the form of heat. (Oxygen and hydrogen are the fuels that power many of the booster rockets used in the U.S. space program.) As the hydrogen burns, H_2 and O_2 react to form water. This is precisely the reverse of the reaction that produced the gases to begin with. The energy given off in this rapid burning of gases was there all along, stored as chemical energy in the bonds between the atoms of hydrogen gas and oxygen gas.

Living systems trap and store energy in chemical form too, by synthesizing (building up) carbohydrates and other energy-rich compounds. The ability of the chemical bond to store energy is what makes life possible on our planet. The rules that govern the storage and release of energy are contained in a branch of science known as thermodynamics. The first and second laws of thermodynamics, which govern the relationship between matter and energy in chemical reactions, will be covered in Chapter 21.

HETEROTROPHS: OBTAINING ENERGY CAPTURED BY PLANTS

If all living organisms were autotrophs, understanding energy would be easy; the chemical energy hoarded in sugars would be stored by primary producers to be used as needed. But many organisms are **heterotrophs;** that is, they depend, directly or indirectly, on primary producers for energy. The term *heterotrophs* comes from *hetero* ("other") and *troph* ("feeding").

Heterotrophs can *store* energy by synthesizing and accumulating a variety of complex molecules such as fats, oils, and proteins. But they must *obtain* that energy by eating energy-rich molecules contained in other organisms. Because energy flow through ecosystems affords us an excellent opportunity to see how chemical reactions affect the biosphere, we will briefly examine a few simple chemical concepts relevant to ecology.

Carbohydrates: Energy Carriers in the Biosphere

Sugars and edible carbohydrates   Just about every molecule found in one organism can be used as food by some other organism. (That is one of the things that can make life dangerous.) But there are no molecules more important as a food source to living things than carbohydrates.

The monosaccharide sugars are important both because they are good molecules in which to store energy

Figure 3.4 In this diagrammatic representation of simple electrolysis, solar energy produces an electric current that breaks the bonds between the hydrogen and oxygen molecules in water and forms new bonds between pairs of oxygen atoms (in oxygen molecules) and pairs of hydrogen atoms (in hydrogen molecules).

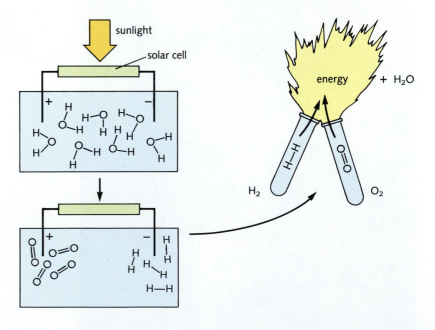

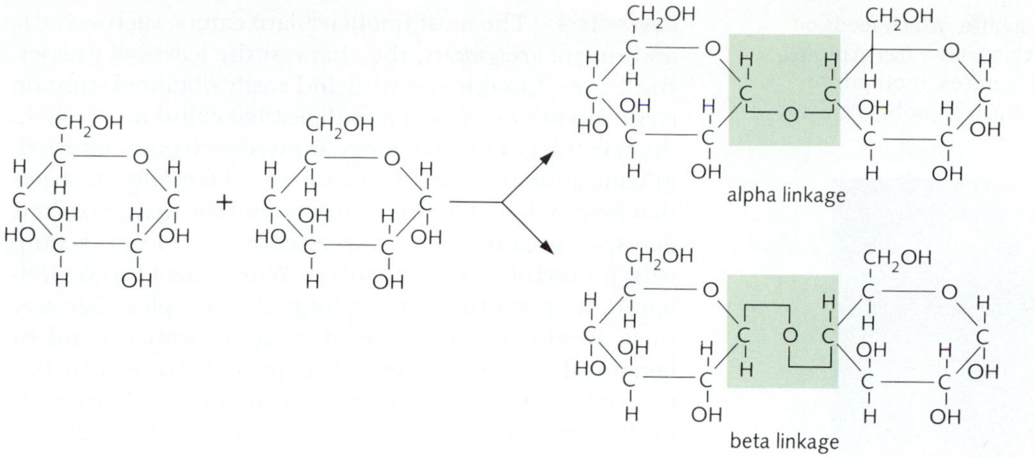

Figure 3.5 Two simple sugars may be joined by a chemical bond. Two of the possible bonds, the alpha linkage and the beta linkage, are shown. While most organisms can break (and digest) the alpha linkage, very few can break the beta linkage.

alpha linkage

beta linkage

and because they are chemically quite versatile. Simple sugar molecules can easily be joined together into more complex carbohydrates such as *starch* and *cellulose*. Before cells can use the energy stored in complex carbohydrates, however, those large molecules must be broken down—or digested—into their component sugars. Shown in Figure 3.5, for example, is a chemical reaction in which two glucose molecules are joined by a chemical bond. The bond that joins the two glucose molecules in Figure 3.5 is important. As can be seen from the figure, the oxygen attached to the carbon atom can be in either of two positions when the bond is made. And although the two bonds do not look very different in the diagram, there is a world of difference between them in living systems.

Plants store much of their energy in the form of a carbohydrate called 'starch. Many other organisms (including humans) can break the chemical bonds in starch very easily, so starch serves as an excellent source of food. In fact, the starches found in potatoes, corn, rice, wheat, and barley are the major sources of food energy for the human species.

Cellulose Many plants use another carbohydrate for an entirely different purpose. By linking thousands of glucose molecules in a different way, plants form a substance known as cellulose, which they use to build tough, flexible cell walls to support and protect their cells. In larger plants, cellulose fibers form the basis of a much stronger material: wood.

In terms of chemical components, cellulose is very similar to starch, but the beta linkages in cellulose make it very difficult for other organisms to digest. As we will learn in more detail in Chapter 34, many steps in digestion depend on very specific *digestive enzymes*. These enzymes are substances that break up large molecules,

such as complex carbohydrates, into small molecules, such as glucose.

The beta linkages between the glucose molecules in cellulose can be broken only by the enzyme cellulase. However, most animals cannot manufacture cellulase; only a handful of snails, some clams, and a few species of microorganisms produce this enzyme and so can digest cellulose. Actually, we can be thankful that the cellulose bond is so difficult to break. Large plants such as trees could not survive long in a world where every bug, fungus, and mammal could use them as easy meals.

ECOLOGICAL STRATEGIES OF HETEROTROPHS

Heterotrophs obtain the energy they need in many ways, most of which can be divided into three main strategies: eating plants (herbivory), eating animals (carnivory), and decomposing the remains of other organisms and their waste products.

Herbivores: The Diverse Plant Eaters

Herbivores (*herb* means "plant"), such as rabbits and deer, eat plants that have stored energy in the forms of sugars, starches, and other complex carbohydrates. Because they are the first organisms to consume the energy and carbon that is "fixed" by primary producers, herbivores are also called **primary consumers.** The enormous diversity of plants has spawned an equally diverse set of herbivores, who survive by selectively attacking different parts of different plants (Figure 3.6).

Figure 3.6 Herbivore diversity. (top) A giraffe, which feeds on the leaves of tall trees. (center) A sea urchin uses a five-part jaw to rasp encrusting marine algae off hard surfaces. (bottom) An Achilles tang uses needle-sharp teeth to graze on both encrusting and leafy green algae.

Leaf eaters The most familiar plant eaters, such as cattle and sheep, are *grazers;* they harvest the leaves of grasses. But leaves, though plentiful and easily obtained, contain large quantities of nearly indigestible cellulose. To help them extract as much energy as possible from their food, grazing animals depend on a variety of feeding strategies that involve both feeding behavior and the design of their digestive organs. First, grazers chew leaves into a pulp with powerful grinding molars. When this pulp is swallowed, it enters the grazer's long and complex digestive tract, where cellulose is broken apart with the aid of beneficial microorganisms (Figure 3.7). Even with this microbial assistance, however, grazers can extract relatively little energy from each mouthful of food and must spend most of their waking hours eating.

Seed, fruit, and berry pickers Many other herbivorous animals, including a variety of fishes, birds, and mammals, feed on seeds and fruits. Some, such as many finches, specialize in eating seeds of particular sizes and types. Others, such as South American toucans, use their long, sharp bills to pluck small berries or to chop large fruits into bite-sized pieces (Figure 3.8).

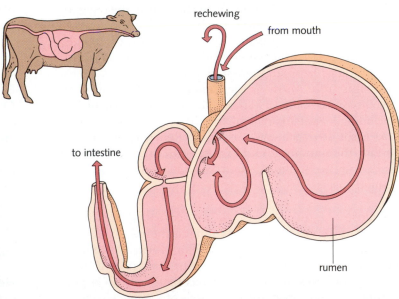

Figure 3.7 Cows have four stomach chambers. In the rumen, the first and largest chamber, bacteria and other microorganisms produce the enzyme cellulase, which combines with the cow's digestive enzymes to attack leaf pulp. Cows and some other grazers regularly regurgitate the pulp back into their mouths, chew it a second time, grind it into an even finer paste, and add more digestive enzymes from saliva before swallowing it again. This is what cows are doing when they "chew their cud." After the food has been processed again in the rumen, it passes through the other stomach chambers and into the intestines.

Figure 3.8 The toucan is a colorful herbivore that feeds on berries and fruits.

Figure 3.9 Carnivore diversity. (top) Sparrow hawks use aerial agility and sharp talons and beaks to prey upon small birds. (center) Chameleons snare insects, using long, sticky-tipped tongues that shoot out with lightning speed. (bottom) Lions hunt in groups and generally share their prey, usually large herbivores like this unfortunate zebra.

Humans also depend heavily on plant seeds and fruits as food. Most of the world's human population, in fact, lives on the seeds of a single plant family, the grasses, which includes rice, corn, wheat, oats, and barley.

Carnivores: The Meat Eaters

Carnivores, such as cats and hyenas, eat other animals that have stored energy in the form of fats, oils, and proteins. Animals that eat herbivores are called **secondary consumers.** Carnivores that eat other carnivores are called **tertiary consumers.**

Many meat eaters expend more energy than grazers in actually obtaining their food; stalking and running down a zebra or subduing a wildebeest is no easy matter. Other carnivores, such as chameleons, lie in wait to ambush their prey; these animals may spend much of their time waiting for unsuspecting meals to wander by (Figure 3.9).

Table 3.1 *Comparison of Birds' Intestines Showing Longer Length with Poorer Diet*

	Quail	Woodpigeon	Starling
Type of Diet			
Poor diet	artificial	brassica	plant
Rich diet	artificial	grain	animal
Intestine Length (cm)			
Poor diet	51	220	33
Rich diet	46	157	27
Poor/rich	1.1	1.4	1.2
Cecum Length (cm)			
Poor diet	17.0	—	0.8
Rich diet	14.5	—	0.6
Poor/rich	1.2	—	1.3

SOURCE: Sibly, 1981.

Figure 3.10 Fungi are vital decomposers that recycle the nutrients locked up in the complex molecules of dead animals and plants.

The ease with which meat is digested is reflected in several aspects of carnivore anatomy and behavior. Even tame carnivores, such as pet cats, hardly bother to chew their food. Their razor-sharp teeth quickly chop food into pieces small enough to be swallowed. Because these chunks of meat yield their energy so readily, carnivore intestines are much shorter on average than those of herbivores. (There is more than circumstantial evidence linking intestine length to food quality. Table 3.1 records experimental evidence that birds fed on poorer diets grow longer guts in as short a time as 25 days!) Many meat eaters need to feed only occasionally, and some—such as the big cats of Africa and the boa constrictors of South America—spend much of their time sleeping or relaxing.

Omnivores, such as humans and several other primates, eat both plant and animal material. Some omnivores, including certain fishes of the Amazon river system, vary their diets according to the seasonal availability of preferred foods. Others eat both meat and vegetable matter year round. As you might expect, the teeth and digestive systems of these animals often (though not always) combine characteristics of both carnivores and herbivores.

Decomposers: Organisms of Decay

Saprophytes, such as many fungi and bacteria, obtain energy by breaking down, or *decomposing,* the complex molecules in the decaying tissues of plants and animals (Figure 3.10). These organisms of decay are vital to the health of all ecosystems, for they play indispensable roles in the cycling of most important nutrients. Without these invisible, yet vital, heterotrophs, bodies of dead animals would litter the landscape and the nutrients those organisms accumulated during their lifetimes would be lost to the ecosystem. We will give specific examples of how decomposers function later in this chapter.

Other Feeding Techniques

Many feeding techniques, however, do not fit neatly into these familiar categories. Some herbivores and carnivores are **liquid feeders** that feed only on liquids extracted from their prey (Figure 3.11). Aphids and other sucking insects pierce plant stems to suck out the sugary sap. Mosquitoes and ticks attack animals in the same way, using syringe-like mouthparts to drink the blood of birds and mammals.

Another group is made up of the **waste feeders.** Animals such as Egypt's famed scarab beetles, for example, feed on the dung of other animals.

Many other animals feed primarily on **detritus**, a combination of bits of decaying organisms and dung containing partially digested food. On land, detritus mixes with soil particles and is eaten primarily by such animals as earthworms and some soil-dwelling insects. Worms swallow large amounts of soil and digest the detritus particles as they pass through the worm's gut. A large number and variety of marine animals, from corals to sea cucumbers, either filter floating detritus from the water or pick it up off the bottom after it settles.

Aquatic environments offer several food sources not found on land. In addition to floating detritus, water carries many dissolved organic molecules. Both fresh and salt water also harbor tiny floating or swimming plants and animals—the **phytoplankton** (*phyto* means "plant"; *plankton* means "drifter"; *phytoplankton* means "drifting plants") and **zooplankton** (*zoo* means "animals"; *zooplankton* means "drifting animals"). To exploit these foods, many marine species have become *filter feeders*, with mouthparts, gills, and specially designed limbs that filter food from the water around them (Figure 3.12).

Figure 3.11 Spiders paralyze prey trapped in their webs, inject digestive enzymes into the prey's body, and suck out the partially digested tissues as they liquefy.

ENERGY AND LIFE

Because all heterotrophs depend totally on preformed organic molecules they cannot make themselves, their lives are structured around the need to obtain those molecules. Herbivores must live within reach of growing plants. Carnivores survive only where there is prey, and saprophytes where there are dead plants or animals. For these reasons, *any conditions that control the growth of primary producers in an ecosystem indirectly control the lives of all other organisms as well.*

Although many factors affect plant growth both on land and in the sea, the availability of solar energy for photosynthesis is the single most important factor controlling marine plant life. This dependence has profound effects on both life in the sea and on human interactions with that life, because most of the earth's surface is under water.

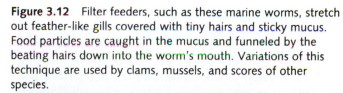

Figure 3.12 Filter feeders, such as these marine worms, stretch out feather-like gills covered with tiny hairs and sticky mucus. Food particles are caught in the mucus and funneled by the beating hairs down into the worm's mouth. Variations of this technique are used by clams, mussels, and scores of other species.

Energy and Life in the Sea

On land, sunlight is virtually the same everywhere. Wildflowers in the high Rockies, grasses on the plains, and cacti in Death Valley all live in the same kind of light. There are seasonal variations, but plenty of energy for photosynthesis makes its way to most terrestrial environments.

However, over two-thirds of the earth's surface is covered by water, and more than half of the photosynthesis

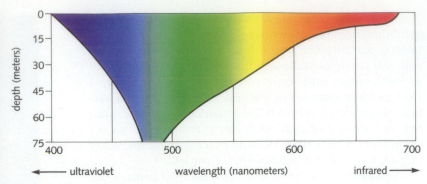

Figure 3.13 (left) Accessory pigments endow many algae with a range of striking colors. (right) As light passes through seawater, most of the longwave (red) light and much of the extreme shortwave (ultraviolet) light is absorbed. The light remaining in deep water is restricted primarily to the blue region of the spectrum.

on our planet occurs in the sea. And from the standpoint of solar energy, the open sea is very different from terrestrial environments because seawater absorbs much more light than the atmosphere does, particularly in the shortwave (ultraviolet) and long-wave (infrared) light regions of the spectrum. Coastal water also absorbs blue light because of the dissolved compounds carried into the sea by rivers and streams.

This means that the farther beneath the surface a marine organism lives, the less light reaches it and the more strongly colored the light is. In the open sea, light much below 10 meters (33 feet) is a vibrant, turquoise blue; closer to shore it has a more yellow-green hue (Figure 3.13a). Such dramatic lighting changes have profound effects on marine plants because chlorophyll absorbs and harnesses most efficiently light in the violet and red regions of the spectrum—precisely those colors removed by seawater.

Algae and plants have evolved several different kinds of chlorophyll, each of which captures energy from slightly different parts of the spectrum. But even these chlorophylls can't carry on photosynthesis deeper than a few meters beneath the surface. So marine algae use other compounds, called *accessory pigments,* to collect more energy by absorbing light of wavelengths that chlorophyll does not absorb (Figure 3.13b). This extra energy is then passed on to the machinery of photosynthesis, enabling the plants or algae to photosynthesize, and to grow, deeper than they could otherwise.

Even so, marine plants can only grow close to the ocean's surface. The region in which photosynthesis can occur—called the photic zone—may be as shallow as 30 meters in the turbid waters of the North Atlantic or as deep as 200 meters in the crystalline tropical Pacific. Only in this uppermost part of the ocean can phytoplankton grow actively. Bottom-growing plants and algae can grow only where the ocean floor is within this depth along

continental margins. Below the photic zone are vast volumes of water in which darkness prevents plants from growing.

To imagine a similar situation on land is nearly impossible. If the atmosphere absorbed light as seawater does, terrestrial plants could carry on photosynthesis only on the world's highest mountaintops. The vast plains and green, rolling hills of our continents would be as dark and devoid of plant life as a pharaoh's tomb.

FOOD CHAINS AND FOOD WEBS

In every ecosystem, plants and animals are linked together by feeding relationships into food chains and food webs. Looking at a few species at a time, we can imagine a primary producer, an herbivore, and a carnivore forming a simple **food chain.** For example, in a temperate woodland, a grass–eating mouse becomes prey for a predatory owl:

and in tropical Africa, impala feed on range grasses and are fed upon by the big cats:

But if we observe all the animals and plants in most ecosystems, we see immediately that feeding relationships usually weave numerous organisms into large, complex, and dynamic networks called **food webs,** in which many animals eat several different kinds of food (Figure 3.14). Furthermore, branches of most food webs interconnect with those of adjacent areas; thus food webs in

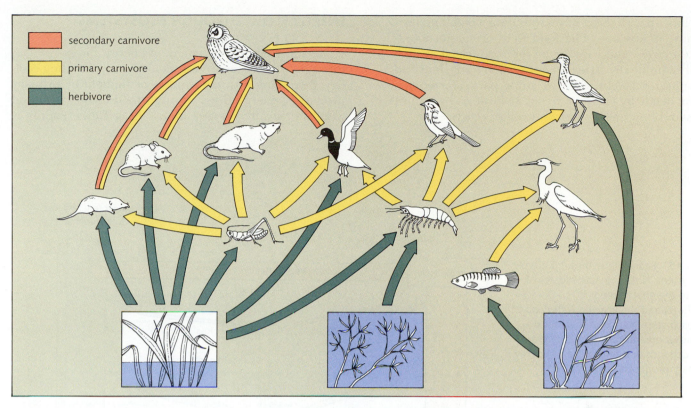

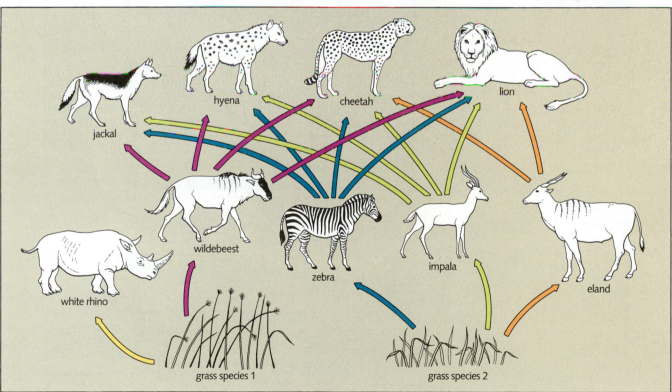

Figure 3.14 These two food webs give some idea of the complexity of feeding relationships in most environments. (top) A simplified food web from a California salt marsh. Arrows are color-coded to indicate trophic levels. (bottom) Feeding relationships among selected organisms of the African savannah, still enormously simplified. Arrows are color-coded to highlight simpler food chains within the web.

ponds link up with those of surrounding woods and feed into those of streams. The food webs of coastal woodlands, estuaries, and bays are all interconnected as well. We will examine the structures of several important and interesting food webs when we study particular ecosystems in detail in Chapter 6. First, however, we must understand how energy moves through ecosystems.

ENERGY FLOW THROUGH ECOSYSTEMS

Understanding the movement of energy, the *energy flow*, through living systems is essential to understanding ecology. Beginning with the capture of sunlight by primary producers, energy flows through food chains and food webs in a steady, *one-way stream*. As it flows, energy is

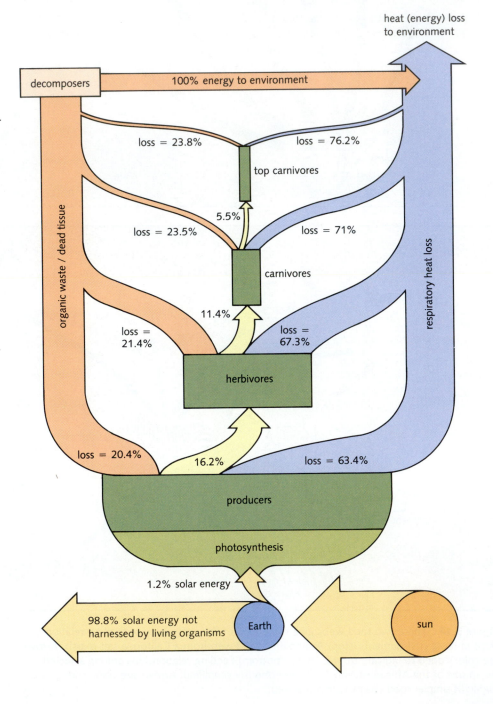

Figure 3.15 Energy flow through a typical ecosystem. Note the major losses of energy to the environment through respiration (heat) and the loss through dead tissue at each step. Note also that this diagram shows dramatically how the entire system depends on the capture of energy by primary producers.

heat (energy) loss to environment

decomposers — 100% energy to environment

loss = 23.8% loss = 76.2%

top carnivores

5.5%

loss = 23.5% loss = 71%

carnivores

11.4%

loss = 21.4% loss = 67.3%

herbivores

loss = 20.4% 16.2% loss = 63.4%

organic waste / dead tissue

respiratory heat loss

producers

photosynthesis

1.2% solar energy

98.8% solar energy not harnessed by living organisms

Earth

sun

alternately stored and used to power the life processes of animals through which it moves, much as the potential energy of a river stored behind a dam can be used to drive a mill wheel. The actual use of energy in living systems, of course, is more complex than that simple analogy, because several important processes can take place at each step along life's energy chain.

Trophic Levels

The energy captured by producers and consumers is temporarily stored until one organism eats another. Each of these storage steps along a food chain or food web is called a **trophic level.** The primary producers represent the first trophic level, herbivores occupy the second, carnivores that eat herbivores form the third trophic level, and so on.

Although there is theoretically no limit to the number of trophic levels, in reality there are practical limitations. Every time one organism eats another, only a small fraction of the energy present in the lower trophic level is stored in the next higher level. Finding, catching, and ingesting food uses up energy, as does digesting food. And very few animals even come close to extracting all the energy in their diet. In addition, life processes such as cellular respiration convert a great deal of chemical energy into heat, and much of that heat (nearly all of it in some organisms) is lost to the environment (Figure 3.15).

Estimates of energy loss vary widely, but on the average, only about 10 percent of the energy fixed by plants is ultimately stored by herbivores. Only 10 percent of the energy herbivores accumulate ends up being stored in the living tissues of the carnivores that eat them. And only 10 percent of *that* energy is successfully converted into living tissue by carnivores on the third trophic level. This is known as the ecological **rule of 10** or the *10 percent rule* (although the actual efficiency of energy transfer in food webs varies widely).

The rule of 10 is important because the less energy available at any trophic level, the less living tissue that trophic level can support. Inefficient energy chains thus create what are called **ecological pyramids** (Figure 3.16). In an ecological pyramid, each trophic level contains only one-tenth as much living tissue as the layer beneath it. Often this relationship also produces a *pyramid of numbers,* which means that each successive trophic level contains fewer individual organisms than the level below.

In some cases, however, the consumers are much smaller than the organisms they consume, and the pyramid of numbers is turned upside down. (Thousands of insects may graze on a single tree, for example, and countless mosquitoes can feed off a few unfortunate deer or humans.) But even in situations where the pyramid of numbers is reversed, the pyramid of biomass still applies.

These ecological rules have dramatic social and ecological consequences. For example, if a human chooses to eat only red meat to gain a kilogram of weight, that person must eat 10 kilograms of beef. The cattle that

Figure 3.16 Ecological pyramids and pyramids of numbers, biomass, and energy. (left) Numbers of animals of different sizes in a tropical forest. (center) Dry biomass of organisms at Silver Springs, Florida. (right) Percent of total ecosystem metabolism for different groups of organisms in a temperate meadow.

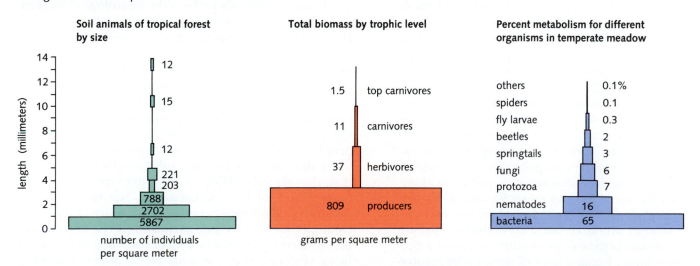

produced that 10 kilograms of flesh must have originally eaten at least 10 times that weight, or 100 kilograms, of fodder. For us to obtain 1 unit of energy from beef thus requires the storage of at least 100 units of energy in cattle feed. (And this calculation ignores the energy spent to process, store, transport, and sell food.)

That's why it makes good ecological sense that the largest land animals—such as elephants—are vegetarians and that the largest marine animals—such as whales—are plankton feeders. Such large animals need prodigious quantities of food to build their own living tissue. Were they to feed on large animals, those already on the fourth or fifth trophic levels, the total amount of primary production needed to support them would be inconceivable. A given ecosystem could support very few, if any, of them. But by feeding at or near the base of the ecological pyramid, whales and elephants make much more efficient use of energy so that a given ecosystem can support many more of them.

THE CYCLING OF NUTRIENTS

We have seen that animals acquire the energy essential to life by ingesting organic molecules carrying that energy. At the same time, animals obtain from food the essential organic nutrients they need to construct their own body tissues.

For proper growth, most organisms require roughly 17 chemical elements or *essential nutrients*. Six of these—carbon, hydrogen, oxygen, nitrogen, phosphorus, and potassium—are required in large amounts. Calcium, magnesium, sulfur, iron, and manganese are required in lesser quantities. Still smaller amounts of sodium, boron, molybdenum, copper, zinc, and chlorine are needed. Other elements, including vanadium, cobalt, iodine, selenium, silicon, fluorine, and barium, are required in very small quantities (*trace* amounts) by some organisms. Several members of these last two groups, though essential in minute amounts to ensure proper growth, may be toxic to both plants and animals in high doses.

Interestingly, though both energy and nutrients are passed from one organism to another by the same complex molecules, the paths that energy and nutrients ultimately take through ecosystems are quite different.

Energy, as we have seen, flows through ecosystems. It arrives steadily from the sun, passes from one trophic level to the next, and dissipates in the environment along the way. But the earth receives no such continuous supply of chemical elements from space. These nutrient elements are neither produced nor used up but are passed around from one organism to another in closed loops called *nutrient cycles*. Because most of these cycles involve the passage of elements through both living organisms and geological features of the globe, they are often referred to as **biogeochemical cycles.**

Nutrient cycling and recycling have been going on since life began, and the nutrient elements circulating today have seen many previous incarnations. Atoms of carbon that reside in your body for the time being may once have been part of rocks on the ocean floor, of single-celled organisms floating in the central Pacific, or of the tail of a long-extinct dinosaur.

The Hydrological Cycle

The cycling of water through clouds, rain and snow, and rivers is the most familiar and visible of the biogeochemical cycles, and its fundamental processes (evaporation, condensation, precipitation, and runoff back to the sea) are commonly understood (Figure 3.17).

Global air movements often carry moisture for hundreds or even thousands of kilometers before it condenses and falls as rain. It is tempting, therefore, to view the **hydrological cycle** as a strictly physical phenomenon that *affects* life but is itself little affected *by* life. It has recently become clear, however, that water often cycles much more locally and that rainfall patterns are often strongly affected by living organisms. We now know, for example, that trees in a tropical rain forest return a great deal of water to the atmosphere through transpiration from their leaves. Much of that moisture feeds the heavy local rainstorms that keep the forest so well watered. We will see in Chapters 6 and 7 that removing large areas of rain forest interrupts this cycle and can cause major, long-lasting changes in local climate. As humans continue to change the face of the biosphere, we must understand how the changes we make may affect the hydrological cycle.

Human activity also affects the hydrological cycle by redistributing surface and underground water supplies. Humans use enormous quantities of fresh water for irrigating farm crops, mining mineral resources, producing steel, and converting shale and coal to liquid fuels. Much of the world's current water supply is pumped from underground reservoirs called *aquifers*. Some aquifers are continuously replenished by percolation of rainfall through the ground. Other aquifers—especially in desert and prairie areas—were established thousands of years ago when those regions received substantially more rainfall than they do today. Water in the latter type of aquifer is often called *fossil water,* because it has been trapped underground for long periods of time. Much of the water currently used to irrigate farmland in the central and midwestern United States is fossil water and hence is not being replenished at the same rate at which it is being

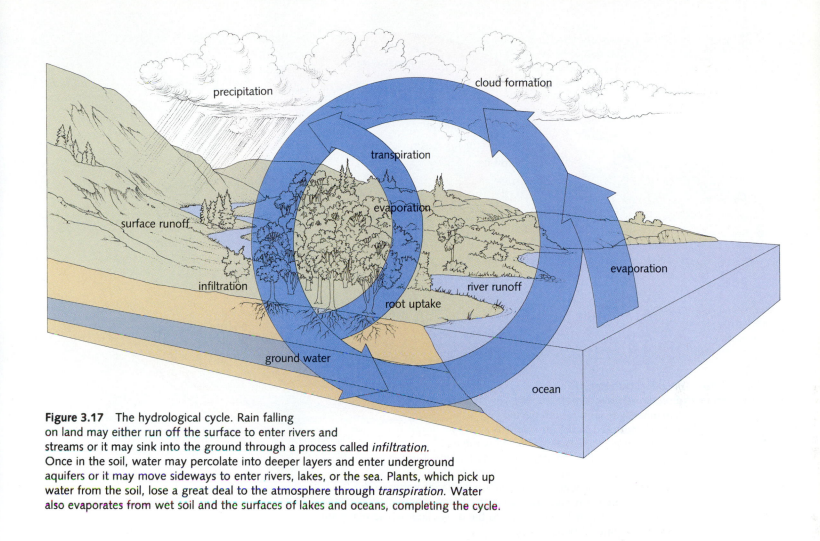

Figure 3.17 The hydrological cycle. Rain falling on land may either run off the surface to enter rivers and streams or it may sink into the ground through a process called *infiltration*. Once in the soil, water may percolate into deeper layers and enter underground aquifers or it may move sideways to enter rivers, lakes, or the sea. Plants, which pick up water from the soil, lose a great deal to the atmosphere through *transpiration*. Water also evaporates from wet soil and the surfaces of lakes and oceans, completing the cycle.

consumed. This situation requires careful attention if we are to avoid serious problems in the future (see Theory in Action, The Hidden Water Crisis, p. 49.) We will discuss the additional problem of water pollution in Chapter 7.

The Carbon Cycle

Carbon is an element of many faces. Bonded to hydrogen and oxygen, carbon is the backbone of the organic molecules on which life is based. It is also an important component of carbonate rocks. And as we noted in Chapter 2, carbon in atmospheric carbon dioxide helps regulate the earth's temperature.

Not surprisingly, the movement of carbon through the biosphere is of paramount importance to life on earth. Scientists have understood the basics of the **carbon cycle** for some time, but many questions still remain about how

much carbon travels along which part of the pathway at any particular time (Figure 3.18).

Almost all organisms consist of 49 percent carbon by dry weight. This means that organic compounds—both in organisms alive today and in the fossil remains of ancient organisms (such as coal, oil, and natural gas)—are important carbon storage sites in the biosphere.

Before human civilization arose, the biosphere gradually converted more and more carbon from gaseous form into carbohydrate, bone, and the shells of marine organisms. This lowered the carbon dioxide content of the atmosphere and kept the earth's temperature constant as the sun grew slowly warmer (Chapter 2). But during the last two centuries, humans have begun to alter the carbon cycle in two ways. First, we are cutting down the world's great forests, particularly in tropical countries where many of the trees are burned as fuel. Cutting trees destroys living tissue that fixes atmospheric CO_2, and burning these trees immediately returns the carbon they contain to the

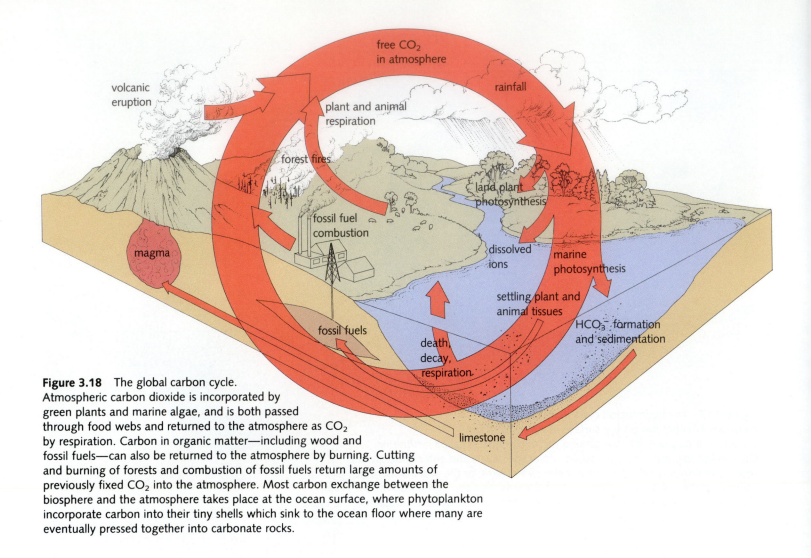

Figure 3.18 The global carbon cycle. Atmospheric carbon dioxide is incorporated by green plants and marine algae, and is both passed through food webs and returned to the atmosphere as CO_2 by respiration. Carbon in organic matter—including wood and fossil fuels—can also be returned to the atmosphere by burning. Cutting and burning of forests and combustion of fossil fuels return large amounts of previously fixed CO_2 into the atmosphere. Most carbon exchange between the biosphere and the atmosphere takes place at the ocean surface, where phytoplankton incorporate carbon into their tiny shells which sink to the ocean floor where many are eventually pressed together into carbonate rocks.

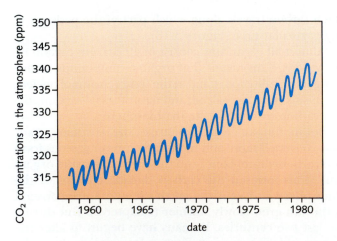

Figure 3.19 Atmospheric researchers began taking careful records of atmospheric CO_2 in 1957 at Mauna Loa in Hawaii. Since that time, the amount of CO_2 has been increasing steadily. The annual fluctuations are caused mostly by the seasonal fixing of CO_2 by green plants in the northern hemisphere.

atmosphere. Second, we are burning large quantities of coal, oil, and natural gas. These two forms of combustion return enough carbon dioxide to the atmosphere to raise atmospheric CO_2 concentrations measurably (Figure 3.19).

Because CO_2 is a major factor in controlling the earth's temperature, this steady increase may cause a steady rise in global temperatures, and could affect the biosphere in a number of ways. But many questions about the importance of increased atmospheric CO_2 remain unanswered, as we will see in Chapter 7.

The Nitrogen Cycle

From primary production to consumers Primary producers absorb nitrogen in a simple form—usually either ammonia (NH_3), ammonium (NH_4^+), or nitrate (NO_3^-).

The Hidden Water Crisis

Both in Hawaii and in the midwest, water essential for both human consumption and agriculture is being removed from aquifers more rapidly than it is being replenished by rainfall.

Though all the Hawaiian islands are blessed with ample rainfall, most rainwater is allowed to run off, unused, into the sea, while groundwater supplies are used for both drinking and irrigation. Because stream water in tropical places like Hawaii can carry a variety of diseases, it makes sense to reserve the islands' underground water supply—some of the purest fresh water in the world—for drinking. But most Hawaiian plantations consume prodigious amounts of that precious groundwater even though the enormous quantities of water needed to irrigate Hawaii's sugar cane fields do not need to be pure enough to drink. During the unusually dry years of the early 1980s, strong demand for water seriously depleted Oahu's groundwater, threatening to force the island either to treat and drink its surface water or to spend huge sums to desalinate seawater. In the middle of the decade, fresh rains and decreased demand for sugar relieved the drain on Oahu's water supply—for a time. Similar situations face populated coastal and island regions from Cape Cod to the Florida Keys.

A related problem involves large areas of Kansas, Arkansas, and other midwestern states, which for years have based their economies on enormous farms irrigated with groundwater. Unfortunately, the groundwater supplies of many midwestern states consist of fossil water. Until very recently, this groundwater was being pumped much faster than it is being replenished, and water levels in these aquifers had dropped between 23 and 40 percent by 1980.

Facing dire predictions that even their deepest wells would soon run dry, farmers responded. Some now use advanced irrigation techniques to conserve water, and others have planted some cropland in native grasses to allow the aquifer to replenish itself. The water table in many places is now dropping only one-third as rapidly as it was in the 1970s, but the long-term outlook for irrigation in the area remains uncertain.

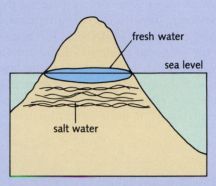

This schematic shows an island with its freshwater lens "floating" above salt water beneath it.

This map shows the distribution of the 134,000 square-mile Ogalalla Aquifer, a major fossil aquifer that varies in thickness from a few meters to more than 200 meters.

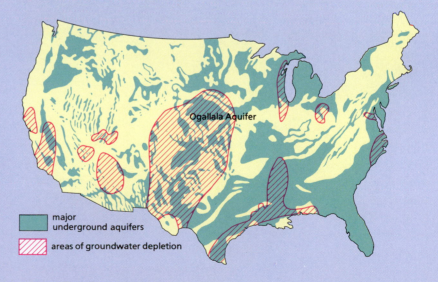

Plants and algae use energy from photosynthesis to concentrate nitrogen in their tissues and then use more solar energy to fuel the assembly of nitrogen, hydrogen, carbon, and other elements into proteins and amino acids, the complex molecules used as building blocks in living systems.

From organisms to the environment When animals eat plants or other animals, they digest proteins, breaking them down into their component nitrogen-containing amino acids. Some of these amino acids are reassembled into the personal proteins of the animal that has eaten them. Others are broken down to liberate the energy they contain, releasing nitrogen in the form of ammonia (NH_3).

Because ammonia is toxic to many cells even in low concentrations, it must be eliminated from body fluids or changed to a less poisonous form immediately. Most aquatic animals eliminate ammonia continuously into the water around them. Terrestrial animals often convert ammonia into *urea* and concentrate it in urine before eliminating it. Other strategies for eliminating waste are described in Chapter 35.

Microorganisms in the environment Nitrogen wastes in any form don't persist in the environment for long; bacteria go right to work on them. One group, the *Nitrosomonas* bacteria, combine ammonia with oxygen and convert it into *nitrite* (NO_2^-). The nitrite may then be converted into *nitrate* (NO_3^-) by bacteria called *Nitrobacter*. These two processes together are called **nitrification.**

Although they rarely appear in food web diagrams, the bacteria and fungi of decay are critical to the **nitrogen**

35

Figure 3.20 The terrestrial nitrogen cycle as described in the text.

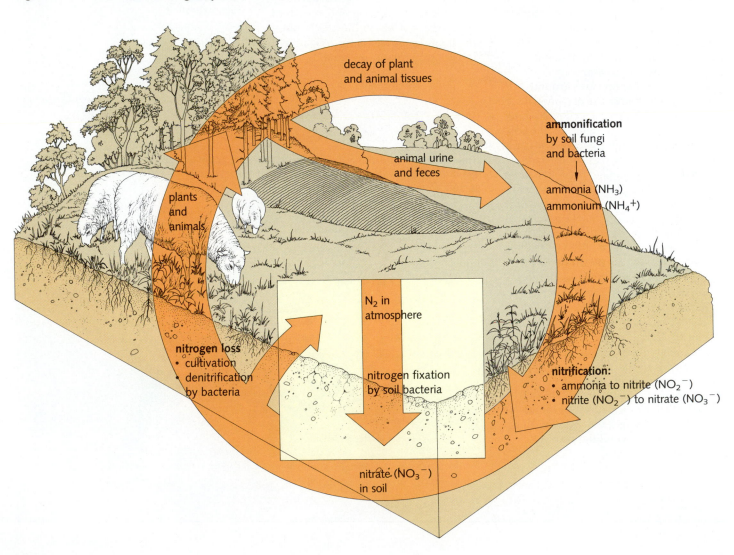

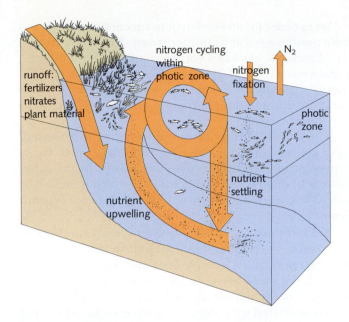

Figure 3.21 The oceanic nitrogen cycle. Note the recycling of nitrogen within the photic zone, the input from land runoff, the steady sedimentation of organic matter out of the photic zone into the depths, and the periodic return of those nutrients to the photic zone in areas of upwelling.

cycle; nitrogen locked up in dead animals and plants is useless to primary producers. Decomposers disassemble the organic molecules in animal and plant carcasses and release their component elements in simpler form. This process is called **ammonification,** because bacteria and fungi usually release nitrogen in the form of ammonia.

Completing the cycle When nitrate, nitrite, and ammonia are released into the soil of a healthy forest or into shallow water filled with growing algae, they may be quickly reabsorbed by the primary producers. Different primary producers preferentially absorb nitrogen in different forms (Figure 3.20). Many terrestrial plants absorb nitrate best, which is one reason why most fertilizers contain a lot of nitrogen in nitrate form. Some terrestrial plants and many algae, however, pick up ammonia more readily than nitrate.

In the open sea, dead marine organisms and their nutrient-rich solid wastes sink rapidly out of the *photic zone.* Large quantities of nutrients, therefore, end up in slowly moving currents near the ocean floor, far out of reach of photosynthetic primary producers. In certain places around the world, particular combinations of winds and ocean currents pull large quantities of this nutrient-laden water back up from the ocean floor into the photic zone, a phenomenon known as *upwelling* (Figure 3.21).

In areas of steady upwelling, phytoplankton thrive and grow at rapid rates, providing the basis for a vigorous and productive food web that may include anchovies, herring, lobster, cod, hake, bluefish, tuna, and many other commercially important kinds of seafood. The productivity of upwelling areas is staggering; up to 50 percent of the worldwide fish catch comes from upwelling areas that comprise a mere tenth of 1 percent of the total ocean surface. In Chapters 6 and 7, we will discuss the importance to oceanic ecosystems and to humans of upwellings such as those off the coasts of California and Peru.

Earth's nitrogen reserve: the atmosphere While aquatic and terrestrial primary producers scramble for available nitrogen, an enormous reservoir of gaseous nitrogen floats out of reach in the atmosphere. Unfortunately, the paired atoms in nitrogen gas are held together by a powerful chemical bond that cannot be broken by most organisms. In fact, only a few bacteria can break that bond and incorporate atmospheric nitrogen into living tissue, a process known as **nitrogen fixation.** The most familiar of these are the **nitrogen-fixing bacteria** that live on the roots of such terrestrial plants as peas, soybeans, and many other members of the legume family. The most important nitrogen fixers in the sea are photosynthetic cyanobacteria. These bacteria are often incorrectly called blue-green algae.

Denitrifying bacteria carry out precisely the reverse reaction; they convert organic nitrogen back into nitrogen gas and release it to the atmosphere. Ironically, the constant addition of nitrogen fertilizers to farm soil can encourage the growth of denitrifying soil bacteria that pump significant amounts of nitrogen out of the soil and back into the atmosphere!

Nutrient Limitation

Major and minor nutrients all revolve through their own complex biogeochemical cycles. Understanding how all the cycles work together is a formidable task, because as long as atoms are bound together in a single organic compound, they must travel together through the ecosystem. For example, protein molecules contain carbon, nitrogen, phosphorus, sulfur, iron, magnesium, and other elements.

Figure 3.22 This schematic representation of interlocking nutrient cycles demonstrates metaphorically how the movement of each nutrient through ecosystems is dependent upon the movements of all others. If even a minor nutrient is present in limiting quantities, it can slow down the entire assemblage.

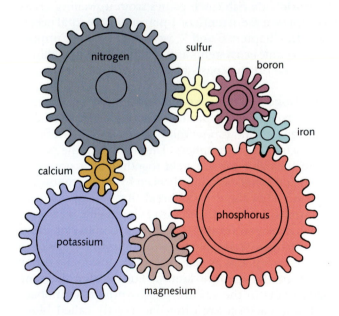

Yet in order for an ecosystem to function, each nutrient cycle must revolve at its proper speed. The whole assembly of cycles works like a complicated set of interlocking cogwheels (Figure 3.22). The steady, one-way flow of energy through the system provides the power to drive the gears through their perpetual revolutions, each at its own specific rate.

The **law of limiting factors** states that the growth of an organism will be limited if any essential growth factor is present in insufficient quantity relative to the other factors. In deserts, plant and animal growth is limited by lack of water. In the deep sea, the limiting factor is insufficient light.

The productivity of an entire ecosystem can thus be curtailed by a single nutrient present in short supply, a phenomenon called *nutrient limitation*. In the cogwheel analogy, the speed of the entire assembly is controlled by the speed of the slowest wheel. When any single nutrient cycle is interfered with—when any wheel sticks—the whole system must slow down or stop.

Nutrient limitation in the abstract is neither a "good" phenomenon nor a "bad" one; it is simply a fact of life and an important part of the dynamic balance of nature. In some situations, nutrient limitation keeps ecosystems functioning in ways we find attractive and desirable, while in other cases it curtails the productivity of ecosystems we would like to exploit more efficiently.

In ponds and lakes, for example, we often find ample supplies of nitrogen, iron, magnesium, and several other nutrients that could support much higher rates of plant growth than usually occur. The primary producers in these ecosystems—and hence the system as a whole—are held back by a lack of phosphorus. If phosphorus is suddenly added in large amounts, the phosphorus wheel becomes unstuck, and in no time there begins the prodigious growth of algae called an algal bloom that can cover ponds and rivers, choking out life beneath it.

On land and in the sea, on the other hand, phosphorus is often plentiful, and nitrogen is commonly the limiting nutrient. For this reason, farmers use extra nitrogen, either in natural forms such as "green manure" (plants and other organic matter added to the soil) or in manufactured chemical fertilizers, to increase crop growth. Similarly, the addition of nitrogen-rich sewage and agricultural runoff to coastal waters can greatly increase the productivity of marine systems. But massive discharges of untreated sewage from coastal towns can seriously upset the local ecological balance, and in some cases they are thought to foster blooms of undesirable—even poisonous—algae.

In the next two chapters we will see how the availability of energy combines with the storage and cycling of nutrients to control the growth of animal and plant populations, the structures of natural ecosystems, and the response of natural communities to human activities.

SUMMARY

Autotrophs, also called primary producers, obtain energy from sunlight and create living tissue from nonliving matter. Photosynthesis is the process that most autotrophs carry on to store energy as carbohydrates. Heterotrophs obtain both energy and nutrients by eating other organisms and digesting the complex organic molecules in living tissue. Heterotrophs obtain energy by various ecological strategies including herbivory, carnivory, and decomposing the decaying tissues of plants and animals.

In every ecosystem, feeding relationships link plants and animals together into food chains and food webs. Energy flows through food chains and food webs in a one-way stream from lower to higher trophic levels. In contrast, nutrients—such as carbon and nitrogen—cycle continuously.

Limiting factors—the lack of any single requirement for growth—can constrain productivity in different living systems throughout the biosphere. Any conditions that control the growth of primary producers in an ecosystem indirectly control the lives of all other organisms as well.

STUDY FOCUS

After studying this chapter, you should be able to:

■ Explain how energy flows through collections of plants and animals.
■ Describe how the principles of energy flow affect the structure of ecosystems.
■ Demonstrate how nutrients cycle, both within and between ecosystems.
■ Show how nutrients and energy connect organisms and ecosystems to each other.
■ Explain the law of limiting factors.

SELECTED TERMS

autotroph *p. 34*
primary producer *p. 34*
photosynthesis *p. 34*
carbohydrate *p. 34*
heterotroph *p. 36*
herbivore *p. 37*
primary consumer *p. 37*
carnivore *p. 39*
secondary consumer *p. 39*
tertiary consumer *p. 39*
food chain *p. 42*
food web *p. 42*
trophic level *p. 45*

rule of 10 *p. 45*
ecological pyramid *p. 45*
hydrological cycle *p. 46*
carbon cycle *p. 47*
nitrogen cycle *p. 50*
law of limiting factors *p. 52*

REVIEW

Discussion Questions

1. Many aquatic organisms use feeding techniques that are not used by terrestrial creatures. Give two examples, and explain how these feeding techniques help aquatic organisms obtain nutrients.

2. What is the photic zone? How do the existence of the photic zone and the movements of nitrogen in the sea combine to cause nutrient limitation of phytoplankton in the open sea?

3. What are the three steps in the nitrification process? What organisms perform these transfers of nutrients, and why are they important?

4. Why is it difficult for grazing animals to extract energy from their feed? What physical and behavioral characteristics of a grazing animal increase the efficiency of its digestion?

5. What are the differences between the ways in which energy and nutrient elements move through living systems? What is energy flow? Why is it possible for energy to flow through ecosystems whereas nutrients must cycle?

6. Assume that a particular human lives entirely on the meat of a bird that in turn eats only herbivorous insects. If the human gains 1 kilogram on this diet, how many kilograms of plant material has the human eaten indirectly?

Objective Questions (Answers in Appendix)

7. The energy for most communities comes from
   (a) the sun.
   (b) decomposition of organisms.
   (c) the oceans and lakes.
   (d) secondary consumers.

8. Which of the following statements about ecosystems is *false*?
   (a) All the energy entering the ecosystem is passed on to the decomposers.
   (b) The primary producers control the lives of herbivores and carnivores.
   (c) The efficiency of energy transfer in food webs varies widely.
   (d) Ecosystems must start with the capture of energy by autotrophs.

9. Organisms that synthesize their own food are called
   (a) microorganisms.   (c) heterotrophs.
   (b) autotrophs.   (d) secondary consumers.

10. Which of the following nutrients is often a limiting factor in ponds and lakes?
    (a) nitrogen   (c) magnesium
    (b) phosphorus   (d) iron

11. Which process causes a loss of nitrogen from the soil because of bacterial activity?
    (a) upwelling   (c) nitrification
    (b) ammonification   (d) denitrification

CHAPTER **4**

Organisms, Environments, and Environmental Variables

Rarely are two adjacent environments as different as they are in the Middle East, where the turquoise waters of the Red Sea sparkle between the parched sands of Africa and Arabia (Figure 4.1). On land, mountains and sand dunes are nearly devoid of visible life. The desert air is hot and dry. Plant life is rare outside oases; here an acacia tree spreads its gnarled branches, there a patch of succulent plants clings to a crevice in the rocks. An occasional sand lizard chases desert flies who are themselves searching constantly for moisture. Now and again a lone camel wanders by in search of forage. The unbroken silence is both unfamiliar and unsettling to those from more forested regions: no insects chirp, no songbirds sing, and the wind encounters no leaves to rustle as it whips the sand into swirling dust devils.

But where desert rocks plunge into the sea, the scene changes dramatically. The coral reef teems with life. Around and above the corals swarm scores of multicolored fishes. On the surface of the reef, crabs scuttle for shelter, worms undulate in and out of their holes, and shrimp scramble between coral branches, making loud snapping sounds with their claws. Everywhere, plants and animals compete fiercely for every inch of available space.

These are environmental extremes. At one end of the spectrum, the Sinai desert is an example of a *physically controlled* environment in which physical stresses such as daytime heat, nighttime cold, intense sunlight, and lack of water act as *limiting factors* to restrict animal and plant growth. At the other end of the spectrum, growth of reef animals and plants is largely *biologically controlled,* for these organisms are limited as much by interactions with each other as they are by physical factors.

THE IMPORTANCE OF UNDERSTANDING ENVIRONMENTAL VARIABLES

Studies of the harsh Sinai desert and the exotic Red Sea coral reef are interesting in their own right, but can also teach us general lessons about the effects of environmental conditions on living systems. First, environmental conditions vary widely: Tropical deserts are hot and dry; temperate rain forests are cool and wet. Second, plants and animals "divide up" the biosphere: Different species live in different places, in part because environmental conditions vary.

It is easy to see why camels don't live on coral reefs and why sharks wouldn't get along very well in the desert. But other environmental variables exert more subtle influences. It is often difficult, therefore, to pinpoint the factors that determine why organisms live where they do.

Yet this information is more important today than ever before. Earth's human population is constantly interacting with the environment and initiating changes in ecosystems around us. We need to predict which systems will respond positively to human intervention, which will tolerate our influence, and which will collapse if we disturb them. We also need to understand the critical (and often hidden) links between ecosystems.

Making useful predictions about the impact of environmental change, however, requires more than a feeling that water *seems* to be in short supply in the desert, or that reef animals *seem* to depend on each other. Rather, ecologists must *quantify,* or describe in numerical terms, the ways in which environmental conditions affect the ability of organisms to survive and reproduce. For example, what is the minimum amount of water a particular species of desert plant needs to grow, flower, and produce seeds? After learning how individual organisms respond to environmental conditions, ecologists can try to predict how groups of organisms in complex ecosystems will respond to alterations in those conditions. Although we will not discuss the detailed mathematical models ecologists use, it is worthwhile to understand the more important theories and experiments that have shaped our understanding of environments and their effects on organisms.

ENVIRONMENTS, HABITATS, AND NICHES

The term **environment** refers to all the conditions surrounding an organism. Those conditions are often divided into two subgroups: the physical, or *abiotic,* environment and the biological, or *biotic,* environment. The **physical environment** includes all the conditions created by the nonliving components of the organism's surroundings: sunlight, heat, moisture, the speed of wind and water currents, the size of sand or sediment grains, and so on. The **biotic environment** consists of the living organisms in the habitat—that is, other species that may serve as food, parasites, predators, or competitors.

In practice, however, it is often difficult to separate biotic and abiotic factors. Earthworms, for example, live within the organic soils and moldy leaves that accumulate in woods and forests. The decaying vegetation around the worm absorbs rainwater and stays moist far longer than clean sand does. The plant material itself decays because it is home to the myriad fungi, bacteria, and protozoans that inhabit the forest floor. Forest soils are held in place by plant roots, shaded from direct sun, and protected from strong winds by the leafy forest canopy. The environment experienced by the worm, therefore, is created by a combination of physical and biotic factors.

Habitat: Where an Organism Lives

The actual location an organism inhabits within its environment—its address, in effect—is its **habitat.** Just as we describe our addresses by referring to street names, ecologists can characterize habitats by referring to physical characteristics or dominant plant forms.

But ecologists must be careful to analyze habitats from the perspective of the organisms they wish to study, for humans perceive and understand most easily the sorts of environmental conditions encountered by animals their own size. Yet within every habitat are scores of distinct, miniature habitats called **microenvironments** or *microhabitats* that are not always obvious to human observers (Figures 4.2 and 4.3). From the earthworm's tunnel to beetle burrows under the bark of trees, each microhabitat has its own temperature, humidity, and wind speed—conditions referred to collectively as **microclimate.** A wide range of microenvironments can exist within short

Figure 4.1   (top) Bitterly cold winter nights and scorching summer days combine with lack of water to make the Sinai desert one of the world's harshest terrestrial environments. (bottom) Ample light and even temperatures help make the coral reef a hospitable environment for many animals. Corals create an intricate three-dimensional habitat shared by hundreds of fish and invertebrate species.

Figure 4.2 Microclimates within a temperate woodland. These are but a few of the organisms sheltering in various microclimates in and around a single deciduous tree. Clockwise, from top: A moth and its larvae in the uppermost branches, a bird parasite crawling through the down feathers of its host, a bark beetle drilling cavities beneath the bark, nitrifying bacteria beneath the soil surface, an earthworm in the leaf mold within the topsoil, termites within the decaying log, fungi in the process of decomposing moist wood, a salamander beneath a rotting log, and a gall wasp.

distances of each other. For newly germinating seeds, small insects, or bacteria living on tree roots, microclimates are important because environmental conditions even a meter away are of little consequence. All that matters are the amounts of light, nutrients, temperature, and moisture available within a few millimeters of their location.

Niche: How an Organism Makes Its Living

The way each species reacts to its microenvironment, and the ways in which plants and animals interact with other species around them, divide habitats into distinct **niches.** An organism's niche should not be confused with either its habitat, or its microhabitat. If an organism's habitat can be viewed as its "address," its niche can be likened to its "profession."

A full description of an organism's niche includes three important sets of factors pertaining both to where an organism lives and the way it "makes a living:"

- The range of physical factors in which the organism can survive and reproduce: temperature, humidity, salinity, pH, grain size of the soil, and other such variables.

- The biological factors with which the organism interacts: predators, prey, parasites, organisms that provide shelter, and those that compete for the same limiting resources.

- The organism's behavior: when, where, and upon what it feeds, its social organization, and its behavioral interactions with other organisms.

It is not possible to provide a picture or a complete graphical representation of a niche, because pictures and graphs cannot represent so many different factors at once. We can, however, look at two or three aspects of a niche at a time as we try to determine which factors are most important in the ecology of a particular species.

Physical aspects of the niche   One view of niches concentrates on the effects of physical factors in determining where particular organisms can live. Figure 4.4, for example, shows how several different moss species divide up available habitats according to the strength of incoming light and the acidity or alkalinity of the soil.

Biological aspects of the niche   Another perspective on the niche focuses on the way organisms subdivide habitats by dividing up available resources. For example, similar species that live in the same area often prefer slightly different types of food, obtain that food in different places,

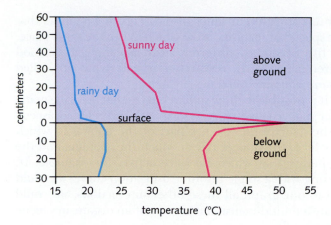

Figure 4.3   Air and soil temperatures near the surface of a sand dune during typical rainy and sunny days. Note the dramatic temperature differences dune organisms would encounter across a vertical distance of only a few centimeters.

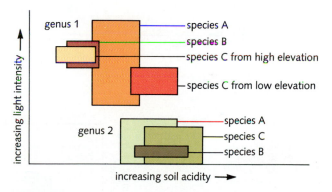

Figure 4.4   The response of several moss species to two physical factors: light and soil acidity. Each species prefers a different light intensity and different range of pH values. Because of these preferences, these species inhabit different microhabitats within the same forest. For example, genus 1 species A tends to live in a less acidic and more intensely lighted habitat than genus 2 species C.

or hunt at different times of day. Figure 4.5 shows that five species of seabirds, all of which live in the same place, eat fish of slightly different sizes. To show those food preferences clearly, researchers create what are called *resource utilization curves*.

It is possible to draw similar curves that describe many other interactions between organisms and selected aspects of their surroundings: the part of the habitat in which they feed, the time of day at which they forage, and the temperature at which they are active, for example. If, starting with a single resource utilization curve, we could add to our graph all these other interactions, we could create a multidimensional picture of an organism's niche (Figure 4.6).

The broadest of all possible niches an organism is capable of occupying is called its *fundamental niche*. However, both positive and negative interactions with other organisms often restrict a species to a smaller niche, called the *realized niche*. We will see how interactions among organisms affect realized niches in the next chapter. Here we will concentrate on the ways environmental conditions control the distribution of both plants and animals.

ENVIRONMENTAL VARIABLES

Sunlight

Sunlight powers the photosynthesis that supplies energy to nearly all life on earth. For that reason, dim light may act as an ecological limiting factor on primary producers and may ultimately limit the productivity of an entire ecosystem. Sunlight is also essential for vision, which many animals rely on for catching food, spotting predators, and communicating with each other (Figure 4.7).

Too much sunlight can be equally dangerous, however. Trees and cacti that live in full sunlight must protect their delicate tissues from damaging ultraviolet radiation. Organisms used to protective shade have as difficult a time tolerating brilliant sunshine as open field organisms would have surviving in dense forest.

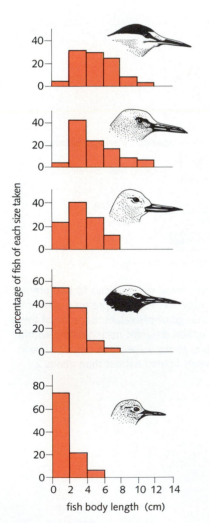

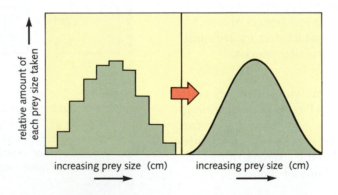

Figure 4.5 (left) Beak length and food choice in seabirds. Each of these tern species from Christmas Island in the Pacific has a different size bill. Each chooses fish of various sizes in different proportions, as shown by the bar graphs with each species. (above) Ecologists translate these bar graphs into curves called *resource utilization curves*. Resource curves make it easy to compare the feeding habits of different species.

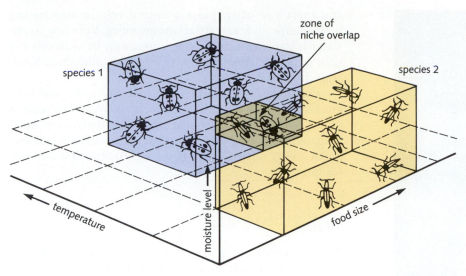

Figure 4.6 Niches visualized as three-dimensional spaces created by environmental conditions and food characteristics. Shown here is a three-dimensional version of Figure 4.3 for two imaginary insect species, each of which has different preferences for temperature, moisture, and food size. Each insect has part of its niche space to itself and shares a portion with the other species. Where overlap occurs, competition is likely to take place (as we will see in Chapter 5).

Temperature

Most organisms can survive only within a specific, limited range of temperatures. If their body temperatures either rise above or fall below that range, the critical chemical reactions in their tissues get "out of sync" with one another, and metabolic chaos results. Extremes of either cold or heat can be equally deadly.

Water and Dissolved Salts

All organisms must maintain a precise balance of water, dissolved salts, and organic molecules in their body fluids to keep their cells alive. For this reason, both the availability of water and the concentration of dissolved salts in the immediate vicinity are important environmental variables.

Many plants and animals cannot survive in deserts, for example, because they cannot acquire and store water under such dry conditions. But an equally large number of organisms cannot live in swamps and marshes because they cannot survive where there is so much water in the soil.

Salinity, the concentration of dissolved inorganic salts, affects organisms' abilities to control their water balance

Figure 4.7 These plants use their large, thin leaves to catch sunlight in the shade beneath the canopy of tropical rain forests.

Figure 4.8 This diving seal's ability to hold its breath and regulate its body's demand for oxygen allows it to feed underwater for extended periods of time.

and therefore often plays a critical role in structuring both aquatic and terrestrial communities. Very few freshwater organisms can survive immersion in seawater for very long, and few marine plants and animals can tolerate fresh water.

The ability of crop plants to tolerate salty soil is of great importance to world agriculture today. In dry regions of the world, especially places where fields have been irrigated for a long time, salts can accumulate in soil and make it very difficult for plant roots to absorb water (see Chapter 7).

Oxygen

The concentration of available oxygen can be an important limiting factor in a variety of environments. Most multicellular organisms are **aerobes** that require oxygen for respiration. But many bacteria are **anaerobes** that thrive only in the absence of oxygen. Too much oxygen in the surroundings of anaerobic organisms can be just as fatal to them as a lack of oxygen is to aerobes. In aquatic environments and in wet soil or mud, oxygen concentrations can easily fall too low to support aerobic life. Under these conditions, anaerobic organisms thrive. Air-breathing animals that spend time under water and aquatic animals that must periodically survive exposure to drying air have to conserve oxygen when it is not readily available (Figure 4.8). We will learn more about the ways organisms obtain oxygen and deliver it to their tissues in Chapter 33.

Metabolic Wastes

All organisms produce metabolic waste products. As we noted in Chapter 3, animals exhale carbon dioxide and excrete nitrogen. Plants release oxygen by day, give off carbon dioxide by night, and often discard leaves and stems on a seasonal basis. In most cases, waste products such as these enter biogeochemical cycles in which they are broken down or carried away. But in certain microenvironments, wastes can accumulate. In mud, for example, noxious gases such as hydrogen sulfide (responsible for the odor of rotten eggs) and methane (commonly called marsh gas) may build up, making the immediate environment unsuitable for many organisms.

Nutrients

The distribution of nutrients is also critical in determining where organisms can grow and where they cannot. Soils formed from volcanic rock, for example, may contain far more magnesium than soil formed from sand-stone or shale, and may therefore support very different plant communities. In the sea, the distribution of nutrients by ocean currents is a major factor in determining where marine primary producers can grow.

THE IMPACT OF ENVIRONMENTAL VARIABLES ON ORGANISMS

Keeping Internal Conditions Constant

Many organisms are able to keep their internal conditions—temperature, water and salt balance, availability of nutrients, and so on—within certain limits through a sort of physiological balancing act called *homeostasis* (*homeo* means "similar"; *sta* means "stand or remain"). Although behaviors and physiological mechanisms that permit homeostasis offer many advantages, each has its costs, as we can see by examining various methods of **thermo-regulation,** the control of body temperature.

Mammals, for example, generate heat within their tissues and use several techniques to conserve that heat and maintain high body temperatures. Such organisms are called **endotherms** (*endo* means "internal"; *therm* means "heat"). Other animals, such as turtles, lizards, and snakes, bask in the sun to warm themselves up during the day and shelter in below-ground burrows to conserve heat during cold nights. Because these animals obtain most of their body heat from their surroundings, they are called **ectotherms** (*ecto* means "external").

 We will discuss techniques of thermoregulation in more detail in Chapter 35. For the time being, it is sufficient to realize from this example that temperature regulation under extreme conditions can be energetically quite expensive. During cold weather, endotherms require large amounts of food to stoke their metabolic furnaces, and ectotherms must spend a great deal of precious time basking. During hot weather, animals that sweat require more water, and both endo- and ectotherms must often shelter in the shade (Figure 4.9). For these reasons, *the greater the environmental stress on an organism's homeostatic systems, the more work it has to do just to stay alive, and the less energy it has left over for growth and reproduction.*

Law of Tolerance

Given a range of environmental conditions and the opportunity to move as they choose, animals actively select the best conditions for them from those available (see Theory in Action, Where Are the Bass?, p. 62). However, forced to occupy unfavorable habitats, animals either die

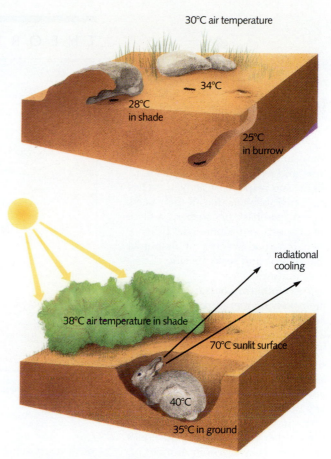

Figure 4.9 (top) Body temperatures of an ectothermic desert insect in different locations at noon in the desert. The body temperatures of this ectothermic animal match those of its microhabitat. (bottom) Many endothermic animals take advantage of microclimate differences as well. Here, a jackrabbit shelters by digging a hole shaded by vegetation to help keep cool in a hot environment.

Where Are the Bass?

Many animals sense light, temperature, and oxygen in their environment, and respond by actively "selecting" microhabitats that fit their needs. The striped bass, or "striper," a favorite of sports and commercial fishermen, is a good example of an animal that detects and seeks specific temperatures.

As ectotherms, bass cannot control their body temperatures independently of the temperature of the water around them. Yet their metabolic processes function best at specific temperatures that vary with the age of the fish. Adult striped bass normally live in coastal ocean waters and migrate up rivers to spawn in the early spring. The eggs hatch in the frigid water of early spring, juveniles grow as streams warm in summer, and adults return to the cooler sea. Over time, bass physiology has adapted to these different temperature regimes. Newly hatched fish survive best near 20°C (68°F), young fish grow fastest between 24°C and 26°C (75°F–79°F), and adults grow best at around 22°C (72°F). Laboratory and field experiments have shown that bass of every age sense the temperature best suited to their needs and seek out water near that temperature.

The natural tendency of bass to seek out a specific temperature at each stage in their life cycle serves the fish well in nature, but can frustrate biologists' attempts to cultivate bass in certain environments. After researchers found that stripers can survive in fresh water, for example, they stocked many rivers and reservoirs around the country with bass. The fish thrive in deep, cool reservoirs, but in others, such as Cherokee Reservoir in Tennessee, scores of adults died mysteri-

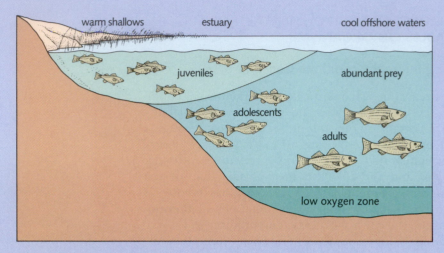

Separation of striped bass by age and thermal preference in a healthy estuary.

ously every summer. Autopsies revealed empty stomachs (despite abundant prey near the surface) and body sores caused by disease. The figure at the right shows how temperature-seeking behavior provides a probable explanation for this situation.

Understanding the role of temperature in bass biology may help fishery managers create conditions more acceptable to bass in a variety of habitats. Fishery personnel could, for example, provide cool water for adult

bass in artificial lakes by controlling the release of water from dams. They might also be able to increase the oxygen content of the bottom water by cutting down on the inflow of organic material. Elsewhere, in the polluted river systems of New York harbor and Chesapeake Bay, striped bass stocks have been declining drastically. Understanding the relationship between bass behavior, low oxygen concentrations in bottom water, and the effects of toxic pollutants may some day help save these fish as well.

Bass behavior in Cherokee Reservoir. Adult bass need water temperatures below 25°C (77°F) and an oxygen concentration of at least 2 milligrams per liter. In the spring, conditions are fine. Problems develop during the hot Tennessee summer, when decomposing organic matter uses up much of the oxygen in the bottom water, and the sun heats the surface water. The adults are squeezed in between these areas of unsuitable conditions (early summer). Then, as juveniles mature and change preferences to cooler water, they crowd the acceptable microhabitats (late summer). Adults overflow into water warmer than they can cope with, and die. ▶

SOURCE: Based on material from Charles C. Coutant, August 1986, Thermal niches of striped bass, *Scientific American*, 98.

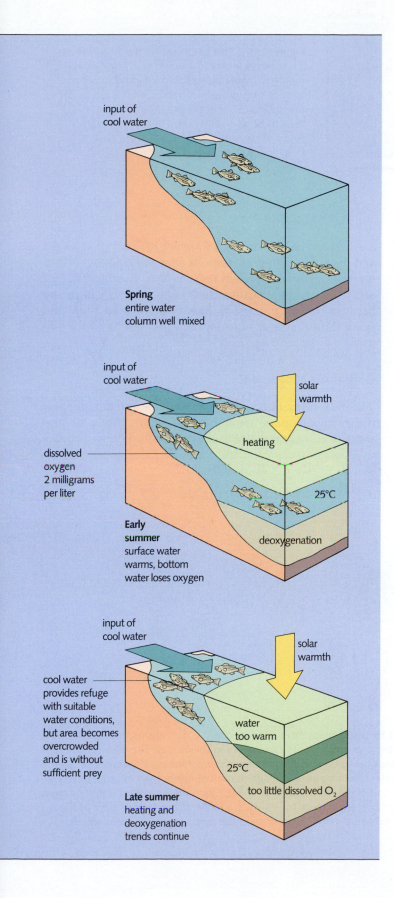

Spring
entire water
column well mixed

input of
cool water

input of
cool water

solar
warmth

heating

dissolved
oxygen
2 milligrams
per liter

25°C

**Early
summer**
surface water
warms, bottom
water loses oxygen

deoxygenation

input of
cool water

solar
warmth

cool water
provides refuge
with suitable
water conditions,
but area becomes
overcrowded
and is without
sufficient prey

water
too warm

25°C

too little dissolved O_2

Late summer
heating and
deoxygenation
trends continue

or fail to reproduce. Similarly, plants grow and flower best under certain conditions. When seeds are spread over a large area, only those that fall in favorable microhabitats survive.

Those conditions of temperature, moisture, sunlight, oxygen concentration, or nutrient availability under which an organism grows and reproduces most vigorously comprise the organism's optimum range for that particular variable. Unless other factors are unfavorable, the organism should be able to survive and reproduce within this optimum range. When conditions fall outside the optimum range, the organism's chances for survival are described by the ecological law of tolerance. The *law of tolerance* states that the growth of a species is limited by that environmental factor for which the organism has the narrowest range of tolerance.

Figure 4.10 shows the response of a typical organism to a gradient of some environmental variable. Above or below the optimum range, too much or too little of the factor in question forces the organism to work harder to maintain homeostasis. These regions along the scale are referred to as *zones of stress*. Under low stress, organisms usually survive. Individual organisms may be able to live under somewhat higher stress as well, but they may be forced to expend so much energy to maintain homeostasis that they do not have enough energy left to reproduce. Under these conditions, though individuals survive, populations cannot. Still higher and lower on the scale are ranges beyond which organisms simply cannot live; these are the *zones of intolerance*.

GLOBAL CLIMATE AND LIFE

Global climate patterns ultimately govern all life on the earth by creating predictable variations in temperature, rainfall, and humidity in different regions of the globe. Because climate and weather are so important, we need to understand them and the factors that determine the conditions under which life must exist around the world.

Climate Versus Weather

Climate consists of the *means* or *averages* of temperature, rainfall, hours of sunlight, wind speed, and so on. **Weather** comprises the day-by-day (or even moment-by-moment) changes in those same factors. Weather therefore includes descriptions of *extremes* in environmental conditions.

Climate data, such as average annual temperature and rainfall, enable biologists to determine whether the averages for particular environmental variables fall within an organism's optimal range. It is on the basis of such averages that certain areas are said to have "wet" or "dry,"

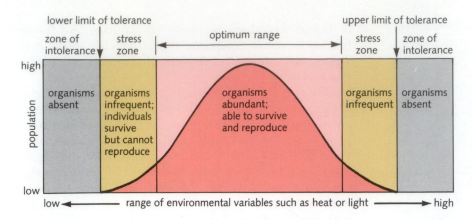

Figure 4.10 This graph shows the response of a typical organism to a range of a single environmental variable such as sunlight, temperature, or oxygen concentration. At the center of the optimum zone, the organisms are potentially most abundant. They become rarer in zones of physiological stress and are absent from zones of intolerance.

"temperate," "subtropical," or "tropical" climates. But only weather data describe the full range of environmental conditions that test organisms' absolute tolerance limits and therefore determine where those organisms can survive (Figure 4.11).

Solar Energy and Temperature

The earth's major climate zones are formed by relatively simple phenomena: unequal heating of the planet's atmosphere, land masses and oceans, and the rotation of the earth around its axis. Although every region on the

earth receives the same *total number of hours* of daylight and darkness each year, regions near the equator receive far more solar energy than regions near the poles (Figure 4.12). The unequal heating that results is further increased because the polar regions radiate more heat back into space than do the equatorial regions.

Global Air Movement

This differential heating between poles and equator sets the earth's climatic machinery in action. Both air and water become less dense as they warm and denser as they

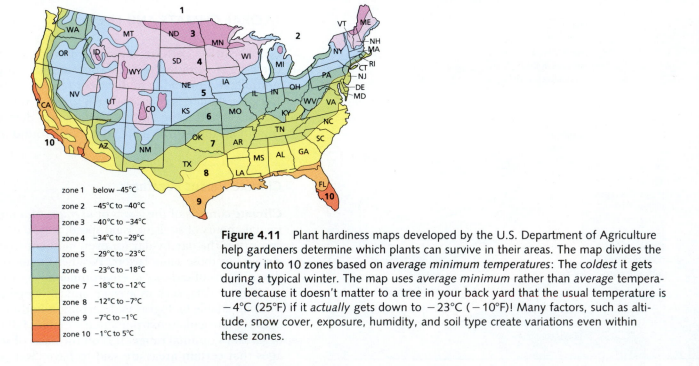

zone 1   below −45°C
zone 2   −45°C to −40°C
zone 3   −40°C to −34°C
zone 4   −34°C to −29°C
zone 5   −29°C to −23°C
zone 6   −23°C to −18°C
zone 7   −18°C to −12°C
zone 8   −12°C to −7°C
zone 9   −7°C to −1°C
zone 10   −1°C to 5°C

Figure 4.11 Plant hardiness maps developed by the U.S. Department of Agriculture help gardeners determine which plants can survive in their areas. The map divides the country into 10 zones based on *average minimum temperatures*: The *coldest* it gets during a typical winter. The map uses *average minimum* rather than *average* temperature because it doesn't matter to a tree in your back yard that the usual temperature is −4°C (25°F) if it *actually* gets down to −23°C (−10°F)! Many factors, such as altitude, snow cover, exposure, humidity, and soil type create variations even within these zones.

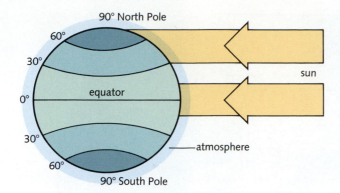

Figure 4.12 Unequal heating of the earth's surface caused by two imaginary beams of solar energy from the sun. The same amount of energy is spread over a larger area at the poles than at the equator, so each square kilometer near the poles receives only a fraction of the energy received by the same area at the equator. Furthermore, the energy reaching the equator travels straight through the earth's atmosphere, while that hitting the poles travels at an angle, passing through more atmospheric gases, water vapor, clouds, and suspended dust that remove energy.

cool (until water reaches 4°C). As a result, warm air and water masses near the equator rise, while cooler air and water near the poles tend to sink. Because the earth is so large, however, the behavior of its atmosphere and oceans is complicated. Figure 4.13 shows that as warm equatorial air rises and moves toward the poles, it loses heat, cools, and becomes more dense. By the time equatorial air reaches 30° North and South latitudes, it cools sufficiently to sink toward the ground. In the meantime, cold air that sinks at the poles moves southward near the earth's surface, picking up heat from the ground as it flows. By the time this polar air reaches 60° North and South latitudes, therefore, it becomes warm enough to rise.

Whenever air rises or sinks, nearby air moves in sideways to replace it, so near the earth's surface air is constantly moving from areas of sinking air to areas of rising air. This horizontal air movement forms prevailing winds such as the *trade winds* that blow steadily and predictably on either side of the equator. Because the earth spins on its axis from west to east, any air currents traveling north or south are deflected sideways. In the northern hemisphere, the deflection is to the right, in the southern hemisphere, to the left. This phenomenon gives rise to the circular air movements familiar to anyone who watches television weather forecasts.

Winds and oceanic surface currents The earth's prevailing winds act steadily on the ocean's surface, pushing water from east to west near the equator, from west to east further north, and from east to west again close to the poles. This influence, combined with the shapes and locations of the continents and certain other factors, generates the surface currents that govern marine ecosystems the way air currents shape life on land (Figure 4.14). Several important currents throughout the world have profound effects on both terrestrial and marine life over vast regions of the globe (see Theory in Action, *El Niño*, p. 69).

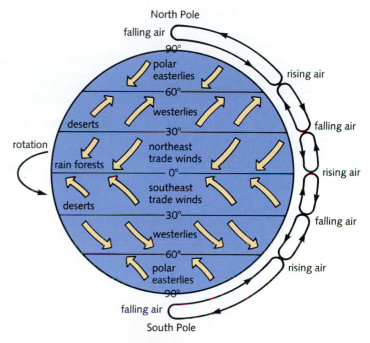

Figure 4.13 This schematic view of the globe shows the symmetrical patterns of atmospheric circulation in each hemisphere and the surface winds they cause.

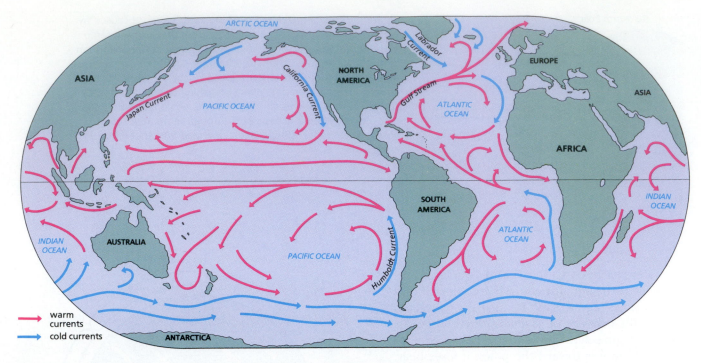

Figure 4.14 Major oceanic surface currents. Note the warm Gulf Stream circulating northward along our Atlantic coast and the cold California current running southward along the Pacific coast. The Gulf Stream transports somewhere around 30 billion gallons every second—more than 65 times the amount of water carried by all the rivers of the world combined.

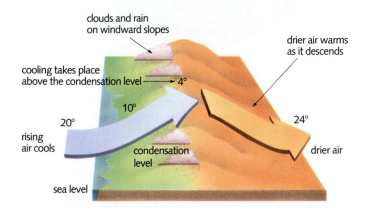

Figure 4.15 The effect of mountain ranges on wind and rainfall.

Winds, mountains, and rainfall Whenever air containing water vapor rises and cools, that water vapor condenses and falls as rain, leaving the air drier than it was before. For this reason, whenever warm, moist air at sea level is pushed up over a mountain range, it drops most of the water it carries on the windward side of those mountains. As this drier air sinks downwind of the mountain range, it warms and dries even more. Therefore, wherever mountain ranges abut the sea, heavy rains drench the windward slopes and desert conditions prevail in the lee, a phenomenon called a "rain shadow" (Figure 4.15 and Figure 4.16).

Lakes and oceans: climate moderators It takes substantially more energy to raise the temperature of water than it does to raise the temperature of air, soil, or rock by the same number of degrees. Thus sunlight warms land and air much faster than bodies of water. Conversely, land and air cool much more rapidly than water.

Because the oceans warm more slowly than continents in the spring and cool more slowly in the fall, they help keep summers cooler and winters warmer over nearby land areas. On sunny summer days the land warms rapidly, heating the air immediately above it. That warm air rises, pulling cool air in off the water to create refreshing sea breezes. In winter, on the other hand, oceans retain more heat than the land, so winds off the water are warmer than those that blow from a land mass.

Effects of Vegetation on Climate

Ecologists have long known that vegetation strongly affects microclimate, but evidence now exists that vegetation influences major climate patterns as well. Plants (especially dense forests) affect the amount of solar radiation a land area absorbs or reflects, the amount of water retained in the soil after each rain, the speed at which that water

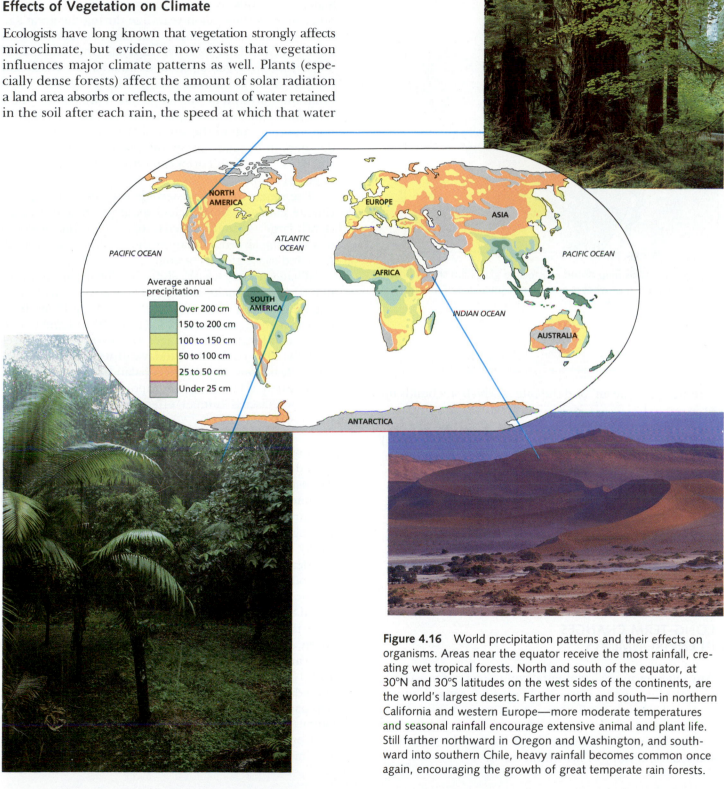

Average annual precipitation

Over 200 cm
150 to 200 cm
100 to 150 cm
50 to 100 cm
25 to 50 cm
Under 25 cm

Figure 4.16 World precipitation patterns and their effects on organisms. Areas near the equator receive the most rainfall, creating wet tropical forests. North and south of the equator, at 30°N and 30°S latitudes on the west sides of the continents, are the world's largest deserts. Farther north and south—in northern California and western Europe—more moderate temperatures and seasonal rainfall encourage extensive animal and plant life. Still farther northward in Oregon and Washington, and southward into southern Chile, heavy rainfall becomes common once again, encouraging the growth of great temperate rain forests.

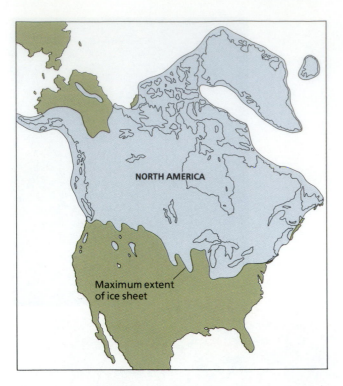

Figure 4.17 This map of the Pleistocene glaciation shows how ice covered much of what is now North America and Europe.

is returned to the air, and the rate at which soil builds up or is eroded.

Furthermore, up to three-fourths of the rain that falls on the great rain forests of the Amazon basin is returned to the atmosphere by either evaporation or transpiration (the loss of water from leaf tissues) from living foliage. Thus, much of the water in the area is continuously recycled. When forests are cleared, this critical cycle is altered, and when too much forest is destroyed, the cycle is broken. Local rainfall decreases dramatically, and, as we will see in Chapters 6 and 7, the result can be either the expansion of existing deserts or the creation of new ones.

LONG-TERM CHANGES IN ENVIRONMENTS

Any changes in environmental conditions can drastically alter the face of an ecosystem. As we discussed in Chapter 2, global climatic conditions have remained remarkably stable over most of the earth's history. Both globally and locally, however, deviations from this stable state have shaped the history of life on the earth.

Global Climatic Changes

Astronomical causes   At least four times over the last 3 million years, decreases in world temperatures have triggered ice ages that gripped the entire planet. Vast ice sheets called glaciers covered much of the northern United States as recently as 12,000 years ago (Figure 4.17). At other times—100 million years ago during the great age of the dinosaurs, for example—the earth became so warm that even Greenland and Antarctica had tropical climates.

Astronomers, geologists, and climatologists think that two classes of astronomical events—changes in the earth's orbit and encounters with other heavenly bodies—may be behind some of these phenomena. The earth's position in orbit around the sun is a critical determinant of the amount of solar energy the planet receives, so instabilities in the earth's orbit can cause long-term changes in global temperature.

The impact of large asteroids may also cause global climate changes. Some scientists argue that if a sufficiently large asteroid crashed onto land, the dust it would toss into the air would blanket the earth, block the sun's rays, and cause global cooling.

Whatever their cause, shifts in global temperatures that triggered climate change were relatively small. During the dinosaur "heat wave," the earth's average temperatures were only 6 to 12°C (10 to 20°F) higher than they are today, and the ice ages saw world temperatures only 5°C (9°F) cooler than today's. But these apparently minor deviations in global heat balance changed the face of the earth, a fact it will be important to keep in mind when we discuss human effects on global climate in Chapter 7.

Geological causes   In 1915 a German scientist named Alfred Wegener published a book entitled *The Origin of the Continents and Oceans.* In it he proposed that the earth's continents were not stationary but moved slowly around the surface of the planet, sometimes colliding, sometimes pulling apart. Wegener collected evidence from geography, botany, and geology to support his views, but had no idea of the forces that could *cause* such movement.

Today, Wegener's notions have been expanded and elaborated into the current theory of **plate tectonics,** also called the *theory of continental drift.* According to plate tectonic theory, the earth's crust is divided into large, irregularly shaped pieces called *plates* (Figure 4.18). Forces in the earth's mantle create powerful currents of molten rock, causing new crust to be created at some of the boundaries between plates. At other boundaries, plates slip past one another or are forced over and under each other (Figure 4.18).

Continental movements have been extremely important in shaping both local and global climate for several

El Niño

In 1983 world climate seemed to be going haywire. Floods devastated Ecuador and Peru, and unusually heavy rains drenched California. Fishes and seabirds died by the thousands off the coast of South America, while seals, whales, and dolphins took off for parts unknown. Record drought fueled brushfires in Australia and devastated farmlands throughout Indonesia.

These seemingly unconnected events, spanning half the globe, were in fact caused by a single set of interrelated changes in global wind and ocean current patterns. Scientists studying the phenomenon soon adopted the name Peruvian farmers had been using to describe it for years—El Niño, or "The Child"—after its regular occurrence around Christmas. Normally, El Niño appears as a short-term, local phenomenon: a slight warming of the surface waters off the coast of South America that Peruvians associate with a spell of bad

fishing. But during 1982–1983 El Niño was much stronger than usual, and the warming of the Eastern Pacific was combined with a change in air pressures over the Western Pacific called the Southern Oscillation. The worldwide effects of this joint phenomenon sparked an international investigation into its causes.

The story of El Niño—Southern Oscillation (or ENSO) is complex. For most of the year, steady winds blow from east to west across the equatorial Pacific. These easterly trades, as they are called, pull surface water away from the west coast of South America and bring nutrient-rich deeper water to the surface. As we noted in Chapter 3, this upwelling supports a rich and productive marine food web. Half a world away in the Western Pacific, the same easterly winds interact with local winds to control the seasonal monsoon rains.

At the start of an ENSO, the trade winds slack off and then reverse. The

upwelling of deep water off South America stops and is replaced by a southward flow of warm water from the equator. Deprived of the nutrient-rich deep water, phytoplankton growth stops, and the coastal food web off Peru collapses. Related changes in ocean temperatures and wind patterns caused drought in the west and heavy rains in the east.

Such phenomena seem to occur about every 4–5 years, but their timing and intensity vary. Exactly what sets this series of events in motion is not known, and it is not clear what causes the effect to be stronger in some years than in others. Ocean surface temperatures, unexplained changes in wind speed and direction, habitat alteration on land, and even undersea earthquakes and major volcanic eruptions may be involved. Though their causes remain a puzzle, ENSOs are clear reminders of the interconnectedness of the earth's ecosystems.

These false-color satellite images illustrate the differences in sea surface temperatures in the tropical Pacific during January 1984, a normal winter (left), and 1983, during an El Niño year. (right) Note the "finger" of colder water stretching westward from the Pacific coast of South America in the normal pattern and the solid band of warm water in 1983.

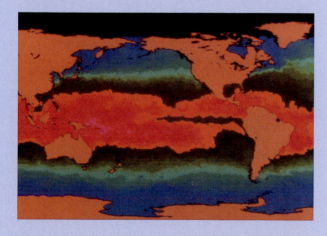

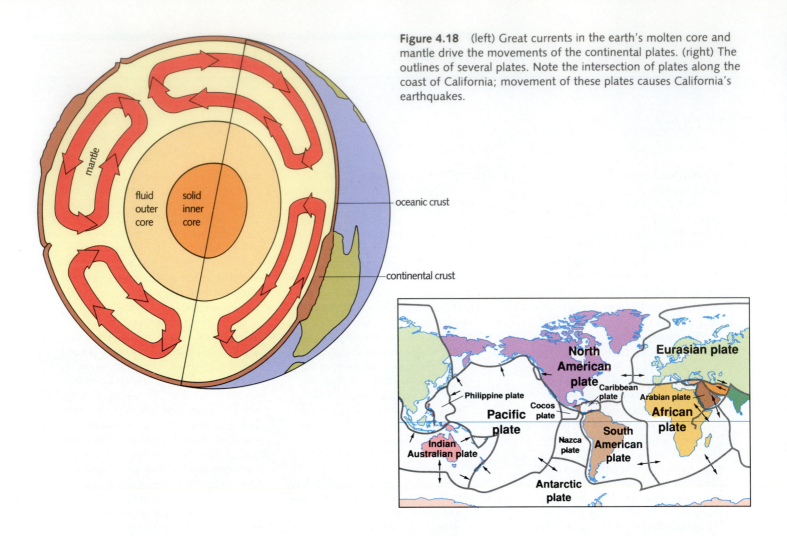

Figure 4.18 (left) Great currents in the earth's molten core and mantle drive the movements of the continental plates. (right) The outlines of several plates. Note the intersection of plates along the coast of California; movement of these plates causes California's earthquakes.

reasons. First, continents have not always been where they are today. Continents that drift from equator to pole, as Greenland and Antarctica seem to have done, carry their resident organisms from tropical to arctic climates. Second, the earth's land masses have repeatedly joined and separated from one another, at times forming "supercontinents" and at other times splitting up into many smaller pieces. The interiors of large continents experience greater fluctuations in weather than the interiors of smaller land masses or areas near major inland seas. Third, when previously separate continents join, they divide the oceans that once flowed between them, dramatically altering the ocean current patterns. Those currents, in turn, have powerful effects on global rainfall and temperature patterns.

The planetwide geological forces driving all this activity themselves influence climate through their effects on the biogeochemical cycles we discussed in Chapters 2 and 3. For reasons yet unknown, the forces that drive continental drift seem to fluctuate over millions of years. At certain times, major episodes of volcanic activity have

expelled large quantities of carbon dioxide from the earth's interior into the atmosphere. Until this "extra" heat-retaining greenhouse gas was removed by either geochemical or biological means, it could have caused the earth's temperature to rise. Some researchers believe this mechanism was responsible for warm spells like that of the dinosaur era. We will discuss both the historical and ongoing effects of continental drift on the evolving biosphere in Chapters 14–18.

Biological Causes of Environmental Change: Succession

Because organisms affect their environments so powerfully, any major changes in the collections of plants and animals living in an ecosystem can have significant effects on local environmental conditions. Long ago, ecologists noticed that the collection of plants and animals living in a given area often changes slowly but steadily over time. They gave the name **succession** to this gradual process of environmental change and species replacement. In many cases, succession is an orderly and predictable process that ends with the establishment of a final, stable ecological association called a **climax community.**

Primary succession   When succession occurs for the first time in newly formed habitats such as lava flows, deep bogs or lakes, or sand dunes, it is called **primary succession.** The earliest arrivals in newly formed habitats are organisms that can tolerate relatively harsh physical conditions. Many species that appear in the first stages of succession grow rapidly, mature quickly, and reproduce in large numbers.

As soon as colonizing species start to grow, however, they begin to change conditions around them. As grasses and hardy herbs grow on a newly formed dune, for example, they hold the sand down, shade the surface, and add organic matter in the form of dead leaves and roots year after year. These changes affect the stability of the sand, its temperature on sunny days, the amount of moisture the soil holds, and the availability of nutrients within it. Similarly, as the water-loving mosses and other plants at the edges of a bog grow and die, they build up layers of peat that gradually fill in the bog, change its pH, and alter its nutrient content.

These biologically caused environmental changes often allow other plants, less tolerant of the original harsh con-

ditions, to survive. On the dune, shrubs and pine trees move in, shading and gradually replacing the pioneering grasses and herbs. As these taller plants grow, they change the environment further, encouraging the growth of still other plants that ultimately replace them. Pine woods give way to oak, and in some places oaks are ultimately replaced by a beech–maple forest. In the bog, azaleas and tamarack trees follow the mosses, to be replaced in turn by birches and spruce (Figure 4.19).

Changes in dominant plants are followed by changes in both the animals and the other plants that can live in an area; forest worms would die quickly in an open sand dune, and birds that eat grass seed would find very little food in a beech–maple forest.

Secondary succession   The process of **secondary succession** occurs in places where either human activity or natural events destroy or disturb ecosystems. Secondary succession occurs, for example, when a plowed field is abandoned. As the field lies fallow, it is colonized first by a mixture of annual and perennial grasses and small herbs. Over the years, its flora progresses through small shrubs and trees to larger trees and, ultimately, mature forest.

Naturally caused episodes of secondary succession are often attributable to disturbances such as fires, severe storms, or outbreaks of disease. Any of these events can destroy mature forest trees, for example, creating

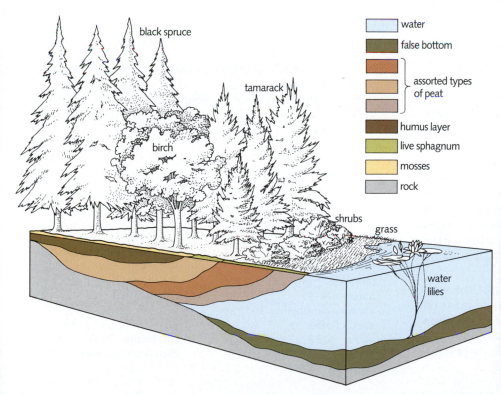

water	
false bottom	
	assorted types of peat
humus layer	
live sphagnum	
mosses	
rock	

Figure 4.19   (left) This bog in cross section shows a view of primary succession "frozen in time." The open bog is to the right, the forest to the left. Over many years, the open area will slowly fill with peat and vegetation, and the floating mosses and shrubs will encroach on the open water. As the bog fills in, larger shrubs and trees will follow. (right) This typical northern peat bog shows a number of plants ranging from floating sphagnum moss and small shrubs, to spruce and other trees in the background.

openings in the canopy and allowing sunlight to reach the normally shady forest floor. Along rocky coastlines, storm-driven waves can scour the shore of its normal cover of algae or mussels, exposing bare rock.

In some types of secondary succession, disturbance knocks the system back to a relatively pristine state. In such cases, secondary succession proceeds by the same mechanisms as primary succession; early colonizers pave the way for later arrivals. In other cases, such as the re-colonization of old fields, the seeds of both early and late colonizers may be present and capable of growing. In these situations, the first plants to appear may simply be those that grow fastest. Their early advantage is ulti-mately lost, however, as slower-growing but taller shrubs and trees overgrow and shade the field.

Climax Communities: Stability and Replacement

For years, many ecologists agreed that succession in any particular area always proceeded in a predictable fashion toward a specific climax community that remained stable over long periods of time. Succession in one region, for example, would always lead to pine woods; in another area, the climax community would invariably be a beech–maple forest.

More recently, however, it has become clear that chance plays a major role in many forms of succession. Different organisms reproduce at different times of year, for exam-ple, and the seeds or eggs of one colonizing species may "win out" over those of another just because they hap-pened to arrive in a disturbed area first. Whether a storm-scoured rocky shore ends up covered with barnacles or mussels may depend on the season of the storm.

Furthermore, many so-called climax communities are disturbed so often that they can't really be called stable. Lightning sporadically sets forest fires that sweep across large and small areas of forests and grasslands, wiping out some species and encouraging the growth of others. Similarly, periodic hurricanes severely damage coral reefs. These fairly common, localized disturbances create such a patchwork of early, middle, and late successionary stages that some ecologists have begun to question whether the concept of climax has any real validity. In biology, it seems, the only constant is change itself.

In this chapter, we have examined many of the vari-ables that shape the conditions under which life exists on the earth. We have seen that global climate determines many aspects of the environments that support life—and that one of the major factors affecting the nature of those environments is the presence of life itself.

SUMMARY

An organism's ability to survive in a given area is determined both by the environmental conditions it encounters and by its ability to cope with those conditions. Both animals and plants must keep conditions inside their bodies within certain limits in order to grow and reproduce.

Environmental conditions encountered by organisms fall into two broad categories: (1) The strictly physical, or abiotic characteristics of a particular region—the means and extremes of sunlight, temperature, rainfall, and humidity—are deter-mined both by global meteorological phenomena, such as wind and ocean currents, and by local geographic features, such as the presence of a lake or mountain range. (2) Biological, or biotic, conditions an organism encounters are determined by the presence of other organisms that can provide food or shel-ter or may act as predators.

All organisms strongly affect the local environmental con-ditions around them, creating microhabitats with microclimates that often determine which organisms can survive in any par-ticular environment. Because living systems are adapted to pre-vailing environmental conditions, changes in those condi-tions—from either natural or human-caused events—can dramatically reshape many ecosystems.

STUDY FOCUS

After studying this chapter, you should be able to:

- Define the concepts of environment, habitat, and niche.
- Describe environmental variables and explain their impor-tance to plants and animals.
- Illustrate the role of organisms themselves in shaping their environments.
- Explain the importance of global climate patterns.
- Show how environmental changes can affect plant and ani-mal growth.

SELECTED TERMS

environment p. 55
physical environment p. 55
biotic environment p. 55
habitat p. 55
microenvironment p. 55
microclimate p. 55
niche p. 57
aerobe p. 60
anaerobe p. 60

thermoregulation p. 61
endotherm p. 61
ectotherm p. 61
climate p. 63
weather p. 63
plate tectonics p. 68
succession p. 70
climax community p. 71

REVIEW

Discussion Questions

1. What does the term *environment* mean? What conditions determine the effects a given environment has on organisms living in it?

2. What is a habitat? Why is it important to consider micro-habitats when looking for the effects of environmental conditions on small organisms?

3. What is a niche? What are some ways organisms divide their habitats into niches?

4. How do global air circulation and ocean currents determine climate and weather where you live now?

5. What is succession? How does it exemplify biologically caused, local environmental change?

6. What is plate tectonics and how does it help to explain global climate changes over geological time?

Objective Questions (Answers in Appendix)

7. Which of the following describes primary succession?
   (a) natural reforestation of a burned out forest
   (b) may go more rapidly because favorable conditions are already established
   (c) mosses growing on newly hardened lava rock
   (d) plants growing are less tolerant of harsh conditions

8. Which of the following describes secondary succession?
   (a) farmers abandon a field and move to the city
   (b) organisms that can tolerate harsh conditions
   (c) organisms growing on a lake formed by a recent earthquake
   (d) mosses growing on newly hardened lava rock

9. A climax community
   (a) is self-sustaining.
   (b) never changes.
   (c) is one in which growth proceeds in a predictable pattern.
   (d) is not likely to be disturbed by localized climatic changes.

10. Which of the following variables does not determine the distribution of plants and animals?
    (a) the size of the habitat
    (b) incoming solar radiation
    (c) salinity of the waters
    (d) available nutrients

11. According to the theory of plate tectonics, the plate boundaries
    (a) are the focus of strong geologic activity.
    (b) separate the continents from each other.
    (c) continue to move away from each other.
    (d) continue to move toward the magnetic poles.

CHAPTER 5

Population Growth and Control

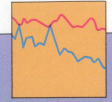

Across the eastern United States, the caterpillars of native North American butterflies and moths go about their business, consuming roughly as many leaves as they always have. Yet only decades after the accidental release of a few European gypsy moths in Massachusetts, the species periodically multiplies out of control, invading woodlands by the hundreds of thousands (Figure 5.1). These pests can defoliate thousands of acres of forests, and—during years when their numbers peak—can literally cover cars and driveways with their bodies and feces.

Similarly, in environments where they are normally found, rabbits are hardly threats to the environment. Yet in Australia, where an enterprising (though foolish) farmer released twelve pairs of rabbits in 1859, and in England, where the conquering Normans (who knew no better) brought them more than 800 years ago, rabbits are pernicious pests whose populations fluctuate wildly. In Australia and New Zealand, periodic plagues of grass-gobbling rabbits threaten not only to drive certain kangaroos and other native herbivores to extinction, but to outcompete sheep for pasture as well. And in Britain, where an estimated 100 million rabbits roamed free in 1953, rabbits inflict more than 140 million pounds (sterling) in damages on food crops every year. There are still some parts of Britain today in which crops can flourish only when surrounded by rabbit-proof enclosures.

DYNAMICS OF POPULATIONS

Gypsy moths and rabbits are classic examples of apparently harmless organisms that have turned into "plagues" when introduced to new habitats. But why? Gypsy moths don't multiply out of control in their native Europe. And rabbits don't lay waste the countryside in their native habitats either. As a rule, plant and animal populations do not fluctuate wildly; rather they are kept under control by interactions with each other and with their environments.

Functional Units of Ecology

To ecologists and geneticists, the word *population* has a very specific meaning, although populations can differ enormously, both in size and in the amount of territory they cover. A **population** is a group of individuals of a single species that interact and, most important, interbreed with each other.

Because, by definition, individuals within a population must be able to breed with one another, single populations may cover large areas or be restricted to small ones, depending on the organisms' ability to encounter and mate with one another. For aquatic animals such as fishes, a strip of land just a few yards wide between two adjacent ponds may form an impenetrable barrier that separates the individuals in those ponds into two distinct populations. On the other hand, a long chain of lakes connected by a river might host only one or two populations of each species because the individual fishes can swim freely back and forth and interbreed along the entire waterway. The area covered by plant populations can be determined by several factors. Because many plants depend upon insect pollinators, the distance over which

Figure 5.1 Gypsy moth caterpillars, an introduced exotic species, have multiplied out of control. These pests regularly defoliate thousands of acres of forests.

they can exchange genetic material may be controlled by the movements of those insects. Additionally, because many plants rely on birds or mammals to spread their seeds, plants' "movements" may be directly tied to the activities of particular seed- or fruit-eating animals.

Exponential Growth

Nearly any population of organisms whose nutritional needs are met and that is free of predators and disease soon begins to grow rapidly.

Take the case of a single bacterium, which is able to divide to form 2 new cells every 20 minutes. At the end of the first 40 minutes there are 4 individuals. After the first full hour, each of those individuals divides to produce 2 × 4, or 8, cells. Twenty minutes later, each of *those* divides, producing 16 individuals. As long as local conditions remain favorable, the size of this population doubles every 20 minutes.

Graphing the size of this population over time produces a curve that rises slowly at first and then accelerates rapidly as the population doubles from 2 to 4 to 8 to 16 to 32, 64, 128, and so on. This is known as **exponential growth:** The larger the population is, the faster it grows. Exponential growth curves are also sometimes called **J-shaped** growth curves (Figure 5.2).

Organisms growing in this uninhibited way exhibit what is called their **intrinsic rate of growth.** The rate at which a population grows is determined by the rate at which new individuals are produced, the **birth rate** b, and the rate at which other individuals die, the **death rate** d. The intrinsic rate of population growth (or r_0), therefore, is simply defined as the birth rate minus the death rate $(b - d)$:

$$r_0 = b - d$$

Exponential population growth can thus be described by the following simple equation:

$$\begin{pmatrix} \text{rate of} \\ \text{increase in} \\ \text{number of} \\ \text{individuals} \end{pmatrix} = \left[\begin{pmatrix} \text{birth rate per} \\ \text{individual} \end{pmatrix} - \begin{pmatrix} \text{death rate per} \\ \text{individual} \end{pmatrix} \right] \times \begin{pmatrix} \text{number of} \\ \text{individuals} \end{pmatrix}$$

That is,

$$I = (b - d)N$$

or

$$I = (r_0)N$$

where

I = rate of change in the number of individuals in the population over time
r_0 = rate of population growth
N = size of the population

Note that N is constantly increasing, so that I grows faster and faster with each generation. Clearly, natural populations cannot grow exponentially for very long.

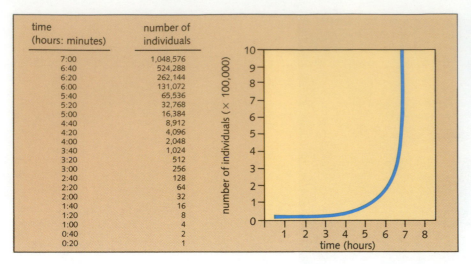

time (hours: minutes)	number of individuals
7:00	1,048,576
6:40	524,288
6:20	262,144
6:00	131,072
5:40	65,536
5:20	32,768
5:00	16,384
4:40	8,912
4:20	4,096
4:00	2,048
3:40	1,024
3:20	512
3:00	256
2:40	128
2:20	64
2:00	32
1:40	16
1:20	8
1:00	4
0:40	2
0:20	1

Figure 5.2 If provided with an ideal environment, continually supplied with food, and protected from the buildup of waste products, bacteria—like many other organisms—can grow exponentially. The larger the population becomes, the faster it increases in size.

Under certain circumstances, however, organisms can grow exponentially for a short time. For example, pioneering species in secondary succession and species introduced by humans to favorable new environments can exhibit short-term exponential growth. But J-shaped curves rarely last long in natural populations. Why does exponential growth stop when it does, and what happens next? Much of ecological theory attempts to explain these very questions and to understand why natural populations grow the way they do.

Logistic Growth

If a laboratory population of bacteria is grown with a limited amount of space and nutrients, its growth can be graphed with an **S-shaped,** or **logistic growth,** curve (Figure 5.3). At the top of this curve, population growth is zero; although new individuals are continually being born, others are dying at the same rate so $(b - d)$ equals zero. The total number of individuals therefore remains constant, and the population exists in a *steady state,* or **equilibrium.**

The Concept of Carrying Capacity

Ecologists say that a population in equilibrium contains as many individuals as its particular environment can sustain for extended periods of time. In other words, the population has reached the **carrying capacity** of the

Figure 5.3 The S-shaped, or logistic growth, curve has four parts. In the *lag phase,* the population is established and begins to grow. In the *acceleration* phase, births greatly exceed deaths, and the population grows exponentially. During the *deceleration* phase, the rate of population growth slows down. Ultimately, the rate continues to decrease until the population reaches a constant, maximum size.

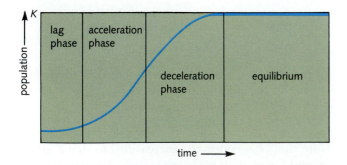

environment for that species, a number often represented by K. The exponential growth equation can be modified to describe logistic growth; the term K is simply added in a way that simulates what happens as the population approaches the carrying capacity of its environment. Thus

$$I = r_0 N\left(\frac{K - N}{K}\right)$$

where, as before,

I = rate of increase in the population over time
r_0 = intrinsic rate of population growth
N = size of the population
and K = carrying capacity of the environment

Consider what this equation means for populations of different sizes (Table 5.1). When the population is far below the carrying capacity (when N is far less than K), the term $(K - N)/K$ is close to 1 and has little effect on population growth. But as the population approaches the carrying capacity (as N approaches K), the term $(K - N)/K$ multiplies r_0N by a smaller and smaller fraction, slowing down the increase in population size. When the population size reaches K, N equals K, $(K - N)/K = 0$, and population growth stops altogether.

In practice, population growth may slow down for any of three reasons. The birth rate may drop, the death rate may rise, or birth rates may drop and death rates may increase at the same time. When birth rate and death rate precisely balance each other, the population reaches a point of *zero population growth*. Note also in Table 5.1 that if N ever exceeds K, $(K - N)/K$ becomes *negative*, indicating that deaths exceed births. Under these conditions, of course, the population decreases in size.

Factors determining carrying capacity   The logistic growth equation elegantly describes the inhibitory effect of carrying capacity on populations' tendency to grow exponentially. But what factors in a real environment determine its carrying capacity for a particular species? And how do those factors alter birth rates and death rates to slow population growth?

Nearly any of the ecological limiting factors we encountered in Chapters 3 and 4 can limit population size (Figure 5.4). Insufficient sunlight or a deficiency in any essential nutrient can limit plant growth and reproduction. A shortage of suitable prey can limit animal populations. Too much or too little water, heat, or humidity can affect both plants and animals.

Although we have treated carrying capacity as constant, in reality the carrying capacity of a specific environment for a particular species is often anything *but*

Table 5.1   *Effect of Carrying Capacity on Population Growth*

N	$(K - N)/K$	I	
1	999/1000	$0.999r_0N$	
10	990/1000	$0.99r_0N$	
100	900/1000	$0.9r_0N$	
500	500/1000	$0.5r_0N$	Growth
900	100/1000	$0.1r_0N$	slows
990	10/1000	$0.01r_0N$	
999	1/1000	$0.001r_0N$	
1000	0/1000	0.0	Growth stops: equilibrium
1010	$-10/1000$	$-0.01r_0N$	Population
1100	$-100/1000$	$-0.1r_0N$	declines

K for this population is 1000.

Figure 5.4   This graph represents the relationship between the theoretical potential of a species to grow exponentially and the effect of environmental limiting factors on actual population growth. The shaded area between the two curves represents the growth-inhibiting effects of density-dependent regulating factors represented by the term $(K - N)/K$.

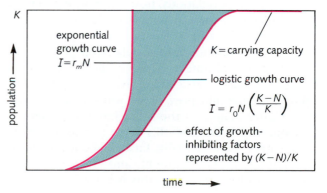

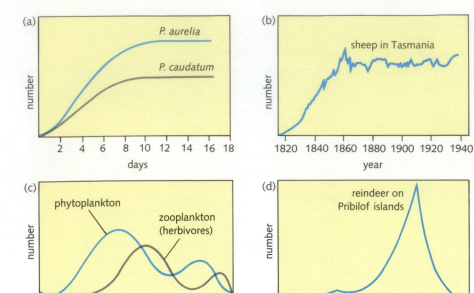

Figure 5.5 Patterns of population growth. (a) Logistic growth of two species of the single-celled *Paramecium* in a laboratory culture. Notice *P. aurelia* has a higher *K* under these conditions than *P. caudatum*. (b) Near-logistic growth of sheep when first introduced to the island of Tasmania. (c) Seasonal fluctuations in marine phytoplankton and zooplankton in temperate coastal waters. (d) "Boom and bust" growth curve of reindeer introduced to one of Alaska's Pribilof islands. The 26 original animals grew exponentially and overgrazed their food supply. The population then plummeted to 18 in less than a decade.

constant. Physical conditions (weather) change from season to season, and biological conditions (interactions with predators, prey, and parasites) also vary. And because the carrying capacity (*K*) is determined by these factors, it often changes with them.

Population Growth in Nature

The formula for logistic growth accurately predicts growth in certain laboratory populations and for certain species under specific conditions in nature. The graphs in Figure 5.5 (*a*) and (*b*) show laboratory and field populations whose growth is reasonably well described by the logistic growth curve.

Other natural populations, however, may exhibit any of several different growth patterns. Figure 5.5(*c*), for example, shows the annual fluctuations in populations of marine phytoplankton and zooplankton populations in temperate coastal waters. These cyclical changes are typical of interactions between herbivores and plants and between predators and prey, as we will see later in this chapter. Figure 5.5(*d*), on the other hand, represents a wild "boom and bust" growth curve. In this situation, organisms grow exponentially, far overshoot the carrying capacity of the environment, and crash instead of reaching a steady state. Often this is because they have seriously degraded their environment.

DEMOGRAPHY: THE STUDY OF POPULATIONS

So far we have been concerned only with the total number of individuals present in a population over time. But to fully understand populations and to predict their growth, ecologists must examine individuals' ages, their reproductive abilities, and their chance of survival at each age. This type of study is called **demography** (*demos* means "the people"; *graphos* means "measurement").

Demography is a crucial component of the biology of all species. Because extremely accurate and detailed population data are available for humans, and because knowledge of human population growth is vital to understanding global ecology, we will use data on humans as we discuss demography. All the principles that emerge, however, are just as valid for frogs, flies, and oak trees as they are for *Homo sapiens*.

Population Age Structure

The **age distribution** of a population refers to the relative number of individuals of each age, from newborns to the oldest survivors present.

Mortality and survivorship   Individuals of different ages have different chances of living for the same length of

Table 5.2 *Life Tables*

	MALES					FEMALES					
Age	Population	Deaths	Probability of Dying Within Five Years	Survivorship per Population of 100,000 at Birth	Life Expectancy	Population	Deaths	Births	Probability of Dying Within Five Years	Survivorship per Population of 100,000 at Birth	Life Expectancy
0	414700	8706	0.020609	100000	68.646	393600	6274	0	0.015716	100000	74.831
1	1716199	1489	0.003465	97939	69.088	1631199	1193	0	0.002923	98428	75.024
5	1980499	868	0.002191	97600	65.323	1882499	572	0	0.001520	98141	71.239
10	1715999	676	0.001977	97386	60.461	1633099	430	222	0.001319	97992	66.344
15	1734799	1544	0.004445	97193	55.575	1681799	624	81853	0.001854	97862	61.428
20	1857199	1711	0.004598	96761	50.812	1838899	739	295946	0.002011	97681	56.538
25	1538699	1346	0.004370	96316	46.035	1500799	793	240807	0.002651	97484	51.646
30	1504099	1553	0.005159	95896	41.226	1426099	1076	125316	0.003776	97226	46.777
35	1508799	2446	0.008080	95401	36.426	1443299	1738	58083	0.006009	96859	41.944
40	1540799	4652	0.014997	94630	31.700	1522099	3152	15904	0.010309	96277	37.181
45	1626399	8274	0.025168	93211	27.141	1644899	5568	1140	0.016797	95284	32.540
50	1411599	13022	0.045200	90865	22.771	1488799	7847	1	0.026037	93684	28.050
55	1458299	23367	0.077259	86758	18.720	1585299	12600	0	0.039025	91245	23.728
60	1292799	34784	0.126664	80055	15.063	1483799	19195	0	0.062862	87684	19.583
65	991000	43709	0.200187	69915	11.863	1280499	27232	0	0.101552	82172	15.717
70	631100	43425	0.295219	55919	9.183	1033400	37935	0	0.169360	73827	12.191
75	399200	42010	0.415613	39411	6.970	753600	47723	0	0.275469	61324	9.138
80	206600	32674	0.558305	23031	5.182	464300	49948	0	0.424249	44431	6.632
85	101100	27036	1.000000	10173	3.739	275100	58912	0	1.000000	25581	4.670

Source: Data from N. Keyfitz and Wilhelm Flieger, 1971. *Population: Facts and Methods of Demography*, W. H. Freeman, San Francisco, pp. 154–157.
Note: Data collected from populations in England and Wales, 1968.

time into the future. For example, a 90-year-old person is less likely to survive another 20 years than a teenager is. The chances that individuals of a particular age will live or die in a finite amount of time are expressed in **life tables** (Table 5.2). Life tables reflect two complementary sets of data: the percentage of the population that can expect to live to a given age, or its **survivorship,** and the population's age-related death rate, or its **mortality.**

Different species have different patterns of mortality and survivorship. Demographers recognize three classic patterns, although there are many intermediates (Figure 5.6). The human data shown in Table 5.2 translate into what is called a Type I survivorship curve.

Fertility There is another reason why age structure influences a population's ability to grow. Individuals of different ages have different **fertility rates;** they do not produce offspring at the same rate. At any given time, some are too young to reproduce, others too old. Even those capable of reproduction differ in *fertility* according to age. The peak biological reproductive years for human females, for example, fall between the ages of 15 and 44. But ethnic traditions and contemporary trends in different societies interact with biological potential to influence actual human fertility rates very strongly. Consider

Figure 5.6 Three survivorship curves. The human curve, Type I, shows that humans who survive their early years usually live to a ripe old age. Hydra exhibits a Type II curve; these animals have a fairly equal chance of dying throughout life. Many organisms, such as fishes and oysters, exhibit a Type III curve. Type III organisms produce tremendous numbers of offspring, most of which die at a very early age. Once these organisms reach a certain age, however, they are likely to live a long time.

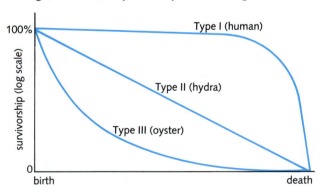

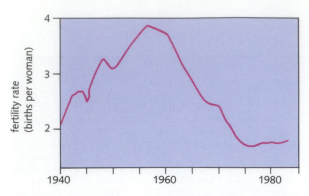

Figure 5.7 This graph of the total U.S. fertility rate from 1940 to 1985 shows its dramatic rise during the post-war baby boom and subsequent return to more normal levels.

the total fertility rate in the United States: From a post–World War II baby-boom peak of 3.8 births per woman, our current birth rate has dropped to 1.8, which is slightly below the level of 2.0 required for zero population growth (Figure 5.7). This shift reflects a social change, rather than a biological or medical change.

Age structure and population growth When demographers know the age structure of a population, they can predict its future by combining information on the number of females, their ages, and the fertility rate for each age class. Once they have all this information, they can describe the population as belonging to one of four basic types of population age structures:

- stable
- slowly expanding
- rapidly growing
- gradually declining

Rapidly declining populations can be recognized too, but they don't last very long in nature.

Figure 5.8 illustrates three of these situations in the human populations of different countries. Sweden's age structure reflects a low, stable birth rate and an even distribution of individuals across all age classes. This sort of population usually remains stable because the same number of females are constantly entering and leaving their peak reproductive years. Mexico, by contrast, exhibits

Figure 5.8 Age structure diagrams of human populations; males are represented on the left-hand side of each figure, and females on the right. In Sweden, a country with zero population growth, many individuals are far beyond primary reproductive age. This population structure reflects a low, stable birth rate. In Mexico, a country with rapid population growth, a large percentage of the population will soon reach prime reproductive age. This structure reflects a high birth rate that will continue into the future. The United States is a country with relatively stable population. Note the bulge in the population of 15- to 40-year-olds representing our post-World War II baby-boom generation.

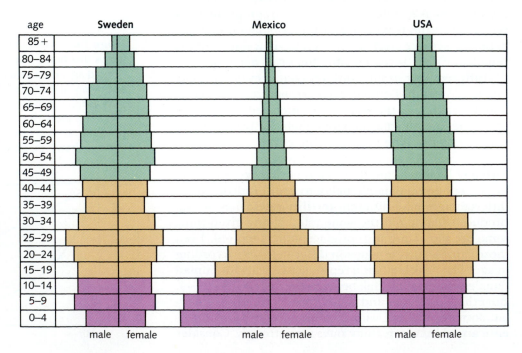

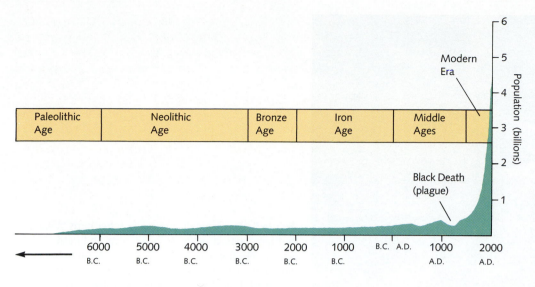

Figure 5.9 Growth curve for the global population of *Homo sapiens*. After a long lag period—lasting until the industrial revolution—human populations entered exponential growth that was maintained through the middle of this century.

a distribution that clearly indicates very rapid growth, because each year more and more females are entering their reproductive period. The American age structure as it was in 1985 has been interpreted by various experts as growing slowly, as heading toward zero population growth, and as declining slowly.

HUMAN POPULATION GROWTH

Many biologists are worried about global human demographic data, because our species' growth to date describes a classic J-shaped curve that cannot continue indefinitely (Figure 5.9). It is to be hoped that human population growth will slow down, taking on the S-shape of a population in equilibrium with the carrying capacity of the global environment. If it doesn't, our species may be faced either with the sort of "boom and bust" growth that is characteristic of all species that continue to exhibit J-shaped growth curves, or with other unpleasant circumstances that will limit growth for us.

One of the first to recognize the problems of unchecked human population growth was the eighteenth-century English economist Thomas Malthus. Malthus observed that human populations of his day were growing exponentially but that society's ability to increase food production was growing much more slowly. If exponential human population growth continued, Malthus reasoned, sooner or later we would run out of food and space. Malthus believed that the runaway population growth of humans could ultimately be checked only by food short-

ages, plagues, and wars. These less-than-optimistic observations are called the *Malthusian Doctrine*.

To Malthus's gloomy predictions, modern ecologists have added questions about the effects of the earth's human population and pollution on the biosphere. There is spirited debate being held among ecologists, human demographers, and social scientists about what will happen to the world's human population in the next few decades (see Current Controversies, Human Population Growth, p. 84).

POPULATION-REGULATING FACTORS

Ecologists have attempted to divide natural population-regulating factors into two main categories. The first type exerts stronger effects when population density is high than when it is low. Because their influence varies with population size, these are called **density-dependent factors.** The classic example of a density-dependent factor is competition: Organisms vie with each other for limiting resources such as food and water. The higher the population density, the more organisms compete with one another and the more the population is affected.

Density-independent factors, on the other hand, influence population growth to the same extent whatever the population density. Sudden, unpredictable changes in the physical environment often act to reduce plant and animal populations in a density-independent manner. A sudden frost in Florida, for example, kills a certain proportion of the orange trees it hits, regardless of how many trees are growing in the orchard.

Figure 5.10 Seagulls and many other birds stake out territories in which to build nests and raise young. Nesting space is in short supply on many breeding grounds, and gulls fight fiercely to defend their turf. Pairs that do not win suitable pieces of ground simply cannot breed.

Competition

Competition occurs when organisms of the same or different species require common resources, such as sunlight, food, water, or space, that are present in limited supply. Alternatively, if the resources are not actually in short supply, competition can occur when the animals that utilize those resources harm one another in the process anyway. To many ecologists, competition is the single most important density-dependent population-regulating factor, and a potent process that also directs the course of evolution.

Ecologists recognize two major kinds of competition. In **intraspecific competition,** members of the same species compete among themselves (*intra* means "inside or within"). In **interspecific competition** (*inter* means "between"), members of two or more species compete with each other.

Plants and animals may compete for a variety of important resources. In crowded plant communities, inorganic nutrients, water, and light may become scarce enough to limit growth. Animal populations may compete for water, edible plants, suitable prey, or space. Many animal species, from seagulls to lions, need various kinds and amounts of space in which to feed, breed, and/or hide from predators (Figure 5.10). When population densities rise, such spaces, or territories, may function as limiting resources.

Competing for resources either intraspecifically or interspecifically takes both time and energy. For this reason, individuals engaged in competition have less energy left for growth and reproduction than individuals with no competitors. The stress caused by competition may decrease the birth rate, increase the death rate, or both. When competition is intense enough, it can slow down population growth or halt it altogether.

Competitive exclusion In 1934, Soviet ecologist G. F. Gause performed a classic series of experiments involving two species of the single-celled animal *Paramecium.* Gause grew each species in culture under two sets of conditions—first alone and then together with the other species. In both cases he found that one species drove the other species to extinction (Figure 5.11).

These and similar experiments led ecologist G. Hardin to propose **the principle of competitive exclusion.** This principle states that two species cannot live together in the same place at the same time if they are both limited by one or more of the same limiting resources. Under those conditions, the principle states, one species always drives the other to extinction locally.

Competition and the niche To understand fully the importance of competition in ecological and evolutionary theory, we must look once again at the niche, for the competitive exclusion principle is another way of saying

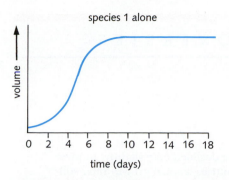
species 1 alone

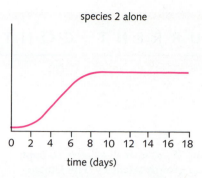
species 2 alone

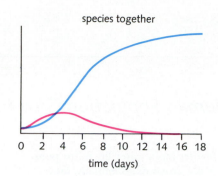
species together

Figure 5.11 Competition in the laboratory. The two species Gause studied have similar requirements. When grown in culture alone, both exhibit logistic growth, reach the carrying capacity of their container, and persist indefinitely. When grown *together*, however, species 1 drives species 2 to extinction.

that no two organisms can occupy the same niche in the same place at the same time. Recall that the niche describes not only where an organism lives but also what it does to "make its living." In the last chapter we found that an organism's use of food or living space can be represented by a resource utilization curve. Such curves can be used to illustrate the extent of competition between two species with similar ways of life, as shown in Figure 5.12.

Because competition between two similar species affects their birth and death rates, it can force the less efficient competitor out of any habitats the two share. Thus competition can restrict organisms to a part of their fundamental niche (a phenomenon we encountered in Chapter 4).

Competition in nature   It is easy to demonstrate competition and competitive exclusion under simple laboratory conditions. Even in the laboratory, however, the outcome of competition between two species can be strongly affected by the response of competing species to physical factors in the environment. This warns us that predicting the outcome of competition in nature can be difficult indeed.

For example, in Gause's experiments, it turned out that the successful competitor won out not because it was better at grabbing the available food, but because it was more resistant to the chemical waste products that accumulated in the densely populated culture medium. The competitive outcome could be completely reversed if the culture medium was changed frequently to limit the buildup of those waste products.

These and other experiments made it clear that any change in the physical environment can alter the outcome of competition. That observation in turn led ecologists to predict that differences in physical conditions

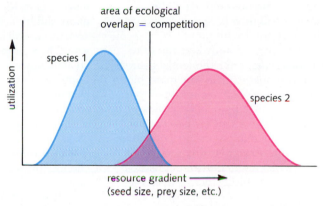

Figure 5.12   The resource utilization curves of two similar species whose use of a common resource overlap. The extent of competition between the species is indicated by the shaded area on the graph.

Human Population Growth

A farmer has a water lily in his pond. The lily grows exponentially, doubling in size every day. If nothing interferes with it, the lily can cover the water's surface in 30 days, suffocating all other life in the pond. But the farmer is busy, and he decides not to bother with the water lily until it covers half the pond. Now that may sound reasonable, but think about it. The lily won't cover half the pond until the twenty-ninth day. The farmer will then have only 24 hours in which to save his pond.

This parable aptly illustrates the hidden danger of exponential growth, and it should explain why many biologists are worried about human population. World population growth now averages about 1.7 percent per year, a rate that causes our population to double about every 41 years.

Why is earth's human population growing so quickly today? To answer that question, we must consider the phenomenon called *demographic transition*—a drop first in death rates, and then in birth rates, often considered the most crucial event in the history of human populations.

Remember that *r*, the rate of population growth, is equal to the birth rate minus the death rate. The death rate falls as countries begin to develop because of improvements in health care, nutrition, and sanitation. As a result, (*b* - *d*) is small or zero, so population growth stabilizes. The United States, Europe, and Japan completed the demographic transition between 1850 and 1950.

Researchers attribute the decline in birth rates in fully developed societies to factors associated with changes in socioeconomic conditions: (1) As urbanization proceeds, overcrowding and migration from farms to cities make it hard for families to house many children. Also, extra hands aren't so useful in the city as they would have been on a farm. (2) Increased educational and vocational opportunities offer new social and economic alternatives for women in societies where they once spent most of their lives producing and raising children. (3) The availability of birth control technology enables couples to choose how many children they will have.

Whatever the reasons, across the Western world (and in Japan), birth rates and death rates are both low, (*b* − *d*) is small, and population growth averages between 0.5 and 1 percent per year.

More than two-thirds of the world's population, however, lives in the "Third World" countries of Central and South America, Asia, and Africa. There death rates have dropped as they did in the West, but birth rates

have declined only slightly. In these countries, annual growth rates still average between 1.2 percent (in the South Pacific) and a staggering 4.1 percent in parts of Africa.

Few question the harsh realities of life in the Third World today. Disagreements arise immediately, however, when discussion shifts to the importance of population growth in development. For this reason, we will return to this question in Chapter 7.

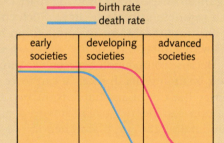

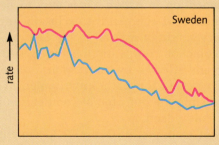

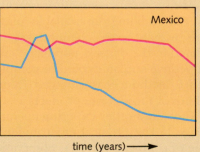

(top) Birth and death rates in societies at three stages: "early," "developing," and "developed." In early societies, high birth rates are matched by high death rates, so population growth is slow. In societies that complete the demographic transition, both birth rates and death rates fall, as they did in Sweden during the nineteenth and early twentieth centuries (center). In such societies population growth is also slow. (bottom) In Mexico, a country in the midst of the transition, death rates have fallen, while birth rates have just begun to drop. This population is growing rapidly.

Figure 5.13 This photograph shows the sharp line dividing several species that inhabit rocky, temperate seacoasts. The experiments of Joseph Connell sought to explain the factors behind this sort of rigid zonation.

might cause one competitor to win out in one microhabitat and another to prevail somewhere else. The first clear demonstration that competition in nature can work that way was a study of two barnacle species by J. H. Connell from the University of California.

Biologists had known for years that plants and animals on rocky seashores are arranged in distinct, horizontal layers. These layers form a predictable pattern stretching from a perpetually wet zone below the lowest tide line to a splash zone above the highest high tides (Figure 5.13). Working on the rocky coast of Scotland, Connell noticed that two species of barnacles, *Balanus* and *Chthamalus*, were found in adjacent, but separate layers. Adult *Chthamalus* lived only on the highest rocks, just above normal high tides. Below a specific height relative to the tides, adult *Chthamalus* virtually disappeared and were replaced by dense communities of *Balanus*. Connell wanted to find out why this was so.

Because adult barnacles are small, are sedentary, and live at high densities, Connell could perform several experiments that are difficult or impossible to do with other animals. On rocks treated as controls, he watched the barnacles' natural growth and survival from year to year. At the same time, he performed a series of experiments on other barnacle-covered rocks. Some he transplanted to different heights; others he cleared of adult barnacles to see which species would settle and grow. Connell found that *Chthamalus* did better higher up on the rocky shore because it had a higher tolerance than *Balanus* for the physical stresses of heat and drying out. But farther down, on rocks exposed less often to the air, *Balanus* grew much faster than *Chthamalus*, and became a superior competitor for a vital limiting resource: space on the crowded rocks. Under these conditions, *Balanus* could either overgrow or undercut its slower-growing neighbors—starving individuals or prying them off the rocks. Thus an interplay between the effects of physical stress and competition on *Balanus* and *Chthamalus* results in a remarkably clear division of the rocky habitat between these species.

Such data indicate that interspecific competition often works together with physical environmental factors (such as heat, desiccation, stress, cold, or exposure) to regulate plant and animal populations, and therefore to mold organisms' distributions in ecosystems around the world.

Predation

Predation is another powerful biological interaction that shapes natural communities, because herbivores and carnivores affect plant and animal populations in a variety

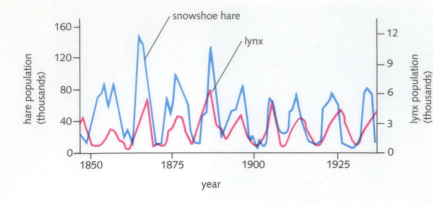

Figure 5.14 Lynx and hare population fluctuations. These data, provided by records of pelts sold by fur trappers, have been used for years to tell a simple story. Whenever hare populations increase, lynx populations can increase as well. At some point, the predators devour the hares faster than the prey can reproduce, and the hare population declines rapidly. As food becomes scarce, the lynx population plummets, releasing hares from predation pressure, and the cycle begins again. For more information on the real complexity of the situation, see text.

of ways. For example, wandering moose colonized the island called Isle Royale in Lake Superior around the turn of the century, at a time when the island supported no large predators. The moose population grew exponentially, exceeded the island's carrying capacity, and severely overgrazed the vegetation. The result was a dramatic decline, or crash, in the moose population in 1930. With the moose population reduced, the vegetation recovered. But the moose population quickly rebounded, only to crash once more in the 1940s. This is the sort of "boom and bust" growth cycle we saw in Figure 5.5, a pattern commonly observed when population size is not controlled in a density-dependent manner.

On Isle Royale, however, the situation changed when a few timber wolves crossed the winter ice to the island in 1948. The wolf population grew to about 24 individuals, and the moose population decreased to just under 1000. Since then the wolves and moose have reached a dynamic equilibrium, in which both populations remain at relatively constant levels.

Predation may thus either increase the prey's death rate, decrease its birth rate, or both. In simple laboratory environments, predators may completely eliminate their prey. In nature, predators are usually only able to harvest their prey to a certain point, because the physical complexity of the natural environment often offers hiding places for prey.

Predator–prey oscillations One classic example of a predator–prey interaction is the **oscillation** of lynx and snowshoe hare populations (Figure 5.14). These regular population fluctuations, in which prey and predator alternately rise and fall in numbers, are common in nature. It is easy to interpret such data simplistically; one can assume that available prey are the major limiting factor for lynx and that predation is the major factor limiting hare population size. But these data are obtained by observation and not experiment; they lack the controls and experimental manipulation necessary to prove the point and exclude competing hypotheses.

Recently ecologists have tried to determine whether factors other than predator–prey interactions could be responsible for these oscillations. Observations on Anticosti Island in the Gulf of St. Lawrence showed that hare populations oscillate even where no lynx or other large predators are around, suggesting that early interpretations focused on the wrong predator–prey interaction. On Anticosti Island, the real population-limiting interaction may be between the hares and their plant prey! Elsewhere, it may well be a combination of predation, food limitation, and weather conditions—rather than any single phenomenon alone—that explains cycles in hare populations.

Some of the same arguments may apply to the wolves and moose. In that situation, it seems reasonable to assume that the wolf population is limited by food and the moose population by predation. Some ecologists disagree, however, arguing that the moose population is primarily controlled even today by food availability, disease, and other factors. In any case, it is now clear that in most ecological interactions in nature, populations and distribution patterns are controlled by several interacting factors rather than by one alone.

Parasitism

Parasites can also act as density-dependent population regulators. In laboratory settings, as well as on farms and in zoos, parasites may completely eliminate their hosts, but this rarely happens in nature. Rather, parasites and their hosts reach the same sort of balance as predators and prey. Australia's experience with rabbits is an excellent example of host–parasite interactions.

Rabbits were introduced to Australia intentionally as a source of meat and game. In the absence of all their

Competition, Predation, and the Complexity of Natural Ecosystems

The complex relationship between predation, competition, and environmental conditions in nature is illustrated by a recent study in Africa's Etosha National Park. There the excavation of gravel pits during road construction, a seemingly harmless change in the park ecosystem, had unexpected results: Wildebeest and zebra populations crashed; cheetah, brown hyena, and eland populations declined, and lion, spotted hyena, and springbok populations increased! Biologist Hugh Berry pieced together the following explanation.

Stagnant rainwater in the gravel pits, which were used as watering holes by many animals, became living reservoirs for deadly anthrax bacteria. Wildebeest and zebra became sick in large numbers, and lions and spotted hyenas (immune to anthrax) easily captured their usually elusive prey.

As lion and spotted hyena populations increased, they captured larger numbers of eland. (Eland are immune to anthrax.) Concurrently, lions and spotted hyenas began to outcompete the cheetahs and brown hyenas, the next largest carnivores in the area. As cheetah and brown hyena numbers dropped and the number of carcasses continued to rise, the next largest carnivore/scavenger, the black-backed jackal, increased in numbers.

Finally, the declines in wildebeest and zebra populations freed the springbok from competition for food. Cheetahs (now in decline) had been the springbok's major predator. With fewer adversaries and fewer competitors for food, the springbok population grew rapidly.

This story demonstrates several important principles about predation and competition as population-regulating factors. First, a variety of ecologically similar organisms, such as Africa's many grazing animals, often coexist. At times, similar species may coexist in a stable fashion only because predators remove enough to prevent any one species from forcing the other to extinction. Anything that shifts the balance of predators in such a system can also change the relative numbers of their prey. The resulting imbalance may lead to the extinction of one or more competitors.

Second, when a change in the environment aids one or two of several competing species, increased numbers of those species can overpower their competitors.

Third, when one or more of a group of competing species is hit hard by an environmental change, its competitors benefit from the end of the competitive struggle.

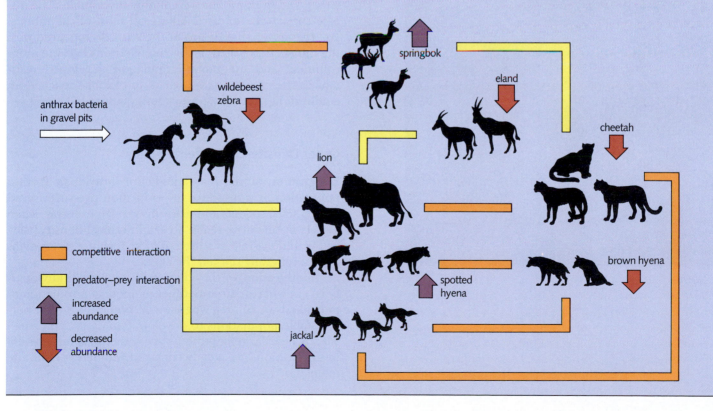

springbok

wildebeest
zebra

eland

anthrax bacteria
in gravel pits

cheetah

lion

brown hyena

competitive interaction

predator–prey interaction

increased
abundance

decreased
abundance

spotted
hyena

jackal

Figure 5.15 Commensal relationships in nature. (right) Commensal anemone shrimp on its host. (below) Orchids clasp trees with their roots and gather nutrients from decaying leaves and rainwater. The orchids obtain support and a place in the sun from their host tree, which is neither helped nor harmed.

natural population-limiting factors, such as predators and disease, the rabbit population grew exponentially and devastated the countryside. In a desperate effort to control the rabbits, biologists introduced the viral disease myxomatosis. At first, nearly all strains of the virus were lethal, and most rabbits were highly susceptible. The disease spread like wildfire, killing off most of the rabbits. But the most lethal strains of the virus disappeared with the rabbits they killed, less virulent strains became more common, and the rabbit population developed resistance. After several years, virus and rabbits reached an equilibrium that allowed host and parasite to coexist.

Other Density-Dependent Factors

Emigration, or the mass exodus of individuals from a population, is a response to population pressure found among some rodents and many insects. For example, when locusts in one area reach a certain critical density, they gather and migrate in search of food, forming vast swarms that can literally darken the sky.

Immigration, or the addition to an area of individuals born elsewhere, can sometimes maintain a population of individuals in an area where local conditions do not permit stable populations to develop. Populations on small islands near large continents are sometimes maintained in this fashion by a steady trickle of wanderers from the mainland.

Cannibalism, or the eating of members of one's own species, increases in some animal species under crowded conditions. Many kinds of stress, including overcrowding, have strong effects on both the behavior and the physiology of animals. Either male or female rodents, for example, may eat their own babies when population densities become too high.

Waste products released by animal or plant activity may accumulate to the point of toxicity in dense populations. Wastes may either affect the survival of all organisms, including the species that creates them, or they may be more toxic to other species, thus affecting the outcome of interspecific competition. Yeasts grown in culture, for example, produce alcohol as a waste product. The toxicity of this alcohol can act both to stem the population growth of the strain that produces it and to suppress the viability of competing strains of yeast.

Positive Interactions Between Organisms

Nature is not ruled only by negative relationships; there are many cases in which the presence of one organism helps rather than hinders another. These are classic examples of beneficial **symbiosis** (*sym* means "together"; *bios* means "life") in which two or more species form close relationships. Many organisms involved in symbioses depend utterly on their partner and cannot exist without them. For that reason, any environmental factors that affect the distribution of one member of a symbiotic pair also affect the other. Note that parasitism, because it involves a close relationship between two species, could technically be considered a form of symbiosis. However, the term *symbiosis* is most often used to refer to positive associations among organisms.

Commensalism   A common type of symbiosis is **commensalism,** in which one member of the association benefits while the other is neither helped nor harmed. Commensal relationships are abundant on coral reefs and in tropical rain forests (Figure 5.15).

Mutualism   In **mutualism,** both partners benefit from the association. On the coral reef, clownfishes live within the stinging tentacles of the sea anemone (Figure 5.16). The clownfish's body is covered with mucus that protects it from the anemone's sting. The anemone offers the fish protection from some of its enemies. The clownfish, in turn, defends its living home against other fish species that like to eat the anemone's tentacles.

Mutualistic relationships are common on land as well, even within human bodies. Our intestines, for example, contain beneficial bacteria that aid in the digestion of food. We provide the bacteria with a warm, moist envi-

Figure 5.16   A mated pair of symbiotic clownfish in the tentacles of their anemone host.

ronment and with food, and they in turn synthesize certain vitamins our bodies could neither obtain nor produce otherwise. Animals from termites to cows and sheep rely on symbiotic protozoa and bacteria to help them digest the tough cellulose in wood and plant leaves (Chapter 3).

Density-Independent Population Regulation

Sudden, extreme, or unpredictable changes in environmental conditions (such as droughts, floods, cold spells, or heat waves) periodically wipe out large numbers of organisms. Because such events may influence populations in an area without any regard to population density, they are said to operate in a *density-independent* fashion.

In reality, even temperature changes rarely act in a way that is completely unrelated to population density. A hard frost, for example, may kill only those insects, lizards, or birds left out in the cold after other members of their species have occupied all suitable shelters. The higher the population density, the higher the proportion of "homeless" individuals—and the more organisms killed by the cold.

In the final analysis, all natural populations are regulated by the combined action of *all* potential population-regulating factors, most of which act most of the time in a density-dependent manner.

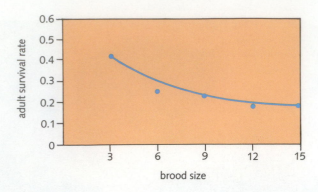

Figure 5.17 Survival of female songbirds in the year after breeding, plotted against experimentally manipulated brood size. Females with large broods had lower survival rates than females raising fewer young.

THE EFFECT OF ENVIRONMENTAL PREDICTABILITY

In some environments, such as tropical rain forests and coral reefs, physical conditions are stable and predictable throughout the year. In others, such as mountain highlands and shifting sand dunes, physical conditions may be widely variable, less predictable, or continually disturbed. These different environments make different demands on organisms and affect their population biology and ecological interactions in a variety of ways.

The optimal reproductive strategy for an organism, for example, is strongly dependent on the sort of environment it faces. Under certain circumstances, organisms that reproduce early in life and produce many offspring are favored in the long term. That is not always the case, however, because reproduction "costs" organisms energy they might otherwise put into growth, protection, and maintenance. Producing lots of offspring may leave parents weakened in the face of competition or harsh environmental conditions (Figure 5.17). This is an evolutionarily acceptable tradeoff if and only if most of the offspring produced survive to reproduce. For this reason the nature of the local environment determines when high rates of reproduction are favorable.

Unpredictable Environments

In physically difficult or unpredictable habitats, environmental upsets often kill organisms in large numbers, thereby opening up spaces that are available to any newcomers. It is usually advantageous for species living in such environments to reproduce as quickly as possible, because adults cannot survive for long and because many of their offspring can survive for brief periods. Many bacteria, protozoans, and "weeds" have evolved a reproductive strategy geared to unstable environments: These organisms typically reproduce at a very young age, produce large numbers of offspring, devote little or no care to their brood, and may die after reproducing. Such spe-

Table 5.3 *Comparison of Opportunistic and Equilibrium Species*

	Opportunistic	Equilibrium
Environmental conditions	Variable and/or unpredictable; uncertain	Fairly constant and/or predictable; more certain
Survivorship	Often type III	Usually types I and II
Population size	Variable in time; nonequilibrium; periodic ecological vacuums allow recolonization	Fairly constant in time; equilibrium at or near carrying capacity of the environment; stable, ecologically saturated communities
Intraspecific/ interspecific competition	Variable, often low	Usually intense
Typical characteristics	1. Rapid growth 2. High *r* 3. Early reproduction 4. Small body size	1. Slower growth 2. Greater competitive ability 3. Delayed reproduction 4. Larger body size
Length of life	Short, usually less than 1 year	Longer, usually more than 1 year

SOURCE: Adapted from Pianka, 1970, On *r* − 1 and *K*-selection, *American Naturalist,* 104:592–597.

cies are often called **r-selected,** because they maximize r_0, their intrinsic rate of reproduction. These organisms are also called *opportunistic species:* Their reproductive strategy allows them to grow maximally during brief periods of environmental benevolence. Such species are often first to invade new or disturbed environments in the early stages of succession.

Predictable Environments

In stable, predictable environments, on the other hand, competition, predation, and parasitism are often the major sources of mortality. Under these conditions, organisms gain little by stretching their resources to produce large numbers of offspring, because most of those offspring will not survive: Food, nesting sites, or other factors limit the size of the adult population and make it difficult for newcomers to get established.

Organisms in predictable environments, therefore, usually mature and reproduce later in life. Although they may reproduce periodically, they produce far fewer offspring than opportunistic species. They also spend more time and/or energy nurturing and protecting their young. In this way they can produce larger, stronger, more developed offspring that are more likely to survive in a highly competitive environment. Such organisms (they include many tropical mammals and birds) are often called **K-selected** species because they seem to maximize their carrying capacity *K*. These organisms, also called *equilibrium species,* are typically found late in the process of succession and make up a large proportion of the species in climax communities. Table 5.3 compares opportunistic and equilibrium species on several dimensions.

CAN WE PREDICT (AND CONTROL) OUR OWN IMPACT?

Humans are affecting environments around the world in many ways. In order to predict the effects of our own population on the environment, ecologists must understand the population biology of natural populations. We now appreciate the significance of competition, predator–prey interactions, and host–parasite interactions in controlling natural populations, but in most cases our knowledge works only "backwards." We can explain why things go awry *after* the fact, but we can seldom predict outcomes in advance. Our knowledge has helped us reverse the ill effects of tampering with some natural ecosystems (Figure 5.18), but it has proved useless in other situations. To improve that record, we must learn a good deal more about population growth and ecological interactions than we know today.

Figure 5.18 Correcting an ecological error. (top) The prickly pear cactus, introduced into Australia from Latin America, grew wildly and uncontrollably, and covered vast areas with tangled, thorny growth. (bottom) After careful research, *Cactoblastis*, a ravenous, cactus-eating moth from the prickly pear's native haunt was released in Australia. The plant's natural predator soon helped bring the cactus under control.

Populations are collections of individuals, belonging to a single species, that regularly interbreed with one another. Under ideal conditions in the laboratory, most populations grow exponentially, increasing in size more rapidly as they get larger. In nature, one or more environmental factors place a ceiling on population size, either by decreasing the birth rate, increasing the death rate, or both.

Demography, the study of individuals' ages, their reproductive abilities, and their chance of survival at each age, is important in predicting the future growth of a population. When population-regulating factors in the environment operate in a density-dependent fashion, population growth follows an S-shaped or logistic growth pattern leveling off as the population reaches the environmental carrying capacity, or K. The most important density-dependent population-regulating factors are competition, predation, and parasitism. But nature is not ruled only by negative relationships; intimate, positive biological relationships among organisms may also affect population growth and the carrying capacity of local environments.

Populations that face major environmental perturbations, such as hard frosts, droughts, or unpredictable environmental fluctuations of other kinds, often fluctuate between exponential growth and calamitous declines.

After studying this chapter, you should be able to:

- Explain how and why populations grow.
- Discuss the factors that control populations in nature.
- Explain competition, predation, and symbiosis.
- Trace some of the ecological effects of interactions among organisms.
- Appreciate the complexity of natural ecosystems.

population *p. 74*
exponential growth *p. 75*
logarithmic growth *p. 75*
logistic growth *p. 76*
equilibrium *p. 76*
carrying capacity *p. 76*
demography *p. 78*
survivorship *p. 79*
mortality *p. 79*
fertility rate *p. 79*
competition *p. 82*
intraspecific competition *p. 82*

interspecific competition *p. 82*
competitive exclusion principle *p. 82*
emigration *p. 88*
immigration *p. 88*
cannibalism *p. 89*
symbiosis *p. 89*
commensalism *p. 89*
mutualism *p. 89*
r-selected species *p. 91*
K-selected species *p. 91*

Discussion Questions

1. What is a population? How do organisms' physical abilities and environmental conditions combine either to limit populations to small areas or enable them to cover large regions?

2. What is exponential growth? Under what conditions in nature is exponential growth likely to occur? What would happen if exponential growth continued unchecked?

3. What are the four stages of growth in a population that follows a logistic growth curve?

4. What is the difference between density-dependent and density-independent population-regulating factors? Which are most likely to result in a stable equilibrium? Why?

5. Compare and contrast opportunistic and equilibrium organisms and give examples of each.

6. How can interactions between predator and prey result in density-dependent control of both populations?

7. How can interactions among organisms amplify apparently minor environmental changes into major shifts in population size?

Objective Questions   (Answers in Appendix)

8. The birth rate minus the death rate determines a population's
   (a) reproductive rate.
   (b) intrinsic rate of growth.
   (c) logistic growth.
   (d) carrying capacity.

9. When a species overshoots its carrying capacity in the environment,
   (a) the population will plummet.
   (b) population growth will not be affected.
   (c) the population growth will reach a steady state.
   (d) none of the above.

10. A J-shaped curve represents a population that
   (a) has a limited supply of food.
   (b) has reached the carrying capacity of its environment.
   (c) is growing exponentially.
   (d) was growing exponentially and then crashed.

11. Which of the following represents a density-independent factor regulating population growth?
   (a) competition for food and raw materials
   (b) activities of predators
   (c) parasites
   (d) climate

12. Which of the following descriptions represents a "K-selected" strategy?
   (a) large number of young produced
   (b) little or no care devoted to young
   (c) large investment of energy in reproduction
   (d) advantageous to reproduce as quickly as possible

CHAPTER 6

Ecosystems: The Working Units of the Biosphere

Across southern New England, rich farms produce crops year after year in places once covered by deciduous forests. In the Midwest, corn and wheat fields blanket what were once vast prairies. In both areas—and in similar places around the world—intensive human activity has replaced natural assemblages of plants and animals. But western-style mass agriculture is not ill-suited to these areas; the soil remains productive for decades, and abandoned farms return, after many years, to ecosystems not too dissimilar from those that covered the region in the past. Much of "wild" New England today, for example, is actually what is called second-growth forest, a collection of organisms that have covered areas cleared for farming more than a century ago.

But in certain parts of the world, efforts to turn natural habitat into farmland have ended in ecological and economic disasters. In Africa, sheep and goats forage in ever-wider circles over desert that was once dry forest. And in Brazil, huge areas of lush rain forest, cleared for farming less than five years ago, are already reduced to wastelands covered with sterile soil so hard it can break a plow. Crops no longer grow, and the soil—once stripped of the rain forest that held and protected it—becomes so compacted that original vegetation cannot return.

Is there any way to predict which human interactions with nature will work and which will turn sour? To answer this question properly, we must first recognize the importance of biological diversity and the factors that govern the response of ecosystems to human activity.

THE IMPORTANCE OF BIOLOGICAL AND ECOLOGICAL DIVERSITY

The term **biological diversity** refers to the variety and variability in living organisms and the ecological systems in which they live. Diversity exists on several levels: **Ecosystem diversity** is the variety of ecosystems in a region. An area with patches of farmland, forests, ponds, marshes, and grasslands has higher diversity than a landscape devoted exclusively to farming. **Species diversity** refers to the myriad kinds of organisms alive today. To date, 1.7 million species of plants and animals have been identified, but experts estimate that our planet may harbor up to 30 million species. **Genetic diversity** consists of the heritable variability among individual members of a single species. As we will discover when we study evolution, genetic diversity can be essential to a species' survival.

This wide assortment of organisms and ecosystems presents fascinating ecological and physiological phenomena for us, as students of biology, to explore. These organisms also constitute an invaluable reservoir of new crops, new drugs, and other products that have important potential applications in industry, medicine, and agriculture.

The Resilience of Natural Systems

It is tempting to portray the living world as a delicately stacked house of cards: Push one too hard and the whole structure tumbles down around our heads. Luckily, however, that is not the case. Organisms can respond to change in their surroundings. Indeed, as we noted while studying population growth and succession, populations and ecosystems are constantly changing. The biosphere, too, is a living network and can repair itself following certain kinds of disturbance.

But ecosystems are far from invulnerable. Some are more susceptible to disturbance than others, and certain types of disturbance have more serious effects than others. Today, human activity is changing the biosphere and destroying many of the world's habitats and species at record rates. As you will see in Chapter 7, the situation is critical; if those species and ecosystems are to survive, we need to start understanding and preserving them now. In this chapter, therefore, we will survey the world's major living systems, to marvel at their beauty and diversity, to appreciate the ways they contribute to human life, and to understand how their various components react to stress—both natural and human-caused.

Biomes

Ecologists divide the continents and oceans into major zones called **biomes** according to their physical environmental characteristics. Each biome is home to certain communities of plants and animals (Figure 6.1). Though physical conditions are similar throughout a single biome, differences in local temperature, soil, or rainfall can encourage the growth of different living communities. We will conduct our investigation of earth's living systems by examining first the major terrestrial ecosystems and then the aquatic.

LIFE IN TERRESTRIAL ECOSYSTEMS

The Importance of Soil

Because all life on land depends on green plants, terrestrial ecosystems are strongly affected by the soil that sustains those plants. The basic inorganic components of soil are partially weathered rocks and minerals of varying size and composition. Sand is composed of large grains often made of quartz. Clays are formed from much finer

Figure 6.1 Major biomes of the world, identified by the ecosystem that existed in each before human intervention. Note that on this scale, nearly the entire eastern United States is classified as a single biome entitled "temperate forest."

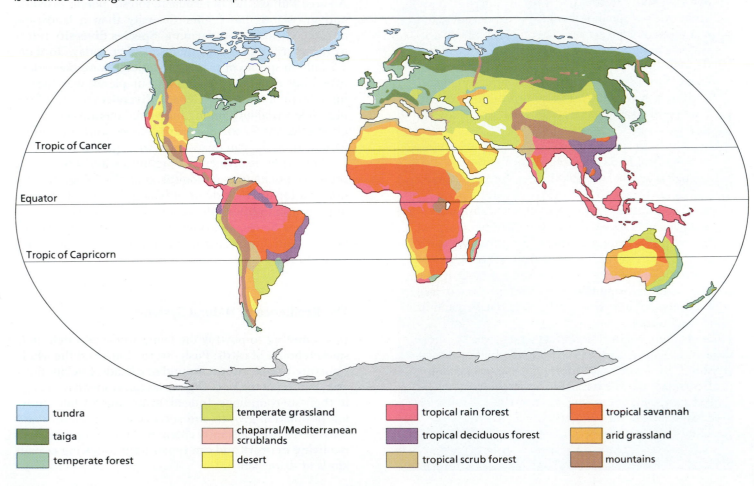

tundra	temperate grassland	tropical rain forest	tropical savannah
taiga	chaparral/Mediterranean scrublands	tropical deciduous forest	arid grassland
temperate forest	desert	tropical scrub forest	mountains

Figure 6.2 Averages and extremes of temperature, rainfall, and soil type shape the world's biomes.

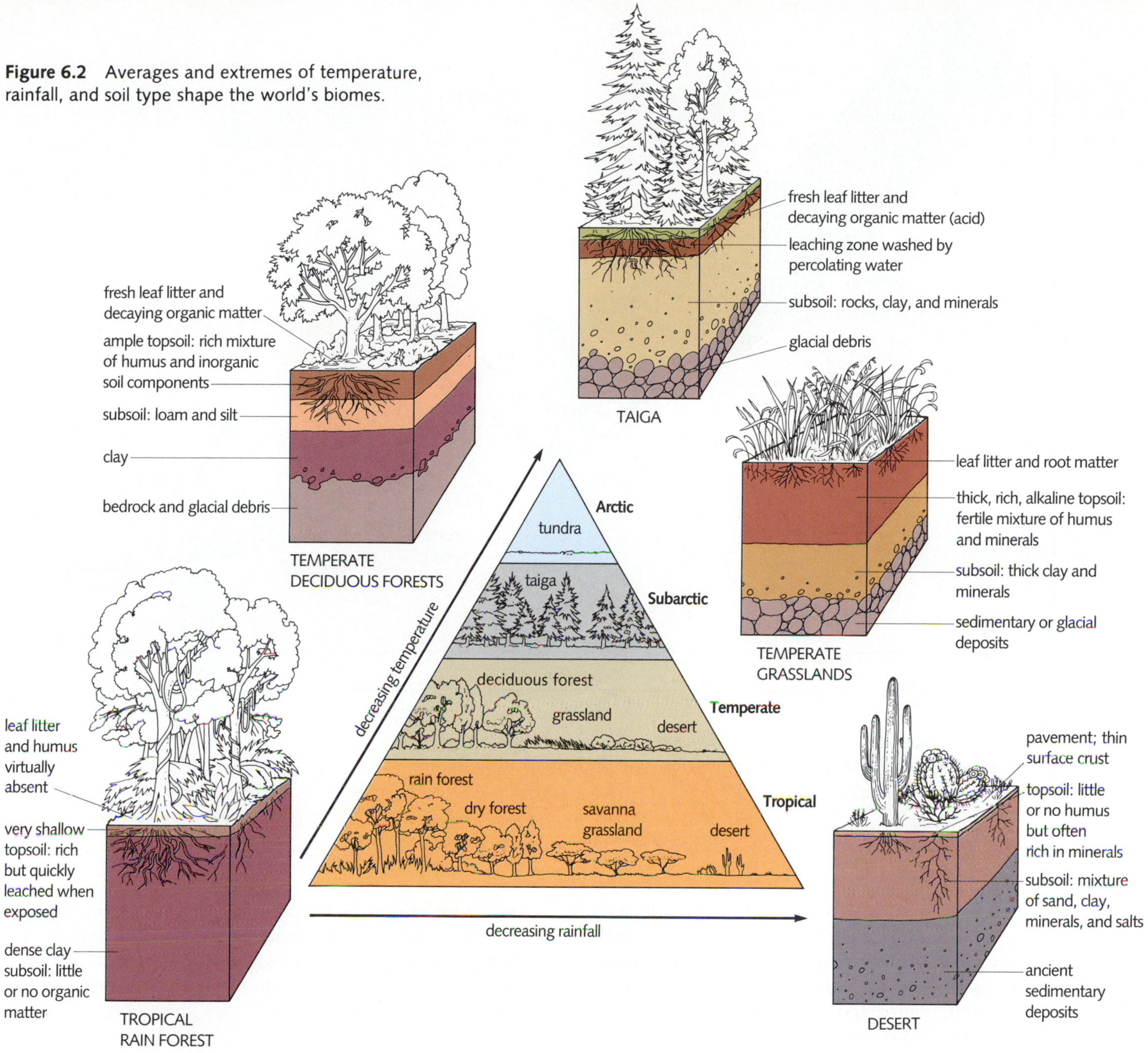

TAIGA

fresh leaf litter and decaying organic matter (acid)

leaching zone washed by percolating water

subsoil: rocks, clay, and minerals

glacial debris

TEMPERATE DECIDUOUS FORESTS

fresh leaf litter and decaying organic matter

ample topsoil: rich mixture of humus and inorganic soil components

subsoil: loam and silt

clay

bedrock and glacial debris

TEMPERATE GRASSLANDS

leaf litter and root matter

thick, rich, alkaline topsoil: fertile mixture of humus and minerals

subsoil: thick clay and minerals

sedimentary or glacial deposits

decreasing temperature

Arctic — tundra

Subarctic — taiga

Temperate — deciduous forest / grassland / desert

Tropical — rain forest / dry forest / savanna grassland / desert

decreasing rainfall

TROPICAL RAIN FOREST

leaf litter and humus virtually absent

very shallow topsoil: rich but quickly leached when exposed

dense clay subsoil: little or no organic matter

DESERT

pavement; thin surface crust

topsoil: little or no humus but often rich in minerals

subsoil: mixture of sand, clay, minerals, and salts

ancient sedimentary deposits

grains. These inorganic materials combine with varying amounts of organic matter to form productive soils.

The physical and biological characteristics of a region not only determine which plants and animals can live there but also affect the use to which humans can put an area (Figure 6.2). In the grasslands that once covered the midwestern United States, for example, the roots of native plants penetrated 4 to 10 feet into the ground, forming over time a dense sod that held moisture and prevented erosion. Because both leaves and roots of grass plants die back each year, organic matter built up and created rich, dark soil. Across New England and the Mid-Atlantic states,

an equally fortuitous combination of rainfall, temperature, and soil structure encouraged the growth of lush temperate forests. The dense roots of these plant communities held soil in place, and the annual fall of leaves contributed the organic materials that decomposed to form **humus,** a spongy bed of matter. The ecological history of both these regions, therefore, has endowed them with highly productive, organically rich soils, well suited to the demands of long-term, intensive agriculture.

In tropical rain forests, on the other hand, dense clay subsoils, heavy rainfall, and high temperature create poor, shallow soils. Here dead organic material breaks down

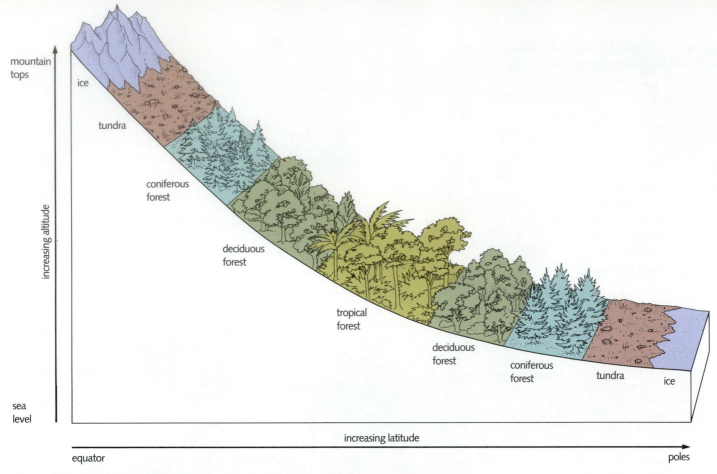

Figure 6.3 Because temperature decreases with increasing altitude, ecological zonation on mountains mirrors the pattern seen when moving from the equator toward the poles.

rapidly, so the humus layer is very thin. Soluble soil nutrients not bound up in living tissue are quickly washed away by heavy rains. When stripped of forest cover, such soils support agriculture for only a few years before they become exhausted, compacted, and useless.

Desert soils hold little organic matter but, because of low rainfall, do not lose their nutrients. With proper irrigation and fertilization, some desert areas (such as California's Imperial Valley) can become productive farmland. But because water in these soils evaporates so quickly, and because the salts left behind are not washed away by rain, soil in irrigated deserts around the world is slowly but steadily becoming too salty for agriculture.

Environmental Variables and Elevation

As you learned in Chapters 4 and 5, climatic variables, including temperature, rainfall, humidity, and winds, control plant and animal growth. These factors vary not only with latitude but also with elevation. At every lati-

tude, locations at higher altitudes experience conditions similar to those at higher latitudes. For this reason, the zonation of life from river valleys to mountaintops mirrors the zonation of life from the tropics to the arctic or the antarctic (Figure 6.3).

TERRESTRIAL BIOMES

Tundra

Near the North Pole and the South Pole, winter is a way of life; average monthly temperatures rarely exceed 10°C. The top few feet of soil thaw during the six-to-eight-week summer, but the subsoil stays permanently frozen in a state called **permafrost.**

This is the **tundra,** a physically controlled environment with predictable but extremely harsh conditions. Most precipitation here falls as snow that remains frozen and unavailable to plants as liquid water for most of the

Figure 6.4 Plants of the tundra, predominantly mosses and quick-growing annuals, make the most of the region's brief summers to grow and reproduce.

Figure 6.5 The taiga supports a diverse and complex community, including evergreens, deciduous trees, and a wide variety of animals.

year. This makes winter on the tundra as inhospitable as summer in the desert. When summer arrives, continuous daylight warms the soil surface, but that warmth does not penetrate far. Any water that melts is trapped at the surface by the permafrost below, forming bogs and ponds.

Microclimates are very important on the tundra, where most plants and animals live just at, just above, or just below ground level. During a typical midsummer day, the permafrost retreats to a depth of 35 centimeters below the surface, the temperature of the soil surface is about 38°C (100°F), air temperature just above the ground hovers around 15.5°C (60°F), air temperature a meter above the ground drops to 4.7°C (40.5°F), and air temperature above that level may be below freezing.

Most trees can't grow on the tundra, where summers are too short and too cool to support the growth of any but a few dwarf willows. The dominant plants are sphagnum mosses and lichens. The relatively few species native to the tundra make the most of the short summers, reproducing quickly and in large numbers (Figure 6.4). As the surface ice thaws, flies and mosquitoes form vast

swarms. Herbs and grasses grow and flower as soon as liquid water becomes available. Flocks of migratory birds feed on abundant seeds and insects and then breed before heading south again in the fall. Because no small area produces enough vegetation to support large herbivore herds for very long, reindeer range far and wide, and caribou migrate south to places where food is more plentiful. Smaller herbivores such as arctic hares and lemmings burrow and hibernate for much of the winter. Carnivores include arctic foxes, snowy owls, arctic wolves, weasels, and (near the coast) polar bears.

Taiga

South of the tundra lies the belt of evergreen forest called the **taiga** (Figure 6.5). During midwinter the climate of the taiga is similar to that of the tundra. Summers, however, are warmer and longer, soil thaws completely, and water is liquid for longer periods. The growing season is still short, and temperature fluctuates wildly.

Figure 6.6 (left) Temperate deciduous ecosystems such as this oak-hickory forest once covered most of eastern North America, Europe, and China. Light reaches the ground in these forests, encouraging the growth of shrubs and herbs. (right) Moist temperate coniferous forests, such as this stand of Douglas fir, may also be dominated by conifers such as redwoods and spruces. These areas receive both ample rainfall and lots of fog.

Evergreen, cone-bearing trees such as spruce, fir, pine, and larch are the most important here, although deciduous trees such as beech, aspen, willow, and ash appear. Evergreens predominate for several reasons. Their tough, needle-like leaves retain water when the soil is frozen. The leaves remain on the stems through the winter, ready to carry on photosynthesis as soon as temperatures moderate in the spring. The reproductive habits of evergreens are also suited to short growing seasons. Pine cones do not release mature seeds until nearly a year after pollination, giving the seeds much longer than a single growing season to mature.

A host of plant-eating insects live here, along with such familiar herbivorous mammals as rodents, porcupines, rabbits and hares, moose, elk, and deer. Carnivores range from weasel, mink, and polecat to lynx, wolves, and bears. Birds are everywhere in summer; seed-eating thrushes, finches, buntings, grosbeaks, nuthatches, and jays are joined by such insect-eating migrants as sandpipers, oystercatchers, and ducks that frequent the taiga's bogs and streams. Eagles, falcons, and buzzards are the most common birds of prey.

Temperate Forest

South of the taiga, milder winters, longer growing seasons, and ample, year-round rainfall support the great deciduous forests of the United States, Europe, and Asia. Often considered a single biome, the **temperate forest** actually comprises at least three different communities (Figure 6.6): the temperate deciduous, the moist-temperate coniferous, and the broad-leaved evergreen forests.

In all these forests, fallen leaves produce a deep, rich layer of decomposing plant material that builds up over many years. Held in place by plant roots and protected from direct sun by the forest canopy, this spongy humus layer is a dependable reservoir of both water and nutrients. Within the humus layer, decomposers such as bacteria and fungi break down the cellulose in fallen leaves and branches, playing vital roles in nutrient cycles.

Temperate forests provide enough buds, leaves, seeds, and fruits to support a wide variety of herbivores and predators. Insects are abundant, in terms of both species diversity and absolute numbers. In spring, summer, and fall, temperate forests provide food for scores of bird species. Some birds live there year-round; others migrate back and forth between temperate breeding grounds and tropical or subtropical wintering grounds. We in North America think of migratory songbirds and shorebirds as northern species that winter in the south. Not surprisingly, residents of Mexico and the Caribbean view them as tropical birds that head north to breed. Actually, they are neither and both. In many cases, their two-biome habits make them vulnerable to habitat destruction in both summer and winter homes.

Mammals, once present in great numbers and diversity, included deer, boar, foxes, wildcats, martens, and lynx coexisting with elk, moose, and caribou. But farming and hunting displaced or killed off the larger predators, and herbivore populations reached new equilibria in their absence. Deer and moose are now managed by wildlife authorities who control their numbers by regulating hunting. But recently, under the protection of law, wolves, coyotes, and other native predators have been making a comeback.

Desert

Deserts are places where annual rainfall is less than 25.5 centimeters. But beyond that unifying factor, deserts are extraordinarily diverse (Figure 6.7). Some deserts, like that of the Sinai, are nearly barren year-round. Others, like North America's Sonoran Desert and Israel's Negev, appear dead for most of the year but actually harbor many drought-tolerant plants that burst into life quickly after seasonal rains. Many of these are annuals that sprout, grow, mature, flower, and die within weeks of the first rain.

Mean temperature records from deserts are nearly meaningless because both daily and seasonal fluctuations can be enormous. African deserts, such as the Kalahari, sometimes experience temperature fluctuations of up to 38°C (from just below freezing to 99°F) in the space of only 24 hours.

Desert plants called *xerophytes* (*xeric* means "dry"; *phytos* means "plant") gather and conserve water in many different ways. The root systems of desert plants may spread horizontally, like those of the saguaro, or plunge 50 meters into the ground, like those of the mesquite. To conserve water, some plants, such as cacti, have reduced their leaves to spines and carry on photosynthesis in thick, waxy stems. Plants that bear leaves often lose them during long dry spells, whereas others (called succulents) rely on thick, fleshy leaves to store water.

Desert animals exhibit remarkable adaptations to extreme environments (Figure 6.8). Numerous insects and other small animals escape the extremes of desert life by burrowing to create hospitable microclimates

Figure 6.7 Cacti, native to the Americas, are premier examples of plants adapted to life in arid environments.

Figure 6.8 Large animals, such as camels that cannot burrow for shelter, cope with desert life through physiological adaptations that permit them to survive the loss of far more of their body water than most other vertebrates. Contrary to myth, the hump does not store water; it stores energy in the form of fat. By concentrating fat in one place, this mammal allows metabolic heat to escape more readily from the rest of its body.

Figure 6.9 In their pristine state, temperate and tropical grasslands—such as the Serengeti seen here—can support vast herds of several herbivores.

beneath the sand (see Chapter 5). Desert reptiles, small mammals, spiders, and scorpions generally remain inactive below ground during the day and emerge to feed only at night.

Grassland

Both temperate and tropical regions of the world have large areas where rain is more abundant than in deserts but is either too sparse or too seasonally concentrated to support forest growth. These areas, which have played crucial roles in human history, are the **grasslands** or **savannahs.** In the United States they are called prairies; in South America they are known as pampas; in South Africa they are referred to as veldt. Intermediate between wooded areas and deserts, grasslands are both fertile and fragile. As the name implies, the dominant plants here are grasses (Figure 6.9).

Grasses are among the world's most efficient plants at converting sunlight and inorganic nutrients into living tissue. They have widely spreading, fibrous root systems that hold soil firmly and capture water from the wet season's often-violent storms. Grasses can grow with extraordinary speed when conditions are right, and they can enter dormancy during dry seasons. During dry times, the above-ground portions of grass plants die, leaving only underground stems alive.

Grasslands are among the world's most important food-producing areas, because grasses directly or indirectly provide food for most of the world's human population. Corn, wheat, rye, oats, and barley are just a few of the grasses American farmers grow, and only three grass species—maize (corn), wheat, and rice—support more than half the world's human population. Grasslands can be farmed for years if treated properly, but carelessness has turned many formerly fertile grasslands into deserts or dust bowls (Figure 6.10).

Tropical Rain Forest

The world's **tropical rain forests** are the most complicated of all terrestrial environments, and the diversity of species that they support is staggering. Rain forests' consistently warm temperatures, high humidity, and year-round rainfall create the perfect biologically controlled environment. Their organisms are classic equilibrium species; they are ecologically specialized, long-lived, and slower to reproduce than related forms elsewhere. Tropical rain forests harbor more species than any other terrestrial environment, although each species is represented by fewer individuals. A single 25-acre plot in a rain forest in Borneo is home to more than 700 species of trees alone!

The trees in a tropical rain forest create a complex, three-dimensional canopy of up to five distinct layers (Figure 6.11). On these trees grow climbers, or lianas, which may be as thin as a clothesline or as thick as tree trunks in the temperate zone. In clumps on the branches grow epiphytes (*epi* means "on top of"), including ferns, mosses, orchids, and bromeliads. Most obtain nutrients from decaying leaves that accumulate around their roots.

Humus, live vegetation, seeds, and fruits are scanty at ground level, so ground-dwelling animals (except the numerous ants and termites) are fairly rare. Most rep-

Figure 6.10 Careless replacement of native grasses with less hardy agricultural crops created the great American dust bowl of the 1930s. Much of Texas and parts of Oklahoma, once fertile prairies, are wastelands today because of overgrazing by domestic cattle.

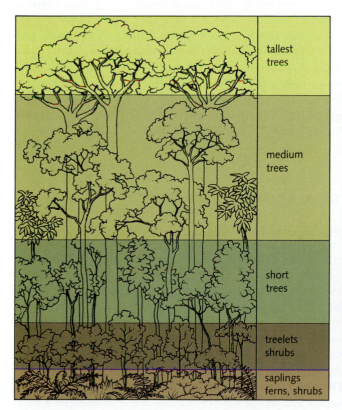

tallest trees

medium trees

short trees

treelets shrubs

saplings ferns, shrubs

Figure 6.11 (left) The multilayered canopy of the tropical rain forest creates a complex, three-dimensional habitat that houses thousands of animal and plant species. (below) The rain forest canopy is home to animals, such as this slow loris, that are normally seen by humans only when trees are cut down.

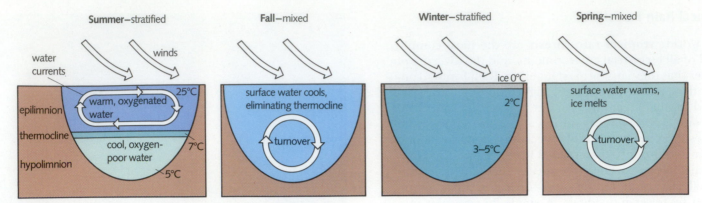

Figure 6.12 Water currents and temperature stratification in a deep lake. In summer, surface waters—the *epilimnion*—are warm and saturated with oxygen, but may be depleted of nutrients for plant growth. Bottom waters—the *hypolimnion*—may be dark, cold, and nearly anaerobic. Low light levels make plant growth difficult, while sinking fecal material and decomposing plants and animals encourage the growth of bacteria that simultaneously use up oxygen and return nutrients to the water. In fall, as air cools (and winds pick up), surface waters cool until the thermocline nearly vanishes and winds mix the lake from top to bottom. This *turnover* redistributes oxygen and nutrients. In winter, ice coats the surface, and mild stratification occurs, this time with the coldest water on top. In spring, ice melts and turnover occurs once again. Some lakes turn over only in the spring, while others mix both spring and fall.

tiles, birds, mammals, and amphibians live in the rain forest canopy hundreds of feet above the ground, where flowers, seeds, and insects are plentiful. The canopy is home to scores of fruit eaters ranging from birds such as parrots, toucans, and hornbills to mammals such as bats, apes, and monkeys.

Despite the rain forests' lushness and robust appearance, their poor soils make them highly vulnerable to disturbance. High rainfall, high temperatures, and high humidity encourage such rapid growth of microorganisms that the soil humus layer and topsoil remain very thin. Nearly all the system's nutrients are therefore tied up in living plant and animal tissue. When that tissue is removed to clear land for farming, heavy rains quickly leach away nutrients and erode humus and topsoil. The poor soil that remains is soon exhausted and worthless. Local climate changes, crops can't grow, the jungle can't grow back, and even the fishes in surrounding rivers disappear (see Theory in Action, The Fishes and the Forest, p. 106). Unfortunately, as you will learn in Chapter 7, many of the world's rain forests are in grave danger.

LIFE IN AQUATIC ECOSYSTEMS

The Importance of Light and Temperature

Life under water is often controlled by different environmental factors than life on land. As we noted in Chapter 4, the availability of light is a critical limiting factor for plants in deep water.

Water temperature is also important in structuring aquatic ecosystems, both because it has a direct influence on organisms and because temperature differences between water layers can cause thermal stratification. **Thermal stratification** is a condition in which surface waters and deeper water masses are separated by a **thermocline,** or zone of dramatic temperature change (Figure 6.12). Because the density of water varies with its temperature, a thermocline often creates two distinctly different environments in a single body of water.

Because many plants and animals float in water without expending energy, aquatic environments (particularly those in the sea) can be more three-dimensional than terrestrial habitats. The most three-dimensional terrestrial habitat, the tropical rain forest, for example, ranges in height from 30 to 60 meters. The photic zone in marine habitats, on the other hand, is regularly deeper than that, and animal life in the sea ranges down to a depth of several kilometers.

Freshwater Ecosystems

Lakes and *ponds* are open bodies of fresh water with aerobic surface layers. Clear lakes with low nutrient content and relatively little phytoplankton growth are called **oligotrophic** (Figure 6.13a). Lakes with heavy phytoplankton and attached plant growth and high nutrient concentrations are said to be **eutrophic.**

Most large lakes in nature are oligotrophic, because rivers flowing into them and groundwater seeping through the soil around them carry relatively few nutrients. Over long periods of time, as lakes slowly fill with sediment and decaying plants, they tend to become eutrophic (Figure 6.13b). This process is speeded up dramatically when lakes are overloaded with nutrients by runoff water from fertilized fields or sewage from cities and towns.

Bogs are small, deep, and usually stagnant bodies of water in which mosses such as *Sphagnum* grow densely (Figure 6.13c). *Sphagnum* mosses acidify the bog waters they populate. Few aquatic animals can tolerate this acidity, which also makes it difficult for plants to take up the nitrogen they need for growth.

On the surface, *swamps* and *marshes* look similar. The distinction between them is that water moves (however slowly) through marshes, whereas it just sits in swamps. In the northern United States, freshwater swamps and marshes usually shelter a variety of tall grasses, although some are dominated by cedar trees. In the South, especially in Louisiana and Mississippi, they may contain either grasses and bog plants or cypress.

Brackish-Water Environments

Estuaries   An **estuary** occurs where a partially enclosed arm of the sea is fed by freshwater runoff from the land, creating a brackish zone of intermediate salinity. In part because of this mixing of fresh and salt water, estuaries are highly productive. Freshwater runoff usually contains nitrogen and silica, but it may lack phosphorus. Surface seawater often carries phosphorus but tends to

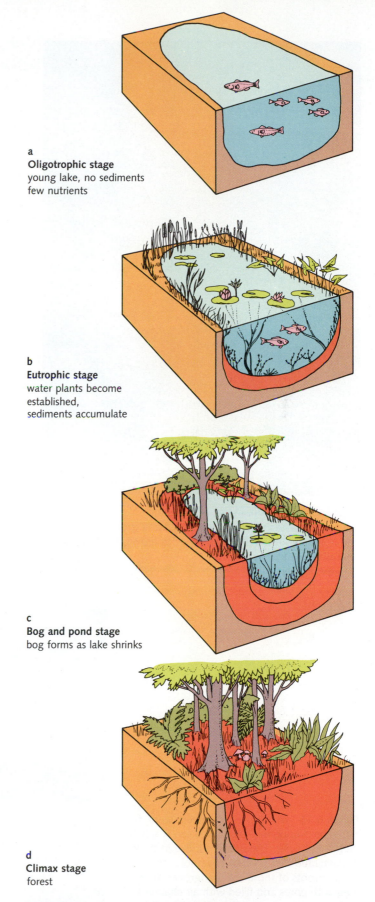

a
Oligotrophic stage
young lake, no sediments
few nutrients

b
Eutrophic stage
water plants become
established,
sediments accumulate

c
Bog and pond stage
bog forms as lake shrinks

d
Climax stage
forest

Figure 6.13   Stages in lake eutrophication.

Figure 6.14 This salt marsh at Monomoy National Wildlife Refuge in Massachusetts serves as feeding ground, nursery, and breeding ground for many species of fish, shellfish, and migratory birds.

be deficient in nitrogen and silica. Each type of water can thus support limited plant growth, because each lacks one or more essential nutrients. When combined and mixed in shallow estuaries with ample sunlight, their nutrients complement each other and encourage high productivity.

Estuaries present their inhabitants with extremely variable conditions, from full-strength salt water on the ocean side to completely fresh water upstream. Thus estuaries are physically controlled habitats where a few ecologically broad, opportunistic species grow in large numbers.

Temperate zone salt marshes Many temperate estuaries host the highly productive community known as the **salt marsh** (Figure 6.14), which is both created and dominated by one or more species of the marsh grass *Spartina*. *Spartina's* tightly woven roots stabilize sediments, providing stability and protection from the ravages of hurricanes.

Mangroves In the tropics and subtropics, *Spartina* grasses give way to salt-tolerant shrubs and trees called **mangroves,** which fringe tropical islands from the Bahamas to Indonesia and straddle the deltas of tropical rivers such as the Mekong, Amazon, Congo, and Ganges. The peculiar root systems of red and black mangroves, the keys to success in their difficult environments, are also their Achilles' heels. Increases in suspended sediments, long-lasting increases in water level, or oil spills can damage those roots enough to harm or even kill the entire forest (Figure 6.15).

Figure 6.15 Mangroves create estuarine forests whose uppermost branches can reach 30 meters in height. Their roots survive in thick organic mud that is often anaerobic. The arching roots of these red mangroves are covered with pores and filled with air channels that conduct oxygen to buried roots.

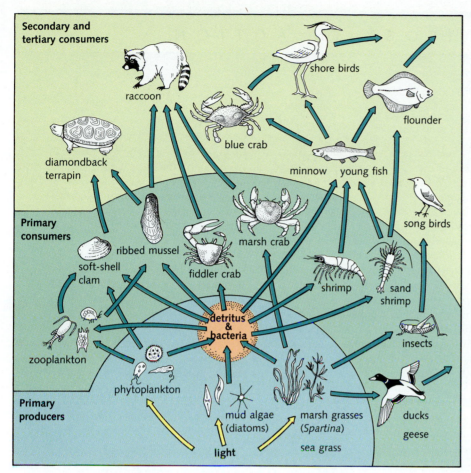

Figure 6.16 This simplified schematic of an estuarine food web shows the central position of detritus in these ecosystems. Scores of small fish, zooplankton, and fiddler crabs ingest detritus particles, digesting epiphytes and decomposers growing on them. The nitrogen these animals release back into the water is quickly resorbed and recycled by primary producers.

Figure 6.17 Estuaries serve as both nursery grounds (as shown here) and as spawning grounds by many fishes and invertebrates that spend their adult lives elsewhere.

Marsh and mangrove ecology   Marshes and mangrove swamps provide shelter for many other primary producers, such as nitrogen-fixing blue-green bacteria and diatoms. Shallow saltwater ponds often harbor one or more of the totally aquatic, bottom-dwelling plants called seagrasses. Seagrass meadows are so productive that they rival even cultivated tropical plantations.

Most energy and nutrients in both salt marshes and mangrove swamp food chains flow through a mixture of dead plant and animal parts called *detritus*. Bacteria and fungi attack these dead plant parts, breaking down the plants' cellulose (Figure 6.16).

Spawning grounds   Many fishes and shellfish either use estuaries as spawning grounds or prey upon species that do (Figure 6.17). Commercially valuable pink shrimp, spiny lobsters, blue crabs, bluefish, spotted sea trout, and groupers all enter mangrove lagoons periodically to feed on abundant small prey. Because both temperate and tropical estuaries are so critical as spawning grounds and nurseries, sizable, healthy marshes must be left at regular intervals along the shore to ensure the health of vital commercial fisheries offshore.

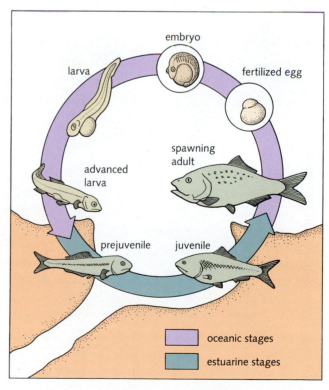

The Fishes and the Forest

It begins with the first drops of rain in the cloud-forests of the Andes mountains. It builds as mountain streams swell and rush toward the Brazilian lowlands. And it reaches its peak as seasonal storms drop so much water that even the world's mightiest river cannot contain it. It is the annual flood of the Amazon river, and it wields the power not only to shape a continent, but also to create an ecosystem of unrivaled size, scope, and diversity. If the Amazon river is, as many have described it, like a thing alive, this seasonal pulse of floodwaters is its heartbeat. Each year the flood submerges more than 70,000 square kilometers of jungle, creating the unique habitat known as the flooded forest, or igapo.

As the waters rise, so do the animals of the river, swarming over the riverbanks to feed and breed among the leaves and trunks of tropical plants. Over time, the river's fish species have evolved bizarre shapes and behaviors and fascinating ecological relationships with animals and plants of the forest. Piranhas, notorious for their supposed attacks on humans and cattle, actually feed for most of the year on fruits and flower buds. But large, predatory fishes do snatch birds off low-lying branches, and scores of species feast on ants and other insects that fall into the water.

Throughout the Amazon basin, the larger cousins of many popular aquarium fishes (such as discus, angelfish, neon tetras, and catfish) form the basis of a giant commercial freshwater fishery, providing protein in the diets of millions of people.

But the Amazon river, its fishes, and its forests are in trouble. Steadily expanding pressure on fish stocks has begun to take its toll, and large-scale deforestation threatens the entire ecosystem. A host of local and international economic problems have forced Brazil to look upon rain forest development as a source of vitally needed jobs and capital. Unfortunately, Brazilians may inadvertently be destroying their only real hope for self-sufficiency. To help prevent this, ecologists all over the tropics are working on studies the results of which may help regional planners save the igapo and the other forests of the region.

The Amazon may shelter as many as 3000 fish species, ranging from retiring leaf-eaters to voracious carnivores, bloodsuckers, and parasites; from tiny, iridescent neon tetras to the giant arapaima shown here and the 100-kilogram catfish. Only recently has the extent and importance of the relationship between the Amazon fauna and the life of the flooded forests been understood.

The Open Sea

Out in the open sea, all higher life depends on the sea's diverse assemblage of zooplankton and phytoplankton. The floating world of the plankton is a dynamic, multi-layered ecosystem that covers the oceans, reaching down into the depths for hundreds, or even thousands, of feet (Figure 6.18).

The surface waters of tropical seas offer intense sunlight but carry very few nutrients. Only tiny primary producers, growing little larger than 20 microns, can thrive there. In areas along the coasts of major continents, however, freshwater runoff from the land provides enough nutrients for larger phytoplankton, including *dinoflagellates* and *diatoms*. And where winds and ocean currents produce upwellings, such as above Georges Bank in the

Figure 6.18 Provinces of life in the ocean. (left) Organisms that live on the bottom are called *benthic,* while those that live in open water are called *pelagic.* Animals and plants that spend their time floating in the water column are called *plankton* or "drifters." Plant plankton are called *phytoplankton;* animal plankton are called *zooplankton.* Animals that actively swim through the water are called *nekton.* (right) Concentration of organisms is highest in the photic zone and at the sea bottom.

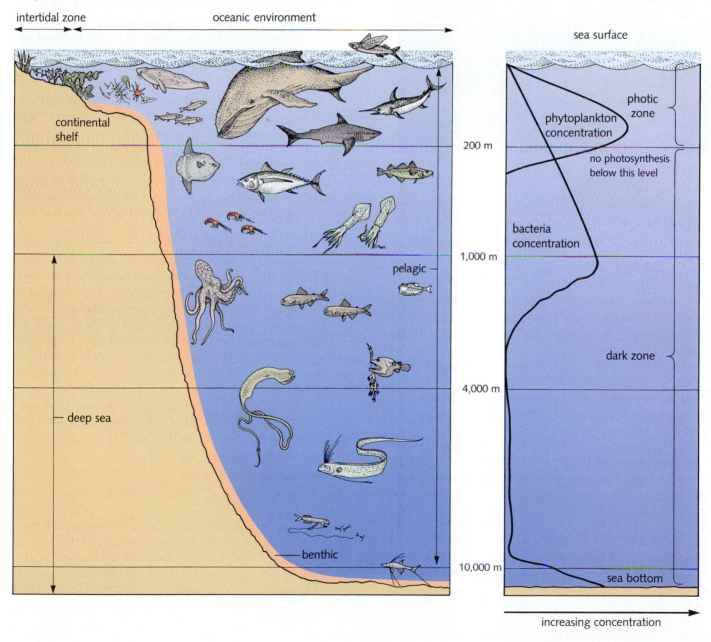

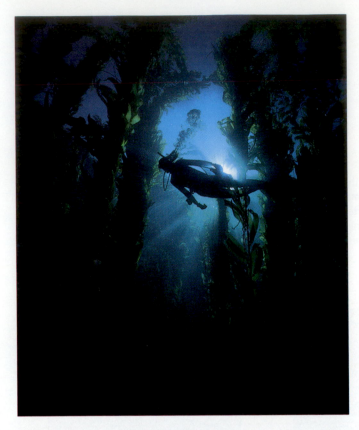

Figure 6.19 In the Pacific, the tall, narrow fronds of *Macrocystis* create open, roomy forests.

North Atlantic and off the coast of Peru in the Pacific, nutrients brought up into the sunlit photic zone from the ocean floor support dense populations of the largest phytoplankton, chain-forming diatoms.

This difference in nutrient availability and its effect on primary producers creates very different food chains in different areas. The minuscule primary producers of the open sea are suitable food only for very small zooplankton, creating a food chain with many levels between primary producers and sizable carnivores. As we saw in Chapter 3, food chains that contain many trophic levels can support very little living tissue at their upper end with a given weight of primary producers. Accordingly, open-ocean food chains are very inefficient at producing large organisms.

In coastal waters, larger phytoplankton can be either eaten directly by good-sized zooplankton or filtered from the water by clams, mussels, worms, and other bottom-dwelling filter feeders. Here, higher primary productivity and a shorter food chain combine to support a greater mass of high-level consumers.

Upwelling areas combine even higher productivity and shorter food chains to support the largest amount of large animal tissue of any planktonic system—up to 36,000 times the final productivity of the open sea. Chain-forming diatoms are eaten directly, not only by such large zooplankton as krill, but also by fishes such as anchovies. Krill and anchovies are food for seabirds, seals, whales, and even humans.

Between the Tides: Life on Rocky Shores

From the rocky promontories of northern Maine to the cliffs of Washington, Oregon, and northern California, wherever hard surfaces meet the sea a thriving community lives in the space bounded by high and low tides. The plants and animals of all intertidal areas are arranged in distinct horizontal bands, a pattern called **vertical zonation.** The upper and lower edges of these bands can be related predictably to the highest high tide and the lowest low tide.

The principles behind zonation uncovered by Connell in his studies of barnacles (Chapter 5) seem to hold true for many other intertidal animals as well. The upper edge of many species' ranges is determined by the physical stresses (largely desiccation and overheating) caused by living high up on the shore. The lower edge of many species' ranges, on the other hand, is often determined by biological factors such as interspecific competition and predation.

Predation is crucial in structuring many intertidal communities. For example, carnivorous starfish dominate both temperate and subtropical rocky shores, where they choose among several prey species. Top carnivores such as these are often called **keystone predators** because their presence is like the keystone that holds up an arch; remove them and the whole structure of the system may change dramatically. When experimenters removed the starfish known as *Pisaster* from an area, for example, its former home was soon overwhelmed by its favorite food, the mussel *Mytilus*. In this situation, the mussel is a superior competitor for space that is normally kept in check through heavy predation by the starfish.

Kelp Forests

Kelp are giant, cold-water algae that grow along the Pacific coasts of North America (as far south as northern California), Japan, Siberia, and South America, and off the Atlantic coasts of Canada, New England, South Africa, Great Britain, and Scandinavia. These remarkable plants can grow up to 50 cm per day and, when fully grown, can extend 20 to 40 meters between the rocky bottom and the surface of the water (Figure 6.19).

Kelp begin as tiny plantlets that germinate from microscopic spores. At this stage, any algae eaters that happen by can devour them. For this reason, herbivore density strongly influences the establishment of new plants. Adult kelp, while continuously growing, are also constantly eroding, producing a steady stream of detritus that supports a food web like those of salt marshes and mangrove swamps. Abalone (large, flattened, snail-like animals) are common in Pacific kelp habitats, whereas American lobsters, *Homarus americanus*, favor Atlantic kelp beds.

Hunting and fishing have upset the ecological balance in many kelp forests. The Pacific sea otter, once common on the West Coast, was nearly exterminated by fur traders in the nineteenth century (Figure 6.20). And on the Atlantic Coast, lobsters have been fished down to a fraction of their former population size. Because both otter and lobster prey upon sea urchins, they normally help keep kelp-eating sea urchins under control. But where both otter and lobster populations have been reduced, their abilities to keep kelp-eating sea urchins under control have been curtailed, and urchin populations have exploded. Huge herds of these voracious grazers carpet the bottom, devouring immature kelp and damaging the holdfasts of mature plants until storms uproot them and wash them ashore.

The future of kelp forests on both coasts is still uncertain. However, where sea otter populations have recovered (around Amchitka island in the Aleutians, for example), urchin populations have fallen and kelp grow densely once again. In the Northeast, an epidemic struck sea urchins recently, causing a crash in the urchin population and giving the kelp some chance to grow back.

Coral Reefs

Coral reefs, which support an abundance of life in nutrient-poor tropical seas, are both productive and ecologically diverse for two reasons.

First, reef-building corals are actually two organisms living as one—coral animals engaged in a tightly knit symbiotic relationship with algae called *zooxanthellae*. At the heart of this symbiosis, which enables both partners to grow more efficiently than either could alone, is a steady exchange of nutrients between the corals and their photosynthetic guests. The corals generate carbon dioxide and ammonia as waste products. Though toxic to the corals themselves, these wastes are perfect nutrients for the zooxanthellae. Because the algae grow inside the ani-

mal tissues, they can absorb those compounds with minimal effort. The nutrients don't get diluted in seawater, so the plants don't have to work to concentrate them again. In return, the algae facilitate the growth of corals, both by providing their hosts with certain complex products of photosynthesis and by somehow helping the corals lay down their calcium carbonate skeletons. This combination of primary and secondary producers enables both algae and animals to obtain essential nutrients more efficiently than either could alone.

Second, like trees of the tropical rain forest and giant kelp, corals create a complex, three-dimensional environment in which other organisms live. With nearly constant, hospitable environmental conditions, coral reefs in the western Pacific shelter more species, that exhibit more complex ecological interconnections among them, than most other habitats on earth.

Many reef animals are active only by day and seek shelter by night (Figure 6.21). Others wander only after dark and hide from sunrise to sunset. These animals alternate occupancy in coral caves so regularly that one could describe the reef as a collection of boardinghouses where the same rooms are occupied day and night by different guests.

Figure 6.20 Sea otters are carnivores that seem to have important roles in controlling populations of abalone and sea urchins.

Figure 6.21 (below) By day, huge schools of scarlet *Anthias* roam the waters of the coral reef, as do many other diurnal species. (right) The change in the reef at sunset is striking, as these and many other daytime fishes seek shelter in coral nooks and crannies. At the same time, nocturnal animals, hidden throughout the day, emerge to prowl the darkening waters.

INTERACTIONS AMONG ECOSYSTEMS

In the last two chapters we have seen how primary producers, herbivores, predators, and parasites live in dynamic balance with one another in different ecosystems. It is not so obvious that ecosystems, too, interact with one another.

As we first mentioned in Chapter 1, the earth's atmosphere and oceans flow over and between ecosystems, carrying water and other nutrients from one place to another. Global meteorological phenomena such as El Niño cause biologically important events on opposite sides of the globe. And migratory animals continually cross the boundaries of ecosystems in their travels to feeding and breeding grounds. Therefore no ecosystem exists in isolation.

Similar connections between terrestrial and aquatic habitats, and between coastal and open-water marine communities, are much newer to science but are just as strong. A coral reef, for example, produces calcium carbonate faster than it is eroded by the surf, keeping the crest of the reef near the turbulent surface. Reefs thus act as living breakwaters, taming waves and making it possible for more delicate mangroves and seagrasses to grow in their lee. For their part, mangrove and seagrass systems export both dissolved organic matter and detritus, which serve as food for corals and other filter-feeding reef organisms. Reef predators stalk small fishes in mangrove nursery grounds. Osprey, herons, and pelicans eat in one habitat, deposit their nutrient-bearing feces in another, and raise their young in still a third. Back and forth, day and night, animals and currents carry energy and nutrients among these systems, binding their inhabitants together.

The same is true of all earth's living systems. If we think of the biosphere as a living fabric, all life forms are woven into it. From its diversity of threads, that fabric acquires strength and resilience, characteristics vital to survival in a dynamic world.

Into that same cloth are woven the threads of human existence. From a biological point of view, it is not merely arrogant, but foolish to imagine that we can keep that cloth intact without preserving the biological and ecological diversity that has evolved over the earth's 4.5-billion-year history. Although the interdependence of all forms of life and the importance of species diversity in ecosystems have been common knowledge among ecologists for decades, many individuals, developers, and governments around the world continue to ignore their significance. Some results of this oversight and some possible remedies will be discussed in Chapter 7.

Biological diversity refers to the variety of ecosystems and organisms in the biosphere. A substantial portion of that diversity is important for the stability and resilience of the earth's living systems. Ecologists divide the world into major habitat groups called biomes, characterized by similar physical environmental characteristics. Each biome, however, may house several different living communities.

Certain ecosystems, because of both their physical and biological characteristics, are able to withstand considerable pressure from human use, while others cannot. Grasslands, when properly treated, can support intensive, large-scale farming of corn, wheat, and other plants for many years; tropical rain forests cannot.

Though the world's ecosystems are named and described as though they were independent, self-contained units, they are not; all the earth's living systems are connected by winds, ocean currents, nutrient cycles, and animal migrations. Understanding both the biological riches and the limitations of the world's ecosystems is critical to the future of all life on the earth—including our own.

STUDY FOCUS

After studying this chapter, you should be able to:

- Define biological diversity and explain its importance to natural systems.
- Describe the several kinds of terrestrial and aquatic environments, and list some of their resident species.
- Explain how environments influence plant and animal communities.
- Demonstrate the operation of principles of ecology in the context of living communities.
- Provide examples of both positive and negative interactions between humans and ecosystems.

SELECTED TERMS

biological diversity *p. 93*	thermocline *p. 102*
biome *p. 94*	oligotrophic *p. 103*
humus *p. 95*	eutrophic *p. 103*
tundra *p. 96*	estuary *p. 103*
taiga *p. 97*	salt marsh *p. 104*
temperate forest *p. 98*	mangrove *p. 104*
desert *p. 99*	vertical zonation *p. 108*
grassland/savannah *p. 100*	keystone predator *p. 108*
tropical rain forest *p. 101*	kelp *p. 108*
thermal stratification *p. 102*	coral reef *p. 109*

REVIEW

Discussion Questions

1. Using examples from natural ecosystems, how would you support or refute the idea that harsh physical conditions support less species diversity than more benign areas?

2. Why does a hiker, climbing from the foot of a tropical mountain to its summit, encounter ecological zones similar to those found across great distances from the equator to the poles?

3. Why are tropical rain forests so interesting to scientists and so potentially valuable to human society?

4. Why are upwelling areas able to produce so much more useful food for humans and other large animals than the open sea?

5. What is an estuary? Why is life there so abundant?

6. What is biological diversity? Why is it so important? Why is the world's biological diversity threatened?

Objective Questions (Answers in Appendix)

7. The biome that can have enormous fluctuations of daily and seasonal temperatures is the
   (a) temperature deciduous forest.   (c) tundra.
   (b) savannah.   (d) desert.

8. In tropical rain forests
   (a) there is an intense competition for available sunlight.
   (b) diversity of species is limited.
   (c) the soil is fertile.
   (d) the bulk of mineral nutrients is in the soil.

9. The difference between eutrophic and oligotrophic lakes is that the oligotrophic lake
   (a) has little phytoplankton growth.
   (b) has decaying plants on the bottom.
   (c) contains many nutrients.
   (d) tends to be shrinking in size.

10. The _____ biome would be best suited for cropland if cleared.
    (a) tropical rain forest   (c) taiga
    (b) temperate forest   (d) tundra

11. In a tropical rain forest, the majority of the nutrients are present in
    (a) plants.
    (b) animals.
    (c) decaying humus, seeds, and fruits at ground level.
    (d) soil.

12. In marine ecosystems, the critical limiting factor for plants in deep water is the
    (a) intensity of solar radiation.
    (b) availability of nutrients.
    (c) salinity of the water.
    (d) amount of fertilizer run-off from adjacent farm lands.

CHAPTER 7

Human Ecology

Bali. Easter Island. Even in today's small world these names whisper of mystery and beauty. The soft, sea air that sweeps over these islands and the magical light that floods their landscapes rouse even the most jaded visitors to romantic flights of fancy. Yet these isolated scraps of land, in addition to inspiring the imagination, can teach us practical lessons about the relationship between human societies and the living world.

In theory, islands should convey to their inhabitants a sense of physical limits; the oceans that surround them place undeniable (and easily visible) boundaries on habitable turf. Yet the responses of dissimilar cultures to both the blessings and the limitations of island life have led to dramatically different human histories on different bits of rock.

Both Bali and Easter Island, for example, were colonized long ago. Bali was heavily populated by 300 B.C., and Easter Island was settled by 400 A.D. On both islands, populations grew for centuries, and sophisticated cultures flourished. On both islands, human activity totally transformed the landscape, leaving virtually no natural habitat untouched. Yet today the circumstances of these islands' human populations could hardly be less alike.

ECOLOGY AND CIVILIZATION

To be fair, Bali is much larger than Easter Island, has a more equitable climate, and is blessed with fertile volcanic soil (Figure 7.1). So it is not easy to make direct comparisons between these islands' ability to support humans. Yet the success of the Balinese is undeniably due, at least in part, to their skill in managing limited resources. Their farming techniques are based on water conservation, nutrient recycling, and soil preservation. As a result, the very same plots on the island's terraced hillsides have supported continuous crops for at least 1000 years. This reliable food supply has maintained both the island's people and its extraordinary culture.

Easter Islanders, in contrast, made fatal errors in handling their environment. By completely clearing the island of its forests, they exhausted the supply of logs that served both as fishing canoes and as rollers for moving their monolithic statues from quarry to final pedestals. Without proper land management, the island's once-fertile topsoil washed into the sea, and crop yields fell. Deprived of both offshore fish stocks and sufficient food crops, Easter Islanders faced starvation. The once-proud culture disintegrated in a tragic storm of warfare, cannibalism, and slavery. Soon, islanders were living in caves for protection.

Island Earth

Humanity at large has a great deal to learn from the experience of these island people. How? Recall the earth as astronauts have seen it; an inviting island of life in the desolate void of space. It is an enormous island, to be sure, but it is an island still—an island teeming with burgeoning populations of the remarkable species known as *Homo sapiens.*

Figure 7.1 (left) These rice terraces on the island of Bali have supported continuous, intensive agriculture for hundreds of years. (right) These abandoned giant statues on Easter Island bear mute witness to the collapse of the ecosystem that once supported a sophisticated culture.

Thus far, by all but the most cynical definitions, our species has triumphed. We have transformed Earth's landscapes by clearing forests and grasslands for our crops. Over the last 2000 years, as technological and medical advancements have allowed us to live longer and support more offspring, world population has soared from around 130 million to well over 5 billion.

But if we take the planet-as-island metaphor seriously, we have some sober thinking and many difficult decisions facing us in the near future. For just under 2 million years, small populations of humans exploited their surroundings to provide life's necessities. Some cultures—such as those of the Balinese and many New World native peoples—either lived in harmony with nature or learned to exploit natural environments in a sustainable fashion. Others—such as Easter Islanders and the great civilizations of the formerly fertile Tigris-Euphrates region—miscalculated their impact on their immediate surroundings and were destroyed by ecological collapse. In the last several decades, problems such as toxic waste dumps and pockets of polluted air have bedeviled cities ranging from Niagara Falls to Los Angeles. But through it all, most environmental issues were seen as local concerns—caused by local problems and affecting local people.

More recently, however, the exponential growth of human populations (and the impact of our expanding technology) have had effects well beyond the local level. Human activity has begun to affect the air we breathe, the water we drink, the soil we depend on, and the climate we must live in on a planet-wide scale. As noted by Soviet Foreign Minister Eduard Schevarnadze in 1988, ". . . man's so-called peaceful constructive activity is turning into a global aggression against the very foundation of life on Earth."

THE STATE OF THE WORLD

What is the status of human ecology today? What do our interactions with the biosphere portend for our future? These are complex questions, and experts disagree strongly on answers and possible solutions. In large part, these disagreements stem from the disparate outlooks of economists and ecologists discussed in Chapter 2, although political perspectives also contribute to the divergence of opinions. In the early 1980s, for example, two groups of world-class analysts made diametrically opposite predictions about humanity's future.

The *Global 2000* report prepared and funded by the Council on Environmental Quality and the Department of State, 1982:

> If present trends continue, the world in 2000 will be more crowded, more polluted, less stable ecologically, and more vulnerable to disruption than the world we live in now. Serious stresses involving population, resources, and the environment are clearly visible ahead. Despite greater material output, the world's people will be poorer in many ways than they are today.
>
> For hundreds of millions of the desperately poor, the outlook for food and other necessities of life will be no better. For many it will be worse. Barring revolutionary advances in technology, life for most people on earth will be more precarious in 2000 than it is now—unless the nations of the earth act decisively to alter current trends.

The Resourceful Earth: A Response to Global 2000, by Julian Simon and Herman Kahn; financed by the Heritage Foundation, 1984:

> If present trends continue, the world in 2000 will be *less* crowded (though more populated), *less* polluted, *more* stable ecologically, and *less* vulnerable to resource-supply disruption than the world we live in now. Stresses involving population, resources, and environment will be *less in the future than they are now.* . . . The world's people will be *richer* in most ways than they are today. . . . The outlook for food and other necessities of life will be *better* [and] life for most people on earth will be *less* precarious economically than it is now.

In biological terms, these perspectives reflect differing opinions about the earth's ultimate carrying capacity (K) for humans, the challenges we face as we approach K, and the dangers we risk if we overshoot it.

The *Global 2000* report presents what some call an unnecessarily gloomy and others call a biologically realistic picture. It states that the earth has some finite K for humans that places ultimate limits on human population size and on certain kinds of economic growth. *Global 2000* recommends centralized national and international control of population growth, resource use, and pollution. Throughout the early 1980s, however, this perspective was criticized for being overly pessimistic and naive. For although ecologists could point out ecological problems and suggest ideal solutions, they had few economically and politically viable recommendations. And, thankfully,

several extreme worldwide food and energy shortages that had been predicted did not come to pass.

The *Resourceful Earth* report, based on purely economic models, proposes that technological and agricultural advancements, driven by market forces of supply and demand, will enable us to raise our K before we overshoot it. These analysts argue that falling birth rates, new inventions, and more efficient means of pollution control will enable us to solve today's problems as they have those of the past. Environmentalists criticize these economic models as biologically uninformed and point to serious local and global ecological problems that have resulted from careless exploitation of the biosphere.

Each of these views represents an extreme position on a complicated situation. In order to appreciate these arguments fully, we must examine human impact on the biosphere to date.

HUMAN IMPACT ON THE BIOSPHERE

Water Pollution

Roughly half the United States receives sufficient rainfall to supply rivers and lakes with drinking water. The other half taps **groundwater** (underground) reserves held in geological formations called **aquifers.** Though Americans assume that there will always be enough water to go around, in many regions (particularly in California and the Southwest), domestic, industrial, and agricultural uses are making demands that neither surface water nor aquifers can satisfy indefinitely. The problem in some Third World countries is far worse; in 1980, 1.32 billion people were without safe drinking water.

Chronic surface-water pollution Pressure on water supplies is compounded by pollution from both industrial and domestic wastes. Some water pollution problems clear up within a few years if the contamination is stopped, but certain types of pollution have serious, long-term effects.

Industrial water pollutants Numerous industrial and agricultural chemicals, such as the insecticide **DDT,** do

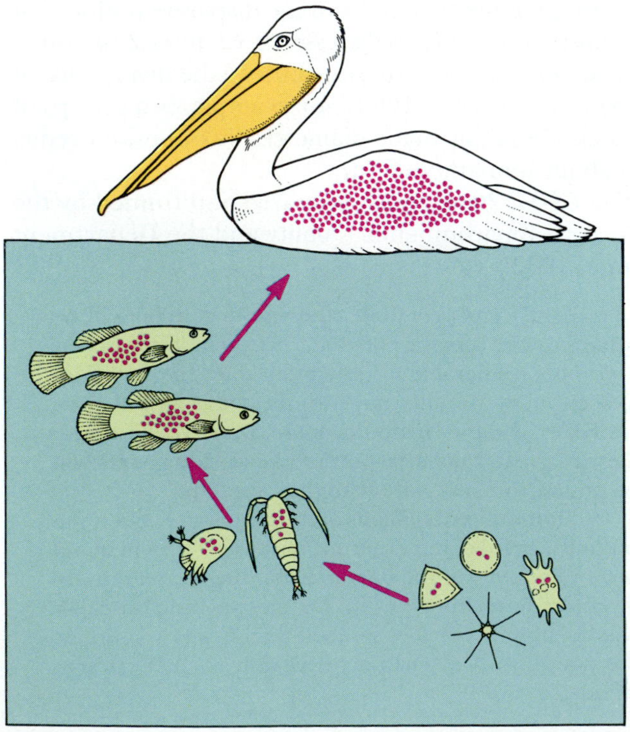

Figure 7.2 Concentration of toxic compounds through biological magnification. Some, including pesticides such as DDT, and heavy metals such as lead and mercury, are neither eliminated nor broken down by animals that eat them. Though present in the environment in small amounts, these pollutants are concentrated as they move through food webs. Though they rarely kill animals outright, some compounds damage reproductive organs and cause sterility, while others cause cancer.

Figure 7.3 Mercury released from a chemical plant in Minamata Bay, Japan, accumulated in the marine food chain and produced tragic debilitation and death in over 100 local residents who consumed large quantities of contaminated fish. Once the problem was recognized, residents warned, and mercury sources eliminated, "Minamata disease" disappeared.

not break down in nature and are toxic to animals and humans. DDT and several other toxic waste products—including **PCBs** (polychlorinated biphenyls), compounds used in the manufacture of electronics parts—are odorless, colorless, and tasteless, and in the environment for a long time. They are picked up by plants and passed through the food chain, where they accumulate in the fatty tissues of fishes, shellfish, birds, and mammals through a process called **biological magnification** (Figure 7.2).

It is currently difficult or impossible to remove DDT, PCBs, and similar compounds from the environment. One promising line of research involves the discovery in the mid-1980s of bacteria that can break down certain PCBs and similar compounds, much as the microorganisms involved in nutrient cycles process animal wastes. Laboratory and field tests with these organisms show great promise, but no commercially viable applications have been developed yet.

Other industrial water pollutants are heavy metals such as cadmium, lead, mercury, and zinc. Like DDT, heavy metals can be concentrated in the food chain and can pose serious threats to human health. Even at very low concentrations, lead causes neurological problems in young children, and mercury can cause serious brain damage in children and adults (Figure 7.3).

Residential sewage Humans, like all animals, produce organic wastes containing nitrogen and phosphorus.

These biological wastes by themselves are not toxic; they are normally processed by bacteria involved in nutrient cycles. In small quantities—such as those produced from small numbers of humans living in scattered settlements—these nutrients can be absorbed into natural systems and, in fact, increase local primary productivity. But in large towns and cities, human wastes are combined with other materials flushed into drains and sewers to produce large amounts of *sewage*. Sewage, which is eventually dumped into rivers, lakes, or coastal waters after varying degrees of treatment, can have serious environmental impacts.

From a public health standpoint, untreated sewage carries large numbers of bacteria and viruses (such as the Hepatitis B virus) that can survive prolonged exposure to both fresh and salt water in the environment. Fresh water containing these disease-causing organisms is a major health hazard in Third World countries; cholera and several types of potentially fatal diarrhea, for example, are spread largely by contaminated water. Furthermore, marine filter feeders (such as clams and mussels) tend to concentrate bacteria and viruses in their guts. For this reason, sewage contamination of coastal areas often forces the closing of thousands of acres of clam and mussel beds each year.

Other components of sewage compound the problem. Because homeowners in industrialized countries use a wide range of detergents, cleansers, paints, and oils, and because water and sewage pipes in many cities often

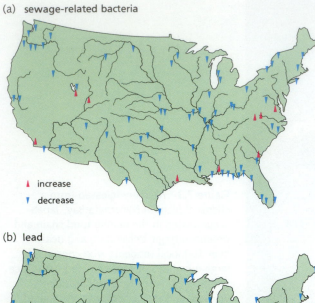

(a) sewage-related bacteria

▲ increase
▼ decrease

(b) lead

(c) nitrates

Figure 7.4 Trends in water contaminants at sampling stations from 1974 to 1981. Upward pointing triangles indicate increase; downward pointing triangles indicate decrease. Map A: sewage-related bacteria; Map B: lead; Map C: nitrate.

contain several metals, even residential sewage usually carries many potentially dangerous compounds more often associated with industrial wastes. Residential sewage from a small town on Cape Cod, for example, was shown to carry significant quantities of every metal in the periodic table from Antimony to Zinc. Many of these elements and other compounds can be concentrated by biological magnification to cause significant problems.

Complete sewage treatment, which releases as an end product water that is safe into the environment, is both costly and difficult, and involves several steps. *Primary treatment* employs settling tanks and screens to filter out solid wastes. *Secondary treatment* uses giant, controlled cultures of bacteria to break down organic compounds, releasing nutrients in forms that algae and plants can utilize. The intense heat generated by metabolic activity in these bacterial cultures can sometimes be harnessed to kill at least some dangerous bacteria and viruses. *Tertiary treatment* uses algae or plants to "scrub" nutrients from wastewater before its release.

Today, most major towns and cities use some combination of primary and secondary treatment. Though tertiary treatment is both possible and ecologically desirable, its expense has severely restricted its use. In the absence of tertiary treatment, the enormous quantities of organic wastes in sewage can cause two types of problems. Sewage that has experienced only primary treatment stimulates "blooms" of bacteria that rob water bodies of oxygen, turning them into anaerobic wastelands where no organisms can grow. Secondarily treated sewage avoids this problem, but contains sufficient nutrients to initiate algal blooms, eutrophication, and the death of native organisms. Often, chlorine is added to kill many (though not all) harmful microorganisms just before the sewage is released. Unfortunately, chlorine itself is toxic to many aquatic organisms. Additionally, chlorine combines with certain organic components of sewage to produce compounds that are even more toxic to aquatic life.

With careful attention to the presence of heavy metals and infectious microorganisms, secondarily treated sewage, often called "night soil," can be used to fertilize farm crops or ecosystems such as estuaries whose productivity is limited by their supplies of those nutrients. In Asia, night soil has been used to fertilize both vegetable crops and fish farms for centuries. The disposal of partially treated sewage has been a major problem for large American cities for many years, and the situation is getting worse. Near New York City, for example, beaches are now regularly closed because of a hideous mixture of sewage, other wastes, and ocean muds called "the black mayonnaise" that washes up after certain storms.

Contributing factors A 1987 study by a New York State environmental research group confirmed that drainage from storm sewers and agricultural land contributes more

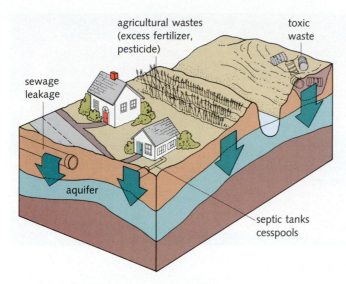

Figure 7.5 Potential sources of pollution affecting underground aquifers: Groundwater contaminants range from inorganic fertilizer salts, such as chloride and nitrate, to complex organic compounds and radioactive isotopes. Particularly insidious is the groundwater pollution that occurs when industrial chemicals such as TCE leak out of chemical waste dumps.

of certain pollutants to the Hudson River than either sewage plants or factories. Farmers and suburban gardeners use chemical fertilizers and pesticides in large amounts, and heavy rains flush these compounds into streams and street drains that empty into lakes and rivers.

Airborne contamination, which can be washed into water supplies by rainstorms, also represents a major source of surface-water contamination. Acid rain brings nitrate ions, the burning of leaded automobile fuel adds lead, and the salting of highways in winter adds both sodium and chlorine ions.

But recent research has shown that environmental legislation and pollution-control expenditures can have direct, positive effects on purity of surface water (Figure 7.4). One study confirmed that the billions of dollars spent on improved sewage treatment had lowered both the levels of sewage-associated bacteria and the oxygen requirements of bodies of water receiving the runoff. The use of unleaded automobile fuel has had positive effects as well; lead levels had dropped in nearly all rivers surveyed. On the other hand, increased application of chemical fertilizers to farmlands and use of salt on highways in winter had raised the concentrations of nitrates and chloride in those rivers.

Groundwater contamination Pollution of underground aquifers has become frighteningly common; in the mid-1980s it was discovered that many towns across the nation were pumping water from contaminated wells (Figure

7.5). For example, **TCE** (trichloroethylene), a suspected carcinogen, was discovered in 39 wells that supplied water to 13 cities in California's San Gabriel Valley. TCE contamination also forced the closing of municipal wells in several Massachusetts towns.

Air Pollution

Our atmosphere, as you learned in Chapter 2, is never static; it is a dynamic mix of gases that constantly interacts with plants, animals, and microorganisms in the soil and oceans. As you will learn in Chapter 14, the atmosphere has changed dramatically during Earth's history, in large part through the actions of living organisms. Today, however, human activity is changing the atmosphere in several ways, faster than it has ever changed in the history of life on Earth.

Because global environments are complicated, making precise predictions about the long-term effects of these changes is difficult. We are, in a very real sense, making ourselves and our environment the subjects of a global experiment in atmospheric chemistry (Figure 7.6). As usual, biologists, economists, and politicians have different interpretations regarding the severity of various types of air pollution, and widely divergent views on how to improve the current situation. Although most

Figure 7.6 A single pollution source can have dramatic effects on air quality for kilometers—or even hundreds of kilometers—downwind. This copper smelting plant in Arizona was the largest single emitter of sulfur dioxide in the United States for nearly half a century. The plant was closed in 1987 as a result of pressure from environmentalists fighting acid rain, to which airborne sulfur dioxide is a major contributor.

Figure 7.7 (left) Healthy lake trout from unpolluted waters. (right) A lake trout suffering from the effects of acid rain (pH 5.0 in lake).

forms of air pollution are linked to a few common sources, we can divide them up for the sake of discussion into several categories.

Smog Scientists have known for decades about the health hazards of several local forms of air pollution. One of the most notorious of these is *smog,* a catch-all term that is often used to describe a noxious collection of airborne dust, smoke, particles produced by internal combustion engines, and a variety of gases. One particularly nasty component of certain types of smog is ozone. Ozone is vital where it normally occurs in the upper atmosphere, as you will see shortly, but at ground level it can pose serious health hazards both to humans and to agricultural crops.

Carbon dioxide and global climate Carbon dioxide, as you learned in Chapter 1, plays a major role in keeping Earth's temperature constant. But the burning of fossil fuels such as oil and coal and the global destruction of the world's forests are returning carbon dioxide to the atmosphere faster than plants are removing it. Many scientists predict that increased CO_2 levels will raise the earth's average temperature, causing significant changes in climate. If that happens, some of today's most agriculturally productive areas will become too dry or too hot to support the crops now grown there. Sea levels could also rise by several meters, flooding coastal cities from Miami to Bangladesh. Such predictions are quite controversial and impossible to test, and other scientists argue that a slight warming will produce more clouds that will *prevent* excessive warming by reflecting heat out of the atmosphere.

Acid rain The burning of fossil fuels by automobiles and industries also releases large quantities of sulfur dioxide and nitrous oxides into the atmosphere. When these compounds dissolve in fog or raindrops, they produce **acid rain,** precipitation that contains dangerously high levels of sulfuric acid and nitric acid. Acid rain causes numerous problems. Combined with other airborne pollutants, it damages the leaves of plants and places those plants under stress that can slowly kill them. Acid rain can also harm roots by releasing aluminum and other metals from some soils, and it can interfere with bacterial decay in topsoil and humus. Acidification of surface water kills aquatic organisms from algae to fishes (Figure 7.7). Many lakes and streams in New England, eastern Canada, Scandinavia, and central Europe are already classified as "dead" because of acid rain, and 14,000 Canadian lakes are showing serious damage.

Where did today's widespread problems originate? Removing sulfuric and nitric oxides from smoke requires expensive technology. So in an effort to improve local air quality at minimal expense, industries and utility companies built giant smokestacks to carry pollutants aloft. This "solution" doesn't eliminate pollutants; it simply sends them high into the atmosphere so that winds carry them elsewhere. As a result, pollution generated in the midwestern United States falls on New England and eastern Canada, and smoke from Great Britain creates acid rain in Scandinavia (Figure 7.8). Nearly half of central Europe's forests already show damage so serious that Germans have coined a new word to describe it—*Waldsterben,* or "forest death."

Because acid rain falls far from its sources, and because removing pollutants at those sources requires billion-dollar investments, the politics and economics of cleanup are complicated. Acid rain generated primarily in the Ohio valley, for example, falls on New England and eastern Canada, where it threatens forestry, the maple sugar industry, fisheries, agriculture, and tourism. Industries and consumers in the Midwest, however, object to paying for cleaning up someone else's environment. These conflicting interests make it likely that pollution control measures will be delayed until economic pressure justifies their cost.

Changes in the ozone layer Evidence is accumulating that the atmosphere's vital **ozone** layer is thinning out. Ozone is produced in the atmosphere when oxygen molecules are split by the ultraviolet radiation in sunlight.

Ozone then continues to absorb ultraviolet light, preventing much of that potentially dangerous radiation from reaching the earth's surface.

Several chemicals we release into the atmosphere combine with ozone and break it down. These compounds include chlorofluorocarbons (CFCs) such as freon, which was once used in aerosol spray cans and is still widely used in refrigerators, air conditioners, and fire extinguishers. Oxides of nitrogen released by automobile and airplane exhaust can also destroy ozone. Research flights over the Arctic and Antarctica have shown that CFCs combine with ice crystals in frigid air in a manner that allows sunlight to destroy ozone rapidly (Figure 7.9). It is not yet known how quickly and how extensively this process occurs elsewhere.

Figure 7.8 Airborne pollutants. (**a**) While particulate pollutants generally settle close to their source, gases such as sulfuric and nitric oxides and chlorofluorocarbons often travel great distances. (**b**) Some regions are more seriously affected by acid rain because of wind and rainfall patterns and because of soil composition factors. Pollutants released in the Midwest, for example, are carried by winds in the upper atmosphere to Canada and the northeastern United States. Maps (**c**) and (**d**) show the increase in severity of acid rain since the 1950s.

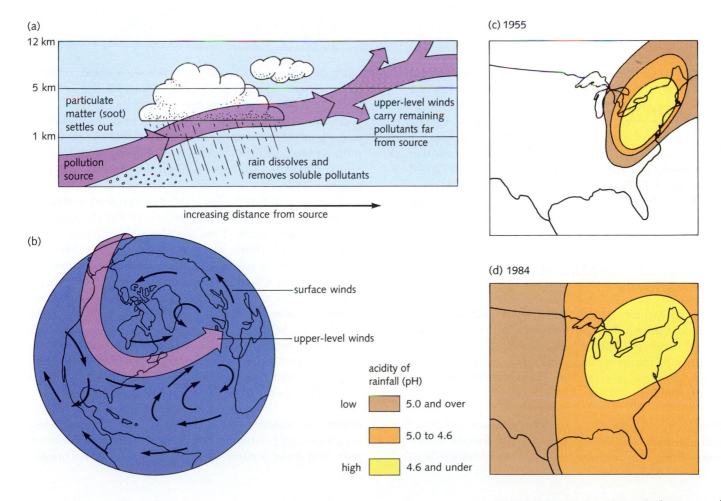

Figure 7.9 Recent satellite data, represented in this false-color image, show major "holes" in the ozone layer over north and south poles. If those holes spread over populated areas, many new cases of skin cancer and immune system damage could result, along with crop damage and an increase in global temperature.

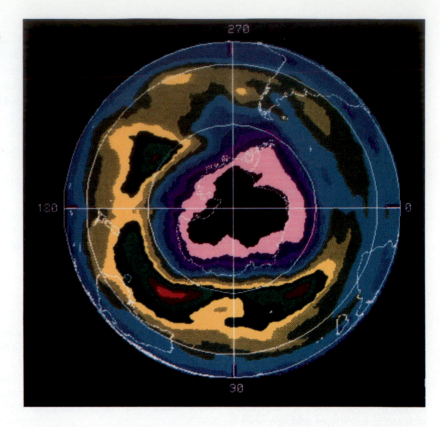

A treaty to first limit and then slowly roll back the use of ozone-destroying chemicals in 43 nations was signed in 1987. Though the treaty failed to limit CFC use as strictly as some environmental analysts think necessary, it was a landmark step in international legal action on environmental issues. The treaty will be costly to enforce, because the development of safe alternatives to CFCs will cost billions of dollars. The long-term benefits of protecting the world's ozone layer, however, will far outweigh its expense.

Habitat Destruction and Loss of Biological Diversity

Exponential human population growth and economic development have led to the dramatic alteration of many terrestrial and aquatic environments. Some habitat changes are of minor ecological significance, while others have serious negative effects on biological diversity and ecosystem stability.

Deforestation A process called **deforestation** that some say will eliminate the world's tropical rain forests before the end of this century, has several causes. Many subsis-

tence farmers in the tropics practice *slash-and-burn* agriculture, clearing and burning rain forest for use as farmland. Because cleared rain forest is fertile for only a few years, these farmers must constantly relocate and clear more land. Additionally, the demand for cheap paper pulp and hamburger in developed nations provides economic incentives to transform tropical forests into tree plantations and cattle ranches.

Extensive deforestation, however, can lead to **desertification,** the creation of new deserts or the extension of existing desert areas. This has already occurred in the Sahel region of Africa, where climate change and habitat destruction have made Ethiopia far more sensitive to droughts than it was in the past (Figure 7.10). Deforestation in mountainous areas destroys soil fertility and increases these regions' susceptibility to both droughts and floods. Without the protection of vegetation, rich and porous topsoil washes away and subsoil hardens, preventing later rainfall from penetrating. Instead of adding to local groundwater supplies, rainwater therefore rushes downhill, causing floods and erosion downstream.

Marine and estuarine habitat destruction Human activities have had relatively little adverse effect on the open seas. But most oceanic primary productivity is concentrated in narrow bands along the coastlines called the **continental shelves**—precisely the areas most affected by human activity. In a few states, wetlands and estuaries are protected, but elsewhere these vital breeding and nursery grounds are being filled to build marinas, hotels, and condominium developments or are being polluted by garbage and sewage from coastal cities and towns. The loss of critical coastal habitats has already hurt flounder and striped bass fisheries in the north and has reduced the yield of spiny lobsters and shrimp in the south.

Estuaries are also highly vulnerable to oil spills because of their location between the tides. Immediate effects, such as helpless seabirds dripping oil, hit public sensibilities the hardest, but the long-term effects of petroleum on estuarine life are far more significant. They include the death of mangroves and seagrasses, the smothering of bottom-dwelling food species, and the incorporation of toxic compounds into the food web.

Loss of species Pollution and habitat destruction are forcing animal and plant species into extinction at an ever-increasing rate. Before the year 2000, nearly 1 million of the earth's species will have vanished forever; between one-third and one-half the world's species will

Figure 7.10 Destructive farming practices in Ethiopia have caused the Sahara Desert to grow. These false-color Landsat photos show the retreat of vegetation from sub-Saharan Africa.

Figure 7.11 The rosy periwinkle, a plant native to Madagascar, contains two compounds now used in chemotherapy for Hodgkin's disease and several other forms of cancer including acute lymphocytic leukemia. Worldwide sales of these drugs now top $100 million each year.

have disappeared within 500 years. No extinction of this magnitude has occurred for 65 million years.

Human society gains much from the diversity of the earth's species. In their struggle for existence, many organisms have evolved chemicals that protect them from enemies, and humans have long used these compounds in food, medicine, and industry. For example, pine trees contain turpentine, which is used in solvents, and foxglove contains digitalis, used to treat heart ailments. In fact, more than half of all the drugs in use today were originally discovered in wild plant species (Figure 7.11). No one knows what treasures might still be locked away within the plant kingdom, and as every unstudied specimen disappears forever from the face of the earth, its potential applications for human welfare vanish also.

Biodiversity and agriculture   Modern food crops have been derived from wild species through centuries of selective breeding for improved yield, strength, and ease of cultivation. Yet the success of agricultural hybrids carries hidden dangers, for the best hybrid strains are both widely planted and highly inbred. An estimated 70 percent of

Figure 7.12 New disease strains find uniform fields easy pickings. The corn blight that struck the United States in 1970 proved resistant to every defense modern agriculture could muster. Once a field was infected, nothing could eradicate the disease, and roughly 15 percent of the crop was destroyed. Future blights could be worse.

American corn, for example, is derived from no more than five hybrid lines. Because crop plants covering hundreds of square miles are nearly identical genetically, any new diseases can spread rapidly (Figure 7.12).

To combat this problem, crop geneticists must introduce new, disease-resistant strains every 10 to 15 years. But disease-resistant genes do not simply appear when we need them; they must be discovered in wild strains and either bred into the hybrid stock or transferred by genetic engineering techniques. For that reason, the wild relatives of domestic food crops, though seldom grown for food today, are invaluable to the future of agriculture. Yet our available genetic crop diversity is shrinking fast. Habitat destruction is making those wild populations harder to find every year. And tropical farmers who once grew crops from their own genetically diverse seed stock are switching to higher-yielding hybrid seeds. The situation is not unique to corn; identical trends threaten wild relatives of rice in Southeast Asia, cereals in Turkey, and barley in Ethiopia.

Sustainability: Living Within Our Means

Faced with these ecological dilemmas, ecologists and economists are searching for a new style of interacting with the global environment, a style they describe as *sustainable development*. An ideally sustainable system is characterized by stability, resilience, use of appropriate technology, efficiency, and satisfactory productivity.

A sustainable system must be *stable*; it must operate in such a way that it neither upsets ecological systems nor overexploits living organisms (Figure 7.13). Sustainable management ensures that natural resources will be managed in a renewable fashion and that environments will be *resilient* enough to survive inevitable floods, droughts, heat waves, and cold winters. *Appropriate technology* refers to equipment and practices suited both to the local ecosystem and to the abilities, training, and economic system of the people using it. Finally, a sustainable system needs to be both *efficient* and *productive*. It should provide not only the means for survival but also enough profit to improve the local economy.

Presented in the abstract, sustainable development is an ideal goal. In various cultures around the world, sustainable systems have indeed been developed and have survived for centuries (see Theory in Action, The Goddess and the Computer, p. 126). In contrast, wherever ecological principles have been ignored, the long-term costs have eventually become obvious. Why, then, has sustainable development not been the rule around the globe, and why is it not yet the first priority of every culture? To answer those questions, we must examine the complex dilemmas behind the environmental issues.

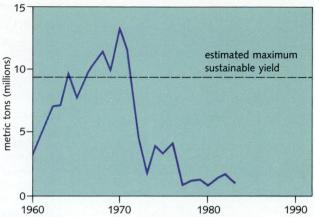

Figure 7.13 Costs of nonsustainable management. This graph of the Peruvian anchovy catch shows how the most productive fishery in the world was destroyed by overexploitation. Before the fishery began (and during its early years), the upwelling ecosystem off Peru was resilient enough to survive the disruption caused by El Niño events. At the fishery's peak, however, the anchovy population was so seriously overfished that the entire food web collapsed during a major El Niño. Twenty years later, the system has not yet recovered.

Table 7.1 *Projected Population Size at Stabilization*

Country	Population in 1986 (million)	Annual Rate of Population Growth	Population at Stabilization (million)	Change from 1986
Slow Growth Countries				
China	1,050	1.0%	1,571	+ 50%
Soviet Union	280	0.9	377	+ 35
United States	241	0.7	289	+ 20
Japan	121	0.7	128	+ 6
United Kingdom	56	0.2	59	+ 5
West Germany	61	− 0.2	52	− 15
Rapid Growth Countries				
Kenya	20	4.2	111	+455
Nigeria	105	3.0	532	+406
Ethiopia	42	2.1	204	+386
Iran	47	2.9	166	+253
Pakistan	102	2.8	330	+223
Bangladesh	104	2.7	310	+198
Egypt	46	2.6	126	+174
Mexico	82	2.6	199	+143
Turkey	48	2.5	109	+127
Indonesia	168	2.1	368	+119
India	785	2.3	1,700	+116
Brazil	143	2.3	298	+108

SOURCE: World Bank, 1985, *World Development Report 1985*, Oxford University Press, New York.

ENVIRONMENTAL PROBLEMS: CAUSES AND SOLUTIONS

Human Population Growth

For the last several decades, the world's human population has grown exponentially (see Chapter 5). Disagreements exist, however, about current trends in population growth and about the role that population size plays in causing ecological problems.

Some analysts view population growth as the cause of many global problems and contend that exponential growth is continuing in many developing countries. They further argue that population growth in several countries is causing ecological and economic conditions to deteriorate, and they fear that this ecological impoverishment will block the economic and social changes that could lower birth rates. Such countries could become "stalled" in the middle of the demographic transition, caught in a "demographic trap" from which they cannot escape. Once that happens, they may be caught in a downward ecological–economic spiral.

Proponents of this view urge active family planning programs throughout the Third World. Table 7.1 summarizes population sizes and growth rates for selected countries and projects the ultimate sizes of those countries when growth stabilizes. If these projections are correct, Nigeria will ultimately house more people than all of Africa in 1986, and Kenya and Ethiopia will quintuple in size. Table 7.2 shows population growth and change in per capita income for selected countries between 1980 and 1986. Recall that the drop in birth rates that completes the demographic transition has historically accompanied an increase in the per capita income. Note that in eight countries currently "stalled" in that transition, per capita income is declining.

Other analysts believe that population growth has little to do with sustainable development and insist that the real problems are political and economic. In their view, population growth is slowing down on its own, and hunger and poverty are created by unbalanced political and economic systems in developing countries. They therefore feel that money and effort spent on international family planning programs are largely wasted.

There is no simple truth to this matter. Population growth and inequities in land distribution in some countries result in serious ecological problems. In others, economic systems, government policies, and wars are more at fault. Meanwhile, a number of Third World nations are addressing their own population problems, in different ways and with varying degrees of success (Figure 7.14).

World Food Production

Faced with an increasing number of mouths to feed during the 1950s, governments around the world joined forces with researchers to increase global food production. The result was the **Green Revolution,** an unprecedented increase in world agricultural output. Between 1950 and 1970, wheat production in Mexico increased tenfold, while China and India became self-sufficient in food production. In 1983, farmers worldwide produced nearly 1.5 *billion* tons of grain, an increase of 900 million tons over world output in 1950. In most parts of the world, agricultural production is increasing, and prices for grain, rice, and other commodities fell during the mid-1980s.

Yet food shortages persist. According to the World Bank, there are more poor and hungry people in the world today than ever before. The data are grim: Twenty-four people, eighteen of them children less than five years old, die of starvation every minute. More human beings starved to death between 1980 and 1986 than were killed

Table 7.2 *Changes in Population and Per Capita Income*

Country	Rate of Population Growth	Change in Per Capita Income, 1980–1986
Rising Incomes		
China	1.0%	+58%
South Korea	1.6	+34
Japan	0.7	+21
India	2.1	+14
West Germany	−0.2	+10
United States	0.7	+10
United Kingdom	0.2	+12
France	0.4	+ 3
Declining Incomes		
Nigeria	3.0	−28
Argentina	1.6	−21
Philippines	2.5	−16
Peru	2.5	−11
Kenya	4.2	− 8
Mexico	2.6	− 7
Sudan	2.9	− 7
Brazil	2.3	− 6

SOURCE: Reprinted from *State of the World*, 1987, by Lester R. Brown, William V. Chandler, Christopher Flavin, Jodi Jacobson, Cynthia Pollock, Sandra Postel, Linda Starke, and Edward C. Wolf, by permission of W. W. Norton & Company, Inc. Copyright © 1987 by the WorldWatch Institute.

Figure 7.14 China's powerful government has instituted the world's strictest population control measures. While encouraging families to limit themselves voluntarily to a single child, the government strictly prohibits larger families. In many areas women are coerced (psychologically) into sterilization or abortion after bearing their first child. These extremely harsh measures have slowed China's growth rate to 1.0 percent and helped agricultural production catch up with national food needs.

The Goddess and the Computer

It's afternoon on the Indonesian island of Bali. A high priest invokes the blessing of Dewi Danu, the water goddess, as he assigns water allotments, irrigation schedules, and planting dates to the island's assembled farmers.

Daily life and religion are inseparable for the Balinese. Nature gods are everywhere, and natural rhythms of growth, maturation, and decay shape their rituals and their complex calendar. So no one disobeys the priest's agricultural instructions; to plant at the wrong time or to take too much water from communal irrigation ditches would be disrespectful to Dewi Danu. Besides, following the oracle has resulted in a highly productive, sustainable agricultural system. The Balinese have been growing rice on the same terraces for at least 1000 years with neither chemical fertilizers nor pesticides.

Then, in an effort to increase yields, agriculturists introduced a dwarf, high-yielding rice strain that grows fast enough to permit two or three crops each year. Dewi Danu adjusted her advice to accommodate this opportunity, and farmers began to double- and triple-crop the rice. Then the unexpected happened.

The fast-growing rice needed more nitrogen and needed it more quickly than the nitrogen-fixing bacteria could provide, so farmers had to start using chemical fertilizers. With genetically identical rice growing everywhere all the time, pest populations skyrocketed. Ducks and other natural predators couldn't keep up with them, so farmers had to start using chemical pesticides. Although these brought the pests under control, they also either killed the eels and frogs that were part of the ecosystem or accumulated in their flesh, making them unfit for human con-

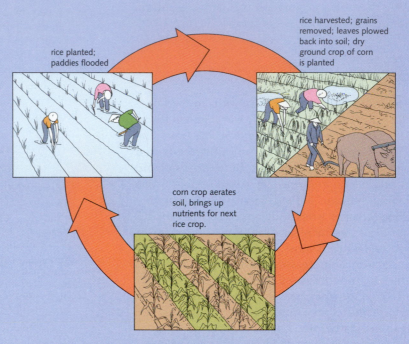

rice planted; paddies flooded

rice harvested; grains removed; leaves plowed back into soil; dry ground crop of corn is planted

corn crop aerates soil, brings up nutrients for next rice crop.

In a typical rice paddy, a community of plants and animals form a self-perpetuating system that thrives under Balinese management practices. According to tradition, after harvest, rice plants are plowed into the soil. The paddy is planted with a dry-ground crop whose deep roots aerate the soil and bring up nutrients flushed down by flooding. This crop rotation, practiced in patchwork fashion across the island, conserves water, preserves soil nutrients and texture, and helps control plant pests and diseases.

sumption. And with pesticides in the paddies, farmers could no longer safely graze ducks there.

Finally, even with chemical fertilizers, rice yields declined after a few years. Paddy soil hardened and became difficult to cultivate. Weighing actual yields against the added cost of chemicals, and factoring in the Balinese preference for more flavorful native rice, farmers began returning to the old ways.

But Dewi Danu's priests and farmers, like most other Balinese, are both educated and resourceful and far from ready to ignore the possible benefits of modern agricultural tech-

nology. Firmly grounded in, and proud of, their time-tested agricultural techniques, they are eager to blend the best of their traditional knowledge with the best the twentieth century has to offer. The day after the ceremony invoking the goddess's blessing, the priest takes time to help American anthropologists and ecologists construct a model of his irrigation system on their portable computer. With luck, the hybrid system they devise together will produce higher sustainable yields than either method alone.

Rumor has it that Dewi Danu is pleased.

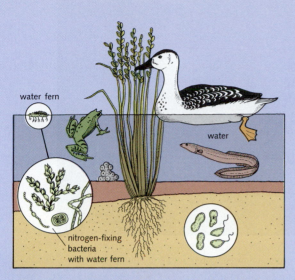

Nitrogen-fixing bacteria growing in association with water ferns supply a constant trickle of fertilizer. Frogs, ducks, and freshwater eels in the paddy provide protein in the Balinese diet, feed on harmful insects, and leave nutrient-rich droppings.

In the traditional Balinese rice-paddy ecosystem, rice plants grow in flooded paddies, maturing in about 210 days.

in all the wars, revolutions, and murders of the last century and a half. In 1987, 730 million human beings lived on the edge of starvation.

What does the future hold for the affluent United States and for the world's poor and hungry? Once again, ecologists and economists are deeply divided. We will now examine the ecological bases for a mixture of cautious optimism and concern about the future of world food production.

The Green Revolution: successes and problems   The Green Revolution techniques that dramatically increased world food production originated primarily in the fertile, temperate farmlands and economic climate of the United States. The central principles are simple: Plow large fields, plant those fields in a single high-yielding crop, maintain soil fertility by using chemical fertilizers, and control pests by applying chemical insecticides. Because these techniques succeeded brilliantly here, they were exported with the best intentions to developing countries. New land was brought under cultivation, new strains of high-yielding crops were bred, and chemical-intensive agricultural techniques were introduced.

The benefits of this strategy were immediate: Yields per harvest increased and more crops could be grown each year. In some countries, such as the United States, grain *surpluses* rather than shortages became an economic problem. The Green Revolution was hailed as an unqualified success. Yet many Green Revolution techniques have not turned out to be sustainable in the sense we have defined it.

Monoculture and chemicals   The classic Green Revolution strategy of planting large areas in genetically uniform, single-species crops is called **monoculture**. Monoculture requires the use of agricultural chemicals for two reasons. Planting the same crops year after year exhausts the soil, resulting in the need for added fertilizer. And large populations of identical plants invite attack by fungi, bacteria, and plant-eating insects that are most easily controlled by chemical pesticides.

When transplanted to tropical ecosystems, these techniques, though spectacularly successful in the short term, cause long-term problems. Fertilizers and pesticides are expensive, and their production consumes fossil fuels such as coal and oil. Increased food production often requires enormously increased energy use; rice production in the United States requires 10 times as much energy as it does in the Philippines, and wheat farmers here require 1000 times as much energy as Indian farmers to produce a unit of grain. Adoption of this "modern" technology can thus necessitate major changes in the economy of developing countries, and it can end up requiring government subsidies because it costs more than the resulting food is worth.

Figure 7.15 Lush tropical rain forests such as this one (right) are best left in their natural state and exploited in a sustainable fashion for their valuable plant and animal products. (below) The result of deforestation.

The use of fertilizers and pesticides in developing countries also exposes their ecosystems to such problems as eutrophication of rivers and streams, pollution of drinking water, and accumulation of toxic compounds in local food chains. Equally alarming is the ability of pests to develop immunity to pesticides.

Soil depletion Soil erosion is a natural process by which soil is removed from an area by wind and water. Normally, soil removed from an ecosystem is more or less replaced by the creation of new soil and humus. But intensive, Green Revolution-style agriculture often leaves large areas of soil barren and open to wind and water erosion between crops, causing soil erosion far in excess of natural rates. In Third World countries, crowding and economic incentives push farmers into clearing and farming hilly areas where rains remove any soil not held down by plant roots (Figure 7.15). Even in the relatively flat American Midwest, dominant forms of plowing and harvesting encourage erosion.

The loss of soil at a rate faster than it is being formed is causing steady deterioration of croplands covering nearly 35 percent of the earth's land surface. Fields on the high plains of the American Midwest, for example, lose an average of 20.9 tons of topsoil per acre per year to erosion. Corn fields in the mountains of Guatemala are in even worse shape; some lose 770 tons per acre per year. Ultimately, even heavy fertilization will not compensate for the loss of fertile topsoil, a grave threat to the sustainability of global agriculture.

Irrigation and salinization Only about one-seventh of the farmland in the United States depends on irrigation, but that land provides a disproportionately large share of our annual harvest. In many areas, from Africa through the Middle East and Asia, irrigation is the only source of water for agriculture. Virtually all water supplies contain some dissolved salts, usually in low enough concentration to be harmless. In arid regions, however, extensive evaporation from water and soil surfaces leaves those mineral salts behind to build up in the soil, a process called **salinization.**

Salinization is not a new problem; it turned the once-fertile valleys of the Tigris and Euphrates rivers into deserts and contributed to the collapse of ancient Mesopotamian civilization. Today agriculturists recognize salinization as a serious problem in arid southern California and in much of the Middle East. As a result, researchers are developing new irrigation techniques, both to conserve water and to decrease the rate of salt build-up. One such technique, *drip irrigation,* uses pipes to provide water slowly, steadily, and directly to tap roots.

World agriculture: trends and prospects The agricultural problems described above, combined with population growth, have created some unsettling trends. Though grain production per person has risen steadily in Western Europe, it is dropping slowly in Africa and is falling rapidly in countries with rapid population growth (Figure 7.16). These trends are slowly eroding the grain self-sufficiency of several countries, including Egypt (Figure 7.17).

Toward a sustainable agriculture The urgently needed reevaluation of world agricultural practices indicated by these data has finally begun. In 1988, for example, the United States government created a $3.9 million "seed fund" to finance further research into sustainable agricultural technology. Ideally, this fund will foster cooperation and combine the best techniques developed in all sectors of the agricultural economy. Organic gardening experts, long neglected by the agricultural establishment, can offer knowledge of ecological principles that regenerate degraded soils and rely on insect predators to control pests without toxic chemicals. Giant agricultural research firms can shift their efforts from the creation of new pesticides to the development of pest-resistant crops. And the benefits of classical Green Revolution technology can be combined with farming methods tailored to local climates and soil types.

In tropical countries, innumerable variations on modern rice and wheat monoculture are possible because so few of the hundreds of potential food and forage crops

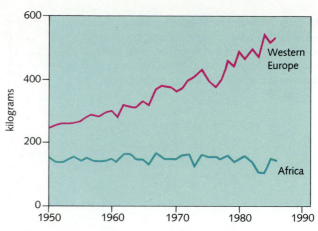

Figure 7.16 Per capita grain production in Western Europe (upper line) and Africa (lower line) from 1950–1986. *Total grain production during this period increased by 164 percent in Europe and by 129 percent in Africa; during the same period, Europe's population increased by about 1/5, while Africa's population doubled.*

Figure 7.17 Grain self-sufficiency in Egypt, 1960–1986. Almost 100 percent self-sufficient in grain production in 1960, Egypt now imports over 50 percent of the grain it consumes.

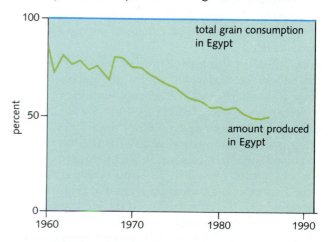

Figure 7.18 In Western countries (such as the United States), intensive, repetitive monoculture is the rule on large farms but does not need to be. *Crop rotation* involving nitrogen-fixing plants such as alfalfa improves soil fertility by adding both organic content and nitrogen.

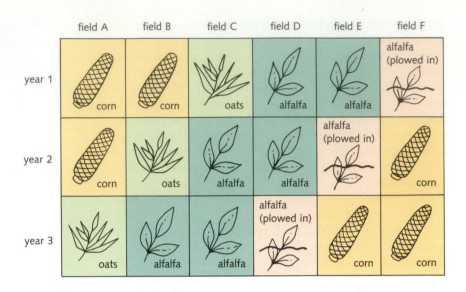

Fisheries

In 1983 the world fish harvest totaled 74 million tons and supplied 23 percent of the animal protein consumed. In several Third World countries (and in a few developed ones), fishes provide nearly the only source of animal protein. From 1950 to 1970, the annual world fish catch more than tripled, from 21 to 66 million metric tons. This spectacular growth led many to believe that the oceans harbored an inexhaustible supply of food. But by the 1970s, the first signs of trouble appeared; more and more effort was required to catch the same amount of fish. By 1983 the world fish catch not only leveled off but began to drop.

Today, intense fishing pressure is harvesting numerous fish stocks beyond their ability to replenish their numbers, a condition called **overexploitation.** Overexploitation can be fatal to a fishery, as shown by the history of the Peruvian anchovy catch. Before the fishery began operating (and during its early years), the natural system that nurtured the anchovies was resilient enough to survive the disruption caused by El Niño events (Chapter 4). At the fishery's peak, however, overharvesting stressed the system so much that it collapsed during a major El Niño. The fishery has not yet recovered, nearly two decades later (see Figure 7.13, p. 123). Several other impor-

tant species—including halibut, cod, salmon, Atlantic herring, pilchard, and Alaskan king crab—are showing signs of overexploitation today. If not protected, these fisheries may go the way of the Peruvian anchovy.

Unfortunately, it is very difficult to estimate how many fishes there are in the sea. Scientists associated with fisheries do their best to predict stock sizes and reproductive rates, and in some cases they have enough data to calculate what quantity of fish can be caught without depleting the stock. But these recommendations are speculative.

Aquaculture On land, human civilizations switched from hunting and gathering to herding and farming thousands of years ago; yet at sea, we are still predominantly hunter–gatherers. Though **aquaculture,** the farming of aquatic animals, has been practiced in China since 1100 B.C. and in Egypt since 2000 B.C., it produces less than one-sixth of the aquatic organisms consumed today.

Commercial aquaculture in Western countries usually involves raising fishes in tanks or ponds and feeding them prepared foods. Some fish species convert vegetable matter to meat far more efficiently than farm animals. Catfish, for example, require 1.7 pounds of grain to build a pound of fish, whereas beef cattle require 7.5 pounds of grain to produce the same weight of meat. Though many technical problems remain to be solved, aquaculture is growing rapidly: American catfish farmers alone increased production from 2600 metric tons in 1970 to 62,400 tons in 1983.

But as Western aquaculture technology has grown, researchers have rediscovered many remarkably efficient culture techniques known in Egypt and Asia for millennia. A single agriculture–aquaculture farm in Thailand, for example, can support several acres of crops; thousands of ducks, chickens, and pigs; and over a million fishes (Figure 7.19). The secret is intensive internal recycling; crops are fed to animals whose wastes fertilize pond water to grow algae and other fish food. Water from fish ponds is used to water and fertilize crops simultaneously, and other wastes produced by the system are either recycled into fertilizer or digested by bacteria to produce methane gas for use as fuel.

ENERGY

Developed countries consume enormous quantities of energy today, and as Third World countries industrialize, their energy demands will rapidly outstrip ours. This trend is important because most forms of energy have become significantly more expensive in the last half-century and will continue to increase in price. The reasons for this are simple. In the early days of the Industrial Revolution, low energy prices reflected only the low costs of production. Now we must factor in not only increased costs of finding, extracting, and transporting energy supplies but also the expenses of controlling pollution.

Nonrenewable Energy

Fossil fuels Today more than 80 percent of the world's energy comes from **nonrenewable energy** sources, which include finite deposits of oil, coal, and natural gas. The fact that these reserves are finite and dwindling cannot be emphasized strongly enough. No matter how much remains today, and no matter how efficiently we extract and conserve it, our supply will run out sooner or later. Analysts warn, for example, that little of the world's petroleum will be left by the middle of the twenty-first century. The transition from fossil fuels to other energy sources will be much smoother if we prepare for it before current supplies run low.

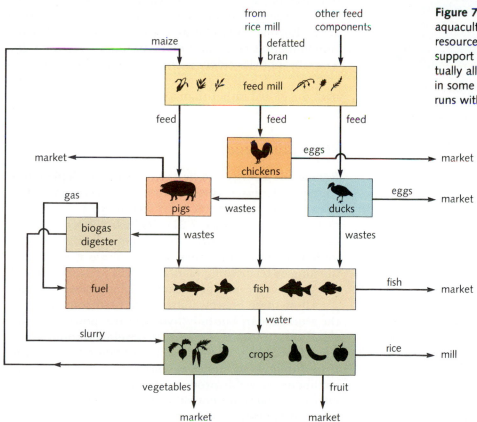

Figure 7.19 This schematic of a Thai energy-aquaculture-agriculture facility shows how resource conservation and nutrient recycling can support high-yielding food production units. Virtually all plant and animal "wastes" are recycled in some way, creating a productive system that runs with minimal inputs of chemical fertilizers.

Figure 7.20 Reindeer are important sources of protein in the Lapp diet and central elements in traditional Lapp culture. Following the Chernobyl accident, however, many reindeer became unfit to eat. Radioactivity from the damaged reactor settled over the tundra, contaminated the abundant lichens, and became concentrated in the herbivorous reindeer.

In theory, America's coal reserves alone could supply the world's energy needs for hundreds of years. But much of that coal is laden with impurities, and the burning of coal using current techniques is a major source of air pollution worldwide. Extracting coal from strip mines leaves massive scars on the face of the land and adversely affects both surface and groundwater supplies. A great deal of investment in research and development will be necessary before coal can supply the energy we want as cleanly and efficiently as we need.

Nuclear power   In 1984 roughly 13 percent of the world's electricity was supplied by nuclear power plants, which use heat generated by the controlled fission of uranium-235 to produce steam. By the mid-1980s, however, falling energy prices and public concern over safety caused a virtual moratorium on new plant construction in the United States.

These safety concerns were legitimate. In 1979 an accident at the Three Mile Island reactor in Pennsylvania released radioactive steam into the atmosphere. Then in 1986, a nuclear reactor in the town of Chernobyl in the Soviet Union experienced a partial *meltdown,* an accident in which the plant's radioactive core overheated out of control. Over the next several weeks, 31 people died of direct radiation poisoning, and radiation released during that accident was carried by winds that blew north and westward over Europe. That radioactive fallout settled across Europe and as far away as the tundra of Finland, contaminating crops and dairy products (Figure 7.20). The Chernobyl accident reminded the world that planet-wide air and water movements make nuclear contamination a global rather than a local issue.

Renewable Energy

Renewable energy can be tapped without depleting its supply. *Solar energy* can be captured directly and either used to generate electricity or channeled to heat water or living space (Figure 7.21). *Hydroelectric power* is generated as water falls through turbines, which are usually located in large dams. *Power from biomass* is generated by burning or fermenting plant materials to produce methane gas. *Wind power* is generated by wind-driven turbines.

All of these energy sources are used somewhere in the world today, but only hydroelectric power provides significant amounts of energy in developed countries. In some cases, such as the direct generation of electricity from sunlight, additional research is needed to produce commercially viable products. In other cases, alternative energy supplies are not yet compatible with established power networks.

Heavy Third World dependence on energy from biomass has its disadvantages. When many trees are felled

Figure 7.21 A solar energy power plant in California harnesses our most plentiful renewable energy source.

for firewood, the loss of topsoil begins. When firewood runs out, dung and crop residues are burned as fuel instead of being returned to the soil as fertilizer, so the topsoil is depleted as the fuel shortage worsens.

TACKLING THE DILEMMA

The Impact of Economics on Ecology

Population growth and food production account for only part of humanity's impact on the biosphere. Equally important is the developed world's desire for *economic* growth even after *population* growth has slowed. Until fairly recently, many industries in both Western and Eastern bloc countries have kept costs down and profit margins high by consuming resources without regard to future supply and by disposing of waste products cheaply, without regard to adverse effects.

In the past, when pollution created only minor, local problems, government absorbed the long-term cleanup costs that resulted from such activities. But now that pollution is global in scale and cost, analysts emphasize that resource depletion and pollution control must be included in figuring the real costs of production. That means consumers must share a larger portion of those real production costs by paying more for goods.

Passing laws designed to control industrial pollution is often an uphill struggle. But the blame for our pollution problems does not rest solely with corporations. To balance the economic and political power of polluting industries, citizens' groups must monitor their environment and be willing to take legal action when necessary. In the mid-1980s, for example, voters in California's Silicon Valley region battled state government and businesses to protect their water supply. The result was a referendum requiring stringent controls on potentially carcinogenic wastes.

Individual ecological responsibility, however, goes much further. Residents of cities and towns, for example, are often not willing to pay additional taxes to improve sewage treatment facilities. And most of us contribute our share of pollution by failing to participate in recycling efforts, to keep our auto engines well tuned, or to minimize our use of caustic household chemicals, fertilizers, and pesticides. There is much room for improvement in individual environmental responsibility.

The Impact of Ecology on Economics

Ecological changes, often driven by economic necessity, can have long-term economic impact. The Brazilian government, for example, while trying to repay loans to Western banks during the 1970s and 1980s, encouraged farmers to clear rain forests and grow cash crops for export. That sort of "development" was encouraged by funding from the World Bank with little regard for its ecological impact.

But though such activities yielded badly needed short-term profits, they damaged soil and water resources, undermining Brazil's abilities to feed and house its

Ecology, Engineering, Banks, and Politics

The Panama Canal is a 51-mile waterway that enables ships to pass from the Atlantic to the Pacific without having to circumnavigate South America. American engineers devised a series of locks that lift ships 85 feet to the height of two inland lakes and lower them back to sea level on the other side. These locks require no active pumping of water. Instead, the canal's engineers harnessed the renewable energy of the area's hydrological cycle. Well-placed dams catch and store runoff from Panama's lush rain forests in artificial Gatun Lake, the highest point in the canal. Between 2 and 3 billion gallons of that water is channeled through the locks every day, where it raises and lowers the ships as it flows toward the sea.

Gatun Lake's rain-forest watershed is a resilient, self-sustaining system. It not only generates much of its own rainfall but also acts like a giant sponge, absorbing and storing water in the rainy season and releasing it slowly in the dry season. For most of this century, that natural system kept the canal's water supply constant, even during exceptionally dry years.

But though the forest around the canal is theoretically protected as a national park, it is rapidly being cleared and transformed into pastureland for cattle ranching. This deforestation is already having its effect; rainfall in the area is decreasing by about an inch a year, water input to the canal is dropping steadily, and increasing volumes of silt are flowing into the waterway. During the dry years of 1982–1983, water levels in the canal were so low that some boats had to unload their cargo for transport across the isthmus by rail; only empty boats could pass through parts of the canal safely.

There is no clear villain in Panama. The ecologically destructive events are directed by a chain of well-meaning individuals and groups, all of whom consider their motives admirable, and none of whom understands the entire situation. Panamanian farmers, though clearing land within protected forest, are breaking no laws, are encouraged by their government, are backed by development banks, and know nothing of the threat to the canal. International development bank officers in Washington insist that they merely respond to the requests

of their Panamanian bank partners. And local agricultural banks are proud of their projects; the farmers they support live well, generate food for domestic consumption, and export meat that generates badly needed foreign currency.

Clearly, decision makers in Panama need to understand the ecological repercussions of development around the Panama Canal region. For if destruction of the rain forest continues, the canal itself, foundation of the Panamanian economy, will soon be in serious trouble.

people over the long term. Brazilian environmentalists argue, for example, that properly managed and harvested rain forests can produce up to 500 pounds per acre per year of tropical fruits, nuts, game, and fishes. If the same rain forest is carved into pastureland for cattle, however, it produces only 45 pounds of meat per acre for the few years before the soil becomes worthless.

Advocates of sustainable development point out that this sort of ecological mismanagement fosters poverty among peasant farmers and that poverty fuels economic crises and sparks social and political unrest. International economic repercussions can be severe; loan defaults by Third World countries have seriously weakened several American banks. Several similar environmental threats to national and international industries can be found around the world. (See Current Controversies, Ecology, Engineering, Banks, and Politics)

Changing perspectives Thankfully, a long overdue dialogue between economic and environmental interests has begun. In 1986, for example, the World Bank began to require projects it funds to incorporate sound resource conservation measures. Over the longer term, World Bank analysts are realizing that the way to direct truly *sustainable* development is to plan and back only projects that maintain or improve the world's resource base. And ecologists are coming up with development projects of their own that encourage local people to participate in projects that benefit them economically and ecologically in both the short and long term. In Bolivia, for example, an innovative "debt-for-nature" swap involving Conservation International, Citicorp Bank, and a private conservation foundation allowed the government to trade $650,000 of its foreign debt in return for guaranteed protection of a tract of virgin Amazon forest.

Figure 7.22 Our "disposable economy" in action in New York City.

ENVIRONMENTAL POLICY: ETHICS AND REALITY

As you have probably realized from reading this chapter and the associated material in Chapter 2, working toward a sustainable world economy will be a long and difficult process. The steps that should be taken, the order in which they should be taken, and even the process by which decisions on such matters should be made are far too complicated for us to adequately address them here.

If there is any lesson to be learned from the impact of economy on ecology to date, it is that legislating compliance through regulations is at best a stop-gap measure. Such legislation can levy stiff fines on offenders—whether those offenders be corporations, municipalities, or individuals—that can help mitigate any damage that

transgressions have caused. But given the reality that ecologically sound activities are often more expensive in the short term, new rounds of laws invariably stimulate the hatching of even more clever schemes for avoiding detection and liability. Still, certain regulations are unarguably effective; efficiency standards for American appliances that were adopted in 1986 will save $28 billion in energy costs and keep 342 million tons of carbon dioxide out of the atmosphere by the turn of the century.

It is also true, however, that our dynamic and flexible social system responds extremely well to economic incentives. In the early 1970s, for example, the United States experienced its first major "energy crisis," largely as the result of an oil embargo by a cartel of oil-producing nations. Prices for petroleum products of all sorts skyrocketed. As a result, the United States increased its efficiency of energy use dramatically over the following decade. In a similar fashion, governmental incentives that encourage further energy conservation and discourage our "disposable economy" could push our society in the direction needed to ease many current problems. There is no reason, for example, why a typical resident of New York City *must* generate nearly two kilograms of solid wastes each day; residents of Rome and Hamburg throw out only half as much. By far the majority of garbage generated by Americans today consists of materials that could be recycled in one way or another. They are discarded because there is little or no economic incentive to do otherwise (Figure 7.22).

In addition to the necessary hard-nosed approaches to specific environmental issues, however, there is the question of our society's overall attitudes toward our environment. As you may recall from Chapter 2, several modern environmental philosophers, echoing the writings of Thoreau and others, suggest that our environmental dilemmas reflect fundamental problems in our ethical and moral relationship with the biosphere. According to these philosophers, modern society has little regard for life in other than human form. We also fail to include in our definition of humanity a relationship with the rest of the biosphere. It is these attitudes, coupled with economic necessity, that often lead us to sacrifice the eternal for the expedient.

Our best hope for developing a sustainable global ecosystem lies with approaches that incorporate economic and political reality, the resourcefulness and creativity of the human mind, and a healthy respect for the living world around us. Many so-called primitive cultures had systems of environmental ethics from which our modern society can learn a great deal. There is much in the way of both warning and encouragement in the following excerpt from a letter written by an American Indian chief to the "Great Chief in Washington" over a century ago:

> We know that the white man does not understand our ways. One portion of the land is the same for him as the next, for he is a stranger who comes in the night and takes from the land whatever he needs. The earth is not his brother, but his enemy, and when he has conquered it, he moves on. . . .
>
> One thing we know which the white man may one day discover. Our God is the same God. . . . [and] . . . This earth is precious to him. And to harm the earth is to heap contempt on its creator. . . . Continue to contaminate your bed, and you will one day suffocate in your own waste. . . .
>
> When the buffalo are all slaughtered, the wild horses all tamed, the secret corners of the forest heavy with the scent of many men, and the view of the ripe hills blotted by the talking wires, where is the thicket? Gone. Where is the eagle? Gone. And what is it to say good-bye to the swift and the hunt? . . . [It is] the end of living and the beginning of survival.

> —"This Earth is Sacred"
> Letter from Chief Sealth (Seattle)
> to President Franklin Pierce—1885

SUMMARY

During the mid-twentieth century, the public became more aware of the environmental effects of human activities on global ecosystems. Still, environmental damage persists. Water pollution from industrial and agricultural chemicals and residential sewage endanger both surface water and underground aquifers; atmospheric pollution may raise the earth's temperature through increased levels of carbon dioxide and cause widespread damage to streams and forests through acid rain. Pollution and habitat destruction result in the extinction of plant and animal species and the loss of genetic diversity. Faced with these ecological dilemmas, ecologists and economists are searching for a new style of interacting with the global environment—a system of sustainable development that provides for human needs in ways that do not adversely affect the biosphere.

Population growth, flawed economic systems, and short-sighted government policies contribute to the problems of hunger and poverty throughout the world. Our best hope for developing a sustainable global ecosystem lies with approaches that incorporate economic and political reality, the resourcefulness and creativity of the human mind, sound knowledge, and a healthy respect for the living world around us.

STUDY FOCUS

After studying this chapter, you should be able to:

- Outline the present interaction between humans and the biosphere.
- Describe the consequences of ecologically imprudent human actions.
- Explain the concept of sustainability.
- Cite various successes and failures in ecosystem management.
- Discuss the connection between global economics and local ecology.

SELECTED TERMS

groundwater *p. 114*
aquifer *p. 114*
DDT *p. 114*
PCB *p. 115*
biological magnification *p. 115*
TCE *p. 117*
acid rain *p. 118*
ozone *p. 119*
deforestation *p. 120*

desertification *p. 120*
continental shelves *p. 121*
Green Revolution *p. 125*
monoculture *p. 127*
salinization *p. 129*
overexploitation *p. 130*
aquaculture *p. 130*
nonrenewable energy *p. 131*
renewable energy *p. 132*

Discussion Questions

1. What economic arguments are used to justify industrial and municipal pollution in your area? In developing countries in the tropics? Why are most of these arguments unconvincing in the long term?

2. What common practices in modern agriculture are detrimental to farmland? What remedial measures can be taken to halt or reverse the damage?

3. Why is slash-and-burn agriculture in the wet tropics ultimately unproductive both to the ecosystem and to its human inhabitants? Why does it continue?

4. What is happening to the concentration of carbon dioxide (CO_2) in the atmosphere? Why? Why is that change important?

5. When a large midwestern industrial facility spews out large quantities of sulfur dioxide and nitrous oxides into the atmosphere, what are the environmental effects and where are these effects felt?

6. If renewable energy such as solar and wind power can be tapped without depleting their source, why are we not using this renewable energy exclusively?

Objective Questions   (Answers in Appendix)

7. The least common type of sewage treatment used is
   (a) primary treatment.
   (b) secondary treatment.
   (c) tertiary treatment.
   (d) a combination of primary and secondary treatment.

8. Ozone
   (a) is a highly poisonous gas that combines with oxygen in the lungs, causing breathing difficulties.
   (b) is a gas that absorbs ultraviolet light and prevents it from reaching the earth's surface.
   (c) is one of several gases that combines with sulfur and oxygen to produce acid rain.
   (d) combines with carbon in the atmosphere to produce the air pollutant carbon monoxide.

9. Nonrenewable energy sources do not include
   (a) oil.
   (b) natural gas.
   (c) hydroelectric power.
   (d) coal.

10. Secondary sewage treatment involves the
    (a) biological breakdown of organic material by bacteria.
    (b) settling out of solid waste.
    (c) chemical removal of toxic compounds.
    (d) heating of toxic compounds using blast furnaces to decompose them.

11. In which organism would DDT most likely be in the heaviest concentration once it had been introduced into the ecosystem?
    (a) grasshopper
    (b) toad
    (c) snake
    (d) seal

PART THREE

Evolution and Mendelian Genetics

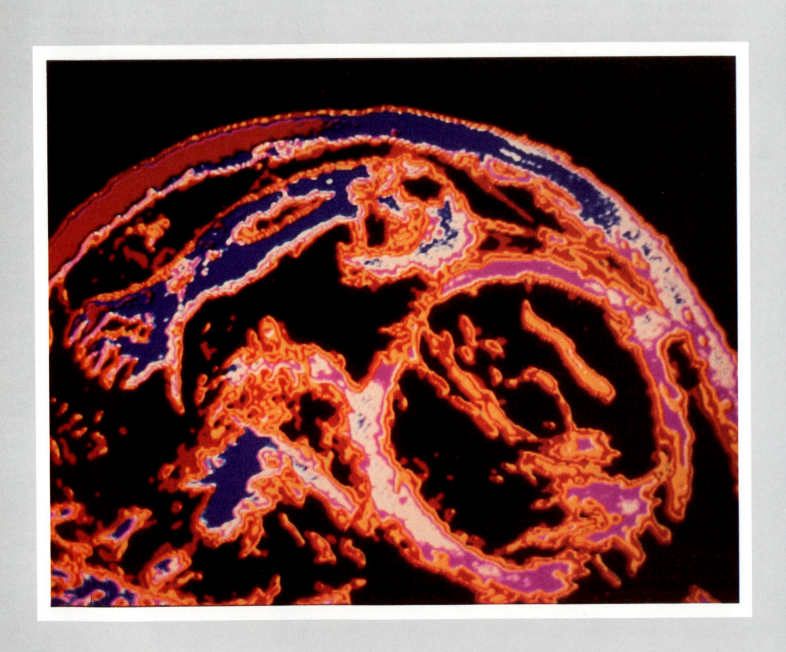

Biologists have accumulated evidence that proves, beyond a shadow of a doubt, that animals and plants have *evolved*, or changed over time. This certainty is perhaps the single most important fact in all of biological science today. But knowing that a complex process such as evolution *has* occurred is very different from understanding just *how* evolution proceeds. Following the lead of Charles Darwin, some of the greatest minds in biology have struggled not just to *describe* the evolutionary process but also to *explain* it in scientific terms. The great body of observations, experiments, and hypotheses they have produced constitutes what is called *evolutionary theory*.

Why is evolutionary theory so powerful and important? Why has evolution always been so controversial? To understand fully the revolutionary nature of Darwin's work and the impact it has had on modern thought, this section reviews the Western world view of the mid-eighteenth century, examines the mysteries that inspired Darwin and his contemporaries, presents Darwin's original theory of evolution, and traces the development of evolutionary theory to the present day. In the process, it places evolutionary theory in context in science and society.

Of course, evolutionary theory raises a central and profound question: How is it that like begets like and yet life evolves? This question and others equally important to a coherent view of life over time are addressed by the field of genetics, founded by the work of Gregor Mendel. Because understanding the principles of Mendelian genetics is essential to a coherent perspective on evolution, and because evolutionary thought provides both context and substance to genetic principles, this section considers these disciplines together in an integrated treatment.

◀ This living fetus within the womb, dramatically revealed by the non-invasive technique of ultrasound imaging, is developing according to instructions carried in the genes it received from its parents.

CHAPTER 8

Darwin's Dilemma: The Birth of Evolutionary Theory

The word **evolution** literally means "unrolling or unfolding," though in common usage it simply means "change." A theory of evolution is nothing more (or less) than a theory of biological change. Yet *On the Origin of Species*, Darwin's first collection of evolutionary facts and theories, has been described as "the book that shook the world." Because our species is among the organisms that have evolved, facts and theories about evolution have profound practical and philosophical implications for humanity. It should come as no surprise, therefore, that evolutionary thought has guided, inspired, frightened, and infuriated scientists, philosophers, religious authorities, and lay people from the nineteenth century to the present. A century after Darwin published *On the Origin of Species*, philosopher J. Collins asserted that "there are no living sciences, human attitudes, or institutional powers that remain unaffected by the ideas . . . catalytically released by Darwin's work." To understand why Darwin's work was so important, we must first place his theory within the context of eighteenth- and nineteenth-century Western scientific philosophy.

THE WESTERN WORLD VIEW BEFORE DARWIN

From the time of the Greeks, most Western philosophers viewed the material world as rigid, static, and innately flawed. To Plato and Aristotle, both living organisms and inanimate objects were inferior mimics of perfect models called **ideal types.** Ideal types, found only in the transcendent world of ideas, were perfect and unchanging. But their imperfect copies on earth were full of flaws that human naturalists saw as variations among members of plant and animal species.

Western philosophers combined many Platonic and Aristotelian ideas with Christian thought to create a world view that encompassed religion, science, and society. Two beliefs in particular constrained the natural sciences. First, theologians believed that because God was perfect, all His work had to be perfect. Second, philosophers asserted that perfection *necessarily* implied stability; things that were divine and perfect should not change.

For those reasons, orthodox Christian philosophy taught that after God created the first ideal types, species were fixed for all time. Imperfections appeared in living things because the material world, unlike the spiritual world, is corrupt and imperfect. But because the original Creation was complete and perfect, no organisms had appeared or disappeared, and the "type" for each species did not change.

Furthermore, each living species had a permanent place in the divine order of things called *the Great Chain of Being.* Derived from Aristotle's *Scala Naturae,* the Great Chain of Being stretched from nonliving matter, through lower forms of life, to humans at the top of the earthly chain. Most important, nearly all European scientists shared the belief that because humans were created in God's image, our species was unique and essentially different from all other forms of life.

Stability in Biology and Society

The idea of the unchanging type permeated biology into the nineteenth century. Biologists sought to look beyond the visible flaws of earthly organisms—to ignore individual variation—and to reveal God's master plan by studying the ideal types those organisms represented. When a new species was discovered, a specimen thought to resemble its ideal type most closely was deposited in a museum collection, where it represented its species as the "type" specimen. It was this philosophy that guided Swedish naturalist Carolus Linnaeus (1707–1778) when he established the system still used today for naming and classifying organisms (Chapter 14).

This view of a divinely ordered and stable world governed not only the natural world but social systems as well. Just as the human race had dominion over creation, kings ruled over humanity by "divine right." The rigid system of upper and lower social classes was thus seen as an extension of the immutable world order represented by the Great Chain of Being. Talk of change was immoral; such change was unthinkable, whether in the human social order or the natural world.

DARWIN'S TIME: A WORLD IN FLUX

Charles Darwin (1809–1892) was born and raised in a privileged family within a society growing uncomfortable with the rigid status quo (Figure 8.1). Social and technological change was sweeping through Europe. The merchant class created by the Industrial Revolution was not satisfied with the hereditary social structure and struggled to change it, while philosophers searched for a new world view that could accommodate the emerging, competitive social order. At the same time, some of the greatest scientists Europe ever produced were making major contributions to the emerging scientific view of the world (Figure 8.2).

By the time Darwin was born, nearly every branch of science except biology had challenged the established view of a static, divinely ordered world. Astronomers argued that the earth was not the center of the universe, as religious dogma maintained. Newton provided mathematical explanations both for the previously mysterious orbits of planets and for the movements of objects on earth.

Then, during the eighteenth and early nineteenth centuries, global explorers made discoveries that defied traditional biological perspectives. They found that Asia, Africa, and the New World harbored hundreds of exotic plant and animal species unknown in Europe and that many of these animals and plants lived only in certain parts of the world. If all living things had been recently

Figure 8.1 Charles Darwin was born into a wealthy family immersed in the social change of the nineteenth century. Both his maternal grandfather, Josiah Wedgewood, and his paternal grandfather, Erasmus Darwin, rose from poverty to become industrial magnates. These self-made men didn't fit into the static, hereditary social structure; they and their peers wanted society to change and wanted to affirm upward mobility based on individual achievement.

created in the same place at the same time and had later been released from Noah's ark, what could explain these distribution patterns? Biology and geology were ripe for revolution.

Fossils and Catastrophism

Fossils presented another problem for biologists. As more and more preserved remains of plants and animals were found, it became impossible for biologists to deny that fossils represented extinct organisms (Figure 8.3). Where did these fossils come from? And why did the animals they represented die out?

The French anatomist Cuvier was so overwhelmed by fossil diversity that he suggested there had been not one but six separate creations! Other scientists suggested that there had been several successive creations followed by

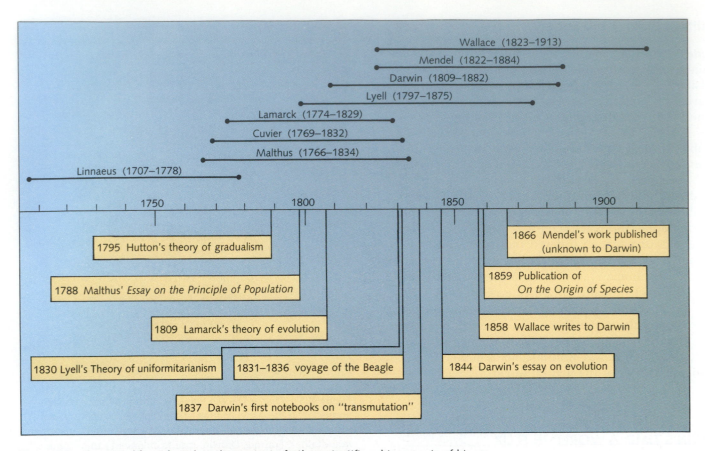

Figure 8.2 Darwin's life and work in the context of other scientific achievements of his era.

Wallace (1823–1913)
Mendel (1822–1884)
Darwin (1809–1882)
Lyell (1797–1875)
Lamarck (1774–1829)
Cuvier (1769–1832)
Malthus (1766–1834)
Linnaeus (1707–1778)

1750   1800   1850   1900

1795 Hutton's theory of gradualism

1788 Malthus' *Essay on the Principle of Population*

1809 Lamarck's theory of evolution

1830 Lyell's Theory of uniformitarianism

1831–1836 voyage of the Beagle

1837 Darwin's first notebooks on "transmutation"

1866 Mendel's work published (unknown to Darwin)

1859 Publication of *On the Origin of Species*

1858 Wallace writes to Darwin

1844 Darwin's essay on evolution

Figure 8.3 This fossil sampler shows examples of the sort of fossils unearthed before and during Darwin's life. The diversity of these fossils and their abundance posed serious problems for creationists of the Victorian era.

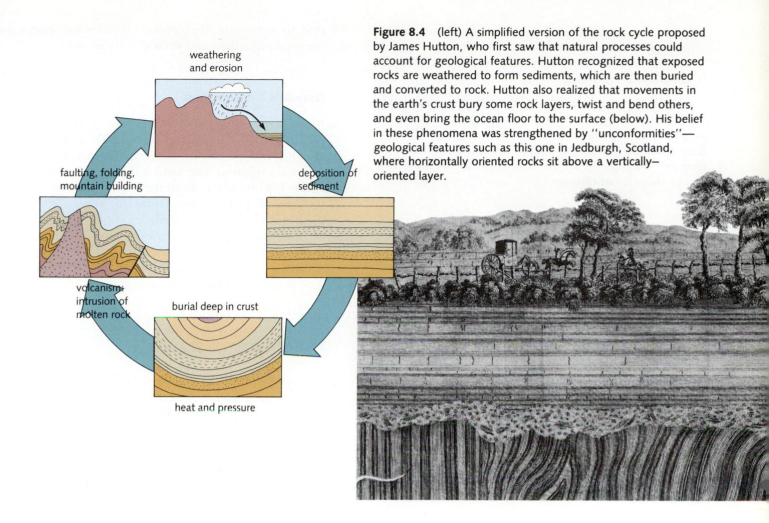

Figure 8.4 (left) A simplified version of the rock cycle proposed by James Hutton, who first saw that natural processes could account for geological features. Hutton recognized that exposed rocks are weathered to form sediments, which are then buried and converted to rock. Hutton also realized that movements in the earth's crust bury some rock layers, twist and bend others, and even bring the ocean floor to the surface (below). His belief in these phenomena was strengthened by "unconformities"—geological features such as this one in Jedburgh, Scotland, where horizontally oriented rocks sit above a vertically—oriented layer.

(diagram labels)
weathering and erosion
deposition of sediment
burial deep in crust
heat and pressure
volcanism: intrusion of molten rock
faulting, folding, mountain building

floods or other catastrophes, a doctrine that became known as **catastrophism.** Note that catastrophism did not really challenge the dominant philosophy; although catastrophists suggested that the earth was far older than scriptures implied, they still assumed that the world and its inhabitants were stable and had been specially created. Every now and then, the state of the world changed, but those changes were rare events that occurred only by divine decree.

GEOLOGISTS CHALLENGE THE STATIC EARTH

Geologist James Hutton produced the first scientific challenge to this concept of a static world in his revolutionary *Theory of the Earth,* published in 1788. Hutton proposed that the same geological principles in action today—weathering and erosion, deposition of sediment, and volcanism—had shaped the world over time (Figure 8.4). In order for these processes to have occurred, of course, the earth had to be *very* old. Hutton was the first to introduce to both geology and biology the critical concept of *deep time.* According to Hutton, "The result . . . of our present enquiry is, that we find no vestige of a beginning—no prospect of an end."

The man best known for championing geological change was Charles Lyell, whose *Principles of Geology* was published in 1830. Lyell's book expanded on Hutton's work, putting forth three principles that constitute the theory of **uniformitarianism.** (The first two principles must apply not only to geology but to all scientific inquiry.)

1. *Natural laws are constant in space and time.* As Harvard biologist Stephen Jay Gould notes, this is a statement of method that any scientist must make to analyze the past. For "if the past is capricious, if God violates natural law at will, then science cannot unravel history."

2. *Scientists should attempt to explain events of the past through the same sorts of natural processes that we can observe directly today.*

3. *Most geological change occurs slowly and gradually, not through sudden, catastrophic events.* This rule of uniformity of rate was an extremely important influence on Darwin's thinking.

By 1830 most scientists believed that the earth was old (though they didn't agree on *how* old) and that the structure of the earth had changed substantially over time (though they argued over the causes and rates of change). Biology, of all the sciences, was the slowest to accept theories of change.

BIOLOGICAL SCIENCE AND CHANGE

Mid-nineteenth-century intellectuals were intensely interested in theories of change in nature. Darwin's grandfather, Erasmus Darwin, discussed the origin and evolution of life as early as 1794. And in 1857, two years before Darwin published *On the Origin of Species*, the English philosopher Herbert Spencer argued that life *must* have evolved, because change was universal in all other domains. But though people suggested that organisms *could* evolve, no one could explain *how* and *why* organisms changed over time.

Lamarck's Theory of Evolution

The first to propose a mechanism for evolutionary change was the French scientist Jean Baptiste de Lamarck. Lamarck's theory of **evolution through inheritance of acquired characteristics** was published in 1809, the year Darwin was born. Though he has often been maligned, Lamarck was an innovative theoretician who combined evolutionary and ecological thinking. Lamarck, who believed in divine creation and the Great Chain of Being, actually conceived of the chain as more like an escalator; he felt that each species was created with a God-given drive toward perfection that, combined with environmental factors, impelled it along a relentless journey up the chain.

Lamarck made two basic assumptions that we know now to be incorrect. First, his theory was **teleological;** he believed that evolution had a goal, or directed purpose, and that species changed over time because they "wanted" to "better" themselves. Second, he believed that characteristics acquired during the life of an organism could be passed on to its offspring—a belief we now know to be generally false. To Lamarck's lasting credit, however, he championed biological evolution during a time when such ideas were not at all popular. He was also the first to propose a truly *scientific* theory of change, as well as the

first to recognize that evolution involved interaction between organisms and their environments.

Darwin's Contribution

When Darwin and his contemporary, Alfred Russel Wallace (see Theory in Action, Darwin's Delay and Wallace's Insight), independently suggested a scientific explanation for evolution, the last of the pieces of the puzzle finally fell into place. Both the scientific community and the general public were primed to accept a biological theory of change. The publication of Darwin's book in 1859 was not just a scientific event but also a public sensation. The first printing sold out the day it was released, and many other printings followed before the year ended. The world hasn't been the same since.

DARWIN ON THE *BEAGLE*

Just after Christmas in 1831, Charles Robert Darwin, an educated gentleman of 22, embarked on HMS *Beagle* for a global voyage of exploration (Figure 8.5). Darwin had always been interested in natural history, but he had obtained his university degree in theology from Cambridge University. Darwin was, in fact, a pious man; he wrote home midway through his voyage that he could see himself spending the rest of his days as a country preacher. But during the *Beagle*'s voyage, Darwin's intellect, his naturalist's eye, and some extraordinary luck led him far from the commonly accepted religious views of nature.

Geological Observations

Darwin carried with him a gift from his mentor Professor J. S. Henslow—the first volume of Lyell's *Principles of Geology.* Darwin was so impressed by Lyell's work that he had subsequent volumes delivered to him en route.

With Lyell's ideas in mind, Darwin had the good fortune to witness the forces of geology in action. In January of 1835, he saw the eruption of a volcano in Chile and later learned that another volcano 480 miles away had blown its top the same night. Just over a month later, he experienced an earthquake that lifted beds of marine mussels "still adhering to the rocks, ten feet above the high water mark." Still later, he observed beds of fossil mussels a thousand feet and more above sea level. Whereas others were simply awed by these phenomena, Darwin realized that he had glimpsed the geological processes that were gradually building the Andes mountains. His belief in geological uniformitarianism was established.

Darwin's Delay and Wallace's Insight

When Darwin returned from his voyage on the *Beagle* in 1836, his theories were still in the formative stages. But though he had completed most of his important work by 1844, he chose not to publish it. Instead he put it aside, instructed his wife to publish the work in the event of his death, and turned for more than a decade to a study of worms and barnacles.

Several of Darwin's colleagues repeatedly urged him to publish his ideas before someone else beat him to it. Admonished, Darwin started assembling his thoughts, chapter by chapter, in 1856. Meanwhile, halfway around the world in Malaysia, an English naturalist named Alfred Russel Wallace took ill. Wallace, who had also worked in the tropics and had also read Malthus's essay, awoke in a fever with an inspiration. Writing with the same fever that had inspired

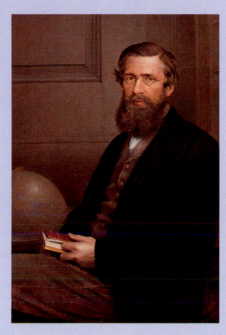

Alfred Russel Wallace

him, Wallace outlined a theory of evolution nearly identical to Darwin's and mailed it to Darwin asking for his opinions.

Darwin was downcast as he conveyed the paper to Lyell for public presentation. All his work was for nought; Wallace would receive credit for publishing his ideas first. But Lyell presented both Wallace's paper and excerpts from Darwin's earlier, though still unpublished, essay. Darwin worked furiously to publish *On the Origin of Species,* and everyone (including Wallace) agreed that Darwin's exhaustive research and documentation of evolutionary phenomena entitled him to the lion's share of the credit.

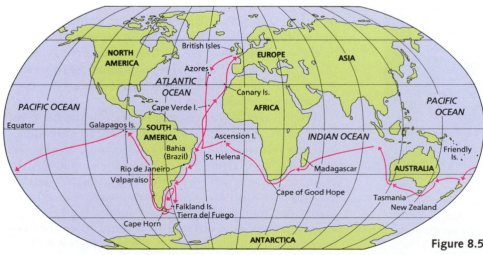

Figure 8.5 The voyage of the *Beagle*, 1831–1836.

Figure 8.6 Darwin explored as far inland as time allowed. In Argentina, he discovered exotic fossil animals, including giant ground sloths, peculiar horses, and this bizarre, extinct relative of the armadillo called Glyptodon. As numerous as Darwin found living organisms to be, he was soon convinced that they were far outnumbered by forms no longer living.

Figure 8.7 These photographs show two remarkably similar, yet unrelated, animals—a marsupial mouse (top) from Australia and a common wood mouse (bottom). Occupying similar niches, they exhibit strikingly similar forms and behaviors.

Biological Diversity

Darwin was staggered by the variety of animals and plants he encountered during his voyage. Everywhere he looked, he saw dozens of new and oddly shaped trees, hundreds of exotically colored flowers and birds, and beetles and other insects almost beyond counting. But Darwin quickly realized that the diversity of living organisms was only part of the mystery of life, for he found even greater numbers of fossil species. In Argentina, he discovered fossil armadillos, giant ground sloths, peculiar horses, and creatures that reminded him of the hippopotamus.

As numerous as living organisms were, Darwin was soon convinced that they were vastly outnumbered by extinct forms (Figure 8.6). That extraordinary diversity of fossil species convinced Darwin that extinction and the appearance of new species were real phenomena that had to be explained. (Biologists today estimate, in fact, that more than 99.9 percent of the species that have lived on the earth are now extinct. Because current estimates place the number of living species as somewhere between 2 and 30 million, over 2 billion species must have come and gone since life began.)

Darwin also discovered that both flora and fauna differed markedly from continent to continent and on opposite sides of natural barriers such as mountains, deserts, and large rivers. The Argentinean pampas, for example, supported very different animals from the grasslands in Australia. And Darwin noticed that although plants and animals in ecologically similar but geographically separate areas differed from each other, they possessed similar structures and behaviors (Figure 8.7).

Figure 8.8 Fitness for animals includes the design of legs, wings, and claws. Fitness for plants includes the features of leaves, stems, roots, and flowers. (above) This mole has no use for vision but obtains a great deal of information through a highly developed sense of smell. Its limbs and digits are modified into efficient tools for digging tunnels. (right) Many large tropical plants, such as this banyan tree (a species of fig), develop extensive systems of prop roots that both gather nutrients and serve as extra trunks to support the spreading canopy.

Fitness

Throughout his journey, Darwin marveled at the "perfection of structure" that made it possible for organisms to do whatever they needed to do to stay alive and produce offspring. Darwin called this perfection of structure **fitness,** by which he meant the combination of all traits that help organisms survive and reproduce in their environment (Figure 8.8). As he observed, Darwin also began to question what process had "fit" these organisms to their physical environments and to each other.

The Galapagos Islands

Off the west coast of South America, the *Beagle* visited a cluster of tiny, rocky islands called the Galapagos after the Spanish word for "tortoise" (Figure 8.9). Darwin noted that these islands were inhabited by a surprising number of bizarre and often beautiful plant and animal species. He surmised correctly that many of these species were *endemic,* which means they are found nowhere else (Figure 8.10). One group of Galapagos birds has since been named Darwin's finches in his honor.

These islands and their inhabitants made lasting impressions on Darwin, and his notes and collections served him well in later years. In his first *Journal of Researches*, published in 1837, he noted that the Galapagos Archipelago was like "a little world within itself" that was "very remarkable" in its organic productions. But not

Figure 8.9 Map of the Galapagos islands drawn by the officers of the *Beagle* in 1835.

Figure 8.10 (left) Darwin found that each of the Galapagos islands had its own peculiar type of giant tortoise. These tortoises are from the islands of Pinta (top), Hood (center), and Isabella (bottom). Although Darwin learned of these differing forms during his visit, he did not recognize their evolutionary significance until much later. (above) Four species of Galapagos finches, each with head and beak sizes suited to different diets. So different in appearance and habits were some of these birds that Darwin erroneously placed them in separate subfamilies.

until the second edition of his *Journal,* published in 1845, did Darwin begin to recognize the full significance of what he had seen there. In one of his most famous passages, he wrote

> Considering the small size of these islands, we feel the more astonished at the number of their aboriginal beings, and their confined range. Seeing every height crowned with its crater, and the boundaries of most lava-streams still distinct, we are led to believe that within a period geologically recent the unbroken ocean was here spread out. Hence both in space and time, we seem to be brought somewhat near to that great fact—that mystery of mysteries—the first appearance of new beings on this earth.

But it took Darwin nearly a quarter of a century after his Galapagos sojourn to publish his solution to that great mystery. On his return to England, he discussed his discoveries with prominent botanists and zoologists. Later, in 1837, he started his first notebook on "transmutation" (changing to a different form). There, and in other notebooks to follow, Darwin documented his growing belief in evolution and his search for a mechanism that could explain it. Then in October of 1838, he read "for amusement," a 40-year-old essay that crystallized his thinking, the *Essays on the Principle of Population* by mathematician Thomas Malthus (see Chapter 5).

DARWIN'S THEORY OF EVOLUTION BY NATURAL SELECTION

Malthus, you will recall, believed that without some external control, human population growth would invariably outstrip our ability to feed ourselves. Darwin realized that if this were true for humans, who usually had fewer than ten children, it was doubly true for plants and animals that produced hundreds or even thousands of offspring. Darwin calculated, for example, that if all the descendants of even a slowly reproducing pair of elephants survived, that pair would have 19 million descendants in 750 years. After a great deal of work, Darwin finally devised a scientific explanation for why this didn't happen.

Variation

Darwin began by abandoning the idea of species as perfect and unchanging. Instead of viewing differences among members of a species as imperfections or deviations from an ideal type, Darwin realized that these **variations** were a basic fact of all life (Figure 8.11).

From plant and animal breeders, Darwin learned that seed from a single ear of corn produces both plants with ears that are larger than average and plants with ears that are smaller. Among crops of garden flowers, flocks of pigeons, and packs of dogs, some individuals always differ from the norm in color and size. These breeders knew that much of this variation could be passed on from parent to offspring—in other words, it was **heritable.**

Figure 8.11 The variation in color patterns found in "ladybug" beetles across their range. We now know that in addition to such visible variation, organisms also exhibit "invisible" variation in body chemistry, physiology, and behavior.

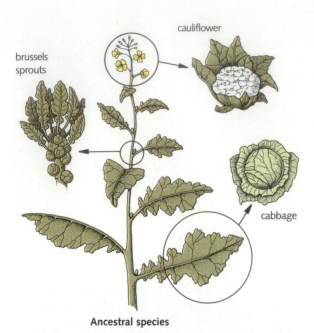

brussels
sprouts

cauliflower

cabbage

Ancestral species

Figure 8.12 (above) Cabbage, cauliflower, and brussels sprouts are all descendants of a single ancestral species, shown at the center here. The modern vegetables acquired their forms through generations of artificial selection favoring individuals with larger leaves, denser heads of flower buds, or larger leaf buds. (below) Darwin was a pigeon fancier, and he knew that the fancy breeds popular in his day were descended from the rock dove, a species resembling common city pigeons. All these breeds were created through artificial selection.

From his field observations, Darwin knew that wild species varied in nearly every characteristic he could observe and measure. Of course, he had no idea how variations arose or how they were inherited. (That understanding had to await the emergence of modern genetics.) But Darwin did realize that heritable variation—in both domestic and wild organisms—is *random, purposeless, and in no way subject to control.* Farmers cannot *cause* any particular kind of variation to appear among their crop plants, any more than wild animals can will their necks to grow longer or their fur to grow darker.

Artificial Selection

Darwin built his argument brilliantly. His first chapter, "Variation Under Domestication," details how farmers take advantage of random variation in crops and livestock. In a process Darwin called **artificial selection,** farmers choose the most desirable cows, sheep, or tomato and corn plants for breeding. Over several generations, this selective process produces individuals that differ markedly from their forebears (Figure 8.12). In Darwin's words, "The key is man's power of accumulative selection: nature gives him successive variations; man adds them up in certain directions useful to him."

Natural Selection

Darwin's next insight was to see in nature an analogue to the farmer as selective agent. Here again, Darwin took a giant philosophical step. Rather than postulating a mysterious or supernatural force behind change, as his predecessors had done, Darwin searched for a material force, or a scientific mechanism, to drive evolution. Darwin called that material force **natural selection** and described it as a process that favors the survival and reproduction of those organisms exhibiting variations best suited to their environment.

> As many more individuals of each species are born than can possibly survive, and as, consequently, there is a frequently recurring struggle for existence, it follows that any being, if it vary however slightly in any manner profitable to itself under the complex and sometimes varying conditions of life, will have a better chance of surviving, and thus be *naturally selected*. From the strong principle of inheritance, any selected variety will tend to propagate its new and modified form.

But though Darwin compared natural and artificial selection, he emphasized that natural selection operates without the foresight and purpose of the farmer. Darwinian evolution is **nonteleological,** which means that the process of natural selection operates without any ultimate goal of perfection. Natural selection operates only in the here and now for each organism, selecting those variants that are most fit to survive and reproduce *under local environmental conditions.*

Note that the forces of natural selection and farmers practicing artificial selection share a major handicap: Neither has the ability to *cause* desirable variations. Particular variants either arise, or do not arise, by chance. Only after such variations arise can they be favored by either artificial or natural selection.

Adaptation

According to Darwin, organisms are "selected," or molded over many generations by natural selection to become better suited or "fitted" to their environment in a process called **adaptation.** Thus it is through the process of adaptation that organisms acquire fitness. The word *adaptation* is also used to describe any characteristic of an organism that increases its fitness.

Adaptation is a complex process that involves all parts of an organism's anatomy, physiology, and behavior. Woodpeckers, for example, have evolved a suite of adaptations, all of which work together to enable these birds to feed on insects that live in the bark of trees (Figure 8.13).

feet with toes that can grip bark

stiffened tail feathers provide support

Figure 8.13 Woodpeckers are superb examples of evolutionary adaptation of a basic bird body plan that suits a specific niche—in this case, eating insects that live in and beneath the bark of trees. Woodpecker adaptations include a powerful, chisel-tipped beak; strong neck muscles for hammering; a sturdy skull with extra padding that protects the brain from impact; a flexible, protrusile tongue that snags insects in crevices; feet adapted for grasping onto tree trunks; and stiff tail feathers that support the bird's body against the tree.

SUMMARY OF NATURAL SELECTION AS PRESENTED BY DARWIN

The evolutionary process as Darwin envisioned it can be summarized as follows:

1. Organisms alive today were not specifically created as we see them but have descended from species that lived before them. This concept of **common descent** links plants or animals together into groups descended from ancestors they share.

2. More organisms are produced than can possibly survive, most die before reaching sexual maturity, and many that do survive fail to reproduce. Individual organisms are constantly struggling against each other, and often against hostile environmental conditions, for the necessities of life.

3. The physical characteristics of individual members of each species vary a great deal, and much of this variation can be inherited.

4. Some variants in each generation are better suited to life in their environment—that is, better adapted—than others.

5. Better-adapted individuals are more likely than others to survive and reproduce; hence the phrase "survival of the fittest."

6. Over time, natural selection can both produce changes in existing species and create new species from preexisting ones.

Scientific and Philosophical Significance

Evolutionary theory has profound practical and philosophical repercussions that make it essential for *all* educated people to understand the essentials of Darwinian thought.

Philosophical ramifications   Darwin knew that accepting his theory required believing in *philosophical materialism,* the conviction that matter is the stuff of all existence and that all mental and spiritual phenomena are its by-products. Darwinian evolution was not only purposeless but also heartless—a process in which the rigors of nature ruthlessly eliminate the unfit.

Suddenly, humanity was reduced to just one more species in a world that cared nothing for us. The great human mind was no more than a mass of evolving neurons. Worst of all, there was no divine plan to guide us. These realizations troubled Darwin deeply, for in his day, materialism was even more outrageous than evolution (Figure 8.14). Some scholars speculate that fear of being branded a heretic for his materialism contributed to Dar-

Figure 8.14   "I see no good reason why the views given in this volume should shock the religious feelings of anyone," Darwin wrote earnestly. But contemporary clergymen quickly condemned his implying that humans had descended from ape-like ancestors, and cartoonists responded much as they would today.

win's 21-year delay in publishing his theory. The same antimaterialistic reasoning also drives much modern-day opposition to evolutionary thought.

Yet as pointed out by evolutionary scholar Douglas Futuyma, seldom do the detractors of the Darwinian world view take note of its positive implications. In Darwin's world we are not helpless prisoners of a static world order, but rather masters of our own fate in a universe where human action can change the future. And from a strictly scientific point of view, rejecting evolution is no different from rejecting other natural phenomena such as electricity and gravity.

Darwin remained to the end a devout, if somewhat unorthodox, Christian. "I see no good reason why the views given in this volume should shock the religious feelings of anyone," he wrote. Like religious scientists of many faiths today, he found no less wonder in a god that directed the laws of nature than in one that circumvented them.

Darwin and politics   Political theorists have always had a field day with Darwin's materialistic world view, although different individuals have interpreted and extended its message in diametrically opposite directions. Karl Marx and Friedrich Engels, for example, saw in evolution both justification for the overthrow of the aristocratic order and proof of the inevitability of the class struggle. Yet Henry Ford, America's preeminent capitalist, found in Darwinism the perfect rationale for the free-enterprise system.

Herbert Spencer championed the twisted logic of *social Darwinism,* which had nothing to do with Darwin himself.

According to social Darwinism, society should operate according to the same principle of "survival of the fittest" as nature does. Although social Darwinism itself is now a discredited philosophy, its underlying principles have had considerable impact in modern history. For example, the ideology that led Adolph Hitler into power in Germany was largely founded on the ideas of social Darwinists.

Evolution and other sciences Evolutionary biology has exerted powerful influences on anthropology, the social sciences, and psychology. By drawing attention to evolutionary trends and similarities in the behaviors of animals and humans, Darwin laid the groundwork for comparative psychology and animal behavior. Both Freud's theories of the id and sexual development and Jung's notions of the collective unconscious were influenced by evolutionary thought.

Biological significance Last, but most certainly not least, evolutionary theory provides the scientific underpinning of all the biological sciences and medicine. Because evolution from shared ancestors has shaped organs and physiological processes that operate in similar ways from worms to humans, we can learn about our own bodies and cells by studying bacteria, dogs, and monkeys. Because ecological assemblages of plants and animals—herbivores and plants, parasites and hosts, predators and prey—have evolved together over time, many features of their anatomy and physiology are intimately interconnected in ways that we ignore to our peril and can understand to our advantage.

DATA SUPPORTING THE FACT OF EVOLUTIONARY CHANGE

Darwin's success at proving the *fact* of evolutionary change was admirable; evolution as a fact of natural history was widely accepted by the time of his death in 1882. The honored position that Darwin's work earned him in the world of ideas is evidenced by his burial next to Sir Isaac Newton in Westminster Abbey.

Why was evolution accepted so rapidly? Part of the reason was that lay people and scientists were ready for a theory of change in nature. But much of the credit goes to Darwin himself for his impressive skills in gathering data, both during his voyage of discovery and from observations in his own back yard. Darwin was able, for example, to point out how well his concept of evolution by common descent explained a variety of puzzling biological phenomena that had been documented over the years. More recent discoveries enable us to understand many of those phenomena far more thoroughly than Darwin could, but the basis of our understanding is still rooted in his original observations.

Similarities in Anatomy and Development

Darwin and his contemporaries knew that early embryos of many animals look nearly identical and that the earliest stages of development in "lower" animals seem to be repeated in the early development of "higher" animals such as ourselves (Figure 8.15). Darwin realized that the

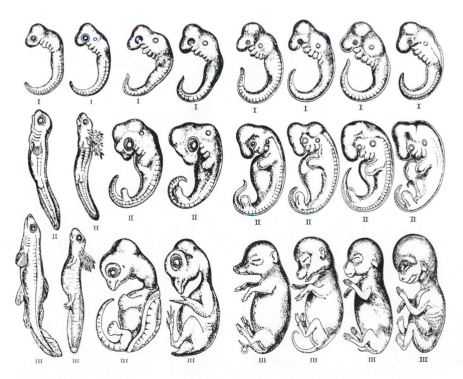

Figure 8.15 The early embryos of humans and other vertebrates look so similar that it takes an expert to tell them apart. During the earliest stages of development, all these embryos have gill pouches and a tail—remnants of structures needed by our aquatic ancestors.

similar developmental paths followed by animal embryos make sense if all of us evolved from common ancestors through a long series of evolutionary changes.

These striking embryological similarities led some of Darwin's contemporaries (though apparently not Darwin himself) to believe that the embryological development of an individual repeats its species's evolutionary history. Biologists have known for years that this is not the case.

Why, then, should the embryos of related organisms retain similar features when adults of their species look quite different? The cells and tissues of the earliest embryological stages of any organism are like the bottom levels in a house of cards. The final form of the organism is built upon them, and even a small change in their characteristics can result in disaster later. It would hardly be adaptive for a bird to grow a longer beak, for example, if it lost its tongue in the process.

The earliest stages of the embryo's life, therefore, are essentially "locked in," whereas cells and tissues that are produced later can change more freely without harming the organism. As species with common ancestors evolve over time, divergent sets of successful evolutionary changes accumulate as development proceeds, but early embryos stick more closely to their original appearance.

Homologous structures and adaptive radiation Darwin knew of many remarkable similarities among body parts of radically different animals. "What can be more curi-ous," Darwin wrote, "than that the hand of man, formed for grasping, . . . the leg of a horse, . . . the paddle of the porpoise, and the wing of the bat should all be constructed on the same pattern, and should include similar bones in the same relative positions?" Figure 8.16 shows how the components of several such limbs are all derived from the same structures in developing embryos. These structures are said to be **homologous** (*homo* means "same").

Cuvier and other creationist–catastrophists found it difficult to explain such obviously homologous structures. Why, they were forced to ask, would the creator have chosen to patch together such different animals out of the same mix-and-match bag of parts instead of using unique structures best suited to each?

Darwin's evolutionary theory, on the other hand, provided a satisfying explanation for homology. Sometimes a population of a species or populations of several related species may develop a new, advantageous adaptation or migrate to a new environment that offers many ecological opportunities. Under these circumstances, Darwin hypothesized, organisms would rapidly evolve new adaptations that enabled them to occupy those different niches. We call this evolutionary "spreading out," or divergence of organisms from a common heritage, an **adaptive radiation.**

During an adaptive radiation, evolution shapes new body parts for flying, digging, hunting, or hopping, not by continually reinventing the wheel, but by modifying

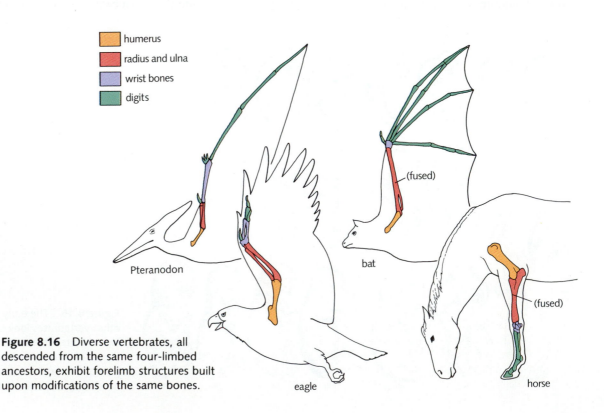

- humerus
- radius and ulna
- wrist bones
- digits

Pteranodon

(fused)

bat

(fused)

eagle

horse

Figure 8.16 Diverse vertebrates, all descended from the same four-limbed ancestors, exhibit forelimb structures built upon modifications of the same bones.

Figure 8.17 Homologous and analogous flight structures. The wings of bats and birds, though different in certain particulars, are derived from homologous limbs. The wings of butterflies, on the other hand, though they serve the same function, are strictly analogous.

existing structures to serve new functions. Bats, for example, did not evolve from mouse-like ancestors by sprouting a new set of wings and losing their front legs. Instead, the digits of their forelimbs elongated and developed a membrane of skin stretched between them (Figure 8.17).

Analogous structures and convergent evolution There is another kind of similarity in nature: the resemblance between

a bird's wing and a butterfly's wing and among a whale's flukes, a fish's fins, and a penguin's webbed feet (Figure 8.18). Such structures, which look and function similarly but are built of embryologically *un*related parts, are called **analogous.**

Evolutionary theory explains analogous structures as a result of **convergent evolution,** which occurs when different organisms face similar environmental demands (such

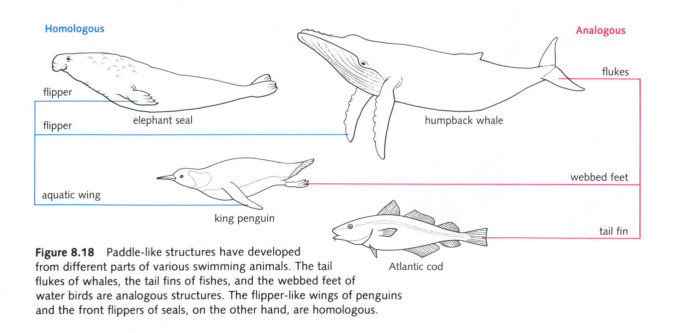

Figure 8.18 Paddle-like structures have developed from different parts of various swimming animals. The tail flukes of whales, the tail fins of fishes, and the webbed feet of water birds are analogous structures. The flipper-like wings of penguins and the front flippers of seals, on the other hand, are homologous.

striped opossum

honeycreeper

aye-aye

woodpecker finch

Figure 8.19 Different animals have adapted to the "woodpecker" niche in habitats lacking woodpeckers. The New Guinea striped opossum and the Madagascar aye-aye both have one elongated finger for digging through bark and spearing insects. The Akiapolaau, one of the Hawaiian honeycreepers, uses its lower beak to dig and its upper beak for impaling prey. The Galapagos woodpecker finch holds a cactus spine in its beak to accomplish the same feat.

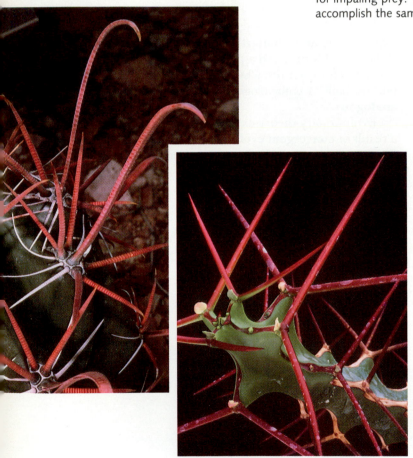

Figure 8.20 Deserts in the New World have spawned our familiar cacti (family Cactaceae) (left), while similar conditions in Africa have given rise to remarkably cactus-like plants in a completely different family (the Euphorbiaceae), which includes this relative of the familiar "crown of thorns" (right).

as flying and swimming) but start out with quite different "raw materials" for evolution to work with. Natural selection then molds these divergent structures along similar lines. Note that when two groups of organisms undergo adaptive radiations in separate, but ecologically similar, environments, convergent evolution often produces remarkably similar-looking—though unrelated—organisms (Figures 8.19 and 8.20).

Vestigial structures Darwin found further evidence for evolution in apparently useless structures, called **vestigial structures.** He recognized that perfection of form was acceptable evidence of evolution if one believed in evolution at the outset. But because creationists argued that perfection in nature reflected the inspired engineering of the divine architect, the mere existence of perfect creatures would not shake their views.

Useless structures, on the other hand, made no sense from the creationist viewpoint. Why would the creator have burdened humans not only with the troublesome appendix but also with the useless remains of an embryonic tail that never completely disappears? These sorts of structures make sense only as evolutionary remnants of organs that were useful to the organism's ancestors. Such remnants, Darwin argued, are proof that today's species were not created in their present form but evolved through time.

Darwin and Fossils: Possibilities and Problems

Darwin knew that the fossilized remains of extinct organisms provide critical evidence that life has evolved over time. But he had trouble reconciling the fossil record as he knew it, in the late nineteenth century, with certain aspects of his theory.

First, Darwin worried about the incompleteness of the fossil record, for he believed that biological evolution, like Lyell's geological evolution, proceeded slowly and gradually. "Nature," Darwin asserted, "does not take leaps." Thus Darwin felt it should be possible to find a long line of fossils documenting gradual transitions between one species and another. In Darwin's time those examples were not forthcoming.

Second, primitive members of most major groups of organisms appeared abruptly at the beginning of the geological period known as the Cambrian (Figure 8.21). Of the simplest, earlier life forms there was no record.

Relative Time Span	Era	Period	Epoch	Began (millions of years ago)	Length (millions of years)
Cenozoic	Cenozoic	Quaternary	Recent	0.01	
			Pleistocene	1.5	
Mesozoic		Tertiary	Pliocene	12	10
			Miocene	25	13
Paleozoic			Oligocene	34	9
			Eocene	56	22
			Paleocene	63	7
	Mesozoic	Cretaceous		135	70
		Jurassic		180	45
		Triassic		225	45
	Paleozoic	Permian		280	55
		Pennsylvanian (Carboniferous)		310	30
Precambrian		Mississippian (Carboniferous)		350	40
		Devonian		400	50
		Silurian		430	30
		Ordovician		500	70
		Cambrian		570–600	70–100
	Precambrian	Proterozoic		2500	2000
		Archaezoic		4600	2000

Figure 8.21 (above) The Grand Canyon, where the Colorado River has eroded through sedimentary rock formed over millions of years. (left) By observing such slices of Earth's history, geologists assembled the geological time scale.

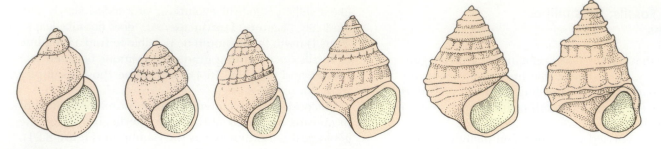

Figure 8.22 These snails illustrate a smooth series of intermediate fossils from the oldest known ancestors to present-day forms. The transitions along the way are gradual, often making it difficult to divide the lineage into distinct species.

This "Cambrian explosion," as it was called, looked to many of Darwin's contemporaries like the sudden population of the earth by a Creator, rather than the outcome of gradual evolution.

Furthermore, nineteenth-century scientists could not prove precisely *how* old any fossils were. No one argued that ancient sedimentary rocks had formed as sediments accumulated in the sea and that younger rocks had formed on top of them. Thus, as long as fossil-bearing rocks under study had not been disturbed, fossils found in lower layers were clearly older than those found in upper layers. But although these straightforward techniques could arrange fossil series in chronological order, they could not assign an absolute age to any specimen. Thus, in an absolute sense, they could neither prove fossils' antiquity nor provide evidence on rates of evolutionary change.

Darwin spent two chapters in *On the Origin of Species* wrestling with these problems. He pointed out that soft-bodied organisms and nonwoody plants do not fossilize unless they are buried in the right types of sediments before they decompose or are torn to shreds by predators. He also realized that even hard body parts, such as bones and shells, could be destroyed or scattered before fossilization occurred. And he noted that many habitats—such as deserts and mountain ranges—don't offer proper conditions for fossilization. In conclusion, he argued that the fossil record was "extremely imperfect" and allowed that anyone who did not accept that imperfection could "rightly reject my whole theory."

Paleontology and Evolution Today

As biologists applied new insights and new techniques to evolutionary biology and paleontology, they learned that Darwin was overly pessimistic. On the one hand, we will see in Chapter 13 that evolution may not always proceed

at the slow, steady pace on which Darwin insisted. On the other hand, although the fossil record is undeniably patchy, thousands of fossil finds have filled in many of the gaps in the evolutionary record over which Darwin agonized.

We now know of many examples of transitional forms in evolutionary sequences, and even in cases where some pieces are missing, striking intermediate creatures document the transition between one group and another (Figure 8.22). Paleontologists have recently uncovered single-celled organisms that date back much further than the Cambrian explosion (Chapter 14). And several theories involving the physiology and evolution of the first modern cells help explain why this explosive increase in new species occurred when it did.

Furthermore, we can now date fossils by examining their ratios of various **radioisotopes,** unstable atoms that break down to form other atoms. Solar radiation, for example, generates radioactive carbon (carbon-14 or ^{14}C) in the atmosphere. Plants take in that isotope along with normal carbon (^{12}C) as they photosynthesize, and animals pick it up from the plants they eat. Once an organism dies, the ^{14}C in its tissues decays steadily to ^{14}N (nitrogen-14). Thus the older a fossil is, the smaller its ratio of ^{14}C to normal carbon (^{12}C).

Carbon-14 has a half-life of roughly 5700 years, which means that half the molecules of ^{14}C that were present at the organism's death decay to ^{14}N over that length of time. A fossil sample that has a ratio of ^{14}C to ^{12}C that is one-half of the ratio of ^{14}C to ^{12}C in a living specimen will be about 5700 years old. Because ^{14}C has a relatively short half-life, other radioisotopes are used to date really ancient samples. One long-term method measures the decay of radioactive potassium (^{40}K) to argon (^{40}Ar) over a half-life of about 1.3 million years. And the clock provided by the decay of uranium into lead can be read as far back as 4 billion years (Table 8.1).

Table 8.1 *Radioactive Isotopes and Half-Lives*

Radioactive Isotope	Half-Life (years)	Useful Range (years)
Carbon-14	5730	0 –60,000
Potassium-40	1.25 billion	> 100,000
Uranium-235	704 million	> 100 million
Thorium-232	14 billion	> 200 million

Coevolution: Support from Plant and Animal Relationships

Some of the most fascinating pieces of living evidence for evolutionary change are the products of a phenomenon called **coevolution.** When we say that two species have coevolved, we mean that they have evolved in a tightly knit, reciprocal fashion, each evolutionary change in one occurring as a response to a change or changes in the other. Coevolution has occurred innumerable times among animals, among plants, and between plants and animals. Because these relationships are seen most clearly in the context of evolutionary history, you will more fully appreciate the significance of coevolution during our discussion of the history of life on earth in Chapters 14–18. It is worthwhile, however, to examine selected dramatic examples here.

Plant–pollinator coevolution   Darwin, a master at recognizing the significance of common phenomena in nature, observed that most flowering plants require pollen from another plant to fertilize their flowers, and that many insects spend hours collecting nectar, pollen, or both for food. Darwin correctly hypothesized that flowers and pollinating insects have evolved together in ways generally beneficial to all parties.

After studying the many intricate and unusual flower parts of many orchids, for example, Darwin predicted—although he did not directly observe—one of the most extreme examples of flower–pollinator coevolution. In Madagascar, Darwin discovered an orchid (*Angraecum sesquipedale*) whose flower sports a foot-long spur with nectar at its base. He correctly predicted the existence of a moth with a proboscis long enough to sip that nectar (Figure 8.23).

Plant–herbivore coevolution   Over time, many plants have evolved tough leaves, armored spines, indigestible compounds such as tannins, and potent poisons such as alkaloids. (Pyrethrum, one of the most powerful insecticides known, was first extracted from a species of *Chrysanthe-*

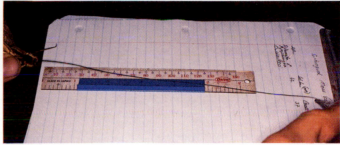

Figure 8.23   This orchid, *Angraecum orchidglade* (top), carries a long spur with nectar at its base. Darwin, aware that this unusual nectary must have evolved with a specific pollinator able to reach it, predicted the existence of a moth with a long proboscis (bottom). The moth was discovered years later.

mum.) Sometimes these toxic compounds enable plants to live free from insect predators. But sooner or later, one insect species or another evolves either resistance to a particular herbivore deterrent or an enzyme that inactivates it. From then on, plant and insect are locked in a coevolutionary "arms race."

Fighting the deterrents of several plants is apparently not feasible for an insect. For this reason, it is not unusual to find that closely related species of insects have specialized to feed only on closely related plant species. The larvae of butterflies in the genus *Heliconius,* for example, feed on either passion flower vines of the genus *Passiflora* or on related species in the same plant family. These plants all contain similar poisonous compounds, and these larvae are among the few insects that can ingest these compounds without harming themselves. In response, several *Passiflora* species have evolved stiff, hooked hairs that trap, injure, and kill caterpillars (Figure 8.24).

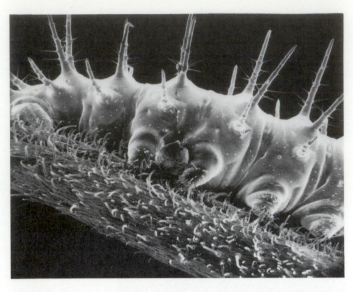

Figure 8.24 This scanning electron micrograph of a caterpillar impaled on the spines of a *Passiflora* leaf shows a striking example of a plant that "bites back" at its predators. Although the caterpillar has evolved resistance to toxins present in the leaves, it must still contend with these physical defenses.

Symbiosis: Life Together

As we noted in Chapter 5, many organisms engage in symbiotic relationships that have powerful effects on their ability to survive and reproduce. Each type of symbiosis—parasitism, commensalism, and mutualism—provides examples of animal and plant species that have evolved intricate and often essential relationships with one another.

Parasitism Technically, *parasitism* is a form of symbiosis in which one partner benefits from the association while the other partner is harmed. Although parasites can cause serious illness or death, no evolutionarily successful parasite could be rapidly lethal to all its hosts or it would soon be out of business! As we saw in the case of rabbits and myxomatosis in Chapter 5, parasite virulence and host resistance evolve together in a continually shifting balance.

Commensalism In a form of symbiosis called *commensalism*, one of the participants benefits but the other is neither helped nor harmed. Orchids and Spanish moss, for example, may look like parasites but they are not; they attach harmlessly to the branches of host trees. They obtain their nutrients from rainwater and decaying leaves that collect around their roots, not from the host plant.

Mutualism In *mutualism*, both members of the association benefit. In fact, the participants in many mutualistic

relationships cannot survive without their partners. We have already encountered such examples of mutualistic symbioses as wood-digesting microorganisms in the guts of herbivores, clownfish in the tentacles of sea anemones, and lichens. In fact, one hypothesis (discussed in detail in Chapter 14) proposes that the very cells of all higher organisms are symbiotic associations between two simpler cell types.

Mimicry

Mimicry, in which two or more species resemble each other closely, demonstrates intimate evolutionary relationships between species but may or may not fall precisely in the category of coevolution. In **Batesian mimicry** one organism evolves to resemble another that has a defense against a common predator. In **Müllerian mimicry** two species with similar defense mechanisms evolve to resemble each other. Note that no individual (or species) consciously "learns" to "trick" another; the evolution of mimicry is a long, random, evolutionary process guided by natural selection.

The best-known group of North American animal mimics includes monarch, viceroy, and queen butterflies (Figure 8.25). At the center of the group is the conspicuously colored monarch, whose noxious taste protects both itself and its tasty mimics (Figure 8.26). Mimicry also occurs in plants, where the selective advantage it confers may enhance either pollination or protection from predators (Chapter 28).

One example of coevolutionary mimicry occurs in the plant–herbivore "arms race" between *Passiflora* and *Heliconius*. Several *Passiflora* species have evolved a remarkable defense that specifically exploits the egg-laying behavior of their insect enemies. The larvae of *Heliconius* butterflies—which compete for food on small patches of *Passiflora*—are often aggressive and cannibalistic. For that reason, larvae emerging on leaves that already host young caterpillars are less likely to survive than those emerging on vacant leaves. Research has shown that female butterflies tend not to lay eggs on leaves that already have eggs on them. In a remarkable defensive adaptation, several *Passiflora* species have evolved structures that mimic the shape and color of the *Heliconius* eggs, ostensibly to "fool" females into avoiding them!

Support from Comparative Biochemistry

Just as anatomists have found structural homologies among related animals, biochemists have found chemical homologies among related species. Living organisms share many biologically important chemical compounds, and the more closely related two species are, the more closely

Figure 8.25 Poisons, resistance, and mimicry in butterflies. Monarch caterpillars pick up toxic compounds from the milkweed plants they eat. The monarchs not only tolerate those compounds but also store them in their tissues and retain them after metamorphosing into adult butterflies (above left). Any predator unfortunate enough to swallow a monarch containing those compounds becomes violently sick to its stomach. The similar-looking queen butterfly (above, right) also tastes bad to birds and so is classified as a Müllerian mimic. The viceroy butterfly (below, right) contains no such poison but gains protection by mimicking the monarch; viceroys are therefore considered Batesian mimics.

Figure 8.26 Once a bird has been made sick by eating a monarch, it avoids anything that resembles the distasteful species.

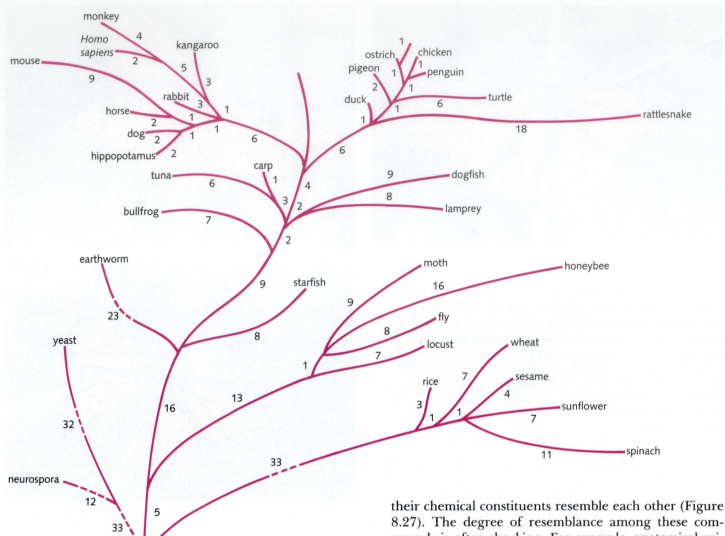

Figure 8.27 It is possible to construct evolutionary trees based on similarities and differences in organic molecules shared by all living organisms. This evolutionary tree is based on similarities in the molecule called cytochrome *c*, which retains its function but differs slightly in all these organisms.

their chemical constituents resemble each other (Figure 8.27). The degree of resemblance among these compounds is often shocking. For example, anatomical evidence convinced biologists long ago that humans and chimpanzees are close relatives. But it is fascinating to discover that we share 98 percent of our DNA with chimps! We will discuss chemical homologies in more detail in Chapter 10.

DARWINIAN THEORY EVOLVES

Despite his ability to convince the world that life has evolved (and continues to evolve), Darwin was initially less successful in promoting his theory of natural selection, the process he credited with directing evolutionary change. But though natural selection as Darwin envisioned it has never been universally accepted, most of its basic principles have been reinforced and extended as biological knowledge has grown over the last century. Part of the strength of Darwinian theory lies in its ability to absorb information and itself evolve as our knowledge of the world evolves.

One major weakness of the theory in Darwin's time was that neither Darwin nor his colleagues understood how heritable traits are controlled and passed from generation to generation. In Darwin's day, it was believed that parents' traits blended in their offspring, somewhat like mixing paints: Blend a strong color and a weak one, and you get a mixture of intermediate hue. Everyone knew, for example, that crossing large animals with small ones produced offspring of intermediate size and that crossing red flowers with white ones often yielded pink progeny. The idea made sense, but no one knew why things worked out that way.

Darwin realized that this view of inheritance could spell trouble for natural selection. When an organism appeared with a new, favorable characteristic, it would have to mate with an organism which probably did not have the new characteristic to pass that adaptation to the next generation. But if all its characteristics were "blended" with those of typical members of the species, they would be blended toward the species' average and would diminish in strength generation after generation. Any new trait, no matter how useful, would disappear so quickly that evolutionary change would be impossible.

This problem concerned Darwin. In later editions of *On the Origin of Species,* he suggested that the environment might somehow cause variation and that characteristics acquired in this way might be inherited. Without a better understanding of inheritance, Darwin was up against a formidable obstacle.

The Contribution of Genetics

Ironically, even as Darwin was grappling with the problem, ground-breaking studies on the nature of inheritance were being conducted by an Austrian priest named Gregor Mendel. But Mendel's elegant work, which set the stage for a radically new way of thinking about heredity, was ignored by the scientific community for over 30 years, so Darwin may never have learned about it.

Then, around the beginning of the twentieth century, several independent researchers obtained results similar to Mendel's, and Mendelian genetics was born. The maturation of genetics as a discipline since that time has revolutionized evolutionary theory. Because genetics is fundamental to modern evolutionary theory, we will spend the next three chapters examining the principles of Mendelian genetics. We will then explore the ways in which genetics, ecology, and population biology are woven together in the exciting and ever-changing world of evolutionary biology today.

SUMMARY

Darwinian evolutionary theory has profoundly influenced scientists, philosophers, religious leaders, and lay people since the day it was published. Today evolutionary theory provides the basis for understanding all the biological sciences and medicine, and over the years it has affected the development of the social sciences, anthropology, psychology, and politics.

Though Darwin was not the first to propose that life has evolved, he provided the first plausible scientific mechanism by which evolution could occur. His thinking was influenced both by the accumulating evidence for gradual geological change and by his travels and reading. Darwin's theory of evolution by natural selection explains evolutionary change on the basis of the differential survival and reproduction of those individual organisms best adapted to their local environments. The variation in living species that makes such change possible is random, purposeless, and not subject to control.

Many lines of evidence prove that evolution has occurred. Fossils document extinct transitional forms between living groups of organisms. Similarities among diverse living organisms—ranging from the structure of bones and muscles to similarities among proteins and DNA—prove that modern organisms share common ancestors. And the phenomena of coevolution, symbiosis, and mimicry demonstrate intimate, long-term evolutionary relationships between species.

STUDY FOCUS

After studying this chapter, you should be able to:

- Outline the nineteenth-century scientific world view, and explain the developments in science and society that made the world view indefensible.
- Place Charles Darwin and his ideas in the context of nineteenth-century society.
- Describe the events that led Darwin to formulate his theory.
- Explain classical Darwinian evolutionary theory.
- Appreciate the importance of evolutionary theory in the biological and social sciences.

SELECTED TERMS

evolution *p. 140*	variations *p. 149*
catastrophism *p. 143*	heritable *p. 149*
uniformitarianism *p. 143*	artificial selection *p. 150*
teleological *p. 144*	natural selection *p. 151*
fitness *p. 147*	adaptation *p. 151*

common descent *p. 152*   vestigial structures *p. 156*
homologous *p. 154*   radioisotopes *p. 158*
adaptive radiation *p. 154*   coevolution *p. 159*
analogous *p. 155*   Batesian mimicry *p. 160*
convergent evolution *p. 155*   Müllerian mimicry *p. 160*

R E V I E W

Discussion Questions

1. How do natural selection and artificial selection differ? Can pressure from either natural or artificial selection cause specific variations in organisms?

2. What was the Platonic view of variation in nature, and how did Darwin's view differ from it?

3. How did the Malthusian doctrine inspire Darwin?

4. What changes in the scientific world view were necessary before Darwin could even begin to think about biological evolution?

5. What three principles constituted Lyell's theory of uniformitarianism?

6. Darwin's theory dispensed with the teleological thinking of Lamarck and his colleagues. Using descriptions of variation and natural selection, explain why Darwinian evolution can have no ultimate purpose.

7. Why were Darwin's ideas both welcomed and feared by nineteenth-century Europeans?

8. Give three examples of homologous structures in familiar animals.

9. Give three examples of analogous structures in either animals or plants.

10. Why did Darwin so enthusiastically advance vestigial structures as proof of his theory?

Objective Questions (Answers in Appendix)

11. Evolution can be described as
    (a) a continuing process.
    (b) a catastrophic event in the past.
    (c) static.
    (d) the attaining of an ideal type.

12. During Darwin's time, inheritance was considered to result from
    (a) the passing of genes from generation to generation.
    (b) the passing of chromosomes from generation to generation.
    (c) dominant maternal traits.
    (d) the blending of parents' traits.

13. Of the greatest importance in evolution is the number of individuals who, under local environmental conditions,
    (a) are born and survive.
    (b) survive and reproduce.
    (c) undergo coevolution.
    (d) undergo adaptive radiation.

14. Which of the following is *not* an example of coevolution?
    (a) flowers and their pollinators
    (b) predators become more effective hunters, and their prey evolve a better means of escape
    (c) plants and specialized plant eaters
    (d) plants that can be eaten by a variety of insect species

15. A major weakness in Darwin's theories was that
    (a) there was no explanation of how characteristics are transmitted from parents to offspring.
    (b) the concept of survival of the fittest was omitted.
    (c) the concept of natural selection was omitted.
    (d) he did not explain how the theory of special creation supported his observations on the Galapagos Islands.

CHAPTER **9**

Continuity of Life: Cellular Reproduction

*Death, be not proud, though some have called thee
Mighty and dreadful, for thou art not so;
For, those whom thou think'st thou dost overthrow,
Die not, poor Death, nor canst thou kill me.*

—John Donne (1633)

Of all things that are certain in life, death is the surest. Every living thing will ultimately die, ourselves included. Yet despite this sad fact, life itself seems almost immortal. For 3 billion years, the thread of life on this planet has been spun into a wide variety of organisms, each of which is linked to the whole by common ancestry as well as by the common properties of life itself.

Although the triumph of death over the biological individual is inevitable, death, even as John Donne noted, does not triumph over life. Why not? The ability of cells and organisms to reproduce, to pass life from one generation to the next, has ensured that life itself outlasts the death of any one individual.

For simple organisms, death is *not* inevitable. Think about that for a moment. If a single-celled organism divides to produce two identical cells, which is parent and which is offspring? The answer is that for two *identical* cells, that question has no meaning. In a sense, each of the two cells has the same identity as the one that gave rise to them. We could trace each cell backward through time in an unbroken series of cell divisions that reach as far into the past as we care to go. As long as such cells are able to divide, they do not age in the sense that we do. One might almost think of them as immortal.

The ability to outlast death via reproduction is a property of the individual cell. Ultimately, life *does* triumph over death. It does this not by magic, but by the simple act of growth and cell division.

THE DIVERSITY OF CELLS

The invention of the microscope in the seventeenth century led directly to the discovery that living organisms are composed of cells. Although all cells share certain similarities, in many ways the most striking thing about the cells that make up a living organism is their diversity. It is commonly estimated that the human body is made up of at least 85 completely different types of cells, each distinguishable from the other—and that is just for our species alone. When we examine the larger world, we find that cells exist in an almost infinite variety of sizes, shapes, colors, and structures. It is still tempting to generalize, of course, and try to describe a typical cell. Such a cell would be about 20 μm (2.0×10^{-5} m) in diameter; it would have a prominent, centrally placed structure about 6 μm in diameter that we call a **nucleus** (after a Latin word meaning "kernel," because it looks like a seed in the center of a fruit); it would be surrounded by a cell membrane that separates it from its surroundings, and it would contain a number of tiny structures that might be thought of as miniature organs: **organelles.** We might be bold enough to draw such typical cells, and it would indeed be possible to find a few examples of cells that bear a remarkable resemblance to our drawings (Figure 9.1).

Generalizations are always useful, but we must not allow them to keep us from appreciating the enormous number of cases in which the generalization does not apply. Let's consider, for example, our generalization about cell *size*. Cells are small and that's why the use of the microscope preceded the development of the cell theory. (Microscopes are covered in detail in the discussion on cell organization in Chapter 24.) The smallest living cells are a kind of organism known as *mycoplasma*, a form of bacteria. They are less than 0.2 μm across and can be observed only in an electron microscope. But many cells

165

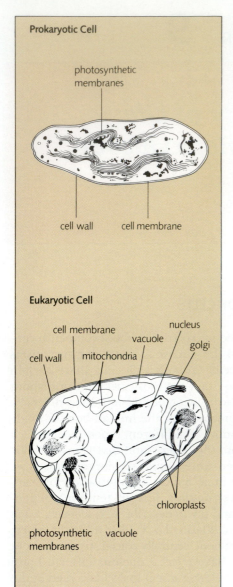

Prokaryotic Cell

photosynthetic membranes

cell wall   cell membrane

Eukaryotic Cell

cell membrane   nucleus
vacuole   golgi
cell wall   mitochondria
chloroplasts
photosynthetic   vacuole
membranes

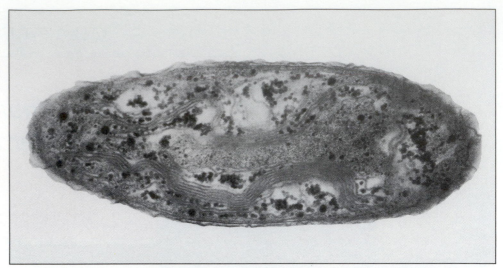

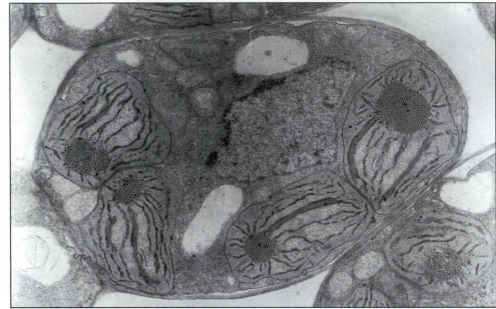

Figure 9.1 These cell diagrams and electron micrographs of two photosynthetic organisms provide a basis for comparing the structures of eukaryotic and prokaryotic cells. Eukaryotic cells possess nuclei, while prokaryotic cells do not. Each cell type possesses different photosynthetic membranes. In the eukaryotic cell, these membranes are enclosed in an organelle called the chloroplast. In the prokaryotic cells, these membranes are free in the cytoplasm.

are large enough to be seen with the unaided eye. In oceans throughout the world, tiny plant-like organisms grow on the surfaces of submerged rocks. A particularly beautiful one is *Acetabularia mediterranea* which grows in the Mediterranean. Individual *Acetabularia* measure up to 2 or 3 cm, yet each one is a single cell (Figure 9.2). Even larger cells are found elsewhere in nature. The nerve cells carrying the impulses that control the mus-

cles in your lower limbs, for example, can be as long as 1.5 m.

Types of cells Living cells can be divided into two distinctly different classes. Cells with nuclei are commonly known as **eukaryotic** (*eu* means "true"; *karyon* means "kernel" or "nucleus") cells, whereas living cells that lack nuclei are called **prokaryotic** (*pro* means "before") cells.

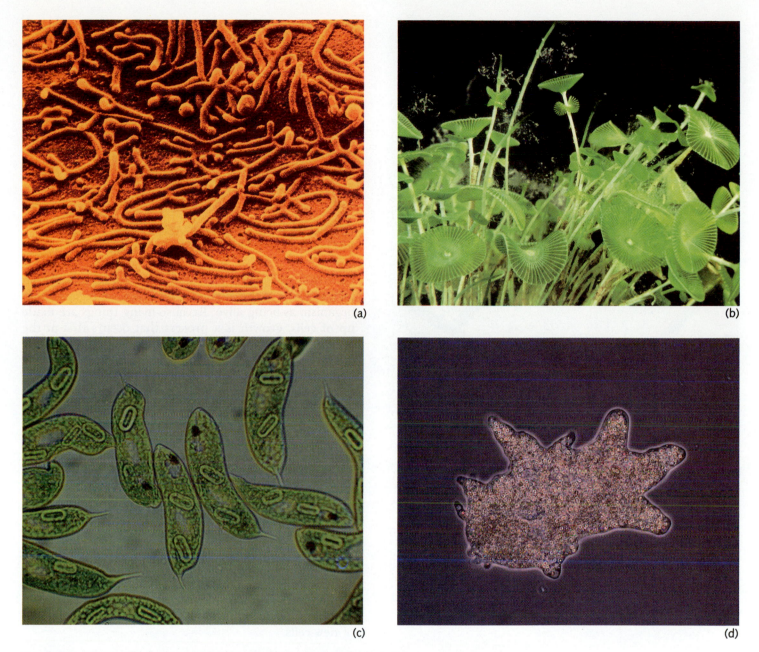

(a)

(b)

(c)

(d)

Figure 9.2 Living cells display a wide diversity of structures and adaptations.
(a) *Mycoplasma:* The cells of this small and simple prokaryote grow, often undetected, in a variety of environments (scanning electron micrograph). (b) *Acetabularia:* The mushroom-shaped cells of this green alga are among the largest in nature. Many species are 1–3 centimeters in length, like these specimens, photographed beneath the waters of the Mediterranean. (c) *Euglena:* These motile cells swim rapidly, propelled by two whip-like flagella. (d) *Amoeba:* These large and flexible cells move by streaming their cytoplasm into false feet, or pseudopodia.

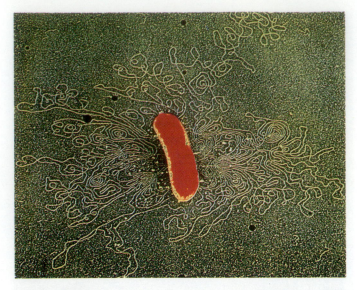

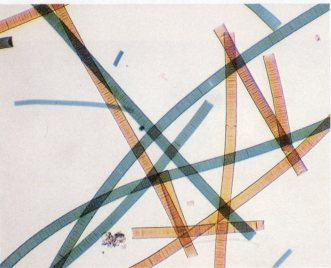

Figure 9.3 *Escherichia coli* (top) and *Oscillatoria* (bottom) are prokaryotes, cells that do not contain nuclei.

Typical prokaryotic cells are bacteria such as *Escherichia coli* and blue-green algae such as *Oscillatoria* (Figure 9.3).

Although prokaryotic cells may contain internal membranes, as a rule they are less complicated internally than eukaryotic cells. Eukaryotic cells generally contain a large number of specialized, membrane-enclosed organelles that enable the cell to carry out a variety of activities in different compartments. In other words, eukaryotic cells seem to display a greater degree of internal specialization. The fundamental distinction between prokaryotes and eukaryotes, of course, remains the presence or absence of a nucleus.

We will use some terms that are related to the earlier terminology, however. In eukaryotic cells we might think of the cell as divided into two large sections: the nucleus and the cytoplasm. The word **cytoplasm** ("cell fluid") is widely used by biologists to refer to the major *compartment* in most living cells. In this way, we can quickly describe molecules found outside the nucleus as "cytoplasmic" and can speak of specific organelles as being found in the cytoplasm (as opposed to the nucleus). In prokaryotic cells, where there is no nuclear compartment to describe, we can still speak of the cytoplasm as the large compartment bounded by the cell membrane.

GROWTH AND DEVELOPMENT: THE PROBLEM OF REPRODUCING A CELL

Growth is one of the principal characteristics of life; in fact, it is one of the attributes by which we recognize an organism as being alive. Because living things are made up of cells, growth is a process that occurs first at the level of the cell. Within limits, cells may increase in size, but no process of growth can be sustained for very long without involving **cellular reproduction**—the formation of new cells.

The basic patterns of cellular reproduction are deceptively simple: A cell passes through a phase in which it increases in size, then it divides and two new "daughter" cells are formed. In many respects, each of these daughter cells is identical to the single cell that produced it. Therefore, as we examine the process of cellular reproduction, we should expect to find two things: (1) a process of *duplication,* so that each daughter cell will possess a complete set of cellular structures, including the information coded in DNA, and (2) a process of *separation,* so that cell division carefully and precisely parcels structure and information between the two daughter cells. In this chapter we will examine these two processes, which are at the heart of life's most essential mystery, the formation of new cells.

Cellular Growth

As a cell increases in size, its needs increase as well. It requires larger amounts of nutrients and other materials from its environment, and it produces larger amounts of waste products. These increasing demands place an effective limit on how big a cell can be. The mathematics of size and shape dictate that when a cell doubles in diameter, its surface area increases by a factor of 4. Its volume, however, increases by a factor of 8. The simplest solution for the problem of increasing cell size is the process of cell division, the formation of two cells by the splitting of a preexisting one.

In single-celled organisms, such as the bacterium *Escherichia coli,* cell division is often the means by which members of the species *reproduce,* because the production of two new cells means the production of two completely new individuals (Figure 9.4). Except in special cases, the process of cell division produces two cells, each of which has all the characteristics of the single cell that produced them.

The Cellular Life Cycle

The process of cell growth cannot be continuous and unbroken. A cell is not filled with a nondescript molecular soup. Rather, each cell contains within it the necessary information, in biochemical terms, to synthesize the molecules it needs for life. In order for cell division to be successful, this essential information must be duplicated and then carefully divided between the two daughter cells.

The most important form of biochemical information is carried from one generation to the next in the molecule known as **DNA** (deoxyribonucleic acid). DNA molecules contain instructions for synthesizing proteins, and this enables them to carry the information needed to build nearly all of the important molecules of the cell. Therefore, one of the critical points in the time between one cell division and the next is the time when the duplication of DNA molecules occurs (Figure 9.5).

When essential information has been duplicated, the process of cell division must then ensure that each daughter cell receives a complete set of that vital information.

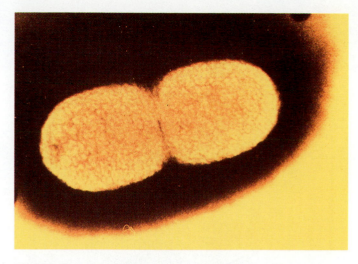

Figure 9.4 "Binary fission" aptly describes the process of cell division in most prokaryotes. When it has completely copied its genetic material in the form of DNA, the cell splits in two.

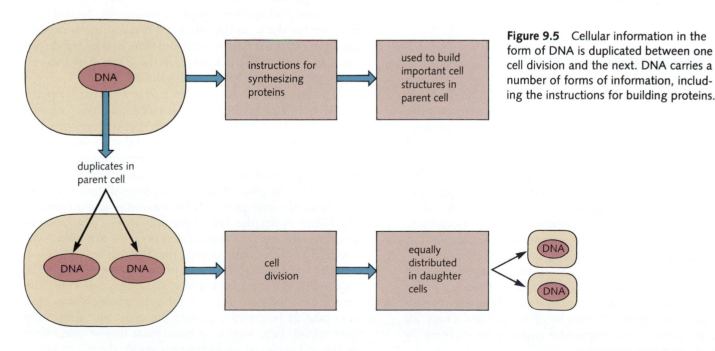

Figure 9.5 Cellular information in the form of DNA is duplicated between one cell division and the next. DNA carries a number of forms of information, including the instructions for building proteins.

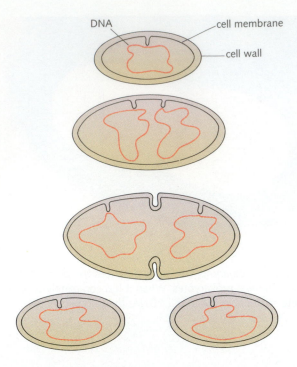

DNA — cell membrane
— cell wall

Figure 9.6 In many prokaryotes, attachments to the cell membrane ensure that each cell receives a complete DNA molecule.

Mechanisms must exist to distribute DNA molecules between the two daughter cells and, finally, to break what had been a single cell into two cells independent of each other. The cycle of changes involved in cellular reproduction occurs repeatedly as living cells progress from one generation to the next.

CELL DIVISION IN PROKARYOTES

Prokaryotic cells—cells without nuclei—are able to divide in a relatively simple way. After the cell has grown to the point where division is possible and has duplicated its DNA molecule(s), a small fissure begins to grow inward from the cell membrane. Prokaryotic DNA molecules are attached at several points to the cell membrane, and these DNA molecules can be separated at their membrane attachment points, as shown in Figure 9.6. As the DNA molecules are gradually separated, the cell membrane and cell wall continue to grow inward until the point at which the two cells will divide becomes apparent.

This process resembles a simple splitting of one cell to form two, and it is often called **binary fission** (*fission* means "splitting"). Bacteria placed in rich nutrient broth can carry out binary fission as often as once every 20 minutes, which gives them the capacity to increase in numbers at a fantastic rate.

CELL DIVISION IN EUKARYOTES

The process of cell division in eukaryotes is known as **mitosis.** Mitosis differs somewhat from one cell type to the next, but its general features are the same in most eukaryotic cells. Eukaryotic cells have an additional feature that makes cell division more complex: The majority of their DNA molecules are collected in structures known as **chromosomes,** which are found within the nucleus. Chromosomes (*chromo* means "colored"; *soma* means "body") contain the genetic information of the cell.

Chromosomes were originally discovered in eukaryotic cells because of their distinctive appearance during mitosis. Chromosomes are composed of a coiled web of protein and DNA, which is dispersed within the nucleus during most of a cell's life cycle. During cell division, however, each chromosome condenses into a distinct structure as the cell enters mitosis (Figure 9.7).

The Eukaryotic Cell Cycle

The life cycle of a eukaryotic cell can be divided into two general phases: **mitosis** (the process of cell division) and **interphase** (the time between cell divisions). Because many cells proceed quickly from one mitosis to the next, we can visualize some of the repeating events as laid out graphically in a circular pattern that is known as the **cell cycle** (Figure 9.8).

The term *interphase* is misleading because it gives the impression that little is happening in the time between cell divisions. However, this is the period when a cell is rapidly growing, increasing in size, developing new structures, and synthesizing new molecules. In order to prepare for the next round of cell division, each rapidly growing cell must have a period of time in which it duplicates its DNA. This is known as the *S phase* of the cell cycle, where S stands for the synthesis of DNA.

The fact that DNA is synthesized at a particular point in the cell cycle enables us to break interphase down a bit further. The G_1 *phase* is the "gap" between mitosis and the beginning of DNA synthesis, and the G_2 *phase* is the time between the S phase and the beginning of mitosis.

Scientists studying the cell cycle have found that certain events take place during each phase of the cycle. The synthesis of some proteins, for example, is confined to G_1. Other molecules are made only in G_2, just prior to mitosis, whereas DNA-associated proteins are often synthesized preferentially during the S phase.

The lengths of the various phases of the cell cycle are extremely variable. In rapidly growing yeast cells, the entire cycle takes as little as 2 hours. Human skin cells may take 24 hours to complete the cycle, and some cells may spend days, weeks, or even years in just one phase of the cell cycle. In an important sense, the G_1 phase of

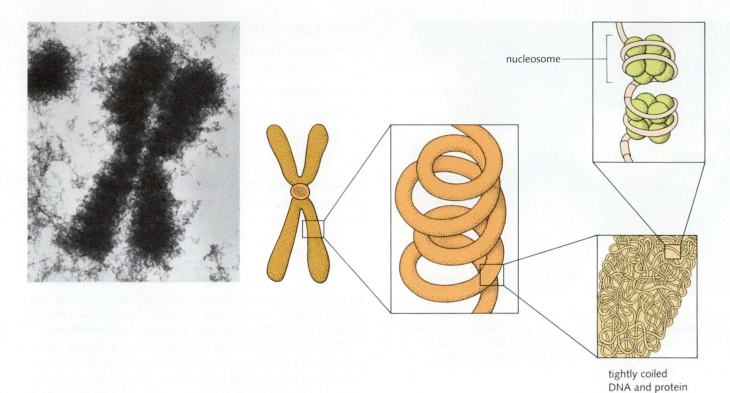

nucleosome

tightly coiled
DNA and protein

Figure 9.7 In most eukaryotic cells, chromosomes condense to form distinct structures during mitosis. The electron micrograph shows a single mitotic chromosome. Protein and DNA make up the structure of a chromosome. DNA is wound around protein cores and folded into a complex pattern.

the cycle is the "decision" period for each cell. When a cell completes the G_1 phase and enters the S phase, it has made the "decision" to divide—there is no point in beginning DNA replication unless mitosis will take place. This is why G_1 is the phase in which nongrowing cells find themselves locked. For example, sugar-conducting cells of a woody plant may spend decades in the G_1 phase of the cycle.

Cellular changes during the cycle Cells go through a series of dynamic changes during the cell cycle, some of which are reflected in changes at the cell surface. Although there are slight differences in cellular appearance during the G_1, S, and G_2 phases, the most profound changes occur as a cell enters mitosis. Cells grown in culture "round up" as they begin to enter mitosis, withdraw from the substrate upon which they have been growing, and mobilize their resources for the dramatic changes of mitosis. Scanning electron microscope images of cells in various phases of the cycle confirm this (Figure 9.9).

Mitosis: The Beginning of Division

Like life, mitosis is a continuous process. And just as there are no clear-cut distinctions between adolescence and adulthood, two phases of human life, there are few obvious points at which the process of cell division can be divided into stages. Nonetheless, since the end of the nineteenth century, scientists have broken mitosis into four distinct stages or phases to make studying the process a little

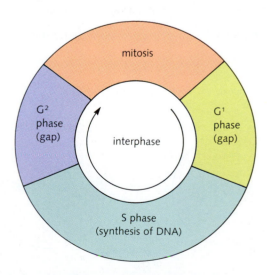

Figure 9.8 The eukaryotic cell cycle. DNA synthesis occurs during the S phase, which is separated from mitosis by two "gaps," G_1 and G_2.

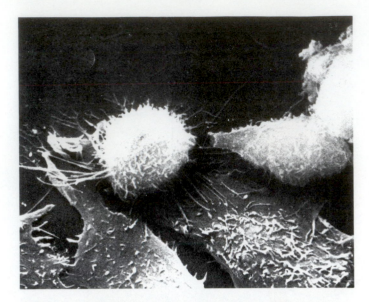

Figure 9.9 These cells in culture change shape as they pass through the phases of the cell cycle. The cell in the center of the field has rounded up and is beginning mitosis. The two cells to its right are in late G₂ and have begun to separate from the substrate.

easier (Figure 9.10). The four stages are

- **Prophase.** Preparation for cell division. Chromosomes condense, and the nuclear envelope disappears.

- **Metaphase.** Chromosomes align along a central plane.

- **Anaphase.** Chromosomes separate and move.

- **Telophase.** Mitosis ends and the two daughter cells separate.

Prophase After **interphase** (Figure 9.11), the cell begins the **prophase** stage of mitosis. In animal cells, the first hint that mitosis is about to take place is the duplication of two small structures known as **centrioles** (Figure 9.12). The centrioles are found just outside the nuclear envelope, and after duplication, they slowly move to either side of the nucleus. Gradually the **mitotic spindle,** a structure of fine filaments, develops near each of the centrioles and envelops the nucleus. The mitotic spindle is made up of tiny structures known as **microtubules.**

During prophase, the nuclear envelope slowly disintegrates (see Theory in Action, Where Does the Nuclear Envelope Go?, p. 176), and dense, thread-like chromosomes begin to appear (Figure 9.12, Figure 9.13, and Figure 9.14). Gradually these chromosomes untangle themselves and gather near the center of the cell. With a good microscope, it is possible to see that each chromosome consists of two identical strands called **chromatids** (Figure 9.13, inset) joined at a region that biologists have named the **centromere** ("central zone"). The point at which each chromatid is attached to the mitotic spindle is known as the **kinetochore.**

Figure 9.10 The four stages of mitosis.

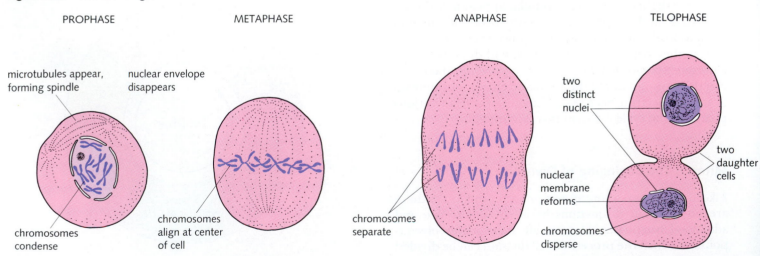

PROPHASE

microtubules appear, forming spindle

nuclear envelope disappears

chromosomes condense

METAPHASE

chromosomes align at center of cell

ANAPHASE

chromosomes separate

TELOPHASE

two distinct nuclei

nuclear membrane reforms

chromosomes disperse

two daughter cells

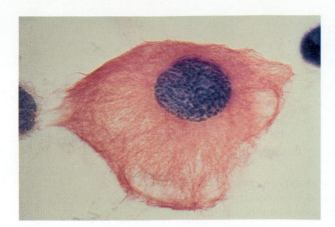

Figure 9.11 A typical cell late in G₂ contains a nucleus with dispersed chromatin, as shown in this light micrograph of an interphase cell from an African flower of the genus *Haemanthus*.

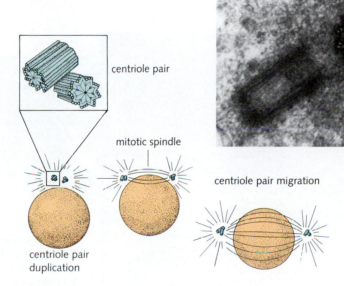

centriole pair

mitotic spindle

centriole pair migration

centriole pair duplication

Figure 9.12 In most animal cells, one of the first clues that mitosis is about to begin is the replication of centrioles. Following their replication, a pair of centrioles moves to either side of the nucleus (right).

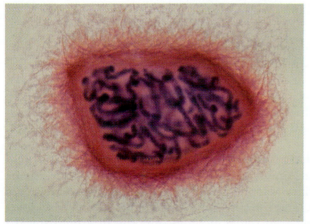

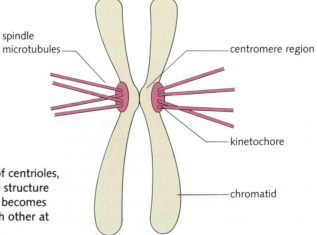

spindle microtubules

centromere region

kinetochore

chromatid

Figure 9.13 Early prophase (left) is marked by the separation of centrioles, the appearance of individual chromosomes, and a change in the structure of the cytoskeleton. During prophase (right), each chromosome becomes visible as a pair of chromatids. The chromatids are joined to each other at the centromere and to the mitotic spindle at the kinetochore.

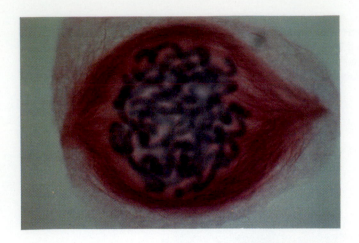

Figure 9.14 In late prophase, the nuclear envelope has disintegrated and fully condensed chromosomes are visible.

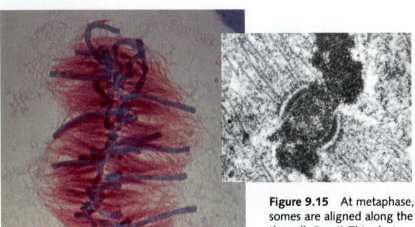

Figure 9.15 At metaphase, the chromosomes are aligned along the central plane of the cell. (inset) This electron micrograph of the centromere region shows both kinetochores.

Metaphase When the centromeres of each chromosome are aligned in a single plane at the center of the cell, **metaphase** begins (Figure 9.15). At metaphase, the entire set of chromosomes can be seen across the center of the cell. The mitotic spindle extends from one "pole" of the cell to the other (Figure 9.15, inset).

As we have seen in animal cells, the poles of the spindle are oriented around tiny structures in the cytoplasm known as centrioles. Plant cells lack centrioles, but the poles of their mitotic spindles have a similar shape. The spindle is a dynamic structure. Tension between spindle fibers extending to each pole seems to be responsible for moving chromosomes toward the center of the cell. In what can be described as a gentle tug-of-war, each side pulls with roughly the same pressure, moving the chromosomes toward the center of the plant cell.

Metaphase, often the briefest stage of mitosis, lasts only until the chromosomes begin to separate (just a few minutes in many cells). But metaphase is also the stage at which many chromosomes can be observed most clearly, and the fact that each chromosome is composed of two identical chromatids is often apparent at metaphase.

Anaphase The most dramatic phase of mitosis is **anaphase.** Almost as if in response to a signal, attachments that have held the paired chromatids at the centromeres break, and the single *chromatids* (now called *chromosomes* in their own right) move in opposite directions toward the regions where two new nuclei will be organized. Each chromosome comes apart in this fashion, contributing one chromatid to each of the new cells. Anaphase concludes when each set of chromosomes has arrived at its pole (Figure 9.16).

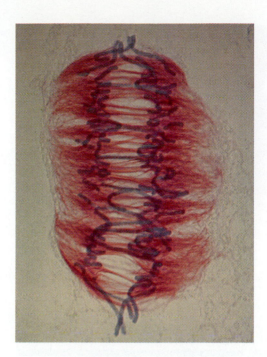

Figure 9.16 At anaphase, the chromatids separate and begin to move to the poles of the spindle.

A great deal of scientific effort has gone into studying the mechanism of anaphase movement. It is clear that microtubules play a principal role in moving the chromosomes to opposite poles of the cell, but the actual nature of the force-generating mechanism is still not clear.

Telophase When chromosome movement is complete, the two separated groups of chromosomes enter **telophase,** during which they start to organize into two cell nuclei. The condensed chromosomes begin to disperse, the nuclear envelope reforms, the mitotic spindle disappears, and the formation of two complete nuclei has been accomplished. Mitosis is complete when the two nuclear envelopes form once more (Figure 9.17).

Cytokinesis

In most cells, the cytoplasm of the cell divides and produces two distinct cells, a process known as **cytokinesis** (*cyto* means "cell"; *kinesis* means "movement"). In animal cells, cytokinesis begins with the formation of a **cleavage furrow,** a progressive constriction of the cytoplasm between the two nuclei (Figure 9.18). Contractile proteins in the cytoplasm gradually pull the furrow inward until the cell is broken in two. In many plant cells, small vesicles and vacuoles gather to form a structure known as the *phragmoplast.* The vesicles of the phragmoplast gradually fuse to form a **cell plate** that separates the cytoplasm into two distinct cells (Figure 9.19).

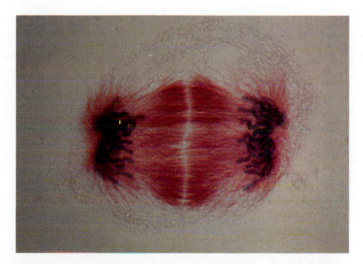

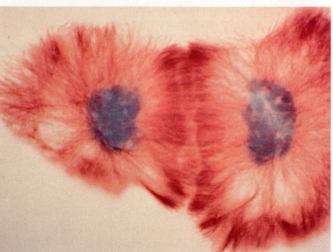

Figure 9.17 Telophase (top) begins when anaphase movement is complete. Two new nuclei begin to organize around the two sets of chromosomes. Late telophase (bottom) marks the beginning of cytokinesis, the division of the cytoplasm.

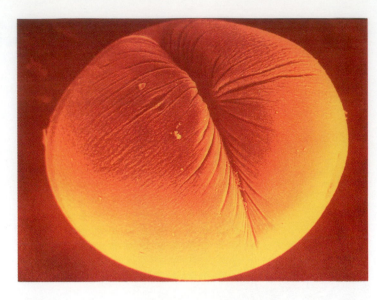

Figure 9.18 Cytokinesis in this dividing animal cell is a dramatic event, as the cell membrane forms a large cleavage furrow to complete cell division.

Where Does the Nuclear Envelope Go?

Near the end of the nineteenth century, biologists studying cell division began to wonder about one of the very first events of prophase, the apparent dissolution of the nuclear envelope. Even under the highest-power light microscope, the envelope, which is composed of two distinct membranes, seems simply to disappear. Just as magically, two nuclear envelopes re-form at the conclusion of telophase. Where do the nuclear membranes go during mitosis? Until very recently, this was just one of many unsolved mysteries of cell division.

Peter Hepler, a scientist at the University of Massachusetts, seems to have found the answer to the riddle of the disappearing membrane. By using the electron microscope to study cells just after the disappearance of the membranes, he has identified a large number of small vesicles, bounded by membranes, that appear

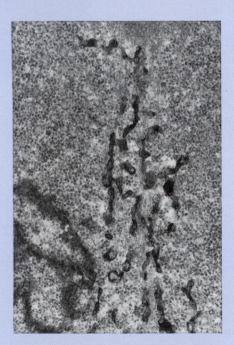

These small vesicles, seen in an electron micrograph of a plant cell in prophase, are the remnants of the nuclear membranes.

late in prophase of mitosis. Late in anaphase, these vesicles disappear. It seems that the nuclear membranes don't dissolve after all. Instead, the material that makes up these membranes becomes rearranged into a series of smaller structures. At the end of mitosis, these smaller structures are brought back together to assemble the nuclear membranes of each daughter cell. At the level of the light microscope, which is usually how we look at mitosis, the membranes seem to disappear, but the fact is that they simply change into a form that keeps them out of the way during cell division.

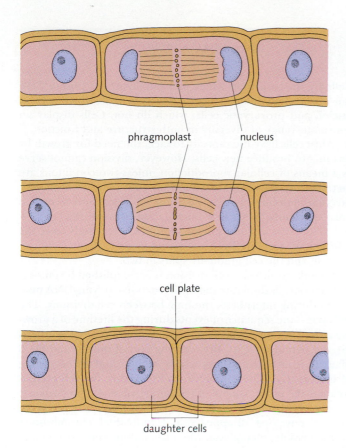

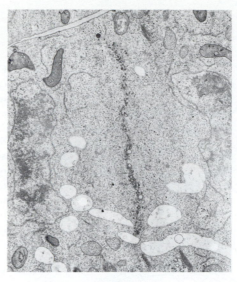

phragmoplast   nucleus

cell plate

daughter cells

Figure 9.19 A diagrammatic view (left) of phragmoplast and cell plate formation in plant cells. The phragmoplast (right) forms from hundreds of small vesicles. These vesicles fuse to form a cell plate, thereby completing the process of cytokinesis.

Control of Cell Division

No large organism could survive without strict controls over the rates of cell division. Some cells, such as those in the nervous and muscular systems of individuals, never divide after the embryonic stage. Others, such as those lining the digestive system, divide regularly to replace cells lost to wear and tear. Still others maintain low rates of cell division until they are stimulated. When part of the liver is destroyed by surgery or disease, healthy cells in the remaining portions divide rapidly until the liver returns to its normal size. The rate of cell division then slows down to its usual level. How is the rate of cell division controlled?

Contact inhibition   One of the key factors in regulating cell division is the physical contact of neighboring cells. The importance of cell contact can be illustrated by a simple laboratory experiment. Mammalian cells in a culture dish grow until they form a single layer, as shown in Figure 9.20. Cell division stops when the layer is only one cell thick. However, if cells are removed from the center of the dish, the cells surrounding the empty area enter the cell cycle and divide rapidly until that empty space is filled. Cell division stops when cells on both sides of the empty space make contact with each other. This contact-dependent regulation of cell growth and division is known as **contact inhibition.** The cellular signals that produce contact inhibition seem to be activated by the binding of protein molecules on the surface of one cell to receptor molecules on its neighbor. If these molecules are disrupted, another round of cell division takes place.

External controls on cell division   The simple experiment outlined in Figure 9.20 can be compared to the healing of a small wound in the skin. Cells on the edges of the wound are stimulated to divide, and rapid growth fills the wound area. When the healing is complete, cell division slows down and returns to its normal rate. Biologists have discovered that chemical factors known as **growth factors** or **mitogens** ("*mitosis-gen*erators") are responsible for such rapid changes in the rate of cell division. A host of different growth factors exist; some affect skin cells involved in the healing of wounds, some regulate the growth and development of nerve cells, and others control the rate of cell division in blood-forming cells. Still other factors, which are less well-defined, inhibit cell division and slow down the rate of cell growth.

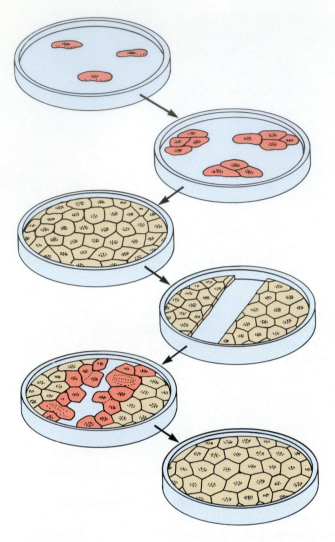

Figure 9.20 When cells are grown in culture, they often stop dividing when they make contact with each other. This is known as contact inhibition.

Cells are the basic units which make up living organisms. Cells can be classified into two types: eukaryotic cells, which contain **nuclei,** and prokaryotic cells, which do not. Cells display an enormous range of diversity in both structure and function.

Most cells have the capacity to meet the need for growth by dividing to produce new cells. However, division cannot serve as a means of cellular reproduction unless two conditions are met: First, a cell must make a duplicate copy of the information-carrying molecules that its offspring will need, and second, it must efficiently divide that information between the "daughter" cells. After cellular information has been duplicated, the information-carrying DNA molecules move to the two ends of the cell, and the cell gradually splits in two.

In eukaryotic cells, cell division is accomplished by mitosis. Eukaryotic cells duplicate their information-carrying DNA molecules during interphase, the time between cell divisions. The cell cycle—the sequence of events during the lifetime of a growing cell—has four principal phases: mitosis (M), DNA synthesis (S), the period or "gap" between the end of mitosis and the S phase (G_1), and the gap between the S phase and the beginning of mitosis (G_2). The timing of events during the cell cycle is critical to the proper control of cellular growth and reproduction.

Mitosis also has four distinct phases: prophase, when the nuclear envelope disappears and individual chromosomes become visible; metaphase, when the mitotic spindle arranges chromosomes along the central plane of the cell; anaphase, when the chromatids separate from each other and move toward the poles of the cell; and telophase, when the chromosomes reorganize into two clusters and two distinct nuclei begin to form. Following these four phases is cytokinesis, the process by which the cytoplasm of the cell separates into two, completing the process of cell division.

The rate of cell division in a multicellular organism is closely regulated. Cell division can be stimulated by growth factors, which may be released when rapid cell division is necessary to heal a wound or complete a process of development. If control over cell division is lost, cells may proliferate wildly, resulting in a disease known as cancer.

Cancer: when control is lost The importance of maintaining control over the rate of cell division cannot be overemphasized. When that control fails, a disease known as **cancer** results. Normal body cells become cancer cells when they lose the ability to respond to the normal controls that regulate cell division. Cancer cells do not grow or divide more rapidly than normal cells. The problem is that cancer cells do not know when to stop. By ignoring inhibitory factors, cancer cells form large masses known as **tumors,** which damage the body by absorbing nutrients and interfering with the functions of vital organs. Cancer is a leading cause of death in developed countries, and we will discuss this disease in detail in Chapter 38.

After studying this chapter, you should be able to:

- Explain what must be accomplished in the precise formation of new cells.
- Describe the simple process of cell division in prokaryotes.
- Outline the phases of mitosis, and discuss the events that occur within the cell during mitosis.
- Describe how the rate of cell division is controlled.

nucleus *p. 165*
binary fission *p. 170*
mitosis *p. 170*
chromosomes *p. 170*
interphase *p. 170*
centriole *p. 172*
mitotic spindle *p. 172*
chromatids *p. 172*
centromere *p. 172*
prophase *p. 172*

metaphase *p. 174*
anaphase *p. 174*
telophase *p. 175*
cytokinesis *p. 175*
cleavage furrow *p. 175*
cell plate *p. 175*
contact inhibition *p. 177*
mitogens *p. 177*
cancer *p. 178*
tumors *p. 178*

REVIEW

Discussion Questions

1. What are some of the differences between cell division in eukaryotes and in prokaryotes?

2. What are the four phases of the cell cycle and the four phases of mitosis? Draw a sketch of a unified cell cycle that displays each of these stages.

3. Distinguish between mitosis and cytokinesis. Why is the process of cell division sometimes not completed at the end of telophase?

4. How does the process of cytokinesis differ between plants and animals? What difference between a typical plant cell and a typical animal cell does this distinction point out?

Objective Questions (Answers in Appendix)

5. DNA is replicated
   (a) during interphase.
   (b) immediately before telophase.
   (c) immediately after telophase.
   (d) as part of cytokinesis.

6. During the division of cytoplasm, a cell plate appears in
   (a) plant cells.
   (b) animal cells.
   (c) both plant and animal cells.
   (d) neither plant nor animal cells

7. The mitotic spindle
   (a) holds the cells together.
   (b) helps in the movement of proteins from the cytoplasm to the nucleus.
   (c) attaches to the cell plate and the cell membrane.
   (d) is a network of fibers to which the chromosomes are attached.

8. The number of chromatids per chromosome is
   (a) one.   (c) three.
   (b) two.   (d) four.

9. The nuclear envelope disintegrates and the contents of the nucleus are released into the cytoplasm
   (a) at the end of the S phase.
   (b) at the end of prophase.
   (c) during metaphase.
   (d) at the beginning of telophase.

CHAPTER 10

Genetics: The Science of Inheritance

For thousands of years people have selectively bred plants and animals. The idea of selective breeding is simple. One chooses a few organisms with desirable characteristics to be the parents of a new generation. When that generation is grown, the hardiest plants, the fastest racehorses, or the most obedient dogs are selected for the next round of breeding. It's simple in practice and simple in theory. Offspring tend to resemble their parents, so by patiently choosing parents with the proper characteristics, the breeder can accentuate those characteristics.

Is inheritance really that simple? For hundreds of years, people regarded inheritance as a *blending*—the characteristics of both parents were thought to blend to produce offspring. In most cases, the blending explanation seems to make sense. Most of us look a little like our mothers and a little like our fathers. A cross between a large dog and a small dog produces medium-sized dogs. Everything seems to make sense. Well, almost everything.

INHERITANCE: A PROBLEM

Many people enjoy the common parakeet *(Melopsittacus undulatus)*, an attractive bird native to Australia. Because you have a preference for birds with green wings, you select a matched pair of such birds and place them in a breeding cage at home. The birds get along, conditions are right, and before long they are the parents of five little parakeets. But something is wrong. The new generation is not a simple blend of the two green parents. In fact, some of the birds don't look anything like their parents at all.

Two of the new birds are green, one is blue, one is bright yellow, and one is completely white (Figure 10.1)! Where did the new colors come from? And why don't all of the new birds look like their parents? More to the

Figure 10.1 These five parakeets are the offspring of two green-colored parents.

point, what should you do now? You might rush back to the pet store, complaining that the green parakeets do not "breed true." Or you could wonder whether another parakeet had slipped into the cage while you were away.

However, there is something else that you could do. Puzzled by the results of this cross, you might *experiment*. You could conduct more breedings and try to learn the rules by which parakeets inherit their color. You would need a much larger cage, some extra time, and a lot of birdseed. But if you had done this 150 years ago with patience and care and insight, you just might have founded the science of **genetics,** the science of inheritance. As luck would have it, someone else got there first.

MENDEL AND THE BIRTH OF GENETICS

As we saw in Chapter 8, Darwin realized that the blending theory of inheritance presented special problems for evolution. Specifically, he wondered whether favorable characteristics might be "blended away" before natural selection had a chance to increase their frequency in a population. Ironically, and unknown to Darwin, this problem was being solved just as *On the Origin of Species* was being published. This ground-breaking work, performed by an Austrian priest named Gregor Mendel, set the stage for a radically new way of thinking about heredity. But because Mendel's work did not receive widespread recognition until the early twentieth century, Darwin probably was unaware of it.

Gregor Mendel was born in 1822 in the town of Heinzendorf. A bright student, he entered the Augustinian monastery at Brünn, which, like his birthplace, is now part of Czechoslovakia (Figure 10.2). Mendel became a priest when he was 25. He had a special interest in natural science and took an exam to qualify for a science teaching certificate. Although he failed that exam, his superiors at the monastery thought enough of the young priest to send him to the University of Vienna for two years to study science and mathematics. When he returned to the monastery he was given two assignments: to teach physics and biology in a local high school and to supervise the monastery gardens. We do not know how well he fulfilled the first assignment. But we know a great deal about his work in the garden.

Mendel took a special interest in the garden's peas. The other gardeners helped Mendel isolate several strains of plants with distinct characteristics: One always produced tall plants, another only short plants; one always produced purple flowers, another only white flowers. Mendel selected plants that bred true—that is, each variety produced seeds that grew into plants identical to the parent.

Figure 10.2 Gregor Mendel (1822-1884)

Mendel knew that peas are self-fertilizing. Because of the structure of the pea flower, the pollen that produces the male reproductive cells usually fertilizes female reproductive cells in the very same flower. To cross one plant with another, Mendel opened the pea flowers and removed the pollen-producing anthers. He then dusted the female portions of the flower with pollen from another plant (Figure 10.3). In this way, he could carry out controlled pollination from one plant to another—a process known as *crossing*. Because each plant produced many seeds, Mendel could obtain reliable statistics on the outcome of each cross.

Mendel's First Experiments

In one experiment, Mendel fertilized the flowers of a strain that produced purple flowers with pollen from plants that produced white flowers. He also reversed the process by using pollen from purple-flowered plants to fertilize the white-flowered strain. One of his first observations was that it did not matter which plant contributed the pollen and which received it.

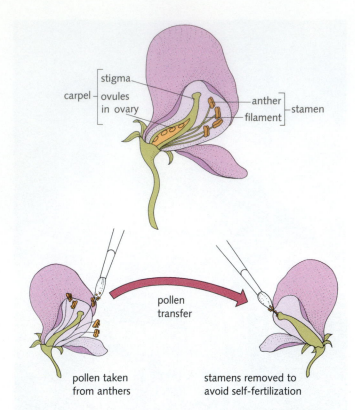

Figure 10.3 Although pea plants are normally self-fertilizing, Mendel carried out crossbreeding by brushing the pollen of one plant on the stigma of another.

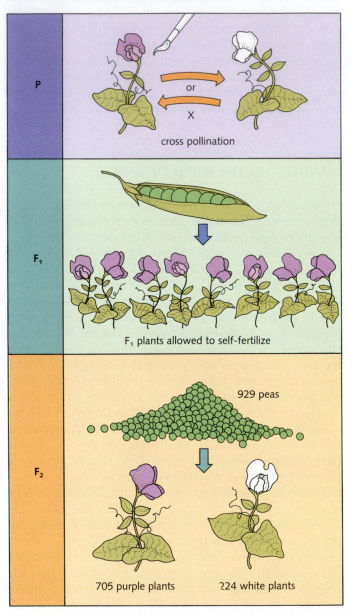

Figure 10.4 Mendel's first experiments involved the crossing of parental (**P** generation) plants with white flowers and plants with purple flowers. The seeds produced by this cross (**F₁** generation) grew into plants which produced purple flowers. When these plants were allowed to cross naturally, both white and purple flowers were produced in the **F₂** generation.

The original plants involved in the cross are known as the **parental,** or P, generation. The seeds produced from this first cross Mendel called the **first filial,** or F₁, generation (*filial* comes from the Latin root for "son"). All of these seeds produced plants that bore purple flowers, so it seemed that the white-flower trait had disappeared. Mendel took the seeds from this cross, planted them, and allowed the resulting plants to self-fertilize. This F₁ × F₁ cross produced the **second filial,** or F₂, generation. Mendel collected these seeds and planted 929 of them. To Mendel's surprise, the white flowers reappeared in some of these plants; 224 plants bore white flowers and 705 had purple flowers, a ratio of purple to white flowers of 3.1 to 1 (Figure 10.4).

A Theory of Particulate Inheritance

At first, these results must have seemed every bit as confusing as the blue and yellow parakeets mentioned in the opening pages of this chapter. But Mendel was able to explain them. He proposed a system of **particulate inheritance** in which heritable characteristics were controlled by individual "units." Mendel assumed (correctly, as it turned out) that each plant had two such units for each trait. Mendel called each unit a "Merkmal," the German word for "character." Today we call these units **genes,** and we know that the cells of the pea plant carry two genes for most characteristics. If we represent the units with symbols, we might use *P* for Mendel's purple-flower character and *p* for the white-flower character.

P and *p* are known as **alleles,** which are alternate forms of a single gene, in this case the gene for flower color. In Mendel's first cross, the original, true-breeding purple-flowered parent would have been *PP*, and the original white-flowered parent would have been *pp*.

One of Mendel's most extraordinary insights was the realization that when organisms produce their reproductive cells, or **gametes,** each gamete carries only one allele for each gene. Therefore, the two alleles for each gene are **segregated** from each other when gametes are formed. In Mendel's experiment, for example, the F_1 generation received an allele for purple flowers from one parent and an allele for white flowers from the other parent. In this way, each parent makes an individual genetic contribution to its offspring.

The Punnett square We can visualize the alleles involved in each cross by making a diagram known as a **Punnett square** (Figure 10.5). On one side of the square we write down all the *gametes* that can be formed by one parent, and on the other side all the gametes that can be formed by the other parent. Then we use the blocks within the square to represent all the possible *combinations* of alleles that may occur in the offspring. When two plants in the F_1 generation are crossed, as shown, the result is that, on average, 1/4 of the F_2 offspring have the *PP* alleles for flower color, 1/4 are *pp*, and 1/2 are *Pp*. The Punnett square is a powerful predictive tool. In this case, the square predicts that 3/4 of the offspring from the $F_1 \times F_1$ cross will be either *PP* or *Pp*.

Dominance

One of the remarkable results of the crosses shown in Figures 10.4 and 10.5 is the fact that plants with contrasting alleles (*P* and *p*) produce purple flowers. The purple allele of the flower-color gene is **dominant** over the white allele. Mendel carried out crosses with seven other characteristics, including plant size, flower color, and seed shape (Figure 10.6). In each case, one characteristic was "dominant" in the F_1 generation. The characteristic that seemed to disappear in the F_1 generation and to reappear in the F_2 generation he called a **recessive** trait. Mendel explained this phenomenon by proposing that whenever a dominant allele and a recessive allele

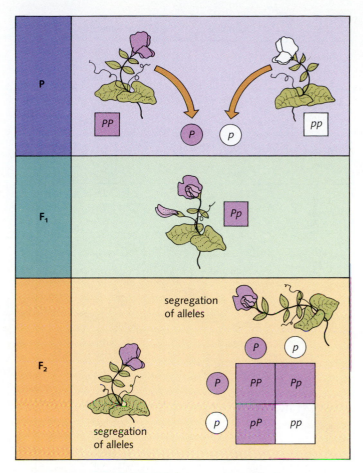

Figure 10.5 Mendel explained the results of his experiment by assuming that each pea plant carried two "characters," or alleles, for flower color. In the F_1 generation, each plant inherits an allele for purple (*P*) and an allele for white (*p*). When the F_2 generation is formed, these alleles are segregated as gametes are produced. By analyzing the possible combinations of these gametes using a Punnett square, we can predict the allele combinations that the F_2 generation will inherit.

were found together in the same organism, the dominant allele alone controlled the appearance of the plant. Only when two recessive alleles occurred together did the recessive characteristic, or "character," emerge.

Now we can look at the results of the crosses in Figure 10.4 and appreciate the value of Mendel's data. Because both *PP* and *Pp* plants produce purple flowers, the Punnett square predicts a 3:1 ratio of purple to white flowers in the F_2 generation. Mendel's 3.1:1 ratio was a close match.

The same 3:1 ratio held for crosses involving the seven other traits. For example, when Mendel crossed plants producing round seeds (alleles: *RR*) with those producing wrinkled seeds (alleles: *rr*), all the F_1 plants produced

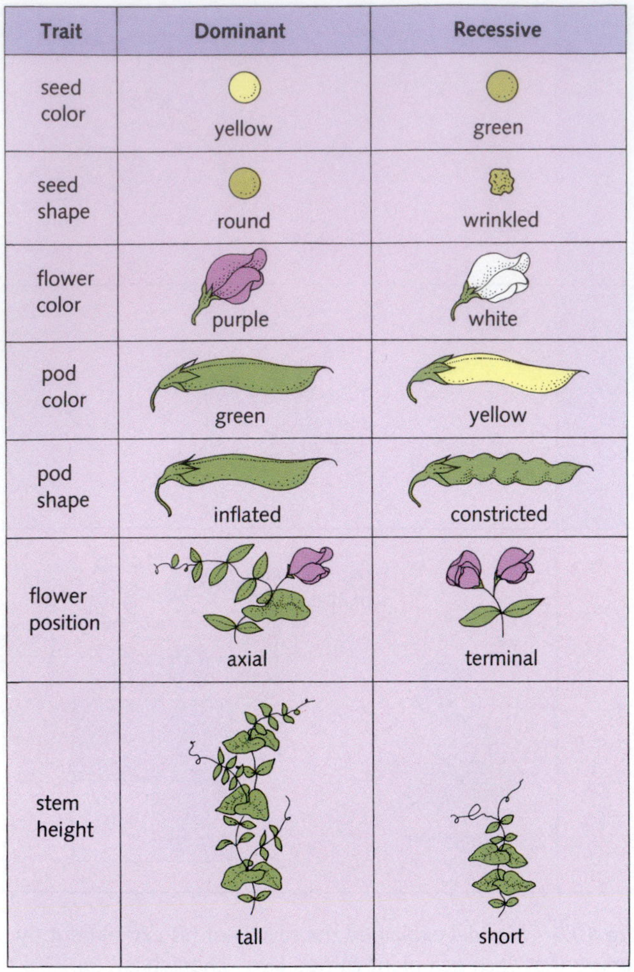

Trait	Dominant	Recessive
seed color	yellow	green
seed shape	round	wrinkled
flower color	purple	white
pod color	green	yellow
pod shape	inflated	constricted
flower position	axial	terminal
stem height	tall	short

Figure 10.6 The seven contrasting traits investigated by Mendel.

round seeds, showing that the *R* allele is dominant over the *r* allele. When the *Rr* F₁ plants were crossed with themselves, the F₂ generation included 5474 plants that bore round seeds and 1850 plants with wrinkled seeds, a ratio of 2.96:1. Obviously, Mendel's patience in counting seeds knew no bounds!

Mendel's First Principles

It is hard for us today to realize just how revolutionary Mendel's theory of particulate inheritance was. With only the evidence of his garden pea crosses, he established the first two of three important principles on which the science of genetics is founded:

- The characteristics of an organism are determined by individual units of heredity called genes. Each adult

organism has two alleles for each gene, one from each parent. These alleles are *segregated* (separated) from each other when reproductive cells are formed. This is known as the principle of **segregation.**

- In an organism with contrasting alleles for the same gene, one allele may be *dominant* over another (as round is dominant over wrinkled for seed shape in the garden pea). This is known as the **principle of dominance.**

Although Darwin could not take Mendel's principles into account, it is clear they solved the problems presented by a blending theory of inheritance. When a new and beneficial allele appears in a population, the characteristic that allele produces won't disappear by repeated "blending." Instead, the allele is preserved from one generation to the next. Even a recessive characteristic can appear in a later generation through the changing gene combinations generated by sexual reproduction.

Genotype and Phenotype

A second major contribution of Mendel's experiments was the idea that every organism has a *genetic* makeup called its **genotype.** The actual characteristics an organism exhibits are called its **phenotype.** The genotype is inherited, whereas the phenotype is produced under the influences of the environment and the genotype. Plants may have the same phenotype and different genotypes. For example, about 2/3 of the 705 purple-flowered plants of the F₂ generation in Figure 10.4 have the *Pp* genotype, and 1/3 have the *PP* genotype.

Plants with two identical alleles for the same gene (*PP* or *pp*) are said to be **homozygous** for that gene. Plants with two contrasting alleles for a gene (*Pp*) are **heterozygous** for that gene. How can we determine which of the 705 purple-flowered plants are homozygous for the purple allele of the flower-color gene and which are heterozygous? Mendel developed a simple technique known as a **test cross** that enabled him to determine the genotype of any plant. The plant in question is crossed with another plant that is homozygous for the recessive version of the gene in question. The phenotypes of the offspring from the test cross then reveal the genotypes of the parents.

Figure 10.7 shows how this works in the case of plants with purple flowers. One test cross produces only plants with purple flowers, proving that the plant used for that cross had the *PP* genotype. The second test cross produces 9 plants with purple flowers and 11 with white flowers (roughly a 1:1 ratio), proving that the original genotype was *Pp*. The test cross is a powerful technique that can be used to determine genotype in animals as

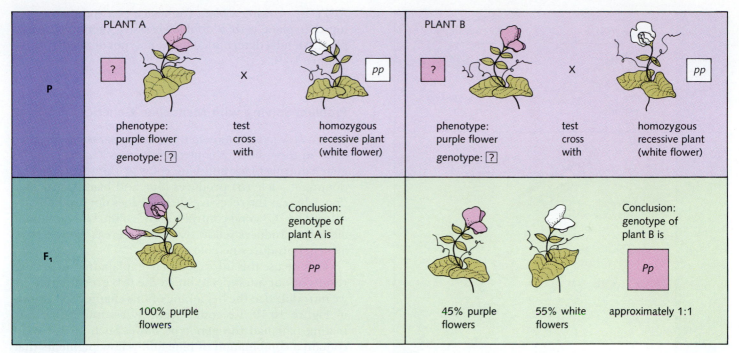

Figure 10.7 Genotypes of plants can be determined in a test cross with plants containing homozygous recessive alleles. The characteristics of the offspring of a test cross reveal the genotypes of the parents.

well as plants. Can you imagine a test cross that might be useful in the case of our parakeets? We'll come back to this problem later.

Independent Assortment

Knowing that Mendel had seven different traits available for study, each of which had two contrasting alleles, what do you think his next experiment might have been? If you guessed that he might have tried to follow the alleles for two different genes at the same time, then you guessed right.

Mendel wondered whether the alleles for different genes would segregate in the same pattern as alleles of a single gene did. He carried out a series of experiments like those shown in Figure 10.8. A parental line of plants with yellow-colored and round seeds was crossed with a line that produced green and wrinkled seeds. The seeds in the F_1 generation were all round and yellow, showing that those two alleles are dominant over the alleles for wrinkled and for green seeds.

If the alleles for the seed coat and seed color genes were able to assort independently during gamete formation, then plants grown from the F_1 generation seeds

could produce four different kinds of gametes:

$$RrYy$$
$$\downarrow$$
$$RY \quad Ry \quad rY \quad ry$$

But these four types of gametes could be produced only if the alleles for the two genes were free to segregate and to assort independently of each other. **Independent assortment** means that which allele a gamete receives for the seed coat gene (**R** or **r**) has no effect on which allele it receives for the seed color gene (**Y** or **y**). The Punnett square for this cross shows that the 16 possible offspring of an F_1 cross display 4 different phenotypes and that these plants are present in a 9:3:3:1 ratio. Mendel's breeding experiments produced 556 seeds whose phenotypes matched these ratios almost exactly. These results show that the R and Y genes assort independently.

■ Each of the seven genes that Mendel investigated obeys the third of Mendel's three important principles, the concept of **independent assortment.**

It's interesting to note that Mendel confirmed the genotypes of each of his F_2 plants by conducting test crosses. A round, yellow pea from the F_2 generation might have any of 4 possible genotypes: *RRYY, RrYY, RrYy,* or

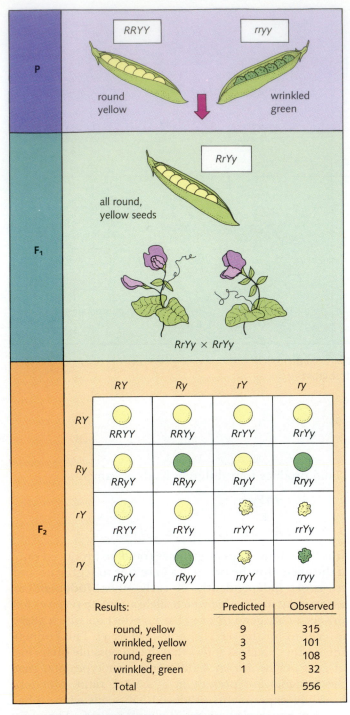

	RY	Ry	rY	ry
RY	RRYY	RRYy	RrYY	RrYy
Ry	RRyY	RRyy	RryY	Rryy
rY	rRYY	rRYy	rrYY	rrYy
ry	rRyY	rRyy	rryY	rryy

Results:	Predicted	Observed
round, yellow	9	315
wrinkled, yellow	3	101
round, green	3	108
wrinkled, green	1	32
Total		556

Figure 10.8 A two-factor cross carried out by Mendel. The dominant characteristics of the parents appear in the F₁ generation. When the F₁ plants are crossed among themselves, independent assortment of the alleles for both characteristics produces four distinct phenotypes in the F₂ generation with a 9:3:3:1 ratio.

RRYy. By growing a plant from each pea and then crossing that plant with a double-recessive (genotype: *rryy*) plant, Mendel determined the genotypes of the offspring (Figure 10.9).

Problem Solving with Mendelian Genetics

We are now in a position to think about parakeets. Feather color in these birds is controlled by two genes. The "B" gene controls black and blue color in the feathers. The dominant allele (*B*) produces blue and black pigmentation, whereas the recessive allele (*b*) does not produce any color. The "C" gene controls yellow color. The dominant allele (*C*) produces yellow feathers, whereas the recessive allele (*c*) produces no color.

Can we use this information to explain the surprising results of our attempts to breed the two green parakeets we introduced at the beginning of this chapter? As shown in Figure 10.10, we could begin by assuming that our mating pair had the genotype **BbCc.** Such birds would contain a combination of blue and yellow pigments that would appear green. By using a Punnett square to analyze the gametes that such birds would produce, we can predict the outcome of a cross between our two parents. As you can see, we would expect about 1/16 of the offspring to be white (genotype: **bbcc**), so the single white bird in our cross can be explained as a result of Mendelian genetics. We would also expect green, yellow, and blue birds in the offspring, and that prediction came true. Just for fun, you might plan a way to test our explanation of the parental genotypes with a test cross. One of the 5 parakeets that resulted from crossing the two parental green parakeets is the ideal bird to use for such a test. Can you identify it? (See Appendix for the answer.)

Genetics and the Cell

Mendel's three important principles, of *dominance, segregation,* and *independent assortment,* place certain restrictions on the way genes and their alleles behave. For example, the fact that individuals inherit one allele for each gene from each parent implies that the cells of an adult organism contain two sets of alleles. Similarly, because segregation requires that the two alleles for each gene be separated when gametes are formed, there must be a mechanism that accomplishes this separation. What an opportunity! To Mendel, the "gene" was a purely hypothetical factor that controlled heredity. But if genes were real, then we should be able to find physical structures within the cell that behave in accordance with Mendel's three principles.

Unfortunately, no one was ready to seize upon Mendel's work when it was published. It was, after all, just

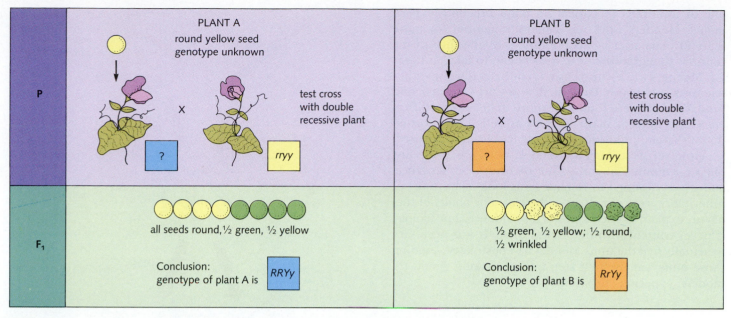

Figure 10.9 Test crosses with plants homozygous recessive for two characteristics reveal the genotypes of the parent plants.

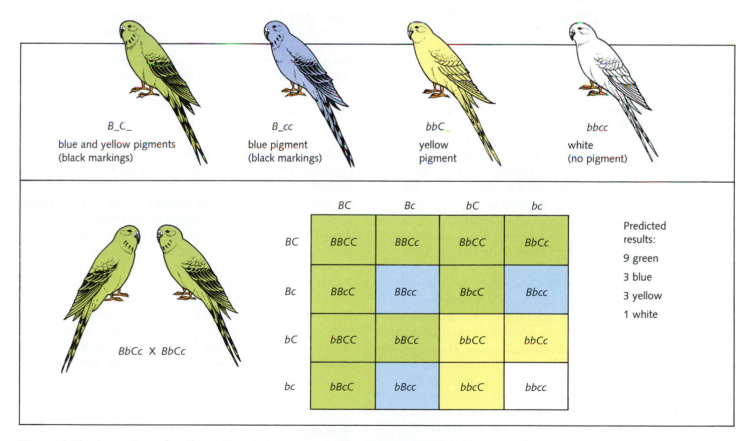

Figure 10.10 A genetic explanation using a Punnett square of how two green-colored parakeets could produce offspring with four different color patterns. Each of the green parents is heterozygous for the two genes that control feather color. The Punnett square shows how birds with four different colorations may be produced by the cross.

one of hundreds of papers on plant breeding experiments. Mendel's work was generally ignored for more than 30 years after it was published. By the turn of the century, though, other scientists began to take an interest. Mendel's work with peas was confirmed by two other scientists (Erich von Tschermak and Carl Correns), and similar results were found for more than a dozen different plants and for many animals as well. There was no doubt that an important principle had been discovered.

However, as striking as Mendel's work was, it was just one of a number of competing theories of inheritance, and it was destined to remain so as long as his units of inheritance were strictly hypothetical. The ability of theory to explain events is powerful evidence, but biology is never content with theoretical units. It was necessary to actually *find* the units Mendel had postulated. Finding those units was the first step toward making genetics a modern experimental science.

MEIOSIS: FORMING A NEW ORGANISM

As we saw in Chapter 9, during mitosis each chromosome of the cell appears as a pair of *chromatids*. As mitosis proceeds, the chromatids of each chromosome separate and two daughter cells are formed, each containing the same number of chromosomes as the original cell. To many biologists, the fact that chromosomes were duplicated just prior to mitosis and then carefully separated into the two daughter cells suggested that chromosomes were important structures. Some even suggested that they contained hereditary information.

The critical evidence, however, came from studies of what happened to chromosomes during the formation of reproductive cells, or *gametes*. In sexually reproducing organisms, a new individual is formed by the fusion of male and female gametes to form a single cell known as a **zygote.** The zygote then goes through a process of development and growth to become an independent organism.

Gametes contain only *half* the number of chromosomes found in other cells of the body. The sperm and egg cells of cats, for example, contain 19 chromosomes each. A fertilized zygote, about to develop into a kitten, contains a total of 38 (19 + 19) chromosomes. How does this work? Does the sperm contribute chromosomes *A, B, C,* and *D* while the egg contains *E, F, G,* and *H*? No. In fact, with one exception that we will encounter shortly, the two sets of chromosomes are very similar to each other. A better way to describe it would be to say that the egg contributes *A, B, C,* and *D* while the sperm donates *A', B', C',* and *D'.* For each chromosome contributed by one parent, a corresponding chromosome, or **homologous chromosome,** is contributed by the other parent.

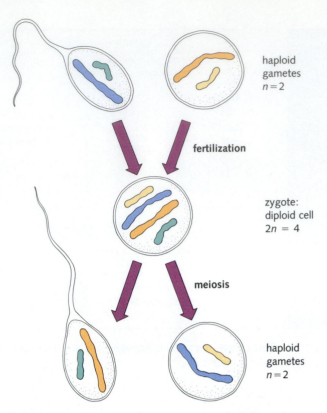

Figure 10.11 A diploid organism contains two complete sets of chromosomes, one from each parent. The contributions of each parent are illustrated in the formation of a fertilized zygote in this figure. When the diploid organism reproduces, it too will produce haploid gametes.

Each cell of an adult cat contains two sets of homologous chromosomes; one set is derived from the male parent and one from the female.

Biologists use the term **diploid** to describe a cell that contains two sets of homologous chromosomes. The *somatic* (body) cells of peas, cats, and humans are diploid. Their gametes, however, which contain only a single set of chromosomes, are **haploid** (Figure 10.11). We use the symbol n to represent a single set of chromosomes. Diploid cells, therefore, are $2n$ and haploid cells are n. (The term *haploid* does not mean that such cells have *half* a set of chromosomes. In fact, haploid cells have a full, single set. A better term for n might be *monoploid*, but *haploid* and *diploid* are now firmly entrenched in the scientific vocabulary.)

Reduction Division

Haploid gamete cells are formed by a special process that is often called *reduction division* because it reduces the chromosome number of cells that are developing into sperm or eggs. Reduction division is commonly known as **meiosis,** and it occurs in all organisms that reproduce sexually. Meiosis generally takes place over the course of two rounds of cell division. It may resemble mitosis, but there are big differences in how chromosomes behave in the two processes.

Meiosis is a two-stage process that produces four haploid cells from a single diploid cell. Homologous chromosomes are segregated from each other in the **first meiotic division.** The **second meiotic division** is not preceded by an *S* phase of the cell cycle, so DNA and chromosome duplication do not occur between the two divisions. This results in a reduction of chromosome number. Figure 10.12 illustrates the features of meiosis.

One of the first organisms in which meiosis was intensively studied was the ordinary fruit fly, *Drosophila melanogaster* (*Drosophila* means "dew lover" and *melanogaster* means "dark belly"). Diploid *Drosophila* cells have eight chromosomes ($2n = 8$). Four of these chromosomes were originally provided by the fly's mother and four by its father. We shall use a female *Drosophila* to trace the details of meiosis (Figure 10.13).

The first meiotic division   Prior to meiosis, each of the eight chromosomes is duplicated, so we now have eight chromosomes, each consisting of two paired, identical chromatids. This duplication is no different from that which occurs before an ordinary cell division. However, as the process continues, something interesting happens: The chromosomes *pair* to form bundles known as **tetrads** (the same structure is sometimes referred to as a *bivalent*) composed of four chromatids each. The pairing of homologous chromosomes, also known as **synapsis,** produces an appearance at metaphase that is very different from a mitotic metaphase. As shown in Figure 10.13, a *Drosophila* cell at metaphase I of meiosis has four tetrads.

Which chromosomes pair to form the tetrads? The *homologous* chromosomes derived from both parents. Chromosome *A* from the mother pairs with *A'* from the father, *B* with *B'*, and so on. At metaphase, all four tetrads are aligned across the center of the cell.

Anaphase begins as the tetrads draw apart and the homologous chromosomes are separated. Each daughter cell receives one chromosome from each pair. The alignment of each chromosome pair is random with respect to the others. In other words, the fact that the maternal copy of one chromosome pair moves to a particular cell does not affect the way the maternal copy of a different chromosome pair goes. The chromosomes exhibit *independent assortment*!

The first meiotic division (meiosis I) produces a random shuffling of the genetic deck: Each cell has an equal chance of getting either the maternal or the paternal copy of any individual chromosome. As you know, Mendel's principle of *segregation* states that the maternal and

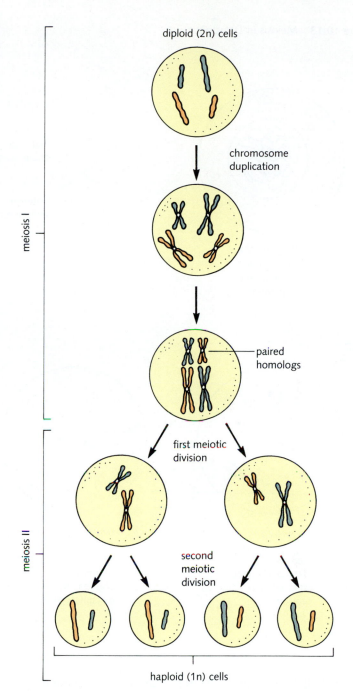

Figure 10.12   An overview of meiosis in an organism with a diploid chromosome number of 4.

Figure 10.13 Meiosis in *Drosophila*.

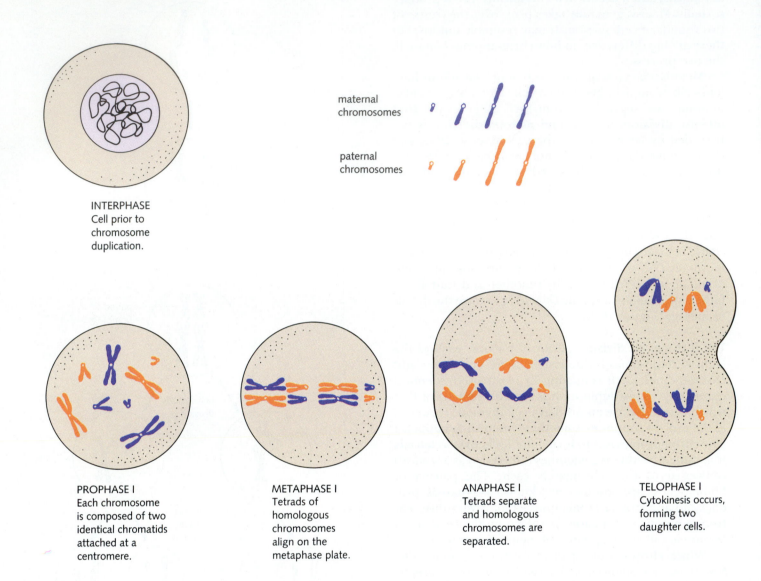

maternal chromosomes

paternal chromosomes

INTERPHASE
Cell prior to chromosome duplication.

PROPHASE I
Each chromosome is composed of two identical chromatids attached at a centromere.

METAPHASE I
Tetrads of homologous chromosomes align on the metaphase plate.

ANAPHASE I
Tetrads separate and homologous chromosomes are separated.

TELOPHASE I
Cytokinesis occurs, forming two daughter cells.

paternal alleles of a gene are segregated during gamete formation. The pairing of homologous chromosomes during the first meiotic division makes that possible. As we observe anaphase of the first meiotic division, we are watching the microscopic events that underlie the segregation of alleles. Mendel never worked with a microscope, but microscopic observation confirms that chromosomes obey the principles of independent assortment and segregation.

The second meiotic division   Each of the two cells produced by the first meiotic division now enters a second meiotic division (meiosis II). The cell cycle that occurs

between the first and second meiotic divisions is exceptional, because the chromosomes are *not* duplicated during it. Each of our two *Drosophila* cells received four chromosomes (with two chromatids each) from the first division, and these two cells now enter another round of division with exactly the same number of chromatids. In metaphase, four chromosomes are visible in each cell, and anaphase separates these four chromosomes into two new cells, each of which now receives four single chromatids. The four cells formed by the second meiotic division are true haploid cells, each containing four distinct chromosomes ($n = 4$). Although the cellular details of meiosis differ in male and female animals (a sperm cell

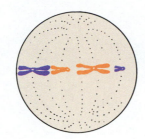

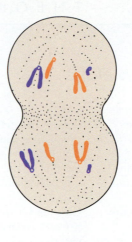

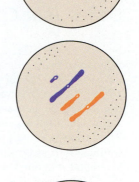

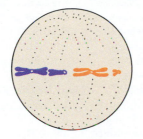

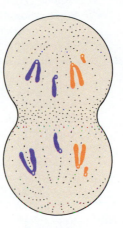

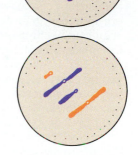

PROPHASE II
Each daughter cell
has one chromosome
from each pair;
chromosomes migrate
toward metaphase plate.

METAPHASE II
Chromosomes align
on metaphase plate.

ANAPHASE II
Centromeres of
chromatids separate,
and individual
chromosomes move
toward opposite
poles of each cell.

TELOPHASE II
Cytokinesis occurs,
forming four daughter
cells, each with
haploid number of
chromosomes.

is produced quite differently from an egg cell), the nuclear details of meiosis are similar in both sexes and in plants as well as animals.

Meiosis and the Rise of Mendelian Genetics

An American graduate student, Walter Sutton, and the German biologist Theodor Boveri were the first to appreciate the fact that meiotic chromosomes behave *exactly* as one would expect for Mendel's genes. In 1902 they were bold enough to say that Mendel's genes "are on chromosomes." They proposed a model in which genes were located on chromosomes like beads on a string. And they were right. Mendelian genetics was rapidly accepted because studies on chromosome behavior during cell division and meiosis showed that the Mendelian model was essential to account for the inheritance of genetic units located on chromosomes and passed from one generation to the next by means of haploid reproductive cells.

Mendelian genetics was also accepted because of its ability to *predict* the results of experimental crosses. In the remainder of this chapter, we will investigate some aspects of Mendelian genetics and examine how the laws of genetics can be used to solve problems in inheritance—and even to predict genetic disorders.

Genes That Cheat at Meiosis

Why do so many organisms use sexual forms of reproduction? The usual answer is that sex is a method of reshuffling the genes in a population and results in new gene combinations. The production of as many gene combinations as possible increases the likelihood that one will be formed that will be favored by natural selection. However, sexual reproduction depends on the "fairness" of the mechanisms of gene sorting when gametes are formed.

We have seen that during meiosis, homologous chromosomes of an organism pair to form tetrads. The alignment of chromosomes during metaphase is random, which gives each cell roughly a 50–50 chance of getting either the maternal or the paternal homologue when the first meiotic division occurs. Natural selection can then act on the organisms whose gene composition is decided by the chance mechanisms of meiosis. All of this is true—assuming that nobody's *cheating*. Is it possible for genes to "cheat" during meiosis?

A gene that cheated during meiosis would have some way of making sure that the sorting of genes was not random. A cheating gene might be able to make certain that it was *always* included in reproductive cells, ensuring that it, and not its homologue, was passed along to succeeding generations. How could it do this? James F. Crow and his associates at the University of Wisconsin have discovered exactly such a system in *Drosophila*.

The genetic system is known as **segregation distorter,** or *sd* for short. The genes that produce *sd* are capable of producing a kind of genetic warfare during meiosis—one which ensures that chromosomes carrying the *sd* genes, and *only* the *sd* genes, are passed to the next generation. When *sd* is present in the heterozygous form, it actually sabotages the homologous chromosome, the one that doesn't have *sd*. The sabotage seems to occur at the beginning of meiosis, and the results are evident a few days later when sperm cells begin to develop. The accompanying photo shows the effect of the sabotage. *Drosophila* sperm are normally produced in packets of 64 cells each. The electron micrograph shows *sd* "cheating" in action: 32 of the cells have begun normal development, but 32 are fail-

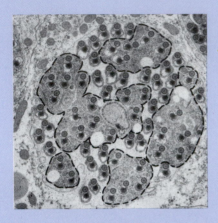

ing to develop. Which 32? The ones that lack *sd*. Somehow, the *sd* gene has destroyed every cell containing its competition in the next generation! It actually breaks the rules of meiosis; it cheats!

Despite its seemingly devious behavior, the *sd* gene is not very common in wild *Drosophila* populations. Why? Because it causes a number of destructive changes in other genes that reduce the fitness of the flies that contain it. Its cheating behavior may be the only thing that protects it from the rigors of natural selection in the natural population.

Based on an article by James F. Crow, "Genes that Violate Mendel's Rules," *Scientific American,* February 1979.

Gene Linkage and Crossing Over

Although one of Mendel's important principles was the law of independent assortment, geneticists soon discovered that not all genes assorted independently. Some were inherited in groups, as though they were *linked* together. The explanation for *gene linkage* turned out to be a simple one. Genes that are located on the same chromosome are linked because whole chromosomes separated together during meiosis.

One of the first examples of linked genes appeared in an experiment done in 1905 by William Bateson and R. C. Punnett (yes, the inventor of the Punnett square) in Cambridge, England. Their experiments were conducted on peas, and they had developed two lines of plants whose genetic composition they had determined. One line contained a gene for purple flowers (*P*) that was dominant over its allele for red flowers (*p*); it also contained a gene to produce long pollen grains (*L*) that was dominant over round pollen grains (*l*). Their first cross was a mating of plants that were homozygous for both alleles:

$$PPLL \quad \times \quad ppll$$
$$\text{(purple, long)} \quad \text{(red, round)}$$

The cross produced a crop of heterozygotes ($PpLl$) as its F_1 generation. All of these plants produced purple flowers and long pollen grains. When the F_1 plants were crossed among themselves, an F_2 generation was produced containing 381 plants with the following phenotypes:

284	purple, long	21	red, long
21	purple, round	55	red, round

This does not match the 9:3:3:1 ratio we would have expected for such a cross. In fact, the expected and observed numbers of offspring are quite different (Figure 10.14).

Phenotype	Number Observed	Percent Observed	Percent Expected
Purple, long	284	74%	56% (9)
Purple, round	21	6%	19% (3)
Red, long	21	6%	19% (3)
Red, round	55	14%	6% (1)

Why should there be *more* of the purple, long plants and more of the red, round plants than expected? And why *fewer* than expected of the purple, round and of the red, long? Bateson and Punnett did not have an immediate explanation. Neither did scientists wrestling with data on other systems which, like these, suggested that Mendel's ratios did not correspond to reality. They entertained the notion that Mendel had been wrong. Can you think of an explanation for these results?

The most striking thing about the data from this cross is that the phenotypes present in larger-than-expected amounts were **"parental"** phenotypes, in the sense that they resembled the original *P* generation parents. How could this have happened? Suppose that the two genes under study had been "linked" together in some way? Those genes would then not assort independently, and they would be inherited "together."

As shown in Figure 10.15, the gametes produced by the parental plants must contain the linked *P* and *L* genes or the linked *p* and *l* genes. The F_1 organisms, therefore, contain a pair of genes linked in this fashion. What will happen when the F_1 plants produce their gametes? If the genes are unlinked, we would expect four different kinds of haploid gametes to be formed in equal proportions:

PL   Pl   pL   pl

However, if the genes are linked, then only 2 combinations are possible:

PL   and   *pl*

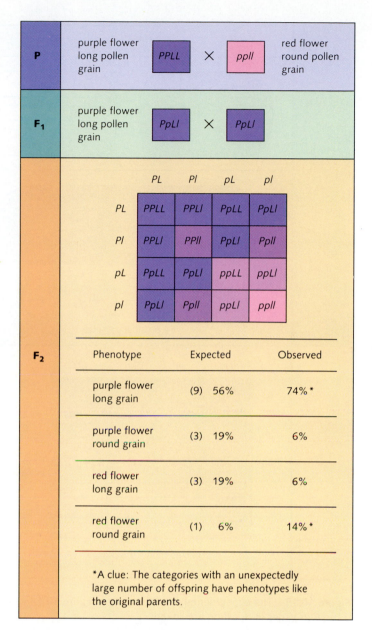

Figure 10.14 An experimental cross of peas showing evidence of gene linkage. If the two genes were unlinked, independent assortment would produce a 9:3:3:1 ratio in the F_2 offspring. Instead, the F_2 offspring have an unexpectedly large number displaying parental phenotypes.

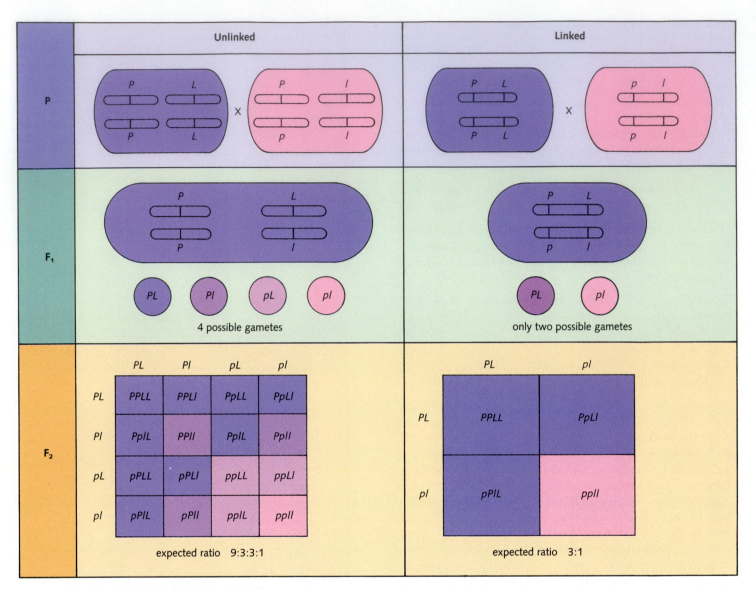

Figure 10.15 The experimental results shown in Figure 10.14 can be partly explained if the two genes in question are linked.

If only two kinds of gametes can be produced, then the possibilities for offspring are severely limited (Figure 10.15). But the simple explanation that these genes might be linked to each other by virtue of being located on the same chromosome was not sufficient. The real data conform to *neither* model:

Phenotype	*Unlinked*	**Real Data**	*Linked*
Purple, long	56%	**74%**	75%
Purple, round	19%	**6%**	—
Red, long	19%	**6%**	—
Red, round	6%	**14%**	25%

Why would these genes behave as though they were partly linked, but not completely? We now know that the answer can be found by watching how chromosomes behave during meiosis. During meiosis, the homologous chromosomes pair to form tetrads.

While the homologues are paired, microscopists have observed exchanges of material between one chromosome and its homologue (Figure 10.16). These exchanges allow genes to cross over from one chromosome to its homologue.

Exchanges like these can occur anywhere along a pair of homologues in synapsis, and the process is sometimes known as **crossing over.** Because the existing genetic

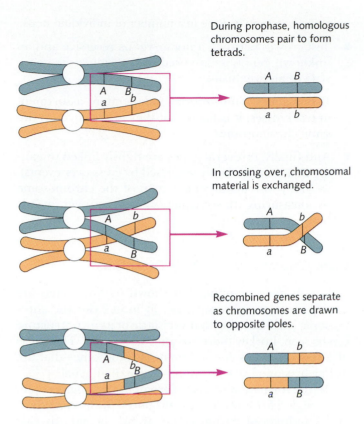

During prophase, homologous chromosomes pair to form tetrads.

In crossing over, chromosomal material is exchanged.

Recombined genes separate as chromosomes are drawn to opposite poles.

Figure 10.16 Linked genes can still be separated if a crossover event occurs between them. Crossing-over results in the formation of new combinations of linked genes. Here a cross-over event occurs between genes A and B during meiosis.

chromosomes. These results, therefore, do not invalidate the Mendelian model for inheritance but rather add another dimension to it.

GENETIC MAPS

Alfred Sturtevant, while still an undergraduate working in a genetics research lab at Columbia University, noticed that although some genes were tightly linked, others were not. The tightness (or the looseness) of the coupling was reproducible from one experiment to the next. It occurred to Sturtevant that the degree of linkage could be explained in a very straightforward way. If the genes were arranged on the chromosomes and if cross-over events occurred at random, then the closer together two genes were, the less likely they were to be separated by cross-over events. Random cross-over events are relatively unlikely to separate two genes that are right next to each other, whereas they are almost certain to separate genes that are at opposite ends of a chromosome (Figure 10.17).

Geneticists quickly realized that the frequency of cross-over events between two genes could be used to measure how far apart two genes were on a chromosome. In other words, they could construct a **genetic map** showing the location of genes on a chromosome (Figure 10.18). Gene

material is *recombined* as a result of crossing over, the process is also called **genetic recombination.** Now we have a mechanism that can be used to explain the results from our genetic cross.

If two genes are located on the same chromosome, then they are **linked** and will be inherited together *unless* a cross-over event occurs between them. As shown in Figure 10.16, if a cross-over event occurs between two linked genes, the genes will be separated onto different chromosomes. Thus linkage can be variable, just as Bateson and Punnett's breeding experiment with peas suggested. The reason why the actual data are intermediate between what we might expect for linked and for unlinked genes is the occurrence of occasional cross-over events that separate the genes onto separate chromosomes.

Crossing over requires us to modify the law of independent assortment. As we have just seen, genes on the same chromosome do not assort independently; they are linked. However, the law of independent assortment does apply to chromosomes themselves, so we may still refer to the independent assortment of genes on different

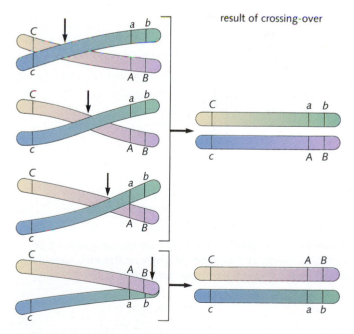

result of crossing-over

Figure 10.17 Because cross-over events occur at random, two genes that are located close together (genes A and B) are less likely to be separated by a cross-over than two genes that are some distance apart (genes C and A for example).

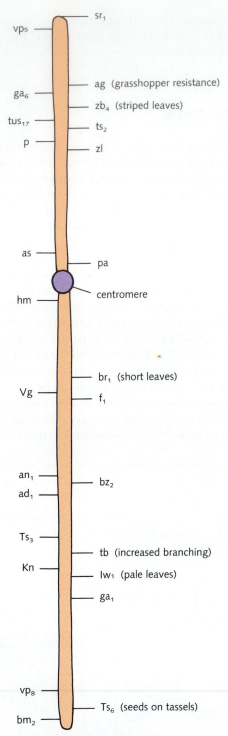

Figure 10.18 A gene linkage map from chromosome #1 of corn (*Zea mays*). The locations of genes on this map were determined by the frequency of cross-over events between them.

sr₁
vp₅
ag (grasshopper resistance)
ga₆
zb₄ (striped leaves)
tus₁₇
ts₂
p
zl
as
pa
centromere
hm
br₁ (short leaves)
Vg
f₁
an₁
bz₂
ad₁
Ts₃
tb (increased branching)
Kn
lw₁ (pale leaves)
ga₁
vp₈
Ts₆ (seeds on tassels)
bm₂

mapping is generally done in a number of individual steps:

- Test crosses between a homozygous recessive and an unknown genotype are used to determine the genotype of an organism.

- If two genes fail to display independent assortment in test crosses, it is likely that they are located on the same chromosome.

- And finally, if several genes are tightly linked (meaning that they are rarely separated by cross-over events), we can construct a genetic map of the chromosome by measuring the frequency of cross-over events between pairs of genes.

Giant Chromosomes in Tiny Organisms

Although gene mapping has shown us how genes are arranged on the chromosome, in most cases the chromosome itself is so tiny that very little of its own structure can be seen. Luckily there are some exceptions. The nuclei of a few cell types contain giant chromosomes that are as much as one hundred times larger than usual.

In the formation of giant chromosomes, many copies of a single chromosome are produced without mitosis, and hundreds of chromosome strands lie side by side (Figure 10.19). Because of the way they are formed, these chromosomes are called **polytene,** which means "many-stranded." Once polytene chromosomes have formed in a cell, the cell never divides again. The cells containing them, therefore, represent a final stage of development. By a great stroke of luck, *Drosophila*—the very same organism that had been so useful in studies of genetics—was discovered to have polytene chromosomes in its salivary glands.

A comparison of polytene chromosomes with normal metaphase chromosomes illustrates why these structures have been so useful in cellular genetics. The tiny individual bands enable observers to recognize different regions of the chromosome, and that has given geneticists an important opportunity. Each band marks the location of an individual gene on the chromosome, and genetic mapping techniques have made it possible to identify a great many of the bands with particular genes.

Regions of genetic activity can also be seen on the chromosomes. When a gene becomes active, the band corresponding to it sometimes swells to form a distinct **chromosome puff** (a puff appears in Figure 10.19). The presence of puffs in certain genes enables geneticists to determine which genes are active at particular stages of development. For more than half a century, these giant chromosomes have given biologists an unparalleled opportunity: the chance to "see" the positions of genes on chromosomes and to "watch" genes in the process of being turned on and off.

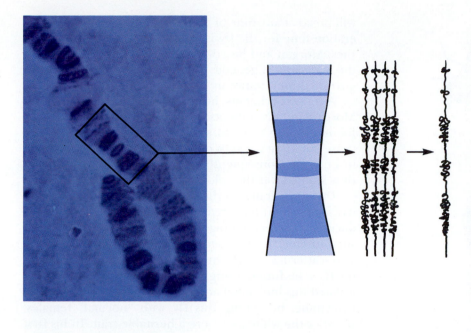

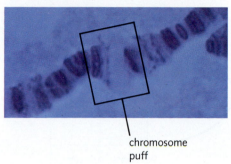

Figure 10.19 (left) A polytene chromosome from salivary gland cells of *Drosophila melanogaster*. Polytene chromosomes are formed by repeated rounds of chromosome replication. Instead of being separated, the identical chromosome strands are laid out side-by-side to produce a polytene chromosome which contains hundreds or thousands of copies of the single chromosome. (below) A chromosome puff.

chromosome puff

THE INHERITANCE OF SEX

Mendel's simple principles account for many types of inheritance, but the reality of nature is more complex. One of the first areas to require modification involved differences in genetics between the sexes. These differences have their origin in the fact that in most organisms, the sexual identity of an individual is determined by special chromosomes known as the *sex chromosomes*. When we examine each of the eight chromosomes found in a diploid female *Drosophila* cell, for example, we can easily group them into four pairs of homologous chromosomes. In a cell taken from a male fly, however, the story is a little different. We can arrange six of the chromosomes in pairs identical to those found in the female cell, but two dissimilar chromosomes are left over (Figure 10.20). One of these unmatched chromosomes is identical to the last pair in the female cell, but one is unlike any other chromosome. This is the exception (which we referred to earlier) to the rule that nearly all chromosomes occur as one member of a homologous pair. We call these two sex-related chromosomes *X* and *Y*. Between them, they determine the sex of an organism.

Inheritance of the Sex Chromosomes

The *X* and *Y* chromosomes are passed from one cell generation to the next like any other chromosomes, except during the process of meiosis. Female cells have two *X* chromosomes. These constitute a homologous pair that in the sex cells produces a tetrad along with the other chromosome pairs during the first meiotic division. Male

Figure 10.20 Chromosomes of male and female *Drosophila melanogaster*.

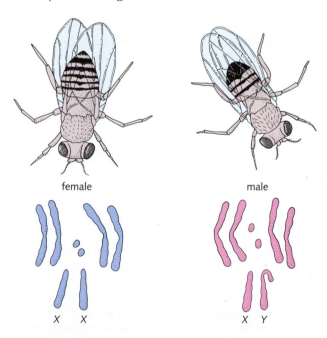

female

male

X   X

X   Y

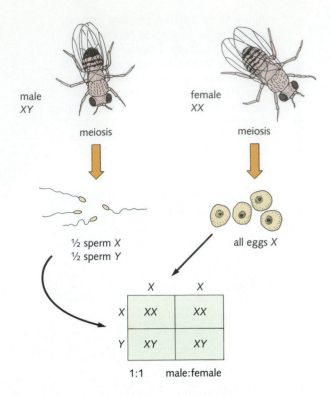

Figure 10.21 A Punnett square shows why male and female *Drosophila* are produced in roughly equal numbers.

In the figure:

male
XY

meiosis

female
XX

meiosis

½ sperm X
½ sperm Y

all eggs X

	X	X
X	XX	XX
Y	XY	XY

1:1   male:female

cells, however, have both an *X* and a *Y* chromosome (and no tetrad forms). During meiosis, each gamete receives either an *X* or a *Y* chromosome. Therefore, about half of the sperm cells produced by a male fly contain the *X* chromosome and about half contain the *Y*. Because all eggs contain an *X* chromosome, the sex of the zygote is determined by the sex chromosome carried by the sperm. If the sperm cell donates an *X* chromosome, the zygote will be female; if it donates a *Y* chromosome, the zygote will be male.

The *X-Y* mechanism, which is also how sex is determined in humans, produces a 1:1 ratio of males to females (Figure 10.21). But what happens to a gene located on one of the sex chromosomes? The answer, as first discovered by the American geneticist T. H. Morgan, is that inheritance can be *sex-linked.*

Sex Linkage

Thomas Hunt Morgan was a professor of biology who did his first work at Columbia University beginning in 1904. Morgan pioneered the use of the fruit fly, *Drosophila melanogaster,* in genetics research. These organisms offer several advantages over Mendel's peas. Fruit flies can be grown in small bottles on a laboratory shelf, they

will breed at any time of the year, and above all, the generation time for the fly is only 10–14 days. Using these flies, Morgan and his co-workers were able to show that the basic laws of genetics Mendel had discovered in peas governed inheritance in flies as well.

A few experiments, however, yielded baffling results. Morgan and his students were constantly on the lookout for interesting traits to study, and they examined first those traits that could be observed quickly and easily even in a tiny fly. Consequently, one of the first characteristics they studied was the color of its eyes.

Drosophila found in the wild (wild-type flies, as geneticists call them) have reddish-colored eyes, a color produced by a screening pigment that is located in the cells surrounding the light-sensitive cells and prevents excessive scattering of light in the eye. As Morgan examined his flies, he found a single male fly with white eyes. He isolated this individual and decided that he would try to determine, by mating this fly with "normal" females, whether the white eyes were a heritable trait. In his first cross, he produced an F_1 generation made up entirely of red-eyed flies (Figure 10.22).

It looked as though the gene for red eye color was dominant over the gene for white eye color. The best way to test this idea was to cross members of the F_1 generation with each other. Morgan carried out an $F_1 \times F_1$ cross and got the 3:1 ratio of red to white eyes in the F_2 generation, just as one would expect if the gene for red eyes were dominant over the gene for white eyes. However, he noticed something puzzling about the offspring. *All of the white-eyed flies were male!* In other words, none of the female flies showed white eyes, and about half of the males did. Why were the white eyes concentrated in the male flies?

Drawing on the observations of Nettie Stevens (see Theory in Action, Nettie Stevens and the *Y* Chromosome, p. 200) and Edmund Wilson, Morgan produced a simple explanation for his results. *Drosophila* females had four chromosome pairs, whereas males had three pairs and a nonmatching *XY* pair. Assuming that the females were *XX*, Morgan suggested that the gene for white eye color was located on the *X* chromosome. If we use the symbol *W* for the red-eye gene and *w* for the white-eye gene, his original white-eyed male must have been X_wY. The wild-type females were X_WX_W. No white-eyed flies appeared in the first filial (F_1) generation, because all flies in that generation, whether male or female, had an *X* chromosome with the *W* (red-eye) gene. In the second filial (F_2) generation, only one combination of gametes could produce a white-eyed fly: a sperm cell carrying the *Y* chromosome and an egg cell carrying the X_w chromosome. Although about half the females in the F_2 generation *carried* the *w* gene, the fact that they also carried a dominant *W* gene meant that they had red eyes (Figure 10.23).

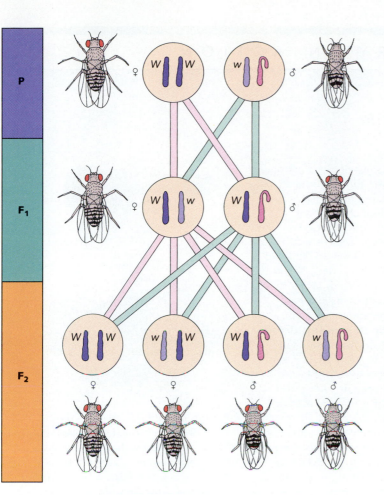

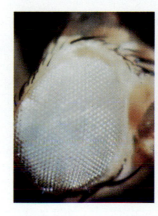

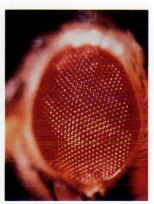

Figure 10.22 The results of two generations of crosses involving eye color in *Drosophila*. The original breeding couple are a white-eyed male and a red-eyed female. In the F₁ generation all flies have red eyes, but in the F₂ generation 1/2 of the male flies are white-eyed.

females, being *XX*, must have two copies of the white eye gene (*ww*) to display the white eye phenotype

female

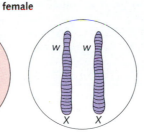

heterozygous = red eyes

homozygous recessive = white eyes

males, carrying only one *X* chromosome, will display the white eye phenotype with just one *w* gene

male

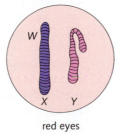

red eyes

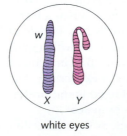

white eyes

Figure 10.23 Sex-linked inheritance explains the appearance of the white eye color in half of the F₂ males in Figure 10.22.

Nettie Stevens and the Y Chromosome

The history of genetics, like that of any science, is filled with wrong turns, mistakes, and errors in judgment that time and the experimental method have corrected. The identification of the *X* chromosome is a case in point. The first suggestion that sex could be a chromosomal characteristic was made in 1901 by Clarence E. McClung, who suggested that a so-called *accessory chromosome* in insects determined maleness. The term had been coined 10 years earlier by an investigator named H. Henking. Henking noticed this chromosome in studies on the cells of *male* insects. Unlike the other chromosomes, it did not have a paired chromosome, so he called it an accessory. He correctly noted that it divided in only one of the two meiotic divisions, which showed that it was present in only two of the four sperm cells. This unusual behavior made many biologists doubt that it even *was* a chromosome, and the designation *X* reflects that doubt. Because McClung found the unpaired accessory chromosome only in male cells, he concluded that the *X* chromosome determined maleness.

We can look back on the work of Henking and McClung with the benefit of hindsight, but only because a more patient observer followed them—one who noticed an even smaller structure in these insect cells, a structure they had missed. That scientist was Nettie M. Stevens. Nettie Stevens was born in rural Vermont, studied in Massachusetts at a state teachers' college, and then taught school for nearly 10 years in order to save enough money to attend Stanford University to study science. From Stanford she went to Bryn Mawr College in Pennsylvania in 1900 and began a series of studies focused on the way in which the sex of an animal is determined. While many other investigators believed that sex was *not* genetically determined, Stevens was convinced that McClung was on the right track.

Stevens made a series of very careful observations with *Tenebrio molitor*, the common mealworm. She found that female cells contained 20 chromosomes, whereas male cells contained 19 large chromosomes and 1 small one. She correctly concluded

that this small difference was the result of chromosomal sex determination! Nettie Stevens had discovered the *Y* chromosome. The determinant of maleness was not the *X*, for females had two of those, but the small *Y* chromosome, which was found only in males. Stevens's discovery of the *Y* chromosome was a pivotal event in the development of genetics, because it made possible for the first time a correct explanation of sex-linked inheritance.

Morgan was the first to show that sex-linked inheritance could be explained if the gene in question were located on the *X* chromosome. The same pattern is seen for other genes located on the *X* chromosome. It is important to appreciate why a gene found on the *X* chromosome produces sex-linked inheritance. When a recessive gene is located on one of the two *X*s in a female, it may or may not be expressed, depending on which gene is present on the other *X* chromosome. But in a male, *any* gene on the single *X* chromosome is expressed. The tip-off that a particular gene is sex-linked is the preferential expression of a recessive gene in males.

Other Patterns of Sex Determination

Humans and fruit flies happen to follow one pattern of sex determination, but a great many organisms follow different patterns. In some insects there is no *Y* chromosome. Males in such species are *XO* and females are *XX*. (The letter *O* represents the absence of a chromosome.) In birds and moths, the males possess a pair of identical sex chromosomes referred to as *ZZ*, whereas the females are *ZW* or *ZO*. (The letters *Z* and *W* are used to prevent confusion with the *XY* system we have become familiar with.)

In bees and ants, sex determination occurs by means of a system called **haplo-diploidy.** In haplo-diploidy there are no sex chromosomes as such. The males develop from unfertilized eggs and are *haploid*, whereas the females develop from fertilized eggs and are *diploid*.

OTHER FORMS OF GENETIC VARIATION

Thus far we have considered only traits in which one gene is recessive and one gene is dominant. Such situations are known as **simple dominance** or **complete dominance.** However, not all genes behave quite so neatly. Flower color in one strain of snapdragons, for example, is controlled by a system in which genes exist for red flowers (*R*) and white flowers (*r*). When both genes for a particular trait are identical, we say that an organism is homozygous for that trait. In snapdragons, the homozygous organisms have either red flowers (*RR*) or white flowers (*rr*). The heterozygous organism, of course, has different genes for the same trait. Which gene is dominant in the heterozygote?

Incomplete Dominance

In snapdragons *neither* gene is dominant. Instead, the flowers produced by *Rr* plants are pink, a blend of red and white. Such a situation is called **incomplete dominance,** because neither gene is clearly dominant. In incomplete dominance, the characteristics of the heterozygote are not the same as those of either homozygous organism. In some respects, one might consider this a form of "blending" inheritance, in which the offspring look a little bit like one parent and a little bit like the other. However, incomplete dominance still follows the basic ratios of Mendelian inheritance. When an F_1 generation of pink snapdragons is crossed with itself, the ratio of flower color in the F_2 offspring is 1:2:1 (red:pink:white) (Figure 10.24). This ratio is identical to the 3:1 ratio that would be obtained for simple dominance in such a cross *except for one thing*: The heterozygote has different characteristics from the homozygous dominant (it is pink rather than red).

Snapdragon flower color is not the only example of incomplete dominance. It is observed in a host of different genes, including a gene that causes disrupted feathers (known as frizzle) in chickens. A homozygous chicken with the frizzle gene (*ff*) shows "extreme frizzle" in its feathers. The heterozygous cross of such a chicken with a normal chicken (*FF*) displays "mild frizzle" as a result of its *Ff* gene composition.

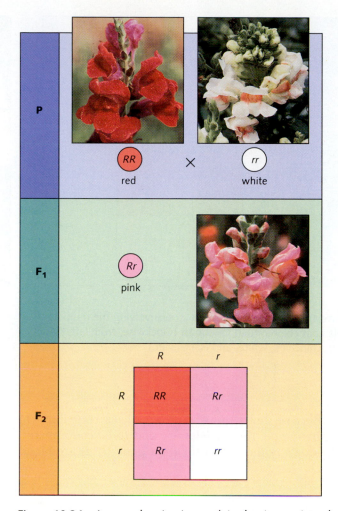

Figure 10.24 A cross showing incomplete dominance in red, white, and pink snapdragons. The pink color results from incomplete dominance of the red allele over the white allele.

Codominance

When each of two different alleles contributes to the phenotype, the alleles are said to be **codominant.** In such situations, it is not possible to say that one allele is dominant over the other. One of the best-known examples of codominance occurs in humans, where it determines the ABO blood group. The gene locus for this blood group system has three possible alleles: I^A, I^B, and i. The I^A and I^B alleles determine carbohydrates present on the surfaces of blood cells. An individual with only the I^A allele has A-type carbohydrates and is said to be blood type A. An individual with only the I^B allele has B-type carbohydrates and is blood type B. I^A / I^B heterozygotes have *both* A-type and B-type carbohydrates and are blood type

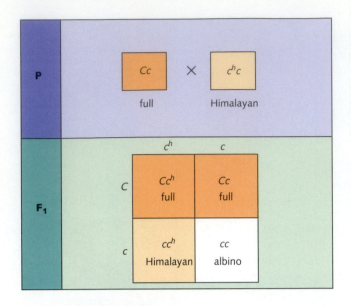

full $C\_$

chinchilla $c^{ch}c^{h}$ or $c^{ch}c$

Himalayan $c^{h}c$

albino cc

Figure 10.25 An experimental cross involving the multiple allele coat color system in rabbits. (Clockwise, from left: full color, chinchilla, white, and Himalayan.)

AB. Because both the I^A and I^B alleles contribute to the phenotype, these alleles are said to be codominant. Individuals homozygous for the i gene (i / i) are blood type O.

Multiple Alleles

Although many of the genes we have examined may have two different alleles, not all genes do. In fact, for many genetically determined characteristics, three or four or even ten different varieties of the same gene are found in a population. The seven characteristics investigated by Mendel were all examples of two-allele systems. Only two types of each gene were present in his breeding population. However, Mendel was either very lucky or very clever in his choice of genes for study. Multiple alleles are not at all uncommon.

One of the most familiar examples of a multiple allele system is found in the rabbit (Figure 10.25). The coat color of rabbits is controlled by a series of four alleles for the same gene:

C	full color
c^{ch}	chinchilla
c^{h}	Himalayan
c	albino (white)

These four genes show a pattern of simple dominance:

$$C > c^{ch} > c^{h} > c$$

This means that the heterozygote cc^{ch} would have a chinchilla coat, Cc would have a full-color coat, and $c^{h}c^{ch}$ would also have a chinchilla coat. There are many other examples of genes with multiple alleles (including the ABO blood-group genes in humans), as we will see in the next chapter.

Traits Controlled by Gene Interaction

The apparent simplicity of the principles that govern the inheritance and expression of some traits should not be taken to suggest that all characters are controlled as simply as the color of a flower or the fur of a rabbit. In reality, many phenotypic differences that seem clear-cut are controlled by gene interaction.

An interesting one is a two-gene system that seems to control coat color in Labrador retrievers (Figure 10.26). These dogs commonly come in both black and yellow varieties, and they occasionally exhibit a rich brown color called chocolate. The genetics of coat color in these dogs has been investigated at breeding colonies where several generations of controlled matings have been carried out. It turns out that their coat color is controlled by genes found at two **loci** (singular, *locus*)—that is, at two different positions in the genetic system.

The two loci are known as the B locus and the E locus. Each locus has two alleles. At the E locus the double-recessive genotype e/e produces yellow fur regardless of the situation at the B locus. But if the E locus contains the dominant E allele E/E or E/e), then control shifts to the B locus. Two alleles are possible at the B locus: a dominant B allele, which produces black coat color, and a recessive b allele, which produces chocolate color if it is in the homozygous form (b/b). A system like this, in which the genotype at one locus controls the expression of genes at another, is an example of **epistasis.** The E gene is said to be epistatic to the B gene, which literally means that it "stands above" the B locus.

In Labrador retrievers an interesting twist enables us to determine the condition of genes at the other locus. Look closely at Figure 10.26. Both dogs are yellow, which means that they are genotype e/e at one locus. However, they have different pigmentation around the nose and lips. Breeding experiments show that dogs with black pigmentation around the nose and lips are genotype B/B or B/b in the B locus and that dogs with pale noses and

lips are genotype b/b. Therefore, coat color in Labrador retrievers is governed by an epistatic two-locus system, but one in which it is sometimes possible to determine the status of both loci by looking closely at the phenotype.

A question. The coat color system in parakeets is also a two-locus system, as we have seen. Can you determine whether this coat color system is epistatic? (See Appendix for the answer.)

Polygenic Systems

Systems wherein a single trait is controlled by more than two genes are said to be **polygenic** ("many genes"). Many polygenic systems, particularly those that control such traits as shape and form, are so complex that we are years away from understanding them fully. However, we do know the details of a few simple systems.

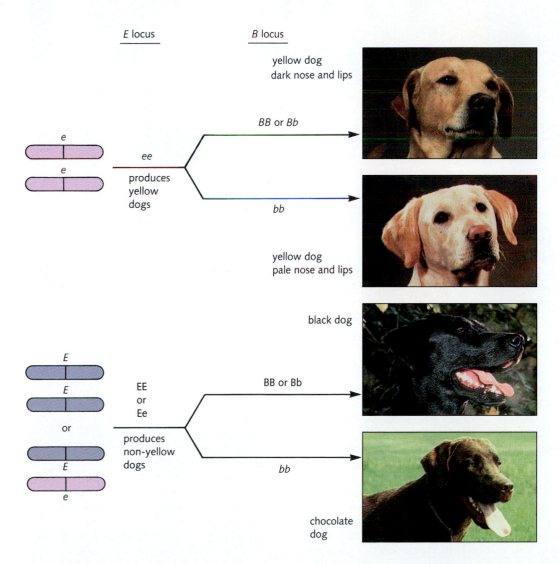

Figure 10.26 Coat color in Labrador retrievers is controlled by four alleles found at two different loci. Note how the alleles present at the E locus affect the expression of the alleles at the B locus. (right) Labrador retrievers showing the four possible coat and skin color combinations.

E locus

B locus

yellow dog
dark nose and lips

BB or Bb

e
e

ee
produces
yellow
dogs

bb

yellow dog
pale nose and lips

black dog

E
E

EE
or
Ee

BB or Bb

or

produces
non-yellow
dogs

E
e

bb

chocolate
dog

Using Genetics to Solve Problems

Part of the power of genetics as a scientific tool stems from the ability it confers on us to analyze and solve problems and, often, to predict the results of crosses between two organisms. Let's look at one example of a genetics problem.

Two *Drosophila* flies, each of which seemed to be phenotypically wild type, were crossed. Their offspring (the F_1 generation) were then collected, and each one was test-crossed with a fly displaying purple eye color, a recessive trait. Of these crosses (the F_2 generation), half produced flies with wild-type eye color, and the other half produced 50 percent flies with wild-type eyes and 50 percent flies with purple eyes. *Determine the genotype of the original parents.*

By reasoning backwards from the last cross, we can solve the problem. Purple is a recessive trait, so each of the purple-eyed flies must have been genotype *ww*. Because half of the F_1 flies that were mated to the purple flies produced offspring with 50 percent purple eyes, these F_1 flies must have had a copy of the *w* gene. Their own eye color was wild type, so they must also have had a copy of the wild-type gene, making them genotype *Ww*. The other half of the F_1 generation did not produce any purple-eyed flies in their F_2 offspring. This means that their genotype must have been *WW*. What would the genotypes of the two original fly parents have to have been to produce an F_1 generation that was half *WW* and half *Ww*? The answer is that they must have been *WW* and *Ww*.

It is also possible to use genetics to predict future events. You will remember that Mendel discovered that tall is dominant over short in garden peas. Two tall peas are crossed and 100 seeds are collected. Two seeds are planted and grown under identical conditions. One produces a tall plant and the other a short plant. On the basis of these results, *determine*

P		Both parents have wild-type eye color.
F_1		All of F_1 offspring have wild-type eye color.
		All of F_1 offspring are crossed with double recessive purple–eyed flies.
F_2		Half of F_1 generation produces F_2 generation of all wild-type eye color.
		Half of F_1 generation produces F_2 generation made up of 1/2 purple color and 1/2 wild-type eye color.

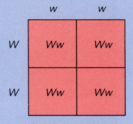

F_1 flies that produced all wild-type flies when mated with purple flies must have been genotype *WW*.

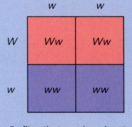

F_1 flies that produced 1/2 purple flies and 1/2 wild-type flies when crossed with purple flies must have been genotype *Ww*.

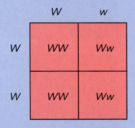

To produce these two types of F_1 flies the parents must have been genotype *WW* and *Ww*.

how many of the remaining 98 plants can be expected to grow into tall plants.

Remember that we were not told the genotypes of the original parents—only their phenotypes (they were both tall). However, because they were both tall, we know that each had at least one of the dominant genes for tallness (*T*). The single seed of the first filial generation that grew into a short plant had to be genotype *tt*, because the *t* gene is recessive. The

only way in which a short plant could have been produced was for each of the parents to have had at least one *t* gene. Therefore the two parents were both genotype *Tt*. Simple genetic analysis tells us that 25 percent of the seeds in a cross between two such plants will grow to display the short phenotype. Because 75 percent of 98 = 73.5, 73 or 74 of the remaining seedlings can be expected to be tall.

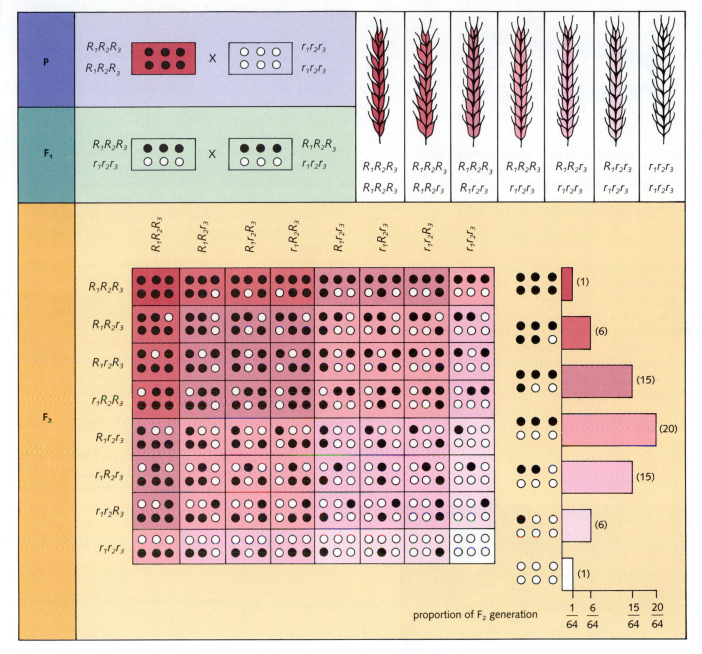

Figure 10.27 Many complex traits are controlled by polygenic systems. Kernel color in wheat is determined by three loci which produce a range of phenotypes from white to dark red.

The reddish color of wheat kernels is controlled by a three-locus system. There are two contrasting alleles at each gene locus, a red allele (R_n) and a white allele (r_n). (The n designates a particular locus in the three-locus system, as shown in Figure 10.27.) The red alleles show incomplete dominance over the white alleles, and each locus makes a contribution to the color of the kernel. Therefore, the overall kernel color is determined by the total number of red and white alleles. Kernels with six red alleles are bright red, those with six white alleles are white, and those with intermediate numbers are somewhere between red and white.

Figure 10.27 illustrates the expected result of crossing a white-kernel plant with a red-kernel plant. The F_1 kernels would be medium-red in color. When the F_1 plants are crossed, 64 possible gene combinations result. Only one of these would produce a white kernel, and only one would produce a bright red kernel. The remaining 62 kernels would be intermediate between these two colors. It's easy to see how the kernels themselves would give the

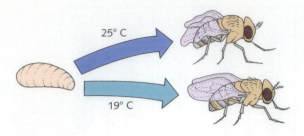

Figure 10.28 The expression of some genes is affected by the environment. The expression of the "curly-wing" allele in fruit flies is dependent on the temperature at which the flies develop.

impression that the white and red colors of the original kernels had *blended* to produce the kernels of the F₂ generation.

Polygenic traits sometimes give the impression that **continuous variation**—a full range of phenotypes between two extremes—exists in a population. The traits that Mendel investigated involved **discontinuous variation;** all phenotypes fell into a few well-separated categories. But you would be hard–pressed to sort each of the F₂ kernels into seven categories of pigmentation. As the number of gene loci involved in a polygenic trait increases, the range of variation becomes almost completely continuous.

Figure 10.29 The expression of coat color genes in Siamese cats varies with temperatures. Black pigment is produced only in those areas of the skin which are lowest in temperature, such as the ears and tail. This variation produces the typical Siamese markings.

Environmental Effects on Gene Expression

Earlier in the chapter, when we made the distinction between genotype and phenotype, we were careful to say that the phenotype of an organism develops under the influence of its genotype *and* its environment. Some of the effects of the environment are obvious. As Mendel learned, the height of pea plants is genetically determined. One of his seven traits was determined by a height gene, for which there were two alleles, tall and short. The actual size of each plant, however, was influenced by environmental factors, including the amount of water, sunlight, and nutrients that each plant received. The effect of environment means that even a group of plants with identical genotypes displays variability in phenotype.

The *curly* allele is a dominant gene that is found in *Drosophila* and influences the shape of the wing. The effect of this allele is controlled by the temperature of the environment. If the flies develop at 19°C, most wings are normal. If they develop at 25°C, however, the wings are curly (Figure 10.28).

Temperature also influences gene expression in the Siamese cat (Figure 10.29). Siamese cats are pale in color, with gradually darkening fur near the tips of the tail, feet, ears, and nose. This coloration is produced only in cats that are homozygous for a particular recessive allele that produces the black pigment melanin. Unlike the normal melanin allele, the recessive allele produces melanin only at temperatures just below the usual body temperature. Black pigment, therefore, is produced only in the cooler regions of the skin, creating the characteristic Siamese markings.

SUMMARY

The science of genetics traces its origins to the work of Gregor Mendel, who determined that a set of characters, or genes, was passed from one generation to the next by reproductive cells, or gametes. Mendel's three important principles of dominance, segregation, and independent assortment describe the way the genotype is determined. The genotype and the organism's environment determine its phenotype, the sum total of all its expressed characteristics.

Although Mendel was not aware of their existence, other biologists had discovered that cell nuclei contained chromosomes, visible during mitosis. Cells that contain one maternal and one paternal set of chromosomes are said to have the diploid number (2*n*) of chromosomes. During meiosis, the chromosome number is reduced from diploid to haploid (*n*).

Genes are located in distinct positions on chromosomes. Genes located on the same chromosome are said to be "linked," since they are inherited together. This linkage is not absolute, however, and linked genes may be separated by cross-over events which take place during meiosis. Because random cross-over events are more likely to separate distant genes than nearby ones, the frequency of crossing over between genes provides geneticists with the information they need to construct a map of genes on a chromosome.

The existence of giant polytene chromosomes has enabled geneticists to visualize chromosome regions that are associated with specific genes, and even to watch genes in the process of being activated.

In many organisms, sex determination occurs by means of sex chromosomes such as the *XY* system, in which individuals of genotype *XX* are female and those of genotype *XY* are male. Genes carried on either sex chromosome display a sex-linked pattern of inheritance in which the phenotypes they determine are preferentially expressed in one sex or the other. Living organisms exhibit a wide range of genetic systems involved in the control of different characteristics. These include incomplete dominance, codominance, multiple alleles for a single gene, interactions between different genes, and polygenic control of a single characteristic.

STUDY FOCUS

After studying this chapter, you should be able to:

- Explain how the concept of a gene originated and the science of genetics developed.
- Apply basic Mendelian genetics to real-life situations in the breeding of animals and plants.*
- Describe the connection between one generation and the next. Give evidence that genes are real structures that occupy definite positions on chromosomes.

*Note: See Appendix for Genetics Problems, answers to Genetics Problems, and answers to in-text questions.

SELECTED TERMS

genes *p. 183*
dominant *p. 183*
recessive *p. 183*
alleles *p. 183*
homozygous *p. 184*
heterozygous *p. 184*
test cross *p. 184*
genotype *p. 184*
phenotype *p. 184*
principle of dominance *p. 184*
segregation *p. 184*
independent assortment *p. 185*

zygote *p. 188*
homologous
   chromosomes *p. 188*
diploid *p. 188*
haploid *p. 188*
meiosis *p. 189*
synapsis *p. 189*
genetic map *p. 195*
incomplete
   dominance *p. 201*

Discussion Questions

1. How did Mendel's discoveries elucidate Darwin's theory of evolution?

2. In what respect does the behavior of *chromosomes* during mitosis and meiosis resemble the behavior of *genes* during the same two processes?

3. Why do we need a two-factor cross to demonstrate the principle of independent assortment?

4. The existence of chromosomes requires an important modification of the principle of independent assortment. What is it? Why is the process of crossing over an exception?

5. Is a 50:50 balance of males and females maintained in a population wherein sex determination is based on the number of *X* chromosomes present: *XX* for females and *XO* for males? How about a population wherein *ZZs* are males and *ZWs* are females?

Objective Questions (Answers in Appendix)

6. In a test cross of one trait, the unknown genotype was found to be Pp. The phenotypic ratio in the F_1 generation must be
   (a) 9:3:3:1.
   (b) 3:1.
   (c) 2:2:1.
   (d) 1:1.

7. The outward appearance of an organism is called its
   (a) phenotype.
   (b) genotype.
   (c) allele.
   (d) chromosome.

8. The genetic makeup of an organism is called its
   (a) phenotype.
   (b) genotype.
   (c) allele.
   (d) chromosome.

9. The principle of independent assortment assumes that
   (a) gametes combine in an ordered fashion.
   (b) genes segregate freely during mitosis.
   (c) genes segregate freely during gamete formation.
   (d) genes are linked together in the chromosome.

10. For a diploid species, fertilization of the gametes will
    (a) reduce the diploid number by one–fourth.
    (b) reduce the diploid number by one–half.
    (c) restore the diploid number.
    (d) result in a haploid number in the zygote.

CHAPTER 11

Human Genetics

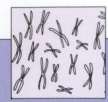

"Know thyself." The philosopher's first commandment is an important one, for it stresses the strengths and limitations of human inquiry. More to the point, the desire to know about ourselves is one of the things that motivates many of us to study biology in the first place. That is particularly true of the science of genetics. Therefore, it is appropriate that we spend a good deal of time reviewing some of what we have learned about ourselves as genetic organisms.

Except for the fact that we find the species interesting, *Homo sapiens* is a most inappropriate organism for the study of genetics. It is large and complex and cannot be maintained in the laboratory. Its generation time is very long—about 20 years in most societies. It usually produces only a single offspring after mating, and most individuals produce no more than three or four offspring in a lifetime. Finally, we cannot perform experimental crosses, as we can for other species, so we must depend on the organisms to tell us their own life histories. Knowing that any sensible person would study fruit flies, roundworms, or bacteria, we plunge ahead anyway.

THE HUMAN GENETIC SYSTEM

Humans are multicellular organisms that reproduce sexually (Figure 11.1). An analysis of human genetics begins with individual cells. A diploid human cell contains 46 chromosomes, and nearly all of its genetic content, or *genome*, is contained on these chromosomes. Two of these are the **sex chromosomes** (*XX* in females and *XY* in males), and the other 44 are referred to as **autosomal chromosomes.** The autosomes consist of 22 pairs of homologous chromosomes, one member of each pair inherited from each parent. In order to describe human chromosomal composition quickly, we may write 46 *XX* for a normal female and 46 *XY* for a normal male. This shorthand

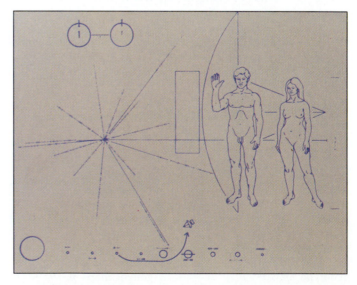

Figure 11.1 The Voyager spacecraft was fitted with this panel describing the human species to extraterrestrials. Studies of the genetics of our species, once so difficult to carry out, are now advancing rapidly.

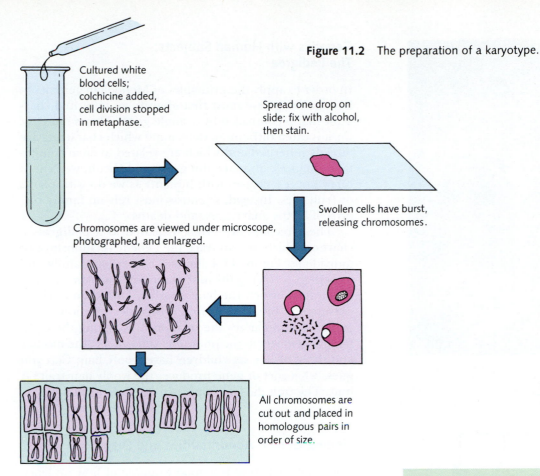

Figure 11.2 The preparation of a karyotype.

Cultured white blood cells; colchicine added, cell division stopped in metaphase.

Spread one drop on slide; fix with alcohol, then stain.

Swollen cells have burst, releasing chromosomes.

Chromosomes are viewed under microscope, photographed, and enlarged.

All chromosomes are cut out and placed in homologous pairs in order of size.

indicates the *total* number of chromosomes and the nature of the sex chromosomes.

We can get a good look at human chromosomes via a technique known as **karyotyping** (*karyon* means "nucleus"). Karyotyping is a standard procedure in genetics and medicine and is used to diagnose a "chromosomal abnormality." A few white blood cells are removed from a blood sample and grown in culture (Figure 11.2). Then *colchicine*, a chemical that causes the microtubules making up the mitotic spindle to disassemble, is added to the cells. This enables the cells to enter mitosis but prevents the chromosome separation that normally occurs during anaphase. The cells are now locked into metaphase, and each chromosome is visible as a pair of identical chromatids joined by a single centromere. The cells are then fixed on a slide and stained to make it easier to visualize the chromosomes released from cells that have burst.

Photomicrographs of these metaphase chromosomes can be analyzed in a very simple way. We take a pair of scissors and cut out each of the chromosomes from the photograph. When we sort through the loose pictures, we find a match for each chromosome in our pile, and they can be arranged in order of size (Figure 11.3). A male karyotype contains two chromosomes for which no match can be made: the *X* and the *Y*.

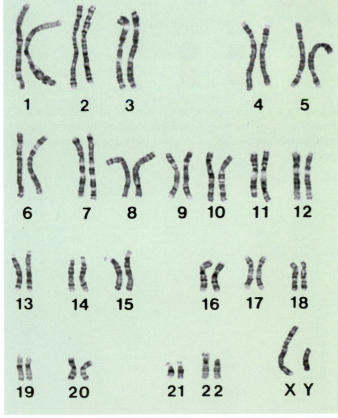

Figure 11.3 Karyotype of a human male.

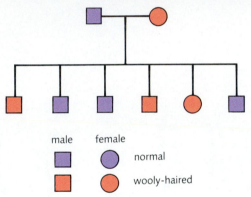

male   female

■ ● normal

■ ● wooly-haired

Figure 11.4 A Norwegian family, some of whose members have the wooly hair trait. A pedigree chart of the family (bottom) shows the passage of this trait from one generation to the next.

Genetics with Human Subjects: The Pedigree

In order to apply the principles of Mendelian genetics to human beings, we must first identify an inherited character that is controlled by a single gene. This is not easy, for it is often difficult to determine which characters are directly inherited and which are related to environmental influences. To make things more difficult, we cannot carry out test crosses with humans as we do with plants or fruit flies. Instead, scientists must rely on family records of births, marriages, and deaths.

These records are often displayed as a **pedigree,** a chart on which we can trace the genetic relationships of individuals. Figure 11.4 shows a Norwegian family and its pedigree, tracing the inheritance of *wooly hair,* an unusual trait in which the hair is tightly kinked and brittle, causing it to break off before it becomes very long. The pedigree summarizes the relationships among the eight family members in the photo. As you can see, the mother and three of her six children have wooly hair. Can you guess what sort of gene produces the wooly hair trait? A hint: The trait did not appear in the father's family.

Finding Genes: Abnormalities and Disorders

The wooly hair trait is a good example of how we learn about the existence of a human gene. It is only because the wooly hair trait is *abnormal*—uncommon in the population—that we can recognize it as a distinct characteristic. In that sense, it is a **genetic abnormality,** a characteristic very different from the norm, or average.

Some genetic abnormalities produce medical problems. A **genetic disease** is a serious disorder caused by a gene or group of genes. The alleles that cause such diseases may be either dominant or recessive. A great many genetic diseases have been described in the past two centuries, and our acquaintance with these diseases makes up a large part of our knowledge of the human genome. This is not due to any inclination on the part of geneticists to be fascinated with disease; rather, it is because the presence of a genetic disorder highlights the "missing" function that the gene normally carries out.

HUMAN GENES

More than 4000 human genes and their alleles have been described, and many of these have been mapped to specific chromosomes. **Sex-linked genes** are carried on either the *X* or the *Y* chromosome. Genes located on any of the other 44 chromosomes, or *autosomes,* are known as **autosomal genes.**

Autosomal Recessive Inheritance

At least 600 known human traits are produced by recessive autosomal alleles. Such alleles do not affect the phenotype unless an individual is homozygous for them.

Albinism   The genetic disorder known as **albinism** is caused by a recessive allele found on an autosomal chromosome and present at low frequencies in all human population groups (Figure 11.5). Because the allele is recessive, individuals who are heterozygous for the trait express their normal skin color, so the presence of the allele is "hidden" by the dominance of the normal allele. Albinos are unable to synthesize **melanin,** the pigment molecule responsible for most human skin coloring. This makes their skin and eyes extremely sensitive to light, and they must avoid exposure to bright sunlight.

What happens when an albino and a normally pigmented person have children? We can represent the genotype of the normal individual as A/A and that of the albino as a/a. A simple genetic analysis reveals that their children will all carry the genotype A/a. But these children will not be albinos; they will be normally pigmented. Now what happens when two people who carry the genotype A/a have children? We can analyze this with a simple Punnett square in which the possible gamete combinations are enumerated (Figure 11.5).

Our result is the classic 3:1 ratio for a heterozygote cross. What this means is that the probability that a child of this couple will display the normal phenotype for this gene is three in four, or 75%. However, chances are two out of three (66%) that a normally pigmented child will carry a copy of the defective allele. The passage of an inherited disorder from one generation to the next can be predicted by Mendelian genetics. Because it can be used to help individuals who carry a genetic disorder determine the odds of their having healthy children, the science of genetics is a useful tool for family counseling (see Theory in Action, Genetic Counseling: Knowing the Odds, p. 212).

Tay–Sachs disease   In the 1880s two physicians, Warren Tay of Great Britain and Bernard Sachs of the United States, described a fatal disease that took the lives of many of their young patients. At about six months of age, babies with this disease developed a reddish patch on the retina of their eyes. Gradually they became blind and deaf, suffered convulsions, and eventually died by the age of 3 or 4. Both physicians noted that the disease occurred in Jewish families of eastern European ancestry.

Today it is recognized that **Tay–Sachs disease** is caused by a recessive allele. Therefore, each parent of a Tay–Sachs child carries a single copy of the Tay–Sachs allele. The Tay–Sachs allele is a defective version of a normal allele that helps cells in the nervous system dispose of fatty

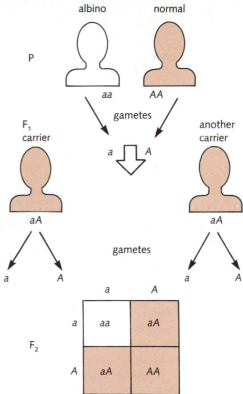

Figure 11.5   (top) An albino child with his family. (bottom) The inheritance of the recessive albinism allele.

molecules known as *gangliosides*. A child who is homozygous for the defective allele cannot dispose of these compounds. They gradually build up within the cells of the nervous system, leading to mental degeneration and death. There is a simple test that can detect heterozygotes who carry the Tay–Sachs allele, warning them that they are at risk of giving birth to children with the disease.

Cystic fibrosis   The most common fatal genetic disease in the United States is **cystic fibrosis,** which affects 1 child in 1600. Victims of the disease suffer from digestive disorders and produce a thick, heavy mucus that clogs their lungs. With medical intervention the breathing passages can be kept clear, and some victims survive to young adulthood, but most die as children. The cystic fibrosis allele is an autosomal recessive gene that may be carried in the heterozygous form by as many as 3 percent of the white population in the United States.

Other autosomal recessives   A host of other genetic diseases are produced by alleles that, like Tay–Sachs, are defective versions of normal alleles. These include **galactosemia,** a disorder that makes it impossible for babies to digest lactose (milk sugar), and **sickle-cell anemia,** a sometimes fatal disorder of the oxygen-carrying proteins of the blood. We will examine the molecular basis of sickle-cell anemia in detail in Chapter 23.

Autosomal Dominant Inheritance

Autosomal dominant alleles affect the phenotype even when they are present in the heterozygous condition with a normal allele. Therefore, children showing an autosomal dominant phenotype may have inherited the allele from either parent.

Darwin tubercle   A small thickening of cartilage near the upper rim of the ear is called the **Darwin tubercle** (Figure 11.6). This trait is caused by an autosomal dominant allele. Therefore, if you have it, one of your parents is likely to have it as well.

THEORY IN ACTION

Genetic Counseling: Knowing the Odds

One of the most interesting aspects of genetics is its predictive power. Genetics can be used to predict the future course of events. It is possible, for example, to examine the genetic system by which human sex is determined and predict that male and female babies will continue to be born in roughly a 1:1 ratio. The statistical power of genetics is useful in animal and plant breeding and even in predicting the possibilities of genetic diseases in human offspring.

At the very heart of genetics is the principle that the assortment of individual genes is a matter of chance. **Genetic counseling** is a discipline which makes use of the principles of genetics to advise potential parents of the risk that they will have children who may suffer from a genetic disease. Parents who have supported one child in his or her struggle against a genetic disease must know the odds of facing the same battle with another child.

Consider a healthy couple whose first child was a boy with hemophilia. If they have a second child, what are the chances that it will suffer from the same disease? This is a question you should be able to answer yourself. One of the mother's two X chromosomes carries the hemophilia allele. If their second child is a girl, she cannot suffer from hemophilia, because one of her two X chromosomes will be the normal X chromosome from her father. If their second child is a boy, he will have a 50:50 chance of inheriting his mother's X chromosome with the hemophilia allele. Once they know the odds, a couple can make an informed decision.

As another example, say a healthy couple gives birth to a boy who develops Duchenne muscular dystrophy. Can you calculate the odds of their having a second child suffering from the disease? If your father were to develop Huntington's disease, what are the chances that you would develop it sooner or later?

Achondroplasia This disorder affects the conversion of cartilage to bone, which is a normal part of the growth process in young people. Individuals with a defective allele for this gene are unable to make the conversion rapidly enough, and this slows down the rate at which their long bones grow, producing dwarfism. Growth and development in other tissues of the body and development of the nervous system are unaffected by the allele.

Huntington's disease This rare disease takes its name from the physician (George Huntington) who first showed that the disease has a genetic basis. Most individuals with **Huntington's disease** have no symptoms until their late thirties or forties, when they begin to lose some control over their voluntary muscles. Early symptoms include muscle twitches and convulsions. Later the nervous system itself begins to degenerate, and the patient usually dies within 12 years after symptoms first appear.

Huntington's disease is controlled by a dominant allele carried on chromosome number 4. Therefore, individuals who are *either* homozygous or heterozygous for the gene suffer from Huntington's disease. Because this disease does not manifest itself until later in life, many individuals had to decide whether or not to have children *before* they knew whether they carried the allele. As we will see in Chapter 23, all of this has now changed. Molecular biologists have now devised a test for the presence of the Huntington's allele, though the test itself has raised new issues and new problems.

Polydactyly A dominant allele produces **polydactyly,** a condition that results in extra fingers and toes (Figure 11.7). This allele is found throughout the world, although it is especially common in the Ukraine republic of the Soviet Union. Polydactyly is variable in its influence on phenotype, sometimes producing extra fingers, sometimes extra toes, and sometimes both. Polydactyly also displays **partial penetrance,** meaning that sometimes the gene has no effect, and an individual carrying it has the normal number of fingers and toes.

Multiple Alleles

Many human genes display more than two alleles, and this can complicate inheritance considerably. One important gene that falls into this category is the blood group gene that is involved in blood *transfusion* (the replacement of lost blood with blood from a donor). When the first recorded human blood transfusions were carried out in 1818 by James Blendell, the results were mixed. His patients were women who were in danger of bleeding to death after childbirth. Although he was able to save the lives of three patients by transfusion, he lost four. These patients' reactions to the transfused blood were fatal.

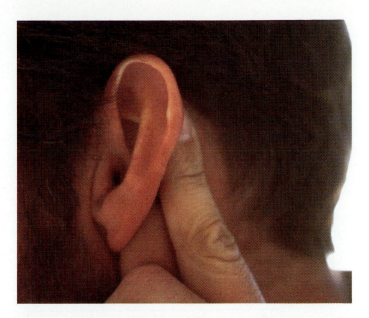

Figure 11.6 This thickening of cartilage near the upper rim of the ear is a Darwin tubercle, caused by an autosomal dominant allele.

Figure 11.7 Polydactyly, the presence of extra fingers or toes, is caused by a dominant allele.

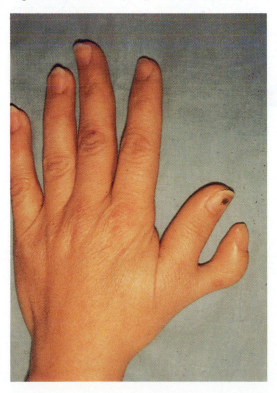

Table 11.1 *Blood Group Genotypes and Corresponding Phenotypes*

Genotype	Phenotype
$I^A I^A$	A
$I^B I^B$	B
$I^A I^B$	AB
ii	O
$I^A i$	A
$I^B i$	B

A physician named Carl Landsteiner discovered that human blood could be classified into four different groups that made it possible to predict whether a transfusion would be successful. The four groups, often referred to as blood "types," are determined by molecules known as *antigens* that occur at the surface of the red blood cells. An **antigen** is a molecule that the body's immune system recognizes as foreign. Violent reactions against transfused blood occur when the new blood brings with it an antigen that the body recognizes as foreign. Success in a transfusion, therefore, depends on making sure that the blood being used does not contain such antigens.

There are four blood groups in the system discovered by Landsteiner. The **A** blood group contains a molecule on the surface of red cells known as the *A antigen,* **B** contains the *B antigen,* **AB** contains both *A and B antigens,* and **O** contains *neither antigen.* As previously described in Chapter 10, these four different phenotypes are determined by the alleles found at a single gene locus, the **immunoglobulin (I)** locus, or gene. This gene has three different alleles (I^A, I^B, and i) which determine the antigens (Table 11.1).

Blood type O, lacking both A and B antigens, is sometimes known as the "universal donor," because it was thought that any of the ABO blood types could safely receive O blood (Figure 11.8). Similarly, blood type AB is sometimes known as "universal recipient," because it was thought that an individual with this blood type could safely receive any blood type in transfusion. In actual practices, physicians insist on an exact match of blood types for transfusion. As we noted in Chapter 10, the ABO blood group system is an example of **codominance;** that is, neither of two alleles is clearly dominant and both affect the phenotype. The alleles, I^A and I^B, are *codominant;* the combination produces blood type AB, a phenotype different from that produced by either I^A or I^B alone, and the allele i is recessive to both.

It is worth noting that the ABO blood grouping is not the only example of human blood typing. There is another important blood antigen known as the **Rh factor,** named for the Rhesus monkey, in which it was discovered *before* it was found in humans. Individuals are classified as Rh-positive when the factor is present and as Rh-negative when it is absent. In medical work it is important to match both blood groups, if possible, so blood is generally classified according to both sets of antigens: O^+ for type O blood, Rh positive; B^- for type-B blood, Rh negative; and so on.

The Rh blood group is particularly important during pregnancy. When an Rh-negative mother carries an Rh-positive fetus, a bit of the fetus's blood may leak into her circulation during birth and "sensitize" her immune system to Rh-positive blood. Generally this does not cause a problem with a first child. But if a second pregnancy

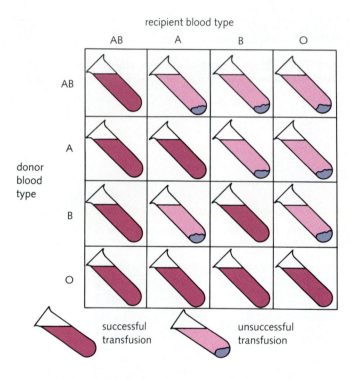

recipient blood type

AB   A   B   O

donor blood type

AB

A

B

O

successful transfusion   unsuccessful transfusion

Figure 11.8 Compatibility of blood transfusions in the ABO blood groups.

with Rh-positive blood occurs, medical attention is required to prevent her immune system from attacking the blood of the "foreign" Rh$^+$ fetus, causing damage to both mother and fetus.

Fortunately, there is a routine medical treatment to lessen Rh compatibility problems. A substance known as *rhogam* can be added to the mother's blood shortly after her Rh$^+$ baby is born. This substance is a form of **antibody,** a protein that binds directly to the Rh antigen. The binding of antibody to antigen prevents the mother's own immune system from being exposed to the Rh antigen, and strong reactions that might develop against the next child are avoided. We will discuss this effect and the immune system more thoroughly in Chapter 39.

Polygenic Traits or "You've Got Your Uncle's Nose!"

Size and shape We can speak with precision about the inheritance of well-defined traits such as blood type and Darwin's tubercle, but what about the other characteristics that common experience suggests we inherit from our parents: a tendency to be short or tall, the shape of our ears, a tendency to be athletic, and our very faces, which often are similar enough to enable total strangers to recognize brothers and sisters? Do genetic systems operate in these cases, too?

In most respects, it is clear that these characteristics are genetically determined. Nothing illustrates this more clearly than so-called identical twins (Figure 11.9). Such twins are properly called **monozygotic,** because they have developed from a single cell that divided at an early stage to form a pair of genetically identical embryos. The resemblance between such twins, even those who have been raised apart, is uncanny. It is clear that complex traits such as the appearance of the face and the shape of the hands and feet are inherited. However, it is also apparent that the genetics of the systems that determine *morphology* (size and shape) is extremely complicated.

Fingerprints S. B. Holt carried out a detailed analysis of a simple morphological system in humans, the shape of our fingerprints (Figure 11.10). The patterns of fingerprints are genetically controlled, and the fingerprints themselves are formed very early in embryonic development. Using a system that law enforcement officers developed to analyze fingerprint patterns, Holt tested the fingerprints of members of various families for similarity. He found that identical twins, who are genetically identical, had roughly 95 percent of their fingerprint features in common. Parent and child had about 48 percent similarity, which is consistent with the fact that parent and child share an average of 50 percent of their

Figure 11.9 Identical twins.

Figure 11.10 Analysis of fingerprint patterns indicates that their major features are inherited as polygenic traits.

Table 11.2 *Correlation Between Relatives for Total Dermal Ridge Count* *

Relationship	Observed Correlation	Expected	Comments
Parent–child	0.48	0.50	Indicates 50% of genes in common
Father–mother	0.05	0.00	Indicates no relationship
Sibling–sibling	0.50	0.50	Indicates 50% of genes in common
Identical twins	0.95	1.00	Indicates 100% of genes in common
Fraternal twins	0.49	0.50	Indicates 50% of genes in common

*The "total dermal ridge count" is a way of quantifying one of the key features of the fingerprint pattern. It is the number of ridges crossed by a line drawn between the center of the pattern and the "triradial point," a place where three groups of ridges meet to produce a small triangle.

SOURCE: S. B. Holt, 1961, Quantitative genetics of fingerprint patterns, *British Medical Bulletin*, 17:247–250.

genes. Siblings also exhibit roughly a 50 percent similarity and can be expected to have about 50 percent of their genes in common. These figures, Holt emphasizes, do not mean that 50 percent of the *fingerprints* of siblings are identical. In fact, generally, *none* of their fingerprints are identical. Their individual fingerprints merely show about 50 percent similarity in the patterns of lines and ridges that occur on corresponding fingers (Table 11.2).

In genetic terms, this means that a fingerprint cannot be the product of a single gene in the same way that ABO blood type is. Instead, such traits are **polygenic:** They are specified by more than one gene. The similarity of fingerprints among members of the same family does not mean that the police will ever confuse a father's fingerprints with his son's. Instead, it reflects the fact that shared individual genes produce similar features that make up each fingerprint, generating many regions of similarity rather than identical patterns.

Skin color   Our species is remarkably diverse. We inhabit every corner of the globe, have successfully invaded the extreme environments on the planet, and display an extraordinary range of genetic diversity. One measure of this diversity is human skin color. The coloration of human skin is caused by a dark pigment called *melanin* that is found in a number of cells, including specialized cells near the surface of the skin known as *melanocytes* (Figure 11.11). More than 30 different shades of human skin color have been described, and they follow a general adaptive trend: Native populations with darker skin color occur in areas closest to the equator, and populations with lighter skin color occur in regions toward the north that receive less sunlight. There are many exceptions to that rule, however, and these have prevented us from developing a complete explanation for the origin of differences in skin color.

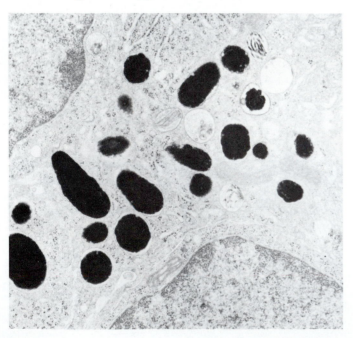

Figure 11.11   A human melanocyte, containing the dark, melanin-rich granules that produce skin color.

Human Intelligence: Are There Genes for IQ?

Every person is an individual with unique capacities, abilities, strengths, and weaknesses. Just as individuals differ in physical characteristics, they differ in mental characteristics too. In 1903 Alfred Binet, a French social scientist, introduced a test to measure intelligence—the *IQ test*. The premise of the test was simple. Just as all people have a physical age, Binet reasoned, they have a mental age. He also assumed that an *intelligence quotient (IQ)* could be computed by dividing the mental age by the physical age and multiplying by 100. A child of 10 with a mental age of 9 would have an IQ of 90, for example, whereas a child of the same age with a mental age of 11 would score 110 on the test.

Intelligence quotient scores differ widely among the human population. A typical distribution is shown above. However, the claim that IQ tests measure intelligence has been challenged by many investigators, some of whom have charged that the tests are flawed by hidden racial and ethnic bias. We will leave those questions unanswered and instead will ask a question that is a bit more biological. Are individual differences in IQ scores the result of genetic differences between individuals or the result of differences in the individual environments that children experience in their formative years? In short, are there genes for IQ?

The answer to this question is emotionally charged for a number of reasons. One of them is the finding that average IQ scores of American blacks are about 15 points lower than average IQ scores of American whites. Some scholars, including Arthur Jensen and William Shockley (one of the inventors of the transistor) have cited this as proof that differences in aver-

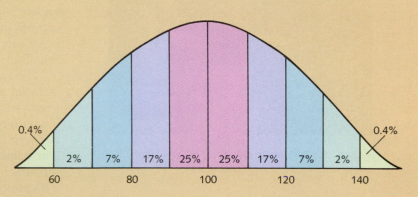

The distribution of IQ scores in an idealized human population.

age academic performance between white and black children are biological and cannot be eliminated by improved schooling. There is substantial evidence that the ability to score well on an IQ test may indeed be inherited to some degree. Identical twins, whether they are reared together or apart, are more likely to have similar IQ scores than nonidentical siblings reared together. Unrelated children who are adopted into the same home, by contrast, do not show any strong similarity of scores.

However, the situation is not as simple as these statistics may make it appear. Studies of children adopted from orphanages have shown that adoption itself may raise the adopted child's IQ score by as much as 10 points, and there is a very high correlation of IQ level with social and economic status. Individual IQ is now known to be variable, which is to say that it can be changed by study and a positive learning environment, and it may be dramatically affected by self-image (what a child

thinks of himself or herself). Therefore, many other social scientists have cautioned that the prevalence of social and economic inequality is not entirely the *result* of differences in IQ scores between different racial groups but rather acts as one of the *causes* of such differences. At this point, there is little doubt that intelligence is shaped by both genetic and environmental factors, but there is little scientific support for the notion that biological differences will undermine the positive effects of improved schooling for any children.

Figure 11.12 Human skin color is highly variable, suggesting that it is a polygenic trait.

Human skin color is a *polygenic* trait. If a single gene with two alleles governed color, humans would come in three colors at most: dark-skinned, light-skinned, and an intermediate skin tone if the alleles displayed incomplete dominance. But the many different shades of human skin color tell us differently (Figure 11.12).

Calculations based on the degree of variation in human skin color suggest that at least four different genes, each with several alleles, govern skin pigmentation. Such complexity is just one of many reasons why the traditional concept of a limited number of well-defined human races is not tenable. There are no clear categories into which the observed human phenotypes can be neatly fit.

The same is quite probably true for such subtle traits as facial appearance. Each of these characters is polygenic—the result of interactions among many genes—so it would not be appropriate to expect Mendelian ratios in the inheritance of a cute earlobe or a noble profile. The science of genetics, then, offers us no help in deciding which side of the family the baby most resembles. This important task must still be left in the capable hands of grandparents.

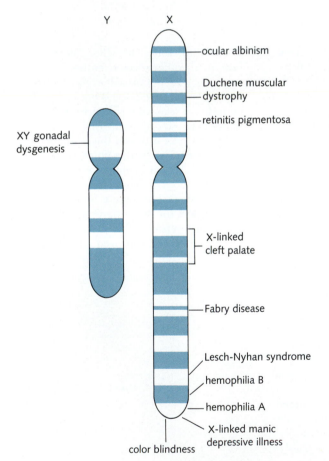

Figure 11.13 Genetic maps of the human *X* and *Y* chromosomes. The markers indicate the positions of genes that have been mapped to the chromosomes by conventional genetic techniques.

SEX-LINKED HUMAN INHERITANCE

Genes located on the *X* and *Y* chromosomes are inherited in a **sex-linked** pattern. As we have already seen, females are diploid with respect to the sex chromosomes (they have two *X*'s, whereas males are functionally haploid for both the *X* and the *Y*).

The human *X* chromosome contains a large number of important genes. The complexity of the *X* chromosome is particularly striking when the large number of *X*-linked genes is compared to the miniscule number traced to the *Y* chromosome (Figure 11.13). Because males have a single copy of the *X* chromosome, they are particularly susceptible to *X*-linked gene defects. The two *X* chromosomes of females, however, mean that they must have two copies of the recessive allele before being affected by them. Therefore, a gene defect on one of the *X* chromosomes of a female is no more likely to be expressed than a gene defect on any other chromosome, except in cases of Lyonization (as we will discuss on p. 225). But because males have only a single *X* chromosome, *any* defective gene on that chromosome is expressed.

The mathematics of sex-linked disorders are revealing. If 1 percent of the *X* chromosomes in a population contain a particular recessive gene defect, that genetic disorder is expressed in 1 percent of the males but in only 0.01 percent of the females (because the chances of *both* X chromosomes having the defect are $0.01 \times 0.01 = 0.0001$). Thus 1 in 100 males, but only 1 in 10,000 females, will express the disorder.

Sex-Linked Disorders

Hemophilia "Bleeders' disease," or **hemophilia,** results from a defect in one of the genes required for the normal clotting of blood. The two most common forms of hemophilia are both *X*-linked recessive disorders. When a normal allele for one of the genes is not present, a **clotting factor,** one of the proteins needed for normal blood clotting, is missing (Figure 11.14). Even a small wound can be serious for a hemophiliac because it is so hard to stop the bleeding. Rough physical activity often has to be avoided because of the possibility that a bump or bruise will cause uncontrollable internal bleeding. Although the disease cannot be cured, it can be successfully treated with regular administration of clotting factors prepared from the blood of healthy donors.

As is true of the victims of other human sex-linked disorders, the overwhelming majority of hemophiliacs are male. For a female to suffer from the disease, *both* of her *X* chromosomes must carry the recessive allele for hemophilia. This means that she must be the daughter of a man who suffered from hemophilia himself and of a woman who carried at least one gene for the disease. Because the hemophilia allele is relatively rare, this seldom happens (Figure 11.15). However, the presence of the gene on the *X* chromosome of the male has some interesting consequences that apply to other sex-linked traits as well.

For example, although the allele for the disease is expressed in males, it cannot be passed from father to son. Every boy carries the *Y* chromosome he got from his father, but his *X* chromosome is derived from his mother. Therefore, fathers can pass the gene only to their daughters.

Wounded tissue of broken blood vessel activates a blood protein (Factor VIII): This factor is missing in an individual with hemophilia A. It is sometimes called anti-hemophiliac factor.

Factor IX is always present in normal blood. This factor (also known as the Christmas factor) is missing in an individual with hemophilia B.

These two clotting factors change prothrombin, an inactive enzyme, into thrombin, its active form.

Thrombin changes fibrinogen, an abundant blood protein, into fibrin, a thread-like protein that is able to stick to itself and the blood vessel walls to produce a blood clot.

Figure 11.14 Hemophilia is an inherited disorder that affects the protein factors needed for normal blood clotting.

Figure 11.15 The inheritance of hemophilia in human families.

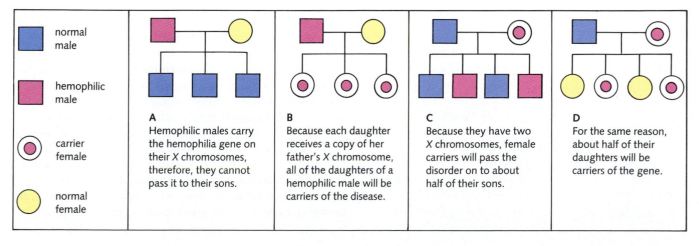

normal male

hemophilic male

carrier female

normal female

A Hemophilic males carry the hemophilia gene on their *X* chromosomes, therefore, they cannot pass it to their sons.

B Because each daughter receives a copy of her father's *X* chromosome, all of the daughters of a hemophilic male will be carriers of the disease.

C Because they have two *X* chromosomes, female carriers will pass the disorder on to about half of their sons.

D For the same reason, about half of their daughters will be carriers of the gene.

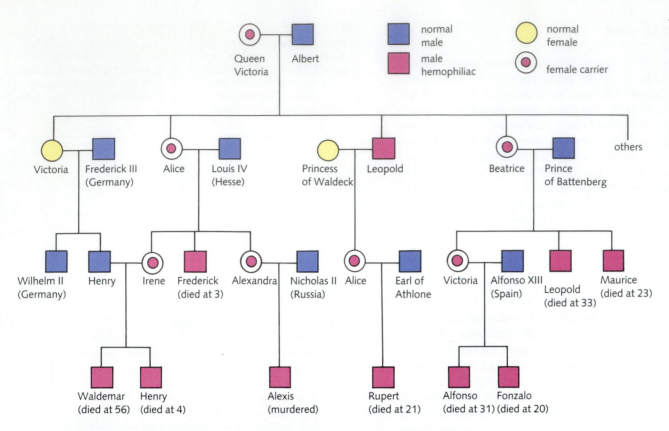

Figure 11.16 The inheritance of hemophilia in a famous human family, that of Queen Victoria of Great Britain. Only a few of her children are shown, to simplify the diagram. There was no known incidence of this disorder in the royal family before Victoria, so that it is possible that a mutation in her germ line was responsible for the disease in her offspring.

Furthermore, although women rarely suffer from sex-linked diseases, they are the only parents capable of passing such a disease on to their sons. Because every boy receives his *X* chromosome from his mother, the birth of a son with a sex-linked disease to a normal woman indicates that such a woman is a carrier for the sex-linked genetic disease. A healthy woman who gives birth to a hemophiliac son, for example, must have the hemophilia gene present on one of her *X* chromosomes. Because her other *X* chromosome must contain the normal gene, the odds of her having a second son with the disease are 50:50 (Figure 11.15).

Hemophilia has played an important role in world history. Queen Victoria (1819–1901) of Great Britain had nine children during her reign. One son, Leopold, suffered from hemophilia, and at least two daughters, Alice and Beatrice, carried a single copy of the hemophilia gene (Figure 11.16). Alice passed the gene along to her daughter Alexandra, who married Nicholas II, the tsar of Russia. Their only son, Alexis, suffered from he-

mophilia. Alexis's hemophilia, and his parents' desperation in seeking spiritual and medical treatment, preoccupied the tsar to the extent that it limited his ability to deal with issues of state. What happened during this time? The Russian Revolution deposed the tsar, and the whole royal family was executed.

Colorblindness One of the most common *X*-linked recessive traits is **colorblindness.** About 10 percent of the male population of the United States suffers from one form or another of colorblindness, which is caused by a defective allele for one of three *X*-linked genes. In the most common form of colorblindness, individuals are unable to distinguish pale shades of red from green.

Check the color test pattern reproduced in Figure 11.17. The chances are good that one of your male friends (assuming that you have at least 10 male friends) will be unable to read the pattern correctly. (There are several types of colorblindness. Total colorblindness is a very rare disorder caused by an autosomal recessive gene and is known as *achromatopsia*.) When a colorblind male has children, is the gene for colorblindness passed on to his sons or his daughters?

Lesch–Nyhan syndrome First recognized in 1965, **Lesch–Nyhan syndrome** is a rare and devastating *X*-linked genetic disease. Boys suffering from this disease have chemical imbalances in the body, some of which can be successfully treated with diet and medication. However, in some unknown way Lesch–Nyhan affects the nervous system, producing jerky, uncoordinated movements, violent behavior, and self-mutilation including biting of the hands and arms. Victims of the disease rarely live through childhood.

Retinitis pigmentosa Tristan da Cunha is a small group of islands in the Atlantic Ocean between Africa and South America. No more than 300 people live on the islands. For years they have known a strange form of progressive blindness that affects mostly young men. At first, victims of this disease must twist their heads to see clearly. Growing patches of pigmented tissue cover portions of their retinas, the light-sensitive tissue at the back of the eye. As these pigmented tissues merge, the victims lose their sight completely. **Retinitis pigmentosa,** as this condition is called, is a rare *X*-linked recessive disorder. The few women who have suffered from this disease were daughters of men who also had the disease. These islands were first settled in 1814 by 15 British colonists, one of whom may have carried this allele, explaining why this tiny population has the highest incidence of retinitis pigmentosa in the world.

Duchenne muscular dystrophy One boy in 3000 in the United States begins to experience a sudden weakness of skeletal muscles between the ages of 3 and 6. **Muscular dystrophy,** as this disorder is known, produces crippling paralysis and eventually leads to death, usually before the early twenties. The most common form is **Duchenne muscular dystrophy,** caused by a recessive *X*-linked allele. The normal version of this allele is responsible for a protein found in small amounts in muscle cells. There is no treatment or cure for the disease, but it is the object of intensive medical research (Figure 11.18).

Hypophosphatemia Not all sex-linked disorders are recessive. Several are caused by dominant alleles carried on the *X* chromosome. One such disorder, **hypophosphatemia,** causes a severe deficiency of phosphates in the blood. Because this disease is a dominant *X*-linked disorder, males with the disorder transmit it to *all* of their daughters but to *none* of their sons. Conversely, females with the disorder transmit it to about *half* of their children, regardless of the children's sex.

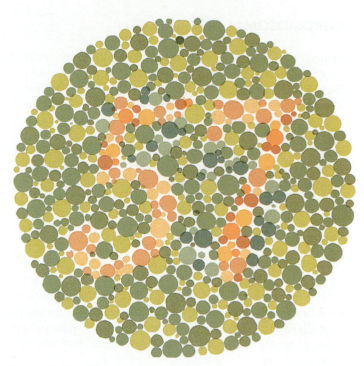

Figure 11.17 This pattern tests for red-green colorblindness, although difficulties in maintaining precise colors during the printing process make it less accurate than a test administered by a physician.

Figure 11.18 A young girl with muscular dystrophy. Females suffering from muscular dystrophy are very rare, because their fathers must have the disorder and their mothers must be carriers.

CHROMOSOMAL INHERITANCE

Human reproduction begins with the cellular events that produce sperm and egg (Figure 11.19). Meiosis, tetrad formation, crossing over, and the segregation of homologues occur just as they do in other species. When meiosis is completed, the male and female gametes contain 23 chromosomes each: 22 autosomes and 1 sex chromosome. When **fertilization** occurs, the haploid male and female gametes fuse with each other to form a diploid **zygote** with a full complement of 46 chromosomes.

Disjunction Abnormalities

Normally, each chromosome pair separates during the first meiotic division. However, in some cases the separating mechanism fails, and a pair of homologous chromosomes ends up in one of the two cells produced by that division. The failure of a chromosome pair to separate properly is known as **nondisjunction** (literally, "not coming apart").

When nondisjunction involves the sex chromosomes, zygotes may be formed with abnormal combinations of sex chromosomes (Figure 11.20). The X chromosome contains a great many genes that are indispensable to human development. Zygotes that contain only a Y chromosome (they are abbreviated OY) fail to develop. However, cells containing other unusual combinations of sex chromosomes can and do develop to maturity.

Turner syndrome People with only a single X chromosome suffer from **Turner syndrome.** Genetically, they are 45 XO (44 autosomes and 1 X chromosome; the O is included to draw our attention to the missing sex chromosome). Even though these individuals are females, their ovaries do not develop fully, they are infertile, and they do not reach normal sexual maturity. Individuals with Turner syndrome tend to be shorter than average and to have a number of characteristic features, including a slight webbing of the skin at the back of the neck, enlarged feet when they are babies, slight abnormalities in the aorta (the main blood vessel leading from the heart), and a slightly lowered hairline. Turner syndrome has been thought to involve a degree of mental retardation, although this may not be the case. In fact, many individuals who are inflicted with the syndrome are able to lead normal lives.

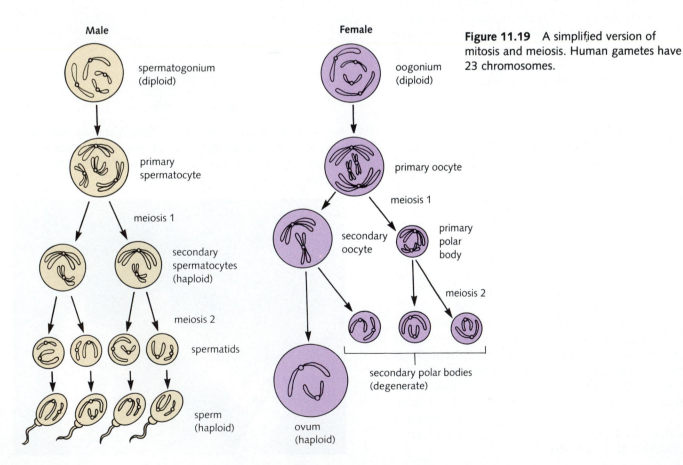

Figure 11.19 A simplified version of mitosis and meiosis. Human gametes have 23 chromosomes.

Klinefelter syndrome Individuals who are genetically 47 *XXY* suffer from **Klinefelter syndrome,** named after the physician who first recognized and described it. These individuals are sterile males. They tend not to develop the sex-related body structure of most males (wide shoulders and narrow hips), and they tend to be unusually tall. At puberty about half of them develop breast tissue that is female in appearance. Males with Klinefelter syndrome tend to be mentally retarded, although this varies greatly from one individual to the next. There are more severe variants of Klinefelter syndrome in which individuals have a genetic makeup of 48 *XXXY* or 49 *XXXXY*. In these cases, mental retardation is pronounced.

Other sex chromosome abnormalities There are also people whose chromosomal compositions are 47 *XXX* and 47 *XYY*. As you might expect, the former are females and the latter males. Although there have been some controversies in each case, no clear pattern of abnormalities is associated with either of these conditions.

Autosomal Trisomy

We have already seen how nondisjunction can cause abnormalities in the number of sex chromosomes. What happens when nondisjunction occurs in one of the autosomal chromosomes? One of the most common genetic abnormalities is known as **trisomy 21,** in which a child is born with *three* copies of chromosome number 21. It might seem that having an extra copy of a particular chromosome would cause no great harm, but this is not the case. Trisomy 21 is also known as **Down syndrome** and is associated with mental retardation, reduced resistance to infection, and a greatly lowered life expectancy. The degree of mental retardation associated with Down syndrome varies from one individual to the next, and Down children are often warm, loving individuals who are capable of developing strong relationships with those around them (Figure 11.21; See Theory in Action, Prenatal Genetics, p. 224).

Why is an extra copy of chromosome 21 so serious? The real answer may be that trisomy 21 is the *least* serious of the many possible trisomies. In fact, there are medical reports of at least two other trisomies (chromosome numbers 18 and 13), each of which results in a more serious set of abnormalities than Down syndrome. Many other trisomies are so serious that fetuses affected by them do not develop to term. Ironically, we may see so many children with trisomy 21 (about 15 per 10,000 births) precisely because the effects of Down syndrome are so *mild*.

The incidence of Down syndrome increases quite dramatically in mothers over the age of 35 (Figure 11.22).

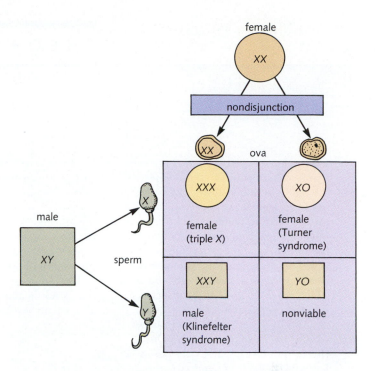

Figure 11.20 When nondisjunction, a failure of chromosomes to separate during meiosis, occurs in the sex chromosomes, the zygote may contain an abnormal number of sex chromosomes.

Figure 11.21 A child with Down syndrome (left). Partial karyotype of a child with Down syndrome (right). Note the presence of an extra copy of chromosome #21.

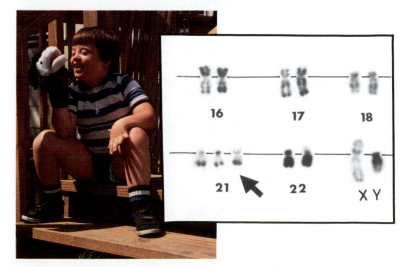

Prenatal Genetics

Until relatively recently, the only way that a woman could find out whether the baby she was carrying was suffering from a genetic disease was to wait for the child to be born. But it is now possible to routinely diagnose a number of genetic disorders while a child is still developing within the uterus. *Amniocentesis* is a medical procedure in which a needle is carefully inserted into the uterus to withdraw a small amount of the amniotic fluid that surrounds the developing fetus. This fluid contains cells that were released from the developing embryo and are genetically identical to it. The cells can be grown in culture and examined by a number of methods in the lab (see the accompanying figure).

In recent years a new technique, *chorionic villi biopsy,* has become common. A thin tube is inserted through the vagina into the tissue surrounding the placenta, and cells from the chorion (a tissue surrounding the fetus) are removed. These cells are derived from the developing fetus, and they can be examined directly for abnormalities. This method has the advantage that the cells collected can be examined immediately. With amniocentesis it often takes three or four weeks to grow enough cells for analysis. Chorionic villi biopsy can also be performed earlier in a pregnancy than amniocentesis.

The number of defects that can be detected by these techniques is growing larger every year. A few disorders, such as Down syndrome and Turner syndrome, are easily discovered in karyotypes of the fetal cells. Karyotypes can also inform prospective parents whether their child is a girl or a boy. Chemical tests on the fetal cells can detect a host of other genetic disorders. Some of these are diseases that result in abnormally high or abnormally low levels of specific carbohydrates, lipids, or proteins in the developing embryo. Tay–Sachs disease and sickle-cell anemia can be detected by such tests. The pathways by which these molecules are produced are genetically determined, and more than 250 such diseases can now be detected before birth, some in time for effective treatment if the prospective parents decide not to terminate the pregnancy.

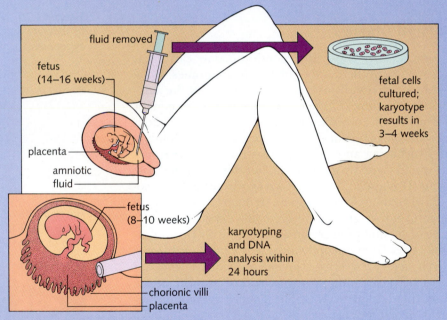

Amniocentesis and chorionic villi biopsy are two methods used to obtain cell samples from a developing fetus.

Mothers aged 20 to 30 have about 1 chance in 1000 of having a child with Down syndrome. Mothers over 40 years of age have better than 1 chance in 100. Why should maternal age make such a difference? On the day that a girl is born, every egg cell that she will release during her lifetime is already in prophase of the first meiotic division. Each egg cell remains in the first meiotic prophase until just before it is released. This long period of inactivity increases the chance that a nondisjunction error will occur when the egg cell finally completes meiosis.

X-Chromosome Inactivation

Females have two *X* chromosomes and males have one. Mary Lyon, a British geneticist, proposed that only one *X* chromosome is active in a female cell. Her model, called **Lyonization,** suggests that one of the two *X* chromosomes is inactivated during the process of embryonic development. The other *X* chromosome remains active and genes on that chromosome are expressed. Because this occurs early in development, females are actually "mosaics," composed of small patches of tissue in which alternating *X* chromosomes are active.

The most obvious examples of female mosaicism are calico cats. Many domestic cats have dark patches of fur on a white background. In males, the dark patches may be either yellow or black, depending on which allele is found on the single *X* chromosome. However, when a female is heterozygous for coat color, the patches of skin in which one *X* chromosome is active produce orange fur, and those patches in which the other *X* is active produce black fur. The result is the striking three-color calico (Figure 11.23). In females, the patches of fur on a calico cat occur because of random inactivation of the *X* chromosome in the cells that produce coat color. Calico cats are almost always female, and the few males that have been reported have been sterile, implying that they may be *XXY*. (On the theory that a male, if found, would be so rare as to be very valuable, calico cats are sometimes called "money cats.")

Lyon's ideas were also supported by the discovery of a dark-staining structure called a **Barr body** (Figure 11.23) within the nucleus of female cells. A Barr body is actually a single condensed, inactivated *X* chromosome. The active *X* chromosome remains dispersed in the nucleus and is not visible during interphase.

Lyonization occurs in humans, too. Patches of color-insensitive cells have been demonstrated on the retinas of women who carry one copy of a colorblindness allele, although enough of the retina responds normally to light to produce normal color vision. *Anhidriotic dysplasia*, a disorder that prevents the formation of sweat glands, also shows a mosaic pattern of development. Males with the allele have no sweat glands. Heterozygous females have patches of skin with normal sweat glands and patches of skin without sweat glands (Figure 11.24).

CHROMOSOME MAPPING

Mapping Human Genes

We saw in Chapter 10 how experimental crosses in other species can be used to determine recombination frequencies between two genes located on the same chromosome. We can do the same thing with human genes,

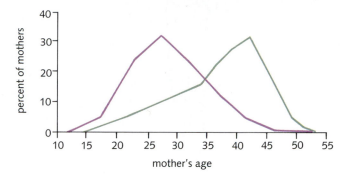

Figure 11.22 Down syndrome shows a strong correlation with maternal age.

Figure 11.23 (top) The three-color coat of a calico cat is the result of *X*-chromosome inactivation. (bottom) The inactivation of an *X*-chromosome produces a dense structure in the nucleus known as a Barr body. Barr bodies are formed in cells with 2 or more *X* chromosomes, including normal females (*XX*) and males with Klinefelter's syndrome (*XXY*).

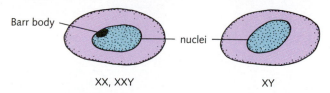

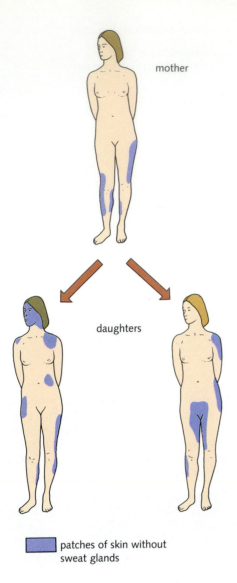

mother

daughters

patches of skin without
sweat glands

Figure 11.24 Mosaic women showing patches of skin without
sweat glands.

but only up to a point. Although classical genetics has
produced a limited amount of information regarding the
placement of linked genes (especially on the X chromosome), it has not been possible to use standard genetic
techniques to construct human gene maps in the same
detail as has been achieved for *Drosophila* or even for
laboratory mice.

You may remember that our ability to calculate recombination frequencies depends on having large numbers of offspring available to generate reliable statistics.
This is usually not possible with humans. It also is not
possible, for obvious reasons, to generate stocks of organisms that are homozygous recessive and use them in
experimental crosses with heterozygous organisms. As
we noted earlier, *Homo sapiens* is not a very good organism
for studies of genetics. As we will see in Chapter 23, however, the new tools of molecular biology have superseded
classical genetics and have begun to supply a wealth of
information about the organization of the human genome.

Gene Mapping by Cell Fusion and Culture

The ability to grow human cells in culture has made it
possible to determine the chromosomal location of genes
in a new way, using a technique known as **somatic cell
hybridization.** One of the first applications of this technique was in the search for the chromosome containing
the gene for a protein called **thymidine kinase (TK),**
which is needed to synthesize thymidine, a molecule
essential to cell growth. Special techniques are used to
fuse in culture a human cell (which contains the TK gene
on one of its chromosomes) with a mouse cell that lacks
its own TK gene. The fused **hybrid cells,** containing a
mixture of mouse and human chromosomes, are then
grown in culture for many generations.

Now something interesting happens. One at a time,
the human chromosomes are lost from the hybrid cells.
After many rounds of cell division, only a few of the
human chromosomes are left (Figure 11.25). If this is
done in a medium that lacks thymidine, any cell that
discards the chromosome carrying the TK gene dies.
Investigators then simply identify each of the human
chromosomes still present in the cells. In this experiment, they found that every living cell contained human
chromosome number 17. The TK gene had to be on
chromosome 17!

Chromosome Deletions and Translocations

For quite some time, biologists have studied chromosomes as though the genes they contain were locked in
place. However, it has recently been revealed that this is

not necessarily so and that changes in chromosome organization and structure are common. The several well-known kinds of changes include **deletions** (part of a chromosome is missing), **inversions** (part of a chromosome is inverted with respect to the rest of the chromosome), and **translocations** (a portion of one chromosome is broken off and attached to another chromosome).

We can use these events to map gene locations on human chromosomes. Figure 11.26 shows a human chromosome *deletion* that occurred in a young boy with crippling birth defects. Researchers were able to grow a few of the boy's cells in culture and to determine that a part of chromosome number 2 had been deleted (Figure 11.26). Chemical studies of the cells confirmed that the gene for *acid phosphatase*, an important enzyme, had been lost as a result of the deletion. Therefore, the gene must have been on that part of chromosome number 2.

Translocations have made it possible to determine the portion of chromosome 21 that is responsible for Down syndrome. Researchers discovered a small fraction of Down syndrome patients whose karyotypes were 46 *XY* instead of the usual 47 *XY*. Soon it was discovered that these patients had a *translocation* in which a tiny portion of chromosome 21 had become attached to one of the other autosomal chromosomes. Because they also had two normal copies of chromosome 21, they actually had *three* copies of the portion involved in the translocation. Although the translocated fragments were different sizes in different patients, each contained one critical region of chromosome 21. Researchers are now trying to analyze the genes in this segment of the chromosome to understand the cause of the disorder.

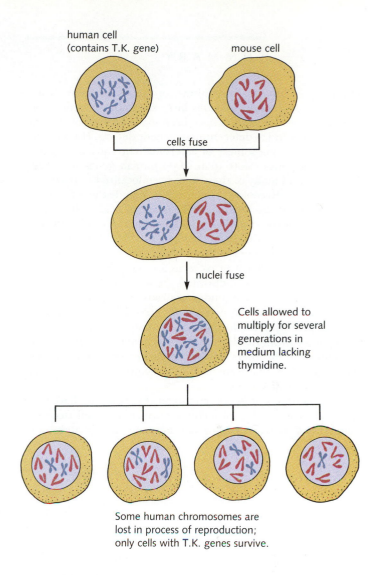

Figure 11.25 Somatic Cell Hybridization, a technique that has been used to localize genes to particular chromosomes.

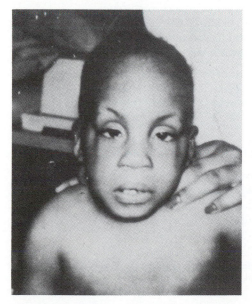

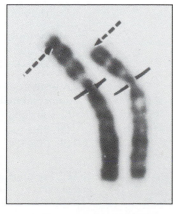

Figure 11.26 A young boy with crippling birth defects caused by chromosomal deletions (left). The deletion took place in the terminal portion of chromosome #2. The dotted arrows point to the position of the deletion (right).

Turner syndrome *p. 222*
Klinefelter syndrome *p. 223*
trisomy 21,
  or Down syndrome *p. 223*
Lyonization *p. 225*

Barr bodies *p. 225*
deletions *p. 227*
inversions *p. 227*
translocations *p. 227*

SUMMARY

Humans are not ideal organisms for genetics studies. Nonetheless, because of our obvious interest in human genes, considerable scientific efforts have been made to develop human genetic maps and to detect human genetic defects. Human cells contain 46 chromosomes, 44 autosomal chromosomes, and 2 sex chromosomes. More than 4,000 human genes have been described, and many of these have been localized to particular chromosomes. Recessive human alleles include those for albinism, Tay–Sachs disease, cystic fibrosis, and sickle-cell anemia. Dominant human alleles include those that cause achondroplasia, Huntington's disease, and polydactyly. Many human traits are produced by genes with multiple alleles, and others, like skin color and fingerprint patterns are polygenic.

A number of human disorders are sex-linked, and the most common sex-linked conditions are caused by genes on the X chromosome, including hemophilia, colorblindness, and muscular dystrophy. Non-disjunction, the failure of chromosomes to separate properly during meiosis, results in a number of chromosomal disorders, including Turner syndrome (45 *XO*) and Klinefelter syndrome (47 *XXY*). Down syndrome results from a nondisjunction event.

The positions of genes on human chromosomes have been mapped by a number of techniques, including cell fusion in culture and the analysis of chromosome breaks and translocations.

STUDY FOCUS

After studying this chapter, you should be able to:

■ Show how human inheritance can be analyzed by techniques used in genetics.

■ Describe special problems, diseases, and disorders caused by defective genes.

■ Explain the techniques used in chromosome mapping.

■ Discuss the heritability of eye color, facial appearance, and other traits.

SELECTED TERMS

sex chromosomes *p. 208*
autosomal
  chromosomes *p. 208*
karyotyping *p. 209*
pedigree *p. 210*
albinism *p. 211*
Tay–Sachs disease *p. 211*

Huntington's disease *p. 213*
antigen *p. 214*
codominance *p. 214*
Rh factor *p. 214*
antibody *p. 215*
hemophilia *p. 218*
nondisjunction *p. 222*

REVIEW

Discussion Questions

1. What are some of the reasons why the fruit fly (*Drosophila melanogaster*) is a better organism than the human (*Homo sapiens*) as a subject in studies of genetics?

2. At which stage of meiosis are tetrads formed in human gamete formation? At which stage does crossing over occur?

3. Carriers of Huntington's disease are generally unaware until late adulthood that they carry the gene. Why?

4. What evidence suggests that fingerprints are determined by a polygenic system, rather than by a system of one or two genes with multiple alleles?

5. What information would you give the same couple if their first child were afflicted with Down syndrome? How would the ages of the two parents be significant?

Objective Questions (Answers in Appendix)

6. A karyotype is used to
   (a) determine the stage of mitosis.
   (b) determine the number of chromosomes.
   (c) substitute a normal allele for a defective allele.
   (d) determine the stage of meiosis.

7. A chromosome map is used to determine
   (a) the order of genes on a chromosome.
   (b) the type of gene on a chromosome.
   (c) if Lyonization has occurred.
   (d) if nondisjunction has occurred.

8. Which genetic disorder is caused by a recessive allele?
   (a) Tay–Sachs disease　　(c) Huntington's disease
   (b) Darwin tubercle　　(d) Polydactyly

9. An example of polygenic inheritance is
   (a) blood type.　　(c) fingerprints.
   (b) Darwin tubercle.　　(d) albinism.

10. A woman who is heterozygous for color blindness marries a normal male. What is the probability that their child will be colorblind?
    (a) 25%　　(c) 75%
    (b) 50%　　(d) 100%

NOTE: See Appendix for Genetics problems, answers to the problems, and answers to in-text questions.

CHAPTER 12

Darwinian Theory Evolves

During the early twentieth century, geneticists explained the nature of heritability and demonstrated several sources of variation among individuals in populations. Then, as fields related to Mendelian genetics matured, mathematicians began to build on the simple models of gene segregation and recombination represented by Punnett squares. Soon they began modeling the more complicated behavior of genes in plant and animal populations, creating the field of *population genetics* that transformed evolutionary theory.

The 1930s and 1940s were filled with intense debate and creative thinking among evolutionary biologists. During that time, researchers combined the essence of Darwin's original theory with insights afforded by new developments in Mendelian and population genetics, paleontology, and natural history. The result was a body of theory known as the modern evolutionary synthesis—the core of the biological sciences today.

Because the tenets of the modern synthesis are so important, we will spend most of the next two chapters discussing them. It is important to remember, however, that evolutionary theory, like all scientific theory and the organisms whose nature it seeks to explain, is always subject to change. The birth of molecular biology and recent advances in population genetics that began in the 1960s and 1970s (fueled in part by the computer revolution) have initiated another transformation in evolutionary theory that is still in progress today. We will mention some of the more fundamental revelations of molecular evolution in these chapters, but will reserve detailed treatment for Chapter 23.

SPECIES AND FITNESS: GENETIC DEFINITIONS

Because genetics is now the keystone of evolutionary biology, we will begin our discussion of evolutionary theory by redefining several important Darwinian concepts in genetic terms.

The Species

Biologists have historically recognized and classified species strictly according to physical characteristics, or what we can now call phenotype (see Chapter 10). But genetics and population biology suggest that species can and should be defined in terms of genes and the reproductive behaviors that affect the movements of genes.

From a genetic perspective, a **species** is a group of natural populations that can (at least potentially) interbreed among themselves to produce fertile offspring and are reproductively isolated from other such groups. Because mating within and between populations of a species means that members mix genes among their offspring, a species is also a genetic unit whose members share a group of genes called a **gene pool.**

This genetic definition is important. It means that when a genetic change occurs in some individuals of a species, that change can spread through the gene pool of that species but not to other gene pools (Figure 12.1). Thus each species evolves as a separate unit. You will understand the importance of this genetic definition of species more clearly after we discuss, in the next chapter, the way new species originate.

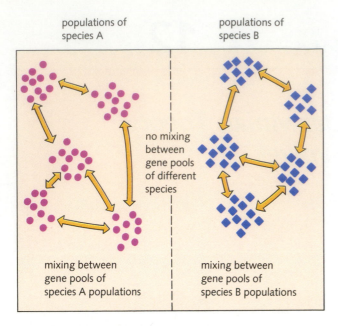

populations of species A populations of species B

no mixing between gene pools of different species

mixing between gene pools of species A populations

mixing between gene pools of species B populations

Figure 12.1 Genes mix among interbreeding populations of a species, but because of reproductive isolation, they do not mix between species.

less fit   more fit

Figure 12.2 Evolutionary fitness is ultimately determined, not by any particular characteristics of its physiology, but by the number of copies of its genes an organism successfully passes on to the next generation.

Evolutionary Fitness and Adaptation

Genetics also provides a concise definition of evolutionary or Darwinian fitness: An individual's **evolutionary fitness** is defined by the size of its probable genetic contribution to the next generation. In other words, an organism that produces many offspring, carrying many copies of its genes, has high fitness (Figure 12.2).

An **evolutionary adaptation** is defined as any genetically controlled characteristic that increases an organism's fitness. Larger muscles that are built up through exercise (and thus are an acquired characteristic) are not considered an evolutionary adaptation. A gene that causes an individual's muscles to grow faster in response to exercise, on the other hand, could be considered an adaptation—*if* it increased the individual's fitness.

EVOLUTION AND NATURAL SELECTION

Defined in genetic terms, **evolution** is any change over successive generations in the relative frequencies of different genes in the gene pool of a population. We can see the logic in that definition and examine the process of evolution by natural selection by looking at any heritable, adaptive characteristic in a population of organisms.

Observable Variation in Organisms

Every characteristic of organisms in a population has a frequency distribution (Figure 12.3). That is, if you look at any heritable trait in a large enough population, you will see that it has some average value and some degree of variation or deviation from that average value. Much of this variation represents the constant reshuffling of genes that occurs each time an organism reproduces sexually, but it also includes the much rarer occurrence of new mutations in existing genes. To the extent that this phenotypic variation reflects differences in genotype, it can be affected by evolutionary change. And to the extent that this variation affects the fitness of organisms in the population, it can provide the raw material on which natural selection operates.

Remember that genetic variation is random. It does not occur because an organism *needs* or *wants* to evolve. (That's close to what Lamarck thought.) There is no way for an organism to cause variation in its genes, and there is no way to prevent it. Similarly, specific environmental conditions do not give rise to specific sorts of mutations or other variations in genotype that are useful to organisms. In other words, genes do not "know" precisely how to mutate or recombine in order to benefit their organism.

The occurrence of a genotype that endows an organism with resistance to an environmental poison, for example, does not occur *because* the poison is present in the environment. In other words, the poison in the environment neither produces new genes that confer resistance to poison nor directs the reorganization of existing alleles to confer resistance. Instead, the existence of the poison simply increases the fitness of those individuals that *already* carry resistance genes or gene combinations. This was demonstrated by a series of elegant experiments examining the resistance of bacteria to penicillin and was later shown to hold true for the development of resistance to DDT in *Drosophila* (Figure 12.4).

Note that humans can increase the *speed* at which new variants appear in a population by exposing organisms to radiation or chemicals that cause mutations to occur more often. And as you will learn in Chapter 23, genetic engineers can move certain genes and groups of genes from one organism to another, dramatically increasing the power of artificial selection. But even the most sophisticated laboratory techniques cannot yet cause specific sorts of useful mutations in existing genes, and we still have no idea how to create new genes "from scratch."

Figure 12.3 Military recruits arranged in groups by height. Notice that this grouping produces a distribution much like those represented in the bar graphs and histograms you have seen for characteristics in animals and plants.

Figure 12.4 An elegant experiment that uses a technique called replica plating to demonstrate that mutations conferring resistance to penicillin are not *caused* by the presence of penicillin in the environment. The same number of resistant colonies are found in both cases; the presence of penicillin does not affect the appearance of the resistant mutation.

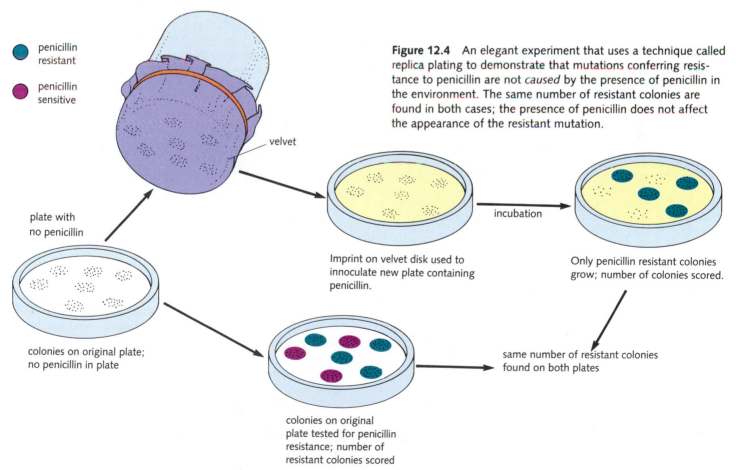

● penicillin resistant

● penicillin sensitive

velvet

plate with no penicillin

colonies on original plate; no penicillin in plate

Imprint on velvet disk used to innoculate new plate containing penicillin.

incubation

Only penicillin resistant colonies grow; number of colonies scored.

colonies on original plate tested for penicillin resistance; number of resistant colonies scored

same number of resistant colonies found on both plates

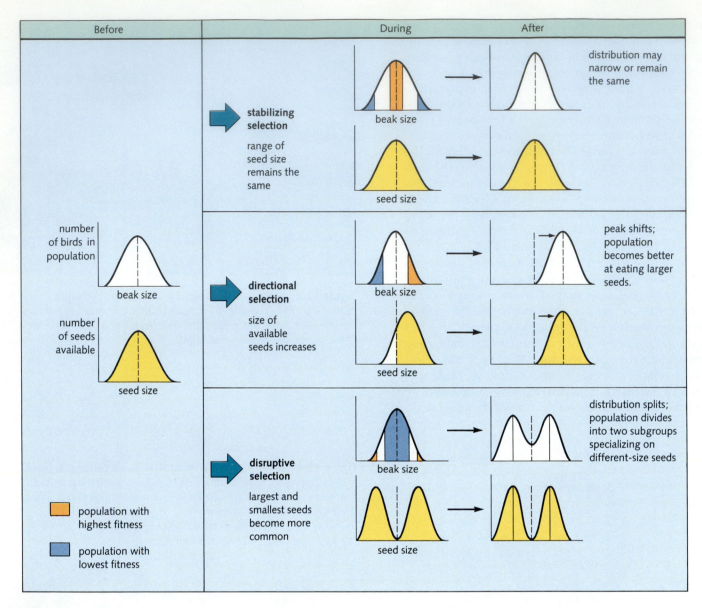

Before	During	After

stabilizing selection

range of seed size remains the same

beak size

seed size

distribution may narrow or remain the same

number of birds in population

beak size

number of seeds available

seed size

directional selection

size of available seeds increases

beak size

seed size

peak shifts; population becomes better at eating larger seeds.

disruptive selection

largest and smallest seeds become more common

beak size

seed size

distribution splits; population divides into two subgroups specializing on different-size seeds

☐ population with highest fitness

☐ population with lowest fitness

Figure 12.5 Distribution of beak length in a hypothetical population of birds and the effect of changes in seed availability on that distribution. In each case, the distribution of beak length in the population changes over time, because some individuals in the population (blue) have lower fitness and others (orange) higher fitness under the new conditions. Stabilizing selection: Average-sized seeds become more common, and the birds become more specialized with around the same (average) beak length. Directional selection: Larger seeds become more common, and the bird population evolves larger beaks. Disruptive selection: Average-sized seeds become less common, and larger and smaller seeds become more common. In response, the bird population splits into two subgroups specializing in eating the larger and smaller seeds, respectively.

Natural Selection: Effects on Phenotype

Natural selection, however, never touches genes themselves. Rather, it operates only through the effects that variations in genotype have on organisms' phenotypes. Selection can influence the distribution of phenotypic characters in a population in three basic ways: *stabilizing selection*, which tends to maintain the genetic status quo, *directional selection*, which tends to shift the mean of the frequency distribution in one direction or another, and *disruptive selection*, which tends to split a population into two divergent forms.

We will examine these three categories of natural selection in two ways. We will first propose a series of hypotheses about the effects of different environmental conditions on a bird population modeled after Darwin's finches. We will then examine data from controlled observations and experiments that test our hypotheses on other organisms.

The hypothetical population   Suppose a population of seed-eating birds exhibits substantial variation in beak size, a genetically controlled characteristic. Suppose also that this variation in beak size affects the birds' ability to feed on seeds of different types. Birds with small beaks can feed most effectively on smaller seeds, whereas birds with bigger, stronger beaks, can handle larger, thicker, harder seeds more easily. The frequency distribution of beak size in our hypothetical population at the outset is shown in Figure 12.5.

Stabilizing selection   If the sizes of seeds available to these birds remain unchanged, natural selection will favor individuals with beaks best suited to average-sized seeds. **Stabilizing selection** will tend to eliminate individuals with beaks very much larger or much smaller than average, because those individuals will be able to handle most efficiently only some of the available seeds and so will have lower fitness (Figure 12.5, *top*). Because sexual reproduction tends to maintain the original degree of genotypic variation in each generation, the frequency distribution of beak size will not necessarily narrow, but it is not likely to broaden either. Stabilizing selection discourages evolution away from the average condition.

Directional selection   If, however, the population of small-seeded plants declines for some reason, or if the population of large-seeded plants increases, the relative fitness of birds with the smallest beaks will fall (Figure 12.5, *center*). After a few generations, this type of **directional selection** will cause the average beak size in the population to increase. The resulting bird population will be better adapted to feeding on larger seeds.

Disruptive selection   Now suppose instead that the supply of intermediate-sized seeds suddenly decreases. (In nature this could happen either because plants producing those seeds suddenly decline in numbers or because a new bird species appears that eats large numbers of mid-sized seeds.) In this case, the relative fitness of individuals near the middle of the frequency distribution would fall relative to that of those individuals at either extreme (Figure 12.5, *bottom*). The resulting **disruptive selection** would tend to split the population into two groups, one with larger beaks for large seeds and one with smaller beaks for small seeds.

VARIATION AND SELECTION: A CASE STUDY

The preceding discussion makes straightforward predictions from evolutionary theory, but it is entirely hypothetical. Can we test any of its predictions on a real population in nature? Thanks to a long series of observations and experiments on the fascinating birds from the Galapagos Islands called Darwin's finches, we can. These 14 species are so ecologically diverse that Darwin first classified them as members of several separate subfamilies, including finches, orioles, and warblers. Only much later, after showing his specimens to ornithologists in England, did he realize that they were, in fact, all finches. His confusion was understandable; some of these birds eat insects, others eat fruit, several eat seeds, and there is even a "vampire finch" that drinks the blood of larger seabirds (Figure 12.6). You can readily see that beak size and shape vary extensively among these species (Figure 12.7).

Figure 12.6   These photos show the remarkable diversity of beak shapes found among Darwin's finches. (top, left) A woodpecker finch, *Cactospiza pallida*. (bottom, left) A medium ground finch, *Geospiza fortis*. (right) A sharp-beaked ground finch, *Geospiza difficilis*.

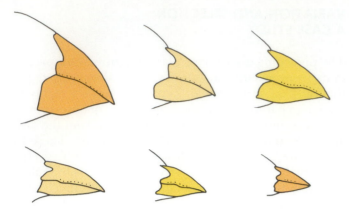

Figure 12.7 Here you can see clearly the variations in beak size among species that characterize Darwin's finches. Clockwise from top left, *magnirostris, fortis, conirostrus, fuliginosa, difficilis,* and *scandens.* [From Grant (1986) after Abbott *et al.* (1977).]

It is easy to accept that these major variations in beak size and shape *between* species—along with related variations in the size and shape of muscles that control the beak—are related to the ways in which birds use those beaks in feeding (Figure 12.8). Long, pointed bills work well for probing flowers or woody plant tissues. Massive, deep beaks are well suited to crushing large seeds. Curved beaks with sharp tips are good for biting and grasping small insects. Thus we can hypothesize (if we agree with the premise of natural selection in advance) that evolutionary pressure to avoid competition has somehow "pushed" these birds into different ecological niches.

But the charge of evolutionary biologists is to *test* such hypotheses rigorously, not to accept them on faith. And to test an hypothesis about the way natural selection has acted on finch populations over time, we must make observations at a finer level of detail. We must look for variation *within* species and search for evidence that natural selection can, in fact, cause differential survival based on that variation.

Because these birds, quite apart from their place in the history of evolutionary thought, are biologically fascinating, they have been studied extensively over the last century. The results of those studies fill several books, and we cannot do them all justice here. In this chapter, to give an example of modern evolutionary biology in action, we will concentrate on experiments examining variation and directional selection in seed-eating finch species. These studies, summarized, organized, and significantly expanded by Peter and Rosemary Grant of Princeton University, are among the most elegant examinations ever performed into the nature of natural selection. In the next chapter, we'll expand our observations to consider how this group of species arose.

Figure 12.8 A relatively simplistic—yet revealing—comparison between selected finch beak types and certain human tools that apply force in different ways for different ends. [From Grant (1986) after Bowman (1963).]

Platyspiza	*Geospiza*	*Pinaroloxias*	*Certhidea*
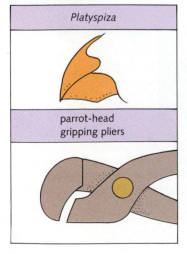	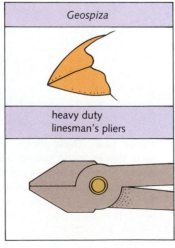	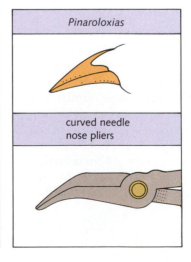	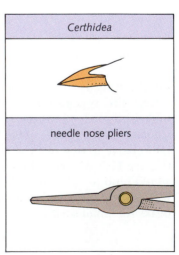
parrot-head gripping pliers	heavy duty linesman's pliers	curved needle nose pliers	needle nose pliers

Variation in Beak Size: Raw Material

Measurements of both living birds and preserved specimens in museums make it clear that there is substantial variation in beak size among individuals of each of several species of seed-eating Galapagos finches. This sort of variation, evident in graphs such as those shown in Figure 12.9, agrees with the first premise in the hypothetical case outlined previously.

But are these heritable differences in beak size of any adaptive significance to the birds? In other words, is the difference in fitness caused by these variations sufficient to cause differential survival of these birds in their natural environment? Early workers in the Galapagos guessed "yes," but they had no quantitative data to support their hypothesis. The Grants and other workers since that time have provided solid evidence that beak size is, in fact, of real importance.

Beak size and seed-handling ability   By observing birds feeding in their natural habitats, several researchers found that individuals with different beak sizes fed on different-sized seeds. Birds seen feeding on the smallest and softest seeds had smaller beaks than those seen feeding on larger, harder seeds that are tougher to crack.

This relationship is true both within and between species; small-beaked species eat a higher percentage of smaller seeds than large-beaked species, and smaller-beaked individuals within each species prefer smaller seeds on average than their larger-beaked fellows. These data were bolstered by studies showing that larger, tougher seeds were more readily accepted by larger-beaked birds (Figure 12.10). Smaller-beaked birds either would not or could not utilize these seeds.

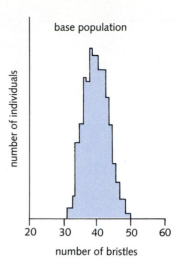

Figure 12.9   This graph, showing the frequency distribution of *Drosophila* individuals with various numbers of abdominal bristles, demonstrates the classic pattern of variation found in many heritable characteristics of organisms. The same distribution occurs in final beak sizes, as you will see in Chapter 13.

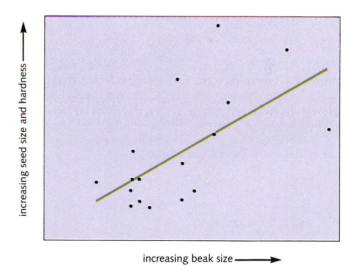

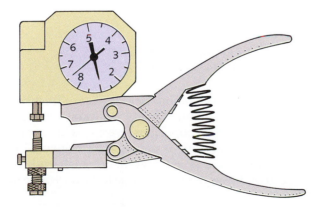

Figure 12.10   (left) This device, used to measure seed hardness, shows the ingenuity often required in the design of field experiments in evolutionary biology. To test the hypothesis that beak size affects the birds' ability to handle large, hard seeds, the experimenters had to do more than guess at seed toughness; this device provided replicable, quantitative measurements of seed hardness. (above) This figure shows the positive correlation between beak size in ground finches and an index that represents the relative size and hardness of seeds in their diets. [From Abbott et al. (1977).]

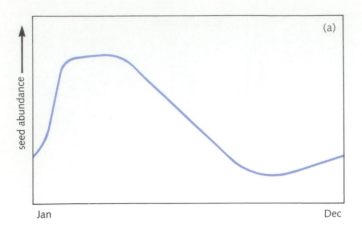

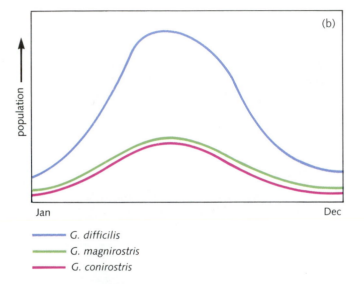

G. difficilis
G. magnirostris
G. conirostris

Figure 12.11 Relationship between seed availability and finch populations in the highly seasonal Galapagos climate. (top) Annual fluctuations in seed abundance. (bottom) Annual fluctuations in the numbers of three species of ground finches. The rise in bird populations represents the onset of the breeding season. The subsequent fall occurs as individuals starve when food is scarce. [From Grant and Grant (1980).]

The importance of food limitation It has become clear that in most years, food is a limiting resource for these finches. The Galapagos Islands have extremely seasonal climate; rains, during which plants grow and set seed, are followed by dry months during which many plants enter dormancy. The supply of seeds thus rises and falls markedly (Figure 12.11, *top*). Finch populations also fluctuate seasonally, rising as new individuals hatch and falling as many of them fail to survive food shortages during the dry season (Figure 12.11, *bottom*).

Such seasonal fluctuations, it turns out, are part of a larger cycle during which wetter years alternate with drier ones. These long-term cycles, too, affect finch population size. During one exceptionally wet period during the years 1982–1983, plants grew more abundantly, and finches bred far more prolifically than usual. And during a prolonged drought, documented by the Grants from mid-1976 through the end of 1977, seed production fell markedly and finch populations plummeted from around 1200 individuals to no more than 180 (Figure 12.12*a,b*).

Response to Drought: Directional Selection

Interestingly, the birds that survived the seed shortage of 1977 were *not* simply a representative sample of the birds present before the drought began. Population surveys taken before, during, and after the drought showed clearly that large-beaked birds survived the famine in greater numbers than small-beaked members of the same species.

Why might this have been the case? Both small and large seeds were available at the beginning of the drought, offering food for both large- and small-beaked birds. But during the drought, as the seed supply ran low, the supply of small seeds was exhausted. By early 1977 the birds had to depend almost exclusively on large, hard-to-crack varieties. Large-beaked birds, which could exploit this food, survived to reproduce later on; small-beaked individuals starved. The result, a classic demonstration of directional selection in action, was an increase in the average beak size of surviving birds (Figure 12.13*a,b*).

OTHER EXPERIMENTAL STUDIES OF NATURAL SELECTION

In addition to ambitious field studies of finches and other organisms in nature, many laboratory studies of both plant and animal populations have sought to test the effects of natural selection or to search for evidence of selection in action. Here we will present one such study examining each of the modes of natural selection we have described.

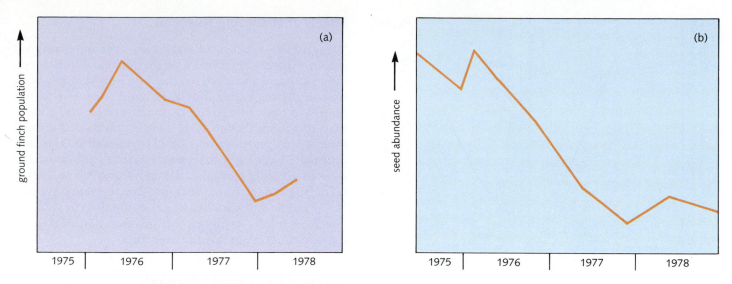

Figure 12.12 During a particularly severe drought in 1977, both seed abundance (a) and ground finch populations (b) dropped steadily. [From Boag and Grant (1981).]

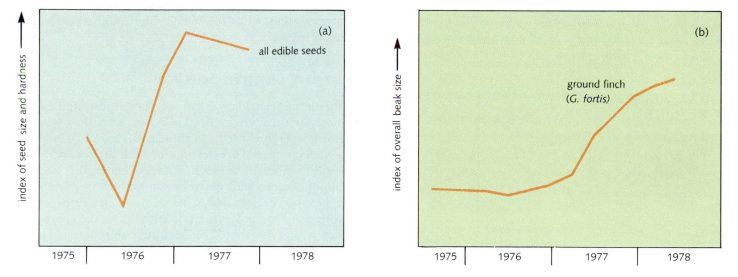

Figure 12.13 (a) During the first several months of the drought, small, easy-to-crack seeds were consumed in large numbers. As a result, the average size and hardness of remaining seeds rose steadily. (b) Forced to feed on large, hard seeds, the population of *G. fortis* was subjected to intense natural selection for large beaks. This graph shows a steady increase in an index related to both beak size and body size; the line on the graph represents the average for the remaining living birds at each point in time. [From Boag and Grant (1981).]

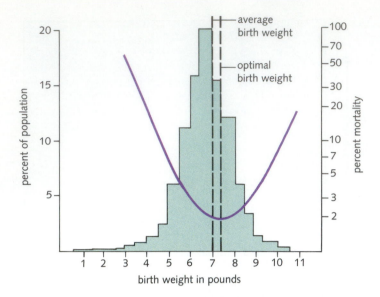

Figure 12.14 Stabilizing selection and human birth weight. The curve on this graph shows that human infants have the best chance of surviving the trials of birth if they weigh between 7 and 8 pounds; at higher and lower weights, mortality is higher. The histogram shows that average birth weight matches that optimal weight quite closely. [From Cavalli-Sforza and Bodmer (1971).]

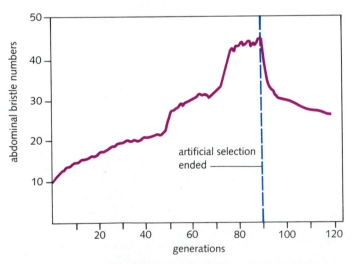

Figure 12.15 This graph shows the response of a laboratory *Drosophila* population to artificial selection for increased numbers of abdominal bristles. In generations 1 through 90, only those flies with the highest numbers of bristles were allowed to mate. Note that when this artificial selection ended after 90 generations, the average bristle number began to decline. [From Yoo (1980).]

Effects of Stabilizing Selection

We can find evidence of stabilizing selection acting on our own species by looking at the relationship between infant survival and birth weight. The weight of newborn human infants varies widely, ranging from less than a pound in cases of extremely premature births to more than 10 pounds (Figure 12.14). The survival rate for those infants also varies widely; both very small infants and unusually large ones have much higher early mortality rates than babies averaging between 6 and 8.5 pounds at birth. You can see from the figure that the actual average birth weight and the theoretical optimal birth rate are nearly identical—evidence for the operation of stabilizing selection on our species over time.

Effects of Directional Selection

The effects of directional selection have been demonstrated in the laboratory through artificial selection in a number of organisms. One variable character in laboratory populations of *Drosophila*, for example, is the number of bristles the flies carry on their abdomens. It is a simple matter to count these bristles and to select for breeding from a "wild-type" population those individuals with higher or lower numbers of bristles. Figure 12.15 shows the steady increase in bristle numbers in a laboratory fly population selected for high numbers of abdominal bristles.

Effects of Disruptive Selection

Disruptive selection has also been demonstrated experimentally through artificial selection with *Drosophila*, as shown in Figure 12.16. In these experiments, a population was subjected to selection for both higher and lower bristle numbers. The result, after 35 generations, was two subpopulations with mean bristle numbers significantly higher than normal and lower than normal, respectively. Note that Figure 12.16 lends support to the claim that even strong directional or disruptive selection—like stabilizing selection—does not necessarily decrease variation in the population.

THE HARDY–WEINBERG LAW

Once biologists defined evolution as change in the frequencies of genes in populations, they could examine the circumstances in which such genetic change occurs. In 1908 two independent investigators, George H. Hardy of Cambridge University and Wilhelm Weinberg, a German physician, realized that although the segregation

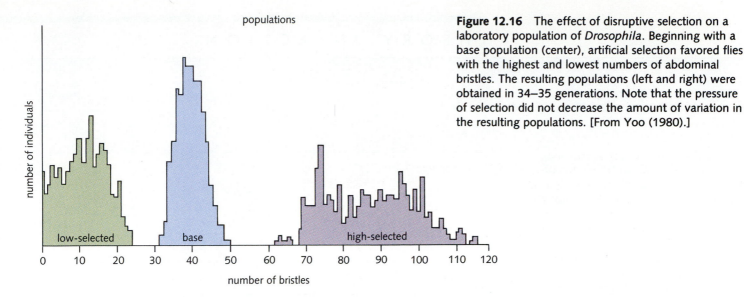

Figure 12.16 The effect of disruptive selection on a laboratory population of *Drosophila*. Beginning with a base population (center), artificial selection favored flies with the highest and lowest numbers of abdominal bristles. The resulting populations (left and right) were obtained in 34–35 generations. Note that the pressure of selection did not decrease the amount of variation in the resulting populations. [From Yoo (1980).]

and recombination of genes during mating provide genetic variability among offspring, sexual reproduction by itself would not change relative gene frequencies. They went on to describe a restricted set of conditions under which the relative frequencies of genes in a population remain the same indefinitely.

When Evolution Will Occur and When It Will Not

According to what is now called the **Hardy–Weinberg law,** the relative gene frequencies in a population remain constant from generation to generation—in other words, evolution does not occur—*as long as*:

1. Chance events do not affect the genetic frequencies in the population.

2. Mutations do not occur, or, if they do occur, they balance each other out.

3. All genotypes have equal reproductive success; in other words, there is no natural selection operating on the population.

4. There is no net flow of genes into or out of the gene pool. This means that there is no net migration of organisms into or out of the population.

5. All mating in the population is completely at random. In other words, no genotype in the population can have a preference for mating with any other particular genotype.

If and only if all these conditions are met, the gene pool of a population remains in a state of genetic equilibrium. But only rarely do all these conditions exist in nature for all of an organism's alleles at the same time. Chance *can* affect genetic frequencies in small populations. Mutations *do* occur, and it is impossible for them to be completely balanced by opposite mutations. Organisms are *constantly* entering and leaving most populations, though it is only the net genetic change that counts. And mating in natural populations is *rarely* completely random; certain phenotypes (and hence certain genotypes) are almost always preferred over others as mating partners.

Why is the Hardy–Weinberg law important if its conditions are seldom met? First, the law provides equations that predict the frequencies of different genotypes in a stable gene pool. If the relative proportions of genotypes in a population match those predicted by the equations that express the Hardy–Weinberg law, that population's gene pool is stable and the particular locus in question is not under strong selective pressure. If the genotype ratios are different from those the equations predict, the gene pool is under pressure, from natural selection or some other factor, to change.

Second, the Hardy–Weinberg law clearly defines the set of conditions under which evolution does *not* occur. Biologists who look at natural populations find that these conditions rarely all prevail at once for prolonged periods of time. We therefore know that many natural populations do not remain in genetic equilibrium for long, and we have additional evidence that change is a fact of life on earth.

The Hardy–Weinberg Equilibrium

Imagine that you are a geneticist studying a trait controlled by two alleles, A and a, that follow the rules of simple dominance at a single locus. When you survey a natural population of organisms for this trait, you discover that only 4 percent of the population exhibits the phenotype representing genotype aa. Fully 96 percent of the population is either AA or Aa and thus exhibits the dominant phenotype. Not surprisingly, you assume that over many generations, the dominant allele will somehow displace the recessive allele, not because it confers any adaptive advantage, but just because of its greater apparent frequency in the population. Your assumption, however, would be wrong, for reasons demonstrated by the Hardy–Weinberg principle.

The Hardy–Weinberg principle, simplified for a discussion of a single locus with only two possible alleles, A and a, can be demonstrated as follows:

Let us represent the frequency of the A allele as p and the frequency of the a allele as q. Because all "places" at this locus in the population must be occupied by one allele or the other, the sum total of all occurrences of p and q must always equal 100 percent of those loci. Stated in another way, $(p + q) = 1$.

In any cross involving these alleles, there are three possible genotypes: AA, Aa, and aa.

As you can see from the accompanying diagram, one generation of random mating among the three genotypes creates a binomial distribution of those genotypes in the next generation. This occurs because segregation of these alleles produces gametes that carry the alleles in the same relative frequencies at which those alleles occur in the population. Thus the rel-

Phenotype			
Genotype	AA	Aa	aa
Frequency of genotype in population	0.64 ($p^3 = 0.64$)	0.32 ($2pq = 0.32$)	0.04 ($q^2 = 0.2$)

Frequency of each gamete type

A → 0.32 + A → 0.32 + A → 0.16 a → 0.16 + a → 0.02 + a → 0.02

$A = p = 0.8$ $a = q = 0.2$

Filial genotypes

sperm A $p = 0.8$ A $p = 0.8$ egg

AA $p^2 = 0.64$

a $q = 0.2$ Aa $pq = 0.16$ Aa $pq = 0.16$ a $q = 0.2$

aa $q^2 = 0.04$

ative frequency of A-carrying eggs is p, and the relative frequency of a-carrying eggs is q. The same is true for sperm. Accordingly, the three types of resulting zygotes are produced in the same relative numbers as the individuals in the Punnett squares you met earlier:

$$(p + q)^2 = p^2 + 2pq + q^2$$

Note that because the sum of all genotypes in the population must also equal 100 percent, $(p^2 + 2pq + q^2) = 1$.

If 4 percent of the population exhibit the aa genotype, $q^2 = 0.04$ and q therefore is 0.2. If $q = 0.2$, then $p = 0.8$. In each generation, there-

fore, the relative frequency of AA homozygotes will be p^2, or 0.64; the number of aa homozygotes will be q^2, or 0.04; and the number of Aa heterozygotes will be $2pq$, or 0.32. As long as the necessary conditions continue to be met, neither the frequency of the genotypes nor the frequencies of the alleles (p and q) will ever change from generation to generation.

You can demonstrate this stability over time to yourself by modeling this situation through several generations. To do that, begin with an imaginary population of 1000 organisms in which the first generation consists of 640 AA individuals (p^2), 320 Aa individuals (2*pq*), and 40 *aa* individuals (*q*²).

You can also see how population geneticists and evolutionary biologists actually use this seemingly abstract relationship by trying your hand at the following problems.

1. In a particular human population, roughly 1 in 5000 individuals is afflicted with a recessive genetic disorder that causes mental degeneration and death, but usually not until after age 55. What percentage of this population would you expect to be heterozygous for this gene?

Hint: Assume that this population meets all the criteria needed to satisfy the Hardy–Weinberg principle. Remember that those afflicted with the disease are homozygous recessives *(aa)* and that the total number of these individuals is q^2.

2. A researcher collects individuals from a natural population of butterflies that displays the three phenotypes indicated below. Is this particular locus at or near Hardy–Weinberg equilibrium?

Hint: Set up a diagram for this population similar to the figure included in the box, but use as your numbers the numbers of individuals of the three genotypes instead of frequencies. Calculate *p* and *q* using the binomial equation $(p^2 + 2pq + q^2)$ to predict the percentage of individuals of each genotype in the population.

Phenotype	Genotype	Number of Individuals
Heavy white spotting	(AA)	1469
Moderate spotting	(AA')	138
Little spotting	(A'A')	5
	Total	1612

As important as natural selection is under certain circumstances, both field and theoretical studies in population genetics have shown that it is not the source of all change in gene frequencies in natural populations. It is possible for a mutation to become common in a population entirely by accident through a phenomenon called **genetic drift.** Even a harmful mutation can become established, and can remain in a gene pool, as long as a single copy of it is not lethal.

Operating by chance, genetic drift works most strongly in either small populations or populations of dramatically rising and falling size. When a nonlethal mutation occurs in such a population, it is possible for the organism carrying that mutation to have more offspring than its neighbors *not* because it is better adapted, but just by chance.

Also, a sudden change in the environment of that population may just happen to wipe out many of the individuals that do not carry that mutation. When either of these situations arises, a new mutation can appear in most members of the population after only a few generations (Figure 12.17). In large, stable populations, this is much less likely to happen by accident.

Figure 12.17 This demonstration of genetic drift shows how evolution can occur by chance. The population begins with equal numbers of A and B genotypes. Each type produces identical offspring. In each generation, half of the population dies *at random*, regardless of genotype. The relative numbers of A and B genotypes vary randomly until A becomes extinct not because it is poorly adapted, but by chance alone.

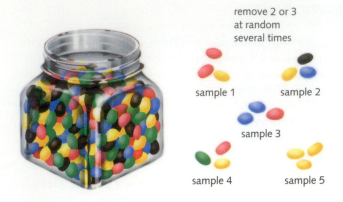

remove 2 or 3
at random
several times

sample 1

sample 2

sample 3

sample 4

sample 5

Figure 12.18 A schematic demonstration of the founder effect. Imagine a jar of 1000 jellybeans, containing equal numbers of colors. If you poured out half of the jellybeans, you would probably find roughly equal numbers of all five colors in your sample. But if you picked out only two or three beans at random, your sample would be missing at least two of the colors present in the main population, and you could conceivably end up with three yellow ones. The same thing can happen with the genes in a small founding population.

The Founder Effect

If a small number of individuals migrate off to found a new population, they may carry genes in different relative frequencies from the main population just by chance. This phenomenon, known as the **founder effect,** is an extreme case of genetic drift. Figure 12.18 offers an illustration of how the founder effect might work.

The founder effect can be especially important in places such as the Hawaiian or Galapagos Islands, where populations of mainland species may be established by a handful of stranded wanderers. The tiny gene pool of those individuals may not be at all representative of the gene pool of the main population from which they came. Founding populations, therefore, can start out with peculiar collections of genes.

There are several well-documented cases of the founder effect among small groups of humans whose practices isolate their gene pool from that of the surrounding community. One example is the Amish community of Pennsylvania, an orthodox religious group that was founded by a very small number of immigrants and allows no marriage to outsiders. Among the founders of the American Amish community was the family of a Mr. and Mrs. Samuel King. The Kings happened to carry a rare gene that, when inherited from both parents, causes dwarfism, the growth of extra fingers, and heart defects.

Those who are afflicted with this condition usually die at a very early age. In the outside world, the frequency of this gene is less than 1 in 1000, but the Kings and their descendants had larger families than the others in their Amish community. Thus, through a combination of the founder effect and genetic drift, this gene occurs in the Amish population today at the extraordinarily high frequency of about 1 in 14.

PLEIOTROPY AND HETEROZYGOUS ADVANTAGE

One unexpected finding of modern genetics is that a single gene often has *several* important effects on the phenotype of an organism. This phenomenon is called **pleiotropy,** which means "many effects." The same gene can have several very different, simultaneous effects, some of which may be beneficial while the others are harmful. The selective advantage of one trait controlled by such a gene may be powerful enough to make the gene common in a population, even though its ill effects would lower the organisms' fitness if they occurred alone.

One example of pleiotropy in human populations is the sickle-cell anemia gene that causes a hemoglobin defect in blacks of African descent. In people homozygous for the sickle-cell gene (those who have inherited copies from both parents), red blood cells become distorted into rigid, sickle-shaped forms. These cells lose some of their oxygen-carrying ability, and get stuck in tiny blood vessels throughout the body. This causes a number of serious health problems (Figure 12.19). These homozygotes become very ill, and usually die before puberty. Even the red cells of heterozygotes, who inherit only a single copy of the gene from one parent, may "sickle" under certain conditions.

Why, then, has the sickle-cell gene never been eliminated by natural selection? In addition to all its negative effects, this gene has two additional, unexpected, beneficial effects on phenotype. Individuals who are heterozygous for the sickle-cell anemia gene are endowed with increased resistance to malaria and with slightly enhanced fertility. For this reason, in areas of western Africa where malaria is a common and often lethal disease, the sickle-cell gene significantly increases the fitness of individuals who carry a single copy of it.

Such a situation, in which an individual with one copy of a particular gene has an advantage over an individual with either two copies or none at all, is called *heterozygote advantage.* Note that in malaria-free environments (such as the United States) this advantage disappears. Here, as expected, the fitness of the heterozygote is lower than that of individuals who carry no sickle-cell gene, so the frequency of that gene has decreased significantly.

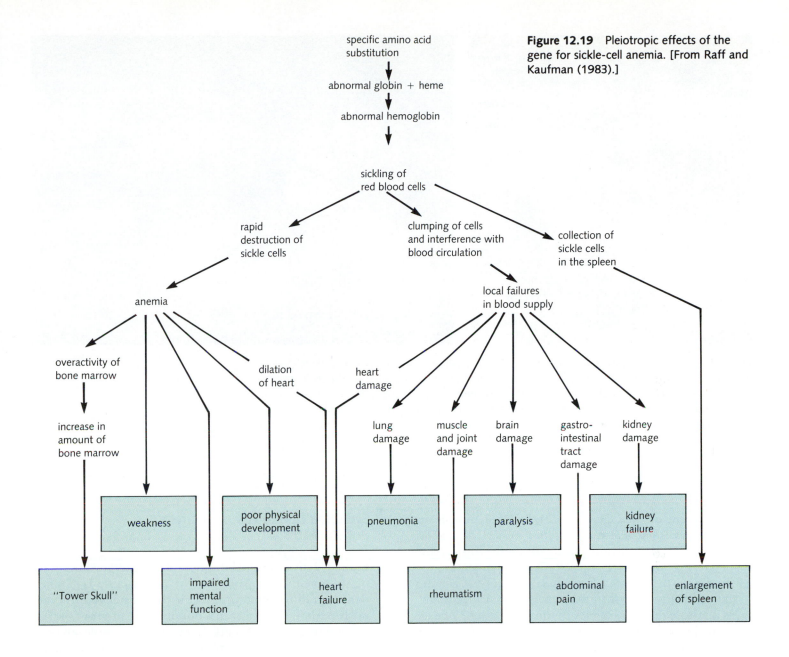

Figure 12.19 Pleiotropic effects of the gene for sickle-cell anemia. [From Raff and Kaufman (1983).]

MODERN STUDIES OF GENETIC VARIATION

Note that all our models of evolutionary change assume that sufficient genetic variation exists in nature to provide raw material on which natural selection can operate. In order to understand why some species evolve rapidly while others stagnate or become extinct, evolutionary biologists are trying to test that assumption in a variety of species. To do so, they must answer two related questions: How many different alleles of each gene are typically present in a population? And how much do members of a population differ from one another in their genetic makeup? In other words, what is the degree of genetic **polymorphism** in the population? (By definition, a population is polymorphic for a given trait whenever more than one allele of a given gene is present. Mendel's peas, for example, were polymorphic for both seed color and seed shape.)

Darwin observed many heritable differences among the organisms he studied. But a lot of genetic variation cannot be discerned simply by looking at an organism's physical characteristics. Molecular biology has provided the tools that evolutionary biologists needed to study this previously undetectable genetic variation, and they have uncovered more polymorphism than anyone ever thought existed. In the fruit fly *Drosophila,* for example, average individual flies carried more than one allele for as many

Figure 12.20 The adaptive significance of industrial melanism in *Biston betularia*. (left) On normally colored tree trunks, the light-colored form is nearly invisible, whereas the dark form can be seen clearly. (right) On tree trunks darkened by soot, the light form is clearly seen—and hence more vulnerable to bird predators—whereas the dark form is well camouflaged.

as 12 percent of their genes, a level of polymorphism that turns out to be about average for insects. Plants, on the whole, have a polymorphism of about 18 percent, while average vertebrates exhibit about 7 percent.

In many species, therefore, there is plenty of genetic variation for natural selection to work on. But what maintains all those different alleles in the population? Are all of them adaptive? Is polymorphism itself adaptive? In at least some cases the answer seems to be "yes." We have seen, in the case of sickle-cell anemia, how heterozygote advantage can maintain polymorphism, even when one of the alleles is lethal in the homozygous condition.

The *neutral theory* suggests that in other cases of polymorphism, natural selection neither favors nor opposes the spread of alternative alleles. In such cases, adherents of the neutral theory argue, polymorphism may be maintained entirely by the action of genetic drift.

But environments in nature change regularly. Under different environmental conditions, an originally neutral trait—such as a different flower color or the ability to digest a certain chemical compound—might become advantageous. The existence of neutral variation in a population, therefore, can prove useful in the long term.

Evolution Due to Human Activity: British Moths

Human activity in the biosphere has caused evolution to occur in many organisms over very brief periods of time, demonstrating dramatically how the existence of previously neutral variability in a population can prove useful in changing environments. In one such example, an ecological side effect of the Industrial Revolution in Britain around Darwin's time caused several British moth species to change color.

Before Britain's industrialization, most of these moths were lightly mottled and blended superbly with the tree trunks on which they rested, motionless, for much of the day (Figure 12.20). Darker individuals were always present, but they never constituted more than a small percentage of the population. By 1880, however, many decidedly darker moths were being observed in such industrial areas as Manchester and London. Before the end of the century, more than 98 percent of the individuals of *Biston betularia* caught in Manchester were of the darker form. What had caused this remarkable shift in coloration?

E. B. Ford of Oxford University proposed that the moth populations had shifted toward darker color because of natural selection. The moths' major predators, Ford reasoned, were birds that hunted by sight. In pre-industrial England, the light-colored moths were so well camouflaged as they clung to normally colored tree trunks that those birds had trouble locating them. Industrial burning of coal, however, noticeably blackened the tree trunks around major cities. Ford proposed that darkened tree trunks had conferred a selective advantage on dark-colored moths and that natural selection had caused the black form to take over the population. The phenomenon thus described was called *industrial melanism* after the pigment, melanin, that darkened the moths' wings.

Later experimental studies by H. B. D. Kettlewell confirmed that moth-eating birds hunt primarily by sight and that those moths that most resemble their backgrounds in color have the highest chance of escaping predation. Here was an example of rapid evolutionary change occurring practically in Darwin's backyard!

Note, however, that the patchy nature of industrial and rural environments across the English countryside has maintained polymorphism in the moth population, rather than favoring a permanent switch to the darker form. Darker moths have higher fitness only on soot-blackened trees; elsewhere, lighter forms are still better camouflaged.

Evolution Today: Out of Control

Not all examples of human-caused evolution are quite so innocuous. Efforts to eradicate weeds from cultivated areas and to eliminate disease and parasites from both humans and domestic plants and animals have imposed new selective pressures on wild plant and animal populations. The hand weeding of cultivated fields, for example, creates selective pressures in favor of weeds that resemble crop plants. Over successive generations, many weeds do converge in appearance with crops, making later weeding much more difficult.

Additionally, humans now use antibiotics to control bacterial infections, pesticides to kill insects on farms and around human dwellings, and herbicides to kill weeds. All three of these uses have important positive effects on human health and well-being. But our regular use of those compounds has made them part of the environment of insects and microorganisms. As it turns out, many natural populations of insects and bacteria contain enough genetic variation for a few of them to be resistant to any given toxic compound. By killing susceptible individuals, these insecticides dramatically increase the relative fitness of resistant variants.

Already, DDT and several other insecticides no longer kill many of the insect species they controlled a few years ago. Even more worrisome is the fact that many bacteria and other parasitic microorganisms are rapidly evolving resistance to penicillin and other antibiotics. In the tropics, for example, the organism that causes malaria has evolved resistance to the drug that once effectively controlled the disease.

Note, however, that resistance to poisons is rarely a "free ride" for either insects or other organisms, because the selective trade-offs imposed by pleiotropy often maintain polymorphism either within or between populations of a species. Some populations of Norway rats, for example, have evolved resistance to the rat poison *warfarin*. Where the poison is in widespread use, homozygotes for the allele that confers resistance are common. But that allele also lowers rats' ability to synthesize vitamin K, a compound essential in allowing blood to clot. For that reason, in places where warfarin is not used, individuals homozygous for this allele are at as much as a 54 percent selective *dis*advantage compared to "wild-type" rats, and the allele is far less common. The same sort of phenomenon has been demonstrated for the alleles that confer resistance to DDT and to dieldrin on mosquitoes.

SUMMARY

Evolutionary biology today, which has absorbed significant contributions from Mendelian genetics, population genetics, and ecology, deals both with the fact of evolutionary change and with theories about how and why evolution takes place.

The gradual accumulation of evolutionary change within populations can often be explained by the differential action of selective pressures on different phenotypes that represent different genotypes. Recent advances in molecular biology have revealed unexpected riches of genetic variation in natural populations. Some of these variations are sufficient to provide the raw material necessary to operate; others may be neutral, or selectively meaningless.

Natural selection is not necessarily responsible for all evolutionary change. Additional mechanisms, including genetic drift, operate more according to chance than in terms of selective advantage.

After studying this chapter, you should be able to:

- Define *adaptation*, *selection*, and *fitness* in genetic terms.
- Explain how Mendel's principles of genetics affected evolutionary biology.
- Understand the significance of the Hardy–Weinberg principle in population genetics.
- Describe the way natural selection can affect heritable characteristics in a population.
- Appreciate that natural selection is not necessarily responsible for all evolutionary change.

SELECTED TERMS

species *p. 229*
gene pool *p. 229*
evolutionary fitness *p. 230*
evolutionary adaptation *p. 230*
evolution *p. 230*
stabilizing selection *p. 233*
directional selection *p. 233*
disruptive selection *p. 233*
Hardy–Weinberg law *p. 239*
genetic drift *p. 241*
founder effect *p. 242*
pleiotropy *p. 242*
polymorphism *p. 243*

REVIEW

Discussion Questions

1. How did genetic theory provide biologists with a new way to define species?

2. What is genetic drift? How can genetic drift cause changes in gene frequencies in a small population, even in the absence of natural selection?

3. What is the founder effect? How did it cause unusual gene frequencies among the American Amish?

4. How does pleiotropy help to maintain the existence of the sickle-cell gene in Africa? How does it oppose the spread of the gene for warfarin resistance in Norway rats?

5. What do antibiotic resistance in bacteria and pesticide resistance in insects have in common? Why are both cause for concern?

Objective Questions (Answers in Appendix)

6. The critical source of variability within a population is
   (a) sexual reproduction.
   (b) recombination of genes.
   (c) environmental changes.
   (d) mutations.

7. When the mean of the frequency distribution of a trait in a population shifts in one direction or another, this describes
   (a) directional selection.
   (b) disruptive selection.
   (c) pleiotropy.
   (d) stabilizing selection.

8. For a mutation to be important in the evolution of a species, it must be found in
   (a) the reproductive cells.
   (b) all of the body cells.
   (c) random body cells.
   (d) those cells that keep an organism healthy.

9. Evolutionary fitness is defined as
   (a) producing just enough offspring to be supported by the existing food supplies.
   (b) being physically fit for the age and type of species.
   (c) adapting to changes in the environment.
   (d) contributing to the gene pool of the next generation.

10. A type of natural selection in which two extreme phenotypes take hold in a population is known as
    (a) divergent selection.
    (b) environmental selection.
    (c) disruptive selection.
    (d) stabilizing selection.

Abbott, I., L.K. Abbott, and **P.R. Grant** (1977). Comparative Ecology of Galapogos Ground Finches (*Geospiza* Gould): Evaluation of the Importance of Floristic Diversity of Interspecific Competition. *Ecological Monographs, 47:* 151-184.

Boag, P.T. and **P.R. Grant** (1981). Intense Natural Selection in a Population of Darwin's Finches: (*Geospizinae*) in the Galapagos Finches. *Science, 214:* 82-85.

Bowman, R.I. (1963). Evolutionary Patterns in Darwin's Finches. *Zoology, 58:* 1-302.

Cavalli-Sforza, L.C. and **W.F. Bodmer** (1971). The Genetics of Human Populations. Freeman, San Francisco.

Grant, P.R. (1986). Ecology and Evolution of Darwin's Finches. Princeton University Press, New Jersey.

Grant P.R. and **B.R. Grant** (1980). The Breeding and Feeding Characteristics of Darwin's Finches on Isla Genovesa, Galapagos. *Ecological Monographs, 50:* 381-410.

Raff, R.A. and **T.C. Kaufman** (1983). Embryos, Genes, and Evolution: The Developmental Genetic Basis of Evolutionary Change. Macmillan, New York.

Yoo, B.H. (1980). Long-Term Selection for a Quantitative Characters in Large Replicate Populations of *Drosphilia melanogaster.* I Response to Selection. *Genetic Research, 35:* 19-31.

CHAPTER **13**

Evolution of Species

If you ever read *On the Origin of Species*, you will notice a rather paradoxical omission; despite its title, the book doesn't really say much about the origin of species! Throughout the book, Darwin discusses his ideas about how natural selection, operating on heritable variations within species, produces steady, gradual change in *existing* species. Today biologists often use the term *microevolution* to describe these sorts of small changes, such as shifts in beak size in finches and the darkening of wing color in moths.

But though Darwin wrote elegantly about the way microevolutionary changes enable species to adapt to changing environments, he wrote very little about the multiplication of one species into two or more—the process we call **speciation.** The accumulation of genotypic and phenotypic changes large enough to create new species, genera, and higher taxonomic categories (Chapter 14) is often called *macroevolution*. In this chapter we will discuss several hypotheses about the conditions under which new species are formed, and we will examine the rates at which existing species may evolve.

THE ORIGIN OF SPECIES

In Chapter 12 we defined a species as a population of physically similar, interbreeding organisms that are *reproductively isolated* from other such groups. Thus, in genetic terms, speciation occurs when two or more populations diverge in their heritable characteristics in such a way and to such an extent that breeding between them no longer takes place and, with a few exceptions, is no longer possible.

What sorts of mechanisms could cause reproductive isolation? Organisms can be prevented from breeding with one another by any of several **reproductive isolating mechanisms**—characteristics that prevent them from interbreeding. Depending on whether these mechanisms have their effect before or after mating, they are separated into two main categories: *prezygotic* and *postzygotic isolating mechanisms*.

Prezygotic Isolating Mechanisms

Prezygotic isolating mechanisms prevent eggs and sperm from ever coming together to form fertilized eggs (*pre* means "before"; *zygote* means "fertilized egg"). This type of isolation can occur in several ways.

Mechanical isolation   In some species of animals and plants, males and females have reproductive organs that have coevolved for so long that they are uniquely suited to one another. Among insect species, for example, male reproductive organs vary enormously in size and shape, preventing the effective transfer of sperm to females of other species. Such species are said to exhibit **mechanical isolation** from one another.

Similarly, plant–pollinator coevolution in certain groups of plants ensures that interspecific pollination

247

Figure 13.1 Behavioral isolating mechanisms. Courtship displays of birds, such as these great frigate birds, are species-specific and greatly reduce the likelihood of accidental interspecific mating.

never happens in nature. In many cases, a particular plant species is pollinated by a single insect species that is adapted both physically and behaviorally to deal with that plant's unique flower characteristics. Such specific pollinators either do not or cannot pollinate the flowers of other species. Numerous orchids are reproductively isolated in nature in this manner; although flowers can be hand-pollinated to produce fertile interspecific hybrids in cultivation, such hybrids are rarely (if ever) seen in the wild.

Behavioral isolation   Animals of many species engage in elaborate courtship rituals before mating (Figure 13.1). During such rituals, males may display brightly colored body parts, serenade females with species-specific songs, or release unique chemicals as odor signals. Differences in particular aspects of these rituals enable the female member of the pair to be certain that her prospective mate belongs to her own species. If courtship begins between two closely related yet different species, the male soon gives the wrong signals and the female refuses to mate with him. Such species illustrate **behavioral isolation,** which we will discuss in more detail in Chapter 43.

Behavior not directly involved in mating can also lead to reproductive isolation. For example, the closely related lions and tigers once lived in broadly overlapping ranges throughout India, but their behavioral ecology kept them separated. Lions are highly social creatures that prefer to hunt in open grasslands. Tigers, on the other hand, are solitary and choose forest over open space. Thus, although these animals can interbreed and produce fertile hybrids in zoos, there is no record of hybrids ever having occurred in nature.

Temporal isolation   If two species breed at different times in nature, they do not exchange gametes even if they could interbreed in captivity. This is called **temporal** (time-related) **isolation.** Many species of potentially interfertile plants, for example, flower at different seasons, so even if they can be pollinated by the same insects, their flowers never open at the same time. Closely related animals, too, may be reproductively isolated from each other because they breed at different seasons or at different times of day.

Gamete incompatibility   Even if pollen is passed from one flower to another, or if sperm from one animal reaches the reproductive tract of another, successful fertilization may not occur. In some cases, the female reproductive tract is inhospitable to sperm from different species; in other cases, egg and sperm simply do not fuse. These are cases of **gamete incompatibility.**

Postzygotic Isolating Mechanisms

Even if mating and fertilization occur between members of different species, the zygotes that result may not give rise to functioning individuals. Genetic and physiological characteristics that keep hybrid zygotes from developing properly, or from becoming established in nature if they do survive past birth, are called **postzygotic isolating mechanisms** (*post* means "after").

Often, genetic differences between species are so great that the mismatched sets of genes cannot function together to direct the development of the embryo. This is a situation known as **hybrid inviability.**

In other cases, interspecifically fertilized embryos do develop into viable organisms. But though they are able to survive as individuals, these hybrid plants and animals cannot reproduce. In some cases, they are simply too weak or stunted to reproduce. In other cases, such as the mules produced by crossing horses and donkeys, hybrids are healthy, vigorous animals but are sterile. Because of this **hybrid infertility,** a self-sustaining mule population could never develop; mules must always be produced by repeating the original cross.

Ecological isolation   Sometimes two species that can produce viable and fertile hybrids under laboratory conditions do not do so in nature because they are adapted to habitats that may be adjacent to one another but are ecologically quite different. For example, *Quercus lobata* and *Q. dumosa*, two species of oak trees found in the western United States, respectively inhabit fertile valley grasslands and drier, less fertile chaparral habitats on steeper slopes (Figure 13.2). These species experience **ecological isolation,** a phenomenon that can act as both a prezygotic and a postzygotic isolating mechanism.

Adaptations to distinct habitat types such as these can discourage hybridization in two ways. First, the different ecological conditions the plants prefer may minimize the number of places where they occur close enough to one another to exchange gametes. Second, when viable hybrids do occur, they are poorly suited to either parent's habitat (and thus fail to compete successfully for resources there) and have no intermediate environment in which they can thrive and reproduce.

MECHANISMS OF SPECIATION: HOW REPRODUCTIVE ISOLATION DEVELOPS

Because reproductive isolation is a critical part of the definition of a species, any explanation of how populations of one species evolve into two or more species must explain how those populations become reproductively isolated from one another. There are several mechanisms that describe how the sorts of reproductive isolating mechanisms we have discussed can arise, and each explains certain cases of speciation observed in nature.

Allopatric Speciation

Allopatric speciation (*allos* means "different"; *patria* means "native land") occurs when two or more populations are separated from one another by a geographic barrier that may be as large as a mountain range or as small as a

Figure 13.2   *Q. dumosa* (top) and *Quercus lobata* (bottom) in their respective habitats.

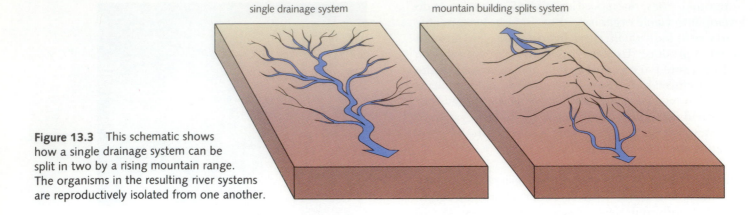

single drainage system   mountain building splits system

Figure 13.3 This schematic shows how a single drainage system can be split in two by a rising mountain range. The organisms in the resulting river systems are reproductively isolated from one another.

narrow stream. Any barrier, large or small, can isolate populations of a species if that barrier creates conditions in which individuals of that species cannot survive or across which they will not travel. Once populations are isolated in this way, their members cannot interbreed, and gene flow between them is blocked. This sort of separation may occur in several ways, some of which depend on geological events and others of which result from the actions or habits of the organisms themselves.

Wide-ranging populations may be separated from one another by events in the physical world around them. As we will see in Chapters 14–18, the major continents of the world have united, drifted apart, and collided again, while global climate has warmed and cooled during and between ice ages. Although major geological or climatic changes and long-distance migrations are not necessarily the most common causes of allopatric speciation, they have produced some of the most spectacular adaptive radiations known.

Less drastic geological change can also isolate populations. Changes in continental topography may cause major river systems to change course, separating once-contiguous populations of freshwater organisms (Figure 13.3). Over time, those streams may carve valleys or create new lakes that can also harbor isolated aquatic populations. And those very same valleys can isolate populations of certain (though not all) terrestrial organisms living on either side of them. A chasm the width and depth of the Grand Canyon, for example, is a formidable barrier to small mammals such as squirrels, although it is trivially easy for large birds to traverse.

On an even smaller scale, patches of one sort of habitat are often separated from one another by bands or patches of other habitat types. Mountains in the tropics, for example, may be capped with snow-covered alpine habitats while valleys at their feet shelter tropical rain forests. Not surprisingly, high-altitude populations of plants and animals are often unable to survive in tropical valleys. And organisms adapted to wet tropical forests may find cool mountain ranges and alpine deserts equally inhospitable. Thus these patches of similar habitat are home to local populations of organisms that only rarely interbreed with neighboring populations.

Occasional individuals, however, may cross such barriers. When they do, they may find new homes in previously vacant patches of suitable habitat on the other side of the ecologically hostile zone. These infrequent colonizers can establish populations that are fairly isolated from one another even if they are separated by relatively short distances. In other cases, highly mobile animals and seeds traverse major geographic barriers to establish populations in truly remote locations.

Effects of separation Geographic isolation of populations is often the first step in the evolutionary processes that eventually give rise to other isolating mechanisms. Why should this be the case? Once two populations are physically separated, gene flow between them stops. From then on, all factors that affect gene frequencies, such as genetic drift and natural selection, can operate on those populations independently. Especially when separated populations experience different environmental conditions, natural selection may favor changes in gene frequency between them over time.

The English oak, *Quercus robur*, for example, easily hybridizes with either of the two American oak species mentioned earlier if they are planted together in a botanical garden. Under natural conditions, however, the Atlantic Ocean and North America separate them (Figure 13.4). Though their ancestors may have been identical, these isolated populations have evolved independently, diverging in several characteristics. Although they have reached a point where they *look* sufficiently different to have convinced morphologists to classify them as different species, they have not yet developed intrinsic reproductive isolation mechanisms.

Sooner or later, however, separated populations such as these may diverge in such traits as color, habitat preference, breeding behavior, or breeding (or flowering) season that act as reproductive isolating mechanisms. Once that happens, the populations can no longer interbreed, even if the physical barrier between them is removed or if they jointly colonize a common area. Those populations have, in effect, become separate species.

There are in nature numerous examples of allopatric animal populations that seem to be in the process of diverging into separate species. Often these populations form what is called a "ring species" in which a series of slightly dissimilar populations are arranged in the shape of a ring. Adjacent populations within the ring—which can and do interbreed with their immediate neighbors on either side—are frequently assigned the status of subspecies. But typically, populations on either end of the ring do not interbreed and are thus classified as distinct species (Figure 13.5).

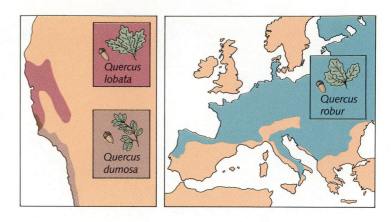

Figure 13.4 Distribution maps of three oak species, one native to Europe, the other two native to the west coast of the United States.

Figure 13.5 This series of closely related gull species rings the north pole today. It appears that the ancestral species first evolved in Siberia and then spread around the northern continents in both directions. Genetic divergence among those populations has resulted in differences among adjacent populations. At opposite ends of this "ring," those differences are enough to have produced two populations that interbreed so rarely that they are considered distinct species. [From *Illustrated Origin of Species* (1979).]

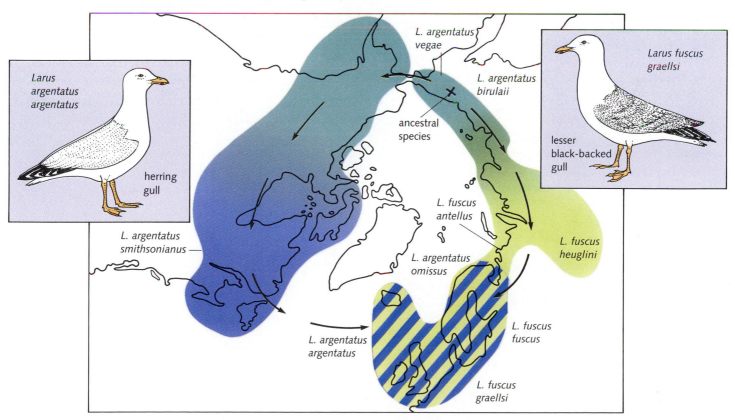

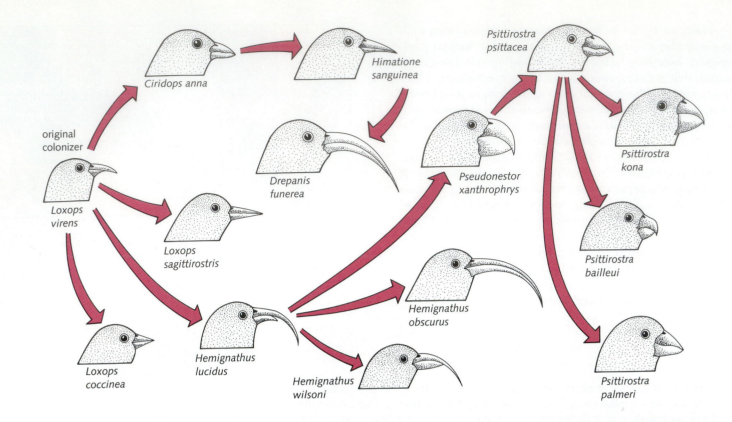

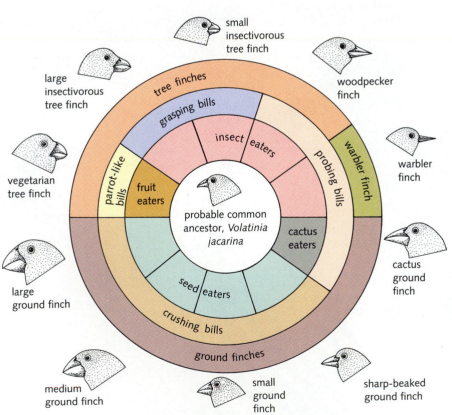

Figure 13.6 (top) Hawaiian honeycreepers form a remarkable adaptive radiation. Founded by a small number of ancestral birds that wandered far from land, these species evolved to fill the many vacant niches for birds on the islands. (bottom) In the process two groups, specialized to feed on seeds and insects, respectively, converged in beak size and shape with several ecologically similar Galapagos finches.

Many of the most spectacular cases of allopatric speciation have occurred on various kinds of islands: terrestrial islands in the sea, aquatic "islands" in lakes and streams, and mountainous islands in tropical lowlands (see Theory in Action, Worlds Apart: Island Mountains and Inland Seas, below). There, in seclusion from the rest of the world, a few stranded travelers have evolved into **endemic species**—organisms found nowhere else on earth. Two of the best-studied cases of allopatric speciation in island faunas involve the honeycreepers of the Hawaiian archipelago and Darwin's finches (Figure 13.6). Both cases are explained by similar logic; we will examine in detail here a plausible scenario for the finches' adaptive radiation in the Galapagos.

THEORY IN ACTION

Worlds Apart: Island Mountains and Inland Seas

When most people hear the word *island*, they envision islands of land in bodies of water. But in biological terms, an island is any habitat that is isolated in some way from other, similar habitats. Islands of all sorts are living laboratories for the study of evolution in isolation.

Towering 9000 feet above the steamy rain forests of Venezuela, the craggy, fog-shrouded plateau called Neblina is the image of a world lost in time. In fact Neblina, like other Venezuelan table-topped mountains, is one surviving piece of a giant plateau more than 1.5 billion years old. Over time, the soft sediments that made up most of the plateau eroded away, leaving isolated outcrops of harder rock, like Neblina, behind.

Environmental conditions at the mountain's cool, misty summit differ dramatically from those in the surrounding lowland jungle. The jungle, therefore, isolates Neblina's plants and animals from those of the other scattered plateaus as effectively as any ocean could. This ecological isolation has spawned scores of endemic insects and amphibians, peculiar giant earthworms, unusual carnivorous pitcher plants, and a whole genus of plants called *Neblinaria* whose members grow nowhere else.

Allopatric speciation does not necessarily require separation by major geological features. In East Africa's Lake Malawi, for example, hundreds of species of cichlid fishes have evolved from a few common ancestors within 30,000 to 40,000 years—a mere instant in geological time. Biologists who study these species believe that allopatric speciation has occurred repeatedly in at least one group, the Mbuna, or rock dwellers.

The shallow regions of Lake Malawi are dotted with rock piles separated by long stretches of sandy bottom. The Mbuna really are rock dwellers; they hover above those rock piles and rarely swim out over the sand flats. Each rock pile thus serves as a tiny "island" whose isolated population has minimal genetic contact with other populations. Given the right combinations of chance migrations, the founder effect, genetic drift, and natural selection, these fish populations could have evolved as Darwin's finches did in the Galapagos archipelago.

This peculiar fish, discovered on Neblina, is believed to be a kind of catfish. It is so new to science that its temporary name, *Neblinichthyes*, may well have been changed by the time you read this.

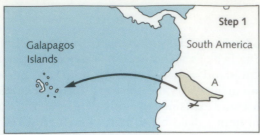

Small mainland population (A) is blown to archipelago. Individuals colonize adjacent islands.

Two populations on different islands are isolated by distance. As they adapt to their different environments, they become behaviorally isolated from one another, and become distinct species, (B) and (C).

Following reproductive isolation, a small population of species (B) migrates to the island of species (C). The two species enter into ecological competition with each other.

Directional selection driven by competition leads to further change in (B). This population becomes reproductively isolated from both species (C) and its parent population (B), and becomes a new species (D).

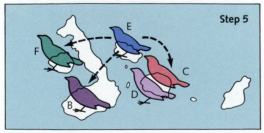

Repetition of this process and several variations of colonization, divergence, and competition can create many species on an archipelago.

A Case Study in Allopatric Speciation: Galapagos Finches

The Galapagos Islands arose, devoid of life, in the Pacific Ocean between 3 and 5 million years ago. More than 1000 kilometers west of Ecuador, these islands have never been connected either to the mainland or to each other. Somehow, more than a dozen of the finch species we met in the last chapter have evolved on this archipelago from a small founding population that strayed from the mainland (Figure 13.7). Why and how could that single population have given rise to so many distinctively different kinds of birds? Let us reconstruct a plausible scenario.

Colonization Assume that a small finch population (it could even have been a single impregnated female) was blown to the archipelago in a storm (Figure 13.7, *Step 1*). Once a breeding population was established on one island, wandering individuals could occasionally colonize adjacent islands. But finches seldom stray far from their nesting sites and do not like flying over open water. Because the islands are separated by enough water to make island-hopping uncomfortable, their finch populations are isolated not only from the mainland but also from each other.

Genetic divergence in allopatry Ecological conditions across the archipelago vary significantly, so the various islands offer different kinds and quantities of food. Under those conditions, small, reproductively isolated bird populations would be expected to evolve different feeding characteristics—a supposition confirmed by studies published in the early 1980s. Species on different islands have evolved beak sizes and feeding behaviors that enable them to exploit efficiently the seeds, fruits, and insects present on their particular islands. Knowing for certain whether that divergence was originally or primarily driven by natural selection or whether it occurred through genetic drift is not critical for us here. Recall from Chapter 12, however, that natural selection, particularly during droughts, has been shown to cause measurable changes in heritable characteristics related to feeding.

Reproductive isolation At some point, the gene pools of these isolated populations diverged from one another enough so that their members would no longer interbreed, even if they lived in the same place. At that point, the divergent island forms could legitimately be called different species (*Step 2*).

Figure 13.7 Hypothetical mechanism for the evolution of Galapagos finch species through repeated episodes of allopatric speciation.

Why Shouldn't Species Interbreed?

Why is reproductive isolation so important? Why, in other words, is it evolutionarily worthwhile for species to invest time and energy in complex, species-specific breeding behaviors and other prezygotic isolating mechanisms to prevent hybridization? There is no simple general answer to these questions, but in certain cases, at least, interspecific hybridization can result in lowered fitness.

In southern Germany, for example, the ranges of two species of European house mice (*Mus musculus* and *M. domesticus*) overlap in a region about 20 miles wide. That limited area harbors mice rarely seen elsewhere; they are hybrids between the two dominant species. While studying these animals, Richard Sage of the University of California noticed that the hybrid mice had more fleas than either parent species. Further investigation showed that the hybrids also had many more parasitic worms and more different kinds of worm parasites than either parent. Although these infestations are not fatal, they can damage the host's internal organs and increase its susceptibility to other diseases.

There is evidence that resistance to parasites has a genetic basis and can therefore be affected by natural selection. The two mouse species in question seem to have specific genes that control resistance to the bacterium *Salmonella*. Furthermore, resistance to parasitic worms in these species acts like a simple, dominant trait; it can be passed from generation to generation in the sorts of ratios Mendel found in his peas.

It is not clear precisely how interspecific hybridization decreases resistance to parasites in these species. But whatever the genetic cause, its effect is to lower the fitness of the hybrids. For that reason, natural selection would favor any isolating mechanism that reduces the likelihood of a mismatch.

Source: From a letter by R. D. Sage *et al.*, *Nature*, November 6, 1986, 324.

Recent studies have shown that, except in rare cases, finches reliably choose mates of their own species. The birds apparently discriminate among prospective mates by using a combination of physical cues (such as beak size) and behavioral cues (such as song patterns learned from their parents). Thus, as the isolated island populations evolved different-sized beaks adapted to feeding on different foods, and as they diverged (possibly at random) in their song patterns, they became behaviorally isolated from each other.

Competition and further divergence After behavioral reproductive isolation occurred, birds from one island that occasionally colonized adjacent islands were still reproductively isolated from the local population (*Step 3*). Under these circumstances, the two species might have entered into competition with one another for food. Such competition could favor the survival of those individuals in each species that were most different in feeding habits from their competitors (*Step 4*). This kind of divergent evolution should push the competitors ecologically farther apart from one another.

This part of the hypothesis sounds satisfying, but are there any quantitative data to back it up? Indeed there are, because several finch species occur on several islands, and because each island has its own assortment of species. It is thus possible to compare, for example, beak sizes between two populations of the same species of seed-eating finch, one of which occurs with a similar competing species and the other of which has no competitors. In several such cases, beak measurements show clearly that these birds differ more when the species occur together than when they occur separately (Figure 13.8). It is thus reasonable to assume that competition has, in fact, caused directional selection.

Further speciation Theoretically, divergent evolution, either driven by this sort of competition or allowed by genetic drift, could cause sufficient changes in beak size to lead to the formation of yet another species (*Step 5*). Repeated over time and over the entire archipelago, this process could have produced the 13 living species and the several extinct forms that we know once existed in the Galapagos Islands.

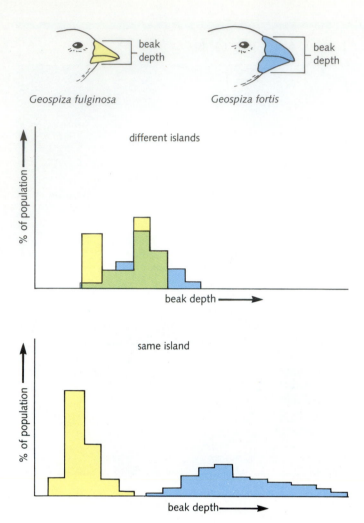

Figure 13.8 Quantitative evidence for the effects of interspecific competition on beak sizes in Galapagos ground finches. When populations of *Geospiza fortis* and *G. fuliginosa* occur on separate islands, their range of beak sizes is fairly similar. When populations of those same species occur together on the same island, however, their average beak sizes shift markedly in a manner that reduces the overlap (and hence the competition) between them. [From Futuyma (1979) after Lack (1947).]

Parapatric Speciation: Life on the Fringe

In **parapatric speciation** (*para* means "next to"; hence *parapatric* means "next to the native land"), a small population on the fringe of a larger population diverges, often despite the absence of a major geographic boundary and despite the existence of some gene flow between it and the main group.

Why might this occur? Fringe populations of a species are regularly subjected to different and often more stressful conditions, and they may be exposed to different food and predator species than populations in the center of the species' range. Under these circumstances, and if interbreeding with individuals from other populations is rare enough, the combination of natural selection and genetic drift might cause these fringe populations to diverge sufficiently to become distinct species. Because both natural selection and genetic drift operate more rapidly in small populations than in large ones, some biologists believe that small populations on the fringe of large populations' ranges are the primary sites of macroevolutionary changes.

Researchers disagree about whether or not several cases of variant populations of organisms recorded in nature represent parapatric speciation in progress. In several industrial areas, for example, heavy metals that are normally toxic to plants have entered the soil. In some, populations of grass species have slowly evolved tolerance for those toxins and are thus able to grow in patches of habitat where tainted soil excludes other members of their species. Some of those populations appear to be diverging from wild-type populations around them in both flowering time and gamete compatibility. This partial reproductive isolation may be an example of parapatric speciation in progress. The pair of American oak species mentioned earlier may represent a similar situation, in which nearly complete reproductive isolation has evolved.

Sympatric Speciation: Alone in a Crowd

Under certain conditions, it is possible for new species to arise in the midst of their parent populations. Such speciation, which occurs "together in the native land," is called **sympatric speciation.**

There is convincing evidence that sympatric speciation has occurred repeatedly, at least among plants. It is not uncommon for individual plants to appear with double the normal amount of genetic material, either through errors in cell division in plant buds or through errors in the production of pollen and eggs. These tetraploid individuals can perpetuate themselves only by breeding with one another, for they are instantaneously reproductively isolated from other members of what had been their species. (Although such genetically doubled individuals may be able to cross with their parents and produce viable offspring, those offspring are usually sterile and are thus an evolutionary dead end.)

Many horticulturally and agriculturally important plant species, including several strains of wheat, have more than the normal amount of genetic material and may well have arisen sympatrically through this sort of genetic event. Some botanists estimate that more than half of today's flowering plant species may have arisen as the result of chromosome doubling.

Several laboratory experiments have shown that strong selection can lead to reproductive isolation within controlled populations. For example, disruptive selection for high and low bristle numbers in *Drosophila* (Chapter 12), unexpectedly created subpopulations of few-bristled and many-bristled flies that preferred to mate with others of their own kind. Similarly, one researcher working with corn planted two genetically marked strains together but planted in each generation only seed from plants that carried the smallest number of hybrid kernels. In just a few generations, the rate of fertile crosses between the strains dropped from 40 percent to less than 5 percent (Figure 13.9).

Models of sympatric speciation in natural populations, however, are highly controversial. Although several theories have been proposed to explain how reproductive isolation could arise gradually within a natural population of animals, none has yet been conclusively proved. And although some biologists believe that sympatric speciation has been clearly demonstrated in the case of several insects, others insist that those cases are actually examples of parapatric speciation.

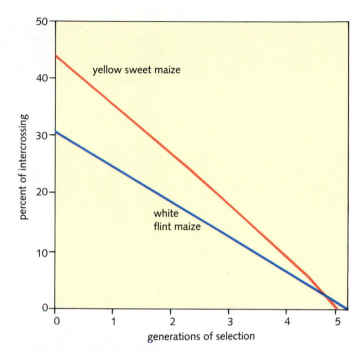

Figure 13.9 When two genetically marked strains of corn are planted together, cross fertilization invariably occurs. After several generations of artificial selection against hybrids (selecting for plants less likely to interbreed) the amount of intercrossing drops markedly, as shown here. Experiments such as this one support the contention that, under certain conditions, reproductive isolation can arise within sympatric populations. [From Futuyma (1979).]

CURRENT DEBATE ON EVOLUTIONARY THEORY

Recent advances in genetics, molecular biology, and paleontology have called into question certain assumptions abut evolutionary change. Is natural selection the *only* force behind genetic change? Are all evolutionary changes adaptive? Does evolution really have to progress slowly and gradually? Spirited debate on these issues continues today.

In an honest attempt to interpret these debates for the public, the popular science press has featured articles implying that the very concept of evolutionary change is in serious doubt. Nothing could be further from the truth. Certain aspects of Darwinian theory *have* come under scrutiny. But as Stephen Jay Gould eloquently stated, "Facts don't disappear while scientists debate theories.... Einstein's theory of gravitation replaced Newton's, but apples did not suspend themselves in mid-air pending

the outcome." Similarly, the fact of evolution remains, regardless of whether scientists agree on how evolution proceeds.

Yet these debates are informative, both for what they tell us about evolution and for what they have to say about the process of science. To give you some insight into contemporary debates on evolutionary theory, we will recall two basic pillars of Darwinian thought: the nature of species and the pace of evolutionary change.

The Darwinian View of Species

Recall that Darwin abandoned the idea of species as fixed entities because in his day "fixed" meant forever, and immutable ideal types left no opportunity for change over time. In what may have been overreaction, Darwin went to the other extreme, essentially denying species any real existence over time. As paleontologist Niles Eldredge wrote, Darwin came to view species as nothing more than "momentary collections of similar and inter-

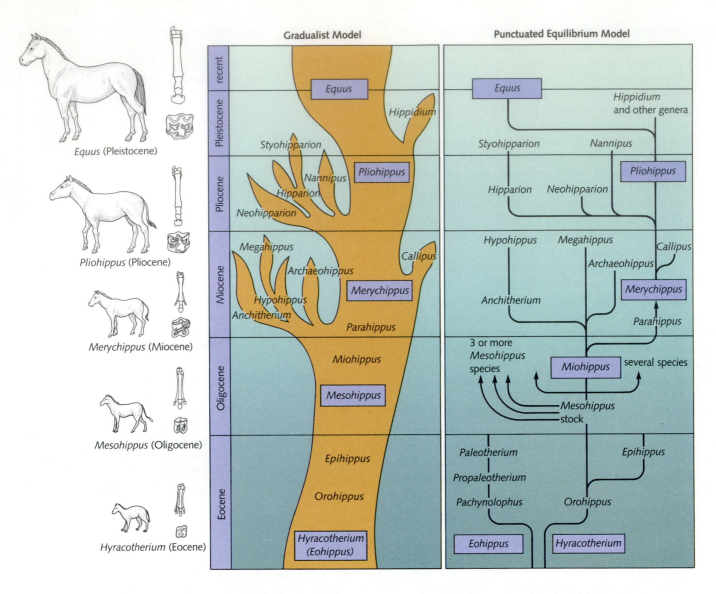

Figure 13.10 Fossil remains of the ancestors of modern horses provide one of the most nearly complete records of evolutionary change for any lineage of mammals. Evolution and speciation in equine lineages are shown at the left. Several trends are obvious: increase in size, loss of side toes to form a single, central hoof, and changes in tooth structure. This figure illustrates the older idea of evolution in animal lineages—-that a single, dominant species gradually changes in response to natural selection. But where in this constantly changing series of ancestors does one species end and another begin? The definitions are necessarily arbitrary. At the right is shown the evolution of the modern horse as reconstructed according to the modern, expanded fossil record and arranged to more closely reflect the ideas of punctuated equilibrium theory. This theory relies on the same fossil history as the gradualist model with two critical differences. First, we now know of scores of fossil horses, only a few of which are shown here. These species represent a highly branched evolutionary "bush" rather than a tree with a straight central trunk. Modern horses are merely one twig off a side branch of this bush that happened to survive. Second, this view attributes most evolutionary change to brief periods of speciation, rather than to long periods of gradual evolution.

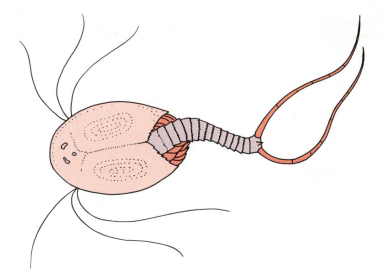

Figure 13.11 Visible characteristics of this small shrimp species have remained unchanged since it first appeared in the fossil record over 180 million years ago. [From Kaestner 1970]

breeding organisms that looked somewhat different in the not-too-distant past, and are destined to become modified in the not-too-distant geological future."

Species and Gradual Change

Following in Darwin's footsteps, the current mainstream of evolutionary thought stresses the importance of slow, continuous transformation of one species into another. This view, known as **gradualism,** holds that natural selection, genetic drift, and other processes lead to the accumulation of small (microevolutionary) changes within species. Large-scale (macroevolutionary) events result from the accumulation of these smaller changes over time.

This view, though easily explained, causes some problems with definitions of species over time. Anyone who studies mammals, for example, can give you a precise definition of the species we call the modern horse. But that clearly defined entity loses its identifiable characteristics when we try to follow it back over geological time. Remember, too, that Darwin's often unsuccessful search for the intermediate fossils this sort of change should produce led him to criticize the fossil record as grossly incomplete.

Punctuated Equilibrium

In 1972 Niles Eldredge and Stephen Jay Gould challenged the doctrine of gradualism, adjusting their view of evolutionary change to accommodate their greater respect for the record in the rocks. They say the fossil record shows that most species remain "virtually unchanged" from the time they appear to the time they disappear. They also argue that new species, when they do arise, appear more quickly than Darwinian gradualism allows.

Gould and Eldredge propose that evolution proceeds in a manner they call **punctuated equilibrium.** According to this theory, very little microevolutionary change occurs over a species' lifetime, because each species is a stable, relatively unchanging entity. This stability, or "equilibrium," is periodically broken, or "punctuated," by brief periods of change during which new species arise from preexisting ones. The new species evolve for a time but then, like their predecessors, remain intact for the rest of their geological lives, perhaps as a result of stabilizing selection (Figure 13.10).

There are some extreme examples of species that have remained in equilibrium since they first appeared in the fossil record, despite the fact that, around them, other species have evolved considerably while countless more have suffered extinction. A few living species, such as the coelacanth (see page 330) closely resemble fossils over 250 million years old. And the living tadpole shrimp, *Triops cancriformis,* is so close in morphology to a well-preserved series of 180-million-year-old fossils that both living and fossil animals are placed in the same species. Steven Stanley, another participant in the punctuated equilibrium debate, suggests that this small shrimp is the oldest known living animal species (Figure 13.11).

Several mechanisms have been proposed to explain punctuated equilibrium. Some researchers point out that changes in the gene pool of large populations tend to occur relatively slowly. In small populations, on the other hand, such changes can accumulate much faster. For this reason, periodic parapatric speciation events could occur

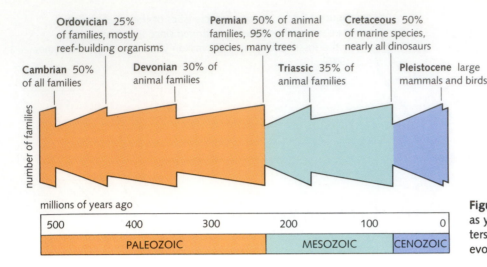

Figure 13.12 Periodic mass extinctions, as you will learn in the next several chapters, have been important episodes in the evolution of life on earth.

rapidly enough to produce changes during periods very much shorter than the lifetime of the more stable, parent species. Additionally, sympatric speciation (in plants) through polyploidy can produce new species literally in an instant. And the colonization of new habitats (such as the Galapagos Islands) by either animals or plants can result in rapid adaptive radiations.

There have also been several times in the earth's history when large numbers of species have disappeared during what are called **mass extinction events** (Figure 13.12). During such episodes, global ecosystems have been disrupted and many previously filled ecological niches have been opened up. The best-known mass extinction occurred at the end of the Cretaceous period, eliminating the ruling dinosaurs and paving the way for the earliest mammals (Chapter 17). Although such mass extinctions are not necessary for rapid speciation to occur, it is clear that they have created several major "punctuation marks" in the history of life on earth.

But more significant than these views on rates of change is the new vitality this theory infuses into the concept of species. For in this view of evolution, species are not constantly shifting biological will-o'-the-wisps but distinct individuals from their birth at speciation to their death at extinction. That critical change in perspective opens up an entirely new series of questions about the ways in which natural selection operates. If punctuated equilibrium continues to gain acceptance and comes to dominate evolutionary thinking, it may set the stage for yet another period in the evolution of evolutionary thought.

SUMMARY

Speciation—often called macroevolution—is the process by which populations of a single species diverge evolutionarily to produce more species. The first step in speciation is usually the creation of at least a partial barrier to gene flow between two populations of an existing species. Speciation is complete when the two populations are reproductively isolated by any of several reproductive isolating mechanisms: mechanical, behavioral, temporal, or ecological isolation, gamete imcompatibility, hybrid inviability, or hybrid infertility.

Allopatric speciation, the most widely accepted mechanism for speciation among animals, presumes that an initial barrier to gene flow occurs because of a physical barrier between populations. This isolation may be caused by major geographic features or through the colonization of patches of habitat separated by environments hostile to the organism in question. Parapatric speciation involves the isolation of small populations at the fringes of a species's range. Subsequent genetic divergence in both cases occurs because of both natural selection and genetic drift.

Sympatric speciation, the probable mechanism for nearly half the speciation events among flowering plants, occurs principally as a result of genetic changes that reproductively isolate certain individuals from surrounding members of their parent population.

Classical Darwinian theory holds that speciation occurs gradually, as small changes accumulate over long periods of time. The theory of punctuated equilibrium, on the other hand, holds that most species remain relatively unchanged for most of their existence and that speciation occurs during relatively brief periods.

After studying this chapter, you should be able to:

- Discuss modern theories on the evolution of new species.
- Define the term *reproductive isolation* and explain how populations of a species may be reproductively isolated from one another.
- Explain the theories of speciation.
- Outline the modern evolutionary theory of punctuated equilibrium.

SELECTED TERMS

speciation *p. 247*
reproductive isolating mechanism *p. 247*
prezygotic isolating mechanism *p. 247*
mechanical isolation *p. 247*
postzygotic isolating mechanism *p. 248*
behavioral isolation *p. 248*
temporal isolation *p. 248*
gamete incompatibility *p. 248*
hybrid inviability *p. 248*
hybrid infertility *p. 249*
ecological isolation *p. 249*
allopatric speciation *p. 249*
endemic species *p. 253*
parapatric speciation *p. 256*
sympatric speciation *p. 256*
gradualism *p. 259*
punctuated equilibrium *p. 259*
mass extinction events *p. 260*

REVIEW

Discussion Questions

1. Why is reproductive isolation so important in the process of speciation?

2. What are two ways in which the behaviors of animals can result in reproductive isolation?

3. What is the difference between hybrid inviability and hybrid infertility?

4. How does the punctuated equilibrium theory challenge Darwin's view of the fossil record?

5. How can geographic or ecological isolation eventually lead to the evolution of isolating mechanisms that keep new species separate even when the external barriers are removed?

Objective Questions   (Answers in Appendix)

6. A postzygotic barrier prevents
   (a) the union of sperm and egg.
   (b) the development of the embryo.
   (c) fertilization.
   (d) difficulties during meiosis.

7. The theory of punctuated equilibrium
   (a) can account for sudden appearance and disappearance of species.
   (b) explains that evolution occurs during periods of slow, gradual change.
   (c) assumes that speciation is an on-going process.
   (d) cannot be substantiated using fossil records.

8. Speciation cannot occur unless there is _____ isolation.
   (a) prezygotic
   (b) genetic
   (c) divergent
   (d) convergent

9. Sympatric speciation
   (a) most likely involves sudden genetic changes.
   (b) most likely involves gradual genetic changes.
   (c) is dependent on allopatric speciation.
   (d) requires geographical barriers.

10. Hybrids, if they live, are
    (a) always sterile.
    (b) never sterile.
    (c) usually sterile.
    (d) unusually fertile.

Darwin. C. (1979). Illustrated Origin of Species, Abridged and Introduced by Richard E. Leakey. Hill and Wang, New York.

Futuyma, D.J. (1979). Evolutionary Biology. Sinauer Associates, Sunderland, Mass.

Kaestner, A. (1970). Invertebrate Zoology. Volume III. Translated by H. Levi and C.R. Levi. Wiley, New York.

Lack, D. (1947). Darwin's Finches. Cambridge University Press, New York.

PART FOUR

The Diversity of Life

It now remains to speak of animals and their nature. So far as we can, we will not exclude any of them, no matter how mean; for though there are animals which have no attraction for the senses, yet for the eye of science, for the student who is naturally of a philosophic spirit, and who can discern the causes of things, nature which fashioned them provides joys that cannot be measured.

—Aristotle (ca. 330 B.C.)

Earth's millions of living species are products of between 3.5 and 4 billion years of genetic change, natural selection, and chance. The history of each species can be viewed as an evolutionary experiment that constantly tests the fitness of a particular combination of body structures, physiological processes, and styles of reproduction under specific ecological conditions. If the species passes the test, it survives; failure dooms it to extinction.

The ecological conditions that define life's evolutionary trials never remain static for long. Movements of the earth's crust create continents, tear them apart, and smash them into one another. Mountains grow and erode, and rivers change course. Sea levels rise and fall. Recurrent ice ages blanket continents with ice and snow. These changes in global environments drive the processes of natural selection that shape new species and influence their subsequent evolution. And periodically, episodes of major environmental change initiate mass extinctions, wiping out scores of species and making room for adaptive radiations among the survivors.

For this reason, knowledge of the earth's environmental history is vital to understanding the diversity of life, and paleontologists use a variety of techniques to reconstruct our planet's past. Some clues come strictly from geology; glaciers, for example, carry sand and stones that scour the rocks beneath them as they move, leaving unmistakable evidence of ice ages. Other clues come from comparisons of fossil and living organisms ranging from mammals to phytoplankton.

Remember also that organisms that have disappeared over the eons were not "inferior" in any absolute sense to those that survived. The survivors were just coincidentally better adapted to life under the prevailing environmental conditions. At any step along the way, chance could have produced a dramatically different earth. Sixty-five million years ago, for example, a mass extinction cleared the earth of dinosaurs. Had those giant reptiles not disappeared, the adaptive radiation of mammals might never have occurred, and our own ancestors might still be nothing more than small, shrew-like creatures scurrying to keep out of harm's way in a lizard's world.

◄ Displaying the symmetry typical of many living things, this colony of the fungus *Penicillium* symbolizes the importance of life's diversity to human society. The yellowish liquid on the surface is a metabolic waste product whose presence indicates that the colony is producing another by-product—the antibiotic penicillin.

CHAPTER 14

Biological Classification and the Origin of Life

In the evolutionary narrative in this section, you will see both successes and failures in organisms' adaptations to the earth's physical and biological environments. The individual stories illustrate three main themes.

1. *Key evolutionary innovations set the stage for adaptive radiations.* Innovations in body structure and life cycle provide the raw materials that fuel adaptive radiations in major new groups of organisms.

2. *Diversity is subject to geology.* Continental drift and changes in climate create opportunities for adaptive radiations by isolating once continuous populations (as continents split apart), by opening up new territory (as new oceans are created and seas invade the land), and by triggering changes in temperature or humidity that cause extinctions.

3. *No organism evolves in a vacuum.* The evolutionary history of each species is shaped by its interactions with contemporary plants and animals.

FINDING ORDER IN DIVERSITY TAXONOMY: THE SCIENCE OF NAMES

To study the vast numbers of living and fossil organisms, biologists must name them and arrange them into scientifically logical groupings. **Taxonomy,** the science of naming organisms, dates back to Aristotle (Figure 14.1).

As more organisms were discovered through the years, finding suitable names for them became a problem. Common names were vague and often confusing. "June bug," for example, refers to at least a dozen species of beetles in the United States alone, and the name "bluebell" is used for several different plants (Figure 14.2). Conversely, a single species of American wildcat is known in different areas as the puma, the cougar, the catamount, the American panther, and the mountain lion.

Early scholars tried to standardize the system by giving each organism a name in Latin, the universal language of science. Unfortunately, these Latin names were difficult to use; they were changed regularly and could be up to 15 words long. One carnation-like flower, for example, bore the cumbersome name *Dianthus floribus solitariis, squamis calycinis subovatis brevissimus, corollis crenatis.* Biologists clearly needed a single, stable, concise way of naming things.

The System of Binomial Nomenclature

Swedish botanist Carolus Linnaeus (1707–1778) devised a shorthand method of naming organisms called the system of **binomial nomenclature** (*bi* means "two"; *nome* means "name"; hence *binomial* means "two names"). The Linnaean system, which is now used around the world, gives each species a unique, two-part name. For the plant whose nine-word name is given above, for example, Linnaeus used the name *Dianthus caryophyllus.*

Figure 14.1 The process of naming and describing organisms inspired scientific art during the Middle Ages, when drawings of useful or interesting organisms were copied and collected into magnificently illustrated volumes. Bestiaries, such as this fourteenth-century French example, depicted both real and imaginary animals.

The first part of each name is a single word, usually a proper noun derived from either Latin or Greek, called the **genus** name (plural, *genera*). Because Latin was the language of science, many genera represent the common names used by the Romans. Similar cats were placed in the genus *Felis*, for example; dogs in the genus *Canis;* and horses in the genus *Equus*. Note that generic names begin with a capital letter and are italicized in print.

The second part of the name, usually a Greek or Latin adjective, describes some significant characteristic of the species. The plant *Campanula rotundifolia*, for example, is a bellflower with rounded leaves (*campana* means "bell"; *rotund* means "round"; *folia* means "leaves"). Similarly, the name for our own species is *Homo sapiens* (*homo* means "man"; *sapiens* means "wise"). The second part of the scientific name is italicized but not capitalized.

Figure 14.2 These three plants are only a sample of those commonly called "blue-bells." Their scientific names and families are (left) *Mertensia ciliata* (Boraginaceae), (center) *Campanula rotundifolia* (Campanulaceae), and (right) *Clematis hirsutissima* (Ranunculaceae).

Higher Taxonomic Categories

Scientists also find it useful to group plants and animals into larger taxonomic categories or **taxa** (singular, *taxon*), each defined by a certain set of important characteristics. Animals belonging to our own class, Mammalia, for example, are identified in part by having body hair, being "warm-blooded," bearing young alive (instead of laying eggs), and nourishing newborns with milk from mothers' mammary glands.

Biological taxonomy uses a *hierarchical* system; that is, each category is nested within the category above it and contains lower categories within it. You use hierarchical systems to group objects all the time. To fill out a check in a store, for example, you ask for a "pen." Although there are many types of pens, members of another class of writing utensils, namely pencils, are clearly excluded from your request.

Biologists identify organisms using a hierarchy that ranges from the largest divisions, called *Kingdoms,* down to species:

Kingdom
  Phylum (for animals) or Division (for plants)
    Class
      Order
        Family
          Genus
            Species

(A simple mnemonic device to help you remember these main divisions in the taxonomic hierarchy is the nonsense sentence "**K**ings **P**lay **C**hess **O**n **F**ine **G**reen **S**and.") In some cases, however, systematists have decided that these categories alone are not sufficient to demonstrate natural groupings among organisms and have created additional categories. Phyla, for example, are sometimes divided into subphyla, orders are occasionally grouped into superorders or divided into suborders, and so on. Table 14.1 gives the classification of three familiar organisms.

Table 14.1 *Classification of Common Organisms*

Taxon	Human	Fire Ant	Sunflower
Kingdom	Animalia	Animalia	Plantae
Phylum	Chordata	Arthropoda	Anthophyta
Class	Mammalia	Insecta	Dicotyledones
Order	Primates	Hymenoptera	Asterales
Family	Hominidae	Formicidae	Compositae
Genus	*Homo*	*Solenopsis*	*Helianthus*
Species	*sapiens*	*saevissima*	*annuus*

SYSTEMATICS: THE SCIENCE OF CLASSIFICATION

Working alongside taxonomy, the discipline of **systematics** tries to group organisms into higher taxa in ways that provide the most biologically relevant information. In Linnaeus's time, kingdoms and families were used simply to "pigeonhole" similar-looking organisms into groups. But the recognition that living species have evolved from earlier organisms posed a new challenge: to group organisms into taxa that represent lineages or lines of evolutionary descent. Thus, ideally, higher taxa should represent major "branches" on the evolutionary tree of life. By this reasoning, species within a single genus resemble each other because they share a recent common ancestor. Similarly, members of a family represent a larger evolutionary lineage descended from common stock in the more remote past.

Creating taxa that accurately represent evolutionary history, however, is difficult. Until recently, most systematists aimed to group together those organisms that most closely resembled one another. To this end, systematists following a school of thought called *phenetics* examined as many anatomical and physiological characteristics as possible. Many characteristics—such as bone structure, tooth arrangement, body size, and claw size—were given equal weight. Other characteristics were considered more important. Those organisms that shared the greatest number of similar characteristics were assumed to be most closely related to one another (Figure 14.3a,b). But the phenomena of adaptive radiation and convergent evolution often make it difficult to distinguish organisms that look alike because they are, in fact, closely related from those that look alike because they have adapted to similar niches. For this reason, classifications that rely exclusively on structural similarities do not always reflect evolutionary history.

Another school, *cladistics,* aimed specifically to create taxonomic groupings that more accurately reflected organisms' evolutionary histories. In this method, characteristics were separated into two clearly defined groups: those shared by living organisms because they had evolved from recent common ancestors, and those shared by larger groups of organisms because an ancient common ancestor had them. By treating recently evolved characteristics differently from very old characteristics, and by relying heavily on the fossil record, this scheme emphasized evolutionary branching patterns while downplaying shared structural similarities (Figure 14.3c,d).

Many modern evolutionary systematists attempt to combine the best features of both phenetic and cladistic approaches. This effort is fraught with conflict, because considerations of overall similarity and evolutionary relatedness often lead to the sorts of dramatically different conclusions shown in Figure 14.3. Perhaps the fairest

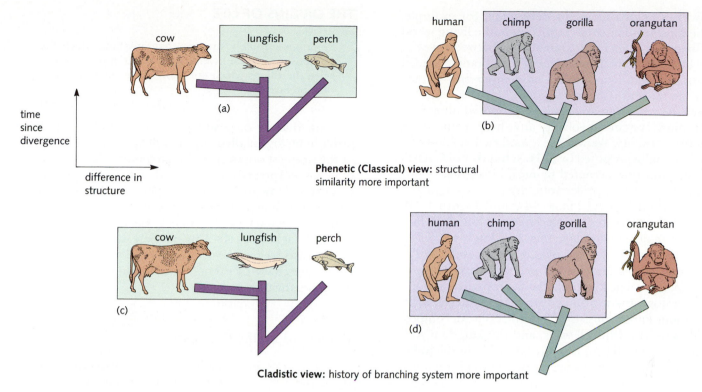

difference in structure

(a)

(b)

Phenetic (Classical) view: structural similarity more important

(c)

(d)

Cladistic view: history of branching system more important

Figure 14.3 **(a)** Though the lineage leading to lungfish separated from other fishes hundreds of millions of years ago, both still live in water and look like ''fish,'' so the phenetic (classical) approach groups them together. **(b)** Chimpanzees, gorillas, and orangutans retain many characteristics of early primates, whereas humans have undergone extensive changes in anatomy and behavior. Pheneticists therefore group apes in the family Pongidae and humans in their own family, the Hominidae. **(c)** Though cows and lungfish look different because they have adapted to different environments, they are more closely related to each other than either is to a perch. Cladists thus group cows and lungfish separately from other fishes. **(d)** Humans, chimps, and gorillas share recent common ancestors, whereas orangutans branched off in the more distant past. Cladists therefore group humans with our closest living relatives.

way to summarize the situation is to admit that, as in all types of classification schemes, the usefulness of a particular method depends on its specific application. To an ecologist, a lungfish is indeed more similar to a perch than to a cow. To a physiologist, however, lungfishes bear certain striking resemblances to mammals.

Molecular Systematics

Recent advances in molecular biology have provided an entirely new set of systematic tools; we can now judge similarities and differences among organisms on the molecular level. Evolution by common descent ensures that many organisms share many important biological molecules ranging from DNA to hemoglobin (Chapter 20) and chlorophyll (Chapter 26). New techniques in

molecular biology enable systematists to compare similarities and differences among these molecules and among a host of important body proteins. The importance of molecular techniques in evolutionary biology will be discussed in detail in Chapter 23.

THE FIVE KINGDOMS

In Aristotle's time, the largest taxonomic categories (the kingdoms) were obvious: Most things were clearly Animal, Vegetable, or Mineral. A few odd organisms, such as bread molds and sponges, didn't fit neatly into any of those groups, but the dividing lines were generally apparent. As biological knowledge grew, however, it became clear that additional kingdoms were needed.

Microscopes helped biologists determine, for example, that the most fundamental distinction in the living world is not between plants and animals but between *eukaryotes*—whose cells contain a variety of membrane-bounded structures, including a nucleus—and *prokaryotes*—which contain no nucleus (Chapter 9).

Over the years, researchers have proposed numerous classification systems involving anywhere from three kingdoms to twenty. We will use a modified version of a system originally proposed by R. H. Whittaker of Cornell University and now accepted by most biologists. In our five-kingdom system, prokaryotic organisms are placed in the kingdom Monera (Table 14.2 and Figure 14.4). Single-celled eukaryotes are placed in the kingdom Protista (some multicellular algae that are closely related to single-celled forms are often placed here as well). Multicellular eukaryotes are divided primarily according to nutritional type: The kingdom Plantae contains predominantly photosynthetic primary producers with cell walls, the kingdom Fungi consists of heterotrophic decomposers and parasites with cell walls, and the kingdom Animalia includes a variety of heterotrophs without cell walls.

Although these groupings appear straightforward, some living things defy easy classification. Certain single-celled organisms can act as either plants or animals, depending on environmental conditions. Other organisms straddle the line between single-celled and multicellular life forms. And viruses, which many biologists do not even consider organisms, fall outside this classification scheme altogether.

THE ORIGINS OF LIFE

. . . life is infinitely stranger than anything which the mind of man could invent. We would not dare to conceive the things which are really mere commonplaces of existence.

—A. Conan Doyle (1891)

Now that we have some idea of how biologists find order in biological diversity, we will approach the biology of the simplest organisms with an inquiry into a question that has perplexed and inspired the human mind for centuries: How did life first appear on the primordial earth? And how did those earliest life forms begin the long evolutionary process that has led to the biosphere today? That inquiry is still incomplete, but geologists and biochemists have come up with several plausible theories about the chain of events that culminated in the first living organisms.

Conditions on the Early Earth

As the cosmic gas that formed the earth solidified to form the first solid rocks about 4.5 billion years ago (Chapter 1), volcanic activity produced an atmosphere that consisted mainly of water vapor (H_2O), carbon monoxide and carbon dioxide (CO and CO_2), nitrogen (N_2), hydrogen sulfide (H_2S), and hydrogen cyanide (HCN). Water existed only as a gas because the earth's surface was so hot that any rain that fell boiled right back into

Table 14.2 *Major Characteristics of the Five Kingdoms*

Monera	Protista	Fungi	Plantae	Animalia
Prokaryotic; no membrane-bound intracellular structures	Eukaryotic; nucleus, mitochondria, some have chloroplasts	Eukaryotic; nucleus, mitochondria, but no chloroplasts; cell wall of chitin	Eukaryotic; nucleus, mitochondria, chloroplasts; cell wall of cellulose	Eukaryotic; nucleus, mitochondria, but no chloroplasts; no cell wall
Solitary, filamentous, or colonial	Mostly solitary, some colonial or multicellular	Some unicellular, most grow in thread-like branches	Multicellular, most sedentary on land	Multicellular, most motile
Aerobic or anaerobic	Mostly aerobic	Mostly aerobic	Strictly aerobic	Strictly aerobic
Autotrophic or heterotrophic	Autotrophic or heterotrophic	Mostly saphrophytic	Mostly photosynthetic autotrophs	Heterotrophic
Bacteria of diverse types	*Amoeba*, paramecia, seaweeds	Yeasts, molds, mushrooms	Mosses, ferns, flowering plants	Sponges, worms, snails, insects, mammals

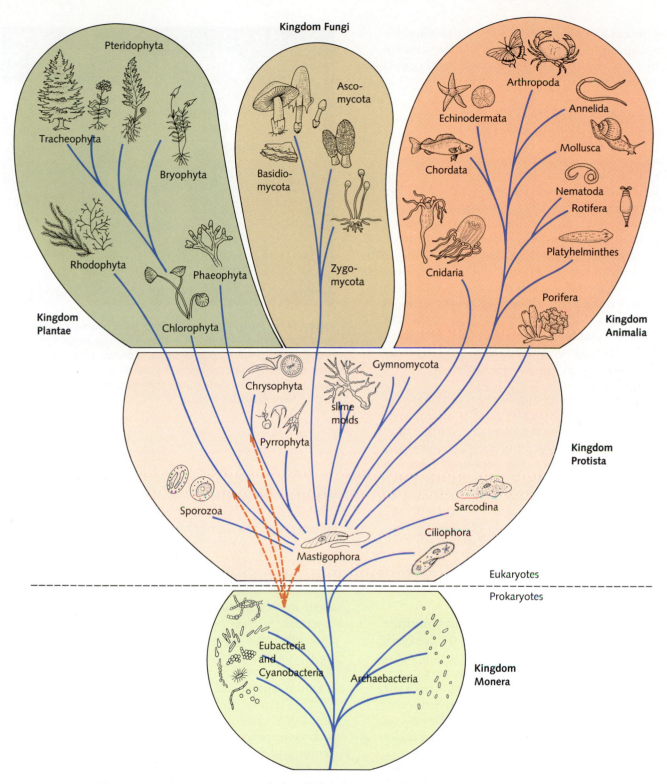

Kingdom Fungi

Kingdom Plantae
Pteridophyta
Tracheophyta
Bryophyta
Rhodophyta
Phaeophyta
Chlorophyta

Ascomycota
Basidiomycota
Zygomycota

Kingdom Animalia
Echinodermata
Arthropoda
Annelida
Chordata
Mollusca
Nematoda
Rotifera
Cnidaria
Platyhelminthes
Porifera

Kingdom Protista
Chrysophyta
Gymnomycota
slime molds
Pyrrophyta
Sporozoa
Sarcodina
Ciliophora
Mastigophora

Eukaryotes
Prokaryotes

Kingdom Monera
Eubacteria and Cyanobacteria
Archaebacteria

Figure 14.4 This schematic of the five-kingdom system depicts hypothetical evolutionary relationships. All prokaryotes are placed in the kingdom Monera. Unicellular eukaryotes are placed in the Protista, and multicellular eukaryotes are divided among the Fungi, Plantae, and Animalia. Note that though this system keeps all prokaryotes within the Monera, recent discoveries have led to the subdivision of the Monera into the Archaebacteria and Eubacteria. Some systematists feel that these divisions should be separate kingdoms.

the atmosphere. Once the planet cooled sufficiently for water to remain on the surface as a liquid (around 3.8 billion years ago), vast thunderstorms raged and oceans began to form.

The Organic Soup: Life from Non-Life

Russian biochemist A. I. Oparin began theorizing about the origin of life as early as 1924, but it was not until 1953 that an American graduate student, Stanley L. Miller, tested Oparin's ideas by simulating conditions on the primordial earth in a laboratory experiment. Miller built a flask in which he enclosed water, some inorganic matter, and several gases thought to have been present in the earth's early atmosphere. Miller sterilized the equipment to kill all microorganisms, excluded all traces of oxygen, and simulated lightning by generating sparks between tungsten electrodes. Incredibly, within a week the raw materials of life assembled themselves like enchanted building blocks, creating the ordered structures of amino acids—the building blocks of proteins—where before there had been only chaos.

Recent experiments using different gas mixtures produced results similar to Miller's, forming several important organic molecules in the total absence of life. Other scientists believe that the intense heat, pressure, and steady supply of high-energy inorganic compounds found in certain locations on the sea floor make those environments equally likely sites for similar critical reactions, even today (see Theory in Action, Submarine Hot Springs: Ancient Bacteria, Symbiosis, and the Origin of Life, p. 272).

From Molecules to Life

Organic compounds alone, however, do not constitute living cells, and we can only conjecture how the first cells appeared. One line of reasoning points out that if the organic "soups" created in these experiments are exposed to clay surfaces, many molecules stick to the electrically charged clay minerals. Because clay crystals have layered, repeating structures, molecules that stick to clay end up being arranged in a nonrandom order. Furthermore, clay crystals are interspersed with atoms of metals such as zinc and iron that encourage certain chemical reactions. Held close to one another and to metal atoms in this way, small organic molecules often join together spontaneously to form larger molecules.

Guided by the structure of clay and agitated by waves or currents, large organic molecules may organize themselves into different classes of spherical objects called

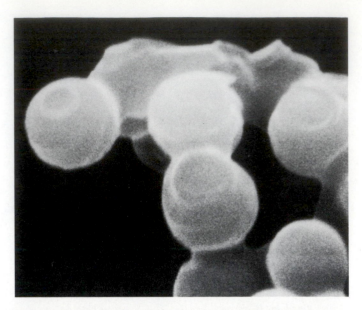

Figure 14.5 Proteinoid microspheres form spontaneously under conditions similar to those present on the early earth. Some microspheres act as natural chemical factories, encouraging reactions that incorporate materials from their surroundings. When they grow beyond a certain size, they become unstable and may split to form daughter spheres.

coacervate droplets and *proteinoid microspheres* (Figure 14.5). Both of these structures exhibit some properties of life: Their structure is nonrandom, they are surrounded by a two-layered membrane, they tend to accumulate and incorporate organic substances from their surroundings, and they have the ability to "grow." Although these collections of molecules are not truly alive, many of them facilitate certain biochemical reactions in their ordered interiors. Several hypotheses link these "protocells" or "precells" with more organized structures able to direct more complicated biochemical reactions, culminating in the various pathways of cellular metabolism you will learn about in Chapter 25.

One gap in the theory remains: the origin of RNA and DNA, the complex molecules around which all life today is organized. One hypothesis, offered by G. Cairns-Smith and J. Bernal as an alternative to the protocell theories, suggests that clay crystals might have served as templates or blueprints not only for the first proteins, but for these vital giant molecules as well. Recent evidence indicates that once molecules resembling RNA were formed, they could have been able to catalyze their own reproduction. From that point on, evolutionary processes took over.

THE EARLIEST LIFE FORMS

Regardless of how the earliest living cells evolved, they must have been anaerobic—that is, they must have lived in the absence of oxygen—because there was no free oxygen on the early earth. They must also have been heterotrophs, depending on organic compounds formed around them, because they had not yet evolved ways to capture energy in any other manner. Finally, they were unicellular and prokaryotic, like many modern-day bacteria. Researchers have found fossils of these microscopic organisms in rocks that formed 3.5 billion years ago, barely a billion years after the earth solidified. This discovery surprised even the most avid searchers, for it pushed the origin of life much farther back in time than anyone expected.

The Road to Autotrophy

At first, life for these primitive bacteria would have been simple, for the oceans were well stocked with the complex organic molecules those cells needed to survive. But as early cells multiplied, they would have used up their chemical "foods," and there would soon have been intense natural selection favoring individuals that were able to manufacture complicated molecules from simpler ones. The elaborate series of organic reactions you will learn about later would have evolved one step at a time.

Many metabolic pathways used by bacteria today probably evolved during this early period of the earth's history. Numerous living bacteria, in fact, still depend on *anaerobic fermentation* pathways to break sugars apart into alcohol and carbon dioxide or similar products in the absence of oxygen. We will examine the details of such processes in Chapter 25.

As these early heterotrophs exhausted the finite supply of abiotically produced complex molecules, they also encountered another problem—a shortage of available energy to fuel their metabolic processes. The result of selection pressure to harness new energy sources led to the evolution of a way to capture and utilize solar energy. What resulted were the earliest forms of bacterial photosynthesis—solar-powered reactions quite different from those carried on by modern plants. One type of bacterial photosynthesis, for example, takes in hydrogen sulfide (H_2S) instead of water (H_2O) and releases free sulfur instead of oxygen gas. These early photosynthesizers spread quickly. By around 3.5 billion years ago, several different types were growing together in layered, mat-like formations called **stromatolites** (*stroma* means "layer"; Figure 14.6). Modern purple sulfur bacteria still use this kind of photosynthetic pathway today.

Figure 14.6 (top) These fossils of ancient stromatolites are more than 3 billion years old. (bottom) These modern communities of photosynthetic bacteria exist only in extreme environments, such as these saline lagoons in Australia. Grazing pressure excludes these organisms from environments where plant-eating animals can survive.

Submarine Hot Springs:
Ancient Bacteria, Symbiosis, and the Origin of Life

Much of the deep ocean floor is nearly devoid of multicellular life. Water pressures reach several tons, and temperatures hover around freezing. The deep sea is pitch dark, so photosynthesis is impossible. Yet in this inhospitable environment, scientists have uncovered both a new class of ecosystem and clues to the origin of life.

In 1977, geologists aboard a deep-sea submersible near the Galapagos

Islands stumbled upon a rich animal community that thrives around submarine hot springs. Here geysers of mineral-laden water as hot as the core of a nuclear reactor erupt from the sea floor. Abundant animal life grows densely round the vents. No one had dreamed such a profusion of life could exist on the sea floor, and at first no one had any idea what energy source nourished these animals in the absence of sunlight.

We know now that this entire community is powered by archaebacteria that take in hot hydrogen sulfide (deadly poison to most eukaryotes) and use its chemical energy to produce complex organic compounds. These chemosynthetic bacteria, among the most ancient organisms alive today, thus replace photosynthetic plants as the base of the energy pyramid. The tube worms, clams, and mussels survive because they have

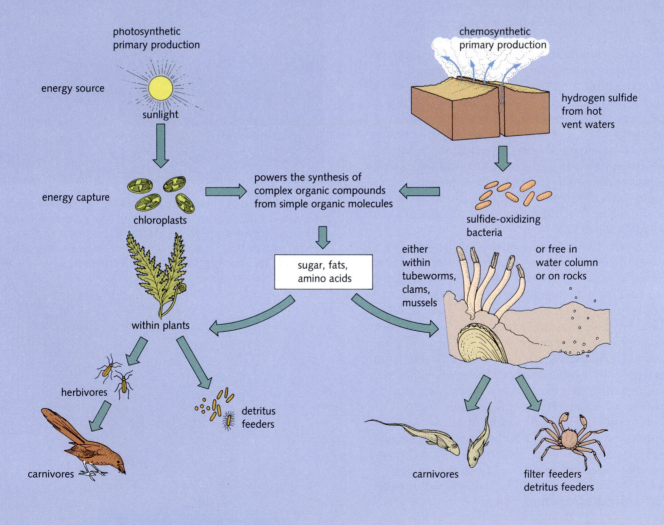

photosynthetic primary production

energy source

sunlight

energy capture

chloroplasts

within plants

herbivores

carnivores

powers the synthesis of complex organic compounds from simple organic molecules

sugar, fats, amino acids

detritus feeders

chemosynthetic primary production

hydrogen sulfide from hot vent waters

sulfide-oxidizing bacteria

either within tubeworms, clams, mussels

or free in water column or on rocks

carnivores

filter feeders detritus feeders

(above) The deep-sea hot springs community includes scarlet tube worms a meter long, clams 30 centimeters in length, clusters of mussels, and scores of shrimps, crabs, and fishes. (left) This schematic compares energy flow in a system fueled by photosynthesis and one powered by chemosynthesis.

evolved symbiotic relationships with bacteria in their gills. Water passing over the animals' gills supplies oxygen, carbon dioxide, and hydrogen sulfide, and the bacteria provide the necessary organic carbon.

Several scientists note that all the conditions required for the chemical reactions leading to the origin of life exist at these vents and have existed there since the oceans were formed: high temperature, the necessary chemical menu, turbulent currents, and clay deposits. By duplicating these conditions, researchers have produced not only amino acids but also short stretches of RNA.

Given millions of years, these building blocks of life could well have assembled first into protocells and ultimately into the earliest living prokaryotes. That the oldest living direct descendants of those first cells, the chemosynthetic archaebacteria, inhabit the hot springs today lends additional support to this fascinating, though highly controversial, hypothesis.

Photosynthesizers Change the World

About 3 billion years ago, a new group of photosynthetic prokaryotes evolved and completely reshaped the entire planet. These organisms began to photosynthesize the way plants do today, releasing oxygen as a by-product. Oxygen-producing photosynthesis evolved for very sound biochemical reasons that you will discover in Chapter 26. This striking metabolic innovation had profound consequences for the history of life on earth.

It is difficult to convey the revolutionary importance of oxygen production. Because we depend on oxygen, we think of that gas as essential to life. In fact, oxygen is *not* essential to most life processes, but only to the particular metabolic pathways used by eukaryotes today. On the contrary, oxygen is a highly reactive gas that is *poisonous* to many forms of life because it causes the destructive oxidation of organic compounds. Oxygen will kill any cell not protected by shielding compounds!

So imagine the scene 3 billion years ago: In an anaerobic world, one group of organisms started churning out this deadly gas in large quantities. For a time, geochemical activities shielded life from this toxic waste, because free oxygen rapidly combined with iron ions dissolved in the world's oceans. Over several million years, the resulting insoluble ferric oxide—rust—precipitated out as layers on the ocean floor. But once earth's reactive minerals were all oxidized, free oxygen began to accumulate in the atmosphere.

Anaerobes that did not evolve protection from toxic oxygen either perished or were excluded forever from the earth's surface and from most of the oceans. These physiologically archaic organisms survive today only in places where oxygen cannot penetrate, such as hot sulfur springs and deep mud. This was the first time that wastes produced by one group of organisms poisoned the biosphere for others.

Evidence of an Aerobic World

When oxygen reached a concentration equal to 1 percent of its modern-day levels, the ionizing effects of ultraviolet radiation would have produced the beginnings of an ozone layer. As that layer began shielding the earth from those harmful rays, organisms could live safely near the sea's surface. Then, about 2 billion years ago, a new sort of organism appeared, and its fossils offer the first evidence of adaptation to higher oxygen levels. These organisms formed chains of small cells punctuated at intervals with larger, thicker-walled cells. The resulting colonies resemble the modern prokaryotes called **cyanobacteria** or *blue-green bacteria*. The thick-walled cells, or **heterocysts** (*hetero* means "different"; *cystis* means "bladder or pouch";

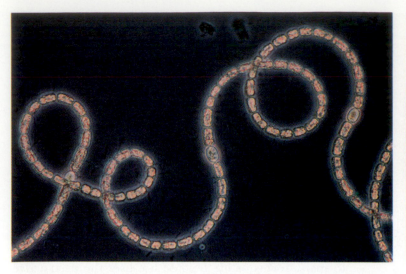
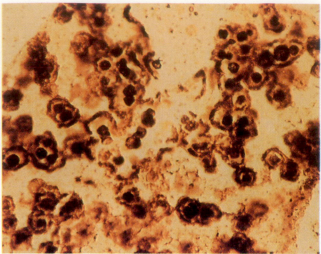

Figure 14.7 (left) This blue-green bacterium shows the enlarged structures called heterocysts that protect its anaerobic, nitrogen-fixing enzymes from oxygen. (right) This blue-green fossil from the ancient Gunflint Formation is evidence that prokaryotic organisms were evolving ways to tolerate increasing free oxygen levels in the atmosphere.

hence *heterocyst* means "different-looking cell") of modern blue-greens shield nitrogen-fixing enzymes from oxygen, enabling those anaerobic pathways to function in an aerobic world (Figure 14.7).

THE RISE OF EUKARYOTES

Between 1.4 and 2.5 billion years ago, a series of events occurred that set the pace for the rest of earth's history. First, as oxygen concentrations continued to rise, organisms evolved that not only *tolerated* oxygen but also *depended* on it. Over time, new forms of aerobic metabolism evolved, ultimately developing in ways that extracted 18 times as much energy from each sugar molecule as the older, anaerobic metabolism.

Second, the first eukaryotes evolved—organisms whose cells contain a nucleus surrounded by a protective double membrane and other classes of membrane-bound structures. The transition from prokaryotic to eukaryotic structure probably occurred among floating marine microorganisms whose remains comprise a class of microfossils called *Acritarchs* (from a Greek phrase meaning "of uncertain origin"; Figure 14.8). Fossils of presumed eukaryotic cells have been dated at between 1.4 and 1.6 billion years of age, which means that eukaryotic cells first evolved in, and were fully adapted to, an aerobic environment.

Third, within 2 or 3 hundred million years, **sexual reproduction** occurred among some eukaryotes and

Figure 14.8 This microfossil Acritarch is thought to exemplify the first eukaryotic cells.

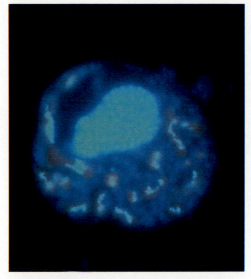
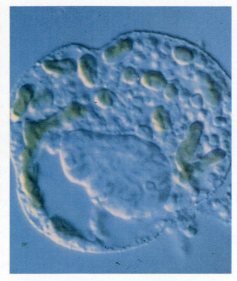
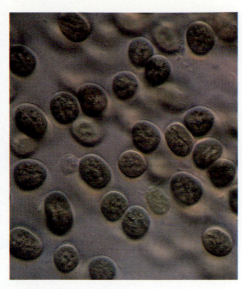

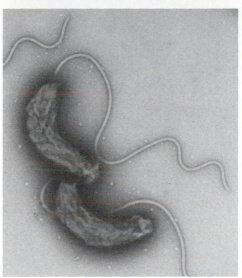

Figure 14.9 These micrographs show why many researchers believe that chloroplasts and mitochondria originated as free-living organisms. Like the grin of the Cheshire Cat, the DNA of these symbionts lingers as a clue to their independent origins. (top left) An alga, *Gyrodinium*, photographed in such a way that all its DNA fluoresces red. Why does this cell have DNA in both its nucleus and its chloroplasts? (center) The same alga, photographed in natural light to highlight its green chloroplasts. (right) This blue-green bacterium, a descendant of the earliest photosynthesizers, looks much like the proposed free-living ancestors of modern chloroplasts. (bottom left) This bacterium, *Bdelovibrio*, is thought to resemble the free-living ancestors of modern mitochondria.

accelerated the process of evolution to speeds never seen before. Prokaryotic cells, with some exceptions, reproduce by simple **binary fission,** duplicating their genetic material and sending half to each daughter cell, which is therefore a precise replica of its parent. Although this is an effective means of cell division, it restricts genetic variation to mutations and errors in DNA replication. The recombination and reshuffling of genetic material during sexual reproduction provide many more chances for existing genes to combine in new ways and thus vastly increase the genetic variation on which natural selection can operate. Once eukaryotes adopted sexual reproduction, therefore, new species began to appear far more rapidly than they had before. The first multicellular organisms evolved within a few hundred million years, and not long after that, a host of exotic multicellular creatures appear in the fossil record.

The Symbiotic Theory of Eukaryote Origins

Biologists have long pondered the evolutionary origin of *mitochondria* and *chloroplasts*, the complex, membrane-bound structures found in eukaryotic cells. Mitochondria and chloroplasts are puzzling because they contain their own DNA and reproduce on their own while the cell in which they reside divides. These organelles also bear an uncanny physical and biochemical resemblance to certain living prokaryotes such as the alga *Gyrodinium* and the bacterium *Bdelovibrio* (Figure 14.9).

It now appears that the evolution of eukaryotes from prokaryotes involved several distinct steps whose timing and order are yet unknown. At some point, the ancestors of all eukaryotes somehow evolved a membranous envelope around their nuclei. Just how or why this happened is not understood.

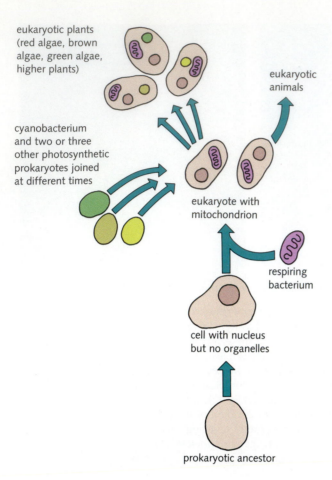

eukaryotic plants
(red algae, brown
algae, green algae,
higher plants)

eukaryotic
animals

cyanobacterium
and two or three
other photosynthetic
prokaryotes joined
at different times

eukaryote with
mitochondrion

respiring
bacterium

cell with nucleus
but no organelles

prokaryotic ancestor

Figure 14.10 The symbiotic origin of eukaryotic cells. First, a host cell that had already evolved a nucleus combined with a bacterium to form an animal-like cell with mitochondria. This duplex cell (or another like it) was later joined by a photosynthetic prokaryote to form a unicellular alga with chloroplasts. These latter events probably happened several separate times.

Then, according to a widely accepted theory championed by Lynn Margulis, eukaryotic evolution continued in several distinct lines as two or more simple organisms joined to form lasting symbiotic associations. According to Margulis, mitochondria and chloroplasts don't just *look* like prokaryotes; they once *were* prokaryotes that took up residence inside other cells and lost their independence. In light of the chemical differences among living algae and higher plants, it is likely that these events happened not just once, but two or three separate times, giving rise to several lines of photosynthetic eukaryotes (Figure 14.10). This fascinating story of intracellular symbiosis reminds us that even the most basic components of our bodies—our cells—are collections of once-independent parts.

Figure 14.11 offers a review of key milestones in the history of the earth up to the appearance of eukaryotic cells.

Living Prokaryotes and Single-Celled Eukaryotes

Although there is ample microfossil evidence of early life, many important steps in early evolution were chemical events that left no fossil traces. Biologists, therefore, try to trace the links between ancient prokaryotes and their descendants by making biochemical comparisons among living forms. The advances in molecular biology over the last 20 years have led to extensive revisions of the classification of unicellular organisms.

BACTERIA: KINGDOM MONERA

Organisms similar to today's living bacteria were the earth's first life forms, and living members of the kingdom Monera are the most abundant. A handful of soil contains billions of bacteria, and the bacteria in your mouth alone outnumber all the humans who have ever lived. Bacteria owe their success to an extraordinary metabolic diversity. Though all plants use basically the same photosynthetic pathways, for example, bacteria exhibit several. Bacteria reproduce rapidly by binary fission. The intestinal bacterium *Escherichia coli*, when grown in the laboratory, matures and is ready to divide in less than 15 minutes. This short generation time creates genetic diversity in the absence of true sexual reproduction, helping bacteria adapt to changing environments (Figure 14.12).

The higher-level classification of monerans is both complex and under dispute. Many experts now believe that the ancient prokaryotes split early in their evolution

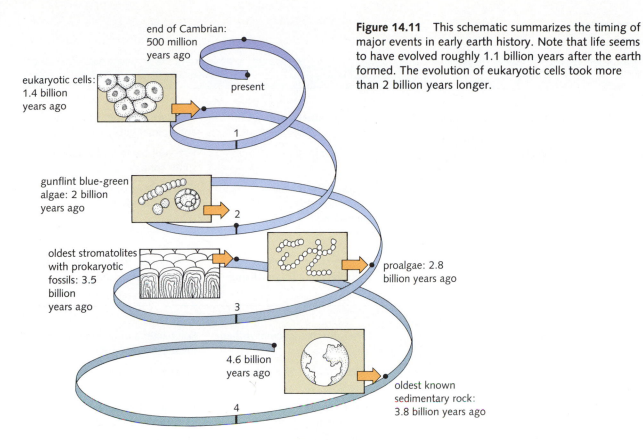

Figure 14.11   This schematic summarizes the timing of major events in early earth history. Note that life seems to have evolved roughly 1.1 billion years after the earth formed. The evolution of eukaryotic cells took more than 2 billion years longer.

end of Cambrian: 500 million years ago

present

eukaryotic cells: 1.4 billion years ago

1

gunflint blue-green algae: 2 billion years ago

2

oldest stromatolites with prokaryotic fossils: 3.5 billion years ago

3

proalgae: 2.8 billion years ago

4.6 billion years ago

4

oldest known sedimentary rock: 3.8 billion years ago

into at least three separate lines or divisions: the *Archaebacteria,* the *Eubacteria,* and the *Cyanobacteria.* Some microbiologists, however, place the cyanobacteria within the eubacteria, and others favor giving the archaebacteria their own kingdom (Table 14.3).

"Ancient" Bacteria: Division Archaebacteria

The **archaebacteria** are called "ancient" bacteria (*archaic* means "old") because they seem to be quite similar to the first prokaryotes on the earth. Many bacteriologists place the archaebacteria in their own kingdom because they are chemically unique in several important ways. Archaebacteria have different RNA and chemically different cell walls from other bacteria.

Some archaebacteria, called **methanogens,** live in the anaerobic mud of swamps and marshes, where they produce methane gas as a metabolic by-product. Other methanogens live inside the intestinal tracts of many organisms (including sheep, cows, and other mammals), where the methane they produce accumulates until it exits from the end of the gut. Some researchers even feel that the methane these bacteria produce in the guts of

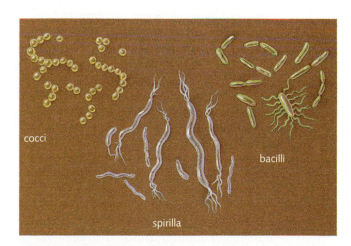

cocci

bacilli

spirilla

Figure 14.12   Moneran shapes: spherical cocci (*singular*: coccus), rod-shaped bacilli (*singular*: bacillus), spiral-shaped spirilla (*singular*: spirillum), and chain-forming streptobacilli and streptococci.

Table 14.3 *Bacterial Diversity*

Division	Shape	Nutrition	Example
Archaebacteria			
Methanogens	Bacilli, cocci, or spirilli	Chemosynthetic or photosynthetic	Cellulose digesters in human and animal guts
Thermoacidophiles	Bacilli	Chemosynthetic	Rift-vent bacteria; bacteria of sulfer springs
Halophiles	Bacilli	Photosynthetic (using bacteriorhodopsin, not chlorophyll)	Pink bacteria of the Dead Sea and salt ponds
Cyanobacteria (some place these within the Eubacteria)	Cocci, bacilli, streptobacilli	Photosynthetic (using chlorophyll); nitrogen-fixing	Common in fresh water, brackish water, and marine habitats
Eubacteria			
Pseudomonads	Bacilli	Heterotrophic; chemosynthetic	Soil bacteria of decomposition; invaders of hot tubs
Spirochetes	Spirilli	Heterotrophic; free-living or parasitic	Parasite that causes syphilis
Enteric bacteria	Bacilli	Heterotrophic symbionts; in mammalian guts may be mutualistic, commensal, or parasitic	Normal gut symbionts; serious parasites causing sometimes-lethal diarrhea and typhoid
Chlamydias	Very tiny spheroids	Heterotrophic; parasitic	Urogenital infections in humans
Actinomycetes	Bacilli, streptobacilli	Heterotrophic; free-living or parasitic	Soil bacteria that produce streptomycin; nitrogen-fixing symbionts with plant roots; parasites that cause tuberculosis and leprosy

farm animals contributes significantly to the greenhouse effect! The members of another group, the **thermoacidophiles,** inhabit extremely hot, acid habitats such as sulfur springs and deep-sea volcanic vents (*thermal* means "heat"; *philos* means "to love"; hence *thermoacidophile* means "heat and acid lover"; see Theory in Action, p. 272). The **halophiles** (*halo* means "salt"; hence *halophile* means "salt lover"), are photosynthetic but thrive only in places such as the Dead Sea, where the water is so salty that it is lethal to virtually all other organisms.

The "Blue-Greens": Division Cyanobacteria

The cyanobacteria, or "blue-greens," are common worldwide. Cyanobacteria carry on the same sort of photosynthesis as green plants, and they fix nitrogen in the conspicuous, specialized cells called *heterocysts* (see Figure 14.7; p. 274). Some cyanobacteria are solitary, floating forms; others form clusters; and still others live in mats, forming stromatolites similar to those their ancestors made more than 3 billion years ago.

"True" Bacteria: Division Eubacteria

The **eubacteria,** or "true" bacteria, are both numerous and ecologically diverse. Some are aerobic, others strictly anaerobic. Some are photosynthetic, others depend mostly on fermentation for their energy, and still others are dangerous parasites of multicellular eukaryotes. Major groups of eubacteria are shown in Table 14.3

Many keys that aid in the identification of eubacteria rely on a staining color test known as the Gram stain, developed in the late 1800s by the Danish physician Hans Gram. Depending on the structure of their cell walls, bacteria appear either violet, in which case they are called *Gram-positive*, or red, in which case they are called *Gram-negative*.

Bacteria in Nature and Technology

Most bacteria are not only harmless to humans but also indispensable in the nutrient cycles of both natural and artificial ecosystems. Recall that bacteria oxidize the nitrogen-containing wastes of animals and keep the nitrogen cycle turning in both terrestrial and aquatic habitats. Furthermore, nitrogen-fixing bacteria are the biosphere's only link to the vast reservoir of nitrogen in the atmosphere around us.

Many useful bacteria have been "domesticated" for centuries. Bacteria that ferment milk sugar (lactose) into lactic acid produce milk products such as cheeses and yogurt. The bacteria of the nitrogen cycle are used to break down waste products in artificial systems ranging from compost heaps to sewage treatment plants, aquaculture systems, and even your home fish tank. Bacteria are used in the medical industry as well, producing such modern antibiotics as erythromycin and streptomycin. Much more recently, our skills in manipulating the genetic material of bacteria has turned them into the new workhorses of biotechnology. Bacteria are now being used to produce scores of important biological compounds, including insulin and other hormones that are difficult or impossible to synthesize in the laboratory.

Bacteria and the human body Our intestines are filled with symbiotic eubacteria, most of which are either commensal forms (living without causing harm) or mutualistic forms (benefiting both themselves and their hosts). Some members of this "gut flora" help us digest such foods as fats; others synthesize and release vitamin K and a few of the B vitamins. (Recall the importance of gut bacteria to herbivores noted in Chapter 3.)

The human digestive system becomes accustomed to the normal activities of these microbial boarders. Anything that disrupts their activity or population, such as treatment with certain powerful, broad-spectrum antibiotics, invariably causes intestinal distress. Travelers' diarrhea occurs when unfamiliar strains of bacteria are introduced to the gut in drinking water, incompletely cooked food, or unwashed raw vegetables and fruits.

There are relatively few parasitic bacteria, but they cause some of the most deadly diseases known: bubonic plague, cholera, tuberculosis, bacterial pneumonia, diphtheria, and tetanus, to name just a few. Most bacterially caused diseases have been under control for years in industrialized nations, though they linger under unsanitary conditions in less-developed countries. Diarrhea brought on by cholera bacteria is still one of the major causes of death among Third World infants. Bacteria are also responsible for several of the most common *venereal* or *sexually transmitted diseases* (STDs, in medical parlance), including gonorrhea (*Neisseria gonorrhea*), syphilis (*Treponema pallidum*) and chlamydial infections (*Chlamydium trachomonas*).

UNICELLULAR EUKARYOTES: KINGDOM PROTISTA

The Protista is by far the most diverse of the eukaryotic kingdoms; its members range from free-floating plankton to deadly parasites in mammals. The classification scheme we have adopted for this book defines the protists as unicellular, eukaryotic organisms. It further divides protists into three broad groups according to their nutritional habits: animal-like heterotrophs, fungus-like saprophytes, and plant-like autotrophs. Among the ancestors of these groups, we might expect also to find the ancestors of the three multicellular kingdoms.

The evolutionary systematics of the kingdom Protista is under some dispute. Because of differences between protists and higher plants and animals, it is not certain whether living protists and multicellular organisms share a common ancestor or evolved along completely separate lines. Because of differences among the protists themselves, it is not even clear that all living protists evolved from a single ancestor, as classic evolutionary trees imply (Figure 14.13).

Protists, though unicellular, are far from simple. Their single-celled condition hasn't kept them from evolving intricate relationships with their environment and with more recently evolved organisms. Whereas multicellular organisms have evolved specialized organs and organ systems, protists have evolved specialized structures ranging from complex organelles to locomotor structures protruding through their outer membranes.

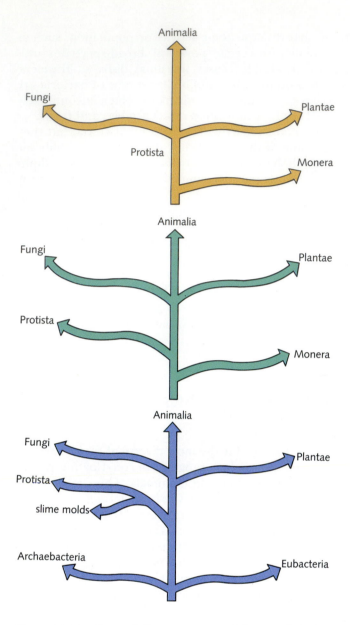

Figure 14.13 The evolutionary position of the unicellular kingdoms is still in dispute. The Protista are viewed by some as ancestors of all eukaryotes and by others as a separate line. Still others believe that slime molds, archaebacteria, and eubacteria should all be treated as belonging to separate kingdoms.

Animal-Like Protists

Phylum Mastigophora The *mastigophora*, also known as **flagellates,** are among the simplest in appearance of the living protozoa (Figure 14.14). Despite their structural simplicity, however, many modern flagellates are highly specialized. Some live symbiotically in the guts of termites, where they help these insects break down the complex molecules of cellulose (Chapter 3). Other mastigophorans are deadly parasites of humans and other animals. The genus *Trypanosoma,* for example, causes African sleeping sickness, a disease serious enough even today to make much of Africa uninhabitable for both humans and livestock (Chapter 39).

Phylum Sarcodina Members of the phylum Sarcodina, commonly called **amoebae,** are simple in appearance but ecologically diverse. Many, such as members of the genus *Chaos,* wander freely in wet habitats. Radiolarians and foraminiferans are abundant and ecologically important members of free-floating plankton communities, whose delicate shells are common in marine sediments. Some amoebae, such as *Entamoeba histolytica,* are parasitic in mammals (including humans), where they can cause amoebic dysentery. After entering the body in contaminated food or water, these parasites live in the lining of the intestine, where the histolytic enzymes for which they are named cause major tissue damage (*histo* means "cell"; *lyse* means "break open"; hence *histolytica* means "breaker or destroyer of cells"). In severe cases they can spread through the bloodstream to other tissues in the liver, the lungs, the skin, and even the nervous system.

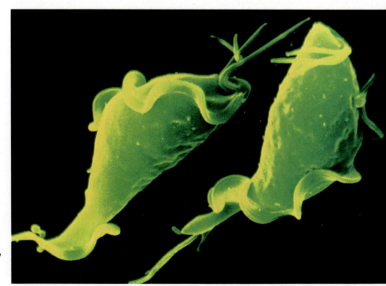

Figure 14.14 These typical mastigophorans bear a conspicuous flagellum, with which they propel themselves through their liquid environment.

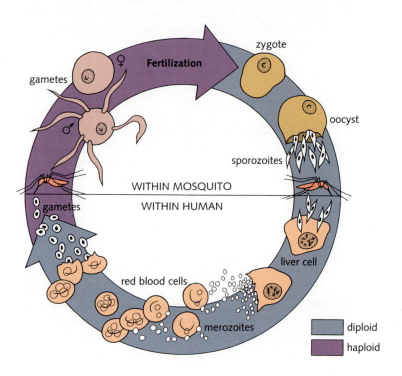

Figure 14.15 The complex life cycle of the malaria parasite demonstrates the coevolution of the parasite with both host and insect carriers. Starting with the left side of the cycle: The female mosquito bites an infected person and takes in protozoan gametes along with blood. Within the mosquito digestive tract, gametes unite and form a zygote. Structures called *oocysts* develop from zygotes. Thousands of small, spindle-shaped *sporozoites* develop within oocysts and migrate to mosquito salivary glands when the oocysts burst. The mosquito then bites another person and infects that person with sporozoites, which enter liver cells and multiply to form *merozoites*. The liver cells burst; merozoites are released and enter red blood cells, where they multiply. Merozoites break out of red blood cells and cause fever in the victim. Some of these merozoites become gametes. The female mosquito then bites another infected person, and the cycle continues.

Phylum Sporozoa Organisms belonging to the phylum Sporozoa have extremely complex life cycles, often involving two hosts and usually including some sort of spore-like phase. The best known of these **sporozoa** are the members of the genus *Plasmodium,* which causes malaria in humans (Figure 14.15).

Phylum Ciliata The phylum Ciliata contains another one of the best-known protozoans, the genus *Paramecium.* Covered with countless rapidly beating hairs, or *cilia,* its members may be free-swimming or sedentary (Figure 14.16). **Ciliates** such as *Paramecium* exhibit the most sophisticated organelles of all protozoans. Some are detritus feeders, others voracious predators.

Fungus-Like Protists

The phyla grouped among the fungus-like protists contain a variety of unicellular organisms with fungus-like characteristics (Figure 14.17). They may be either saprophytic or parasitic on plants or animals. Among the most interesting of the fungus-like protists are the cellular slime molds. These curious organisms spend part of their lives as independent, free-living, amoeboid cells. When food becomes scarce, however, they secrete a chemical attractant that causes hundreds of cells to swarm

Figure 14.16 These marine ciliates clearly show the hairs, or cilia, that give the group its name.

Figure 14.17 These fruiting bodies, or sporangia, are the spore-producing stage of a slime mold.

together and form a multicellular, slug-like *plasmodium.* The plasmodium crawls around on its own for some time and then produces a tall fruiting body that releases spores when conditions become favorable.

Plant-Like Protists

Phylum Euglenophyta The phylum Euglenophyta contains *Euglena,* a genus that defies firm classification (Figure 14.18). Some individual **euglenoids** have chloroplasts and photosynthesize as plants do; others lack chloroplasts and absorb complex molecules to survive as heterotrophs or absorb nutrients from their environment like fungi. By manipulating green species in the lab, researchers can produce colorless forms lacking chloroplasts, thus crossing the traditional boundary between plant and animal kingdoms.

Phylum Chrysophyta The **chrysophytes,** which make up the phylum Chrysophyta, include the single-celled algae commonly called *diatoms, golden-brown algae,* and *yellow-green algae.* Many are solitary, single-celled members of both freshwater and saltwater phytoplankton communities. However, some form long floating chains, and others grow attached to larger objects underwater. Diatoms are among the most important aquatic primary producers. They often grow densely; a five-gallon pail of seawater can hold several million diatoms. Many diatoms have spectacularly beautiful shells made of silica (Figure 14.19).

Figure 14.18 *Euglena,* a genus that blurs the boundary between autotroph and heterotroph.

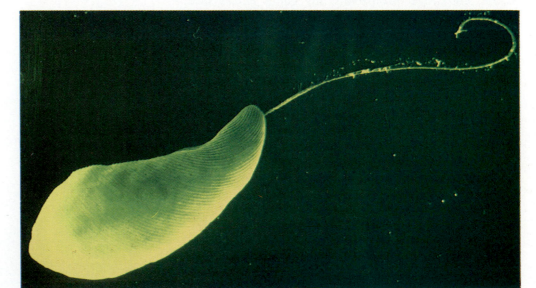

Phylum Pyrrophyta The **pyrrophytes** are commonly called **dinoflagellates,** or "armored" flagellates (*dino* means "powerful"), because of the tough cellulose plates that encase their bodies. These protists swim through the water using a pair of long, whip-like flagella.

Dinoflagellates are second only to diatoms as aquatic primary producers, and they come in a wide variety of sizes and shapes. Some give off a bright, bluish light when disturbed in the water. The famous phosphorescent bay in Puerto Rico glows because of its dinoflagellate population. Other dinoflagellates live symbiotically within the tissues of many marine organisms, from corals to giant clams.

Several dinoflagellates produce powerful toxins that have the potential to cause illness or even death in higher organisms. *Red tides* are caused by the sudden, explosive growth or *bloom* of certain free-swimming dinoflagellates. A virulent form of seafood poisoning called ciguatera is caused by dinoflagellates that grow on the blades of bottom-dwelling algae in the tropics. Poisons manufactured by the dinoflagellates are picked up by herbivores and accumulate in the food chain through biological magnification.

THE VIRUSES

Somewhere between living cells and inanimate crystals exist the **viruses,** whose name means "poison" in Latin. Viruses, so small they are invisible under even the most powerful light microscope, are life stripped down to its barest essentials: a central core of nucleic acids surrounded by a coat of protein. Viruses are not "alive" the way other organisms are. Though other specialized parasites can *reproduce* only inside the cells of their hosts, a naked virus can carry on *no* metabolic processes of any kind on its own. Viruses "come alive" only by taking over the metabolic machinery of their hosts' living cells.

Viruses were discovered in the early 1900s by the Russian biologist Dimitri Iwanowski, who was studying a disease of tobacco plants. He discovered that he could transmit the disease by dripping juice from infected leaves onto healthy plants. Even when he passed the juice through a filter so that not even a tiny cell fragment could pass through it, the filtered liquid could still transmit the disease. He concluded that an infectious organism (later called the tobacco mosaic virus) smaller than any yet known to science must be causing the disease.

Figure 14.19 Chrysophytes, such as these diatoms, are vitally important primary producers in aquatic food chains.

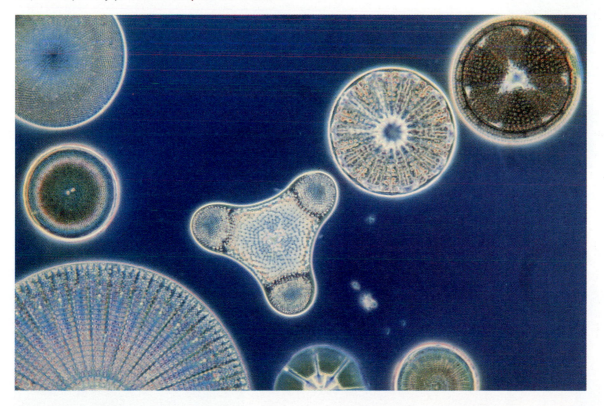

Viral Diversity

Viruses exist that can infect any kind of cell—animal or plant, prokaryote or eukaryote. The electron microscope has made viruses visible to us, and these small particles often have exotic and beautiful structures (Figure 14.20). There are many "families" of viruses, all classified according to the nature of their nucleic acid core; some contain DNA, and others contain RNA (Table 14.4). In either case, these macromolecules carry just enough information to enable the viruses to take over their host cells and reproduce themselves. Viral diseases range from mild infections, such as the common cold, to debilitating and fatal diseases, such as AIDS. You will learn more about viruses and diseases in Chapter 39.

There are several competing theories about the evolutionary origin of viruses, but we know for certain that despite their simplicity, they cannot be the oldest forms of life. Because they can reproduce only with the aid of living cells, they must have originated well after the first prokaryotes.

Table 14.4 *Important Groups of Viruses*

Virus Type	Results of Infection
DNA Viruses	
Adenovirus	Respiratory illness; some animal tumors
Herpesvirus	Herpes (both oral cold sores and genital herpes); infectious mononucleosis; Burkett's lymphoma; chicken pox; shingles
Papavovirus	Warts; cervical cancer; tumors in monkeys
Parvovirus	In joint infection with adenoviruses: gastric and intestinal distress; possibly hepatitis A in humans
Poxvirus	Smallpox; cowpox
RNA Viruses	
Orthomyxovirus	Flu
Paramyxovirus	Mumps and measles
Picornavirus	Polio; common cold; intestinal ailments
Reovirus	Diarrhea
Retrovirus	Certain cancerous tumors and leukemias; AIDS
Togavirus	Yellow fever; equine encephalitis; German measles

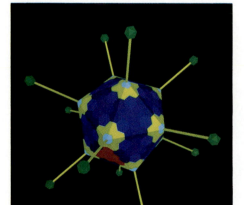

Figure 14.20 The crystalline structure of viruses is readily apparent whether reconstructed using computer graphics techniques (common cold virus top left), or visualized in electron micrographs (influenza virus below left, polyoma virus at center, and pox virus at right).

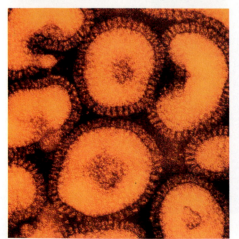
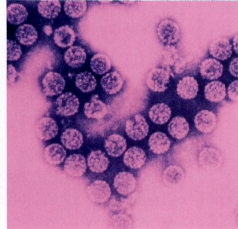

ciliates *p. 281*
euglenoids *p. 282*
chrysophytes *p. 282*

pyrrophytes *p. 283*
dinoflagellates *p. 283*
viruses *p. 283*

SUMMARY

The sciences of taxonomy and systematics work together in assigning names to organisms and grouping them into useful categories. Evolutionary systematics attempts to group organisms according to common ancestors they share.

Life began on earth as, over billions of years, complex organic molecules formed abiotically and became organized in progressively more complex ways. The evolution of photosynthesis liberated free oxygen and radically changed the composition of earth's atmosphere.

The prokaryotic kingdom Monera is divided into three distinct evolutionary lines: the divisions archaebacteria, eubacteria, and cyanobacteria. Many bacteria are essential participants in global nutrient cycles; others are commensal or mutualistic symbionts with higher organisms.

The unicellular eukaryotic kingdom Protista contains many diverse organisms; some resemble plants, others resemble animals, and still others resemble fungi. Many species are free-living and occupy vital positions at or near the bases of aquatic food chains. Others are dangerous parasites.

The viruses, nature's genetic engineers, can survive and reproduce only by using their genetic material to take over the metabolic machinery of a host cell. Viruses parasitize organisms ranging from bacteria to humans and cause diseases ranging from the common cold to AIDS.

STUDY FOCUS

After studying this chapter, you should be able to:

- Explain the scientific system for classifying and naming organisms.
- Show how different taxonomic approaches lead to different groupings of organisms.
- Outline the five-kingdom taxonomic system.
- Cite various hypotheses about the origin and early evolution of life.
- Describe the kingdoms Monera and Protista and the viruses.

SELECTED TERMS

taxonomy *p. 264*
binomial nomenclature *p. 264*
genus *p. 265*
taxa *p. 266*
systematics *p. 266*
stromatolites *p. 271*
cyanobacteria *p. 273*
heterocysts *p. 273*
sexual reproduction *p. 274*

binary fission *p. 275*
archaebacteria *p. 277*
methanogens *p. 277*
thermoacidophiles *p. 278*
halophiles *p. 278*
eubacteria *p. 279*
flagellates *p. 280*
amoebae *p. 280*
sporozoa *p. 281*

REVIEW

Discussion Questions

1. What was earth's earliest atmosphere like? What secondary atmosphere soon replaced it?

2. How could organic molecules have been formed abiotically early in the earth's history?

3. Why must the first organisms, or proto-organisms, have been heterotrophs? How could autotrophy have evolved?

4. Explain the symbiotic theory of the origin of eukaryotic cells.

5. Why do viruses fall outside all five kingdoms of life?

6. Devise two classification systems for the contents of a typical home workshop: tools, construction supplies, and various types of nuts, bolts, nails, screws, and fasteners. Devise one system based exclusively on the physical appearances of objects and a second system based solely on their function. What are the benefits and disadvantages of each system?

Objective Questions (Answers in Appendix)

7. The classification of organisms into meaningful groups is the science of
   (a) taxonomy.
   (b) systemics.
   (c) evolutionary biology.
   (d) organismic biology.

8. The cladistic taxonomist would study _____ among organisms.
   (a) molecular differences
   (b) evolutionary history
   (c) similarity of function
   (d) similarity of appearances

9. The five kingdom system classifies organisms according to _____ and _____.
   (a) plants / animals
   (b) cellular organization / nutritional type
   (c) biochemical differences / nutrition
   (d) photosynthetic / heterotrophic characteristics

10. The earliest living cells
    (a) lived in aerobic conditions.
    (b) could get energy from other organic compounds.
    (c) were eukaryotes.
    (d) were autotrophs.

11. Which group of bacteria are in the division archaebacteria?
    (a) bacteria that produce methane gas
    (b) bacteria that have cells called heterocysts
    (c) bacteria that fix nitrogen
    (d) bacteria that live in aerobic conditions

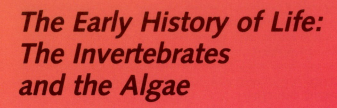

CHAPTER 15

The Early History of Life: The Invertebrates and the Algae

Feature boxes appear throughout the text serving as valuable study aids designed to enrich your understanding of each chapter. You may wish to refer to the following feature boxes as you read Chapter 15.

The Proterozoic Era	p. 287
The Cambrian Period	p. 288
Phylum Porifera	p. 290
Phylum Cnidaria	p. 291
Phylum Platyhelminthes	p. 293
Phylum Nematoda	p. 294
Phylum Mollusca	p. 296
Phylum Annelida	p. 298
Phylum Arthropoda	p. 300
Phylum Echinodermata	p. 302
The Ordovician Period	p. 304
Division Chlorophyta	p. 305

LATE PROTEROZOIC (OR PRECAMBRIAN) ERA

The Proterozoic era, stretching from about 2.5 billion years ago to about 570 million years ago (see p. 287) was one of the longest periods in earth's history. For most of that time, life evolved slowly; it took 2 billion years for the first eukaryotic cells to appear. But toward the end of that era, about 650 million years ago, two global trends sparked the evolution of the first multicellular organisms: a change in global climate and an increase in the amount of free oxygen in the atmosphere.

The Dawn of Multicellular Life

The mid-Proterozoic had seen a rapid diversification of single-celled plankton—the first great *adaptive radiation* of eukaryotes. But around 650 million years ago, a global cold spell triggered an ice age that wiped out many early plankton species in the world's first mass extinction. When the ice retreated and the climate warmed, a second adaptive radiation of plankton began, and this time a variety of multicellular animals called *metazoa* were among them.

This was the first era in earth's history that multicellular, aerobic organisms *could* have evolved, because by this time photosynthesis had increased the atmosphere's free oxygen to 10 percent of present levels. This oxygen concentration was important; because the earliest metazoa had not yet evolved respiratory and circulatory systems, they had to depend on simple diffusion to supply oxygen to their cells. And because oxygen diffuses through tissues very slowly at the low concentrations that prevailed during the early Proterozoic, it would have been impossible for sizable clusters of aerobic cells to obtain the oxygen they required.

Most of what we know about the first metazoa comes from excavations in the Ediacaran Hills of Australia. These hills have yielded a rich fauna of soft-bodied, multicellular animals dated at 680–700 million years of age. Some paleontologists believe that this *Ediacaran fauna* contained the forebears of modern jellyfish, segmented worms, and their kin. Other researchers disagree, arguing that the Ediacaran fauna represents a series of ultimately unsuccessful evolutionary experiments that left no descendants. In any case, these organisms probably lived in somewhat the same manner as many corals and jellyfish do today, depending on photosynthetic symbionts for energy in nutrient-poor seas.

By the end of the Proterozoic, about 570 million years ago, a few of the first fossil shells appeared, along with remnants of the first multicellular plants. From this time, too, date the first complex **trace fossils:** petrified tracks and burrows in the sediment left behind by animals that were not themselves preserved. These early shells and trace fossils indicate that other animals too were evolving—hard-shelled forms that would soon dominate the seas. For most of this time the continents were barren; life was still an aquatic phenomenon. But by the end of the era, bacteria may have begun to colonize the soil, and primitive fungi and multicellular green algae may have ventured out onto moist beaches around shallow bays.

The Proterozoic Era

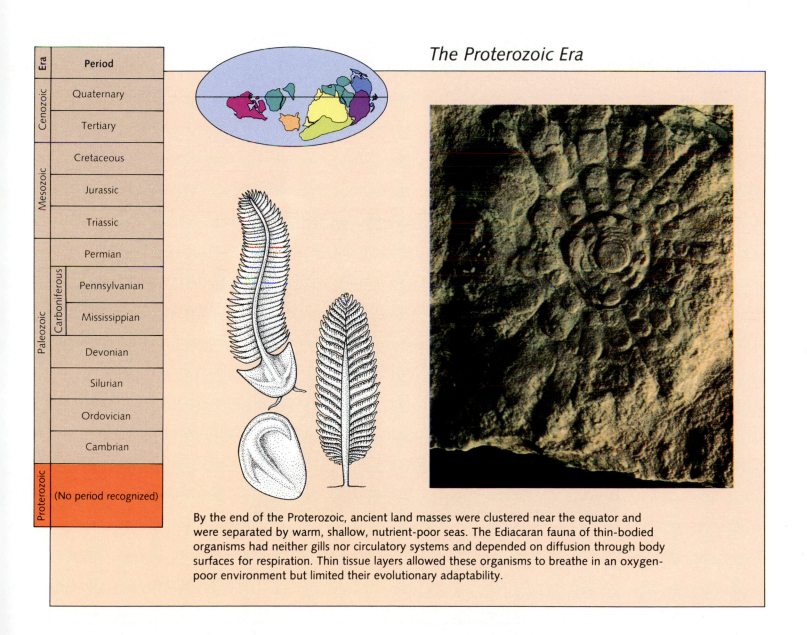

Era	Period	
Cenozoic	Quaternary	
	Tertiary	
Mesozoic	Cretaceous	
	Jurassic	
	Triassic	
Paleozoic	Permian	
	Carboniferous	Pennsylvanian
		Mississippian
	Devonian	
	Silurian	
	Ordovician	
	Cambrian	
Proterozoic	(No period recognized)	

By the end of the Proterozoic, ancient land masses were clustered near the equator and were separated by warm, shallow, nutrient-poor seas. The Ediacaran fauna of thin-bodied organisms had neither gills nor circulatory systems and depended on diffusion through body surfaces for respiration. Thin tissue layers allowed these organisms to breathe in an oxygen-poor environment but limited their evolutionary adaptability.

PALEOZOIC ERA: THE CAMBRIAN PERIOD

During the geologically active Cambrian period, the giant supercontinent began to fragment and pull apart, creating many shallow seas and thousands of miles of shoreline (see below). Atmospheric oxygen concentrations were still rising and soon reached 20 percent of current levels. That increased oxygen availability enabled multicellular organisms to grow thicker tissue layers with greatly increased evolutionary flexibility.

Under the pressures of natural selection, different cell groups within those thicker tissues began to specialize, adapting in structure and function to perform a variety of tasks impossible for generalized cells. This biological division of labor led to the evolution of complex organ systems. Some excitable cells evolved into nerve cells, which in turn combined into *nervous systems* that facilitated rapid communication between body parts. Other excitable cells evolved the ability to contract, forming *muscles* that made complicated movement possible. *Gills* served as specialized respiratory structures, and *circulatory systems* distributed oxygen and food throughout the body.

This increased structural complexity and diversity fueled the Cambrian explosion, life's second series of experiments in body design. Some Cambrian body plans, as fantastic as any imaginary space creatures, were failures that rapidly vanished. Other Cambrian period experiments, no less unusual, proved evolutionarily successful and adaptable. These organisms founded several major phyla, many representatives of which are alive and abundant today.

The Cambrian Period

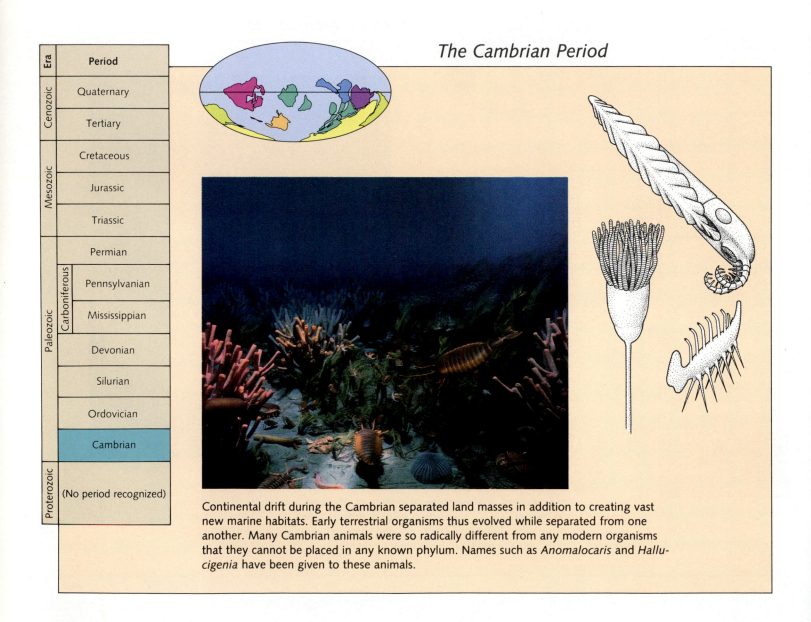

Era		Period
Cenozoic		Quaternary
Cenozoic		Tertiary
Mesozoic		Cretaceous
Mesozoic		Jurassic
Mesozoic		Triassic
Paleozoic		Permian
Paleozoic	Carboniferous	Pennsylvanian
Paleozoic	Carboniferous	Mississippian
Paleozoic		Devonian
Paleozoic		Silurian
Paleozoic		Ordovician
Paleozoic		Cambrian
Proterozoic		(No period recognized)

Continental drift during the Cambrian separated land masses in addition to creating vast new marine habitats. Early terrestrial organisms thus evolved while separated from one another. Many Cambrian animals were so radically different from any modern organisms that they cannot be placed in any known phylum. Names such as *Anomalocaris* and *Hallucigenia* have been given to these animals.

Ecological Diversity

Before the Cambrian, most animals had lived either by scavenging dead organic matter or by relying for energy on symbioses with photosynthetic organisms. The rise in oceanic nutrient levels during the Cambrian, however, supported abundant growth of both unicellular and multicellular algae, which encouraged the evolution of herbivores with new and varied ways of feeding. We know, for example, that there was a dramatic decline in the diversity of stromatolites during the Cambrian, probably because the first grazing herbivores had appeared.

As the earliest members of modern phyla emerged, they took their places in the world's first complex food webs. Many species assumed the roles of predators and began feeding on other animals. The simultaneous emergence of anatomical complexity and ecological interaction snowballed, as organisms evolved and occupied new niches and adjusted to the emerging pressures of competition and predation. Freed by gills and circulatory systems from the need to breathe through body surfaces, for example, many animal groups evolved armor for protection against the first predators.

Modern Phyla First Known From the Cambrian

Many of the major living phyla probably trace their roots back to yet-unknown, presumably soft-bodied ancestors that first evolved during the Proterozoic era. The dearth of fossils from that time, however, still shrouds these origins in mystery. By the beginning of the Cambrian period, several of these major lineages began leaving recognizable fossils, although modern members of these groups did not evolve for millions of years.

In some ways, the basic body plan of each phylum represented a sort of evolutionary blueprint or archetype for the organization of body tissues and appendages. Within each phylum, variations on those blueprints could form new classes or orders, each of which could diverge in adapting to new ways of life. The first of those variations were tried and tested in the Cambrian and Ordovician periods. Whether they subsequently evolved was determined by their success or failure then.

THE INVERTEBRATES

Many phyla of **invertebrates,** animals without backbones, developed during the Cambrian (*in* means "without"; *vertebrae* means "skeletal elements of the backbone"; hence *invertebrate* means "without a backbone"). Ecologically and evolutionarily, their body plans have been enormously successful; they outclass members of our own phylum (Chordata) in length of presence on the earth, number of living species, and number of living individuals.

The Sponges: Phylum Porifera

Like their ancestors, modern-day sponges have very simple body plans. Sponges, although multicellular, are just barely so; they have neither organs nor well-differentiated tissues. They do have several different cell types, including wandering *amoebocytes* and flagellated *collar cells* that cooperate to assemble the sponge structure (see p. 290). Sponges are so different from other metazoans in their organization that they probably evolved multicellularity independently. (Some taxonomists place the sponges in a separate subkingdom—the Parazoa—to emphasize their uniqueness.) Like many other primitive animals, sponges can reproduce either sexually, using eggs and sperm, or asexually, by budding.

Jellyfish and Sea Anemones: Phylum Cnidaria

The phylum Cnidaria was the first to contain specialized tissues, including a simple nervous system called a *nerve net.* Cnidarians also have specialized sensory cells, muscles, and cells for catching and digesting food (Figure 15.1). Their body plans exhibit **radial symmetry:** The body parts are arranged like the spokes of a bicycle wheel around a central mouth. Cnidarians don't have a diges-

Figure 15.1 Jellyfish spend most of their lives as medusae; their polyp stage is short-lived and serves only to produce new medusae asexually.

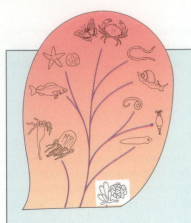

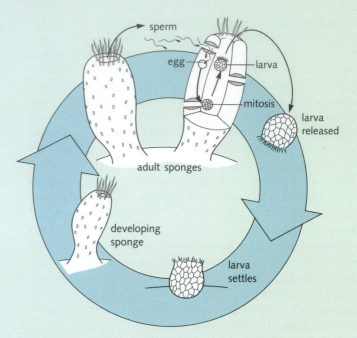

sperm

egg

larva

mitosis

larva released

adult sponges

developing sponge

larva settles

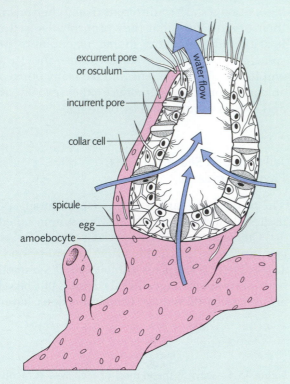

excurrent pore or osculum

water flow

incurrent pore

collar cell

spicule

egg

amoebocyte

Life cycle: Sponges reproduce either asexually (by budding) or sexually. In sexual reproduction, sperm released by other sponges are drawn through the incurrent pores, and fertilization occurs internally. Zygotes grow into multicellular larvae that swim to a point of attachment and metamorphose into miniature adults.

Anatomy: Sponges propel water currents through their bodies using flagellated *collar cells.* Water enters through incurrent pores and exits through the *excurrent pore,* or *osculum.* This current allows sponges to breathe and to filter feed. Needle-like *spicules* (''little spikes'') of calcium carbonate offer skeletal support to many sponges. Tough elastic fibers of *spongin* serve that purpose in other species.

A vase sponge.

Phylum Cnidaria: Jellyfish, Corals, and Sea Anemones

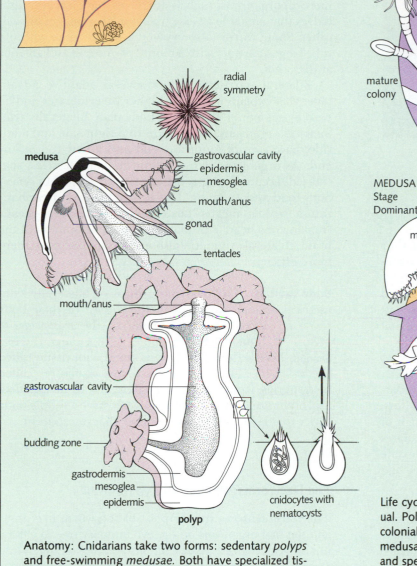

radial symmetry

medusa
- gastrovascular cavity
- epidermis
- mesoglea
- mouth/anus
- gonad
- tentacles

mouth/anus

gastrovascular cavity

budding zone

gastrodermis
mesoglea
epidermis

polyp

cnidocytes with nematocysts

POLYP Stage Dominant

female
male
medusae
feeding polyp
egg   sperm   **Fertilization**
reproductive polyp
zygote
mature colony
branching
planula
polyp forming

MEDUSA Stage Dominant

medusa
zygote
planula
Fertilization
polyp

sexual

asexual

Anatomy: Cnidarians take two forms: sedentary *polyps* and free-swimming *medusae.* Both have specialized tissues, including an *epidermis* for protection, a *nerve net* for motor coordination, and a *gastrodermis* for digestion. Both also exhibit radial symmetry. The jelly-like mesoglea separates the epidermis and the gastrodermis. The *gastrovascular cavity* has a single opening that serves as both mouth and anus. Tentacles of both body forms carry stinging *nematocysts.*

Life cycle: Reproduction can be both sexual *and* asexual. Polyps reproduce asexually by budding. Some colonial forms have *reproductive polyps* that bud off medusae. Medusae reproduce sexually, producing *eggs* and sperm that unite to form a *zygote.* The zygote grows into a *planula* larva, which metamorphoses into a polyp. Corals grow as either colonial or solitary polyps; some have small, short-lived medusae, while others produce eggs and sperm that form new polyps.

Figure 15.2 A colonial coral.

tive tract with separate mouth and anus. Instead they have a *gastrovascular cavity* with a single opening that serves for both taking in food and expelling wastes.

Cnidarians may be either solitary, like many sea anemones and jellyfish, or colonial, like hard and soft corals (Figure 15.2). Many cnidarians are predators that use stinging structures called **nematocysts** to paralyze or ensnare prey ranging from small shrimp to sizable fishes (*nema* means "thread"; *kystis* means "bag"; hence *nematocyst* means "thread in a bag"; see p. 291). Other cnidarians fall somewhere between predators and filter feeders, spreading web-like arrays of individual polyps, each of which contributes its tentacles to a network that snares passing plankton. Many cnidarians exhibit a complex life cycle that alternates between sexual reproduction and asexual budding.

The cnidarian body plan is simple yet successful. Though most numerous and diverse in the tropics, cnidarians are common in temperate seas and are represented in freshwater habitats by *Hydra*. Hard corals, which secrete calcium carbonate skeletons, build coral reefs, the largest biologically formed structures in the world and one of the most important marine ecosystems.

Worm-Like Phyla

Modern worm-like animals actually belong to several very different phyla, including two simple ones that probably evolved by the middle of the Cambrian: the platyhelminthes, or flatworms, and the nematodes, or round-worms. The earliest members of these groups were free-living; they lived independently as herbivores and carnivores. Modern representatives range from such free-living forms to highly specialized parasites whose lives are irrevocably linked to those of other organisms.

Flatworms and roundworms exhibit **bilateral symmetry,** a body plan in which similar body parts occur on either side of a central dividing line. All bilaterally symmetrical organisms, such as ourselves, have left and right halves that are (at least externally) mirror images of each other. Some worms also evolved a linear "tube within a tube" digestive tract with a mouth at one end and an anus at the other.

Bilateral symmetry offers several adaptive advantages that explain why most higher invertebrates and all vertebrates exhibit this sort of body plan. Bilaterally symmetrical organisms, in addition to a right side and a left side, have a front (anterior) end and a rear (posterior) end. As organisms evolved more mobile lifestyles, specialized, accumulating sense organs developed to detect both food and enemies. The steady process of **cephalization** (*cephalos* means "head"; hence *cephalization* means "making of a head") ultimately led to development of the brain. Bilateral symmetry also makes directed movement more efficient.

Body cavities The evolving groups of bilaterally symmetrical animals diverged in the design of their body cavity or *coelom*. The most primitive, the *acoelomates* (*a* means "without"), have no body cavity. Others, the *coelomates*, have a fluid-filled cavity lined with tissue called the *mesoderm*. This mesoderm forms structures called mesenteries, from which internal organs are suspended. Still other organisms, *pseudocoelomates,* have a simple body cavity that is lined with tissue other than mesoderm.

The Flatworms: Phylum Platyhelminthes

The flatworms are the simplest of the worm-like phyla and the first organisms on the evolutionary tree to exhibit not only specialized tissues, but also tissues arranged into *organs*. Most have neither respiratory nor circulatory systems, but flatworms do have primitive nervous systems, excretory systems for getting rid of wastes, and reproductive systems with separate male and female organs (see opposite page). Today's free-living flatworms may be either scavengers or active predators in fresh water, salt water, or very wet terrestrial habitats. Two classes within this phylum contain the **flukes** and the **tapeworms,** exclusively parasitic forms with complex life cycles. Infestations of parasitic flatworms can cause devastating damage in humans.

Phylum Platyhelminthes: Flatworms

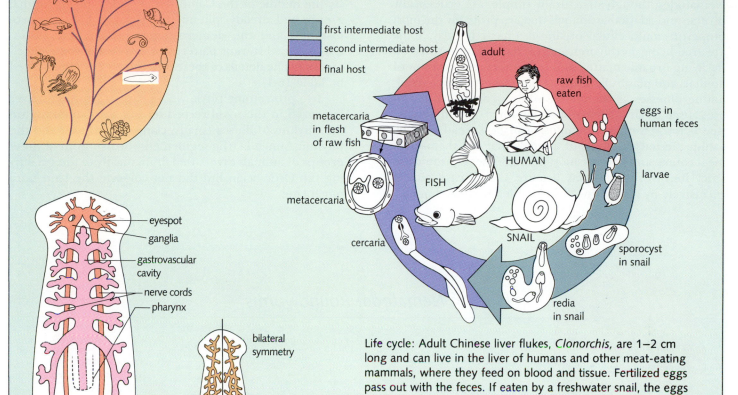

first intermediate host
second intermediate host
final host

adult

raw fish eaten

HUMAN

eggs in human feces

FISH

larvae

metacercaria in flesh of raw fish

metacercaria

SNAIL

sporocyst in snail

cercaria

redia in snail

eyespot
ganglia
gastrovascular cavity
nerve cords
pharynx

bilateral symmetry

protonephridia

Life cycle: Adult Chinese liver flukes, *Clonorchis*, are 1–2 cm long and can live in the liver of humans and other meat-eating mammals, where they feed on blood and tissue. Fertilized eggs pass out with the feces. If eaten by a freshwater snail, the eggs hatch within the snail. They reproduce asexually, forming thousands of free-swimming *cercarias*. These cercarias break out of the snail and burrow into the flesh of the next host, usually a freshwater fish. There they transform into *encysted metacercaria*, which can remain alive for years. If a human eats infected fish raw, the cyst is digested. The young worms are released into the intestine and migrate to the liver, where they can live for up to 30 years.

Anatomy: *Dugesia,* a free-living flatworm, has a *gastrovascular cavity* with a single opening, and several branches with dead-end sacs. The mouth is at the end of the *pharynx* that emerges from the animal's underside. The nervous system is organized into two *ventral nerve cords* that meet at a pair of simple "brains" called *ganglia.* Rudimentary *eyespots* can detect only light and shadow. Primitive excretory organs called *protonephridia* control water and salt content. Flatworms, like many higher organisms, exhibit bilateral symmetry.

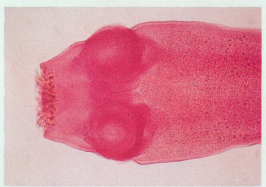

The head of a tapeworm (above), a parasite of mammals, shows suckers and hooks that cling to the host's intestinal lining. (left) A colorful, free-living flatworm.

The Roundworms: Phylum Nematoda

Nematodes, though structurally simple, are ecologically diverse and adaptable animals that are probably the most numerous multicellular organisms on earth today. Some are free-living herbivores and predators, others live on decaying organic matter, and still others are parasites of animals and plants. Parasitic nematodes cause such debilitating illnesses as trichinosis, whose painful symptoms are caused by larval nematodes that form cysts in muscles (see below). Because trichinosis can be spread only by eating the muscle tissue of animals that are themselves meat eaters, and because most mammals we consume are strict herbivores, humans acquire the disease almost exclusively by eating raw or incompletely cooked pork.

THE HIGHER INVERTEBRATES

By the middle of the Cambrian period, evolving animal groups had already split into two separate lines, the **protostomes** and the **deuterostomes,** which were distinguished by different patterns of early embryonic development. The details of these differences are summarized in Table 15.1.

Two protostome phyla whose earliest representatives appeared in the fossil record during the Cambrian are the phylum Mollusca (snails, clams, and squid) and the phylum Annelida (the segmented worms). Both of these groups of animals diversified dramatically in the sea during the Ordovician but did not colonize the land until much later.

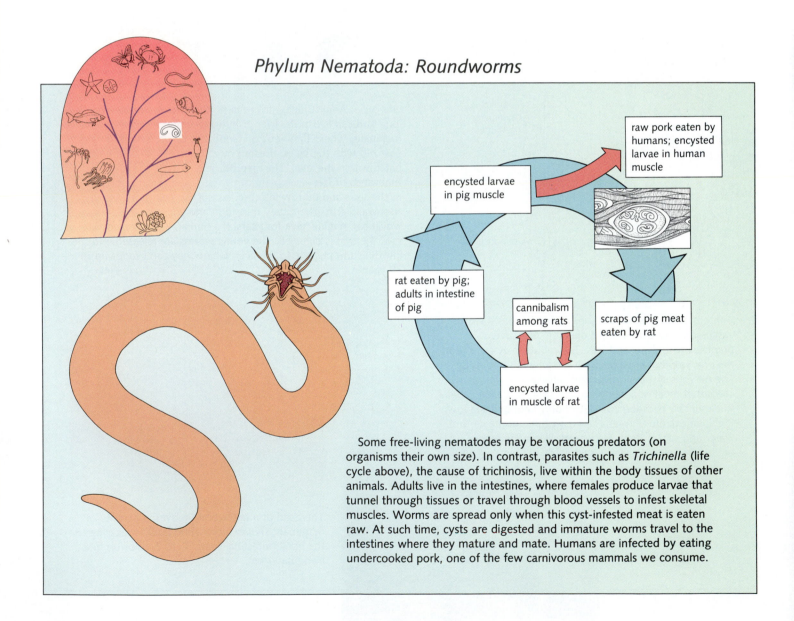

Phylum Nematoda: Roundworms

raw pork eaten by humans; encysted larvae in human muscle

encysted larvae in pig muscle

rat eaten by pig; adults in intestine of pig

cannibalism among rats

scraps of pig meat eaten by rat

encysted larvae in muscle of rat

Some free-living nematodes may be voracious predators (on organisms their own size). In contrast, parasites such as *Trichinella* (life cycle above), the cause of trichinosis, live within the body tissues of other animals. Adults live in the intestines, where females produce larvae that tunnel through tissues or travel through blood vessels to infest skeletal muscles. Worms are spread only when this cyst-infested meat is eaten raw. At such time, cysts are digested and immature worms travel to the intestines where they mature and mate. Humans are infected by eating undercooked pork, one of the few carnivorous mammals we consume.

Table 15.1 *Protostomes and Deuterostomes**

Protostomes	Deuterostomes

Cleavage: Division of Early Cells

In protostomes, cells divide at an angle to one another, producing a *spiral* arrangement of cells. The ultimate fate of each cell is determined early in development.

In deuterostomes, the early cells divide vertically, producing a *radial* arrangement of cells. The fate of embryonic cells is not determined until much later in development. Each of these early cells can become part of almost any tissue.

Formation of the Coelom

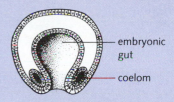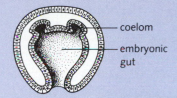

Protostomes are said to be *schizocoelous* (*schizo* means "split"). Here, solid masses of mesoderm tissue *split* to form the coelom.

Deuterostomes are said to be *enterocoelous* (*enteric* means "intestine"; hence *enterocoelous* means "intestine coelom"). The coelom is formed by buds that branch out and separate from the embryonic gut.

Development of the Gut

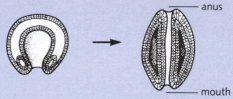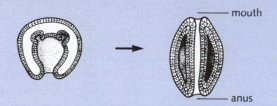

Protostome means "mouth first" (*proto* means "first"; *stoma* means "mouth"). The embryonic gut opening becomes the animal's *mouth*, and a new opening forms the anus.

Deuterostome means "mouth second" (*deutero* means "second"). The embryonic gut opening becomes the *anus*, and a new opening forms the mouth.

*Protostomes and deuterostomes are separated by differences in embryonic development. Because these events happen early in development and are constant across all members of a phylum, they are considered fundamental characteristics dividing groups of phyla.

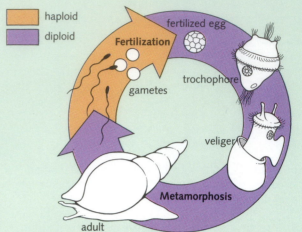

Life cycle: Molluscs reproduce sexually. In most forms, gametes are shed into the water. Fertilized eggs develop into larvae that undergo one or more metamorphoses.

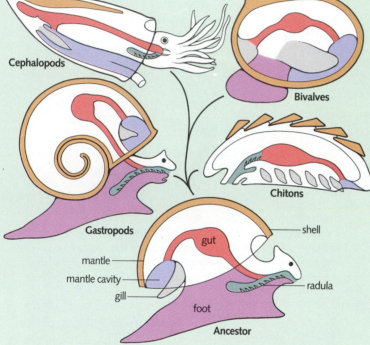

Chambered nautilus

Anatomy: The body plan of ancestral molluscs has been modified in several ways to produce modern forms. The digestive tract is a tube with both mouth and anus. The heart forces blood through open spaces to bathe internal organs. A soft *mantle* surrounds part of the body and often lines protective shells. Most aquatic forms have gills, whereas terrestrial snails have mantles modified to extract oxygen from moist air. The radula contains rows of teeth, and the foot serves in both locomotion and attachment.

The Snails, Clams, and Squid: Phylum Mollusca

The **molluscs** were the first animals to evolve hard external coverings, an invaluable evolutionary advantage that helped early molluscs discourage predators and enabled recent forms to colonize the land. Members of this phylum ultimately diversified into several classes, each utilizing a variation on the phylum's basic body plan (see previous page). Because their shells are easily preserved, molluscan fossils are abundant.

Chitons: Class Polyplacophora

The **chitons** are ancient molluscs whose living members resemble their Cambrian ancestors (Figure 15.3). They are slow-moving herbivores that spend most of their time clamped firmly to rocks, where they scrape off encrusting algae with a sandpapery tongue, or *radula*. Organisms with feeding mechanisms such as this undoubtedly led to the downfall of stromatolites, which are now found only where herbivorous animals cannot reach them.

Snails and slugs: Class Gastropoda

The **gastropods** include both forms with shells (snails) and forms lacking external shells (slugs). The class name means "feeding foot" and refers to the large, fleshy foot and the mouth and radula used in feeding. Many gastropods are herbivores that use their radulae to scrape food from the surfaces over which they crawl. A number of modern gastropods, however, are predators; some bore through the shells of their prey, and others paralyze victims with poison-tipped harpoons.

Although one would hardly expect to find great thinkers among the gastropods, even the lowliest sea snails are capable of mastering primitive learning tasks in laboratory situations.

Clams, oysters, and scallops: Class Bivalvia

The **bivalves,** as their name implies, have two shells or "valves" held together by powerful muscles. Most bivalves are filter feeders that sift plankton from water passed over comblike gills. Some bivalves, such as the tropical giant clams, can grow to more than a meter across (Figure 15.3). Many are attached in one place for life, although scallops can move about by rapidly clapping their shells together.

Octopi, squid, and nautilus: Class Cephalopoda

The **cephalopods** (or "head-footed" molluscs), in many ways the most advanced of the molluscs, are also among the oldest. Fossilized cephalopods bear a striking resemblance to the living chambered nautilus. Today's cephalopods are agile, fast-moving predators with remarkably advanced nervous systems (Figure 15.3). Octopi, among the most intelligent of all invertebrates, can learn a number of complex tasks once thought to be the exclusive province of the vertebrate brain. Cephalopod eyes are also among the most advanced in the world, operating in much the same way as ours to produce clear images of the environment.

◁41

Figure 15.3 Living molluscs: The most common early molluscs were ancestors of the modern cephalopods—octopi, squid, and the chambered nautilus shown on the facing page. Chitons (top left) live like snails, attached to submerged rocks from which they scrape algae. (above) *Tridacna*, the giant clam, filter feeds and houses symbiotic algae in its mantle. (bottom left) Octopi are the most intelligent predators among the cephalopods.

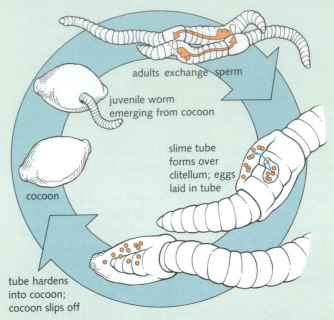

adults exchange sperm

juvenile worm
emerging from cocoon

slime tube
forms over
clitellum; eggs
laid in tube

cocoon

tube hardens
into cocoon;
cocoon slips off

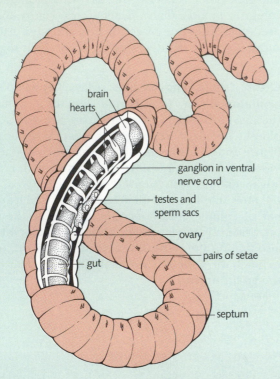

brain
hearts

ganglion in ventral
nerve cord

testes and
sperm sacs

ovary

pairs of setae

gut

septum

Life cycle: Annelids reproduce sexually. Terrestrial species, such as earthworms, function as males and females at the same time. During mating, they exchange sperm. The clitellum secretes mucus that protects the sperm during the exchange and later secretes a slime tube that becomes the cocoon. After mating, worms separate and lay eggs, releasing stored sperm at that time. These eggs develop directly into adults. In marine forms, gametes are shed into the water; free-swimming larvae live in plankton before metamorphosing into adult form.

Anatomy: Each body segment is separated from the next by an internal partition. Most body parts, including hearts, blood vessels, and nerves, are repeated in many segments.

Living annelids: Some leeches (left) are parasites that drink the blood of fishes and mammals. Many free-swimming marine annelids, such as this *Glycera* (above), are predators.

The Segmented Worms: Phylum Annelida

The **annelids,** which had probably been in existence for millions of years, diversified extensively in the sea during the Cambrian period, but did not invade freshwater habitats until millions of years later. Annelid bodies are composed of many identical segments, each of which bears bristles known as *setae.* In aquatic forms, many of these segments have appendages modified to serve as gills. Some annelids are active predators, others are harmless scavengers, and still others use delicate gills to filter food from the water as they breathe.

The coelom of each annelid body segment is sealed off from adjacent segments by divisions, or *septa* (see previous page). These sealed segments enable the worms to use a combination of muscle power and hydraulic power to create what is called a *hydrostatic skeleton.* By alternately contracting different sets of muscles, annelids use the fluid yet incompressible contents of their segments to generate power for crawling, swimming, or burrowing, even though they lack any sort of hard skeletal support.

The First Arthropods: Phylum Arthropoda

Perhaps the best-known Cambrian fossils are the **trilobites,** the first representatives of the phylum Arthropoda or "jointed leg" animals. **Arthropods** have enjoyed one of the longest-running and most spectacularly successful adaptive radiations in Earth's history. This phylum, which includes the crustaceans, spiders, insects, and a variety of other animals many people call "bugs," contains more species than any other group of organisms on Earth today. Between 800,000 and 900,000 arthropod species have already been described, and estimates of the number of species yet undiscovered range from 2 million to as high as 30 million. The arthropods' basic body plan contains several key features retained from their ancestors and several innovations responsible for the group's success.

Like annelids, arthropods have bodies divided into a number of segments. In some, such as centipedes, those segments are as readily visible externally as they are in annelids. In others, such as crabs and beetles, a number of embryonic segments fuse to form larger body regions (see p. 300). Each segment carries a pair of the jointed appendages that give the phylum its name, and whenever several body segments have combined, the appendages associated with those segments have become specialized, serving a variety of functions. Through the course of evolution, arthropods have developed mouth parts for chewing, flippers and legs for walking and swimming, and powerful claws for seizing and dismembering prey.

All arthropods have hardened external skeletons, or *exoskeletons,* that are jointed to enable legs and body segments to move. Exoskeletons are made from a mixture of proteins, a substance called *chitin,* and varying amounts of calcium salts. They bestowed on the early arthropods many advantages, including protection, support, and hard structures to which their muscles could attach. Millions of years later, when the phylum invaded land, the exoskeleton also served as a useful barrier to loss of water through evaporation.

Exoskeletons do have their limitations, however; arthropods can't grow inside such inflexible coverings. As a result, all arthropods must periodically shed their old skeletons and grow new ones, a traumatic process called *ecdysis,* or molting. During molting, the animal is completely immobilized, and for some time afterward the new shell is too soft either to support the animal or to protect it. The animals behave accordingly; as the time for ecdysis approaches, arthropods seek secluded hiding places in burrows, under stones, or in piles of debris.

Arthropods as a group have well-developed nervous systems and an elaborate set of sensory cells. The sophisticated compound eyes of many species enable them to see in color, and extraordinarily sensitive taste buds on the feet and antennae of others can detect vanishingly small concentrations of chemicals given off by food. Arthropods also have well-developed respiratory organs, circulatory systems, and excretory systems.

Zoologists believe that annelids, molluscs, and arthropods are closely related and evolved from a common protostome ancestor. Although modern members of these three great phyla differ widely from one another, they share enough larval and developmental characteristics to indicate close ancestral ties. Furthermore, chance and natural selection have kept alive several plausible descendants of a "missing link" between these phyla—animals called Onychophorans (see Theory in Action, *Peripatus:* The Worm That Time Forgot, p. 303).

The extinct trilobites: Subphylum Trilobita The trilobites, the first animals to display all the arthropod characteristics we have noted, rapidly took over the early Cambrian seas. At first their speed, agility, keen vision, and armor made them deadly and invincible predators (Figure 15.4). By the middle of the Cambrian, however, predators such as *Anomalocaris* and giant sea scorpions (see the description of chelicerate arthropods that follows) began to take their toll on the trilobite population.

Chelicerates: Subphylum Chelicerata Both living and extinct members of this subphylum are classified together because all possess mouthparts called *chelicerae,* often

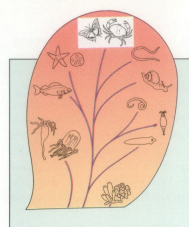

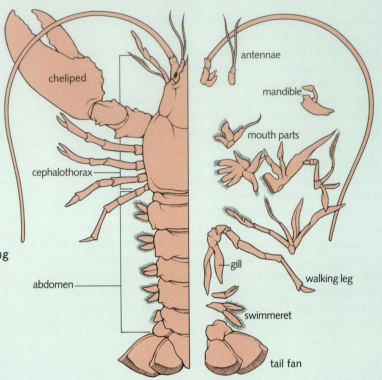

antennae

mandible

cheliped

mouth parts

cephalothorax

gill

walking leg

abdomen

swimmeret

tail fan

Anatomy: In crustaceans, such as lobsters, numerous body segments have fused, forming the cephalothorax (*cephalo* means "head"; *thorax* means "breastplate") and abdomen. The numerous appendages have specialized and perform various functions.

Arthropod body plans show the fusion of primitive segments during development. In centipedes (left), most visible body segments carry one pair of legs. The daddy-longlegs (center) and the scorpion (right) are examples of organisms whose embryonic segments have fused to form larger body regions.

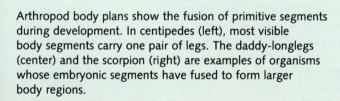

shaped into poisonous fangs or pincers, that evolved from the frontmost pair of appendages of the early arthropods. **Chelicerates** have another pair of modified legs called *pedipalps* that are used in some species for feeding and in others only for grasping the partner during copulation. In contrast to the three pairs of walking legs evolved by insects 200 million years later, chelicerates have four pairs.

The first chelicerates evolved soon after the trilobites. The earliest marine representatives were the *Eurypterids,* the giant sea scorpions. Another ancient chelicerate still alive today is the common horseshoe crab or king crab, *Limulus* (Figure 15.4), whose larvae look so much like trilobites that they are named after them.

Neither trilobites nor their predators survive today; many died out by the end of the Cambrian. Other annelid-like Cambrian animals that also developed exoskeletons gave rise millions of years later to other arthropods, including scorpions, spiders, crabs, shrimps, and insects.

Starfish, Sea Lilies, and Sea Cucumbers: Phylum Echinodermata

The **echinoderms,** whose name means "spiny-skinned," are deuterostome animals that evolved during the Cambrian (see p. 302). Most echinoderms are radially symmetrical, although they seem to have evolved from bilaterally symmetrical ancestors. Echinoderms have never spread into freshwater environments but have always been abundant in the sea, where they live as filter feeders, detritus feeders, or carnivores.

The most peculiar characteristic of this phylum is the unique *water-vascular system,* a network of tubes connected to muscular, extensible suction cups called *tube feet.* Echinoderms use muscles in parts of the tube network to move water around inside, effectively employing hydraulic power to extend or retract their tube feet. Like hydraulic lifts used in garages, the water-vascular system can develop tremendous force over long periods of time. This sustainable power enables starfish to prey on clams and mussels by attaching scores of tube feet to the bivalves and pulling relentlessly until the shells gape open. The starfish then turns its stomach inside out into the mussel shell, spilling out digestive enzymes that digest the prey.

Although modern echinoderms have been evolving since the Cambrian period, many (particularly the filter-feeding "sea lilies" or "feather stars") look remarkably similar to their fossilized forebears (Figure 15.5). Sea cucumbers are detritus feeders found in shallow marine habitats around the world and in "herds" of thousands on the ocean floor.

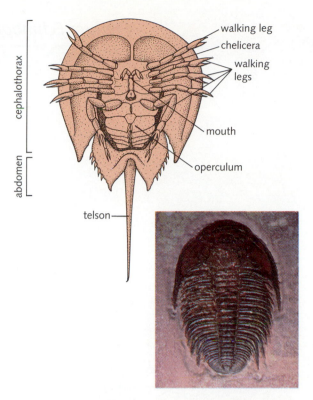

Figure 15.4 (top) *Limulus,* the horseshoe crab, a living chelicerate arthropod. (bottom) Trilobite fossils are the most common remains of primitive arthropods.

Figure 15.5 A feather star from the phylum Echinodermata.

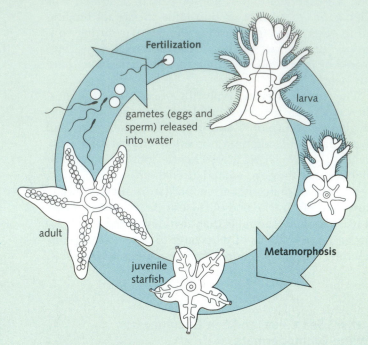

Fertilization

gametes (eggs and sperm) released into water

larva

adult

juvenile starfish

Metamorphosis

Life cycle: Echinoderms reproduce sexually, releasing eggs and sperm into the water. Free-swimming larvae go through complicated metamorphoses before settling down.

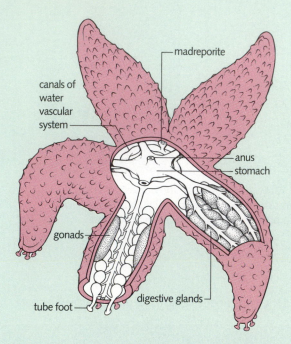

madreporite

canals of water vascular system

anus

stomach

gonads

tube foot

digestive glands

Anatomy: The water vascular system includes a *sieve plate* or *madreporite*, a series of canals, and tube feet. The arms contain digestive glands and reproductive organs.

Living echinoderms: A sea cucumber (left) is a detritus feeder while most starfish (right) are carnivores that prey on bivalve molluscs. Urchins (center) are grazers that scrape algae from submerged rocks.

Peripatus: *The Worm That Time Forgot*

In 1826 the Reverend Lansdown Guilding published a description of an animal he couldn't quite place in the established scheme of things. It looked rather like a slug, yet it had many body segments and several dozen pairs of tiny feet with claws at their tips. It fed by shooting a sticky, gluey substance that entangled prey. At the base of each antenna it had a simple eye that much more closely resembled an annelid eye than the compound eyes of arthropods. Most of its body was soft and flexible, but the tips of its feet were encased in the rudiments of an exoskeleton.

Guilding called this organism *Peripatus* and created a new class for it among the molluscs, where he thought it belonged. Other zoologists thought it was an annelid; still others were certain it was a primitive arthropod. Today most taxonomists place *Peripatus* and its relatives in their own phylum, Onychophora.

The onychophorans have survived since at least the Ordovician, and they were sufficiently widespread on the supercontinent to have left survivors in widely separate parts of the tropics. Their host of transitional characteristics bridge the gaps between annelids and arthropods. Because onychophorans have survived nearly unchanged for millions of years, they—like *Limulus*—are often called living fossils. Yet in their other, far more advanced features, the onychophorans remind us that evolutionary time never stands still. One pair of limbs near the head has evolved into jaws. And though some onychophorans lay eggs, others bear young alive after a gestation period of 12 to 15 months.

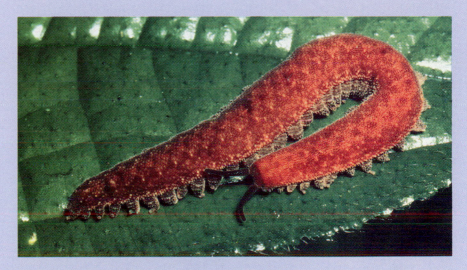

The onychophoran *Peripatus*, a possible "missing link" between molluscs, annelids, and arthropods.

PALEOZOIC ERA: THE ORDOVICIAN PERIOD

The closing days of the Cambrian ushered in one of the largest mass extinctions on record; nearly 50 percent of the existing animal families, including many trilobites, became extinct. The Ordovician period that followed (see p. 304) saw dramatic adaptive radiations among survivors of that extinction. The Ordovician adaptive radiation generated diversity within each of the major phyla. Molluscs, particularly cephalopods, diversified extensively, as did the echinoderms and aquatic arthropods. One group of early cnidarians (not the corals we know today) formed the first reefs.

THE ALGAE

The Ordovician saw substantial evolution among multicellular marine algae, whose major divisions were probably established by this time: the chlorophytes, or green algae; the rhodophytes, or red algae; and the phaeophytes, or brown algae (see p. 306). By the Ordovician, too, the ancestors of green algae and terrestrial plants had taken separate evolutionary paths, the latter entering shallow, freshwater environments. But because physiological characteristics (such as photosynthetic pigments) do not leave fossil traces, and because modern algae have changed extensively since the Paleozoic, the family tree of the plant kingdom is full of question marks.

There are so many evolutionary connections among the unicellular green algae, the multicellular green algae, and the higher plants that their taxonomy is problematic. Many biologists, for example, like to classify unicellular green algae in the kingdom Protista. But because unicellular and multicellular algae are similar in so many ways, such a strategy would force us to add multicellular algae to that unicellular kingdom. In this book, therefore, we place all of the green, red, and brown algae in the plant kingdom with higher plants.

Each division of modern algae contains a different set of chlorophylls and *accessory pigments* that help the algae capture light for photosynthesis. These chemical characteristics indicate that the three algal divisions arose independently from the symbiotic union of three different pairs of ancestral prokaryotes.

The Green Algae: Division Chlorophyta

The more than 7000 species of living green algae vary enormously in size and complexity, from single-celled *Chlamydomonas* and its relatives through colonial forms such as *Volvox* to truly multicellular species such as *Ulva*,

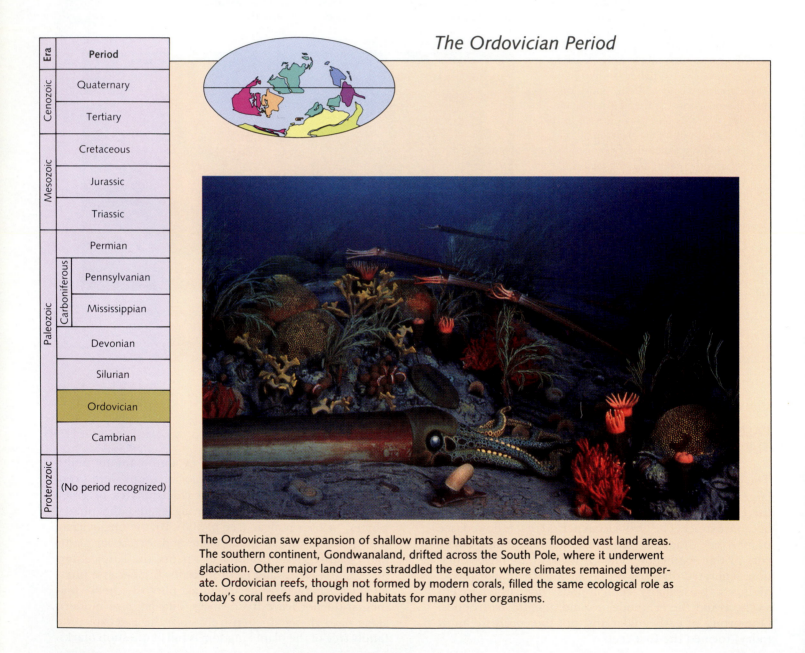

The Ordovician Period

Era	Period	
Cenozoic	Quaternary	
Cenozoic	Tertiary	
Mesozoic	Cretaceous	
Mesozoic	Jurassic	
Mesozoic	Triassic	
Paleozoic	Permian	
Paleozoic	Carboniferous	Pennsylvanian
Paleozoic	Carboniferous	Mississippian
Paleozoic	Devonian	
Paleozoic	Silurian	
Paleozoic	Ordovician	
Paleozoic	Cambrian	
Proterozoic	(No period recognized)	

The Ordovician saw expansion of shallow marine habitats as oceans flooded vast land areas. The southern continent, Gondwanaland, drifted across the South Pole, where it underwent glaciation. Other major land masses straddled the equator where climates remained temperate. Ordovician reefs, though not formed by modern corals, filled the same ecological role as today's coral reefs and provided habitats for many other organisms.

the common sea lettuce (see below). Although most chlorophytes are aquatic, a few can be called "semiterrestrial"; they can grow on land in very wet soil or on moist tree trunks. Like the higher plants, green algae contain two types of chlorophylls—chlorophyll *a* and chlorophyll *b*—and a variety of accessory pigments called *carotenoids*.

28

The ancestors of modern chlorophytes and higher plants were probably much like *Chlamydomonas*: single-celled, motile, and exhibiting a primitive reproductive pattern. *Chlamydomonas* provides the simplest example of **alternation of generations,** in which a sexually reproducing stage that produces gametes alternates with an asexual stage that produces spores. In *Chlamydomonas* the stage that produces spores, called the **sporophyte,** has two complete sets of chromosomes and is called **diploid.** The stage that produces gametes, called the **gametophyte,** has half that number of chromosomes and is called **haploid.** In Chlamydomonas both stages are single-celled, but in *Ulva* and many other plants, both stages are multicellular.

In algae and in many lower plants, male gametes must swim to join the female gametes. Because swimming requires water, plants employing this style of reproduction can grow only in places that are covered in at least

Division Chlorophyta: The Green Algae

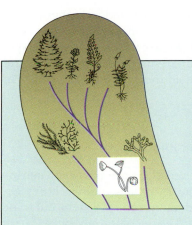

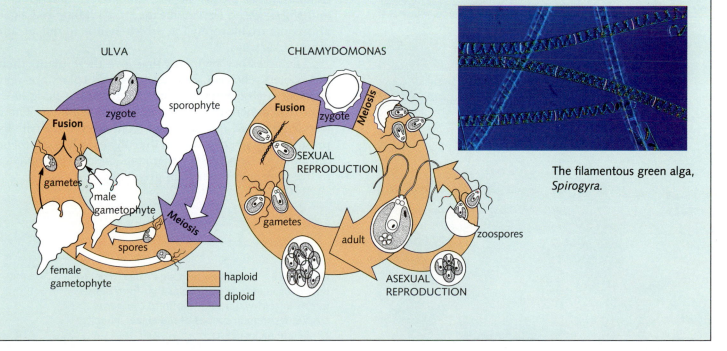

Ulva life cycle: Both haploid and diploid stages of the life cycle are multicellular. The diploid plant that produces haploid spores is called the *sporophyte.* These spores grow into multicellular, haploid plants called *gametophytes,* because they produce gametes. *Ulva*'s gametes are different from one another; larger ones are arbitrarily called "female," and smaller ones, "male." The condition of having different-looking gametes is called *heterogamy.* Fertilization forms a diploid zygote that grows into a multicellular sporophyte.

Chlamydomonas life cycle: This unicellular green alga is capable of sexual or asexual reproduction. Many individuals are haploid and reproduce asexually by cell division. Sometimes a mature haploid cell will produce identical haploid gametes, which fuse to form a diploid zygote. This zygote does not grow, but either develops a protective covering and rests, or divides to release haploid cells. In *Chlamydomonas,* the gametes look identical, a condition called *isogamy* (*iso* means "equal"; *gamy* refers to gametes; *isogamy* means "similar gametes").

The filamentous green alga, *Spirogyra.*

ULVA

CHLAMYDOMONAS

zygote
sporophyte
Fusion
gametes
male gametophyte
Meiosis
spores
female gametophyte

haploid
diploid

Fusion
zygote
Meiosis
SEXUAL REPRODUCTION
gametes
adult
zoospores
ASEXUAL REPRODUCTION

a thin film of water during the plants' reproductive periods. We will soon see that adaptive changes in these primitive plant life cycles were crucial in enabling plants to colonize terrestrial habitats.

The Brown Algae: Division Phaeophyta

The brown algae, which obtain their color from accessory pigments called *xanthophylls* and *fucoxanthins,* include most of the plants known as seaweeds: the common "rockweeds" of the intertidal zone, the giant kelps of deeper water, and the floating "sargassum weeds." Each of these major groups creates a complex, three-dimensional habitat for many other organisms.

Although clearly not closely related to higher plants, the phaeophytes are among the most structurally complex of all marine algae (Figure 15.6). Many have specialized *holdfasts* that secure them to a rocky bottom and bladders, or floats, that hold their long stems upright in the water.

The Red Algae: Division Rhodophyta

The 4000-odd species of red algae are primarily marine and very diverse. Many have extremely complex life cycles. Red algae are most abundant in tropical waters, but some species (such as *Chondrus crispus,* commonly called "Irish moss") are found below the low tide line along rocky temperate shores (Figure 15.6). All red algae possess chlorophyll *a* and varied accessory pigments called *phycocyanins* (*phyco* means "algae"; *cyan* means "blue-green") and *phycoerythrins* (*erythro* means "red"). Many red algae also contain chlorophyll *d,* which is not found in any other eukaryotes.

The accessory pigments that give these plants their red, violet, yellowish, or even brownish colors are very efficient at absorbing the blue and green wavelengths of light that penetrate to great depths in seawater. These wavelengths are not absorbed well by chlorophyll. By passing this energy on to the algae's photosynthetic machinery, the accessory pigments enable rhodophytes to live in deeper water than green algae.

Figure 15.6 (top) The green alga, *Ulva.* (center) Brown algae include giant kelps that form dense forests in cold-water seas. (bottom) *Chondrus crispus,* a common species of red algae, inhabits temperate rocky coasts.

alternation of generations *p. 305*
sporophyte *p. 305*
diploid *p. 305*
gametophyte *p. 305*
haploid *p. 305*

SUMMARY

The early history of life on earth, from the Precambrian through the end of the Silurian, set the stage for all changes to come. The great adaptive radiations in the Proterozoic, Cambrian, and Ordovician were evolutionary "experiments" during which many body plans and life cycles appeared. The Cambrian saw the appearance of the major invertebrate groups, some of which successfully diversified to found major animal phyla. Others became extinct due to competition and/or environmental change. Those phyla that remain, including the Porifera, Cnidaria, Nematoda, Platyhelminthes, Echinodermata, and Arthropoda, are defined by physical structure, developmental histories, and life cycles.

The Cambrian also saw the first abundant growth of algae. By the Ordovician, all major groups of modern algae, the Chlorophyta, Rhodophyta, and Phaeophyta, were established and separated from the ancestors of higher plants. Most of these algae have life cycles with alternation of generations between a diploid sporophyte and a haploid gametophyte. This type of life cycle is found (with modifications) throughout the plant kingdom.

The success of algae permitted the evolution of varied herbivores. Herbivores, in turn, became the victims of the first predators as the first complex food webs were established.

STUDY FOCUS

After studying this chapter, you should be able to:

- Using a geological time scale, trace the evolution of the first multicellular organisms.
- Describe the evolutionary events of the Cambrian and the Ordovician periods.
- Name and describe the phyla of the lower and higher invertebrates, and outline their life cycles.
- Name and describe the divisions of multicellular algae. Outline the life cycle of typical green algae.

SELECTED TERMS

invertebrates *p. 289*
radial symmetry *p. 289*
bilateral symmetry *p. 292*
cephalization *p. 292*
protostomes *p. 294*
deuterostomes *p. 294*
molluscs *p. 297*
chitons *p. 297*

gastropods *p. 297*
bivalves *p. 297*
cephalopods *p. 297*
annelids *p. 299*
trilobites *p. 299*
arthropods *p. 299*
chelicerates *p. 301*
echinoderms *p. 301*

REVIEW

Discussion Questions

1. Why was the evolution of multicellular life tied to the production of oxygen by photosynthesis?

2. What were the key innovations that made the arthropod body plan such an evolutionary success?

3. What makes the organization of sponges so different from that of other metazoans? Why is the sponge considered just barely multicellular?

4. Why are the nematodes probably the most numerous multicellular organisms on earth today?

5. What characteristics distinguish the three divisions of modern algae from each other?

Objective Questions   (Answers in Appendix)

6. Sponges differ from other multicellular animals because they
   (a) can only reproduce asexually.
   (b) have complex body plans.
   (c) do not reproduce sexually.
   (d) lack organs and tissues.

7. The nervous system of Cnidaria contains
   (a) a nerve net and sensory cells.
   (b) a nerve net only.
   (c) specialized gametes.
   (d) many-celled sense organs.

8. Bilateral symmetry is a characteristic of
   (a) flatworms.
   (b) cnidarians.
   (c) sponges.
   (d) Porifera.

9. Which of the features of embryonic development is characteristic of protostomes?
   (a) radial arrangement of cells in embryo
   (b) the fate of embryonic cells is not determined until late in development
   (c) coelom is formed by the splitting of the mesoderm
   (d) the anus forms first, then the mouth

10. In the species of *Chlamydomonas* that exhibits isogamy, the gametes
    (a) are produced by the sporophyte.
    (b) are diploid.
    (c) are the same size.
    (d) have the same number of chromosomes as the spores.

CHAPTER **16**

The Silurian Period:
The First Lower Plants
and Jawed Vertebrates

Feature boxes appear throughout the text serving as valuable study aids designed to enrich your understanding of each chapter. You may wish to refer to the following feature boxes as you read Chapter 16.

The Silurian Period	p. 309
Division Bryophyta	p. 311
Division Pterophyta	p. 313
Division Ascomycota & Division Basidiomycota	p. 317
Phylum Arthropoda	p. 319
Phylum Chordata: Tunicates	p. 321
Phylum Chordata: Lampreys	p. 322

PALEOZOIC ERA: THE SILURIAN PERIOD

Global climate and geological events during the Silurian period set the stage for two crucial evolutionary events: the colonization of the land by plants and the evolution of the first jawed aquatic vertebrates. A global cold spell that ended the Ordovician and caused the extinction of many marine organisms gave way to a warm, moist climate that lasted through the Silurian. Continental drift carried the continent of Gondwanaland northward, toward other land masses near the equator. Plentiful swamp-like, freshwater habitats served as cradles for the evolution of both land plants and freshwater fishes. During the Silurian, several important groups of marine animals that had appeared much earlier evolved hard body parts and left clear fossil records for the first time.

ORIGINS OF LIFE ON LAND

As life in the sea continued to evolve, competition and predation became fierce. Because terrestrial habitats were still devoid of competing life, any organism that could survive out of water would have enjoyed major selective advantages. But because life evolved under water, both plants and animals encountered similar obstacles in adapting to terrestrial life. Plants were the first to colonize the continents successfully, so we will examine the problems they faced and the evolutionary adaptations that made the transition to land possible.

Desiccation, the loss of body water to surrounding air, is the foremost obstacle to terrestrial life. Plants could not colonize the land, therefore, without evolving some way to retain water. Most land plants have evolved waxy,

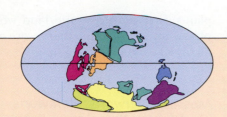

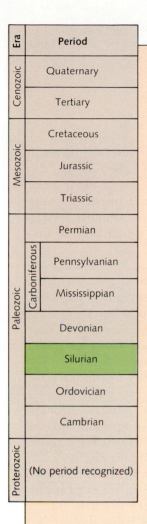

Era		Period
Cenozoic		Quaternary
		Tertiary
Mesozoic		Cretaceous
		Jurassic
		Triassic
Paleozoic		Permian
	Carboniferous	Pennsylvanian
		Mississippian
		Devonian
		Silurian
		Ordovician
		Cambrian
Proterozoic		(No period recognized)

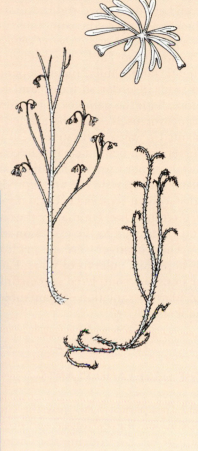

During the Silurian, many land areas were uplifted, pushing back shallow seas and creating moist, tropical, freshwater habitats. Predatory arthropods reached 2 meters in length, and numerous other strange animals inhabited the seas. During this period, the first multicellular plants emerged onto land.

waterproof **cuticles** that cover their leaf surfaces and retain water in tissues surrounded by dry air. This evaporation barrier is interrupted by openings called **stomata,** which open or close as needed to allow exchange of carbon dioxide and oxygen for respiration and photosynthesis.

Transport of body fluids is another problem for terrestrial plants. Because aquatic plants can absorb water and nutrients through the leaf-like tissues they use for photosynthesis, all their cells have ready access to the requirements for life. Terrestrial plants, on the other hand, stretch their leaves up to the sun and probe underground with their roots for water and nutrients. Because roots cannot carry on photosynthesis, and because most leaves cannot absorb water, land plants require sophisticated transport

systems to move these essential materials through their bodies.

This function is served in modern land plants by two types of plant **vascular tissues,** which are tissues specialized to conduct fluids. Water and inorganic nutrients are transported through the **xylem,** and the organic products of photosynthesis are carried through the **phloem.** These tissues evolved slowly as various groups of plants adapted to progressively drier environments. (We will have more to say about all these structures in Chapter 27.)

Physical support for body parts is another requirement that terrestrial organisms must meet. Because living tissues are generally either lighter than water or only slightly denser, they are buoyed up by the medium around them. Aquatic plants can thus simply *extend* photosynthetic surfaces without having to hold them up. Land plants, however, have developed rigid supporting structures to hold their leaves up to the sun (Table 16.1).

Reproductive and dispersal strategies must also adapt to conditions on land. The sexual reproduction of early aquatic multicellular plants depended on sperm that had to swim in order to locate and fertilize eggs. Furthermore, both the gametes of aquatic plants and the young plants themselves can be carried far from their parent plants by waves and currents. But on land, normal air currents can neither carry gametes effectively over great distances nor disperse mature plants. For these reasons, plants could not survive and reproduce far from the water's edge without reproductive and dispersal strategies adapted to terrestrial life. Ultimately, the combination of necessity and chance led to the evolution of flowering plants.

These changes were not made quickly; they occurred in steps over millions of years. Each time a step or series of steps was completed successfully, however, it initiated an adaptive radiation of plants able to colonize ever-drier habitats.

THE FIRST LAND PLANTS

The first plants to colonize the land undoubtedly resembled modern green algae and probably began their invasion of the land during the Ordovician. By the Silurian they had already given rise to the two main branches of the plant kingdom: the single surviving nonvascular division Bryophyta and the divisions of vascular plants. The

plants' first ventures onto land would have gone completely unopposed, for there were as yet no herbivorous terrestrial animals.

Mosses, Liverworts, Hornworts: The Division Bryophyta

Bryophytes are at once an ancient and successful division of more than 16,500 species and a collection of "also rans" that never achieved the diversity of other plant groups. Viewed by some botanists as a primitive group close to the ancestors of "higher plants," they are seen by others as highly specialized forms that have become simplified during their long history.

Bryophytes, like all other plants, have a life cycle involving the alternation of generations, as shown on the next page. Dust-like spores, easily spread by the wind, germinate on moist surfaces and grow into thread-like *protonemata* that look remarkably like green algae. These protonema ultimately give rise to the leafy green shoots we know as moss plants, the separate male and female gametophytes that reproduce sexually to complete the life cycle.

Though well suited for life in wet terrestrial habitats, bryophytes have no true roots to take up water, no vascular tissue to transport fluids, and no cuticles to retain water. For those reasons, mosses cannot grow very tall; most are no more than 2 or 3 centimeters in height. Although bryophytes require standing water to reproduce, many species can enter a dormant state during droughts or subfreezing temperatures. Thus, though they probably evolved in the wet tropics, mosses are today the most abundant plants both in the Arctic and in Antarctica.

Table 16.1 *Evolutionary Adaptions from Aquatic to Terrestrial Environment*

Multicellular Algae	Requirement	Terrestrial Plants
Medium supportive; no specialized support tissues	Support of plant tissue	Medium nonsupportive; specialized structural tissues
Occurs in most cells	Photosynthetic areas	Confined to portions above ground
Direct access to environmental water and minerals; no specialized transport system	Transport of water and dissolved nutrients	Elevated parts not in direct contact with water and minerals; transport of water and minerals required
By water	Transportation of gametes	By wind, water, animals
By water	Seed/spore dispersal	By wind, water, animals

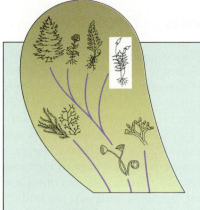

Mosses, Liverworts, and Hornworts: Division Bryophyta

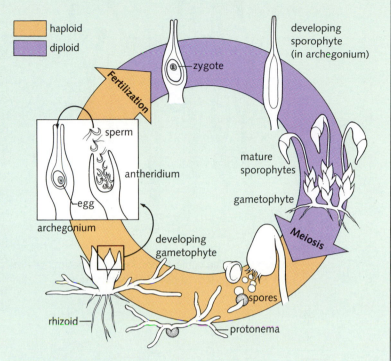

haploid
diploid

zygote

developing sporophyte (in archegonium)

Fertilization

sperm

antheridium

egg

archegonium

mature sporophytes

gametophyte

Meiosis

developing gametophyte

spores

rhizoid

protonema

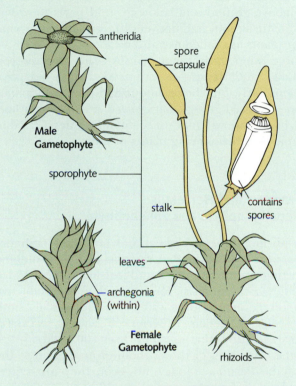

antheridia

Male Gametophyte

spore capsule

sporophyte

stalk

contains spores

leaves

archegonia (within)

Female Gametophyte

rhizoids

Life cycle: Sperm must swim to the archegonium to fertilize the egg. The diploid sporophyte grows from the archegonium, producing a spore capsule on a slender stalk. Within this capsule, dust-like, haploid spores are produced and released during dry, windy weather. Spores landing in favorable locations grow into algae-like *protonemata*, from which gametophytes sprout.

Anatomy: The haploid moss gametophyte has short stems, primitive vascular tissue, root-like rhizoids, and leaves one cell layer thick. Reproductive organs grow at the tips of these shoots. The male *antheridium* produces sperm, and each female *archegonium* produces a single egg.

(center) Many of the green moss gametophytes support the stalk and spore capsule of the sporophyte generation. Liverworts (top) and hornworts (left) are other bryophytes common in moist environments.

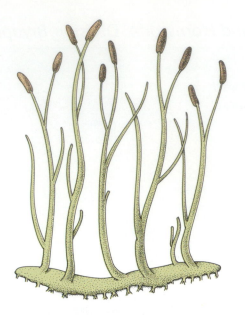

Figure 16.1 The earliest vascular plant species, such as *Rhynia,* lacked both true leaves and true roots.

Figure 16.2 Living sphenophytes, though neither as large nor as widespread as their ancestors, retain their division's characteristic shape.

EARLY VASCULAR PLANTS

The earliest vascular plants appeared around the beginning of the Silurian, about 430 million years ago. The stepwise development of vascular tissue, true roots, and water-retaining leaves rapidly led to the emergence of several early plant divisions, each with a distinct body plan. Like the Ediacaran animal fauna (Chapter 15), some of these early plant groups were evolutionary "experiments" that failed to survive environmental changes and became extinct. The division Rhyniophyta was one such group; it contains the first vascular plants that are known well from the fragmentary fossil record (Figure 16.1). Their above-ground and below-ground parts were virtually identical; they had not yet differentiated into what in later plants are called *root* and *shoot.*

Three other early plant groups—the club mosses, horsetails, and ferns—experienced major adaptive radiations early, but they relinquished their dominance to more advanced groups that evolved later. Another line, whose history we will pick up again in Chapter 17, led to the conifers and flowering plants of today.

Club Mosses and Horsetails: Divisions Lycophyta and Sphenophyta

The club mosses (**lycophytes**) and horsetails (**sphenophytes**) were among the first vascular plants to combine conducting tissue, true roots, and leaf-like structures for efficient photosynthesis. However, their reproductive cycles still require standing water. After a slow start, both mosses and horsetails experienced wildly successful adaptive radiations that lasted for over 100 million years, from the beginning of the Carboniferous period to the end of the Permian. Today, however, they are represented by only 5 genera and about 1000 species worldwide (Figure 16.2).

Ferns: Division Pterophyta

Like bryophytes, ferns (**pterophytes**) can reproduce only with the aid of standing water (see next page). But fern sporophytes, with their better-developed vascular tissue, thicker and more resilient leaves, and true roots, can survive under somewhat drier conditions and may live for many years. Like club mosses and horsetails, ferns were enormously successful in the Carboniferous and Permian when their tree-sized representatives formed huge forests. Because they were better adapted to survive changing conditions, however, ferns are still both diverse (some 12,000 species) and widespread today.

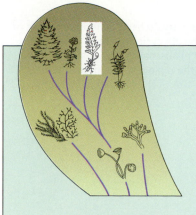

Ferns: Division Pterophyta

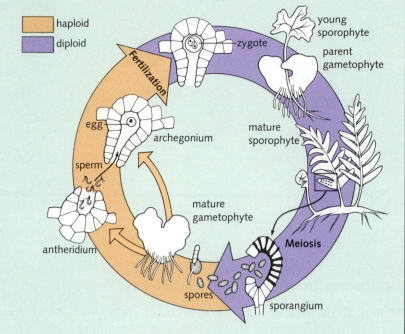

haploid
diploid

Fertilization

zygote

young
sporophyte

parent
gametophyte

egg

archegonium

sperm

mature
sporophyte

antheridium

mature
gametophyte

Meiosis

spores

sporangium

Life cycle: Haploid spores are released from sporangia to be carried on the wind. The spores germinate into small, haploid gameto-phytes. Sperm produced in antheridia swim to eggs in archegonia; the zygote produces a diploid sporophyte that grows out of and soon dwarfs its parent.

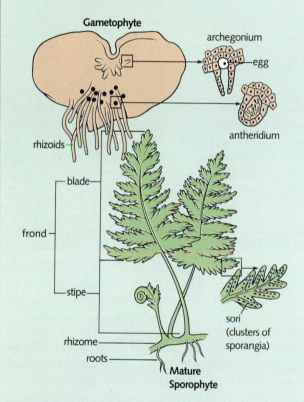

Gametophyte

archegonium

egg

antheridium

rhizoids

blade

frond

stipe

rhizome

roots

sori
(clusters of
sporangia)

**Mature
Sporophyte**

Anatomy: Leafy fronds, composed of *blade* and *stipe*, belong to diploid fern sporophytes, which are larger and more independent than those of mosses. Fronds unfurl from creeping *rhizomes* that grow underground or along the surface. Clusters of spor-angia called *sori* form on the undersides of mature fronds. Well-developed roots extend into the soil.

Living ferns range from small species common in moist-temperate environments to giant tropical tree ferns.

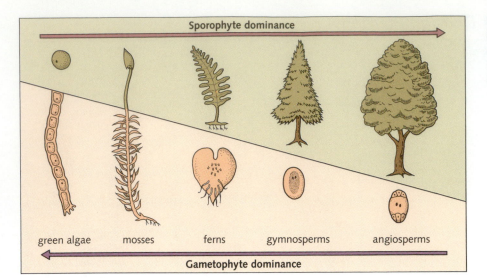

Figure 16.3 This schematic representation of trends in the life cycles of algae, lower plants, and higher plants shows the steadily increasing size of the diploid sporophyte and the decreasing size of the haploid gametophyte.

Evolutionary Trends in Plant Reproduction

Though all modern higher plants and many algae share reproductive cycles based on alternation of generations, a trend emerges as we survey plant evolution. Among single-celled forms, haploid gametophyte and diploid sporophyte are of roughly equal size and importance. As plants become increasingly adapted to terrestrial life, however, the sporophyte becomes dominant and the gametophyte becomes progressively smaller (Figure 16.3). This early trend came to fruition with the evolution of seed plants.

KINGDOM FUNGI

With the plants (or perhaps long before them), ancient members of another kingdom, the **fungi,** made their way onto land. Fungi are among the most important, diverse, and curious organisms on earth. Ecologically, some fungi are indispensable decomposers (or saprophytes) and some are vital symbiotic partners of green plants. Other fungi are predators that capture nematodes, and still others are parasites that infest both animals and plants, causing serious diseases and millions of dollars worth of crop damage annually.

Some fungi, such as yeasts, are single-celled, but most are called "multicellular." Most fungi are not, however, multicellular in the same way as plants and animals, because, as a rule, individual fungal "cells" are not separated by complete cell walls, and many of the cell-like compartments have more than one nucleus. The pre-

dominant fungal body form is the **mycelium,** a tangled, cottony mass composed of branched filaments called **hyphae** (from the Greek *hyphe,* meaning "web"). The familiar structures we recognize as "mushrooms" are actually the fruiting bodies produced by much larger concealed masses of mycelium.

Fungi as Decomposers

Saprophytic fungi, which live on dead organic matter, are the most important decomposers of plant tissue in terrestrial environments, where they break down such complex organic molecules as cellulose and allow their constituent elements to be recycled. The hyphae of saprophytic fungi penetrate dead wood, leaves, and animal tissues and secrete digestive enzymes into the substrate around them. The hyphae then absorb the products of this digestion, which takes place outside their bodies.

Some saprophytic fungi live by digesting organic matter that is important to humans: Many species spoil bread and fruit or attack and decompose paper, wooden structures, and natural fabrics. Humans use numerous fungi to produce beer and wine, make bread rise, flavor many cheeses, and produce penicillin and other antibiotics. A number of fungi are also edible and tasty.

Fungi as Symbionts

Symbiotic fungi enter into relationships with primary producers. Those associations enable the primary producers either to live in habitats they could never colonize alone or to function more efficiently in hospitable locations.

Lichens The 25,000 species of organisms we commonly call **lichens** are actually symbiotic associations in which cells of either photosynthetic cyanobacteria or green algae are woven into a dense mat of fungal hyphae. Those hyphae are occasionally vividly colored, a trait that probably evolved because colored hyphae shield the photosynthetic partner from too much sun (Figure 16.4). Often, specialized hyphae penetrate the algal or bacterial cell to pick up nutrients more efficiently. The photosynthetic cells respond to their fungal hosts by producing certain substances they do not manufacture when growing alone.

This intimate association equips lichens to live in habitats where virtually nothing else can grow, such as on bare rock, on mountaintops above the timberline, and on the tundra. In such places, lichens are often the first colonists during succession, breaking down and enriching sterile rock and sand, thus paving the way for other, less hardy plants. When the photosynthetic partner is a cyanobacterium, the association can fix nitrogen, some of which is released to the immediate environment in organic form. For this reason, some researchers propose that lichens may have been among the very first forms of multicellular terrestrial life. Because terrestrial life did not exist before the Silurian, early soils would have been totally devoid of the organic, nitrogen-containing compounds found in today's rich soils. In such inorganic habitats, nitrogen-fixing lichens would have been ideal pioneers.

Figure 16.4 (left) Vividly colored lichens form a patchwork over the rock surface. (right) Lichens with fruiting bodies.

Mycorrhizae Numerous fungi called **mycorrhizae** (*myco* means "fungus"; *rhizae* means "roots") enter into symbiotic associations with the roots of as many as 80 percent of all higher plants (well over 200,000 species). In most mycorrhizal associations, fungal hyphae penetrate the outer cells of delicate plant roots. Part of the fungal mycelium then forms within the root cells, and the rest spreads out into the surrounding soil. Other mycorrhizae encase roots with mycelia but do not actually penetrate root tissue (Figure 16.5).

Although these associations might seem parasitic, they are highly beneficial for both partners. Mycorrhizae help host plants take up copper, phosphorus, zinc, water, and other nutrients from the soil. In return, the plants provide the fungi with organic carbon. Mycorrhizal symbioses are widespread today; according to some estimates, these symbiotic fungi account for up to 15 percent of the weight of plant roots round the world. Plants with mycorrhizae seem to be more resistant than others to drought, nutrient shortages, cold, and possibly even the effects of acid rain. Mycorrhizae have been found on the earliest fossil plant roots, leading to speculation that they may have helped vascular plants to colonize inorganic Silurian soils.

Figure 16.5 Many mycorrhizal fungi form a thick mantle of hyphae around plant root hairs.

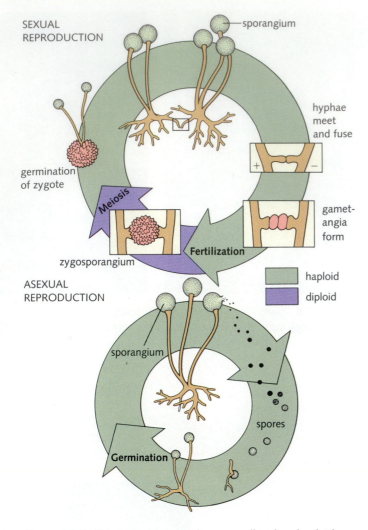

SEXUAL
REPRODUCTION

sporangium

hyphae
meet
and fuse

germination
of zygote

Meiosis

gamet-
angia
form

Fertilization

zygosporangium

ASEXUAL
REPRODUCTION

haploid

diploid

sporangium

spores

Germination

Figure 16.6 Zygomycetes reproduce sexually when haploid hyphae of different mating types (called simply "+" and "−") meet. One cell from each hypha fuses to form the diploid zygospore. Upon germination, the zygote immediately produces a sporangium that releases haploid spores that grow into the next generation of hyphae. Zygomycetes can also reproduce asexually, producing several generations of haploid organisms.

MAIN GROUPS OF FUNGI

Common Molds: Division Zygomycota

The **zygomycetes** include bread molds, saprophytes that live in soil and water, and parasites of animals, plants, and even other fungi. The common bread mold, *Rhizopus*, is a zygomycete that can reproduce either sexually or asexually (Figure 16.6). During sexual reproduction, zygomycetes produce a diploid zygote inside a thick-walled *zygosporangium* that can survive inhospitable conditions by remaining dormant for months.

Sac Fungi and Yeasts: Division Ascomycota

The **ascomycetes,** also called the sac fungi because their spores are carried inside a sac, or *ascus* (see next page), include edible saprophytes such as truffles and pathogens such as those that cause Dutch Elm disease. The single-celled yeasts, although dramatically different in appearance from other ascomycetes, are placed in this group because under certain circumstances they can form spores internally. The fungal components of most lichens are ascomycete fungi. In the course of competitive evolutionary interactions with other decomposers, many ascomycetes have evolved chemicals that inhibit the growth of bacteria around them. Cultured varieties of these fungi are widely used in the production of antibiotics.

Mushrooms: Division Basidiomycota

The **basidiomycetes** are what most of us think of as mushrooms. The familiar shape, however, is only the organism's fruiting body; the bulk of the mushroom is the mass of hyphae slowly digesting its way through the organic matter beneath the surface. Many basidiomycetes are edible, although many others (such as *Amanita*) are hallucinogenic, deadly poisonous, or both. It is therefore most unwise to hunt mushrooms to eat unless you are extremely knowledgeable. Furthermore, among many potentially hallucinogenic species, the difference between a hallucinogenic dose and a lethal one is very small.

The Water Molds: Division Oomycota

The **oomycetes,** commonly called water molds, grow mainly in water or in wet soil. Some taxonomists classify these organisms with fungi, while others consider them more closely related to algae. Many are parasites that cause mildew and blights on beets, grapes, tomatoes, and

Division Ascomycota and Division Basidiomycota

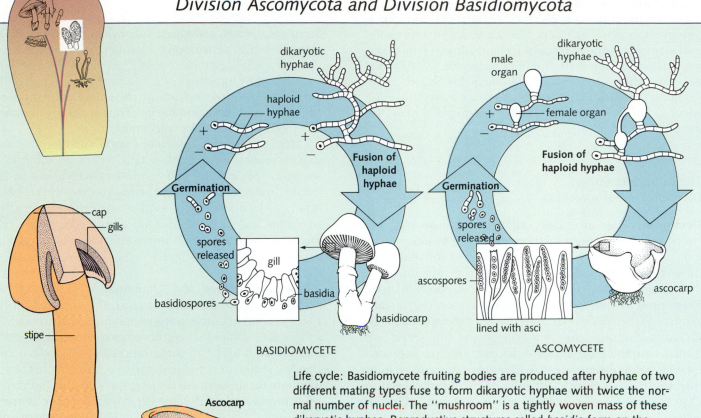

dikaryotic hyphae

haploid hyphae

+

−

Germination

spores released

basidiospores

gill

basidia

Fusion of haploid hyphae

basidiocarp

BASIDIOMYCETE

dikaryotic hyphae

male organ

+

−

female organ

Germination

spores released

ascospores

lined with asci

Fusion of haploid hyphae

ascocarp

ASCOMYCETE

cap

gills

stipe

Basidiocarp

mycelium

Ascocarp

Anatomy: Familiar "mushrooms" are the fruiting bodies of basidiomycete fungi. A cap covers the reproductive structures located on the gills underneath. The stalk of the mushroom is called the *stipe*. A typical ascomycete fungus consists of a conspicuous cap or *ascocarp*. Both fruiting bodies are small compared with the *mycelium*, the much larger underground portion of the fungus.

Life cycle: Basidiomycete fruiting bodies are produced after hyphae of two different mating types fuse to form dikaryotic hyphae with twice the normal number of nuclei. The "mushroom" is a tightly woven mass of these dikaryotic hyphae. Reproductive structures called *basidia* form on the undersurface of the cap. Ascomycete fruiting bodies are formed after hyphae of two mating types meet. Some of these hyphae fuse to create *dikaryotic hyphae*. All three classes of hyphae (+, −, and dikaryotic) build the ascocarp. Across cells of the ascocarp surface, nuclei fuse to create the spore sacs, or asci, which produces haploid spores.

The basidiomycete *Amanita* (above left) is poisonous to mammals; many (but not all) species of the genus *Russula* (above right) are edible. (left) The fruiting bodies of an ascomycete.

other crops. One oomycete, *Phytophora infestans,* which causes late blight in potatoes, changed the course of human history. Between 1845 and 1860, a series of unusually cool, damp growing seasons in Ireland allowed *P. infestans* populations to devastate the potato crop on which Irish peasant farmers depended for their livelihood. The resulting famine and outbreaks of disease forced unprecedented numbers of Irish families to emigrate, many to the United States.

The Imperfect Fungi: Division Deuteromycota

Deuteromycetes are also called *fungi imperfecti,* or "imperfect fungi," because they seem to be lacking a sexual stage in their life cycle. As a matter of convenience, these 25,000-odd species are lumped together into this division even though they are probably not closely related. From time to time, a sexual stage is discovered in this group and the species is moved to one of the other fungal divisions. Among the deuteromycetes important to humans are the *Penicillium* species that produce antibiotics. Other deuteromycetes are parasites that cause serious diseases in many plants and animals, including humans. One species, *Candida albicans,* causes a throat infection called candidiasis—"thrush." This infection strikes people with impaired immune systems, for example in people with AIDS.

THE FIRST LAND ANIMALS

Once plants and fungi colonized the continents, animals could follow. But not all aquatic animals were equally good candidates for a terrestrial existence, because emerging land animals had to solve the same problems faced earlier by plants: desiccation, the need to support their bodies, and the need for well-developed systems for the internal transport of water and essential nutrients.

By those criteria, the body plan of cnidarians, though eminently successful in the sea, was useless; it provided natural selection with no raw material from which to shape the necessary terrestrial adaptations. **Arthropods,** on the other hand, had evolved exoskeletons that, with minor evolutionary modifications, provided both waterproof body coverings and stiff, jointed appendages for walking, burrowing, or flying. It should come as no surprise, therefore, that among the first land animals were members of the same arthropod subphylum that contained the early sea scorpions and horseshoe crabs.

Arthropods

Scorpions, spiders, and mites: Subphylum Chelicerata The first terrestrial arthropods, members of the chelicerate class Arachnida (see next page), predated the first land vertebrates by hundreds of millions of years. Some of the earliest animals to crawl among the emerging plants were the detritus-feeding and herbivorous ancestors of today's scorpions, spiders, ticks, and mites.

Modern terrestrial scorpions are venomous predators that carry their stings on their tails. There are about 1200 scorpion species alive today, most of which live in the tropics, in the subtropics, and in deserts everywhere. Many other modern arachnids, such as spiders and mites, are predators and parasites that have evolved intricate ecological relationships with more recently evolved groups of plants and animals.

Subphyla Crustacea and Uniramia While chelicerate arthropods were evolving on land, the ancestors of the two most important groups of modern arthropods, the subphyla Crustacea (ancestors of modern shrimps, crabs, and lobsters) and Uniramia (ancestors of modern centipedes, millipedes, and insects), were evolving in the sea. Though the history of these groups in the Silurian is undistinguished, both later experienced major adaptive radiations. These subphyla combined well-developed nervous systems, adaptable exoskeletons, efficient respiratory systems (gills in crustaceans, air-filled tubes called *tracheae* in uniramians), and effective circulatory systems to produce a body plan that has proved adaptable to an infinite variety of ecological conditions.

Over the course of their evolution, crustaceans and uniramians developed innumerable specialized versions of the adaptable appendages that characterize all arthropods. It is on the basis of those appendages that members of these two subphyla are distinguished from one another. Both groups have jaws called **mandibles** that evolved from the appendages of the second or third body segment (rather than from the first, as did the chelicerae of the Chelicerata). Beyond that similarity, however, lies a fundamental difference. Crustacean limbs have two branches and are called *bi*ramous, or two-branched. *Uni*ramian appendages, on the other hand, have only one branch.

Shrimps, crabs, and lobsters: Subphylum Crustacea The **crustaceans,** which include the shrimps, crabs, lobsters, copepods, and barnacles, evolved primarily in the sea, where they quickly became the ultimate aquatic arthropods. Living crustaceans include more than 26,000 species, the vast majority of which live in marine and freshwater habitats. Ecologically they can be detritus feeders, filter feeders, scavengers, herbivores, carnivores, or omnivores.

Spiders, Crustaceans, and Insects: Phylum Arthropoda

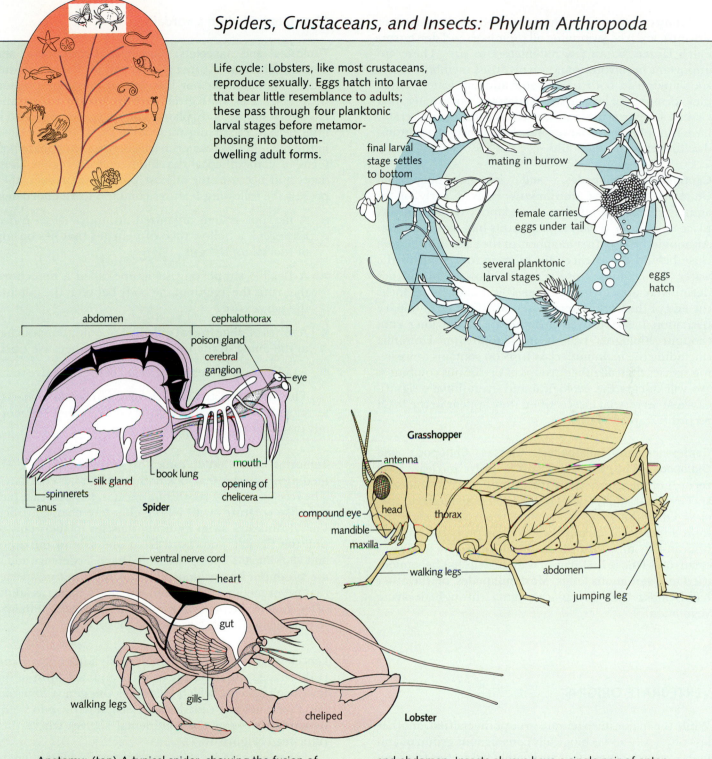

Life cycle: Lobsters, like most crustaceans, reproduce sexually. Eggs hatch into larvae that bear little resemblance to adults; these pass through four planktonic larval stages before metamorphosing into bottom-dwelling adult forms.

final larval stage settles to bottom

mating in burrow

female carries eggs under tail

several planktonic larval stages

eggs hatch

abdomen

cephalothorax

poison gland

cerebral ganglion

eye

silk gland

book lung

mouth

opening of chelicera

spinnerets

anus

Spider

Grasshopper

antenna

compound eye

head

thorax

mandible

maxilla

walking legs

abdomen

jumping leg

ventral nerve cord

heart

gut

walking legs

gills

cheliped

Lobster

Anatomy: (top) A typical spider, showing the fusion of body segments into cephalothorax and abdomen. Arachnid nervous systems have a large *cerebral ganglion* and several compound eyes. Silk for weaving webs or other traps is produced by glands called silk glands. Arachnids breathe primarily through respiratory organs called *book lungs,* whose blood-filled folds resemble the pages of a book. Spiders do not swallow prey whole; they inject digestive enzymes through openings made with their chelicerae and suck up the resulting liquid. (center) Most insects have three main body parts: head, thorax, and abdomen. Insects always have a single pair of antennae and three pairs of uniramous walking legs, but they may be wingless or have either one or two pairs of wings. All insects have dorsal hearts and ventral nerve cords. (bottom) Most crustaceans have two main body parts, the cephalothorax and abdomen. All appendages are two-branched, or *biramous.* They carry two pairs of antennae, a pair of grasping *chelipeds,* and four pairs of walking legs. The ventral nerve cord, with its numerous ganglia, shows evidence of the animal's segmented ancestry.

Many crustaceans, such as copepods and a variety of shrimps, are free-swimming all their lives and don't grow much larger than a few millimeters in size. These are important herbivorous and carnivorous members of the zooplankton in both fresh water and salt water, where they form critical links in many complex, open-water food chains. Larger crustaceans, such as lobsters and crabs, have tiny, planktonic larvae but settle down to a bottom-dwelling existence as adults.

Centipedes, millipedes, and insects: Subphylum Uniramia The first **uniramians** evolved early on in aquatic habitats, but their major adaptive radiations took place much later, on land, after plants had paved the way. Although the very first members of the class Insecta were close behind the chelicerates in their forays out of the water, for example, insects as we know them did not definitively appear for another 100 million years (near the end of the Carboniferous period). That evolutionary hesitation hasn't proved a handicap in the long run, though: Biologists believe the subphylum Uniramia (including the mammoth class Insecta) contains more living species than all other groups of organisms combined. We will discuss the ecological and evolutionary significance of terrestrial insects when they appear in the fossil record (Chapter 17).

Centipedes and millipedes: Classes Chilopoda and Diplopoda The earliest uniramia to evolve and leave fossil records of themselves were the centipedes (2500 modern species) and millipedes (10,000–50,000 modern species). Of all the living arthropods, these classes show the most superficial resemblance to their probable onychophoran-like ancestors. Modern centipedes are voracious carnivores whose first pair of walking legs has been modified into venomous fangs. Most millipedes, on the other hand, make their living scavenging through decaying vegetation.

VERTEBRATE ORIGINS

While the major invertebrate groups diversified, another less physically and ecologically conspicuous group of animals evolved—those that ultimately led to the first animals with backbones. The fossil record of those organisms is virtually nonexistent; they seem to have lacked any hard body parts, and, as important as they are to vertebrate history, they don't seem to have played a major role in early ecosystems. Nevertheless, by the Silurian, the vertebrates had come into their own and had founded a dynasty that would ultimately control most of the earth.

Chordates: Phylum Chordata

Tunicates and lancelets: Subphyla Urochordata and Cephalochordata At first glance, you probably wouldn't place either tunicates or lancelets in the same phylum with vertebrates. Both groups are exclusively marine and have very simple body plans. The sparse fossil record is little help in classifying these animals. They are placed in the phylum *Chordata* primarily because they share patterns of embryological development with higher chordates and because, for at least part of their lives, they possess the following features common to all **chordates:**

- A nervous system organized around a *dorsal hollow nerve cord,* a hollow tube of neurons that lies on top of (dorsal to) the gut.

- A flexible supporting structure called a **notochord** that runs the length of the body between the gut and the nerve cord.

- A series of **pharyngeal gill slits:** openings or pouches in the lining of the upper part of the digestive tract.

- A tail that continues past the end of the digestive tract.

The last three items on this list of chordate characteristics often apply only to the larvae or embryos of more advanced forms, such as our own species.

Tunicates Most adult tunicates are stationary, filter-feeding organisms that look nothing like other chordates. The free-swimming larvae of tunicates, however, are tadpole-like creatures with all the chordate characteristics. Larval tunicates swim by using muscles in their tails to pull the flexible notochord from one side to the other and allowing it to spring back into shape. Sedentary adults are produced by metamorphosis. A few tunicate species that never metamorphose into sedentary forms as adults offer clues to the paths that early chordate evolution may have taken as far back as the Cambrian.

Lancelets Lancelets are small (usually under 5 cm), torpedo-shaped, superficially fish-like animals. Most of the few living species, which belong to the genus *Branchiostoma* (erroneously called *Amphioxus*), burrow into sandy or muddy areas of the continental shelves where they filter-feed (Figure 16.7).

Lancelets have many characteristics that link them to tunicate "tadpoles," but they have also evolved a few more advanced traits that link them to higher vertebrates. Like most higher chordates, lancelets have segmental muscles organized into chevron-shaped units on either side of the notochord, a distinct advance over the muscular organization of tunicates. By contracting units on opposite sides of the body alternately against the flexible but incompressible notochord, lancelets flex their bodies during swimming much the way fishes do. But because lancelets lack the paired fins of fishes (Chapter 17), they have poor directional control.

Subphylum Vertebrata The ancestors of vertebrates, which are likely to have separated from the other lines of chordate evolution as early as the lower Cambrian, were probably shallow-water marine organisms. They had evolved a flexible series of cartilaginous units that wrapped around the nerve cord to protect it and interlocked with one another to provide a rudimentary backbone.

The first fossil traces of vertebrates are tiny fragments of scale-like and tooth-like hard parts from the late Cambrian and Ordovician periods. From this evidence we infer that the earliest vertebrates were heavily armored, a trait that offered protection from such formidable predators as the eurypterids. These fragments of armor,

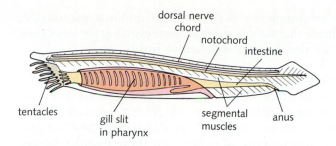

Figure 16.7 The lancelet, a simple, filter-feeding chordate, is a poor swimmer that spends most of its time partially buried in sediment.

Tunicates: Phylum Chordata

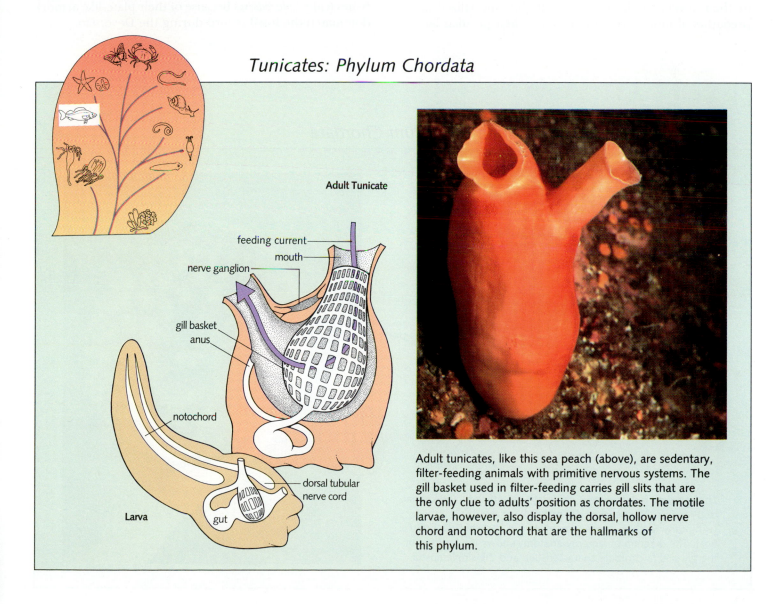

Adult tunicates, like this sea peach (above), are sedentary, filter-feeding animals with primitive nervous systems. The gill basket used in filter-feeding carries gill slits that are the only clue to adults' position as chordates. The motile larvae, however, also display the dorsal, hollow nerve chord and notochord that are the hallmarks of this phylum.

whose complexity is clearly seen under a microscope, indicate that vertebrates had been evolving for some time before they left any substantive fossils.

Jawless fishes: The first vertebrates The first vertebrates of which we have any reliable record are a curious collection of fishes without jaws, commonly called **agnathans** (*a* means "without"; *gnathos* means "jaw"). Agnathans had mouths, of course, but those mouths did not have jaws: the reinforced, hinged structures that serve the rest of us in feeding. The first agnathans became established in the Ordovician and diversified further in the Silurian. These **ostracoderms** (so named because of bony plates in their skin) exhibited a variety of odd shapes. Although jawless fishes diversified well into the Devonian, only about 45 species of lampreys and hagfishes remain alive today.

Lampreys, though they retain many primitive features of their forebears, have become highly specialized as predators of more recent organisms. Their peculiar lar-

vae remind us of the connections between invertebrate and vertebrate chordates. Lamprey larvae are pink worm-like organisms that look like crosses between fishes, lancelets, and tunicates. They bury themselves in the mud of rivers and streams and filter-feed for several years before metamorphosing into adults.

Hagfishes, too, are evolutionary relics. They burrow through the mud of the continental shelves, where they hunt annelid worms and scavenge pieces of dead and dying vertebrates.

Placoderms: The first jawed fishes Sometime during the middle of the Silurian, one group of primitive fishes evolved true, hinged jaws, apparently from the most anterior of the structures that had previously supported gills (Figure 16.8). Jawed fishes diversified quickly and appeared in the fossil record fully developed. Beginning their diversification in the mid-Silurian, the first jawed fishes (called *placoderms* because of their plate-like armor) dominated the fossil record during the Devonian.

Lampreys: Phylum Chordata

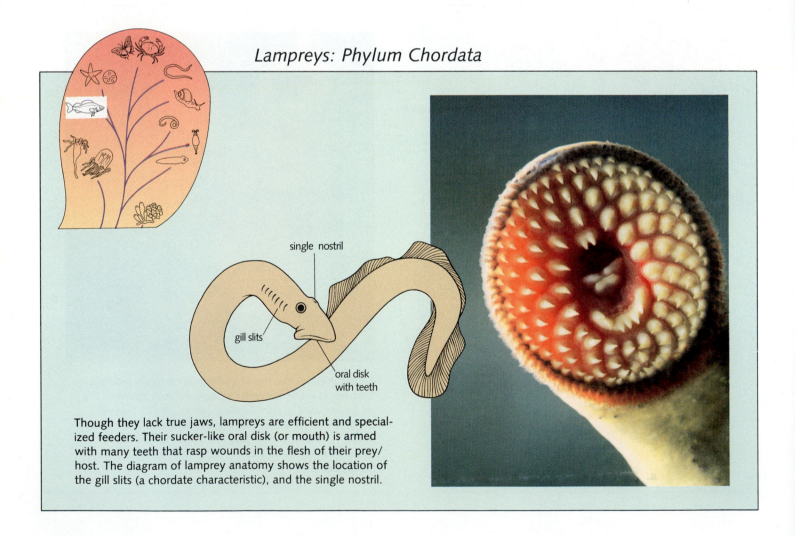

single nostril

gill slits

oral disk
with teeth

Though they lack true jaws, lampreys are efficient and specialized feeders. Their sucker-like oral disk (or mouth) is armed with many teeth that rasp wounds in the flesh of their prey/ host. The diagram of lamprey anatomy shows the location of the gill slits (a chordate characteristic), and the single nostril.

Evolutionary Innovations and Adaptive Radiations

Many popular accounts of evolution imply that advanced groups of modern organisms evolved either *from* more primitive living forms, or at least *after* those primitive forms underwent their major adaptive radiation. One might get the impression, for example, that vascular plants evolved from modern green algae or that jawed fishes evolved from living hagfishes. This is virtually never the case.

Throughout evolutionary history, adaptive innovations that lead to advanced groups have occurred in an evolving line *before* the primitive members of that line diversified very much, as Figure 16.9 shows. In the case of the earliest fishes, for example, the ancestors of placoderms evolved jaws before the agnathans diversified. And, as you will learn in the next chapter, the very first sharks and bony fishes appeared well before the placoderms reached their prime.

Although there are several possible explanations for this situation, the powerful ecological principle of competition may be sufficient to account for it. It is likely that under certain circumstances, adaptation and diversification are stimulated by competition. If so, then competition with the very first jawed fishes excluded agnathans from some niches and stimulated them to diversify and exploit other niches where they temporarily became equal or superior competitors. In the long term, however, the advances in body plan of jaw-bearing fishes provided them with an adaptive superiority that, together with the stresses of changing environments, ultimately spelled doom for their predecessors. The placoderms in turn, after flourishing in the Devonian, fell victim to a mass extinction, clearing the way for the higher fishes whose ancestors had emerged as early as the late Silurian.

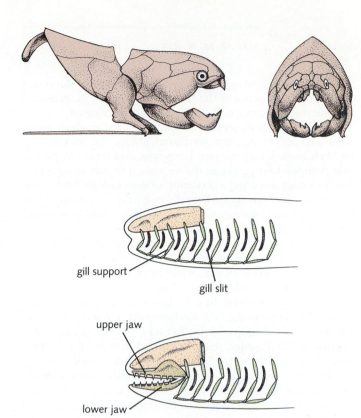

Figure 16.8 (top) Many of the early jawed fishes were fearsome predators; this one reached a length of nearly three meters. (bottom) Through many steps over millions of years, several of the anterior gill supports fused and changed shape to create the highly adaptable jaws that characterize most fishes and higher vertebrates today.

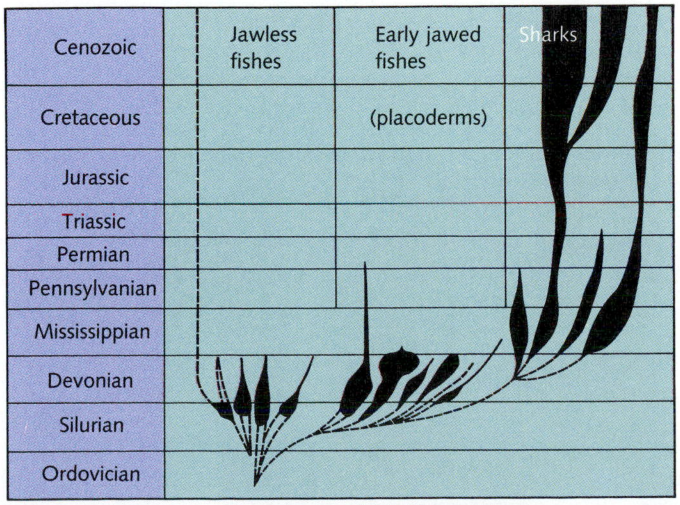

	Jawless fishes	Early jawed fishes	Sharks
Cenozoic			
Cretaceous		(placoderms)	
Jurassic			
Triassic			
Permian			
Pennsylvanian			
Mississippian			
Devonian			
Silurian			
Ordovician			

three classes of fish-like vertebrates

Figure 16.9 This chart illustrates the relative abundance (signified by the width of the bands) of early fishes in the fossil record. For an explanation, see the text.

SUMMARY

The Silurian period witnessed two momentous evolutionary events: the emergence of the first plants adapted to terrestrial life and the appearance of the first jawed vertebrates.

The line of multicellular plants that sired modern green algae also led to the first terrestrial plants, the ancestors of modern mosses, club mosses, horsetails, and ferns. As part of that common heritage, most modern plant species share reproductive cycles involving alternation of generations. Over time, as certain plant lineages adapted more fully to terrestrial life, the sporophyte stage steadily increased in size and importance while the gametophyte stages became smaller.

The diverse kingdom Fungi contains organisms that function as decomposers in many ecosystems and as vital mutualistic symbionts with many plants. Some fungi, however, are pernicious parasites on both plants and animals.

Arthropods evolved first in the sea, but some lineages emerged to become the first land animals. Their descendants, which include shrimps, crabs, lobsters, insects, spiders, centipedes, and millipedes, have diversified to make the Arthropoda the most diverse phylum of multicellular organisms.

The first vertebrates, who share common roots with the invertebrate ancestors of starfish and tunicates, left virtually no fossil record. Their descendants, first the jawless fishes, and then the jawed fishes and sharks, began to fill the seas by the end of the Silurian.

STUDY FOCUS

After studying this chapter, you should be able to:

- Explain the adaptations that enabled aquatic plants to colonize the land successfully.
- Describe the lower plants and their life cycles.
- Trace the successful movement of arthropods onto land and the early evolution of the vertebrates.

SELECTED TERMS

bryophytes *p. 310*	ascomycetes *p. 316*
lycophytes *p. 312*	basidiomycetes *p. 316*
sphenophytes *p. 312*	arthropods *p. 318*
pterophytes *p. 312*	mandibles *p. 318*
fungi *p. 314*	chordates *p. 320*
mycelium *p. 314*	notochord *p. 320*
hyphae *p. 314*	pharyngeal gill slits *p. 320*
lichens *p. 315*	agnathans *p. 322*
mycorrhizae *p. 315*	ostracoderms *p. 322*
zygomycetes *p. 316*	

REVIEW

Discussion Questions

1. What trend emerged in the relative size of sporophyte and gametophyte as plants became increasingly well adapted to terrestrial life?

2. Why are fungi among the most important, diverse, and curious organisms on earth? How is multicellularity in fungi different from that in plants and animals?

3. What characteristics contributed to the success of arthropods as the first terrestrial animals?

4. What shared features place tunicates and lancelets in the phylum Chordata?

5. What is one explanation for the fact that jawed fishes appeared before the diversification of the agnathans (the jawless fishes) in the fossil record?

Objective Questions (Answers in Appendix)

6. Mosses and liverworts are not considered to be completely adapted to land because they
   - (a) require water to reproduce.
   - (b) do not grow in soil.
   - (c) have alternation of generations as part of their life cycle.
   - (d) require warm, moist temperatures to grow and reproduce.

7. As plants became more adapted to land,
   - (a) the haploid gametophyte predominated.
   - (b) the diploid sporophyte predominated.
   - (c) the haploid gametophyte and diploid sporophyte predominated.
   - (d) alteration of generations ended completely.

8. Most fungi produce cellular filaments called
   - (a) hyphae.
   - (b) mycelia.
   - (c) saprophytes.
   - (d) ascomycetes.

9. Chordates share a common ancestor with the
   - (a) centipede.
   - (b) lobster.
   - (c) mollusc.
   - (d) tunicate.

10. Jawless, predatory, extinct fishes with bony plates in their skin are classified as
    - (a) invertebrates.
    - (b) agnathans.
    - (c) urochordates.
    - (d) uniramians.

CHAPTER 17

Emergence of the Modern World

Feature boxes appear throughout the text serving as valuable study aids designed to enrich your understanding of each chapter. You may wish to refer to the following feature boxes as you read Chapter 17.

The Devonian Period	p. 326
Class Chondrichthyes	p. 327
Class Osteichthyes	p. 329
Class Amphibia	p. 332
The Carboniferous Period	p. 333
The Permian Period	p. 335
Division Coniferophyta	p. 336
The Mesozoic Era	p. 339
Division Anthophyta	p. 343

THE DEVONIAN PERIOD: THE REIGN OF VERTEBRATES BEGINS

The Devonian period saw rapid evolutionary radiation of vertebrates in the sea and extensive colonization of the land by plants and animals. During this geologically active time, the earth's continents once more drifted together, forming a single, vast land mass called *Pangaea* (see p. 326). This supercontinent became the stage for the evolution of all major lines of terrestrial plants and animals, for it was to persist for 200 million years, into the late Jurassic. All across Pangaea, Devonian climates were moist and comfortable, though cool, and ample rainfall created freshwater basins that cradled many new aquatic and soon-to-be-terrestrial organisms.

The Devonian is often called the "age of fishes" because of the explosive radiation of aquatic vertebrates during this period. As the jawed *placoderms* reached the peak of their adaptive radiation, four key innovations appeared among different lines: jaws, paired appendages, lungs, and the two sets of limbs that characterize higher vertebrates. As each of these innovations appeared, it established a new body plan and sparked the adaptive radiation of one or more classes of vertebrates (see Figure 17.1 on page 328).

The Cartilaginous Fishes: Class Chondrichthyes

The first innovation was the development of *paired fins*. The earliest fishes had only single *dorsal* or *ventral* fins, which acted as stabilizers, and tail fins that propelled them forward (see p. 327). As a result agnathans and most placoderms couldn't steer very well, either to chase prey or to escape predators. Paired fins, by adding significant ability to steer, bestowed an enormous selective advantage. The first vertebrates to possess both jaws and paired fins were the **cartilaginous fishes,** so called because their skeletons were made of firm but resilient cartilage. The scientific name for this group is class Chondrichthyes (*chondro* means "cartilage"; *ichthyes* means "fishes").

The first adaptive radiation of cartilaginous fishes produced a host of species, most of which became extinct fairly rapidly. The class survived to radiate once again during the Cretaceous period, however, producing the modern sharks, skates, and rays. Though they have retained such primitive features as bulky fins and a cartilaginous skeleton, sharks and their relatives have been

evolving for nearly 400 million years. Armed with reasonably good eyesight and a keen sense of smell, they are among the planet's most efficient and sophisticated predators. In addition, many can detect the minute electric currents generated by the muscles of distant potential prey (Chapter 41).

The Bony Fishes: Class Osteichthyes

Another line of early placoderms combined two sets of paired fins with strong, lightweight bone to found the lineage of **bony fishes,** known as the class Osteichthyes

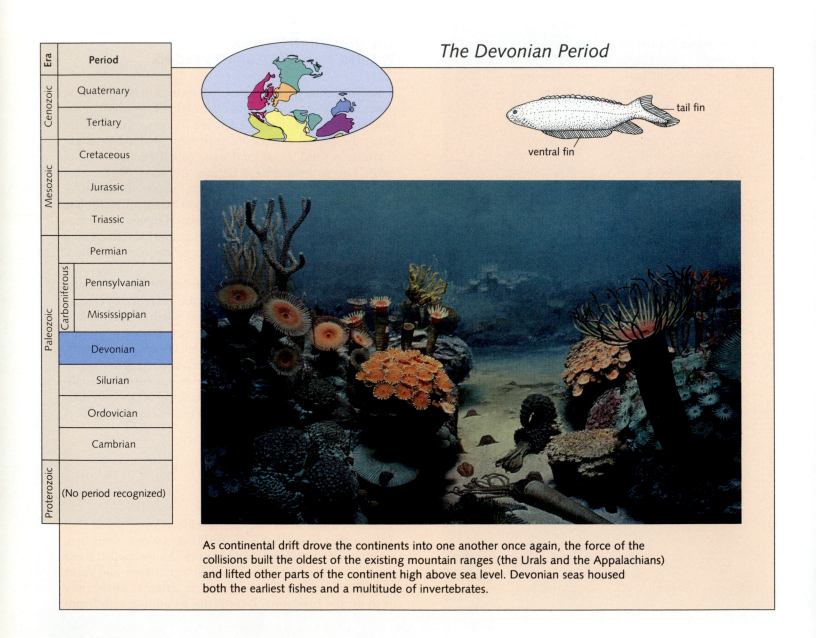

The Devonian Period

tail fin

ventral fin

Era	Period	
Cenozoic	Quaternary	
	Tertiary	
Mesozoic	Cretaceous	
	Jurassic	
	Triassic	
Paleozoic		Permian
	Carboniferous	Pennsylvanian
		Mississippian
		Devonian
	Silurian	
	Ordovician	
	Cambrian	
Proterozoic	(No period recognized)	

As continental drift drove the continents into one another once again, the force of the collisions built the oldest of the existing mountain ranges (the Urals and the Appalachians) and lifted other parts of the continent high above sea level. Devonian seas housed both the earliest fishes and a multitude of invertebrates.

(*osteo* means "bone") (see p. 329). By the time they appeared in the mid-Devonian, the bony fishes had split into two distinctive lines, the *ray-finned fishes,* the most diverse group of vertebrates alive today, and the *fleshy-finned fishes,* among whom the ancestors of land vertebrates arose.

The ray-finned fishes   The relatively modest radiation of ray-finned fishes during the Devonian has left behind only a few primitive species, such as gars and sturgeons (Figure 17.2). Two hundred million years later, however, the main line of ray-finned fishes finally began the spectacular adaptive radiation that produced the more than 30,000 species alive today. Many modern fishes are extraordinarily agile swimmers that can dart, hover, pivot, swim in reverse, and stop on a dime. They use this agility in feeding, avoiding predators, and navigating around such complex structures as coral reefs.

The fleshy-finned fishes   There are only a few species of fleshy-finned fishes surviving today as evidence of the final early vertebrate innovations. These living fossils are virtually identical to their Devonian forebears.

Early *lungfishes* probably evolved in shallow Devonian swamps where the water contained little oxygen. In these fishes, a pouch connected to the upper digestive tract

Sharks, Skates, and Rays: Class Chondrichthyes

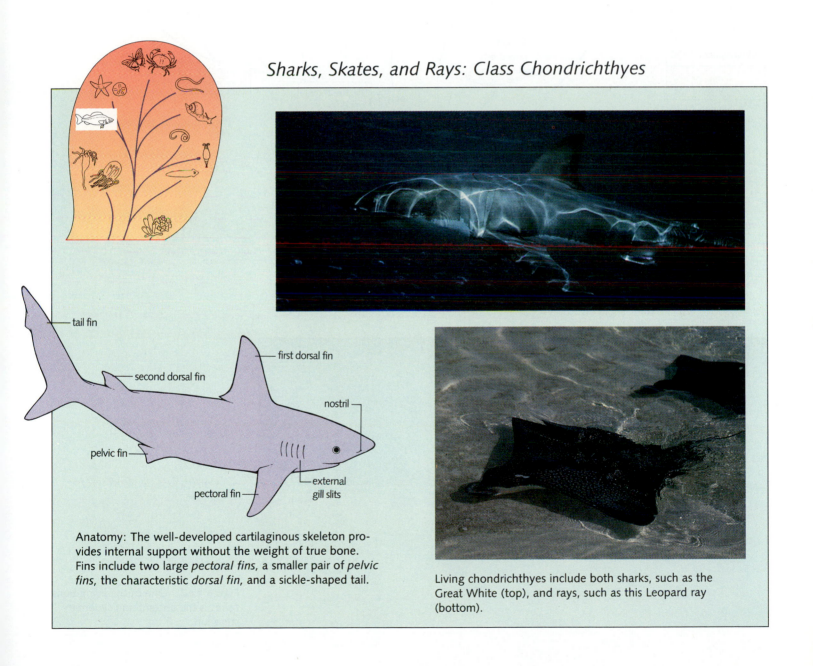

tail fin
first dorsal fin
second dorsal fin
nostril
pelvic fin
pectoral fin
external gill slits

Anatomy: The well-developed cartilaginous skeleton provides internal support without the weight of true bone. Fins include two large *pectoral fins,* a smaller pair of *pelvic fins,* the characteristic *dorsal fin,* and a sickle-shaped tail.

Living chondrichthyes include both sharks, such as the Great White (top), and rays, such as this Leopard ray (bottom).

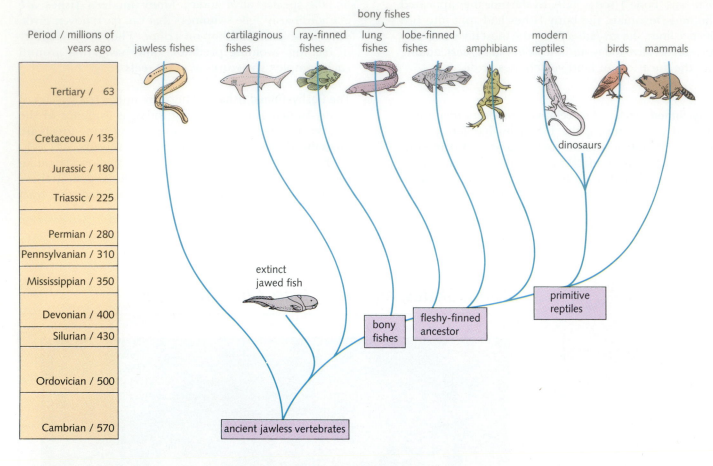

Period / millions of years ago

| Tertiary / 63 |
| Cretaceous / 135 |
| Jurassic / 180 |
| Triassic / 225 |
| Permian / 280 |
| Pennsylvanian / 310 |
| Mississippian / 350 |
| Devonian / 400 |
| Silurian / 430 |
| Ordovician / 500 |
| Cambrian / 570 |

jawless fishes

cartilaginous fishes

bony fishes
ray-finned fishes   lung fishes   lobe-finned fishes

amphibians

modern reptiles   birds   mammals

dinosaurs

extinct jawed fish

bony fishes

fleshy-finned ancestor

primitive reptiles

ancient jawless vertebrates

Figure 17.1 The family tree of the vertebrates. Note that both the earliest and the most numerous vertebrates are aquatic. Note also the sequence of evolutionary innovations in the vertebrate line that fueled the series of vertebrate adaptive radiations.

Figure 17.2 One of the living bony fishes is the ancient and sedentary sturgeon.

Bony Fish: Class Osteichthyes

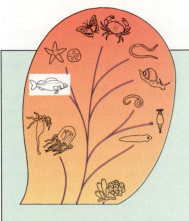

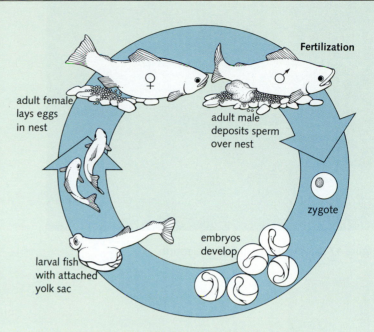

Life cycle: In most fishes, sexes are separate, fertilization is external, and eggs hatch into free-swimming larvae. There are also many species with internal fertilization that bear their young alive.

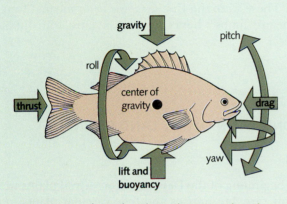

Maneuvering through a three-dimensional medium such as water is no easy matter; to capture prey efficiently, fishes must control roll, pitch, and yaw, just as airplane pilots do. The single fins of the earliest fishes provided only rudimentary steering ability; paired fins in more advanced species permit precision maneuvering.

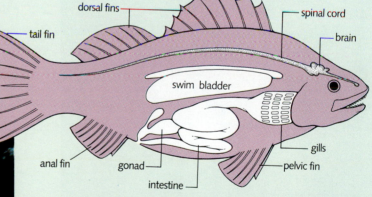

Anatomy: Ray-finned fishes have three or four *unpaired fins* and two sets of highly mobile *paired fins*. The Nassau grouper (left) clearly shows the highly mobile paired pelvic fins and lateral pectoral fins (not depicted in anatomy diagram). Most fishes of this class also have gas-filled *swim-bladders* that counter the weight of bone and muscle and allow the fish to hover effortlessly. Fish anatomy varies greatly with ecological specialization. This carnivorous species has jaws armed with sharp teeth and a short digestive tract. Herbivorous fishes have jaws specialized for rasping and long intestines to digest plant tissue.

Figure 17.3 The fleshy-finned coelacanth.

evolved into a primitive lung that enabled its bearers to breathe air. Though common and diverse at first, lung-fishes dwindled to a handful of species in parts of Australia, Africa, and South America.

The living **coelacanth** was thought to be extinct, until a single living species was captured in the Pacific Ocean. Since then, several dozen specimens have been captured or sighted from submersibles. The coelacanth is the sole survivor of a primitive group of fleshy-finned fishes that crawled around Devonian swamps on four stubby, fleshy fins (Figure 17.3). X-rays reveal that these pudgy versions of normal fish fins were supported by bones homologous to the bones in the limbs of all terrestrial vertebrates. These were the first *tetrapods* (*tetra* means "four"; *poda* means "limbs"; hence *tetrapods* means "four-limbed animals").

COLONIZATION OF THE LAND BEGINS

Plants

At the beginning of the Devonian, terrestrial plants were probably rare, but mosses and true vascular plants soon became common (Figure 17.4). One or more of these lines that led ultimately to higher plants continued to evolve refinements in their vascular tissue, enabling them to grow taller and thicker stems. But as we have seen, both mosses and primitive vascular plants could reproduce only in very moist environments.

Then, sometime during the Devonian, a new suite of adaptations ushered in a more fully terrestrial way of life. The most important single innovation was the evolution of the **seed,** a self-contained, drought-resistant reproductive package that houses a dormant plant embryo and an ample supply of food within a protective seed coat. The seed coat protects the infant plant from desiccation, enabling it to survive for weeks, months, or even years until conditions are right for germination. The stored food nourishes the embryo until it develops its own functioning roots and leaves. Of all plant reproductive strategies, those involving seeds are best adapted to terrestrial life.

Seed plants remained rare for some time before giving rise to the major groups of modern plants that display several additional changes in their life cycles. The haploid female gametophytes of seed plants were reduced to little more than a few cells living obligatorily on the diploid sporophyte. Male gametophytes dwindled to dust-like **pollen grains** carried by wind or by animals to the female reproductive organs where fertilization eventually occurred.

Figure 17.4 By the mid-late Devonian, tall horsetails and lycopod trees were abundant, forming the world's first dense forests.

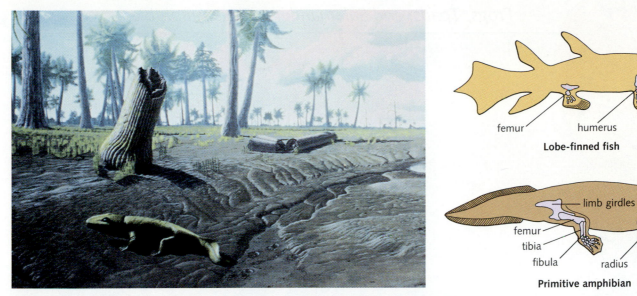

Lobe-finned fish

limb girdles

Primitive amphibian

Figure 17.5 (above) *Ichthyostega*, one of the earliest amphibians. (right) Evolutionary changes in limb girdles between fleshy-finned fishes and amphibians. Note the homology between coelacanth fin bones and the limbs of more recent tetrapods.

Early Terrestrial Vertebrates: Class Amphibia

The first terrestrial vertebrates evolved from ancestors much like the early fleshy-finned fishes through a range of intermediate forms seen in the fossil record (Figure 17.5). Unfortunately, none of these can be unambiguously labeled *the* single ancestor of all land vertebrates. It is certain, however, that the tetrapod limb evolved among the fleshy-finned fishes and that the limb bones of the first **amphibians** were modified from those fins' original bony supports.

The first amphibians—ancestors of frogs, toads, and salamanders—climbed out of shallow swamps into moist terrestrial environments in tropical parts of Pangaea (Figure 17.6). There, constant temperatures and high humidity made the transition from water to land less stressful than it would have been in colder or drier habitats. Possibly as a result of the value of being able to waddle from pond to pond in search of food and open water, or possibly as an adaptive response to competition or predation in shallow water, their descendants became progressively better adapted to terrestrial conditions.

These first terrestrial vertebrates faced the same problems as the first land plants: providing body support and avoiding desiccation. And the first amphibians—like the first mosses and lycopods—developed some, but not all, of the adaptations necessary for a fully terrestrial existence. The fleshy fins of their ancestors lengthened and strengthened into stocky limbs, each with five digits,

Figure 17.6 The midland mud salamander is representative of the class Amphibia.

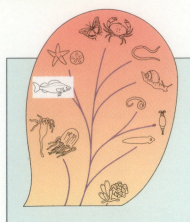

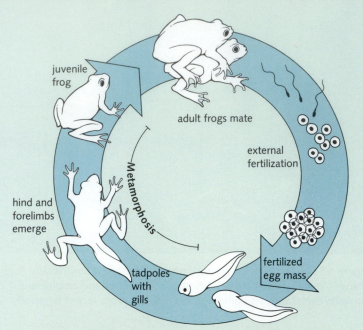

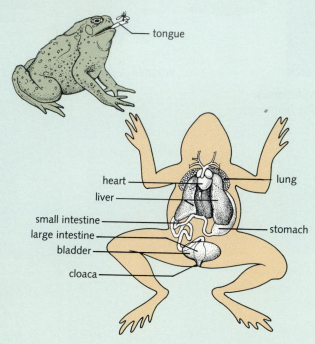

tongue

heart
liver
small intestine
large intestine
bladder
cloaca
lung
stomach

Anatomy: Modern amphibians show many of the internal organs common to higher terrestrial vertebrates.

juvenile frog
adult frogs mate
external fertilization
Metamorphosis
hind and forelimbs emerge
tadpoles with gills
fertilized egg mass

Life cycle: The typical cycle includes a fully aquatic larval stage, or *tadpole.* Frog and toad tadpoles are herbivorous filter feeders whose internal anatomy changes as profoundly as their external anatomy as they metamorphose into carnivorous adults.

Many modern amphibians have become highly specialized in adapting to specific habitats. Eggs of the Surinam toad (above) sink into the protective skin of their parent's back, where they develop. Spadefoot toads (left) have turned water-permeable skin into an advantage in desert habitats. They burrow deeply at the end of rainy season and stay buried for up to 10 months, using their skin to extract water from moist soil as the roots of plants do.

and two skeletal "girdles" evolved allowing the attachment of those limbs to the backbone. These major adaptations initiated the slow, steady amphibian radiation that spawned a complex and fascinating ancient fauna over a period that lasted more than 130 million years. Most early amphibians seem to have been carnivores that fed on the early land-dwelling scorpions and spiders and on the very first insects that appeared at this time.

Like lower plants, however, most amphibians lack a waterproof barrier to prevent the loss of body fluids, and therefore they lose water rapidly through their skin in dry air. Furthermore, all primitive amphibians (and many modern forms) had to lay their eggs in water and spend at least part of their lives as fully aquatic larvae. For that reason, although numerous adult amphibians can survive on land for limited periods of time, most can survive and reproduce over the long term only in wet (or at least seasonally wet) habitats (see previous page).

THE CARBONIFEROUS PERIOD

During the Carboniferous period, Pangaea remained intact while mountain building created a great diversity of terrestrial environments ranging from moist, swampy lowlands to cooler and drier upland areas. Toward the end of the period, several major glaciations affected the northern and southernmost land areas.

Although the Devonian period closed with a mass extinction that claimed most jawless fishes, some placoderms, and many trilobites, the major lines of higher plants and animals continued to evolve. The first forests, whose fossilized remains form most of our coal deposits and give the period its name, spread across Pangaea (see below). In and around those swampy forests, the amphibian adaptive radiation continued, encouraged both by the expansion of the great, moist forests and by the emergence of a new type of prey in abundance: insects.

The Carboniferous Period

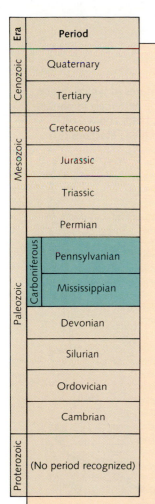

Era	Period	
Cenozoic	Quaternary	
	Tertiary	
Mesozoic	Cretaceous	
	Jurassic	
	Triassic	
Paleozoic	Permian	
	Carboniferous	Pennsylvanian
		Mississippian
	Devonian	
	Silurian	
	Ordovician	
	Cambrian	
Proterozoic	(No period recognized)	

During the Carboniferous, extensive mountain building created a great diversity of terrestrial environments, ranging from moist, swampy lowlands to cooler and drier upland areas. Toward the end of the period, several major glaciations affected the northern and southernmost land areas. Carboniferous forests were dominated by giant, spore-bearing lycopods and sphenopsids (giant club mosses and horsetails 12–15 meters high). Closer to the ground, ferns competed for space with early seed plants.

Order	Main Characteristics
Odonata	Mouthparts (biting); wings (2 pairs); metamorphosis incomplete *Examples:* dragonflies, damselflies
Orthoptera	Adult mouthparts (biting and chewing); membranous wings (2 pairs); metamorphosis incomplete *Examples:* grasshoppers, locusts
Isoptera	Mouthparts (chewing); wings (2 pairs) or wingless; social; employs division of labor; metamorphosis incomplete *Example:* termites
Hemiptera	Mouthparts (piercing, sucking); horny, membranous wings (2 pairs); metamorphosis incomplete *Example:* bedbugs
Coleoptera	Mouthparts (biting and chewing); membranous wings (2 pairs); armored exoskeleton; metamorphosis complete *Example:* beetles
Diptera	Mouthparts (sucking, piercing, lapping); wings (1 pair); metamorphosis complete *Examples:* flies, mosquitoes
Siphonoptera	Mouthparts (piercing, sucking); wingless; legs (jumping); metamorphosis complete *Example:* fleas
Lepidoptera	Tongue (long, coiled, for sucking); hairy body; wings (2 pairs) or wingless; metamorphosis complete *Examples:* butterflies, moths
Hymenoptera	Head mobile; eyes developed; mouthparts (chewing, sucking); membranous wings (2 pairs); metamorphosis complete *Examples:* bees, wasps, ants

Insects: The Arthropod Class Insecta

The great quantity and wide diversity of Carboniferous land plants provided countless new ecological opportunities for **insects,** whose first representatives had appeared much earlier. Although not as old as either the arachnids or the crustaceans, insects proved more successful, partly because of their ability to fly. Insects were, in fact, the only flying organisms for over 100 million years.

During the Carboniferous, insects diversified rapidly, founding most major orders: silverfish, grasshoppers, roaches, dragonflies, stone flies, and beetles (Figure 17.7). The success of the insects had both positive and negative repercussions for other organisms. To plants, the first insects were predators that quickly became the earth's principal herbivores. To early carnivorous vertebrates, insects were prey that provided abundant, bite-sized animal food for the first time in history.

Today, insects are both numerous and ecologically important: Ants, bees, and wasps alone comprise fully

Figure 17.7 (below) Representative features of typical insects. (left) Selected insect orders. Note that the extremely important order Hymenoptera containing the ants, bees, and wasps, though shown here, did not evolve until the Cretaceous.

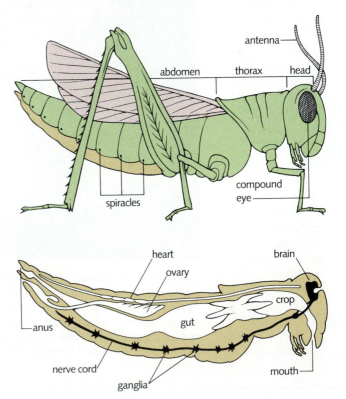

30 percent of the total animal matter in tropical rain forests. More than 900,000 species of insects have been described, and estimates of yet-unknown species range from 2 million to 30 million or more.

The First Reptiles: Class Reptilia

While amphibians diversified, **reptiles,** which first appeared during the Devonian, began their long-lived and enormously successful reign. The reptiles' adaptive radiation did not get under way until long after their amphibian predecessors had diversified, but the group radiated explosively from the second half of the Carboniferous into the Permian.

Like amphibians, reptiles quickly seized upon insects as their primary food source and took advantage of much stronger jaw muscles that enabled them to *bite*, rather than just snap at prey. Early reptiles also evolved more agile limbs and stronger limb girdles. But the key inno-

vation that spurred the reptilian radiation was the **amniotic egg,** a watertight egg produced by internal fertilization and wrapped in three protective membranes. Amniotic eggs serve land vertebrates much the same way seeds serve land plants—by making it easier for them to reproduce without standing water. Amniotic eggs protect developing embryos from desiccation, nourish them as they grow, provide a space for the storage and ultimate disposal of waste products, and enable the egg to exchange respiratory gases with the surrounding air.

THE PERMIAN PERIOD

The Permian period was a time of great environmental stress and innovation for all plants and animals, as geological changes in Pangaea produced cooler, drier climates. A mass extinction claimed more than 50 percent of all terrestrial animal families and more than 95

The Permian Period

Era		Period
Cenozoic		Quaternary
		Tertiary
Mesozoic		Cretaceous
		Jurassic
		Triassic
Paleozoic		Permian
	Carboniferous	Pennsylvanian
		Mississippian
		Devonian
		Silurian
		Ordovician
		Cambrian
Proterozoic		(No period recognized)

During the Permian, the plates composing Pangaea continued to grind against one another, lifting larger and larger areas above sea level. The climate became cooler and drier, the great continental seas retreated, and many swamps dried up completely. Two plant relics from this period are ginkgoes (left) and cycads (right). The ginkgo, thought by Western scientists to have been extinct for millions of years, was rediscovered in China where it had been maintained in horticulture.

Conifers: Division Coniferophyta

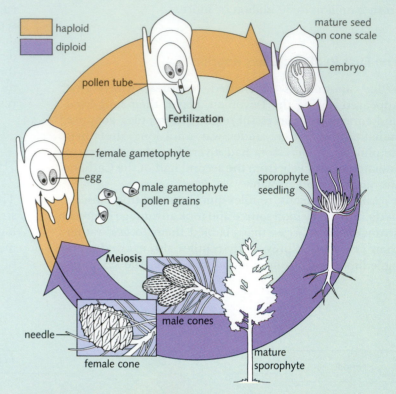

haploid

diploid

mature seed on cone scale

embryo

pollen tube

Fertilization

female gametophyte

egg

male gametophyte pollen grains

sporophyte seedling

Meiosis

needle

male cones

female cone

mature sporophyte

Life cycle: The pine tree, the diploid sporophyte, grows male and female cones that produce spores. Those spores produce the tiny gametophyte generation: pollen in male cones, and the egg and associated cells in female cones. Pollen is wafted by the wind to the female cones, where it is caught and held by a sticky secretion as the pollen tube grows slowly toward the egg. The winged seeds grow naked on the scales of the female cone.

A representative conifer, the Jeffrey pine, seen here in Yosemite National Park (top). A coniferous forest of Douglas fir (bottom).

percent of all marine species. Among the victims were the last trilobites, many amphibians, and nearly all non-seed-bearing plants. At the beginning of the period, spore-bearing plants and amphibians were masters of the land; by its end, seed plants and reptiles held sway.

The First Higher Plants: Gymnosperms

Because the Permian's cooler, drier conditions spelled hard times for spore-bearing plants, they opened up new ecological opportunities for seed bearers by eliminating competition. By the end of the Permian, early seed-bearing plants had diversified into numerous lines, most of which are now extinct. Their living descendants are often called **gymnosperms,** or "naked seed" plants, because their seeds are not enclosed in well-developed fruits as are the seeds of the flowering plants that evolved later on. In all these plants, the haploid gametophytes are reduced to a few cells that are totally dependent on the sporophyte for sustenance.

Cycads and ginkgoes: Divisions Cycadophyta and Ginkgophyta   Two "naked seed" groups diversified successfully during the Permian period but have left few living representatives. **Cycads** are slow-growing, tropical and subtropical, palm-like plants with thick trunks that are represented today by only about 100 species. **Ginkgoes** are represented by a single "living fossil," *Ginkgo biloba.*

Conifers: Division Coniferophyta   Another gymnosperm group, the **conifers,** has had a long and successful reign and is represented today by about 550 species. Following their first radiation in the Permian, conifers remained the dominant land plants for nearly 200 million years. Though there are only a fraction as many species of modern conifers as there are species of flowering plants, vast coniferous forests still cover much of the north temperate high-altitude regions of the world.

In conifers, male and female gametophytes are produced on separate, specially modified reproductive structures called **cones.** The processes of fertilization and seed maturation in conifers proceed very slowly. Seeds are usually not released from the cone for nearly two years after pollination occurs.

The Rise of the Reptiles

Driven by changes in climate and encouraged by diversifying terrestrial ecosystems, animal evolution proceeded more and more rapidly. Reptiles made the most impressive progress, as the ancestors of modern reptiles, dinosaurs, and mammals appeared in quick succession.

By the end of the Permian, terrestrial ecosystems were dominated by the **therapsids,** a collection of curious animals that roamed across Pangaea and diversified into both carnivorous and herbivorous niches (Figure 17.8). These animals were clearly intermediates in the evolution of mammals from reptiles. In fact, paleontologists

Figure 17.8   The curious therapsids combined reptilian and mammalian features.

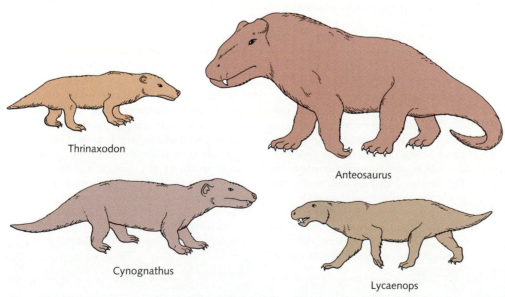

Thrinaxodon

Anteosaurus

Cynognathus

Lycaenops

differ as to whether they should be called transitional mammal-like reptiles or reptile-like early mammals. In any case, because therapsids clearly maintained high internal body temperatures by metabolic means, they are described as **endotherms** (*endo* means "inside or interior"; *therm* means "heat"). In contrast, living reptiles and amphibians, all invertebrates, and most lower vertebrates are called **ectotherms,** because their body temperatures rise and fall with the temperature of their external environment (*ecto* means "outside"). Endothermy is a useful—though energetically expensive—strategy that ultimately held the key to the success of mammals (see below).

THE MESOZOIC ERA

The Mesozoic era, often called the "age of dinosaurs," spanned 183 million years and encompassed three periods: the Triassic, the Jurassic, and the Cretaceous (see next page). Following the mass extinctions of the late Permian, the Mesozoic saw major adaptive radiations among molluscs, crustaceans, echinoderms, insects, bony fishes, reptiles, and land plants. At the same time, the first true mammals appeared. Because the evolutionary histories of all these groups are closely intertwined, we will discuss the Mesozoic as a single unit.

CURRENT CONTROVERSIES

Hot-Blooded Dinosaurs?

Many body tissues, including muscles, operate well when body temperature is within a narrow range. The ability to move rapidly, either to capture prey or to avoid becoming prey, thus depends on a constant, optimal body temperature.

Living reptiles are all ectotherms and make active use of environmental conditions to regulate their body temperatures. To get warm, they bask in the sun. To avoid overheating, they shelter in the shade. This strategy works well in warm climates, but in very cold weather, many ectotherms must become dormant.

Endothermic mammals and birds use insulation such as feathers, fat, and fur to retain metabolic heat. When it gets cold, endotherms burn more food to keep warm. When it gets too hot, they sweat or pant to get rid of extra heat. Endotherms can perform muscular work for longer periods and under more widely varying environmental conditions than ectotherms. They can also grow faster. This vitality is expensive; an

endotherm needs *up to fifty times* more food than a reptile of the same size.

Some paleontologists, led by Robert Bakker of the University of Colorado Museum, argue that dinosaurs were endotherms that regulated their body temperatures as modern mammals do. This theory would account for dinosaurs' adaptive superiority and for how they grew so rapidly, and it would explain how birds became endothermic.

Other paleontologists point out that dinosaurs were an extremely diverse group and that not all were necessarily endothermic. It is not unlikely, say researchers like Yale's John Ostrom, that several small, carnivorous dinosaurs were endothermic. Indeed, endothermy among the small, active dinosaurs that gave rise to birds would explain the role of the first feathers. But Ostrom and others point out that mammalian-style endothermy in larger herbivorous dinosaurs is unlikely for practical reasons; they could never have eaten enough to survive. Elephants spend

up to 15 out of every 24 hours stuffing nearly 300 pounds of leaves and branches into their huge mouths. Could the tiny-headed apatosaurs have gathered the daily *2000* pounds of conifers and lower plants needed to maintain their endothermic metabolism?

Bakker's supporters point out that elephants use their large mouths for chewing and grinding. Dinosaurs, on the other hand, used their jaws only for *taking in* food; grinding was accomplished by a gizzard-like arrangement in which stones that were swallowed were rubbed together by muscles inside the digestive tract.

Many researchers agree that dinosaurs had body temperatures higher than those of modern lizards. The debate over the maintenance of those temperatures will doubtless continue for some time. For extinction, as John Ostrom says, ensures a certain amount of ambiguity. "When the dinosaurs died," he writes, "they took many of their secrets with them."

The Mesozoic was a time of great geological changes that profoundly influenced the evolution of terrestrial organisms. The supercontinent of Pangaea remained intact through the Triassic but began to break up during the Jurassic. As the land mass fragmented, its pieces carried once-unified plant and animal populations into isolation and set the stage for new adaptive radiations.

Climatically, the Mesozoic was stable and warm; the dry warmth of the Triassic was followed by a warm, moist period that lasted into the Cretaceous. For reasons not completely understood, episodes of mass extinction terminated all three periods. The extinction at the end of the Triassic, for example, wiped out 35 percent of all animal families, including many primitive reptiles on land and molluscs in the sea. Toward the end of the Cretaceous period, the rise of the Rocky Mountains and of the Andes drained the inland seas and cooled continental climates worldwide, with dramatic consequences for both plants and animals.

The Ruling Reptiles

The reptilian adaptive radiation reached its peak during the Mesozoic, and reptiles dominated both land and sea. The reptilian lineage, which had diverged back in the

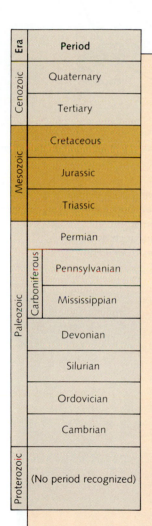

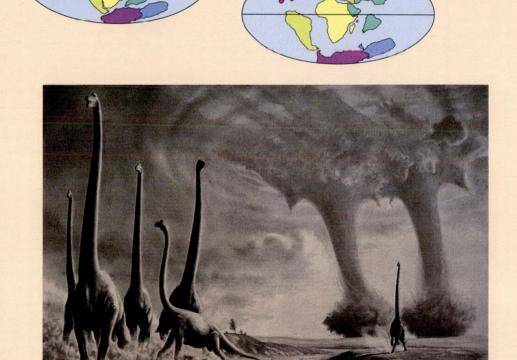

The Mesozoic Era

The Mesozoic era consisted of Triassic, Jurassic, and Cretaceous periods. (left globe) During the Triassic, Pangaea remained in one piece but seems to have become drier. Many areas were covered by vast deserts, while the single world ocean seems to have had relatively few shallow seas. (right globe) During the Jurassic, Pangaea split into a single northern continent, Laurasia, and a southern continent, Gondwanaland. Across both these land masses, the climate was warm, was moist, and exhibited very little seasonal variation, resembling conditions in many tropical countries today. Laurasia and Gondwanaland broke apart into recognizable modern continents, though they were not yet in their present-day positions.

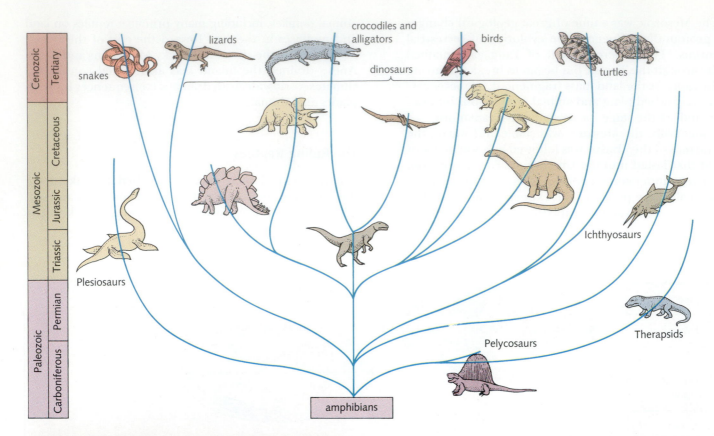

snakes
lizards
crocodiles and alligators
birds
turtles
dinosaurs
Plesiosaurs
Ichthyosaurs
Therapsids
Pelycosaurs
amphibians

Cenozoic — Tertiary
Mesozoic — Cretaceous, Jurassic, Triassic
Paleozoic — Permian, Carboniferous

Figure 17.9 (above) The reptilian radiation, showing both extinct and living forms. Iguanas (left center) and crocodiles (left bottom), two living reptiles, are ectotherms that depend on their environment to maintain adequate body temperature. (below) Dinosaurs, such as these hypsilophodonts, may have used several techniques to maintain dependable body temperatures.

Permian, spawned two extinct aquatic lines (ichthyosaurs and plesiosaurs); the forebears of modern lizards, snakes, and turtles; the ancestors of dinosaurs, crocodiles, and birds; and therapsids, the ancestors of mammals (Figure 17.9).

Dinosaurs Dinosaurs were extraordinary creatures. More than half were as large as or larger than any living land mammal. Evolving in an ecological arena already populated by therapsids and the first mammals, they quickly outstripped the competition to take over an incredible variety of niches. Dinosaurs ranged from chicken-sized carnivores preying on even smaller animals to giant flesh eaters and scores of herbivores, large and small.

The most famous of Jurassic reptiles were *Allosaurus*, *Ichthyosaurus*, *Apatosaurus* (formerly called *Brontosaurus*), and *Stegosaurus*. Their Cretaceous successors were the latest and greatest of the dinosaurs: *Tyrannosaurus rex*, the king of the carnivores; *Triceratops* and the ankylosaurs, the heavily armored herbivores; *Pteranodon*, the great flying reptile, and the duck-billed dinosaurs.

Until the mid-1960s, most scientists envisioned dinosaurs as oversized versions of modern reptiles: plodding, pea-brained evolutionary misfits that survived simply because they were bigger than everybody else. Apatosaurs, estimated to have tipped the scales at 25 to 35 tons, were depicted slogging through swamps submerged up to their bellies, because it was assumed that their legs could barely support their weight.

But during the past 25 years, paleontologists have learned that accepted reconstructions of dinosaur skeletons were often in error. Most dinosaur skeletons in museums, for example, show the animals' legs splayed out sideways like those of living reptiles and amphibians. In that configuration their bones would have splintered under their massive bulk. More careful and extensive comparison of fossil bone and joint structure has shown that dinosaurs held their legs erect beneath them, as mammals do.

Furthermore, fossil footprints tell us that even the largest dinosaurs moved at respectable speeds and that some lived in social groups like those of today's large, herbivorous mammals. One set of Cretaceous tracks shows a herd of 30 individuals traveling in such a way that larger animals on the edges of the group protected juveniles in the center. Also, dinosaur nests preserved with well-developed babies still inside suggest that certain dinosaurs incubated their eggs and cared for their young.

Birds In 1861 the discovery of a truly remarkable fossil caused remarkably little excitement. The fossil, named *Archaeopteryx*, had feathers and a wishbone, but it also had teeth in its beak and a long, bony tail. The animal was presumed to be an early bird that had evolved from

the ancestors of dinosaurs and other reptiles at the beginning of the Mesozoic.

In the early 1970s, however, John Ostrom examined several *Archaeopteryx* fossils and noticed that the animals' teeth, shoulders, forelimbs, pelvis, and tail looked much more like parts of a dinosaur than those of a modern bird. Ostrom concluded that birds had evolved from a group of small, running, carnivorous dinosaurs. More recently Sankar Chaterjee discovered another, older fossil he named *Protoavis*, whose characteristics were also intermediate between those of a small dinosaur and a bird. The question of how these first birds evolved the ability to fly is the subject of considerable debate and experimentation (see Current Controversies, How Does Half a Bird Fly?, p. 344).

Such finds are valuable because they demonstrate a phenomenon known as **mosaic evolution,** in which different sets of traits evolve at different rates as organisms exploit new ecological opportunities. In-between forms such as these clearly document evolutionary links between extinct reptiles and living birds, and they refute the arguments of "creation scientists" who specifically deny the existence of evolutionary intermediates between major groups.

The First Mammals

The promising radiation of the therapsids was cut short during the Triassic as these mammal-like reptiles were replaced by the ruling reptiles. The only survivors of the therapsid line were a group of small, insect-eating, furry, endothermic creatures: the first true mammals (Figure 17.10). They were probably nocturnal, feeding at night

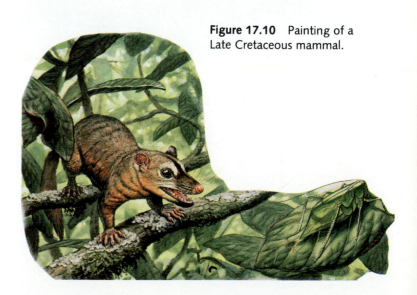

Figure 17.10 Painting of a Late Cretaceous mammal.

Figure 17.11 (top) These wild barley grasses are representative monocots. (bottom) The dogwood is a representative dicot.

when small dinosaurs might have been sluggish. While the dinosaurs ruled the earth, these mammals seem to have cowered in the underbrush. But from these small beginnings came the entire mammalian radiation.

The Rise of Seed Plants

With all the attention lavished on dinosaurs, it is easy to forget that other important organisms were evolving along with them. Among the most important of those were the still-diversifying land plants. For most of the Mesozoic, the dominant plants were gymnosperms; dinosaurs and the first mammals lived in habitats dominated by conifers, cycads, ginkgoes, and a few extinct seed plants. Then during the Cretaceous, the **angiosperms,** the true flowering plants, began to dominate the land. The first definitive angiosperm fossils are found near the beginning of the period, although the very first angiosperms probably evolved well before that time.

Angiosperms radiated explosively during the mid-Cretaceous, their success fueled by their reproductive innovations. By the end of the Cretaceous, the coast and river valleys along North America's inland sea were carpeted with dense vegetation that probably resembled undisturbed areas along the coast of Louisiana today. Redwoods, cypresses, and a variety of broad-leaved trees flourished in the wet, subtropical climate.

Flowering plants: Division Anthophyta—Classes Monocotyledonae and Dicotyledonae Two major groups of angiosperms emerged during the Cretaceous. The *monocotyledons* include the lilies and the grasses and today comprise around 65,000 species (*mono* means "one"; *cotyledon* means "seed leaf"; hence *monocotyledon* means "bearer of a single seed leaf") (Figure 17.11). The *dicotyledons* (*di* means "two"; hence *dicotyledon* means "bearer of two seed leaves") include most common trees and shrubs (Figure 17.11). Today they number about 160,000 species.

In angiosperms, the trends that we have noted in plant reproduction are carried even further. Not only are the male and female gametophytes reduced to a few cells living on the sporophytes, but they complete their functions much more rapidly than gametophytes of most lower land plants. Many angiosperms can produce seeds in a few weeks. This ability to reproduce rapidly may have given early angiosperms a great advantage under the intense grazing pressure, first from herbivorous dinosaurs and later from plant-eating insects and mammals. We will examine angiosperm structures and their functions in detail in Chapters 27–30.

To understand the full significance of higher plant evolution, we must consider the animals with which the evolving plants shared the continents. The first modern

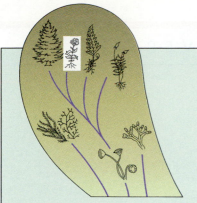

Flowering Plants: Division Anthophyta

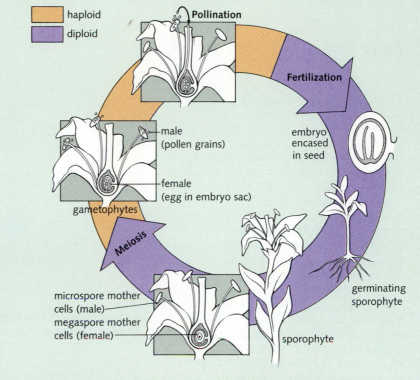

haploid
diploid

Pollination

Fertilization

embryo encased in seed

male (pollen grains)

female (egg in embryo sac)

gametophytes

germinating sporophyte

Meiosis

microspore mother cells (male)

megaspore mother cells (female)

sporophyte

Life cycle: Flowering plants completed plants' adaptation to a terrestrial existence in a world populated by insects, birds, and mammals.

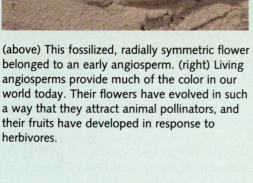

(above) This fossilized, radially symmetric flower belonged to an early angiosperm. (right) Living angiosperms provide much of the color in our world today. Their flowers have evolved in such a way that they attract animal pollinators, and their fruits have developed in response to herbivores.

How Does Half a Bird Fly?

The fossilized remains of both *Archaeopteryx* and *Protoavis* raise two thorny questions. Neither possessed the combination of muscle, bone, and feather development necessary for full flight. But if these creatures couldn't fly, why did they have feathers? Furthermore, as their descendants lurched along the evolutionary road to full-fledged birdhood, how did they get off the ground? What good is half a bird that can almost—but not quite—fly?

The presence of feathers is relatively easy to explain. Long before they were far enough developed to use in flight, feathers could have provided efficient insulation for small, endothermic animals that had to keep warm at night. But two groups of investigators hold conflicting views about the way the first flying vertebrates left the ground.

In 1974 Ostrom suggested that the earliest birds, such as *Protoavis* and *Archaeopteryx,* were swift runners whose feathery forelimbs gradually evolved into net-like insect-catching devices. Then, once they began to resemble wings, the forelimbs would have enabled the animal to spring off the ground and maneuver to catch

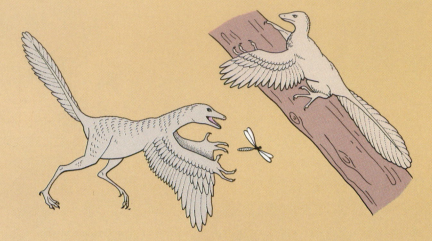

Reconstructions of Archaeopteryx illustrating two views of the evolution of flight in primitive birds.

insects. From that point on, the animals could have begun taking longer and longer leaps and spending more and more time in the air. Ostrom recently declared that the insect-net part of the idea was probably wrong, but he still favors a "ground-up" origin of flight.

Others argue for a "tree-down" evolution of flight. According to this idea, the earliest birds were tree dwellers that first became airborne by

gliding from tree to tree. Sooner or later they learned to extend their range by flapping their wings.

Currently, these theories are being tested both through sophisticated aerodynamic analyses worked out on computers and by physical models based on a combination of fossil evidence and modern calculations. Here, perhaps, is a paleontological question that can be solved by experiment. May the best bird win!

gymnosperms and the ancestors of flowering plants appeared toward the end of the Jurassic, just as reptilian herbivores were reaching their peak. Concurrently, the second adaptive radiation of terrestrial insects produced earwigs, caddis flies, flies, mosquitoes, and the orders Isoptera and Hymenoptera—termites, ants, and bees. Appearance of these animals made them a part of the environment to which the angiosperms adapted, and they became plants' partners in two major classes of coevolutionary relationships.

First, intense herbivory (eating of plants) created powerful selection for the evolution of mechanisms to discourage plant eaters. Recent gymnosperms and many angiosperms evolved ways of "biting back" at their predators; these plants synthesize and concentrate poisonous or distasteful chemicals in their leaves.

Equally important, the reproductive habits of modern flowering plants coevolved with insects and mammals. Whereas gymnosperms are almost exclusively wind pollinated, angiosperms as a group are pollinated by animals that carry pollen from flower to flower. Bees are among the most important pollinators today, but moths, beetles, bats, and even hummingbirds also play that role. The flowers that we admire look and smell as they do because they have evolved as lures for pollinating animals.

And while flowers were evolving with pollinators, seeds evolved with birds and mammals that both ate and dispersed them. Angiosperms wrap their seeds inside **fruits,** structures that are often both brightly colored and tasty and thus are attractive to many birds and mammals. At least some of the hard-shelled seeds inside survive passage through the animals' digestive tracts and are scattered far from the parent plant.

These coevolutionary relationships provided flowering plants with two major advantages over their predecessors. Plants fertilized by animal pollinators, even if they were separated by large distances, could exchange genetic material without having to produce pounds of wind-carried pollen. And plants whose seeds were spread by animals could invest substantial energy in provisioning their embryos, even though all that extra food and protection produced seeds far too heavy to be dispersed by wind like the spores of lower plants.

THE GREAT CRETACEOUS EXTINCTION

The end of the Cretaceous was literally the end of an era; the Mesozoic closed with one of the most dramatic mass extinctions in the history of life on land. Dinosaurs, many archaic birds, and most gymnosperms vanished. Interest in Cretaceous extinction runs high, not only because dinosaurs are fascinating but also because if the dinosaurs had not been wiped out, mammals might never have undergone their adaptive radiation, and *Homo sapiens* might never have evolved!

Many explanations for the great dying are nearly as interesting as dinosaurs themselves. One modern proposal suggests that a giant meteor (or several meteors) struck the earth, raised enormous clouds of dust into the atmosphere, blocked solar radiation, and triggered a sudden, planet-wide cold spell. But although dinosaurs and many gymnosperms did disappear, mammals, insects, and many nonreptilian marine animals survived. And scrutiny of the fossil record has shown that even groups that did become extinct died out very slowly, over millions of years. Rather than suggesting a sudden, global, catastrophic event, the fossil record paints a picture of prolonged changes in the global environment punctuated by a brief period of more rapid change.

Many researchers maintain that the Cretaceous extinction can be explained by long-term changes in climate caused by global geological events. For as continents rose, inland seas retreated, taking with them their tem-

pering effects on weather. Continental climates became both cooler and more variable; the hottest days became hotter and the colder days grew *much* colder. Those changes might have been accelerated by dust clouds cast up either by a long period of intense volcanic activity or by the impact of a large meteorite, but many researchers argue that such extreme mechanisms are not necessary.

Whatever problems the end of the Cretaceous posed for dinosaurs, however, it offered mammals and birds conditions under which their strategy of endothermic temperature regulation finally paid off. Endotherms spend more than 90 percent of the energy they take in just to maintain their body temperatures, and while climates were warm and comfortable, this was an expensive strategy with few selective advantages. When temperatures dropped world-wide, however, endothermy became a great advantage, as we will see in the next chapter.

SUMMARY

During the Devonian, the evolution of jaws, paired appendages, and two sets of limbs sparked the adaptive radiation of higher vertebrates. The evolving bony fishes split into two groups, the ray-finned fishes (most modern fishes) and the fleshy-finned fishes (lungfishes and coelacanths). The ancestors of the fleshy-finned stock also gave rise to the earliest terrestrial tetrapods, four-limbed amphibians that emerged onto land.

This period also saw the emergence of the first seed plants, whose reproductive methods were adapted to drier environments. The seed—a self-contained, drought-resistant package containing a plant embryo and stored nutrients—allowed plants to colonize an ever-widening variety of terrestrial habitats.

The Carboniferous period saw a major expansion in terrestrial animal species. Insects diversified rapidly, feeding on the formerly untouched land plants. On the heels of the insects, carnivorous amphibians diversified extensively, and the first reptiles emerged. The evolution of the amniotic egg among reptiles served much the same purpose that seeds served for plants; it allowed those organisms to reproduce without standing water.

During the Mesozoic era, reptile diversity reached its peak, dinosaurs ruled, and the first birds and mammals appeared. During this time, too, the first flowering plants appeared, and they have been coevolving with insects, birds, and mammals ever since. The era was ended by a mass extinction that wiped out the dinosaurs, setting the stage for the modern world.

After studying this chapter, you should be able to:

- Explain the early diversification of vertebrates, referring to the geological time scale.
- Describe the major classes of fishes, amphibians, and reptiles and explain the emergence of mammals.
- Trace the evolution of higher plants and explain their co-evolution with modern insects and other pollinators.

SELECTED TERMS

class Chondrichthyes (cartilaginous fishes) *p. 326*
class Osteichthyes (bony fishes) *p. 326*
coelacanth *p. 330*
seed *p. 330*
pollen grain *p. 330*
amphibians (class Amphibia) *p. 331*
insects (class Insecta) *p. 334*
reptiles (class Reptilia) *p. 335*
amniotic egg *p. 335*
gymnosperms *p. 337*
conifers *p. 337*
therapsids *p. 337*
endotherms *p. 338*
ectotherms *p. 338*
mosaic evolution *p. 341*
angiosperms *p. 342*

REVIEW

Discussion Questions

1. How did animals and plants of the Mesozoic era influence each other's evolution? Name two classes of coevolutionary relationships that began during this time.

2. Why might cool global temperatures seriously jeopardize the survival of the large dinosaurs?

3. What problems did the first terrestrial vertebrates face when colonizing land? What adaptations evolved as a result of these pressures?

4. Why was the evolution of the seed such an important innovation in terrestrial plants?

Objective Questions   (Answers in Appendix)

5. The Devonian Period is called the Age of
   (a) Fishes.
   (b) Amphibians.
   (c) Reptiles.
   (d) Birds.

6. The supercontinent from which major evolutionary lines originated is called
   (a) Mesozoic.
   (b) Devonian.
   (c) Pangaea.
   (d) Chondrichthyes.

7. Mammals and birds originated during the _____ Period from reptile groups.
   (a) Devonian
   (b) Carboniferous
   (c) Permian
   (d) Triassic

8. Dinosaurs became extinct during the
   (a) Triassic Period.
   (b) Jurassic Period.
   (c) Cretaceous Period.
   (d) Permian Period.

9. Flowering plants, called angiosperms, began to dominate over gymnosperms during which of the following periods?
   (a) Devonian
   (b) Cretaceous
   (c) Permian
   (d) Triassic

CHAPTER 18

Evolution of Mammals and the Ascent of Homo sapiens

Features appear throughout the text serving as valuable study aids designed to enrich your understanding of each chapter. You may wish to refer to the following features as you read Chapter 18.

The Cenozoic Era   p. 348
Primate Family Tree   p. 352
Models of Hominid Evolution   p. 355

THE CENOZOIC ERA

During the Cenozoic era, which spans the last 65 million years, the fragments of Pangaea acquired their modern continental shapes and drifted into their current positions. As the continents separated, ocean currents as we know them today began to flow (Figure 18.1), exerting profound effects on global climate. The first part of the Cenozoic was warm, but its middle period, from 50 to 20 million years ago, turned cooler and drier. The tropical and subtropical forests of earlier times were replaced in North and South America, Africa, and Asia, first by temperate forests and then by vast plains. The period climaxed with several major glaciations, the last of which ended only 10,000 years ago.

These climate changes had important effects on the era's rapidly evolving flora and fauna. The mass extinction that ended the Cretaceous period had wiped out the dinosaurs and decimated the lower plants, offering ecological and evolutionary opportunities to the mammals, birds, and flowering plants that survived.

Angiosperms dominated the terrestrial flora by the end of the Mesozoic. One of the most important families, the grasses, appeared early in the Cenozoic era but did not diversify and spread widely until the major cooling and drying trend began. Then, as the tropical forests gave way to drier plains and savannahs, grasses and other quick-growing, rapidly reproducing plants flourished. Such plants, examples of ecological opportunists described in Chapter 5, quickly colonized the great plains and became major food sources for evolving mammals.

Mammals responded to the elimination of their reptilian competitors by diversifying dramatically toward the end of the Cretaceous period. The earliest mammals spread widely while Pangaea was still intact, though they remained small and ecologically inconsequential. Mammalian evolution thus began on the Mesozoic supercontinent and continued slowly for about 80 million years on the great land masses of Laurasia (the northern continent) and Gondwanaland (the southern continent). By the time the mammalian adaptive radiation finally began in force, at the beginning of the Cenozoic, all the continents had separated except Australia and Antarctica.

MAMMALS: CLASS MAMMALIA

The 4100 species of modern **mammals** share several important characteristics. Mammals possess four-chambered hearts and a diaphragm to aid in breathing. Mammals are also *endothermic*, generating body heat through metabolism and conserving that heat via insulating layers of subcutaneous fat and body hair. With two exceptions, they (or rather we) bear live young and feed newborns on milk secreted by the mother. Mammalian nervous systems are the best developed in the animal kingdom; large brains and excellent muscular coordination have helped this class invade most of the earth's terrestrial habitats and many of its marine environments.

Because the early ancestors of mammals were isolated from one another on diverging continents, three simultaneous adaptive radiations of mammals occurred—one in Laurasia, one in South America, and one in Australia. Separated for nearly 100 million years, each of these radiations produced similar-looking, though unrelated, animals through *convergent evolution* (Figure 18.2). On the ancient continent of Gondwanaland, for example,

The Cenozoic Era

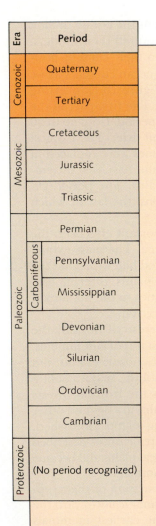

Era		Period
Cenozoic		Quaternary
		Tertiary
Mesozoic		Cretaceous
		Jurassic
		Triassic
Paleozoic		Permian
	Carboniferous	Pennsylvanian
		Mississippian
		Devonian
		Silurian
		Ordovician
		Cambrian
Proterozoic		(No period recognized)

As Cenozoic continents drifted to their modern positions, the supercontinent of Gondwanaland completed the breakup it began in the Jurassic. During the Cenozoic, however, new connections between continents—called land bridges—periodically appeared between Alaska and Siberia (across the Bering straits) and between eastern North America and Europe (through Greenland). Ultimately, Africa collided with Europe, and North and South America were joined through Central America.

the group of mammals known as marsupials radiated to produce a spectacular fauna including many species adapted to the same niches as their familiar North American or African counterparts. The remnants of that radiation survive today, for the most part, only in Australia, where they have been preserved by that continent's isolation from other continents.

Egg-Laying Mammals: Subclass Prototheria

In Australia and parts of New Guinea live two of the most peculiar animals on the face of the earth: the duck-billed platypus (Figure 18.3) and the echidna, or spiny anteater. These strange creatures represent a fascinating mixture of mammalian and reptilian traits. **Prototherians** are clearly endothermic, but they do not regulate their body temperature as effectively as other mammals. Most peculiarly, they lay eggs but suckle their young after the eggs hatch.

Pouched Mammals: Subclass Theria, Infraclass Metatheria

Isolated on the continents formed from Gondwanaland, the first **metatherians, marsupials** such as kangaroos, evolved along a path distinct from, yet parallel to, the main mammalian line (Figure 18.4). These pouched

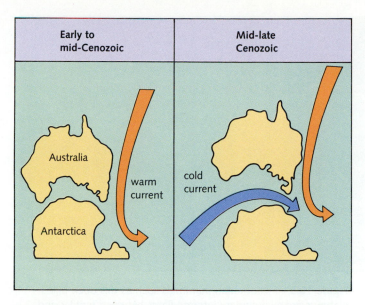

Figure 18.1 Continental drift profoundly affected climate, in part by changing ocean currents. When Australia and Antarctica were connected, a warm current reached Antarctica, moderating its temperatures. As the continents separated, a cold current flowed between them and deflected the warm stream, resulting in the present-day frigid temperatures in Antarctica.

North America

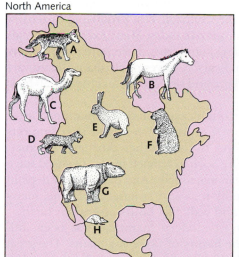

South America

Australia

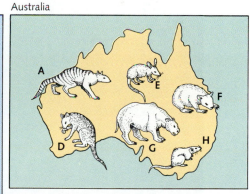

Figure 18.2 Cenozoic mammalian radiations. Mammalian ancestors dispersed across Pangaea were isolated on wandering continents. Convergent evolution among the independently evolving faunas of Laurasia, South America, and Australia produced both remarkable ecological look-alikes and bizarre, one-of-a-kind animals.

Figure 18.3 Prototherian mammals such as this duck-billed platypus are evolutionary intermediates between reptiles and mammals. The platypus does not represent the ancestors of modern forms, but it undoubtedly preserves several characteristics of those ancestors.

Figure 18.4 Marsupials, such as this red kangaroo, look much like familiar placental mammals but differ from our closer relatives in reproductive physiology and metabolic rate. Marsupials are "born" as little more than embryos that complete their development in their mother's pouch.

mammals were once common across all the southern continents that had formed Gondwanaland, but significant numbers of species survive today only in Australia. Nearly all the marsupials that roamed South America during the early Cenozoic became extinct when that continent ultimately collided with North America. Marsupials do not fare well in competition with placental animals, and when North American mammals entered South America via the Central American land bridge, most local mammals—including many fossil species discovered by Darwin—became extinct. The only South American marsupials to make the trip north successfully were the opossums.

Marsupials are better at endothermic temperature regulation than prototherians, but they have somewhat lower metabolic rates than other mammals. Their young are born alive but emerge from the womb at an extraordinarily premature stage of development. Kangaroos, for example, which ultimately grow to the size of an adult human, emerge from the birth canal while still little more than embryos only 2 to 3 centimeters long. Each marsupial infant crawls blindly and instinctively up its mother's belly and down into her pouch. There it attaches to a teat that then swells in its mouth so the baby cannot drop off. The infant hangs on for weeks or months (depending on the species) until it is ready to begin independent life.

Placental Mammals: Infraclass Eutheria

The **eutherian,** or "true," mammals diversified rapidly at the beginning of the Cenozoic era, producing dozens of orders, 17 of which survive today. Many extinct mammals were giants by today's standards; there were once beavers as large as modern bears, relatives of elephants that dwarf living species, and ground sloths 6 meters (nearly 20 feet) tall!

Eutherians are often called **placental** mammals after the highly developed reproductive organ that connects the developing fetus to the mother during pregnancy. Eutherian young are born in a much more advanced stage of development than marsupials. And though they depend on the mother's milk for varying lengths of time, many are able to stand and walk just hours after their birth.

The evolution of modern mammals has been intimately linked to the history of flowering plants. The great herds of bison, buffalo, wildebeest, gazelle, and other mammalian herbivores evolved in direct response to the widespread success of grasses during the Cenozoic. Among the changes seen in the evolution of modern horses (Chapter 13), for example, was the development of molar teeth that grind the tough leaves of grasses into digestible pulp. Our own order, the Primates, owes most of its basic character to millions of years spent evolving in the branches of angiosperm forests. And, as we will see shortly, our immediate ancestors probably emerged from the trees in response to ecological opportunities in the spreading grassy savannahs of Africa.

The biology of eutherian mammals could easily fill several books; mammalian physiology will feature prominently in Chapters 31–43. Here we will concentrate on the evolution of the remarkable species known as *Homo sapiens*, a mammal that has had dramatic effects on earth's flora and fauna since early in its evolution.

PRIMATE AND HUMAN EVOLUTION

Human beings, *Homo sapiens* (*homo* means "man"; *sapiens* means "wise"), are members of the order **Primates.** Our closest living relatives are the chimpanzees and gorillas of Africa. We can trace our line of descent back through a host of intermediate forms to tree-dwelling ancestors that lived in the forests of Africa some 65 million years ago. Because no other statements of evolutionary fact have caused such an uproar from Darwin's time to the present, we will examine the events in human evolution closely.

The earliest primates were small, insect-eating, tree-dwelling mammals that probably looked a great deal like today's tiny tree shrews (Figure 18.5). Even though a number of primates ultimately abandoned the trees for a ground-dwelling existence, many of the characteristics we consider quintessentially human first evolved as adaptations to life in trees.

1. Highly mobile digits (fingers and toes), strong, grasping, flexible hands, and freely movable arms.
2. A complex visual system involving large, mobile, forward-pointing eyes connected to enlarged visual processing centers in the brain. This combination (eyes with overlapping fields of view and sophisticated information processing in the brain) endows primates with excellent *binocular vision* (*bi* means "two"; *ocular* means "eyes") that permits precise perception of depth and distance.
3. Complex central nervous systems based on enlarged brains that can take in and evaluate large amounts of environmental information quickly and efficiently. Information flow to and from the brain is channeled largely through a well-developed spinal cord protected by the backbone.
4. A trend toward bearing only one offspring per pregnancy. This trend is reflected in the reduction in the number of nipples throughout the order. Primitive primates have several pairs, as do most other mammals; humans and other higher primates have only a single pair.

Figure 18.5 The earliest primates, like other primitive mammals, seem to have resembled the diminutive living tree shrew.

5. A complex yet flexible social system in which parents and relatives care for infants and juveniles for long periods of time. As you will see in Chapter 43, higher primates often exhibit a number of strikingly human behavior patterns.

Division of the Primate Line

Soon after the primate line appeared, it split into three groups (Figure 18.6). There were two groups of **prosimians** (*pro* means "before"; *simians* means "monkey"), or "lower primates," including the relatively primitive tarsiers and lemurs, pottos, and galagos. These are the primates whose origins lie closest to the base of the family tree.

The third group, **anthropoids** (*anthro* means "human"; *anthropoid* means "human-like"), or "higher primates," first appeared in the fossil record between 40 and 36 million years ago. Their origins among the lower primates are still uncertain, but anthropoids clearly split into two groups as the continents of Africa and South America drifted away from one another. The *New World monkeys* of Central and South America have *prehensile* (long and grasping) tails that they can use much like a fifth hand for grasping branches. The *Old World monkeys* and *great apes* often lack tails, and when tails do occur, they are not prehensile.

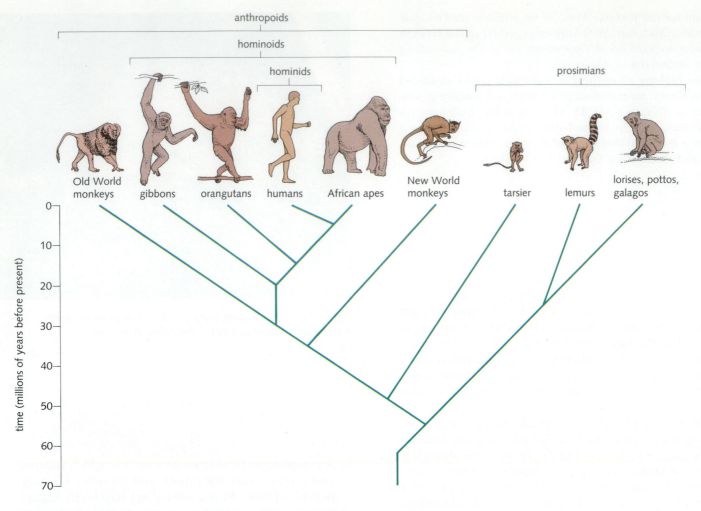

Figure 18.6 The primate family tree shows how the order divided first to produce the prosimians and anthropoids, and then again to produce the New and Old World monkeys and great apes. Paleontologists still argue over which group of lower primates contained the ancestors of the anthropoids.

The Hominoid Lines

Humans, extinct human ancestors (the family Hominidae), and the living apes (the family Pongidae) together make up a group of anthropoids called the **hominoids.** Early hominoids quickly evolved into several species that spread out over most of Africa, Europe, and Asia by about 26 million years ago. Between 10 and 20 million years ago, two early branches off this line led to the gibbons and orangutans of Asia (Figure 18.7). These apes have remained in the trees that sheltered their ancestors. Other ancient hominoids—the ancestors of both the African apes and humans—began to spend more and more time on the ground as the Cenozoic climate changed and Afri-

ca's tropical forests gave way to broad savannahs. Unfortunately, this stage of hominoid evolution occurred in parts of Africa where researchers have not yet found good fossil sites from this time period. For that reason, there is little fossil evidence to date the final splitting of the human line from that of the apes. As we mentioned in Chapter 9, however, biochemists can estimate the date of that split using the "molecular clock" approach. By examining similarities and differences among the genes and blood proteins of chimpanzees, gorillas, and humans, they have calculated that the hominid–pongid (living ape) split occurred somewhere between 4 and 9 million years ago. Many agree that humans and chimps diverged from human ancestors no more than 5 million years ago.

Hominid radiation Precisely what happened next to the hominid line is not clear, but the fossil record reveals that human-like organisms appeared with increasing frequency between 2.5 and 3.5 million years ago. When the very first of these specimens were found more than 20 years ago, they were assembled into a sort of evolutionary "ladder" along which ape-like species led to progressively more human-like forms with larger brains and more erect posture. You may recall that this is much the same view that people once had of horse evolution.

But beginning in the 1970s, a series of exciting discoveries threw this simple scheme into disarray. These finds made it clear that between 1.5 and 2.5 million years ago there was a whole cluster of hominid species, several of which must have lived virtually side by side. Four of the more primitive species from this time period are usually classified in the genus *Australopithecus* (Figure 18.8), whereas the two more advanced forms are placed in our own genus, *Homo*. The new wealth of hominid fossils had turned the simple evolutionary ladder into a bush with several branches, just as similar finds had done for the story of the horse.

Figure 18.7 Though the orangutan may look more "human" than several other apes, its shared ancestor with *Homo sapiens* was farther back than the ancestor we share with African apes.

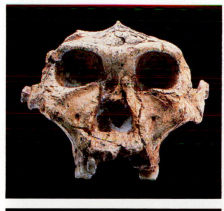

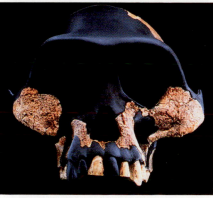

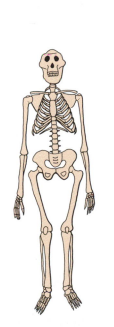

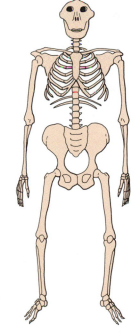

Figure 18.8 All these species of australopithecines lived on the ground, stood between 1 and 1.5 meters tall, and weighed around 20 kilograms. Clockwise from top , the skulls of *Australopithecus robustus*, *A. afarensis*, and *A. boisei*. (right) Representative skeletal structures of *A. afarensis* and *A. boisei*.

Australopithecus afarensis   *Australopithecus boisei*

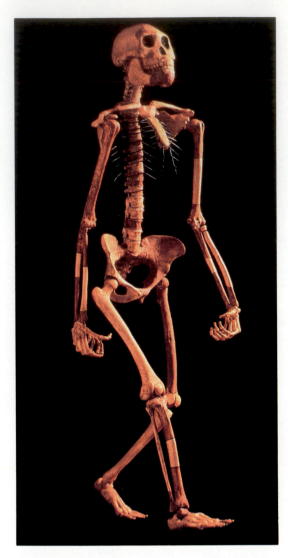

Figure 18.9 This dramatic reconstruction of Lucy, one of the most nearly complete fossil hominids ever discovered, has been assigned to the species *A. afarensis*. This skeleton is between 3 and 3.5 million years old.

The Australopithecines

In 1974 Donald Johansen discovered the skeleton of a female hominid dated at between 3 and 3.5 million years old (Figure 18.9). The find was spectacular. In a field where most specimens are reconstructed from isolated teeth and skull fragments, this skeleton was nearly 40 percent complete and included enough bones to yield a great deal of information about this woman's appearance. Johansen and his co-workers named the skeleton "Lucy" because the night of the discovery they held a party at which they played the Beatles tune "Lucy in the Sky with Diamonds."

Lucy is an excellent example of a transitional organism between apes and humans. From the shape of her pelvis and leg bones, we know that she walked *bipedally*—that is, erect on two legs (Figure 18.10). But her stance was intermediate between the knock-kneed, uneven shuffle of chimpanzees and the more straight-legged and graceful modern human walk.

Lucy was assigned to the species *Australopithecus afarensis*, the species many researchers believe to be the common ancestor of all other hominids. But the relationships between the other three species of *Australopithecus* (*A. boisei*, *A. africanus*, and *A. robustus*) and the early members of the genus *Homo* (*H. habilis* and *H. erectus*) are unclear. The situation was further confused when Alan Walker of the Johns Hopkins University medical school found a fairly intact skull—now referred to as KNM-WT 17000—in a fossil bed dated 2.5 to 2.6 million years of age. The discovery of this skull provoked suggestions that there may have been a sudden adaptive radiation of the hominid line between 3 and 3.5 million years ago, producing

Figure 18.10 Differences and similarities in skull shape and general skeletal structure between humans and chimpanzees.

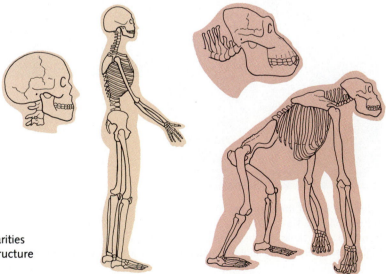

either three or four parallel branches. No one knows what evolutionary or ecological event might have sparked this sudden proliferation, and experts are still arguing over who gave rise to whom (Figure 18.11). (Even the names of these hominids are still in dispute; the fossils we refer to as *Australopithecus robustus*, for example, are assigned by some authorities to the genus *Paranthropus*.) Because several of these species existed at the same time, however, one obviously could not have evolved into another.

Though much is often made of the "human-ness" of **australopithecines,** there is ample evidence that they

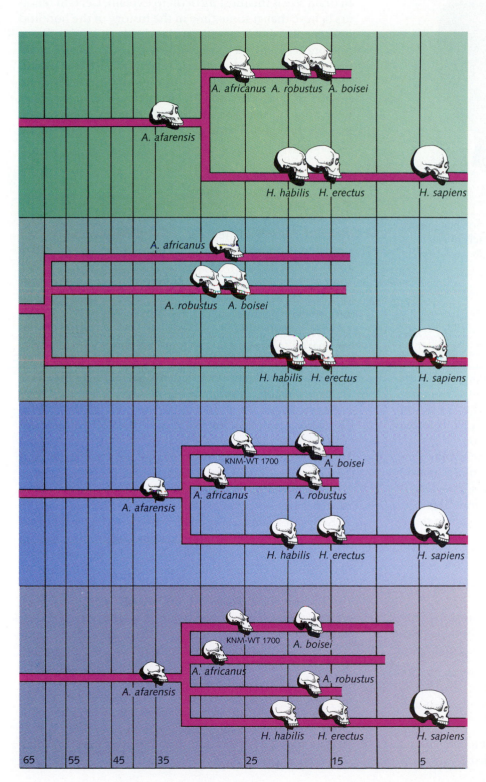

Figure 18.11 Three alternative models of early hominid evolution, all of which can be supported, and none of which can be conclusively disproved, through existing fossil evidence. Adapted from *Discover* magazine, September 1986.

Figure 18.12 *Homo habilis* may have been the first hominid to use stone tools. These flaked stone tools are artifacts of Neanderthal culture found from southern Africa through western Europe.

actually occupied the same sort of ecological niche in African savannahs that baboons do today. Australopithecines had not yet learned to hunt; rather, they were still the prey of the carnivorous mammals with whom they shared the grasslands. South African paleontologist C. K. Brian has found holes in 2-to-3-million-year-old hominid skulls that match precisely with the bite of a leopard's canine teeth. By 1 million years ago, these sorts of injuries vanish, but it is not certain whether hominids had learned to hunt on their own or had simply grown more adept at avoiding other hunting species.

Evolutionary Trends in Hominids

Although their evolutionary interrelationships are uncertain, the early hominids exhibit clearly the evolutionary trends that separated them from the great apes:

- A continued *lack* of specialization in jaw and tooth structure, indicating an omnivorous diet. Studies of wear patterns in fossil teeth show that both meat and vegetables were consumed.

- The continual modification of lower spine, pelvis, and leg bones to permit more efficient bipedal walking.

- A phenomenal increase in brain size, from 280–400 cubic centimeters in chimpanzees to 440–530 cubic centimeters in australopithecines, 775–1100 cubic centimeters in *Homo erectus,* and 1300–1400 cubic centimeters in modern *Homo sapiens.*

Of all these changes, the extraordinarily rapid increase in brain size is the most difficult to explain. Certain scholars believe that at some time in the history of the hominid line, mental development in the slowly enlarging hominid brain reached a crucial milestone and made possible the development of a primitive culture. At that point, mental ability and social organization themselves became influential factors in hominid life, fueling ever-more-rapid evolutionary increases in mental abilities. This is the process biologist Edward O. Wilson calls *gene–culture coevolution.*

Genus *Homo*

Around 1.9 million years ago, the first member of our own genus appeared in the form of *Homo habilis* (*habilis* means "able"). Many (though not all) authorities believe that *H. habilis* was the first hominid to use stone tools (Figure 18.12). Despite this ability and despite a broad range covering much of eastern and southern Africa, this species lasted only about 500,000 years, a short span compared to the life of the hominid line.

Judging by the fossil record, *Homo habilis* was replaced around 1.5 million years ago by *H. erectus.* This more modern-looking, more intelligent species not only used tools but regularly employed fire as well. *H. erectus* was also a great traveler; the species migrated far from Africa into Europe and across India into Asia. Two of the first fossil hominids ever discovered in Asia, "Java man" and "Peking man," belong to this genus.

We know a great deal about this species because of a nearly complete specimen of a young *H. erectus* male found in Kenya by a team led by Alan Walker and Richard Leakey (Figure 18.13). Contrary to previous assumptions that all early hominids were small, this skeleton shows that *H. erectus* could have stood nearly 2 meters tall. At 880 cubic centimeters, his brain was significantly larger than that of australopithecines, but his spinal cord was so much smaller that some researchers feel it would have seriously limited neuromuscular control and coordination.

This information is consistent with what we know about *H. erectus* culture. Members of the species apparently lived mostly by gathering food and scavenging meat from carcasses left by larger carnivores, although they may have done some hunting on their own. They quickly developed the hand axe but never went any further in tool development, even though they survived for more than a million years.

Neanderthals: The first *Homo sapiens* The earliest fossils of our own species indicate that *Homo sapiens* originated in Africa about 500,000 years ago and spread to Europe around 70,000 years ago. Because the first of these European fossils were found in Germany's Neander Valley, they were called Neanderthal man. They are now generally recognized as a subspecies of *Homo sapiens: Homo sapiens neanderthalensis.*

Neanderthals were mentally similar to modern humans, though by modern standards they looked rather brutish. Their skulls were massive, bony ridges jutting above their eyes, and they were heavyset and muscular, though no more so than modern-day weight lifters. Despite their physically primitive appearance, Neanderthals had clearly begun to develop a sophisticated and multidimensional culture. They buried their dead provisioned with weapons, food, and flowers, a behavior usually taken to signal belief in life after death. Neanderthal skeletons showing extensive healing of serious wounds give evidence that these early humans cared for their sick and injured. Whether they did so because they relied on one another for survival or because they were emotionally attached to each other is not clear.

Even so, most authorities believe that Neanderthals, like several australopithecines before them, turned out to be a 300,000-year evolutionary dead end. They were replaced by a wave of modern humans that swept out of Africa into Europe and Asia in the Neanderthals' footsteps.

Cro-Magnons The first skeletally modern humans—*Homo sapiens sapiens,* or **Cro-Magnon** man—appeared in Africa about 100,000 years ago. What appear to have been small bands of Cro-Magnon wanderers headed into Europe and Asia, where they quickly replaced the resident Neanderthals about 35,000 years ago. Some students of the human molecular clock, in fact, believe that the entire human race descended from no more than a handful of Cro-Magnons. These scholars use their data to remind us how closely related all humans are today.

The Cro-Magnons were physically weaker than Neanderthals but had a much more advanced culture. Whether they interbred with the Neanderthals or wiped them out cannot be determined from the fossil record. The Cro-Magnons made a wide variety of tools out of stone, bone, and ivory. The paintings they left behind in caves are remarkable for the use of color, form, and texture (Figure 18.14). These paintings, together with what look like ritual implements, suggest sophisticated religious practices related to hunting. The Cro-Magnons were skilled hunters who survived the great ice ages of the late Cenozoic by feeding on game as large as 3-ton ground sloths, mammoths, mastodons, and woolly rhinoceroses.

Most of this early human history occurred in the Old World. Although Cro-Magnon people spread rapidly throughout Europe and Asia, there were none in the

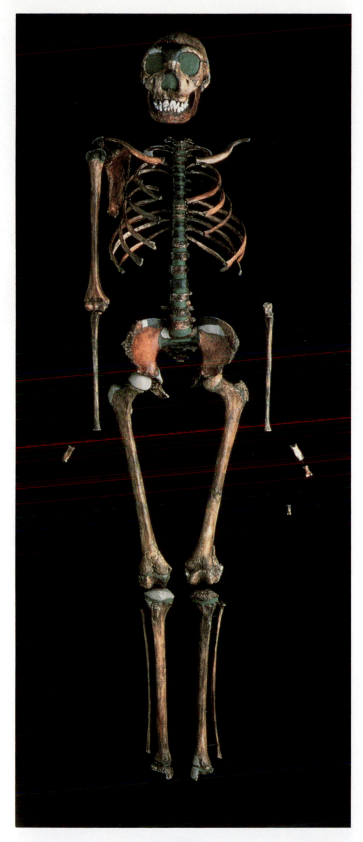

Figure 18.13 This nearly complete specimen of a young *Homo erectus,* often called the Nariokotome boy, shows that his species could have stood nearly 2 meters tall.

Figure 18.14 The Cro-Magnon culture left behind many cave paintings, such as this image of a bison. These paintings are interpreted as artifacts used in religious rituals aimed at appeasing animal spirits and ensuring success in dangerous hunts for large game.

Figure 18.15 Fossils such as this one, showing a Clovis spear point between the ribs of a bison, are taken by some as evidence that early humans were responsible for the disappearance of many large mammals, even in prehistoric times.

Americas until about 12,000 years ago. At that time, an ice-free period gave travelers the opportunity to cross over a land bridge from Siberia through western Canada southward into the American plains. These ancestors of the American Indians were called the Clovis people because their stone tools were first discovered near the town of Clovis, New Mexico.

Hunting with distinctive stone blades and building homes of mammoth skin and bones, the Clovis people quickly spread across the Americas. Within 1000 years they had colonized North America and had expanded from Canada all the way to Tierra del Fuego.

Just at the time the Clovis people were expanding their range, 75 – 80 percent of the large mammals in the New World abruptly disappeared. Many biologists believe that the skillful slaughter of large mammals by the Clovis people caused this wave of extinction about 11,000 years ago, though the "overkill" hypothesis cannot be conclusively proved. Opponents point out that modern "primitive" peoples such as the Australian aborigines are not particularly adept at hunting big game and argue that a change in climate was the main culprit in the extinction.

The theory's supporters point to fossils of bison preserved with spear tips between their ribs (Figure 18.15) and to numerous bones at fossil campsites as proof that the Clovis people hunted large animals extensively at precisely the time of the great extinctions. If early humans did contribute to the demise of other mammals, it was the beginning of humanity's impact on the environment and the start of the largest mass extinction since the end of the Cretaceous.

SUMMARY

The Cenozoic era saw major fluctuations in climate; grassy plains replaced forests and several glaciations took place. Mammals, birds, and flowering plants experienced major adaptive radiations as the continents drifted into their current positions.

Mammals are endothermic organisms that have four-chambered hearts, bear live young, and nourish newborns with mother's milk from mammary glands. Three separate mammalian lines evolved: (1) Prototherians suckle their young but lay eggs like reptiles. (2) Marsupials give birth to tiny live young that continue development attached to a nipple in the mother's pouch. (3) Eutherians develop in the womb with the aid of a placenta; many can walk within hours of birth.

Primates, the order that includes *Homo sapiens*, is characterized by mobile digits, flexible appendages, binocular vision, and trends toward fewer offspring per pregnancy, progressively larger brains, and increasingly complex social systems. The primate line is divided into three groups: two prosimian groups (tarsiers and lemurs) and the anthropoids (higher primates). The Hominoids include humans, extinct human ancestors, and living apes, such as gibbons, orangutans, and chimpanzees. The first human ancestors to approach the cultural complexity of modern humans were the Neanderthals, members of a *Homo sapiens* subspecies. The Cro-Magnons, *Homo sapiens sapiens,* were the first members of our own subspecies and had a culture sufficiently advanced to include religious rituals.

STUDY FOCUS

After studying this chapter, you should be able to:

- Describe the climatic events of the Cenozoic and explain how plants and animals responded to these global changes.
- Discuss the mammalian adaptive radiation that took place during the Cenozoic period.
- Trace early primate history and the evolution of the human being, *Homo sapiens.*

SELECTED TERMS

mammals *p. 348*

prototherians *p. 349*

metatherians *p. 349*

marsupials *p. 349*

eutherian *p. 350*

placental *p. 350*

primates *p. 351*

prosimians *p. 351*

anthropoids *p. 351*

hominoids *p. 352*

australopithecines *p. 355*

Neanderthals *p. 357*

Cro-Magnons *p. 357*

REVIEW

Discussion Questions

1. Why is the skeleton "Lucy" considered a transitional organism between apes and humans?

2. About 11,000 years ago, 75–80 percent of the large mammals in the New World disappeared. What are two possible explanations for this extinction?

3. What evidence leads scientists to believe that Cro-Magnons engaged in religious practices?

4. How is the evolution of modern mammals tied to the history of flowering plants?

5. The duck-billed platypus exhibits the reptilian trait of laying eggs. Is the duck-billed platypus a reptile or a mammal? What characteristics determine how it is classified?

Objective Questions   (Answers in Appendix)

6. Mammals share all of the following characteristics *except:*
   (a) they are exothermic
   (b) they have body hair
   (c) they bear live young
   (d) they nurse their young

7. The Cenozoic era began approximately _____ million years ago.
   (a) 5
   (b) 20
   (c) 30
   (d) 65

8. Primate evolution began
   (a) on the savannah.
   (b) near the seacoasts.
   (c) in the forest.
   (d) on the desert.

9. An early primate might have resembled today's
   (a) kangaroo.
   (b) tree shrew.
   (c) platypus.
   (d) chimpanzee.

10. Early hominids exhibited all of the following evolutionary trends except:
    (a) an increase in brain size
    (b) a modification of lower spine, pelvis, and leg bones for efficient bipedal walking
    (c) a continued specialization of jaw and tooth structure
    (d) highly mobile digits

PART FIVE

Molecules of Life

. . . we cannot categorically deny that perhaps we may be able to grind genes in a mortar and cook them in a beaker after all. Must we geneticists become bacteriologists, physical chemists and physiologists simultaneously with being zoologists and botanists? Let us hope so.

—H. J. Muller (1922)

Hermann J. Muller, one of the great pioneers of genetics in the United States, might well have been kidding when he wrote about the prospects for "grinding" and "cooking" genes in the laboratory. To many of his contemporaries, the possibility that one day geneticists might be able to manipulate genes in a chemical sense must have seemed impossibly optimistic. It was a wild idea.

Despite all of the scientific advances of the early 20th century, there was no apparent way to link the physical world with the living world. To be sure, chemists had discovered that the elements found in living organisms were similar to those found in nonliving matter. A few chemists had even succeeded in producing molecules in the test tube which previously had only been found in living cells—a sure sign that there was no fundamental chemical barrier between life and nonlife. Nonetheless the major goal, an explanation of the material basis of life, seemed almost beyond reach. The most ordinary properties of living things, including the abilities to grow, reproduce, and react to the environment, could not even be approached in chemical terms. This inability led many to conclude that there was something mystical about living things, that they contained a *"vitalism"* that transcended the ordinary laws of nature. The idea of *vitalism* was particularly attractive in studies of inheritance, where it was argued that no mere molecule could carry genetic information from one generation to the next.

It was Muller himself who helped to convince biologists that it was possible to think of the gene as a molecule. Muller irradiated fruit flies with X-rays and discovered that the number of mutations detected in the flies was directly proportional to the X-ray dosage. Because X-rays cause atomic changes in matter, Muller guessed that the gene must have a molecular basis. In short, heredity might operate as a molecular code.

As we know today, Muller was correct. Little by little, biologists have been able to do exactly what he imagined: unravel the physical and chemical basis of inheritance, and with it, develop a new understanding of the material nature of living things. In this part, we will follow the pathway that has led to this new understanding. Beginning with biological chemistry, we will examine the key compounds found in living things, investigate their special chemical properties, and follow the thread of experiment and discovery that leads from the physical world to the living world. It has been, as Muller anticipated, a journey that has changed the way in which scientists view the world, and which may ultimately change the world itself.

◄ Crystals of glycine, an amino acid. Amino acids are the building blocks from which proteins are formed.

CHAPTER 19

The Molecules of Life

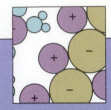

On a cold winter night, the snowy landscape is quiet, even desolate, and sound carries easily through the barren woods. Most trees have lost their leaves, the birds have flown south, and insects have all but disappeared. At times it may seem that life itself has vanished in the cold weather. Despite the cold, we all know that things will change in the spring. The snow will melt, and the songbirds will return. We know that living things are subject to the seasons. But *why* should seasonal changes affect living things so profoundly? Why should the cold of winter cause life to all but disappear from the forest? And why should so many forms of life reappear with the warm winds of spring? The answers to these questions hold the key to one of the most fundamental principles of biology: *Life is chemical.*

The changes we see in the forest during the winter are a reflection of the fact that living organisms, plants, and animals contain water, and when water becomes solid, living organisms must adapt to the change. Living things and nonliving things are made of the very same types of atoms, and living matter is composed of molecules that are subject to the same laws of physics and chemistry that govern nonliving matter. Life really *is* chemical. In this chapter, we will begin to investigate the material organization of life. We will look closely at some of the ways in which the "ordinary" atoms of nonliving things have been put together to produce the "extra-ordinary" phenomenon called *life* (Figure 19.1).

LIFE IS CHEMICAL

One biological discovery after another has been made possible by an ever-more detailed knowledge of chemistry. Naturalists could not appreciate the cycles of carbon, nitrogen, and oxygen that exist on the earth until those elements were discovered, and cell biologists could not understand the nature of cellular membranes until the basic chemistry of *lipids,* one of the principal classes of molecules found in such membranes, was understood. Biologists could not investigate the role of the blood in

Figure 19.1 This medieval engraving portrays the laughable chaos of an alchemist's shop. Alchemists took some of the first steps along the road that led to an understanding of the chemical nature of living things.

transporting oxygen and carbon dioxide until chemists discovered these gases and described their basic characteristics.

First, however, it might be interesting to reconsider a bit of the biology that we have just covered. We have seen how the laws of genetics provide a systematic basis for passing distinct traits from one generation to the next. These laws also form the basis for genetic change that makes evolution possible. One of the ultimate goals of biology is to understand complex processes, including genetics, in chemical terms. By understanding the details of how ordinary matter can be organized into the components of a living organism, we can approach a much deeper understanding of life itself. What we call modern biology has developed because biologists have sought chemical explanations for biological events.

ATOMS AND MOLECULES

The basic unit of chemical structure is the **atom**. Atoms are composed of three different kinds of **subatomic particles: electrons, protons,** and **neutrons.** We represent their positions in the atom by diagrams such as Figure 19.2, which shows the electrons occupying a region of space, known as an **atomic orbital,** around the protons.

Each of these subatomic particles has its own set of physical characteristics. Electrons have very little mass and are negatively charged. Protons and neutrons are much heavier (1836 and 1839 times the mass of an electron, respectively). The proton is positively charged and the neutron is neutral. The electrostatic interactions between these oppositely charged particles help to hold the components of an atom together. The simplest atom is common hydrogen, which contains a single electron and a single proton. The single rapidly moving electron surrounds a lone proton, and attractive forces between these oppositely charged particles keep the electron moving in its specific orbital (Figure 19.2a). The single proton forms the **nucleus** of the atom, a core at the center of the electron orbital.

More complicated atoms—helium for example—have more than one electron arranged in orbitals and have several protons and neutrons clustered together to form their atomic nuclei (Figure 19.2b,c). Atoms are electrically neutral. The number of positively charged protons in the nucleus is equal to the number of negatively charged electrons. The protons and neutrons found in the nucleus are more massive than the electrons that surround it. In fact, more than 99.9 percent of the mass of an atom is found within its nucleus.

We classify atoms according to their **atomic number,** which is defined as the number of protons found in the

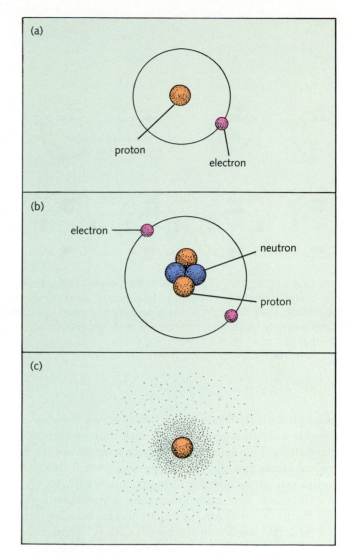

Figure 19.2 (a) The basic unit of an element is the atom. Atoms are composed of three elementary particles: the neutron, the proton, and the electron. This drawing portrays hydrogen, the simplest atom, which consists of a single proton and a single electron. (b) Larger atoms, such as the helium atom shown here, have a complex nucleus composed of protons and neutrons. (c) Any attempt to draw an atom is necessarily inexact, because the positions of subatomic particles at any one time cannot be precisely fixed. Because of this fact, atoms may be represented graphically in different ways, including a probability distribution for electrons.

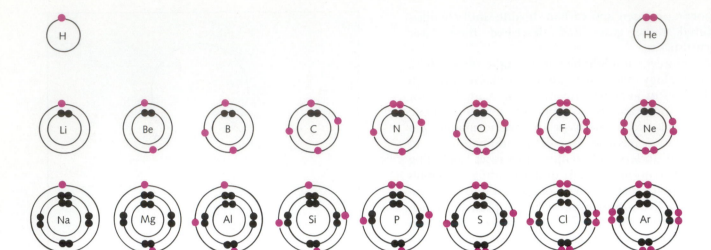

Figure 19.3 In the first 18 elements of the periodic table, atoms in the same column share similar chemical properties by virtue of having the same number of electrons in their outer-most orbitals. Note that helium, neon, and argon possess completed electron outer shells, and this property makes them chemically inert, generally unable to form covalent bonds with other atoms.

atomic nucleus. By this standard, the atomic number of hydrogen is 1, that of helium 2, and that of carbon 6 (Figure 19.3). There are 92 different naturally occurring **chemical elements,** and each of these chemical elements is composed of a different type of atom with its own atomic number. The most important elements in living things (hydrogen, oxygen, nitrogen, and carbon) are very common at the surface of the earth; each is found in the gases of the atmosphere. The heaviest naturally occurring element is uranium, which has an atomic number of 92. Heavier elements with larger atomic numbers can be created experimentally by bombarding other elements with high-energy radiation.

Only about a third of the naturally occurring chemical elements are found in living organisms. Table 19.1 lists the 26 elements that are most commonly found in living things and also indicates their relative abundance in the earth's crust.

Chemical and Physical Properties

The **physical properties** of a material are the characteristics we can see or measure, and they include the color, density, physical state (solid, liquid, or gas) and melting and boiling points of a substance. Each of the 92 chemical elements has different physical properties. Some, such as gold, zinc, and sodium, are shiny solids. Some are gases, such as hydrogen, chlorine, and oxygen, and some,

including bromine and mercury, are liquids at room temperature.

Although the physical properties of the different elements can vary considerably, their **chemical properties**—the ways in which atoms combine with other atoms to form chemical compounds—show certain important similarities among groups of elements. Fluorine and chlorine, both highly reactive gases, are examples of elements with such similarities. These similarities were noticed in the last century by Mendeleev, a Russian chemist. Mendeleev thought that it would be useful to group the elements in a table according to their chemical properties and atomic weight. Today the product of his idea is a chart known as the **periodic table.** Mendeleev's efforts were successful because the chemical properties of an element are determined mainly by the number of electrons in its outer atomic orbital. Two different elements have similar chemical properties if they have the same number of electrons in their outermost orbitals.

Figure 19.3 shows how the periodic table predicts chemical similarities. As the number of electrons in an atom increases, orbitals are occupied by electrons in a regular pattern: The orbital closest to the nucleus can hold only two electrons, the second shell, which as four orbitals, can hold as many as eight, and the third shell also holds eight. The reasons for similarities between fluorine (F) and chlorine (Cl) become clear when we examine the arrangements of electrons in their atoms: Each needs but a single electron to complete its outermost orbital.

Table 19.1 *Elements Important to Life*

Symbol	Element	Atomic Number	Percentage of Earth's Crust by Weight	Percentage of Human Body by Weight
H	Hydrogen	1	0.14	9.5
B	Boron	5	Trace	Trace
C	Carbon	6	0.03	18.5
N	Nitrogen	7	Trace	3.3
O	Oxygen	8	46.6	65.0
F	Fluorine	9	0.07	Trace
Na	Sodium	11	2.8	0.2
Mg	Magnesium	12	2.1	0.1
Si	Silicon	14	27.7	Trace
P	Phosphorus	15	0.07	1.0
S	Sulfur	16	0.03	0.3
Cl	Chlorine	17	0.01	0.2
K	Potassium	19	2.6	0.4
Ca	Calcium	20	3.6	1.5
V	Vanadium	23	0.01	Trace
Cr	Chromium	24	0.01	Trace
Mn	Manganese	25	0.1	Trace
Fe	Iron	26	5.0	Trace
Co	Cobalt	27	Trace	Trace
Ni	Nickel	28	Trace	Trace
Cu	Copper	29	0.01	Trace
Zn	Zinc	30	Trace	Trace
Se	Selenium	34	Trace	Trace
Mo	Molybdenum	42	Trace	Trace
Sn	Tin	50	Trace	Trace
I	Iodine	53	Trace	Trace

Atomic Numbers and Atomic Weights

We have defined the atomic number of an element as the number of protons found in its nucleus. We can also speak of **atomic weight** (more properly, **atomic mass**) as the sum of the weights of the elementary particles in a single atom. The unit of measurement that is used for this purpose is the *atomic mass unit* (amu), which is defined as $1/12$ the atomic mass of the carbon atom. Therefore, ordinary carbon has an atomic mass of 12.0, and hydrogen's atomic mass is 1.0078. For our purposes the proton and neutron have masses that are nearly equal to 1.0, but the electron is so small that its mass ($1/1839$ the mass of the neutron) is often neglected.

Atomic weights are important because they reflect the relationship between numbers of atoms (atoms are difficult to count!) and mass (which we can measure easily with a scale or balance). If we know that the atomic mass of oxygen is 16, we can place 2 grams of hydrogen and 16 grams of oxygen in a reaction vessel and be certain that we have exactly 2 atoms of hydrogen for every atom of oxygen.

Isotopes

The identity of a particular atom is determined by the number of protons in its atomic nucleus. If the nucleus contains 6 protons, the atom is carbon; 7 protons, nitrogen; 8 protons, oxygen. The number of neutrons, however, can vary. For example, the most common form of carbon has a nucleus of 6 protons and 6 neutrons: a total of 12 elementary particles. Therefore we refer to it as carbon-12 (in atomic shorthand: ^{12}C). However, there are other forms of carbon as well. One of the most important, ^{14}C, has 8 neutrons in its nucleus. Carbon-14 is chemically identical to carbon-12, despite the fact that it has a different atomic nucleus, because it has the same

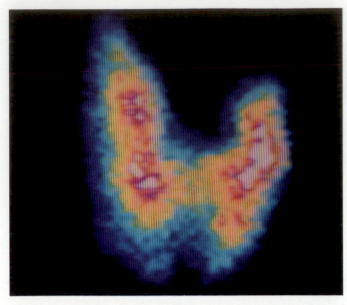

Figure 19.4 The presence of ^{125}I, a radioactive isotope of iodine, reveals the outline of the thyroid gland in this scan. The thyroid concentrates iodine atoms for use in synthesizing the thyroid hormone, thyroxine.

number of electrons. We refer to ^{12}C and ^{14}C as **isotopes** of the element carbon.

Atomic isotopes are nearly identical in chemical behavior because they have the same configuration of outer electron shells. It is the arrangement of electrons that determines chemical properties. However, isotopes differ from each other in one and sometimes two respects:

■ Because their nuclei have different numbers of neutrons, they have different atomic weights. For example, ^{12}C has an atomic weight of 12 (actually 12.0113), whereas ^{14}C has an atomic weight of 14 (actually 14.1023). Therefore, molecules that contain ^{14}C are a bit heavier than ones that contain only ^{12}C.

■ Some chemical isotopes have unstable nuclei, which means that they may break down, or decay, releasing atomic *radiation*. Such isotopes are said to be *radioactive*. For example, ^{14}C emits **beta particles** (fast-moving electrons), a form of radiation, and changes to ^{14}N after the beta particle is released.

These two differences mean that isotopes can be used as *tracer molecules*—they can be used to "label" a molecule and identify compounds that contain the isotope. This enables investigators to trace the movements of specific molecules through cells and tissues (Figure 19.4).

We describe the rate at which a radioactive atom decays by stating its **half-life.** The half-life for ^{14}C is 5700 years, which means that if we were to take 1000 atoms of ^{14}C and wait for 5700 years, roughly *half* of these atoms would have released one beta particle each and changed into ^{14}N by that time. In another 5700 years, half of the 500 remaining particles (or about 250) would do the same, and at the end of another 5700 years only about 125 (⅛) of the original ^{14}C atoms would be left.

Not all chemical isotopes are radioactive. An important isotope of nitrogen is ^{15}N, which has a single neutron more than ^{14}N, the more common form of nitrogen. Neither ^{15}N nor ^{14}N is radioactive, and the only difference between them is in atomic weight.

Isotopes are particularly useful in biology. In order to follow a certain molecule through a chemical reaction, all that is necessary is to find an appropriate isotope of one of its chemical elements. The labeled molecule can be recognized either by virtue of its mass or by its emission of radioactivity.

CHEMICAL COMPOUNDS AND CHEMICAL BONDS

Atoms may be combined in fixed ratios to form **compounds.** Water, for example, is a chemical compound composed of oxygen and hydrogen atoms. Because there are two atoms of hydrogen for each oxygen, we write the chemical formula of water as H_2O. The chemical and physical properties of a compound are different from those of the atoms from which they are formed. Oxygen and hydrogen are highly reactive gases at room temperature, whereas water is a stable liquid.

Just as the atom is the basic unit of an element, the **molecule** is the basic unit of a compound. A single molecule of sodium chloride is composed of one atom of chlorine and one of sodium. We can even consider each molecule to have a *molecular weight* in the same manner that single atoms have atomic weight. How does this work?

We can estimate the molecular weight of a compound simply by adding up the atomic weights of its individual atoms. In the case of NaCl, the molecular weight is found as follows:

$$
\begin{array}{ll}
22.98 & \text{(the atomic weight of Na)} \\
+\,35.45 & \text{(the atomic weight of Cl)} \\
\hline
58.43 & \text{(molecular weight of NaCl)}
\end{array}
$$

Molecules that contain 10, 15, or 20 atoms have much larger molecular weights, and the very large **macromolecules** found in living cells may have weights of a million or more.

The atoms in a compound are linked together by **chemical bonds.** Although there are several types of chemical bonds, we will begin our study of biological chemistry by looking at two of the most important ones: **ionic bonds** and **covalent bonds.**

Ionic Bonds

The atoms we have looked at so far have all been chemically *neutral.* They contain equal numbers of electrons and protons, so that their positive and negative charges balance. However, that is not always the case. Most atoms are able to gain or lose electrons in their outer shell. When this happens the balance of electrons and protons is upset, and the atoms become electrically charged. An atom or molecule that becomes charged as the result of gaining or losing an electron is called an **ion.** Under the right conditions, an electron may transfer from one atom to another. The two atoms involved in such a transfer may then form an **ionic bond.**

Sodium (Na), a shiny metal, and chlorine (Cl), a greenish gas, are two very different and very dangerous materials. Sodium reacts violently when it comes in contact with water, and chlorine is a deadly poison that has been used as a weapon of war. Sodium has 1 electron in its outermost shell; chlorine has 7. These two atoms can undergo a **chemical reaction** in which sodium loses its outermost electron and chlorine gains an electron. This transfer allows each atom to fill its outermost orbit with the maximum complement of 8 electrons (Figure 19.5). Chemically, this is the most stable form in which the electrons can be arranged, and reactions in which an outermost electron shell can be filled are very common. Because the electron has a negative (−) charge, losing an electron makes sodium positively charged (we symbolize it Na^+) and chlorine negatively charged (Cl^-). Both Na^+ and Cl^- are ions; Na^+ is a **cation,** or positively charged ion; and Cl^- is an **anion,** or negatively charged ion.

Electrostatic forces exist between charged chemical groups: Opposite charges attract and like charges repel. Because the sodium and chlorine ions are oppositely charged, they are attracted to each other by electrostatic forces, and an ionic bond exists between them. Earlier we said that the chemical properties of an atom depend on the number of electrons in its outermost shell. That

---

THEORY IN ACTION

Making Chemical Solutions: The Mole

Because the molecular weights of compounds are different, it is often difficult to calculate just how much of a chemical should be added to a solution. For example, suppose we wanted to have equal numbers of sodium chloride ions and glucose molecules in a solution. If we added 100 grams of salt, how much glucose would we have to add? The molecular weight of salt is 58.5, and that of glucose is 180. So we would have to add 100 grams × 58.5/180, or . . . well, you get the picture: It would be complicated. But there's a simpler way to do it.

Scientists generally make up chemical solutions in terms of moles. A *mole* is a *gram molecular weight* (the molecular weight in grams) of a compound. One mole of salt would be 58.5 grams, one mole of sucrose, 180 grams. If we were to add one mole of each to a container, we could be certain (within the limits of our ability to measure) that there were equal numbers of each compound in the mixture. When describing solutions, we take this a step further. We might, for example, make up a one *molar* solution of NaCl (designated 1.0 M NaCl), which would contain exactly 1 mole of NaCl per liter of solution. A 0.1 molar solution would have 0.1 moles/liter, and so forth. By mixing solu-

tions with their contents described in moles, a worker can be sure of the relative amount of each compound in a particular solution, even if the numbers of individual molecules can't be counted.

A flask containing 1.0 M NaCl, 0.5 M KCl, and 0.1 M $MgCl_2$ has exactly two sodium (Na) atoms for every potassium (K) and five potassiums for every magnesium (Mg).

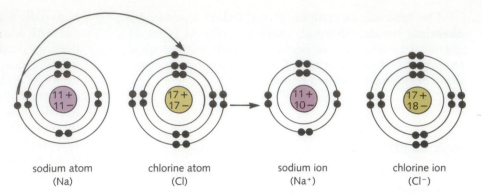

sodium atom
(Na)

chlorine atom
(Cl)

sodium ion
(Na⁺)

chlorine ion
(Cl⁻)

principle applies to ions as well. Therefore, the chemical properties of the two ions are completely different from what they were before they gained or lost electrons. The name for the compound NaCl is sodium chloride. Despite the noxious nature of each of the elements that go into it, sodium chloride (ordinary table salt) is an essential nutrient in the human diet.

Covalent Bonds

Although many compounds are formed by ionic bonds, another kind of chemical bond is equally important. Hydrogen has a single electron in its outermost orbit and carbon has 4. When hydrogen and carbon combine to form methane (CH_4), they actually *share* electrons so that carbon seems to have 8 electrons in its orbital and each

of the four hydrogen atoms seems to have 2. This sharing of electrons produces four *combined* orbitals for the compound different from that formed by either atom alone. The atomic bonds that exist between carbon and hydrogen are said to be **covalent bonds,** because electrons are shared in the combined orbits of the atoms (see Figure 19.6 below).

Each covalent bond results from the sharing of a pair of electrons. When a single pair of electrons is shared, the bond is represented as a single line drawn between the symbols for the atoms. However, it is also possible for 2 or even 3 pairs of electrons to be shared. In carbon dioxide, for example, each oxygen atom shares 4 electrons with the carbon atom, and we represent the compound with a symbol that shows the existence of a **double covalent bond** between the atoms. Hydrogen cyanide contains a **triple covalent bond** (Figure 19.7).

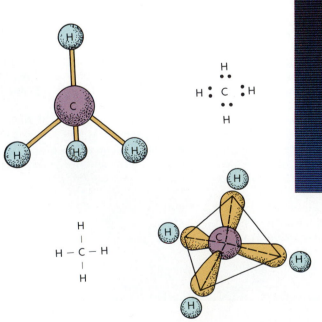

Figure 19.6 Formation of a covalent bond can be represented in many different ways. Here the four covalent bonds in methane (CH_4) are shown in a ball-and-stick model (upper left), as a structure showing paired electrons (upper right), as a molecular diagram drawn in a flat plane (lower left), or as overlapping electron orbitals (lower right). (above) A computer representation of the probability distribution of the two electrons shared in the single covalent bond of a hydrogen molecule (H_2).

The Differences Between Covalent and Ionic Bonds

Although both types of chemical bonds are important, there are clear differences between them (Figure 19.7). Covalent bonds are formed between two distinct atoms and involve the actual *sharing* of electrons. Ionic bonds involve the *exchange* of electrons to form oppositely charged ions. In a salt crystal, for example, sodium and chloride ions are in a regular lattice that has the appearance of a schoolyard jungle gym (Figure 19.8). Each sodium is surrounded by six chlorines. Each chlorine is surrounded by six sodiums. Each atom in the lattice is equally attracted to all six surrounding atoms, but the sodium and chlorine atoms are not distinctly bonded to any of their six neighbors. That is not true for atoms held together by covalent bonds. Even in a crystal of water (ice!) each hydrogen is covalently bonded to *one* and only one oxygen atom.

WATER: SOMETHING SPECIAL

From outer space, the single most striking feature of the planet Earth is its color, the deep blue of its oceans and the striking white puffs of clouds (Figure 19.9). Clouds and oceans, of course, are composed of water. Considering how water dominates the surface of our planet, it is not surprising that the single most common molecule in living organisms is **water.** Living cells are more than 50 percent water by weight, and some are as much as 98 percent water. Without water, the kind of life that we know on earth would not be possible. Water *seems* to be a simple molecule. Two atoms of hydrogen are held by single covalent bonds to a single atom of oxygen. But that simplicity is more apparent than real.

Chemists say that the water molecule is **polar,** which means that different parts of the molecule have different electrical charges. Although the entire water molecule is electrically neutral, the electrons in the shared orbits are not evenly distributed. Instead, the electrons are strongly attracted to the large, positively charged oxygen nucleus, as shown in Figure 19.9. Because of this, the electron cloud is much denser in the region of the oxygen nucleus than it is near the two hydrogen nuclei. When we consider this (along with the fact that the two hydrogen atoms are arranged to one side of the oxygen), it is obvious that the two ends of a water molecule have slightly different charges. In other words, water molecules have two positive poles and one negative pole.

The polarity of water has a number of important consequences. The polar ends of the molecule have a strong attraction for each other, and they also have a strong attraction for other charged molecules. The polar water

$$\ddot{\text{O}} = \text{C} = \ddot{\text{O}}$$

carbon dioxide

$$\text{H} - \text{C} \equiv \text{N}$$

hydrogen cyanide

Figure 19.7 Two double covalent bonds are present in carbon dioxide (CO_2). Hydrogen cyanide (HCN) contains a triple bond linking carbon and nitrogen.

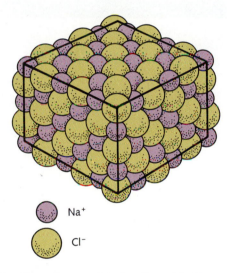

Na⁺

Cl⁻

Figure 19.8 Atoms in a crystal of sodium chloride are arranged in a cubic lattice.

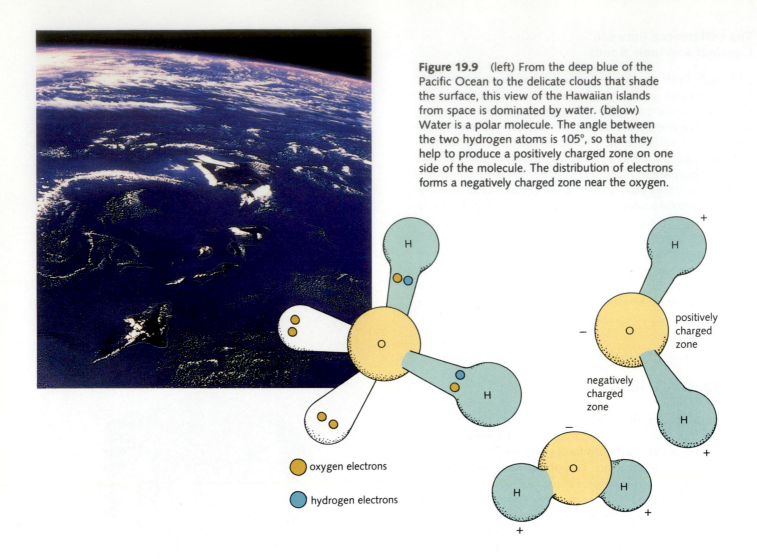

Figure 19.9 (left) From the deep blue of the Pacific Ocean to the delicate clouds that shade the surface, this view of the Hawaiian islands from space is dominated by water. (below) Water is a polar molecule. The angle between the two hydrogen atoms is 105°, so that they help to produce a positively charged zone on one side of the molecule. The distribution of electrons forms a negatively charged zone near the oxygen.

oxygen electrons

hydrogen electrons

positively charged zone

negatively charged zone

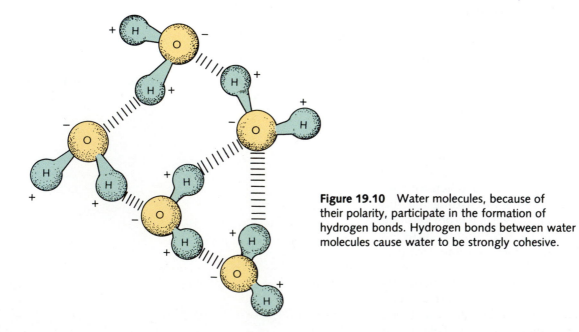

Figure 19.10 Water molecules, because of their polarity, participate in the formation of hydrogen bonds. Hydrogen bonds between water molecules cause water to be strongly cohesive.

molecule forms weak bonds between the hydrogen and oxygen atoms of different molecules, as shown in Figure 19.10. This interaction, which is called a **hydrogen bond,** is not nearly as strong as a covalent bond, but it does cause water molecules to be attracted to each other. Other molecules can form hydrogen bonds, and in the next chapter we will encounter several examples of the importance of these bonds in biological systems.

When other molecules are placed in water, their ability to dissolve is determined by how strongly they are attracted to water molecules. The most **soluble** molecules are those that can interact strongly with polar water molecules. Ions dissolve very well, so that NaCl, which actually dissociates in solution to Na$^+$ and Cl$^-$, dissolves very quickly in water (Figure 19.11). Larger compounds (such as sugars), which are uncharged, can still dissolve well if they too are polar because of uneven electron distributions. In fact, the ability to dissolve an enormous range of substances is precisely what makes the water molecule so special. No other solvent can keep such large amounts of so many chemicals in solution, a property that is at the heart of life itself. Water is sometimes called the "universal solvent."

Hydrophilic ("water-loving") or *polar* molecules are those that interact strongly with water. Of course, many substances do not dissolve in water, and that is important for the existence of life as well. **Hydrophobic** ("water-fearing") or *nonpolar* molecules do not interact with water and are *insoluble* in it, like salad oil in an oil-and-vinegar dressing (Figure 19.11).

Hexane, a molecule commonly found in oil and gasoline, has a molecular structure in which the electrons are evenly distributed; therefore, the hexane molecule has no charged regions to form ionic or hydrogen-bonding interactions with water. Hexane molecules do not dissolve well in water and tend to cluster together, excluded from the bonds that link water molecules to each other. The interactions between hydrophilic and hydrophobic molecules are important in the formation of a number of biological structures, including cellular membranes.

Acids, Bases, and Buffers

Besides its polarity, water has another chemical property that is important in living things. The water molecule can dissociate to form two different ions. The process goes like this:

$$HOH \rightarrow H^+ + OH^-$$

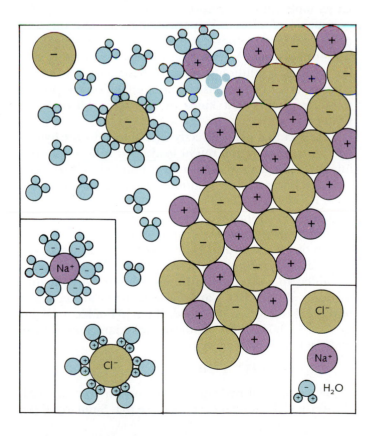

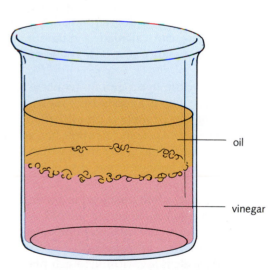

Figure 19.11 (left) Many substances, such as NaCl, interact with the polar water molecules and dissolve very quickly. These materials that interact with the water molecule are said to be hydrophilic ("water-loving"). (right) Hydrophobic ("water-fearing") molecules are those that cannot interact with the polar water molecule.

oil

vinegar

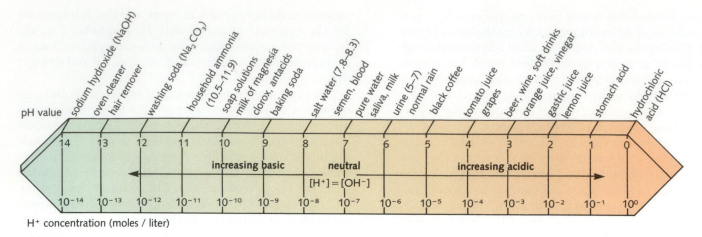

Figure 19.12 The pH scale. The scale measures the concentration of hydrogen ions in a solution. Strong acids produce low pHs, while strong bases produce high pHs.

The **H⁺** ion is called the **hydrogen ion;** it is merely a *proton* and is often referred to as such. The **OH⁻** group is called the **hydroxide ion.**

The chemical properties of these ions require us to pay special attention to their concentrations. A solution that contains excess hydrogen ions is said to be **acidic;** a solution in which hydroxide ions predominate is **basic.** Lemon juice, vinegar, and battery acid are examples of solutions that contain an excess of hydrogen ions and are acidic. Ammonia, hair remover, and oven cleaner are examples of bases. To determine the degree to which a solution is acidic or basic, we can calculate a quantity called **pH,** which tells us at a glance just how acidic or basic a solution is.

The pH is determined by the hydrogen ion concentration. If the hydrogen ion concentration is 10^{-9} molar, the pH is 9; if it is 10^{-2} molar (10 million times higher than 10^{-9} molar), the pH is 2. An *acidic* solution has a pH lower than 7; a *basic* solution has a pH higher than 7. A solution of pH 7 is *neutral* because it has equal concentrations of H⁺ and OH⁻ ions.

An **acid** is a compound that, when dissolved in water, releases hydrogen ions into solution, lowering the pH. A **base** is a compound that raises the pH by removing hydrogen ions from solution.

As you can see, the pH scale is logarithmic: Each pH unit represents an increase or decrease in hydrogen ion concentration by a power of 10 (Figure 19.12). A solution at pH 5 has 100 times as many hydrogen ions per liter as a solution at pH 7, and therefore it is 100 times as acidic.

Clearly, the dissociation of water is of enormous significance. A solution is called acidic simply on the basis of H⁺ ions, ignoring other ions such as K⁺ or Na⁺. We are free to do this because the hydrogen ion, as we've emphasized, is nothing more than a single proton, lacking any electrons. In water, this powerful positive charge is surrounded by an envelope of water molecules. The hydrogen ion is strongly attracted to negatively charged ions and is able to attract electrons from a great many larger molecules. Therefore, the concentration of hydrogen ions in a solution is more important than the concentration of any other ion, and there is good reason for inventing a special terminology to keep track of it. Knowing whether a solution is acidic or basic tells us what the concentration of protons is, and it enables us to predict to some extent what kinds of reactions will occur in solution.

The fact that pH is directly determined by the hydrogen ion concentration might lead us to believe that cellular pH fluctuates whenever large numbers of hydrogen ions are released into solution. To some extent that is true. However, a number of compounds found in living systems tend to stabilize pH at certain levels. Such pH-stabilizing compounds are known as **buffers.** Buffers absorb or release hydrogen ions as the pH of a solution varies and thereby stabilize the pH. Acids or bases can be added to a strongly buffered solution without dramatically altering pH. Carbonic acid and bicarbonate ion buffer pH changes in the blood. If the pH of blood begins to rise, carbonic acid tends to dissociate and release hydrogen and *bicarbonate* ions:

$$H_2CO_3 \rightarrow H^+ + HCO_3^-$$

The release of hydrogen ions tends to lower the pH, "buffering" the tendency to increase pH. Similarly, if the pH of a solution begins to fall, the additional hydrogen ions can be "soaked up" by the bicarbonate ions:

$$H^+ + HCO_3^- \rightarrow H_2CO_3$$

Buffers such as the carbonic acid/bicarbonate ion pair make it possible for living systems to resist rapid changes in pH and to maintain a constant internal environment.

Other Properties of Water

Water has a number of other special properties that affect living systems. It is relatively *transparent*, making it possible for light to penetrate through some living tissues as well as deep into lakes, streams, and oceans. Water expands slightly as it freezes, making solid water (ice) somewhat less dense than liquid water at most temperatures. This causes ice to form at the surfaces of ponds and lakes, insulating many forms of life below the ice against the harsh extremes of cold weather. Water is strongly **cohesive,** a property brought about by hydrogen bonding between water molecules. Water's cohesiveness makes it possible for water molecules to be drawn into small openings or tubes and then to rise against the force of gravity, a phenomenon known as **capillary action** (Figure 19.13).

Hydrogen bonds produce such strong attractions between water molecules at an air–water interface that water displays a measurable **surface tension.** This surface attraction is so great that small animals are able to walk directly on the surface of water, and many (such as water striders) spend their lives skimming the boundary between air and water (Figure 19.13). Finally, water has a high *heat capacity*. A substantial amount of heat energy is required to change the temperature of a body of water by just a few degrees. Even more energy is required for water to move through a *phase transition*: from solid to liquid or from liquid to gas. These properties enable water to act as a "heat buffer," producing stability in the face of rapid changes in air temperature.

THE MOLECULES OF LIVING THINGS

"Organic" Chemistry

As chemistry developed in the nineteenth century, scientists were able to determine the general patterns in which the chemical elements combine to form compounds. They developed a science of chemistry that led directly to our modern understanding of the nature of matter. Not surprisingly, some chemists began to explore the applicability of this knowledge to living things as well as to the materials under study in their labs. At first, such efforts were unsuccessful and led to confusion. Some chemists suggested that living organisms obeyed a different kind of chemistry from nonliving material. In fact, the term *organic chemistry* was originally coined to distinguish the chemistry of a living organism from the *inorganic chemistry* that chemists had been investigating in the lab.

To scientists today, however, **organic chemistry** is the chemistry of carbon compounds. This change of emphasis reflects the fact that carbon compounds dominate the chemistry of living things. Nineteenth-century scientists experienced difficulties in studying the molecules found

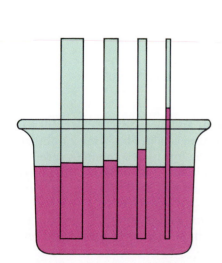

Figure 19.13 Capillary action. The cohesive properties of the water molecule enable water to be drawn upwards into small openings, provided that the walls of the capillary tube also interact with water. The surface tension of water enables small creatures, such as this water strider (right), to live safely on its surface.

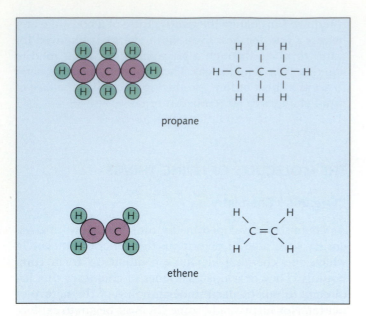

Figure 19.14 Carbon is capable of forming four covalent bonds, and one or more of these bonds may be to other carbon atoms. Carbon's ability to bond to itself in chain-like fashion enables it to play a key role in the formation of the large molecules found in living organisms.

in living organisms because they did not appreciate the extreme complexity of carbon-based molecules.

What's so special about carbon? To begin with, carbon's atomic number is 6, which means that it has four electrons in its outermost shell. Because that second shell can contain a maximum of eight electrons, carbon can form a total of four covalent bonds with other atoms to complete that shell. In addition, carbon readily forms bonds to itself, so it can form long chains that may contain single, double, and even triple carbon-to-carbon covalent bonds. Carbon also forms covalent bonds to the other elements found in living material, including hydrogen, oxygen, and nitrogen (Figure 19.14).

Chemical Groups

Carbon is a versatile element; it can form covalent bonds in many different ways. The most direct way to analyze the kinds of molecules that carbon forms is to look at the **chemical groups** in which carbon can participate. *Chemical groups* are individual clusters of atoms bonded in a certain pattern, and any particular chemical group tends to behave the same way in different molecules.

The **hydroxyl** group is one of the most common. Alcohols contain a hydroxyl (—OH) group and a hydrogen covalently bonded to a carbon atom. The letter R represents an unspecified chemical group that makes up the remainder of the molecule. For example, the *R group* may vary from a single atom to a chain of atoms or a complex ring.

hydroxyl group	— OH
general formula for an alcohol	R — OH
methanol, a simple alcohol	$H-\overset{\displaystyle H}{\underset{\displaystyle H}{C}}-OH$

Alcohols include *methanol* (methyl alcohol, also known as wood alcohol and commonly used as a solvent and cleaner) and *ethanol* (ethyl alcohol, or grain alcohol, found in alcoholic beverages). Hydroxyl groups also form an important part of starches, sugars, and other compounds. The hydroxyl group is polar. This property makes the hydroxyl capable of interacting with water so molecules that contain many hydroxyl groups are usually water-soluble.

A **carboxyl** group is really an organic acid. An acid is a molecule that *releases* protons into solution, and the basic structure of a carboxyl group shows how it is able to do this:

$$R-\overset{\displaystyle OH}{C}=O \;\rightarrow\; R-\overset{\displaystyle O^-}{C}=O+H^+$$

organic acid → dissociated form

Organic acids are polar and often reactive. We will see in the pages ahead that they are the basis for a number of important chemical reactions. We sometimes refer to the carbon–oxygen double bond ($C{=}O$) as a **carbonyl** group. In this case the carbonyl is part of the carboxyl group.

The nitrogen-containing **amino** group is one kind of organic base. Bases are molecules that *remove* protons from solution. The amino group ($-NH_2$) is capable of doing exactly that:

$$R - N \Big\langle \begin{matrix} H \\ H \end{matrix} + H^+ \rightarrow R - N^+ \Big\langle \begin{matrix} H \\ H \\ H \end{matrix}$$

By removing H^+ ions from solutions, amino groups can effectively raise the pH of a solution, making it more basic. An amino group need not be bonded directly to a carbon atom, but in organic molecules that is generally the case.

Aldehydes are compounds that contain a carbonyl group bonded to a single hydrogen atom and an R group:

$$\begin{matrix} H \\ | \\ R - C = O \end{matrix}$$

Aldehydes are strongly reactive chemically, and aldehyde groups are frequently involved in forming covalent bonds between different molecules.

Ketones are molecules that contain a carbonyl in the middle of a longer chain:

$$\begin{matrix} R' \\ | \\ C = O \\ | \\ R \end{matrix}$$

Ketones include common solvents such as acetone.

Phosphates are important chemical groups that are often used to link other organic groups together:

$$\begin{matrix} O^- \\ | \\ -O - P = O \\ | \\ O^- \end{matrix}$$

The negative charges on phosphate groups make many organic compounds containing phosphates very soluble in water. Phosphate groups are partly responsible for the water-solubility of many large molecules, including nucleic acids.

This short list of chemical groups provides a series of generalizations which will enable us to discuss the larger molecules which are important in living systems.

SUMMARY

The basic unit of matter is the atom. Atoms are composed of subatomic particles: protons, electrons, and neutrons. The chemical and physical properties of an atom are determined by the number of subatomic particles in its nucleus and its electron orbitals. Atoms may be joined by chemical bonds to form compounds. The chemical and physical properties of compounds are often very different from the atoms that form them. Chemical bonds may be ionic or covalent, depending on whether electrons are transferred from one atom to another or shared between them.

The most abundant molecule in living things is water. Water has a number of special properties, including an asymmetric arrangement of electrons around the three atomic nuclei that make up the compound. This produces a polarity in the molecule that makes it capable of dissolving other molecules, one of water's most important characteristics. Water molecules may dissociate to produce protons (H^+) and hydroxide ions (OH^-), and the amounts of these ions in a particular solution may be represented on a system known as the pH scale.

The properties of larger molecules can be analyzed in terms of the chemical groups that make them up. Biological molecules contain a number of distinct chemical groups whose basic properties are similar from one molecule to the next. These groups make it possible to describe the chemical properties of complex molecules.

STUDY FOCUS

After studying this chapter, you should be able to:

- Appreciate that chemistry and chemical processes underlie biological systems.
- Describe chemical systems in general terms.
- Explain why biological research increasingly involves chemical techniques and chemistry as a level of analysis.
- Recognize some of the most important biological molecules and chemical groups that are crucial to the study of living things.

SELECTED TERMS

subatomic particles   *p. 363*
atomic orbital   *p. 363*
atomic number   *p. 363*
molecule   *p. 366*
compound   *p. 366*

macromolecules   *p. 366*
isotopes   *p. 366*
beta particles   *p. 366*
half-life   *p. 366*
ionic bonds   *p. 367*

covalent bonds *p. 367*
cation *p. 367*
anion *p. 367*
polar *p. 369*
hydrogen bond *p. 371*
hydrophilic *p. 371*

hydrophobic *p. 371*
pH *p. 372*
buffer *p. 372*
capillary action *p. 373*
surface tension *p. 373*

R E V I E W

Discussion Questions

1. What is the difference between atomic number and atomic mass? Why does adding additional neutrons to an atom not affect its chemical properties, although adding protons does?

2. What are the three most important types of chemical bonds, and how are they different from each other?

3. What is the origin of the term *organic chemistry,* and how does its original meaning differ from its contemporary one?

4. Why is water such an important compound to living things?

5. Which solution has a higher concentration of H^+ ions, one at pH 3 or one at pH 6?

Objective Questions (Answers in Appendix)

6. Atoms of different elements react differently from one another because of
   (a) the number of neutrons in their nuclei.
   (b) the number of electrons in their outer orbitals.
   (c) the number of neutrons in their outer orbitals.
   (d) their atomic weight.

7. Isotopes of an element differ in all the following *except*
   (a) the number of neutrons in their nuclei.
   (b) atomic weight.
   (c) stability of the nuclei.
   (d) chemical behavior.

8. Electrons that are transferred from one atom to another form
   (a) covalent bonds.
   (b) ionic bonds.
   (c) hydrogen bonds.
   (d) beta bonds.

9. Electrons that are shared between elements form
   (a) covalent bonds.
   (b) ionic bonds.
   (c) hydrogen bonds.
   (d) cooperative bonds.

10. What type of bond is found between molecules of water?
    (a) covalent
    (b) hydrogen
    (c) ionic
    (d) nonpolar

11. H^+ and OH^- are
    (a) atoms.
    (b) compounds.
    (c) ions.
    (d) molecules.

CHAPTER 20

Macromolecules

For many years, chemists were baffled by the sheer complexity of living matter. Although living organisms are composed of the same *atoms* found in nonliving material, many of the *compounds* in living organisms are unlike anything found in the nonliving world. The complexity of life, the chemists began to understand, was due to the way in which the chemical elements were joined together in living things. Gradually, chemists began to appreciate that many of the important molecules in living things are very, very large. In the 19th century, chemists had learned how to analyze materials with molecular weights as high as 1,000. Compounds isolated from living things were much larger. We know these compounds as *macromolecules*, and they may have molecular weights of 100,000 or more.

MACROMOLECULES AND POLYMERIZATION

Macromolecules are assembled by a process known as **polymerization** (Figure 20.1a). Polymerization refers to the building of large molecules by hooking together a great many smaller ones. The individual molecules are known as **monomers.** When two monomers are hooked together, a *dimer* is formed; three form a *trimer,* four a *tetramer,* and a large number is said to form a **polymer** (*poly* means "many").

The ability to form polymers is not confined to biological molecules. In fact, polymer science is one of the major branches of chemistry. Synthetic polymers include molecules such as nylon, rayon, teflon, and polyethylene (Figure 20.1b). Unlike natural polymers, many of these synthetic polymers are not biodegradable because cells have not evolved the ability to break them down.

Each of the four major types of biological macromolecules is produced by the process of polymerization. These four classes are **carbohydrates,** compounds such as simple sugars, cellulose, and starch; **proteins,** which are important in forming the structures of the cell and in controlling chemical reactions within it; **lipids,** waxy or oily compounds that include the common fats; and **nucleic acids,** which store and transmit biological information. Except for a few of the lipids and simple sugars, each of these compounds is a polymer.

CARBOHYDRATES

Carbohydrates are organic molecules that exhibit some variation of the general chemical formula $(CH_2O)_n$. Most carbohydrates are constructed from monomers known as *simple sugars*. A sugar molecule can have from three to

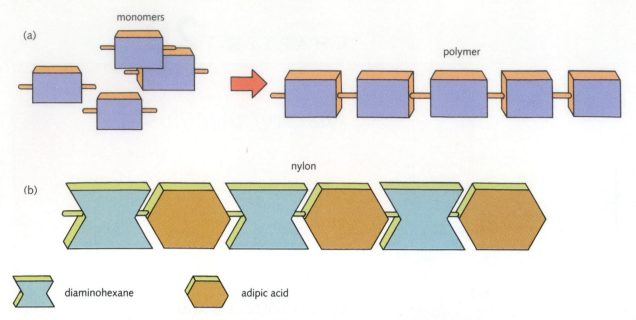

monomers

(a)

polymer

nylon

(b)

diaminohexane

adipic acid

Figure 20.1 (a) Polymerization. This is a general chemical process in which individual compounds (monomers) are joined together to form larger assemblies (polymers) which may take the form of chains, as shown. (b) Nylon is an excellent example of a polymer. Nylon is formed by the polymerization of diaminohexane and adipic acid. (Because the two monomers are chemically different, nylon is properly known as a *heteropolymer. Homopolymers* are compounds in which the monomers are chemically identical.)

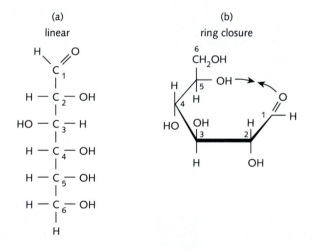

Figure 20.2 Glucose can exist in several forms. Although we sometimes draw it as a linear molecule (a), it is generally in a closed ring form. In solution, the ring can spontaneously open and close with very little loss of energy. (b) Note that the position of the hydroxyl group on the #1 carbon determines whether alpha- or beta-glucose (c) is formed when the ring closes.

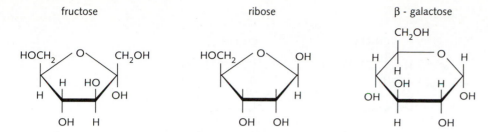

Figure 20.3 Carbohydrates are a diverse group of molecules that fit the general chemical formula of $(CH_2O)n$, where $n \geq 3$. They include five-carbon (pentose) sugars like **ribose**, six-carbon (hexose) sugars such as **glucose** and its structural isomers **fructose** and **galactose.**

seven carbons: A sugar with three carbons is known as a *triose*, and those with five, six, and seven carbons are called *pentose*, *hexose*, and *heptose* sugars respectively. Simple sugars are also known as **monosaccharides,** because each molecule consists of a *single* sugar unit. We can get a general picture of the organization of simple sugars by examining the structure of *glucose*, which is a simple hexose sugar.

Glucose is a common carbohydrate with the molecular formula $C_6H_{12}O_6$. It can be drawn schematically as a linear molecule in which the carbon atoms can be numbered from 1 to 6, beginning with the carbon that forms a carbonyl group (Figure 20.2a). In solution, however, the glucose molecule does not actually exist in this linear configuration. Instead, the molecule undergoes spontaneous structural rearrangement in which a covalent bond forms between the number-1 carbon and the hydroxyl group of the number-5 carbon. The new bond causes glucose to form a closed, six-membered ring (Figure 20.2b). There are actually two forms of glucose, α *glucose* and β *glucose*, which differ in the position of the hydrogen atom attached to the number-1 carbon when the bond that closes the ring is actually formed (Figure 20.2c). This doesn't seem like much of a difference, and in fact, α glucose and β glucose can *spontaneously interconvert* (change from one form to another without assistance) in solution. However, when individual glucose monomers are linked together to form glucose polymers, the position of that hydrogen atom on the number-1 carbon is fixed, and it is locked in either the α or the β position.

Monosaccharides

Many other sugars, which are chemically different from glucose, still share the chemical formula $C_6H_{12}O_6$. Figure 20.3 shows two such molecules, which are known as *structural isomers* of glucose. They include such molecules as *galactose*, which differ only in the orientation of hydroxyl groups, and *fructose*, which has a five-membered ring instead of the six-membered ring found in glucose. Fructose is also a hexose sugar, but it forms a five-membered ring by means of a bond that links the number-2 and the number-5 carbons through an oxygen atom. The pentose sugar, *ribose*, is also shown in Figure 20.3.

Disaccharides

Complex carbohydrates are formed by linking simple sugars together. A single sugar molecule is known as a monosaccharide. When two sugars are joined by a covalent bond, the compound is known as a **disaccharide.**

Some of the compounds we think of as "sugar" in our everyday experience are in fact disaccharides. A common disaccharide is ordinary table sugar, which is formed by linking together glucose and fructose. Other important disaccharides include *maltose, lactose,* and *sucrose,* common table sugar (Figure 20.4a).

Maltose is produced by a reaction that results in the formation of a chemical bond between two glucose molecules. A link is formed between the number-1 carbon of one glucose and the number-4 carbon of the other. Because the hydrogen atom on glucose's number-1 carbon is in the α position, the link between the two sugars is referred to as an *α-1,4 bond.* As you can see, the formation of the bond is accompanied by the loss of a hydrogen atom (H) from one sugar and the loss of a hydroxyl group (—OH) from the other (Figure 20.4a). Because these two ions condense to form a water molecule, this kind of reaction is frequently known as a **condensation reaction** (Figure 20.4b). Condensation reactions release water molecules.

Condensation reactions are common in the formation of biological polymers. The reverse reaction, in which polymers are broken down to monomers, consumes a

(a) important disaccharides

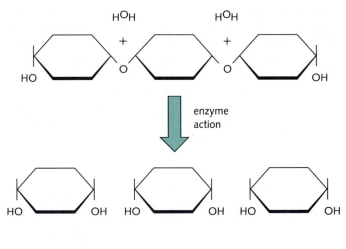

α - glucose + β - glucose → β - maltose + H_2O

Sucrose (α - glucose + β - fructose)

β - lactose (β - galactose + β - glucose)

(b) condensation

enzyme action

+ HOH + HOH

(c) hydrolysis

HOH + HOH

enzyme action

Figure 20.4 (a) The formation of β-maltose, a disaccharide. Maltose is formed in a reaction between alpha- and beta-glucose which releases a water molecule. Sucrose, table sugar, is a disaccharide composed of alpha-glucose and fructose. Lactose, another disaccharide, is composed of beta-galactose and beta-glucose. (b) Formation of a trisaccharide by covalently bonding three monosaccharides releases two molecules of water. (c) These two covalent bonds may be broken by subsequent enzyme action in which hydrolysis, the addition of a water molecule, occurs at each linkage.

water molecule and is known as **hydrolysis** ("water splitting") as water molecule atoms are divided between the two subunits as the bond is broken (Figure 20.4c).

Polysaccharides

Polysaccharides are formed by joining together three or more monosaccharides. An enormous variety of carbohydrates can be formed by linking simple sugars together (Figure 20.5). *Starch* consists of long chains of glucose molecules connected by α-1,4 linkages. Plants store much of their excess carbohydrate in the form of starch. *Glycogen* is a polysaccharide similar to starch, except for the presence of a few α-1,6 linkages that allow the chain to branch. Glycogen is used in animal cells to store carbohydrate in a form that can be quickly broken down to use as a source of energy. *Cellulose* is identical to starch, except that its linkages are in the β-1,4 form. As we saw in Chapter 3, this makes all the difference in the world when the time comes to digest these molecules. Cellulose is the major molecule found in plant cell walls and in wood. *Chitin* is a very important polysaccharide that consists of a sugar called *N*-acetylglucosamine joined by β-1,4 linkages. Chitin is tough and resilient, and it forms the external skeletons of insects and crustaceans.

You might have noticed that we haven't said anything about how long the carbohydrate chains can be. That's because these chains are almost unlimited in size; they may contain more than a million linkages. Starch is a complex polysaccharide consisting of glucose molecules joined by alpha 1,4 bonds. Glycogen is similar to starch, but also includes branching chains with 1,6 linkages. The glucose monomers in cellulose are joined by β-1,4 bonds, while chitin consists of chains of n-acetylglucosamine joined by β-1,4 bonds (Figure 20.5). Figure 20.5 also shows starch granules in a raw potato, clusters of glycogen in a liver cell, cellulose in a bristlecone pine, and chitin constituting the bulk of a crab's exoskeleton.

PROTEINS

Proteins are probably the most diverse class of macromolecules. Proteins are formed by the polymerization of monomers known as **amino acids.** There are structural proteins, such as *keratin* which forms hair, skin, and fingernails; proteins that transmit chemical messages, such as *insulin;* proteins that carry oxygen in the blood, such as *hemoglobin;* and proteins that catalyze chemical reactions, such as *amylase,* an enzyme that is found in saliva and breaks starch down into simple sugars. Proteins are involved in every activity in the cell: They pump molecules across cell membranes, form a cytoskeleton that

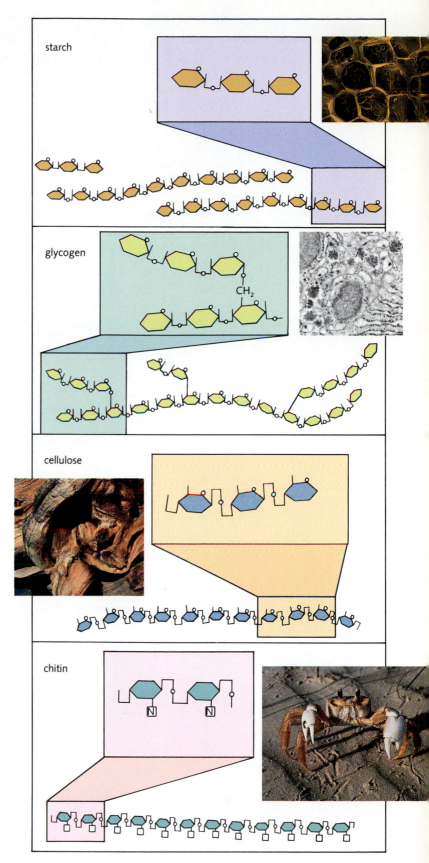

Figure 20.5 The chemical structures of major polysaccharides.

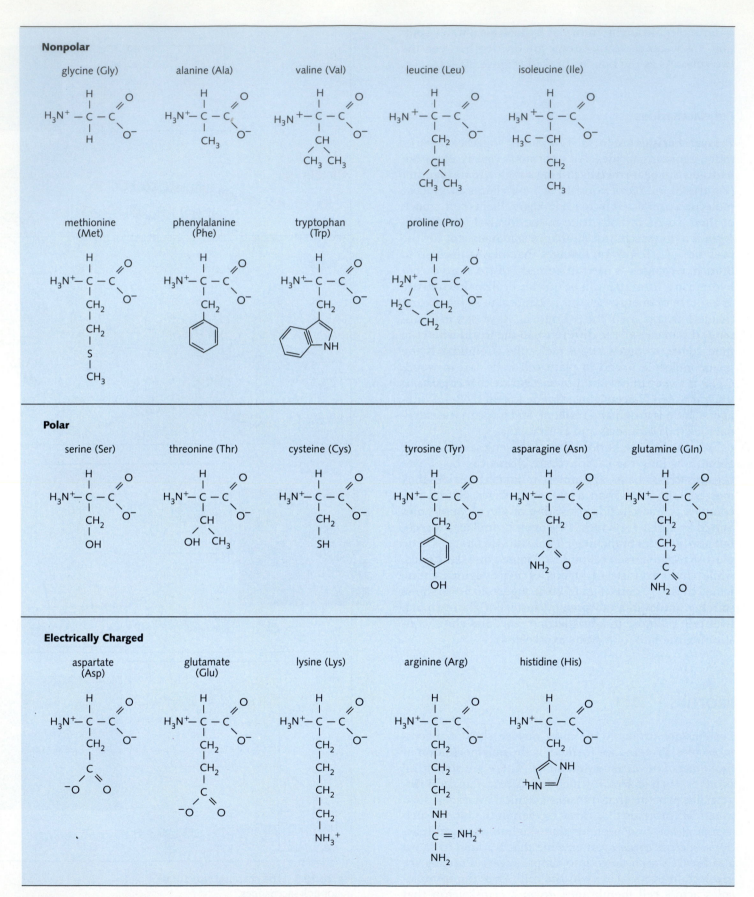

Figure 20.6 Twenty of the most common amino acids found in living organisms. The three groupings (nonpolar, polar, and charged) reflect chemical differences in the R groups of the amino acids.

supports cell movement, synthesize other macromolecules, and help to regulate gene expression and cell recognition.

Amino Acids

Proteins are polymers of amino acids. More than 40 kinds of amino acids are found in living organisms, but only about 20 different types are so common that they are found in all cells.

Amino acids are simple chemical compounds that contain three principal chemical groups: an *amino* group (—NH$_2$), a *carboxyl* group (—COOH) and a third chemical group that differs from one amino acid to the next. This third group is known as an *R group*.

The simplest amino acid is *glycine*, in which the R group is merely a hydrogen atom. The structures of all of the most common amino acids and their R groups are shown in Figure 20.6. Because all amino acids have an amino (base) group and a carboxyl (acid) group, amino acids have some properties of both acids and bases.

The Peptide Bond

Amino acids can be joined one to another in a chemical reaction that links the amino group of one to the carboxyl group of another. This condensation reaction removes a molecule of water, and forms a covalent bond between carbon and nitrogen (Figure 20.7).

Although we draw it as an ordinary covalent bond, the link between the nitrogen and carbon atoms of adjacent amino acids is referred to as a **peptide bond.** Two amino acids joined by such a bond form a *dipeptide.* Three would constitute a *tripeptide,* and a very long chain forms a **polypeptide.** Polypeptides as long as 1000 amino acids are common. Because peptide bonds are always formed between the carboxyl and amino groups of amino acids, one end of the chain has a "free" amino group and one end has a free carboxyl group. We call the carboxyl end the *C–terminus* and the amino end the *N–terminus.* This gives polypeptide chains a directionality; that is, each end is different (Figure 20.8).

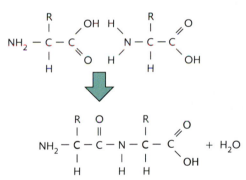

Figure 20.7 The formation of a peptide bond. The bond is formed between the carbon of a carboxyl group and the nitrogen of an amino group. A molecule of water is split off when the bond is formed.

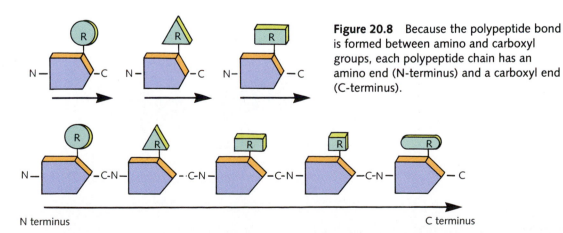

Figure 20.8 Because the polypeptide bond is formed between amino and carboxyl groups, each polypeptide chain has an amino end (N-terminus) and a carboxyl end (C-terminus).

N terminus

C terminus

PROTEIN STRUCTURE

The Possibilities of a Polypeptide

Polypeptides are long chains of amino acids joined by peptide bonds. How many chemical possibilities can there be to polypeptide structure? The answer may surprise you, for we have not yet examined the most important aspect of any polypeptide chain: the R groups. Each amino acid has a different R group or side group (Figure 20.9). Some of them are small hydrocarbon chains (as in leucine and isoleucine). Some have complicated ring structures (as do phenylalanine and tryptophan). Some side groups are actually acids (glutamic acid) or bases (arginine) themselves. One amino acid has a sulfhydryl side group (—SH) that can form covalent bonds on its own. Such diversity allows polypeptides almost unlimited chemical possibilities.

If we were to consider a polypeptide that contained just 30 amino acids, how many different polypeptide molecules could be assembled from 20 amino acids? Because each position could be filled with any one of the 20, the number of possibilities is 20^{30} (20 multiplied by itself 30 times), a very, very large number. Equally important is what the chemical diversity of the side groups means: Our polypeptide could be acidic, basic, or uncharged; it might be neutral at one end and basic at the other; it could contain large bulky R groups or small ones; and these R groups might or might not be able to form covalent bonds. The amino acids represent a kind

of chemical construction set with which it is possible to build a vast number of different molecules. That is the great power of this important class of macromolecule.

Polypeptides and Proteins

You might have noticed that we have yet to say exactly what a "protein" is, although we have defined a *polypeptide* with some precision. There is a reason for such evasiveness. In some cases, a protein actually *is* a single polypeptide, though in others many polypeptides are required to form a protein. What's the distinction? Proteins are polypeptides, but they also are *functional molecules*. A protein is a single molecule that may consist of one or several polypeptides and may also include other components. We will look at some of these subtleties by examining the detailed structure of a single protein. Biochemists recognize four levels of structure in the organization of a protein molecule.

Primary Structure

The first level of structure, **primary structure,** of a protein is nothing more than the sequence of amino acids that compose it. To describe the primary structure of a protein, therefore, we need simply to make a list of its amino acids in the order in which they occur (Figure 20.10).

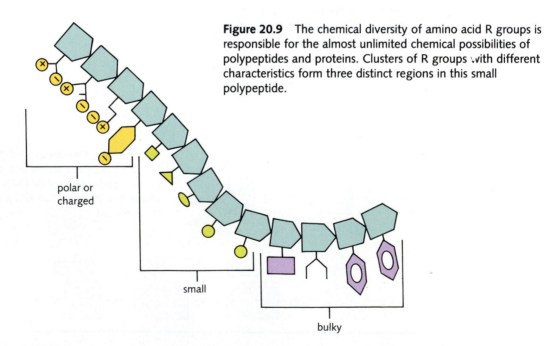

Figure 20.9 The chemical diversity of amino acid R groups is responsible for the almost unlimited chemical possibilities of polypeptides and proteins. Clusters of R groups with different characteristics form three distinct regions in this small polypeptide.

Secondary Structure

Secondary structure is the term given to the localized folding of the polypeptide chain. The interactions that govern this folding usually involve the carbon-nitrogen backbone of the polypeptide. The polypeptide chain can fold into a variety of patterns, so that the secondary structure of a protein can take many forms. However, protein chemists have noticed that certain folding patterns are much more common than others.

The alpha helix One of the most common folding patterns, the **alpha helix,** was first described in 1951 by Linus Pauling, the great American chemist (Figure 20.11a). Pauling discovered the existence of the helix by looking at the X-ray scattering patterns produced by protein crystals. The patterns suggested that many of the polypeptide chains of the crystalized proteins were folded in a helical fashion, not unlike a spring. Pauling toyed with models of polypeptide chains until he found an arrangement that matched the X-ray pattern and made good chemical sense.

The arrangement Pauling hit upon is shown in Figure 20.11(b). When a segment of a polypeptide chain assumes the alpha-helical structure, the basic carbon–nitrogen backbone of the chain twists into a helix with a very specific spacing: Each turn of the helix occupies exactly 3.6 amino acids, and each amino acid advances the helix 1.4 Å (1.4 Å = 0.14 nm) along its axis. Why should a polypeptide assume such a rigid shape? Pauling was puzzled

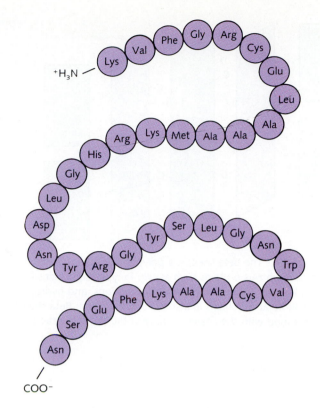

Figure 20.10 The primary structure of a protein is its amino acid sequence.

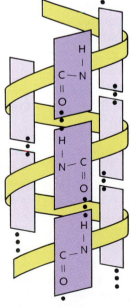

Figure 20.11 Linus Pauling and the alpha helix, one of several potential folding patterns of a polypeptide chain. The helix is stabilized by hydrogen bonds which form between amino acids.

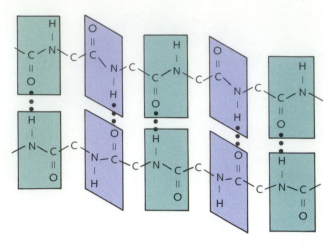

Figure 20.12 The beta sheet is a polypeptide folding pattern common in fibrous proteins such as silk. Polypeptide chains in the sheet run in opposite (antiparallel) directions, and hydrogen bonds link adjacent chains in a stable configuration. Beta sheets are also found with the chains running in the same direction.

by that too, until he noticed that when the chain was bent into the helix configuration, the N—H group of one amino acid was positioned exactly beneath the C=O group of another. Because the oxygen in this position is slightly negative and the H at the other position is slightly positive, a *hydrogen bond* can form between them. This slight electrostatic attraction stabilizes the helical shape.

Although individual hydrogen bonds are weak, dozens or hundreds of hydrogen bonds produce a considerable force and hold the helix in its shape. This helps to explain why the alpha helix is so common in nature. However, the alpha helix is just one kind of chain folding.

The beta sheet The **β sheet** is a type of secondary structure that is very common in fibrous proteins, such as those that make up bird feathers and silk. In the beta-sheet configuration, hydrogen bonds link regions of the polypeptide chain that run in parallel or antiparallel directions (Figure 20.12). When large portions of a protein lie in this structure, the protein assumes a thin, fiber-like structure that is strong, stable, and flexible.

There are other kinds of secondary structure as well, and we have mentioned the alpha and beta configurations only because they are common. In fact, many proteins have large regions of chain folding that do not follow a definite pattern and are completely unique. Some biochemists describe such regions as folded in an open chain pattern, meaning that their structure exhibits no apparent pattern.

Tertiary Structure

Besides the local pattern of chain folding that constitutes secondary structure, proteins show a large-scale folding pattern called **tertiary structure** (Figure 20.13). Tertiary structure is stabilized by a number of bonds that can be formed between the R groups of amino acids. These bonds include hydrogen bonds, ionic and electrostatic interactions, weak interactions between uncharged side groups, and covalent bonds.

The covalent bonds are particularly interesting. They form between two *cysteines*, amino acids that have the —SH (sulfhydryl) side group. Under the right chemical conditions, a covalent bond can form between two cysteines (Figure 20.13), linking together two regions of the polypeptide chain that may be very far apart in the primary sequence of the molecule. This bond, sometimes called a **disulfide bond,** is very important in holding together the shapes of large and complicated proteins.

Disulfide bonds are the basis of some very practical chemistry. A hair permanent, a series of washes and rinses used to produce curls in normally straight hair, works by altering protein tertiary structure (Figure 20.14). First the hair is rinsed with a solution containing a chemical agent that breaks the disulfide bonds in *keratin*, the principal protein found in hair. Then the hair is rolled into the desired shape, and finally it is treated with another chemical that re-forms disulfide bonds. The new bonds help to make the wave permanent—all because of protein chemistry.

How can we distinguish the localized folding of secondary structure from the large-scale folding of tertiary structure? In truth, the distinction is sometimes hard to make. But in general, the difference between the two is a question of scale. Secondary structure is more localized, a bit like the pattern of twisting used to form a rope. Tertiary structure involves details over the length and breadth of a molecule, and it is a bit like the folding of a rope to tie a knot (Figure 20.15).

Quaternary Structure

The fourth level of protein structure, **quaternary structure,** concerns the way a protein is folded when it consists of more than one polypeptide chain. Strictly speaking, a protein that contains only one polypeptide chain does not have a quaternary structure. However, let's consider a protein that has four different subunits, which we can call A, B, C, and D as shown in Figure 20.16. We might arrange these subunits in a host of different ways. In describing the quaternary structure, therefore, we specify the exact arrangement of the different subunits in three dimensions—a critical piece of information for a complex system.

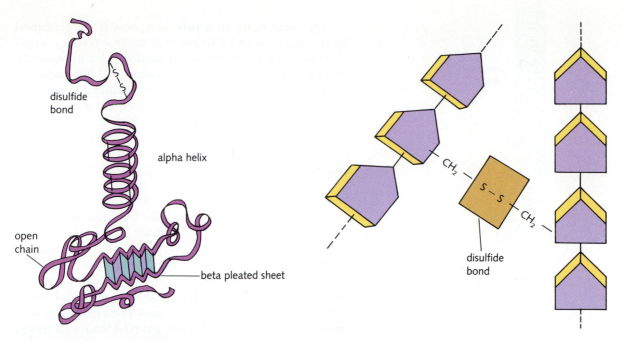

disulfide bond

alpha helix

open chain

beta pleated sheet

disulfide bond

Figure 20.13 (left) The tertiary structure of a polypeptide is the complete three-dimensional pattern into which it folds. Tertiary structure may be stabilized by a variety of interactions between amino acids at different positions in the polypeptide, including covalent disulfide bonds. (right) Disulfide bonds form between the sulfhydryl (—SH) groups of cysteine amino acids. These covalent bonds may link adjacent polypeptide chains.

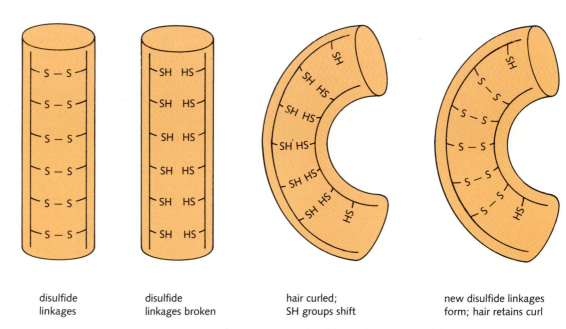

disulfide linkages

disulfide linkages broken

hair curled; SH groups shift

new disulfide linkages form; hair retains curl

Figure 20.14 Permanent waves take advantage of the protein chemistry of hair. First, hair is treated with a chemical which breaks disulfide bonds. Then the hair is folded or rolled into the desired shape. The shaped hair is then treated with a chemical bath which re-forms disulfide bonds, stabilizing each hair in its new position. When the hair is washed clear of the chemicals, new disulfide bonds will hold each hair in its new position, producing a "permanent" wave.

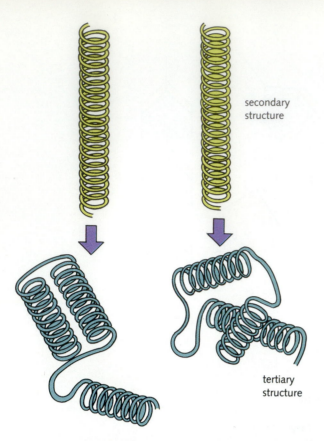

secondary structure

tertiary structure

Figure 20.15 The distinction between secondary structure (localized folding of the polypeptide chain) and tertiary structure (overall folding of the chain) is a subtle one. To help make the distinction, this diagram shows two hypothetical proteins with identical secondary structures (alpha-helix) but different tertiary structures. Note the manner in which tertiary structure involves the interactions of amino acids which are far removed from each other in the primary sequence of the protein.

The quaternary structure of a protein is determined by the same kinds of forces that control tertiary structure: hydrogen bonds, ionic attractions, disulfide bonds, and weak interactions between amino acid R groups. All of these interactions influence the ultimate shape of a protein, and shape determines which chemical groups are exposed on the protein's outer surface.

Prosthetic Groups

Although the tremendous chemical diversity of the amino acids would seem to give a protein the chemical potential to do just about anything, there are some tasks that are best performed by another type of molecule. A good example is *hemoglobin*, a protein molecule that gives blood its deep red color. Hemoglobin consists of four individual polypeptides, two of which are called α *globin* and two of which are called β *globin*. The task of hemoglobin is to bind oxygen molecules. Oxygen gas (O_2) binds to a region in the center of each of the four subunits of the molecule. But it doesn't bind to any of the amino acid side groups. Instead, each subunit contains a molecule known as *heme*, which itself contains an iron atom in its center.

Oxygen binds directly to the iron atom, and this is how oxygen is carried within the bloodstream. Heme is not an amino acid. It is almost a "foreigner" within the folded structure of the polypeptide chain. Heme performs a chemical task that the polypeptide part of the molecule cannot carry out. Heme is called a **prosthetic group**, a colorful name which suggests that it is almost like a *prosthesis*, an artificial limb. Prosthetic groups are non–amino acid portions of proteins, and many proteins contain them (Figure 20.17).

LIPIDS

Lipids are biological compounds that are waxy or oily and that dissolve in organic solvents (an organic solvent is a liquid, such as hexane, acetone, ether, and alcohol that dissolves nonpolar compounds which are generally *hydrophobic*). That is not a very precise chemical definition, and frankly, it's not intended to be. The lipids are a

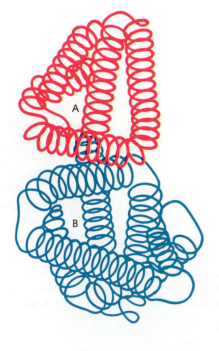

A

B

Figure 20.16 The quarternary structure of a protein is the spatial arrangement of several polypeptides to form a complete protein. Because there are many possible arrangements of subunits, a complete description of quarternary structure pinpoints the position of each subunit in the protein with respect to the others.

very diverse group. Lipids dissolve in organic solvents because a major part of most lipid molecules is *hydrocarbon,* made up solely of hydrogen and carbon:

$$-CH_2-CH_2-CH_2-CH_2-CH_2-$$

The electrons in hydrocarbons are evenly distributed, so hydrocarbons do not contain polar regions and do not interact strongly with water—in other words, they are "oily." We will return to this point shortly.

Many lipids are formed from molecules known as **fatty acids.** A fatty acid consists of a long hydrocarbon chain with a carboxyl group attached to one end. A typical fatty acid is stearic acid, which contains a total of 18 carbons (Figure 20.18).

There are many different kinds of fatty acids, and they differ from each other in two fundamental ways: the number of carbon atoms they contain and their number of carbon-to-carbon double bonds. As you can see from Figure 20.18, stearic acid doesn't have any carbon-

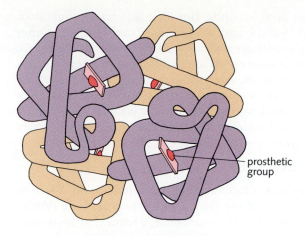

Figure 20.17 A model of hemoglobin shows that each of its four polypeptide subunits contains a single oxygen-binding **heme** group. Heme is a prosthetic group, a non-amino acid portion of the protein.

prosthetic group

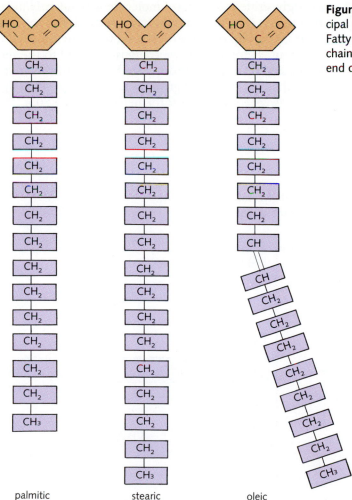

Figure 20.18 Fatty acids are the principal building blocks of many lipids. Fatty acids consist of long hydrocarbon chains with a carboxyl group at one end of the molecule.

palmitic   stearic   oleic

to-carbon double bonds. But oleic acid has a single double bond near the middle of its hydrocarbon chain. Other fatty acids may have two or even three such double bonds.

These double bonds are involved in an interesting bit of jargon that you are probably familiar with. Because the hydrocarbon chain of stearic acid has the maximum number of hydrogens per carbon, it is said to be *saturated*. But when a double bond is inserted, there are two fewer hydrogens in the molecule:

$$-CH_2-CH_2- \rightarrow -CH=CH- + H_2$$

Therefore, a fatty acid with a double bond is said to be *unsaturated*. If the fatty acid has two or three double bonds, we say it is *polyunsaturated*, a term that food companies often use to describe certain kinds of cooking oil and margarine. Unsaturation affects how the body is able to use food containing the fatty acids, and also whether the lipid is liquid or solid. Saturated fats are often solid at room temperature, whereas unsaturated fats are generally liquid.

Triglycerides and Neutral Lipids

We have emphasized the importance of fatty acids because they are the building blocks of most lipids. One such lipid is a type of compound known as a **triglyceride.** As shown in Figure 20.19, a typical triglyceride consists of a single **glycerol** molecule (commonly called glycerine and used

in many commercial products, including shampoos) to which three fatty acids have been covalently joined. The carboxyl group of each fatty acid reacts with one of the three alcohol groups on glycerol, three molecules of water are removed in the process, and a triglyceride results (therefore this process, which is similar to the condensation reactions involved in polysaccharide synthesis, is known as a dehydration synthesis). A triglyceride is a very common form of lipid found in fat cells and used for food storage in both plants and animals. Triglycerides are called **neutral lipids** because they are relatively nonpolar and are uncharged.

Polar Lipids

Many other kinds of lipids can be assembled on this general pattern, using glycerol as a kind of "backbone." When one of the groups that is attached to glycerol is strongly polar, a *polar lipid* is formed. One of the most important of the polar lipids is a class of compounds known as **phospholipids.** Phospholipids are formed along the same lines as triglycerides: A glycerol molecule is covalently linked first to one fatty acid and then to a second. But the third hydroxyl group on glycerol is used for something a little different: Here a *phosphate group* is attached to the molecule, and yet another group can then be attached to the phosphate. If that final group is a choline group, a lipid called *phosphatidyl choline* is formed (Figure 20.20a). That

Figure 20.19 Triglycerides are simple lipids which form from three fatty acids and glycerol. Condensation reactions form covalent bonds between each fatty acid and one of the three carbon atoms of glycerol.

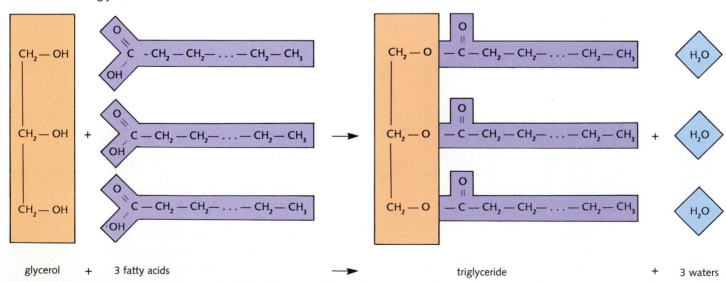

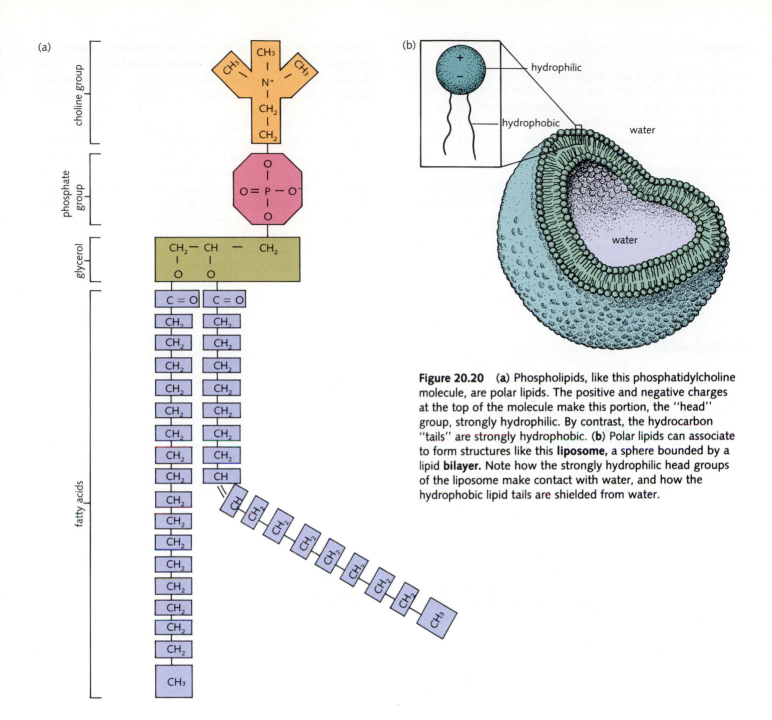

(a)

choline group

phosphate group

glycerol

fatty acids

(b)

hydrophilic

hydrophobic

water

water

Figure 20.20 **(a)** Phospholipids, like this phosphatidylcholine molecule, are polar lipids. The positive and negative charges at the top of the molecule make this portion, the "head" group, strongly hydrophilic. By contrast, the hydrocarbon "tails" are strongly hydrophobic. **(b)** Polar lipids can associate to form structures like this **liposome,** a sphere bounded by a lipid **bilayer.** Note how the strongly hydrophilic head groups of the liposome make contact with water, and how the hydrophobic lipid tails are shielded from water.

complicated name simply indicates that this is a phospholipid molecule that includes a choline group.

Phospholipids are particularly interesting because the opposite ends of each molecule have quite different properties. The hydrocarbon chains derived from their fatty acids are quite oily and do not interact well with water. But the opposite end of the molecule is strongly polar and carries a positive charge, a negative charge, or both. This end interacts very strongly with water and

would be able to dissolve easily were it not for the attachment to the opposite, nonpolar end. We often draw simple diagrams that represent the structure of polar lipids like phospholipids and can be used to emphasize the point that one end of the molecule is hydrophilic, and the other end is hydrophobic (Figure 20.20b). As we will see shortly, this property of polar lipids is the basis for one of their most important functions in living systems: It enables them to form biological membranes.

Steroids

Besides polar lipids and triglycerides, both of which are based on glycerol, there are many other types of lipids, some of which are constructed quite differently. One of these is a class of molecules known as **steroids** (Figure 20.21). Steroids are based on an intricate ring structure that is best exemplified by one of the most common steroids, **cholesterol.** Steroids are important molecules that play a wide variety of roles in living systems. Some of them aid in the assembly of cell membranes, and other steroids function as critical **hormones**—chemical messengers that regulate the activity of cells throughout the body. Later on in the text we will have a chance to examine the functions of still other types of lipids, including *prostaglandins* and *sphingolipids.*

NUCLEIC ACIDS

The final class of macromolecule we will examine in this chapter is the **nucleic acid.** The name *nucleic acid* reflects the fact that they were first discovered in the cell nucleus and their very mildly acidic character. Like proteins, nucleic acids are linear polymers of smaller subunits. Nucleic acids are formed from monomers known as **nucleotides.** The structure of a single nucleotide is shown in Figure 20.22(a). As you can see, a nucleotide consists of three main parts: a *pentose* (5-carbon) sugar, a *phosphate group,* and a nitrogen-containing *base.*

There are two main classes of nucleic acids: **ribonucleic acid (RNA),** which contains ribose as its pentose sugar, and **deoxyribonucleic acid (DNA),** which contains a sugar that is identical to ribose except that it is lacking one oxygen atom. Therefore it is called *deoxy*ribose. Nucleotide polymers are formed by covalent bonds between two or more nucleotides. As shown in Figure 20.22(b), the bonds are produced between the phosphate groups of one nucleotide and the number-5 carbon of another. When two nucleotides are linked by such a bond, a *dinucleotide* is formed. When many nucleotides are attached to form a long chain, we say that a *polynucleotide* has been produced.

Like polypeptide chains, polynucleotide chains have a particular orientation. One end of the chain must always contain a free phosphate group attached to the number-3 carbon of the sugar, and the other end has a free —OH group on the number-5 carbon. Therefore, we can call the two ends of the chain the *3' and 5' ends,* respectively. Nucleic acids are the carriers of biological information. In many of the pages ahead, we will consider the details of their structure and the roles they play in living cells.

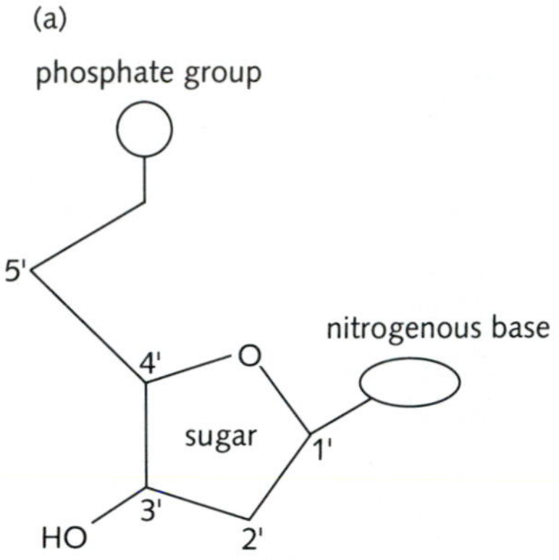

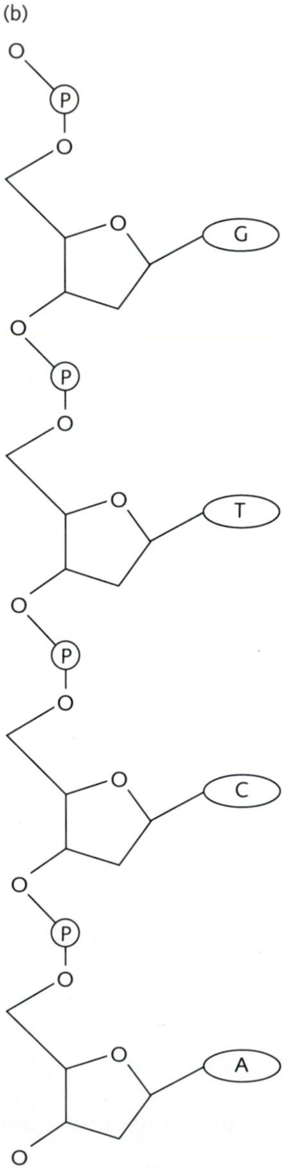

Figure 20.22 **(a)** A nucleotide is formed from a 5-carbon sugar, a nitrogenous base, and a phosphate group. **(b)** Covalent bonds between the sugar and phosphate groups of adjacent nucleotides form nucleic acids.

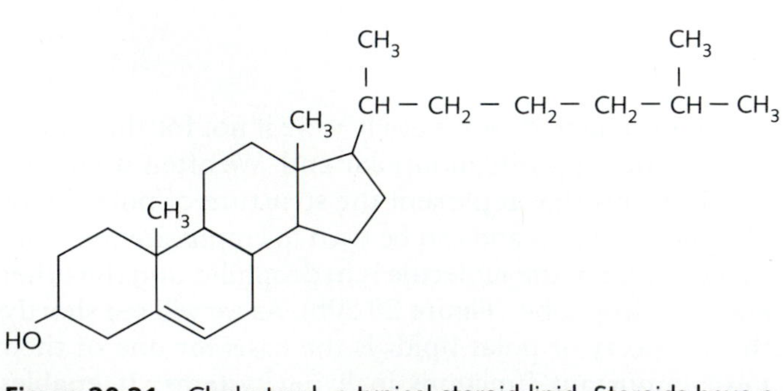

Figure 20.21 Cholesterol, a typical steroid lipid. Steroids have a basic four-ring structure like cholesterol, but differ in the chemical groups which are attached to the rings.

ribonucleic acid (RNA) *p. 392*
deoxyribonucleic acid (DNA) *p. 392*

SUMMARY

Macromolecules, very large molecules often found in living cells, are assembled by a process known as polymerization. Polymerization is the building of large molecules (polymers) by hooking together a great number of smaller ones (monomers). The major biological macromolecules include carbohydrates, nucleic acids, lipids, and proteins. Carbohydrates include monomers (simple sugars) and their polymers (the polysaccharides). Table sugar, starch, and cellulose are examples of polysaccharides.

Proteins, probably the most diverse class of macromolecules, are polymers formed from amino acids. Individual amino acids are joined by covalent bonds called peptide bonds; long chains of amino acids are known as polypeptides. The diversity of amino acid side groups, 20 of which commonly appear in living organisms, allows polypeptides almost unlimited chemical possibilities. Proteins can also assume complex three-dimensional structures; biochemists have defined four levels of structure to help them analyze and describe that complexity.

Lipids are waxy or oily organic compounds that are soluble in organic solvents. Nucleic acids are polymers formed from nucleotides. They play important roles in maintaining and transferring cellular information.

STUDY FOCUS

After studying this chapter, you should be able to:

- Explain the principle of polymerization and emphasize its importance in the synthesis of natural and artificial macromolecules.
- Describe the four major classes of biological macromolecules and emphasize the nature of the chemical bonds that are involved in each of them.
- Illustrate the richness and chemical flexibility of protein structure and emphasize the levels of protein structure.
- Present a number of examples of the importance of certain macromolecules in key cellular structures.

SELECTED TERMS

polymerization *p. 377*	polypeptide *p. 383*
carbohydrate *p. 377*	alpha helix *p. 385*
protein *p. 377*	beta sheet *p. 386*
lipid *p. 377*	disulfide bond *p. 386*
nucleic acid *p. 377*	fatty acid *p. 389*
condensation reaction *p. 379*	triglyceride *p. 390*
hydrolysis *p. 381*	glycerol *p. 390*
polysaccharide *p. 381*	phospholipid *p. 390*
amino acid *p. 381*	cholesterol *p. 392*

REVIEW

Discussion Questions

1. What are the four principal kinds of biological macromolecules? Give an example of each.

2. How is polymerization associated with the formation of each of the four main biological polymers?

3. How do the four levels of structure in proteins differ from one another?

4. What important characteristic of lipids makes it possible for them to self-assemble into larger structures?

Objective Questions (Answers in Appendix)

5. A saturated fat has _____ than an unsaturated fat of the same size.
   - (a) more fatty acid molecules
   - (b) more hydrogens per carbon
   - (c) shorter hydrocarbon chains
   - (d) fewer carbon atoms

6. A lipid that exhibits a ring structure is known as a(n)
   - (a) fatty acid.
   - (b) triglyceride.
   - (c) glycerol.
   - (d) steroid.

7. The process of hydrolysis involves
   - (a) evaporation of excess water upon heating the polymer.
   - (b) breaking down a long chain of carbohydrates by adding water.
   - (c) linking small chains of carbohydrates by removing water.
   - (d) removal of hydrogen ions from the long chain of carbohydrates by heating.

8. Which of the following is a polysaccharide?
   - (a) cellulose
   - (b) glucose
   - (c) hemoglobin
   - (d) keratin

9. Which of the following is an amino group?
   - (a) —COOH
   - (b) H^+
   - (c) OH^-
   - (d) —NH_2

10. The major linkage between amino acids is the
    - (a) hydrogen bond.
    - (b) peptide bond.
    - (c) nucleotide bond.
    - (d) polymer bond.

11. The primary structure of a protein consists of
    - (a) linear sequence of amino acids.
    - (b) beta-sheet configuration.
    - (c) alpha helical pattern of amino acids.
    - (d) more than one polypeptide chain with hydrogen bonds between each.

12. A quaternary protein structure has
    - (a) four different kinds of peptide bonds.
    - (b) more than one polypeptide chain.
    - (c) four in-foldings of the polypeptide chain.
    - (d) four types of amino acids.

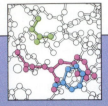

CHAPTER 21

Chemical Reactions and Energy

Imagine a warm summer morning at the beach. The sun has just risen and it casts long shadows across the mounds of sand and grass. Waves rise gently up the shoreline and breezes carry seagulls just above the shallow waters. A few birds poke at passing fish and drifting seaweed, while small clams, exposed by the falling tidal waters, dig into the sand for protection (Figure 21.1). Like every other part of the living world, the organisms at the beach might be described in terms of matter: proteins, carbohydrates, and other chemicals, simple and complex. Yet such a description would be incomplete. The beach is full of movement and sound. Organisms search for food, they react to their environment, and they change. To deal with the real world, we must consider more than the materials of life. We must consider how matter moves and changes from one form to another and, even more important, the cause of those transformations, a factor known as energy.

ENERGY AND LIFE

A physicist defines **energy** as *the capacity to do work*. In everyday usage, work is the expenditure of energy to move something: to lift a bale of hay or throw a ball, each of which requires energy. In molecular terms, work is required to move matter; the movement of material inside a cell is a form of work requiring energy. Energy can exist in several forms, and under the right conditions, energy can be converted from one form to another. **Kinetic energy** is the energy of a moving object, whether a baseball, a planet, or a molecule of carbon dioxide. The kinetic energy of that baseball can be changed into other forms. When a fast-moving baseball strikes a glass window, its

Figure 21.1 The beauty of a sunrise highlights the flow of energy from the sun to the earth.

kinetic energy is reduced as the baseball is slowed down by the impact. Some of the original kinetic energy is converted into *sound,* some into energy used to break the bonds that hold the glass together, and some into *heat* (the glass is slightly hotter as a result of the impact).

Energy can also exist in a form known as **potential energy.** We can produce potential energy by an act as simple as squeezing a balloon. By compressing the balloon, we do work on it. When the balloon is released, some of the work reappears as kinetic energy (Figure 21.2). The compressed balloon has *potential energy.* This is stored energy that can be released under the right conditions. **Chemical energy** is a special form of potential energy. Chemical energy is not contained in the compression of a balloon but in the structure of a molecule itself, and under the right conditions, chemical energy can be released and converted into other forms. When the potential energy in a liter of gasoline is suddenly released, an explosion produces kinetic energy, as well as heat, sound, and light.

As we saw much earlier in the text, nearly all of the energy available on the earth comes from the sun. The amount of energy that sunlight delivers to the United States in one *day* is more than a billion times greater than the amount of electricity generated at all the power stations in the United States in one *year.* Most of this solar energy is immediately transformed into other forms of energy: Sunlight warms the planet, and the uneven distribution of solar energy causes powerful currents in the atmosphere and the seas. A small but significant amount of solar energy, however, is captured by living things and converted into chemical energy. The constant formation of potential energy makes life possible on earth, and we will devote much of this chapter to developing an understanding of how chemical energy is produced, stored, and changed from one form to another.

ENERGY AND CHEMICAL REACTIONS

Chemical reactions involve the making and breaking of chemical bonds, and chemical bonds involve energy. Let's take a look at a very simple reaction: the combination of oxygen and hydrogen to form water. In this reaction, oxygen and hydrogen are referred to as the *reactants* and water is called the *product* of the reaction. Oxygen and hydrogen are both *diatomic* gases, which means they exist as molecules of O_2 and H_2. When oxygen and hydrogen undergo a chemical reaction to form water, the H—H and O=O bonds of the reactants are broken and replaced with 2 H—O bonds. But that's not all that happens.

The chemical energies of the product and the reactants are different: Specifically, there is a great deal less

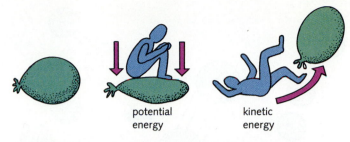

potential energy   kinetic energy

Figure 21.2 Stored energy is potential energy. A compressed balloon contains potential energy which may be released to do work. Chemical energy is a form of potential energy.

energy stored in the 4 O—H chemical bonds present in the products than there was in the 2 H—H bonds and the 1 O=O bond found in the reactants. What happens to all that extra energy? It is released into the environment in the form of heat and light, as shown in Figure 21.3. In the case of this reaction, there is a great deal of energy to be released. Hydrogen gas is highly explosive, and this property brought an abrupt end to zeppelin travel when the Hindenburg, a hydrogen-filled airship, exploded and burned at the end of a transatlantic trip in 1937 (Figure 21.3).

Thermodynamics

The energy changes that occur in the physical world are studied in a branch of science known as **thermodynamics.** Although thermodynamics began originally as part of the physical sciences, its methods and principles apply to biological systems as well. We will not deal with the mathematics of thermodynamic systems, but we will examine some of the principles of this science that have implications for biology.

The **first law of thermodynamics** is also known as the principle of **conservation of energy.** The first law states that "energy can be neither created nor destroyed." In other words, *the total amount of energy available in the universe does not change.* Einstein showed that matter and energy are equivalent, and the first law is sometimes written in a way that points this out. The first law tells us that we must account for all of the matter and energy involved in a chemical reaction.

Let's examine the operation of a gasoline-powered lawn mower. There is a certain amount of chemical energy available in the gasoline that is burned to power the engine. The expansion of gasoline and air as they burn in the cylinder of the engine provides the mechanical force to

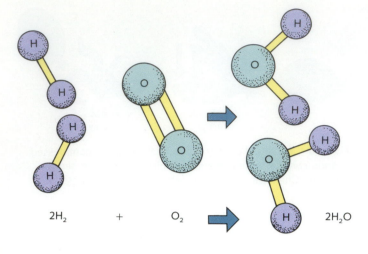

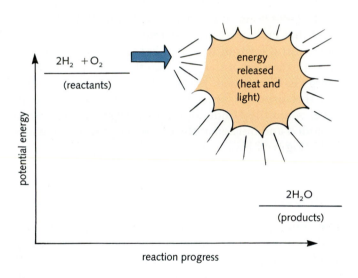

Figure 21.3 (left) The potential energy of products and reactants in a chemical reaction is often different. Hydrogen and oxygen gas have a higher chemical energy level than water, the product of a reaction between them. Therefore, energy will be released when the reaction takes place. (top) The tragic explosion of the Hindenburg airship resulted from the burning of hydrogen gas.

do work—such as spinning a rotary cutting blade at high speeds, which sends blades of grass scattering out the cutting chute. The first law tells us that *none* of the energy from the gasoline is "lost." That does not mean that *all* of the energy from the gasoline is transferred to the blades of grass that fly out of the mower. Most of the chemical energy is converted into the heat released in the hot exhaust, the heat of the engine itself, energy left in unburned gasoline, and even the energy represented by the noise the lawn mower makes (Figure 21.4).

This example introduces us to the **second law of thermodynamics,** which says that *once energy has been used to do work, it becomes less available to do additional work.* There are other ways to phrase this law, but let's not forget the lawn mower. Though all of the energy in the gasoline still exists after the job has been done, it is dispersed in a form much less available once the grass has been mowed.

It is not possible to collect the noise and exhaust heat released during the pumping process, combine them with the kinetic energy of the clippings, and reconcentrate them to produce more fuel. The energy has become dispersed. It is less available. This is the change that is at the heart of the second law.

Chemists use the word **entropy** in connection with the second law of thermodynamics. Entropy is a precise term that is a measure of disorder. When the energy in a system becomes dispersed, we say that the system's entropy (disorder) has increased. The second law says that *in the universe as a whole and in any isolated system, the entropy associated with any chemical change always increases.* Energy becomes more dispersed, less available, and more disorderly, and entropy increases.

Although the second law of thermodynamics seems to say that things must always run "downhill," as energy becomes increasingly dispersed and the entropy of the universe increases, it is important to realize that our statements of the second law have been carefully phrased to apply to closed systems: The universe as a whole is a closed system, but the earth is not. It has a strong and steady input of energy every day in the form of sunlight.

What does this say about the second law and biology? On our planet, there is more than enough energy available for an increase in order and complexity because we have the sun: a source of energy to support the activities of living things (Figure 21.5). The second law does point out a critical fact for biological systems, however. Without an input of energy, living systems are unable to maintain the organization and molecular order characteristic of life. Life depends on energy, and the first two laws of thermodynamics provide a physical explanation of why this is so.

The Chemical Balance of Energy

If living cells require a source of energy, it is only fair to ask how this energy is obtained and how it is transformed from one form to another. Let us begin by considering a molecule that all cells use to store and transform energy. The molecule is **adenosine triphosphate (ATP).** ATP is actually a nucleotide, one of the building blocks of nucleic acids. But it is also used to help store and transform chemical energy in the cell.

ATP can be formed by adding two phosphate groups to **adenosine monophosphate (AMP),** or by adding one phosphate group to **adenosine diphosphate (ADP).** This is an energy-requiring process; that is, energy must be "invested" in the production of ATP (Figure 21.6). The first law of thermodynamics tells us that energy is never lost, so some of the energy invested can be recovered when the two phosphate groups are removed from the molecule to again produce AMP and phosphate.

Calculating the actual amount of energy involved in any chemical reaction is complicated, because we must take into account both the heat change associated with the chemical reaction and the change in entropy (molecular disorder). However, the units we use to measure energy are straightforward enough. Chemical energy is measured in **calories.** A calorie is the *amount of heat required to raise the temperature of 1 cm^3 of water 1°C.* One thousand calories are equal to one kilocalorie. We can use these units to express the amount of energy that is available in a molecule of ATP.

Breaking the bonds that attach the second and third phosphate groups to ATP releases about 7000 calories (7 kilocalories) of energy per mole of ATP for each bond. (Remember that one mole is equal to the molecular weight of a compound in grams and that 7 kcal/mole is a large amount.) Frequently these bonds are described as "high

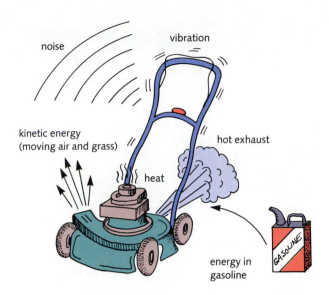

Figure 21.4 Chemical energy released from burning gasoline in a lawn mower is converted into a variety of forms, including noise, motion, and heat. The principle of conservation of energy tells us that energy will not be lost in the conversion processes.

Figure 21.5 Earth is not a closed thermodynamic system. The sun provides a continuous input of energy that makes possible the great diversity of life on Earth.

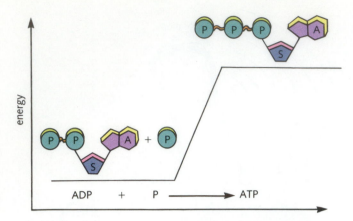

Figure 21.6 ATP (adenosine triphosphate) is formed by the process of *phosphorylation,* in which a third phosphate group is added to ADP (adenosine diphosphate). This is an energy-requiring reaction.

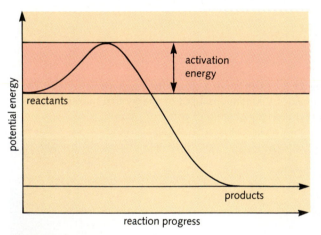

Figure 21.7 Energy changes in a typical chemical reaction. In order to participate in a chemical reaction, the reactants must first be raised to a certain energy level. The energy required to reach this level is known as *activation energy.*

energy" phosphate bonds. Actually, there is nothing special about the chemistry of the bonds that attach the second and third phosphates to ATP. The principal distinction of these chemical bonds is that they are easily made and broken and that chemical energy is involved in each process. This enables the cell to use ATP as a form of temporary chemical storage for energy. It takes energy to attach those two phosphate groups, and breaking them can release energy. Because a molecule of water is used when the bond is broken, we refer to the process as *hydrolysis.* As we continue our study of biology, we will see many examples of biological processes that involve ATP as a carrier of chemical energy.

STARTING A CHEMICAL REACTION

Nearly all chemical reactions involve some sort of an energy change. Those that release energy are known as **exergonic reactions,** and those that require energy are termed **endergonic reactions.** By this definition, the synthesis of ATP is endergonic and the hydrolysis of ATP is exergonic.

Does this mean that all exergonic reactions will proceed automatically, just because they yield energy? Fortunately, no. Flammable material, like paper and wood, will combine with oxygen and release carbon dioxide and water vapor in a reaction that is strongly exergonic. But this doesn't mean that every sheet of paper or stick of wood is in imminent danger of bursting into flames. We all know from common experience that we must *ignite* combustible material to start a fire. A small amount of energy is needed to get the reaction under way.

Activation Energy

There is a reason why a little bit of energy is required to start many chemical reactions, even ones that will eventually release much more energy than is required to get them going. Such reactions are said to require a small amount of energy for *activation.* We can think of this energy in a number of ways. Because chemical reactions occur at the molecular level, heating molecules up increases their thermal motion. As the molecules move faster and faster, they collide with each other with more force and greater frequency. In addition, chemical reactions often require that bonds between some atoms be broken before new bonds can be formed to take their place. In many cases, activation is required to weaken an existing bond before the reaction can actually begin. The quantity of energy required for activation is known, appropriately enough, as **activation energy** (Figure 21.7).

Catalysts

The fact that many chemical reactions require substantial activation energy has important consequences for the world around us. To begin with, it lends some stability to things. If every molecule that is capable of participating in an energy-yielding reaction were to go ahead and enter such a reaction, most of the molecules around us would break down very quickly. Our clothes, most of our homes, our books and papers, and our records and tapes would quickly react with oxygen and burn if the requirement for activation energy were lifted. For a living organism, however, the need for activation energy can be troublesome. It would seem, for example, that an organism would have to heat food molecules to the point of burning in order to release the energy they contain.

Chemists have discovered that there are ways to lower the energy of activation. Observations dating back more than a century suggest that certain materials have the effect of "promoting" or "assisting" chemical reactions. These materials are known as **catalysts,** and they have the effect of lowering the activation energy of a chemical reaction (Figure 21.8).

One classic example of a catalyst is *platinum.* You will remember that hydrogen and oxygen can react to form water, releasing a tremendous amount of energy. However, the reaction requires activation energy—high temperature in the form of a spark or a lighted match—to get the reaction going. Just mixing the two gases together is not enough to start the reaction. If the two gases are passed through a mesh of fine platinum wires, however, they ignite spontaneously and a flame appears. The activation energy is reduced to a level available at room temperature. Apparently the surface structure of platinum metal binds small amounts of each gas in just the right positions to allow the molecules to combine to form water. The platinum is not used up, but it does lower the activation energy just enough for the reaction to proceed. The platinum acts as a catalyst.

There are many other kinds of chemical catalysts. They are extremely important in industry and are used in everything from the curing of plastic to the refining of petroleum. In many automobiles, unburned fuel is removed from the exhaust gases by passage through a device known as a catalytic converter. In the converter, the catalyst lowers the activation energy of the fuel just enough so that it can be broken down into products that will not harm the environment. Although an automobile may seem very different from a living cell, the fact is that cells have a similar problem: They need to be able to lower the activation energy for chemical reactions so that extraordinary temperatures are not required.

The energy that living things use for their day-to-day activities comes from food molecules that can be broken

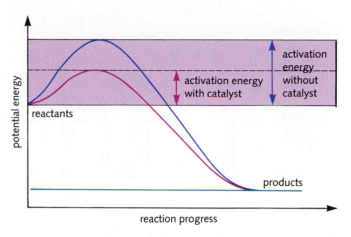

Figure 21.8   Catalysts act by lowering the activation energy of a chemical reaction.

down to release energy. Outside the living cell, raising food molecules above the activation energy usually means bringing them to a temperature so high that they actually begin to burn. This obviously cannot be done within a living cell. The solution, from the cell's point of view, would be to find the right kind of catalyst. Something that, like platinum, would lower the activation energy of a reaction so that it could occur at ordinary temperatures. The name we give to a biological catalyst is **enzyme.**

ENZYMES

Enzymes lower the activation energy of chemical reactions, allowing such reactions to take place at normal cellular temperatures and greatly increasing the rate at which the reaction occurs. Enzymes were first discovered in the nineteenth century, when Eduard and Hans Büchner showed that an extract prepared from broken yeast cells could perform all of the reactions of fermentation. (History lives on in the vocabulary of science: The word *enzyme* comes from *zyme,* the Greek word for "yeast.") As the molecules in the extract were analyzed, it was discovered that the catalytic portion of the extract was made up of proteins.

How does an enzyme lower the activation energy of a reaction? Most enzymes seem to follow a pattern in which they bind the molecules involved in the reactions they catalyze. The *reactants* in an enzyme-catalyzed reaction are referred to as the **substrates** of the reaction. The products? They're still called *products.* The first step in the reaction is the formation of an **enzyme–substrate complex** in which the substrate molecules are tightly

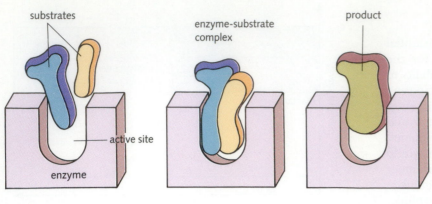

substrates

enzyme-substrate complex

product

active site

enzyme

1. Substrate molecules approach active site.

2. Substrate molecules bind to active site; reaction takes place.

3. Product molecule leaves active site; enzyme is unaltered.

Figure 21.9 Enzymes are chemical catalysts. Enzymes bind substrate molecules to a region known as the active site. The bound substrate then undergoes the chemical reaction, and product is released from the active site. Enzymes are able to speed up reactions by factors as large as a million.

bound to a part of the enzyme called the **active site** (Figure 21.9). The active site is a specialized part of the enzyme that performs at least two functions: (1) It binds the substrates tightly, and (2) it lowers the activation energy of the reaction.

The binding between an enzyme and its substrates is very specific. Many enzymes will bind one, and only one, group of substrate molecules and are able to distinguish these substrates from molecules that are nearly identical.

Carboxypeptidase—A Model Enzyme

We can appreciate of the nature of enzyme–substrate binding by examining an enzyme whose molecular mechanism we understand in detail. *Carboxypeptidase* is an enzyme found in the digestive system. It helps the body digest proteins in food by breaking their peptide bonds to release free amino acids. We know the complete molecular structure of the enzyme, and we understand how it binds its substrate.

Figure 21.10 illustrates the active site of the enzyme. The shape of an enzyme's active site is determined by its tertiary structure—and sometimes by its quaternary structure (in those cases where several subunits form the active site). As you can see, the movement of substrate into the active site causes the enzyme to change its shape just enough to lock the substrate into a precise position. Then a series of chemical groups in the enzyme bind to the amino acid at the end of the substrate polypeptide chain. The R groups of the amino acids in the active site then interact with the protein, weakening and ultimately breaking the peptide bond. An amino acid is then released. Carboxypeptidase enzymes in the digestive system clip

one amino acid after another off long polypeptide chains until the proteins are completely broken down.

The way carboxypeptidase works is typical of the chemical "tricks" used by enzymes to catalyze chemical reactions. The binding of substrates to the active site brings together a host of chemical groups in the side chains of the enzyme that are able to help make or break the appropriate chemical bonds.

As the case of carboxypeptidase illustrates, side groups of the amino acids in the active site often contain chemical groups that form weak bonds to hold the substrates in the best possible positions for the reaction to occur. Some active sites contain chemical groups that donate a proton to the substrate or remove one at just the right moment to assist the reaction, and some literally twist a molecule in order to help break one chemical bond or to form another. Regardless of the mechanism, though, substrate is converted into product in the active site. When the product appears, it is released from the enzyme, and the enzyme is free to bind another substrate and start the process all over again. This three-step process—*binding, catalysis,* and *release*—is characteristic of most enzyme-catalyzed reactions.

THE CHARACTERISTICS OF ENZYMES

With one or two important exceptions (see Theory in Action, The Enzyme That Isn't a Protein, p. 404), all enzymes that have been studied to date are proteins. The tremendous chemical flexibility of proteins has provided evolution with the material to develop enzymes that will catalyze almost any reaction.

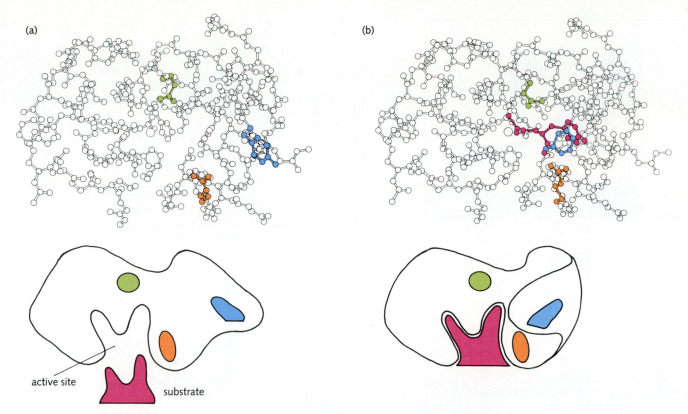

(a)

(b)

active site

substrate

Figure 21.10 The detailed geometry of an enzyme's active site often explains how it catalyzes a chemical reaction. A few enzymes, including carboxypeptidase, have been studied to the point where their action is understood right down to the atomic level. In this diagrammatic view of the action of the carboxypeptidase enzyme, a substrate molecule binds to the active site. The act of binding leads to a conformation change in the enzyme, which closes the cavity and holds the substrate firmly in place.

Enzymes Obey the Laws of Chemistry

As remarkable as they are, even enzymes cannot violate the laws of thermodynamics. Therefore, although enzymes can lower activation energy and dramatically increase the *rate* of a chemical reaction, they do not affect its energy requirements (Figure 21.11). A reaction in which the products are at a higher energy level than the reactants still requires a source of energy to power the reaction.

One of the important principles of chemistry is that *all chemical reactions, at least in principle, are reversible*. This means that a reaction like the burning of methane gas:

$$CH_4 + 2O_2 \xrightarrow{\quad HEAT \quad} 2CO_2 + 2H_2O$$

could also run in the reverse direction:

$$2CO_2 + 2H_2O \xrightarrow{\quad HEAT \ (energy) \quad} CH_4 + 2O_2$$

Common sense tells us that the reverse reaction is less likely than the forward one: The burning of methane *releases* 196 kcal/mole of energy, but the formation of methane *requires* 196 kcal/mole. In energetic terms, the first reaction is downhill and the second is uphill. Nonetheless, in a real situation, where a reaction vessel may contain 10^{20} molecules or more, at least a few molecules will undergo the reverse reaction. Let us suppose that we begin a chemical reaction by filling a container with compound A and compound B. A + B can undergo a chemical reaction in which products C + D are formed, so we can represent the complete reaction as

$$A + B \rightarrow C + D$$

However, once C and D begin to form, the reaction may also run in the reverse direction:

$$A + B \leftarrow C + D$$

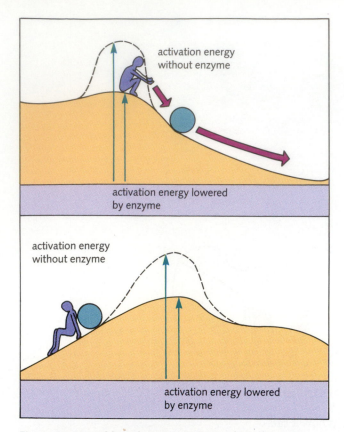

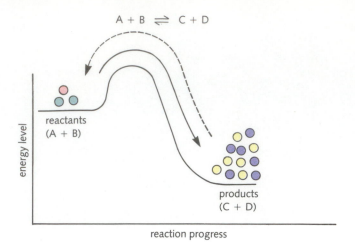

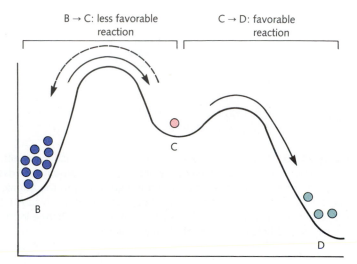

Figure 21.11 Although enzymes may reduce the activation energy associated with a chemical reaction, they do not affect the overall energy balance of the reaction. A reaction that releases energy (top) may occur more rapidly as a result of an enzyme. However, a reaction that requires energy will need a source to provide that energy, even if an enzyme is present to reduce the need for activation energy.

Figure 21.12 (top) At equilibrium, the forward and reverse reaction rates are equal. The fact that the products (C + D) in this reaction are at a lower energy level than the reactants (A + B) means that there will be an excess of product over reactant at equilibrium. (bottom) An energetically favorable reaction (C → D) that is coupled to an unfavorable one (B → C) may help to "pull" the less favorable reaction along.

So we can best represent the complete reaction like this:

$$A + B \rightleftharpoons C + D$$

Eventually the reaction reaches a point called **equilibrium,** where *the rates of the forward and reverse reactions are equal.* When a reaction reaches its equilibrium point, although individual molecules continue to participate in the reaction in both directions, there is no longer any *net* change in the amount of the products or reactants (Figure 21.12). The ratio of products to reactants at equilibrium in a chemical reaction tells us in which direction

the reaction tends to go. Reactions that tend to go in the forward direction, for example, produce an excess of products over reactants when they reach equilibrium.

The ratio of product to reactant is related to the energy change in a reaction. If equal amounts of product and reactant are present at equilibrium, then the energy change associated with the reaction is close to zero. Reactions releasing a large amount of energy usually show an excess of product over reactant. The hydrolysis of ATP to ADP and phosphate (P_i) is an example of such a reaction:

$$ATP + H_2O \rightarrow ADP + P_i + \text{energy released}$$
$$(7 \text{ kcal/mole})$$

At equilibrium, the ratio of product to reactants in this reaction is greater than a million to one; this is typical of reactions that release energy. In contrast, chemical reactions that require energy show an excess of reactants over products at equilibrium—they tend to run in the reverse direction.

Enzymes cannot affect the equilibrium of a chemical reaction. Just as enzymes cannot affect the energy change associated with a reaction (although they can lower energies of activation), they cannot affect the ratios of reactant to product at equilibrium. However, enzymes can dramatically affect the *rate* at which a reaction reaches equilibrium, speeding up a chemical reaction enough to make it useful for living organisms.

Enzymes and the Mass-Action Principle

Because they are governed by the energetics of a reaction and do not affect an equilibrium constant, enzyme-catalyzed reactions are subject to an important chemical principle. *Any reaction at equilibrium, even one that requires energy, can be made to flow in the reverse, or uphill, direction by adding an excess of one of the components.* This is known as the principle of *mass action.*

The principle of mass action has an important consequence for the way in which enzymes work when they catalyze a series of reactions. Say we have a series of reactions that involve the formation of compound D from compound B: $B \rightarrow C \rightarrow D$. As shown in Figure 21.12, let's suppose the reaction that forms C is slightly uphill, so that the equilibrium constant is less than 1: There will be more compound (B) than compound (C) at equilibrium. Does this mean that the whole pathway cannot go forward without a special source of energy for the reaction from B to C? Not necessarily. If the reaction that forms D is strongly downhill, then the small amount of

compound C that forms will be converted to D just as soon as it appears. This means that a very favorable reaction can "pull" or "push" a less favorable reaction simply by removing the product or increasing the amount of substrate. This principle of mass action applies to enzyme-catalyzed reactions, and is important whenever a long series of reactions operate as a single pathway.

Enzyme Regulation

Enzymes can be affected by molecules other than their substrates. For example, molecules that are similar in shape to a substrate molecule may bind to the active site and prevent the enzyme from making contact with its substrate. When this occurs, the enzyme is said to be *inhibited.* Molecules that block the active site in this way are known as **competitive inhibitors,** because they compete with the substrate for the active site (Figure 21.13).

Penicillin, an antibiotic widely used to fight bacterial infections, is a competitive inhibitor of the enzyme *transpeptidase.* This enzyme normally forms chemical crosslinks between the fibrous molecules that make up the bacterial cell wall. When this enzyme's active site is filled with penicillin, these crosslinks cannot be formed and the cell wall is weakened. Gradually, the cell wall swells and bursts under osmotic pressure.

Enzymes can also be affected by molecules that bind outside the active site. These molecules can regulate the activity of the enzyme in a way that either increases or decreases the rate of the enzyme-catalyzed reaction. The molecules that do this are called by many different names, including **effector.** There are both **positive effectors** (which increase the reaction rate) and **negative effectors** (which decrease the reaction rate). Negative effectors are often called **noncompetitive inhibitors,** because they inhibit the reaction but do not compete with substrate for the active site (Figure 21.13).

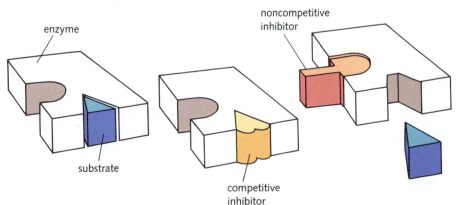

Figure 21.13 Enzyme inhibition. *Competitive inhibitors* compete with substrate for binding to an enzyme's active site. *Noncompetitive inhibitors* bind to other sites on the enzyme and alter its activity by causing a change in the shape of the enzyme molecule which ultimately affects the active site.

The Enzyme That Isn't a Protein

Anyone who studies a biochemical reaction usually expects to find an enzyme at the heart of it. Enzymes are often first purified by trying to produce a cell extract that shows the greatest activity in catalyzing a particular reaction. In that fraction, the investigator expects to find the enzyme. Imagine the surprise, therefore, of Thomas Cech, Paula Grabowski, and Arthur Zaug at the University of Colorado a few years ago. They were studying the processing of one particular form of ribonucleic acid (RNA) in *Tetrahymena*, a protozoan.

This RNA molecule is synthesized as a large precursor and "processed" to its final form: An intervening sequence, or intron, of 413 nucleotides cut out of the precursor molecule. Once the intron is removed, the two ends of the RNA are spliced back together to form the mature RNA. When these researchers maximized the rate of the cutting and splicing reactions, they expected to find a series of proteins that catalyzed the reactions. Instead, they found that the solution contained only the RNA that was being cut and spliced. There was no protein at all in the fractions that had the highest enzymatic activity!

Concerned that they had made a mistake, they set the results aside and continued to study the cutting and splicing reactions. Gradually, they returned to their original results, confirmed them, and confirmed them again. Only one conclusion was possible: *The RNA molecule was acting as an enzyme that catalyzed its own splicing.* The enzyme for the reaction was RNA itself. It became clear that the reaction proceeded in three steps. First, a guanosine base attacks and breaks a linkage in the precursor. Next, the intervening sequence is

removed and the two ends of the mature RNA are spliced together. Finally, a 15-base fragment breaks off the intron, and the remaining portion of the intervening sequence joins to itself to form a small circular RNA. Strictly speaking, the whole RNA molecule cannot be considered a true enzyme because it is altered by the reaction. But further studies have shown that the portion of the RNA molecule that is removed from the

RNA itself can act as a true enzyme, catalyzing nucleotide sequences.

It is now appreciated that several chemical reactions are catalyzed by RNA. To many scientists, the knowledge that RNA can catalyze chemical reactions suggests something else: the possibility that the evolution of life may have begun with small RNA molecules capable of controlling chemical reactions and even making copies of themselves.

An outline of the self-splicing RNA molecule discovered by Cech, Zaug, and Grabowski.

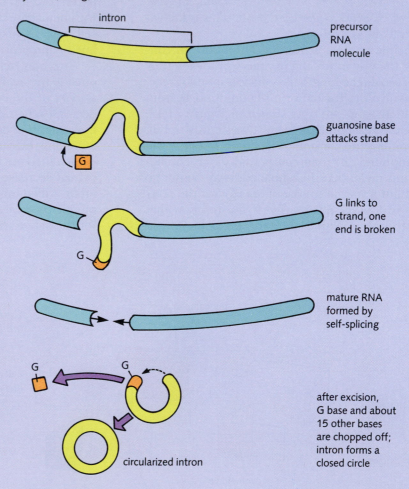

precursor RNA molecule

intron

guanosine base attacks strand

G links to strand, one end is broken

mature RNA formed by self-splicing

after excision, G base and about 15 other bases are chopped off; intron forms a closed circle

circularized intron

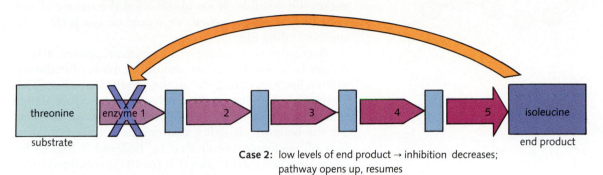

Case 1: high levels of end product → noncompetitive inhibition; pathway shuts off

Case 2: low levels of end product → inhibition decreases; pathway opens up, resumes

Figure 21.14 Feedback inhibition in a biochemical pathway. Isoleucine is an amino acid which is synthesized by a multistep pathway catalyzed by a series of enzymes. Isoleucine, the final product of the pathway, acts as a noncompetitive inhibitor to threonine deaminase, the first enzyme in the pathway. This feedback inhibition allows the pathway to be self-regulating.

How can a molecule that binds outside the active site affect the activity of an enzyme? Most of these molecules act by causing a change in the shape—the **conformation**—of the enzyme after they bind to it. A molecule whose binding affects the shape of an enzyme may change the shape of the active site itself and thereby regulate enzyme activity. The places where regulatory molecules can bind are sometimes called **allosteric sites.** The word *allosteric* means "different shape"; positive and negative effectors often change the shape of an enzyme.

Allosteric Effects and Feedback Pathways

Many enzymes are involved in chemical pathways in which a sequence of reactions synthesizes a molecule step by step from a starting compound. The level of the pathway's end product is often kept constant by a **feedback system** in which the end product *inhibits* the activity of the first enzyme in the pathway by binding to an allosteric site. This allows the pathway to be self-regulating. When the level of the end product is high, noncompetitive inhibition shuts the pathway off; when the level of the end product drops, the number of enzyme molecules that are inhibited also drops, and the pathway starts up again, replenishing the supply. The synthesis of *isoleucine,* an amino acid, is regulated in this way (Figure 21.14). Isoleucine is synthesized in a five-step pathway from another amino acid, *threonine.* The first enzyme in the pathway, *threonine deaminase,* contains an allosteric site that binds the end product, isoleucine. This self-regulating pathway, in which **feedback inhibition** from the end product controls the first reaction, enables the cell to regulate the level of isoleucine needed for protein synthesis.

Cofactors

Many enzyme-catalyzed reactions require more than just an enzyme and its substrates. Some enzymes require a particular ion or even a small molecule. These molecules are called **cofactors.** If a cofactor is an organic molecule, it is called a **coenzyme.** Cofactors are not considered substrates to a reaction, because they are not used up and converted to product, and they may be used over and over again. This means that a cell needs very little of a particular cofactor.

Many of the *vitamins* that we need in our daily diet are actually coenzymes that human cells are unable to synthesize by themselves. Many of the B-type vitamins, including vitamin B_2 (riboflavin), pantothenate, and niacin, are required as coenzymes for the chemical reactions that release energy within the cell. The ill effects caused by deficiencies in these vitamins are the direct results of having a particular set of reactions slowed down by the absence of a coenzyme.

What Can Enzymes Do?

When biochemists began to investigate the properties of enzymes, they came to appreciate the powers of these extraordinary molecules. The color of a flower, the materials we can eat for food, the shape of an eggshell, and the strength of a muscular contraction are all determined by enzymes. It is not unfair to say that the process of life itself is under the control of enzymes. However, we are left with an interesting question. If enzymes can do so many things, where does the information for *their* synthesis come from? As we will see, attempts to find an

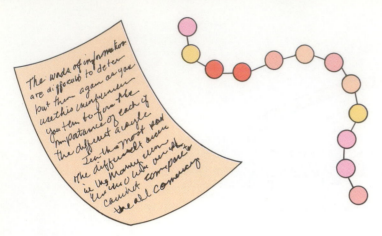

Figure 21.15 Schrödinger's concept of aperiodic crystals can be likened to a page of written text, capable of carrying information. In molecular terms, the closest analogy to a coded message might be a polymer constructed from a sequence of variable subunits.

answer to that question have led to the most important biological discovery of the twentieth century.

As the field of biochemistry developed in the 1930s and 1940s, many scientists joked that the cell could be thought of as nothing more than a "bag of enzymes." Even at the time, scientists realized that this was not quite true, but the sweeping generalization was their way of emphasizing the importance of understanding the chemical pathways that enzymes govern. There may be a lot more to a cell, but enzymes are critically important.

EPILOGUE: A LINK BETWEEN BIOCHEMISTRY AND GENETICS?

As the first half of the twentieth century was drawing to a close in the midst of a tragic world war, a number of scientists thought that the time was right to explain genetics in chemical terms. One of these was Erwin Schrödinger, a famous physicist who fled Germany at the beginning of the war and, early in the 1940s, turned some of his attention away from theoretical physics to indulge his interests in theoretical biology. Schrödinger had paid close attention to the science of genetics, and he was convinced that the nature of living things would be best understood by a plan of study that would integrate genetics with the rapid advances of biochemistry.

In 1944 he gave a series of lectures that were rewritten into a book entitled *What Is Life?* The logic of Schrödinger's approach was powerful and simple. Genetics, he reasoned, held the key to the nature of life. The genes that are passed from one generation to the next do much more than simply determine the colors of flowers and

the shapes of wings. They contain the information that makes life possible. If we understood the nature of the gene, Schrödinger argued, we would be much closer to understanding life.

Because of his training in the physical sciences, Schrödinger believed that genes should be made of ordinary matter, like everything else in the physical universe. The work of biochemists on the mechanism of enzyme action had helped to convince him that this view was correct. Yet the behavior of living matter was *so* different from that of nonliving matter that he believed it had to be organized along principles that (in 1945) remained to be discovered.

How is the *information* that living things require passed from generation to generation? What is the molecular nature of the gene? Schrödinger suggested that the basis of cellular information was something he called an "aperiodic crystal." This confusing term was widely misunderstood, but Schrödinger meant it to imply a molecule in which individual subunits, each of which might be different from its neighbors, held fixed positions (Figure 21.15). Such a molecule could contain information coded at the molecular level and would be able to pass that information from one generation to the next.

New Laws of Physics?

The central tenet of Schrödinger's book was that *new* laws of physics would have to be discovered to explain the properties of living organisms. He put it this way:

> It emerges that living matter, while not eluding the "laws of physics" as established to date, is likely to involve "other laws of physics" hitherto unknown, which, however, once they have been revealed, will form just as integral a part of this science as the former.

More than four decades after Schrödinger's book, phrases like this cause many a scientist to smile. One year *before* his book was published, a group of scientists had shown the molecular nature of the gene. Within ten years, the nature of the genetic code was understood, and within twenty years the code could be translated. In everyday language, many of Schrödinger's speculations were just plain wrong.

Nonetheless, his book is an important one in the history of biology. First, it convinced a number of talented scientists to leave physics and begin to work on biological problems. Second, he pointed with astonishing insight to two ideas that were very important in our coming to understand the molecular nature of the gene: (1) Life is so very different from nonlife that new principles will have to be developed to account for its properties. (2) These new principles will be based on the ordinary laws of physics and chemistry that are familiar to us.

Energy, the ability to do work, is fundamentally important to living things. The vast majority of energy available to living things on the earth is derived from sunlight. Energy is also involved in chemical reactions, where it may be either absorbed or released during the formation and breaking of chemical bonds. Biological systems make use of the ATP molecule as the common currency of energy storage and transfer.

Chemical reactions, even those that release energy, cannot begin until the reactants surpass an energy level known as the activation energy. This activation energy can be supplied by heating the reactants, but the requirement for activation energy can also be met via catalysts or enzymes, which are biological catalysts. Enzymes promote chemical reactions by lowering activation energies, and this can dramatically increase the rate of a particular reaction. Enzyme action usually involves the binding of substrate molecules to the active site of the enzyme, and the molecular details of the active site are important in causing the reaction to go forward.

Enzymes may be regulated by effector molecules and inhibitors. The actions of enzymes, which are protein molecules, directly control the chemical pathways and many of the characteristics of living organisms.

STUDY FOCUS

After studying this chapter, you should be able to:

- Explain the relevance of chemical energy to biology.
- Explain the concept of a chemical reaction and analyze what happens when a reaction takes place.
- Describe catalysis and activation energy.
- Illustrate how enzymes are able to catalyze biological reactions.

SELECTED TERMS

kinetic energy p. 394
potential energy p. 395
chemical energy p. 395
first law of
    thermodynamics p. 395
second law of
    thermodynamics p. 396
entropy p. 396
ATP p. 397
activation energy p. 398
catalysts p. 399
enzyme p. 399
substrates p. 399

active site p. 400
equilibrium p. 402
effector p. 403
competitive inhibitor p. 403
noncompetitive
    inhibitor p. 403
allosteric sites p. 405
cofactor p. 405
coenzyme p. 405

Discussion Questions

1. What happens during a chemical reaction? Are changes in energy always associated with chemical reactions?

2. What do investigators in thermodynamics study? Describe the two principal laws of thermodynamics.

3. What is entropy? How is entropy associated with chemical reactions?

4. What is the role of ATP in energy storage and transformation?

5. Explain the role of activation energy in a chemical reaction. How can the energy of activation be lowered?

6. What is an enzyme? What role does an enzyme play in a chemical reaction? How are enzymes controlled and regulated?

7. Is it possible to have two enzymes catalyzing a reaction, one of which makes the reaction go forward while the other makes it go in the reverse direction?

Objective Questions   (Answers in Appendix)

8. Which of the following statements is not true about enzymes?
   (a) An enzyme can catalyze a variety of reactions.
   (b) An enzyme lowers the activation energy of the chemical reaction.
   (c) An enzyme temporarily binds substrates at an active site.
   (d) An enzyme is usually a protein.

9. Entropy is defined as
   (a) disorder.
   (b) conversion from one state of matter to another.
   (c) coupling.
   (d) a state of equilibrium.

10. When an energy change takes place, the entropy of the universe increases. This statement best describes the
   (a) First Law of Thermodynamics.
   (b) Second Law of Thermodynamics.
   (c) Law of Energy of Activation.
   (d) Third Law of Third Thermodynamics.

11. Energy of activation can be defined as the
   (a) amount of randomness in a system.
   (b) total amount of energy used in biological reactions.
   (c) input needed to start a reaction.
   (d) conversion of chemical to heat energy.

12. By lowering the activation energy of a reaction, enzymes
   (a) become competitive inhibitors.
   (b) compete with substrate for the active site.
   (c) follow the principle of mass action.
   (d) increase the rate at which chemical reactions occur.

CHAPTER 22

Molecules and Genes

Life requires *information*. A living cell contains thousands of different proteins, each with a specific sequence of amino acids and a specific three-dimensional structure. The cell contains a bewildering variety of membranes, organelles, and filaments. Within the cell an interlocking series of biochemical pathways carry out the chemical reactions which make life possible. The organization of the cell requires a small mountain of information, and that information must be passed on from one generation to the next. But how can a cell carry information?

In our everyday experience, we are all familiar with systems that carry information: the alphabet, the dots and dashes of Morse code, and the binary code that represents information on a computer. For a molecule inside a living cell to carry information, it must have a structure as variable as these more familiar information systems. Such a molecule might be structured like a series of symbols strung out to produce a coded message. The cell would have to contain a de-coding system to put that information to useful work, and it would need something else as well: a system to *copy* the code so that a complete copy could be passed to each daughter cell when cell division takes place. If living cells contain a molecular code, the molecule that carries it must be capable of **replication**—of making an exact copy of itself.

In this chapter we will consider how a living cell might carry information, and we will retrace the pathway of one of the greatest discoveries of 20th century science—the molecular nature of the gene.

THE CHEMICAL NATURE OF THE GENE

We have seen how studies of inheritance led to the idea that the characteristics of an organism are controlled by individual elements called *genes*. We've also explored some aspects of biological chemistry and examined the kinds of molecules found in living cells. Is it possible that genes are made of ordinary molecules? Could a molecule actually carry genetic information? In this chapter we will try to answer these questions.

Coding Capacity: Proteins Seem to Be the Best Bet

Could protein be the genetic material? Eukaryotic chromosomes contain very little carbohydrate; they are about 30 percent nucleic acid and from *60 to 75 percent* protein. Most scientists in the first half of the twentieth century agreed that proteins were most likely to be the molecules that carry information. To begin with, protein molecules are composed of 20 amino acids, but nucleic acids are composed of just 4 nucleotides. Therefore, proteins should be better coding molecules (an alphabet with 20 letters would be far more expressive than one with just 4). Proteins also differ a great deal more from one species to another than nucleic acids do. Finally, although scientists at the time were familiar with the ability of protein enzymes to control and regulate chemical reactions, there was no evidence that nucleic acids were able to *do* anything.

The Transforming Principle

As early as the 1920s, a few people were working on projects that were eventually to identify the molecules responsible for inheritance. Frederick Griffith was a British scientist studying the way in which certain types of

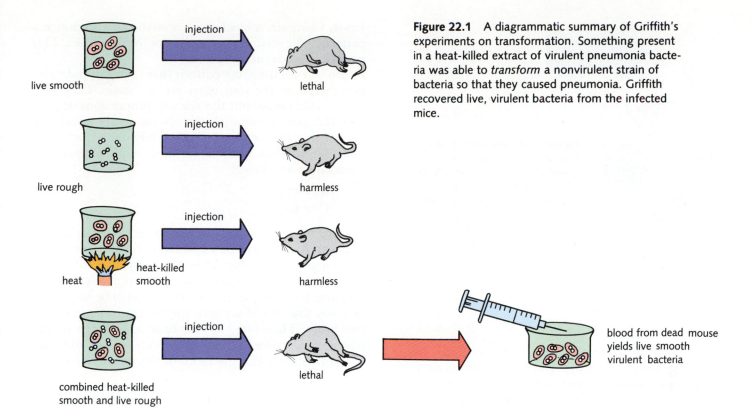

Figure 22.1 A diagrammatic summary of Griffith's experiments on transformation. Something present in a heat-killed extract of virulent pneumonia bacteria was able to *transform* a nonvirulent strain of bacteria so that they caused pneumonia. Griffith recovered live, virulent bacteria from the infected mice.

live smooth — injection — lethal

live rough — injection — harmless

heat → heat-killed smooth — injection — harmless

combined heat-killed smooth and live rough — injection — lethal

blood from dead mouse yields live smooth virulent bacteria

bacteria cause *pneumonia,* a serious and often fatal lung disease.

In 1928 Griffith had in his laboratory two different types—*strains*—of pneumonia bacteria. Both strains grew very well in special culture plates in his lab, but only one of them actually caused pneumonia in mice. Griffith managed to grow both strains successfully in the lab and noticed that he could distinguish one strain from the other simply by its appearance on a culture dish. The **virulent** (disease-causing) bacteria produced a jelly-like coating around themselves that made their colonies look smooth. The **nonvirulent** (harmless) strains made no such coating and instead grew into colonies with rough, jagged edges.

Griffith first tested to determine whether the coating might contain a disease-causing poison. To do this, he killed a culture of virulent cells and injected the dead cells (coatings and all) into the mice. The mice were not harmed by the injections. Griffith concluded that his suspicions about a poison were incorrect. Next he injected both live nonvirulent and killed virulent cells into a mouse (neither one of these injections by itself made any of the mice sick). To Griffith's surprise, when the two types of cells were injected at the same time, the mice developed pneumonia (Figure 22.1).

Today it is hard to appreciate just how startling this result was. The live nonvirulent bacteria never made the mice sick, and neither did the killed virulent ones. Why should the *combination* of these two have an effect that was different from what was seen when the two were administered *separately*?

To confuse matters further, Griffith recovered live bacteria from the animals that had developed the disease. Were these bacteria the same nonvirulent ones he had injected? He grew the bacteria on plates to find out. Now they formed the smooth colonies that were characteristic of the virulent strain. *Griffith's extract had transformed one kind of bacterium into another!* Griffith called the process he had discovered **transformation.**

Griffith speculated that when the live and killed bacteria were mixed together, a "factor" was transferred from the killed cells into the live ones. This factor changed the characteristics of the live cells in a permanent way, so that they henceforth acted like the virulent ones. What had actually happened? *The molecule of inheritance had been transferred, and it had changed the characteristics of the cell that received it.*

The Transforming Principle is DNA

In 1942 Oswald Avery, a scientist at the Rockefeller Institute in New York City, devised a simple research project: He would determine the chemical nature of the material

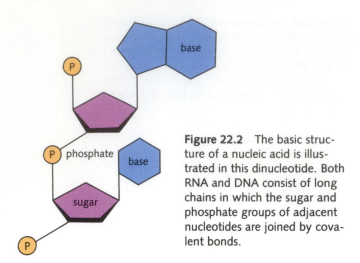

Figure 22.2 The basic structure of a nucleic acid is illustrated in this dinucleotide. Both RNA and DNA consist of long chains in which the sugar and phosphate groups of adjacent nucleotides are joined by covalent bonds.

responsible for the transformation effect in Griffith's experiments. Using Griffith's transformation system, Avery and his co-workers treated the transforming extract in ways that destroyed proteins, carbohydrates, lipids, and a kind of nucleic acid known as ribonucleic acid (RNA). Transformation still occurred. However, when they treated the transforming extract with an enzyme that destroyed another type of nucleic acid, deoxyribonucleic acid (DNA), transformation did not occur. Proteins did not carry genetic information. DNA did.

Avery's results were treated with some skepticism by his colleagues, if only because his answer seemed to make so little sense. Many in the scientific community had not suspected that nucleic acids were capable of playing such an important role.

CLUES TO THE STRUCTURE OF DNA

Base Composition

We have already seen that nucleic acids are polymers of nucleotides. Long chains, **polynucleotides,** can be built by linking nucleotides together (Figure 22.2). By the late 1940s, scientists understood the general chemistry of the polynucleotide. Still there was nothing that might distinguish the nucleic acid from any other biological polymer.

Some of the first clues to the structure of DNA came from experiments in which scientists determined the *composition* of DNA obtained from different organisms. DNA is a nucleic acid containing four nucleotide bases: **adenine, cytosine, guanine,** and **thymine.** Each of these bases is represented by a single letter: **A, C, G,** and **T.**

Erwin Chargaff, a biochemist, was the first to notice a pattern in the relative percentages of the four bases. This pattern is apparent in Table 22.1.

Do you see the same pattern that Chargaff did? The percentages of the four bases are not static. They vary over a wide range, but the relative proportions of guanine (G) and cytosine (C) are always nearly equal; the same is true for the proportions of adenine (A) and thymine (T). In more symbolic form, we might express this observation as follows:

$$[A] = [T]$$
$$[C] = [G]$$

This observation became known as **Chargaff's rule.** Chargaff himself had no explanation for the rule, but it did suggest something about the structure of DNA. Because there were always equal amounts of adenine and thymine, for example, the molecule had to be organized in a way that would account for the equivalence of the two bases. Therefore, to our basic knowledge of the chemical structure of a polynucleotide we can add a bit of information about the ratios of the various bases.

The X-ray Pattern

If a molecule can be crystallized, X-ray diffraction may produce a scattering pattern from the crystal that yields information about the molecule's internal structure. The repeating pattern of atomic bonds that exists within a crystal scatters a beam of X rays in a regular pattern. When researchers realized that DNA might be an interesting and important molecule, a number of X-ray crystallographers began to work on its structure. In most cases, however, the formation of good crystals from DNA turned out to be both a practical and a theoretical problem. Crystals were difficult to produce, and the patterns from successful crystals were difficult to interpret.

Table 22.1 *Base Composition (% of Total DNA Bases)*

Source of DNA	A	T	G	C
E. coli	26.0	23.9	24.9	25.2
Streptococcus p.	29.8	31.6	20.5	18.0
Yeast	31.3	32.9	18.7	17.1
Herring	27.8	27.5	22.2	22.6
Human	30.9	29.4	19.9	19.8

SOURCE: E. Chargaff and J. Davidson, eds., 1955, *The Nucleic Acids,* New York: Academic Press.

A few scientists, however, believed that DNA was such an interesting molecule that X-ray diffraction should be attempted even if perfect crystals couldn't be formed. One such person was Rosalind Franklin, a young scientist working with Maurice Wilkins, a crystallographer in London. Franklin drew a thick suspension of the fiber-like DNA molecules up into a glass capillary tube and used this DNA sample to scatter X rays. In the tube, she hoped, the thick suspension of DNA molecules would be forced to line up so that the molecules were parallel to the tube. Like spaghetti drawn through a straw, the molecules were all arranged in the same direction—not perfect enough to give a crystal-like pattern, but just good enough to yield a few clues about the structure of the DNA molecule.

Interpreting the X-ray Pattern

One of the X-ray patterns produced by Franklin's DNA samples is shown in Figure 22.3. The pattern contains two critical clues to the structure of the DNA molecule (graphically summarized in the figure).

Clue number 1: The two large dark patches at the top and bottom of the figure showed that some structure in the molecule was arranged at a right angle to the long axis of DNA and repeated at a distance of 3.4 Å (see Appendix: The Metric System). In other words, something in the molecule was arranged like the rungs of a ladder.

Clue number 2: The X-like mark in the center of the pattern showed that something in the molecule was

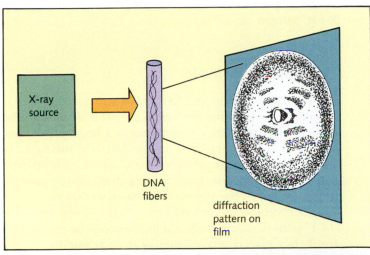

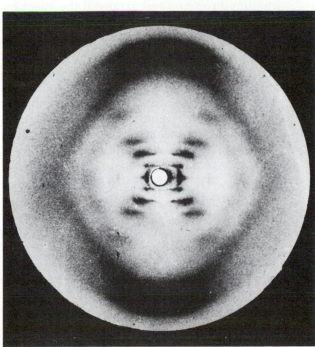

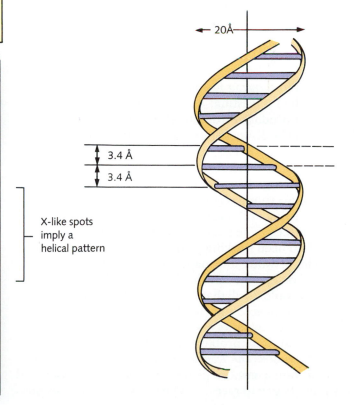

Figure 22.3 (left) DNA fibers were taken up in a thin tube so that most of them were oriented in the same direction. An X-ray diffraction pattern was then recorded on film. (bottom) X-ray diffraction pattern of DNA in the "B" form, as taken by Rosalind Franklin in 1952. Franklin's X-ray pattern contained two important clues to the structure of DNA. The large spots on the top and bottom of the pattern indicate that there is a regular spacing of 3.4 Å along the length of the fiber. The "X"-shaped pattern in the center indicates that there is a zigzag feature in the molecule, which might be consistent with a helix.

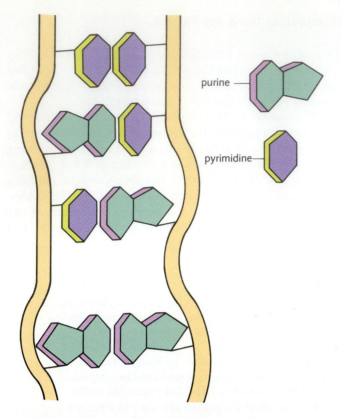

purine

pyrimidine

Figure 22.4 There were many false steps along the road to solving Franklin's X-ray pattern. For example, placing the nitrogenous bases inside the sugar-phosphate chains would give a fiber of the correct average width, but the different sizes of the bases, if there were no restrictions on how they were placed, would produce a "lumpy" fiber.

arranged in a zigzag fashion at a spacing of about 20 Å. If DNA had a twisting, helical configuration, the X-ray pattern it produced from the side would indeed look like that "X." The X-ray pattern suggested that the molecule was a *helix* and that its diameter was about 20 Å.

The scientific challenge remaining was to put all of these clues together to determine the structure of DNA. In 1953 two young scientists working at Cambridge University in England learned about Franklin's remarkable X-ray pattern. The scientists, James Watson and Francis Crick, had been working with molecular models, twisting little bits of wood and paper into various shapes in an effort to determine how nucleotides could form a structure that could do all that DNA was able to do. According to their own accounts of that discovery, one look at Franklin's X-ray pattern was the last bit of evidence they required, and the problem was solved.

In his book *The Double Helix,* James Watson wrote, "The instant I saw the picture my mouth fell open and my pulse began to race. . . . The black cross of reflections

which dominated the picture could arise only from a helical structure."

Watson and Crick had to account for several things. An ideal model for the structure of DNA would:

1. Explain Chargaff's rule.
2. Be able to carry coded information.
3. Be capable of replication.
4. Fit the chemistry known for polynucleotides.
5. Agree with the X-ray pattern's three predictions:
   (a) The molecule is helical.
   (b) One feature of the molecule is stacked at a spacing of 3.4 Å.
   (c) The width of the molecule is about 20 Å.

Because the X-ray data were not detailed enough to determine a structure directly, Watson and Crick hoped to find an arrangement of nucleotide subunits that would be consistent with each piece of the puzzle, including the X-ray pattern.

The Double Helix Model

In early 1953 Watson and Crick believed that they had a sensible structure for the molecule: the **double helix** model. They published their ideas in a brief paper that appeared in April of that year. The details of their model were surprisingly simple. Watson and Crick realized they could account for the 3.4 Å spacing in the X-ray pattern if they arranged the nitrogenous bases so that they were "stacked" on top of each other.

The 20 Å width of the molecule could be accounted for if they placed two *antiparallel* strands side by side in opposite directions and arranged the bases facing each other.

The helical twist that was evident in the diffraction patterns could be accounted for as well. All they had to do was twist the molecule so that the two strands twisted about each other.

At first, however, there were two problems with the model. First, what kinds of forces might hold the two strands together? Second, how could one solve the problems posed by the sizes of the nitrogenous bases? Two of the bases, adenine and guanine, belong to a chemical group known as the **purines.** They have *two* carbon–nitrogen rings in their basic structures. The other two, thymine and cytosine, are **pyrimidines:** They have a single ring, meaning that they are quite a bit smaller than the purines. This would cause a problem in the model. If two pyrimidines were paired, the two strands would have to be much closer than when two purines were paired, making the model "lumpy" (Figure 22.4).

Chargaff's rule showed how a "lumpy" helix could be avoided. If a purine was always paired with a pyrimidine, the helix wouldn't be lumpy. But was it possible for bonds to form between purines and pyrimidines to hold the two strands together? Watson and Crick remembered how hydrogen bonds (weak, noncovalent interactions) seemed to stabilize the structure of the α helix in proteins. To their delight, when James Watson drew a sketch of the bases, they could find perfect places for hydrogen bonds to form between A and T and between G and C (Figure 22.5).

The specific hydrogen bonding between the bases is known as **base pairing.** Watson, by the way, had never done very well in chemistry, and his sketch showed it. He didn't notice that a third hydrogen bond (highlighted in Figure 22.5) existed between G and C. Watson and Crick's insight solved one of the critical problems regarding the *biological* role of DNA. Prior to 1953, no one had been able to come up with a reasonable scheme for how a molecule might be *replicated*. But the structure of DNA itself contained an obvious answer to the riddle. Each strand is a *complementary* "copy" (although not an *exact* copy) of the other, which means that each contains the "information" required to reproduce the other strand. All that is required to copy the molecule is to separate the two strands and then to form a new strand for each original one by using the base-pairing rules suggested by Watson and Crick. James Watson and Francis Crick (Figure 22.6) had put together a three-dimensional model that accounts for the biological properties of DNA (Figure 22.7; see Theory in Action, The Prize, p. 414.)

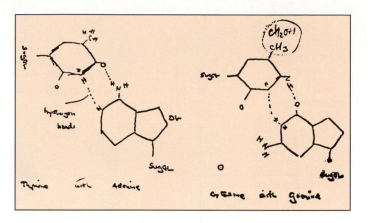

Figure 22.5 The problem of placing nitrogenous bases was solved by James Watson, who drew sketches (top) showing how hydrogen bonds might pair thymine with adenine and guanine with cytosine. The modern representation of A–T and G–C base pairing (bottom) shows *three* hydrogen bonds in the G–C pair, something that was missing from Watson's first sketch.

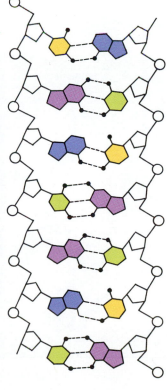

Figure 22.6 (left) James Watson and Francis Crick shortly after the publication of their double helix model for the structure of DNA. (right) The two strands of a DNA double helix are held together by hydrogen bonds between the nitrogenous bases. The strands run in antiparallel directions, forming a helix with a diameter of 20 Å.

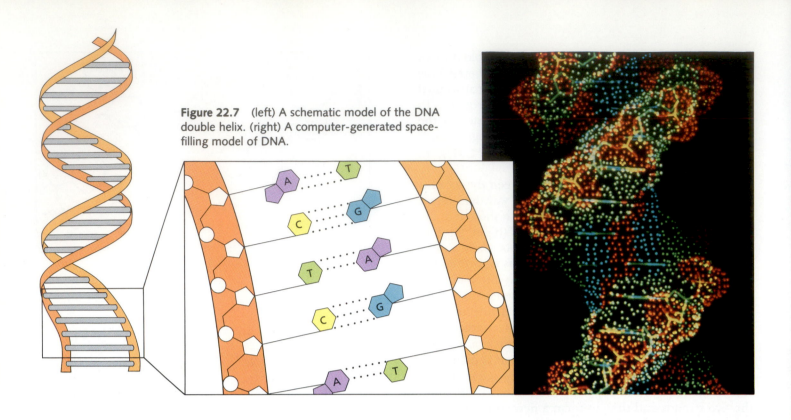

Figure 22.7 (left) A schematic model of the DNA double helix. (right) A computer-generated space-filling model of DNA.

THEORY IN ACTION

The Prize

The X-ray pattern that provided the critical clues for Watson and Crick was made by Rosalind Franklin. Her paper, which contained the pattern, was actually published in the very same issue of *Nature* that contained the double helix paper. Needless to say, she was not exactly overjoyed that a detailed interpretation of her own data was published simultaneously with the data itself.

Watson and Crick have never denied that Franklin's pattern contained the important new information that made their model possible. But some writers have questioned whether their actions, interpreting the data after an advance look, were ethical. One of Franklin's biographers has

argued that Watson and Crick deprived Franklin of the credit she deserved for making one of the fundamental scientific discoveries of the twentieth century.

Unfortunately, Rosalind Franklin lived long enough to see only the first hint of how important her work had been. She died of cancer in 1957, just as DNA was beginning to become a central focus in biological research. A few years later, the Nobel Prize committee decided that the time was right to award the Prize to the group that had developed the double helix model. The Nobel Prize is never given posthumously. It was awarded to Watson and Crick, and Franklin's associate, Maurice Wilkins.

DNA REPLICATION

The DNA molecule, as we see, suggests a method for its own replication. Watson and Crick pointed this out in a paper published shortly after the announcement of the double helix structure.

Figure 22.8 illustrates how the double helix is unwound to enable each strand to serve as the *template* for the synthesis of a new strand. The rules of complementary base pairing help to control the process and ensure that each newly synthesized strand has the appropriate base sequence.

The replication process is said to be **semiconservative,** which means that the two original strands of the helix are separated and that at the end of replication, each strand is paired with one of the newly synthesized strands. In this way, the replication process produces two identical DNA molecules, each composed of one "old" strand and one "new" strand.

DNA Polymerase

It was soon discovered that DNA does not replicate in isolation but rather requires a number of special enzymes to unwind the double helix and synthesize new DNA strands. The most important enzyme in this process is known as **DNA polymerase.** The action of DNA polymerase is illustrated in Figure 22.8. DNA polymerase contains a binding site for attachment to the DNA strand and also a binding site for nucleotides. The nucleotides that attach to the enzyme are *triphosphates:* Like ATP, they have three phosphate groups attached to them. DNA polymerase binds the correct nucleotide to the growing DNA strand and uses the energy from splitting off two of the three phosphates to form a covalent bond linking the nucleotide to the growing chain. As DNA polymerase moves along the chain to attach the next molecule, part of the enzyme "proofreads" the work it has just done by checking the nucleotide pair to ensure that the proper base pairing has taken place. If an incorrect nucleotide has been inserted, this portion of the enzyme swiftly removes the nucleotide from the chain, and the molecule starts work over again.

This intricate process of base pairing, covalent bond formation, and proofreading is at the heart of DNA replication in all organisms. In tiny bacteria, a few DNA polymerase molecules may work to replicate the single DNA molecule that contains all of the cell's genetic information. In larger organisms, thousands of DNA polymerase molecules may work at scattered sites throughout many chromosomes to complete DNA replication in time for cell division to begin.

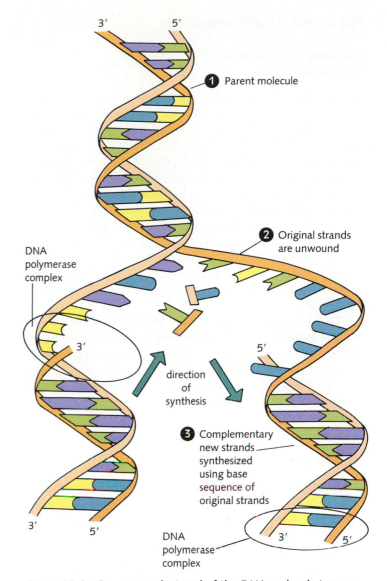

Figure 22.8 Because each strand of the DNA molecule is complementary to the other strand, each may serve as a *template* against which a new strand may be constructed. DNA replicates in semi-conservative fashion. Each strand of the helix serves as a template against which a new strand is assembled, following base-pairing rules.

THE FLOW OF INFORMATION

The Role of DNA as a Coding Molecule

Although the double helix model makes it clear that the sequence of bases in DNA could be copied, it does not explain how DNA could encode information that is important to the cell or how DNA actually works on a day-to-day basis. What, exactly, does DNA do? A quick answer is that *DNA directs the synthesis of proteins.* Why are proteins so important? On a biochemical level, the reason for emphasizing proteins is very simple. The formation of most other biological molecules, and therefore most of the cell, is catalyzed by proteins. Enzymes, nearly all of which are proteins, are responsible for the synthesis of nearly all other macromolecules. If DNA codes for proteins, then the proteins are capable of making everything else.

These discoveries and others laid the foundation for what is generally called **molecular biology,** the study of biology at the level of the molecule. Once the role of DNA was understood, "molecular biology" began to take on a special meaning as it focused on the structure and function of nucleic acids.

Nucleic Acid Templates

A **template** is a mold or model by which something is constructed. A cookie cutter acts as a template, allowing the baker to cut consistent cookie shapes from rolled out dough. A stencil sheet serves as a template for producing lettering. Nucleic acids can act as templates, too, and we have already seen one example of this: DNA replication. The source of this ability is the *base-pairing* mechanism proposed by Watson and Crick. The double-stranded DNA molecule is held together by a series of relatively weak hydrogen bonds. The weakness of these bonds is an important feature of the DNA molecule, because it enables the two strands to be separated when the molecule is replicated.

A single nucleic acid strand contains a sequence of molecules that are ready to form hydrogen bonds with any other molecule that will fit. In molecular terms, that strand forms a surface against which new molecules can be placed and to which they will bind if the molecules "fit." This means that one nucleic acid molecule can serve as a template against which another can be assembled. It also means that *information* in the form of a sequence of bases can be copied from one molecule to another by using the existing molecule as a template.

The Role of RNA

We have said that the sequence of bases in DNA directs the synthesis of proteins. *How* does DNA do this? To begin with, DNA does not directly determine the amino acid sequence of a protein. Instead, the cell makes a complementary copy of one strand of a DNA molecule by synthesizing a single-stranded molecule of **ribonucleic acid (RNA).** RNA (Figure 22.9) differs from DNA in two important respects:

1. It uses a different sugar in its backbone (ribose instead of deoxyribose).

2. The four bases found in RNA are adenine, cytosine, guanine, and **uracil.**

Thymine, which is present in DNA, is replaced in RNA by uracil. (Scientists are divided about *why* RNA and DNA should differ in one of their bases. One theory is that the very slight difference between the shapes of the two bases makes it easier for enzymes to distinguish RNA from DNA.) The process of RNA synthesis is known as **transcription.** A special enzyme known as **RNA polymerase** synthesizes RNA as a complement to one strand of a double-stranded DNA molecule (Figure 22.9). Several types of RNA molecules are produced by transcription, and each of these plays a role in one of the cell's most important activities, the synthesis of proteins.

PROTEIN SYNTHESIS

An RNA molecule used to direct the synthesis of a protein is known as **messenger RNA (mRNA).** The name is appropriate, because mRNA literally carries a message from the DNA molecule to the site of protein synthesis. The message is a set of instructions for how to build a protein, and the language of that message is known as the **genetic code.**

The Genetic Code

Twenty different amino acids are commonly found in living cells. How can a code that consists of only 4 different bases in an mRNA strand specify 20 different components? If each base stood for a single amino acid, only 4 different amino acids could be specified. If *pairs* of bases were used to stand for amino acids, only 16 different amino acids could be represented ($4 \times 4 = 16$ different combinations). But if 3 bases were used for each amino acid, we would have more than enough coding

ribose

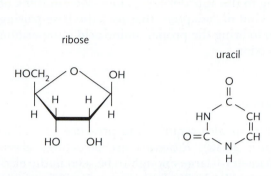

uracil

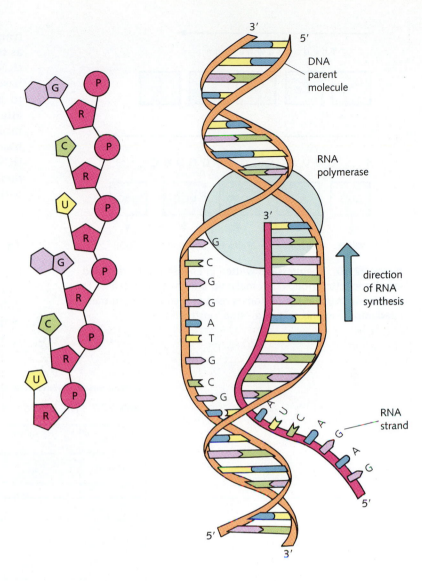

Figure 22.9 The structure of RNA. There are two important differences between RNA and DNA: RNA contains *ribose* (in place of *deoxyribose*) and employs the nitrogenous base *uracil* (in place of *thymine*). (right) Employing the same base-pairing rules as DNA, a complementary RNA strand may be made by the enzyme RNA polymerase.

capacity: $4 \times 4 \times 4 = 64$ different combinations. Remarkably, this little bit of numerical reasoning is borne out in the lab.

The genetic code is a *triplet* system, in which combinations of three bases are "read" by the protein-synthesizing machinery, and one amino acid is brought into position according to each *three-base triplet*, or **codon** (Figure 22.10). In the early 1960s biologists began to unravel the genetic code. One codon at a time, they learned what each three-base combination "stood for." Today we can represent these data in a simple table that shows the precise amino acid specified by each codon (Table 22.2). You will note that there are three codons that don't stand for *any* amino acid. They are **stop codons,** and their occurrence signifies the end of a polypeptide chain.

Most amino acids can be specified by more than one codon, and this gives the genetic code a great deal of flexibility. For example, there are six different codons for the amino acid *arginine* and four for *proline*.

Transfer RNA

The involvement of RNA with protein synthesis does not stop with mRNA. There are two other major forms of RNA—transfer RNA and ribosomal RNA—each of which is directly involved in the synthesis of proteins. Each cell contains a set of special RNA molecules that literally "read" the codons in mRNA and ensure that the proper amino acids are brought into position. These molecules, which

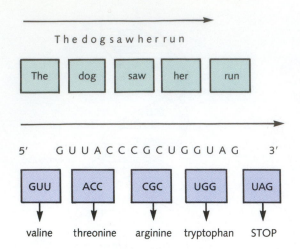

The dog saw her run

| The | dog | saw | her | run |

5′   G U U A C C G C U G G U A G   3′

| GUU | ACC | CGC | UGG | UAG |

valine   threonine   arginine   tryptophan   STOP

Figure 22.10 The message written in English can be decoded into three-letter groups that make up a five-word sentence. The genetic message at the bottom is read as 5 codons, each translated into an amino acid or a "stop" instruction.

transfer the proper amino acids into position, are known as **transfer RNA (tRNA)** molecules (Figure 22.11). Each tRNA molecule contains an **anticodon,** a string of three bases that is complementary to the codon it is designed to help read. At the other end, a special enzyme has attached the proper amino acid. Therefore, the tRNA molecules can line up the proper amino acids simply by base-pairing to the appropriate codons. We can think of tRNA as a kind of "adaptor" that uses the base-pairing mechanism to bring the proper amino acid into position with each codon.

The Ribosome

Protein synthesis also requires the presence of a **ribosome** (Figure 22.12). *Ribosomes* are structures about 250 Å in diameter—large enough to be seen in the electron microscope. Each ribosome is composed of two sub-

Table 22.2 *The Genetic Code*

First Position (5′ end)	Second Position				Third Position (3′ end)
	U	C	A	G	
U	phenylalanine	serine	tyrosine	cysteine	U
U	phenylalanine	serine	tyrosine	cysteine	C
U	leucine	serine	stop	stop	A
U	leucine	serine	stop	tryptophan	G
C	leucine	proline	histidine	arginine	U
C	leucine	proline	histidine	arginine	C
C	leucine	proline	glutamine	arginine	A
C	leucine	proline	glutamine	arginine	G
A	isoleucine	threonine	asparagine	serine	U
A	isoleucine	threonine	asparagine	serine	C
A	isoleucine	threonine	lysine	arginine	A
A	(start) methionine	threonine	lysine	arginine	G
G	valine	alanine	aspartate	glycine	U
G	valine	alanine	aspartate	glycine	C
G	valine	alanine	glutamate	glycine	A
G	valine	alanine	glutamate	glycine	G

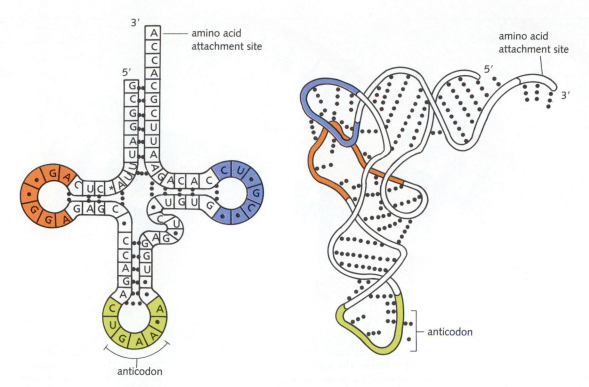

Figure 22.11 Transfer RNA (tRNA), shown in diagrammatic form (left) and in its actual three-dimensional shape (right). tRNA molecules are single strands of RNA which are looped and twisted into a shape somewhat like a cloverleaf. A group of three unpaired bases (the anticodon) is exposed at one end of the molecule. An amino acid is covalently attached at the attachment site, as noted.

units (one large, one small), and each of those subunits is composed of a large number of ribosomal proteins and from one to three RNA molecules. The RNA that makes up the ribosome is called the ribosomal RNA (rRNA). The function of rRNA is not clear. It plays an important role in holding all of the ribosomal proteins together, and there is some evidence that it may help the ribosome to recognize the proper site on an mRNA molecule to start making proteins.

The process of protein synthesis is known as **translation.** Protein synthesis occurs in a number of distinct steps, which are listed below and diagrammed in Figure 22.13.

- A messenger RNA molecule is released into the cytoplasm. Near one end of the mRNA molecule, there is a codon with the base sequence *AUG.* In nearly all mRNA molecules, the *start signal* is *AUG,* which codes for the amino acid *methionine.*

- The *small-ribosomal subunit* (along with several soluble proteins called *initiation factors*) recognizes the end of the mRNA and binds to the AUG codon to form an

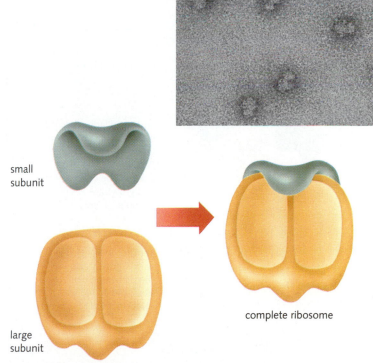

Figure 22.12 (top) Electron micrograph of ribosomes. (bottom) Ribosomes consist of large and small subunits, each of which contains RNA and proteins.

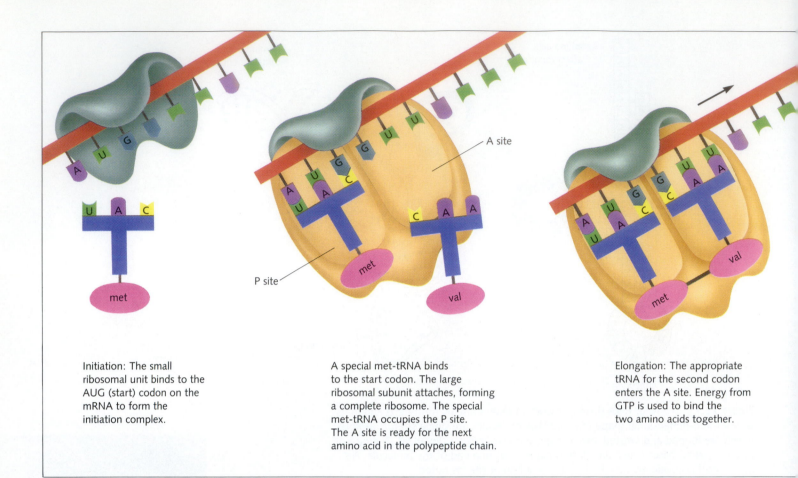

Initiation: The small ribosomal unit binds to the AUG (start) codon on the mRNA to form the initiation complex.

A special met-tRNA binds to the start codon. The large ribosomal subunit attaches, forming a complete ribosome. The special met-tRNA occupies the P site. The A site is ready for the next amino acid in the polypeptide chain.

Elongation: The appropriate tRNA for the second codon enters the A site. Energy from GTP is used to bind the two amino acids together.

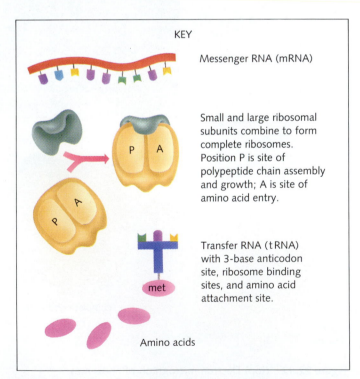

KEY

Messenger RNA (mRNA)

Small and large ribosomal subunits combine to form complete ribosomes. Position P is site of polypeptide chain assembly and growth; A is site of amino acid entry.

Transfer RNA (tRNA) with 3-base anticodon site, ribosome binding sites, and amino acid attachment site.

Amino acids

Figure 22.13 Protein synthesis takes place on ribosomes as a multistep process.

initiation complex. A special version of met-tRNA (tRNA with the amino acid methionine attached to it) enters the small subunit and binds to the start codon, and then the *large subunit* attaches, forming a complete *ribosome.*

■ As far as we can tell, the first tRNA molecule occupies a special site inside the ribosome called the *P site.* ("P" stands for polypeptide, because it is at this site that the growing *p*olypeptide chain is later found.) When the right tRNA for the second codon enters the ribosome at the *A site* (because the *a*mino acids enter there), we have the conditions necessary to begin building a polypeptide.

■ Now part of the ribosome acts like an enzyme, forming a peptide bond between methionine and the second amino acid—in this case, valine. The energy to form the bond comes from the hydrolysis of GTP (guanidine triphosphate, a molecule very similar to ATP). The ribosome briefly contains a dipeptide in the A site.

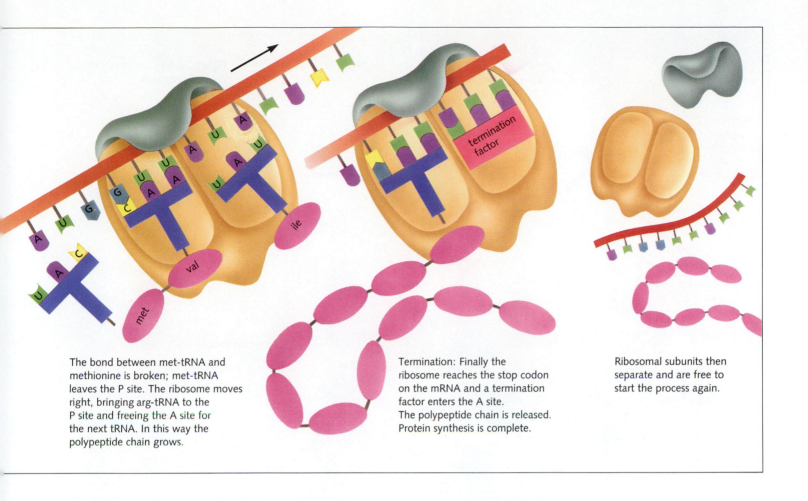

The bond between met-tRNA and methionine is broken; met-tRNA leaves the P site. The ribosome moves right, bringing arg-tRNA to the P site and freeing the A site for the next tRNA. In this way the polypeptide chain grows.

Termination: Finally the ribosome reaches the stop codon on the mRNA and a termination factor enters the A site. The polypeptide chain is released. Protein synthesis is complete.

Ribosomal subunits then separate and are free to start the process again.

■ Next, the tRNA that has lost its amino acid leaves the P site, and the ribosome moves relative to the mRNA so that the second tRNA containing the dipeptide now occupies the P site. The ribosome is now ready for a third tRNA to enter at the A site, allowing the growth of the polypeptide chain to continue.

■ Finally, after 50, 60, or maybe 300 amino acids have been added, the mRNA reaches the point of its stop codon. As we have seen, there is no amino acid corresponding to the stop sequences. Instead, when the stop codon takes up a position within the A site, a special set of molecules known as *termination factors* enter the A site and cause the release of the polypeptide chain. Synthesis of the protein is complete.

■ At termination, not only is the chain released, but the ribosome actually falls apart into large and small subunits. These are released into the cytoplasm and are free to start the process all over again.

The role of the ribosome is to organize the process of protein synthesis: from the recognition of the start codon to the formation of every peptide bond to the termination of the chain (Figure 22.14). The ribosome is really a small factory with a host of different enzymatic functions directed toward the synthesis of a finished product, the protein (see Theory in Action, Seeing Genes at Work, p. 423).

The flow of information from DNA through RNA to protein can be summarized in a simple diagram:

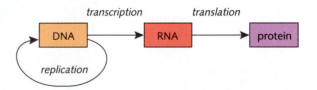

Among molecular biologists this pattern of information flow has become known as the "Central Dogma." The word "dogma" is used almost as a joke, because science must never be dogmatic. As we will see shortly, the central dogma is not universally correct.

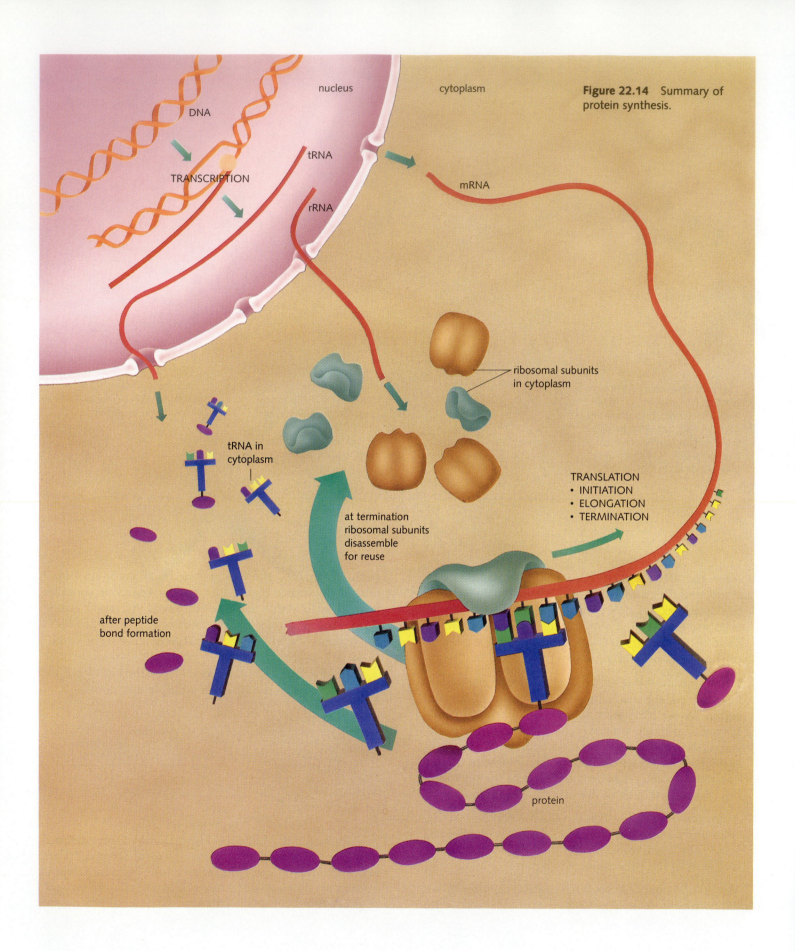

nucleus

cytoplasm

Figure 22.14 Summary of protein synthesis.

DNA

tRNA

mRNA

TRANSCRIPTION

rRNA

ribosomal subunits
in cytoplasm

tRNA in
cytoplasm

TRANSLATION
• INITIATION
• ELONGATION
• TERMINATION

at termination
ribosomal subunits
disassemble
for reuse

after peptide
bond formation

protein

Seeing Genes at Work

Is a picture really worth a thousand words? In some cases, maybe a little more. In the case of molecular biology, sometimes a single picture makes everything clear. Oscar Miller, a scientist at the University of Virginia, has made an art of getting just the right picture. He has helped develop techniques to isolate DNA molecules gently enough so that they can be prepared for electron microscopy without breaking.

One of his most striking electron micrographs is shown here. It was made from nucleic acids isolated from a bacterium that was actively making protein. This one micrograph captures both transcription (RNA synthesis) and translation (protein synthesis). The picture also shows that prokaryotes carry out transcription and translation simultaneously: Ribo-

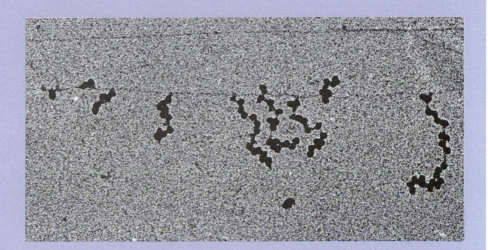

somes actually begin to translate one end of an mRNA before the synthesis of the other end is complete. This would not be possible in eukaryotic cells, in which the nuclear envelope separates transcription, which occurs in the nucleus, from translation, which occurs in the cytoplasm.

THE VIRUSES: A LESSON IN INFORMATION TRANSFER

Our short history of the birth of molecular biology has been, out of necessity, incomplete. Until now we have left out one branch of biology that was very important in the identification of DNA as the molecule of heredity: the study of viruses. The viruses have taught us some important lessons about the role of DNA in living cells.

The word *virus* means "poison" in Latin, and for quite a few years the term was applied to any organism that could cause disease. Viruses as we understand them today were discovered in the early 1900s by the Russian biologist Dimitri Iwanowski.

Iwanowski was interested in a disease that had ruined thousands of acres of tobacco plants. The leaves of the infected plants were covered with large yellow spots, forming a pattern that the farmers called a "mosaic." The spots grew larger and larger, gradually destroying the

leaf and often the whole plant. The scientist discovered that he could transmit the disease from sick to healthy plants simply by crushing some of the diseased leaves and dripping the juice on the leaves of healthy plants. Within a few days the healthy plants were infected with the mosaic disease. Something in the juice must have transmitted the disease to the healthy plants.

Iwanowski passed the juice through a filter so fine that not even a tiny cell fragment could pass through. Under the microscope he could see nothing in the filtered liquid. Nonetheless, the filtered juice *still* caused the mosaic infection! He concluded that an infectious organism so small that it could pass through the finest filter must be causing the disease.

Today we call such tiny, subcellular particles *viruses*, and the organism that Iwanowski encountered is called the *tobacco mosaic virus (TMV)*. **Viruses** *are subcellular particles that are composed of nucleic acid and protein and can infect a living cell.* Viruses exist that can infect virtually

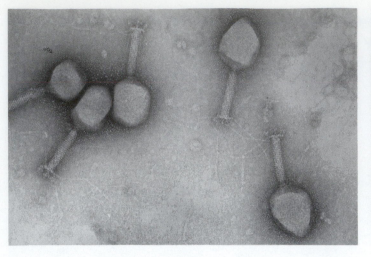

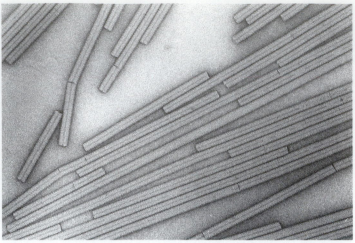

Figure 22.15 Electron micrographs of (left) bacteriophage T4, and (right) tobacco mosaic virus.

any type of cell: animal or plant, large or small, prokaryote or eukaryote. The electron microscope has made viruses "visible" to us, and these small particles often have exotic and beautiful structures (Figure 22.15).

Bacteria Eaters

One important class of viruses attacks bacteria. Because these viruses are easy to grow in the laboratory, we know a great deal about them and they have taught us much about the biological role of DNA. They are known as **bacteriophage,** which means "*bacteria eaters.*" We will take a close look at the life cycle of just one bacteriophage, a virus with the scientific name *T2*, and will glance briefly at bacteriophage T4.

Bacteriophage T2 contains a single long DNA molecule enclosed in a protein capsule (Figure 22.16). Left to itself, a single virus particle is almost *inert*; it does not grow or eat or metabolize or reproduce. But when a T2 particle makes contact with the right kind of bacterium, all of that changes. First, the tail fibers of the protein capsule make contact with the cell wall of the bacterium. The virus changes its shape a bit to allow the head portion of the capsule to attach to the cell wall, and the genetic material of the virus is injected into the bacterium.

In 1952, Alfred Hershey and Martha Chase realized that this system could be used to determine the chemical nature of the infectious viral genetic material. They prepared particles of bacteriophage T2 that contained two radioactive labels: Their DNA was made radioactive with ^{32}P, and their proteins with ^{35}S. They mixed these labeled viruses with bacteria and waited a few minutes for the virus particles to attach to the bacteria and begin the process of infection. Next they shocked the mixture of bacteria and viruses with violent agitation (in a blender),

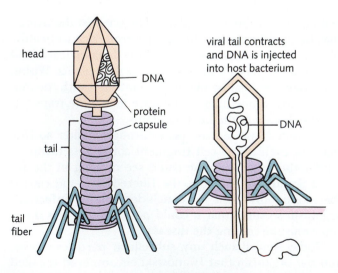

Figure 22.16 Bacteriophage T2. The DNA core of the virus is surrounded by an intricate protein coat (left) that protects the nucleic acid and helps to inject it into a bacterial cell (right).

breaking the virus coats off the surface of the larger bacteria, and separated virus particles from bacteria by centrifugation. They then analyzed the pattern of radioactivity in the bacteria. Hershey and Chase found that most of the ^{35}S label was still with the virus particles, whereas the ^{32}P label had been inserted into the bacteria. The virus particles that eventually burst from the infected bacteria also contained the ^{32}P label in their DNA (Figure 22.17).

This simple experiment showed that the DNA carried the genetic material necessary for a viral infection and furnished powerful proof that the basic molecule of inheritance was DNA.

Once inside a cell, the viral DNA goes to work, and just like a rebel army, begins to take over the cell. To the molecules of the host cell, the viral genes seem no different from the cell's own. A number of viral mRNAs are made and translated into proteins, which begin to shut down the bacterium's own genetic machinery. Then the viral DNA begins to replicate, again using the infected cell's own enzymes, and before long the cell is filled with hundreds of copies of viral DNA.

In the late stages of a T4 infection, genes on the viral DNA molecules coding for the proteins that make up the coat of the virus are activated. As the cell manufactures and accumulates these proteins, complete virus particles begin to assemble inside the cell, until the cell is filled with them. Finally the cell bursts, releasing hundreds of virus particles to the surrounding medium. The whole process takes only about 40 minutes, and all of it is the work of a single DNA molecule from the virus (Figure 22.18).

The power of a single DNA molecule to be capable of destroying a healthy cell tells us quite a bit about the role of DNA as the central molecule in the process of heredity. The viral DNA, after all, contains the genetic information of the virus—information that is capable of shutting down the host cell, replicating its own DNA, and producing new copies of the virus particle, coat proteins and all, that are ready to infect other cells.

Other Viruses

Not all viruses work in the same way as bacteriophage. A great many viruses do not destroy the cells they infect but rather grow inside them and "bud" new virus particles off from the surface of the cell. A few others insert

Figure 22.17 The Hershey–Chase experiments. In order to determine whether DNA or protein carried viral genetic information, Alfred Hershey and Martha Chase labeled viral DNA with ^{32}P and viral protein with ^{35}S. By shearing virus particles off the surface of bacteria, they discovered that ^{32}P (DNA) entered the infected cell.

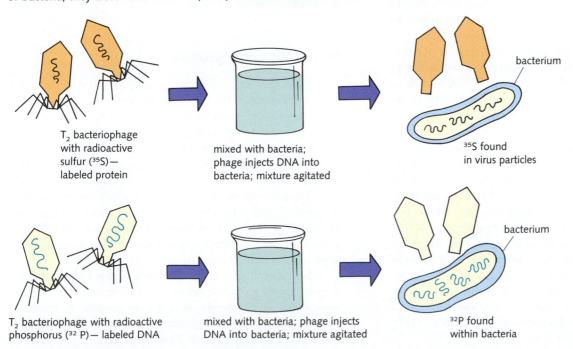

T$_2$ bacteriophage with radioactive sulfur (^{35}S)—labeled protein

mixed with bacteria; phage injects DNA into bacteria; mixture agitated

^{35}S found in virus particles

bacterium

T$_2$ bacteriophage with radioactive phosphorus (^{32}P)—labeled DNA

mixed with bacteria; phage injects DNA into bacteria; mixture agitated

^{32}P found within bacteria

bacterium

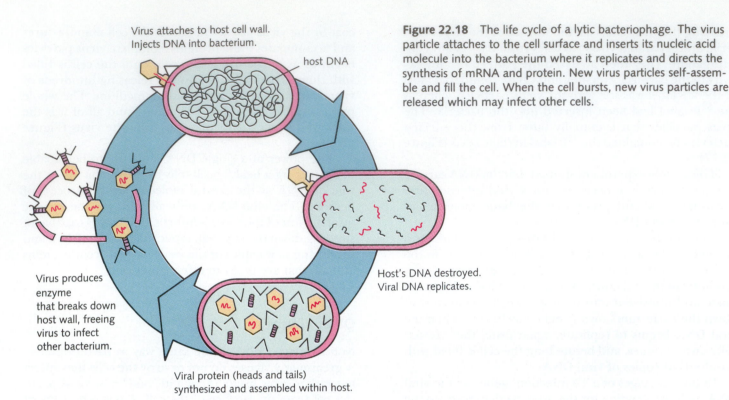

Virus attaches to host cell wall.
Injects DNA into bacterium.

host DNA

Host's DNA destroyed.
Viral DNA replicates.

Virus produces enzyme that breaks down host wall, freeing virus to infect other bacterium.

Viral protein (heads and tails) synthesized and assembled within host.

Figure 22.18 The life cycle of a lytic bacteriophage. The virus particle attaches to the cell surface and inserts its nucleic acid molecule into the bacterium where it replicates and directs the synthesis of mRNA and protein. New virus particles self-assemble and fill the cell. When the cell bursts, new virus particles are released which may infect other cells.

their DNA into the DNA of the host cell, and for a time the viral genes are actually carried in the genome of the infected cell. In addition, not all viruses contain DNA. An important group of viruses contain RNA as their genetic information. At first this puzzled investigators, and many biologists wondered whether these RNA viruses worked by a completely different genetic mechanism. In a sense, they do.

In 1972 biologist Howard Temin suggested that there might be an enzyme, which he termed **reverse transcriptase,** that could synthesize DNA from an RNA template. Temin thought that these RNA viruses might first make a DNA transcript that could then function as an ordinary piece of DNA. Only a few years passed before he was proved right and the enzyme that he postulated was discovered. RNA-containing viruses are now called **retroviruses,** because they seem to work in a *retro*grade (backward) fashion; that is, they make a DNA molecule from RNA. The DNA molecules that are synthesized by retroviruses are sometimes used to produce messenger RNAs that help the virus grow and produce new virus particles. Sometimes these DNA molecules are integrated into the chromosomes of the infected cells. Many of the viruses that cause tumors are retroviruses, as is the AIDS virus. In Chapters 38 and 39 we will examine the biology of retroviruses and their involvement in cancer and AIDS, respectively.

CHROMOSOMES AND DNA

One question central to the study of cell biology is how DNA molecules are organized within a cell. The answer to this question depends on whether the organism involved is a prokaryote or a eukaryote.

Prokaryotic Chromosomes

The DNA molecules in prokaryotes are free in the cytoplasm. Originally, the term *chromosome* was applied only to eukaryotes, because only eukaryotic chromosomes are large enough to be visualized in the light microscope. More recently, however, the term has also been applied to the DNA molecules found in prokaryotic cells. If a prokaryotic cell is broken open, it is possible to visualize the chromosome with the aid of an electron microscope (Figure 22.19). The chromosomes of most prokaryotes are single, circular DNA molecules. In each cell, there is generally only a single chromosome that contains the organism's complete genome.

There's an important exception to this rule, however. A great many prokaryotes contain additional "minichromosomes," or **plasmids.** A cell may contain as many as 30 plasmids, which are generally identical copies of the same short DNA segment. Plasmids contain special DNA sequences that ensure that they are replicated along with the rest of the cell's DNA. Occasionally, plasmids can insert themselves into the cell's chromosome.

Eukaryotic Chromosomes

As we have already seen, eukaryotic cells contain more than a single chromosome, and those chromosomes are enclosed within a special cellular organelle known as the *nucleus.* Chemical analysis shows that when chromosomes are isolated from living cells, they contain more than 65 percent protein and only 30–35 percent DNA (there are also often traces of RNA).

The DNA in a eukaryotic chromosome, unlike the DNA in a prokaryotic chromosome, is tightly bound to protein. The complex of protein and DNA found within the eukaryotic nucleus is known as **chromatin.** There are two categories of proteins found in chromatin. **Histones** are proteins that contain large amounts of the extremely basic amino acids *lysine, arginine,* and *histidine.* **Nonhistone proteins** are distinguished only by the fact that they are not histones and bind very tightly to DNA. The histone proteins have been the easiest to study. They are present in very large amounts, and there are only a few different types of them.

If the nucleus of a single cell is gently disrupted and allowed to spill out on a thin carbon film, it is possible to stain or shadow the chromatin and examine it in the electron microscope. A typical chromatin preparation is shown in Figure 22.20. The unraveling mass of material shows a distinct substructure that looks a bit like beads on a string. The beads were discovered in 1973 by Christopher Woodcock (University of Massachusetts) and Don and Ada Olins (Oak Ridge National Lab). These beads, which are tightly organized structures made up of DNA and histones, are known as **nucleosomes.**

Each nucleosome contains a double coil of DNA wound around eight protein molecules. One of the reasons for the very basic nature of the histone proteins is now clear: The positive charges on the amino groups ($-NH_3^+$) found in lysine, arginine, and histidine help them bind to the DNA strands, which are negatively charged because of their phosphate groups.

What is the function of the nucleosome? That has proved to be a very difficult question for researchers to

Figure 22.19 Electron micrograph of an *E. coli* chromosome. The great length of the single circular DNA molecule in this bacterium is apparent and is attached to fragments of the cell membrane.

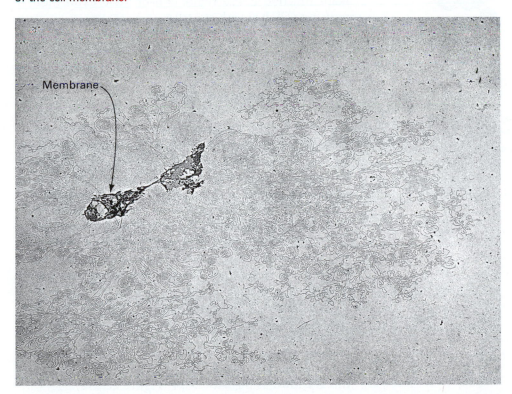

Membrane

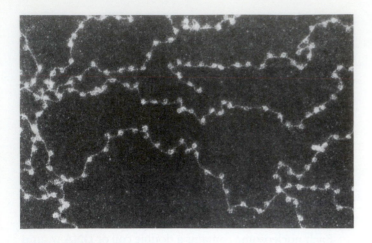

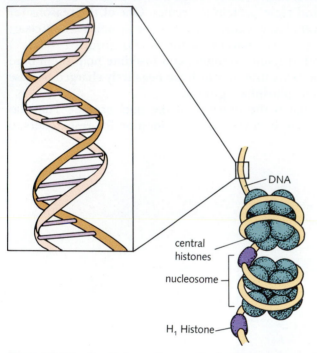

DNA

central
histones

nucleosome

H₁ Histone

Figure 22.20 An electron micrograph of extracted chromatin (top) reveals the *nucleosome* structures which are associated with DNA in most eukaryotic cells. Nucleosomes appear as "beads on a string" and are composed of histone proteins and DNA. Diagram of the structure of a nucleosome (bottom), which contains two copies of each of four different histone proteins.

answer. It was suggested that nucleosomes might play a role in gene regulation, but that does not seem to be the case. Nucleosomes seem to be found everywhere, in both active and inactive genes. The most widely accepted theory is that nucleosomes do "nothing" but that they do it very well. Consider this problem. Each human cell has enough DNA to form a double helix about 10 *meters* in length, but it all has to fit within a nucleus that is generally smaller than 10 *μm* in diameter. This means that the DNA has to be folded by a factor of about 1 million just to fit into the nucleus. How could such folding be accomplished without making the nucleus a hopeless mass of interconnected molecular spaghetti?

The nucleosome may hold part of the answer. Each nucleosome contains two complete loops of DNA (about 150 base pairs) wrapped tightly against the histone proteins of its core. The nucleosomes in turn are clustered to form larger and larger aggregates from which the eukaryotic chromosome is formed. The function of the nucleosome, at least in part, may be to help manage the enormous amount of DNA contained in each of the chromosomes of a eukaryotic cell.

Nonhistone proteins are something else again. They do not form regular structures such as the nucleosome. The nonhistone chromosomal proteins are also much more diverse than the nucleosomes. They vary a great deal from one kind of cell to the next, and between organisms the variation is enormous. The nonhistone proteins seem to include a great many DNA-binding proteins whose main function is the regulation of gene expression, a topic we will deal with shortly.

THE CONTROL OF GENE EXPRESSION IN PROKARYOTES

Up to this point, we have considered the basic mechanisms by which DNA is replicated, how a genetic message is transcribed to RNA, and how the message is translated into the sequence of amino acids in a protein. Genetic studies tell us that even an organism as simple as *E. coli* has several thousand genes. It would be remarkably wasteful for each of these genes to be expressed at the same time. Instead, the products of most genes are adjusted relative to the need for them. One good example is found in the genes that enable *E. coli* to use lactose as a source of food.

Lactose is a form of sugar found in milk, a food that frequently finds its way into *E. coli*'s normal habitat, the human intestine. In order for *E. coli* to be able to utilize **lactose,** it must produce an enzyme that cuts this disaccharide into two simple sugars. The enzyme that does this is known as β-*galactosidase* (Figure 22.21a).

When cells are grown on a medium that doesn't contain any lactose, they have no need of β-galactosidase. Not surprisingly, such cells have only very small amounts of the enzyme. However, when the same cells are transferred to a medium in which a lactose sugar is the only food source, the amount of the enzyme in the cells rises very quickly (Figure 22.21b). It seems almost as though the cell "knows" what kind of sugar is in the medium and then begins to make more of the appropriate enzyme. This phenomenon—a rapid increase in the amount of an enzyme in response to culture conditions—is known as **enzyme induction.** The substance that causes the increase (in this case, the sugar lactose) is known as the **inducer** of the enzyme. How does induction occur? How does the cell "know" that lactose is in the medium? The answers to these questions were provided in the late 1950s by two researchers at the Pasteur Institute in Paris, François Jacob and Jacques Monod. In a brilliant series of genetic studies, they proposed a scheme for the control of the gene. Their ideas were confirmed by a series of subsequent studies and now serve as a model for how many prokaryotic genes are regulated.

The *lac* Operon

The gene for β-galactosidase is found in a cluster of three genes, each of which is related to lactose metabolism: Z is the gene that codes for β-galactosidase itself, and the Y and A genes code for other enzymes that are also involved in lactose metabolism. Because these three genes code directly for the *structure* of proteins, they are known as *structural genes*. On one side of the three genes are two sections of DNA that are termed the **promoter (p)** and **operator (o)** sequences. The o and p regions are regulatory sequences that help control the expression of the three genes. The entire structure of regulatory and structural genes is known as an **operon** because the genes are "operated" together. Because these genes affect the metabolism of lactose, this region of the *E. coli* chromosome is known as the *lac* **operon** (Figure 22.22).

The promoter (p) region is a segment of DNA to which RNA polymerase can bind. If the polymerase does bind to the p region, it can then move to the right until it encounters a special "start transcription" signal at the beginning of the Z gene. The polymerase then produces a messenger RNA that contains a complementary copy of each of the three structural genes, and that mRNA can be translated into proteins: One of them is β galactosidase. If this is the case, what prevents the cell from filling up with the enzyme?

Another gene, known as the *i* gene, codes for a protein known as the *lac* **repressor** (i). The repressor is a DNA-binding protein, and it binds tightly to the operator

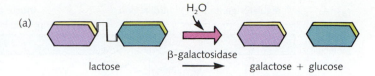

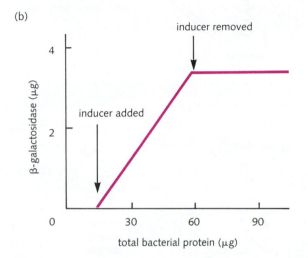

Figure 22.21 **(a)** Beta-galactosidase catalyzes the hydrolysis of lactose to produce galactose and glucose. This is the first step in the chemical pathways which utilize lactose as a source of food energy. **(b)** When lactose is added to a culture of *E. coli*, the amount of beta-galactosidase in the cells quickly rises. This phenomenon is known as enzyme induction.

(o) region of the gene. The effect of repressor binding is dramatic. Because RNA polymerase must pass through the o region to get to the structural genes, the presence of the repressor prevents the structural genes from being transcribed. In effect, the repressor turns off the operon—it represses it.

Now our problem is a bit different. With the repressor around, how does the cell make certain that β-galactosidase can be synthesized when it is needed? The repressor has *two* binding sites. One, of course, is for the o region of the operon. The other binding site is for lactose and similar sugars. When an inducer molecule such as lactose enters the cell, it binds to the repressor and causes a *conformational change* in the structure of the molecule. The repressor–inducer complex is no longer able to bind to the operator, and it falls off the DNA back into solution. Now the operator region is open, and RNA

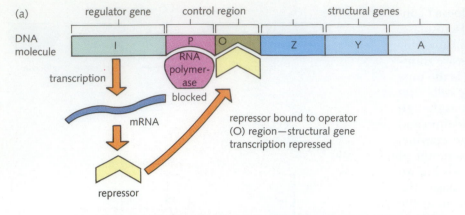

(a)

regulator gene   control region   structural genes

DNA molecule: I   P   O   Z   Y   A

transcription

RNA polymerase blocked

mRNA

repressor

repressor bound to operator (O) region—structural gene transcription repressed

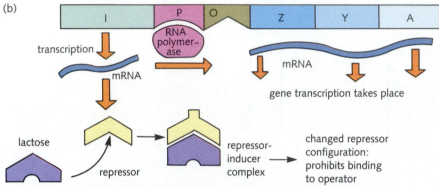

(b)

I   P   O   Z   Y   A

transcription

RNA polymerase

mRNA

mRNA

gene transcription takes place

lactose

repressor

repressor-inducer complex

changed repressor configuration: prohibits binding to operator

Figure 22.22   The *lac* operon. RNA polymerase binds to the promoter (p) region, and transcribes an mRNA molecule which codes for the proteins specified by the z, y, and a genes (the z gene product is beta-galactosidase). The repressor, a protein which binds to the operator (o) region, blocks transcription of the genes, preventing their expression. Lactose induces expression of the operon by binding to the repressor and unblocking the o site.

polymerase is free to move from the promoter across it to allow gene transcription to take place.

The system regulates itself. When lactose is scarce, the repressor binds to the o region of the operon and very little of the enzyme is made. But when lactose is provided as the food source, it diffuses into the cell, binds to the repressor, and "turns on" the genes that enable the cell to use lactose for food. This automatic control system provides the cell with the right amount of lactose-metabolizing enzymes to suit the moment. The *lac* operon also has a positive control system that is sensitive to the overall energy needs of the cell and does not activate the operon if the cell has adequate supplies of other sugars for food, especially glucose.

Not all bacterial genes are like the lac *operon.* Some genes are expressed at low levels regardless of environmental stimuli. A few others are repressed, rather than induced, when specific molecules from the environment are introduced. Other operons are regulated by positive control elements, which enhance transcription rather than restrict it. But the *lac* operon provides a general model for the way in which a gene can be organized in such a way that it responds to changes in the environment in specific ways.

GENE REGULATION IN EUKARYOTES

Gene regulation in eukaryotes is more complex than it is in prokaryotes. There are several reasons for this. First, even a simple eukaryote may have as many as 50,000 genes, making it especially important that genes be expressed only when they are needed. Second, the different cell types in complex organisms require vastly different patterns of gene expression. Cells as different as muscle cells and nerve cells must express very different groups of genes. Finally, the different cell types which are produced during development in a complex organism are themselves the product of differential gene expression. Development requires that gene expression be regulated in *time* and *space* (Figure 22.23).

Regulation at Many Levels

Gene expression can be regulated at many different levels. The *lac* operon was an example of transcriptional regulation. However, gene expression can also be controlled post-transcriptionally. Gene expression can be affected by the manner in which RNA molecules are

Figure 22.23 Position-sensitive expression of genes in the *Drosophila* embryo. A marker gene was placed in four different positions in the genome. Cells expressing the gene can then be stained. Depending on the control elements that surround the gene, it is expressed in (**a**) the trachea and tissues in posterior segments, (**b**) sensory organs, (**c**) muscle, or (**d**) the dorsal ectodermal stripe.

THEORY IN ACTION

Prokaryotes and Eukaryotes: Differences in Style and Substance

Earlier in the text we saw that eukaryotic cells contain a nucleus whereas prokaryotic cells do not. But there are other important differences between prokaryotic and eukaryotic cells.

■ *Messenger RNA:* You'll remember that the mRNA produced by the *lac* operon is a transcription of three different genes (*Z, Y,* and *A*) and that three different proteins are produced from it. Having multiple genes produce a single mRNA is common in prokaryotes. In eukaryotes, mRNAs generally correspond to just one gene sequence.

■ *Chromosomes:* We have already seen that eukaryotic chromosomes contain large amounts of histone proteins in the form of nucleosomes. Prokaryotic chromosomes contain far fewer proteins bound to their DNA and no histone proteins. In addition, eukaryotic cells contain *much* more DNA than the amount found in prokaryotic cells.

■ *Ribosomes:* Eukaryotic ribosomes are a little bigger. They have about 70 different proteins and 4 rRNA molecules, whereas prokaryotes have about 55 proteins and only 3 rRNA molecules.

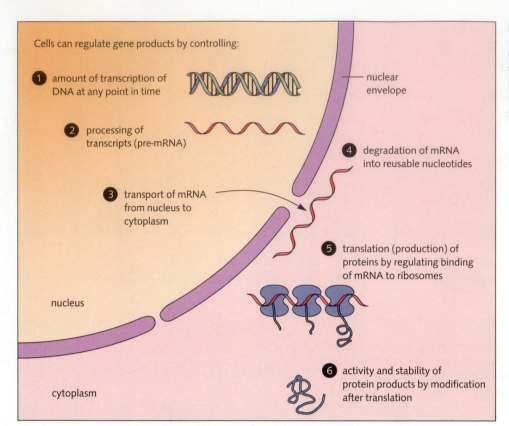

Cells can regulate gene products by controlling:

1. amount of transcription of DNA at any point in time

2. processing of transcripts (pre-mRNA)

3. transport of mRNA from nucleus to cytoplasm

4. degradation of mRNA into reusable nucleotides

5. translation (production) of proteins by regulating binding of mRNA to ribosomes

6. activity and stability of protein products by modification after translation

nuclear envelope

nucleus

cytoplasm

Figure 22.24 Gene expression can be regulated at many levels, including (1) transcription, (2) pre-mRNA processing, (3) transport of RNA from the nucleus, (4) stability of RNA in the cytoplasm, (5) translation, and (6) stability and activity of protein product.

processed before they leave the nucleus, by the stability of mRNAs in the cytoplasm, and by the rate at which mRNAs bind to ribosomes to begin protein synthesis. In fact, there are at least six distinctly different levels at which gene regulation can occur in eukaryotes (Figure 22.24). Researchers have found examples of gene regulation that act at *each* of these possible levels.

Genes that Control the Expression of Other Genes

The complexity of eukaryotic organisms requires that whole groups of genes be turned on and off in groups. We have already seen one example of this in the form of X-chromosome inactivation (Chapter 11). The genes in one of the two X-chromosomes of a female mammal are inactivated as a group. Other genes may be inactivated by repressor molecules which bind to DNA and prevent transcription. In some cases, tightly condensed chromatin structures may prevent RNA polymerase from making contact with DNA, effectively turning off all of the genes within the condensed region.

Naturally, genes may be activated as well as repressed. Researchers have theorized for years that cells might con-

tain "master genes" which activate clusters of genes that control complex developmental patterns (Figure 22.25). It now seems that master genes do exist, and several such systems have been investigated. Immature muscle cells, for example, produce a regulatory protein known as *myoD1*. The expression of myoD1 activates a series of genes which in turn produce a range of muscle-specific proteins required for development.

Researchers have shown that connective tissue cells (fibroblasts), which do not normally express muscle-specific genes, still have the capacity to respond to the myoD1 master gene. When a piece of DNA containing the myoD1 gene was inserted into fibroblast cells, the cells began to produce muscle-specific proteins and to develop into muscle-like cells.

Transcription Factors

Many eukaryotic genes are regulated by *transcription factors,* which bind to specific DNA sequences and enhance gene expression (Figure 22.26). One of the best-studied transcription factors regulates the transcription of a small ribosomal RNA gene in *Xenopus,* the African clawed frog. The binding of this transcription factor increases the rate of DNA transcription of these RNA genes, possibly by making it easier for RNA polymerase to bind to the gene. Remarkably, this transcription factor binds to a region in the middle of the gene, showing that gene regulation need not be confined to the regions "upstream" from the gene itself.

Intervening Sequences and the Control of RNA Processing

In the early 1970s a number of techniques were developed that made it possible to isolate segments of DNA about the size of a single eukaryotic gene. One of the first experiments that several laboratories tried was to compare the DNA segment of a gene with the mRNA it produced. Naturally, scientists expected the DNA and the mRNA to be a perfect match. But a surprise was waiting. One of the first genes to be examined was the DNA segment that codes for *ovalbumin,* the major protein in the "white" of a chicken egg. The gene was discovered to be much larger than the mRNA. When a piece of single-stranded DNA containing the gene was allowed to base-pair with the ovalbumin mRNA, things got even stranger: The gene seemed to be filled with segments that had no matching sequence in the RNA.

The existence of **intervening sequences** in DNA, which had no corresponding segment in mRNA, was a complete surprise. And these **introns,** as they are now called, were found to be a common feature of eukaryotic genes (although they are not universal). The regions that *are*

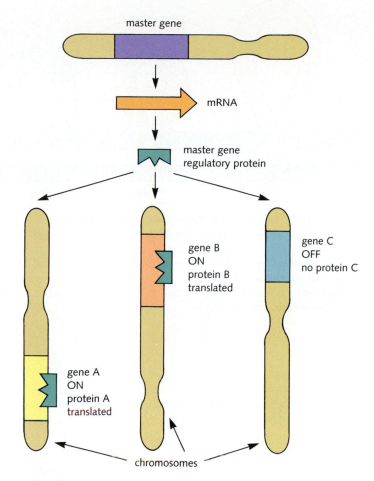

Figure 22.25 The regulatory proteins produced by some genes are able to control the expression of several other genes.

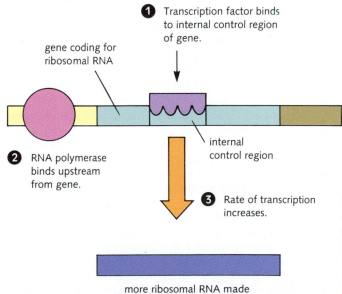

Figure 22.26 Expression of a class of ribosomal RNA genes in *Xenopus* is enhanced by a transcription factor that binds to an internal region of the gene itself. The factor is a DNA binding protein that may fold the gene into a shape that increases the efficiency of RNA polymerase binding.

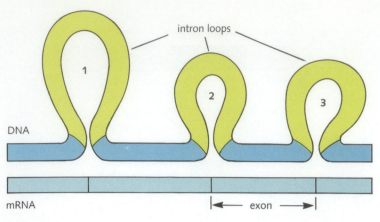

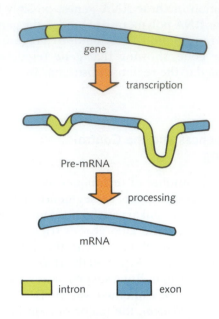

intron   exon

Figure 22.27 (top) Intervening sequences, or introns, were discovered when mRNA was hybridized to the DNA sequences from which it had been transcribed. Although complementary RNA/DNA sequences form a double-stranded hybrid, the regions of DNA that do not match the RNA sequence are visible as large loops. (right) Introns and exons are transcribed into a pre-mRNA. The introns are then removed as the pre-mRNA is processed into mature mRNA.

complementary to mRNA are called **exons** (expressed sequences). What is the function of introns? We now know that most eukaryotic mRNAs are made as "pre-mRNAs" and then processed to remove introns and produce a mature mRNA molecule (Figure 22.27).

It is also apparent that the splicing and processing of pre-mRNA is yet another level at which gene expression may be regulated. Several recent studies have shown that the same pre-mRNA molecules may be spliced in different ways, producing different mRNAs in different tissues.

Genome Structure

In prokaryotes, closely related genes are often found together, like those of the *lac* operon, making it possible for them to be regulated as a group. While this is sometimes true for eukaryotes, it is quite common for related genes to be scattered in different locations around the genome, often on different chromosomes. Scattered genes can still be regulated as a group if they all contain similar regulatory elements which can respond to the same transcription factors, and this is often the case.

Eukaryotic genes are sometimes present in multiple copies. The genes for ribosomal RNA and the histone proteins may be repeated hundreds or even thousands of times, making it possible for cells to produce large amounts of their products in a short period of time. Other genes, including those for contractile and cytoskeletal proteins, are present as members of *multigene families,* closely related groups of genes which serve similar purposes.

SUMMARY

The simultaneous development of biochemistry and genetics led to an expectation that genetics should have a molecular basis. Frederick Griffith discovered that an extract of killed bacterial cells was able to transform a strain of harmless bacteria into pathogenic organisms. Oswald Avery and his co-workers examined the transforming extract and showed that the transforming molecule was DNA, suggesting that DNA was the molecule of genetics. The structure of DNA was determined by James Watson and Francis Crick on the basis of an important X-ray diffraction pattern made by Rosalind Franklin. The Watson–Crick structure is a double helix, in which two polynucleotide strands are twisted around each other in antiparallel fashion. The strands are held together by hydrogen bonds, which produce specific pairs between the nitrogenous bases. The sequence of bases in DNA is copied to mRNA in a process known as transcription. This message is then translated as mRNA directs the synthesis of a protein.

The essential role of nucleic acids in information transfer can be seen in viruses, which infect host cells by inserting either DNA or RNA molecules. These viral genomes then direct the synthesis of new virus particles, disrupting the cells that they infect and releasing new viruses, which may infect other cells. In eukaryotic chromosomes, DNA is tightly bound to several types of proteins, including the histone and nonhistone chromosomal proteins. DNA is replicated in semi-conservative fashion. Because most organisms contain far more genes than they will ever need to express at the same time, nearly all genes are regulated. One well-studied example of gene regulation is found in the bacterium *E. coli,* where three genes are operated as a group known as the *lac* operon. The expression of the operon is controlled by a repressor system.

STUDY FOCUS

After studying this chapter, you should be able to:

- Trace the history of how DNA was discovered as the genetic material and how its structure was deduced.
- Explain the "central dogma" of information transfer: DNA → RNA → protein.
- Give the details of the connection between genetics and the biochemical characteristics of an organism.
- Describe the molecular basis of a number of human genetic diseases.
- Recount some of the latest discoveries in molecular biology and appreciate its exciting, changing nature.

SELECTED TERMS

replication *p. 408*
transformation *p. 409*
purine *p. 412*
pyrimidine *p. 412*
template *p. 416*
transcription *p. 416*
codon *p. 417*
anticodon *p. 418*
ribosome *p. 418*
translation *p. 419*
reverse transcriptase *p. 426*
retrovirus *p. 426*
plasmid *p. 426*
nucleosome *p. 427*
chromatin *p. 427*
histone *p. 427*
lac operon *p. 429*
intron *p. 433*
exon *p. 434*

REVIEW

Discussion Questions

1. What is the molecule of inheritance? How was it discovered?

2. What is Chargaff's rule? What does it imply about the structure of DNA?

3. Describe the major features of the Watson–Crick model for the structure of DNA.

4. What are the basic differences among transcription, translation, and replication?

5. Describe the basic steps associated with the process of translation. Begin with the formation of an initiation complex.

6. What is a virus? What key characteristic distinguishes viruses from other forms of life?

Objective Questions   (Answers in Appendix)

7. Complementary base pairing in DNA happens between
   (a) adenine and guanine.
   (b) adenine and thymine.
   (c) cytosine and adenine.
   (d) thymine and guanine.

8. A strand of DNA serves as a _____ because it can generate a new _____ strand.
   (a) base pair / identical
   (b) biological strand / different helical
   (c) template / complementary
   (d) helix / exact

9. The coded information for synthesizing a protein is carried to the ribosome by
   (a) DNA.   (c) guanine.
   (b) mRNA.   (d) tRNA.

10. The mRNA is found in the
   (a) cytoplasm only.
   (b) ribosome.
   (c) nucleus only.
   (d) cytoplasm and nucleus.

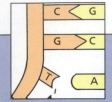

CHAPTER 23

Evolution at the Molecular Level: Natural and Artificial

The development of scientific knowledge does not always proceed along direct paths. We have already seen how Darwin's theory of evolution was produced well before the birth of modern genetics, and we have discussed how genetics was integrated with evolutionary theory. The result was a "new synthesis" of evolutionary theory that combined genetics with evolution. A centerpiece of that theory was an emphasis on mutation as the source of genetic diversity.

In the last chapter, we traced the development of molecular biology and saw how the actions of a gene could be understood in chemical terms. Information coded in DNA is expressed through RNA and translated into proteins. These proteins then help to determine the phenotype of an organism. Molecular biology makes it possible to provide a molecular basis for evolutionary change. As you might expect, molecular biology has changed some of the fundamental ways in which we think about evolution.

MUTATIONS AS A SOURCE OF GENETIC CHANGE

Mutations Are Heritable Changes in DNA

Darwin, Mendel, and the early pioneers of genetics were well aware that sometimes heritable changes occur spontaneously. Animals and plants that displayed sudden changes were sometimes called "sports" by the people who bred them. Such sports were a frequent source of new varieties that were useful for further breeding. Many varieties of ornamental flowers have arisen in this way (Figure 23.1). The term **mutation,** which comes from a Latin word meaning "to change," is applied to such heritable changes. Mutations can be passed on to future generations, they can occur in any gene, and they occur at random.

An early clue to the nature of mutations came from studies on the effects of radiation. In 1927 H. J. Muller irradiated flies with high levels of X rays and found that the radiation caused a dramatic increase in the incidence of mutations. Muller understood almost immediately that radiation could cause physical changes in the molecules of the genes themselves. Later it was discovered that certain chemicals could also act as **mutagens,** agents that cause mutations. Today we know something Muller did not—that genes are made of DNA. Therefore, we define mutation in a different way: *Mutations are heritable changes in the DNA.*

A change in the base sequence of a DNA molecule has the potential to cause changes in RNA molecules and in protein molecules as biological information flows from one molecule to the next. Many of these mutations are the result of errors that occur during the replication of DNA. During replication, the nucleotide bases that form the new strands must be matched with preexisting strands according to the base-pairing mechanism discussed in

Chapter 22. Although the base-pairing mechanism usually ensures that only the correct nucleotide is placed in the new strand, sometimes a mistake is made and an incorrect nucleotide is used. Once a mistake has been made, the replication mechanism can pass the mistake along to succeeding generations of cells, causing a mutation to appear in the cell line.

The Effects of Mutations

Mutations caused by changes in one or two bases are known as **point mutations,** because they exert their effect at a specific *point* in the gene. **Base substitutions** are point mutations that involve a change in one of the nucleotide bases within a gene. What effect can a base substitution have? In some cases it has no effect, but generally the substitution *does* affect the amino acid specified by its codon. Such changes can drastically affect the activity of a protein. A substitution may also change an ordinary codon into a stop codon, causing only the first part of the protein to be made. Many genetic diseases (including sickle-cell anemia, colorblindness, and hemophilia) are the result of a few base changes at critical regions of important genes (Figure 23.2).

Point mutations can also involve the **insertion** or **deletion** of bases. These mutations have far-reaching effects

Figure 23.1 These mutants of black-eyed susan have dramatically enlarged flowers.

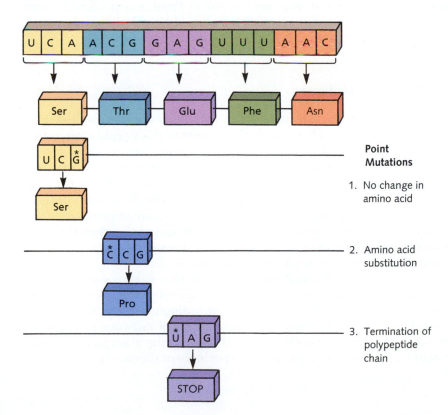

Point Mutations

1. No change in amino acid

2. Amino acid substitution

3. Termination of polypeptide chain

Figure 23.2 The effect of a substitution mutation depends on the location of the base change. In some cases there is no change in the amino acid specified by the codon. In other cases, a different amino acid is specified, or a stop codon that causes premature termination of the polypeptide chain is produced, as shown.

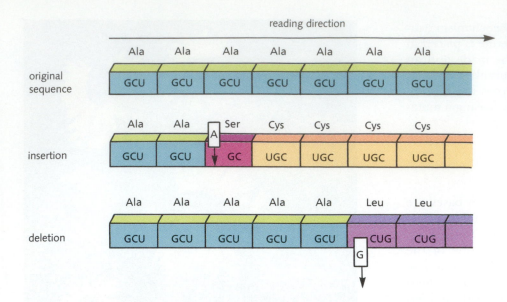

reading direction

	Ala	Ala	Ala	Ala	Ala	Ala	Ala
original sequence	GCU	GCU	GCU	GCU	GCU	GCU	GCU

	Ala	Ala	Ser	Cys	Cys	Cys	Cys
insertion	GCU	GCU	GC A	UGC	UGC	UGC	UGC

	Ala	Ala	Ala	Ala	Ala	Leu	Leu
deletion	GCU	GCU	GCU	GCU	GCU	CUG	CUG

Figure 23.3 The insertion or deletion of a base can result in a frameshift mutation, in which the reading frame of the mRNA is shifted. This has a dramatic effect on the protein produced by the mRNA, changing most of the amino acids that are downstream from the point of the mutation.

on the protein products of the genes in which they occur. Consider a gene sequence in which a strand of 300 nucleotides codes for a polypeptide containing 100 amino acids. The genetic code is read in groups of 3 bases. However, if a single base is added or deleted in the middle of that sequence, the entire message is thrown off. Such changes are said to *shift* the *reading frame* of the message, and they are often known as **frameshift mutations.** A single frameshift mutation disrupts the reading of every codon "downstream" from it and very often produces a useless protein product (Figure 23.3).

Although we have paid particular attention to mutations that occur at the level of individual bases, many mutations involve hundreds and even thousands of bases. These mutations are the result of large-scale movements of whole regions of a chromosome: Large sections of DNA may be *deleted, inverted,* or moved from one region of the genome to another (*transposed*). These mutations occur at the level of the chromosome and are sometimes referred to as **chromosomal mutations.** Many chromosomal mutations are produced by **transposable elements,** short segments of DNA that are removed and then reinserted in the chromosome. These transposable elements are sometimes called "jumping genes," and the name is appropriate.

Transposable elements have been found in bacteria, insects, and mammals (including humans), but they were first discovered in maize (corn) by Barbara McClintock. McClintock's work, which took place in the 1940s and 1950s (well before the era of modern molecular biology),

showed that certain genes were able to jump from one place to another in the maize genome. When these transposable elements were inserted in the middle of a gene controlling kernel color, the gene was inactivated and the color was changed. The movement of transposable elements in maize produces a colorful speckling in the kernels of corn (Figure 23.4). McClintock's work, for which she was awarded the Nobel Prize, has produced a new appreciation among scientists of how flexible and dynamic chromosomes may be.

Mutagens

Although many point mutations are the result of unavoidable errors in the process of DNA replication, others are the result of specific agents known as mutagens. **Chemical mutagens** are molecules that can increase the rate at which mutations occur in DNA. Many chemical mutagens are molecules that can be chemically confused with nucleotides and increase the rate of point mutations dramatically. Other chemical mutagens affect the function of DNA polymerase itself in such a way that the error rate of replication is increased.

Physical mutagens can also exert powerful influences on DNA. The best-understood physical mutagens are forms of radiation, including ultraviolet light, X rays, and gamma rays. These forms of radiation can produce chemical changes in DNA, and the absorption of radiation energy can cause permanent chemical changes (such changes were the cause of Muller's results with X rays). Ultraviolet light, for example, is strongly absorbed by DNA molecules. Energy from UV light causes excitation in the electrons of the nitrogenous bases, and wherever two thymidine molecules are next to each other in the DNA sequence, a covalent bond can be formed between them, producing a **thymine dimer.** A repair mechanism, which operates in most cells, can remove the dimer and replace the damaged bases (Figure 23.5). However, if DNA replication takes place *before* the dimer can be repaired, random bases are placed in the new DNA strand, resulting in a pair of adjacent base substitutions. Some other forms of radiation, including gamma rays, are known as **ionizing radiation** because they are able to knock electrons out of stable molecules, causing the formation of ions. Ionizing radiation causes chemical changes in DNA that affect replication enough to produce serious mutations. Although they work by slightly different mechanisms, the chemical and physical causes of mutations have one feature in common: *They are able to alter the base sequence of DNA molecules.*

Figure 23.4 The mottled colors of these maize kernels are the result of transposable genetic elements which have disrupted the gene coding for pigmentation enzymes. McClintock's studies provided the first proof that DNA segments could move from one position to another within the genome.

Figure 23.5 Thymine dimers are produced by ultraviolet light. A group of DNA repair enzymes in the cell searches for such dimers, removes them from DNA, and replaces the missing bases. If the dimer is not removed by DNA repair enzymes prior to DNA replication, it causes a heritable error in the DNA sequence—a mutation.

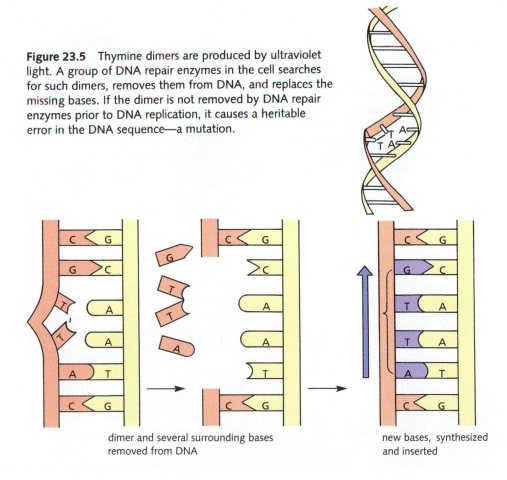

dimer and several surrounding bases removed from DNA

new bases, synthesized and inserted

A Molecular Disease: The Good and Bad Aspects of One Mutation

In 1904 Dr. James Herrick examined a young patient with symptoms so striking that he was moved to write a paper for the *Archives of Internal Medicine* that described the young man's afflictions. He began, "This case is reported because of the unusual blood findings, no duplicate of which I have ever seen described."

His patient had a fever and chills. There were about 20 scars on his legs and thighs, some as big as silver dollars. His heart was enlarged and the heart itself beat with a murmur, a sign that it might be damaged. Herrick's thorough examination found nothing else unusual. But then he made a routine smear of the young man's blood, and he saw something that baffled him: "The shape of the red cells was very irregular, but what especially attracted attention was the large number of thin, elongated, sickle-shaped and crescent-shaped forms."

Herrick was careful enough to do control experiments until he was satisfied that the sickling was not caused by some flaw in how he prepared the smear but was related to a problem in the blood itself. He concluded by saying that his results,

> . . . while suggesting that the chemical composition of the fluid suspending the corpuscles may have something to do with the peculiar formations, perhaps suggest more strongly that some unrecognized change in the composition of the corpuscle itself may be the determining factor.

The final sentence of his paper, 80 years after his patient's visit, rates an A+ for intelligent speculation. Within a few decades, scientists noticed that *sickle-cell anemia,* as the disease was called, runs in families—it is an

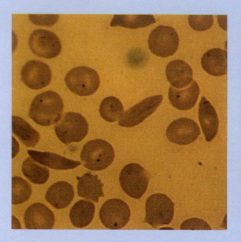 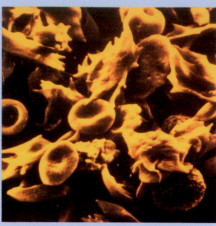

(left) Cells such as these sickled red cells were observed by James Herrick in his initial description of sickle-cell anemia. (right) Electron micrograph of sickled red cells.

inherited disorder. We now understand that hemoglobin, the oxygen-carrying protein found in red blood cells, is chemically different in sickle-cell patients. Specifically, the protein molecules are a bit less soluble. When the blood of a sickle-cell sufferer is low in oxygen, a large proportion of these altered hemoglobin molecules come out of solution and form bundles of tiny fibers within the red cells. These fibers change the shape of the red cell, and they are the culprits that produce the characteristic sickle-cell appearance. As sickle cells pass through the smallest passageways of the circulatory system, they tend to get stuck in tight spots, block the circulation, and clog the flow of blood. Pressure builds up behind the blockage, and as a result small blood vessels can burst. Internal bleeding occurs. Organs with rich blood supplies, such as the spleen, heart, and liver, are repeatedly damaged by

sickle-cell crises in which blood flow is blocked in many places at once.

Scientists now understand sickle-cell anemia at the molecular level. The sickle-cell disease is caused by a defective gene for β-globin, one of the polypeptides in the hemoglobin molecule. Individuals heterozygous for the disease have one copy of the normal β-globin gene and one copy of the sickle-cell version of β-globin. Individuals who are heterozygous are often said to be *carriers* of the disease, because they can pass it along to their offspring. But these people rarely suffer symptoms of the disease: The fact that about half of their hemoglobin molecules are normal seems to provide a degree of protection. They are said to have the sickle-cell *trait.* However, individuals with two copies of the sickle-cell gene are not so lucky. They suffer the full range of problems first noticed by Herrick. Their life expectancy is drastically reduced,

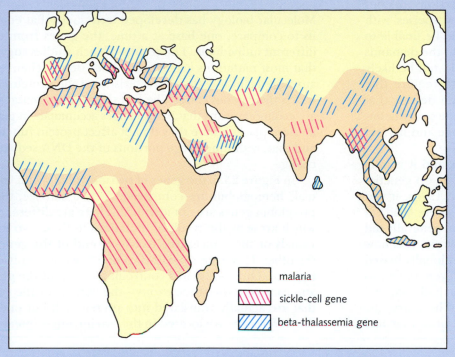

Distribution of the sickle-cell gene in human populations closely parallels the incidence of malaria.

Legend:
- malaria
- sickle-cell gene
- beta-thalassemia gene

as 10 percent in the black population. Why should a simple gene defect be so common in just one population group? A clue can be found in mapping the incidence of the sickle-cell trait worldwide.

The occurrence of the sickle-cell gene is not limited to Africa. In fact, it follows the equatorial region from Africa into Asia in a way that is almost exactly parallel to the incidence of *malaria*. Malaria is caused by a single-celled, mosquito-borne parasite that grows within the red blood cell. Careful experiments have shown that the very process of sickling kills the parasite, and even the cells of a heterozygous carrier of the disease are difficult for the parasite to survive in. An interesting situation: Although sickle cell causes tremendous problems for those who bear two copies of the sickle-cell gene, it confers a great advantage on heterozygous persons who must live in a malaria-filled environment.

Evolutionary biologists look on the occurrence of this gene in the human population as a classic example of how natural selection influences the genotype of a population. Sickle cell is common where malaria exists, because in such environments the positive effects of the gene outweigh its negative effects on survival and reproduction. Sickle cell is rare where malaria is absent, because the negative effects of the gene tend to remove its carriers from the population: Sickle-cell sufferers often die before they are able to reproduce. The high incidence of sickle cell among African–Americans has a simple explanation: The environments from which their ancestors were kidnapped during the slave trade were laden with malaria.

and their lives often end with severe internal bleeding or blood clots in the lungs.

What is the nature of the gene defect in sickle cell? About as small a change as you can imagine: *one* amino acid. At position 6 on the β-globin chain, the amino acid *glutamate* is changed to *valine*. This simple change, from a charged amino acid to the nonpolar valine, seems to be enough to make the whole molecule just a bit less soluble and more likely to form the fibers that produce the sickling phenomenon. What sort of a change would be required to produce this at the level of DNA?

Glutamate codons:
GAA GAG
Valine codons:
GUA GUG GUC GUU

Sickle cell is the result of a point mutation. A single base substitution in either of the possible glutamine codons would be enough to change the codon to one that specified for valine. It is a perfect example of what can go wrong when mutations occur in vital genes. But there's another side to the story. In the United States, sickle-cell anemia is almost exclusively confined to African–Americans, and the incidence of carriers runs as high

MUTATIONS AND EVOLUTION

Although mutations may cause serious genetic problems, mutations are also a source of *variability* in a population, and variability is essential to evolution. A population of organisms in which DNA replication flawlessly produced new generations without a single mutation would face a serious problem. The species would be unable to generate the variation on which natural selection operates to produce evolutionary change. A prey species with absolutely flawless DNA replication would be unable to adapt to changes in its environment and to the demands of evading evolving predators, so it would not be likely to survive very long. Evolution seems to require that mutations be *rare enough* so that the gene defects that mutations sometimes produce don't become overwhelming, but just *common enough* to allow change and natural selection to continue.

Mutations occur constantly in nature, and it is only fair to ask how much of an effect they have on genetics and the process of evolution. Before the 1970s, it was commonly suggested that most mutations were "bad" in the sense that a gene containing a mutation generally did not function as well as one without. However, it is now apparent that this conclusion was unintentionally biased by the science of genetics itself. In classical genetics it was impossible to know for sure that a mutation had occurred unless it caused a change in phenotype. Therefore, the only mutations we knew about were the ones that had observable effects, such as the white eye color or altered wing of *Drosophila* mutants or the inability of a bacterium to survive in its usual growth medium. Now, however, it is possible to discover mutations that do *not* affect phenotype.

Neutral Mutations in the Human Genome

Molecular biology has developed techniques that enable us to compare, one base at a time, the genes from two different individuals. The results of such studies suggest that most mutations are neutral. They do not change in any important way the characteristics of the proteins for which their genes code. The fact that neutral mutations are common means that base changes accumulate in genes in those places where they do not affect protein function and that, with time, large portions of the DNA sequence change.

In Figure 23.6, the genes of two individuals with "normal" hemoglobin are compared. As you can see, these two globin genes are not identical. There are differences, which arose as the result of mutations in these two individuals or their ancestors, from one end of the gene to the other. However, these differences are not evenly distributed. Only one is in a controlling region of the gene, and *none* is in the three exons—the regions of the gene that are actually translated into protein. All but one of the differences are located in the two introns—interven-

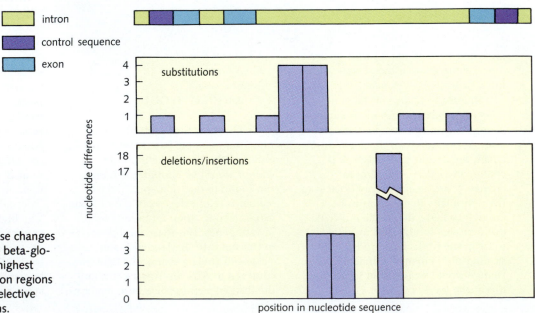

Figure 23.6 The frequencies of base changes in the gene which codes for human beta-globin. As the diagram illustrates, the highest mutation rates are found in the intron regions of the gene, reflecting the lack of selective pressure on changes in these regions.

ing sequences that are removed from mRNA before it is translated. This kind of pattern is seen not only in globin but also in most genes that code for important protein products. What is the significance of the fact that large numbers of mutations are found in the nontranslated introns?

Evolutionary theory has an easy explanation of why more differences are found in intervening sequences. Because the bases that correspond to intervening sequences are "spliced out" of the final messenger RNA, a mutation in this region doesn't affect the protein product. In a sense, it is a *"silent"* or neutral mutation. If mutations can occur randomly, anywhere along the gene, why should so many of them be found in the introns? Probably because the introns are the only places where a base substitution doesn't matter!

A base change in the middle of an intervening sequence has no effect on the protein. Therefore, an individual carrying such a mutation still produces a perfectly normal protein. However, when a mutation occurs within an exon, it usually changes the protein that is produced, and it may cause the new protein to be nonfunctional. If a mutant globin molecule fails to bind oxygen, fails to attach to the other globin polypeptides, or fails to dissolve within the red blood cell, the individual with the mutant may not even survive the process of embryonic development. Therefore, mutations within an exon are serious: Changes in an important part of the molecule, even slight ones, may doom the organism.

An evolutionary biologist might put it this way: *Because intron mutations don't alter phenotype, there is little or no selective pressure on base changes in introns. Therefore, such changes can occur with little effect on an organism's chances for survival.* This is the reason why introns are regions of rapid genetic change. *However, exon mutations do alter phenotype and therefore are subject to strong selective pressure: Substantial changes may be lethal.*

Introns and Evolution

There's an interesting sidelight to all of this. Because biologists have not yet been able to find a clear-cut biological function for intervening sequences, several biologists have suggested that introns are the means by which evolution causes large genetic changes. More specifically, the intron system may provide a way for cells and proteins to evolve whole sequences of protein in a fraction of the time that would be necessary for single mutations and natural selection. Many modern theories of evolution, including punctuated equilibrium, depend on the existence of a mechanism that can produce just such changes. The intron may be an evolutionary tool (Figure 23.7).

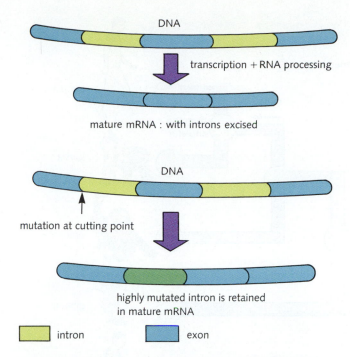

Figure 23.7 Introns may serve as places where mutations can accumulate. When many mutations have altered the base composition of an intron, a final mutation at the border of an intron may alter the site which normally serves as the cutting point for RNA-processing enzymes. This may cause the entire intron to be added to the coding portion of the protein, introducing a substantial change in the amino acid sequence of the protein product.

Tracing the Course of Evolution

It is possible to use the comparison of DNA sequences to study evolutionary relationships. If two organisms once shared a common ancestor, we can estimate the time that has passed since they diverged into two species by analyzing the mutations that have accumulated in their genes.

There is very good agreement between the molecular approach to evolution and the evidence that paleontologists have dug up in the form of fossils. If we examine the differences in amino acid sequence between a group of proteins in two different species and compare the results to the length of time the two species have been separate, we get a very strong correlation. Organisms that diverged only a short time ago (on an evolutionary scale) have few differences between them. Organisms that diverged a long

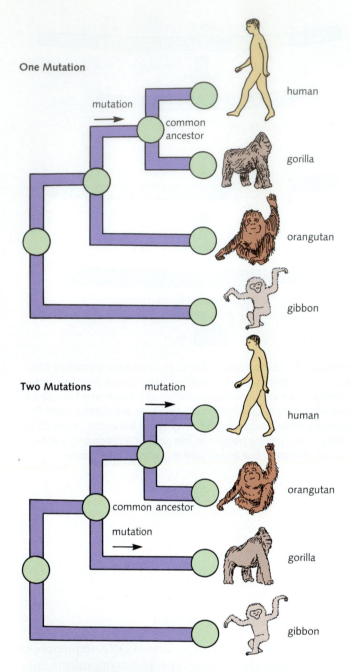

One Mutation

mutation →

common ancestor

human

gorilla

orangutan

gibbon

Two Mutations

mutation →

human

orangutan

common ancestor

mutation →

gorilla

gibbon

Figure 23.8 Phylogenetic trees can be constructed on the basis of DNA sequences of particular genes. Two different schemes can be constructed for human evolution from an analysis of the gene for NAD dehydrogenase 5, a protein involved in energy transduction. One scheme explains the differences as the result of *one* mutation per branch point of the phylogenetic tree, and the other as the result of *two* mutations. The simplest tree (one mutation per branch point) is often held to be most likely.

time ago exhibit many more differences. Not only does molecular biology confirm the basic details of evolution recorded in the fossil record; it also enables us to use a new tool where the fossil evidence is sparse.

Although the human fossil record contains dozens of specimens, we have very few good fossils of the ancestors of contemporary apes, quite possibly because in the past as in the present, the great apes never flourished in very large numbers. As shown in Figure 23.8, scientists have begun to use the tools of molecular biology to study the relationships between humans and our closest primate relatives.

Evolution and Molecular Change

One trend that clearly has taken place during evolution is a dramatic *increase* in the amount of DNA per cell. A typical prokaryotic cell contains about 4 million base pairs of DNA. Simpler eukaryotes such as yeast have between 10 and 20 million base pairs, and typical animals and plants have several billion base pairs in their genome. How has all of this DNA appeared, and what is its function? Interestingly, not all of the DNA is directly associated with genes that code for RNA and protein. In the human genome, there are more than 300,000 copies of a 300-base-pair sequence called *alu*: That's almost 3 percent of all human DNA—and we are not sure that *alu* performs any function at all. It looks as though *alu* may have arisen as a result of random duplication of a particular DNA sequence.

Some functional genes have also arisen through DNA duplication. The organization of a collagen gene in the chicken is a good example of this. In this gene, a basic base sequence of 9 nucleotides is repeated 6 times to produce a 54-base-pair exon. That exon repeats about 50 times in the complete gene. Other genes seem to have developed by duplication of one original gene. Mutations may have modified the extra copies until they diverged into different genes with slightly different functions. The several different globin genes seem to have been formed in this way. Other examples of this phenomenon include the genes for proteins found in muscle cells, and multiple genes for common cellular proteins such as tubulin, which makes up the microtubules found in most cells.

The lesson we can learn from these revelations is that the evolution of gene sequences can proceed by a number of different mechanisms. DNA is a dynamic molecule that can be modified in a number of ways. The accumulation of point mutations, deletions, neutral mutations in introns, and duplicate genes provides a great number of possibilities for the generation of diversity and new gene combinations.

GENETIC ENGINEERING: ARTIFICIAL EVOLUTION

The discovery that genetics has a molecular basis was a pivotal one for biology. Not only has it led us to new levels of understanding of the mechanism of gene expression and the nature of biological information, but it has also given biology an entirely new technology: the ability to manipulate genetic information. Because the sequence of nucleotide bases in the DNA molecule determines the characteristics of a living organism, those characteristics can be changed by changing the DNA. Although biologists have appreciated this possibility ever since the discovery of the double helix, until recently it was not possible actually to make a calculated change in the DNA molecules within a living cell.

This is not to say that the idea of altering plants and animals to suit human needs is a new one. Experiments in plant and animal breeding have been carried out for thousands of years and have resulted in many of our most important crops and farm animals. The common-sense approach of breeding only those individuals that best suited human needs was a kind of genetic screening that gradually selected for desired characteristics. The plant or animal breeder, however, has always been limited by the diversity of genetic material that existed within the population of a species.

In the past 15 years, however, all of that has changed. It is now possible to locate and isolate specific pieces of DNA, read the base sequences on those pieces, alter those sequences, and insert made-to-order DNA molecules into living cells. These developments have important implications for the future of our species.

The Technical Basis of Genetic Engineering

The rapidly advancing field of genetic engineering was made possible by a few basic discoveries and technical advances. Each of these happened during the early 1970s, and collectively they made genetic engineering a reality. We will take a close look at some of these techniques and then see what it has been possible to do with them.

Restriction enzymes We have already seen that DNA molecules are very long. In order to study individual genes, we need a way to cut the DNA molecule up into smaller pieces that can be separated and handled easily. Fortunately, nature has provided a way to do exactly that. Some species of bacteria make special enzymes, called **restriction enzymes,** that *restrict* the kinds of DNA sequences that can survive in the bacteria and help protect them against invading viruses. These restriction enzymes cut DNA at specific sequences that are generally present in the viral DNA. If a virus enters a cell, these enzymes cut the DNA of the viruses to pieces, rendering it harmless.

Bacteria often protect themselves against their own restriction enzymes by attaching methyl groups to the adenine and cytosine nucleotides of their own DNA. This prevents the restriction enzymes from recognizing a cutting site in bacterial DNA. More than 100 different restriction enzymes have been discovered, and each one cuts DNA at a specific base sequence.

One widely used enzyme is *Eco R1,* which got its name because it was the first restriction enzyme discovered in *E. coli* (hence "*E. coli* Restriction Enzyme 1"). An interesting feature of Eco R1 and many restriction enzymes is that they produce *staggered* cuts across the double helix (Figure 23.9). The sequences recognized by such enzymes

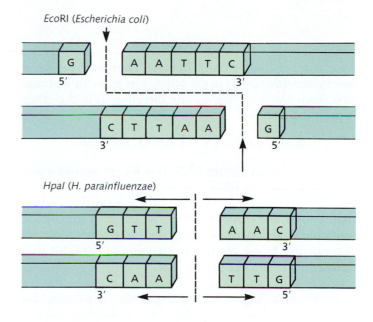

EcoRI (Escherichia coli)

HpaI (H. parainfluenzae)

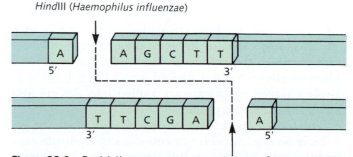

HindIII (Haemophilus influenzae)

Figure 23.9 Restriction enzymes recognize specific sites in DNA and cut both strands of the double helix at those sites. Many of the sites are palindromic, as explained in the text. The cutting sites of a few widely used restriction enzymes are indicated.

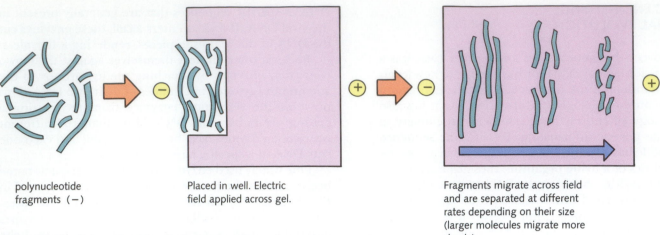

polynucleotide
fragments (−)

Placed in well. Electric
field applied across gel.

Fragments migrate across field
and are separated at different
rates depending on their size
(larger molecules migrate more
slowly).

Figure 23.10 Polyacrylamide gel electrophoresis can be used to separate polynucleotide fragments of different lengths. Smaller polynucleotides move more quickly through the gel and travel farther, while larger molecules move more slowly.

have another interesting property: They read the same way in one direction as in the other. The term for a phrase that reads the same way in one direction as in the other is **palindrome:**

A MAN A PLAN A CANAL PANAMA

As you can see in Figure 23.9, Eco R1 recognizes a *palindromic sequence* in DNA.

The most important feature of restriction enzymes is their exquisite specificity. They always cut at exactly the same place, always recognizing the same base sequence. It is possible to use restriction enzymes to cut DNA into defined fragments of precise size.

Electrophoresis: Separating the fragments Separating the polynucleotide fragments produced by restriction enzymes is actually quite easy. Because polynucleotides are negatively charged, they can be separated in an electric field by a process known as **electrophoresis.** Small samples of a polynucleotide mixture are loaded on top of a porous gel. Microscopically, the gel is a tangle of crossed fibers through which the polynucleotide molecules must pass. When an electric field is applied across the gel, the negatively charged fragments are pulled into the gel and drawn toward the opposite side.

The fragments, however, don't move through the gel at the same rate. Small fragments can slip easily through the meshwork of the gel. But larger molecules get tangled as they pass through, and their movement is much slower. As a result, polynucleotide molecules of different sizes are separated on the basis of size as they pass through the gel. Once the fragments have moved into the gel, the power is shut off and the fragments are located.

Gel electrophoresis is an extremely sensitive procedure. Not only can it be used to separate large molecules as long as 50,000 bases, but it is also so sensitive that two fragments that differ in size by just a single base can be separated into different bands (Figure 23.10).

Reading the sequence For years scientists knew that DNA molecules carried information required to specify protein structures, but they were unable to "read" that sequence. It is one thing to know in principle that the genetic code exists, but it is quite something else to be able to take a DNA molecule and determine, base by base, what the sequence is. In the early 1970s two laboratories, one in England and one in the United States, developed techniques that made it possible actually to "read" the sequence of bases in any DNA molecule.

The two methods, developed by Alan Maxam and Walter Gilbert in the United States and Frederick Sanger in Britain, operate somewhat differently, but both use gel electrophoresis to separate small polynucleotide fragments in a way that reveals the base sequence. Here's how one of the methods, the Sanger technique (Figure 23.11), works:

- Restriction enzymes are used to cut DNA into pieces of manageable size: a bit less than 200 bases. The double helix is separated by heating and the opposite strands are isolated. DNA replication enzymes are then used to make a radioactively labeled complementary copy of one strand. In each of four test tubes used for the reaction, a small percentage of one of the four bases added to the mixture has been chemically modified so that it will terminate the growth of the new DNA strand, as shown in Figure 23.11a.

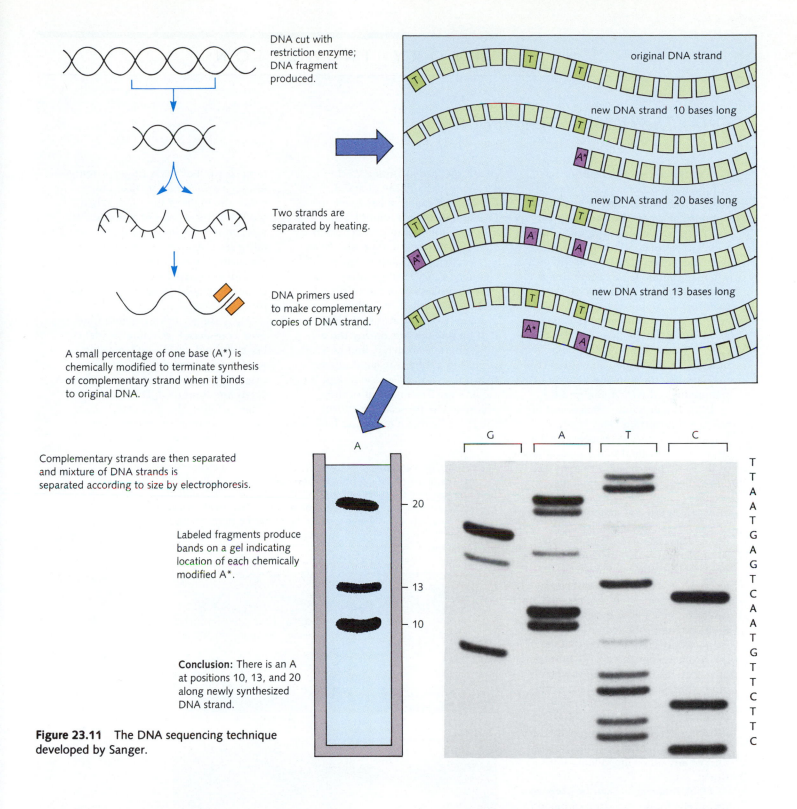

DNA cut with restriction enzyme; DNA fragment produced.

Two strands are separated by heating.

DNA primers used to make complementary copies of DNA strand.

A small percentage of one base (A*) is chemically modified to terminate synthesis of complementary strand when it binds to original DNA.

Complementary strands are then separated and mixture of DNA strands is separated according to size by electrophoresis.

Labeled fragments produce bands on a gel indicating location of each chemically modified A*.

Conclusion: There is an A at positions 10, 13, and 20 along newly synthesized DNA strand.

original DNA strand

new DNA strand 10 bases long

new DNA strand 20 bases long

new DNA strand 13 bases long

Figure 23.11 The DNA sequencing technique developed by Sanger.

■ After replication in each tube is complete, the newly synthesized strands are loaded onto a polyacrylamide gel with four wells, one for each base. Electrophoresis separates the strands according to size, and the banding pattern in each well is visualized by exposing photographic film against the gel.

■ In the "A" well, for example, the presence of bands at 10, 13, and 20 bases in length would mean that the new DNA sequence contained an "A" at each of those positions. When all four wells are visualized side by side, it is possible literally to read the sequence by examining the banding pattern (Figure 23.11b).

DNA: The Ultimate Fingerprint?

Late in the nineteenth century, law enforcement officers began to use fingerprints as a means of identifying criminals. Fingerprints are ideal for this purpose, because they differ so widely from one individual to the next. But fingerprints are useful only when a criminal leaves a number of clean, complete prints that can be matched with police records. In most crimes, fingerprint evidence is not good enough to be used for identification. Today, however, molecular biology has developed a new tool that may become the ultimate weapon of criminology: the *DNA fingerprint*.

DNA fingerprinting makes use of the fact that certain DNA regions between genes are extremely variable from one individual to the next. Many of these "hypervariable" regions contain multiple copies of simple base sequences, and the numbers of copies differ from one individual to the next: One person might have 40 repeats between two genes, another 31, and another 15. By constructing a small DNA fragment known as a *probe,* which is complementary to one of these simple sequences, researchers can examine a gel of total DNA and then use the probe to locate fragments of different sizes on that gel. If the probe recognizes a large number of fragments from around the genome, a single gel can produce a "fingerprint" that is unique for one individual. If two probes are used, the resulting gel pattern is so specific that it can be distinguished from the pattern of any other individual in the world.

DNA fingerprints can be prepared from cells found in a drop of blood or semen and even from fragments of skin caught under the fingernails of a crime victim. In 1988 this technique came to the aid of a 27-year-old com-

puter operator at Disney World in Florida who had been attacked, beaten, and raped in her home in Orlando in 1986. Although she caught a brief glimpse of her assailant's face during the rape, she faced the terrifying prospect of having to convince a jury that her brief eyewitness identification of the suspect, 24-year-old Tommie Lee Andrews, was absolutely certain. Andrews claimed that he was at home during the evening the rape occurred, and he even produced a witness to substantiate that claim. It was a classic case of the victim's word against that of the accused. The prosecuting attorneys sought the aid of a molecular biologist

to perform the "DNA fingerprinting" test on a semen sample taken by police the night of the rape. When the DNA fingerprint of this sample was compared with a blood sample taken from Andrews, the results were conclusive: It was a perfect match. The jury returned a verdict of guilty, resulting in a jail sentence of 22 years for Andrews.

DNA fingerprinting is not, however, considered "absolute" evidence because there are no uniform standards and procedures for the fingerprinting process. When uniform standards are developed, DNA fingerprinting should prove to be the ultimate weapon of criminology.

The first use of DNA fingerprinting in the United States is shown in these photographs. The upper photograph shows two matching bands from DNA obtained from the blood of a suspect and semen left by the rapist. To confirm the result, the test was run with a second restriction enzyme. The result, as seen in the lower photograph, also showed a match between the suspect's blood and semen left by the rapist. The other lanes on the gels are control samples and marker bands.

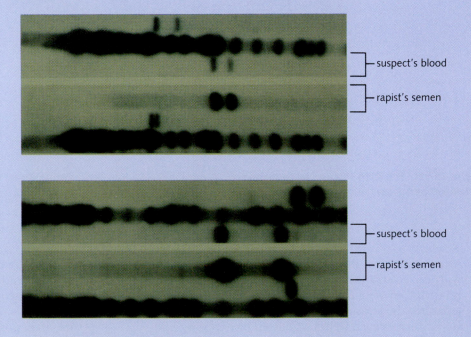

suspect's blood

rapist's semen

suspect's blood

rapist's semen

The development of techniques to read DNA base sequences has been an enormous boon to molecular biology. In the space of a few afternoons, it is possible to take a large DNA fragment, cut it into smaller pieces, read the DNA sequence of each of them, and then reconstruct the complete sequence by "pasting" the results together. Complete DNA sequences for viruses have been determined in this way, and the sequencing of whole chromosomes and genomes is not impossible to imagine.

In the late 1980s a number of scientists suggested that technology was now appropriate for molecular biology to attack its greatest challenge: determining the complete DNA sequence of the human genome. A few scientists challenged this suggestion, pointing out that the most interesting portions of the genome are already in the process of being sequenced and voicing concern that much of the genome will consist of noncoding "junk" DNA. Nonetheless, a major international effort is now underway to sequence the human genome. It is quite clear that such a project is now within the reach of molecular biology and that before long, large portions of the human genome will be completely recorded and available for study.

Plasmids In the early 1950s microbiologists discovered that many bacteria contain small "extra" chromosomes that are separate from the much larger normal chromosomes. These small DNA molecules are known as **plasmids** (Figure 23.12). They often are present in multiple copies, and they frequently contain genes that are necessary for cell viability.

An ideal plasmid for genetic engineering has two characteristics: (1) a single site where a restriction enzyme can cut the molecule, making it easy to manipulate in the lab, and (2) a *marker* gene to help identify cells that contain the plasmid. The best markers are those that make the cell resistant to drugs normally toxic to a bacterium, such as *penicillin* or *tetracycline*. If we've got the right plasmid, the right restriction enzyme, and an interesting piece of DNA, then we are ready to do some genetic engineering: to make that piece of DNA a permanent part of a bacterial cell.

Making the Chimera

There are no technical restrictions on the source of DNA used for cell transformation; it can come from anywhere. Suppose the piece of DNA that we would like to insert into the bacterium was produced by cutting cellular DNA with the restriction enzyme Bam H1. The first step is to mix the DNA fragments with plasmids that have also been cut with Bam H1 (Figure 23.13). Like Eco R1, Bam H1 makes a staggered cut across the two strands of the helix. This exposes four bases at the end of each strand that

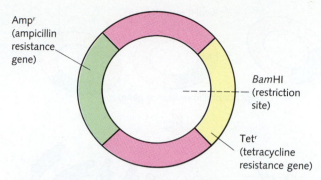

Figure 23.12 A typical plasmid, pBR322, which is widely used for genetic engineering. pBR322 contains a restriction site for Bam H1 as well as a replication origin and two antibiotic resistance genes. The antibiotic resistance genes are useful markers to select cells that contain the plasmid.

are able to base-pair and makes the ends of molecules adhere to each other (they are actually called "**sticky ends**"). Sometimes two DNA molecules pair with each other. But very frequently there forms the combination we have in mind: a plasmid and our DNA fragment. After allowing the strands to "anneal" in this way, an enzyme reattaches the breaks in the strands to form closed molecules.

The combination of plasmid and our DNA fragment is often called a **chimera,** a term that reflects the fact that the two molecules come from different sources. (The word *chimera* in classical mythology refers to a female monster formed by a combination of a lion's head, a goat's body, and a serpent's tail.) Because the technique involves the *recombination* of DNA from different sources, it is also common to refer to the whole procedure as **recombinant-DNA technology.**

Transformation

Once recombinant-DNA molecules are formed, the problem of getting them inside a cell remains. There are several different ways of doing this, depending on what kind of cell will serve as host for the molecules. If the DNA molecules are to be inserted into bacteria, we can use the process of *transformation,* which was discovered by Frederick Griffith, to get the molecules inside.

The recombinant molecules are mixed in with a population of bacteria that contain no plasmids. You might remember that the experiments of Griffith and Avery's group (Chapter 22) showed that DNA molecules could be taken up by bacteria and could transform them permanently. Well, transformation occurs in this case, too.

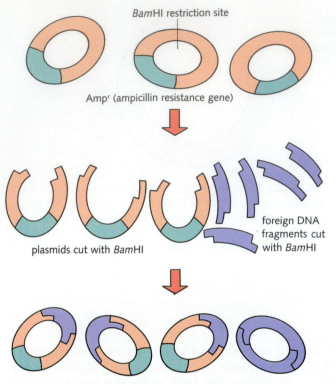

BamHI restriction site

Amp^r (ampicillin resistance gene)

plasmids cut with *BamHI*

foreign DNA fragments cut with *BamHI*

When combined, "sticky ends" of two molecules adhere, forming circular, recombinant plasmid or chimera.

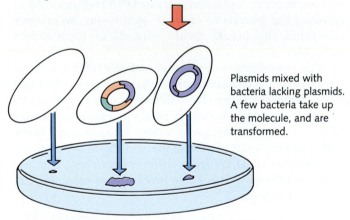

Plasmids mixed with bacteria lacking plasmids. A few bacteria take up the molecule, and are transformed.

Bacteria placed on a medium containing ampicillin. Those that have taken the plasmid with the resistance gene survive, grow into visible colonies, and are identified.

Figure 23.13 Chimeric DNA is formed by mixing the DNA fragment to be cloned with plasmid DNA molecules. After the "sticky ends" have attached, DNA ligase is used to seal the molecules together, producing a *recombinant DNA molecule*. These recombinant DNA molecules may then be taken up by bacteria and antibiotic resistance used to select bacterial clones which contain the recombinant DNA.

When the bacteria and the DNA molecules are mixed together, about 1 bacterium in a million takes up the plasmid molecule and is transformed. The rest are unaffected.

A typical bacterial culture contains as many as 10^8 bacteria per milliliter, which means that rate of 1 in a million could result in the transformation of hundreds or thousands of cells. But how can we separate the 1 bacterium that has a plasmid from the 999,999 that don't? Here's where the resistance gene comes in. Every transformed cell must have a copy of the plasmid, containing at least one gene for antibiotic resistance. We can use the presence of that gene to "find" the cells.

The cells are added to a solution that contains a large dose of the very same antibiotic for which there is a resistance gene on the plasmids used for transformation. The majority of the cells—those that do not contain plasmids—are killed by the drug. But those cells that contain the plasmid are resistant to the treatment, and they survive. In one stroke, then, it is possible to "select" only those cells that have taken up the plasmid.

Once the right cells have been selected, large numbers of them can be grown to produce an almost limitless supply of bacteria containing the recombinant-DNA molecule. This technique is sometimes referred to as "cloning" a DNA molecule. The term **clone** is used because of how the transformed bacteria are handled in the lab: The bacteria are spread on a culture plate in such a way that single cells produce round colonies of bacteria. Each of these colonies is a "clone," a group of cells formed by cell division from one single cell. For this reason, the cells of a clone are genetically identical and contain exactly the same recombinant-DNA molecules.

Why Clone?

There are several reasons why it might be useful to clone DNA molecules into bacteria. Researchers do so in order to get large amounts of a particular DNA fragment for study. If we wanted to learn about the detailed structure of the globin gene, for example, we might try to isolate a restriction fragment that contained the gene and clone that fragment in bacteria. Once we identified a clone of bacteria with the gene, we could keep those cells in storage until we were ready to work with the globin DNA. Then we would be able to grow large amounts of the bacteria, isolate the plasmids after breaking the cells, and separate the globin DNA from the plasmid by using our restriction enzyme. Our gene would then be available as a pure piece of DNA, as much as we wanted, for new experiments.

This is one of the main uses of recombinant-DNA technology in the laboratory. However, there are other reasons for wanting to clone DNA. In some cases, scientists want to use the bacteria as living factories to make large amounts of proteins that cannot be obtained in any other way. When a gene coding for the protein is inserted into a bacterium, along with the proper controlling sequences to make sure that the gene is properly expressed, a culture of the bacteria will synthesize large amounts of the protein. In the last few years, this has become a standard way to synthesize large amounts of important proteins at minimal cost. *Human growth hormone,* once a scarce and precious compound used to treat patients suffering from pituitary dwarfism, is now cheap because it is widely available through the use of a strain of bacteria expressing the human gene.

A bacterium is a complete, living organism. Our techniques for inserting DNA molecules into cells make it possible, in theory, to "design" new organisms. But our knowledge of biology has not yet developed to the point where we are ready to design living organisms from scratch. The sort of genetic design that is possible today is the insertion of a "new" gene into a cell to give it some properties it did not have before. We might imagine developing strains of bacteria with the enzymes necessary for digesting and disposing of harmful organic wastes, other cells capable of infecting corn plants and producing fertilizer that the plants could absorb directly, and even cells to help break down the cellulose fibers in wood pulp so that they could serve as a source of cheap, plentiful food. None of these speculations is far-fetched.

Transforming Other Cells

Bacteria would seem to be "naturals" for recombinant-DNA technology. The existence of bacterial plasmids and the ability to transform cells make the introduction of foreign DNA relatively easy. Can the same technology be applied to other organisms? Little by little an answer to that question has been emerging, and it seems to be yes.

Cloning systems that are very similar to bacteria have been developed in yeast, by the application of a **mini-chromosome** system that resembles the plasmids used for prokaryotic cells. The transformation of yeast cells is now a routine laboratory procedure, and it is possible to carry out genetic engineering in yeast in much the same way as in bacteria.

Why should we be interested in yeast? Although they can be cultured in large quantities, just as bacteria can, yeasts are eukaryotes. Therefore, yeasts are good systems for cloning eukaryotic genes and processing proteins that are unique to eukaryotic cells. Yeasts have now been engineered to make human proteins that may be important in fighting cancer, and one lab has even engineered yeast cells to secrete *rennin,* an enzyme used by cheesemakers. The fact that large amounts of rennin can now be made

cheaply by special yeast strains has been a boon to the cheese industry.

Mammalian cells in culture can also be induced to take up DNA fragments placed in their culture medium. These experiments have not as yet produced any commercial applications, but they have been very useful in research. When certain genes are taken up by normal cells, the cells can be transformed into ones that behave like cancer cells. Not surprisingly, cancer researchers are very much interested in this phenomenon.

Related techniques have also made it possible to insert genes into mammalian ova, and several laboratories have now used such procedures to produce adult mammals that contain and express transplanted genes. Agricultural researchers have begun to experiment with genes that will increase the size and weight of farm animals, and medical researchers have transplanted the genes of pathogens such as the AIDS virus into mice for experimental study. Organisms whose genetic composition has been artificially modified by genetic engineering are known as **transgenic organisms.** Transgenic animals and plants may form the basis of improved breeds of farm animals and more productive crop plants (Figure 23.14).

Figure 23.14 Genetic engineering has been successfully applied to higher organisms. (top) A transgenic pig with higher levels of growth hormone produced the meatier pork chop. (bottom) Transgenic tomato plants containing genes for viral resistance are healthier than those without the resistance genes (right).

Figure 23.15 Frost-damaged citrus trees. Plans for the release of genetically engineered bacteria hope to reduce frost damage.

One of the most useful and widely used transformation systems is one capable of altering plant cells. *Agrobacterium tumefaciens* is a bacterium that infects certain plants and causes tumors known as crown galls. The ability of *Agrobacterium* to cause tumors stems from a plasmid it carries that is known as *Ti* (tumor-inducing). This plasmid enters plant cells and becomes incorporated into the chromosomal DNA within the plant cell nucleus. Scientists have been able to use the Ti plasmid to transform plant cells by inserting new genes into single cells and then growing complete plants from the transformed cells.

IS GENETIC ENGINEERING SAFE?

When several scientists first realized that it would soon be possible to carry out genetic engineering in living organisms, concern began to mount that this new technology should be carefully thought out before it was applied. In 1974 an informal conference of leading scientists from around the world was called to consider the safety question.

The scientists appreciated fully how little we knew about something as complex as the genome of a whole cell. They recommended that genetic engineering experiments should begin with caution and that special care should be taken to ensure that engineered organisms did not escape the laboratory, so that the environment might be protected against the unforeseen consequences of this new technology. Initially, the National Institutes of Health adopted guidelines requiring that genetic engineering be done in specially sealed labs and that the bacteria used for certain experiments be specially weakened so that they could not live outside the laboratory. For several years, these guidelines closely regulated the flow of recombinant-DNA research.

In recent years, however, many of the original rules and guidelines have been relaxed. It has become clear that genetically engineered organisms are usually weaker than their natural counterparts and that there is little danger of creating a "super germ" any worse than the disease-causing microbes that are normally part of the environment. Good evidence has also accumulated that exchanges of DNA molecules occur in nature all the time and that, therefore, the efforts of laboratory scientists have not broken any new barriers in terms of the transfer of DNA between different species.

Nonetheless, concerns about the environmental hazards of genetically engineered organisms persist. As this book is being written, several companies are pressing for permission to use genetically engineered organisms in experiments that would involve releasing them into the environment. In one case, a corporation has taken a genus of bacteria called *Pseudomonas* and removed a gene for a protein that these bacteria normally secrete. This protein serves as a nucleating source for ice crystals, and the *Pseudomonas* bacteria that often grow on the leaves of fruit trees make them susceptible to frost damage. The corporation's idea is to spray crop plants with the engineered bacterium that does *not* make the protein, thereby displacing the other frost-inducing strain and making it more difficult for frost to injure the plant. Because frost damage causes the loss of millions of dollars worth of fruit every year (Figure 23.15), this is an important problem and an interesting solution.

However, opponents of the work argue that we cannot be sure what the ecological consequences of releasing the engineered bacterium will be. If it does penetrate the environment, will it affect the nucleation of ice crystals in the atmosphere and in turn reduce the amount of rainfall in certain areas? Although most scientists don't think that this is very likely, it is difficult to answer such objections with absolute certainty. Controversy over the use of genetic engineering will continue as our ability to manipulate genes and create organisms further develops in the years to come.

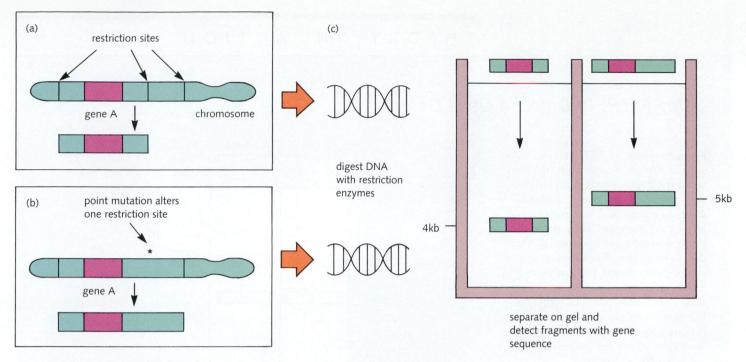

Figure 23.16 A point mutation that alters a restriction cutting site can change the size of a restriction fragment containing a gene of interest. The size change can be detected by gel electrophoresis. The existence of two or more sizes of restriction fragments containing a certain DNA sequence is known as a restriction fragment length polymorphism, or RFLP. RFLPs can serve as genetic markers.

THE NEW HUMAN GENETICS

The powerful techniques of molecular biology have made it possible to carry out genetics experiments in a completely new way. As we have seen, restriction enzymes cut DNA at specific sequences. A hypothetical gene might lie between two restriction sites 4000 bases apart, as shown in Figure 23.16. However, in some individuals, a harmless base change next to the gene might abolish one of those restriction sites so that the gene becomes part of a larger restriction fragment (Figure 23.16).

This change alters the length of the restriction fragment that contains gene A and therefore is known as a **restriction fragment length polymorphism,** or **RFLP** (pronounced "rif-lip"). The presence of a RFLP can be seen in the pattern of DNA fragments produced by restriction enzyme digestion, as shown in Figure 23.16c. When these fragments are separated by gel electrophoresis and the gel is treated to make it possible to detect gene A, the difference in fragment length caused by the RFLP is apparent.

For genetic purposes, a RFLP can be treated like any other trait. It is passed from one generation to the next

and can be detected in DNA isolated from blood samples. Because each RFLP has a definite position, RFLPs on the same chromosome display *linkage,* and the rate of recombination between linked RFLPs can be used to construct genetic maps, as described in Chapter 10. More than 1000 different RFLPs have been discovered in the human genome, and more are added every week. The power of these new genetic markers is so great that they will form the basis of a worldwide effort in the years ahead to map and sequence the entire human genome.

RFLPs and Genetic Disease

Sometimes RFLPs are directly associated with disease-causing genes. This makes it possible to test for the presence of a defective gene simply by probing for a RFLP that is either in the gene or closely linked to it. One of the first such probes was one that detected a RFLP close to the gene for Huntington's disease, a dominant genetic disorder that causes degeneration of the nervous system and death. As we noted in Chapter 11, Huntington's disease does not produce symptoms until middle age, so

Using RFLPs to Detect a Gene Defect

Phenylketonuria, or *PKU*, is a recessive genetic disease. People with two copies of the PKU gene lack an enzyme that is needed to convert phenylalanine, an amino acid, to tyrosine. The lack of this enzyme causes abnormal brain development and mental retardation in infants. Therefore, it is important to detect PKU in newborns and to provide them with a diet that does not contain phenylalanine. In the past it was possible to detect the disease only *after* birth, and there was no way to screen for parents carrying a single copy of the gene.

The presence of some alleles that cause PKU can be detected by probing DNA fragments isolated from both parents and cut with a restriction enzyme. One of the potential parents who make up couple 1 carries a copy of the PKU gene (*), but his wife carries two copies of the normal gene. Therefore, they have no risk of pro-

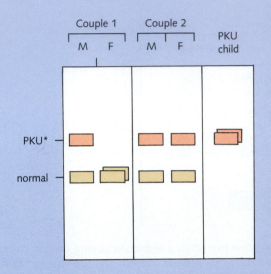

ducing a child with PKU. However, each of the members of couple 2 carries a copy of the PKU gene, giving them a 25 percent chance of having a child with PKU.

The detection of RFLPs linked to PKU provides valuable advance warn-

ing of the possibility of having a child with PKU. Early detection is especially important for PKU, because careful control of the diet can prevent any mental retardation whatsoever and enable people with PKU to lead otherwise normal lives.

---

people are usually unaware that they carry the disease gene until after they have produced children. The test for the Huntington's gene can be done with a small blood sample from which DNA is extracted to test for the telltale RFLP. The availability of this test makes it possible for the children of Huntington's sufferers to learn whether or not they carry the deadly gene (see Current Controversies: Would You Really Want to Know?).

Similar DNA tests are now available for a score of genetic diseases, and more will be developed in the immediate future. Researchers believe that ultimately it may be possible to produce RFLPs that correlate with a *tendency* to develop disorders such as heart disease, diabetes, or cancer. Genetic testing might serve as an early warning marker for these medical problems, aiding in prevention and early detection.

Cystic Fibrosis: New Light on a Mysterious Killer

Cystic fibrosis (CF) has always been one of the most mysterious of all fatal genetic diseases. It affects roughly one child in 1600 in the United States, and most CF patients die as children. For years researchers have tried, unsuccessfully, to find the biochemical basis of CF. CF patients produce an abnormally salty sweat. They are unable to release normal amounts of digestive enzymes from the pancreas, which leads to digestive disorders. Their lungs clog with a thick mucus that makes it difficult for them to breathe and results in serious respiratory infections. Unable to find a single cause for all of these symptoms, many laboratories decided to look for the CF gene itself.

In 1989 a group of researchers including Lap-Chee Tsui of Toronto and Francis Collins of Michigan announced that they had found the gene. They started with DNA probes that were linked to the CF trait and used these probes to locate the gene on the long arm of chromosome 7. By detailed DNA mapping and sequencing, they found a large, complex gene associated with CF. More work needs to be done before we understand CF fully, but researchers believe that the gene itself may explain the nature of the disease.

The protein produced by this gene has the amino acid sequence that is characteristic of a membrane protein, and researchers have already predicted a structure for the protein (Figure 23.17).

This structure suggests that the protein transports ions across cell membranes, and this may explain the symptoms of CF. Improper salt transport could produce the heavy mucus and salty sweat, and it might also explain the digestive problems that victims of CF experience. CF has not been cured, but molecular biology is on the verge of giving us a new understanding of one of our most baffling killers.

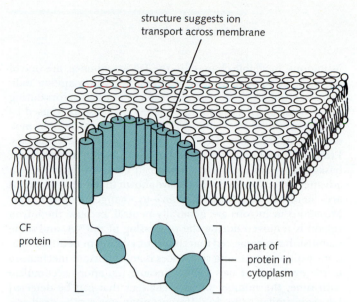

Figure 23.17 The sequence of the cystic fibrosis gene suggests that it may be a membrane protein involved in ion transport. This drawing shows a structural model for this protein in a biological membrane.

CURRENT CONTROVERSIES

Would You Really Want to Know?

The availability of a test for Huntington's disease is a great medical breakthrough. But it confronts the children of Huntington's sufferers with a dilemma: *to test or not to test.* Universal application of the test might make it possible to eliminate the disease. If all potential carriers were to be tested, and if all agreed not to have children if they tested positive, the incidence of the disease might be cut dramatically. Many young adults living in fear and uncertainty would have the relief of learning that they will never develop the disease, and those not so lucky might be grateful for years of advance warning to plan for the onset of the disease.

But many people who may one day succumb to Huntington's have decided that they do *not* want to take the test. Some of these people have said that they prefer uncertainty to the possibility of discovering that they *do* have the disease. Others have said that they will have children regardless of their status on the Huntington's test, seeing nothing tragic about bringing into the world a child who might develop the disease. After all, Huntington's sufferers lead full lives until the disease develops. In fact, about 50 percent of the possible Huntington's carriers who were offered the test in the Boston area in 1989 declined to be tested. Molecular biology may produce powerful new tools with which we can search for genetic diseases, but deciding how to use those tools will remain a matter of individual choice.

SUMMARY

Mutations, heritable changes in the DNA sequence, are one of the most important sources of genetic variation among individuals. Mutations may arise as a result of random errors during the process of DNA replication. They may also be caused by mutagens: chemical or physical agents that affect the DNA molecule. Mutations may affect one base in a gene or several bases. Frameshift mutations are caused by the insertion or deletion of bases and result in drastic changes in gene products. Recent advances in molecular biology have shown that many mutations are "neutral"; they cause no obvious changes in phenotype. Mutations in introns are generally neutral, because the intron region is removed during the processing of mRNA, and some scientists have suggested that the accumulation of neutral mutations within intervening sequences is an important mechanism in the evolution of new genes. Because mutations accumulate with time, the number of base changes that can be detected between similar genes in different organisms can be used as a measure of the length of time that has passed since the two organisms separated in the process of evolution.

Molecular biology has now reached the point where directed genetic change (genetic engineering) is possible. Restriction enzymes, which cut DNA at precisely defined sequences, make it possible to prepare DNA fragments that have been cut at specific points. Other techniques enable technicians to separate those fragments, combine them with bacterial minichromosomes known as plasmids, and use the restriction-fragment–plasmid chimeras to transform bacteria and other types of cells. These technologies make it possible to produce large amounts of specific DNA sequences and also to construct living organisms with gene combinations different from those that occur in nature. The opportunities and challenges posed by this new technology will shape the kind of world that emerges in the next century.

STUDY FOCUS

After studying this chapter, you should be able to:

- Outline the molecular mechanisms that underlie evolution.
- Describe the process of gene mutation, and explain how changes in DNA cause changes in amino acid sequence and ultimately in phenotype.
- Explain some of the tools of modern molecular biology and genetic engineering.
- Ponder some of the related ethical and scientific questions. Should engineered organisms be released into the environment? Is it ethical to attempt human gene therapy? Do humans have the right to alter their own reproductive cells or to control the course of evolution?

SELECTED TERMS

mutation *p. 436*	electrophoresis *p. 446*
mutagen *p. 436*	plasmid *p. 449*
point mutation *p. 437*	chimera *p. 449*
frameshift mutation *p. 438*	recombinant DNA *p. 449*
transposable	clone *p. 450*
elements *p. 438*	transgenic organism *p. 451*
thymine dimer *p. 439*	minichromosome *p. 451*
restriction enzyme *p. 445*	RFLP *p. 453*
palindrome *p. 446*	cystic fibrosis *p. 454*

REVIEW

Discussion Questions

1. What is a mutation? Outline some of the factors that can cause mutations. Are all mutations irreversible?

2. Discuss both the positive and the negative effects of mutations.

3. The position of a point mutation in a gene—intron or exon— helps determine what effect it has on phenotype. Why?

4. Discuss how molecular biology can be used to trace evolution.

5. Describe the use of restriction enzymes to prepare defined sequences of DNA. Would genetic engineering be possible without restriction enzymes? Why or why not?

6. Discuss some of the issues involved in the release of genetically engineered organisms into the environment.

Objective Questions   (Answers in Appendix)

7. A mutation always produces a change in the
   (a) function of DNA.
   (b) nucleotide sequence.
   (c) function of enzyme.
   (d) cytoplasm of the cell.

8. Jumping genes
   (a) regulate protein synthesis.
   (b) are transposable segments of DNA.
   (c) can cause bacterial infection.
   (d) are chemical mutagens.

9. A DNA molecule that has genes from different sources is a
   (a) chimera.
   (b) replicate.
   (c) helix.
   (d) genome.

10. Restriction enzymes are used to
    (a) cut DNA at specific sites.
    (b) inhibit bacteria from infecting humans.
    (c) convert one enzyme to another.
    (d) cut RNA at specific sites.

11. Insertion of foreign DNA into a bacterial plasmid creates a
    (a) minichromosome.
    (b) RFLP.
    (c) chimera.
    (d) clone.

CHAPTER 24

Cell Organization

Some animals know the world through sound. They sense vibrations so precisely that they can fly through the night sky or glide through dark waters, turning and darting through the smallest openings. Others know a chemical world of tastes and scents. They can follow a chemical clue for miles to find food or water, or to locate a mate. Humans have senses of sound and smell, of course, but the very structure of our brains reveals that we humans *see* the world around us. We are visual creatures, and we know the world through our eyes.

For less than four centuries, we have extended our powers of sight with an invention that has opened a new world to human eyes—this startling invention is the microscope.

THE MICROSCOPE: EXTENDING THE SENSES

The construction of a microscope requires painstaking analysis of the interaction between light and matter. Even so, the basic principles of a microscope are simple and easy to describe. If special care is taken to design a glass surface that is not flat, interesting things can occur to light rays passing through it. A thin piece of glass that has been ground down into a lens so that each surface forms part of a nearly spherical curve will focus parallel light rays to a single point on the opposite side of the lens (Figure 24.1). This point is called the *focal point*. When we use a magnifying glass to start a fire, we are taking advantage of this property to focus the nearly parallel

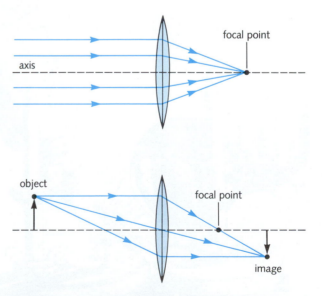

Figure 24.1 An ideal converging lens refracts parallel rays of light so that they converge at a single focal point.

rays of sunlight to illuminate a single spot so intensely that it bursts into flames.

Figure 24.1 also shows what happens when the same lens is used to focus light from an object that is not at the focal point: An *image* of the object is formed at some distance from the lens. Depending on the placement of the original object and the power of the lens, this image can be either magnified or reduced in size from that of the original object. The ability to change the apparent sizes of objects with glass lenses means, quite simply, that magnified images of objects can be made that reveal detail too fine to be seen without assistance. A lens enables us to see things that are either very small or very far away.

The Invention of the Microscope

History does not tell us precisely who invented the microscope, although certain parts of the story are clear. The thriving merchants of early seventeenth-century Holland did a brisk business in cloth and fabric. In order to judge the quality of woven cloth they needed to be able to examine the individual threads of a fabric. The human eye is not able to perceive as separate two structures closer together than 0.1 millimeter. In other words, some important details of woven cloth are just too small to be seen, except by a merchant using a glass lens to produce a magnified image.

Here, as it often does, commercial technology provided an opportunity for science. Anton van Leeuwenhoek (1632–1723) was a young man in Holland when he was apprenticed to a dry goods store by his family. During his apprenticeship, he had used the glass magnifying lenses that his tradesmen had pioneered, but van Leeuwenhoek was interested in other things. As far as we know, he never went very far in the dry goods trade. Instead, history records him as one of the first to use the microscope for biology.

Van Leeuwenhoek's microscope was a simple affair. A single lens mounted in a metal plate enabled him to observe things that no one had ever seen before. He recorded his observations in letters and drawings that were sent to the Royal Society in London. Van Leeuwenhoek's world contained tiny animals and plants with fantastic structures. They inhabited the very water his neighbors swam in and drank, and they were more

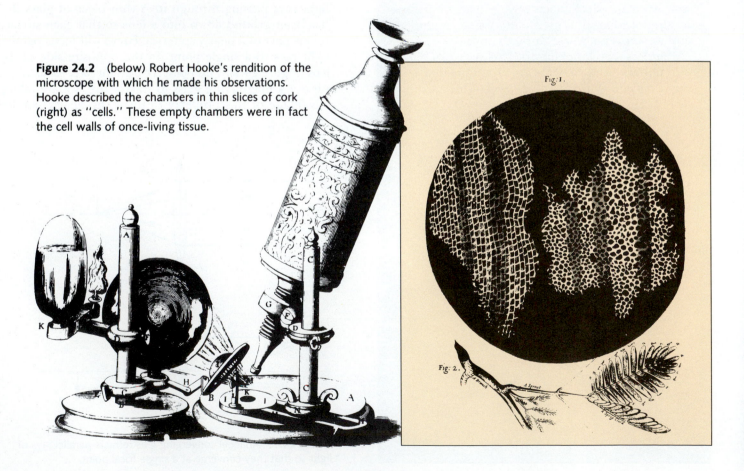

Figure 24.2 (below) Robert Hooke's rendition of the microscope with which he made his observations. Hooke described the chambers in thin slices of cork (right) as "cells." These empty chambers were in fact the cell walls of once-living tissue.

numerous than the birds of the air. Others added to his observations; one of these was an Englishman named Robert Hooke (1635–1703).

While describing a thin slice of cork Hooke wrote that the structure of the material was like a "honeycomb" and that the air within the spongy cork "is perfectly enclosed within little boxes or cells distinct from one another" (Figure 24.2). This observation would be repeated, not just in dry, dead tissue like cork, but also in specimen after specimen of living material. The substructures that both van Leeuwenhoek and Hooke saw have, indeed, come to be called **cells,** and we now realize that cells are the basic structures that make up all living things. The first people to claim that all living things were composed of cells were Matthias Schleiden (1838) for plants and Theodore Schwann (1839) for animals. Today we often consider Schleiden and Schwann the authors of the **cell theory,** a premise that all living organisms are composed of individual, self-reproducing, living structures known as cells. Cells are the basic units of life.

We will now discuss two main types of microscopes used in biology, the light microscope and the electron microscope.

The Light Microscope

A typical **light microscope** contains two main lenses: an *objective lens,* which forms an image of the specimen within the barrel of the microscope, and a *viewing lens* (or *ocular*), which produces an image that an observer can see (Figure 24.3). The reason structures as small as living cells are visible in a microscope is simple: The lenses are organized in such a way that the images they produce are much larger than the objects under observation. The enlargement of small regions in a sample enables us to see fine detail. When we use more powerful lenses, the image is bigger still and the detail finer and finer. But only up to a point.

As microscopes became more refined in the late nineteenth century, some lens manufacturers believed that if lenses could be made better and better, there was no limit to the amount of detail that they would reveal. However, the light microscope is subject to an important limitation: the nature of light itself.

Because light radiation is partly wave-like, as light passes between smaller and smaller structures in living cells, the light is scattered in a process known as *diffraction*. The

Figure 24.3 The optical pathways of a modern brightfield light microscope.

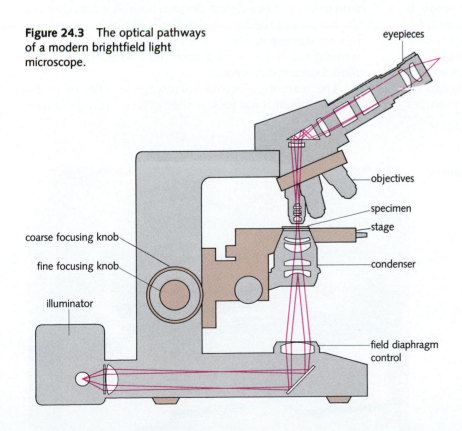

eyepieces

objectives

specimen

stage

condenser

coarse focusing knob

fine focusing knob

illuminator

field diaphragm control

scattering of light caused by diffraction limits the amount of detail that can be distinguished in the microscope. We describe that detail in terms of resolution: the ability to "resolve," or perceive as separate, two or more structures a certain distance apart. Under the very best conditions, the **resolution limit** of a light microscope is about 0.2 μm. This means that the light microscope enables us to see about 500 times the detail that we can see without it (0.1 mm = 500 × 0.2 μm). Because living cells are much larger than this resolution limit (a typical cell is about 30 μm in diameter), we can see them easily in the light microscope. We can also see a great many of the structures within the cell. However, we are not able to see the smallest structures within the cell, nor can we view viruses, tiny disease-causing particles, or individual molecules, all of which are below the resolution limit of the instrument.

The Electron Microscope

Because our eyes are sensitive to visible light, it seems natural to consider first a microscope that uses light to operate. However, there is no reason why *any* kind of radiation cannot be used to construct a microscope, including electrons. Beams of electrons can be focused by magnetic fields in the same way that glass lenses focus light (Figures 24.4 and 24.5).

The principal advantage of an **electron microscope** over the light microscope is the greater resolution it affords. Because the wavelength of an electron beam is much shorter than the wavelength of visible light, very small objects diffract the radiation much less. In fact, the electron microscopes used for biological research have resolution limits in the vicinity of 0.2 nm, almost exactly 1000 times smaller than those the best light microscopes

offer. This increased resolution has opened up a whole new area for biologists, who now study structures within the cell that were never visible before.

However, the opportunity has come at a price. Electrons do not pass through air very easily. (They collide with gas molecules and get scattered about.) Therefore, any material to be observed in an electron microscope must be placed in a vacuum. Samples are first treated with a chemical preservative, or fixative, to preserve their structure. Then they are dehydrated (all of the water removed) so that there is no water to boil in the vacuum of the microscope. Finally, because electrons do not pass through matter easily, biologists must slice samples into very thin slices (usually about 50 nm thick) in order to enable electrons to penetrate them. Despite these limitations, biologists have used electron microscopes to gain a great deal of knowledge about the organization of living things.

In a **transmission electron microscope,** the image is formed on a special screen at the bottom of the microscope column. When large numbers of electrons strike the screen, it glows, and the user of the microscope can "see" an image of the sample.

In the **scanning electron microscope,** a narrow beam of electrons is focused on one spot at the surface of a sample. When electrons strike the surface of the sample, their energy dislodges other electrons in the sample. These "secondary" electrons can be measured by a special electron detector. The original electron beam is scanned across the surface of the sample. As the beam moves, continuous measurements are made of the secondary electrons striking the detector, and these measurements are displayed on a video screen.

The scanning electron microscope enables us to get a three-dimensional look at the surfaces of objects with

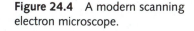

Figure 24.4 A modern scanning electron microscope.

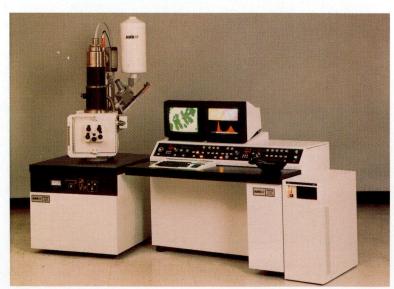

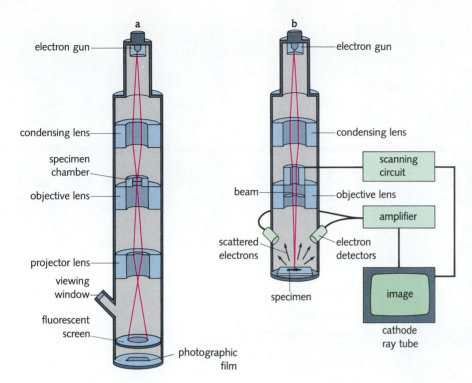

Figure 24.5   In transmission electron microscopes (**a**), a series of electromagnetic lenses focuses a beam of electrons on a thin sample inserted into the specimen chamber. Electrons that pass through the specimen are focused into an image by another series of lenses.

In scanning electron microscopes (**b**), a narrow beam of electrons is scanned across the surface of a specimen in a two-dimensional pattern. Electrons or other radiation that this beam produces at the surface of the sample are collected by a series of detectors and then shown on a television-like display which scans in steps with the electron beam. Samples observed in each type of electron microscope must be placed within the vacuum of the microscope column.

great accuracy. Its resolution limit, generally about 3.0 nm, is considerably better than that of the light microscope, although it does not quite match the resolution of the transmission electron microscope that allows us to see internal cell structure. The scanning electron microscope suffers from some of the same limitations as the transmission electron microscope, including the fact that its specimens must be placed in a vacuum chamber. This remarkable instrument produces dramatic images of cells and tissues that have been very useful in helping us to analyze the detailed organization of biological systems.

Figure 24.6 shows different views of a single cell type as it is seen through the light microscope, the scanning electron microscope, and the transmission electron microscope. The differences between the three views in terms of detail and resolution are startling.

Figure 24.6   Three different views of a single cell type produced in the (left) light microscope, (right) scanning electron microscope, and (middle) transmission electron microscope.

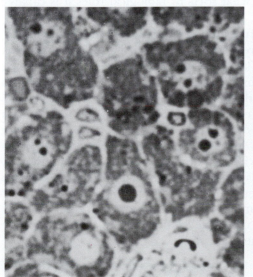
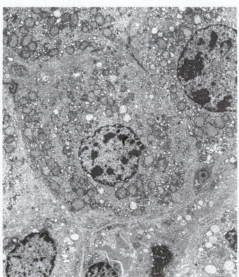
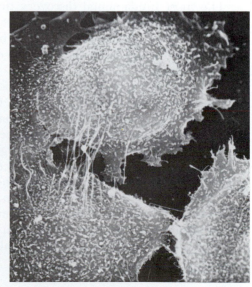

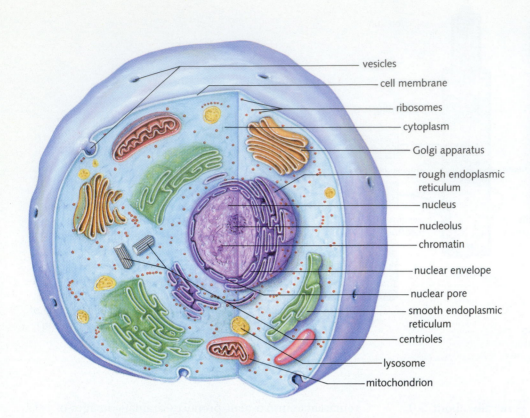

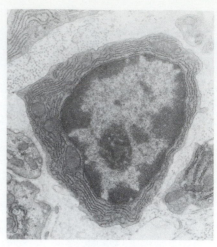

vesicles

cell membrane

ribosomes

cytoplasm

Golgi apparatus

rough endoplasmic reticulum

nucleus

nucleolus

chromatin

nuclear envelope

nuclear pore

smooth endoplasmic reticulum

centrioles

lysosome

mitochondrion

Figure 24.7 Diagrams and electron micrographs showing the basic organization of an animal cell (above) and a plant cell (opposite page). Although these models do not do justice to the tremendous diversity of cellular structure in animals and plants, they do show some of the common features of many cell types. The electron micrographs show a fibroblast cell from rat connective tissue and a cell from a bean seedling leaf.

THE CELL: A DETAILED LOOK

The most fundamental discovery that microscopy has yielded in 350 years is still the cell. It is very difficult to imagine how a modern science of biology could have developed without the discovery of cells, the basis of all living things.

If we take a close look at a typical eukaryotic cell, many things stand out (Figure 24.7). One of these is the complexity of the cell. There are so many different structures within the cell that it should be clear that sorting them all out would be quite a challenge. Nonetheless, we must start somewhere, and the perfect place to begin is with the thin *membranes* that divide the cell into so many compartments. A membrane known as the **cell membrane** separates the cell from its surroundings and deserves special attention. In fact, the electron microscope image of the cell, whether plant or animal, is dominated by

membranes (Figure 24.7). Further in this chapter, we will consider what a biological membrane is, how it is constructed, and what sorts of biological roles it is capable of playing.

When the first anatomists dissected the bodies of animals, they were struck by the fact that living organisms are composed largely of specialized *organs,* each of which seems to perform a specialized task for the organism as a whole. Could the same thing be true of a single cell? Could the cell contain "little organs" that do specific jobs within it?

Before the end of the nineteenth century, cell biologists were convinced that many structures within the cell were analogous to organs and coined the term **organelles** to apply to them. In the second part of this chapter, we will look at some of the most important cellular organelles and will try to develop a picture of how the cell depends on their activities.

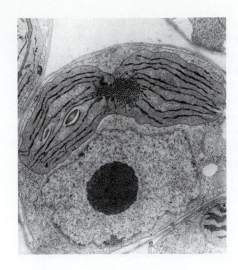

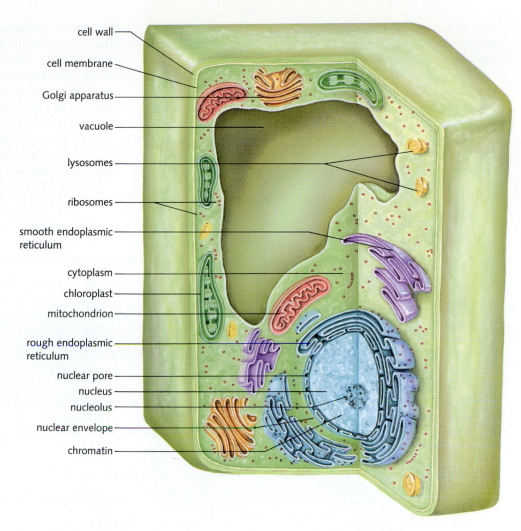

cell wall

cell membrane

Golgi apparatus

vacuole

lysosomes

ribosomes

smooth endoplasmic reticulum

cytoplasm

chloroplast

mitochondrion

rough endoplasmic reticulum

nuclear pore

nucleus

nucleolus

nuclear envelope

chromatin

BIOLOGICAL MEMBRANES

For many years, biologists wondered about the nature of the barrier that separated cells from their environment. A few actually thought that there was *no* membrane surrounding the cell and speculated that the cell was separated from its surroundings by a "phase boundary" like an oil droplet in water. Before long, however, overwhelming evidence emerged that the cell membrane was a genuine structure, and scientists began to work on understanding its composition and form.

The Basic Structure of Membranes

The lipid bilayer   The structural backbone of a biological membrane is a **lipid bilayer.** We have already seen how the basic chemical structure of many lipid molecules

enables them to form lipid bilayers in which individual molecules are arranged with the most polar portions of the molecules facing the water that surrounds the bilayer (Figure 24.8). Lipid bilayers can form spontaneously in mixtures of lipid and water, and they are relatively stable. This is because the oily, or *hydrophobic* ("water-fearing"), hydrocarbon chains are gathered together in the middle part of the bilayer, whereas the polar, or *hydrophilic* ("water-loving"), parts of the lipid molecules are exposed to the water at the surface of the bilayer.

When early studies on biological membranes showed that they contained lipid, one of the first questions scientists tried to answer was whether there was enough lipid in a cell membrane to form a lipid bilayer. In the 1930s, two scientists named Gorter and Grendel did an important experiment the results of which suggested that there was (see Theory in Action, Getting Lucky with Mistakes: How Science Sometimes Works, p. 464). Gorter and Grendel calculated that each cell contained just enough lipid to cover its surface area with *two* monolayers—in other words, with a **bilayer.**

Many other experimental lines of evidence indicate that most of the lipids in biological membranes are in the form of a bilayer. Isolated membranes can be studied via X-ray diffraction, and they scatter X rays in a way that can be explained only by assuming that the lipids are in the form of a bilayer. But biological membranes contain proteins, too. How are the proteins arranged?

Membrane proteins For years many scientists believed that the proteins associated with biological membranes were only found at the surfaces of the lipid bilayer. But more recent work has shown that proteins can actually span the lipid bilayer and make contact with both surfaces of the membrane.

New techniques used in electron microscopy have made it possible to visualize membrane proteins within the lipid bilayer. In a procedure known as freeze-etching, cell or tissue samples are rapidly frozen and then placed in a

---

Getting Lucky with Mistakes: How Science Sometimes Works

In order to carry out their key experiment, Gorter and Grendel took a suspension of red blood cells, extracted all of the lipid in their membranes by using an organic solvent called acetone (you will remember that lipids are soluble in such solvents), and then tried to figure out how much surface area could be covered by a monolayer of the extracted lipid.

The apparatus they used to do this enabled them to pour their lipids on the surface of a water-filled trough and compress them gently until they packed close enough to form a tight monolayer. Next they measured the surface area of the lipid monolayer and used that number to figure out what the surface area of the lipids was in a single red cell. Then all they had to do was to divide the surface area in the trough by the number of cells in the original suspension. Their answer suggested that there was enough lipid to cover the cell surface twice.

An interesting aside to the Gorter and Grendel experiments deserves comment. The two scientists made serious mistakes in their experiment! (1) They made an incorrect estimate for the surface area of a single red cell. (2) The solvent they used to extract membrane lipids was acetone, in which a few of the red cell lipids are not soluble, so they didn't get all of the membrane lipids out. (3) They squeezed too hard in the trough and compressed the monolayer more than they should have, leading them to underestimate the surface area their lipids occupied. In fact, they made only one accurate measurement in the whole experiment: counting the number of red cells in their starting suspension. With so many mistakes, why is their experiment remembered today? There are really two reasons. The first is historical: Their experiment convinced many scientists of the reality of lipid bilayers in cell membranes, a concept that has proved to be correct. The second is sheer, unadulterated luck: Their three errors were in opposite directions and canceled each other out, so that their final results were just about right.

Gorter and Grendel used a trough of water to help measure the surface area occupied by lipid extracted from isolated red cell membranes.

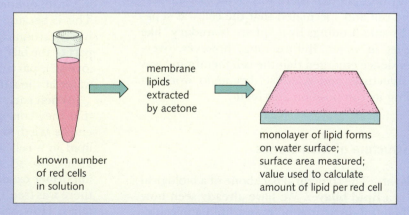

known number of red cells in solution

membrane lipids extracted by acetone

monolayer of lipid forms on water surface; surface area measured; value used to calculate amount of lipid per red cell

---

special vacuum chamber where they can be split in half at low temperatures. A metal film, or replica, is then cast on the fractured surface, and that film can be examined in the electron microscope (Figure 24.9).

When the frozen material fractures near a biological membrane, the fracture tends to split the membrane open between the two halves of the lipid bilayer (this is because the hydrophobic bonds that hold the tails of lipids together are weak at low temperatures). This process reveals the inner region of the membrane, and complexes of transmembrane proteins are visible as distinct particles in the replica. One biological membrane that has been widely studied by means of this technique is the membrane of the red blood cell (Figure 24.9). Mammalian red cells are among the simplest of cells. They lose their nuclei and internal organelles as they develop, so a mature red cell consists of a single membrane surrounding a cytoplasm consisting of hemoglobin and other soluble proteins. The red cell membrane can be isolated for analysis, and we know that it contains large amounts of both lipid and protein and also a significant amount of carbohydrate.

Freeze-etching and other techniques have produced a picture of biological membranes in which proteins are directly associated with the lipid bilayer. However, things are not that simple. In many membranes, including the red cell membrane, certain proteins are attached to

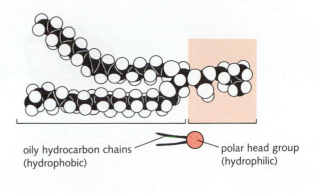

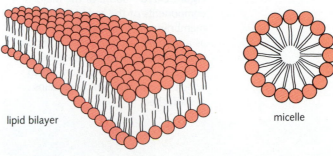

oily hydrocarbon chains (hydrophobic) — polar head group (hydrophilic)

lipid bilayer — micelle

Figure 24.8 Lipids dispersed in water may aggregate to form a number of different structures, including spherical or cylindrical *micelles*, or sheet-like *bilayers*. In each structure, the polar head groups are in direct contact with water, while the hydrocarbon chains are shielded from water.

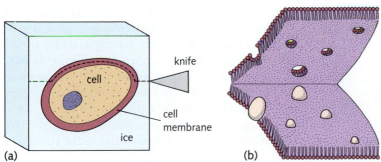

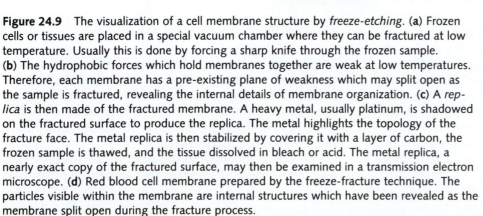

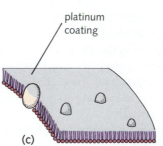

knife

cell

cell membrane

ice

(a)

(b)

platinum coating

(c)

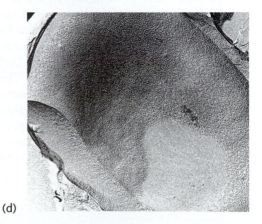

Figure 24.9 The visualization of a cell membrane structure by *freeze-etching*. (a) Frozen cells or tissues are placed in a special vacuum chamber where they can be fractured at low temperature. Usually this is done by forcing a sharp knife through the frozen sample. (b) The hydrophobic forces which hold membranes together are weak at low temperatures. Therefore, each membrane has a pre-existing plane of weakness which may split open as the sample is fractured, revealing the internal details of membrane organization. (c) A *replica* is then made of the fractured membrane. A heavy metal, usually platinum, is shadowed on the fractured surface to produce the replica. The metal highlights the topology of the fracture face. The metal replica is then stabilized by covering it with a layer of carbon, the frozen sample is thawed, and the tissue dissolved in bleach or acid. The metal replica, a nearly exact copy of the fractured surface, may then be examined in a transmission electron microscope. (d) Red blood cell membrane prepared by the freeze-fracture technique. The particles visible within the membrane are internal structures which have been revealed as the membrane split open during the fracture process.

(d)

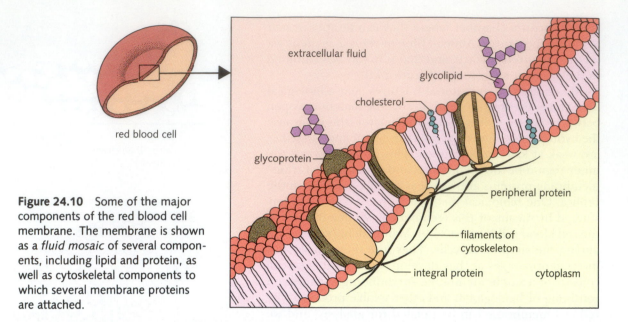

Figure 24.10 Some of the major components of the red blood cell membrane. The membrane is shown as a *fluid mosaic* of several components, including lipid and protein, as well as cytoskeletal components to which several membrane proteins are attached.

Labels in figure: extracellular fluid; glycolipid; cholesterol; glycoprotein; peripheral protein; filaments of cytoskeleton; integral protein; cytoplasm; red blood cell

molecules at the membrane surface (Figure 24.10). One good example is *spectrin,* a protein that is attached at the inner surface of the red blood cell membrane. Because spectrin is bound to the edge, the periphery of the membrane, it is known as a *peripheral* membrane protein. Proteins that are directly associated with the lipid bilayer, in contrast, are known as *integral* membrane proteins. Finally, *carbohydrate* molecules are attached to the membrane in a number of ways. Membrane-associated carbohydrates can be found attached directly to proteins, forming glycoproteins, and they can also be attached to certain lipid molecules to form glycolipids. Protein and glycoprotein molecules exposed at cell surfaces can function as *receptors* for chemical messages, as *markers* that enable cells to identify each other, and as *control points* that regulate cell attachment and cell growth.

All of this may make biological membranes seem like mosaics or conglomerates, which is not far from the truth. In fact, the most widely accepted theory of membrane organization today is known as the **fluid mosaic model.** The name is appropriate. The membrane is a mosaic of many different components, and biological membranes tend to be fluid structures. The "fluidity" of biological membranes reflects the fact that membrane components are free, to some extent, to move about within the plane of the membrane. Both lipid and protein components of the membrane are able to move laterally in the plane of the membrane, and this fluidity is important for many membrane functions.

The Diversity of Cellular Membranes

Although we have emphasized some of the features that most biological membranes have in common, the fact is that the different membranes found within a single cell differ greatly from each other. They vary in lipid composition, carbohydrate content, and the membrane proteins they contain. It is probably accurate to consider the basic fluid mosaic model of membrane structure as a kind of scaffolding. Upon that scaffolding an almost infinite variety of equipment can be installed. Some membranes are specialized for transport, some serve as barriers, and some play important roles in the synthesis of large molecules. A few are as much as 80 percent lipid, and a few are nearly 80 percent protein. In many cases, the different functions of cellular organelles result directly from differences in the membranes that make them up.

The Functions of Biological Membranes

Membranes as barriers The first and foremost task of the biological membrane is to serve as a barrier. Without such a barrier, the components of the cell would diffuse away, and any molecule in the environment would be free to enter. The fluid mosaic plan of organization makes membranes ideal barriers.

Biological membranes can best be described as *selectively permeable,* which means that some things can pass through and some can't. Lipid bilayers, as we have already seen, are held together by the hydrophobic forces that exist between the oily hydrocarbon tails of lipid molecules. Charged molecules (ions such as K^+ and Na^+), which are strongly attracted to water, are not attracted by the hydrocarbon region in the middle of a lipid bilayer, so they are not able to pass through bilayers very easily. Neither are larger, electrically neutral molecules such as glucose, which cannot squeeze between the tightly packed lipid molecules (Figure 24.11).

The molecules that can pass through lipid bilayers most easily are water itself and nonpolar (oily) molecules such as ethanol and propanol, which can "dissolve" in the bilayer as they move through. The fact that membranes are selectively permeable leads to some properties that have important biological consequences. We will examine those properties in the next few pages. But first, we must understand the basic physical process of diffusion.

THE MOVEMENT OF MOLECULES ACROSS MEMBRANES

The Laws of Diffusion

The molecules in living things are in constant motion. Under a light microscope, the movement of water molecules can be seen to cause very small objects to move about in short, sudden jumps—a phenomenon known as **Brownian movement.** This molecular motion is an effect of temperature—the higher the temperature, the greater the rate of movement—and it is random, or undirected. Like marbles jostled about, individual molecules follow a path that depends on their rate of movement and the collisions they experience with other molecules.

The random motion of molecules in a gas or a liquid means that no molecule within a liquid or gas can remain stationary for very long. Consider what happens when we open a bottle of perfume in one corner of a room. The molecules that produce the scent are moving within the liquid perfume (usually a solution of alcohol). As they approach the surface of the liquid, many of the scent molecules pass directly into the air, and the layer of air near the open bottle gradually fills with such molecules. If the movement of air molecules were to stop, the scent would go no further. However, the air molecules continue to move, and their motion bounces the molecules carrying the scent in every direction. Some of them cross back into the liquid and return to solution. But many of them move further into the air, away from the bottle. Little by little, the scent spreads from the bottle until it fills the room. This process is known as **diffusion.**

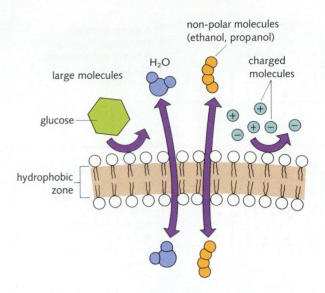

Figure 24.11 Biological membranes are *selectively permeable*. While water and most nonpolar molecules can readily cross such membranes, other molecules cannot. Most biological membranes effectively exclude larger molecules and charged molecules unless the membranes contain specific transport molecules which allow such molecules to cross.

Strictly speaking, *diffusion* is the process by which a substance moves from an area of high concentration to an area of lower concentration, and it commonly occurs in liquids and gases (although in some special cases it can also take place in solids). The gradual movement of scent from our opened bottle of perfume, the dispersal of color from a droplet of dye placed in water, and even the spreading of smoke from a fire are examples of the process of diffusion.

Diffusion occurs in response to a **concentration gradient:** a difference in the concentration of a substance between two areas. This is because diffusion depends only on the random thermal motion of molecules. When two compartments with different concentrations of compound X are prepared, and the barrier between them is opened, diffusion causes a net movement of molecules from one compartment to another until a point is reached at which the concentrations of X in the two compartments are equal. At that point, individual molecules still continue to move from one compartment to the other (molecular motion is random!) but there is no *net* movement. A **diffusion equilibrium** has been reached between the two compartments (Figure 24.12).

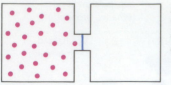

A barrier separates two compartments with different concentrations of a substance.

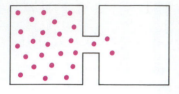

When the barrier is opened, there is a movement of molecules from the region of higher concentration to the region of lower concentration—diffusion.

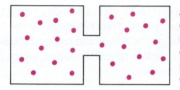

After awhile, a state of equilibrium is reached in which the concentrations of material in each compartment are identical.

Figure 24.12 Diffusion is the movement of molecules between two compartments which have different concentrations of solutes. The removal of the barrier in this example allows the random movement of individual molecules to produce an equilibrium in which the concentration of the solute is uniform throughout.

Diffusion Across a Biological Membrane

We can modify the example of a molecule diffusing from one compartment to another and use it to consider one of the most important kinds of molecular movement in biological systems: diffusion across a biological membrane. Because most biological membranes are selectively permeable, at least a few molecules can cross them. The way in which a particular molecule crosses a biological membrane falls into one of three general categories: simple diffusion, facilitated diffusion, and active transport.

Simple diffusion   As we have just seen, **simple diffusion** causes the movement of molecules from a region of higher concentration to a region of lower concentration (Figure 24.13). Diffusion can take place across a membrane if two conditions are met: (1) The molecule is present on one side of the membrane at a higher concentration than

it is on the other side (a concentration differential). (2) The molecule can actually pass between the molecules that make up the membrane. This second condition requires that the molecule be relatively *small* (big ones can't slip between the molecules of the membrane itself) and *nonpolar* (so that it can pass through the hydrophobic region at the interior of the membrane).

Facilitated diffusion   A special case of diffusion in which molecules cross membranes by passing through pore-like transport molecules that *facilitate* their passage is called **facilitated diffusion.** Facilitated diffusion is similar to simple diffusion in that a concentration differential is required to drive it. But the important difference between this process and simple diffusion is the existence of the pores or carriers that are part of the membrane itself and that enable facilitated diffusion to occur at a faster rate than simple diffusion. Many cells in the human body have, built into their membranes, special transporter proteins that allow glucose to pass through. Excellent effectors of facilitated diffusion, these proteins allow glucose to enter the cell as much as 100 times faster than simple diffusion. A diagram suggesting that such proteins may function like pores is shown in Figure 24.14 (in fact, it's not yet known if they are actually shaped like pores).

Active transport   The use of energy to move molecules across a membrane is known as **active transport.** The energy is derived from active transport molecules which are sometimes referred to as pumps, because they use energy to move material. The most common source of this energy for active transport is ATP, and many membrane proteins use the energy available in ATP to pump molecules across a membrane. Because these molecules can use energy, they are capable of moving material across membranes *against* a concentration gradient (Figure 24.15). This is a key difference between active transport and facilitated diffusion, which depends on a concentration differential for movement of material.

Many cells contain an active transport molecule known as sodium–potassium ATPase. This ATPase is a membrane protein that can pump both sodium and potassium across the cell membrane, using ATP as its source of chemical energy.

The pump works in two directions at once, pumping sodium ions out of the cell and potassium ions into it. Because of the actions of this active transport protein, the cytoplasm of a typical cell contains much more potassium and much less sodium than the fluid surrounding it. Although cell membranes are capable of transporting a number of ions, the concentration gradients involving sodium and potassium are particularly important in nerve and muscle cells, which use these ions to help conduct electrical impulses.

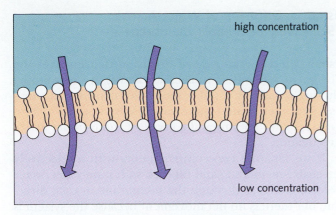

Figure 24.13 Many substances, including water and lipid-soluble organic compounds, can pass through biological membranes by means of simple diffusion.

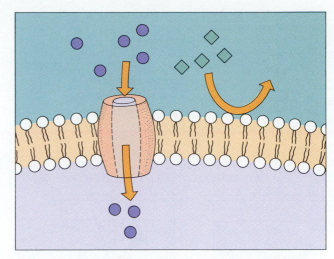

Figure 24.14 Facilitated diffusion across biological membranes is made possible by special pore-like carrier molecules. Many such carriers are integral membrane proteins which form channels across the membrane. These channels can be quite specific, allowing some molecules to pass through, while excluding others.

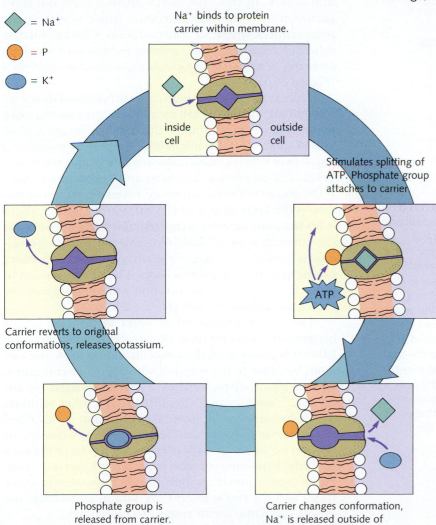

◆ = Na⁺

● = P

● = K⁺

Na⁺ binds to protein carrier within membrane.

inside cell   outside cell

Stimulates splitting of ATP. Phosphate group attaches to carrier

ATP

Carrier reverts to original conformations, releases potassium.

Phosphate group is released from carrier.

Carrier changes conformation, Na⁺ is released outside of cell, K⁺ binds.

Figure 24.15 Active transport involves the use of energy to pump specific substances across a cellular membrane. The sodium-potassium pump, an active transport complex found in many cell membranes, uses the energy of ATP to pump sodium ions out of the cell, while at the same time bringing potassium ions inside the cell. The five stages of this diagram illustrate the cycle of the pump: (Top, clockwise) the binding of sodium; the hydrolysis of ATP and phosphorylation of the pump; the release of sodium and binding of potassium at the cell surface; the release of bound phosphate; and the release of potassium within the cell.

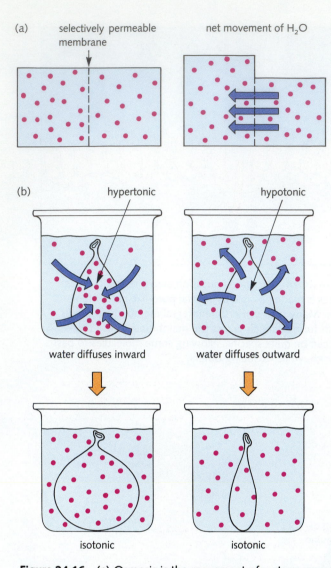

(a) selectively permeable membrane

net movement of H_2O

(b) hypertonic

hypotonic

water diffuses inward

water diffuses outward

isotonic

isotonic

Figure 24.16 **(a)** Osmosis is the movement of water across a selectively permeable membrane. When two compartments contain different concentrations of a solute which cannot pass through a membrane, a net movement of water will occur in the direction of the compartment with a higher solute concentration. The concentration of water in this compartment increases as the result of osmosis. **(b)** The effects of osmosis can be seen in a membrane-like sac placed in two different solutions. If the solution within the sac is more concentrated (hypertonic) than the surrounding solution, osmosis will cause a net movement of water into the sac, and it expands. If the solution within the sac is less concentrated (hypotonic), osmosis will cause a net movement of water out of the sac, and it will shrink. When the concentration of material in the sac is the same as the surrounding solution, net movement of water stops, and the concentration of material in the sac is said to be isotonic with respect to the surrounding solution.

Osmosis: Diffusion of Water Across a Selectively Permeable Membrane

Osmosis is a special case of diffusion that involves the movement of water molecules across a membrane. Although water is a polar molecule, it passes through biological membranes quite easily. Why water is able to do this is still something of a mystery and is under active investigation.

If we separate two compartments with a selectively permeable membrane and fill each with pure water, water molecules diffuse through the membrane in each direction, but there is no net movement because the system is in equilibrium—the solutions on both sides of the membrane are identical. But if we dissolve another molecule—sucrose, for example—in one of the compartments, the situation changes. Sucrose cannot cross the selectively permeable membrane. However, the presence of sucrose in one compartment can be thought of as actually lowering the concentration of water in that compartment. An unequal concentration means that diffusion occurs. In this case, water diffuses from the compartment of higher concentration (pure water) to the compartment where its concentration is lower (sucrose solution). *Osmosis is the movement of water across a selectively permeable membrane in response to a concentration differential* (Figure 24.16a).

Because cells contain high concentrations of dissolved material surrounded by cell membranes, osmosis can cause water to move into or out of the cell. We can get a good idea of what effect this might have on a cell by examining a system in which a small balloon made from a selectively permeable membrane has been suspended in a beaker. If the concentration of dissolved material within the balloon is the same as that of the fluid in the beaker, there is no net movement of water (equilibrium). However, if the concentration of dissolved material within the balloon is greater than that of the surrounding fluid, water begins to diffuse into the balloon. The net movement of water as a result of concentration differences gradually causes the balloon to swell (Figure 24.16b).

The inward movement of water in this case occurs because the concentration of dissolved material in the balloon is *greater* than the concentration outside the balloon. We refer to the solution with a greater concentration of dissolved material as **hypertonic** (*hyper* means "greater"; *tonic* means "strength") to the other solution. The solution with the lower concentration is described as **hypotonic** (*hypo* means "lesser"). The net movement of water across a selectively permeable membrane is always from the hypotonic solution toward the hypertonic one. When equilibrium is reached, the net movement of water stops, and the two solutions are said to be **isotonic** (*iso* means "the same") with respect to each other.

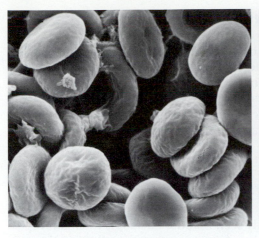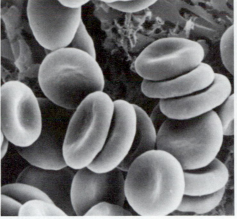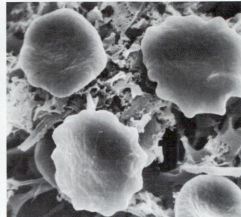

Figure 24.17 Osmosis can cause dramatic changes in the shape of an unprotected cell. These three micrographs show the appearance of red blood cells in hypotonic (left), isotonic (center), and hypertonic (right) solutions.

As we have just seen, a selectively permeable balloon containing a hypertonic solution swelled when osmosis caused water to move into the balloon. The direct cause, of course, is that water moved into the balloon, following the concentration gradient and increasing the balloon's volume. What would happen if the balloon were placed in a wire meshwork so that it could not expand? Although osmosis could not force water to move into the restricted volume of the balloon, the tendency of water molecules to move into the balloon would still generate a force, causing pressure to build up. The force generated by osmosis is known as **osmotic pressure.**

Effects of Osmosis on a Living Cell

The cell membrane is a selectively permeable membrane, and water passes across it easily, although most salts, sugars, and complex molecules do not. Therefore, a single cell suspended in pure water is subjected to a severe osmotic pressure. If the pressure is not counteracted, the cell will swell and burst (Figure 24.17).

Living organisms have evolved three different strategies to contend with osmotic pressure. The first, and most direct, is to solve the problem by changing the environment of the cell. In a large organism, individual cell membranes rarely come into contact with the environment. The organism produces a "microenvironment" around each cell, and that microenvironment includes an isotonic solution. The isotonic environment ensures that cells are not subjected to osmotic pressure, and there is little net movement of water into or out of most cells.

A second strategy is used by a number of single-celled organisms that live in water and cannot alter their immediate environment. These cells contain special organelles known as *contractile vacuoles* that help them cope with osmosis. The details of contractile vacuole function are not completely known. In a general way, however, we do understand that contractile vacuoles help collect the water that osmosis has brought into the cell and periodically release it to the exterior. In a sense, the vacuole works like a bilge pump in a boat, pumping out water as soon as it leaks in. The forces of osmosis are kept in check and the cell maintains its normal shape (Figure 24.18).

For many cells, however, there is an even simpler solution, one that is a lot like our balloon in a cage. Many bacteria, algae, and plants produce tough, rigid **cell walls** that surround their cell membranes (Figure 24.19). The wall is porous enough to allow food molecules to diffuse into, and waste products to diffuse away from, the cell. But it is also tough enough to withstand tremendous pressure. That is important, because the forces of osmosis, which tend to force water into the hypertonic cytoplasm, create a substantial osmotic pressure. The cell wall counteracts that pressure and prevents the cell from literally exploding on account of it.

The great strength of this strategy is that it is simple, quick, and requires a minimum of energy. The disadvantage is that it puts the cell at the mercy of its own cell wall. Any break in the cell wall can cause disaster: Osmotic pressure may cause the flexible membrane to expand through the break until the cell swells to the point of bursting. As we will see later, a number of drugs used to combat bacterial infection exploit the weaknesses of this method of coping with osmotic pressure.

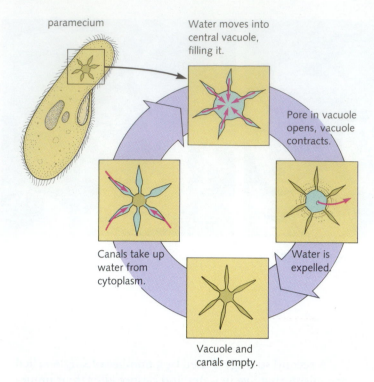

paramecium

Water moves into central vacuole, filling it.

Pore in vacuole opens, vacuole contracts.

Water is expelled.

Vacuole and canals empty.

Canals take up water from cytoplasm.

Figure 24.18 Many cells deal with osmotic pressure by eliminating the excess water through contractile vacuoles. *Paramecium*, a protist, has well-developed contractile vacuoles which take up water from the surrounding cytoplasm, pass it through canals to a central vacuole, and then expel the accumulated water through a pore in the cell surface.

CELL ORGANELLES

Some of the organelles found in eukaryotic cells are involved in the synthesis of molecules needed for cell growth, some in the conversion of energy, still others are part of a microscopic "skeleton" within the cell—the **cytoskeleton** (*cyto* means "cell"). A few organelles do not fit neatly into any category. We will deal with all of them individually.

The Nucleus: The First Organelle

We have already seen that the nucleus is the main repository of genetic information for the cell. The chromosomes are enclosed within it, and the first step in the expression of a gene is the activation of a DNA sequence within the nucleus. The nucleus is surrounded by a **nuclear envelope** consisting of two membranes that are interrupted in several places by **nuclear pores** (Figure 24.20). These pores allow material to pass into and out of the nucleus without passing directly through a biological membrane. The nucleus is a busy place. The activation of gene sequences, the transcription of pre-messenger RNA, and its cutting and splicing into mature messenger RNA all take place within the nucleus. During the S phase of the cell cycle, DNA replication takes place within the nucleus.

The internal structure of the nucleus is difficult to decipher in many cases: The chromosomes that are so obvious during mitosis become dispersed during interphase. However, one can see dark clumps of material

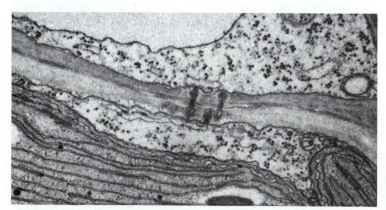

Figure 24.19 Osmotic pressure can be controlled by cell walls. These structures are formed outside of the cell membrane, and provide a semi-rigid casing which prevents excessive cell expansion due to osmotic pressure. This micrograph illustrates the structure of cell walls in a corn seedling. In many plants, cytoplasmic connections known as plasmodesmata (center of micrograph) provide direct contact between adjacent cells despite the presence of the walls.

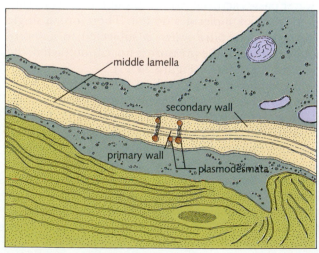

middle lamella

secondary wall

primary wall

plasmodesmata

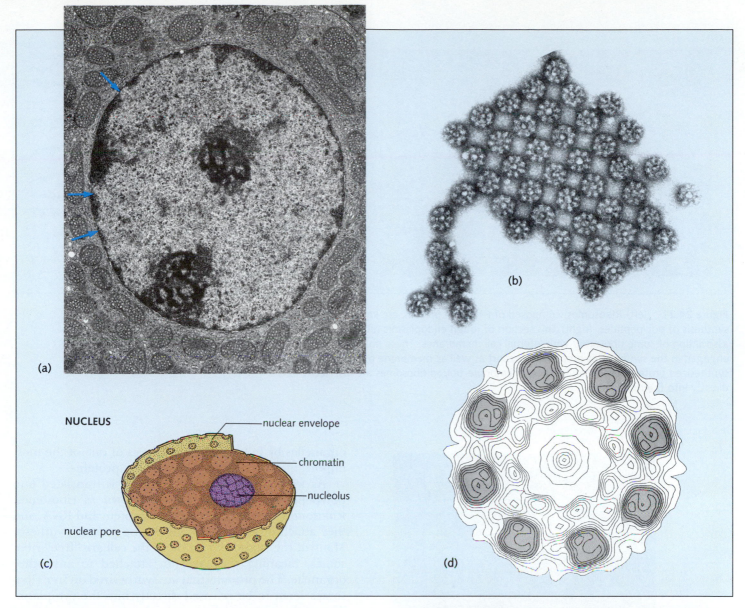

(a)

(b)

NUCLEUS

nuclear envelope

chromatin

nucleolus

nuclear pore

(c)

(d)

Figure 24.20 (a) This nucleus, from the rat adrenal gland, shows two dense nucleolar regions. The darker areas within the nucleus are known as heterochromatin, while the lighter regions are called euchromatin. The nucleus is surrounded by two envelope membranes interrupted by a series of nuclear pores (arrows). (b) A cluster of nuclear pore complexes, isolated by gentle disruption of the nuclear membranes. (c) Nuclear pores are the passageways through which material enters and exits the nucleus. (d) A computer-assisted reconstruction of the basic structure of a nuclear pore, calculated from electron micrograph images like those in Figure b.

known as *heterochromatin* and lighter areas called *euchromatin.* One large clump of heterochromatin is the **nucleolus,** a section of the nucleus that contains ribosomal RNA genes and in which ribosomes are assembled from ribosomal RNA and the appropriate proteins.

The Ribosome

Life is a dynamic process. New molecules that are required for cell growth and for the maintenance and repair of existing cell structures must be synthesized at a rapid rate and brought to their proper places within the cell. A number of cell organelles are so specialized to carry out the synthesis of macromolecules that we can refer to them as *synthetic organelles.* We will examine some of these

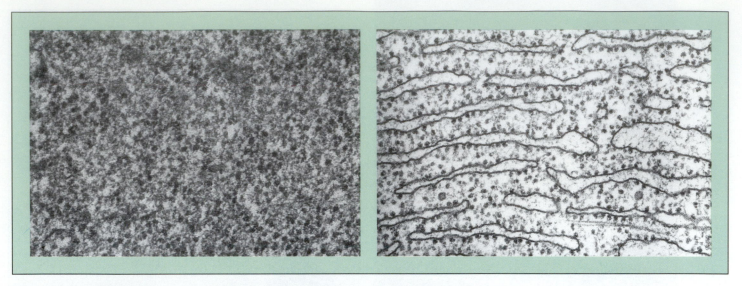

Figure 24.21 (left) Ribosomes, composed of RNA and protein, are key organelles in the synthesis of polypeptides. (right) This section of rough endoplasmic reticulum shows the close association of some ribosomes with internal cell membranes. These ribosomes are engaged in the synthesis of proteins for export as well as membrane proteins. The newly synthesized proteins produced by these membrane-bound ribosomes are inserted directly into the rough ER.

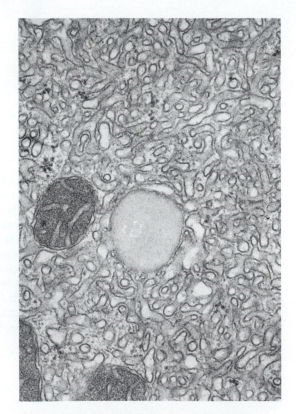

Figure 24.22 Smooth endoplasmic reticulum is involved in a variety of biochemical pathways. The enzymes involved in lipid biosynthesis as well as drug detoxification are contained in smooth ER.

organelles by following the synthesis of one of the most important classes of macromolecules, proteins.

As we saw in Chapter 22, mRNA is translated into polypeptides on small particles known as ribosomes. Ribosomes are composed of both protein and RNA, and they attach directly to mRNA during protein synthesis. Many of the ribosomes found in the cell are "free" ribosomes, meaning they are not attached to any other organelle. The proteins that are synthesized on free ribosomes seem to be released directly into the cytoplasm. These ribosomes can be seen in electron micrographs of the cytoplasm (Figure 24.21).

The Endoplasmic Reticulum

However, ribosomes can also be found in other places. In many cells, a majority of the ribosomes are *membrane-bound;* they seem to be directly attached to internal cellular membranes. The membranes to which ribosomes are most often attached are known as **endoplasmic reticulum (ER)**. Portions of the endoplasmic reticulum that are covered with ribosomes are known as **rough endoplasmic reticulum,** because the scientists who discovered the association thought that the presence of ribosomes made the membrane look "rough," like sandpaper (Figure 24.21). Regions of the ER that do not have ribosomes attached are known as **smooth endoplasmic reticulum** (Figure 24.22).

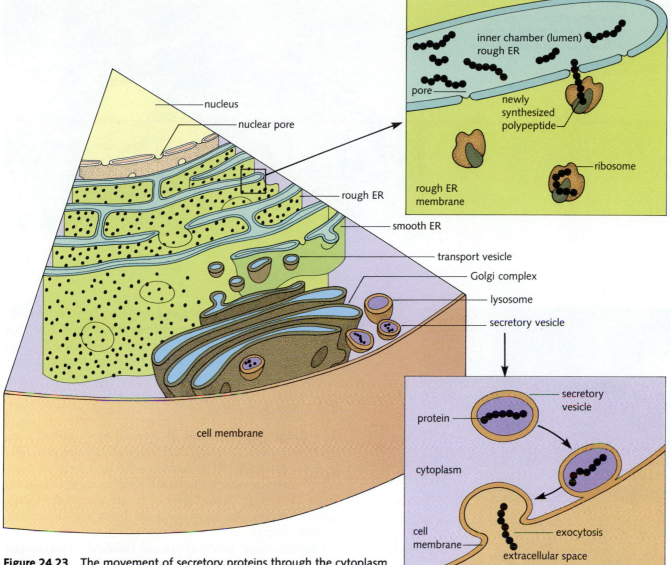

Figure 24.23 The movement of secretory proteins through the cytoplasm of an eukaryotic cell. Newly synthesized proteins are inserted into the rough ER, and then passed to the Golgi apparatus where a variety of chemical modifications may take place, including the attachment of sugars to produce glycoproteins. Secretory vesicles which bud off from the Golgi lead to the cell membrane where their contents are released by exocytosis.

Why are ribosomes attached to the endoplasmic reticulum? Figure 24.23 shows the reason. Ribosomes attached to the rough ER extrude their newly synthesized polypeptides directly through the ER membrane. The proteins synthesized in this way may be released inside the ER, or they may be inserted into the ER membrane. Two types of proteins are synthesized in this way: membrane proteins and proteins that will be released ("secreted") from the cell. In contrast, when a free ribosome completes the synthesis of a protein, the protein is released directly into the cytoplasm. Recent work has shown that the first 15 or 20 amino acids of a newly synthesized polypeptide may serve as a "signal sequence" that helps to direct a growing polypeptide into the endoplasmic reticulum.

Why should proteins destined to leave the cell be inserted into the rough ER? Because this organelle is the first stop on a pathway that leads the newly synthesized protein to its final destination.

The smooth ER is not directly involved in protein synthesis. In many cells, the smooth ER seems to contain a series of enzymes that are responsible for carrying out

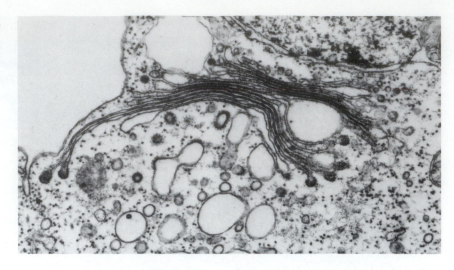

Figure 24.24 The Golgi apparatus, processing site for many secretory proteins, often appears in thin sections as a stack of flattened vesicles.

other types of biochemical jobs that vary from cell to cell. In some, the enzymes responsible for steroid synthesis are found within them. In the liver, smooth ER often contains the enzymes responsible for drug detoxification.

The Golgi Apparatus

The **Golgi apparatus** was discovered by Camillo Golgi, an Italian microscopist. The Golgi apparatus generally appears as a stack of flattened vesicles that is closely associated with the rough ER. Small vesicles seem to bud off from the rough ER and fuse with one side of the Golgi, and other vesicles seem to leave from the other side (Figure 24.24).

A great many proteins are not finished when the last amino acid is put in place and translation is complete. These proteins require a bit more tinkering ("modification") before they are ready for action. The Golgi is the site at which these chemical modifications take place. Depending on the particular protein, sugar molecules must be attached, the polypeptide chain may be cut in a strategic location, or prosthetic groups may be attached— to name just a few possible modifications. In cells that are very active in protein synthesis, the Golgi may be very large and well developed.

Secretory Vesicles

When the synthesis of a protein is complete, that protein can be used by the cell. However, many proteins are released, or secreted, from cells to the exterior. Such proteins pass from the Golgi into **secretory vesicles** (Figure 24.25). In the pancreas, for example, the outer rim of the cell is filled with vesicles that are packed with protein

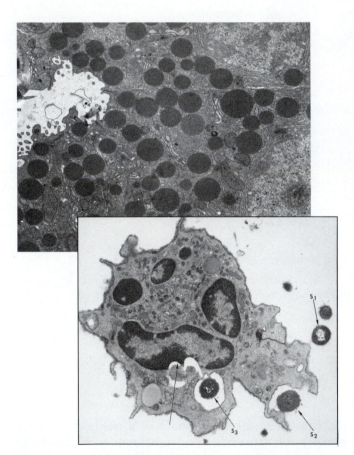

Figure 24.25 (top) These secretory vesicles store zymogen, a collection of the principal secretory products of the exocrine pancreas. The secretory proteins will be released from the cell by exocytosis. (bottom) Endocytosis, which brings material into a cell, is important in a variety of cellular activities. This white blood cell is engaged in the phagocytosis ("cell eating") of several *Streptococcus* bacteria. (S_1) a bacterium remains free; (S_2) the site of a bacterium being engulfed; and (S_3) a bacterium fully surrounded.

products from the Golgi. One of the jobs of the pancreas is to produce enzymes that aid in the digestion of food. These secretory vesicles contain enzymes that were first synthesized on the rough ER, were modified in the Golgi, and were then passed along to the vesicles, where they are stored until they are released from the cell.

The contents of a secretory vesicle are released from the cell in a process known as **exocytosis** (*exo* means "out-side"), in which the vesicle membrane fuses with the cell membrane. Many cells release their secretory contents in response to specific signals, and the link between stimulus and exocytosis is an area of much research interest.

Endocytosis (*endo* means "inside"), the opposite of exocytosis, in which material is brought into the cell enclosed in vesicles, is also an important research topic. Many cell membranes contain protein *receptors* that bind

---

Why Are There So Many Types of Microscopes?

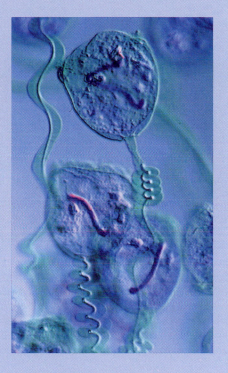

Any microscope is able to magnify the image of a specimen, but magnification alone is not enough to make something visible. In fact, one of the problems we encounter in trying to look at cells under a microscope is that, more often than not, they are invisible. There are two general approaches to solve the problem of **visibility.** One is to *stain* the specimen with chemicals that make it absorb light more efficiently.

But there is another approach. When light passes through a specimen on a microscope slide, there are only three things that can happen to it.

1. The light may pass through unaffected.
2. The light may be absorbed by the specimen.
3. The light may be scattered by the specimen.

A simple light microscope operates under what has come to be known as **brightfield** optics ("brightfield" because an object under the microscope generally looks dark against the background). An object can be seen clearly under brightfield only if it absorbs or scatters enough light to make itself dark against the background. Heavy staining solves that problem and it is usually required because most cells are nearly transparent. But how do we view living cells that would be damaged or killed by staining?

There are a number of special types of light microscopes, each of which is designed to make cells visible by doing something "tricky" with the light scattered by the specimen. The **darkfield** microscope is designed in such a way that *only* scattered light can reach the lens and form the final image. In darkfield optics, objects that scatter a small amount of light stand out brightly against the dark background.

In **phase-contrast** microscopes, the scattered and the transmitted light waves are separated from each other and are passed through different thicknesses of glass on their way through the lenses. This might not seem to make much of a difference. However, when the waves are recombined as the image is formed, the "phase shift" caused by the different optical paths makes the scattered light interfere with the non-scattered light. The result is a very high-contrast image that allows previously "invisible" cells to be seen with great clarity.

Another optical system polarizes the light passing through the specimen to detect differences in the orientation of molecules in the specimen; it is called **polarizing** microscopy. And a very powerful system used for much recent work in cell biology is known as **Nomarski interference** microscopy. Nomarski (named after its inventor) uses a combination of polarizing and phase-contrast effects to produce a striking image with an almost three-dimensional quality as shown above.

None of these different systems increases the **resolution** of the microscope; resolution is a problem that depends on the wavelength of light. These systems help to improve the **contrast** in an image which determines what the viewer is able to see and to study.

Vacuoles

Although most cellular organelles are much more complicated, a few organelles consist of little more than a single large membranous sac. Such organelles are often known as **vacuoles.** Many types of cells contain vacuoles, which may enclose everything from food particles to waste products. In plant cells, however, vacuoles are much more prominent, and they may occupy as much as 90 percent of a cell's total volume. The fluid within these vacuoles serves as a storage reservoir containing water, salts, and sugars. In many plants, certain cells contain large amounts of water-soluble pigments in their vacuoles; these pigments give leaves and flowers their characteristic brilliant colors.

Lysosomes

The small membrane-bounded organelles called **lysosomes** are filled with enzymes (Figure 24.26). Lysosomes are produced by the Golgi apparatus, and the synthetic organelles of the cell direct a flow of very specialized proteins into them. Lysosomes are filled with *lytic* enzymes—enzymes capable of breaking down macromolecules such as proteins, carbohydrates, and nucleic acids into their smaller building blocks. Lysosomes are used for intracellular digestion and destruction.

When endocytosis brings a food particle into the cell, the vesicle containing the ingested food fuses with a lysosome (Figure 24.26). The action of lysosomal enzymes quickly breaks the food molecules down into smaller compounds that can be used by the cell and are easily removed from the lysosome. Lysosomes seem to have other purposes as well. Lysosomes are occasionally seen in the process of destroying a cell's own organelles, and sometimes the rupture of many lysosomes actually causes the destruction of the cell that contains them, a process known as **autolysis** (*auto* means "self"; *lysis* means "destruction"). Why should the lysosome turn on its own cell? These organelles help the cell dispose of damaged or defective organelles, and they also play a creative role in helping destroy cells in strategic locations where a developing organism needs to shape the pattern of growing tissues.

THE CYTOSKELETON

From red blood cells to amoebae to nerve cells, one of the most striking attributes of individual cells is their ability to maintain a characteristic shape. Earlier in the history of biology, the ability of different cell types to maintain different shapes seemed so mysterious that biochemists ignored the problem of cell shape altogether

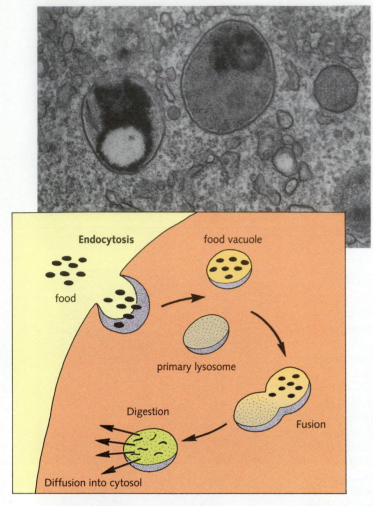

Figure 24.26 (top) Lysosomes are organelles which contain degradative enzymes enclosed within a limiting membrane. They enable cells to degrade and digest endocytosed material. (bottom) Newly endocytosed material is brought into a pathway which involves fusion with lysosomes and the recycling of membrane material to the cell surface. Small molecules and nutrient material are recovered after lysosomal digestion is complete.

to specific molecules. When large numbers of these molecules bind to the surface receptors, the cell membrane turns inward, forming an endocytic vesicle that can then be delivered to a destination within the cell. Liver cells contain receptors for cholesterol-containing lipoproteins, and these enable the liver to remove excess cholesterol from the blood by endocytosis. When very large particles are brought into a cell, the process is known as **phagocytosis** (Figure 24.25).

in order to concentrate on cellular chemistry. That is no longer the case. A host of discoveries about the structure of the cytoplasm have led to the development of an important subtopic within cell biology: the study of the cytoskeleton.

The term *cytoskeleton* really means "cellular support," and the word itself reflects one of the basic discoveries of the past decade and a half: The eukaryotic cell has a skeleton-like substructure. When we think about the skeleton of an animal, we imagine a structure composed of many parts that helps to support the organism and that also aids in movement because muscles are attached to it. The cytoskeleton is similar in both respects.

Microtubules

The first cytoskeletal structure to be appreciated (because it is the largest and most prominent) was the **microtubule.** Microtubules are hollow tube-like structures with a diameter of about 250 Å, and they can easily be spotted in electron micrographs (Figure 24.27a). Microtubules are composed of two proteins, **α tubulin** and **β tubulin,** which form the helical wall of the microtubule itself (Figure 24.27b). Microtubules can *self-assemble* from soluble tubulin under the right conditions, and changes in cellular shape are sometimes associated with the sudden assembly or disassembly of microtubules.

Figure 24.27   (a) Microtubules appear in electron micrographs as hollow tubes roughly 250 Å in diameter. (b) Microtubules are composed of a helical array of dimeric subunits. Each subunit contains one molecule of alpha tubulin and one of beta tubulin. (c) Microtubules are often involved in the maintenance of cellular shape. *Echinosphaerium*, a protist, is surrounded by a stunning series of bristling spikes. These spikes are projections of the cytoplasm, each supported by a double spiral of microtubules. (d) Microtubules help to maintain the shape of many cell types, as fluorescently labeled antibodies to tubulin illustrate in this tissue culture cell.

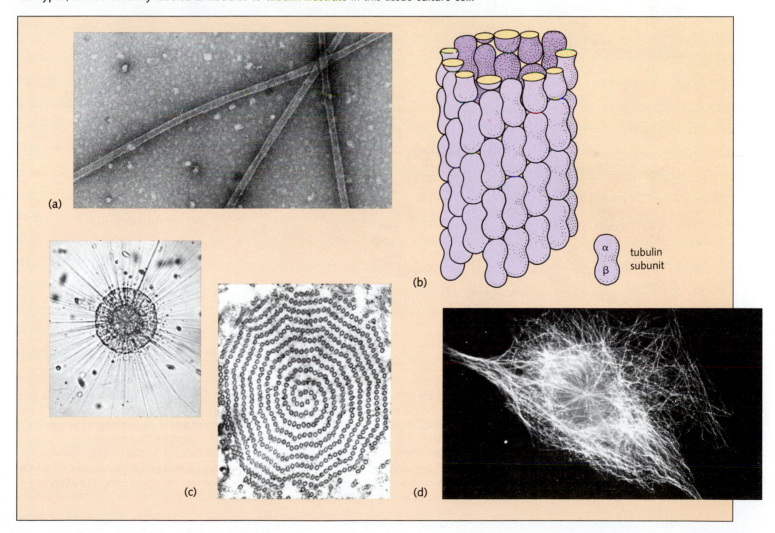

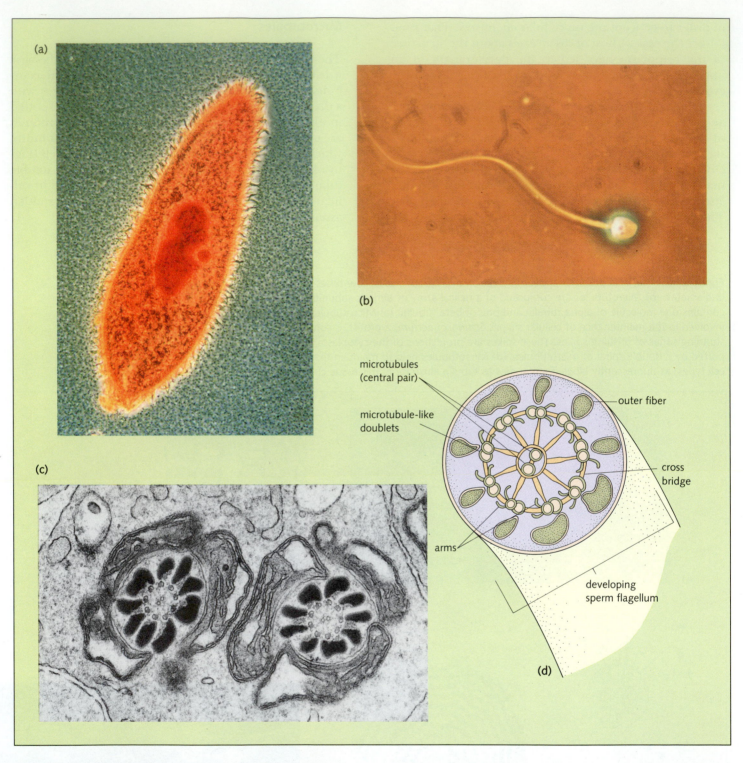

(a)

(b)

(c)

microtubules
(central pair)

microtubule-like
doublets

outer fiber

cross
bridge

arms

developing
sperm flagellum

(d)

Figure 24.28 Cilia (**a**) and flagella (**b**) are projections of the cell surface which are specialized for movement. The cilia of the paramecium beat in a regular pattern which propels the cell through water. The whip-like motions of the sperm's single flagellum enable it to swim quickly. (**c**) These electron micrographs of sperm cells show some of the key structures in both cilia and flagella. Each organelle contains an axoneme, a barrel-shaped arrangement of microtubules with a central pair and nine doublet microtubules arranged in a cyl-inder. Interactions between tubules in the axoneme generate force in cilia and flagella. Sperm flagella often contain additional "outer fibers" as shown in this micrograph of developing rat sperm. (**d**) Force in cilia and flagella is produced by means of connections between neighboring microtubules. The cross bridges between adjacent doublets are made of a protein known as dynein, which uses ATP to provide the chemical energy to bend the axoneme and produce movement.

Microtubules, which are often found in small groups or bundles within the cytoplasm (Figure 24.27c shows *Echinosphaerium* as an extreme example), are sometimes used to provide support for the cell surface (Figure 24.27d). They play a critical role in mitosis and are often associated with cellular movements. Tubulin molecules very similar to those in microtubules are also used to build motile structures in the cell such as cilia.

Cilia and **flagella** are specialized structures that protrude from the cell surface and are used to produce motion (Figure 24.28a,b). (The distinction between cilia and flagella is somewhat arbitrary. The term *cilium* (meaning "hair") is used for structures less than 20 μm in length, and *flagellum* ("whip") for structures 20–100 μm in length.) The internal structure of cilia and flagella in eukaryotes consists of nine microtubule-like doublets that surround a central pair of microtubules (a "9 + 2" arrangement). The individual tubules are held together by a series of cross bridges made up of a protein called *dynein* (Figure 24.28c,d). These cross bridges are capable of generating force, causing the whole structure to whip to one side or the other. Both cilia and flagella are attached to small "rootlets" within the cytoplasm and terminate in structures known as **basal bodies** (Figure 24.29).

Centrioles are structures that are remarkably similar to basal bodies in appearance. Cross sections show that they are composed of nine groups of three tubule-like structures (a "9 + 0" arrangement). A typical animal cell contains two centrioles, which remain closely associated throughout most of the cell cycle (Figure 24.30). Then, just before mitosis, the centrioles are replicated and two "daughter" centrioles are formed. The pairs of centrioles then move to opposite sides of the nucleus as mitosis begins; here they serve as the poles of the mitotic spindle. Surprisingly, plant cells produce perfectly functional mitotic spindles despite the fact that they do not contain centrioles.

Microfilaments and Intermediate Filaments

Two other types of filaments are also components of the cytoskeleton. **Microfilaments** are fibers about 60 Å in diameter that are made up of a protein known as *actin*. Like the microtubule, the microfilament is capable of spontaneous assembly and disassembly. Microfilaments are found in nearly all cells and, like microtubules, help to stabilize cell shape. Actin is also one of the major proteins found in muscle cells, and microfilaments are associated with a number of different forms of cell movement. The streaming cytoplasm of the Amoeba, for example, is produced by the actions of actin filaments and other proteins in the cytoplasm of the cell (Figure 24.31).

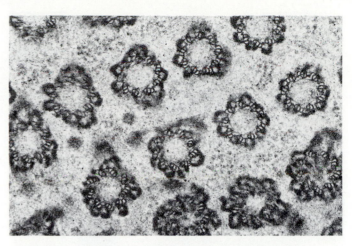

Figure 24.29 These basal bodies, from the lining of the oviduct, are the cytoplasmic anchors to which cilia are attached. Nine groups of three microtubule-like structures are arranged to form this barrel-like structure. Flagella have similar basal bodies.

Figure 24.30 These paired centrioles are surrounded by microtubules.

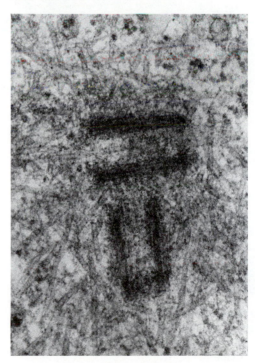

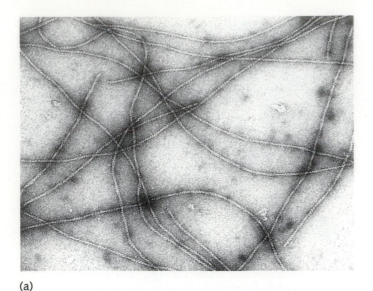

(a)

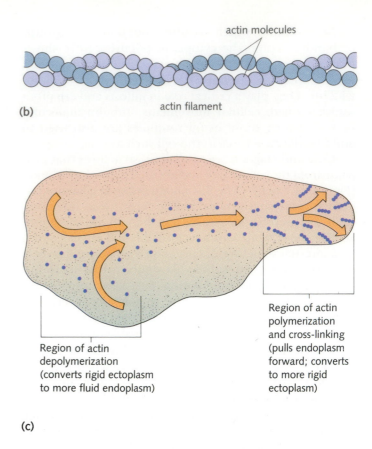

actin molecules

(b)

actin filament

Region of actin
depolymerization
(converts rigid ectoplasm
to more fluid endoplasm)

Region of actin
polymerization
and cross-linking
(pulls endoplasm
forward; converts
to more rigid
ectoplasm)

(c)

Figure 24.31 (a) Microfilaments, which are roughly 60 Å in diameter, are cytoskeletal components made up of actin. Actin filaments are associated with cell movement and with changes in cell shape. (b) Microfilaments are composed of helical arrays of actin proteins. (c) The polymerization and de-polymerization of actin filaments are associated with changes in cytoplasmic structure, including the movement of cells such as the amoeba.

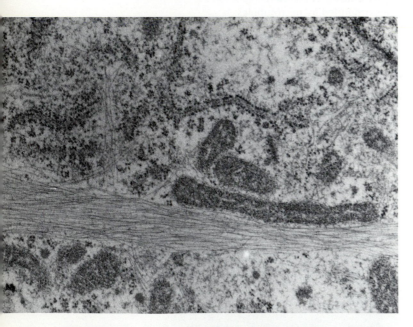

Figure 24.32 The bundle of intermediate filaments in the center of the micrograph is another class of skeletal proteins. They are approximately 100 Å in diameter, and are involved with the maintenance of cell shape.

The third and final class of cytoskeletal proteins is made up of the **intermediate filaments.** These filaments are intermediate in size between microtubules and microfilaments: about 100 Å in diameter. Unlike the other cytoskeletal components, which are made up of actin and tubulin proteins, intermediate filaments are composed of a range of related proteins that vary from one cell type to another. The proteins do seem to have similar properties, however, and scientists are now beginning to make progress in understanding how the intermediate filaments affect cell shape and structure (Figure 24.32).

It is possible to visualize the different cytoskeletal proteins within a single cell by introducing fluorescently labeled compounds that bind to each of them. When this is done, the impression that emerges is that the cytoskeleton is a complex, interconnected structure. We do not yet understand all the relationships between the different protein types the cytoskeleton comprises, and we know only a little about how the cell is able to regulate changes in shape by acting on the cytoskeleton. The answers to these questions will be very important for biology, because such mechanisms help determine the way an organism grows and develops.

ENERGY-PRODUCING ORGANELLES

Mitochondria and Chloroplasts

Besides the organelles we have considered in this chapter, two other organelles are absolutely essential to any discussion of the organization of living cells: **mitochondria** and **chloroplasts** (Figure 24.33). Mitochondria and chloroplasts are energy-transducing organelles: They are involved in the conversion of energy from one form to another. Much earlier in the text we saw that energy could be released from a chemical reaction. We have also seen some of the forms, including the compound ATP (adenosine triphosphate), in which biological energy is used by the cell.

Mitochondria provide most of the energy-rich ATP molecules needed to drive chemical reactions in eukaryotic cells. Cellular respiration, a process in which food molecules such as glucose are broken down to yield water and carbon dioxide, occurs in mitochondria and produces a steady flow of ATP molecules that are available to the rest of the cell.

Chloroplasts are found in plant cells, and they perform *photosynthesis,* a series of reactions in which the energy of sunlight is converted into chemical energy. In the next two chapters we will take a close look at the *process* by which mitochondria and chloroplasts change energy from one form to another and we will see how the membrane molecules in mitochondria and chloroplasts are integral to the function of these organelles.

Unlike other cellular organelles, mitochondria and chloroplasts contain their own DNA molecules. These DNA molecules contain a number of genes that are important for the function of these organelles (Figure 24.33). In addition, both organelles contain ribosomes different from the cytoplasmic ribosomes found in the rest of the cell. A number of important proteins in each organelle, therefore, are coded for by organelle DNA and synthesized on organelle ribosomes.

New mitochondria and chloroplasts seem to arise only by the division of preexisting mitochondria and chloroplasts. Therefore, these organelles are also self-replicating. Does this mean that mitochondria and chloroplasts can be regarded as independent cells within the larger eukaryotic cell? Probably not. Although both mitochondria and chloroplasts synthesize a number of critical proteins on their own ribosomes, most of the proteins found in each organelle are actually synthesized in the cytoplasm and then imported into the organelle. Instead, many biologists believe that the persistence of DNA in mitochondria and chloroplasts may reflect the evolutionary history of each organelle and even of the eukaryotic cell itself—a theory we will explore further in Chapter 26.

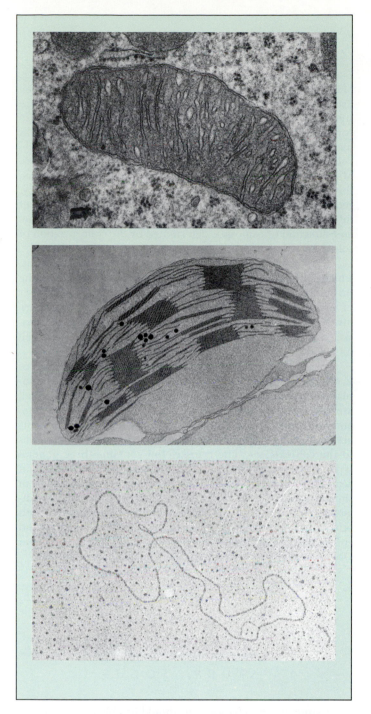

Figure 24.33 (top) Mitochondria are the key energy-releasing organelles of eukaryotic cells. A deeply folded inner membrane contains the electron transport components involved in ATP synthesis. (center) Chloroplasts are organelles that capture light energy and convert it to chemical energy. They are surrounded by two envelope membranes and contain large stacks of internal membranes in which the light reactions of photosynthesis take place. (bottom) Chloroplasts and mitochondria contain their own DNA molecules (this electron micrograph shows mitochondrial DNA) and have the ability to synthesize a few of their own proteins and RNA molecules.

Eukaryotic cells contain a wide variety of complex organelles. Many of these organelles, as well as the boundary of the cell itself, are formed by biological membranes. Membranes are structures that are based on the lipid bilayer, but which also contain proteins and carbohydrates. Membranes can serve as barriers; however, membranes are permeable to some molecules and are able to regulate the passage of these molecules. Material may pass across a membrane by simple diffusion, through special carrier molecules, or by means of an energy-requiring active transport system. Osmosis is a special form of diffusion in which water moves across membranes. The rate and direction of osmosis is determined by the concentration of solute on either side of a membrane.

The primary organelle in any eukaryotic cell is the nucleus, which contains nearly all of the cell's genetic information in the form of DNA. The cytoplasm of the cell contains several classes of organelles, including the synthetic organelles, which are involved with the synthesis of macromolecules, and the cytoskeleton, which helps to maintain cell shape and powers cell movement.

Proteins are synthesized on ribosomes, which may be free in the cytoplasm or bound to the membrane of the rough endoplasmic reticulum. Proteins destined for release from the cell are generally synthesized on the rough endoplasmic reticulum, and are passed to the Golgi apparatus, where many proteins are modified. Finally, they are released from secretory granules to the cell exterior. The cytoskeleton contains three general types of structures: microtubules, intermediate filaments, and actin microfilaments. Each of these structures is assembled from smaller subunits, and two of them, microtubules and microfilaments, can produce cell movement when they interact with other filaments and cross-bridging proteins. Mitochondria and chloroplasts, which are self-replicating and contain their own DNA, are involved with energy transformations in the cell.

STUDY FOCUS

After studying this chapter, you should be able to:

- Be familiar with the structures and functions of the eukaryotic cells.
- Account for some of the structural complexity of the cytoplasm.
- Describe the most important cellular organelles, and explain the ways in which cells grow, divide, and exchange materials with their environments.
- Cite some of the most recent information relating to cell structure and function, including the targeting of proteins to organelles and the molecular basis of cytoplasmic shape and structure.

SELECTED TERMS

organelle p. 462
lipid bilayer p. 463
fluid mosaic model p. 466
diffusion p. 467
facilitated diffusion p. 468
active transport p. 468
osmosis p. 470
cytoskeleton p. 472
nuclear pore p. 472
nuclear envelope p. 472
nucleolus p. 473

rough endoplasmic
  reticulum p. 474
smooth endoplasmic
  reticulum p. 474
Golgi apparatus p. 476
secretory vesicle p. 476
lysosome p. 478
microtubule p. 479
cilia p. 481
flagella p. 481
microfilaments p. 481

REVIEW

Discussion Questions

1. Describe the major organelles of a plant cell. How are these different from the typical organelles found in an animal cell?

2. The amino acid sequences of a large number of integral membrane proteins have now been analyzed in detail. Where such proteins span the lipid bilayer, they tend to have amino acids with side chains (R groups) that are nonpolar. Explain why this observation makes sense.

3. As you have seen, ribosomes active in protein synthesis are either free or membrane-bound. Liver cells synthesize a protein called albumin that is released from the liver into the bloodstream. Would you expect albumin to be synthesized on free or membrane-bound ribosomes? Why? How about tubulin?

Objective Questions   (Answers in Appendix)

4. Which of the following organelles is involved in protein synthesis?
   (a) lysosomes (c) secretory vesicles
   (b) ribosomes (d) vacuoles

5. Molecules can most easily pass through membranes if they are
   (a) water molecules only.
   (b) nonpolar and of a small size.
   (c) polar and of a small size.
   (d) charged molecules and of a small size.

6. Facilitated diffusion is an example of
   (a) osmosis. (c) active transport.
   (b) passive transport. (d) pinocytosis.

7. When a free ribosome completes the synthesis of a protein, the protein is generally
   (a) passed to secretory vesicles.
   (b) released inside the ER.
   (c) inserted into the ER membrane.
   (d) released directly into the cytoplasm.

CHAPTER 25

Cells and Energy

It's surprising to think how many aspects of daily life depend directly on sources of energy. When a car runs out of gas it comes to a sputtering halt. Without fuel, a warm house is at the mercy of the elements. Without electricity, computers, lights, appliances, and mass media such as radio and television grind to a halt. Nearly every thread in the fabric of modern society depends on energy.

Living things depend on energy, too. Nearly everything that happens within a living cell, from cell division to DNA replication to the synthesis of biological membranes, requires energy. We've already seen that autotrophic organisms solve their energy needs by harnessing sunlight and that heterotrophic organisms must find a source of food to obtain chemical energy. On a global scale, the flow of energy from sunlight through autotrophs to heterotrophs affects the flow of carbon, nitrogen, and oxygen throughout the biosphere.

We can see the effects of global energy flow all around us, from winds and storms to the gentle force with which a seedling breaks through the earth. But the critical events involving energy and living organisms, including the *transformation* of one kind of energy into another kind, are often invisible. How do photosynthetic organisms capture the energy of sunlight and transform it into chemical energy? How do heterotrophic organisms release chemical energy from the food they eat? How do they change that energy into a form that can be used for movement and growth? In this chapter we will study the ways in which energy transformation occurs at the cellular level and examine in detail how cells handle the flow of energy.

METABOLISM

In the preceding chapters, we studied a number of biochemical pathways in which macromolecules were synthesized from simpler chemicals. Proteins are assembled from amino acids, and polysaccharides are built from individual sugars. Chemical pathways in which larger molecules are assembled from smaller molecules are said to be **anabolic.** In the majority of energy-producing pathways, larger molecules are broken down into smaller ones. Such pathways are said to be **catabolic.** In every cell, both types of pathways—anabolic and catabolic—are operating at the same time, and the relationship between them is complex (Figure 25.1).

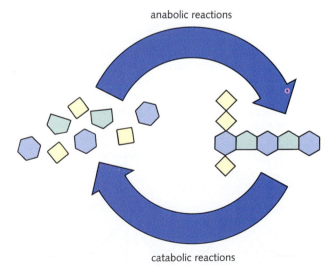

Figure 25.1 Anabolic reactions and catabolic reactions exist within the chemical activity of the cell's metabolism in which the demands for growth and energy production are balanced.

We refer to the sum of all chemical and physical reactions associated with life as **metabolism.** In a way, the metabolism of a cell is nothing more than its internal biochemical economy. The economy of a cell is every bit as varied and complex as the economy of a nation, and studying the metabolism of a cell is just as challenging. The catabolic pathways in which food molecules are broken down and chemical energy is extracted are only one aspect of metabolism, but an important one. They provide the energy that makes the anabolic aspects of metabolism possible.

GLUCOSE: THE ENERGY IN ONE FOOD MOLECULE

Heterotrophs are organisms that require an external source of food, and for nearly all heterotrophs, eating "food" means eating other organisms. What kinds of molecules are present in food? All kinds. Lipids, nucleic acids, carbohydrates, and proteins are found in food sources. Food serves several purposes. Besides containing energy, it provides many of the raw materials that organisms require to synthesize new molecules. In this chapter, however, we will concentrate on how energy is released from food molecules. Nearly all of the thousands of different organic molecules in food can be used as energy sources. Does this mean that we must study each molecule individually to see how energy is obtained from it? Fortunately, no.

Figure 25.2 The basic chemical structure of glucose, one of the key compounds in metabolism, can be represented in several different ways. Only (**a**) is complete, showing every atom of the compound, but each of the other representations (**b, c**) contains enough information for us to be able to recognize the essential features of the sugar.

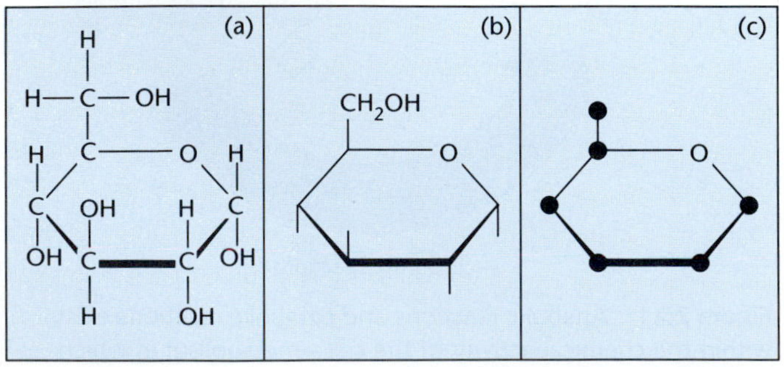

A Logical Molecule to Study

Early studies on cellular metabolism showed that larger molecules are quickly broken down to smaller ones, and at some point the processing of nearly every food molecule involves a series of common chemical pathways. Therefore, we will pay close attention to those common pathways. We will begin our study by examining how **glucose,** a simple sugar molecule introduced in Chapter 3, is broken down to release energy (Figure 25.2). The choice of glucose might seem arbitrary (why not fructose or cholesterol?), but it is not. Glucose is the first molecule used by the most important catabolic pathways, and we can use what happens to glucose to show us how the whole system works. Many foods are first converted to glucose as they enter the pathways.

Oxidation and Reduction

We have already seen that glucose is an important source of chemical energy. One way in which this energy can be released is in the following chemical reaction:

$$C_6H_{12}O_6 + 6O_2 \rightarrow 6CO_2 + 6H_2O + energy$$

The amount of energy released in this reaction is very great: 686 kilocalories per mole of glucose, which means that 180 grams of glucose (one mole) release enough energy in the reaction to raise the temperature of 686 liters of water by 1°C.

Glucose is not the only compound capable of reacting with oxygen. In fact, chemical reactions with oxygen are so common that chemists originally gave such reactions their own terminology, calling them **oxidation reactions.** However, they soon realized that oxidation reactions have one critical feature in common: They involve the *transfer* of electrons from one molecule to another. Therefore we use the term *oxidation* in a more general sense: Any chemical reaction in which electrons are *removed* from a compound is termed **oxidation** whether or not oxygen is involved. When electrons are removed from one compound, they are transferred to another. The *addition* of electrons to a compound is known as **reduction** (Figure 25.3).

When electrons are transferred from one compound to another, both oxidation and reduction reactions take place. The compound losing electrons is oxidized, and the compound receiving them is reduced. In fact, oxidation and reduction reactions occur together so commonly that they are often referred to simply as **redox** reactions ("**red**uction–**ox**idation"). Although the essential feature of a redox reaction is the transfer of one or more electrons, in many cases a proton is transferred

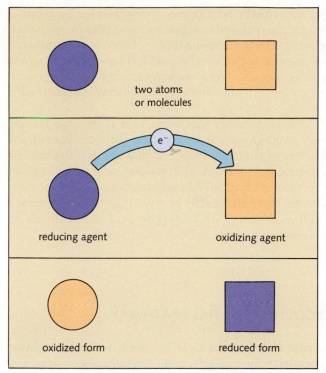

Figure 25.3 Reactions in which electrons are transferred from one compound to another are known as oxidation-reduction reactions. The compound which loses an electron is said to be oxidized, and the compound which receives it is said to be reduced.

along with the electron. An example of this is the oxidation of methane gas:

$$CH_4 + 2O_2 \rightarrow CO_2 + 2H_2O$$

Four electrons are taken from methane and added to oxygen, along with 4 protons (each hydrogen atom has 1 electron and 1 proton). Oxygen is *reduced* to form water, while methane is oxidized to form carbon dioxide. A great deal of energy is released in the reaction—so much that an explosion occurs if enough of the reactants are present.

Twelve Electrons Are Removed During Glucose Oxidation

A glucose molecule contains 12 hydrogen atoms. The electron from each of these 12 hydrogen atoms is transferred to oxygen when glucose is oxidized, and the result is that 6 water molecules ($6H_2O$) are produced. When glucose is oxidized in the chemistry laboratory, both glucose and oxygen must be heated to several hundred degrees, and the reaction occurs in a single step. Glucose burns, releasing a great deal of energy. But living organisms release the energy of glucose at normal temperatures, and they do it in a series of many steps—a little bit of energy at a time. By using so many steps, the cell is able to capture small amounts of energy at some of the steps and use it for work within the cell, rather than losing nearly all of the energy as heat (Figure 25.4).

Energy from Glucose Is Trapped in the Form of ATP

Chemical bonds require energy for their formation, and energy is released when the bonds are broken. The "high-energy" phosphate bonds found in many molecules are particularly well suited for the storage of energy in chemical forms. In Chapter 3 we saw that **ATP** (adenosine triphosphate) can be used to store chemical energy. If we

Figure 25.4 The energy contained in food molecules, like glucose, can be released in a single chemical step. In living cells, the same molecules are broken down in a sequence of far smaller steps, making the release of energy more manageable.

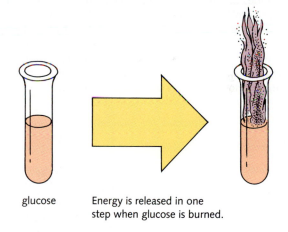

glucose   Energy is released in one
              step when glucose is burned.

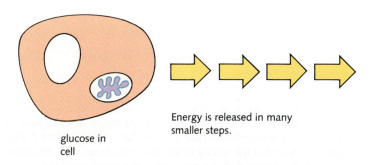

glucose in   Energy is released in many
cell   smaller steps.

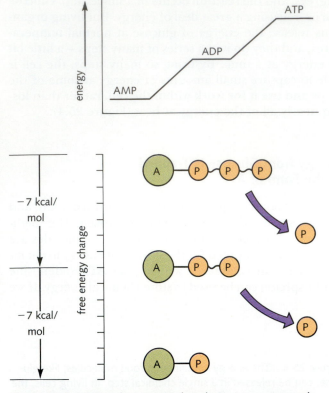

Figure 25.5 ATP, adenosine triphosphate, is a compound widely used by cells as an energy-storing intermediate. Roughly 7 kilocalories of energy per mole is released when a phosphate group is hydrolyzed from ATP to produce ADP (adenosine diphosphate), and another 7 kilocalories when a second phosphate is split off to produce AMP (adenosine monophosphate). The ease with which these phosphates can be removed or re-attached to yield or store energy makes ATP a useful molecule in directing the flow of energy for cellular activities.

begin with **AMP** (adenosine monophosphate), about 7 kcal/mole are required to attach a second phosphate group, producing **ADP** (adenosine diphosphate), and 7 kcal/mole more are needed to produce **ATP**. The reverse reaction can occur, too, allowing a single molecule of ATP to release energy when its phosphate groups are released (Figure 25.5).

The basic strategy that living cells employ to release energy from glucose is very simple. Glucose is oxidized in a series of chemical reactions that provide the energy to synthesize ATP from ADP. By trapping the energy in the form of ATP, a cell can accumulate a supply of ATP molecules that can directly provide the energy necessary for cell movement, ion transport, and the synthesis of macromolecules such as DNA and proteins.

GLYCOLYSIS: THE FIRST PATHWAY

The first chemical pathway involved in the breakdown of glucose is known as **glycolysis.** Literally, glycolysis means "sugar-breaking." The name is well chosen, for the pathway contains an important step in which the sugar molecule is actually broken in two. As we will see, relatively little energy is released in the glycolytic pathway. In the presence of oxygen, glycolysis is only the prelude to other pathways in which a great deal of energy is released (Figure 25.6). However, some organisms are unable to use oxygen, and for them, the glycolytic pathway is the major

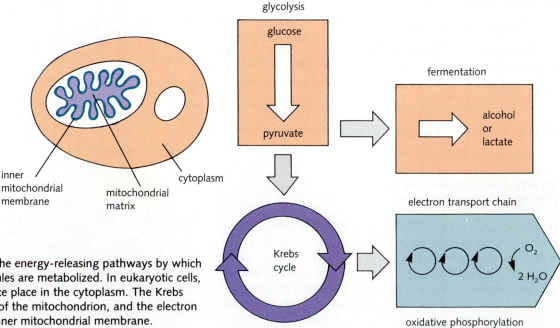

Figure 25.6 An overview of the energy-releasing pathways by which glucose and other food molecules are metabolized. In eukaryotic cells, glycolysis and fermentation take place in the cytoplasm. The Krebs cycle occurs within the matrix of the mitochondrion, and the electron transport chain is part of the inner mitochondrial membrane.

source of chemical energy. In either case, not only is it the first pathway in the process, but it is shared by all organisms, plant or animal, unicellular and multicellular.

Investing Some Energy—
It Takes ATP to Make ATP

The enzyme catalyzing the first reaction of the pathway binds two very different substrates: a molecule of glucose and a molecule of ATP. Once they are in the active site of the enzyme, a phosphate group is transferred from ATP to glucose, producing a new molecule called **glucose-6-phosphate** (the 6 shows that the phosphate group is attached to the number-6 carbon), and releasing ADP (Figure 25.7). At first, this reaction seems to be working backwards. After all, the point of the pathway is to *produce* ATP, not to consume it! However, we can think of the ATP molecule used in this reaction as a kind of investment. The cellular energy represented by the ATP is necessary to "prime the pump" for the flow of energy that will follow.

In step 2 of the pathway, glucose-6-phosphate is rearranged a bit to form **fructose-6-phosphate,** a reaction that requires very little energy. Like every step of glycolysis, this reaction is catalyzed by its own enzyme. In step 3, yet another enzyme adds a second phosphate group to the sugar molecule, using another molecule of ATP as the source of the phosphate. At this point, we have used up two ATP molecules and produced a molecule of **fructose-1,6-diphosphate.** The cell has yet to gain energy, and one might begin to doubt that these chemical reactions have any point to them. In the next few steps of the pathway, however, the investment of ATP begins to pay off.

Figure 25.7 Glycolysis is the energy-yielding pathway by which glucose molecules are first broken down. The pathway consists of distinct steps. In the first four steps, 2 molecules of ATP are required, and therefore there is a net input of cellular energy. In the later reactions, however, a total of four molecules of ATP are synthesized from ADP, providing the cell with a net gain of 2 ATPs per glucose. The electron carriers, NAD^+, which are reduced during glycolysis, can provide more energy if oxygen is present. Pyruvate, the final product of glycolysis, may then be processed further by other chemical pathways.

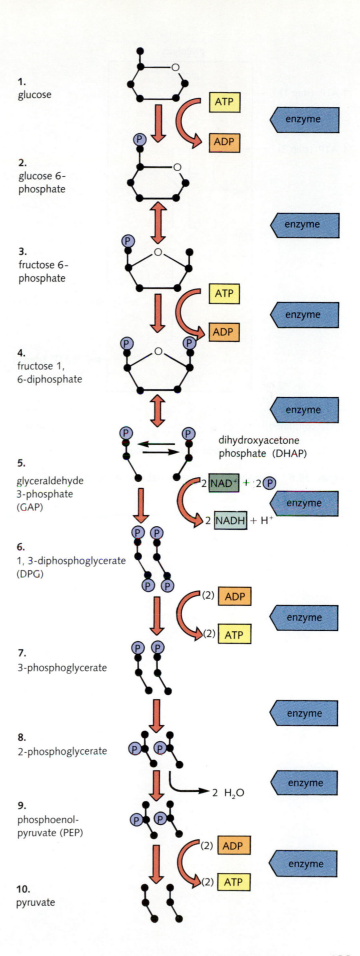

1. glucose
2. glucose 6-phosphate
3. fructose 6-phosphate
4. fructose 1,6-diphosphate

dihydroxyacetone phosphate (DHAP)

5. glyceraldehyde 3-phosphate (GAP)

$2 NAD^+ + 2 P$

$2 NADH + H^+$

6. 1,3-diphosphoglycerate (DPG)

(2) ADP

(2) ATP

7. 3-phosphoglycerate
8. 2-phosphoglycerate

2 H_2O

9. phosphoenol-pyruvate (PEP)

(2) ADP

(2) ATP

10. pyruvate

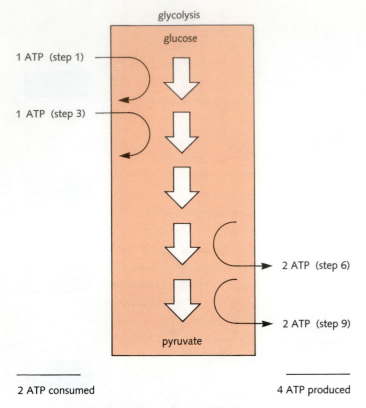

glycolysis

glucose

1 ATP (step 1)

1 ATP (step 3)

2 ATP (step 6)

2 ATP (step 9)

pyruvate

2 ATP consumed

4 ATP produced

Net gain = 2 ATP glucose

Figure 25.8 Glycolysis produces a net energy gain of 2 ATPs per glucose.

Making ATP: Substrate Phosphorylation

In step 4, glycolysis lives up to its name. The 6-carbon sugar molecule is split into two 3-carbon compounds, each of which contains a single phosphate group. In the reactions that follow, two important events take place. The two 3-carbon compounds produced as a result of glycolysis pass through four chemical steps, each step catalyzed by a different enzyme. When the pathway is complete, we are left with two molecules of **pyruvate.** Each pyruvate contains three carbon atoms, so all six carbons of the original glucose are still present. However, two very important events take place in these last few reactions that produce pyruvate.

In two of these reactions, steps 6 and 9, a molecule of ATP is formed, and in each case the phosphate group is transferred directly from one of the substrate molecules in the reaction to form ATP from ADP. Biochemists describe these reactions as **substrate phosphorylation,** because the phosphate groups come directly from the substrates of the reaction. *Phosphorylation* is a general term that simply means the attachment of a high-energy phosphate group to another molecule. Because each of the 3-carbon compounds yields a pair of ATPs in these reactions, we actually obtain 4 molecules of ATP for every molecule of glucose that enters the pathway. We did invest 2 molecules to get things going, so we actually have a *net gain of 2 molecules of ATP per glucose in glycolysis* (Figure 25.8).

In addition to the ATP molecules that are produced, two pairs of electrons are transferred to an electron carrier, a molecule known as **nicotinamide adenine dinucleotide (NAD$^+$).** NAD$^+$ is the oxidized form of the molecule. As Figure 25.9 shows, NAD$^+$ is capable of accepting a pair of high-energy electrons and a proton (H$^+$), converting it to its reduced form, **NADH.** When oxygen is present, NADH can be used to release a great deal more energy.

Glycolysis Releases Only a Small Fraction of Available Energy

Although the cell can make good use of those 2 molecules of ATP per glucose, they don't amount to much when we compare them to the total energy available in glucose. The synthesis of 2 molecules of ATP from ADP accounts for only about 14 kcal of energy per mole, while glucose itself, if completely oxidized to H$_2$O and CO$_2$, would release 686 kcal/mole. The ATP molecules formed directly from glycolysis therefore represent only about 2 percent of the total energy available in glucose itself (Figure 25.10). Where is the rest of that energy?

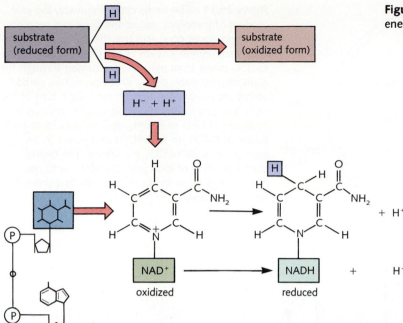

Figure 25.9 NAD$^+$ is reduced by a pair of high-energy electrons to form NADH.

The answer can be found not so much in what glycolysis has done, but in what it has not done. The molecules left over when glycolysis is finished, 2 pyruvates and 2 NADHs, contain a great deal of chemical energy that is still untapped. To release the remaining energy, the cell must find an oxidizing agent to accept the high-energy electrons still present in these compounds. The agent that is used is the most powerful oxidizer of them all: oxygen itself.

THE KREBS CYCLE

Releasing Energy in the Presence of Oxygen

A chemical pathway called the **Krebs cycle** is the next step in breaking down the products of glycolysis. This cycle is known by several other names, including *citric acid cycle* and *tricarboxylic acid cycle,* but *Krebs* is the most common usage, after its discoverer, Hans Krebs. To begin the cycle, pyruvate is oxidized in a reaction in which one of the products, a 2-carbon compound called acetate, is attached to a much larger molecule known as **coenzyme A.** The compound that results is called *acetyl coenzyme A.* One carbon atom from pyruvate is released in the form of carbon dioxide (CO_2), and a pair of electrons is used to reduce NAD$^+$ to NADH.

Figure 25.10 The 2 ATPs made during glycolysis represent only about 2% of the chemical energy available in glucose.

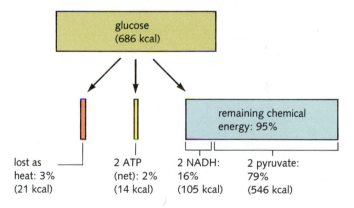

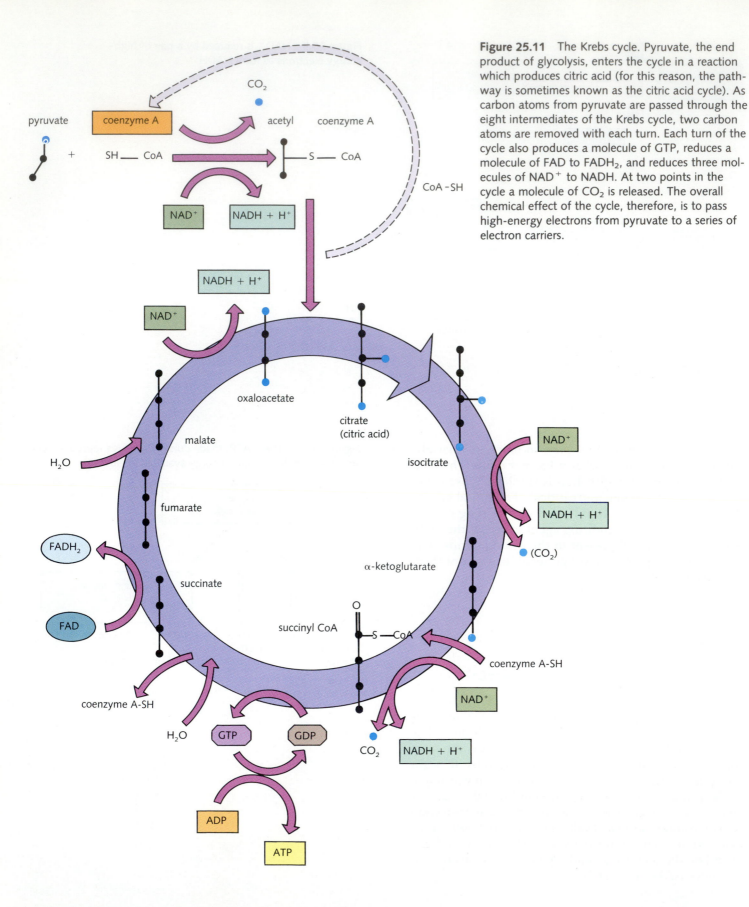

Figure 25.11 The Krebs cycle. Pyruvate, the end product of glycolysis, enters the cycle in a reaction which produces citric acid (for this reason, the pathway is sometimes known as the citric acid cycle). As carbon atoms from pyruvate are passed through the eight intermediates of the Krebs cycle, two carbon atoms are removed with each turn. Each turn of the cycle also produces a molecule of GTP, reduces a molecule of FAD to $FADH_2$, and reduces three molecules of NAD^+ to NADH. At two points in the cycle a molecule of CO_2 is released. The overall chemical effect of the cycle, therefore, is to pass high-energy electrons from pyruvate to a series of electron carriers.

The two carbon atoms from acetyl coenzyme A enter the Krebs cycle by means of a reaction in which they are combined with a 4-carbon compound, *oxaloacetate*. The molecule that results is a 6-carbon compound called **citrate**. Figure 25.11 shows the details of the Krebs cycle. There are 10 intermediate compounds involved in the cycle, and we can look at it in a number of different ways.

A Closer Look at the Krebs Cycle

Carbon At two points in the cycle, a *decarboxylation* reaction occurs in which one carbon atom is removed and released in the form of carbon dioxide. Because of these reactions, by the time the cycle returns to the starting point, only 4 carbons are left, so that with the addition of 2 more carbons from acetyl coenzyme A, citrate (6 carbons) is regenerated. Of the original 6-carbon glucose molecule, 2 carbon atoms are released after glycolysis during the formation of acetyl CoA, and 4 are released during the Krebs cycle. Nearly all of the carbon dioxide that we exhale in the process of breathing comes from CO_2 released in these two pathways.

ATP One of the Krebs cycle reactions is a substrate phosphorylation that produces a molecule of GTP (guanosine triphosphate) by adding a phosphate group to GDP. In terms of energy, GTP is equivalent to ATP, so each turn of the Krebs cycle produces the energy equivalent of a molecule of ATP.

Oxidation–Reduction Four of the Krebs cycle reactions warrant special attention, for they involve the transfer of high-energy electrons to an electron acceptor. In three cases, the acceptor is NAD^+, the same carrier molecule we saw earlier in glycolysis. One of the redox reactions involves FAD (flavine adenine dinucleotide), a carrier that, like NAD^+, can accept a pair of high-energy electrons, reducing it to $FADH_2$. In each turn of the cycle, four pairs of high-energy electrons are removed, producing three NADHs and one $FADH_2$. Because 2 pyruvates are produced from each glucose, a single glucose molecule produces 6 NADHs, and 2 $FADH_2$s in the Krebs cycle (Figure 25.12).

It is important to note that the Krebs cycle is, in a sense, the end of the line for the carbon atoms of the glucose molecule. Sooner or later, every carbon atom of a glucose molecule used by the cycle is released to the environment in the form of carbon dioxide. The cycle is *not* the end of the production of energy, however. As the cycle disassembles carbohydrate molecules, one step at a time, the energy from those molecules is captured in the high-energy electrons used to reduce NAD^+ and FAD. In a final series of pathways, the energy from those electrons is used to produce ATP.

Figure 25.12 A summary of the energy-rich compounds produced during glycolysis and the Krebs cycle.

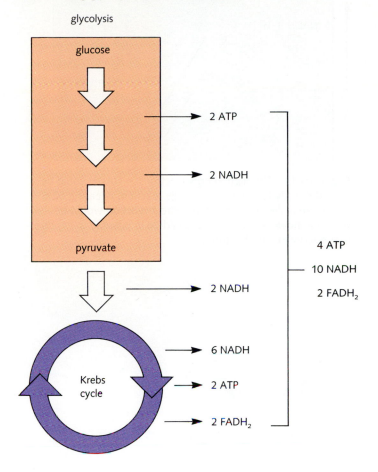

ELECTRON TRANSPORT

Energy from Oxidation and Reduction

So far we have seen a direct gain of only four molecules of ATP per molecule of glucose: 2 from glycolysis and 2 from the Krebs cycle. However, we have seen a great many redox reactions. A careful accounting of the pathways so far shows that the energy in a single molecule of glucose has been used to produce 12 reduced electron carriers: 10 NADHs (2 from glycolysis, 2 from the entry point to the Krebs cycle, and 6 from the cycle itself) and 2 $FADH_2$s. You might also have noticed that although we refer to the Krebs cycle as an **aerobic** (oxygen-requiring) pathway, so far we haven't seen any oxygen involved in any of its reactions. The need for oxygen doesn't become apparent until we begin to consider what happens to these reduced electron carriers. *Oxygen serves as the final acceptor of electrons.*

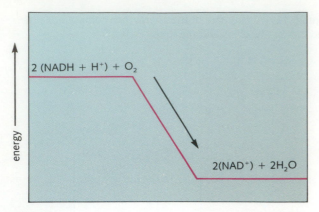

Figure 25.13 The high-energy electrons in NADH could be passed directly to oxygen in a single reaction, releasing 52.6 kcal/mole of free energy. Such a direct reaction would be wasteful, however, because it would be impossible to capture that energy in a usable form.

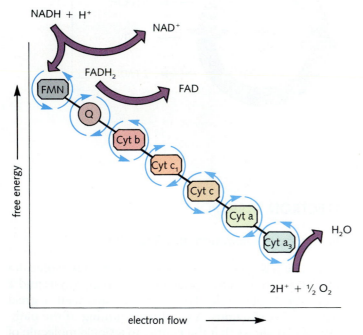

Figure 25.14 The electron transport chain is a series of electron carriers which pass electrons from the Krebs cycle to oxygen, the final electron acceptor. Electrons from the Krebs cycle may enter the pathway from two sources (NADH and $FADH_2$). The energy levels of the electrons drop a bit at each step of the chain. The abbreviations for electron carriers are FMN (Flavine mononucleotide), Q (Coenzyme Q (ubiquinone)), and cyt b, c_1, etc . . . (various cytochromes (electron transport proteins with heme prosthetic groups)).

Although it is possible for oxygen to accept electrons directly from NADH (Figure 25.13), that is not the way it's done in cells. Instead, there are a series of electron carriers involved in the transfer of electrons from NADH to oxygen. In the first reaction, NADH transfers a pair of electrons to a molecule known as *flavoprotein* (FP). The transfer of electrons reduces the flavoprotein, while NADH (having lost electrons) is oxidized back to NAD^+. Electrons flow from the flavoprotein to at least six different carriers. They are passed from one to another until they reach the final carrier of the chain, a protein known as *cytochrome oxidase* (the complex of cytochromes a and a_3). The sequence of electron carriers is known, appropriately enough, as the **electron transport chain** (Figure 25.14).

As you can see, the final acceptor of electrons from the chain is oxygen. The high-energy electrons which enter the chain gradually lose energy as they pass through it, and much of that energy is used to *phosphorylate* ADP to produce ATP. Therefore, the whole process is sometimes known as **oxidative phosphorylation.** Oxidative phosphorylation enables the cell to utilize the chemical energy captured in the form of NADH and $FADH_2$ to produce ATP.

For many years, scientists puzzled over two confusing facts about the chain. First of all, why are there so many steps in the transfer of electrons to oxygen? As Figure 25.14 shows, the energy levels of the carriers in the chain drop a bit lower with each step. Perhaps the steps are necessary to lower electrons just a little at a time so that part of their energy can be captured at each step. Although that explanation seemed to make sense, scientists were unable to find any evidence that the synthesis of ATP or any other high-energy compound occurred during any of the chain's many steps. If the synthesis of ATP was the whole point of the chain, where was it occurring?

Chemiosmosis: Why the Electron Transport Chain Makes Sense

While other scientists were searching for high-energy intermediate compounds formed directly by electron transport, a British scientist named Peter Mitchell believed that important clues could be found by studying *where* electron transport took place. In a eukaryotic cell, glycolysis takes place in the cytoplasm. The pyruvate produced by glycolysis enters the organelles known as mitochondria, and the reactions of the Krebs cycle take place within the matrix of the mitochondria. Other scientists had already shown that the electron transport chain was membrane-bound: All of the carriers of the cycle were actually part of the inner mitochondrial membrane (the

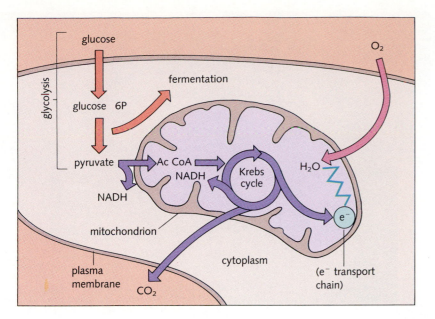

Figure 25.15 The compartmentalization of metabolic pathways. Glycolysis takes place in the cytoplasm. In eukaryotic cells, pyruvate enters the mitochondrion where it enters the Krebs cycle. The electron transport chain, which is located in the inner mitochondrial membrane, accepts electrons from the Krebs cycle, and passes them along to oxygen, reducing it to form water.

cristae) (Figure 25.15). Why should the electron transport chain be membrane-bound? Mitchell and many others believed that the inner mitochondrial membrane must play a pivotal role in the process of ATP synthesis (Figure 25.16).

Mitchell's key experiments were done with isolated inner mitochondrial membranes. The electron transport chain remained attached to vesicles made from the isolated inner membrane, and when oxygen and NADH were added, the chain functioned just as it did within the cell. However, something unexpected happened as electrons began to flow: The pH of the medium containing the vesicles began to rise. You will remember that pH is a measure of the hydrogen ion content of a solution (H^+ means proton). If protons are pumped out of a solution, the pH rises. Protons were being pumped into the vesicles during electron transport. When these isolated membrane vesicles were studied with the electron microscope, scientists noticed that there were large structures

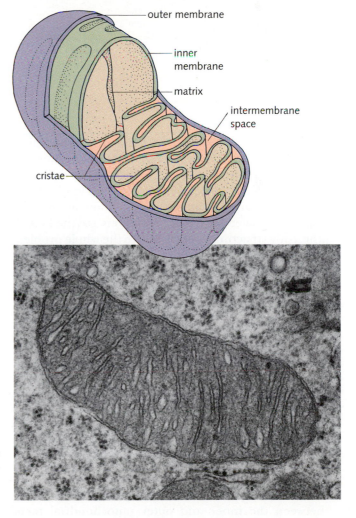

Figure 25.16 The mitochondrion. These energy-transducing organelles are separated from the cytoplasm by an outer membrane. The inner membrane, which contains the electron transport system, is deeply folded in many cells forming structures known as *cristae*, which increase its surface area.

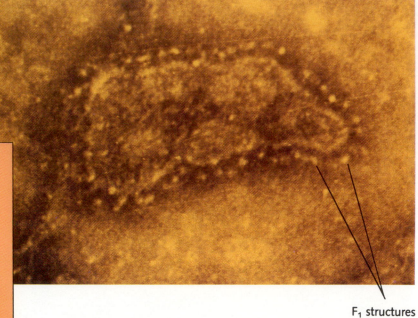

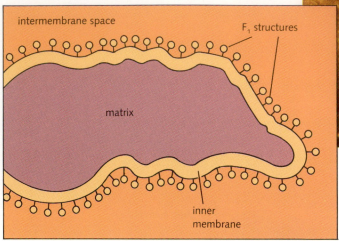

intermembrane space

F₁ structures

matrix

inner membrane

F₁ structures

Figure 25.17 (left) The inner mitochondrial membrane. (right) Electron microscopy of isolated inner membranes shows F_1 structures extending from the membrane surface.

protruding from one surface. A number of experiments showed that these particles were the actual enzymes that synthesized ATP from ADP. In mitochondria these particles were called F_1 *structures* ("factor 1") (Figure 25.17).

Mitchell realized that the combination of electron-transport-associated movement of protons and the F_1 ATP-synthesizing structures was the key to the whole problem. He developed a theory, which is now known as **chemi-osmosis,** to explain the process. The heart of the theory is the notion that the connection between electron transport and ATP synthesis is *indirect*. Here's how the theory works:

1. Carriers in the electron transport chain are built into the inner mitochondrial membrane. When electrons flow through them, part of the energy from the electrons is used to do a special kind of work: to pump protons across the membrane. Because the membrane itself is otherwise impermeable to protons, once they have been pumped from one side to the other, they cannot "leak" back easily (Figure 25.18a).

2. As electron transport continues, more and more protons are pumped across the membrane into the space between the inner and outer mitochondrial mem-

branes (the intermembrane space)—so many that a noticeable change in pH occurs. A gradient of protons has built up across the membrane. The term *gradient* simply refers to a gradual distribution. Because protons are electrically charged, this gradient produces an electrochemical field across the membrane, as shown in Figure 25.18b.

3. The electrochemical gradient can serve as a source of energy—a battery—to power chemical reactions. The F_1 synthetase uses the energy in that gradient to drive the synthesis of ATP from ADP. The energy to attach the third phosphate group comes directly from the electrochemical gradient across the membrane and the flow of protons back across the membrane through the F_1 structures.

The beauty of Mitchell's theory is its simplicity. Electron transport produces a gradient, and the gradient drives the synthesis of ATP. Because there is no direct link between electron transport and ATP synthesis, we often say that the coupling between these two processes is indirect. The complexes within the membrane that are responsible for electron transport do not have to be near those responsible for ATP synthesis.

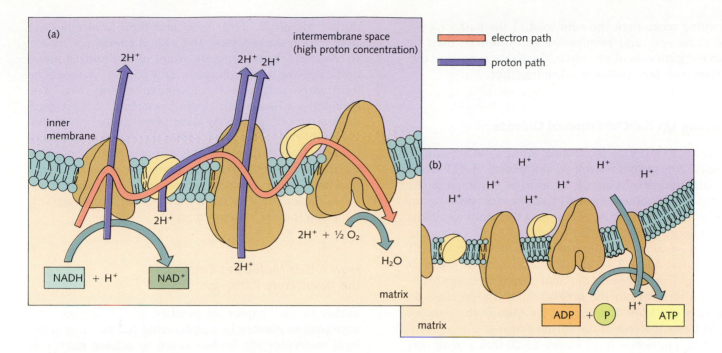

Figure 25.18 **(a)** The electron transport chain. The passage of electrons from one carrier to another results in a movement of protons across the membrane, from the matrix to the intermembrane space. **(b)** ATP synthesis. A proton gradient is produced across the inner mitochondrial membrane by electron transport. The energy in this gradient is sufficient to allow a membrane-bound enzyme, the F_1 ATP-synthetase, to produce ATP from ADP and phosphate.

For each pair of high-energy electrons transferred to the electron transport chain from NADH, enough protons can be moved across the membrane to make a gradient large enough to drive the synthesis of 3 molecules of ATP. The electrons derived from $FADH_2$ are somewhat lower in energy and are able to drive the synthesis of only 2 molecules of ATP (Figure 25.19).

Cellular Respiration

The word *respiration* is often used as a synonym for *breathing*. Organs such as gills and lungs that help to bring oxygen to the tissues are parts of the *respiratory system*. However, we also refer to the oxidation of glucose and of other molecules as "respiration." At the level of the cell, the major need for oxygen is found at the end of the electron transport chain: Oxygen is needed to accept electrons. If the flow of oxygen is stopped, the electron transport chain stops working and the synthesis of ATP cannot continue. (This is why suffocation causes death so quickly. Cells need a constant supply of ATP, and without oxygen, ATP synthesis cannot keep pace with the energy demands of the cell.) Thus, **cellular respiration**

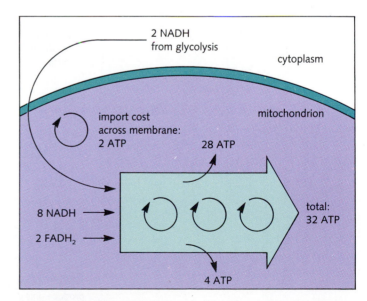

Figure 25.19 The net energy gain from the electron transport chain is 32 ATPs.

is nothing more than the sum total of the pathways we have examined, and **respiration** will often be used to mean the pathways of glycolysis, the Krebs cycle, and the electron transport pathway taken together.

Summing Up the Oxidation of Glucose

Figure 25.20 gives an accounting of the total amount of chemical energy that is trapped in the form of ATP when the oxidation of a molecule of glucose is complete. As you can see, 36 molecules of ATP can be synthesized from each molecule of glucose that enters the pathways. How much energy is captured in the form of ATP? Taking the figure of 7 kcal/mole of the production of ATP from ADP, we find that $36 \times 7 = 252$ kcal are captured for each mole of glucose. Because the total chemical energy available in a mole of glucose is 686 kcal, this represents an efficiency of $252/686 = 38$ percent. Most of the remaining 62 percent is lost in the form of heat.

You might notice from Figure 25.20 that a small correction is applied for the NADH molecules synthesized during glycolysis. This is because these molecules are produced in the *cytoplasm,* and a bit of energy is required to transport them across the outer mitochondrial membrane into the mitochondrion. This points out one of the most important aspects of cellular energy use: The chemical pathways are *compartmentalized.* In a typical eukaryotic cell, glycolysis takes place in the cytoplasm, the Krebs cycle takes place within the matrix of the mitochondrion, and the electron transport chain is bound to the inner mitochondrial membrane. In prokaryotes the situation is a bit different. Prokaryotes don't have mitochondria, and their electron transport chains are actually built into their cell membranes.

How Other Molecules Enter the Respiratory Pathways

Earlier in the chapter we justified all the attention we have paid to glucose by emphasizing the fact that other food molecules are broken down to release energy in much the same way. Figure 25.21 shows how this is done.

Figure 25.20 Roughly 36 molecules of ATP can be produced from a single molecule of glucose if oxygen is available. This represents about 37% of the total free energy in glucose. The remainder is lost as heat.

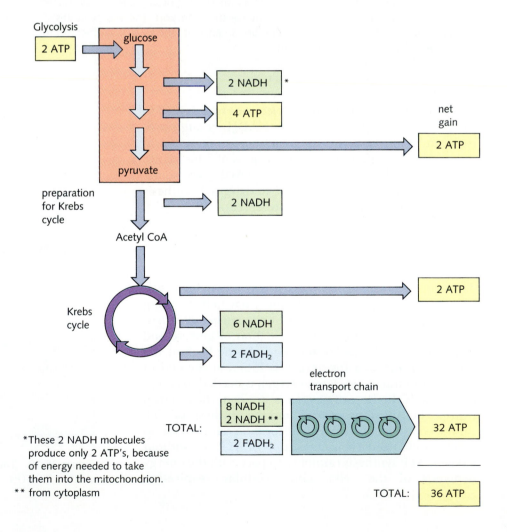

*These 2 NADH molecules produce only 2 ATP's, because of energy needed to take them into the mitochondrion.

** from cytoplasm

Proteins are broken down into individual amino acids, and other enzymes then convert them to compounds that can enter one pathway or another. Carbohydrates are generally broken into simple sugars and converted to glucose. Lipids are broken down to fatty acids and glycerol; these enter the mitochondria where special enzymes cut them up, 2 carbons at a time, to produce acetyl coenzyme A, which enters the Krebs cycle.

FERMENTATION

Oxygen is the living world's best electron acceptor, and that is why oxygen is used in the pathways that follow glycolysis: to accept high-energy electrons from glucose and other food molecules. Does that mean that living organisms are restricted to environments where they can find oxygen? Not at all. You will remember that a small amount of energy (2 ATPs per molecule of glucose) can be trapped just from glycolysis, a pathway that doesn't require oxygen. Therefore, if an organism is able to get

a large supply of food material, or if it has a low metabolic rate, it can use glycolysis to produce enough ATP to get by. This process is known as **fermentation.**

However, an organism that begins to ferment its food has a serious problem to overcome. The glycolytic pathway transfers electrons from glucose to an electron acceptor, NAD^+, which is reduced to form NADH. Without a constant supply of NAD^+, the process would quickly stop. Therefore, in order to keep glycolysis moving, an organism that is fermenting its food must be able to *oxidize* NADH to NAD^+. Living cells have come up with two slightly different ways to do this.

Different Types of Fermentation

Alcohol fermenters A small number of organisms, the best known of which are the **yeasts,** oxidize their NADH in a pathway that reduces *pyruvate,* the end of the glycolytic pathway, and coverts it to **ethanol** (also known as ethyl alcohol, or grain alcohol). One of the carbon atoms

Figure 25.21 Glycolysis and the Krebs cycle are the basic energy-releasing pathways of the cell. Proteins, fats, and carbohydrates may enter the pathways at a variety of different points, enabling cells to draw energy from virtually any food source.

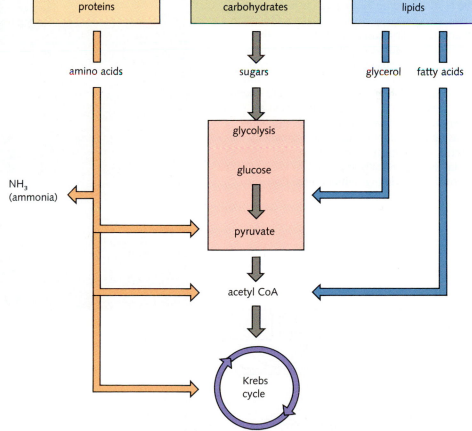

of pyruvate is removed in the process and is released in the form of *carbon dioxide.* **Alcoholic fermentation** is the source of alcoholic beverages, and the bubbles in beer and sparkling wine come from the carbon dioxide released during the fermentation process (Figure 25.22).

The same fermentation process takes place in the manner in which bread is baked. Tiny carbon dioxide bubbles cause the dough to rise, and the alcohol evaporates during baking.

Lactate fermenters Most other organisms are capable of carrying out fermentation in another chemical pathway that reduces pyruvate. The final product of this pathway is **lactate,** or lactic acid, and unlike alcoholic fermentation, lactate fermentation gives off no carbon dioxide. Humans are lactate fermenters, and during brief periods without oxygen, many of the cells in our bodies are capable of generating ATP in this way. The cells best adapted to it, however, are muscle cells, which often need very large supplies of ATP for rapid bursts of activity (Figure 25.23).

The By-products of Fermentation

The use of fermentation as a sole source of energy has a number of drawbacks. For one thing, it is much less efficient: The 2 ATPs formed from each glucose represent only a fraction of the available energy (14/686 = 2 percent). For another, the end products (lactate or ethanol) can produce problems for the organism. Alcohol is toxic to most cells, yeast included, and if fermentation occurs in a closed system then the concentration of alcohol rises until the yeast cells are killed. A similar problem occurs with lactate. During bursts of rapid exercise, lactate accumulates in muscle tissue. The buildup of lactate causes a painful burning sensation that every athlete is familiar with.

Fortunately, there are chemical pathways that can take care of lactate, although first it must be taken to the liver by the circulatory system. These pathways require oxygen, and they are responsible for the out-of-breath sensation, known as *oxygen debt,* that may last for several minutes after only 10 or 20 seconds of intense exertion. In Chapter 42 we will examine this effect more closely.

Figure 25.22 Alcoholic fermentation. (below) The NAD^+ used up in glycolysis is regenerated in a two-step pathway which converts pyruvate to ethanol, releasing carbon dioxide gas in the process. This pathway enables yeasts to obtain energy from glucose in the absence of oxygen. The by-products of this pathway, alcohol and carbon dioxide, help to produce alcoholic beverages. (right) The fermentation process in a winery rapidly converts grape mash into wine, releasing large amounts of carbon dioxide in the process, which must be carefully vented off from the tanks. It is critically important to keep oxygen out of the vats during fermentation. Can you explain why this is the case?

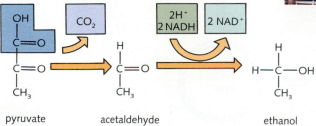

pyruvate   acetaldehyde   ethanol

Energy is central to the activities of living organisms, and the transformation of energy from one form to another is one of the key processes that takes place within the cell. Metabolism, the sum total of the biochemical and physical transformations that take place in living organisms, includes both anabolic reactions (in which molecules are built up from smaller ones) and catabolic reactions (in which larger molecules are broken down).

The oxidation of glucose is a key catabolic pathway that serves as a model for how energy is extracted from organic molecules. Glucose molecules are processed through a series of pathways, beginning with glycolysis, in which the glucose molecule is split into two 3-carbon compounds. Glycolysis results in a small net gain in energy for the cell (2 molecules of ATP per molecule of glucose) and produces pyruvate, a 3-carbon molecule that then, if oxygen is present, enters the Krebs cycle.

In eukaryotic cells, the Krebs cycle takes place within the mitochondrion. The Krebs cycle results in the reduction of a number of electron carrier molecules, which then pass electrons along to a series of electron transport complexes associated with the inner mitochondrial membrane. Electrons move through this transport pathway to oxygen, the final electron acceptor. As they do so, they result in a movement of protons across the inner mitochondrial membrane, producing a chemiosmotic gradient that is used to provide the energy for ATP synthesis, a process known as oxidative phosphorylation.

In the absence of oxygen, cells can capture a small amount of the energy available in glucose by a process known as fermentation, in which electron carriers for glycolysis are regenerated. Yeasts and related organisms carry out alcoholic fermentation, producing carbon dioxide and ethyl alcohol as chemical by-products of the process. Most animals produce lactate, which is released into the bloodstream.

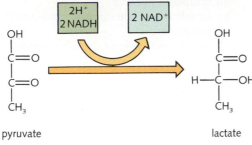

pyruvate   lactate

Figure 25.23 Lactate Fermentation. The NADH used up in glycolysis can be regenerated in a reaction that produces lactate. This process is widely used by animals whenever muscles and other active tissues require more ATP than can be released from oxidative pathways. Olympic runners, beginning the 100 meter dash, will use up all of the oxygen reserves in their large muscles in a matter of 2–3 seconds. For the rest of the sprint, their muscles will produce ATP anaerobically, flooding the bloodstream with lactate. Ten seconds later, this biochemical oxygen "debt" will cause them to pant and breathe heavily for a minute or more as that lactate is removed from the bloodstream.

STUDY FOCUS

After studying this chapter, you should be able to:

- Explain the fundamental importance of energy-producing pathways in biological systems.
- Analyze the steps of the energy-producing pathways needed for all cells to maintain cellular activities.
- Trace the details of respiration at a level that emphasizes its relationship with other chemical pathways in the cell.
- Describe the unique aspects of mitochondria, and explain the chemiosmotic hypothesis.
- Outline some current theories on the evolution of the eukaryotic cell.
- Describe the fermentation process as an energy-producing pathway and its drawbacks.

SELECTED TERMS

anabolic *p. 485*
catabolic *p. 485*
metabolism *p. 486*
glucose *p. 486*
oxidation *p. 486*
reduction *p. 486*
ATP *p. 487*
glycolysis *p. 488*
substrate phosphorylation *p. 490*
Krebs cycle *p. 491*
coenzyme A *p. 491*
aerobic *p. 493*
electron transport chain *p. 494*
chemiosmosis *p. 496*
cellular respiration *p. 497*
fermentation *p. 499*

REVIEW

Discussion Questions

1. Describe the differences between anabolic and catabolic pathways. Is the synthesis of RNA by RNA polymerase an example of a catabolic or an anabolic pathway?

2. Oxidation and reduction reactions always occur together. When hydrogen gas burns in the presence of oxygen, which molecules are reduced? Which are oxidized? Is the product of the reaction an example of oxidation or reduction?

3. Describe the basic elements of Mitchell's chemiosmotic theory. If a preparation of inner mitochondrial membranes actively synthesizing ATP were treated with a substance that made the membranes permeable ("leaky") to protons, would this increase or decrease the rate of ATP formation?

4. You have read that the amount of energy released from aerobic respiration (36 ATPs/glucose) is much greater than the amount released from anaerobic respiration, or fermentation (2 ATPs/glucose). Use this fact to explain the following

observation, which Louis Pasteur made on the making of wine. When yeast cells are added to a sealed vat of grape juice, they use up the sugar in the juice at a very slow rate—until the dissolved oxygen in the juice is gone. When the last trace of oxygen disappears, however, the rate of sugar consumption by the yeast increases dramatically. How would you explain this phenomenon, which is known as the Pasteur effect?

Objective Questions   (Answers in Appendix)

5. A substance that loses electrons is
   (a) an enzyme.
   (b) a catalyst.
   (c) oxidized.
   (d) reduced.

6. As a result of glycolysis, which of the following chemical substances is produced?
   (a) 38 molecules of ATP
   (b) 38 molecules of ADP
   (c) one six-carbon molecule
   (d) two molecules of pyruvate

7. Glycolysis takes place in the _____ of the cell.
   (a) nucleus
   (b) cytoplasm
   (c) plasma membrane
   (d) mitochondria

8. In the Krebs cycle, the most frequent electron acceptor is
   (a) NAD^+.
   (b) coenzyme A.
   (c) ADP.
   (d) pyruvate.

9. In the presence of oxygen, the pyruvates formed at the end of glycolysis will be
   (a) oxidized in the Krebs cycle.
   (b) converted to lactic acid.
   (c) converted to glucose.
   (d) converted to ethyl alcohol.

10. In eukaryotic cells, glycolysis takes place in the _____ and the Krebs cycle takes place in the _____.
    (a) matrix of the mitochondria/cytoplasm
    (b) cytoplasm/nucleus
    (c) membrane of the mitochondria/matrix of the mitochondria
    (d) cytoplasm/matrix of the mitochondria

11. The production of ethanol from glucose occurs as a result of
    (a) aerobic respiration.
    (b) fermentation.
    (c) phosphorylation.
    (d) electron transport.

CHAPTER 26

Photosynthesis

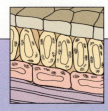

When humans first thought of space travel, their thoughts were outward, toward the moon, the planets, and the stars. As a species, we have long dreamed of leaving our home planet to explore. Given that dream, it is ironic that so many space travelers have said that the most remarkable sight they have seen from space is Earth itself. Its gentle blur of ocean, cloud, and land stands in marked contrast to the harshness of the moon and our planetary neighbors. It seems almost obvious that this great blue planet harbors life.

What makes Earth so hospitable to life? We can find one of the answers in a biochemical pathway. The green plants and microorganisms that dominate the earth's surface have mastered the process of **photosynthesis,** the capture of solar energy and its transformation into energy-rich compounds that can be used by living organisms. Photosynthesis not only provides nearly all of the energy used by living organisms on the earth, but it has also completely changed the atmosphere of the planet by releasing enormous amounts of oxygen.

In the last chapter we saw how cells are able to release the energy stored in glucose and other energy-rich compounds. In this chapter we will see how that chemical energy was trapped in the first place. The ultimate source of energy for life is the sun, and photosynthetic organisms are the crucial links on which other forms of life depend. Without them this planet would be a very different place.

ENERGY FROM SUNLIGHT

In Chapter 25 we saw how chemical energy could be released from glucose in a process known as respiration:

$$C_6H_{12}O_6 + 6O_2 \rightarrow 6CO_2 + 6H_2O + \text{energy}$$

That chemical energy, of course, is trapped initially in the form of ATP, produced at the inner mitochondrial membranes of eukaryotic cells. In many respects, photosynthesis is the opposite of respiration. Energy in the form of sunlight is trapped in a different set of membranes and used to produce glucose from carbon dioxide and water:

$$6CO_2 + 12H_2O \xrightarrow{\text{sunlight}} C_6H_{12}O_6 + 6O_2 + 6H_2O$$

We can visualize the global relationship between respiration and photosynthesis as a biological shuttle system that powers the activities of all living things by passing the energy of sunlight through a series of chemical intermediates as shown below.

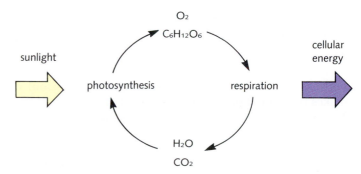

Figure 26.1 The ultimate source of nearly all energy used by living organisms is the sun. Photosynthesis and respiration exist in a global balance.

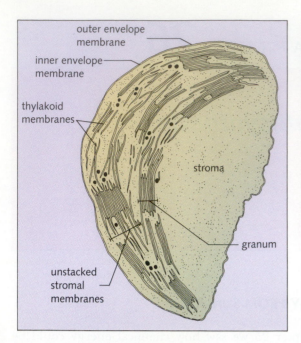
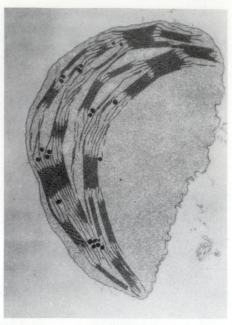

Figure 26.2 The chloroplast. These energy-capturing organelles are surrounded by two enveloping membranes. The chloroplast contains photosynthetic membranes, or thylakoids, which are stacked in several places to form *grana*. The "stroma" region of the chloroplast contains soluble enzymes, ribosomes, and several dark lipid droplets.

In eukaryotic cells, the process of photosynthesis takes place in organelles called **chloroplasts.** The fact that photosynthesis is localized in a membrane-bounded organelle illustrates an important trend in the organization of eukaryotic cells: the tendency to compartmentalize key biochemical pathways. A chloroplast is surrounded by a pair of enveloping membranes and contains stacks of internal membranes known as **thylakoids.** The thylakoids are sac-like membranes that are stacked in some parts of the chloroplasts to form **grana** (Figure 26.2). These membranes contain a number of important proteins and electron carriers, but they also contain the one kind of molecule that is absolutely vital to the photosynthetic process, **chlorophyll.**

Chlorophyll and Sunlight

Light is just one form of **electromagnetic radiation;** it makes up a small part of a larger spectrum that includes X rays, microwaves, and gamma rays. Our eyes are capable of detecting visible light, at wavelengths between 400 nm (violet) and 730 nm (deep red). Ordinary sunlight is actually a mixture of different wavelengths, as passing light through a glass prism demonstrates (Figure 26.3).

Electromagnetic radiation has a dual nature, exhibiting the characteristics of both waves *and* particles. The energy carried in electromagnetic radiation behaves as though it were bundled into particles known as **quanta** (singular; *quantum*). Quanta of light are known as **photons.** When photons strike an object they are either **reflected** or **absorbed.** A surface that absorbs all wavelengths of visible light appears black, whereas one that reflects all wavelengths appears white. A colored object absorbs some wavelengths while reflecting others (Figure 26.3). When light is absorbed by an object, the energy carried in the photons is converted to other forms. In most materials, the energy of absorbed sunlight is converted to heat. Photosynthetic organisms, however, possess a series of molecules that use the absorption of photons to produce chemical energy.

Chlorophyll is a **pigment,** a molecule that absorbs light very efficiently. Like most pigments, it can absorb only part of the visible spectrum of light. Chlorophyll absorbs red light and blue light very well, but it absorbs only slightly in the middle region of the spectrum, where the predominant color is green (Figure 26.4a). Chlorophyll-containing plants appear green to us because the green light is reflected, not absorbed.

Chlorophyll comes in a number of slightly different forms. The most important is *chlorophyll a,* but there are also chlorophylls *b, c,* and *d,* as well as several forms of the pigment that are found only in prokaryotic cells. Chlorophyll molecules consist of a long hydrocarbon tail attached to a complex cyclic structure called a **porphyrin**

ring (Figure 26.4b). The ring has a magnesium atom in its center, and the atoms that make up the ring are linked by an alternating series of single and double bonds. These alternating bonds make the chlorophyll molecule a tremendous absorber of light. Just a few milligrams of chlorophyll in a liter of water will absorb nearly all of the red and blue light passing through it.

When light is absorbed by a chlorophyll molecule, the energy of the light is trapped in the electrons that make up the porphyrin ring. The electrons are raised from their usual "ground" energy state to an excited state (Figure 26.4c). If a chlorophyll molecule is in solution when this happens, the electron usually returns to the ground state by giving off a quantum of light in the form of fluorescence (Figure 26.4d). But if an acceptor molecule is nearby, the electron can be passed from chlorophyll to the acceptor: a classic *redox* reaction. In one sense, the ability of chlorophyll to produce high-energy electrons (by absorbing light) and then to lose them to another molecule in a redox reaction is the secret of photosynthesis. You will remember that the key to the pathways of respiration was the high-energy electrons provided by food molecules such as glucose. In photosynthesis, the high-energy electrons are provided by the interaction of sunlight with chlorophyll.

white light

all wavelengths reflected; none absorbed

red wavelength reflected; others absorbed

green wavelength reflected; others absorbed

Figure 26.3   (left) Refraction of sunlight by a prism. "White" light is actually a mixture of light of many different wavelengths. (right) Reflection. The color of any object, including a leaf, is determined by which wavelengths of light are absorbed and which are reflected.

THE LIGHT-DEPENDENT REACTIONS OF PHOTOSYNTHESIS

Early in the twentieth century, biologists realized that photosynthesis could be separated into two related processes. Today we call these the **light-dependent reactions** and the **light-independent reactions.** The light-dependent reactions require the direct involvement of light energy, and the light-independent reactions do not.

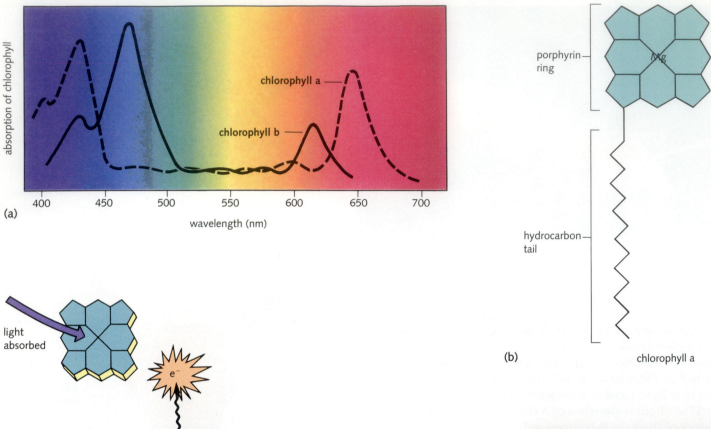

(a)

absorption of chlorophyll

wavelength (nm)

chlorophyll a

chlorophyll b

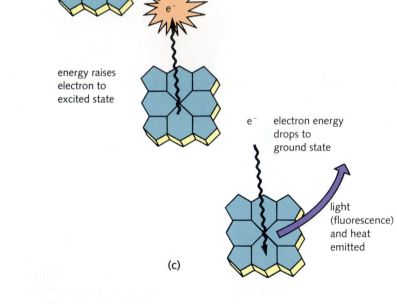

light absorbed

energy raises electron to excited state

e⁻

e⁻ electron energy drops to ground state

light (fluorescence) and heat emitted

(c)

porphyrin ring

hydrocarbon tail

(b) chlorophyll a

(d)

Figure 26.4 (a) Absorption spectra of chlorophylls a and b. These pigments absorb very strongly in the red and blue regions of the spectrum. The poor absorption of these pigments in the middle of the visible spectrum gives chlorophyll its green color. (b) The chlorophyll molecule. The porphyrin ring structure encloses a magnesium atom, and attached to the ring is a long hydrocarbon chain known as the phytol tail. In chlorophyll b the methyl group marked in red is replaced by an aldehyde group (−CHO). (c) Chlorophyll is extremely efficient at absorbing light energy. When a photon strikes chlorophyll, one of the electrons of the pigment is raised to a higher energy level. (d) When this excited electron returns to the ground state, its energy is lost as heat or fluorescence, except in the chloroplast, where this energy is not lost, but is captured in photosynthesis.

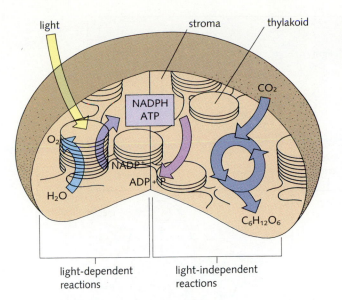

light   stroma   thylakoid

NADPH
ATP

CO₂

O_2

NADP⁺

ADP + P

H_2O

$C_6H_{12}O_6$

light-dependent reactions   light-independent reactions

Figure 26.5 Both the light-dependent and light-independent reactions of photosynthesis take place within the chloroplast. As this schematic illustrates, the light-dependent reactions, which take place in the thylakoid membranes, use the energy of sunlight to produce ATP and to reduce $NADP^+$ to NADPH. Oxygen is released during the light reaction as electrons are removed from water. The light-independent reactions, which take place in the stroma, use the chemical energy available in ATP and NADPH to fix carbon dioxide to produce carbohydrates.

Another way in which we can distinguish between the two sets of reactions is by noting that the light-dependent reactions are directly associated with the *photosynthetic membranes,* or *thylakoids,* of the chloroplast. The light-independent reactions, in contrast, take place in the **stroma,** the semi-fluid matrix within the chloroplast (Figure 26.5).

Figure 26.5 illustrates the interrelationship between the light-dependent and light-independent reactions. The light-dependent reactions trap the energy of sunlight and transform some of that energy into chemical form. The light-dependent reactions produce energy-rich ATP and reduce $NADP^+$ to NADPH. These two compounds can then be used to supply the energy and the reducing power expended by the light-independent reactions to capture carbon dioxide and chemically reduce it to produce sugar in the form of glucose. Starting at the point where light interacts with the chloroplast, we will first consider the light-dependent reactions.

The Photosynthetic Reaction Center

The heart of the light-dependent reactions is the photosynthetic reaction center. As recently as 1985 we knew very little about the organization of these complexes, but a German scientist named Hartmut Michel was able to crystallize a photosynthetic reaction center and determine its structure by using X-ray diffraction (Figure 26.6). The reaction center contains a special pair of chlorophyll molecules surrounded by other chemical groups that are able to accept an excited electron. The reaction center is surrounded by a number of light-harvesting proteins that contain other chlorophyll molecules, as well as "accessory pigments" that absorb light in those regions of the spec-

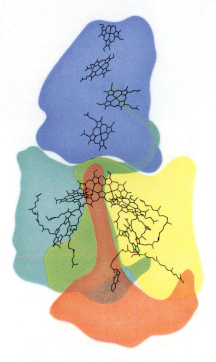

Figure 26.6 The three-dimensional structure of a reaction center from a photosynthetic bacterium. Four polypeptides surround a special pair of bacteriochlorophyll molecules. The absorption of excitation energy by one of these bacteriochlorophylls begins a series of electron transfers which power the light reactions of photosynthesis. The detailed structure of reaction centers from green plants has yet to be determined.

trum where chlorophyll doesn't do such a good job. In higher plants one of the most common accessory pigments is **β carotene,** which gives leaves their reddish-orange color in the autumn after chlorophyll molecules begin to break down, eliminating the green color in the leaves (Figure 26.7).

Figure 26.7 The beautiful colors of autumn leaves come from carotene and other accessory pigments.

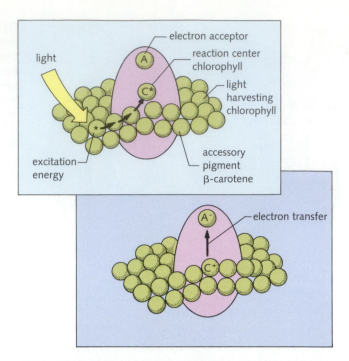

Figure 26.8 Photosynthesis begins with the absorption of light energy by pigment molecules surrounding the reaction center. This excitation energy is passed to the reaction center, which results in the transfer of an excited electron from the central chlorophyll to an electron acceptor.

The reaction center and the light-harvesting proteins associated with it are found within the photosynthetic membranes, or thylakoids, of the chloroplast. When a photon is absorbed by a pigment molecule, the excitation from that photon is passed from molecule to molecule until it arrives at the reaction center chlorophylls, as shown in Figure 26.8. Within the reaction center, the pair of chlorophyll molecules now lose their excited electrons to an electron acceptor. This is the first redox reaction of **photosynthetic electron transport,** and it begins in the reaction center itself.

Photosynthetic Electron Transport: Oxygen Evolution

In plants and many types of algae, there are actually two different photosynthetic reaction centers that work in series. Discovered because they absorb light at slightly different wavelengths, they are called **photosystem I** (700 nm maximum absorbance) and **photosystem II** (680 nm maximum absorbance). Figure 26.9 illustrates how photosystems I and II are organized within the thylakoid membrane of a typical green plant. Historically, photo-

system I was discovered first, but electron transport actually begins at photosystem II. Electrons move from photosystem II to photosystem I.

When photosystem II loses an electron to an electron acceptor, it becomes oxidized. Unless the lost electrons are promptly replaced, all of the reaction center chlorophyll molecules in photosystem II will be permanently oxidized. Fortunately, the photosynthetic membrane has a means of replacing the electrons that chlorophyll has lost. The membrane contains a special set of enzymes known as the **oxygen-evolving apparatus,** which can remove electrons from water. When this happens, the water molecule is actually split (a process called **photolysis**):

$$H_2O \rightarrow 2H^+ + \frac{1}{2}O_2 + 2e^-$$

As you can see, photolysis provides a source of electrons (e^-) to reduce photosystem II, but it also produces oxygen (O_2) and frees a pair of protons (H^+) into the thylakoid sac. The process of photosynthetic oxygen evolution is the major source of oxygen on the planet, and it all derives from a need to evolve a method of reducing

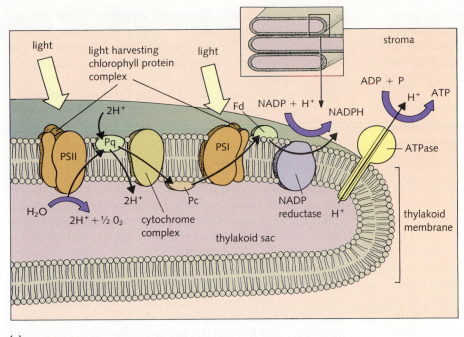

(a)

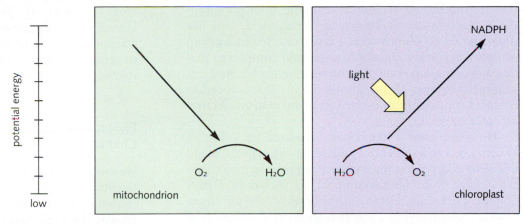

(b)

Figure 26.9 **(a)** The photosynthetic membranes of chloroplasts contain two types of photosynthetic reaction centers as well as an electron transport chain. High-energy electrons produced in the photosystem I and photosystem II reaction centers pass through a series of electron carriers, including plastoquinone (Pq), plastocyanin (Pc), and ferredoxin (Fd), ultimately reducing $NADP^+$, a soluble electron carrier, to produce NADPH. The ultimate source of electrons for this chain is water, and a "water-splitting" apparatus associated with photosystem II releases oxygen when electrons are removed from water. The buildup of protons within the thylakoid membrane, from water-splitting and electron transport, pro-

duces a proton gradient which drives the synthesis of ATP by means of a membrane-bound ATP synthetase known as the coupling factor. **(b)** Electron flow in the photosynthetic membrane is remarkably similar to electron flow in the inner mitochondrial membrane. In most respects, the two chains are opposites: Electrons flow downhill in the mitochondrial system, while light provides the energy for uphill electron flow in the photosynthetic membrane; while each system generates ATP, the direction of electron transfer is *to* water in the mitochondrial system and *from* water in the photosynthetic system.

photosystem II. It is worth noting that there is no absolute requirement that water be used as the source of electrons, and some photosynthetic organisms make use of different compounds. Some photosynthetic bacteria, for example, use hydrogen sulfide as a source of electrons:

$$H_2S \rightarrow 2H^+ + S + 2e^-$$

These bacteria must live near abundant sources of hydrogen sulfide (sulfur springs and volcanic regions are typical habitats), and they produce rich deposits of pure sulfur as they grow.

Photosynthetic Electron Transport: Noncyclic Flow

The energized electron produced by the photosystem II reaction center passes through a series of carrier molecules that are very similar to the electron transport components found on the inner mitochondrial membrane. These carriers include lipids such as *plastoquinone* and proteins such as *cytochrome f* and *plastocyanin*. Eventually the electron transport chain reaches the other photosynthetic reaction center known as *photosystem I*. Just like photosystem II, photosystem I is surrounded by a group of light-harvesting chlorophyll–protein complexes that channel excitation energy to the reaction center. And just as in photosystem II, the photosystem I reaction center is capable of passing an excited electron on to an electron acceptor molecule.

However, once it has become oxidized by losing an electron, photosystem I does not react with water. Instead it is reduced by electrons passed along the electron transport chain from photosystem II, as shown in Figure 26.9(a). Therefore, electron flow between the photosystems is coordinated in a way that allows electrons to flow directly from photosystem II to photosystem I.

The high-energy electrons produced by photosystem I are passed through a new series of electron transport molecules until they are used to reduce a soluble electron carrier called *nicotinamide adenine dinucleotide phosphate* (**NADPH**). NADPH is almost identical to NADH, and it accepts electrons in exactly the same way. It is interesting to note that this pattern of electron flow is the reverse of electron flow in the inner mitochondrial membrane:

chloroplast	$H_2O \rightarrow NADPH$
mitochondrion	$NADH \rightarrow H_2O$

This pattern of electron flow is also known as **noncyclic electron flow**, because the electrons move through a "one-way" pathway from water through chlorophyll to NADPH. In those plants that have been examined, noncyclic electron flow is the major pathway by which photosynthetic reaction centers trap the energy of sunlight.

As Figure 26.9(b) illustrates, electrons in the chloroplast are flowing "uphill" in terms of energy. It takes energy to reduce $NADP^+$ to NADPH. Where does that energy come from? From sunlight, of course, and that is the principal function of the two reaction centers: to trap the energy of sunlight and boost the electrons to an energy level high enough to reduce $NADP^+$ to NADPH.

Cyclic Electron Flow

In some cases, a different pattern of electron flow occurs within the thylakoid membrane, as shown in Figure 26.10. This pattern is known as **cyclic electron transport**, because the pattern resembles a cycle in which excited electrons move out of a photosynthetic reaction center (a cluster of molecules), pass through a series of carrier molecules, and then are returned to reduce the same reaction center from which they originated. Cyclic electron flow may have been the first form of photosynthetic electron transport to evolve, and it is still the predominant form of electron flow in many photosynthetic bacteria. As you can see from Figure 26.10, there are two important differences between cyclic electron transport and noncyclic electron transport. The cyclic pattern of flow does not split water to release oxygen, and it does not reduce the electron carrier $NADP^+$.

Photophosphorylation: Chemiosmosis in the Chloroplast

We saw in Chapter 25 how the pumping of protons across the inner mitochondrial membrane produced an electrochemical gradient that could drive the synthesis of ATP. Exactly the same thing happens in the thylakoid membrane. Both cyclic and noncyclic electron transport result in the movement of protons across the thylakoid membrane. Remember that photolysis also provides a pair of protons every time a single water molecule is split to remove electrons. By increasing the concentration of protons within the enclosed thylakoid sac, photosynthetic electron transport produces a strong electrochemical gradient across the membrane.

This proton gradient is capable of doing work just as it is in the mitochondrion, and an enzyme built into the membrane takes advantage of that capability. This enzyme produces ATP as protons flow out through the membrane back into the stroma. Because this enzyme *couples* electron transport to the synthesis of ATP (**photophosphorylation** in the jargon of photosynthesis), it is generally called the **coupling factor**. The enzyme is almost identical to the F_1 ATP-synthesizing enzyme found in the inner mitochondrial membrane. Photosynthetic electron

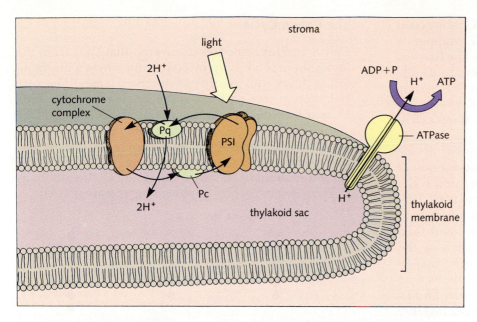

Figure 26.10 Cyclic electron flow. In addition to the non-cyclic pattern of photosynthetic electron transport, electrons may flow in a cyclic pattern. While cyclic electron flow does not reduce NADP$^+$ or release oxygen, it does produce a proton gradient which provides for ATP synthesis.

transport produces ATP and a reduced electron carrier, NADP$^+$. These are then used to provide chemical energy and reducing power to drive the light-independent reactions that produce glucose from CO_2 and H_2O.

Summary of the Light-Dependent Reactions

The light-dependent reactions of photosynthesis, as we have seen, take place within the thylakoid membranes. Photons are absorbed by light-harvesting pigments, and the energy they pass to photosynthetic reaction centers elevates electrons to higher energy levels. These high-energy electrons pass through a series of carrier molecules, eventually resulting in the reduction of the soluble electron carrier, NADPH. Water-splitting enzymes remove electrons from water to replace those that are used up in this pathway. As a result of water splitting and electron flow, protons move into the thylakoid sac, producing a strong electrochemical gradient. The energy from this gradient drives the synthesis of ATP. The products of the light-dependent reactions are NADPH and ATP, both of which will be used in the light-independent reactions.

THE LIGHT-INDEPENDENT REACTIONS OF PHOTOSYNTHESIS

The ATP and NADPH produced in the light-dependent reactions of photosynthesis provide the chloroplast with a light-driven source of chemical energy. These molecules make possible the next steps in the photosynthetic pathway, the light-independent reactions. We can remind ourselves of where we are going by looking at the overall equation of photosynthesis:

$$6CO_2 + 12H_2O \rightarrow C_6H_{12}O_6 + 6O_2 + 6H_2O$$

As you can see, the energy available in ATP and NADPH will be used to produce carbohydrate.

The light-independent reactions take place in the **stroma,** the soluble portion of the chloroplast outside of the thylakoid membranes. These reactions require ATP and NADPH from the light-dependent reactions, CO_2 from the atmosphere, and a complex series of enzymes and cofactors found in the chloroplast stroma.

Biochemists began to explore the pathways by which photosynthetic organisms produce carbohydrate by following the path that the carbon atoms of carbon dioxide took after they entered the cell. As radioisotopes became available at about the time of World War II, radioactive ^{14}C was used in an effort to find out which compounds were formed directly from radioactive CO_2. The first scientists who tried this approach, however, were badly disappointed. Even with a brief pulse of labeled CO_2, nearly every kind of carbohydrate found in the cell was labeled with ^{14}C. The "fixation" of carbon dioxide, as it was called, must be a very fast reaction.

The speed of photosynthesis demanded drastic experiments. In 1946 Melvin Calvin, a young scientist at the University of California at Berkeley, began a series of experiments in which a new approach was tried. Calvin and his co-workers exposed a flat flask of algae to radioactive CO_2 and a few seconds of light, and then opened a valve at the bottom of the flask, allowing the algae to drop into hot alcohol. The cells were instantly killed, and the chemical products of photosynthesis were preserved so that they could be analyzed (Figure 26.11). After many years these scientists were able to put together a complete chemical scheme for how the process of CO_2 fixation takes place.

Figure 26.11 The apparatus used by Calvin and his associates to determine the pathways of carbon fixation in photosynthesis.

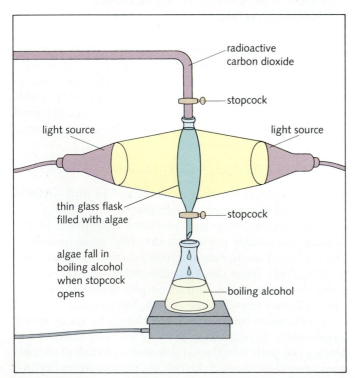

The Calvin Cycle

Calvin and his associates discovered that CO_2 from the atmosphere is taken up by plants and enters a biochemical pathway with nearly two dozen enzymes. The very first step of that pathway is the key reaction, however, because it establishes the first chemical link between CO_2 and an organic molecule, "fixing" the carbon atom. CO_2 is linked to a five-carbon sugar known as **ribulose bisphosphate (RuBP),** briefly forming an unstable six-carbon molecule. This compound quickly breaks apart to form two molecules of **phosphoglycerate (PGA):**

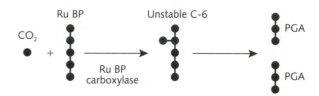

PGA was the first organic molecule to be labeled with radioactive carbon in Calvin's experiments. *Ribulose bisphosphate carboxylase (RuBP carboxylase),* the enzyme that catalyzes this reaction, works very slowly. Nonetheless, the light-independent reactions cannot proceed without the steady supply of new carbon atoms that this reaction provides. Plants respond by producing enormous quantities of the enzyme. In some chloroplasts, as much as 50 percent of the total protein is RuBP carboxylase, making this enzyme the single most abundant protein in the world—eloquent testimony to the importance of carbon fixation for life on Earth.

The two PGA molecules are part of a cycle that produces the carbohydrate used by the plant and regenerates RuBP so that carbon fixation can continue. This pathway is known as the **Calvin cycle,** after its discoverer, or as the **C3 cycle,** in token of the fact that the first stable compounds formed from CO_2 are *three-carbon* compounds.

The phosphoglycerate molecules formed in the first steps of the cycle are used to produce fructose-6-phosphate, a six-carbon sugar. Fructose is formed in a four-step pathway that uses the energy of ATP and NADPH:

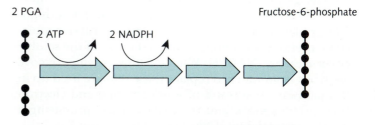

However, if all PGA molecules were converted into fructose, the plant cell would quickly use up the pool of RuBP molecules needed to fix CO_2. Most PGA mole-

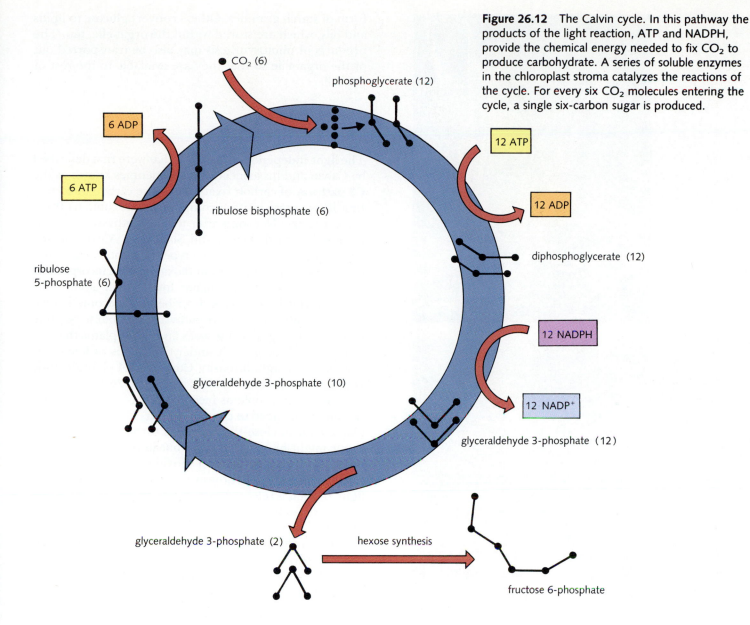

Figure 26.12 The Calvin cycle. In this pathway the products of the light reaction, ATP and NADPH, provide the chemical energy needed to fix CO_2 to produce carbohydrate. A series of soluble enzymes in the chloroplast stroma catalyzes the reactions of the cycle. For every six CO_2 molecules entering the cycle, a single six-carbon sugar is produced.

CO_2 (6)

phosphoglycerate (12)

6 ADP

6 ATP

12 ATP

ribulose bisphosphate (6)

12 ADP

ribulose 5-phosphate (6)

diphosphoglycerate (12)

12 NADPH

glyceraldehyde 3-phosphate (10)

12 NADP$^+$

glyceraldehyde 3-phosphate (12)

glyceraldehyde 3-phosphate (2)

hexose synthesis

fructose 6-phosphate

cules, therefore, enter a pathway that produces more RuBP. This pathway requires the energy of ATP and involves a series of steps with five-carbon, six-carbon, and seven-carbon intermediates. It eventually results in the formation of enough RuBP to keep carbon fixation "in business."

Figure 26.12 is a summary of the light-independent steps of photosynthesis. In order to visualize the synthesis of a six-carbon fructose, we must follow the fates of six CO_2 molecules as they are fixed by RuBP carboxylase.

As shown in Figure 26.12, 6 carbon atoms enter the Calvin cycle from 6 molecules of CO_2. A total of 18 ATPs and 12 NADPHs are required to regenerate RuBP and

produce a single six-carbon fructose molecule. Can we compare the energy required to synthesize a six-carbon sugar with the energy obtained by breaking it down? Here's a quick calculation:

■ The energy of NADPH is roughly equivalent to that of NADH, and as we saw in Chapter 25, 3 ATPs can be synthesized from each NADH in oxidative phosphorylation.

■ Therefore 12 NADPHs are more than equivalent to 36 ATPs.

■ $18 + 36 = 54$ ATPs, compared with the 36 ATPs produced by aerobic respiration.

Figure 26.13 Sugar cane (top) and corn (bottom) are important commercial C4 plants. The special adaptations of C4 plants make them more efficient than C3 plants where light intensity is high and water is limited.

Therefore, the energy required to synthesize a sugar molecule is at least 50 percent greater than the energy obtained by breaking it down! And even this simple calculation understates the energy of synthesis, because 1 NADPH contains more available energy than 3 ATPs.

The carbohydrate that is produced in photosynthesis can be used in a number of ways. Some plants store the products of photosynthesis in the chloroplast itself in the form of starch granules. Others convert glucose to lipids and oils, which are stored within the organelle, too. The products of photosynthesis may also be transported out of the organelle so that they are available to the rest of the cell.

C4 Photosynthesis

The light-independent reactions that were first described by Calvin and his associates are sometimes known as the **C3 pathway** of carbon fixation because carbon dioxide is first incorporated into a compound that contains three carbon atoms. Not long after the C3 pathway was completely described, Hatch and Slack, two Australian scientists, discovered that several types of plants use a different pathway in which carbon dioxide is first incorporated into *oxaloacetate*, which contains four carbon atoms. The pathway Hatch and Slack described has become known as the **C4 pathway.** The C4 pathway turns out to be quite common among desert grasses and other plants that are well adapted for growth under conditions of low moisture and high light intensity. Common C4 plants include corn, sugar cane, and crabgrass (Figure 26.13).

One of the problems faced by plants growing in dry climates is the need to conserve water. The leaves of such plants tend to close the openings through which water vapor might be lost to the air, but in so doing they also make it difficult for carbon dioxide to diffuse into the leaf. Because photosynthesis is rapid in the bright sunlight, oxygen is released from the splitting of water, while the small amount of carbon dioxide within the leaf is quickly reduced by carbon fixation. This leads to a very high O_2-to-CO_2 ratio, and that leads to a problem called **photorespiration.**

RuBP carboxylase is the enzyme that normally catalyzes the first reaction of the C3 cycle:

$$CO_2 + RuBP \rightarrow 2PGA$$

However, when the O_2-to-CO_2 ratio is extremely high, oxygen occupies the active site of the enzyme, and the following reaction takes place:

$$O_2 + RuBP \rightarrow PGA + phosphoglycolate$$

Eventually, phosphoglycolate is broken down elsewhere in the plant cell and CO_2 is released. Because it releases CO_2, photorespiration limits the efficiency of photosynthesis and substantially reduces the productivity of crops grown at high light intensity. Although photorespiration is a common problem among C3 plants, it is almost nonexistent among C4 plants.

Figure 26.14 shows why this is the case. C4 plants carry out the first stages of carbon fixation in leaf tissues known as the mesophyll, passing reduced carbon in the

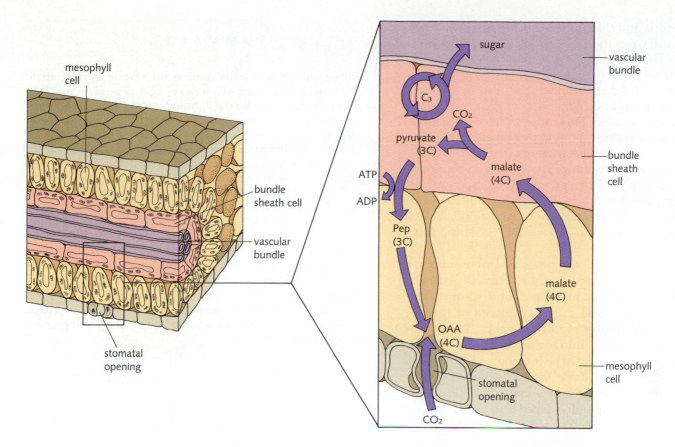

Figure 26.14   C4 plants are so-named because they fix CO_2 to produce oxaloacetate (OAA), a 4-carbon compound. This reaction is the first in a cycle in which malic acid transfers newly fixed carbon to the bundle sheath tissue of the leaf, where the Calvin cycle takes place. In a sense, the mesophyll tissue pumps carbon into the bundle sheath by means of these reactions. Although the C4 cycle requires more chemical energy than the Calvin cycle alone, it can take place at much lower CO_2 concentrations, increasing efficiency at high light intensities.

form of malate (a C4 compound) or oxaloacetate into adjacent tissue known as the *bundle sheath*. Within the bundle sheath, CO_2 is released and used in the ordinary C3 cycle. The C4 pathway within the mesophyll cells uses extra energy in the form of ATP, but it keeps the concentration of CO_2 in the bundle sheath high enough to ensure that photorespiration does not occur. And this, in turn, increases the efficiency of C4 plants in bright, arid environments. Many plant breeders and geneticists have hoped to be able to breed some aspects of C4 photosynthesis into useful C3 plants, such as soybeans, in the hope that higher crop yields will result.

MITOCHONDRIA AND CHLOROPLASTS: REMARKABLY SIMILAR ORGANELLES

Despite their obvious differences, there are striking similarities between mitochondria and chloroplasts. Each organelle is surrounded by a pair of membranes, each one contains a cyclic chemical pathway (the Krebs cycle and the Calvin cycle), each contains a membrane-bounded electron transport system, and each synthesizes ATP in response to an electrochemical gradient established by the process of electron transport. Interestingly, there are other similarities as well.

Mitochondria and chloroplasts contain their own *genetic systems*. In most cases the DNA molecules that they contain are circular, and they are present in multiple copies. These organelles depend on the action of their own DNA, and some very careful experiments have shown that they cannot survive without it. What sorts of genes are found in these DNA molecules? This varies from one organism to another. In most cases, however, organelle DNA contains genes for some of the proteins that make up the electron transport pathways of the inner mitochondrial membrane and the photosynthetic membrane. It may also contain genes for chloroplast rRNA and ribosomal proteins, because chloroplasts and mitochondria also contain their own protein-synthesizing systems, complete with ribosomes.

Does this mean that chloroplasts and ribosomes are genetically independent of the rest of the cell? Actually, no. The majority of the proteins found within these organelles are synthesized in the cytoplasm of the cell and are under the control of nuclear genes. These proteins are then "imported" into the organelles by a mechanism that is just beginning to be investigated. The DNA molecules found within these organelles may be remnants of complete genomes from a time when the organelles were actually independent, free-living cells.

There is no fossil record to tell us exactly when the first mitochondrion or chloroplast appeared. However, scientists have tried to understand how these organelles might have evolved by looking at organisms that are alive today. One suggestion, which has been promoted by Lynn Margulis (see Chapter 11), is that these organelles may have evolved from independent, free-living cells that took up residence in the cytoplasm of early eukaryotes. If this suggestion were correct, it would be easy to understand why each organelle maintains its own DNA and has its own machinery for protein synthesis.

More recently, however, other evidence has emerged to suggest that the story may not be so simple. Molecular biologists have found that some mitochondrial genes contain intervening, nontranslated nucleotide sequences. Such sequences have been found only in eukaryotic cells, so it is not clear why they should be in mitochondria if these organelles are descended from eukaryotic symbionts. Recent work has also shown that some genes from mitochondria are almost identical to genes found in chloroplasts, leading to speculation that DNA may have moved from one organelle to another. Findings like this make it more difficult to construct an evolutionary history for mitochondria and chloroplasts, but only time and more research will tell for sure. In the meantime, it is safe to say that these remarkable organelles may have a few more surprises in store for us.

SUMMARY

Photosynthesis is in some respects the reverse of respiration. Photosynthesis consists of a series of light-dependent reactions that provide energy for the light-independent reactions, which involve the fixation of carbon dioxide to form sugar. The light-dependent reactions begin when electrons in chlorophyll, the primary pigment molecule of green plants, are raised to higher energy levels by the absorption of light energy. These excited electrons are then passed along an electron transport pathway found in the thylakoid membranes of the chloroplast. Electron transport in these membranes produces a chemiosmotic gradient that drives ATP synthesis and also reduces a soluble electron carrier. The light-independent reactions use ATP and the reduced electron carriers to capture atmospheric carbon dioxide and reduce it to form sugar, which then becomes the primary chemical energy source of the plant. Chloroplasts and mitochondria, which carry out these processes, have many features in common. Both possess their own genetic material in the form of DNA molecules, and both have their own protein-synthesizing machinery.

STUDY FOCUS

After studying this chapter, you should be able to:

- Explain how energy-rich food molecules are made by the process of photosynthesis, which uses the sun as its energy source.
- Describe how photosynthesis takes place within the structure of the chloroplast.
- Distinguish between the light-dependent and light-independent reactions, between cyclic and noncyclic electron flow, and between C3 and C4 pathways.
- Cite similarities between respiration and photosynthesis and between chloroplasts and mitochondria.

SELECTED TERMS

photosynthesis   *p. 503*
chloroplast   *p. 504*
chlorophyll   *p. 504*
pigment   *p. 504*
light-dependent reactions   *p. 505*
light-independent reactions   *p. 505*
stroma   *p. 507*
β carotene   *p. 507*

photosynthetic electron transport   *p. 508*
photosystem I   *p. 508*
photosystem II   *p. 508*
oxygen-evolving apparatus   *p. 508*
photolysis   *p. 508*
noncyclic electron flow   *p. 510*
cyclic electron transport   *p. 510*
coupling factor   *p. 510*
Calvin cycle   *p. 512*
C3 pathway   *p. 514*
C4 pathway   *p. 514*
photorespiration   *p. 514*

R E V I E W

Discussion Questions

1. It is sometimes pointed out that the direction of proton movement in the photosynthetic membrane, or thylakoid, is reversed with respect to the inner mitochondrial membrane. Draw a diagram to illustrate what is meant by this.

2. Compare and contrast cyclic and noncyclic electron flow. Which one results in the reduction of a soluble electron carrier?

3. Many C4 plants are grasses found in arid, desert-like regions. Among the most important C4 crop plants is corn. Corn has a remarkable ability to grow rapidly during long, hot summer days with little rainfall. Which aspects of C4 metabolism contribute to this ability?

Objective Questions   (Answers in Appendix)

4. In a hot, dry environment, a plant undergoing C_4 photosynthesis will do better than a plant that can undergo C_3 photosynthesis because
   (a) C_4 plants can take in oxygen more efficiently than C_3 plants.
   (b) C_4 plants can store carbon dioxide more efficiently.
   (c) C_3 plants have a lower rate of energy usage.
   (d) C_4 plants have a higher rate of photorespiration.

5. Chlorophyll is located in the
   (a) thylakoids.
   (b) mitochondria.
   (c) cell membrane.
   (d) cell nucleus.

6. The part of the chloroplast that takes part in the light-independent reaction is the _____
   (a) thylakoids.
   (b) grana.
   (c) stroma.
   (d) photosynthetic membranes.

7. The electrons that pass to the $NADP^+$ during the non-cyclic part of the photosynthetic electron flow come from
   (a) carbon dioxide.
   (b) water.
   (c) sunlight.
   (d) oxygen.

8. In a C_3 plant, carbon dioxide that first enters the Calvin cycle links with
   (a) NADPH.
   (b) PGA.
   (c) RuBP.
   (d) protons.

PART SIX

Plant Systems

In the dooryard fronting an old farm-house near the white-wash'd palings,
Stands the lilac-bush tall-growing with heart-shaped leaves of rich green,
With many a pointed blossom rising delicate, and with perfume strong I love,
With every leaf a miracle—and from this bush in the dooryard,
With delicate-color'd blossoms and heart-shaped leaves of rich green,
A sprig with its flower I break.

—Walt Whitman (1865)

F rom the towering trees of temperate forests to the gentle blades of grass that color the rolling prairie, plants have successfully colonized nearly every corner of the planet. In our effort to study and understand living things, it is all too easy to assign plants a passive role, to relegate them to the background with respect to the ongoing drama of animal life. As animals ourselves, we have a tendency to equate life with movement. The fact that plants are often stationary has led many to the unfortunate conclusion that plants are less than alive. The great irony of this view is that the environment of this planet has been modified by plants and their relatives in a way that has made it hospitable to other organisms, including ourselves.

The atmosphere, one of earth's most precious resources, is the direct result of the process of photosynthesis carried out by plants and their relatives. Plants were essential in forming the biosphere, and they continue to be essential for its maintenance and good health. The central position of plants in the biosphere is reason enough to be curious about plant biology. The utility of plants as providers of food and fiber in our daily lives is another. But there is a third, more subtle reason why we should be interested in plants: They help to expand our understanding of the possibilities of living organisms.

With few exceptions, plants do not gather food, nor do they move about, nor do they struggle directly with predators. They are unable to flee from danger or to strike swiftly when a moment of opportunity arises. Despite these differences from many animals, plants are successful. Indeed, they are *the* most successful form of life on earth. Part of this success arises, naturally enough, from the ability to convert sunlight into chemical energy. But the abilities of plants to respond to challenges, to reproduce under adverse conditions, and to respond to biological and meteorological adversity should prove a powerful reminder that there is more than one road to evolutionary success. In this section we will explore some of the adaptations that have made plants so different from other organisms and yet so successful in the grand scheme of life. Beginning with an analysis of plant structure and function, we will examine the ways in which plants reproduce, obtain nutrients, and respond to their environment.

A study of plants helps us to appreciate the beauty that they lend to our world. It helps us to understand the importance of plants in our daily lives. And it also helps us to learn the most profound lesson of all: that the multiple possibilities of life embrace ways of living very different from our own.

◄ A Madrone tree, an evergreen, sheds bark as well as leaves every year during the heat of the summer. As a result, no moss or epiphytes are able to grow on these trees.

CHAPTER 27

Plant Structure and Function

Imagine that you are asked to write a science fiction novel about the future earth. If you take the course of evolution into account, you might imagine a world in which many new kinds of organisms have evolved and many have become extinct. You might imagine an earth in which reptiles, or birds, or mammals have become extinct. You might even imagine the extinction of human beings. But a world without plants is almost beyond comprehension. Plants are the basis of nearly all life on the land. They provide the material that enters the terrestrial food chain at the bottom, making it possible for so many other forms of life to live on earth.

Plants depend on the sun for energy. Their need for sunlight is so great that it comes to dominate the structure, the life cycle, and even the shape of plants. In this chapter we will explore the basic structure and reproductive patterns of plants, and we will examine some of the ways in which plants reconcile the demands of their environment with the need for sunlight (Figure 27.1).

WHAT IS A PLANT?

We will use the term *plants* to refer to the members of the kingdom **Plantae** and to all **algae,** whether they are classified as Plantae or Protista. These organisms share a number of common features (Figure 27.2). The kingdom Plantae is composed of many divisions of plants that differ from each other in the ways in which they have met environmental challenges. (Plant taxonomists, for historical reasons, use the term *division* rather than *phylum* when they classify plants. For our purposes, however, *division* and *phylum* are interchangeable.) Although we will not examine each division in the kingdom, we can explore some of the ways in which plants are organized by examining representative members of the most important groups.

In earlier chapters we have considered the evolution of plants, and we have seen the vital roles that plants of all types play in the earth's many ecosystems. In this chapter we will begin to examine the organisms themselves and to explore the ways in which plants have adapted to the many environments on the earth.

We discussed the process of photosynthesis in some detail in Chapter 23, but we should also consider the fact that photosynthesis exerts a powerful influence on plant adaptation and evolution. Their dependence on photosynthesis for food and energy means that plants must have access to water and carbon dioxide, and it also requires that as many cells of the plant as possible be directly exposed to sunlight.

The Need for Sunlight

Whether they are found on land or in the water, plants require sunlight to provide the energy needed for photosynthesis. Every species of plant displays some adap-

Figure 27.1 Plants dominate the landscape of the Monteverde cloud forest in Costa Rica, employing a wide variety of survival strategies.

tation shaped by the need to ensure a steady supply of solar energy in its particular environment. Plants living under water are found close to the surface where the light is most intense. Many algae contain organelles that sense the presence of light and direct the cell to swim toward it. Larger underwater plants, including many forms of algae known as seaweeds, produce gas-filled sacs that enable them to float near the surface. Those plants living deeper in the water have special accessory pigments that help them harvest those wavelengths of sunlight that penetrate more deeply.

Plants on the land have body shapes that maximize the amount of light they are able to absorb. The thin, rigid stalks and broad leaves of land plants enable them to gather light efficiently. Trees, the largest plants, grow toward the sky where they have unimpaired access to enormous amounts of sunlight. Plants growing in the shade of these giants increase the amount of light-capturing chlorophyll in their chloroplasts, and they gradually change the orientation of their leaves during the day. This maximizes the amount of light they are able to absorb.

The Need for Water

Plants, like all living things, need water. For the plants that live in water, this problem solves itself, and all plants living on the land have evolved ways to ensure a steady supply of water. Water is the means by which plants absorb mineral nutrients from the soil; plants also require water for internal transport. Plants' need for water is similar to

Figure 27.2 At the cellular level, the common features of plants include cells with nuclei, cell walls, and the green photosynthetic pigments chlorophyll a and b.

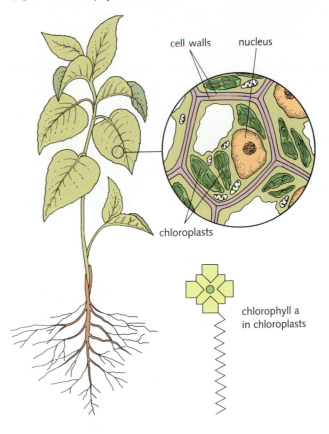

cell walls   nucleus

chloroplasts

chlorophyll a
in chloroplasts

Figure 27.3 The three principal needs of plants are illustrated in the adaptations of these three species: (top, right) some water plants, like the feather boa kelp, stay near the surface of the water with the aid of gas-filled sacs, ensuring that they receive plenty of sunlight; (top, left) the thick, stubby stems and fruits of *Opuntia englemanii* help to conserve water in the Utah desert; (bottom) the airfoil-like wings of these maple seeds increase the chances for reproductive success by spreading these seeds over great distances.

that of animals, with two important differences: For plants, water is one of the raw materials of photosynthesis, so it is used up at a high rate during carbohydrate production. Second, keeping up a steady supply of water is complicated by the need to absorb light energy; sunny places are also places where living tissues can dry out. This presents the plant with an important challenge: how to absorb as much light as possible *and* avoid severe water loss.

The Need to Reproduce

All organisms must ensure that their offspring have a reasonable chance to survive and begin a new generation. Plants, however, face special challenges (Figure 27.3). Plants do not have nervous systems, and they are not able to run away from predators or pests. Because nearly all plants live in fixed positions, they must also manage to find mates without being able to move around. Therefore they have evolved strategies for dealing with these problems that are essentially passive. An important part of such strategies is a reproductive pattern enabling each individual to produce large numbers of offspring.

PLANT STRUCTURE: FROM WATER TO LAND

Plants Restricted to Wet Environments

The simplest plants are the **algae.** Most algae are unicellular like the typical green alga *Chlamydomonas,* shown in Figure 27.4. *Chlamydomonas* is a photosynthetic autotroph, and each cell contains a single, cup-shaped chlo-

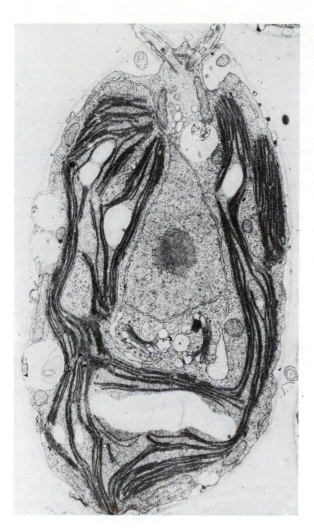

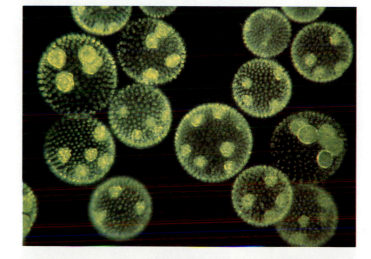

roplast. There are two flagella at one end of the cell, enabling it to swim rapidly through the freshwater ponds and lakes where most species of *Chlamydomonas* live. *Chlamydomonas* is surrounded by a tough cell wall. Under favorable conditions, *Chlamydomonas* cells can grow rapidly by mitosis, and like many algae, they are also capable of sexual reproduction, as we will see in the next chapter.

Other green algae, such as **Volvox**, have individual cells that are similar to *Chlamydomonas*. However, in *Volvox* these cells are associated with each other to form a more complex organism in which a degree of cell specialization develops. More complex algae include **Ulva**, the "sea lettuce." (See Figure 27.5 to compare *Volvox* and *Ulva*.) These plants, which are found along rocky seacoasts, contain a number of specialized cell types, including *holdfast cells* that help attach the plant firmly to the rocks. Many biologists believe that the presence of specialized cells in these organisms suggests that they are similar to the first

Figure 27.5 (top) *Volvox*, a colonial organism closely related to *Chlamydomonas* which displays a limited degree of cellular specialization. Small daughter colonies are seen within each mature colony. (bottom) *Ulva*, a multicellular alga which displays even greater specialization than *Volvox*.

Figure 27.6 Mosses (top) and liverworts (bottom) are bryophytes. The lack of specialized vascular tissue confines these plants to damp environments.

organisms to evolve toward an exclusively multicellular lifestyle.

In a way, the success of *Ulva* in the absence of such specialized tissues points out one of the strengths of the algae. Because algae are small and simple, nearly every cell of an alga is in direct contact with the environment. Therefore, algae have no need to develop specialized tissues to control the flow of water and nutrients. Most algae live in direct contact with water, which means that their exchange of water and other nutrients with the environment occurs on a cell-by-cell basis.

The simplicity of the algae also places severe limits on the places where they are able to live. Although some species (including a few representatives of *Chlamydomonas*) are able to live in the soil, and others (including the algae that associate with fungi to form lichens) are able to live on dry land, for the most part algae must go where the water is. Terrestrial habitats are beyond their reach. Those algae that are able to survive on land do so only in very special environments where the supply of water is ensured. The other types of plants differ from the algae in displaying adaptations that have enabled them to develop a degree of independence from the water. The greater that independence, the more firmly they are committed to the terrestrial lifestyle.

Plant Life on Land

Plants living on land face a completely different set of problems from those that live in salt or fresh water. The problems that confront plants living on land include the following:

- *The lack of buoyant support.* Whereas water supports aquatic plants, the lower density of air requires supporting tissues for all but the smallest land plants.

- *The tendency of land plants to desiccate, or dry out, owing to water loss from evaporation.* Plants exposed to the air face the possibility of disastrous water loss from exposed tissue.

- *Rapid changes in temperature and humidity.* Plants on land face a daily environment that is much less stable than that which water plants face.

- *The extraction of nutrients.* The nutrients used by water plants are taken directly from the external environment, but land plants must acquire nutrients that are present in the soil in solid form.

- *The scarcity of water for reproduction.* Gametes must be exchanged in an environment which often lacks the water that exposed reproductive cells might need to survive.

Land plants have adapted to these problems by evolving a greater degree of complexity in their cells and tissues. The evolutionary patterns of land plants show a clear trend toward establishing independence from water; the details of that evolution are obvious in the very first division of the Plantae, the Bryophytes.

Bryophytes, the mosses, liverworts, and hornworts, are true land plants and have structures very much like stems and leaves (Figure 27.6). They are often covered with a waxy coating, or cuticle, that helps them retain water when the weather is dry, an important adaptation for life on land. Bryophytes are anchored to the soil by thin hair-like structures known as *rhizoids*. Rhizoids are not considered "true roots" because **true roots** contain **vascular tissue,** which is tissue specialized for carrying water. The lack of vascular tissue is one of the most important characteristics separating the bryophytes from "higher" plants.

Water is absorbed through thin cell walls in the rhizoids and then passed from cell to cell by means of osmosis. This cell-to-cell passage of water works well over short distances, but it cannot move water very far against the force of gravity, as vascular tissue can. This is one of the main reasons why these organisms are unable to grow very far from their sources of water—most bryophytes are relatively small. In addition, the sperm cells of bryophytes are propelled by flagella and must swim through water to reach the egg cells. Therefore they are capable of sexual reproduction only when the areas in which they grow are very damp. Their inability to move water efficiently and their need for a moist environment in which to reproduce confines these organisms to places that are very wet. Bryophytes are found on damp forest floors and at the edges of brooks and streams.

The Vascular Plants

Vascular plants, which possess specialized water-conducting tissues, are known as **tracheophytes.** This term comes from specialized cells called **tracheids** that are key components of water-conducting tissue. The existence of specialized water-conducting tissue is just one of the specializations found in tracheophytes. Others include the development of *true roots,* containing vascular tissue, that gather water from the soil and pass it through conducting tissues bundled within the main stalk of the plant. The earliest tracheophytes formed the great forests of the primitive earth, and their ability to move water against the force of gravity enabled them to grow several meters tall. The tracheophytes, including club mosses, ferns, and seed-bearing plants, are true land plants that have decreased their dependence on the presence of water (Figure 27.7). This independence is not complete, for the

Figure 27.7 Two large fronds of *Polypodium scolopendria*, a representative fern. The vascular tissue of ferns and other tracheophytes gives them a degree of independence from water. The structures on the undersides of the fronds are *sori*, clusters of spore-producing tissue.

Plants and Medicine

A young leukemia patient resting peacefully in a cancer ward is thousands of miles away from the exotic island of Madagascar, which lies off the eastern coast of Africa. She and her hopeful parents, who have been told that the odds are good she will recover from the deadly cancer, may not know the location of that island, and they may not have heard of the Madagascar periwinkle *(Catharanthus rosea)*, a plant with white and pink flowers native to the island. Nonetheless, the little girl may owe her life to this flower and to a compound extracted from it known as *vincristine*. Vincristine is a cellular poison that breaks down the mitotic spindle and makes it impossible for cells to complete cell division. Cancer cells, which divide rapidly, are particularly vulnerable to vincristine, and in the last decade the drug has become a powerful weapon in the medical arsenal with which patients are winning battle after battle in the war against childhood cancer.

Vincristine is just one example of a human dependence on plants that is much less obvious than our dependence on plants as sources of food. Plants are one of our most valuable sources of medicines, old and new. In many cases, well-known drugs are responsible for the effectiveness of folk remedies. Hindu folk medicine has made wide use of *Rauwolfia serpentina*, the snakeroot plant, as a treatment for circulatory and nervous disorders. From a careful analysis of the effects of this plant, researchers extracted *reserpine*, a drug absolutely essential to the treatment of high blood pressure. The use of willow bark as a pain remedy is effective because the plant contains a close chemical relative of aspirin. The flowering foxglove contains *digitalis*, one of the most effective

heart medicines known. It is estimated that more than 25 percent of the drugs prescribed for use in the United States contain plant materials as their active ingredients, and that percentage would be even larger if synthetic chemicals first discovered in plants were included in the total.

The possibilities of plants in medicine are almost limitless. For years, biologists have collected rare plant species from around the world with the hope of finding new medicines. The search has been long and costly, but it has also been rewarding. Sometimes success in that search comes from the commonplace: The polyacetylene poisons in marigolds hold great promise as antitumor drugs. But just as frequently, it comes from the exotic, as with *curare*, a potent muscle relaxant derived from the plant poisons used by Amazon natives on their blowgun darts.

Ironically, just as we develop the biochemical techniques to make the

most of these botanical gifts, we may be on the verge of destroying the world's greatest storehouses of plant chemicals. The tropical rain forests, including the Amazon, are vanishing at an astonishing rate. As one acre after another is cleared for the quick gain of timber or for cultivation, the wild forests become ever smaller and the chances for discovery of rare plant species diminish. Sadly, the land beneath the forest is of little use for farming. So much of its nutrient content is tied up in vegetation that once it has been cleared, the land will support farming for no more than a few years. Recent estimates are that more than half the land cleared for the Amazon forest has now been abandoned. The farms and ranches come and go, but once a plant species has been lost, its potential for medical application is gone forever.

Based on "NOVA," April 24, 1988.

Flowering foxglove (*Digitalis purpurea*), the source of digitalis, a powerful medicine used to treat heart disease.

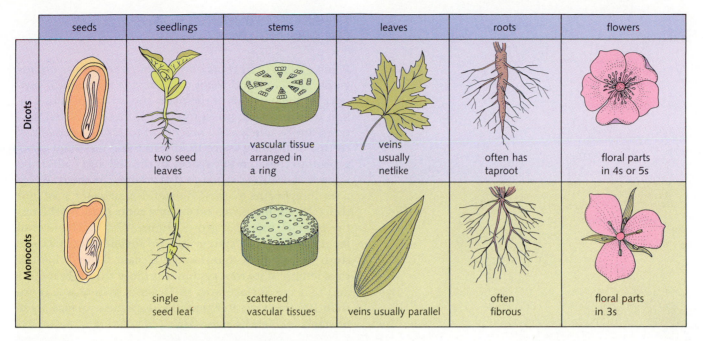

	seeds	seedlings	stems	leaves	roots	flowers
Dicots		two seed leaves	vascular tissue arranged in a ring	veins usually netlike	often has taproot	floral parts in 4s or 5s
Monocots		single seed leaf	scattered vascular tissues	veins usually parallel	often fibrous	floral parts in 3s

Figure 27.8 Some of the principal structural differences between monocots and dicots.

reproductive stages of the life cycle of some members of the group still require the presence of water.

Gymnosperms and Angiosperms

The seed-bearing plants, a subdivision that includes the **Gymnosperms** and the **Anthophyta,** show the greatest organization. All cone-bearing seed plants, including pine and fir trees, are gymnosperms. The Anthophyta, or flowering plants, are sometimes referred to as *Angiosperms,* a name derived from an older classification scheme. Because the angiosperms are the predominant form of plant life on land, we will concentrate our attention on the structure and function of the tissues of flowering plants. These plants have true roots, stems, and leaves, and they have developed an extensive network of vessels that helps to move water, minerals, and nutrients throughout the organism.

The angiosperms themselves are divided into two classes on the basis of the number of *cotyledons,* or seed leaves, they possess. **Monocotyledons** (also known as monocots) contain a single seed leaf, and **dicotyledons** (dicots) contain two. Although this sounds like a trivial distinction, other structural differences between monocots and dicots carry over into the organization of stems, leaves, and flowers, as shown in Figure 27.8.

PLANT ORGANIZATION

The most abundant land plants are the angiosperms; more than 200,000 species are known. As different from each other as many of these species are, we can make some useful generalizations about their structure. The body of a plant contains distinct regions known as **roots, stems,** and **leaves.** The roots are anchored in the soil, providing stability and acting as a conduit for water and mineral nutrients. The stems rise above the ground. They contain supporting tissue to keep the plant rigid as well as conducting tissue to carry water and nutrients from one end of the plant to the other. Leaves are the principal organs of photosynthesis and are commonly arranged in such a way that they are able to capture the maximum amount of sunlight.

Three tissue systems are found throughout the body of the plant (Figure 27.9). **Dermal tissue** is the outer covering that protects the plant from the environment. The dermal layer is continuous from the roots to the leaves. Plants also contain a system of **vascular tissue** specialized for conducting fluid. The vascular system runs throughout the plant and enables its various parts to specialize in different functions. Finally, the vascular system itself is embedded in **ground tissue,** which provides strength and support and also contains most of the cells that are active in photosynthesis.

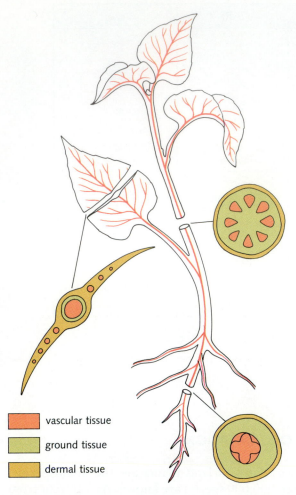

vascular tissue

ground tissue

dermal tissue

Figure 27.9 Plants contain dermal tissue, vascular tissue, and ground tissue. Their general arrangement is illustrated in this diagram.

THE ROOTS: STARTING FROM THE GROUND UP

Plants extend their roots into the soil, where these organs gather minerals and water and provide mechanical support for the portions of the plant above the ground. The best soils for plants are composed of loosely packed particles, and as much as 50 percent of total soil volume may consist of nothing more than water and air. Plant roots grow into these empty spaces in the soil and collect the water and dissolved minerals that accumulate there. When a seedling begins to grow, it extends a **primary root** into the soil, and numerous **secondary roots** branch off from the primary root as it grows (Figure 27.10). The overall length of the root system can be formidable: A classic study of a single 4-month-old rye plant showed that the total length of its root system was more than 600 meters and that the surface area of the root system was more than 100 times greater than the combined surface areas of stems and leaves.

Root Structure

Part of the outer surface of a root is covered with tiny projections called **root hairs,** which extend from the outer surface of the root and make direct contact with the soil. The water and dissolved minerals absorbed into the root are first taken in through the root hairs. The root hairs are part of the outermost layer of cells in the root, the **epidermis.** The epidermis surrounds a spongy layer of cells known as the *cortex,* and near the center of the root is another tissue layer called the *endodermis,* the innermost cells of the cortex (Figure 27.10). The endodermis surrounds the vascular tissue of the root, and the walls of the endodermal cells contain a thin, waxy layer called the **Casparian strip,** which prevents water from moving between the cells of the layer. Therefore, to enter the vascular tissue, water must pass directly through the endodermal cells. Hence the condition of these cells controls the rate at which water enters the root.

Inside the endodermis is a cylinder of cells known as the **pericycle.** During rapid root growth, dividing cells from the pericycle are the sources of new tissue as roots send out lateral branches. Inside the pericycle are two vascular tissues called **xylem** and **phloem.** These tissues are the major fluid-conducting systems in the plant—in effect, they are its circulatory system. The specialization of functions in a plant requires that the movement of material should occur in two directions: Water and dissolved minerals must be brought up from the roots to the rest of the plant, and the products of photosynthesis (sugars and other nutrient molecules) must be trans-

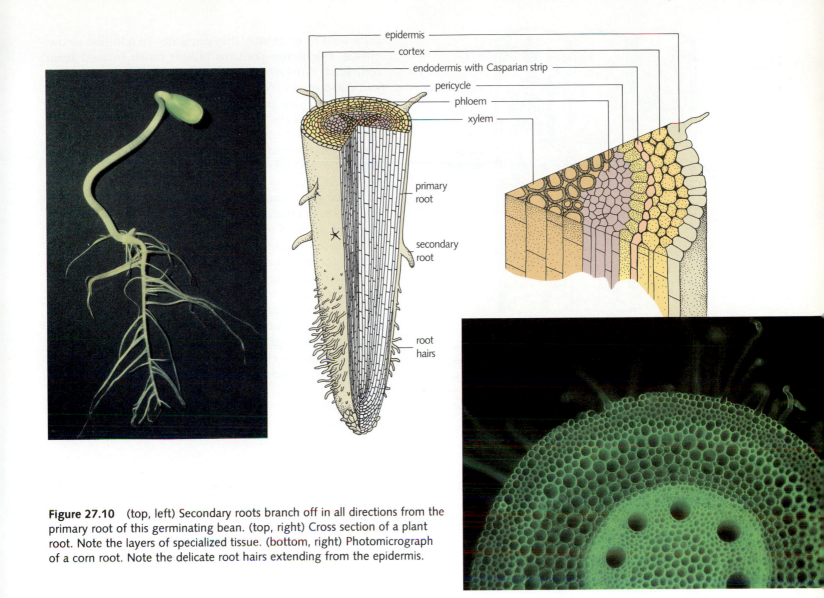

epidermis
cortex
endodermis with Casparian strip
pericycle
phloem
xylem

primary root

secondary root

root hairs

Figure 27.10 (top, left) Secondary roots branch off in all directions from the primary root of this germinating bean. (top, right) Cross section of a plant root. Note the layers of specialized tissue. (bottom, right) Photomicrograph of a corn root. Note the delicate root hairs extending from the epidermis.

ported from their production sites in the leaves to the rest of the plant. *Xylem* is the tissue that conducts water and minerals up from the roots; *phloem* circulates a nutrient-rich sap throughout the plant.

Because of osmosis, water is able to cross biological membranes to enter a cell whenever the concentration of dissolved material is greater within the cell than outside of it (in the terminology of osmosis, whenever the cell is *hypertonic* to its surroundings). In plants, this same situation occurs whenever the soil is damp. Osmosis causes a net movement from moist soil into root hair cells. This movement of water dilutes the cytoplasm of these cells, and water moves by osmosis to adjacent cells in the root.

This transfer of water from one cell to another in the cortex region of the root helps to drive water into the roots and into the vascular system of the plant.

Xylem

Xylem tissue runs from the roots of a plant up through its stem to the leaves and branches. The tissue is composed of *tracheids*, a series of elongated cells that grow as tight bundles and then gradually die, leaving only the interconnected tracheid cell walls. Water can pass from one tracheid cell to another through *pits* that are

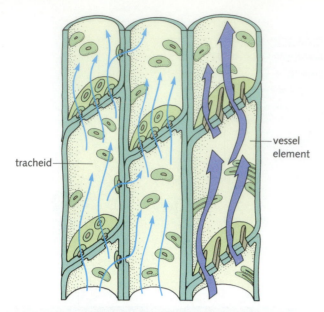

Figure 27.11 The structural organization of xylem. Tracheids are the most basic conducting tissue. Water moves from the empty chamber of one tracheid to the next through *pits,* areas bounded by a thin primary cell wall. Like tracheids, vessel elements are hollow. However, the boundaries between adjacent vessel elements are generally perforated, forming a continuous tubule through which fluid moves easily.

separated by extremely thin cell walls. In addition to tracheids, xylem also contains hollow *vessels* formed from individual *vessel elements* (Figure 27.11). Vessel elements join end-to-end to form a continuous hollow tube through which water can pass easily. In smaller plants, osmotic pressure from the roots is sufficient to force water up through this network of hollow tracheids and vessels. You can see this for yourself if you snip off the end of a healthy blade of grass and watch fluid accumulate on the cut end of the blade.

In larger plants, however, osmotic pressure is not enough to overcome gravity and force water all the way from the ground to the upper branches. There are two other forces in operation that combine to do the job. The first of these is known as **capillary action.** Water molecules have a strong *cohesive* character, meaning that they are attracted to each other (by hydrogen bonding), and water molecules can also be attracted to other molecules, a process called *adhesion.* The combination of cohesion and adhesion helps water to rise in a small tube-like opening. The smaller the diameter of the tube, the greater the capillary action and the higher the level to which water can rise.

The second, and more important, force that helps lift water molecules in larger plants is generated in the leaves

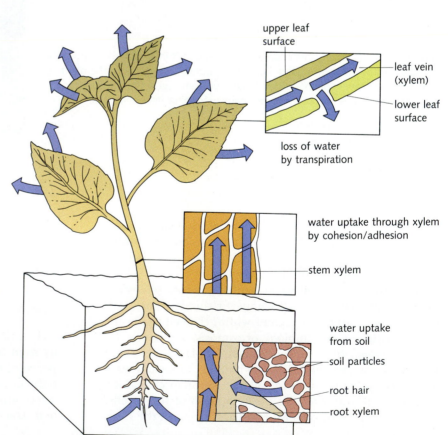

Figure 27.12 The loss of water by transpiration from leaves produces a powerful "transpiration pull" that draws water upward through xylem tissues from the roots.

Figure 27.13 (left) Photomicrograph of a sieve tube member. Nutrients move from cell to cell through the sieve plates which separate phloem members. (below) Companion cells are adjacent to sieve tube members, and perform some of the metabolic tasks required to keep the sieve tube members alive (diagram at left). The contents of phloem vessels can be collected for scientific study by exploiting an insect pest: Aphids are allowed to attach to a stem and begin sucking phloem sap. Then their bodies are sliced off with a razor blade, leaving the mouth parts inserted into the plant. Drops of sap are then collected from the feeding tube.

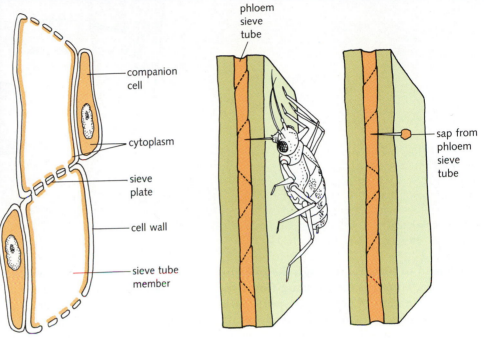

of the plant. A great deal of water can be lost from leaves on warm days by a process known as **transpiration.** A large maple tree on a warm (35°C) day may lose more than 100 liters of water. When water is lost in this way, the water content of the cells surrounding the vascular tissue drops, and osmotic pressure tends to move water out of the vascular tissue into the surrounding cells. This creates an effect known as *transpiration pull,* which draws water into the upper leaves to replace the water lost by transpiration (Figure 27.12). We will soon consider some of the ways in which water loss by transpiration is limited.

Phloem

Phloem tissue is similar to xylem in that it forms a continuous network reaching from the roots to the leaves of the plant. Phloem vessels are sometimes called **sieve tubes,** and individual phloem cells are called *sieve tube members* (Figure 27.13). Unlike xylem, the individual cells within phloem tissue remain alive, and each sieve tube contains

a thin layer of cytoplasm. As they mature, sieve tube members lose ribosomes and a functional nucleus. The roles these organelles play in other cells may be taken over by *companion cells,* which lie next to sieve tube members and are connected to them by *plasmodesmata,* small passageways that permit molecules to flow from the cytoplasm of one cell to that of the next.

Phloem sap, the material that flows through these interconnected sieve tubes, has a completely different composition from the xylem sap. Phloem sap is filled with sugars and other nutrients. Radioactive tracer experiments show that the sugars produced in photosynthesis quickly show up in the phloem sap but not in the xylem sap. These and other experiments suggest that phloem sap is used to move the products of photosynthesis throughout the plant. (See Theory in Action, Another Use for Phloem Sap, page 533.)

The phloem system can move material very quickly in both directions, but this process, known as *translocation,* requires the input of energy. There are a number of theories about what force might be behind the

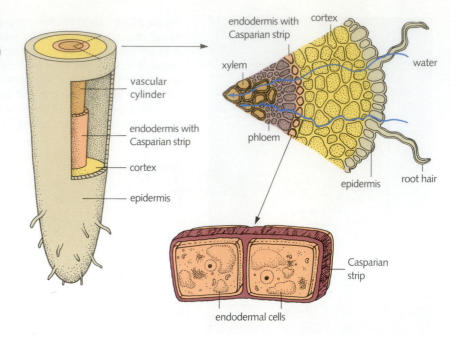

Figure 27.14 The movement of water from root hairs into the vascular cylinder. Endodermal cells are embedded in the waxy Casparian strip, preventing water from moving between the cells. This enables these cells to regulate water movement into the vascular cylinder.

movement of material through the phloem system. One suggestion, known as the **pressure-flow hypothesis,** relies on a combination of active transport and osmotic pressure. If sugars are actively transported into the sap at one location, osmotic pressure produces a flow of water that follows the material. This process helps to transport the sugar. If sugar is then actively absorbed at another location (the fruit, for example), the active uptake of the sugar into the fruit again causes water to follow it. The movement of sugar and water maintains a steady flow between the two points. Although there are some attractive elements to this idea, not enough evidence exists for us to accept it as an explanation for all aspects of phloem transport.

Water Movement into the Roots

If plants had to depend on the presence of enough water around their roots to acquire water by osmosis, they would be able to thrive only in wet, muddy soil. However, the cells of the root cortex help the osmotic process along. These cells pump minerals into the root across their own cell membranes by *active transport*. Mineral ions are pumped into the root cortex cells by protein pumps in the cell membrane that use energy in the form of ATP. As the cells of the root cortex accumulate dissolved nutrients, the concentration of dissolved material within the cells increases. This increase eventually reaches the

point at which the concentration of dissolved material inside the cell is greater than its concentration in the soil, producing an osmotic gradient that causes water to flow into the roots. The flow of water follows the active transport of minerals by the cortex.

After water passes through the root cortex, it moves into the cells of the epidermis, which form a cylinder that encloses the water-transporting vascular tissues of the root. Active transport also takes place in the cells of the endodermal layer. These cells pump minerals into the central region of the root surrounding the vascular layer. As we saw earlier, because the Casparian strip prevents water and minerals from leaking back into the cortical region, the water is "trapped" inside the central region of the root, a zone of water-transporting tissue known as the **vascular cylinder.** No water can leave or enter the vascular layer without passing through the endodermal cells. Hence these cells can regulate the movement of water and dissolved minerals into the vascular tissues of the root. The continuing osmotic pressure forces water into the root system and creates a root pressure that is the starting point for the flow of water through the system (Figure 27.14).

The sensitivity of root hairs to osmotic effects is one of the reasons why saltwater flooding is so severe a threat to crop plants. High concentrations of salt in seawater can cause osmosis to drive water *out of the root hairs* rather than into them. If the water is salty enough and the situation persists, the cells of the root die of water loss,

Another Use for Phloem Sap

The thick, nutrient-rich phloem sap found in the phloem vessels and sieve tubes of plants is their lifeblood. Phloem carries nutrients throughout the organism, and interrupting the flow of sap is one of the surest ways to injure an otherwise healthy plant. As biologists, we can be grateful for another property of phloem sap: the fact that it occasionally leaks from trees and over time can be transformed into a mineralized, gem-like material known as *amber.* The oils and resins found in the sap of pine trees make it a particularly rich source of amber. As droplets of sap dry and gradually oxidize, they acquire the hard texture and yellow-brown color of amber.

Material trapped within the sticky sap is sealed off from the environment as amber forms, and it can be preserved for hundreds or even millions of years. This has made amber particularly useful to biologists and geologists who are trying to learn about the earth's past. Insects trapped on the surfaces of ancient pine trees by large drops of sticky sap more than a hundred million years ago have been locked in those amber gemstones ever since (see accompanying figure). Such specimens have given evolutionary biologists a rare look at ancient living things in a detail unmatched by ordinary fossils.

Besides insects, amber frequently contains small bubbles, and these bubbles have given scientists a chance to determine the composition of the ancient atmosphere. Robert Berner of Yale University and Gary Landis of the U.S. Geological Survey crushed samples of amber that were from 25 to 40 million years old and analyzed the contents of gases released from small bubbles within each stone. The preliminary results of their studies seem to indicate that the ancient atmosphere might have consisted of as much as 30 percent oxygen, considerably more than the 21 percent it exhibits today. More studies are needed to confirm these results, but they illustrate the scientific importance of phloem sap, transformed to amber, as a priceless record of the past.

Amber containing a beautifully preserved midge, so detailed that even a tiny mite on its back is visible.

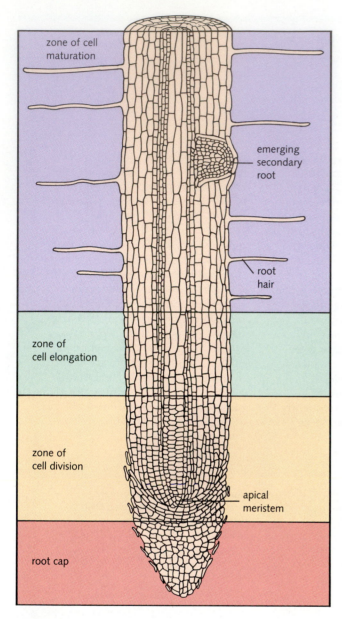

zone of cell maturation

emerging secondary root

root hair

zone of cell elongation

zone of cell division

apical meristem

root cap

Figure 27.15 Growth in roots is localized. Distinct regions, including a *zone of cell division* (where cells are rapidly dividing), a *zone of cell elongation* (where cells increase in size), and a *zone of maturation* (where cell development takes place) can be recognized.

killing the plant. A similar situation occurs in house plants and in fields that have been artificially irrigated for many years. Unlike rainwater, river water and well water contain small quantities of dissolved salts. As irrigation water evaporates, much of this salt is left behind in the soil, making it more and more difficult for plants to grow. This gradual accumulation of salt can ruin soils that have been carelessly irrigated for long periods of time.

Plants that live in salty environments such as salt marshes or coastal dunes are able to endure salt concentrations in the soil that would dehydrate and wilt most other plants. Such plants are known as *halophytes* ("salt plants"). Halophytes are particularly interesting to plant breeders, who would like to transfer their salt tolerance to crop plants, making it possible to grow food in coastal areas where salty soil has limited agriculture. Like other plants, halophytes absorb water via the active transport of salt and other dissolved minerals into their root tissue. When the concentration of dissolved salts is higher there than in the surrounding soil, water follows into the root cortex by osmosis. The key adaptation of halophytes is that their root cortex cells are able to tolerate much higher salt concentrations than the roots of most other plants, enabling them to continue to draw water when other plants have wilted.

Growth Patterns in Roots

Plant growth occurs primarily in regions known as *meristems* in which rapidly dividing unspecialized tissue provides a reservoir of cells that can develop into the more specialized tissues of the plant. Meristematic regions near the tips of roots or stems are known as **apical meristems.**

As a root grows, cells produced in the meristem enlarge and elongate and then begin to develop into the various tissues of the root. This process produces a *zone of elongation,* followed by a region known as the *zone of maturation* (Figure 27.15). At the tip of the root, some of the cells produced by the meristem join the **root cap,** a sheath of protective cells that guards the meristem against injury from the soil particles that it pushes aside as the root grows.

Root growth is not necessarily limited to the apical meristem. In most plants, *secondary roots,* which contain apical meristems of their own, branch off from the main root, increasing root system complexity. Cells in the *pericycle* of the root begin to divide, and these endodermal cells form a new lateral meristematic region that produces cells that push through the endodermis and penetrate the cortex of the existing root. The new root grows through the epidermis of the existing root and forms an independent branch root that is completely connected—xylem, phloem, and epidermis—to the existing root structure.

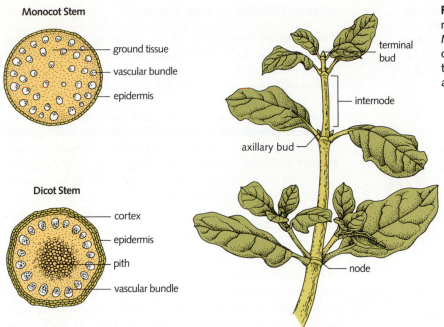

Monocot Stem
- ground tissue
- vascular bundle
- epidermis

Dicot Stem
- cortex
- epidermis
- pith
- vascular bundle

terminal bud

internode

axillary bud

node

Figure 27.16 Many stems produce axillary buds at nodes, which then give rise to lateral branches (right). Monocot and dicot stems (left). In monocots, bundles of vascular tissue are scattered in the ground tissue of the stem. In dicots, the bundles are arranged in a ring around the pith.

THE STEMS: MAINTAINING THE FLOW

In the vascular plants, stems connect the leaves that carry on photosynthesis and the roots that absorb water and nutrients. Stems may run only a few centimeters, or they may extend tens of meters into the air. Either way, they furnish support, water, and nutrients for the other tissues of the plant. Xylem and phloem tissues pass through the stems and directly link all parts of the plant, allowing water and nutrients to be carried from root to leaf or vice versa. In most plants, stems contain distinct **nodes** where leaves are attached and **internode** regions that connect the nodes. Small **axillary buds** are found where leaves attach to the nodes. These nodes contain undeveloped tissue that can produce new stems and leaves (Figure 27.16).

Stems are surrounded by a layer of epidermal cells that have thick cell walls and a waxy coating for protection. In monocots, **vascular bundles** of xylem and phloem are embedded in ground tissue (Figure 27.16). In dicots, these vascular bundles in the stem are arranged in a ring near the center of the stem. Ground tissue within the ring of vascular tissue is known as **pith;** ground tissue outside the vascular bundle forms the **cortex** of the stem.

Growth Patterns in Stems

Like growing roots, growing stems contain active meristematic regions. The apical meristem is found just beneath the apex of the growing stem, and it is followed by zones of elongation and maturation. In some plants, including many common trees, the apical meristem is the only region of the stem capable of increasing in length. This is why tree branches remain at the same height as a plant matures. Other plants, including corn, contain meristematic regions between the leaf nodes, allowing rapid growth as several meristems contribute to stem elongation. The **terminal bud** of a growing woody stem is covered by thick scales shielding the meristem and the leaves that develop around it. **Lateral buds** develop at the points of connection between leaves and the stem, and these serve as the starting points for lateral branches that arise from the stem.

The growth that occurs at apical meristems is known as **primary growth** (Figure 27.17). Primary growth is responsible for elongation of stems and roots. In larger plants, particularly those that live for many years (known as *perennials*), **secondary growth** is also important to plant development. Secondary growth results in a *thickening* of stems and roots in the *lateral* direction. The tissue regions that contain the cells responsible for such growth are known as **lateral meristems,** or **cambium.** Secondary growth is particularly pronounced in the stems of woody plants and trees.

The patterns of primary and secondary growth are important in dicots, but secondary growth does not occur in many monocots. Although few monocots grow large enough to be called trees, those that do (most notably the palm tree) display only primary growth. This is one reason why the trunks of palm trees are of nearly uniform diameter (Figure 27.17).

Figure 27.17 (top) The distinction between primary and secondary growth. Stem elongation is referred to as primary growth. Secondary growth refers to the increase in stem diameter. (bottom) Like most monocots, palm trees display only a limited amount of secondary growth because their growing tissue is localized at the bases of growing scale-like leaves.

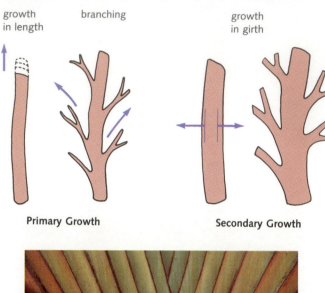

growth in length   branching   growth in girth

Primary Growth   Secondary Growth

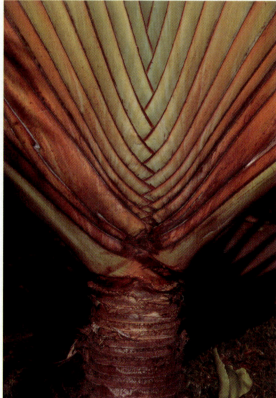

Growth Patterns in Tree Stems

The vascular bundles in dicots are arranged in a ring, phloem tissue near the epidermis and xylem near the center of the stem, as shown in Figure 27.18. As the stem of a woody plant grows, lateral meristems develop at the zone between xylem and phloem in each vascular bundle until they form a layer of **vascular cambium** tissue that completely encircles the stem. This growing layer of cells gives rise to tissues known as **secondary xylem** and **secondary phloem.** As seasons come and go, the cambium continues to produce xylem and phloem cells, causing the stem to become thicker and thicker. This growth presents no special difficulties for the xylem cells; new layers of secondary xylem are simply added to older ones. Continued activity in the cambium, however, pushes the phloem outward, causing the tissue to split and fragment as the diameter of the stem increases. If this process went unchecked, serious breaks in the outer covering of the stem would occur. However, the *cortex* of the stem contains another meristematic tissue known as **cork cambium,** which forms an outer covering of cork that grows to cover places where the phloem tissue has been split by the growth process.

The **bark** of a tree is composed of all the tissues external to the vascular cambium. The **wood** of a tree is formed by the production of new xylem tissue by the cambium, whereas the cambium itself is at the boundary between wood and bark. In other words, *wood is secondary xylem.* In temperate climates the pattern of wood growth follows a yearly cycle. In early spring, when water is generally in good supply and growing conditions are ideal, the cambium forms xylem cells that are quite a bit larger than those formed as the days begin to cool and shorten at the end of the summer. This seasonal variation in the formation of wood produces the **annual tree rings** exposed when you cut through a piece of wood. The annual rings that this process produces reveal more than the age of a tree. By analyzing the thickness and mineral composition of tree rings from many trees, scientists are able to get some idea of what growing conditions were like many hundreds of years ago. Severe weather conditions, including droughts, are clearly reflected in the thinner rings that result from reduced wood growth.

Reaction to Injury

Unlike animals, trees cannot run from predators or take active measures to defend themselves from being eaten. Instead of flight, trees rely on their own growth patterns to react to serious injuries and control predators. When a branch is broken and the wood of a tree is exposed to the air, chemicals in the wood begin to oxidize, producing a series of molecules related to *phenol,* a ringed six-carbon compound toxic to nearly all organisms. The phenol compounds cause a noticeable discoloration in the wood, tinting it in shades of dark blue, red, or green. These compounds help to sanitize the wound, killing fungi and bacteria that might otherwise begin to attack the exposed wood.

Another danger posed by a wound is the loss of fluid: The growing cambium layer produces vascular tissues that carry sap, and were it to continue growing with no change, sap might flow continuously from the wound, weakening the plant. Instead, the cambium at the site of the wound produces far fewer fluid-transporting cells than normal. In their place, the cambium forms a layer of dense-walled cells that cover the wound and are able to resist attack from pathogenic organisms.

Finally, the growth pattern of the tree is adjusted in a way that walls off the injured and infected area. Vascular tissue is plugged on either side of the wound, thick new cell walls surround the injured area, and the tree begins to structure its new growth around the wound. The growth patterns of trees allow them to deal with injury by *compartmentalizing* it: walling it off, limiting damage, and continuing to grow and reproduce.

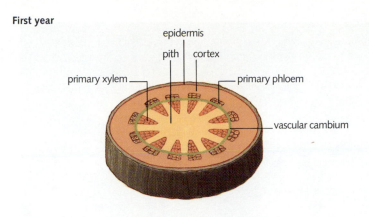

First year

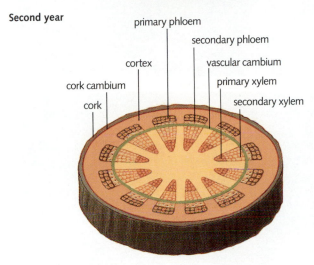

Second year

Figure 27.18 (right) In woody plants, new layers of vascular tissue are produced every year. In the first year of growth (top), xylem and phloem tissues are formed from the vascular cambium. In the second year (center), these primary xylem and phloem tissues are pushed apart by new layers of secondary xylem and secondary phloem produced by the vascular cambium. This process repeats itself every year, producing a series of alternating rings (bottom) as the result of slower growth during the winter months. (below) The spectacular growth of woody trees is recorded in the annual rings of this trunk cross section.

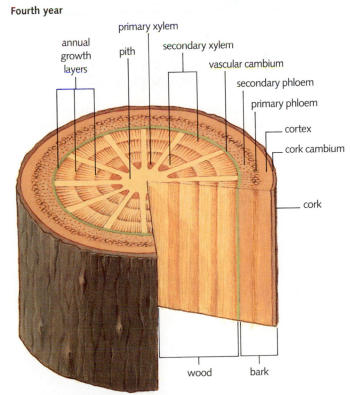

Fourth year

Figure 27.19 The complex branching of the veins of a dicot leaf (top) contrasts sharply with the simpler pattern of a monocot (bottom).

THE LEAVES: HARVESTING SUNLIGHT

The leaves of a plant are the primary organs of photosynthesis. Most leaves consist of long thin *blades* attached to the nodes of the stems by a stalk called a *petiole* (Figure 27.19). Some leaves may be divided and may consist of several interconnected parts, or *leaflets.* Leaves are covered by an outer layer of *epidermal tissue,* which protects them from the environment. The epidermal cells are coated with the **cuticle,** a thick waxy coating that protects the tissue against insect invasion and water loss. Because of their need to exchange gases, leaves cannot be completely sealed off from the environment. Therefore, leaves contain small openings known as **stomata** (singular, *stoma*), most commonly on the leaf's lower surface, which allow gases to flow into the tissue of the leaf itself. The stomata are surrounded by *guard cells* that regulate the passage of gases by opening and closing the opening.

Regulation of Gas Exchange

At first it might seem that leaves would keep their stomata open at all times, to allow carbon dioxide to enter the leaves where it could be used for photosynthesis. But this would cause a problem. Carbon dioxide must be dissolved in water in order to enter the photosynthetic cells of the leaf. Therefore the surfaces of these cells must be kept wet all the time. This means that on a bright sunny day, which is ideal for photosynthesis, transpiration occurs; water evaporates from the cell surfaces within the leaf and is lost through the stomata. We have already seen that transpiration is one of the forces that drives the transport of water from roots to leaves. However, if the stomata were kept open all the time, the water losses due to transpiration would be so great that very few plants would be able to take in enough water to survive. Therefore, plants are faced with a serious problem. They must keep their stomata open just enough to allow photosynthesis to take place, but not so much that they lose an excessive amount of water.

The regulation of the guard cells of the stomata is not entirely understood. It is clear that the stomata respond to fluid levels within leaf cells by closing when fluid levels begin to drop and opening when the amount of water is adequate. One way in which this may occur is represented in Figure 27.20. The cell walls that surround the guard cells are much thicker on the side where the two cells meet than on the other side. When the cells are swollen by water uptake, expansion can occur only on the side with the thinner cell wall. Therefore, when water moves into the guard cells by osmosis, the stomata are forced open. When water is depleted and lost from the

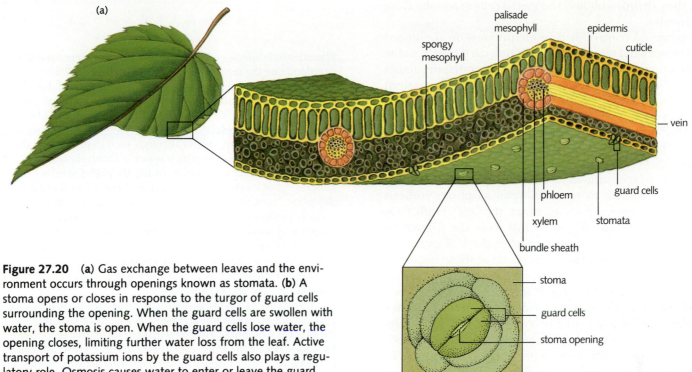

(a)

spongy mesophyll

palisade mesophyll

epidermis

cuticle

vein

guard cells

phloem

stomata

xylem

bundle sheath

stoma

guard cells

stoma opening

Figure 27.20 (a) Gas exchange between leaves and the environment occurs through openings known as stomata. (b) A stoma opens or closes in response to the turgor of guard cells surrounding the opening. When the guard cells are swollen with water, the stoma is open. When the guard cells lose water, the opening closes, limiting further water loss from the leaf. Active transport of potassium ions by the guard cells also plays a regulatory role. Osmosis causes water to enter or leave the guard cells, following the direction of potassium transport. A stoma from a tobacco leaf opens, allowing gas exchange to take place. (c) A scanning electron micrograph showing the internal structure of a turnip leaf. Stomata are present on both surfaces of this leaf, although they are more numerous on the underside.

(c)

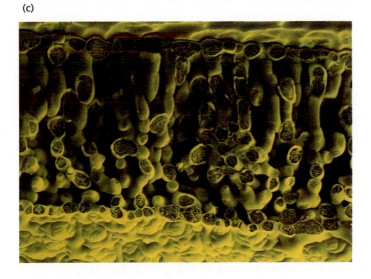

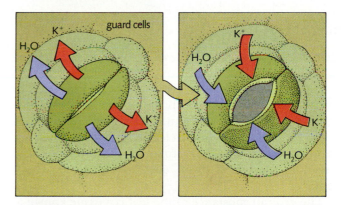

guard cells

K^+

H_2O

K^+

K^+

H_2O

K^+

H_2O

H_2O

(b)

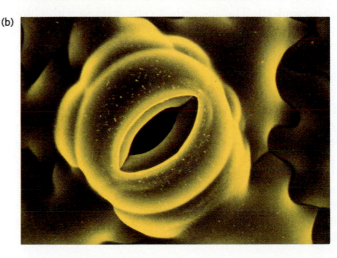

cells, they shrink a bit and the guard cells gradually close the stomata.

The actual control of osmotic pressure in the guard cells is somewhat more complex. The guard cells are able to take up potassium ions (K^+) by active transport across their cell membranes. This accumulation of potassium makes the cytoplasm slightly hypertonic with respect to the surrounding medium, and osmosis drives water into the cells. The resulting increase in *turgor,* or water pressure, within the cells causes them to swell and open the stomata. Several factors seem to affect the active transport of K^+. *Light* stimulates the potassium pump, which causes the stomata to open at dawn as photosynthesis speeds up, allowing CO_2 to enter the leaf. High CO_2 levels within the leaf and water loss cause the stomata to close, whereas low CO_2 levels within the leaf cause the stomata to open. Superimposed on all of these factors in many plants is a 24-hour cycle, or **circadian rhythm,** of stomatal opening and closing. Plants that have been grown in normal day–night cycles and then placed in a dark chamber continue to open their stomata during the daytime and close them at night for several days in the absence of any environmental cues. This internal timing mechanism affects the degree to which the stomata respond to outside influences.

Internal Leaf Tissue

The xylem and phloem tissues found in leaves are direct continuations of the tissues that run through the roots and stems. In leaves these tissues are bound together to form a *vascular bundle* that appears as the *leaf vein* easily seen in most plants. These bundles run more or less parallel to each other in monocot leaves (grasses such as corn are good examples). They follow a more pronounced branching pattern in dicots such as tomato and bean plants.

The ground substance of the leaf is a tissue known as **leaf mesophyll.** The mesophyll cells, packed with chloroplasts, perform most of the plant's photosynthesis. A typical leaf contains an upper layer of *palisade cells,* which form a tightly packed row of cells just below the epidermis. Beneath the palisade cells is a much looser layer of cells known as the *spongy mesophyll.* The air spaces between cells in this region make them especially suitable for gas exchange, and the carbon dioxide that enters through the stomata generally enters these cells first. Moisture is supplied to the mesophyll cells by osmosis from the vascular bundles that run through the mesophyll tissue. C4 plants have a specialized tissue around the vascular bundle known as the *bundle sheath,* and the pathways of carbon fixation in C4 plants are divided between mesophyll and bundle sheath cells.

SUMMARY

Plants include single-celled algae as well as the terrestrial plants that dominate the temperate regions of the earth. Plants require sunlight, water, and carbon dioxide, and the need for these influences their structure and lifestyle. The simplest plants, algae, absorb water directly from the environment. The vascular plants have evolved specialized water-conducting tissues, and the ability to move water through such tissues has enabled them to colonize areas in which water is relatively scarce.

The structural organization of higher plants involves three general tissue layers found in each of three general regions of the plant: root, stems, and leaves. The primary function of roots is to absorb water from the soil. Three different forces combine to draw water into the stems and leaves against the force of gravity: osmosis, capillary action, and transpiration pull. The roots contain another conducting tissue, known as phloem, which serves as a transport system that passes nutrients throughout the organism.

The stems of plants contain the same two conducting tissues, xylem and phloem, arranged in a pattern that identifies the plant as a monocot (one seed leaf) or dicot (two seed leaves). In woody plants the most rapid rates of cell growth are found in the vascular cambium, a tissue that forms between the xylem and phloem, laying down tissue known as secondary xylem and secondary phloem.

The leaves, attached to the stems, are a plant's primary organ of photosynthesis. A rapid rate of photosynthesis can be maintained, because special guard cells on the underside of the leaf control openings known as stomata, which regulate the passage of air into the space within the leaf.

STUDY FOCUS

After studying this chapter, you should be able to:

■ Describe the problems plants face in a terrestrial habitat, and indicate some survival strategies unique to plants.
■ Outline the classification of major plants and plant-like groups.
■ Explain the growth processes of plants and the processes that produce specialized plant tissues.

SELECTED TERMS

vascular tissue *p. 525*
tracheids *p. 525*
dermal tissue *p. 527*
ground tissue *p. 527*
primary root *p. 528*

secondary root   *p. 528*
root hair   *p. 528*
Casparian strip   *p. 528*
pericycle   *p. 528*
xylem   *p. 528*
phloem   *p. 528*
capillary action   *p. 530*
transpiration   *p. 531*
sieve tubes   *p. 531*
apical meristem   *p. 534*
vascular bundle   *p. 535*
primary growth   *p. 535*
secondary growth   *p. 535*
cambium   *p. 535*
stomata   *p. 538*

R E V I E W

Discussion Questions

1. Plants are autotrophic for carbon but not for another important element, as discussed earlier in the text. What is this important nutrient? How do plants generally obtain it?

2. What were some of the special challenges faced by plants as they evolved from an aqueous environment to a terrestrial one? What rewards were available to the first plants to inhabit the land successfully?

3. The vascular plants are sometimes referred to as the "higher" plants. Is this nomenclature justifiable? In what sense are they "higher" than algae and bryophytes? What is the difference between a true root and a rhizoid?

4. Although the high salt concentration found near the seashore is toxic to most plants, a few species thrive in such environments. How does the root epidermis differ between salt-tolerant and typical plants in terms of cellular activities and transport mechanisms?

5. How does transpiration pull function? Does it work in concert with capillary action or antagonistically to it?

6. The loss of water from the leaves of a plant on a hot day causes the cells to lose water pressure and shrink slightly. What direct effect does this have on the guard cells of the stomata? Does this change increase or decrease the rate of water loss?

Objective Questions   (Answers in Appendix)

7. Mosses have not completely adapted to land because they
   (a) need water in order to survive.
   (b) need water for reproduction.
   (c) have true roots.
   (d) every cell is in direct contact with soil nutrients.

8. Which of the following is a seedless, vascular plant?
   (a) fern
   (b) pine tree
   (c) flowering plant
   (d) liverwort

9. Flowering plants called angiosperms are divided into
   (a) gymnosperms and monocots.
   (b) bryophytes and gymnosperms.
   (c) monocots and dicots.
   (d) bryophytes and anthophyta.

10. Plant cells increase in size by
    (a) conduction.
    (b) meiosis.
    (c) differentiating into specialized tissues.
    (d) elongation.

11. The main photosynthetic area of the leaf is composed of
    (a) xylem.
    (b) phloem.
    (c) stomata.
    (d) mesophyll.

CHAPTER **28**

Reproduction in Flowering Plants

The evolution of flowering plants is a relatively recent event. Although we might tend to think of the last 100 million years as the "Age of Mammals," or even the "Age of Humans" (if we confined ourselves to the last 4 million years), we would do better to call it the "Age of Flowering Plants." The rise of the angiosperms, the flowering plants (division Anthophyta), has been one of the most important events to occur in the last 100 million years of the earth's history (Figure 28.1). From their first appearance in the Cretaceous, flowering plants have risen to ecological dominance. With the exception of isolated conifer forests, flowering plants cover nearly every square centimeter of the planet where plant growth on land is possible.

The rapid rise of these plants has led evolutionary biologists to wonder what features made them such fearsome competitors to the gymnosperms that had dominated the earth's landscape for the previous 100 million years. Flowering plants are not significantly different from gymnosperms in terms of their ability to draw water from the soil or to produce large stems and efficient leaves, and each type of seed-bearing plant is able to reproduce well in both dry and wet environments. Even today, evolutionary biologists are divided as to exactly why the flowering plants were so quick to replace the gymnosperms in many environments.

ASEXUAL AND SEXUAL REPRODUCTION

Many biologists believe that the essential advantage enjoyed by the flowering plants over the gymnosperms is the characteristic that separates them: Unlike the gymnosperms (the "naked seed" plants), the flowering plants surround their seeds with a layer of tissue known as **fruit,** formed from the reproductive tissues of the flower itself.

Why should the formation of fruit be such a key characteristic in the ability of flowering plants to profit from natural selection? A hint may be found in the **flowers,** the plant reproductive organs that add so much beauty to our world. Fruit and flowers serve as lures for insects, birds, and mammals. The adaptations by which fruit and flowers interact with animals solve two problems that plants face: finding mates and dispersing their offspring. Being rooted in place, plants cannot move about to exchange reproductive cells with others of the same species, and they face enormous problems in dispersing the thousands of seeds they are capable of producing. Associations with animals, which can carry pollen *and* disperse seeds over great distances, provide a way to solve both of these problems. These associations with animals may have given flowering plants the critical advantage that enabled them to cover the planet.

Asexual Reproduction

Vegetative reproduction in nature Angiosperms, the "highest" of the "higher" plants, are distinguished from the most advanced animals, the vertebrates, in many ways. One of the most interesting of these is the ability they share with other plants to reproduce *asexually.* This process is known as **vegetative reproduction,** and its results are familiar to any gardener. Many plants can grow from a detached leaf, stem, or bundle of roots. Little by little, the growing meristematic cells of the detached tissue form new tissues that replace the parts from which the cutting was taken, and eventually a completely new plant, genetically identical to the original, is formed.

The formal name for this type of asexual reproduction is **fragmentation,** because the "parent" plant is fragmented to produce the vegetative offspring. In nature, vegetative reproduction is an important means of propagation in a number of species. One famous example is the spider plant (*Chlorophytum sp.*), which produces *runners,* slender lateral shoots capable of developing buds of their own. If the *adventitious* roots (roots that appear in an unusual or unexpected place, such as the end of a lateral shoot) of the buds find soil, they grow into an independent organism that can be detached from the original plant. Strawberries can propagate in much the same way, and thick beds of strawberries are produced by carefully allowing a few plants to spread their runners in every direction to establish new plants (Figure 28.2).

Many grasses also propagate by sending runners along the surface of the soil, and thick clumps of dune and desert grasses that are formed as lateral runners take root and develop into separate plants. Because plants produced by vegetative processes are genetically identical to their parents, clusters of plants formed in this way are *clones,* groups of organisms produced by cell division from a single ancestral cell—in these cases, by division of the single zygote cell from which the original plant was formed.

Vegetative reproduction in horticulture   The capacity of plants for vegetative reproduction has been widely used by humans to propagate the plants we grow for our own purposes. Many common houseplants, such as African violets (Figure 28.2), can be propagated from single leaves, and fruit trees are often produced from stem cuttings. As you may know, potatoes produce small buds known as "eyes," which can grow into complete potato plants if they are planted.

Many fruit trees, including apples and pears, are propagated almost exclusively from cuttings of preexisting stock that have been rooted in nutrient solutions and grown into small saplings before they are transplanted to the orchard. The ability of plants to grow under such conditions even makes it possible for two different plants to be grafted together. The careful joining of buds from one plant to the rootstock (stems and roots) of another makes it possible for a single apple tree to yield five or six varieties of apples (the stems of several varieties are grafted to a single tree) and also enables farmers to combine the best aspects of two completely different varieties of plants. The ability to combine characteristics has been

Figure 28.1   The history of flowering plants stretches back more than 100 million years. This ancient multistemmed fossil flower discovered in the clay mine of a Tennessee brick company is an early example of what is now the dominant form of plant life on land.

particularly important to the wine industry. In the late nineteenth century a combination of diseases destroyed thousands of acres of vineyards in France, threatening to wipe out wine production. Fortunately, it was discovered that the Concord grape, which is native to North America, was immune to these diseases and flourished in the soil of France. The fruit of the Concord grape, however, does not make a very good wine. Therefore winegrowers attempted to graft the shoots of European grape varieties onto North American rootstock. The attempt was successful, and today most of the grapes grown for wine throughout the world are combinations of European and American grape species grafted together to produce the best possible plant for all purposes.

Figure 28.2 Strawberries (left) and African violets (right) are plants which can be propagated by vegetative reproduction.

In recent years, tissue culture techniques have been widely used to produce plants from single cells. We will discuss plant cell culture more fully in Chapter 30, and we have already seen how it may be used in plant genetic engineering (Chapter 23).

Sexual Reproduction

Sexual reproduction in the angiosperms takes place in the flowers, which are specialized shoots containing reproductive organs. Like sexual processes in other organisms, reproduction in the flowering plants involves the formation of *gametes* (specialized reproductive cells) and the exchange of genetic information when gametes fuse to form a new organism. Beyond that point, however, diversity is the hallmark of angiosperm reproduction. One of the major themes in the evolution of higher plants has been the progressive modification of reproductive organs and of the reproductive process itself. We often tend to think of evolutionary success in terms of survival. Catch phrases like "survival of the fittest" contribute to a predisposition to view natural selection in

terms of the ability of a group of organisms to compete successfully for survival against predators. Yet evolutionary success is measured most directly by the number of offspring that an organism is able to pass along to the next generation. One powerful strategy for assessing evolutionary success, then, is to focus on the reproductive process itself: the efficient distribution of gametes and the placement of seeds into environments where they are likely to grow. The sexual processes of flowering plants are case studies in the success of this evolutionary strategy.

ALTERNATION OF GENERATIONS

The Principle of Alternation

When we examined the life cycles of other plant groups earlier in the text (Chapter 15), we saw a pattern of alternating haploid (*n*) and diploid (*2n*) generations. This *alternation of generations*, in its simplest form, involves a **sporophyte,** a diploid plant that produces haploid *spores* by means of *meiosis*, and a **gametophyte,** a haploid plant that produces haploid gametes by *mitosis. Fertilization* occurs when two gametes fuse to form a diploid *zygote,* which then grows to form a new sporophyte. The most basic type of alternation of generations—one wherein the gametophyte and sporophyte are both distinct, independent plants—is shown in Figure 28.3. In some plant groups, notably the mosses, the haploid gametophyte is the larger and more prominent phase of the life cycle. In flowering plants, however, the sporophyte dominates, and the gametophyte has become a tiny organism that lives briefly within the tissue of the larger plant.

Alternation of Generations in the Angiosperms

The spores of flowering plants are often produced within the same flower. In Figure 28.3, which illustrates the basic life cycle of a flowering plant, the male and female reproductive tissues are drawn separately for clarity. As illustrated in the diagram, meiosis within the diploid sporo-

phyte produces two types of haploid spores, a large **megaspore** and smaller **microspores.** Each spore type is produced from a diploid **spore mother cell,** which undergoes meiosis to produce four haploid spores. These spore cells then go through several rounds of cell division to produce the female and male *gametophytes.* Fertilization occurs when a haploid nucleus from the male gametophyte and one from the female gametophyte fuse to produce a diploid zygote. This zygote then develops into an embryo that is encased in a seed, awaiting further development into a complete diploid sporophyte plant.

The details of this life cycle are complex; this description is only a simple blueprint. It is necessary to fill in many details to complete a picture of reproduction in the flowering plants. Nonetheless, it is important to appreciate the extent to which alternation of generations has been modified in the evolution of higher plants. Most dramatic has been the reduction in the role of the gametophyte. It is an almost invisible stage of the life cycle that is completely hidden within the plant's reproductive organs.

Figure 28.3 The alternation of generations in plant reproductive cycles. In principle, the sporophyte and gametophyte of a life cycle may be represented by free-living plants, as shown in the idealized cycle at the left side of the diagram. However, in flowering plants (right), the sporophyte stage of the cycle is the dominant form of the plant, and the male and female gametophytes exist only as small groups of cells enclosed within the reproductive organs.

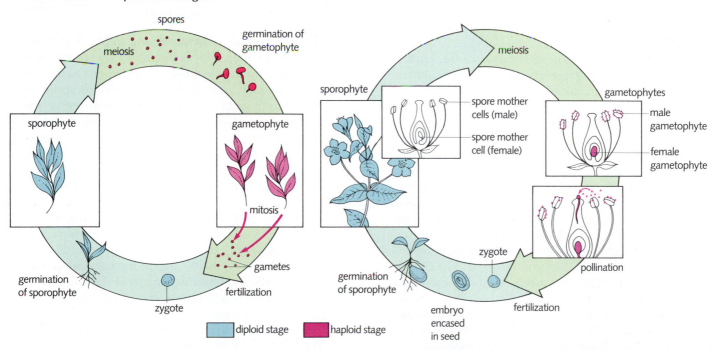

Figure 28.4 An idealized flower, showing the four principal structures of a complete flower: sepals, petals, stamens, and carpels.

anther

filament

stamen

petal

stigma

style

carpel

ovule

ovary

sepal

THE DETAILS OF PLANT REPRODUCTION

The Organs of Reproduction

The structures within a flower are actually modified leaves, and a flower can be viewed as a small, truncated shoot with four distinct leaf layers. We generally do not think of flowers as leaves, an indication of how greatly they have been modified. The four distinct parts of a typical flower are **sepals, petals, stamens,** and **carpels** (Figure 28.4).

In many flowers, the *sepals* are truly leaf-like, enclosing the flower bud during its early stages of development and retaining the green color of chlorophyll throughout the life of the flower. Taken together, the sepals form the *calyx,* the outermost enclosure of the flower. The *petals* of the flower may be brightly pigmented, giving the flower its characteristic color and helping to attract insects, birds, and other animals that may aid in the reproductive process. The actual reproductive organs are found in the two remaining portions of the flower. The male portions of the flower are the *stamens,* which produce **pollen.** The stamens consist of thin filaments that emerge from the stem of the plant and terminate in the **anthers,** enlarged sacs where pollen is developed and released. The female portions of the flower are the *carpels.* A single flower may have one or several carpels, each of which has a broad base containing an **ovary,** within which are **ovules** containing the female gametophyte. The diameter of the carpel narrows as it leads upward through a slender portion known as the **style** and terminates in a sticky **stigma.**

In a sense, the flower shown in Figure 28.4 is an ideal one. It contains the four principal flower parts, and its male and female tissues are perfectly developed, making it a complete flower. The flowers of many plants are not complete, and we will consider them shortly.

Formation of Spores and Gametes

The female tissues: megaspores and eggs The formation of female megaspores begins in the ovary of the flower, within the swollen base of the carpel. On the inner wall of the ovary, one or several clusters of cells have developed into *ovules.* As the ovule matures, a small chamber develops within it, surrounded by a cell layer known as the *nucellus.* The nucellus is in turn surrounded by a protective *integument* that leaves a single opening, the *micropyle,* through which fertilization will occur.

One large cell known as the *megaspore mother cell* appears within the nucellus. This diploid cell undergoes meiosis to produce four haploid megaspores. Three of these megaspores disintegrate in most species, leaving a single haploid megaspore. In lower plants, the mature spore would now begin to divide and to lead an independent existence as the gametophyte stage of the life cycle. In the seed-bearing angiosperms and gymnosperms, however, the gametophyte stage consists of a small group of cells derived from this single surviving megaspore.

Typically, the megaspore, which is now more properly called the *megagametophyte,* passes through three rounds of mitosis, producing first two, then four, and finally eight nuclei that are not separated by cytokinesis (Figure 28.5). The megaspore now contains eight nuclei—four at the end of the cell closest to the micropyle and four at the opposite end. One nucleus from each group moves to the center of the cell, and these two *polar nuclei* are separated from the rest of the cell by cytokinesis to form a binucleate cell. The six remaining nuclei are also split up by cytokinesis to produce a total of seven cells, the complete megagametophyte. Anticipating the role it will soon play in the reproductive process, we call the seven-cell gametophyte the **embryo sac.** The three cells closest to the micropyle form an **egg cell** and two *synergids,* the cell containing the two polar nuclei is known as the *central cell* or *endosperm mother cell,* and the three cells at the opposite end are called the *antipodals.*

The male tissues: microspores and pollen *Microspores* are the smaller spores, and they develop within the anthers of the flower from diploid *microspore mother cells.* It's a bit confusing to speak of both the "male" and "female" spores as deriving from "mother" cells, but bear in mind that the production of spores is *not* a sexual process. The microspore mother cells are contained in chambers within the anther known as **pollen sacs,** which may contain as many as 1000 other mother cells. Each microspore mother cell passes through meiosis and produces a group of four haploid microspores (Figure 28.5).

Each of these microspores then develops into a **pollen grain.** A mature pollen grain is surrounded by a series of thick carbohydrate and protein coats, and the details of pollen wall construction vary so widely that it is possible to identify most plants from their pollen grains alone! Within the developing pollen grain, a single round of mitosis produces two nuclei. One of these, the *generative cell,* will produce two *sperm cells* to fertilize the female gamete; the other, the *tube nucleus,* will form the *pollen tube,* a structure that enables the generative nucleus to penetrate the female tissues of the flower. The binucleate pollen grain is the male version of the gametophyte stage of the plant life cycle.

Pollination and Fertilization

Mature pollen grains are released from the anthers as the pollen sacs split open, allowing the loose pollen grains to be scattered on the surface of the anther (Figure 28.6). From the anthers the pollen grains are spread to the stigma by a variety of methods, including insects, birds, and air currents. If the pollen grains are delayed in finding their way to the stigma, their thick coating allows the cells within the pollen grains to survive many days of dry weather and high temperature, and their ability to sur-

vive outside the sporophyte for prolonged periods is one of the key elements of the independence of seed plants from water. However, pollen cells cannot survive indefinitely, and in many species, successful pollination is a matter of delicate timing between the release of pollen and the development of the female gamete.

Initial events: the growth of the pollen tube The surface of the stigma is covered with a moist, sugar-containing secretion that activates and nourishes the pollen cells that fall upon it. The cells within the pollen grain break through the wall of the grain and form a pollen tube, which begins to grow into the tissue of the stigma and down through the style (Figure 28.5). The generative nucleus divides to form two sperm nuclei, and these two nuclei follow the tube nucleus down the stigma toward the ovule. The three nuclei of the growing pollen tube represent the complete microgametophyte stage of the angiosperm cycle. The pollen tube passes through the micropyle to enter the ovule through one of the synergids, where the two sperm nuclei are released from the pollen tube.

True fertilization occurs when one of these nuclei fuses with the *egg cell* to produce a diploid zygote, the first cell of a new sporophyte plant. Strictly speaking, the female gametophyte produces a single gamete cell, the egg cell of the embryo sac (megagametophyte), and the male gametophyte produces two gametes, the sperm nuclei of the pollen tube (microgametophyte).

Double fertilization—embryo and endosperm In plants and animals that reproduce sexually, the fusion of male and female gametes produces a diploid zygote that then develops into a new organism. This is the case in flowering plants. However, flowering plants also undergo a unique *double fertilization* event in which the second sperm nucleus of the pollen tube fuses with the binucleate endosperm mother cell to form a **triploid (3n) endosperm** cell (Figure 28.5). The formation of endosperm tissue is unique to the flowering plants, and it is one of the most important evolutionary novelties of the angiosperms. The diploid zygote will develop into an embryo that will produce a mature plant. The triploid endosperm cell will develop into a rich, nutrient-laden, living endosperm tissue that will nourish the embryo during its early growth.

Embryonic Development

Immediately after the double fertilization event produces diploid and triploid nuclei within the embryo sac, the process of seed formation begins. Although seeds are complex structures, each of the major tissues of the seed is formed from cells that are already present in the ovary at the time of fertilization. The *zygote* develops into an

Reproductive Tissues of the Flower

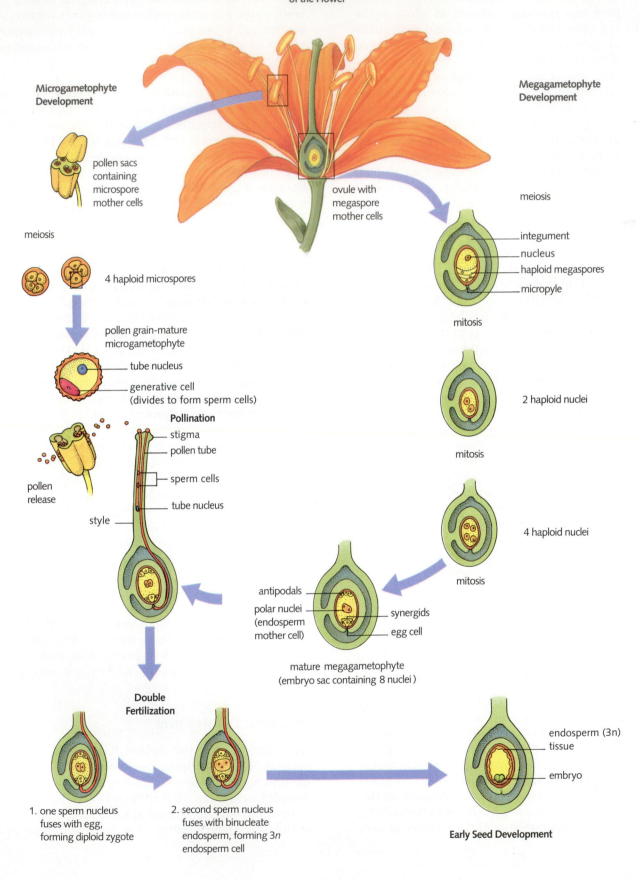

Microgametophyte Development

pollen sacs containing microspore mother cells

meiosis

4 haploid microspores

pollen grain-mature microgametophyte

tube nucleus

generative cell (divides to form sperm cells)

Pollination

stigma

pollen tube

sperm cells

tube nucleus

pollen release

style

Double Fertilization

1. one sperm nucleus fuses with egg, forming diploid zygote

2. second sperm nucleus fuses with binucleate endosperm, forming 3n endosperm cell

ovule with megaspore mother cells

Megagametophyte Development

meiosis

integument

nucleus

haploid megaspores

micropyle

mitosis

2 haploid nuclei

mitosis

4 haploid nuclei

mitosis

antipodals

polar nuclei (endosperm mother cell)

synergids

egg cell

mature megagametophyte (embryo sac containing 8 nuclei)

endosperm (3n) tissue

embryo

Early Seed Development

◄ **Figure 28.5** Details of reproduction in a representative angiosperm. Pollen develops within pollen sacs located at the tips of the anthers. The embryo sac develops within the ovule at the base of the carpel. A double fertilization event occurs in angiosperms, producing a diploid zygote and a triploid endosperm mother cell. Both cells form critical tissues in the seed.

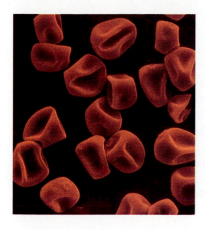

Figure 28.6 Pollen grains of timothy, a common grass. The protective coating around pollen grains enables them to survive until they make contact with a receptive flower.

embryo, the *endosperm mother cell* produces nutrient-storing *endosperm,* the wall of the *ovary* develops into *fruit,* and the *integument* that surrounded the nucellus and embryo sac develops into a *seed coat.*

The first cell division of the zygote produces two cells. One develops into the embryo itself, and the other forms a structure known as the *suspensor,* which attaches the embryo to the *micropyle.* As mitosis continues and the embryo increases in size, the primary tissues of the plant are produced, and recognizable vascular tissue appears as the embryo begins to take shape. Gradually the first leaves of the plant, its "seed leaves" or cotyledons, appear. You will recall that the two major groups of the flowering plants, the *monocots* and *dicots,* are distinguished on the basis of the number of cotyledons that appear in the seed. Dicot embryos, which produce two cotyledons, quickly assume a heart-shaped appearance as the cotyledons emerge from the embryo (Figure 28.7). The apical meristem takes shape between the cotyledons, and the enlarging embryo may gradually bend to fit within the confines of the seed coat. Monocot embryos assume an elongated, cigar-like shape that may also become bent to fit within the seed.

Figure 28.7 The formation of embryonic tissues in dicots (top) and monocots (bottom). The basic pattern of seed leaves is established early in development, and is obvious well before seed development is complete.

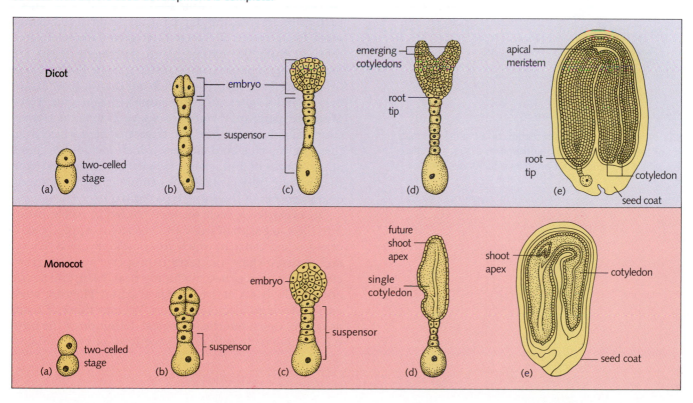

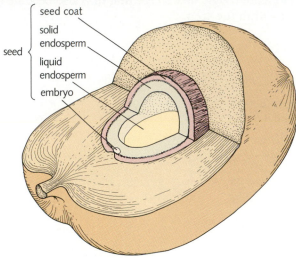

seed {
  seed coat
  solid endosperm
  liquid endosperm
  embryo
}

Figure 28.8 Coconuts and a cross section of a typical coconut. The "milk" of the coconut is liquid endosperm tissue. The thick protective coating of the coconut enables it to survive harsh treatment.

As the tissues of the embryo mature, nutrients from the plant flow into the developing seed, filling the cotyledons and the endosperm with proteins, carbohydrates, and lipids that will be available for the growth of the young plant. When embryonic development is complete, water is removed from the seed until the remaining moisture constitutes less than 10 percent of the seed mass. The coating of the desiccated seed hardens, and it is ready for release into a potentially hostile environment (Figure 28.8).

Fruits

Angiosperm seeds are not borne exposed on the surface of the reproductive tissues, as gymnosperm seeds are, but are enclosed within the protective tissues of the ovary. As the seed matures, the ovary walls surrounding the seed go through a series of transformations until they ripen to form a fruit. Fruits differ widely from one species to another, and the size, type, and structure of the fruit surrounding a seed play a key role in the process of *seed dispersal.*

Fruit types **Simple fruits** are those produced from a single carpel or from several carpels that fuse as the fruit is formed. The ovary wall surrounding a simple fruit may be fleshy, like that of the grape or the tomato, or tough and dry, like a bean pod. In some fruits, such as the peach or cherry, the inner wall of the ovary is rigidly attached to the surface of the seed (Figure 28.9). In others, such as the maple, the dry fruit forms an aerodynamic surface that helps the seed float gracefully when it is released from the parent plant. **Aggregate fruits** are formed from multiple ovaries within a single flower. These ovaries are held tightly together but do not fuse. Strawberries and raspberries are examples of aggregate fruits. **Multiple fruits** are formed from several separate flowers clustered tightly together. The fruits that are formed from these tightly clustered flowers, which include the pineapple, consist of distinct, individual sections derived from the separate flowers. Table 28.1 summarizes the defining characteristics of the three types of fruits.

In everyday speech we tend to divide edible plants into the categories of *fruits* and *vegetables.* Apples, pears, oranges, and bananas, which are commonly called "fruits," are all cases where the term is properly applied. However, many "vegetables," including peas, corn, and beans, are actually fruits in the biological sense: They are seeds enclosed within a ripened ovary. We may think of fruits as tasting sweet, but the biological definition of the term has nothing to do with taste, sweetness, or even edibility.

Figure 28.9 The peach is a simple fruit. Aggregate fruits, like the raspberry, are produced by multiple ovaries in a single flower. Multiple fruits, including the pineapple, are formed from clusters of separate flowers.

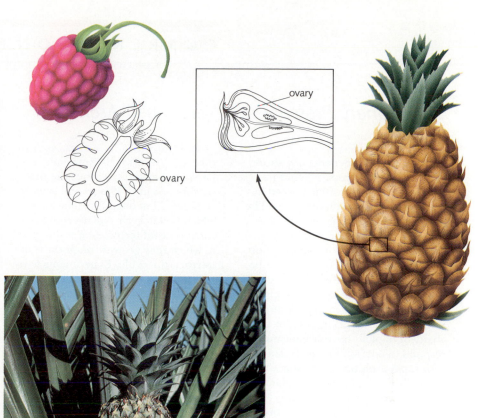

Table 28.1 *Types of Angiosperm Fruits*

Fruit Type	Definition	Examples
Simple	Single ovary or fused ovaries from a single flower	Pea, bean, cherry, grape, banana, peach
Aggregate	Many unfused ovaries from a single flower	Strawberry, raspberry
Multiple	Separate ovaries from adjacent flowers	Fig, pineapple

Evolutionary Pressure in a Very Small Arena

As all students of genetics are aware, the presence of recessive genes in a diploid organism is often masked by the presence of dominant genes. The effect of dominance is presumed to slow down the rate of evolutionary change, because a recessive mutation, no matter how beneficial it might be, cannot be "tried" in the phenotype when it is present in a single copy. This is not the case in haploid organisms, which accounts in part for the ability of haploid microorganisms to respond to environmental pressures with rapid evolutionary change. We normally think of flowering plants as diploid, and we have repeatedly emphasized that the diploid sporophyte phase of the plant life cycle is completely dominant over the haploid gametophyte phase. But is it possible that we are overlooking something in dismissing the gametophyte stage as playing a role in plant evolution?

David and Gabriella Mulcahy, at the University of Massachusetts, believe that we are. Concentrating on the pollen tube, they have examined the conditions under which the male gametophyte grows through the female tissues of the flower. Two conditions are necessary for the haploid gametophyte to be subject to the same kind of rapid genetic change that affects haploid microorganisms: (1) The characteristics of the pollen tube must be partly determined by genes expressed in the haploid stage. (2) The fertilization event must be competitive enough to allow natural selection to favor increased fitness in pollen. These researchers are now convinced that both conditions exist.

Sensitive experiments have shown that key enzymes active in the pollen tube are coded for by genes in the haploid gametophyte; thus the first condition is met. The second is more subjective, but one of the key ways in which pollen grains might compete for successful fertilization is the speed at which they grow through the style to reach the ovule. Experiments have revealed that the rate of pollen tube growth is heritable; rapid growth is a selective characteristic that can be passed on to the next generation. This satisfies the second condition. Flowering plants may face natural selection not just in the diploid sporophyte stage, but in the haploid gametophyte stage as well. Pollination in gymnosperms does not present the same opportunities for competition. The pollen grain is deposited much closer to the developing egg than angiosperm pollen, and it does not have to penetrate the closed carpel of an angiosperm flower.

The implications of this analysis may be significant. One of the most important events of recent natural history has been the rise of the angiosperms, which came to dominate the land in a remarkably short period of time. How did they do it?

The Mulcahys suggest that pollen competition became a "laboratory" in which genetic novelties could be tried out in haploid systems. By acquiring the ability to evolve as haploid systems, flowering plants took hold of a rapid evolutionary engine that enabled them to produce new species and novel genes at such a rate that other organisms could not compete. There is no proof for such a theory, but it does shed a new light on the quiet, invisible growth of tiny pollen tubes. The idea that an evolutionary dynamo is at work within the recesses of each flower is an intriguing concept indeed.

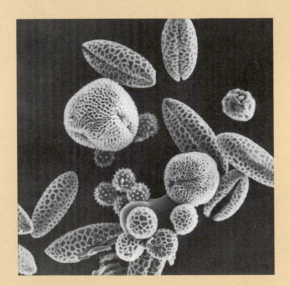

Pollen grains. Because pollen grains are haploid, they represent a potential for the expression of novel genes, which may increase the efficiency of pollination.

Based on D. Mulcahy and G. Mulcahy, Jan.–Feb. 1987, Competition in pollen tubes, *American Scientist*, 75:44.

Figure 28.10 Many plants produce specialized flowers. Corn (left) produces separate male and female flowers. The stamens of the Scarlet passion flower (center) are fused into a single structure. The sunflower (right) is a composite structure, actually consisting of hundreds of smaller flowers.

PLANTS AND ANIMALS—FLOWERS, POLLEN, AND SEEDS

Early in this chapter we stressed the fact that the evolutionary history of flowering plants, as well as their basic structure and organization, reflects a pattern of association with animals. Many plants depend on animal assistance in two of the most important tasks of the reproductive process: pollination and seed dispersal. Evolutionary theory tells us that we cannot expect animals to aid in plant reproduction for altruistic reasons. Evolutionary pressures imply that animal assistance will be available only if plants have developed adaptations that make playing a role in pollination and seed dispersal beneficial to the animals. That is, it must improve their chances for survival, as well as the plants'.

Flower Specialization

The basic flower structure illustrated in Figure 28.5 is the exception rather than the rule for many species of flowering plants. Many plants produce *incomplete flowers* in which either the carpel or the stamens are missing (Figure 28.10). Such flowers are usually described as *staminate* (if only the stamens are present) or *carpellate* (if only the carpel is present). In *monoecious* organisms, staminate and carpellate flowers are found in different portions of the same plant. This is the case in corn, wherein pollen is released from the staminate flowers (the tassels of the plant) to fertilize the carpels (the silk), which then bear aggregate fruits (ears of corn). *Dioecious* organisms are those that produce staminate and carpellate flowers on separate plants, such as the American holly or buckthorn.

Flowers, complete or incomplete, differ from each other in a wide variety of ways. Flowers are classified according to the positions of the stamen and anthers, the number of stamens per bud, the symmetry of petals and sepals surrounding the reproductive structures, and a host of physical characteristics including color, scent, and shape. As we will see, many of these differences can be understood in terms of the process of pollination.

Techniques of Pollination

Wind pollination The simplest technique of pollination is dispersal by the wind. Gymnosperms rely on this technique, and they produce enormous quantities of pollen

Figure 28.11 Silverweed, seen in normal sunlight (left) and ultraviolet light (right), demonstrating the features which serve to guide insects to the center of the flower.

to compensate for the slim chance that any given pollen grain will fall upon a receptive female flower. As you might expect, wind pollination is most effective when a large group of plants from a single species grow side by side, increasing the density of the pollen released and, hence, the chances for success. Corn is a wind-pollinated angiosperm, and gardeners appreciate the fact that a successful corn crop cannot be grown from just a few plants. Corn does well only when a large number of plants are grown side by side in a pattern that maximizes the likelihood of successful wind-aided pollination. Many grasses (corn has evolved from a tropical grass) and a few large trees, including the oak, are wind-pollinated.

Insect pollination Evolutionary biologists have long speculated that the dependence of many plants on insects for pollination may have come about as a natural consequence of adaptations that increased the success of wind pollination. One of the key problems for a wind-pollinated plant is ensuring that the stigma (the tip of the carpel) is sticky enough to catch and hold the pollen grains that chance to fall upon it. These sticky secretions may have been just sweet enough to attract the interest of foraging insects. As the insects traveled from flower to flower to find more of the sticky fluid, they may have inadvertently bumped into the anthers of several flowers, covering their bodies with powdery pollen. In this way, insects may have begun to scatter pollen from one flower to another. As chancy as this process may sound, the fact that insects tend to visit one flower right after another made it much more likely that pollen would arrive by means of their wanderings than via the random stir-rings of the wind. This process would have produced a selective pressure for plants to develop more effective means of luring insects to their flowers. Plants possessing new adaptations ensuring that insects could not visit a flower *without* scattering pollen would have a tremendous advantage in terms of reproductive efficiency—exactly the sort of situation in which natural selection acts most effectively.

The direct spread of pollen from one plant to another by an animal is called **vector pollination** (the insect is the *vector,* the carrying agent). Because hardworking honeybees move quickly from one flower to another, pollen picked up by a bee has a good chance of being deposited directly in another flower. Much less pollen is wasted than in wind-aided dispersal.

Bees are the most important insect pollinators. (See Theory in Action, The Grand Masquerade: Plants as Mimics, page 555.) Worker bees visit flowers repeatedly, gathering sweet secretions known as *nectar* as well as protein-rich pollen. Bees have sharp, well-developed visual systems. They are capable of distinguishing color in the short-wavelength range of the spectrum, meaning that they are particularly sensitive to yellow, green, blue, and violet colors. Their range of sensitivity extends well into the ultraviolet, a region where our own eyes cannot detect light (Figure 28.11). Flowers that attract bees are often strongly scented and brightly colored, enabling bees to locate and visit even widely scattered flowers regularly. The long funnel-like openings of many bee-pollinated flowers ensure that the insect must brush against the anthers to gather nectar and thereby force the animal to scatter pollen on the stigma as it feeds.

The Grand Masquerade: Plants as Mimics

Some plants are able to employ subtle adaptations in flower structure, even to the point where they mimic the flowers of other species. The normal relationship between plant and pollinator is *mutualism*: Each organism benefits, and each organism makes a contribution. But could a plant exploit the insect by luring it without providing nectar as its contribution to the mutual relationship? Ordinarily, one might say no, because that would produce an evolutionary pressure to avoid the nectarless flower (and avoid starving to death!). However, if a species developed that looked enough like a genuine nectar-containing flower, insects might be unable to tell the two species apart.

Such a situation seems to have developed with two common plants in Western Europe—the bellflower (*Campanula*), which produces nectar, and the red orchid (*Cephalanthera rubra*), which does not. Despite differences in shape and color, the reddish orchid and the violet bellflower attract the same species of bees. Each flower is pollinated by the bees, and this first generated confusion among scientists who could not understand why the bees could not distinguish one flower from the other. L. Anders Nilsson of Sweden solved this problem by showing that the flowers do indeed resemble each other in the ultraviolet region of the spectrum. The orchid mimics the bellflower's color in the ultraviolet, where the bee's visual system is the most sensitive. Thus the mimic gets a "free ride" on the mutualistic relationship between the bees and the bellflower.

A few plants have even developed adaptations that mimic means to satisfy drives other than hunger. One genus of orchids (*Ophrys*) has a flower structure that looks like the abdomen of a female bee and releases a fragrance nearly identical to the sexual attractant she produces! When the males emerge in the springtime, they attempt to mate with the flower. Such attempted breedings, known as *pseudocopulation*, do not produce offspring for the bees, but they spread pollen between flowers as male bees engage in one frustrating courtship after another.

Mimicry in plants. The markings of this orchid flower resemble a female bee, attracting male bees in a way that ensures pollination.

Based on Spencer C. H. Barrett, September 1987, Mimicry in plants, *Scientific American*, pp. 76–83.

Figure 28.12 The ruby-throated hummingbird, an active bird pollinator.

Figure 28.13 A cave-dwelling nectar-eating bat, pollinating a banana flower.

There are more than 15,000 known species of bees, and many bee species are specific pollinators of only certain types of plants. For such insects, there is strong evolutionary pressure toward efficient pollination. An adaptation that enables a bee species to pollinate its flowers more efficiently increases the number of offspring of that plant species, thereby increasing the potential food source for the bee species.

Bird pollination Many flowers are specifically adapted to pollination by birds. In North America, the most common bird pollinators are the many members of the hummingbird family (Figure 28.12). The metabolic demands of hummingbirds are so great that they must obtain a steady supply of water and sugary nutrients in order to survive, and bird-pollinated flowers satisfy these needs effectively. The straw-like bills of hummingbirds enable them to drink large amounts of watery nectar from cavities deep within the flower. Meanwhile, anthers brush against their heads and feathers, loading them down with pollen that will be transferred to the next flower on their feeding tour. The storage of nectar in deep, narrow cavities prevents bees from drinking it, ensuring that pollination will be species-specific. Hummingbirds do not have well-developed senses of smell, and therefore bird-pollinated flowers generally do not have the lovely scents that we associate with bee-pollinated flowers. However, birds can see the red end of the spectrum much better than insects, so many bird-pollinated flowers are extremely colorful.

Pollination by other animals Moths and butterflies are also involved in pollination relationships with a series of plants. These flowers attract their insects via a combination of scents and colors not unlike those of bee-pollinated flowers. However, their nectar is generally found at the base of a long, thin tube, where bees cannot reach it. Moths and butterflies insert their long, thin tongues through these tubes and gather the nectar as they hover over the flower.

A few flowers are pollinated by small, pollen-eating bats (Figure 28.13). The poor sense of sight of bats requires that these flowers, which include several species of cactus, be powerfully scented so as to attract the animals when they emerge at night. Flower-feeding bats may lack teeth and have long, brush-like tongues that quickly gather nectar and pollen.

Pollination and self-incompatibility Many plants are capable of self-pollination, and a few have flowers that are structured in a way that makes self-pollination almost unavoidable. In some species, however, there are barriers to self-pollination, even in cases where stamen and carpel are found in the same flower. The most interesting of these are genetic systems that induce **self-incompatibility,** meaning that one plant is unable to pollinate either itself or genetically similar plants. The barriers to self-fertilization are not completely understood, but they seem to act on markers present on the surface of the pollen grain. If the pollen grains carry marker proteins that interact with similar recognition proteins on the stigma, the result is that the entry of the pollen into the flower is blocked. Only pollen grains with different markers are allowed to enter the flower and complete fertilization. Incompatibility systems ensure that the full potential of sexual reproduction, including the ability to produce new combinations of genetic information, is utilized in plant breeding.

Techniques of Seed Dispersal

Seed dispersal raises problems that are similar to those of pollination. In each case, the plant must develop a way to release a small object and increase the odds that that small object will find its way to a distant goal—a stigma of the same species, or a fertile spot to grow—all without the ability to intervene directly itself.

Wind dispersal   The most direct agent of seed dispersal is the wind. Small seeds can be easily scattered in the wind. Large seeds are sometimes encased in a wing-like structure that causes them to spin and twirl, gaining more distance, as they float to the ground. Others, such as dandelions and milkweed seeds, contain feather-like attachments that allow them to float on the slightest breeze, scattering seed and fruit for miles on a moderately windy day (Figure 28.14). Westerners are familiar with the tumbleweed (*Salsola kali*) plant, whose bushes break their attachments to the roots as the plant dies, causing them to tumble along the dry western plains in stiff winds. The plant retains its seeds until it breaks loose; they are then scattered as the dried bushes are blown aimlessly along the prairie.

Water dispersal   Plants that grow in lakes and streams, including the water lily and water hyacinth, produce seeds that trap small air pockets, enabling them to float after they are shed. These floating seeds are carried by currents and surface winds, which disperse the seeds throughout the wet environment. One of the most remarkable seeds is the coconut. This monocot seed contains a *liquid* endosperm layer (the "milk" of the coconut), and it is light enough to float in seawater within its protective coating for many weeks. Volcanic islands in the Pacific, which may form in a matter of days, are quickly

Figure 28.14   Milkweed seeds are dispersed by the wind and can be spread for miles under the right conditions.

Figure 28.15 Fruits play an essential role in seed dispersal. The eastern fox squirrel (below) assists in the dispersal of apple seeds after eating the fruit of the apple. The domestic sheep (right) carries away the burrs of Bidi-bio.

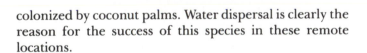

colonized by coconut palms. Water dispersal is clearly the reason for the success of this species in these remote locations.

Dispersal by animals The success of flowering plants in recruiting animals for pollination is mirrored by similar successes in seed dispersal (Figure 28.15). In one sense, these two adaptations are related to the same structure, the flower. Many evolutionary biologists speculate that the development of an ovary, which tightly encloses the female gametophyte, was necessary to prevent the ovule tissue from being eaten by insects in search of nectar. After a successful fertilization, that ovary was available for further modification. As we have seen, ripened ovaries are the fruits that enclose the mature seeds of flowering plants.

When a seed is enclosed within a rich, tasty fruit, it immediately becomes an object of interest to plant-eating animals. Apples, pears, grapes, and cherries are eaten by a wide variety of animals, and although the fruits provide these animals with rich sources of nourishment, the seeds within them pass through the animals' digestive systems unharmed. In many cases, in fact, the coatings of seeds are so tough that they will not allow germination until the seeds have been weakened by digestive enzymes.

Chewing helps prepare for germination the seeds of the common grass timothy, which is widely used as hay for farm animals. Timothy seeds are bundled in a tight cob at the tip of the plant, and the seed surfaces are *scarified* (roughed up) as the hay is eaten. The manure of horses provides a rich source of potential fertilizer for the scarified seeds, and farmers appreciate the fact that spreading such manure on barren fields is an excellent way to plant them with prefertilized timothy.

When fruits are eaten by an animal, the ability of the animal to move guarantees that the seed will be deposited some distance from the parent plant. The bright colors produced by such fruits when they are fully ripe can be seen as "signals" to animals that they are ready to be eaten, helping to make certain that seeds are not dispersed until their development is completed. (See Theory in Action, Is Sex Doomed? Apomixis, page 559.)

Many plants produce seeds that are covered with tiny hairs or barbs, allowing them to stick to feathers or fur. Eventually, animals stop and take the time to remove these burrs from their bodies, but by the time they do, the seeds may have been carried for miles. The saliva that moistens such seeds as the animals pull them out may even help to activate the germination process.

Is Sex Doomed? Apomixis

One of the most widely known flowering plants is *Taraxacum officinale*. This organism thrives in areas of high light intensity, including frequently mowed lawns. It is exceptionally hardy, blooms early in the spring, and produces hundreds of seeds just a few days after blooming. The dandelion, as *Taraxacum officinale* is more commonly known, is one of a handful of plants that have adapted well to twentieth-century suburban life. It has survived even the most vigorous attempts to exterminate it. Could the dandelion be an example of how sexual recombination shuffles genes to achieve the greatest degree of fitness in a population? Not really. Beneath the fields of glistening yellow flowers, the humble, successful dandelion conceals a dark secret: It is one of a handful of plants that have abandoned sex.

For years, evolutionary biologists have been puzzled about why sexual reproduction is so widespread. Although sexual recombination does produce new combinations of genes, the constant and random mixing also makes it more difficult for a beneficial gene to be passed on to the next generation. Asexually reproducing organisms face no such problems; offspring are identical to parents, so beneficial genes take hold in a population quickly. The dandelion seems to have taken a first step along the road to producing identical offspring by abandoning the sexual process.

Although dandelions seem to be typical composite flowers, the pollen they produce is sterile. Furthermore, the female megagametophyte mother cell does not undergo meiosis to produce a haploid gametophyte. Instead, the diploid egg cell begins to develop into an embryo and quickly produces seeds that are genetically identical to the mother plant. The technical name for this process in plants is *apomixis*. Interestingly, although fertilization of the egg does not occur, pollination is still important to development in dandelions: The application of pollen to the carpel seems to serve as a signal that development should proceed. Thus the dandelion has discarded the gametophyte stage of the plant life cycle along with sex. The significance of a "higher" plant going asexual has not been lost on biologists. One might also say that dandelions have discarded our concept of *species* as well. Our definition of this term depends in part on the ability of organisms to interbreed. Because dandelions do not interbreed under any circumstances, they have led biologists to wonder whether the species concept can still be applied to them.

It is possible that dandelions are onto a good thing. They may have discovered that sex is too much trouble, too chancy, too random, and too conservative in the evolutionary sense. They retain the appearance of normal flowers and the need for a pollination signal as evolutionary remnants of processes that they have not yet completely discarded, suggesting that their not-too-distant ancestors at one time reproduced by means of a typical sexual cycle.

Only time will tell whether the dandelions are just an interesting experiment or the harbinger of things to come. Certainly we need not yet conclude that the sexual processes of flowering plants, which have filled the world with such beauty and provided so much inspiration, are about to be pushed off the scene. Just further motivation, perhaps, for those homeowners who resolutely set the lawn mower blades a little lower and cut off every little yellow flower they can reach.

Dandelions in a suburban lawn.

SUMMARY

Flowering plants are the dominant form of plant life on land, primarily because they produce seeds encased within an ovary wall. Although flowering plants can reproduce asexually or sexually, the basic reproductive cycle of flowering plants involves a complex alternation of generations in which a diploid sporophyte plant produces a haploid gametophyte. The male gametophyte, the pollen grain, is formed within organs known as anthers. The female portion of the flower, the carpel, gives rise to the seven-cell female gametophyte, the embryo sac. Pollen released from the anthers falls on the stigma at the tip of the carpel and forms a pollen tube that grows through the female tissue of the carpel to reach the female gametophyte. A double fertilization event in the ovary produces a diploid zygote cell, which produces the new plant embryo, and a triploid cell that forms the food-storing endosperm tissue of the seed. The embryo then develops into a form in which it is prepared for germination, endosperm grows and is packed with stored food material, and the seed is encased in the mature wall of the ovary to form a fruit.

Many flowering plants are dependent on animals for pollination and seed dispersal. Although pollination may occur simply by wind dispersal, pollination by animal vectors, such as bees, birds, and bats, is widespread and of great economic importance. Seed dispersal can occur after an animal has eaten the fruit. As the seeds pass through the animal's digestive system, they are prepared for germination and dispersed by the movement of the animal.

STUDY FOCUS

After studying this chapter, you should be able to:

- Explain the relationship between plant reproductive processes and the life cycles of individual species.
- Trace how the alternation of generations is accomplished in various species, and identify the gametophyte and sporophyte stages of each cycle.
- Describe the details of flower structure, seed formation, and the reproductive cycle in a number of flowering plants.
- Explain the phenomenon of coevolution of plants and animals, especially as it occurred among the flowering plants.

SELECTED TERMS

fruit p. 542
flower p. 542
vegetative reproduction p. 542
sporophyte p. 544
gametophyte p. 544
sepal p. 546
petal p. 546
stamen p. 546
carpel p. 546
pollen p. 546
ovary p. 546
embryo sac p. 547
egg cell p. 547
pollen sac p. 547
pollen grain p. 547
endosperm p. 547
simple fruit p. 550
aggregate fruit p. 550
multiple fruit p. 550
vector pollination p. 554

REVIEW

Discussion Questions

1. How do sexual and asexual reproduction differ? What are the major forms of asexual reproduction in flowering plants?

2. What are the major parts of the flower? What are the differences between complete and incomplete flowers?

3. What are the general features of the angiosperm life cycle? Can pollen grains properly be considered gametes? Why or why not?

4. Sketch the formation of the female gametophyte from a single megaspore cell. Which cell(s) of the gametophyte can properly be considered a gamete and why?

Objective Questions (Answers in Appendix)

5. A gametophyte is
   (a) haploid.
   (b) the plant produced when gametes join.
   (c) produced from zygotes.
   (d) diploid.

6. The female parts of the flower are called the
   (a) sepals.
   (b) petals.
   (c) stamens.
   (d) carpels.

7. Which of the following will develop into the male gametophyte?
   (a) anther
   (b) microspore
   (c) carpel
   (d) ovary

8. In addition to the fusion of male and female gametes, which produces a zygote, flowering plants have a second sperm nucleus that
   (a) fuses with a second haploid female which develops into a second zygote.
   (b) is ejected from the plants once the first fertilization occurs.
   (c) becomes the protective covering of the seed.
   (d) fuses with the diploid endosperm cell to form a triploid endosperm cell.

CHAPTER 29

Plant Growth and Nutrition

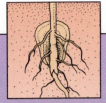

If an alien civilization were to study our planet by means of highly detailed photographs taken from space, it would quickly note that one form of life covers more of the planet's surface than any other. That form of life does not have a nervous system and, in most cases, does not even have the ability to move about. The dominant form of life on land is the flowering plant. All too often we humans suffer from a curious prejudice whereby we fail to consider plants as living organisms. It is all too easy to ascribe the characteristics of animal life to life in general and to forget that in many ways, plants represent the most successful evolutionary experiment of all.

In this chapter we will explore some of the ways in which plants cope with a fundamental need of all living things: to obtain materials for growth and nutrition. Because plants lack a nervous system, we have a tendency to assume they are unable to respond to their environment—after all, that's what a nervous system is for. But we must remember that plants are involved in playing out an entirely different kind of survival strategy than animals are. Because most plants are rooted in place, they cannot search about for nutrients. Plants must obtain nutrients from the soils in which they find themselves. Their success is an important lesson in the diversity of living things.

GERMINATION AND DEVELOPMENT

The seed of a flowering plant consists of an outer coating that encloses and protects a plant embryo arrested at a critical point of development, and endosperm tissue that serves as a storage reservoir of food for the developing plant. In most monocots, the endosperm tissue of the seed remains separate from the embryo and serves as stored food waiting for the embryo to begin rapid growth. In many dicots, however, nearly all of the endosperm is taken up by the seed leaves of the developing embryo, and very little endosperm is left by the time the seed is mature. In such plants the stored food is concentrated in the cotyledons themselves (Figure 29.1).

Dormancy

Most seeds lie dormant for anywhere from a week to a year after they are produced. Cells within the seed have been dehydrated (seed water content is very low: 5–20 percent of the total weight of the seed), and most cellular processes have stopped. Although the cells within a seed are alive, the dehydration that removed nearly all water from the seed has left the cells of the embryo almost in a state of suspended animation. For chemicals to react, they first must come into contact with each other. The drastic reduction that occurs in the amount of free water when a seed develops makes it nearly impossible for chemicals to move around, and it all but stops the normal metabolic processes of the cells in the seed.

Many seeds can survive extended storage. Lotus seeds more than 2000 years old have been sprouted successfully, and seeds of the arctic tundra lupine (*Lupinus arcticus*) recovered in the Yukon, and estimated by radioactive dating to be at least 10,000 years old, have also been

561

Figure 29.1 Seed germination in dicots and monocots. The coleoptile, shown in the diagram of monocot germination, is a protective sheath covering the growing tip of the plant.

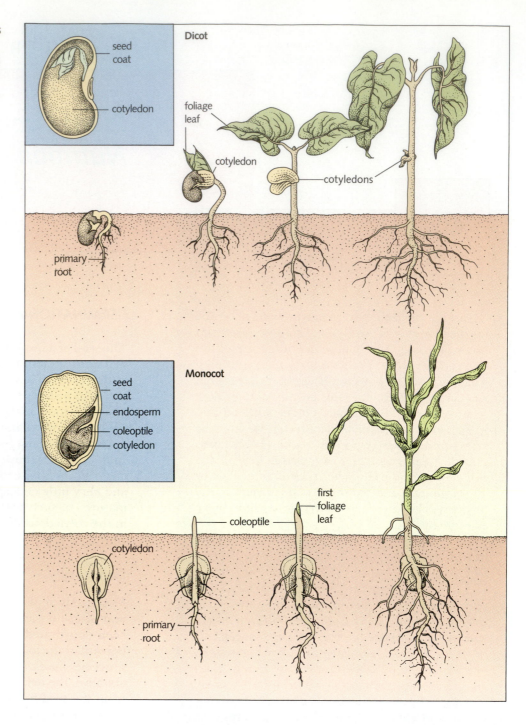

germinated successfully. Despite such remarkable examples, very few seeds can survive indefinitely. Most seeds can last for only a few years without special types of storage, and viable seeds that are more than 100 years old are very rare. The seed is alive, but it cannot last forever in "suspended animation."

In many species, seeds cannot sprout unless they have undergone an extensive dormant period during which normal metabolic processes virtually stop. Other seeds cannot sprout unless they have been nicked or scraped or have had their tough outer coatings damaged in a fire. There are several reasons why seed dormancy is so common. Dormancy aids in the dispersal of seeds, and makes it more likely that the offspring will find a place to develop where they need not compete with the established parent plant. In the temperate regions of the world, extended dormancy can help ensure that a seed sprouts only in the spring. In seeds that are likely to be eaten by animals, a tough outer coating makes it more likely that the seeds will sprout only after the digestive system has weakened their tough outer coating. By that time the seed is likely to have passed out of the animal and to have been deposited in rich manure, which will help to provide nutrients for the early stages of its growth.

Germination

Germination is the reactivation of growth in a plant embryo that leads to the plant's sprouting from its seed. Many factors control germination, and they vary with the species, but three factors are important in the germination of nearly all seeds: water, oxygen, and temperature.

In order to begin using the food stored within it, the seed must absorb a large amount of water, a process known as **imbibition** (literally, "drinking"). The rapid influx of water causes the seed to swell, and the coat of the seed may actually burst under the pressure this swelling creates (Figure 29.2). As cells are activated by the presence of water, they begin to metabolize food stored within the seed, and this creates a demand for oxygen. The seeds require a supply of oxygen, which is plentiful in loose, damp soil. Finally, many species begin germination only at certain temperatures. Seeds from temperate areas of the world often germinate at temperatures as low as 2°C or 3°C, whereas plants from warmer climates may require temperatures no lower than 20°C.

Figure 29.2 Seed imbibition. The lima bean seed at the top has been soaked in water for 12 hours. The rapid intake of water causes the seed to swell and initiates the process of germination.

The Beginning of Growth

The first structure to emerge from the seed is the primary root. The root quickly penetrates the soil and begins to absorb the water the developing embryo needs. As this occurs, the seed is rapidly appropriating and using the food that was stored in endosperm or in the cotyledons. Seeds carry a store of extracellular digestive enzymes that are activated in the early stages of germination and help to break down the complex molecules making up the stored food. The process of digestion makes a brief flood of carbohydrates and amino acids available for plant growth. In grains such as wheat and barley the *aleurone layer,* a single layer of cells surrounding the endosperm, releases these digestive enzymes and makes the food energy stored in the endosperm available for germination. The covering of the seed and the aleurone layer together are known as *bran;* they are removed when white flour is prepared from wheat seeds. The embryo is usually removed as well, although whole wheat flour contains the embryo, which is rich in proteins and oils.

The developing plant quickly establishes directional growth under the soil: Its roots are directed downward, and it begins to send a shoot toward the surface of the ground. We have seen that root apical meristems are protected with a root cap. In monocot grasses, the apical meristem of the first shoot is protected by the coleoptile as it pushes through the soil, and the first vegetative, or *plumular,* leaves, distinct from the cotyledons, do not emerge until after the stem has left the soil. In dicots, a number of different mechanisms are employed to protect the meristem. In some the cotyledons remain closed around the first true leaf until ground is broken. In others the stem bends into a loop that penetrates the soil and drags the top of the stem up into the air after it (see Figure 29.1). Once the growing region of the stem, the meristem, has been moved above the soil, the stem straightens out and the plant begins to grow upward toward the sunlight.

NUTRIENT REQUIREMENTS

Like all plants, a growing seedling requires carbon dioxide and water to support the process of photosynthesis. As we have noted, these two components are available in the atmosphere and the soil, and plants have developed efficient ways of gathering them as they grow. Photosynthesis makes it possible to harness the energy of the sun to produce high-energy carbon compounds, satisfying the major requirements of most plants for a steady source of energy.

Both plants and animals, however, have nutritional needs that carbohydrates alone cannot satisfy. As they grow, all organisms must find the chemicals necessary to produce new proteins and nucleic acids. The synthesis of these compounds requires a source of reduced nitrogen compounds and other chemical constituents. You will recall that many proteins also contain sulfur atoms, and the prosthetic groups of many enzymes include metals such as zinc, iron, and manganese. The chlorophyll molecule contains a magnesium atom. Nucleic acids include phosphorus. Calcium is used within many cells to control the rates of biochemical pathways. These elements are just a few of the essential nutrients that plants must obtain in order to grow and reproduce.

Table 29.1 *Chemical Composition of Typical Crop Plants*

Element	Form in Which Element Is Available	Percent of Dry Weight in Tissue
Carbon	CO_2	45
Oxygen	O_2, H_2O, CO_2	45
Hydrogen	H_2O	46
Nitrogen	NO_3^-, NH_4^+	6
Potassium	K^+	1.5
Calcium	Ca^{2+}	1.0
Magnesium	Mg^{2+}	0.5
Phosphorus	$H_2PO_4^-$, HPO_4^{2-}	0.2, 0.2
Sulfur	SO_4^{2-}	0.1
Chlorine	Cl^-	0.010
Iron	Fe^{3+}, Fe^{2+}	0.010
Boron	H_3BO_3	0.002
Manganese	Mn^{2+}	0.005
Zinc	Zn^{2+}	0.0020
Copper	Cu^+, Cu^{2+}	0.0006
Molybdenum	MO_4^{2-}	0.00001

Plant Chemical Composition

The first step in determining which nutrients are required for plant growth is to analyze the chemical composition of healthy plants. In nonwoody plants, anywhere from 70 to 95 percent of the total weight is water. Most of the remaining dry weight is attributable to organic compounds. Carbon, hydrogen, and oxygen account for 95 percent of dry weight, but 13 other elements are present in amounts large enough to measure.

The average amounts of 16 elements found in significant quantity in typical crop plants are shown in Table 29.1. A list such as this is useful, but it represents only an average. There is tremendous variation within individual plants: Calcium averages about 1.7 percent of dry weight in alfalfa but only 0.4 percent in corn. A white oak leaf contains as much as 10 times the dry weight percent of manganese as does a blade of alfalfa. Furthermore, individual species vary with the type of soil in which they have been raised. Plants grown in the iron-rich soils near mining operations contain much more iron than plants of the same species grown in more typical soils. Finally, the chemical composition of a plant tells us only which elements are present, not which ones are essential for the growth of the plant.

Plant Nutrients

Early in the twentieth century, plant scientists attempted to determine which elements were essential for plant nutrition by growing large numbers of plants in nutrient-poor soil and then watering the plants with solutions containing different combinations of mineral nutrients. This technique identified a number of nutrient elements, including copper, zinc, and manganese, whose importance had not been suspected before that time. More recently, plant biologists have used a liquid growth technique known as **hydroponic culture** to determine a plant's needs for small amounts of nutrients (Figure 29.3). The roots of plants grown this way are immersed in defined nutrient solutions. Because plant root cells need oxygen for respiration, the culture medium must be bubbled with oxygen.

Studies with nutrient cultures have helped to show that a few essential elements must be present in relatively large amounts for plants to thrive. They include oxygen, carbon, hydrogen, nitrogen, potassium, calcium, magnesium, phosphorus, and sulfur. These nine elements, which are known as **macronutrients,** are common in plant tissue. Potassium and sodium, for example, are the major salts found in intracellular and extracellular fluids, respectively, in plants. Calcium is required for the production of cellulose-containing cell walls. The remaining

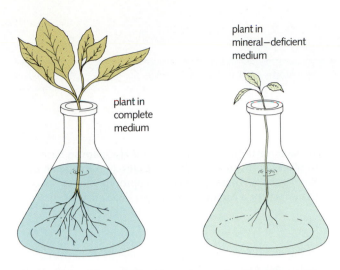

plant in complete medium

plant in mineral—deficient medium

Figure 29.3 The nutritional requirements of plants can be determined by hydroponic culture, growth in a defined liquid medium.

essential nutrient elements are known as **micronutrients.** The roles played by micronutrients, which may be present in plants in amounts less than 10 parts per million of dry weight, are less obvious. The micronutrients include boron, chlorine, copper, iron, manganese, molybdenum, and zinc. Many of these elements are required in key cellular enzymes (manganese is needed for the water-splitting apparatus of photosynthesis), although the actual cellular roles of some of the nutrients remain obscure.

Deficiencies in Mineral Nutrients

It is very rare for any soil to lack a certain nutrient completely. Plants may thrive even in nutrient-poor soils, in part because plant roots contain specific transport systems for many nutrients. One indication of this is the fact that root tissue grown in nutrient solutions may contain 50 to 1000 times the concentrations of sodium, potassium, and calcium found in the nutrient solution itself. Active transport processes move these ions across the cell membranes of root hairs and into the general circulation of water flow into the plant. Remember that in order to enter the xylem, each mineral must first pass through the ring of endodermal cells embedded in the impermeable Casparian strip. This process allows the endodermal cells to control which ions are added to the general circulation of the plant. Cells in other plant tissues are able to concentrate ions by a similar mechanism, ensuring that each cell has an adequate supply of nutrient ions.

However, even the most efficient transport system has its limitations, and in many cases the soil content of

a few nutrients is so low that the plant is unable to concentrate them in its root system. A persistent deficiency of any of the major nutrients eventually causes serious problems for the plant (Figure 29.4). Magnesium deficiencies, for example, limit the ability of the plant to synthesize chlorophyll, because one magnesium atom is needed for each chlorophyll molecule. Magnesium-deficient plants develop **chlorosis**, a yellowing of the leaves, in the older portions of the plants because damaged chlorophyll molecules cannot be replaced quickly enough to maintain the normal green color. By contrast, plants lacking copper synthesize chlorophyll in normal amounts, but they can synthesize neither the electron transport components in mitochondria nor the chloroplasts, which require copper atoms. This places several limitations on their ability to produce and use energy, and copper-deficient plants are severely stunted in growth.

Table 29.2 summarizes the major functions of the essential nutrients in plants and the symptoms that result from a prolonged deficiency of those nutrients.

FULFILLING THE NUTRIENT NEEDS OF PLANTS

Soil and Its Nutrients

Plants obtain nearly all of their nutrients from the soil in which they grow. **Soils** vary widely in their chemical composition as well as in their ability to hold water, and this accounts in part for the great differences in native plant

Figure 29.4 Symptoms of magnesium deficiency in tomato leaves.

Table 29.2 *Essential Nutrients in Plants*

Nutrient	Chemical Form	Need in Plant	Symptoms of Nutrient Deficiency
Macronutrients			
Carbon	CO_2	Organic molecules throughout plant	
Oxygen	O_2, H_2O	Organic and inorganic molecules throughout plant	These three nutrients, CO_2, O_2, and H_2O, are so basic to plant life that major deficiencies result in the death of the organism. Therefore it is not appropriate to speak of the effects of a carbon or hydrogen deficiency.
Hydrogen	H_2O	Organic and inorganic molecules throughout plant	
Nitrogen	NO_3^-, NH_4^+	Found in proteins and nucleic acids	Paling and loss of green color from leaves (chlorosis), which may develop red or purple color
Potassium	K^+	Required for enzyme activities and regulation of cell ionic balance	Chlorosis; brown spots along edges of leaves; general weakening of plant
Calcium	Ca^{2+}	Required for cell wall synthesis, as an enzyme cofactor, and for cell division	Stunting of growth in meristemic (growing) regions; most soils have so much Ca^{2+} that deficiency is rare
Magnesium	Mg^{2+}	Part of the chlorophyll molecule; required for protein synthesis and as an enzyme cofactor	Chlorosis develops between the veins; patches of red or purple color appear on older leaves
Phosphorus	$H_2PO_4^-$, HPO_4^{2-}	Forms part of nucleic acids and phospholipids, ATP, and other energy-storing intermediates	Stunted growth; deep green color develops in older leaves
Sulfur	SO_4^{2-}	Required in many proteins and key compounds	General chlorosis and yellowing of leaves
Micronutrients			
Chlorine	Cl^-	Important in maintaining cellular osmotic balance and in some reactions of photosynthesis	Wilting; stunting of root growth; reduced fruit
Iron	Fe^{3+}, Fe^{2+}	Important part of many enzymes, including the cytochromes; required for chlorophyll synthesis	Chlorosis of younger leaves
Boron	H_3BO_3	Important in nutrient transport within plant	Thickened darker leaves; stunted meristemic growth
Manganese	Mn^{2+}	Required in enzymes of the Krebs cycle and for oxygen release during photosynthesis	Small, speckled dead (necrotic) spots on leaves
Zinc	Zn^{2+}	Required for auxin synthesis and protein synthesis	Dramatically reduced leaf size and chlorosis between veins; reduced flowering
Copper	Cu^+, Cu^{2+}	Enzyme cofactor; required for photosynthetic electron transport	Withered leaf tips and darkened leaf color
Molybdenum	MO_4^{2-}	Required for nitrogen fixation	Chlorosis, including mottling and wilting of leaves

growth in various parts of the world (Figure 29.5). The surface of the earth contains all of the naturally occurring chemical elements. Most of these are locked in the rocks that make up the majority of the earth's crust. The action of wind and water breaks many of these rocks up into smaller particles, and the soluble minerals within the rocks are dissolved by rainwater and washed away. This weathering process gradually produces the small inorganic particles from which soil is formed.

As we saw in Chapter 6, soil contains organic materials as well, and these are formed first from microorganisms, including bacteria, protozoa, fungi, and simple plants. Many of these organisms contribute to the breaking up of solid rocks, as pressures from their growing roots and rhizoids split the largest particles of the crust into smaller and smaller fragments. Gradually, the combination of *physical* and *biological* processes produces a complex soil containing organic and inorganic molecules, as well as a mixture of minerals that may include essential plant nutrients.

Soil particles vary in size and composition. Sandy soils contain quartz (SiO_2) particles that are classified as fine (20–200 μm). (Refer to the Appendix for a discussion of metric units.) Smaller (2–20 μm) quartz particles are known as *silt*; soils containing a mixture of sand, silt, and clay particles are known as **loam.** Soils that are too sandy

are unable to hold water, whereas soils that are mostly clay do not allow water to penetrate. The best loamy soils contain a mixture of sand and clay that allows rainfall to penetrate and then holds much of the water in the moist soil for many days.

Clay particles are covered with negative (−) charges, and this is one of the reasons why clay is able to hold the highly polar water molecule and to bind large amounts of positively charged ions (+). The importance of this property can also be seen in Figure 29.6. Many of the essential elements needed for nutrition are cations (positively charged), and clay particles hold these elements until they are absorbed by root cells.

Soil Fertilization and Management

In some regions of the world, soils are so rich that they contain ideal nutrient mixtures for a wide variety of crop plants. This is true in much of the American Midwest, one of the most productive farming regions of the world. In many places, however, the soil may lack one or two essential nutrients. For thousands of years, farmers have sought to improve the condition of their soils by adding essential nutrients in the form of **fertilizers.** Manure—human and animal—is one of the oldest and most widely

Figure 29.5 Soils vary widely in their ability to support plant life. Rich layers of humus, soil packed with decaying organic matter, are major nutritional sources for plants in many environments.

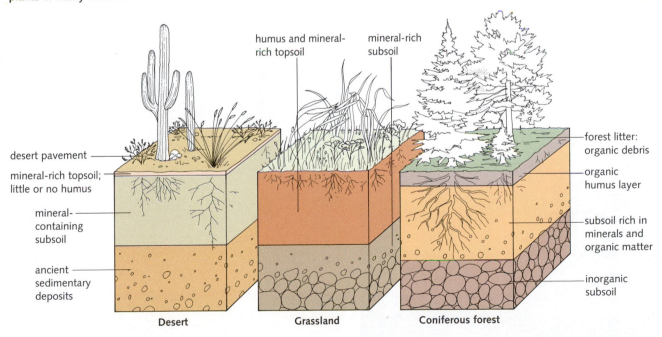

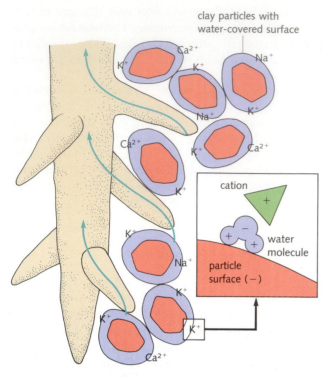

clay particles with
water-covered surface

Ca²⁺

K⁺   K⁺   Na⁺

          Na⁺   K⁺

Ca²⁺   K⁺   Ca²⁺

        K⁺

                    cation
                      +

K⁺   +   water
                − +   molecule
              Na⁺

particle   +
surface (−)

K⁺

    K⁺

          Ca²⁺

Figure 29.6 The surfaces of clay particles hold negative charges, and this enables them to bind positively charged mineral ions as well as water. These properties enable clay-containing soil to bind moisture and nutrients.

Figure 29.7 The proper use (and overuse) of chemical fertilizers. (left) The fertilization of tomatoes contributes to the growth of this valuable cash crop. (right) The overuse of fertilizer may inhibit plant growth and lead to barren, wind-eroded fields.

used forms of fertilizer. Manure is rich in organic and inorganic nutrients. By adding animal manure to heavily farmed fields, it is possible to return to the soil many of the micronutrients and macronutrients that plants remove during their growth process.

Many plants require large quantities of macronutrients. Harvesting these plants may permanently remove macronutrients from the soil and deplete its nutritional value, reducing crop yields in following years. One of the most important food crops, corn, removes large amounts of nitrogen, phosphorus, and potassium from farmland. To use the same land for producing corn year after year requires that the soil be carefully managed and that most nutrients be replaced when the next crop is planted.

One way to replace such nutrients is by using artificial fertilizers (Figure 29.7). Some of the elements in fertilizers are obtained from the mining of mineral-rich rock formations, and others are synthesized by the chemical industry. Many fertilizers used in the United States are particularly rich in the macronutrients nitrogen, phosphorus, and potassium. Commercial fertilizers, including those used for lawn and garden applications, are commonly labeled with the percentage content of these three nutrients. A bag of garden fertilizer labeled "20-10-5" contains 20 percent nitrogen, 10 percent phosphorus, and 5 percent potassium.

These fertilizers are useful in replacing the nutrients that crop plants remove, but they must be used with great care. Overfertilizing, a mistake many backyard gardeners make, can kill crop plants by putting too high a concentration of salts into the soil. The intensive use of fertilizers can also affect the groundwater. When large amounts of nitrogen- and phosphate-containing fertil-

Solar-Powered Marine Animals

The nutritional relationships between plants and animals are so obvious that they scarcely need to be mentioned. Plants are the ultimate food sources for nearly all forms of animal life. In most cases this means that, directly or indirectly, animals eat plants and extract chemical energy from them. As humans, we may believe that the invention of farming represents a fundamental advance, made possible by human intelligence and the development of culture, in harnessing plants for our own uses. As important as farming is, however, we were not the first species to develop it. Corals, for example, have developed associations in which algae are trapped in small chambers within the coral animals. Nutrients released from the algae give the corals much of their food supply and provide much of the energy required to build great coral reefs throughout the world. Does this mean that "farming animals" must lead stationary lives, in the manner of coral, in order to be successful? Not at all.

Consider, for example, the sea slugs, a group of invertebrates known by the more scientific name of nudibranchs. These flattened, shell-less animals are found in warm waters throughout the world feeding on corals, hydroids, and sea anemones. But

as William Rudman, a researcher in Sydney, Australia, has shown, a few nudibranchs have something else going for them: They have developed large colonies of algae that produce their food. One species, the "blue dragon" (*Pteraeolidia ianthina*), is found off the coast of Australia. Early

The blue dragon (*Pteraeolidia ianthina*) gets its color and much of its nourishment from colonies of algae that live within it.

in its life the blue dragon grazes on hydroids that themselves contain colonies of algae. As the blue dragon grows, it gradually develops colonies of its own, possibly by trapping some of the algae from its food in pouches within the nudibranch gut. Algae find points within the gut in which they can remain and thrive, releasing nutrients into the animal's digestive system. The blue dragon is remarkably well adapted to taking advantage of the crop that it grows within itself. Large outgrowths known as *cerata* protrude from its body. These cerata increase the animal's surface area, and thin, algae-filled networks of tubes linked to the animal's gut run just beneath the surface of its skin. Given the photosynthetic resources of a plant, the blue dragon assumes at least some of the structural characteristics of a plant. The delicate clusters of cerata, which furnish nutrients to the animal almost as leaves do to a plant, are arranged so that they seldom shade each other. For its part, the blue dragon makes the most of this association. Once it has acquired a colony of algae, an individual does not seem to return to normal feeding but appears content to bask its nearly transparent body in the sunshine and enjoy the nourishment that comes from within.

Based on William B. Rudman, October 1987, Solar-powered marine animals, *Natural History*, 96:50–53.

izer are used near wetlands and streams, runoff from the fields may contaminate the water with large amounts of these two nutrients. Because the growth of algae in fresh water is often limited by the availability of these nutrients, the sudden increase in nutrients may cause a "bloom"— the sudden increase in the population of algae—that upsets the natural balance of the freshwater ecosystem.

Symbiosis and Plant Nutrition

Mycorrhizae Plants live by their leaves and by their roots. The leaves, of course, are the main reasons why we consider plants autotrophs. Leaves are the major sites of the photosynthetic activity that provides plants with an independent source of energy. Many plants, perhaps as many

Figure 29.8 (left) The roots of many plants are covered with mycorrhizae fungi, which perform many of the functions of root hairs. Mycorrhizae are essential for the growth of several species of plants. (right) This cross section of an orchid root shows the filamentous processes of mycorrhizae fungi inside the cells of the cortex.

 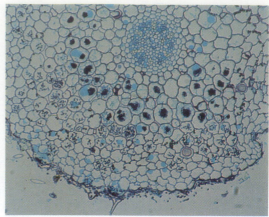

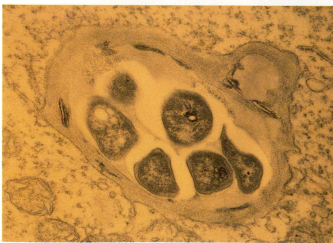

Figure 29.9 (top) Nodules containing nitrogen-fixing bacteria on the roots of peas. (bottom) This thin section through the root system of a pea plant shows portions of six nitrogen-fixing *Rhizobium* bacteria.

as 80% of terrestrial plants, have root systems that are dependent on other organisms. The most widespread example of this is a symbiotic relationship with the fungi known as **mycorrhizae** (*myco* means "fungus," *rhizae* means "roots"; see Figure 29.8).

In many common forest trees, including oaks and pines, these fungi grow in close association with root tips and penetrate the internal tissues of the roots. They take over the functions of root hairs and provide the plant with moisture and mineral nutrients. In a few cases, there is good evidence that the mycorrhizae also manufacture hormones that affect the plant itself and stimulate the growth of new root tissue. Many mushrooms that spring up in shaded areas on the forest floor are merely the reproductive structures of these complex fungi, which cover the roots of the very trees that shade them.

In many types of orchids, the dependence on mycorrhizae has become extreme. Unless their seeds are infected by fungi of the genus *Rhizoctonia*, the orchids cannot germinate. The forest trees that commonly associate with mycorrhizae often cannot be grown in the soil of the plains and prairie, not because the plants themselves have any difficulty with the environment but because their mycorrhizal associates cannot adapt to the soil conditions. The relationship between mycorrhizae and the plants they infect is a perfect example of *mutualism*, the form of symbiosis in which both species benefit. The fungi seem to be more efficient at gathering mineral nutrients from the soil than most plant roots are, and the plant in turn provides the fungus with organic nutrients.

Nitrogen fixation Earlier in the text we considered another important example of symbiosis involving root tissue. The nitrogen-fixing bacteria *Rhizobium* live in nodules in the roots of legumes such as peas and soybeans (Figure 29.9). The roots provide these bacteria with a stable protective environment in which they are supplied with food, and the bacteria in turn use their ability to "fix" nitrogen gas from the atmosphere (N_2) into reduced nitrogen compounds such as ammonia (NH_3). Plants must have reduced nitrogen in order to synthesize proteins, and as we have seen, it often must be provided by adding nitrogen-rich fertilizers to the soil. Plants that can associate with nitrogen-fixing organisms, however, can make their own useful nitrogen fertilizer. Its relationship with these bacteria has made the soybean one of the world's leading protein-producing crops. As you might expect, some of the major research efforts in plant science are directed toward finding ways to produce nitrogen-fixing relationships in such other important crop plants as wheat, corn, and rice. This will not be easy work, but the potential benefits of this research are very great.

Meat Eaters

Though all higher plants are photosynthetic autotrophs, a few species find ways to supplement their diets. Many of these plants are found in nitrogen-poor soil, so the occasional capture of small organisms enables them to obtain badly needed nitrogen compounds. The prey of choice for such plants is small insects that can be lured and trapped by specially modified leaves (Figure 29.10).

The best-known **insectivorous** ("insect eating") plant is the Venus-flytrap, which is native to North America and is found primarily in the Carolinas. The bi-lobed leaves of the Venus-flytrap are an open invitation for flying insects to land and sample the juices that moisten their inner surfaces. When an insect is unfortunate enough to step on the leaves, however, they snap shut in a reaction so fast that the fly is caught inside. The response of the plant is so swift that it almost seems to have a nervous system. Actually, the steps of the fly on the leaves stimulate hair cells that activate a very rapid proton-pumping system in the cells within the leaf. The sudden pH change that the pumping produces causes a rapid movement of water into the tissues where the two lobes of the leaf are joined. As these cells expand with water, they force the jaws of the plant's trap shut. The animal trapped inside is then digested by enzyme-rich solutions secreted from the epidermis of the leaf, and the breakdown products are absorbed directly into the plant.

Few plants can match the mechanical perfection of the Venus-flytrap, but there is more than one way to trap a bug. The pitcher plant, also a North American native, has a large curved leaf covered with tiny inward-pointing bristles. When a fly enters the plant to sip the juices that moisten its inner surface, the direction in which the bristles lie forces the fly to move deeper and deeper. The close quarters of the leaf make it impossible for most insects to fly away, and when they finally reach the bottom of the pitcher-shaped leaf, they are digested in a pool of enzymes and absorbed into the leaf. These plants illustrate not only the remarkable adaptations that plants have evolved but also one of nature's great ironies. Many plants suffer the fate of being eaten by insects, but many others have been able to put insects to use for their own benefit. The most benign example is insect pollination, but the insectivorous plants have added a new chapter to the story: They have invited the bugs to dinner and then served them as the main course.

Figure 29.10 Insectivorous Plants. The Venus-flytrap (left) is about to digest an insect which has become trapped in its clamp-like leaf. The sundew (center) has captured a number of insects in its sticky droplets. The pitcher plant (right) has caught a mosquito in its cup-shaped leaf.

SUMMARY

The seeds of higher plants are structures that are specialized to suit the reproductive patterns of their particular species. In many plants, a pattern of seed dormancy is essential to survival. Dormancy is ensured by the dehydrating of seed tissue, enabling seeds to survive for months or years without apparent changes. Food is stored within the seed, either as endosperm or in the seed leaves of the plant embryo. Germination, the activation of a seed, is often prompted by imbibition. As it begins to grow, the developing plant sends out a shoot toward the surface of the ground as well as a series of roots that push through the soil in search of water and nutrients.

Although they obtain much of their chemical energy from the products of photosynthesis, plants have important nutritional needs in terms of the chemicals, organic and inorganic, that they must obtain from the environment. The exact chemical needs of plants can be determined by several techniques, including hydroponic culture. Deficiencies in any essential nutrients may affect plant growth and cause such conditions as chlorosis, a loss of leaf color. The ability of plants to absorb nutrients is highly dependent on the nature of the soil in which plants grow. The best soils are mixtures of clay and sand that retain some water while allowing excess water to drain. Soils that are lacking in certain nutrients may be supplemented by the addition of fertilizers, although these bring with them their own set of problems and concerns. Many plant nutritional needs are met by close associations with other types of organisms, including the fungi known as mycorrhizae and nodules of nitrogen-fixing bacteria.

STUDY FOCUS

After studying this chapter, you should be able to:

- Cite some basic physiological principles that govern plant behavior, growth, and development.
- Explain the relationship between soil composition and plant growth and development.
- Describe the close relationship between plants and other organisms that provide plants with nutrients and moisture.
- Explain the demands that growing commercial crops places on soil resources, and explore some of the ways in which these resources are replenished.

SELECTED TERMS

germination *p. 563*
imbibition *p. 563*
hydroponic culture *p. 564*
macronutrients *p. 564*
micronutrients *p. 565*

chlorosis *p. 565*
soil *p. 565*
fertilizers *p. 567*
mycorrhizae *p. 569*
insectivorous *p. 571*

REVIEW

Discussion Questions

1. What critical events precede the germination of a seed? Why is dehydration effective in increasing the dormant period of most seeds?

2. What are some of the macronutrients necessary to plant growth? Which of these are provided by the raw materials used for photosynthesis? Which are not?

3. Why might an excess of nitrogen-rich fertilizer cause "bulging" of plant roots (excessive water loss and death of root hair cells)?

4. Carnivorous plants are often able to grow on very poor soil that has serious nutrient deficiencies. How might the ability to trap and digest insects be useful under such conditions?

Objective Questions (Answers in Appendix)

5. A dormant embryo is called a(n)
   (a) endosperm tissue.
   (b) cotyledon.
   (c) seed.
   (d) gametophyte.

6. The type of soil most likely to help plants grow contains
   (a) many negatively charged particles.
   (b) a mixture of sand and silt.
   (c) few air spaces and is tightly packed around the roots.
   (d) mostly silt.

7. Mycorrhizae are
   (a) vascular bundles of phloem.
   (b) plants that have a mutually beneficial association among legumes and nitrogen-fixing bacteria.
   (c) structures on roots that have nitrogen-fixing bacteria.
   (d) fungi that have a mutually beneficial association with a root.

8. A symbiotic nitrogen-fixing relationship is exemplified by
   (a) rhizobium.
   (b) micorrhizae.
   (c) orchids.
   (d) fungi.

9. The folding of the leaf lobes of the Venus-flytrap is a response to changing
   (a) light patterns.
   (b) water pressure.
   (c) wind speed.
   (d) wind direction.

CHAPTER 30

Control of Plant Growth and Development

In many cultures throughout the world, plants are not considered living things. Consciously or subconsciously, we sometimes regard plants as part of the background, the scenery, a landscape that supports life but is distinctly passive. One of the reasons for this belief may be that plants do not seem to display the rapid, coordinated responses to their environment that animals do. If we spend a few quiet hours in a deep forest, our attention at first may be attracted to the great diversity and activity of animal life. It is impossible not to notice the rapid movements of insects, the characteristic songs of birds, or the quiet stealth of mammals as they seek their prey. Everywhere animals seem to possess the ability to respond to their surroundings, quickly and decisively. Behind this activity, the trees and lawn beneath may seem inanimate and unresponsive. Yet just as surely as animals, plants can and do respond to their environment.

If we neglect the ways in which plants react to their surroundings, it may be only because the pace and scale of that response are more subtle than in many animals. As we shall see, this does not make that response any less effective in ensuring the organism's survival, and it does not make it any less important to other organisms, ourselves included. In this chapter we will examine some of the ways in which higher plants produce an organized and coordinated response to their surroundings and examine characteristics that have made these organisms so successful in colonizing the land surface of the planet.

CONTROL OF PLANT GROWTH PATTERNS

When a plant germinates and begins to develop, an obvious need arises to coordinate activities in different parts of the growing organism. The tissue in the root must grow quickly enough to provide moisture, support, and nutrition—but not so quickly that all the resources of the plant are used up in root tissue growth. Similarly, the tissues that produce leaves must grow quickly enough to ensure a rate of photosynthesis that will enable carbohydrate to flow back into the roots. In order to do so, the leaves and stems that carry out photosynthesis must bend toward the sunlight; in this way the plant will find as much solar energy as possible. The plant must respond to the time of day, so that growth and metabolic cycles can be regulated efficiently; it must respond to the time of year, so that reproduction and growth can be completed during the most favorable weather; and it must survive the ravages of predators such as insects that eat its most productive tissues.

In animals, many of these problems are solved by the *nervous system,* a group of tissues and organs specialized to carry messages from one end of an organism to the other. Plants lack nervous systems, yet they perform many of these functions as efficiently as animals.

The Nature of Plant Growth

The growth of a plant is *indeterminate*—that is, there are no absolute restrictions on the precise size a plant will reach or the shape it will assume. This does not mean, however, that plant growth is random. In fact, plant growth is generally subject to a number of controls and restrictions, some of which are so precise that they produce floral and leaf structures identical in all members of a species. These controls are the effects of influences that

Figure 30.1 Although the growth of a plant is indeterminate, it is not random. The growth patterns of trees are often so specific that it is possible to distinguish two species merely by their branching patterns.

Figure 30.2 Geotropism, the ability to respond to the force of gravity, enables plants to orient themselves correctly even in the most difficult terrain.

regulate where and to what extent growth can occur. They are the reason why it is possible to tell, even from a great distance, whether a large tree is an oak or a maple (Figure 30.1). Controls on growth rates and patterns are important to survival because they help mold the shape of the organism to suit important factors in the environment such as light, gravity, wind direction, and moisture.

Responses to the Environment

For thousands of years humans have marveled at the ability of plants to respond to stimuli from the environment. These responses of plants to factors in the environment are known as **tropisms,** a term derived from a Greek word that means "to turn."

Geotropism   Every seedling has the ability to sense and respond to the force of gravity, a property known as **geotropism** (Figure 30.2). Geotropism affects stems and roots differently, directing root growth toward the source of gravitational attraction and stem growth away from it. The differences between root growth and stem growth provide an example of how tropisms can be either positive or negative. Emerging roots exhibit *positive geotropism,* stems *negative geotropism* (they turn against the force of gravity). Besides helping a developing seedling find soil and air, geotropism also helps larger plants respond to major disruptions such as storms. Thanks to geotropism, small trees blown over in wind storms are able to return slowly to a vertical growth pattern.

Phototropism   Plants respond to light, a property known as **phototropism.** When the amount of light hitting one side of a plant is greater than the amount hitting the other side, the plant turns toward the light. If the light source is fixed, the plant gradually orients itself such that its leaves are perpendicular to the angle of illumination. The phototropic response is so quick that young seedlings turn toward a source of light in a matter of just a few hours (Figure 30.3).

Thigmotropism   Plants are also affected by touching solid objects, a response known as **thigmotropism.** Climbing plants, including ivy and pole beans, are able to sense when their growing shoots make contact with a solid object (Figure 30.4). When they do, the rate of growth on the side making contact slows down and the shoot curls around the solid object, attaching to it tightly. This tropism enables these plants to find solid objects and cling to them for support.

Besides these well-defined tropisms, plants can respond to a host of other environmental factors, including the length of the day and the season of the year. Many plants, including such well-known species as the morning glory, open their flowers only during the daytime. Others, such as the cereus cactus of the American Southwest, open only at night. Many plants respond so precisely to seasonal changes that they are said to contain *biological clocks* that adjust biochemical processes going on within them to the length of day and the time of year.

PLANT HORMONES

One of the ways in which plants solve the problems of survival is by using chemical messengers that carry signals from one part of the organism to another (see Theory in Action, Chemical Defense Mechanisms: How Plants Fight Back, p. 576). These messengers ensure that all tissues of the organism can act in concert, enabling even the largest plants to respond directly to changing conditions in the environment. The messengers that perform this task are chemicals known as **hormones** (Figure 30.5). A hormone is a substance that is produced, usually in small quantities, in one part of an organism and affects the physiology of another part. Hormones are accurately described as "chemical messengers." Plant hormones can be produced by several different tissues, including the cells found in leaves, fruits, and seeds.

Figure 30.3 These corn seedlings, bending towards a light source, display positive phototropism.

Figure 30.4 Thigmotropism, the ability to respond to contact, enables the tendrils of this squash plant to obtain support from a solid object.

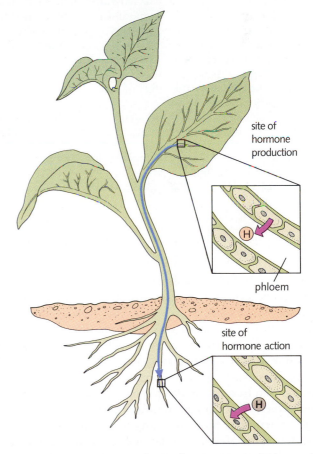

Figure 30.5 Hormones are chemical messengers which may be produced in one region of a plant and affect cells in another region.

Chemical Defense Mechanisms: How Plants Fight Back

Most insects eat plants. The relationship between insects and plants, therefore, would seem to be mostly one-sided. The bugs eat green plants as fast as they can, and the plants, unable to swat or squirm or run away, seem to be unable to do anything to stop them. But plants have evolved a number of weapons that fight back.

Plants produce a whole range of chemicals that are deposited in their tissues seemingly to poison the insects that might otherwise eat them. Plants of the mustard family (Cruciferae) produce a chemical known as *sinigrin*, which contains the toxic compound lylisothiocyanate. By removing this compound from mustard plants (which few insects will even attempt to eat) and infusing it into celery leaves, we can study the effect of the chemical on some insects that normally eat celery plants. One such insect is *Papilo polyxenes,* the black swallowtail butterfly. Larvae of the butterfly that are placed on treated celery leaves grow very slowly when the dose of the chemical is low and are killed outright when the dose is raised to the level found in mustard.

Some scientists have suggested that chemicals that are produced by one species and are able to affect the growth, development, or population levels of another species should be called *allochemicals*. Sinigrin is just one example of a toxic allochemical. There are many more. But plants can be subtle in their defenses as well, and in many cases they can achieve their ends without directly poisoning their insect predators. Insects generally go through several distinct cycles of growth. Most butterflies, for example, hatch first to a series of larval stages that do most of the feeding; the final larval stage forms a pupa; and the pupa develops into the mature butterfly. Each stage of the cycle is triggered by *hormones,* chemical messengers released from special organs within the insect's body.

Recently, Isao Kubo and other scientists were investigating the effects of a devastating plague of locusts in Kenya when they discovered that one plant species, a bugleweed *(Ajuga remota)* had survived the devastation. To figure out why, they fed extracts of the plant to a number of insects. When the time came for the insects to develop into pupae, they all grew several extra head capsules, blocking the development of their mouthparts; they starved and died. The plant extract caused this by producing an allochemical that was a slightly modified version of the insect's own development hormones. The plants had sabotaged the inner workings of their predators!

Based on Gerald A. Rosenthal, January 1986, The chemical defenses of higher plants, *Scientific American,* pp. 94–99.

Plants are able to use hormones to regulate their rate and direction of growth, to control the time at which they produce flowers and drop leaves, and even to coordinate the functions associated with germination. A hormone does not necessarily affect every cell of an organism in the same way. In fact, many cells cannot respond to a hormone message at all. In order to respond to the message carried by a particular hormone, a cell must contain a **receptor** for that hormone. Receptors are molecules to which hormones bind, forming a *receptor–hormone complex* that then affects cellular metabolism. Cells cannot respond to a hormone unless they contain the proper receptor. Those cells that do contain the receptor are known as *target cells,* and it is to such cells that the hormonal message is directed. The nature of the response depends on the amount of hormone that reaches the target cell, and it may also be influenced by the presence of other hormones that affect the same cell.

Auxin

In the 1880s Charles Darwin and his son Francis carried out some simple experiments on phototropism. They were interested in understanding why growing plants seemed to bend toward the light and in whether there was a spe-

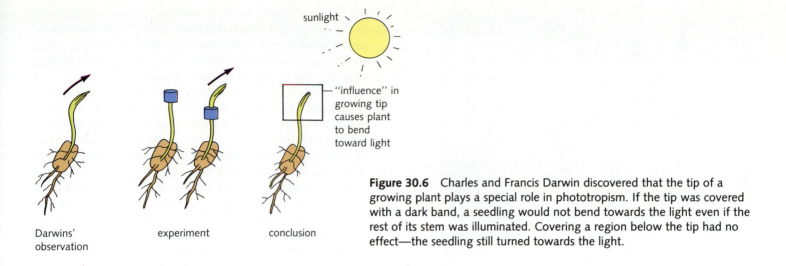

sunlight

"influence" in growing tip causes plant to bend toward light

Darwins' observation

experiment

conclusion

Figure 30.6 Charles and Francis Darwin discovered that the tip of a growing plant plays a special role in phototropism. If the tip was covered with a dark band, a seedling would not bend towards the light even if the rest of its stem was illuminated. Covering a region below the tip had no effect—the seedling still turned towards the light.

cific cause for phototropic growth. The Darwins' experiment showed that the tip of the growing plant was the key to phototropic behavior (Figure 30.6). If the tip of the plant was covered, the plant did not bend toward the light even if the rest of the stem was illuminated. Covering a region below the tip had no effect. Could the tip be releasing a substance (the Darwins called it an "influence") that retarded growth on the illuminated side and accelerated growth on the side away from the light?

In the late 1920s Frits Went showed that the "influence" was a diffusable chemical. His first hunch was that the *coleoptiles* (thin sheaths that cover the tips of rapidly growing grasses) might contain a chemical influence that affected plant growth. To test this hypothesis, he stripped the tips off growing oat seedlings and set them on small blocks of agar to allow fluid from the tips to diffuse into the agar (Figure 30.7). Then he placed one of the agar blocks next to the tip of a seedling whose own tip had been cut off and placed the seedling in the dark. Within a few hours, the growing seedling began to bend away from the agar block as if responding to a light source. The agar was carrying a growth-promoting substance that it had picked up from the oat seedlings. Went named this substance **auxin,** a term derived from a Greek word meaning "to increase," because auxin increased the rate of plant growth. Before long, Went's auxin was identified as *indoleacetic acid* (abbreviated IAA).

Auxin acts on different receptors at different concentrations to induce several different effects. We will examine some of the more important.

Auxin promotes cell growth. It is synthesized in the apical bud, near the tip of the growing seedling, and transported

Figure 30.7 An experiment performed by Frits Went on the nature of the phototropic response. Coleoptile tips were placed on agar blocks so that material from the tips could diffuse into the blocks. When such a block was then placed next to a growing shoot, the shoot bent away from the side touching the block, suggesting that the block had absorbed a growth-promoting substance from the coleoptile tip.

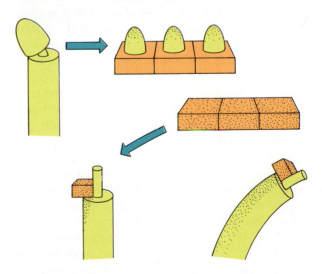

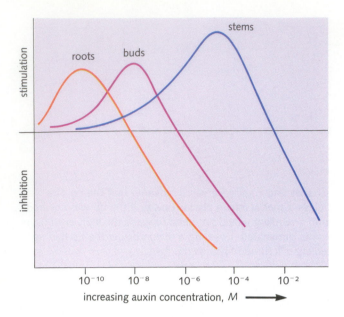

Figure 30.8 The effects of different auxin concentrations on growth in roots, buds, and stems. Note that the same concentration of auxins may stimulate stem growth and inhibit root growth.

down the stem. As Went's experiments showed, a difference in light intensity across the stem results in an unequal distribution of auxin. The accumulation of auxin on the side of the stem away from the light causes the cells on that side to elongate more quickly than those on the side the light shines on. This rapid elongation causes the stem to bend toward the light: the response we recognize as phototropism. We are not sure exactly how the difference in auxin concentration comes about, but it is clear that the apical meristem itself is the source of the difference, as shown by the Darwins' original experiment. The actual effects of auxin are subtle, and they depend on the concentration of the hormone as well as on the tissue responding to it. As shown in Figure 30.8, the same concentration of auxin has different effects on roots, buds, and stems, and very high concentrations of auxin can actually inhibit growth.

Auxin is also the hormone that produces geotropism. When a plant is tilted on its side, auxin accumulates at high concentrations in the bottom side of the main stem. As a result, cells on this side begin to grow more quickly, righting the plant and allowing it to grow in a normal upward pattern (Figure 30.9). Incidentally, auxin does not accumulate on the lower side of the stem because gravity "pulls" it down. (Auxin is a small molecule, and

the force of gravity is not great enough to concentrate it in the lower part of a horizontal stem.) As in the case of phototropism, we still have not determined by what mechanism the plant produces the unequal distribution of auxin that accounts for geotropism.

The effect of auxin on roots provides an interesting extension of this story. Note in Figure 30.8 that the same concentrations of auxin that *stimulate* growth in stems *inhibit* growth in roots. Roots exhibit a behavior exactly opposite to that of stems. If placed horizontally, roots turn away from light and downward toward the force of gravity. Measurements of auxin concentration, however, show that it gathers on the *same* sides of both roots and stems: toward the force of gravity and away from the light. Auxin in the roots is released from the root apical meristem. The pattern of hormone release is the same in root and stem, but the patterns of *target cell response* differ (Figure 30.10).

High concentrations of auxin inhibit the growth of lateral buds. For quite a long time, gardeners have known that the growing tip of a plant inhibits the development of lateral branches but that when the tip is cut off, rapid lateral branching occurs. Horticulturalists call this effect **apical dominance:** The tip, or apex, of the branch tends to dominate the growth pattern (Figure 30.11). If the tip of a growing branch is cut off, lateral buds begin to grow rapidly.

Figure 30.9 Germinating corn seeds show the effects of positive geotropism as they orient root growth in the direction of gravity.

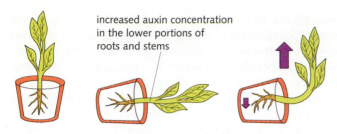

increased auxin concentration in the lower portions of roots and stems

Figure 30.10 The effects of geotropism are different on roots and stems. This is due to the differing sensitivities of roots and stems towards auxin.

A carefully controlled concentration of auxin can promote root growth. This discovery is exploited when cuttings are used to root new plants. When a little bit of auxin is carefully applied to a plant cutting, it promotes root formation and increases the chances that the cutting will be successful. Commercially available rooting and propagation products enable home gardeners to take advantage of this effect.

More than 200 different compounds have been discovered that mimic auxin's effects. Molecules that are chemically similar to IAA have many of the same effects as auxin. Some of these are much stronger than IAA itself, and the most potent compounds affect plants in an unexpected way. Because high concentrations of auxin can inhibit growth, the most potent synthetic auxin-like molecules are *toxic* to fast-growing plants. These compounds have been used as weed killers, or **herbicides,** since the late 1940s. Herbicides are very useful in agriculture, because killing off plants that compete with a desired crop can increase crop yields. One of the most widely used herbicides is a compound known as 2,4-D (2,4-dichlorophenoxyacetic acid). Because 2,4-D is more effective against dicots than against monocots, farmers can spray high levels of the herbicide on monocot crops such as corn and, without badly damaging the corn, eliminate a fair share of the dicot weeds that would otherwise compete with the corn seedlings.

But herbicides are not without their dangers. A mixture of 2,4-D and another auxin-like chemical was used as *agent orange,* a chemical defoliant widely sprayed in Vietnam to deny enemy troops the use of jungle foliage as cover (Figure 30.12). One of the chemical by-products of synthesizing these compounds is dioxin, and dioxin was present in trace amounts in agent orange. Dioxin is a potent carcinogen (cancer-causing chemical), and its presence in agent orange has been a continuing concern

Figure 30.11 Apical dominance controls the shape of these spruce trees, limiting the rate of growth in branches closest to the apical meristem.

Figure 30.12 Agent orange, the chemical defoliant used in Vietnam, contained a mixture of chemicals mimicking the effects of IAA (indolacetic acid).

to the Vietnamese population as well as to American servicemen who fought in Vietnam.

Table 30.1 summarizes the functions, origins, and targets of auxin and the other plant hormones we consider here.

Gibberellin

For many years, rice farmers in Japan had been aware of a disease that weakened their crops by causing the rice plants to grow unusually tall. The giant plants were fragile and often failed to produce rice. They called the disease the "foolish seedling" disease. Eiichi Kurosawa, a Japanese biologist, showed that the rapid growth was due to a fungus called *Gibberella* that grew on the plants. The fungus produced a soluble compound that Kurosawa named **gibberellin.** Gibberellins are actually a family of about 60 closely related compounds. They are similar in chemical structure to the steroid lipids we described in Chapter 20. Gibberellins are also made by higher plants (including rice), and the effect of the fungus was due to its ability to synthesize a chemical that mimics the actions of a class of naturally occurring plant hormones.

Gibberellin seems to regulate the rate at which internode regions of plant stems elongate. Applying gibberellin to a growing plant can result in a rapid increase in the length of stems between branch nodes (Figure 30.13). Gibberellin can cause dramatic increases in size among dwarf plants, often allowing them to reach near-normal sizes. Gibberellins are also important in seed germination in some plants. In barley, for example, as the seed takes up water, special tissues release gibberellin, which diffuses throughout the seed. A target tissue in the seed responds to the gibberellin message by releasing proteolytic enzymes, which begin to break down the stored endosperm tissue in the seed—something the embryo requires for a fast start on growth. Gibberellins, therefore, can cause sprouting in barley. The brewers of beer have noticed this: Barley seeds are sprayed with gibberellin so that they will all sprout at once to produce the uniform barley malt that is the basis of the best beers.

Cytokinins

Cytokinins are plant hormones that were discovered in attempts to improve the growth of plant cells in culture. Carlos Miller, a researcher in the laboratory of Folke Skoog at the University of Wisconsin, was testing the ability of nucleotide-containing mixtures to stimulate growth when he took a bit of herring sperm DNA off the shelf and added it to his cultures. It caused a rapid increase in the rate of plant cell division, but pure DNA had no such effect. Instead, Miller and Skoog showed that one of the breakdown products of DNA was a potent plant hormone. The compound these investigators discovered belonged to a class now known as cytokinins (because of their effect on cell division). Their chemistry is similar to that of *adenine,* one of the bases in DNA.

Cytokinins seem to complement many of the effects of auxin. Cytokinin applied to a lateral bud causes the bud to grow,

Table 30.1 *Major Classes of Plant Hormones and Their Effects on Plant Cells and Tissues*

Hormone	Functions	Origin	Target
Auxins (including IAA)	Stimulate cell growth and stem and root elongation; control geotropism and phototropism	Meristems in roots and shoots	Cells in roots, stems, and leaves
Gibberellins	Stimulate stem elongation and bud and fruit development	Chloroplasts in leaf tissue	Stem cells
Cytokinins	Stimulate cell division and growth	Several plant tissues	Growing tissue in roots, stems, and leaves
Ethylene	Stimulates ripening of fruit; affects (+ or −) development of leaves and fruit	Root tissues; aging leaves	Fruits and flowers
Abscisic acid	Inhibits growth, promotes abscission of leaves in autumn	Leaves	Stem tissue and buds

Figure 30.13 Gibberellin dramatically increased the rate of leaf elongation in the bird's nest fern at the right.

even in the presence of a high concentration of auxin (which would normally inhibit growth of the bud because of apical dominance). Recent experiments show that the ratio between auxin and cytokinin concentrations determines the rate of tissue growth. This is one of the main reasons why the apex itself is not subject to the inhibitory influences of auxin. When applied to clumps of cells in culture, cytokinins tend to cause the development of stems, whereas auxin promotes the growth of roots. When the balance between the two is just right, a mixture of roots and stems is formed. Xylem sap contains cytokinins that are manufactured in the roots and transported upward to the rest of the plant. The balance between cytokinins and auxin seems to be one of the major elements that controls the shape and growth of a plant. This precise control of cellular behavior by the interplay of different hormones is something we will see again when we discuss animal hormones in Chapter 36.

Abscisic Acid

Although many plant hormones seem to be geared toward the stimulation of plant growth, there are cases in which it is important for a plant to limit or restrict its rate of growth. Plants that live in temperate areas of the world must prepare for a winter season in which temperatures are low and light levels reduced. Their survival requires that they conserve material resources gained during the warm seasons of the year and protect them against the ravages of winter weather. **Abscisic acid,** a hormone produced in a variety of plant tissues, has a striking ability to inhibit growth. Abscisic acid is released from cells within a limb bud in early fall. It slows growth in the meristematic tissues of the bud and promotes the development of thick, scaly coverings that will serve as protection for the bud throughout the winter.

In woody tissues, abscisic acid slows down the rates of both primary and secondary growth, preparing the plant for a reduction in the amount of photosynthetic product that occurs as the leaves are shed. Abscisic acid also plays a role in the arresting of cell growth that occurs during seed formation. Although the seeds of many plants are formed by rapid growth of the embryo and ovary tissue, once the seed is mature, abscisic acid slows down the rate of cell growth and prepares the seed for the extended dormant period that may be required for germination.

Ethylene

In the nineteenth century the major source of indoor lighting was illuminating gas. Use of the gas seemed to have some important effects on indoor plants: Growth was stunted, stems swelled, and leaves fell off. Fruit on the indoor plants began to ripen much more quickly than normal. Gradually the effect was traced not to the gas itself but to one of its minor components, **ethylene.** Ethylene is produced by plants themselves, and ethylene production is under the control of auxin. In fact, many of the effects of auxin are actually brought about by the ethylene that auxin causes to be produced. Ethylene is

Figure 30.14 Tomatoes are generally harvested while still green, and then ripened by the application of ethylene gas.

Figure 30.15 Photomorphogenesis, the influence of light on plant form, is dramatically illustrated in this photograph of bean seedlings raised in darkness (left) and light (right).

produced as a chemical trigger during fruit formation, and this explains its ability to control the ripening process.

Commercial producers of fruit exploit these properties. Many commercial crops, including lemons and tomatoes, are generally picked unripe and then packed for shipment under conditions wherein circulating air removes any ethylene the cells of the fruit release. This prevents the fruit from ripening during shipment. Then, just before delivery to stores, the fruit is treated with synthetic ethylene and ripens rapidly (Figure 30.14). Although this treatment has commercial value, it doesn't always duplicate the taste and quality of vine-ripened fruit, in part because light absorption also plays a role in natural ripening for which ethylene does not compensate.

PHYTOCHROME: THE REGULATION OF PLANT GROWTH

The major events in the life of a plant include germination, flowering and reproduction, and senescence. In nature these events are coordinated with the seasons of the year. Given the central role that *light* plays in the life of a plant, it would be surprising if light did not play a major role in regulating the timing of these events in plant development. The fact that all plants, tropical or temperate, experience daily cycles of light and darkness provides a powerful physiological reason why plants regulate their own cycles of activity to make the best possible use of sunlight hours. Not surprisingly, most plants have developed mechanisms by which they can respond to the length of the day as well.

The Discovery of Phytochrome

Light affects the *form* of a plant as well as the direction of its growth. If a bean seedling is allowed to sprout in darkness, its stem grows very quickly and its leaves remain folded and undeveloped. When the seedling is later exposed to light, it rapidly unfolds and enlarges its leaves and slows down its rate of stem elongation to assume a more normal appearance (Figure 30.15). Light actually serves two purposes in this situation. First, light is required for the conversion of chemical intermediates to chlorophyll. Second, light causes a change in the shape and growth pattern of the plant, a process known as **photomorphogenesis.** (In animals the growth and shaping of body parts may have nothing to do with exposure to light and are called *morphogenesis*.)

Photomorphogenesis is not caused by photosynthesis. In 1954 two scientists, Hendricks and Borthwick, who worked at the U.S. Department of Agriculture Experimental Station in Beltsville, Maryland, performed experiments to determine which wavelengths of light were the most effective in inducing photomorphogenesis. The very best wavelengths were in the red region of the spectrum (about 660 nm). Blue light, which drives photosynthesis very effectively, had absolutely no ability to induce photomorphogenesis. The two scientists did other experiments that produced a puzzling result: Far-red light could completely reverse the effects of red or white light. The wavelength that was most effective at this reversal was 730 nm—so far into the red region that most people cannot even sense its presence.

The scientists realized that these experiments could be explained if a single pigment molecule were responsible for photomorphogenesis. The absorption of light in the red region (at 660 nm) should convert the pigment to an active form that triggers photomorphogenesis. However, this new, activated form should absorb light in the far-red region (730 nm), converting the pigment back from the active form to the inactive form and explaining the reversal effect. They proposed that the pigment, which was still theoretical, be called **phytochrome** (*phyto* means "plant"; *chrome* means "color"). It would work like a switch, red light (660 nm) switching it on and far-red light (730 nm) turning it off (Figure 30.16).

These scientists then learned that nearly 10 years earlier two other scientists in the same lab had shown that red light (660 nm) was most effective in causing lettuce seeds to germinate. They had also shown that the effects of this light could be reversed by far-red light (730 nm). Hendricks and Borthwick began to believe that they had discovered one of the fundamental mechanisms by which plants respond to light. And they were right.

Phytochrome has now been isolated and closely studied, although the links between phytochrome and development are still obscure. Phytochrome is a pigment–protein complex with a molecular weight of 120,000, and it does indeed undergo a shift between two forms when it absorbs light, as shown in Figure 30.17.

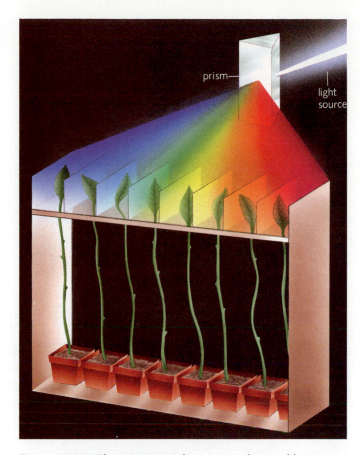

Figure 30.16 The experimental apparatus designed by Hendricks and Borthwick to study photoperiodism enabled them to determine that a pigment absorbing in the far-red region of the spectrum was responsible for the effect. This pigment is now known as phytochrome.

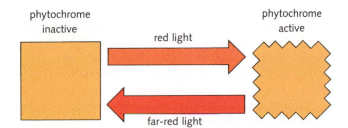

Figure 30.17 Phytochrome is converted into an active form by red light, and can be converted back to its inactive form by far-red light.

Figure 30.18 (top) Wheat *(Triticum aestivum)*, a long-day plant. (bottom) Sierra primrose *(Primula suffrutescens)*, a short-day plant.

Photoperiodism

Besides the germination of seeds and the control of photomorphogenesis, phytochrome is involved in a host of plant responses to light. Phytochrome controls the growth and development of chloroplasts, the synthesis of chlorophyll, the release of hormones such as gibberellin, and even the repositioning of leaves as the angle of sunlight changes. But perhaps the most important response the phytochromes control is **photoperiodism,** a physiological response of a plant to changes in the *photoperiod,* or length of the day. Accurate measurement of the photoperiod is particularly important to plants, for it is the best clue to the passing of the seasons.

Plants are able to time their flowering with some precision, and a stroll through the woods will remind you that some plants flower exclusively in the spring, others only in the fall. Different species clearly have some sort of mechanism to sense the length of the day and adjust their growth and flowering pattern accordingly. In fact, experiments reveal that plants fall into three groups in terms of photoperiodism (Figure 30.18). **Long-day plants** flower only when daylight lasts longer than a certain minimum; they include plants that flower in late spring and early summer, such as hollyhocks and wheat. **Short-day plants** flower only when the daylight period is less than a certain amount; they include soybeans, poinsettias, and ragweed, which flower in the late summer or early fall. A great many plants are unaffected by day length, and they are known as **day-neutral plants.**

At first, you might be tempted to think that plants would respond to changes in the amount of daylight to which they are exposed. However, experiments in controlled growth chambers yielded a surprising result: It was actually the length of the *dark* period that was critical. What the "long-day" plants actually responded to was a "short-night" period of darkness (Figure 30.19). It might be more appropriate to use the terms *long-night plants* and *short-night plants,* but established terminology doesn't yield easily to new discoveries. In any case, the crucial question is how a plant senses the length of the day and responds by flowering.

Phytochrome seems to be at the heart of the photoperiodic response. Cockleburs flower when they are exposed to a dark period of more than 9 hours. However, that dark period must be uninterrupted. If even a flash of light occurs in the middle of the dark period, flowering does not occur. The wavelength of the flash of light that is most effective in preventing flowering is 660 nm. And when the flash of white or red light is followed by a flash of far-red light (730 nm), the plant flowers normally. This is powerful evidence that phytochrome controls the photoperiodic response.

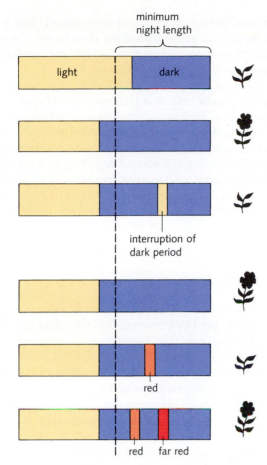

interruption of
dark period

red

red   far red

Figure 30.19 A summary of experiments that showed the involvement of phytochrome in flower production. Cockleburs flower when they are exposed to a dark period longer than 9 hours. If that dark period is interrupted by a flash of red light, flowering does not occur. However, if that flash is followed by a period of far-red illumination, flowering occurs. The ability of far-red light to reverse the effects of the flash are indicative of the involvement of phytochrome.

Florigen

If different parts of the plant are covered when photoperiodic experiments are carried out, it soon becomes evident that the leaves are the source of the response. When even a single leaf of a cocklebur is covered so that it receives a sufficient dark period, the whole plant flowers. This seems to implicate a hormone—one that is produced in the leaf under phytochrome control and then travels throughout the plant system to induce flowering. Botanists have suggested the name *florigen* ("flower maker") for the hormone, and there is additional evi-

dence that florigen exists. To date, however, florigen has not been isolated, and therefore it is not yet possible to say with certainty that the flowering response is caused by a single phytochrome-controlled hormone.

AUTUMN: A CASE STUDY

In the temperate areas of the world, as summer ends the days begin to grow shorter and shorter. The change in photoperiod is sensed by the *phytochrome* system, and several processes are set in motion by the response to activated phytochrome. The *auxins* that were produced in the leaves begin to diminish, and the plant increases production of *ethylene* and *abscisic acid.* Phytochrome turns off the pathways for chlorophyll synthesis in the leaves. Because no new chlorophyll is synthesized, existing chlorophyll molecules are gradually bleached and destroyed by the bright fall sun, leaving only the accessory pigments in the leaves. These pigments, including yellow and orange *carotenes* from the chloroplast and the reddish *anthocyanins* found in cellular vacuoles, become visible as chlorophyll fades, and the lovely colors of autumn are revealed (Figure 30.20).

Figure 30.20 Changes in pigment biosynthesis are responsible for the brilliant colors of autumn, illustrated in this photograph of a sugar maple forest near the Montreal River in Ontario.

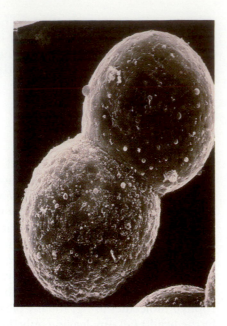

Figure 30.21 The fusion of plant protoplasts, shown in this electron micrograph, is an essential part of techniques for plant gene manipulation.

In time the appearance of *hydrolytic enzymes,* which break down thin cell walls, causes a weakening of the point where the leaf is attached to the stem. This zone of weakness is called the **abscission layer.** The growth of stems begins to slow down, and the apical meristem develops a thick, waxy *terminal bud* that will help it to survive the winter. Little by little, hormones shut down the metabolic activities of the leaf until it breaks off at the weakened abscission layer and drifts to the ground. This pattern is called leaf **senescence,** and it is vital to the plant. Senescence enables the plant to shed tissue that would be difficult to protect through a long cold season and allows the plant to conserve its resources for the winter. Senescence has been a difficult process to investigate, but the programmed senescence of various tissues, including leaves, fruit, and flowers, is an important feature of the normal life cycle of flowering plants.

TOMORROW'S PLANTS

Plant Cell Culture

Today's botanists are experimenting with a number of techniques that promise an unprecedented ability to develop new varieties of plants. In recent years scientists have discovered how to produce complete plants from a few cells taken from another plant. Isolated cells are grown in a broth rich in nutrients and hormones, and groups of cells gradually form a small clump known as a *callus.* Each callus is then transferred to the surface of sterile

agar (also rich in nutrients and hormones), and a small *plantlet* begins to grow with fully developed tissues. As the new plant increases in size, it can be transferred to soil and eventually grown outdoors.

In addition, proper growth conditions can produce plant **protoplasts,** cells that lack cell walls (Figure 30.21). Scientists have developed techniques whereby protoplasts from different plants or even different species can be fused, forming hybrid cells with new combinations of genes and bypassing the natural breeding process entirely. The ability to grow and manipulate plant cells in this way gives scientists an opportunity to alter the genes of a single cell and then grow a complete organism from that altered cell.

Plant Genetic Engineering

In order to manipulate the plant genetic system, one first needs a way to get DNA molecules inside plant cells. There are several methods that scientists have begun to use to insert DNA molecules into plant cells, and it is now clear that there are no fundamental obstacles to carrying out genetic engineering in plant systems.

Several labs have experimented with the use of plant viruses to insert DNA sequences into plant cells. Two main classes of DNA viruses infect plant cells, and each holds some promise as a tool for genetic engineering. At the present time, however, the organism of greatest interest to plant molecular biologists is the bacterium *Agrobacterium tumefaciens. Agrobacterium* is a soil bacterium that infects several types of plants and produces large tumor-like growths of plant tissue known as *galls.* During an *Agrobacterium* infection a small piece of DNA, a *plasmid,* from the bacterium enters the plant cell and transforms it into a gall cell.

This plasmid, known as *Ti,* contains about 200,000 base pairs of DNA. Recombinant-DNA techniques can be used to splice Ti DNA to another segment containing a gene that is to be inserted into plant cells. This altered Ti plasmid can then be used to infect plant cells with the DNA segment. Different strains of Ti can be constructed so that they do not cause galls, and this allows Ti to be used as a vehicle for inserting new genes into plants. These techniques make it possible to isolate a particular gene, insert it into a single plant cell, and then use that cell to grow a complete organism containing the genetic modification (Figure 30.22). Plant scientists are eager to try this technique to endow crop plants with genes for nitrogen fixation, herbicide resistance, and high-efficiency photosynthesis.

The power of these techniques is apparent in a series of experiments in which the Ti plasmid was used to carry the luciferase gene into tobacco cells. Luciferase, an enzyme found in fireflies, produces light from the energy available in ATP. After infecting tobacco cells with the plasmid, Steven Howell and his co-workers in San Diego were able to grow mature tobacco plants from the transformed cells. These plants expressed the luciferase gene, giving them a most un-plant-like quality: They glowed in the dark like fireflies (Figure 30.23)!

These experiments were used because the luciferase enzyme is an excellent "reporter" gene—one that shows an observer that the gene has indeed been inserted into the plant cells and is functioning. However, the same techniques clearly can be used to insert other genes into plant cells. It is still too early to begin designing plants to order, but it may well be possible to insert genes for specific traits, such as nitrogen fixation and high-efficiency photosynthesis, into existing plants. One example that has many agricultural companies interested is the insertion of genes that would make crop plants

Figure 30.22 (right) These corn seedlings, growing in sterile agar, have each been cloned from single cells in culture. (below) Foreign genes can be introduced into plant cells by genetic engineering. One useful technique involves the Ti plasmid from *Agrobacterium tumefaciens*. DNA can be joined to the Ti plasmid, used to transform *Agrobacterium tumefaciens*, which in turn may be used to transform plant protoplast cells in culture. Cells containing the DNA may be used to regenerate a complete plant, as shown.

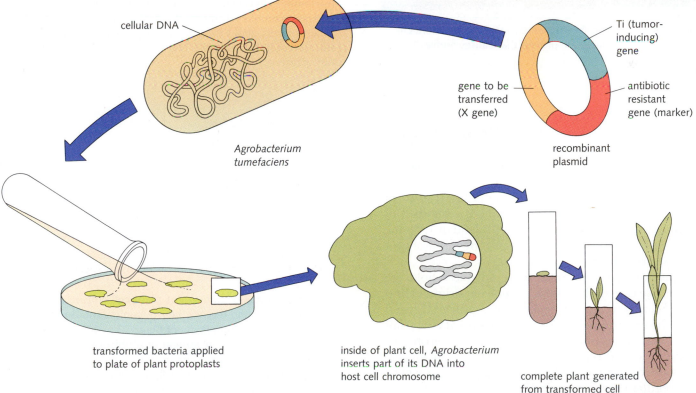

cellular DNA

Agrobacterium tumefaciens

Ti (tumor-inducing) gene

gene to be transferred (X gene)

antibiotic resistant gene (marker)

recombinant plasmid

transformed bacteria applied to plate of plant protoplasts

inside of plant cell, *Agrobacterium* inserts part of its DNA into host cell chromosome

complete plant generated from transformed cell

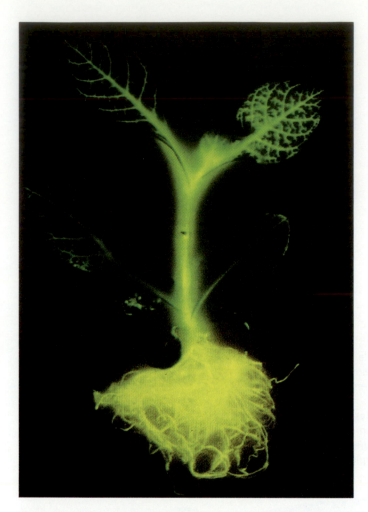

Figure 30.23 A genetically engineered tobacco plant containing the gene for luciferase, the protein that causes fireflies to glow.

SUMMARY

The ability of a plant to survive depends on the coordination of activities in its many cells and tissues. In order for an organism to respond to changing external conditions, there must be a system that controls and coordinates events in different parts of the organism. Many of these responses are known as tropisms. Common tropisms include geotropism, phototropism, and thigmotropism—the ability to respond to gravity, light, and touch, respectively. Tropisms are caused by hormones, substances that are made in one region of the plant and affect cellular behavior in another part. The first plant hormones to be discovered were the auxins. Auxins affect the rate of cell growth, and the accumulation of auxins in different amounts on either side of a stem is the major cause of responses to light and gravity. Chemicals similar to auxins have been used as herbicides (weed killers). Other plant hormones include gibberellins, which regulate the rate of stem elongation and also play an important part in seed germination; cytokinins, which help to regulate cell division; abscisic acid, which inhibits growth under certain conditions; and ethylene, which is important in the ripening of fruit.

Phytochrome is a pigment–protein complex that helps to regulate the timing of important light-sensitive events such as flowering and photomorphogenesis. Phytochrome conversions between active and inactive forms are used by plants to sense the length of the day and to control seasonal responses such as flowering and reproduction.

Emerging techniques, including plant cell culture, depend on the informed use of plant hormones to stimulate cell growth and to promote the development of new tissue in plants grown by cloning cells isolated from other individuals. The application of new techniques in molecular biology has now made it possible to insert genes into plant cells, altering the genetic constitution of individual cells, and then use those cells to produce new plants.

resistant to *herbicides* (chemical weed killers). If the plants were resistant to the chemicals, then herbicides could be used more effectively to control weeds without damaging valuable crops.

Because *Agrobacterium tumefaciens* does not transform monocot cells, scientists have developed a number of techniques to insert DNA into monocots, and in 1988 the first successful expression of a foreign gene in a monocot (maize) was reported.

Despite the great promise these advances hold, we must exercise care in the choice of such experiments and in our attempts to release genetically engineered material into the environment. The construction of new plant varieties may have unforeseen biological and economic consequences, and one of the key questions facing scientists and nonscientists alike in the coming years will be how to regulate such technologies.

STUDY FOCUS

After studying this chapter, you should be able to:

- Define the term *hormone*, and explain how hormones act in controlling plant growth and development.
- Cite some methods and results of experimental plant physiology.
- Describe some techniques of plant cell culture and genetic engineering, and discuss their possible implications.
- Appreciate the complexity of plants as living organisms that respond to their environment in sophisticated and complex ways.
- Explain the central role that light plays in the regulation of plant growth.

SELECTED TERMS

tropism *p. 574*
geotropism *p. 574*
phototropism *p. 574*
thigmotropism *p. 574*
hormones *p. 575*
receptor *p. 576*
auxin *p. 577*
apical dominance *p. 578*
herbicide *p. 579*
gibberellin *p. 580*
cytokinin *p. 580*
ethylene *p. 581*
photomorphogenesis *p. 582*
phytochrome *p. 583*
photoperiodism *p. 584*
long-day plants *p. 584*
short-day plants *p. 584*
day-neutral plants *p. 584*
abscission layer *p. 586*
senescence *p. 586*

REVIEW

Discussion Questions

1. What is a hormone? Why is it justified to classify auxin as a hormone?

2. To many plant biologists, Went's experiments with agar blocks proved the existence of plant hormones (in general) and of auxin (in particular). What else did the experiments show about the nature of the chemical message?

3. Is phytochrome a hormone? What are some of the major differences in actions and effects between phytochrome and auxin?

4. Why would it perhaps be more appropriate to refer to "long-day plants" as "short-night plants"?

Objective Questions (Answers in Appendix)

5. When a seedling rights itself, this is characterized as a _____ response.
   (a) righting
   (b) phototropic
   (c) geotropic
   (d) thigmotropic

6. The processes that allow plants to respond to changing conditions in the environment are most directly influenced by the
   (a) concentration and type of minerals in the soil.
   (b) plant species.
   (c) concentration of hormones.
   (d) amount of shade.

7. Darwin and his son showed that
   (a) plant hormones regulate growth.
   (b) plants demonstrate a phototropic response.
   (c) plants are affected by the force of gravity.
   (d) plants exhibit geotropism.

8. Auxins are active in producing
   (a) cell elongation.
   (b) seeds.
   (c) growth of lateral buds.
   (d) fruit ripening.

9. Auxins, gibberellins, and cytokinins are all
   (a) hormones.
   (b) types of flowering plants.
   (c) types of soil nutrients.
   (d) types of organic fertilizers.

10. A plant that is day-neutral
    (a) will only flower when the amount of daylight exceeds the amount of darkness.
    (b) will only flower when the amount of darkness exceeds the amount of daylight.
    (c) is unaffected by photoperiodism.
    (d) will not flower.

PART SEVEN

Animal Systems

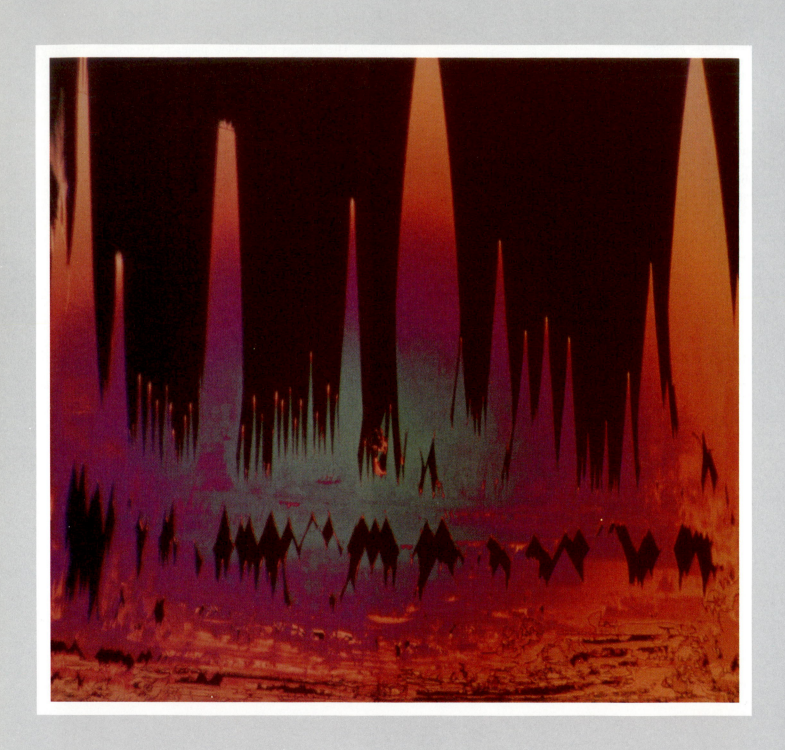

> *We must not conceal from ourselves the fact that the causal investigation of the organism is one of the most difficult, if not the most difficult problem which the human intellect has attempted to solve, and that this investigation, like every causal science, can never reach completeness, since every new cause ascertained only gives rise to fresh questions concerning the cause of this cause.*
>
> —Wilhelm Roux (1894)

Roux's famous statement echoes a complaint that has been heard from many generations of students: "The more we learn the less we seem to know!" To be sure, many scientists would take issue with the tone of this quotation. Although it is often true that answering one question results in our posing two or three new ones, it is also true that each question that science is able to answer enriches our understanding of nature. A scientist works with the inner belief that all knowledge is worth having and therefore that the newer, deeper questions that our studies raise will be that much more profound and satisfying to answer.

Anatomy and physiology—fields devoted to the structure and function of the systems of living organisms—are excellent examples of the sources of Roux's complaint. Thousands of years ago, students of biology recognized that it was appropriate to consider a living organism as a number of systems that could be studied separately. As time passed, the classic Roman and Greek texts were updated by workers in the Middle Ages who compared the writings of the ancients with what they were able to discover on their own. With few exceptions, however, it was impossible to discover a link between any of these systems and the chemicals of which living things are made.

Even so, there were hints of the processes at work within the organism. The electrical activity of nerve and muscle tissue, as well as the acids found in the digestive system, suggested that basic principles of physics and chemistry were at work within living organisms. Engineers pointed out that hydraulics could be used to explain the flow of fluid in the circulatory system, and, little by little, the field of physiology evolved in an effort to link each action of a particular system with a series of cellular and molecular events.

Today, the science of physiology has fully developed the connections with chemistry and physics that were only suspected a century ago. Each of the chapters in this section explores one or more of the systems of the body. Where appropriate, we will examine how similar systems in other animals function and how each system has been shaped by evolution. You may find that Wilhelm Roux's complaint is still quite valid today—that we have learned a great deal only to confront new questions. But we believe that you will also find that such questions have now advanced to the point where our very ability to pose them has made the last few years of the twentieth century a very exciting time to be alive and a very exciting time to study science.

◀ The dual role of the compound epinephrine, captured here in crystalline form, symbolizes the complexity found in regulatory systems of humans and other vertebrates. Epinephrine facilitates communication throughout the body, serving as both a neurotransmitter and a hormone.

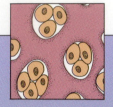

CHAPTER 31

Multicellular Organization and Homeostasis

Imagine that you are visiting Africa's Serengeti Plains in late May. The grasslands are parched and brown, and more than a million wildebeest are on the move, plodding toward greener pastures in Kenya.

Imagine now that you are watching a single animal in this huge herd. He moves mechanically, as if in a trance. Internally, unconsciously, his actions and metabolic processes are geared to scarcity. His measured steps expend the least possible energy. With no food in his gut, his body mobilizes and burns stored energy from fat deposits, distributing it to organs and muscles. Lacking water on today's long march, he retains body fluids by producing as little urine as possible.

Suddenly, a lioness springs from the underbrush. Alarmed, the wildebeest rouses from his stupor, all senses on alert. Energy in the form of glucose pours into his muscles. He gallops away with a grace and speed startling for an animal his size. The lioness strikes and tears his skin with her claws but fails to knock him down. She cannot keep up with his frantic pace.

Several minutes later, once out of danger, the wildebeest relaxes. His metabolism returns to normal, and he lapses into the steady pace of his migratory style. Already, blood in damaged tissues around his wound has thickened enough to stop dripping. Already, bacteria that entered the wound are being surrounded and destroyed by his body's defenses. And already, he has resumed his 200-mile journey toward the northwest.

MAINTAINING THE BALANCE

If you could somehow have watched these events from the perspective of individual cells, you would have seen dozens of remarkable acts of communication, coordination, and control among cells belonging to different tissues. Those acts are neither more nor less remarkable than similar phenomena that occur within your own body or, for that matter, within the body of an ant.

The cells of every organism are at once independent entities and interdependent parts of a larger whole. Each and every cell, whether a nerve cell, a muscle cell, or a skin cell, must respond to changes in its immediate environment within the body. Each must successfully balance the inputs and outputs on which its metabolic activity depends. Yet despite cells' individual requirements, they must also function together as members of the greater cellular community that is the entire organism.

One of the remarkable accomplishments of multicellular organisms is the creation of an internal environment that is carefully controlled. Cells in the wildebeest's brain, for example, must be kept at constant temperature, supplied with energy in the form of glucose, bathed in fluid with a constant concentration of water, and cleansed of their waste products. Those conditions must not change, regardless of droughts, floods, famines, heat waves, or cold snaps. Failure at any of these tasks, even for a few minutes, would lead to permanent injury or death of the entire organism.

Maintaining constant internal conditions in the face of rapidly changing external environments is no small accomplishment. The major systems of the body must rely on a foolproof system of checks and balances that monitors and guides the activities of every cell.

Starting in this chapter, we will examine how the major body systems of animals succeed at this daunting task.

We will begin here by providing a brief overview of the tissues and major body systems of multicellular animals. Earlier in this text (Chapters 15–18), we examined some of the basic differences in structure and physiology in the many diverse groups of the animal kingdom. Although we will include a number of different species in our treatment of animal physiology here, we will most often focus on that species of particular interest to us, *Homo sapiens* (Figure 31.1).

THE RANGE OF CELLS AND TISSUES

The Diversity of Cells

Humans begin life as a single cell, a zygote formed by the fusion of sperm and egg. Over the course of a few weeks, that single cell divides into thousands and ultimately millions of cells in the developing embryo. The embryo develops three cell layers (called **germ layers**), the **ectoderm, mesoderm,** and **endoderm,** and from these the rest of the body is produced. Gradually, the cells in these three layers undergo a process called *differentiation* ("becoming different") in which they are transformed into the astonishing variety of cells found in the adult.

- **Ectoderm** develops into the cells of skin and the nervous system.

- **Mesoderm** develops into the cells of the organs found in between ectoderm and endoderm, including muscles, blood vessels, the reproductive tract, and the kidneys. The mesoderm also provides the outer layers of the organs of the digestive and respiratory systems.

- **Endoderm** produces the cells that form the lining of the digestive and respiratory systems and the glands (liver and pancreas) that are outgrowths of these systems.

As different cell types are produced, they do not take up random locations within the embryo. Rather, they become organized into groups that perform specialized functions. These groups of similar cells, which are called **tissues,** are found throughout the body, and they represent a basic level of organization just above the cell itself.

Traditionally, anatomists (biologists who study the organization of the body) have divided tissues into four types: **epithelial, connective, nervous,** and **muscle.** These categories are based on the characteristics of the tissue itself, not on the germ layer from which it was formed. Within each of these four categories there is a great range of diversity. Cells as diverse as blood cells and bone cells, for example, are considered different types of connective tissue.

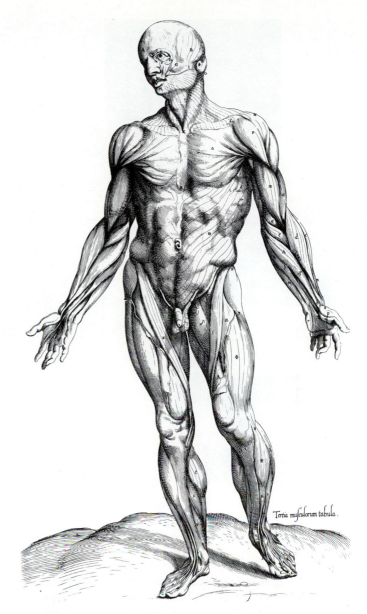

Figure 31.1 A remarkable rendering of the human muscular system, by Vesalius.

Epithelial Tissue

Epithelial tissues form surfaces around and within the body and often act as barriers that divide the body into distinct compartments. The most obvious epithelial tissue is *skin*, which covers the body and serves as its first line of defense against the outside world. Epithelial tissues that are derived from endoderm form the linings of the digestive and respiratory systems, while mesoderm produces the epithelial tissues that cover body organs and line body cavities and blood vessels. *Secretory glands,*

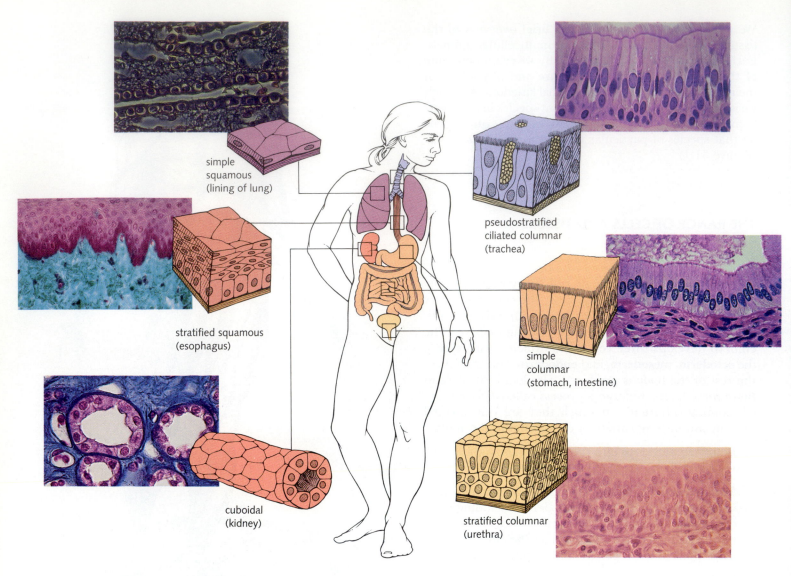

simple
squamous
(lining of lung)

pseudostratified
ciliated columnar
(trachea)

stratified squamous
(esophagus)

simple
columnar
(stomach, intestine)

cuboidal
(kidney)

stratified columnar
(urethra)

Figure 31.2 Epithelial tissues are found throughout the body. They are placed into categories based on the structural organization of the cells of which they are made. Photomicrographs of six major types of epithelial tissue. Clockwise (from the upper left), simple squamous epithelium (from the lung), pseudostratified columnar epithelium (trachea), simple columnar epithelium (stomach), stratified columnar epithelium (urethra), simple cuboidal epithelium (from kidney), and stratified squamous epithelium (esophagus).

found in tissues derived from ectoderm, endoderm, and mesoderm, are usually formed by invaginations (infoldings) of epithelial tissue.

Epithelial tissues are classified according to the shapes of the cells that form them. Figure 31.2 shows examples of *squamous*, *cuboidal*, and *columnar* epithelia. A **simple epithelium** consists of a single layer of cells; a **stratified epithelium** consists of two or more layers of cells. **Pseudostratified epithelium** appears to have layers of cells, yet all the cells actually touch a basement membrane.

Cell junctions In order for tissues to form, the cells that compose them must be held together by some means. The cells of many tissues must also have some means of communication with the cells adjacent to them. **Cell**

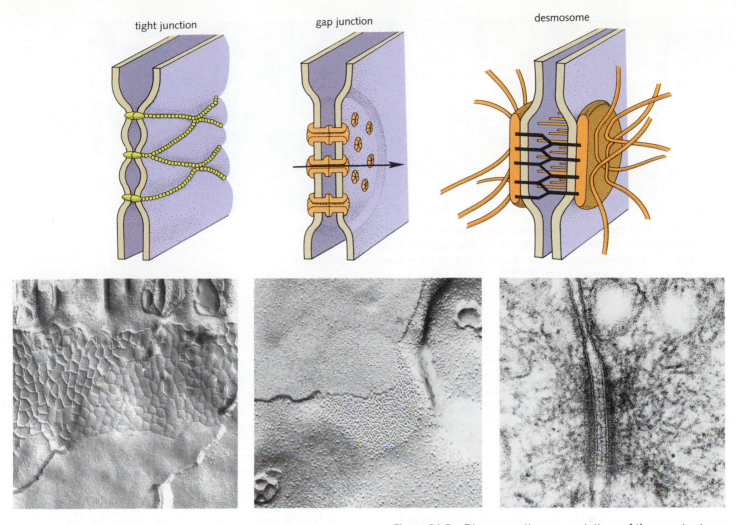

tight junction   gap junction   desmosome

junctions, specific attachments between adjacent cells, perform these tasks. Junctions are found in many non-epithelial tissues as well, and they can be grouped into three general types (Figure 31.3).

Tight junctions seal off the external space between two cells, preventing material from leaking between them. Tight junctions are particularly important in epithelia that line fluid-filled cavities, because they prevent the movement of fluid between cells. Tight junctions between cells of the intestine keep its contents from leaking across the intestinal wall into body cavities.

Gap junctions provide a means of intercellular communication. Gap junctions are small, pore-like protein channels that connect the cytoplasm of one cell to that of its neighbor. Compounds such as salts, sugars, and ions have been shown to pass from one cell to the next through gap junctions. They also allow electrical signals and metabolites to pass between cells, enabling groups of cells to function as a coordinated unit such as heart muscle.

Figure 31.3 Diagrammatic representations of three major types of cell junctions in a simple epithelium. Tight junctions block the passage of materials in the extracellular space between cells, gap junctions form tiny, regulated channels through which small molecules may pass from cell to cell, and desmosomes provide for strong mechanical attachment between adjacent cells. The three major types of cell junctions are shown in the electron micrographs. Tight junctions (left) consist of a web-like series of sealing elements between the membranes of two adjacent cells. Several hundred gap junctions (center) are seen as a cluster of small particles in the center of this micrograph. Each particle is an individual connecting unit which allows material to pass from one cell to the next. A single desmosome (right) consists of extracellular attachments and thickened cell membrane regions anchored to filaments in the cytoskeleton.

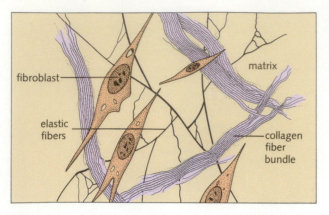
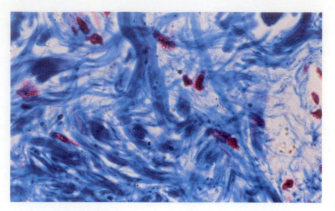

Figure 31.4 Loose connective tissue (left) consists of fibroblasts in a matrix of elastic fibers and tough bundles of collagen. Fibroblasts help to produce this matrix, and are embedded in it. Photomicrograph of fibroblasts (right) in the collagen-rich connective tissue just beneath the human scalp.

Recent studies have shown that gap junctions are quite sophisticated: they may open or close in response to certain cellular signals, including calcium ion and hydrogen ion (pH) concentration. Gap junctions between the contractile cells of the heart are the principal reason why the heart muscle is capable of contracting as a single unit.

Adhering junctions, which are also known as **desmosomes,** are strong mechanical attachments between adjacent cells. These junctions do not block the movement of material between cells. Desmosomes are attached to the *cytoskeleton* of each cell involved in the junction, and this enables the junction to act almost as a "spot weld"—a point at which two cells are cemented together. Desmosomes are particularly common in tissues that are subject to mechanical stress, such as skin, where they prevent the tissues from breaking apart.

Connective Tissue

Connective tissue gets its name from the fact that it makes up the basic support structures of the body; it connects the body's other tissues in a manageable framework. If you place your hand around your upper arm as you flex your elbow, you will feel many of the basic types of connective tissues: bones, ligaments, and tendons.

One of the distinguishing characteristics of this tissue is the fact that cells of connective tissues generally produce an **extracellular matrix,** a layer of material that surrounds the cells. The extracellular matrix may be liquid or solid, loose or dense, flexible or rigid. The unique characteristics of the different types of extracellular matrices are often responsible for the different properties that we associate with various types of connective tissues.

Supporting tissue One fundamental role of connective tissue is providing mechanical support. **Loose connective tissue** surrounds major organs, blood vessels, and epithelial tissues. The support provided by its loose matrix of fibrous proteins is indispensable in forming the structure of an organism. In loose connective tissue the most common cell type is the **fibroblast** (Figure 31.4). The fibroblast produces fiber-like proteins that are woven into a complex extracellular bundle. The most important of these proteins, **collagen,** forms large, tough fibers that are found throughout the body. It is the single most abundant protein in the vertebrate body. In some tissues, the layers of collagen are denser and form the basis for connections between muscles and bones and other points subjected to mechanical stress.

Adipose tissue, or **fat,** is another type of connective tissue widely distributed throughout the body. Each fat cell contains a single large droplet of fat stored for possible future use (Figure 31.5).

Figure 31.5 Adipose tissue is a form of connective tissue.

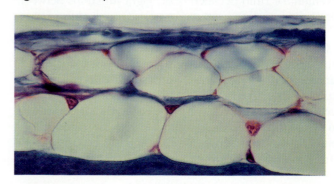

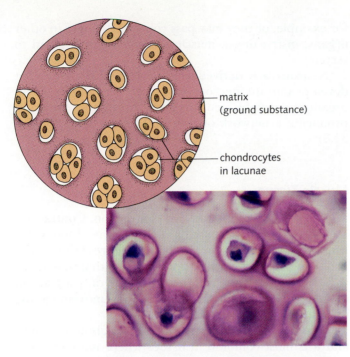

Figure 31.6 Cartilage is produced by chondrocytes, which secrete a tough, flexible extracellular matrix around spaces called lacunae ("holes") in which the chondrocytes remain. The photomicrograph of cartilage shows clusters of chondrocytes confined to isolated lacunae within the tissue matrix.

Cartilage is a form of connective tissue in which an extremely dense matrix containing collagen fibers is laid down in precise orientation (Figure 31.6). Later in development, elastic fibers are laid down with the collagen, and the result is a tough, resilient, springy tissue that can support great weight and still remain flexible. In some vertebrates, notably sharks and rays, the entire skeleton is made of cartilage. In adult humans, cartilage is found at the endpoints of many bones, in the knee, and in flexible structures such as the ear and the tip of the nose.

Bones are strong and rigid, and they provide support and protection for the vertebrate body and its organs. Bone is produced by cells that first lay down a matrix of cartilage and then gradually *mineralize* that matrix by filling it with crystals of calcium phosphate and calcium carbonate. Bone is not a solid matrix, however; it is penetrated by thousands of tiny microscopic canals through which blood vessels flow. The cells that produce the matrix remain alive and active within it, constantly dissolving old matrix and laying down new matrix. This activity enables broken bones to heal rapidly (Figure 31.7).

The inner portion of most large bones, the **bone marrow,** is the principal blood-forming tissue of the body. Most of the cellular components of the circulatory system are produced in bone marrow.

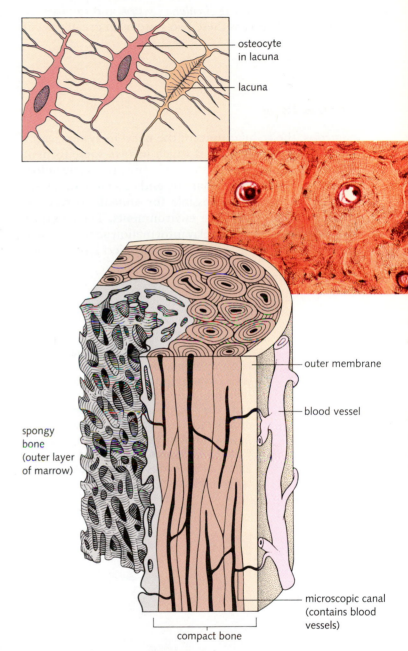

Figure 31.7 Bone is produced by bone cells called osteocytes, which surround their lacunae with a dense, calcium-rich matrix. In spongy bone, the organization of the matrix is loose and open. In compact bone, the tight matrix is penetrated by a series of microscopic canals through which blood vessels and nerves are able to reach the osteocytes. The photomicrograph of human compact bone clearly shows the system of microscopic canals which bring nutrients to osteocytes. It is important to remember that bone is a living tissue, filled with cells that carry on an active metabolism.

Circulating tissue The white and red blood cells that are found in blood are also derived from connective tissue. Unlike the other connective tissues we have described, they have a liquid matrix, and the tissue itself flows throughout the body. Blood cells in their liquid matrix form key tissues of the *circulatory system* and the *immune system*. We will discuss these two systems in greater detail later in the book.

Nervous Tissue

Neurons, the cellular units of the nervous system, carry out a variety of tasks, all of which involve some form of communication. The actions of neurons enable information to pass rapidly from one end of a large organism to another, making it possible for animals to respond rapidly to changes in their environments. The messages carried by neurons may relay information about the environment, informing the body that an object is hot or cold, for example, or they may pass commands along to other organs, instructing a muscle to contract or a gland to secrete.

Neurons are derived from ectodermal tissue that develops into the nervous system of the embryo. As an organism develops, neurons grow throughout the body, producing a network of cells and nerve fibers (Figure 31.8).

Muscle Tissue

Muscle cells are specialized for contraction. Contraction requires energy, and muscular tissue is one of the major users of food energy in the body (Figure 31.9). Their ability to use food energy rapidly means that muscle tissues are major sources of heat. Because only a fraction of the food energy can be converted into movement, the difference is given off as heat.

Smooth muscle tissue is found throughout the body, and it is most common in tissues where muscular con-

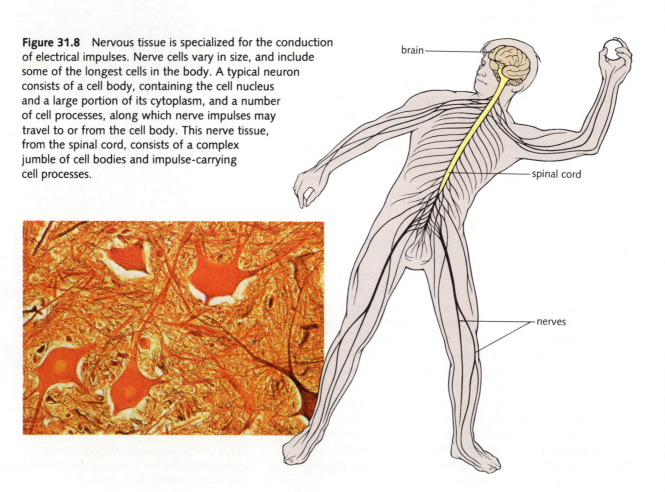

Figure 31.8 Nervous tissue is specialized for the conduction of electrical impulses. Nerve cells vary in size, and include some of the longest cells in the body. A typical neuron consists of a cell body, containing the cell nucleus and a large portion of its cytoplasm, and a number of cell processes, along which nerve impulses may travel to or from the cell body. This nerve tissue, from the spinal cord, consists of a complex jumble of cell bodies and impulse-carrying cell processes.

brain

spinal cord

nerves

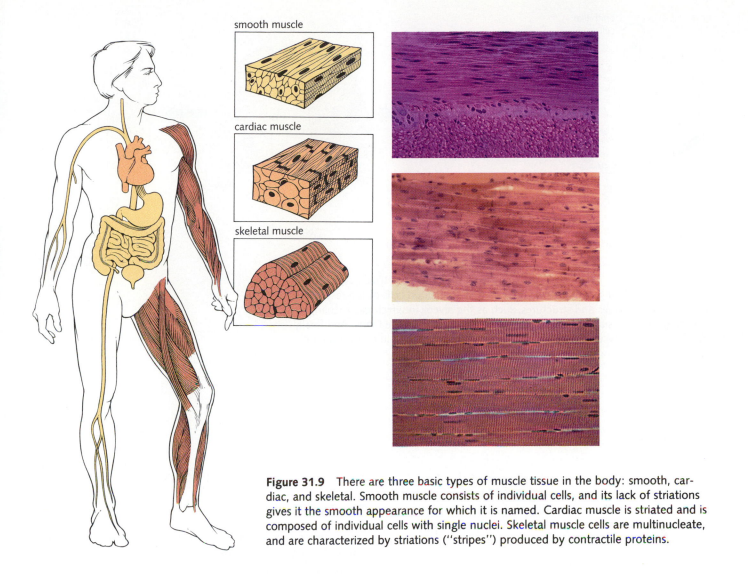

Figure 31.9 There are three basic types of muscle tissue in the body: smooth, cardiac, and skeletal. Smooth muscle consists of individual cells, and its lack of striations gives it the smooth appearance for which it is named. Cardiac muscle is striated and is composed of individual cells with single nuclei. Skeletal muscle cells are multinucleate, and are characterized by striations ("stripes") produced by contractile proteins.

tractions are not under conscious control. Smooth muscle cells surround arteries and veins, as well as the walls of the digestive system and the reproductive tract. Smooth muscle gets its name from the fact that the spindle-shaped cells of this tissue have a cytoplasm that appears smooth and unbroken.

Skeletal muscle tissue forms the large muscles of the limbs and trunk that we often associate with physical fitness and muscular development. **Cardiac muscle** is found in the heart, and it combines many of the properties of smooth muscle and skeletal muscle. Both skeletal and cardiac muscle show a characteristic striped pattern (known as striations) produced by an orderly, repeating array of contractile proteins in the muscle cytoplasm. Although cardiac muscle functions largely without voluntary control, most skeletal muscles are under conscious control.

ORGANS AND ORGAN SYSTEMS

Throughout the body we find examples of different types of tissues organized in ways that enable them to perform a single function or a series of related functions. These groupings of tissues are known as **organs.** Many organs are small, such as the *pituitary,* a gland about twice the size of a pea that is located within the skull and is responsible for the production of important hormones. Other organs are very large, such as the body's most expansive organ, the *skin,* which has a surface area of about 2.0 square meters in the average adult human.

Organs themselves can be grouped into **organ systems,** a higher category comprising groups of organs that are physically or functionally related (Figure 31.10). For example, the *mouth, stomach,* and *large intestine* are

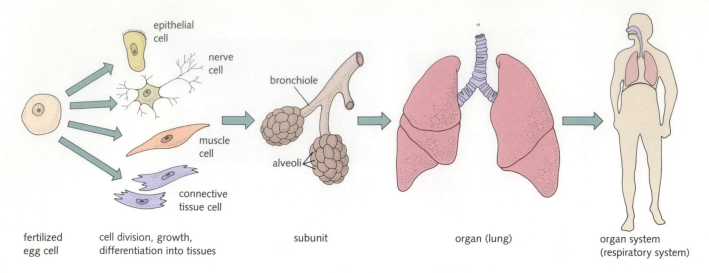

epithelial cell

nerve cell

bronchiole

muscle cell

alveoli

connective tissue cell

| fertilized egg cell | cell division, growth, differentiation into tissues | subunit | organ (lung) | organ system (respiratory system) |

Figure 31.10 Each system of the body develops from the original fertilized zygote. Specialized cell types are formed during development. The specialized cells of the respiratory system form a host of tissues and structures, such as the alveoli in which gas exchange takes place. Many such tissues make up a complete organ, such as the lungs. And finally, the complete respiratory system includes a series of organs that work in concert during breathing.

separate and distinct organs that form part of the *digestive system*, a group of organs that digest food. The many different muscles throughout the body can all be thought of as separate organs, but taken together, they form part of the *muscular system*, the most massive of all organ systems in most vertebrates.

Major Organ Systems

In this book we will group the organs and tissues of the body into 10 major organ systems. These groupings are not absolute, and some organs are functional parts of two or more systems. The diaphragm, or breathing muscle, may be placed in the muscular or the respiratory system, for example, and the mouth serves both the respiratory and the digestive systems. Furthermore, it would be wrong to think of the organ systems as units that function in isolation from each other. Detailed physiological work has shown very clearly that links among the organ systems are important in the daily workings of nearly every organ in the body. Table 31.1 describes and illustrates the 10 major organ systems of the human body.

STUDYING THE HUMAN BODY

The Basic Body Plan

Anatomists use a series of standardized terms to help them describe the positions of tissues and organs. These terms are more precise than our everyday references to "front," "top," and "side," and we will use them in the chapters that follow. Figure 31.11 illustrates how three imaginary planes may be passed through the body to produce the three views (sagittal, frontal, and transverse) that are used to show body structures.

The basic features of the human body are similar to those of other vertebrates. We have an internal skeletal system (endoskeleton) and a dorsal nerve cord (spinal cord). Our bodies exhibit bilateral symmetry, which means that each half is a mirror image of the other half. Even so, that symmetry is not absolute—many internal organs, such as the heart, liver, pancreas, stomach and intestines, are arranged asymmetrically. The human body has three major cavities: dorsal, thoracic, and abdominopelvic (Figure 31.12). We will refer to these cavities frequently to pinpoint the location of organs and systems.

Table 31.1 *The Major Organ Systems and Their Functions*

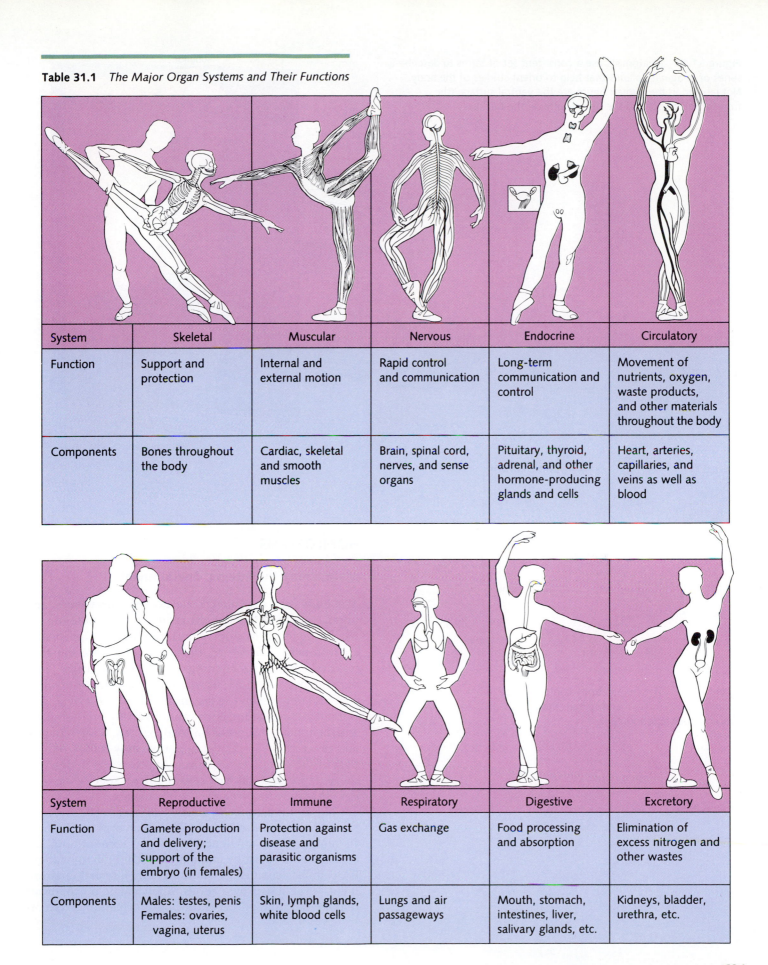

System	Skeletal	Muscular	Nervous	Endocrine	Circulatory
Function	Support and protection	Internal and external motion	Rapid control and communication	Long-term communication and control	Movement of nutrients, oxygen, waste products, and other materials throughout the body
Components	Bones throughout the body	Cardiac, skeletal and smooth muscles	Brain, spinal cord, nerves, and sense organs	Pituitary, thyroid, adrenal, and other hormone-producing glands and cells	Heart, arteries, capillaries, and veins as well as blood

System	Reproductive	Immune	Respiratory	Digestive	Excretory
Function	Gamete production and delivery; support of the embryo (in females)	Protection against disease and parasitic organisms	Gas exchange	Food processing and absorption	Elimination of excess nitrogen and other wastes
Components	Males: testes, penis Females: ovaries, vagina, uterus	Skin, lymph glands, white blood cells	Lungs and air passageways	Mouth, stomach, intestines, liver, salivary glands, etc.	Kidneys, bladder, urethra, etc.

Figure 31.11 Anatomists use a consistent set of terms to describe a series of imaginary planes that help to orient studies of the body. Not labeled in the human figure are the ventral surface (the front of the body) and the dorsal surface (the back).

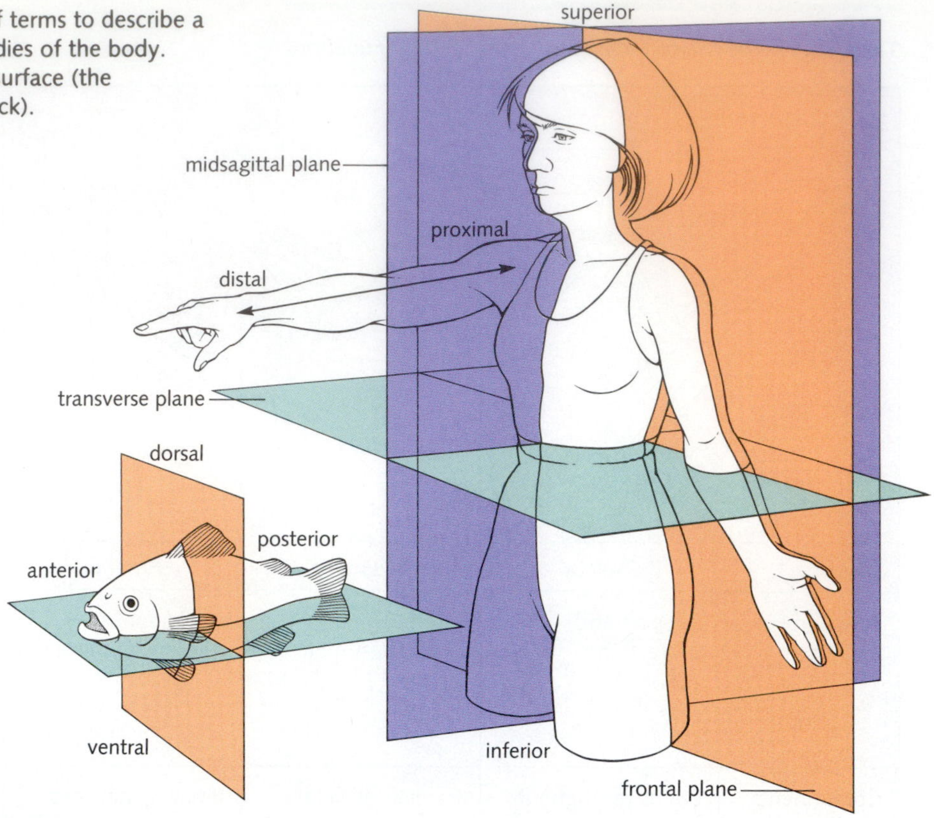

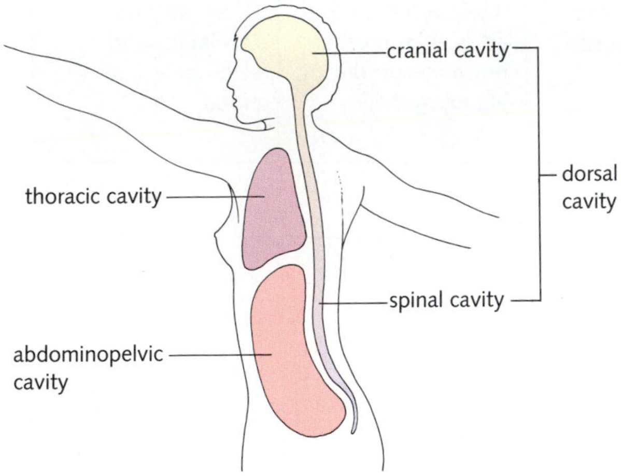

Figure 31.12 The human body contains three major cavities: dorsal (cranial and spinal), thoracic (chest), and abdominopelvic (abdomen).

HOMEOSTASIS

Maintaining the Internal Environment

The body is an organized collection of cells, tissues, and organs whose individual efforts make it possible for the whole organism to survive and function in a particular environment. Therefore, the whole organism depends on the proper functioning of individual cells. This dependence is mutual, of course, because the individual cells also depend on the organism for their survival. The cells and tissues of the body cannot survive on their own, and their individual needs for food, oxygen, waste removal, and protection can be met in nature only if they remain part of the complete organism. Another way to describe the situation is to say that cells require a specific *environment* that only the organism can provide.

The immediate environment of most cells is determined by the *extracellular fluid* in which they are bathed. Its temperature, nutrient and oxygen content, and salt concentration must be maintained within narrow ranges for cellular activities to continue. The internal environment is regulated by a series of automatic mechanisms that are part and parcel of the body's organs and systems

(Figure 31.13). These mechanisms work to counterbalance changes in the environment and to keep the cellular environment consistent.

This regulatory process is called **homeostasis,** which means "keeping things the same." Homeostatic mechanisms work to maintain the constancy of the internal environment. Homeostasis is the central theme of *physiology,* the study of the function of the body's organs and systems. As we will see, it is possible to understand the functions of every system in the body in terms of homeostasis—the maintenance of an acceptable cellular environment.

The Concept of Feedback

Homeostatic systems must respond to changes in the local environment, and then they must counteract those changes by adjusting their own activities. Homeostatic mechanisms are found not only in nature but also in simple machines and everyday appliances. Many of these mechanisms operate by means of a **feedback** principle, in which conditions existing in the environment are "fed back" to the system and used to regulate its activity. At its simplest, a feedback system consists of four basic elements:

1. The *controlled system,* the system regulated to maintain a set level of a particular variable

2. A *receptor* that monitors the level of the regulated variable

3. A *processing center* that integrates information about the set level of the variable and the activity of the controlled system

4. The *output,* the effect or product of the controlled system

Feedback systems may be either positive or negative.

Negative feedback   In **negative feedback** systems, output from the system *inhibits* the system itself. A thermostatically controlled furnace is an excellent example of how negative feedback can control a system.

The heat receptor (thermometer) in the thermostat responds to the household temperature and passes that information along to a processing center, in which the actual temperature is compared to the desired temperature set by the homeowner. If the temperature is at the desired level, no action is taken. If the temperature is too low, an electrical signal is sent to the furnace (the controlled system), which then begins to produce heat (the output).

The furnace burns continually until the temperature in the household rises above the set point in the thermostat. At that point, the receptor "reports" the higher

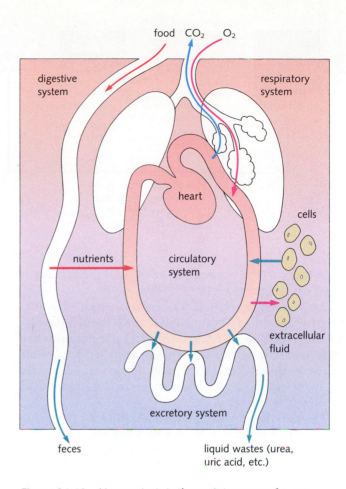

Figure 31.13   Homeostasis is the maintenance of a constant internal environment. Homeostasis demands a variety of systems to regulate temperature, salt and water balance, nutrient flow, and other conditions within an organism.

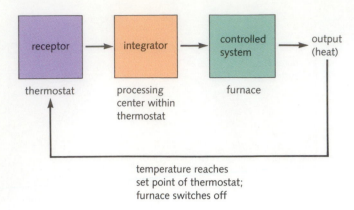

Figure 31.14 The basic elements of a feedback system are found in an ordinary household furnace and thermostat. A feedback system is one that regulates its own output. This block diagram illustrates how *heat*, the output of a furnace, feeds back to stop the production of excess heat, regulating the system to maintain a constant temperature.

temperature to the processing center, and the furnace is turned off (Figure 31.14).

This is a negative feedback control system: The furnace produces an effect (heat) that tends to turn the furnace off over time. Negative feedback mechanisms are widely used to stabilize, or maintain, a particular variable (such as temperature) at a set point. If the variable moves away from the set point, systems are stimulated that tend to return it to the set point.

The body itself regulates temperature by a mechanism that is remarkably similar to a household heating system. The body contains a series of receptors in several organs, especially in the part of the brain called the *hypothalamus,* that monitor the temperature of the blood and compare it to the normal set point of 37°C.

If body temperature drops below this level, cellular activities that use food energy are stimulated and the body produces more heat. One of the most noticeable of these activities is *shivering.* The muscular contractions associated with shivering release heat, causing the body temperature to rise. When temperature receptors in the hypothalamus no longer sense a difference from the set point, they begin to inhibit the production of more heat.

If body temperature rises too far above the set point, *sweating* is stimulated and the evaporation of sweat cools the body surface. This cooling effect lowers body temperature until it returns to the set point and the sweating reaction stops.

Positive feedback Whereas negative feedback acts to stabilize a situation, **positive feedback** tends to intensify a particular effect. During childbirth, for example, the pressure of the baby's head against the wall of the uterus produces a stimulus that increases the contractions of smooth muscles surrounding the uterus. These contractions cause the head to be pushed harder against the wall, and this in turn causes still stronger contractions. The positive feedback in this case (each contraction produces a stimulus that *increases* the strength of the next contrac-

tion) serves an important physiological purpose. It helps force the baby through the uterus into the birth canal. Once birth is completed, the stimuli are gone and the contractions become less intense until they finally stop.

Homeostasis in Action: Eat, Drink, and Be Merry

Water balance Homeostatic systems normally operate so smoothly that we are scarcely aware of their existence. Every action we take has consequences that might affect the internal environment, and it is only because our internal organs regulate that environment so closely that we can carry out everyday functions such as eating and drinking.

When you exercise strenuously on a hot, dry day, for example, you lose body moisture in the form of sweat. If this loss of water continued unabated, your body would soon be in trouble for several reasons. But that doesn't happen, because your body's homeostatic mechanisms swing into action (Figure 31.15).

In addition to the temperature sensors mentioned earlier, the hypothalamus also contains cells that are sensitive to the concentration of water in the blood. As you lose body moisture, causing your blood to become more concentrated, the hypothalamus does two things at once. On one front, it creates in your consciousness a feeling of thirst, so you "know" you ought to take a drink. At the same time, the hypothalamus causes the neighboring pituitary gland to release a substance known as antidiuretic hormone (ADH). ADH molecules pass through the bloodstream to the kidneys, the organs whose activities control the amount of water removed from the blood to be eliminated from the body as urine. ADH *inhibits* the removal of water from the bloodstream, enabling the body to conserve water.

Then, when you finally get around to taking your long-awaited drink, you might take in as much as one or two liters of fluid over the course of an hour. Most of that water is quickly absorbed across the wall of the digestive system into the bloodstream. But two liters of water added to the blood would dilute it so much that the equilibrium between the blood and the cells of the body would be disturbed. Large amounts of water would diffuse across blood vessel walls into the tissues. The cells of the body would swell with the excess water. Capillaries might be blocked by swollen red blood cells, and the skin would become swollen and puffy as water diffused into it from the bloodstream.

Needless to say, this doesn't happen, because the same homeostatic system intervenes. When the water content of the blood rises, as it might after a large drink, the pituitary releases less ADH. With lower ADH levels, the kidneys quickly remove water from the bloodstream, forming large quantities of urine and restoring the blood to its original salt concentration.

Thus the system sets both upper and lower limits for blood water content: A water deficit stimulates the release of ADH, causing the kidneys to conserve water, whereas an overabundance of water causes the kidneys to eliminate the excess (Figure 31.16).

Note that the system works even more quickly when the excess fluid contains alcohol, because alcohol inhibits the production of ADH. For that reason (as you may have noticed), drinking alcohol results very rapidly in a need to urinate.

Figure 31.15 A homeostatic system in action.

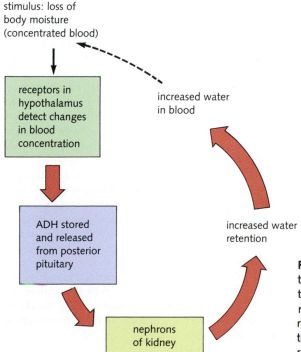

Figure 31.16 Water balance is controlled by the hypothalamus, the pituitary gland, and the kidney. When the blood becomes too concentrated, the hypothalamus causes the pituitary to release ADH (antidiuretic hormone) which *inhibits* water removal by the kidneys. When the water content of the blood is too high, ADH release is inhibited, and the kidneys act to remove excess water. This is a classic negative feedback system.

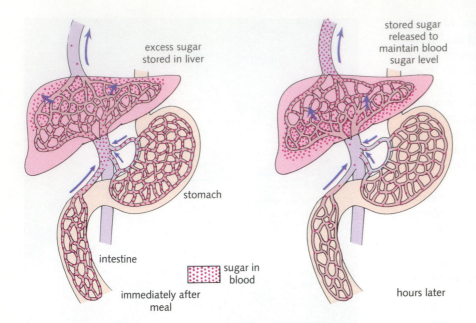

excess sugar
stored in liver

stored sugar
released to
maintain blood
sugar level

stomach

intestine

sugar in
blood

immediately after
meal

hours later

Figure 31.17 Blood sugar level is regulated by the action of the liver. Immediately after a meal, sugar-laden blood from the small intestine passes through the liver, where enough sugar is removed and stored to prevent a sudden rise of sugar in the general circulation. Between meals, stored sugar is released from the liver to maintain a constant sugar level in the blood. The uptake and release of blood sugar is controlled by a series of hormones and will be investigated more fully in Chapter 36.

Sugar balance Without regulation of blood sugar, even the simple act of eating a snack could cause problems for certain cells in the body. Many tissues absorb glucose (sugar) directly from the bloodstream for energy production. For that reason, a sudden surge in blood sugar could completely upset the metabolism of certain cells. Conversely, a decrease in blood sugar between meals would starve other cells—such as brain cells—that depend on blood glucose for nearly all of their energy. This is not an abstract problem. If the body didn't have the means to prevent rapid changes in sugar concentration in the blood, eating a candy bar or drinking a glass of lemonade could double the concentration of blood sugar.

But the body does have a sugar-control mechanism. Blood absorbs glucose and other sugars as it passes through vessels in the small intestine. From the intestine it passes to the liver, where excess sugar is removed from the bloodstream and stored in liver cells, under the control of the hormone *insulin*. The insulin-stimulated absorption of sugar by the liver ensures that sweet foods do not affect blood sugar levels in the general circulation.

As blood sugar levels fall, on the other hand, the liver releases its carbohydrate stores, once again keeping blood sugar levels constant (Figure 31.17). The liver's strategic position in the circulatory system and its ability to regulate blood sugar make it another critical organ of homeostasis. We will explore the regulation of blood glucose in Chapter 36.

SUMMARY

The human body develops in a manner similar to the bodies of other vertebrates. Three cell layers—ectoderm, mesoderm, and endoderm—are produced in the early stages of embryonic development. The wide variety of organs and tissues of the body are formed from these three germ layers. Tissues, groups of cells that perform similar functions, can be grouped into four categories: nervous, muscular, epithelial, and connective. Tissues can be further grouped into distinct organs, and groups of organs in turn form a series of organ systems. Interactions between adjacent cells occur at the level of the cell membrane, and specialized cellular junctions mediate these interactions in a way that assists in the formation of complex, interconnected tissues. The interdependence of the cells in a large organism requires that the internal environment of the organism be closely regulated. Homeostasis, the maintenance of a regulated cellular environment in the face of changing external conditions, is one of the important tasks of nearly every system in the body. Each organ system plays a role in homeostasis. A key part of any homeostatic process is feedback, in which the condition of one aspect of the internal environment directly affects tissues or organs that help to regulate it. The regulation of water balance, in which a series of tissues interact to regulate the concentration of dissolved material in the blood, is a classic example of a feedback system. Similar systems exist to regulate the levels of blood sugar and other compounds within the circulation, as well as such diverse variables as body temperature and hormone levels.

After studying this chapter, you should be able to:

- Explain the cellular organization of tissues and organs in multicellular organisms.
- Describe the major human organ systems.
- Describe the concept of homeostasis and provide some examples of feedback regulation.

SELECTED TERMS

ectoderm *p. 593*
mesoderm *p. 593*
endoderm *p. 593*
epithelial tissue *p. 593*
connective tissue *p. 593*
nervous tissue *p. 593*
muscle tissue *p. 593*
tight junction *p. 595*
gap junction *p. 595*
adhering junction *p. 596*
extracellular matrix *p. 596*

fibroblast *p. 596*
adipose tissue *p. 596*
cartilage *p. 597*
bone *p. 597*
circulating tissue *p. 598*
organs *p. 599*
homeostasis *p. 603*
negative feedback *p. 603*
positive feedback *p. 604*

REVIEW

Discussion Questions

1. Describe the major organ systems of the human body. Which systems seem to have overlapping functions?

2. Do you think that single-celled organisms have homeostatic systems? Why or why not?

3. Describe the major types of tissues. What relationship do they have to the three germ layers that were formed in the embryo?

4. What are the differences between a tissue and an organ? Between an organ and a system?

5. Describe your understanding of the term "homeostasis." Give several examples of homeostatic control mechanisms that exist throughout the body.

6. What critical homeostatic roles are played by the liver and the kidneys when we eat a single large meal and also drink a large amount of liquid?

Objective Questions (Answers in Appendix)

7. Select the proper sequence from the following:
   (a) organs-tissues-cells-systems
   (b) systems-tissues-organs-cells
   (c) organs-cells-tissues-systems
   (d) cells-tissues-organs-systems
   (e) systems-organs-tissues-cells

8. Epithelial tissues are categorized by *function* as
   (a) supportive and fibrous.
   (b) protective and glandular.
   (c) stratified.
   (d) squamous, cuboidal, and columnar.

9. _____ is a flexible tissue that surrounds major organs and contains fibroblasts.
   (a) Epithelial tissue
   (b) Smooth muscle tissue
   (c) Loose connective tissue
   (d) Bone tissue

10. In mammals, the _____ governs control of body temperature and the _____.
    (a) pituitary gland; sugar control mechanism
    (b) hypothalamus, thirst mechanism
    (c) adrenal cortex, hunger mechanism
    (d) pituitary, reproductive mechanism
    (e) adrenal cortex, thirst mechanism

11. The sugar control mechanism in humans is located in the
    (a) liver.
    (b) kidneys.
    (c) large intestine.
    (d) gall bladder.
    (e) small intestine.

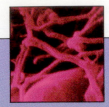

CHAPTER 32

Circulation

"I was about 47 years old at the time. I was at work, and I collapsed one day. There are 22 minutes out of my life that I don't remember. I had gone into cardiac arrest."

That was the experience of one man who survived a heart attack and who—after open-heart surgery—is now leading a healthy, more or less normal life. (He happens to run the Boston Marathon regularly.)

But half a million other Americans who suffer heart attacks each year are not so lucky. They die. In fact, the average American has one chance in four of dying from a heart attack.

The occurrence of heart disease, as we will see shortly, is related to several aspects of modern life, both in the United States and in many other developed nations. But the severe consequences of heart attacks bear grim witness to the importance of this small, muscular organ, about the size of a clenched fist. They also testify to the importance of the circulatory system whose function depends utterly on the heart.

THE IMPORTANCE OF INTERNAL TRANSPORT

To understand why the circulatory system is so important, we must backtrack a bit to consider the requirements of the individual cells of which all multicellular organisms are composed. Each of those cells must have a constant supply of oxygen and nutrients and a method for removing wastes.

When an organism consists of a single cell, this exchange can be carried out by the processes of transport and diffusion across the cell membrane. Some simple multicellular organisms, such as *Hydra,* can also get by without a specialized system to transport nutrients and dissolved gases through their bodies (Figure 32.1). *Hydra* is a simple animal composed of two layers of cells. Food is captured and brought into a large digestive cavity in the center of the organism. Water flows into and out of the digestive cavity, permitting every cell in the organism to come into direct contact with the environment.

But although diffusion is efficient on the microscopic scale, in larger organisms it does not meet the needs of thousands of cells, many of which have no direct contact with the environment. Furthermore, in most multicellular organisms of any size, specialized groups of cells are organized into functional units known as organs. One organ may specialize in the processing of food, another in the exchange of gases with the environment, and another in the elimination of waste material. These organs pose a serious problem: oxygen, carbon dioxide, nutrients, and other substances must be carried from one place to another so that these organs can serve the needs of the whole organism.

A **circulatory system** is a group of organs and tissues that do just that by *circulating* a fluid substance through the body of an organism. The fluid may carry dissolved

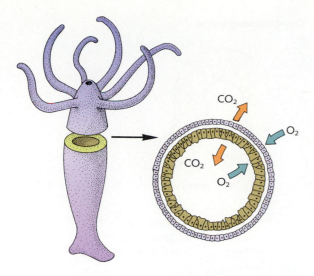

Figure 32.1 In a small organism, gas exchange can take place directly with the environment. *Hydra*, shown here, consists of two cell layers. Oxygen and carbon dioxide diffuse between the environment and each cell layer.

respiratory gases, food, chemical messages, waste materials, and even living cells. The fluid that moves through a circulatory system is generally called **blood,** and the tubules through which it passes are known as **blood vessels.**

Circulatory systems in the animal kingdom can be divided into two broad categories: **open circulatory systems,** in which the circulating fluid percolates throughout the body of an organism and bathes nearly all the cells of the body, and **closed circulatory systems,** in which the circulating fluid is confined to blood vessels.

Open Circulatory Systems

Open circulatory systems are found in most molluscs and arthropods. The blood, or *hemolymph*, that circulates in these organisms moves through open-ended vessels that empty into the body cavity, or **hemocoel.** As blood flows through the hemocoel, it carries cellular waste products to the excretory system and nutrient materials from the digestive system.

One of the best-studied open circulatory systems is found in the common grasshopper. The major organ of the grasshopper circulatory system is a large **heart** that lies along the back of the body (Figure 32.2). This dorsal heart is a muscular tube that collects blood near the rear of the animal and squeezes it toward the front of the animal by a series of regular contractions.

Blood leaving the heart courses through the tissues of the body and gradually flows toward the rear of the

animal, completing the circuit. Blood reenters the heart through small openings called *ostia*, where one-way valves prevent blood from leaking back during a contraction of the heart muscle. The flow of blood carries minerals and nutrients throughout the body and transports waste products to the excretory organs. In insects the blood does not carry oxygen (insects have a special respiratory system in which tubules called *tracheae* carry the air directly to the tissues).

Closed Circulatory Systems

The simplest forms of closed circulatory systems are found in echinoderms and annelids, which include sea cucumbers and earthworms, respectively. More complex systems are found in vertebrates. The circulatory system of the earthworm is a good example of the general plan of a closed system (Figure 32.2). Blood is pumped into large, muscular vessels from a series of five *aortic arches*. These arches are the equivalent of the single pump, or heart, that forces blood through the more advanced vertebrate circulatory systems.

As blood flows from the arches, it passes through smaller and smaller vessels, finally moving into a fine meshwork of **capillaries**—small vessels whose walls are seldom more than one cell thick. Diffusion between the cells of the body and the blood takes place across the capillary walls. It is estimated that no cell in the human body is farther than 0.1 mm from a capillary.

Gradually, blood flows out of the capillary system into larger and larger vessels, which ultimately return the blood to the aortic arches to be pumped through the system again. A closed system produces a higher pressure than an open system does, allowing blood to circulate more rapidly. In many closed systems, including that of the earthworm, blood carries respiratory gases as well as food and waste materials.

THE VERTEBRATE CIRCULATORY SYSTEM

The closed circulatory systems of vertebrates have several important features in common: They all contain a single muscular heart with multiple chambers, and they all contain a number of special features that integrate circulation with gas exchange in the organs of *respiration*.

Vertebrate blood is itself a living tissue, and it contains a range of specialized cells. Its functions include (1) the transportation of nutrients, gases, and wastes; (2) the maintenance of a proper internal environment; and (3) the protection of the organism against disease.

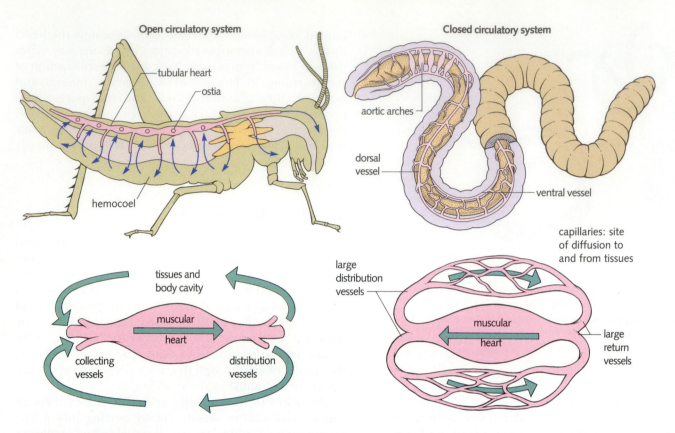

Open circulatory system

tubular heart
ostia
hemocoel

Closed circulatory system

aortic arches
dorsal vessel
ventral vessel

tissues and body cavity
muscular heart
collecting vessels
distribution vessels

capillaries: site of diffusion to and from tissues
large distribution vessels
muscular heart
large return vessels

Figure 32.2   The grasshopper (left) has an *open circulatory system*. Blood is forced through the animal's muscular heart and percolates directly through the tissues. Blood returns to the heart through the *ostia*. The earthworm (right) has a *closed circulatory system*. Blood is forced from the heart and passes into the body by means of vessels and capillaries. The restriction of blood flow to these vessels in a closed system allows the circulatory system to produce a higher pressure than an open system, enabling blood to flow more rapidly.

Vertebrate Blood Vessels

Vertebrate circulatory systems contain three distinctly different types of blood vessels: arteries, veins, and capillaries (Figure 32.3).

Arteries are vessels that carry blood away from the heart. Arteries are thick-walled vessels composed of four layers of tissue: an *endothelium* (a layer of flattened cells that line the artery wall), an elastic layer, a thick layer of smooth muscle, and an outer layer of connective tissue. These four layers are tough and elastic. Because blood is pumped from the heart in powerful spurts, blood in the arteries closest to the heart is under increased pressure. The elasticity of artery walls allows these vessels to expand to accommodate the surges of blood, and by the time blood flow reaches the smallest vessels, it has become smooth and continuous.

As blood moves away from the heart, the arterial system branches into smaller arteries, **arterioles,** which lead

directly to the smallest vessels of the circulatory system, the **capillaries.** Capillaries are blood vessels whose walls, as we have said, are only one cell thick. These tiny vessels form an intricate network that brings blood throughout the tissues of the body, enabling materials to diffuse across the capillary wall in both directions: into the blood and out of it. The interior of a capillary may be as small as 10 μm (just big enough for red blood cells to slip through in single file).

Near the end of the capillary circulation, blood begins to collect into larger vessels called **venules,** which channel blood into larger **veins** that lead eventually to the heart. The walls of veins are similar to arteries: They are composed of three layers—an endothelium, a layer of smooth muscle, and an outer layer that contains both and connective tissue. Because the venous blood has travelled a great distance from the heart, the source of fluid pressure in the system, blood passes through the veins under much less pressure than it does through arteries. Many

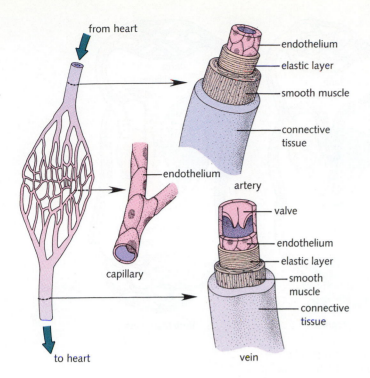

from heart

endothelium
elastic layer
smooth muscle
connective tissue

artery

endothelium

capillary

to heart

valve
endothelium
elastic layer
smooth muscle
connective tissue

vein

Figure 32.3 The basic structure of arteries, capillaries, and veins. Arteries and veins are multilayered structures which include layers of elastic tissue and connective tissue, as well as a layer of smooth muscle. The walls of the smallest capillaries, by contrast, are only one cell thick, bringing blood into close contact with the surrounding tissues.

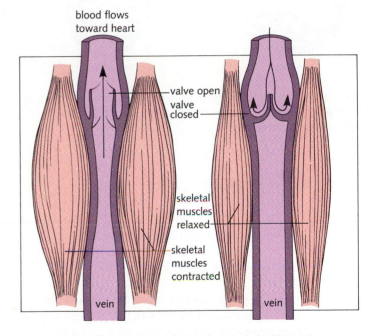

blood flows toward heart

valve open
valve closed

skeletal muscles relaxed

skeletal muscles contracted

vein

vein

Figure 32.4 Many veins are located near skeletal muscles. When these muscles contract, they help to force blood through the vein. Veins contain one-way valves which prevent the backflow of blood.

veins have **valves** that prevent blood from flowing in a reverse direction. In some vertebrates, veins are located near skeletal muscles so that when a muscle contracts, blood is forced through the venous system (Figure 32.4).

The Vertebrate Heart

The center of the vertebrate circulatory system is the heart. Heart structure differs among major vertebrate groups, and comparisons among those variations offer clues to the evolutionary process that gave rise to the mammalian heart.

The simplest vertebrate heart is the *two-chambered heart* found in fish. Blood collects in a single large vein, the *sinus venosus,* and enters the first chamber, the **atrium.** It is pumped into the second chamber, the muscular **ventricle,** which forces it into a single large artery, the *conus arteriosus.* (The words *atrium* and *ventricle* are derived from Latin. The atrium was the room by which one entered a Roman house, and the word *ventricle* is derived from a word that means "to depart.")

The pattern of blood flow in fish is known as **single circulation,** because the blood passes through the entire circulatory system after each pass through the heart. One difficulty with this pattern of blood flow is that blood must pass through two systems of capillaries, one right after another. The first capillary network is in the gills where the blood is oxygenated; and the second is in the

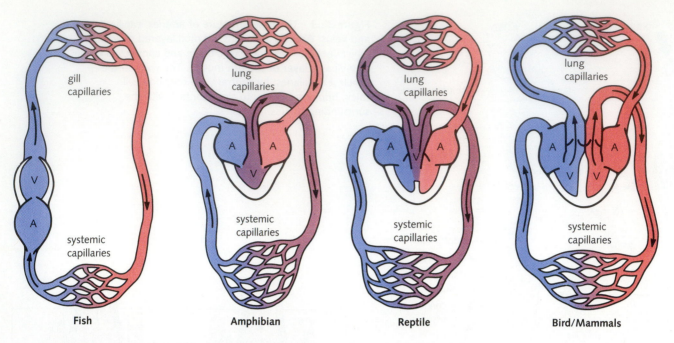

Figure 32.5 Circulatory patterns in fish, amphibians, reptiles, birds, and mammals. Oxygen enters each circulatory system in lung or gill capillaries, and is carried to the tissues in the systemic capillaries.

cells and tissues of the rest of the body. The passage of oxygen-poor blood through the gills reduces fluid pressure, slowing the rate at which oxygen-rich blood moves through the rest of the body, limiting the efficiency of oxygen delivery.

The circulatory systems of higher vertebrates show a trend toward separation of oxygen-rich and oxygen-poor blood, which improves the efficiency of oxygen delivery and helps to support higher metabolic rates in body tissues. Amphibians have a *three-chambered heart* that directs part of the circulatory flow to the *lungs*, making it possible for these animals to use their lungs to exchange gases directly with the air. As shown in Figure 32.5, all the blood that passes through the lungs is returned to the heart, where it is mixed with blood returning from the systemic circulation and pumped back out through a vessel called the **aorta.**

In amphibians, the aorta splits into two main vessels, one that goes to the lungs and one that enters the systemic circulation. Blood flowing in this system moves in a pattern known as **double circulation,** because there are two basic pathways that blood leaving the heart may enter: the **pulmonary circulation** leading to the lungs and the **systemic circulation.**

The circulation of blood in reptiles is similar to that in amphibians, but an important difference illustrates

the beginning of an evolutionary trend. In most reptiles the single ventricle is partially divided by a muscular wall, as shown in Figure 32.5. This wall helps to separate the blood that enters the right and left atria, and experiments have shown that as a result, there is very little mixing of oxygen-rich blood from the lungs with oxygen-poor blood from the systemic veins.

Birds and mammals seem to have completed the evolutionary experiment begun in reptiles. Their ventricles are completely divided into two separate chambers, a right and a left ventricle, so that there is no chance for oxygen-rich blood to mix with oxygen-poor blood. These **four-chambered hearts** can be thought of as two separate pumps. The right atrium collects blood from the veins, and the right ventricle pumps it into the lungs. The left atrium fills with blood from the lungs, and the left ventricle pumps this oxygen-rich blood into the systemic circulation.

This four-chambered heart ensures that (1) *all* the blood that returns to the heart is pumped through the lungs where gas exchange takes place and (2) *all* the oxygen-rich blood returning from the lungs is immediately pumped into the systemic circulation. This double circulation is particularly well suited for animals that are very active and have a high demand for energy, because it pumps fully oxygenated blood into the circulatory system at high pressure and high velocity. Therefore, it is not surprising to find that birds and mammals (warm-blooded animals with high metabolic rates) possess circulatory systems that exhibit this pattern.

The Structure of the Human Heart

The human heart is a cone-shaped muscle a bit larger than a fist (Figure 32.6). It is located in the thoracic cavity, in a space between the lungs, just to the left of the midline of the body. The heart is contained within a loose sac known as the *pericardium* ("around the heart"). A thin film of fluid between the surface of the heart and the pericardium serves to prevent friction during the heartbeat. The wall of the heart muscle is composed of an outer layer, the *epicardium;* a thick middle layer of muscle tissue, the *myocardium;* and an inner lining, the *endocardium.*

The work of the heart is done largely by powerful contractions of the myocardium. In the ventricles, myocardial tissue forms column-like projections known as *papillary muscles.* The myocardium contracts in a regular fashion, providing most of the force that causes blood to flow through the circulatory system.

The heart muscle pumps about 70 ml of blood with each contraction, and it beats about once a second, sleeping or waking, without taking even a few minutes' rest, during an entire lifetime. From the time your heart started beating before birth, to the time it stops, it will have pumped roughly 55 million gallons of blood through your body.

The heart is divided into right and left halves. Each side of the heart consists of two chambers: an atrium through which blood enters the heart and a ventricle through which it leaves. Although blood flows between the two sides of the heart during embryological development, a thick *septum* gradually develops that closes completely at birth, preventing blood flow between the sides. In some infants, the septum does not close properly, allowing blood to flow between the left and right sides of the heart. This produces a mixture of oxygen-rich and oxygen-poor blood that limits the efficiency of the circulatory system. If this problem is not corrected by surgery, these babies are permanently weakened by the delivery of only oxygen-poor supplies of blood to their cells.

Blood Flow Through the Heart

Blood enters the heart through the left and right atria, filling each of these chambers. When the heart muscle contracts, blood moves from the atria into the ventricles (Figure 32.7). Contraction of the ventricles does not force blood back into the atria, because special flaps of tissue form *valves* across the passageways between the atria and the ventricles. The **atrioventricular valves,** or AV valves, are easily pushed open when blood flows through from

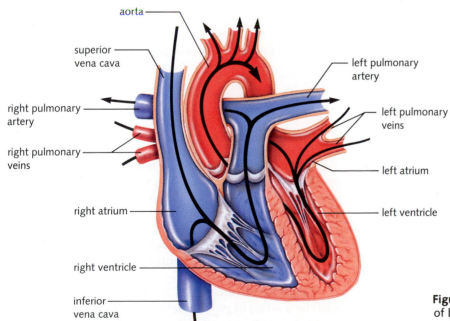

aorta

superior
vena cava

right pulmonary
artery

right pulmonary
veins

right atrium

right ventricle

inferior
vena cava

left pulmonary
artery

left pulmonary
veins

left atrium

left ventricle

Figure 32.6 The basic pattern
of blood flow through the valves
and chambers of the human heart.

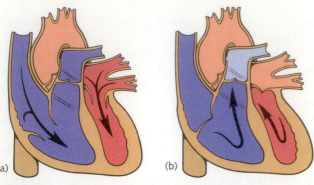
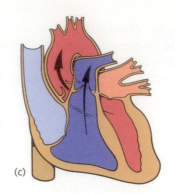

(a) (b) (c)

Figure 32.7 The opening and closing of cardiac valves is crucial to the pumping action of the heart. As the heart muscle relaxes, blood flows through the atria and into the ventricles (**a**). As the heart muscle contracts, the valves between atria and ventricles close (**b**) and blood is forced from the ventricles into the major vessels leading to the lungs and body (**c**).

the atria to the ventricles. But when the ventricles contract and blood is forced in the other direction, these valves are forced shut. Another set of valves, the *pulmonary* and *aortic valves,* are located where blood leaves the ventricles to enter the lungs and the systemic circulation, respectively. (These valves owe their name, *semilunar valves,* to their half-moon shape.)

When blood leaves the right side of the heart, it passes through the **pulmonary arteries** into the lungs, where it releases carbon dioxide and picks up oxygen, a process we will consider in detail in the next chapter. When it leaves the left side of the heart, it passes through a thick, muscular artery known as the aorta. This major vessel branches off to send blood to every branch of the systemic circulation. The two sides of the heart represent the two major circuits of the mammalian circulatory system: The right side of the heart pumps blood through the lungs (the *pulmonary circulation*), while the left side pumps oxygen-rich blood through the body (the *systemic circulation*).

Blood supply to the heart muscle is critical Although an enormous volume of blood passes through the chambers of the heart, the heart muscle itself gets very little nourishment or oxygen from the blood it pumps. Instead, a pair of **coronary arteries** (right and left) branch off from the main systemic circulation and feed a network of tiny capillaries that permeate the heart muscle (Figure 32.8). These vessels are relatively small, and any blockage of the circulation through the coronary arteries can deprive the heart muscle of oxygen. When blockage of the coronary arteries is extensive enough, cardiac muscle cells begin to die, causing a heart attack.

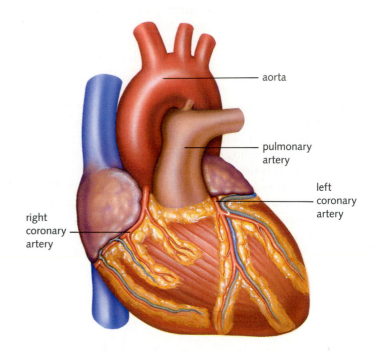

Figure 32.8 Nourishment and oxygen are supplied to the cells of the heart through the coronary arteries.

Control of Heartbeat

Although the heart beats as a single unit, all the cells of the heart do not contract at exactly the same time. Rather, contraction begins in the atria and spreads to the rest of the muscle in a wave-like pattern. Unlike skeletal muscles, the nervous system does not cause each cardiac muscle cell to contract. Instead, individual cardiac muscle cells have the innate ability to contract in a regular, rhythmic cycle.

Each cardiac muscle cell is connected to neighboring cells by a series of *gap junctions* (described in Chapters 24 and 31) that allow small molecules and ions to pass directly from one cell to another (Figure 32.9). When several cardiac muscle cells are grown together in a culture dish, they form gap junctions with each other and gradually begin to beat as a unit. Each contraction is accompanied by the flow of current (in the form of ions) across the muscle cell membrane. Because these ions can also flow through gap junctions, the contraction of one cell begins an ion flow that starts a contraction in the neighboring cell that gradually spreads from cell to cell.

Specialized cardiac muscle cells speed the movement of impulses throughout the heart. These impulses originate from a region of the heart known as the **sinoatrial node** (otherwise known as the **SA node**) located in the upper corner of the right atrium. The sinoatrial node is a source of a wave of contraction in the heart that spreads through both atria via the gap junctions linking cardiac muscle cells. This region is known as the **pacemaker** of the heart. The wave of excitation initiated at the pacemaker causes the two atria to contract together, helping to force blood into the ventricles.

The impulse does not spread directly to the ventricles, however, because there is a region of connective tissue that does not conduct impulses. Instead, the impulse is picked up by a special bundle of conducting fibers (the atrioventricular bundle of His) that originate at an area known as the **atrioventricular node (AV node).** These fibers conduct the impulse to the lower regions of the ventricles, which then contract smoothly to force blood up and out of the ventricles through the semilunar valves. It takes about one-tenth of a second for the impulse to travel from the AV node to the muscular walls of the ventricles, so there is a delay between atrial contractions and ventricular contractions.

In some forms of heart disease, the AV node does not initiate an impulse properly. In its place, other groups of cells may begin to initiate independent contractions, and the usually smooth wave of contraction may become broken up into a series of partial contractions. Blood pumping becomes much less efficient, and the inadequate oxygen supply may cause a feeling of weakness. This condition can be remedied by an *artificial pacemaker,* a tiny electronic device that is surgically installed near the AV node and initiates normal contractions by sending out regular electrical pulses to simulate the pulses of a healthy pacemaker.

At rest, the normal heart rate varies from 60 to 80 beats per minute. During anxiety or vigorous exercise, the heart rate increases; sometimes it approaches 200 beats per minute. Although the nervous system does not initiate each individual heartbeat, it can influence heart rate. Nerves that originate in the central nervous system terminate at the SA node. These nerves release chemical messengers that can increase or decrease heart rate.

Figure 32.9 Electron micrograph of the boundary between two cardiac muscle cells. The arrows point to two regions which contain gap junctions. These junctions allow electrical current to flow directly from one cell into the next, enabling the heart muscle to contract as a unit.

Figure 32.10 The heartbeat begins in a region of the heart muscle known as the *sinoatrial node* (also called the *pacemaker*). The contraction spreads through the heart muscle like a wave. Conducting fibers which originate at the *atrioventricular node* spread the contraction to the ventricular region of the muscle.

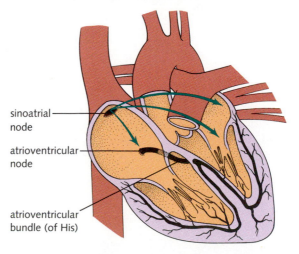

sinoatrial node

atrioventricular node

atrioventricular bundle (of His)

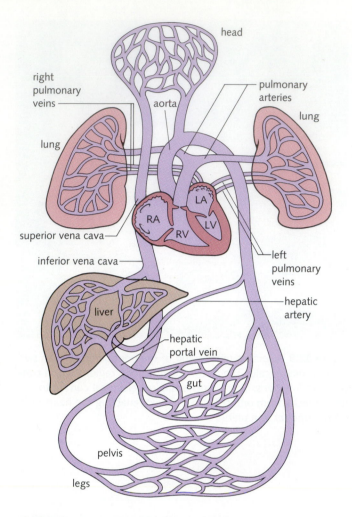

Figure 32.11 A schematic representation of blood flow through the human body. Throughout most of the body, arterial blood passes through a single bed of capillaries before it returns to the venous circulation. However, blood passing out of the capillaries associated with the gut is collected in the *hepatic portal vein* and then passes through a second set of capillaries in the liver. The *hepatic portal circulation* ensures that nutrients absorbed from the gut will pass through the liver before they reach the general circulation.

The Cardiac Cycle

The two-stage cycle pattern of the heartbeat moves blood through the heart's chambers efficiently (the pumping of the atria "primes" the ventricles, so they are full of blood when they contract), and it also produces the characteristic sounds of a heartbeat. If you listen to a heartbeat through a stethoscope, you will notice that it is a double sound: described as "Lub-dub . . . lub-dub."

The sounds are produced by the closing of heart valves. "Lub," the first sound, occurs when the AV valves slam shut at the beginning of a ventricular contraction; "dub" follows when the semilunar valves leading from the ventricles close after blood has been ejected. The heart is completely at rest for about 0.4 sec before a new cycle of contraction begins. Physicians can detect many heart problems, especially those involving the valves, by listening to the heartbeat.

CIRCULATORY PATTERNS IN THE HUMAN BODY

The Pulmonary Circulation

The mammalian circulatory system is organized in a way that enables it to pump the entire blood volume through the lungs before returning it to the rest of the body. This part of the system is known as the pulmonary circulation. Blood passes from the right ventricle to the lungs via the pulmonary arteries and returns to the left atrium via the pulmonary veins (Figure 32.11).

One interesting aspect of this configuration is that it creates an exception to our usual impressions of the circulatory system. Unlike most other arteries and veins, the pulmonary arteries carry *oxygen-poor* blood, and the pulmonary veins carry *oxygen-rich* blood. In the lungs, the pulmonary circulation is routed through a network of tiny capillaries that allow gas exchange to take place between the blood and the air that has been inhaled.

The Systemic Circulation

Blood leaving the heart flows through the largest artery in the body, the aorta, which has an opening about an inch in diameter. The major arteries that lead to the upper and lower parts of the body branch off from the aorta, and they in turn give rise to many smaller vessels that carry blood to individual organs and tissues.

In the organs of the body, arteries branch and rebranch until arterioles lead to capillaries. The capillaries form profuse networks throughout the tissues, where the thin capillary walls facilitate the exchange of oxygen, nutrients, hormones, carbon dioxide, and waste products.

Blood that has passed through the capillary networks is collected by the venules that lead into the larger veins. These veins lead into two major vessels, the **superior vena cava,** which collects blood from the head, neck, and arms, and the **inferior vena cava,** which collects blood from the rest of the body. These large vessels lead directly to the right atrium.

Control of Blood Circulation Patterns

It may be tempting to assume that the circulatory system is just so much passive plumbing through which the blood is pumped like cooling fluid circulating through an engine. It is not. The circulatory system is a flexible, dynamic collection of elements that can quickly accommodate the demands of the body.

Capillaries are often grouped into clusters called *capillary beds* that carry blood from an arteriole to a venule. The rate at which blood may enter the capillary bed is partially controlled by a **precapillary sphincter,** a control region encircled with smooth muscle (Figure 32.12). As long as the smooth muscle is contracted, very little blood can enter the capillary bed. However, other control factors in the body (including the nervous system) may cause the sphincter to open, allowing blood flow to increase through the capillary bed and changing the pattern of circulation.

Several tissues in the body regulate blood flow by means of precapillary sphincters. In many other tissues, blood flow is controlled by contractions of the smooth muscles that surround small arteries and arterioles. Although skeletal muscles have a rich blood supply, when a muscle is called on to work very hard, its capillary beds open and the muscle fills with blood, increasing the rate at which oxygen is carried to the tissue. This increased flow of blood during heavy exercise causes a noticeable increase in the size of the muscle, an effect that weight lifters call "pumping up."

The contractions of smooth muscles surrounding arterioles in the skin produce changes in circulatory patterns that are easy to see. When a light-skinned person is frightened or nervous, these arterioles contract, limiting blood flow to the skin—such a person may literally "pale" in fear. At times when the body temperature is too high, these smooth muscles relax and the face, "blushing" as a result of increased blood flow, becomes reddish. The small intestines, where food absorption takes place, also have a rich blood supply controlled in this way. After a full meal, the capillary beds in the small intestine open, allowing more blood to flow through them so that water and digested food can be absorbed as rapidly as possible. When absorption is complete, the sphincters close again, restricting the blood supply and changing the overall flow pattern of the circulatory system.

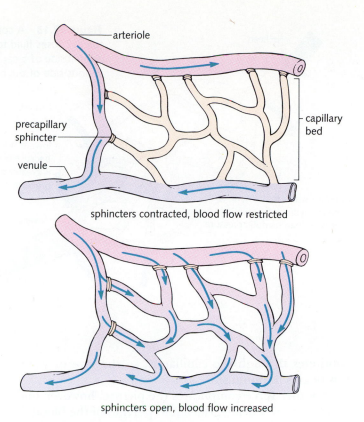

Figure 32.12 Blood flow to muscles and other tissues may be controlled by the opening and closing of precapillary sphincters.

Material Exchange in the Capillaries

The major vessels of the circulatory system—the large arteries and veins—are the ones that most easily attract our attention. However, for most of the cells of the body, the important work of the circulatory system is done across the capillary wall. The extensive network of capillaries makes contact with an enormous surface area of body tissues. This brings the blood into close contact with nearly every cell in the body. The thin wall of the capillary allows materials to leave the blood and also permits molecules in the **interstitial fluid** ("in-between" fluid) that surrounds the cells to enter the blood.

Two important forces act in opposition across the capillary wall (Figure 32.13). As oxygenated blood enters the circulatory system from the heart, it is under significant hydrostatic pressure. This pressure is great enough to force water and small molecules across the membranes that surround arterioles and capillaries, and there is a substantial loss of fluid from the blood into the interstitial fluid. Blood pressure drops dramatically as circulation

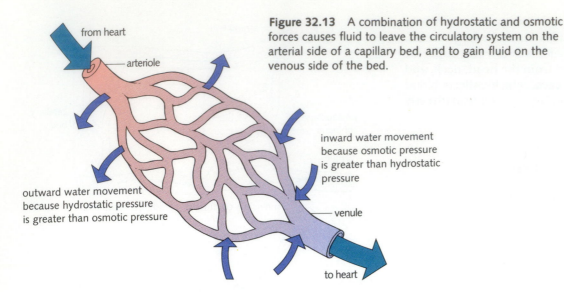

Figure 32.13 A combination of hydrostatic and osmotic forces causes fluid to leave the circulatory system on the arterial side of a capillary bed, and to gain fluid on the venous side of the bed.

continues through the capillaries, so the rate of water loss begins to diminish.

Another force complicates the picture, however, and that is *osmosis*. **Plasma,** the liquid portion of the blood, is a thick liquid containing a rich assortment of dissolved salts, sugars, and proteins. Interstitial fluid contains many of these materials, too, but it lacks the proteins.

Thus plasma is *hypertonic* with respect to interstitial fluid, and the forces of osmosis tend to drive water into the blood. As long as the hydrostatic pressure is higher than the osmotic pressure, water and solutes are lost from the capillaries to the surrounding tissue. But when hydrostatic pressure drops off in the capillaries and venous circulation, osmosis drives fluid back into the plasma.

This movement of fluid out of the circulatory system and back in again has a positive effect: It produces a current of fluid movements that helps move materials between the circulation and the tissues of the body, helping the circulatory system do its work. It also tends to keep the composition of the blood stable. Despite all the movement of water, both ends of the system roughly balance out, so that more than 99 percent of the fluid volume that leaves the heart returns again in the venous circulation.

Blood Pressure

Each beat of the heart exerts a force known as **blood pressure** against the walls of the blood vessels. Blood pressure is a hydraulic force, and it can be studied and analyzed in the same way the forces on other fluids are. Because the heart is not continuously contracted, blood pressure rises and falls with the contraction cycle. The expansion and contraction of arterial vessels with the rise and fall of pressure constitute a **pulse.**

Arterial blood pressure   When the ventricles of the heart contract (a state known as **systole**), blood pressure reaches a maximum in the aorta and the major arteries. When the ventricles relax (a state known as **diastole**), pressure drops to a minimum in these vessels. Blood pressure is therefore expressed as two numbers: systolic pressure and diastolic pressure.

Blood pressure measurements are made with a *sphygmomanometer,* which is connected to a pressure cuff. This device measures blood pressure in units of millimeters of mercury, abbreviated "mm Hg." The pressure of the atmosphere at sea level is 760 mm Hg, equal to a column of mercury 760 mm high. A blood pressure reading of "100" *exceeds* atmospheric pressure by 100 mm Hg.

The person taking a blood pressure reading listens with a stethoscope for the sound of blood flowing through an artery in the arm. The cuff is gradually inflated until the pulse disappears completely. This happens when the pressure of the cuff cuts off the flow of blood through the surface artery. Then the pressure in the cuff is gradually released. When the blood flow is audible again, the pressure of the cuff is noted. This is the **systolic pressure,** the maximum pressure exerted by a heart contraction. However, the blood still flows in spurts—flowing only when it exceeds the pressure of the cuff. The pressure is now lowered still further until a change in sound indicates that the blood is flowing constantly; the blood pressure at this point is the **diastolic pressure.**

Detecting the second sound is difficult and requires considerable experience. Knowing both systolic and diastolic pressure is important, however, because the difference between the two can be critical in diagnosing hypertension. In a young adult, for example, normal blood pressure is roughly 120 mm Hg systolic and 80 mm Hg diastolic; this is expressed as 120/80.

Blood pressure is always measured in the same place (at the inside of the elbow joint with the cuff on the upper arm) because actual blood pressure varies from one part of the circulatory system to another. Blood pressure is highest in the major arteries closest to the heart muscle. That pressure steadily drops as blood moves through arteries, arterioles, and capillaries. Pressure is lowest in the venous part of the system (Figure 32.14).

Factors affecting blood pressure Blood pressure in a healthy individual can be influenced by a number of factors. As the heart rate increases, pressure increases with it, because more blood is being forced into the system. Contraction of the smooth muscles that line the arteries and veins also increases blood pressure by reducing the volume of the circulatory vessels.

Blood pressure is influenced by the volume of blood in the circulatory system. Normally, blood volume remains relatively constant. But when more fluid is added to the blood, its volume increases and blood pressure rises. When fluid is lost, on the other hand, blood volume decreases and blood pressure falls. As you will see in Chapter 36, several organs controlled by hormonal feedback loops regulate the fluid content of blood.

All these mechanisms affecting blood pressure and blood volume normally operate to ensure that blood flow to various parts of the body can be increased or decreased in response to the demands of exercise and physiology. When any of these regulatory systems fails, however, the results can be quite serious.

DISORDERS OF THE CIRCULATORY SYSTEM

Any medical disorder that affects heart muscle can be fatal, for the body cannot survive an interruption in blood flow that lasts longer than a few minutes. *Diseases of the circulatory system are the number-one cause of death* in the United States and several other developed nations.

Hypertension

Hypertension, the technical term for excessively high blood pressure, is a serious medical problem. The heart must pump blood against the diastolic pressure in the arteries, so as that pressure rises, the work load on the heart increases. Heart muscle must work nearly twice as hard to pump the same volume of blood in an individual

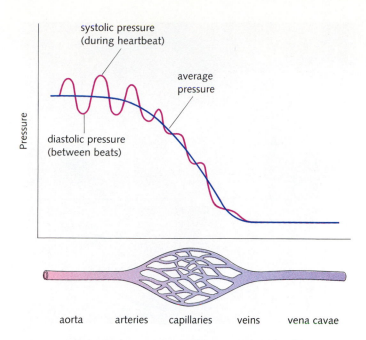

Figure 32.14 Blood pressure drops as the blood moves through the circulatory system.

with a blood pressure of 180/140 as in someone with a blood pressure of 100/70.

Both numbers are important, however. A large increase in *systolic pressure* may mean that the circulatory system has become less elastic. Normally, the major arteries are elastic enough to expand slightly as the ventricles contract, moderating the pressure as blood surges out of the heart. Sometimes, however, arteries "harden" as a result of the buildup of fatty tissue on the arterial inner surface. This condition, known as *atherosclerosis,* is discussed in more detail below.

The extra load that high blood pressure imposes on the heart has its consequences: The heart muscle fatigues easily, can be more easily damaged, and has less reserve capacity to draw on if sudden demands are made on the body. For these reasons, physicians often try to control high blood pressure with a number of treatments. The usual recommendations include moderate exercise and a reduction of salt in the diet.

What has salt to do with blood pressure? Eating too much salt increases the amount of sodium ion in the blood. In some individuals, this increased sodium load causes the body to retain more fluids, increasing both blood volume and blood pressure.

Other recommendations for controlling blood pressure include prescription drugs that help to lower blood pressure by reducing blood volume, and a reduction in fatty foods to reduce atherosclerosis.

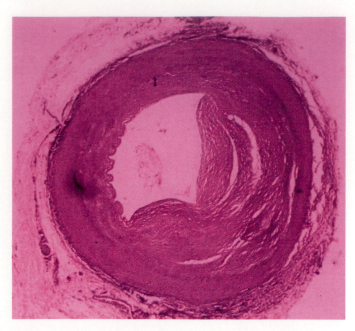

Figure 32.15 Severe atherosclerosis has produced fatty deposits large enough to block more than 60% of the area of this coronary artery.

Atherosclerosis

Many problems in the circulatory system are produced by a disorder known as **atherosclerosis,** or "hardening of the arteries." In atherosclerosis, the normally smooth and unobstructed inner surface of the body's arteries is lined with fatty deposits and clusters of tissue produced by abnormal cell growth (Figure 32.15). We will discuss the nutritional and metabolic factors that *cause* atherosclerosis in Chapter 34; here we will concern ourselves only with its *effects* on the circulatory system.

When atherosclerosis proceeds too far, the obstruction it creates within arteries can have two serious consequences. First, as we have already seen, it can seriously impede the flow of blood and increase blood pressure. Second, it increases the risk that a normally harmless blood clot can block the artery altogether. This condition is serious everywhere, but it is most dangerous in the brain and in the coronary arteries, the vessels that feed blood to the heart muscle.

Heart Attack

Atherosclerosis in coronary arteries is extremely dangerous, because the constantly beating heart muscle requires an equally constant supply of oxygen and nutrients in order to survive. As coronary arteries begin to narrow, blood flow may become insufficient to support high rates of heart muscle activity, and the heart may be seriously weakened and possibly damaged.

When a coronary vessel becomes completely closed, part of the heart muscle may begin to die for lack of blood, causing intense pain and failure of the heart to pump sufficient blood to the rest of the body. This is known as a **myocardial infarction,** or **heart attack.** When a heart attack strikes, *immediate* medical attention is necessary to provide CPR (cardiopulmonary resuscitation) and the anti-clotting medication and drugs that will quickly increase the flow of blood to the heart.

Many heart attacks occur not when atherosclerosis completely seals an artery but when a small blood clot, called a **coronary thrombus,** becomes lodged in a partially obstructed region of the coronary artery. As the clot blocks the flow of blood through that vessel, the heart attack begins. Medical scientists believe that many such clots form elsewhere in the body and drift through the circulatory system until they become trapped in a small vessel or passageway.

Risk factors in heart disease There is no single, simple cause of heart disease. Rather, there are several known risk factors that—especially in combination with one another—increase the likelihood of heart attack. Given the severity and prevalence of heart disease in this country, it is important that everyone understand these risk factors and take appropriate action.

Certain risk factors are genetic; individuals may inherit from either of their parents a predisposition to heart disease in general or to atherosclerosis in particular. Some of these problems can be treated with medication, but to the extent that they are genetically controlled, they persist throughout life.

Several other risk factors, however, are produced by certain patterns of behavior and can therefore be eliminated by appropriate changes in those behaviors.

First, it is now clear that eating too much saturated fat and **cholesterol** leads to atherosclerosis. Second, cigarette smoking enormously increases a person's risk of coronary disease, although the mechanism by which it does so is not completely understood. There is also a high degree of correlation between obesity and lack of exercise and the incidence of heart disease. High blood pressure further compounds the risk, because it increases the amount of work the heart must do and makes even the slightest interruption in coronary blood flow that much more serious.

For these reasons, the best advice physicians can give for avoiding heart attacks are reminders to exercise regularly, to avoid fatty foods and cigarettes, and to control body weight.

EKGs

Every time the heart beats, each of its cardiac muscle cells contracts. The electric currents produced by the impulses that stimulate those contractions are strong enough to be detected at the surface of the body. An *electrocardiogram* is a procedure that produces a recording of the electrical activity of the heart muscle from a series of electrodes placed at the surface of the skin. These electrodes are typically placed at the wrists and on the left ankle.

A normal EKG (below, left) shows five distinct components, or waves, labeled P, Q, R, S, and T. The P wave is produced by the electrical impulse that initiates contraction in the atria—the first event in the contraction cycle. The Q, R, and S waves result from similar impulses in the ventricles. The T wave is generated by postcontraction recovery in the ventricles. The pattern as a whole gives an experienced health worker an overall view of the health of the heart muscle.

A comparison of a normal EKG (right, top) with several abnormal recordings appears below. *Atrial fibrilla-* *tion* (right, middle) is a more serious condition in which uncoordinated impulses spread throughout the atria, and the P wave all but disappears from the EKG. *Ventricular fibrillation* (right, bottom) results from a similar condition in the ventricles. A heart in ventricular fibrillation is ineffective in pumping blood. Because the ventricles do the lion's share of pumping, ventricular fibrillation is fatal unless it is quickly corrected by administration of drugs or electric shock.

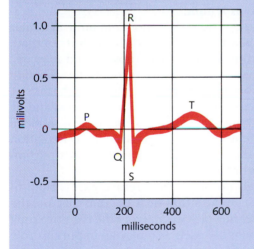

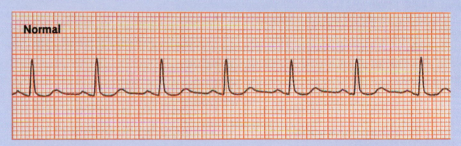

Normal

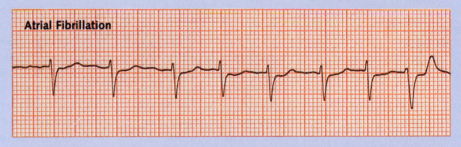

Atrial Fibrillation

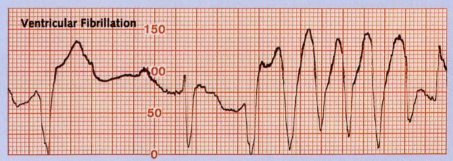

Ventricular Fibrillation

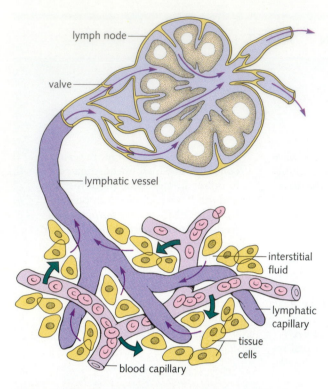

lymph node

valve

lymphatic vessel

interstitial fluid

lymphatic capillary

tissue cells

blood capillary

Figure 32.16 Fluid lost from the circulatory system is collected into the vessels of the lymphatic system. This fluid passes through lymph nodes and is eventually returned to the circulatory system. The lymphatic system plays a vital role in a number of physiological activities including the immune system.

Stroke

A drifting blood clot need not block a coronary artery to cause serious problems. Such a clot may move into the blood vessels leading to the brain, where it may become stuck and block the flow of blood to a group of nerve cells. If the clot manages to block blood flow completely, the brain cells served by the vessel begin to die for lack of oxygen, a condition known as a **stroke.** Strokes can happen without warning, although they are made more likely by the same conditions that have been identified as favoring heart attacks: smoking, atherosclerosis, and high blood pressure.

Strokes vary in their effect on individuals, depending on the region of the brain affected. If the stroke affects brain cells that control movement, a patient may become paralyzed. If blood flow to the visual center of the brain is blocked, blindness may result.

Because strokes are localized, their effects may be confined to one side of the body; temporary or permanent loss of the use of one leg or one arm is not uncommon. Each half of the brain controls motor functions on one side of the body, and the localized nature of a single blood clot accounts for the localized effects of a stroke.

THE LYMPHATIC SYSTEM

We noted earlier that 99 percent of the blood volume returns from capillary beds into the venous system. This implies that about 1 percent of the blood volume *does not* return. What happens to it? The system is set up in such a way that a bit more fluid leaves the capillaries on the arterial side than reenters on the venous side. This extra fluid enters the interstitial fluid, and over time, the volume of interstitial fluid increases. If nothing is done to counteract the flow, large amounts of excess fluid build up in the tissues of the body.

Fortunately, a special system exists to collect the fluid. Thin-walled tubes known as **lymph vessels** absorb the excess fluid and move it slowly through the body. The lymph vessels are part of the **lymphatic system,** and the fluid that moves through the vessels of the lymphatic system is known as **lymph.**

The lymphatic system does not contain a muscular pump like the heart to move fluid through it. Instead, lymph vessels contain valves like those found in veins, which ensure that fluid can move through the system in only one direction. Lymph flows slowly through this system, driven by the osmotic pressure of the blood and by occasional mechanical pressure from nearby skeletal muscles, until it empties back into the general circulation by flowing into the subclavian veins in the upper portion of the chest (Figure 32.16).

Lymph vessels do not merely return excess fluid to the circulation. They also flow near the cells that line the intestines, where they pick up fat from the digestive tract (Chapter 34). Lymph vessels also play an important role in the movement of white blood cells to aid in the immune response. It is essential that the flow of lymph not be blocked. *Edema,* a swelling of the tissues that is due to excess fluid accumulation, can occur when the lymphatic vessels do not function properly.

A parasitic disease caused by a roundworm can destroy and block lymph vessels, resulting in a grotesque swelling known as *elephantiasis.* Another cause of swelling is nutritional. If the diet does not contain enough protein, the protein concentration of the blood falls. Less fluid may return to the circulation by the process of osmosis, increasing the load on the lymphatic system. Beyond a certain point, the system cannot cope with the added work, and the tissues of the body swell—an ironic consequence of prolonged malnutrition.

BLOOD COMPOSITION

For certain purposes, it is convenient to consider blood a fluid and to imagine it flowing through the circulatory system like water through a garden hose. But we must never forget that blood is a tissue, that it contains billions of living cells, and that these cells are absolutely vital to the functioning of the circulatory system.

Plasma

The **plasma,** the fluid in which these cells are found, is 90 percent water. Nearly 7 percent of blood plasma is composed of plasma proteins. There are three major kinds of plasma proteins: **globulins,** which are produced by the immune system and help protect the body against infection; **fibrinogen** and **prothrombin,** which control the clotting process; and **albumins,** which stabilize the plasma by binding fats and other insoluble molecules and, with the other plasma proteins, regulate the osmotic character of the blood. We have already seen that blood is hypertonic with respect to interstitial fluid, and the plasma proteins are the principal reason why.

Blood plasma is slightly basic (its pH is usually about 7.4). If this pH is allowed to vary by as much as a single pH unit, serious problems develop. The solubility characteristics of dissolved proteins, carbohydrates, and salts are all pH-dependent, and the entire blood function of transporting materials can come to a halt if the pH changes rapidly. Vigorous exercise (which increases blood concentrations of carbon dioxide and lactic acid) and hyperventilation (which decreases blood levels of carbon dioxide) have the potential to upset blood pH. Plasma proteins (especially albumin) help to buffer the blood against these and other influences and to maintain its pH within allowable limits.

Salts dissolved in blood plasma include potassium, sodium, calcium, and magnesium (positively charged ions) and chloride, phosphate, carbonate, and sulfate (negatively charged ions). The concentrations of these ions vary within remarkably narrow limits, and large changes in ion concentration are indications of serious medical conditions.

Glucose is present in blood plasma as a result of digestive activities. A great deal of the glucose obtained in a meal is absorbed by the liver and stored in the form of glycogen. As the cells of the body absorb glucose from the circulation, the liver provides more of this sugar to keep the level of blood glucose within a narrow range. Precise control of blood sugar is especially important for cells in the brain, which obtain all of their energy from glucose carried in the bloodstream. The control of blood sugar is a complex process that involves the close coor-

dination of activities in the liver, digestive system, nervous system, and endocrine system. It will be discussed in detail in Chapter 36.

Lipids (fats) are carried in blood plasma bound to proteins, forming **lipoproteins.** One fatty compound of great medical interest is the steroid cholesterol. Cholesterol is carried in two different forms through the blood: in *high-density lipoproteins (HDLs)* and in *low-density lipoproteins (LDLs)*. A certain amount of cholesterol is essential to the body; it is used in the construction of many cell membranes. Excess cholesterol, however, particularly in the form of LDL, leads to atherosclerosis.

Blood Cells

Nearly 45 percent of the volume of human blood is cellular. Blood contains a range of cell types that perform a variety of functions (Figure 32.17). These living components of blood are of three types: erythrocytes (red blood cells; *erythro* means "red"), leukocytes (white blood cells; *leuko* means "white"), and thrombocytes (platelets).

Red blood cells The majority of blood cells, called **red blood cells** or **erythrocytes,** are shaped like biconcave disks, and they lack nuclei, which are extruded during

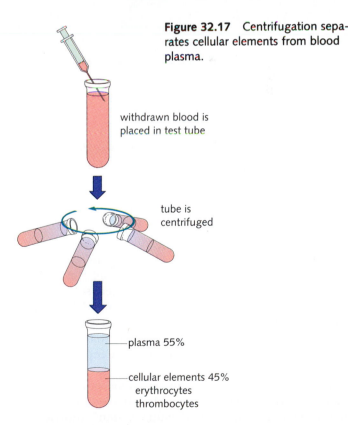

Figure 32.17 Centrifugation separates cellular elements from blood plasma.

withdrawn blood is placed in test tube

tube is centrifuged

plasma 55%

cellular elements 45% erythrocytes thrombocytes

Artificial Hearts

If the heart is nothing more than a pump, then it should be possible to replace it with a mechanical device that pumps blood. This concept has been the inspiration for a number of research teams who have sought to develop an artificial heart to replace worn or damaged human hearts—

and thereby to prolong the lives of patients suffering from severe heart diseases.

Although several labs have developed hearts that have had some success in animal trials, the first successful replacement of a human heart with a mechanical device was done in 1982 by surgeon William DeVries. DeVries removed the heart of Barney B. Clark, a patient who was suffering from acute heart failure, and replaced it with the Jarvik-7, a mechanical heart designed by William Jarvik. The heart was implanted into the chest and was driven by compressed air. The heart included ingenious mechanical devices that automatically compensated for minor variations in blood flow and closely simulated the pumping action of a normal heart.

Initially, the implant into Barney Clark was successful, and he regained consciousness, was able to talk and

move about his hospital room, and clearly survived much longer than he might have without the device. However, Clark suffered a number of seizures and strokes, and his condition eventually deteriorated so seriously that the heart was unable to keep him alive any longer. He died 112 days after insertion of the heart.

The Jarvik-7 and other mechanical hearts were used on more than a dozen patients since then. Although most of these experiments worked well enough to support the potential for designing satisfactory artificial hearts, the incidents of strokes, seizures, and other circulatory problems have been widespread. The Jarvik-7 has now been officially recalled. It is clear that we do not understand every aspect of the heart's relationship to the rest of the body's physiology. A successful artificial heart is possible, but we have not achieved it yet.

---

development (Figure 32.18). Erythrocytes are produced by cells in the bone marrow that undergo rapid cell division. These cells develop into erythrocytes gradually, producing large amounts of **hemoglobin** (the reddish, oxygen-carrying protein) and gradually losing their internal cellular organelles.

As they develop, these cells go through a final, irreversible step: Their nuclei are extruded and destroyed. When released from the bone marrow into the circulation, erythrocytes are really concentrated solutions of hemoglobin and other proteins surrounded by a specialized cell membrane. A new red cell has an average life span of about 4 months, meaning that roughly 1 percent of the body's erythrocytes must be replaced every day.

The number of red cells in the blood does not vary much from day to day, but it can be affected by environmental demands on the circulatory system. Individuals who participate in demanding aerobic exercises (those that stimulate the heart rate and place extra demands on the oxygen-carrying capacity of the blood) produce erythrocytes more rapidly than normal, resulting in a higher concentration of red cells in the blood. People who live at high elevations, where the air is much thinner, show similar changes in their red cell content. When we consider the respiratory system, we will look at the details of the interaction between oxygen and hemoglobin.

Although mammalian erythrocytes lose their nuclei during development, the erythrocytes of most other vertebrates—including birds, reptiles, amphibians, and fish—do not. These cells retain nuclei throughout their useful life span, although they suffer much the same loss of cellular organelles as mammalian erythrocytes do.

White blood cells   Vertebrate blood includes a number of other cell types that do not contain hemoglobin. These are known as **white blood cells** or **leukocytes,** to contrast them with the reddish color of the erythrocytes. There are about 500 erythrocytes for every leukocyte. Like erythrocytes, leukocytes are formed in the bone marrow, and they are released into the bloodstream as fully developed cells with active cytoplasmic organelles. There are five basic types of white blood cells, and they can be distinguished from each other by their size, the shape of their nuclei, and the types of granules visible in their cytoplasm (Figure 32.19).

Leukocytes are important components of the **immune system,** and they function in a number of ways to guard the body against attack from invading bacteria, viruses, and eukaryotic parasites. Some of the cells, including granulocytes and macrophages, deal with foreign cells by engulfing and destroying them, much as an amoeba might dispose of a small cell it happened across. Such cells are aptly called **phagocytes** ("eating cells"). Other cells, known as **lymphocytes** (because they are common in lymph fluid), deal with infection by producing special proteins known

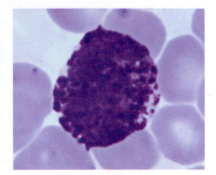

Figure 32.18 Human red blood cells as viewed in the light microscope (left) and scanning electron microscope (right).

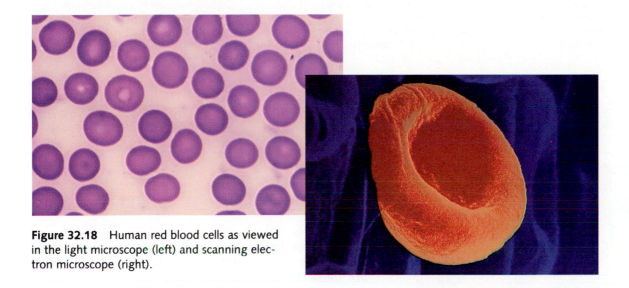

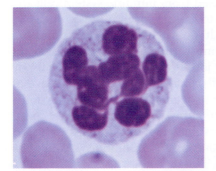

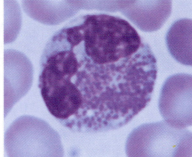

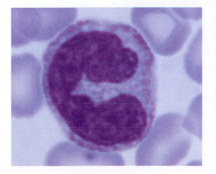

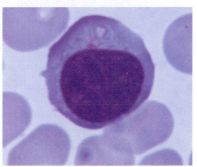

Figure 32.19 Five basic types of *leukocytes*, or white blood cells: (clockwise, from upper left) Neutrophil, eosinophil, basophil, lymphocyte and monocyte.

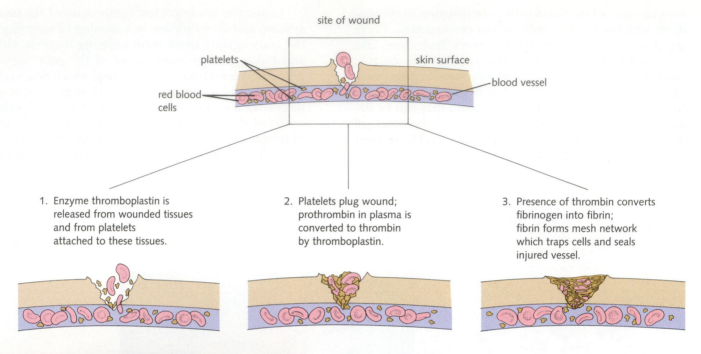

site of wound

platelets

skin surface

red blood
cells

blood vessel

1. Enzyme thromboplastin is released from wounded tissues and from platelets attached to these tissues.

2. Platelets plug wound; prothrombin in plasma is converted to thrombin by thromboplastin.

3. Presence of thrombin converts fibrinogen into fibrin; fibrin forms mesh network which traps cells and seals injured vessel.

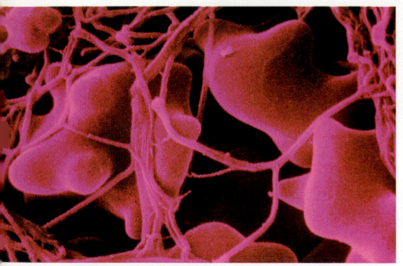

Figure 32.20 Blood clotting is triggered by the release of thromboplastin from broken tissues and platelets. This sets in motion a series of reactions that form the clot, stopping further bleeding and beginning the healing process. The scanning electron micrograph is of red blood cells tangled in the fibrin meshwork of a blood clot.

as antibodies that attach to and help destroy foreign cells. These leukocytes do not occur only in the blood: They are found in much greater numbers in the lymphatic system and circulating through various places in the body. They are also important in variations of the immune reaction, such as allergies, in which special chemicals are released from immune system cells in response to exposure to certain foreign molecules.

Although leukocytes are found in other tissues, circulating with the blood is an ideal way for them to do their job. It allows them to visit all portions of the body rapidly, enabling the blood to serve as a second line of defense against disease and infection. The operation of the immune system is much more complicated (and more interesting) than this suggests, and we will examine the functions of the immune system in Chapter 39.

Platelets and blood clotting As we have seen, blood is more than just a liquid. It can respond to changes in the circulatory system in a sophisticated way. Blood is even able to stop leaks in the circulatory system. We have all had minor cuts and scrapes that bleed for several seconds or minutes and then stop bleeding and begin to heal. The first stage in that process is the formation of a **blood clot,** a fibrous mass of material that forms automatically from molecules in the blood itself and stops the loss of fluid (Figure 32.20).

For many years, physicians believed that blood clotting was caused by exposure to the air. This was a natural enough assumption, given the fact that most visible wounds are open to the air. However, a great deal more is involved. One of the most important molecules in the clotting reaction is **fibrinogen,** a large protein produced by the liver and always present in normal plasma. Fibrinogen polymerizes to produce **fibrin** during clot formation. Fibrin forms a net-like meshwork that traps erythrocytes and platelets and actually creates the clot.

But if fibrinogen is present all the time, what controls its conversion to fibrin? When tissues are torn at the site of a wound, an enzyme called **thromboplastin** is released. This enzyme catalyzes the conversion of a soluble protein called **prothrombin** into another enzyme called **thrombin.** And thrombin, in turn, catalyzes the reaction in which fibrinogen is converted into fibrin.

The clotting process is aided by fragments of cells known as *thrombocytes* or **platelets.** Platelets are small pieces of cells called **megakaryocytes** that are found in the bone marrow. As these cells grow, small pieces of their cytoplasm are broken off and released into the circulation to become **platelets.** Platelets are essential to the clotting reaction. Their surfaces are quite sticky, and they easily attach to the broken tissue at the edges of a wound. There they seem to release thromboplastin, and the platelets serve as anchoring points for the network of fibrin that envelops the blood around a wound.

All of this may give you the impression that blood clotting is complicated. It is indeed. Blood clotting is caused by a sequence of reactions, each of which produces a product that catalyzes the next reaction in a *cascade* of chemical events resulting in the production of large amounts of clot-forming fibrin. The cascade method has a number of advantages. The most obvious is that a few molecules at the beginning of the cascade can ultimately produce millions of fibrin molecules at the end, because each step *catalyzes* the next step. This increases the power of the blood clotting system, as well as its sensitivity to small injuries.

One problem with such a system, however, is the fact that every component of the cascade must be present and must function perfectly if the clotting reaction is to be successful. This makes the system very vulnerable to genetic disorders that affect even one element of the pathway. **Hemophilia** is an example of such a genetic disorder. As we saw earlier in Chapter 11, hemophilia is a human disease that results from a defective protein in the clotting pathway. Individuals who suffer from hemophilia are unable to produce blood clots quickly enough to stop even minor bleeding, and they must take great care to avoid injury. Hemophiliacs are usually treated with extracts of clotting factors prepared from normal whole blood.

SUMMARY

In small animals, the exchange of gases with the environment can occur directly through cell membranes. Nutrients and waste products can diffuse directly from cell to cell and can be eliminated directly into the environment. In larger organisms, however, this is not possible, and specialized systems are required to allow for the internal transport of material.

Circulatory systems generally function by moving a fluid through the body of an organism. In vertebrates a muscular heart moves blood through a closed circulatory system. Fish have a two-chambered heart that pumps oxygen-poor blood into the gills (where gas exchange occurs), from which vessels lead to the rest of the body. Amphibians and reptiles have three-chambered hearts in which there is a partial separation of blood pumped to their gas-exchanging organs (the lungs) and blood pumped to the rest of the body. In birds and mammals this separation is complete: Their four-chambered hearts support completely different pulmonary and systemic circulations. The human heart is a large, hollow muscle that forces blood through atria into paired ventricles. Reverse flow is prevented by valves that separate the chambers, and contractions are under the control of a specialized bundle of cells (the pacemaker) in a region of the muscle known as the SA node.

With each contraction of the heart, a surge of blood under high pressure is pumped into the aorta, the large artery from which every other vessel in the systemic circulation receives blood. Blood pressure is maintained in the arteries, whose thick elastic walls are surrounded by smooth muscle. As blood flows into arterioles and capillaries, its pressure is diminished, and the lowest pressures in the circulatory system are found in the veins that return the blood to the heart. The liquid portion of the blood is known as plasma and consists of a complex mixture of salts, organic compounds, and specific blood proteins. The cellular portion includes erythrocytes (red blood cells), leukocytes (white blood cells), and platelets (subcellular particles that play an important role in the clotting reaction).

STUDY FOCUS

After studying this chapter, you should be able to:

- Analyze the process of diffusion at both the cellular level and the tissue level in order to understand the problems an organism encounters as it exceeds a certain size.
- Explain how organisms in various phyla have evolved ways to deal with the problem of circulation.
- Describe the details of the human circulatory system, its organs, and cells.
- Explain how the circulatory system integrates its function with the larger systems of the body.
- Describe some of the most serious disorders of the circulatory system.

SELECTED TERMS

atrium *p. 611*
ventricle *p. 611*
aorta *p. 612*
atrioventricular valves *p. 613*
sinoatrial (SA) node *p. 615*
atrioventricular (AV) node *p. 615*
systole *p. 618*
diastole *p. 618*
atherosclerosis *p. 620*
myocardial infarction *p. 620*
coronary thrombus *p. 620*
cholesterol *p. 620*
stroke *p. 622*
lymph *p. 622*
plasma *p. 623*
erythrocytes *p. 623*
hemoglobin *p. 624*
leukocytes *p. 625*
platelets *p. 627*

REVIEW

Discussion Questions

1. What are the basic differences between a closed and an open circulatory system? Give an example of an organism with each type of system.

2. What evolutionary trends are apparent in the development of the mammalian heart? What key characteristic of the four-chambered heart enables it to deliver oxygenated blood to the tissues more efficiently than the two-chambered heart?

3. What key role do the valves play in the functioning of the human heart? What problems might result if one of the valves failed to close? How would this affect the heart's ability to pump blood?

4. What is the difference between the pulmonary circulation and the systemic circulation?

5. Why is the level of blood plasma protein important in preventing fluid loss when blood passes through the capillary bed?

Objective Questions   (Answers in Appendix)

6. Which type of circulatory system is found in insects?
   (a) open system
   (b) closed system
   (c) open or closed system depending on the size of the insect
   (d) open system when the insect is immature and closed system when the insect reaches adulthood

7. Which type of blood vessel carries blood away from the heart?
   (a) vein
   (b) venule
   (c) artery
   (d) superior vena cava

8. Which of the following conditions is otherwise known as "hardening of the arteries"?
   (a) atherosclerosis
   (b) myocardial infarction
   (c) stroke
   (d) hypertension

9. Which of the following elements in mammalian blood is a living cell containing a nucleus?
   (a) erythrocyte
   (b) leukocyte
   (c) platelet
   (d) a and b only

10. Which of the following statements correctly describes the human lymphatic system?
    (a) It serves as a conduit between the interstitial fluid and the venous fluid.
    (b) It contains vessels similar to arteries.
    (c) It contains tiny muscular pumps that move fluid through the system.
    (d) All of the above statements are correct.

CHAPTER 33

Respiration

A limp body is pulled from the wreckage of an automobile, and the rescuers who have hurried to the scene wonder whether there is still a life to save. One pulls a tiny mirror from her medical kit and holds it under the victim's nostrils. The mirror fogs—there is hope. With not a moment to lose, they begin to administer first aid and skillfully load the victim through the open ambulance doors.

This situation reflects something we often take for granted: the connection between life and the earth's atmosphere—that is, the process of respiration. **Respiration** is the exchange of gases with the environment: the release of carbon dioxide and the uptake of oxygen.

The process of respiration is so vital that it serves as a universal sign of life, so persistent that it punctuates every other physical activity. The need for an organism to exchange these gases stems from the chemical processes that are at work in living cells. Respiration at the level of the organism, the subject of this chapter, is nothing more than a reflection of cellular respiration, a process we examined in Chapter 25.

THE CELLULAR ROOTS OF RESPIRATION

For hundreds of millions of years, earth's atmosphere has been rich in oxygen. Multicellular organisms evolved in the presence of this gas, and their metabolic pathways use oxygen to generate a constant supply of energy in the form of ATP (Chapter 25). As carbon-containing food molecules are broken down, cells must find acceptors for the electrons released from them. Oxygen is a powerful electron acceptor, and its abundance makes it a logical last step in the electron transport chain of respiration.

So urgent and constant is the need for energy that many cells, especially those of the brain, cannot survive more than a few minutes without oxygen. This cellular demand for oxygen creates a need for oxygen on the part of the whole organism. At the same time, the breakdown of energy-rich carbon compounds in food results in the production of carbon dioxide, a waste product that must be eliminated from the body. Thus all multicellular organisms must constantly exchange oxygen and carbon dioxide with their environment (Figure 33.1).

The term *respiration,* in addition to its meaning at the cellular level, is also applied to this organismal process of *gas exchange* with the environment. In this sense, a **respiratory system** is a group of cells, tissues, and organs that exchanges gases with the environment and helps distribute those gases within the organism.

Respiration in Water and Air

To satisfy the respiratory needs of cells, oxygen must first pass from the gas phase of the atmosphere into the liquid phase that exists within living organisms. Oxygen and carbon dioxide are both soluble in water. For that reason, animals in typical aquatic environments are surrounded by a supply of oxygen that can diffuse directly into cells

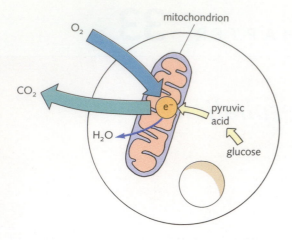

Figure 33.1 Respiration begins at the cellular level. The mitochondria of eukaryotic cells require a steady supply of oxygen to convert the chemical energy in glucose and other foods into biologically useful forms.

to support respiration. Similarly, carbon dioxide can diffuse out of living cells into the surrounding water. Terrestrial animals, which exchange gases directly with the atmosphere, face one additional complication: They must supply a moist surface for gases to dissolve into the liquid phase before these gases can enter living cells.

Many small organisms, whose ratio of surface area to total volume is large, rely on simple diffusion across their body surfaces to meet their cells' needs. This is possible not only for many aquatic animals, but also for those terrestrial species whose habits enable them to keep their outer surfaces moist.

But in many larger organisms—both aquatic and terrestrial—the body cells' demand for gas exchange exceeds the diffusion capacity of the organisms' outer body surfaces. Over time, different groups of animals have evolved various kinds of respiratory systems through which gas exchange with the environment takes place. Not surprisingly, the design of these systems varies with the organism's size, its body structure, its evolutionary history, and the nature of the environment in which it lives. In this chapter we will look at some of the special problems that confront the respiratory systems of various organisms, and we will examine the human respiratory system.

THE VARIETY OF RESPIRATORY SYSTEMS

Every respiratory system has two basic elements: a system to exchange gases with the environment and a system to distribute those gases throughout the organism. Although respiration itself, which occurs at the cellular level, is very much the same from one cell to the next, respiratory systems are not at all the same in different organisms.

Well-developed respiratory systems usually work by bringing either air or water into close contact with moving body fluid. It is not surprising, therefore, that the most advanced respiratory structures are found in animals that have well-developed circulatory systems.

Exchange Through Body Surfaces

The point where gases enter and leave an organism is known as the **respiratory surface.** The simplest respiratory surface is the outer covering of an organism—its skin, or *integument*. Many small aquatic organisms exchange gases through their skins, and so do some terrestrial animals such as the earthworm.

The earthworm illustrates several specializations required in a sizable organism that carries out gas exchange across its skin. First, there is a close coupling between the skin surface and the circulatory system (Figure 33.2).

Figure 33.2 Gas exchange in earthworms takes place directly through the skin. The process is made more efficient by a circulatory system that carries blood close to the surface of the body, and by a thin, moist skin which assists in the diffusion of oxygen and carbon dioxide.

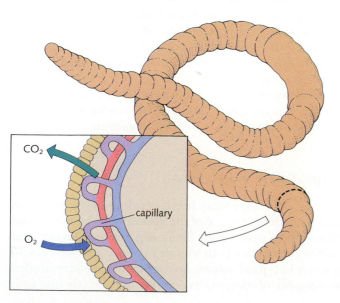

The body of the earthworm is too large for oxygen to diffuse into tissues deep within, and a network of fine capillaries just below the skin allows the oxygen and carbon dioxide that pass through the skin to move directly into or out of the circulation.

Second, the skin must be kept moist to enable oxygen gas to dissolve in it and then diffuse into the body. For this reason, the earthworm's skin contains a series of glands that cover its body with a thick, wet mucus (if you've picked up a worm, you know this is true).

Gas exchange through the integument is not limited to invertebrates. A number of fishes acquire some oxygen directly through their skin, and the thin, moist skin of amphibians is particularly well suited to gas exchange. Frogs fulfill a large fraction of their respiratory needs through the skin.

During one stage of their lives, reptiles and birds carry out gas exchange through their integument as well. The active, growing embryo in an egg needs a constant supply of oxygen, which it obtains through thousands of tiny pores that pass through the eggshell. As the embryo develops, a tissue known as the *chorioallantois* extends from the body of the growing bird and covers the entire inner surface of the shell. Richly endowed with blood vessels, the chorioallantois is separated from the eggshell pores by a thin extracellular membrane that helps to block the entry of bacteria into the egg. Gas exchange occurs across this membrane and completely answers the respiratory needs of the embryo until its lungs become functional, usually just a few days before hatching.

Plants can be considered skin-breathing organisms, too. Although we usually pay more attention to their photosynthesis (a process not found in animals), plants *do* respire. The nonphotosynthetic tissues of a plant derive most of their energy from aerobic respiration, as does the entire plant at night. Gas exchange takes place by means of simple diffusion through the "skin" of the plant.

Exchange Through Gills

Gills are organs specialized for gas exchange in water. Gills may be internal or external, they may function in salt water or fresh water, and they may be exposed at the surface of an organism or protected by shells or parts of the skeleton. All gills, however, solve the problem of gas exchange by bringing oxygen-poor blood into close contact with oxygenated water so that gas exchange can occur between them. The rate of gas exchange depends in part on the size of the surface over which blood and water are in close contact, so gills of all types have one thing in common: thousands of tiny projections that increase their surface area. By doing so, gills enable their bearers to live an active life supported by a high rate of respiration and gas exchange.

The gills of the common sea star, an echinoderm, are delicate stalks of tissue that project from the surface of the animal directly into seawater (Figure 33.3). To protect them from injury, the organism is covered with spine-like projections that extend past the gills. These projections make the surface of the sea star rough to the touch.

Some vertebrates also have external gills. Many have gill clusters on either side of the head that look almost like feathers (Figure 33.3). The tiny structures within the

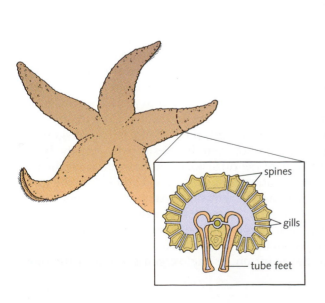

Figure 33.3 (left) Gas exchange in the starfish takes place in feathery gills which exchange oxygen and carbon dioxide with the surrounding seawater. Spiny projections from the surface of the animal provide protection for the gills. (top) The external gills of this tiger salamander larva provide a large surface area for gas exchange.

spines

gills

tube feet

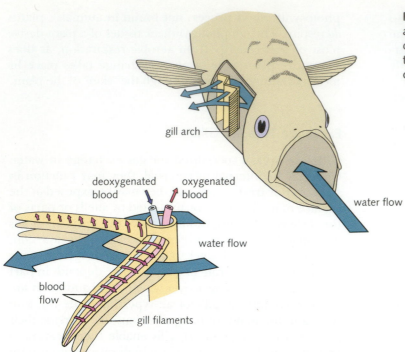

Figure 33.4 The gill systems of fish allow blood and water to come into close contact. Notice that blood flows through the gill filaments in a direction opposite to the flow of water.

gill arch

deoxygenated blood

oxygenated blood

water flow

water flow

blood flow

gill filaments

gills enable the cluster to expose an enormous surface of the circulatory system for gas exchange.

Fishes have two large banks of internal gills that develop from the lining of the throat. These gills are protected behind two bony plates, or gill covers, that also help control the movement of water across the gills. Some fast-swimming bony fishes, such as tunas, simply swim through the water with their mouths and gill covers open wide. As these fishes move, water enters the mouth, flows across the gills, and leaves through the openings behind the gill covers. More sedentary fish species must pump water across their gills. They first suck water in by opening their mouths with the rear ends of their gill covers pressed against the sides of their body. Then they close their mouths and force the water past the gills and out past the relaxed gill covers. Both techniques produce a steady flow of oxygen-rich water across the gill.

Gill tissue has a rich blood supply that comes directly from the heart. This oxygen-poor blood passes into large **gill arches** that support rows of tiny **gill filaments.** Within each filament, there are two vessels—one through which blood enters the gill and another through which it leaves. Between these vessels are dense networks of thin-walled capillaries packed into hundreds of wafer-thin projections called **lamellae.** During the trip through the lamellae, the forces of diffusion cause the blood to release carbon dioxide and to absorb oxygen from the water.

Note that blood is pumped through gill filaments and lamellae in a direction opposite to the flow of water past the gills (Figure 33.4). This pattern is known as **countercurrent flow,** because the blood and the water move in opposite directions.

This countercurrent flow pattern dramatically increases the efficiency of the gill system over that of a configuration in which water and blood move in the same direction. In fact, countercurrent flow is so efficient in this regard that it crops up again and again among many animals and in physiological systems throughout the body.

Why is this arrangement so efficient? Recall that diffusion is a passive, physical process. Molecules diffuse from areas of high concentration to areas of low concentration. In doing so, they move at rates proportional to the difference in concentration between the two areas.

Now imagine the diffusion of oxygen in an imaginary gill through which blood and water flow in the same direction. Here oxygen-poor blood meets oxygen-rich water so, at first, oxygen diffuses rapidly into the blood. But the rate of net diffusion soon slows down as the oxygen concentration rises in the blood and falls in the water.

In a countercurrent flow system, oxygen-poor blood first "encounters" water that has already lost some of its oxygen. But because there is still more oxygen in the water than in the blood, net diffusion occurs anyway. And from that point on, as the oxygen content of the blood

The Ice Fish: Blood Without Hemoglobin

The low solubility of oxygen in water is the principal reason why oxygen-carrying proteins such as hemoglobin are necessary to support the gas-exchange activities of the respiratory system. What might happen if an organism lived in an environment where its own oxygen demands were lower and the solubility of oxygen in water was greater? We need not speak hypothetically in answering this question. An organism known as the ice fish (scientific name, *Mallotus villosus*) lives in the cold waters of the polar regions, where the average water temperature is less than $-4°C$. It is similar to most fish in every respect but

one: Its blood is clear and colorless—it has no hemoglobin. How does this remarkable animal survive with its pale, colorless blood?

Low temperature produces two effects that enable the ice fish to transport oxygen without an oxygen-carrying protein. First, *low* temperatures *lower* the animal's metabolic rate, so its oxygen demands are much less than they would be at higher temperatures. This means that the respiratory system may supply less oxygen. Second, *lower* temperatures *increase* the solubility of gases in water. This is why warm bottles of soda tend to pop or even explode: The CO_2 used to

create the fizz in soda is quite soluble when the drink is cold. But when a warm bottle is opened, the gas comes out of solution so quickly that the drink may foam out of its container. At the temperatures in which the ice fish lives, oxygen solubility is so much greater than normal that dissolved oxygen alone is sufficient to meet the respiratory needs of the organism. Why the ice fish apparently lost the ability to synthesize hemoglobin is a question you might wish to ponder in terms of the selective pressures shaping evolutionary change.

---

increases, it encounters water that has a progressively higher oxygen content (Figure 33.4). The architecture of the gill system thus ensures that concentration differences between blood and water enable gas exchange to occur across the entire length of the capillary.

Exchange Through Tracheae

Terrestrial animals face a different set of problems from those of animals living in water. On the one hand, the atmosphere is a much richer source of oxygen than water. But on the other hand, these animals face the challenge of exchanging gases across a gas–liquid interface. Because oxygen must enter solution to diffuse into the body, terrestrial animals must keep their respiratory surfaces moist. Obviously, such surfaces are potential sites for water loss.

The largest group of terrestrial animals, the arthropods, respire in a unique way (Figure 33.5). Their tough outer shell of chitin is nearly impermeable to gases. Instead, air enters insects and other terrestrial arthropod bodies through a series of openings known as **spiracles.** This air then travels through a series of dead-end

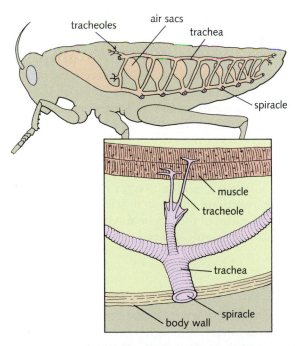

Figure 33.5 The tracheal system of insects allows air to come into close contact with oxygen-utilizing tissues like muscle.

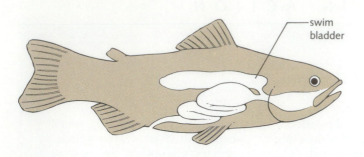

swim
bladder

Figure 33.6 Air-filled swim bladders in fish help to regulate buoyancy. Air sacs connected to the oral cavity might have played a role in the evolution of lungs.

passages known as **tracheae** (singular, *trachea*) that carry oxygen to tissues throughout the body.

Tracheae, in turn, lead into tiny **tracheoles** that make direct contact with individual cells, especially those (such as muscle cells) that have high oxygen demands. Oxygen diffuses directly into the cells from the interior of the tracheoles, and carbon dioxide diffuses in the opposite direction. In smaller insects, the only force that moves air through the system is diffusion, but large insects can assist respiration by moving their abdomen. This helps to pump air back and forth through the system.

Remember that insects have open circulatory systems, which are less efficient than closed ones. The main reason why insects can make do with an open system is that their tracheal systems carry oxygen directly to their tissues. Although this system works well for insects, the nature of its tracheal system places severe restrictions on the size of an organism. As an animal increases in size, it becomes more and more difficult for fresh air to penetrate the deepest passageways of the system. In larger insects, the tracheal system becomes less capable of meeting peak respiratory demands—and this is one of the principal reasons why insects are seldom longer than a few centimeters. Giant insects are common in science fiction movies, but the limitations of the tracheal system have (thankfully) prevented their emergence in the real world.

Lungs

Terrestrial vertebrates carry out gas exchange in internal organs known as **lungs.** Although there is no fossil record of the development of the lung (soft tissues are seldom preserved in fossils), we can get some idea how lungs may have developed by looking closely at present-day fishes. Many fishes have internal air sacs known as *swim bladders* that help them adjust their buoyancy for efficient swimming (Figure 33.6). Some fishes also have sacs in the mouth region that can be filled with air as they swim to the surface. A few fish, including the common goldfish, can trap air in these sacs and extract some of the oxygen from it directly into the bloodstream.

The combination of an oral cavity and an internal sac in ancient fishes might easily have developed into a gas-exchanging system that would have enabled these fishes to survive in shallow, muddy waters where the dissolved oxygen content is low. Developments such as these might have been the first steps in the evolution of the lungs. The development of the lung made it possible for vertebrates to thrive on land.

In most modern terrestrial vertebrates, air enters each lung through a single large tube that branches into thousands of tiny passageways. Those passageways lead to terminal sacs known as **alveoli** (singular, *alveolus*) where gas exchange takes place. Human lungs have more than 300 million alveoli, and their combined surface area is more than 75 square meters—more than 3 times the area of a bowling alley. The flow of blood through a closed circulatory system is routed through the lungs, allowing blood to be used as the oxygen-transporting medium (Figure 33.7).

In most vertebrates, including mammals, lungs are dead-end passageways: Air enters the lungs as the chest cavity expands, and air leaves as the cavity decreases in size. Because of this in-and-out flow pattern, fresh air entering the lungs is always mixed with some residual air that has been depleted of oxygen.

The avian lung—increased efficiency Although soaring seagulls may make flight look easy, flying actually demands tremendous physical effort from all birds. That effort, in turn, requires a steady supply of oxygen. Interestingly, the gas-exchange system in birds is more efficient than that in other terrestrial vertebrates because air flows through the lungs in a one-way cycle.

The efficient gas-exchange system in birds works as follows: Fresh air is first routed to a series of sacs behind the lungs. This fresh air then passes directly into the lungs and through a series of small passageways, from where it is exhaled. Because air flows through the lungs in a one-way cycle, the gas-exchange surfaces in avian lungs are always presented with air containing the maximum oxygen content of 20 percent. The air-sac–lung system of a bird satisfies that demand while minimizing weight to help produce an efficient "flying machine" (see Figure 33.8).

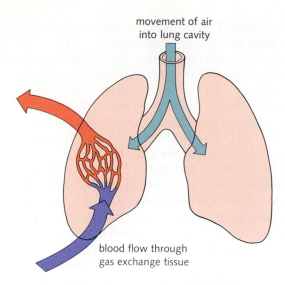

Figure 33.7 The lungs of modern terrestrial vertebrates are sacs into which air is drawn by breathing. The circulatory system absorbs oxygen in the lungs and transports it to the rest of the body.

Conserving Moisture

When dry air moves across the moist surfaces of an animal's respiratory system, some of that moisture evaporates and may be lost. This problem reaches an extreme in desert animals that live in an arid climate characterized by high temperatures and low humidity.

The kangaroo rat illustrates the way in which the respiratory systems of many desert animals are adapted to limit water losses in breathing (Figure 33.9). Its upper air passages are covered with tightly packed hairs moistened with mucus and other secretions from the epithelial tissue beneath them. As the dry desert air moves into the body, it is moistened by water from these passageways, and by the time the air arrives in the lungs, it is humid enough so that the moist membranes across which gas exchange takes place do not dry out.

Every time the animal inhales, evaporation cools the upper passageways of the system. When it exhales, the warm, moist air from the lungs passes through the cooler passageways of the upper part of the system, allowing most of the moisture in the exhaled air to condense on the lining of the upper system. This arrangement limits the loss of moisture, conditions the air that is inhaled, and enables the animal to exist in an environment where water is at a premium.

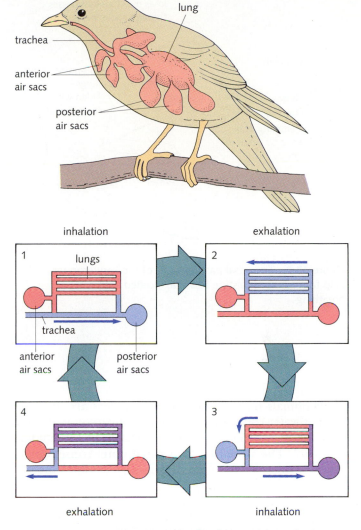

Figure 33.8 Air sacs connected to the lungs of many birds provide for a one-way system of air flow which results in highly efficient breathing.

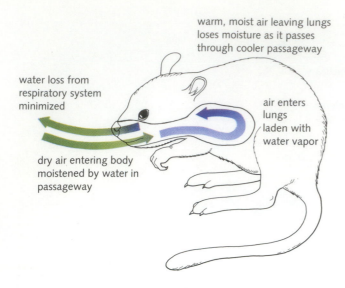

warm, moist air leaving lungs loses moisture as it passes through cooler passageway

water loss from respiratory system minimized

air enters lungs laden with water vapor

dry air entering body moistened by water in passageway

Figure 33.9 The nasal passageways of the kangaroo rat help to minimize moisture losses associated with breathing.

Mammals that are not desert dwellers are less efficient at water conservation than the kangaroo rat, because their nasal passages are not so specialized for holding moisture. The same basic process of water conservation, however, does occur in most organisms, including humans.

THE HUMAN RESPIRATORY SYSTEM

Inhaled air enters the human respiratory system through the mouth and nose (Figure 33.10). The nasal passageways are specially adapted to filter and moisten the air as it passes into the body. Air then passes through the **pharynx** and into the **trachea,** which leads to the lungs. The accidental entry of food or drink into the lungs is prevented by the **epiglottis,** a small flap that closes over the entrance of the trachea during swallowing. The upper part of the air passageway, called the **larynx,** contains the **vocal cords,** which produce sounds by vibrating as air passes between them.

In the chest the trachea divides into two tubes, the **bronchi,** one of which enters each lung. The bronchi are supported by rings of cartilage, which prevent the passageways from collapsing. Within the lungs, the air passageway divides further into smaller bronchi, then into **bronchioles,** and finally into the **alveoli,** where gas

exchange takes place. The rich blood supply of the alveoli makes them ideal sites for gas exchange.

As air moves through the pharynx and larynx, it picks up moisture from *mucus* lining the epithelial tissue of the passageway (Figure 33.10). This helps to keep the gas-exchange surface within the alveoli moist. This mucus also traps inhaled particles of dust or smoke. Such foreign material is forced out of the respiratory system by the movement of *cilia* lining the passageways.

These ciliated cells are among the first to be damaged by environmental assaults such as smoking. Nicotine, one of the components of cigarette smoke, works like an anesthetic on the cilia in the respiratory tract, reducing their efficiency in clearing the lungs. A smoker's morning cough is an attempt to clear the respiratory passages of accumulated mucus—an early warning that the cilia are not working properly and that the respiratory system is headed for trouble.

Breathing

The process of **breathing,** or *ventilation,* is the movement of air into and out of the lungs. The lungs are not directly connected to any muscle system, and the only force that moves air into the lungs is air pressure. The body exploits atmospheric pressure every time we inhale.

The lungs are sealed in the **thoracic cavity.** They lie within two **pleural sacs** that border the **diaphragm,** a large dome-shaped muscle separating the thoracic cavity from the abdominal cavity. The thoracic cavity can be expanded in two ways: by contractions of the diaphragm muscle that expand the lower portion of the cavity, and by contractions of the *external intercostal muscles* between the ribs that raise the rib cage and enlarge the cavity from side to side. Because the thoracic cavity is sealed, this creates a partial vacuum within the pleural cavity, a vacuum that is rapidly filled with inrushing air. Figure 33.11 shows the expansion of the chest that occurs during inhalation, as well as a mechanical model that illustrates the physical principles at work.

In contrast to inhalation, exhalation is largely a passive process: The muscular contractions that expanded the thoracic cavity cease, and the highly elastic components of the lung and the tissue surrounding it retract and return to their original shape. As the thoracic cavity returns to its original position, air is expelled.

This is an efficient system, but it depends on maintaining the thoracic cavity around the lungs intact. A puncture wound in the chest, even if it does not touch the lungs, may admit air into the thoracic cavity and make breathing impossible—one of the reasons why chest wounds are always serious.

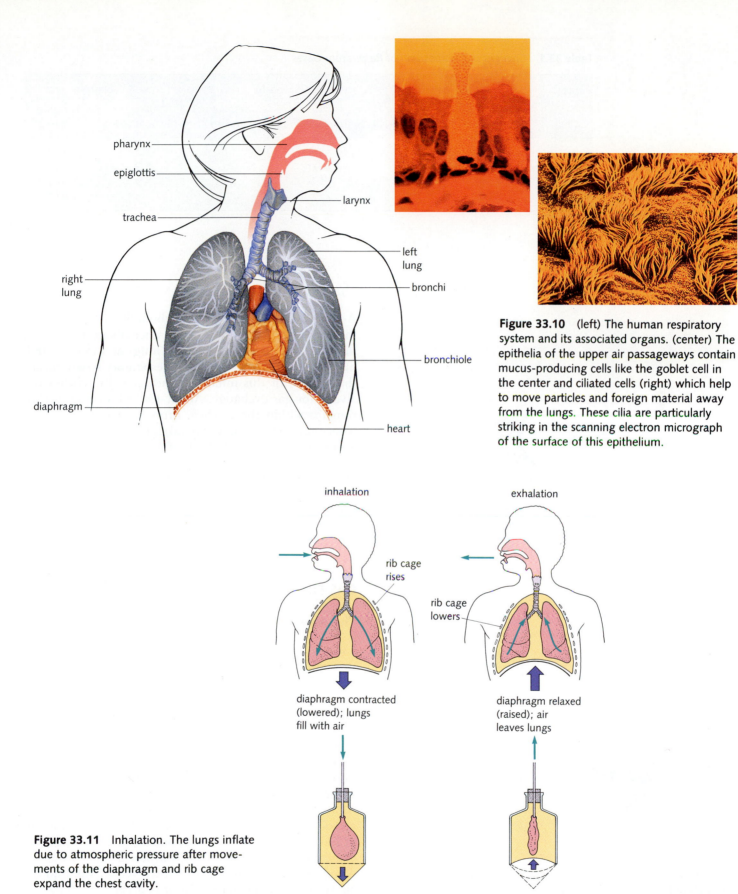

pharynx

epiglottis

larynx

trachea

right lung

left lung

bronchi

bronchiole

diaphragm

heart

Figure 33.10 (left) The human respiratory system and its associated organs. (center) The epithelia of the upper air passageways contain mucus-producing cells like the goblet cell in the center and ciliated cells (right) which help to move particles and foreign material away from the lungs. These cilia are particularly striking in the scanning electron micrograph of the surface of this epithelium.

inhalation

exhalation

rib cage rises

rib cage lowers

diaphragm contracted (lowered); lungs fill with air

diaphragm relaxed (raised); air leaves lungs

Figure 33.11 Inhalation. The lungs inflate due to atmospheric pressure after movements of the diaphragm and rib cage expand the chest cavity.

Table 33.1 *Changes in Composition of Respiratory Gases*

| Gas | Inhaled Air | | Exhaled Air | | Alveolar Air | |
	% (by volume)	Pressure (mm Hg)	% (by volume)	Pressure (mm Hg)	% (by volume)	Pressure (mm Hg)
O_2	20.71	157	14.6	111	13.2	100
CO_2	0.04	0.3	4.0	30	5.3	40
H_2O	1.25	9.5	5.9	45	5.9	45
N_2	78.00	593	75.5	574	75.6	574

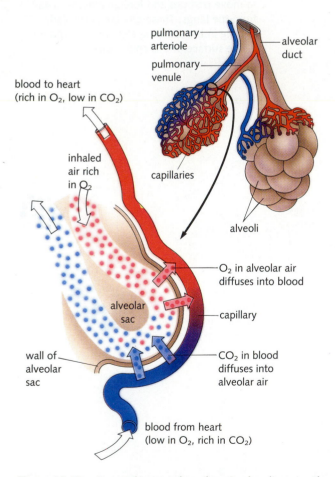

Figure 33.12 Gas exchange takes place in alveoli across the thin walls of the alveoli and capillaries.

The Movement of Air

A normal breath at rest brings about 500 mL of air into the lungs. A person at rest breathes about 10 times a minute, so air is moved into the lungs at a rate of 5000 mL/min. But the anatomy of the respiratory system means that not all of this inhaled breath actually reaches the alveoli for gas exchange. About 150 mL of each breath remains within the trachea, bronchi, and bronchioles, where gas exchange does not take place. This means that, at rest, only the first 350 mL of a breath actually have a chance to reach the alveoli, a condition that reduces the effective respiratory volume. During times of stress or physical exertion, breathing becomes much deeper; a single deep breath can bring in as much as 3500 mL of air.

GAS EXCHANGE

The Alveoli

Table 33.1 illustrates what happens to fresh air as it passes through the respiratory system. By the time it is exhaled, air has lost about 30 percent of its oxygen. In contrast, it contains 100 times as much carbon dioxide. The exchange of these two gases takes place in the alveoli.

The alveoli are extremely thin; their walls are only one cell thick. Each alveolus is surrounded by a delicate network of thin-walled capillaries that bring blood into close contact with the surface of the alveolar wall. Inhaled air enters each alveolus, filling the tiny sac with oxygen. As oxygen-poor blood rushes over the surface of an alveolus, oxygen from within the sac dissolves in the film of liquid coating the alveolus and diffuses through the capillary wall into the blood. Carbon dioxide from the blood, in turn, diffuses across the alveolar wall into the air, which is then exhaled (Figure 33.12).

Bhopal

One of the greatest industrial tragedies in world history occurred in 1985 when a chemical leak developed at the Union Carbide chemical plant in Bhopal, India. A cloud of poisonous methyl isocyanate gas drifted over a residential area near the plant, and within a few minutes more than 2000 people were dead. The accident raised countless legal and ethical issues. To medical scientists, however, one aspect of the accident was particularly confusing: Why did so many people die?

The reason for their confusion was simple enough. Although cyanide compounds are deadly poisons, very few of the victims of the accident had enough cyanide in their bodies to have been killed by the gas. What had they died from? The most probable answer was found by examining the action of cyanide at the cellular level.

Cyanide ions interfere with the last carrier in the mitochondrial electron transport chain, cytochrome oxidase. Cyanide blocks the chain at cytochrome oxidase, and the synthesis of ATP grinds to a halt. A cell poisoned with cyanide is unable to produce enough ATP for normal cellular activities, and it quickly dies.

In Bhopal the cyanide poison was inhaled, so the highest concentrations of cyanide were found in the cells lining the respiratory tract; nearly all of these cells were killed by the gas. Cilia stopped beating, mucus secretion stopped, cells lining the alveoli died, and gas exchange became less and less efficient until the victims literally suffocated. Although the amount of cyanide in their bodies was not large, the fact that it was concentrated in the respiratory system turned a serious accident into a catastrophe.

The movement of oxygen and carbon dioxide across the alveolus is a matter of simple diffusion: *There are no active transport processes associated with gas exchange in the lungs.* By the time it leaves an alveolus, the bloodstream has passed carbon dioxide into the atmosphere in exchange for oxygen. The enriched blood flows back to the heart through the pulmonary veins. This oxygen-rich blood is then pumped through the aorta to the rest of the body.

The Role of Hemoglobin

Although oxygen diffuses rapidly into solution, the oxygen-carrying capacity of most liquids is low. A liter of water in equilibrium with atmospheric pressure contains only about 3 mL of dissolved oxygen. If blood carried only 3 mL O_2 per liter, an enormous volume of blood would be required to meet the respiratory needs of the average human: about 600 mL O_2 per minute. Fortunately, this is not the case. Human arterial blood has a much higher oxygen content: roughly 200 mL O_2 per liter. The increased capacity is possible because only a small amount of the oxygen is actually dissolved in water. More than 98 percent of the oxygen carried by blood is bound to a carrier molecule called **hemoglobin** (Figure 33.13).

Hemoglobin is an oxygen-carrying protein contained within the red blood cell. The hemoglobin molecule consists of four polypeptide chains, and each of these chains contains a **heme** group. Heme is a ring-like, iron-containing prosthetic group (as discussed in Chapter 20) that binds an oxygen molecule. Each hemoglobin can bind a

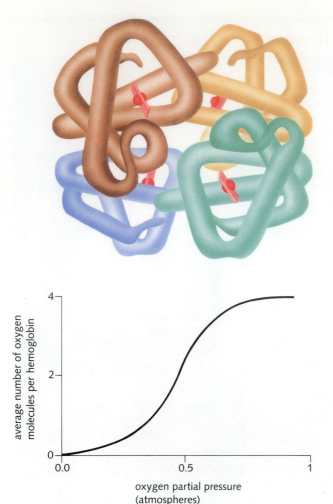

Figure 33.13 Oxygen is carried within red blood cells by hemoglobin (top), a protein with four oxygen-binding sites. Each subunit in this diagram of the hemoglobin molecule contains a single oxygen binding site. The binding of oxygen to hemoglobin follows a sigmoid-shaped curve (bottom), indicating that the affinity of hemoglobin for oxygen increases after one or two oxygen molecules have bound to the protein.

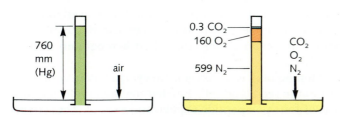

Figure 33.14 Atmospheric pressure at sea level is equivalent to a column of mercury 760 mm high. Oxygen accounts for 160 mm of this pressure, and nitrogen for 599 mm. Therefore, the partial pressure of oxygen in sea level air is said to be 160 mm.

total of four oxygen molecules, one at each of its four heme groups (Figure 33.13).

Hemoglobin accounts for as much as 31 percent of the total weight of the red blood cell. Oxygen binds to hemoglobin molecules as red blood cells pass through the alveoli, and it is released as they pass through capillary networks in body tissues.

The binding of oxygen to hemoglobin causes a slight change in the color of the molecule. Fully oxygenated hemoglobin is bright red, and even partially deoxygenated hemoglobin is a much deeper bluish red. These differences account for the fact that venous blood is a much deeper, more bluish color than the bright red blood that flows through the arteries.

Gas Exchange Between Blood and Tissues

Why does hemoglobin bind oxygen in the alveoli and release it in the tissues? In order to understand this fascinating process, we must first understand a few terms and concepts related to the behavior of gases in mixtures and in solution.

Partial pressure Normal atmospheric pressure is 760 millimeters of mercury, which means that the pressure of the air produces a force equivalent to the weight of a column of mercury 760 millimeters high (Figure 33.14). Only about 21 percent of the atmosphere is oxygen, so only 21 percent of atmospheric pressure—about 160 mm Hg—is due to oxygen. This value is known as the **partial pressure** of oxygen. Partial pressure is the pressure that can be directly attributed to one gas in a mixture of gases.

When a container of water is opened to the air, oxygen from the air diffuses into the water until an *equilibrium* is reached at which the number of oxygen molecules entering the water is matched by the number leaving it. Because the dissolved gas is in equilibrium with the gas found in the air, we speak of the dissolved oxygen as also having a partial pressure of 160 mm Hg.

Hemoglobin and oxygen Hemoglobin serves as a good carrier of oxygen in living systems because of the manner in which it binds oxygen. As it happens, oxygen molecules attach loosely, rather than permanently, to each of the four binding sites in hemoglobin. These loosely-bound oxygen molecules are thus free to dissociate under certain conditions.

The property of hemoglobin that makes it so useful to the body is that it binds oxygen when the O_2 partial pressure is high and releases oxygen when the O_2 partial pressure is low (Figure 33.15). In the alveolus, the O_2 partial pressure is high, so hemoglobin passing through picks up oxygen. In the capillary networks in other body tissues, however, the O_2 partial pressure is low. There oxygen dissociates from hemoglobin, so it is free to diffuse across the capillary wall into the tissues.

The need for oxygen to diffuse into and out of the blood can cause problems under certain circumstances. At high altitudes, for example, atmospheric pressure is lower than at sea level. Therefore, the partial pressure of oxygen is low, and less oxygen is available to diffuse across the alveolus. **Hypoxia,** a deficiency of oxygen in the tissues, may result. To compensate, an individual breathes harder at high altitudes (meaning that a larger volume of fresh air is inhaled). However, at levels above 20,000 ft, diffusion cannot move enough oxygen into the bloodstream to satisfy metabolic demands, and special breathing equipment is usually necessary.

During pregnancy, mammals are confronted with another interesting problem: Oxygen must be transferred from the blood of the mother to the blood of the developing fetus. The circulations of mother and offspring are separate, but they come into intimate contact in the capillary beds of the *placenta,* where oxygen and nutrients diffuse from the maternal circulation to that of the fetus. If the hemoglobin molecules of mother and fetus were identical, the rate of oxygen transfer would be very slow; the best that could be achieved would be a 50/50 equilibrium between the two circulations.

As a result, in most species the red cells of the fetus contain a different form of hemoglobin, which has a higher affinity for oxygen than does adult hemoglobin. This fetal hemoglobin thus picks up oxygen at the same partial pressure at which the maternal hemoglobin releases it. For this reason, transfer of oxygen from mother to fetus is rapid and effective. Shortly before the animal is born and starts to breathe for itself, the body begins to synthesize the adult form of hemoglobin.

Skeletal muscle cells of the body face a similar problem. They need to draw large amounts of oxygen from the blood, and they do so in much the same way. These cells draw oxygen from the circulation and maintain their own stores of oxygen in the tissue. **Myoglobin,** a protein very similar to hemoglobin, is stored in skeletal muscle cytoplasm (Figure 33.16). Each myoglobin molecule binds a single molecule of oxygen, allowing a reserve of oxygen to be built up within the muscle cell. This reserve is released when the partial pressure of oxygen within the cell drops, as it does during vigorous exercise.

The binding curve for myoglobin shows that it, like fetal hemoglobin, has a greater affinity for oxygen than does adult hemoglobin. This ensures that myoglobin will bind to oxygen at partial pressures low enough for hemoglobin to release it. Thus oxygen flows into muscles efficiently. The reddish color of myoglobin, incidentally, is one of the major reasons why the meaty muscle tissue of beef is a deep, bright red. Conversely, the "white meat" of a turkey has very little myoglobin.

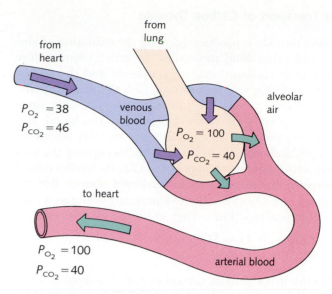

Figure 33.15 Because the lungs are not completely emptied with each breath, oxygen in the alveolus has a partial pressure (PO_2) of 100 mm. The movement of gases between capillaries and alveoli gives blood leaving the alveoli different partial pressures of oxygen and carbon dioxide than the blood that enters the alveoli.

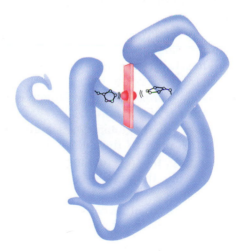

Figure 33.16 Oxygen is stored in muscle and other tissues by binding to myoglobin, a protein containing a single oxygen-binding site.

The Transport of Carbon Dioxide

Carbon dioxide is produced during the oxidation of glucose and other food molecules. In active tissues, large amounts of carbon dioxide are present and the partial pressure of CO_2 rises, prompting CO_2 to diffuse into the bloodstream. As was the case with oxygen, only small amounts of CO_2 can dissolve directly in the blood. But unlike oxygen, only a small amount of CO_2 (about 11 percent) binds directly to hemoglobin, occupying the sites previously taken by oxygen. Some CO_2 molecules (about 8 percent) dissolve directly in the blood plasma. But the majority of CO_2 molecules (81 percent) undergo a two-step chemical reaction as they enter the bloodstream:

$$CO_2 + H_2O \rightarrow H_2CO_3$$
$$H_2CO_3 \rightarrow HCO_3^- + H^+$$

In the first equation, carbon dioxide dissolved in water reacts with a water molecule to form carbonic acid. This chemical reaction is catalyzed by an enzyme known as *carbonic anhydrase* and is found within the red blood cell. In the second equation, carbonic acid dissociates to produce the negatively charged bicarbonate ion and to release a positively charged proton. The proton may then bind to hemoglobin, displacing oxygen which can then be released into the tissues. The binding of protons to hemoglobin helps to release oxygen where it is most needed, and to prevent the blood from becoming highly acidic.

Bicarbonate ion is readily soluble in water, so large amounts of carbon dioxide can be carried by blood returning to the heart. The release of bicarbonate and hydrogen ions slightly lowers the pH of the blood (you will remember that an increase in the hydrogen ion concentration lowers the pH of a solution). However, it turns out that deoxygenated hemoglobin has a high affinity for hydrogen ions, so most of these ions are bound to hemoglobin. Therefore, blood returning to the heart is only slightly more acidic than blood in the arterial circulation. When deoxygenated blood flows through the alveolar capillaries, bicarbonate is converted to CO_2 that rapidly diffuses out of the blood into the alveoli and is exhaled.

CONTROL OF RESPIRATION

Breathing, or ventilation, is caused by muscular contractions under the control of the nervous system. Although breathing occurs automatically, we override that unconscious control when we want to blow up a balloon, play the flute, or hold our breath underwater. This sometimes gives us the mistaken impression that breathing is purely voluntary. It's not. After only a few seconds of hard running, athletes begin to puff to "catch their breath." That adjustment to the increased respiratory demands of the body is automatic. You may be able to hold your breath for a minute or so, but before long your body "forces" you to breathe. How does this happen?

A **breathing center** located in the *medulla* (a portion of the brain just above the spinal cord) controls breathing. Nerves carry impulses from the breathing center to the diaphragm and to muscles of the rib cage. These impulses produce a cycle of contraction and relaxation that ventilates the lungs.

How does the breathing center react to an increased need for oxygen? Cells in the center are particularly sensitive to the partial pressure of CO_2 in the bloodstream. As the content of CO_2 in the blood rises, so does the CO_2 content of the breathing center. The center reacts by stimulating the breathing muscles. If the CO_2 content of the blood reaches critical levels, the impulses from the breathing center become so powerful that we cannot override them, no matter how strong our determination is.

In most cases, the sensitivity of the breathing center serves the body well, because an increased CO_2 partial pressure is a good indication that the body needs oxygen. However, there are some circumstances in which the breathing center can be tricked, and those cases can be fatal. Passengers in hot air balloons at high altitudes often have to be *told* to begin breathing pressurized oxygen, because they do not feel the effects of oxygen deprivation. The thin air at high altitudes starves the body for oxygen, but because CO_2 does not build up in the bloodstream, the breathing center does not react.

A similar problem occurs in divers who "hyperventilate" by rapid breathing before diving underwater. Rapid breathing decreases the carbon dioxide levels of the blood so much that these divers may run out of oxygen, pass out, and then drown underwater before the breathing center senses a buildup of CO_2. Experienced divers avoid hyperventilation!

DAMAGE TO THE RESPIRATORY SYSTEM

Subverting the System: Carbon Monoxide

Few molecules are as dangerous to the respiratory system as **carbon monoxide** (CO), a colorless, odorless gas produced when fuel is burned under conditions wherein oxygen availability is limited. Carbon monoxide is given off by gasoline and diesel engines and also by wood and charcoal stoves that limit air flow.

The carbon monoxide molecule is similar to the oxygen molecule in size and shape, and it attaches very tightly to the oxygen-binding sites of hemoglobin. But unlike

The Bends

Although nitrogen constitutes 79 percent of the atmosphere, our discussion of the respiratory system has all but ignored the existence of this gas. In most respects there is no reason to be concerned with nitrogen; at normal partial pressures it is not known to participate in any physiological process. However, nitrogen does dissolve in the blood, and as its partial pressure increases, the amount of dissolved nitrogen in the blood increases too. This can cause serious problems for deep sea divers, who must breathe pressurized air to keep their lungs from collapsing under the force of scores of meters of water.

Under these unusual pressures, large amounts of nitrogen gas dissolve in the blood. When a diver rises to the surface rapidly, the pressure falls quickly enough to cause nitrogen to come out of solution. The diver's circulatory system is blocked by millions of tiny nitrogen bubbles in the capillaries, causing pain, paralysis, and even death. This serious condition, which is known as *the bends*, can be avoided by rising to the surface slowly, allowing excess nitrogen to leave the blood gradually through the lungs.

Hours of breathing nitrogen-rich air under high partial pressure can affect the nervous system and cause another problem known as *nitrogen narcosis*, in which the diver may seem drunken and silly, losing the ability to make sound judgments. Some of the first divers to experience this condition reported seeing mermaids and other exotic sights, leading some to describe the effect as the "rapture of the deep." The danger of nitrogen narcosis is one of the reasons why professional divers work in pairs and limit the time they spend beneath the water.

oxygen and carbon dioxide, CO does not dissociate easily from the oxygen-carrying protein. This allows a relatively small number of CO molecules to "crowd out" oxygen and dangerously curtails the oxygen-carrying ability of the respiratory system. The tissues of the body are deprived of oxygen, and death may occur within minutes.

Carbon monoxide poisoning is possible whenever the CO concentration of the air reaches significant amounts. Therefore, engines and stoves that produce CO should always be used in open, well-ventilated places where their exhaust fumes cannot accumulate. The first symptoms of carbon monoxide poisoning are drowsiness and disorientation. The best first aid is plenty of fresh air and concentrated oxygen, if available, to increase the O_2 partial pressure.

Because the CO molecule binds directly to the oxygen-binding site of hemoglobin, CO-poisoned blood has the same bright red color as fully oxygenated blood. Emergency squad workers are often struck by the fact that the victims of carbon monoxide poisoning have a flushed "healthy" color; it is caused by this mechanism.

Smoking and Lung Damage

The upper passages of the respiratory system keep foreign particles from entering the delicate alveoli in which gas exchange takes place. Persistent exposure to foreign particles may eventually overwhelm the ability of the system to cleanse itself and may begin to damage the air passageways (Figure 33.17). Remarkably, the major source of such damage is self-inflicted; it is smoking.

When tobacco is burned and the smoke inhaled, a mixture of nicotine, carbon monoxide, ash particles, and dozens of dangerous organic chemicals finds its way into the lungs. Persistent smoking, even in moderate amounts, can lead to *bronchitis*, a chronic inflammation of the bronchi; to *emphysema*, serious damage to the elastic tissue of the lungs; and potentially to the most serious consequence of all, *lung cancer.*

Lung cancer is now the leading cause of death from cancer for both men and women in the United States. Over 140,000 people will develop lung cancer each year in the United States during the 1990s, and very few will

Saving a Life: The Abdominal Thrust

Dozens of people die each year as a result of an unfortunate aspect of the structure of the respiratory system: the fact that air and food enter the body by means of the same passageway. When an individual manages to swallow a piece of food that is just a bit too large, that food may become lodged in the upper part of the throat, blocking the air passageways. In most cases a quick burst of air from the lungs—a cough—dislodges the food, but sometimes the coughing action is not powerful enough to move the food. Air is sealed off from the lungs, and death may follow from suffocation in as little as four minutes. The speed of a "café coronary," as these sudden attacks of suffocation have been called, leaves no time to summon trained medical personnel. Fortunately, a technique for dealing rapidly with the problem has been developed. Originally named the *Heimlich maneuver* after its inventor, this *abdominal thrust* technique takes advantage of the residual air that is always contained within the lungs and the elastic properties of the chest cavity.

The abdominal thrust should be attempted only when there is clear evidence that the air passageways have been completely blocked and the patient is in serious distress. Even if an individual is choking violently, if she or he is able to breathe or speak, the air passageways are not blocked and there is no need for the maneuver. The figure shows how the maneuver is applied. The rescuer reaches around the patient from behind and joins hands together to make a small fist just below the rib cage. The rescuer then drives that fist inward and upward while trying to lift the patient, compressing the thoracic cavity and forcing air against the food fragment. If the force of the maneuver is great enough, the food pops out of the throat, freeing the breathing passage.

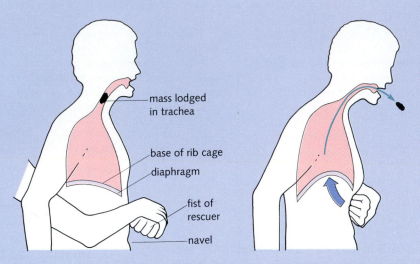

mass lodged in trachea

base of rib cage

diaphragm

fist of rescuer

navel

The abdominal thrust technique depends on a sudden compression of the chest cavity to produce the pressure needed to dislodge an object from the breathing passageways.

survive it; the death rate is greater than 90 percent after 5 years. The cancer cells develop as solid masses that interfere with lung activity. Very quickly, cells break off from the original tumor and spread throughout the body, causing dozens of secondary tumors and often bringing on a painful death.

In addition to its own carcinogenic properties, cigarette smoking also increases the smoker's risk of developing lung cancer as the result of exposure to other environmental hazards. All of us who live in major metropolitan areas continually inhale scores of potentially noxious airborne particles. Normally, the self-cleaning mechanism in the respiratory tract traps these particles and moves them up, out of the lungs and into the pharynx, where they are swallowed. But because cigarette smoke paralyzes this transport system, smokers' lungs cannot clean themselves efficiently and particles accumulate in the lungs and remain for long periods.

The great tragedy of lung cancer is that it is almost completely preventable. Lung cancer in most cases is caused by smoking. If you don't smoke, you greatly diminish your chances of contracting this terrible disease.

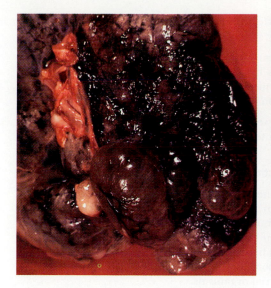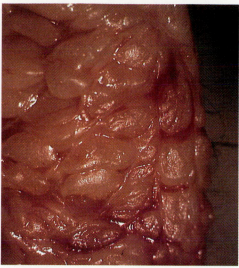

Figure 33.17 Part of the damage done by smoking is shown in this comparison of the lungs of a smoker (left) with the healthy lungs of a non-smoker (right).

Emphysema

The breathing process can be affected by anything that makes it more difficult for air to enter the lungs. **Emphysema** is a condition that often develops in older people who have been exposed to conditions that weaken the respiratory system, including smoke, dust particles, and pollutants.

Recall that the elasticity of healthy lung tissue produces much of the pressure differential that drives air out of the lungs. The lung tissue of an emphysema patient, however, is much looser than normal; when expanded, it does not "snap back" to so small a volume. The lowered elasticity that results from emphysema reduces the volume of air that is exhaled and the amount of air that can be inhaled. In advanced cases of emphysema, the lungs are able to bring in so little air that patients must be given air enriched in oxygen.

Asthma

Asthma is an allergic condition that we will discuss in Chapter 39. The allergies of an asthma patient may trigger an involuntary reaction in which smooth muscle cells around the airways contract, reducing the diameter of the passageways. Because this increases the resistance to air flow (for the same reason that water flows more slowly through a narrow garden hose), less air is able to flow to and from the lungs during a normal breath. To counteract this and allow them to breathe normally, asthma sufferers often take medication that helps to relax the muscles surrounding the air passageways.

SUMMARY

The process of respiration, in which cells oxidize carbon-rich food molecules to obtain chemical energy, requires the presence of oxygen to act as an electron acceptor. In smaller organisms, this oxygen can be obtained directly from the environment by the inward diffusion of oxygen from surrounding water or air, and carbon dioxide is diffused directly outward. In larger organisms, however, diffusion is not adequate to the task, and a specialized respiratory system makes possible gas exchange with the environment. Respiratory exchange may occur through a moistened integument, through gills, through tracheae that conduct air directly to the tissues, or through lungs that allow gas exchange to occur into a circulating blood, which carries oxygen to the tissues.

The human respiratory system is typical of those of terrestrial mammals. Air is inhaled through a series of passageways and enters the lung, driven by a partial vacuum created within the chest cavity when the diaphragm muscle contracts. Air enters the lungs and finds its way into hundreds of thousands of tiny alveoli, where gas exchange takes place between the air and the circulating blood. Oxygen passing across the walls of the alveoli enters the fluid phase of the blood, and most of it is taken up by hemoglobin, the oxygen-carrying protein contained within red blood cells. Oxygen is bound to special sites within the hemoglobin molecule. This binding is loose enough so that when the blood passes into tissue regions where little oxygen is found, oxygen molecules released from hemoglobin are able to diffuse into the surrounding area. The carbon dioxide that is a waste product of respiration in the tissues also enters the blood, and most of it returns to the lungs as bicarbonate ions, where it produces carbon dioxide to be exhaled. The oxygen-carrying system may be subverted by a molecule such as carbon monoxide, which competes with oxygen for binding sites on hemoglobin.

After studying this chapter, you should be able to:

■ Discuss the relationship between respiration at the cellular level and the respiratory system.

■ Explain some of the mechanisms and organs that have evolved to meet the challenges of gas exchange.

■ Trace the operation of the human respiratory system, and describe its connection with the circulatory system.

■ Cite the properties of hemoglobin, and describe how hemoglobin facilitates oxygen transport through the system.

SELECTED   TERMS

gills *p. 631*
countercurrent flow *p. 632*
spiracles *p. 633*
tracheoles *p. 634*
lungs *p. 634*
pharynx *p. 636*
breathing *p. 636*
trachea *p. 636*
epiglottis *p. 636*
larynx *p. 636*

bronchioles *p. 636*
alveoli *p. 636*
thoracic cavity *p. 636*
pleural sacs *p. 636*
diaphragm *p. 636*
hemoglobin *p. 639*
heme *p. 639*
partial pressure *p. 640*
myoglobin *p. 641*
carbon monoxide *p. 642*

REVIEW

Discussion Questions

1. If respiration occurs at the cellular level, why is it necessary for an organism to have a respiratory system?

2. Describe the various types of respiratory structures found in animals. How are they similar? How are they different?

3. Why is an effective respiratory system often associated with a circulatory system?

4. An animal can suffocate in any atmosphere that lacks oxygen. Why is brief exposure to concentrated carbon monoxide much more dangerous than brief exposure to concentrated carbon dioxide?

5. What characteristics of hemoglobin make it an ideal oxygen-carrying protein?

Objective Questions   (Answers in Appendix)

6. Which of the following statements about diffusion is true?
   (a) Oxygen cannot diffuse through the moist external membranes of most animals.
   (b) Oxygen exchange occurs in individual cells by diffusion in both small and large organisms.
   (c) Diffusion does not occur at the cellular level in large terrestrial organisms.
   (d) a and b only

7. All respiratory systems have
   (a) a way to exchange gases by diffusion.
   (b) a tracheal system.
   (c) countercurrent flow.
   (d) a system using lungs.

8. The most efficient gas exchange system is found in
   (a) reptiles.
   (b) mammals.
   (c) humans.
   (d) birds.

9. Which of the following organisms would you expect to have the most finely subdivided lungs?
   (a) frog
   (b) earthworm
   (c) goat
   (d) insect

10. Hemoglobin contains _____ heme units, each of which can combine with _____ of oxygen.
   (a) four/one molecule
   (b) eight/one molecule
   (c) two/one molecule
   (d) two/one atom

CHAPTER 34

Nutrition and Digestion

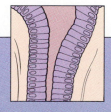

One must eat to live, and not live to eat.

—Molière
Amphitryon (1668)

You can find them in remote rural hamlets and in tumultuous urban ghettos, in the hills of Appalachia and on the plains of Bangladesh: children whose lives will be tragically shortened or permanently impaired by lack of proper energy and nutrients. Yet in an equally universal range of places are children and adults whose lives are in peril from *too much* food: chronically overweight individuals whose hearts, blood vessels, liver, and kidneys struggle daily with problems brought on by misdirected and overzealous eating.

Molière, with his classic combination of wit and insight, captures all too well these extremes of the perpetual human preoccupation with food. Today, more than three centuries after he wrote those lines, the human race is still struggling, not only with the demons of malnutrition and obesity, but also with a host of other medical problems that result from an improper diet.

Even those of us who are healthy are constantly bombarded with a confusing mix of scientific and pseudoscientific advice on nutrition. We are advised by physicians on the basis of scientific evidence to add fiber to our diets and to reduce our intake of fats. We are pressed by various "health food" manufacturers to replace "refined sugar" with "complex carbohydrates" and to swallow megadoses of "essential vitamins and amino acids." And everyone seems to have a new diet plan that promises weight loss with no change in either eating patterns or exercise habits.

A thorough exploration of all these important, daily concerns about nutrition is beyond the scope of this chapter. We can, however, touch on several of them and lay sufficient conceptual groundwork for you to pursue further answers on your own.

NUTRITIONAL NEEDS OF HETEROTROPHS

Organisms need to eat for two main reasons. The *generation of chemical energy* is the principal use to which most food is put. As we saw earlier, carrying on the activities of life requires energy in the form of ATP. That ATP can be produced from the breakdown of glucose and other food molecules in pathways known as *glycolysis*, the *Krebs cycle*, and *oxidative phosphorylation* (Chapter 23). Every cell has energy needs that must be met by a constant supply of food molecules entering these chemical pathways.

Animals' diets must also provide them with numerous *molecules needed for growth and maintenance*. Given only a simple source of chemical energy, such as sugar, cells can synthesize some of these compounds. But there are many other molecules that animal cells cannot synthesize and that must be provided in the diet. These essential nutrients vary from one species to the next, and finding diets that provide all essential nutrients, along with the proper amount of food energy, is a major focus of research in animal nutrition.

Food Energy and Calories

We measure the energy available in a food by measuring the amount of heat that can be released from it when it is completely oxidized. This heat is expressed in units known as calories. A **calorie** is the amount of heat energy needed to raise the temperature of a gram of water 1°C. One thousand calories is a kilocalorie. (Nutritionists usually use the term *Calorie* to represent *1 kilocalorie,* but in this text we will adhere to proper scientific usage to avoid confusion.)

Every organism needs a constant supply of energy just to maintain its basic cellular activities. This background level of activity is known as the **basal metabolic**

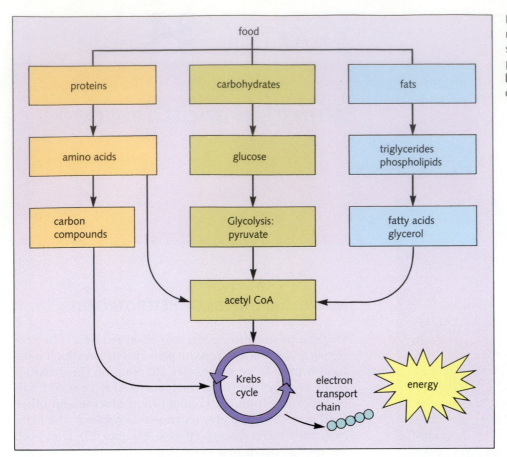

Figure 34.1 Food is a source of raw materials for the body, but it is also a source of energy. Food in the form of proteins, fats, or carbohydrates can be broken down and used as a source of cellular energy.

rate, and it represents the minimum activity required to sustain vital functions such as breathing and nervous system activity. The basal metabolic rate for average-sized humans is about 1600 kcal/day.

Nearly all foods can provide usable energy. In Chapter 21 we traced the chemical pathways followed when glucose was broken down to release energy used in the formation of ATP. We chose glucose because it is the molecule on which the pathways of cellular metabolism are based. However, the same pathways can be used to obtain chemical energy from nearly any food molecules. Fats, proteins, and carbohydrates can all be broken down into simpler molecules, which then enter the same basic chemical pathways (Figure 34.1). In this limited sense, basic energy requirements can be satisfied by any source of calories.

A lack of sufficient calories in the diet is known as **undernourishment.** The effects of chronic undernourishment can be tragic. Starved for sources of energy, the body begins to break down its own macromolecules to provide ATP, causing a loss of muscle and other tissue and a drop in the efficiency of several major organ systems. Prolonged undernourishment is common in areas such as Ethiopia and the Sudan that have suffered from severe famines, but it is also found in parts of many other supposedly wealthy countries, including the United States. Because of its ability to cause permanent damage to the skeleton and the nervous system, long-term undernourishment of children can lead to stunted growth and mental retardation.

Food and Essential Nutrients

Most animals, humans included, cannot manufacture all of the compounds needed to maintain life. Those molecules that an organism cannot synthesize, along with a variety of minerals, are known as *essential nutrients,* and a healthful diet must include them.

Determining which nutrients are essential and the amounts in which they are required is not a simple task, especially in a difficult organism to experiment with, such as the human. However, nutritional scientists now believe that roughly 50 compounds are essential to human life. Given these 50 compounds in proper amounts, and an adequate source of chemical energy, the human body can synthesize each of the thousands of compounds found living in cells and tissues.

However, if one or more of these essential nutrients are missing, the body is said to suffer from **malnourishment,** or malnutrition. Malnourishment cannot be compensated for by excesses of other nutrients; it can be corrected only by a diet that contains proper amounts of the missing nutrients.

Water Water can be thought of as the most important of all essential nutrients. In fact, animals deprived of both food and water die of thirst long before they begin to suffer the ill effects of hunger. More than half of the total body weight of most animals is water—in some animals, that figure is as high as 90 percent. Water is needed in and around every cell of the body, and it makes up the major portion of blood, lymph, and other body fluids. Water is required for processing food in the digestive system, for eliminating waste through the kidneys, and for temperature regulation in animals that sweat or pant when they are hot.

If the total water content of an animal's body drops below the minimum level needed for normal functions, the animal is said to be *dehydrated.* If dehydration is not promptly treated, it can lead to problems with the circulatory, respiratory, and nervous systems. Under typical circumstances, the average human must take in at least a liter of water a day to avoid dehydration.

Carbohydrates The major sources of food energy in the human diet are carbohydrates. **Simple carbohydrates** are the sugars found in fruits and honey and in refined sugar. **Complex carbohydrates** include the starches in potatoes, grains, and vegetables. Enzymes in the digestive system break down complex carbohydrates to release simple sugars. These are absorbed directly into the circulation, making carbohydrates an immediate source of food energy for the body.

Many carbohydrate-rich foods also contain cellulose-based *fibers* that we cannot digest because we lack the enzymes to break down cellulose. However, these fibers add bulk to the food that moves through the digestive system, and a certain amount of fiber is necessary for the digestive system to work properly.

Proteins Proteins are essential nutrients because they are sources of the *amino acids* used by the body to build new proteins. There are 20 different amino acids commonly found in cellular proteins. Plants can synthesize all 20 amino acids as long as they are supplied with chemical energy and a source of nitrogen. Thus plants do not require these compounds as nutrients. Animals, however, can synthesize only about half of these 20 amino acids and must acquire the rest in their food. Those amino acids that cannot be synthesized are called **essential amino acids.**

An average-sized adult needs a minimum of 60 grams of protein per day. However, the need for protein includes *quality* as well as *quantity,* for humans must derive 8 essential amino acids from the protein they eat. If one or more of these essential amino acids are lacking in the diet, **protein deficiency** results. Figure 34.2 shows that when even one of the essential amino acids is lacking in the diet, the overall level of new protein synthesis in the body declines, despite an abundance of other nutrients. For this reason, a diet deficient in *any* of the essential amino acids dramatically reduces the amount of new protein that can be synthesized. Amino acids that cannot be used for protein

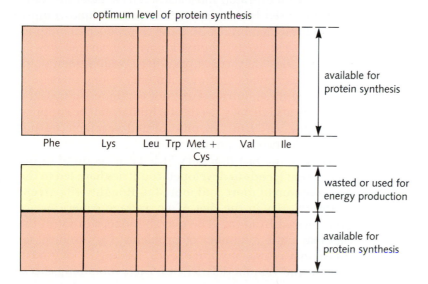

optimum level of protein synthesis

Phe Lys Leu Trp Met + Cys Val Ile

available for protein synthesis

wasted or used for energy production

available for protein synthesis

Figure 34.2 Humans require 8 essential amino acids in the diet. Because each of these amino acids is needed for protein synthesis, a deficiency in any one of them lowers the overall level of protein synthesis and may cause malnutrition, even if the other amino acids are present in excess.

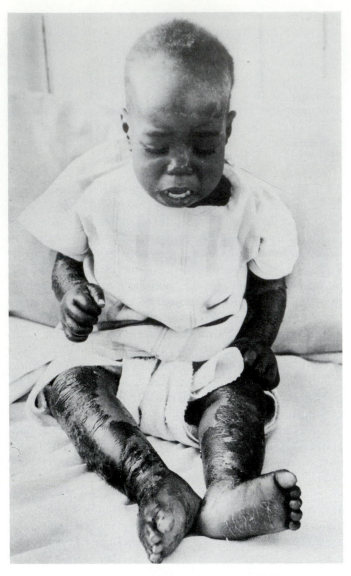

Figure 34.3 This child has suffered from kwashiorkor, a nutritional disorder caused by a diet lacking in one or more of the essential amino acids. The disease is common in parts of the world where diets contain only a single source of protein, such as rice or corn.

synthesis are then used as food energy—a waste of nutrient value in one of the most important types of food.

Protein-rich foods from animal sources, such as eggs, fish, meat, and milk are said to contain *complete* proteins, which means that they include a complete set of essential amino acids.

Many protein-rich plant foods, however, are nutritionally *incomplete* because they lack some of the essential amino acids. Corn, for example, is deficient in *lysine,* an essential amino acid. Beans have plenty of lysine, but they lack adequate amounts of *methionine* and *cysteine,* also essential amino acids. Vegetarians must therefore take care to balance their food sources in order to obtain a complete supply of the essential amino acids. Fortunately, this can be done quite simply. A diet made up exclusively of either beans or corn produces protein deficiency. However, a diet that contains ample portions of both supplies all of the essential amino acids and hence supports normal levels of protein synthesis.

Protein deficiencies produced by diets that lack complete sources of protein cause serious problems in many areas of the world where meat and dairy products are scarce. One such problem is *kwashiorkor,* a severe protein-deficiency disease found in several Third World countries (Figure 34.3). It strikes children (who need large amounts of protein for growth) particularly hard. Kwashiorkor, which is characterized by weakness, stunted growth, skin inflammation, and anemia, is caused by diets based on only a single vegetable source of protein: corn in Africa, rice in Asia. The cure for kwashiorkor is a change in diet to include complementary sources of proteins. The addition of soybean products—such as tofu—to a rice diet can supply the missing nutrients.

Fats Although animals can synthesize many fats and lipids from a few starting compounds, they do require a source of unsaturated fatty acids. These **essential fatty acids** are used both to synthesize certain lipid components of cell membranes and also to produce the important fatty-acid-like hormones known as *prostaglandins.*

Fatty acids that can supply these requirements are found in most plant and animal foods, so long-term deficiencies in these nutrients are not common. In fact, the total fatty acid dietary needs of a human can be supplied by as little as 60 mL of unsaturated vegetable oil per day—about two teaspoons.

Despite the fact that very little fat is required in the diet, recent trends in developed countries have led to the consumption of more and more fat. Roughly 40 percent of the calories in the diet of an average American are now derived from fat. The overabundance of fat in the diet may cause a number of problems. For instance, fat intake may crowd out other foods, leading to long-term protein and vitamin deficiencies. Diets containing excessive amounts of saturated fats also tend to increase the levels of cholesterol and other dangerous forms of fat in the blood, often leading to heart disease (see Theory in Action: Cholesterol and Heart Disease: Research and Changing Perspectives, p. 652). High-fat diets are also associated with increased incidence of breast and colon cancers. Although the ideal level of fat in the diet is hard to determine, most dietary experts believe that the fat content of the typical diet should be reduced to the point where only 20–30 percent of one's daily calories are derived from fat.

Vitamins   Animals also require more than a dozen small organic molecules known as **vitamins.** The term was coined as a contraction of "vital amines," because *thiamine,* the first vitamin to be discovered, belonged to the chemical family known as the amines. Vitamins are usually defined as *complex organic compounds needed in small amounts in the diet.* This distinguishes them from proteins and other nutrients that are needed in larger amounts. Most vitamins are involved in chemical reactions as *coenzymes* or *cofactors,* which help to catalyze chemical reactions. Just a handful of molecules can catalyze hundreds or thousands of such reactions.

Although vitamins are needed in small amounts, vitamin deficiencies can have serious—even fatal—consequences. Vitamin B_1, or thiamine, for example, is present in many meats and grains, including the hulls that surround the grains of natural brown rice. The need for this vitamin was first appreciated in the nineteenth century when prisoners in the Far East fell sick with a disease known as *beriberi.* Prisoners afflicted with beriberi were weak and nervous and suffered from *edema* (swelling in the tissues) as well as heart failure. The fact that none of the jail guards caught the disease suggested that it was not a contagious illness. The jail wardens were also proud to point out that they fed the prisoners the very best rice on the market: polished white rice, from which the brown husks had been removed.

A clue to the source of this mysterious ailment came from an unlikely source: the behavior of chickens kept in the prison compound. These birds suffered from symptoms similar to those of the human prisoners—loss of muscle coordination that caused them to fall repeatedly when they tried to walk. The fact that the chickens had been fed table scraps from the jail led to the idea that the prisoners' food might be responsible for beriberi. In fact, when the ailing chickens were fed table scraps containing ordinary brown rice, they quickly improved.

It became clear that the disease was caused by the absence of thiamine, vitamin B_1, in the polished white rice. Because rice hulls contain plenty of thiamine, all that was necessary to cure the disease was to switch from white to brown rice.

One vitamin after another has been discovered over the years, and at least 14 vitamins are now recognized. Wherever possible, scientists have tried to determine the minimum amount we need of each vitamin for normal functioning. This amount is expressed in terms of a *recommended daily allowance* (RDA), which serves as a guide to developing a daily diet that contains at least the RDA for each vitamin.

As shown in Table 34.1, 10 of the vitamins are water-soluble and 4 are fat-soluble. The body is not able to store any of the water-soluble vitamins; if the diet contains more of these vitamins than can be used, the excess is removed from the blood by the kidneys and excreted in the urine. That's why when you take large quantities of water-soluble vitamins (particularly those in the B-complex), your urine turns dark yellow; you are simply excreting the vitamins your body has not used.

Excess amounts of fat-soluble vitamins, on the other hand, can be stored in the fatty tissues of the body. Although this allows the body to develop storage reserves of these vitamins, it also poses a potential danger. Vitamins A, D, and K are toxic (poisonous) in high concentrations, and the use of excessive numbers of vitamin pills can cause these compounds to build up to the point where they become a serious medical problem. For this reason, the excessive use of vitamin supplements (except under the supervision of a physician) should be avoided.

Minerals   The *inorganic* nutrients known as minerals are usually required in very small amounts. There are at least 18 **essential mineral nutrients** (Table 34.2), many of which are required for the body to build specific tissues. Deficiencies of these and other minerals can result in stunted growth, muscle weakness, or anemia, depending on what is lacking and the severity of the deficiency.

Calcium phosphate, for example, is the major component of bone, and large amounts of both calcium and phosphorus are required in the diet when new bone is being formed.

Iron is needed as a component of a number of important proteins, including hemoglobin and myoglobin, which bind oxygen, and of the cytochromes, which are involved in electron transport. Iron deficiencies are particularly common in women, and often result in chronic anemia. Even college-age women should pay particular attention to the amount of iron in their diets.

Most bodily fluids contain sodium, potassium, and chlorine, and a proper supply of these minerals is important to many excitable tissues, including muscle and nerve.

Sulfur is a component of many proteins, and iodine is used in very small amounts for the synthesis of *thyroxine,* a vital hormone produced by the thyroid.

One of the most critical minerals for animals living on land is sodium. The large amount of sodium in the

Cholesterol and Heart Disease: Research and Changing Perspectives

A 60-year-old man, a veteran of two coronary bypass operations, sits in a chair, connected by tubes from both his arms to an experimental laboratory device. As his blood flows slowly through the apparatus, a chemical filter removes the excess cholesterol that threatens to clog his few functioning coronary arteries. Elsewhere, a 6-year-old girl receives in a single operation both a new heart to replace the one crippled by fatty deposits and a new liver that doctors hope will keep the vessels of her replacement heart intact.

Both these individuals, one 6, the other 60, are victims of *atherosclerosis* (Chapter 32), a disease that contributes to nearly half of all deaths in the United States each year. Theirs are extreme cases, for they carry defective genes that interfere with the homeostatic control mechanism that regulates blood cholesterol levels.

"Normal" blood cholesterol concentrations among Americans range from 100 to 200 milligrams per deciliter (mg/dL). The man described above, who carried a single copy of the defective gene, had cholesterol levels ranging from 300 to 450 mg/dL. The girl, who carried *two* defective genes, had levels in excess of 600 mg/dL. Levels this high are rarely, if ever, seen outside the 1 in 500 individuals with such a defective gene.

Atherosclerosis occurs when fatty deposits, laden with cholesterol, accumulate in the walls of arteries throughout the body. The medical effects of high cholesterol are chillingly clear: In villages in Japan and Yugoslavia where the mean cholesterol level is 160, the heart attack rate is extremely low: fewer than 5 per 1000 men in 10 years. But in Finland, where the mean cholesterol level is 265, the incidence of heart attack is 14 times as high. The incidence of heart attack and the mean cholesterol levels in the United States are midway between these two extremes. It is quite clear that high cholesterol levels—along with other contributing factors such as obesity, lack of exercise, smoking, and stress—lead to elevated risk of heart attacks.

But as we learned in Chapter 32, cholesterol is not simply a "bad" compound; it is an essential part of cell membranes and a vital step in the biosynthetic pathways of several important substances ranging from bile salts to sex hormones. How, then, does excess cholesterol cause such serious problems?

Cholesterol is present in foods derived from animals, including meat, eggs, and dairy products (see table). The body can also manufacture cholesterol, primarily in the liver. Cholesterol itself is insoluble in water, and it therefore circulates in the blood combined with other fats and protein in either of two forms: as *low-density lipoprotein,* or LDL, and as *high-density lipoprotein,* or HDL. It is the LDL form of cholesterol that gets deposited in arterial walls; HDL particles seem to pick up cholesterol from such places and carry it to the liver for disposal.

The workings of the cholesterol regulatory machinery in the body were described by Michael Brown and Joseph Goldstein in a series of elegant experiments that earned them a Nobel Prize. They discovered that the membranes of many cells contain specialized receptor molecules that bind to cholesterol-containing HDLs and LDLs and carry these lipoproteins into the cell. There, cholesterol can be incorporated into the cell membrane, converted into hormones, stored, or metabolized. In healthy individuals, a

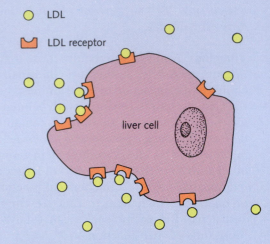

Receptors for low-density lipoproteins (LDLs) on the surface of liver cells help to regulate LDL levels in the bloodstream. When the receptor system does not operate properly, high LDL levels may produce atherosclerotic deposits that threaten the circulatory system.

negative feedback loop within liver cells "informs" the cells' metabolic pathways that enough cholesterol is present, removes excess cholesterol from circulation, and shuts down the synthesis of more cholesterol.

The 60-year-old man and the 6-year-old girl have defective genes for the receptors. Unable to detect and monitor cholesterol in the bloodstream, not only are the liver cells of such individuals unable to *remove* excess cholesterol, but they are also "tricked" into continually producing and releasing more, even if blood cholesterol levels are already high. Thus these individuals cannot control their cholesterol levels with diet alone; their bodies continue to synthesize the compound even when none is taken in. The resulting high levels of blood cholesterol produce life-threatening atherosclerosis.

How is a gene defect in the cholesterol receptor related to cholesterol in the diets of normal individuals? Brown and Goldstein have suggested that a diet rich in cholesterol provides the liver with more cholesterol than it can possibly use. No longer requiring cholesterol for its own metabolism, the liver cell shuts down the synthesis of LDL receptors. Because the liver is subsequently unable to measure and react to the true cholesterol concentration in the blood, blood cholesterol rises. In a subtle but dangerous way, therefore, a diet high in cholesterol causes symptoms that mimic the genetic disease.

Nationwide surveys have shown that as many as 50 percent of Americans have dangerously high cholesterol levels caused mainly by diets containing too much fat. Many—though not all—of these individuals are at increased risk of heart attack and would benefit from lowering their blood cholesterol. There is little question that the American diet contains far too much fat: A typical American consumes more than the equivalent of an entire stick of butter in fat and cholesterol each day. These facts have led researchers to assemble recommendations about cholesterol control that range from diet to drugs:

1. Individuals over 20 years of age should have their cholesterol levels checked once every 5 years.

2. Those with LDL cholesterol levels above 130 mg/dL should be instructed in ways to cut back their dietary intake not only of cholesterol itself, but also of saturated fats that the body converts into cholesterol. Some dietary fats, including those found in olive oil and the "omega-3" polyunsaturated fats found in fish oils, seem to lower LDL and raise HDL in a beneficial way.

3. Those individuals with LDL levels above 160 mg/dL should be considered at high risk and should be monitored closely. If these individuals do not respond to changes in diet, their physicians should consider treatment with one of the recently developed drugs that control cholesterol uptake and synthesis.

Many individuals with high cholesterol levels respond well to fairly simple changes in diet alone. That's because most of us are unaware of just how much cholesterol and saturated fats we eat in many common foods. "Fast foods" in particular are notoriously high in fat, because many of them, from doughnuts to french fries to chicken fillets, are fried in fat containing very high concentrations of beef fat or lard. Others contain palm or coconut oil rich in saturated fats.

Learning from nutritionists how to eat a satisfying meal without all that fat is the first step in controlling high cholesterol. Another helpful step involves eating foods high in soluble fiber, which binds cholesterol-containing bile salts. As a last resort for those whose cholesterol levels are resistant to dietary measures alone, these measures may be supplemented by drug therapy. Combined therapy with several classes of drugs has been shown to lower cholesterol levels by as much as 40 percent, and there is clear evidence that lowering cholesterol reduces the risk of heart attack.

Cholesterol Content of Common Foods

Food	Cholesterol (mg)	Food	Cholesterol (mg)
Fruits/vegetables	0	Creams (1 tbsp)	
Grains	0	Whipped	20
Milks (1-cup serving)		Sour	8
Whole milk	33	Half and half	6
Yogurt (whole		Meats (3-oz serving)	
milk)	29	Veal	86
Low-fat milk	18	Lamb	83
Low-fat yogurt	14	Beef	80
Cheeses (1-oz		Pork	76
serving)		Chicken	39 to 63
Cheddar	30	Eggs (1 large egg)	
Processed		Yolk	274
American	27	White	0
Swiss	26	Fish (3-oz serving)	
Ice cream (1/2 cup)	30	Shrimp	128
Butter (1 tsp)	12	Lobster	72
Margarine, all		Clams	50
vegetable (1 tsp)	0	Oysters	38
		Fish fillet	34 to 75

Source: Adapted from *Cholesterol Information Sheet* (Rosemont, Ill.: National Dairy Council, 1984).

Table 34.1 *Vitamins*

Vitamin	Food Sources	Function	Needed Daily	Results of Vitamin Deficiency
Water-Soluble Vitamins				
Vitamin B₁ (thiamine)	Yeast, liver, grains, legumes	Coenzyme for carboxylase	1.5 mg	Beriberi, general sluggishness, heart damage
Vitamin B₂ (riboflavin)	Milk products, eggs, vegetables	Coenzyme in electron transport (FAD)	1.8 mg	Sores in mouth, sluggishness
Niacin	Red meat, poultry, liver	Coenzyme in electron transport (NAD)	20 mg	Pellagra, skin and intestinal disorders, mental disorders
Vitamin B₆ (pyridoxine)	Dairy products, liver, whole grains	Amino acid metabolism	2 mg	Anemia, stunted growth, muscle twitches and spasms
Pantothenic acid	Liver, meats, eggs, whole grains, and other foods	Forms part of Coenzyme A, needed in Krebs cycle	5–10 mg	Reproductive problems, hormone insufficiencies
Folic acid	Whole grains and legumes, eggs, liver	Coenzyme in biosynthetic pathways	0.4 mg	Anemia, stunted growth, inhibition of white cell formation
Vitamin B₁₂	Meats, milk products, eggs	Required for enzymes in red cell formation	0.003 mg	Pernicious anemia, nervous disorders
Biotin	Liver and yeast, vegetables, provided in small amounts by intestinal bacteria	Coenzymes in a variety of pathways	Unknown	Skin and hair disorders, nervous problems, muscle pains
Vitamin C (ascorbic acid)	Citrus, tomatoes, potatoes, leafy vegetables	Required for collagen synthesis	45 mg	Scurvy: lesions in skin and mouth, hemorrhaging near skin
Choline	Beans, grains, liver, egg yolks	Required for phospholipids and neurotransmitters	>700 mg	Not reported in humans
Fat-Soluble Vitamins				
Vitamin A (retinol)	Fruits and vegetables, milk products, liver	Needed to produce visual pigment	1 mg	Poor eyesight and night blindness
Vitamin D (calciferol)	Dairy products, fish oils, eggs (also sunlight on skin)	Required for cellular absorption of calcium	0.01 mg	Rickets: bone malformations
Vitamin E (tocopherol)	Meats, leafy vegetables, seeds	Prevents oxidation of lipids in cell membranes	15 mg	Slight anemia
Vitamin K (phylloquinone)	Intestinal bacteria, leafy vegetables	Required for synthesis of blood clotting factors	0.03 mg	Problems with blood clotting, internal hemorrhaging

SOURCE: N. S. Scrimshaw and V. R. Young, The requirements of human nutrition, *Scientific American*, 1976.

body is an evolutionary reflection of the abundance of sodium chloride (common salt) in seawater, where life first began. On land, however, sodium is hard to come by. Land plants, which need to conserve water, contain very little sodium. Therefore herbivores often have chronic sodium deficiencies, and this can influence their behavior in the wild. Farmers often provide blocks of salt for their horses and cattle, and hunters know that a block of salt placed in the woods will draw wild animals from miles around. For example, salt blocks are so effective in drawing deer that their use in hunting has been outlawed in most states.

Humans have the same need for salt as other land animals, and ancient civilizations valued salt highly. The discovery that salt could be obtained by boiling seawater predates civilization, and many cultures have incorpo-

Table 34.2 *Essential Mineral Nutrients*

Mineral	Food Sources	Function	Needed Daily	Results of Mineral Deficiency
Calcium	Milk, cheese, legumes, dark green vegetables	Bone formation, blood clotting reactions, nerve and muscle function	800 mg	Stunted growth, weakened bones, muscle spasms
Phosphorus	Milk products, eggs, meats	Bones and teeth, ATP and related nucleotides	800 mg	Loss of bone minerals
Sulfur	Most foods containing proteins (derived from sulfur-containing amino acids)	Used in formation of cartilage and tendons	Provided by sulfur-containing amino acids in diet	Deficiency of sulfur-related amino acids
Potassium	Most foods	Acid–base balance, nerve and muscle function	2500 mg	Muscular weakness, heart problems, death
Chlorine	Salt	Acid–base balance, nerve and muscle function, water balance	2000 mg	Intestinal problems, vomiting
Sodium	Salt	Acid–base balance, nerve and muscle function, water balance	2500 mg	Weakness, diarrhea, muscle spasms
Magnesium	Green vegetables	Enzyme cofactors, protein synthesis	350 mg	Muscle spasms, stunted growth, irregular heartbeat
Iron	Eggs, leafy vegetables, meat, whole grains	Hemoglobin, electron transport enzymes	10 mg	Anemia, skin lesions
Fluorine	Drinking water, seafood	Structural maintenance of bones and teeth	2 mg	Tooth decay, bone weakness
Zinc	Most foods	Enzyme cofactor	15 mg	Fever, vomiting, diarrhea, nausea
Copper	Meats, drinking water	Required for hemoglobin-synthesizing enzymes	2 mg	Anemia
Manganese	Whole grains, egg yolks, green vegetables	Required for several enzymes	3 mg	No reported effects in humans
Iodine	Seafood, milk products, iodized salt	Thyroid hormone	0.14 mg	Goiter (enlarged thyroid)
Cobalt	Meat, liver, milk products	Part of vitamin B_{12}	Required in the form of vitamin B_{12}	No reported effects in humans

SOURCE: N. S. Scrimshaw and V. R. Young, The requirements of human nutrition, *Scientific American*, 1976.

rated the refining and use of salt into their cultural and religious ceremonies.

The successes of human culture in developing means to purify salt as a mineral supplement, however, have had a serious side effect. Probably because our mammalian ancestors needed salt, our taste buds find it eminently desirable. Salt enhances the flavor of dozens of foods, and this has led to an excess of sodium in the typical diet in some countries. The average American takes in roughly 25 times the amount of sodium needed for a healthy diet, and this extra sodium can cause several problems.

The immediate effect of sodium is to raise blood pressure and increase stress on the kidneys, which remove excess sodium from the blood. Over the long term, these effects increase the likelihood of heart disease and stroke. A high level of dietary sodium is one of the major causes

Table 34.3 *What It Takes to Burn Calories*

Food	Activity	Cal/min	Time to Burn
Cheeseburger 470 Cal.	Resting	1.1	427 min.
	Walking	5.5	85 min.
	Swimming	10.9	43 min.
	Running	14.7	32 min.
Milkshake 318 Cal.	Resting	1.1	289 min.
	Walking	5.5	58 min.
	Swimming	10.9	29 min.
	Running	14.7	22 min.
Corn, 2 pats butter 170 Cal.	Resting	1.1	155 min.
	Walking	5.5	31 min.
	Swimming	10.9	16 min.
	Running	14.7	12 min.
Corn, unbuttered 70 Cal.	Resting	1.1	63 min.
	Walking	5.5	13 min.
	Swimming	10.9	6 min.
	Running	14.7	5 min.

of circulatory disease in the United States and other industrialized countries.

Excesses of other minerals can also cause serious problems. An excess of iodine in the diet can depress thyroid activity, large amounts of iron can produce liver damage, and high levels of magnesium in water supplies can cause chronic diarrhea. As with the other major classes of nutrients, a healthful diet contains adequate, but not excessive, amounts of each of the essential minerals.

Balancing Nutrient Input and Requirements

As we have seen, the basal metabolic rate for an average-sized human is about 1600 kcal per day. This figure represents no more than the basic energy needed to remain alive with minimal activity. Another way of thinking about the basal metabolic rate is to note that it is roughly the amount of energy one would burn while spending the whole day in bed.

Physical activity, however, requires energy in *excess* of the basal rate. The normal daily activities of walking, lifting, and doing household chores may consume several hundred kilocalories in addition to the basal rate; vigorous activity may add more than a thousand kilocalories. Swimmers and long-distance runners may need more than 4000 kcal a day. Table 34.3 presents the amount of energy required for a number of physical activities in terms of common foods.

THEORY IN ACTION

The Hundred-Watt Bulbs

The basal metabolic rate for humans ranges from 1300 to 1800 kcal per day. Ultimately, whether it is used for internal movement, breathing, circulation, or philosophical thought, nearly all of this energy is lost as heat. This gives the average human an energy output equal to that of a 100-watt lightbulb. This may not sound like much, but architects realize how important those heat sources can be.

When building a 1000-seat theater, for example, the designer must deal with the fact that 1000 humans produce a steady heat output of 100 megawatts! That's roughly equivalent to the heat output of a large barbecue fire roaring in the middle of the theater. When the audience laughs or cries, the energy output is even higher, because emotional involvement increases basal metabolism. Air

conditioning may be a product of our modern society, but it is made necessary by one of the oldest forces—the metabolic fire of life itself.

Obesity Naturally, it is a rare day when the body takes in exactly enough food to match precisely the number of calories burned. When more food is eaten than can be converted into energy, the body stores the excess as glycogen in the liver and muscles. This ready reserve of carbohydrate can be used quickly when the body demands more energy. However, if the body's reserves of glycogen are already full when still more food is made available, the body converts that food energy into an efficient form for long-term storage: fat.

The energy required to produce fat is provided by ATP, so we should not delude ourselves into thinking that it is possible to gain weight only by eating fatty foods. Fat is produced whenever the amount of food taken in is greater than the amount of energy the body is capable of using. Therefore we can gain weight by eating to excess any food, including proteins, simple sugars, complex carbohydrates, and fat.

Individuals are said to be **obese** when their weight exceeds the ideal weight for someone of their size by more than 20 percent (Figure 34.4). Obesity is a serious health problem. In addition to limiting physical activity, extra weight places an unnecessary strain on the heart and circulatory system.

Losing weight The causes of obesity seem simple. Too much food is taken in, and too little is burned up in daily exercise. The cure for obesity seems equally simple: more exercise and better control of diet. When a person's daily activity *exceeds* his or her food intake, the first sources of extra energy to be called into action are glycogen reserves stored in liver and muscle cells. When these reserves are depleted, the body takes fat molecules out of storage and uses them for energy. If activity exceeds food intake on a regular basis, a slow but steady reduction in fat reserves takes place and weight is lost.

There is, of course, a lot more to it. Many people, despite their best efforts, try and try again and fail to lose weight. Nutritional scientists have noticed that an individual's appetite for food and personal energy level seem to help maintain a *set point* of body weight. When food intake is restricted, basal metabolic rate slows down

Figure 34.4 Ideal weights for individuals of various heights and ages. The "ideal" weight is a range that varies with body type and muscular development, as well as bone structure.

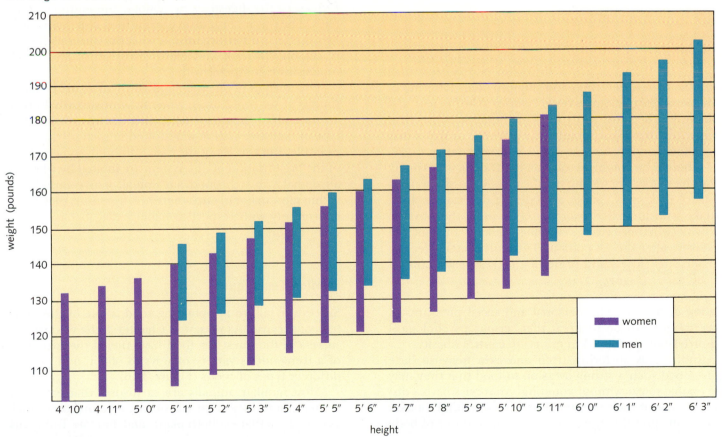

Figure 34.5 Proper exercise is an essential element to a program of weight control. Exercise also strengthens the circulatory and respiratory systems.

and the dieter feels tired and sluggish. This lowering of the metabolic rate causes fewer calories to be burned, and weight loss becomes more difficult. Some scientists have suggested that the set point may be genetic or that it may be determined by feeding habits in early infancy.

There is no complete solution to the difficulties many of us have in losing weight, but the most successful weight-control programs couple a moderate decrease in food intake with increased exercise. Over time, the physical activity (whether cycling, running, walking, or dancing) seems to alter the set point and give the dieter a feeling of personal fitness and pride that makes weight control much easier (Figure 34.5).

Eating disorders The personal and social pressures to lose weight can be so great that serious psychological disorders are produced in efforts to control obesity.

Individuals suffering from **anorexia nervosa** have developed such a strong aversion to food that they may be unable to eat without vomiting. Anorexics have a distorted view of their own bodies, which sometimes makes them interpret the presence of normal amounts of body tissue as a symptom of obesity. Such images may lead these individuals to lose weight to the point where their healthy tissue begins to break down and their very lives are threatened.

A related eating disorder is **bulimia,** in which individuals force themselves to vomit after eating high-calorie foods. In the early stages, sufferers may rationalize their behavior as helping to control weight, but this rationalization masks serious health problems. Vomiting the acidic contents of the stomach gradually destroys tooth enamel, causes a loss of salts from the body (which may produce heart failure), and may rupture the esophagus, causing death. There is no doubt that these eating disorders are partly the result of social pressures that place a premium on a slim appearance. They require immediate medical and psychological attention.

OBTAINING FOOD: THE UNIVERSAL PREOCCUPATION

The need for food affects every aspect of the life of an organism. The need to ensure an adequate supply of food affects physiology; it influences behavior and reproduction, helps determine the structure of the nervous system and the skeletal system, and defines the tasks performed by the digestive, circulatory, and endocrine systems. The way an organism obtains food also determines how it affects other organisms, what pressures its species places on the local and global ecosystems, and which organisms it competes with.

Our analysis of digestion and nutrition would be much simpler if we could make generalizations that held true across each of the major taxonomic groups—if just one type of digestive system were found in each phylum, for example. But this is not the case. Although some aspects of the organization of any system are limited by whether an animal is an arthropod, a mammal, or an echinoderm, within most major groups there are organisms that differ enormously with respect to habitat, diet, behavior, and other factors.

Feeding Techniques

Success or failure in obtaining food is one of the great tests of survival. As we saw in Chapter 3, the rewards for success in the quest for food have fueled the evolution of a tremendous variety of ways to obtain and process different types of foods (Figure 34.6).

Most familiar animals are carnivores that eat other animals, herbivores that eat plants, or omnivores (such as ourselves) that eat both plants and animals. But many aquatic organisms, such as clams, are filter feeders that trap food particles as they strain enormous volumes of water. Others, such as mosquitoes, are fluid feeders that consume either the blood of an animal or the sap of a plant. And many ecologically important organisms are detritus feeders that eat decaying organic material.

Because of this great diversity in feeding strategies, we cannot carry out a phylum-by-phylum analysis. We can, however, illustrate a few common "themes" that emerge in a study of digestive systems.

Cellular Digestion

Digestion is such a basic process that it is found in organisms that consist of nothing more than a single cell. A number of protozoans, including *Paramecium* and *Amoeba*, take in food particles directly from the environment. *Amoeba* engulfs particles directly, enclosing them in part of the cell membrane to form a food vacuole. *Paramecium* uses a ciliated groove to sweep particles into a specialized region of the cell membrane known as the *cytopharynx*, where food vacuoles are formed (Figure 34.7).

Once inside the cell, food particles must be broken down into usable form. Food vacuoles fuse with *lysosomes* to form digestive vacuoles. Lysosomes are organelles laden with enzymes capable of breaking down many large molecules (Chapter 24). These enzymes split polymers in food into monomers that are released, along with other small molecules, into the cytoplasm where they can be utilized by the rest of the cell. (Note that lysosomes are not restricted to protozoans; certain cells in multicellular organisms, such as the white blood cells of humans, handle large food particles in the same way.)

Material that cannot be broken down in lysosomes is released from the cell. In *Paramecium*, the release of indigestible material occurs at a specialized region known as the *anal pore*.

Figure 34.6 There's more than one way for an animal to acquire food, as demonstrated here by (Clockwise, from top left) a brown pelican with fish; a mosquito sucking blood; a spiny sun star attacking a green sea urchin; and a yellow rat snake devouring its prey.

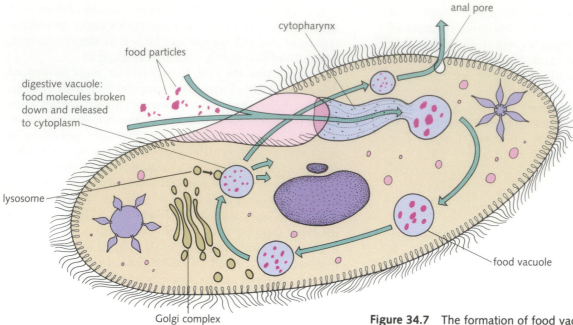

Figure 34.7 The formation of food vacuoles in *Paramecium*. The movement of food particles through this organism is an example of the pathways followed by food material in many other cells. The primary work of chemical digestion is done by lysosomal enzymes.

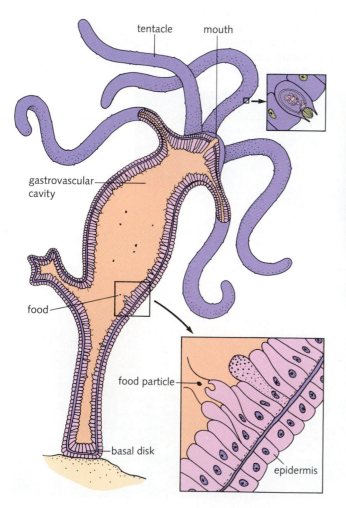

Figure 34.8 Digestion in *Hydra*. (left) Food is broken down in the gastrovascular cavity, and digestion is completed within the cells lining the cavity. (right) The extensive gastrovascular cavity of *Planaria* is stained red in this photomicrograph. The intricate branchings of this cavity allows food to diffuse to all parts of the animal.

Animals with Gastrovascular Cavities

The simplest digestive systems have a single opening to the exterior. Food is taken into the cavity, it is broken down by enzymes released from cells lining the cavity, and the breakdown products are absorbed into the tissues of the organism. Indigestible material is then released from the cavity via the same opening through which it entered.

One organism that digests food in this way is *Hydra,* shown in Figure 34.8. *Hydra* disables its food by stinging it with *nematocyst* cells on its tentacles. Then it pushes its paralyzed prey through its mouth opening into its **gastrovascular cavity,** where digestive enzymes break the food down. This *extracellular digestion* begins the breakdown of food, and *intracellular digestion* completes it, as the partly digested material is taken up by the cells lining the cavity. Material that cannot be used for food, including the mineral skeletons of some small organisms, is ejected from the cavity through the mouth.

A gastrovascular cavity is also found in flatworms such as *Planaria,* and it functions in much the same way (Figure 34.8). The cavity is highly branched, extends throughout the length of the organism, and enables food to diffuse to all parts of the animal. This is why the word *vascular* is used to describe the cavity.

Animals with Alimentary Canals

Many organisms have evolved an **alimentary canal,** or **gut,** a tubular passageway that is open at both ends and thus allows a one-way flow of material to take place. Food enters at the **mouth** and is digested, and the products of digestion are absorbed through the canal. Undigested material is expelled from the terminus of the gut, the **anus** (Figure 34.9).

This one-way movement of food through the alimentary canal allows for a degree of specialization that is not possible in digestive systems with only one opening. Specialized organs have evolved along the length of the canal, forming a kind of "dis-assembly line" in which food is broken down in an orderly and systematic fashion. As food enters the canal, for example, specialized sharp mouth parts, bills, or teeth can break the food into small pieces. In some animals, especially birds, this process continues in a muscular compartment called the *gizzard.*

The early stages of digestion occur in another specialized portion of the canal called the *stomach,* in which food may remain for several minutes or hours. The final chemical breakdown and absorption of nutrients occurs in the *intestines.* In most organisms this is the longest portion of the canal, reflecting the high priority placed on

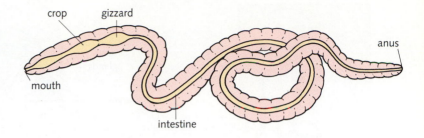

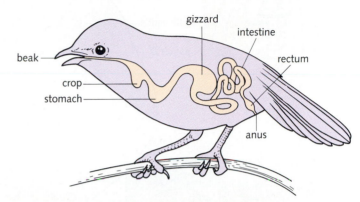

Figure 34.9 The digestive systems of many animals are organized around an alimentary canal that allows one-way movement of food from a mouth to an anus. The alimentary canals of the earthworm and a bird contain a number of specialized organs that process the food as it moves through the digestive system.

absorbing every possible nutrient molecule. Finally, undigested material is collected in the *rectum* and is periodically expelled through the *anus.*

Associated with the alimentary canal are a series of glands that produce the enzymes needed to break down the macromolecules found in food. As food passes through the canal, these enzymes are released in sequence, breaking down food in stepwise fashion and preparing it for the absorption that occurs in the intestines.

Alimentary canals in many organisms have valve-like rings of smooth muscle, called *sphincter muscles,* between different compartments. These valves manage the flow of material through the system, ensuring that food is kept in each compartment long enough for chemical processes to be completed before the food enters the next portion of the canal.

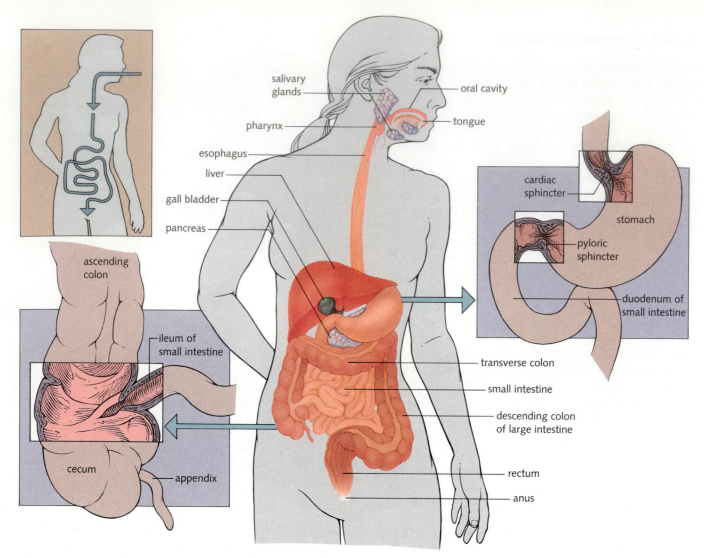

salivary glands
oral cavity
pharynx
tongue
esophagus
liver
gall bladder
pancreas

cardiac sphincter
stomach
pyloric sphincter
duodenum of small intestine

ascending colon
ileum of small intestine
cecum
appendix

transverse colon
small intestine
descending colon of large intestine
rectum
anus

Figure 34.10 The human digestive system. After being ground and chewed in the mouth, food passes down the esophagus to the stomach where partial digestion occurs over a period lasting from 1 to 5 hours. The final stages of digestion and absorption take place in the small intestine, and excess water is resorbed into the body in the large intestine. A number of organs, including the liver and pancreas, produce secretions that are essential for digestion.

THE HUMAN DIGESTIVE SYSTEM

The overall plan of the human digestive system reflects the fact that we are omnivores. This system, called the **gastrointestinal tract,** is illustrated in diagrammatic fashion in Figure 34.10. It is essentially a tube between 6 and 10 meters in length and composed of specialized digestive organs.

A cross section of the tract shows that its wall is composed of four distinct layers (Figure 34.11). The **mucosa** lines the interior of the tube. It consists of a layer of epithelial cells, some of which release mucus, digestive enzymes, ions, and water into the tract; a thin layer of connective tissue; and an equally thin layer of muscle.

Beneath the mucosa is the layer of **submucosa,** containing major blood and lymph vessels and a network of nerve fibers.

Next is the **muscularis externa,** a layer of muscle fibers oriented in two directions. Most of the muscle fibers are arranged in a circular pattern around the tract such that their contractions squeeze the tube-like lumen, making it thinner. A few of the fibers are oriented at right angles to the circular fibers; their contractions shorten the tube. Contractions of these two types of muscle squeeze food through the tract.

Finally, the gastrointestinal tract is surrounded by an outermost layer of connective tissue known as the **serosa.** This pattern varies a bit from one region of the tract to another, but the same four tissue layers are found throughout the system.

Over the next few pages we will trace the pathway followed by food as it moves through the digestive system, and we will examine the ways in which each major organ of the system contributes to the overall absorption of nutrients.

The Oral Cavity

Food processing begins in the **mouth (oral cavity).** The lips are covered with thin, almost transparent skin that allows the color of capillaries to show through. This gives them their reddish color but also makes them susceptible to moisture loss during dry weather—one of the reasons why the lips may become dry and cracked. Food is not absorbed in the mouth, so the inner lining of cells in the oral cavity is not specialized for absorption. The upper portion of the mouth is the **palate.** The palate is hard and bony near the front of the cavity, but the **soft palate** near the rear is a flexible covering that is important in separating breathing and swallowing activities.

Teeth are anchored in special sockets in the bones of the jaw and are connected to the underlying bone by a network of blood vessels and nerve fibers that enter through the root. Teeth are living tissue despite the tough coating of mineralized enamel on their surfaces.

Several basic types of teeth can be distinguished: *incisors,* whose sharp edges are capable of cutting directly through meat; *cuspids* and *bicuspids,* which have one or two points (cusps) to grasp and tear food; and *molars,* whose large flat surfaces make them ideal for grinding food into fine particles. Human tooth structure is midway between that of an herbivore (in which molars predominate) and a carnivore (in which incisors and cuspids are most common), reflecting the mixed human diet of both meat and vegetable material (Figure 34.12).

The oral cavity contains three pairs of large **salivary glands** that produce *saliva,* a fluid that initiates the process of chemical digestion of carbohydrates. The release of saliva is under the control of the nervous system, and

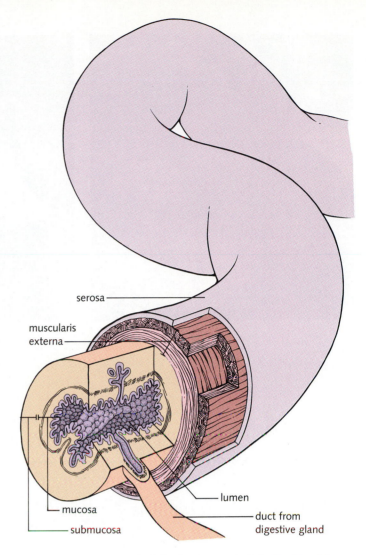

Figure 34.11 The wall of the human gastrointestinal tract is composed of four layers: An outermost *serosa;* a layer of smooth muscle fibers; a *submucosa,* which contains blood and lymph vessels; and finally an inner *mucosa.*

Figure 34.12 The teeth of an animal reflect its diet. The flattened surfaces of an herbivore's teeth are used for grinding plant material (left). The sharp surfaces of a carnivore's teeth (right) are useful in cutting flesh.

THEORY IN ACTION

Intestinal Bacteria: The Essential Dinner Guests

The gastrointestinal tract is inhabited by dozens of different species of bacteria. One of them is the famous laboratory bacterium *Escherichia coli,* whose species name implies that its normal place of residence is the colon. These species of bacteria come from the environment. Much of the food we eat, no matter how clean or well-cooked or carefully prepared, contains bacteria.

A few of these organisms are destroyed by the enzymes of the digestive system. A few pass right through the gastrointestinal tract. And a very few can cause serious diseases and intestinal disorders. But many bacteria are able to grow within the gastrointestinal tract and thrive in the dark, warm, moist, and nutrient-laden environment they find there.

Confined within the gastrointestinal tract, these bacteria are not attacked by the immune system, and they generally do not spread beyond the digestive tract. Some such bacteria help our digestive processes by breaking down food molecules that could not otherwise be broken down and absorbed by the digestive system. They even release vitamins into the gastrointestinal tract, and these vitamins are absorbed by the cells lining the walls of the tract. The major natural sources of vitamin K, in fact, seem to be intestinal bacteria.

Other mammals derive even greater benefits from their intestinal "flora," as these bacteria are called. Symbiotic bacteria in the gastrointestinal tracts of cows and other ruminants release enzymes that break

down the chemical bonds in cellulose—bonds that mammals are unable to digest on their own. This enables grazing animals to subsist on diets of hay and grass that would yield few nutrients to other mammals, such as humans, that are unable to harbor the same intestinal flora.

often just the scent of food increases the production of saliva (this is the biological basis of the expression "mouth-watering"). An adult produces from 1 to 2 liters of saliva in a day, and this fluid serves a critical function in beginning the digestive process.

Buffering chemicals, including *bicarbonate ions*, in saliva maintain the pH of the oral cavity near neutral, even when acidic food is eaten. The enzyme *salivary amylase* breaks the bonds in starch, releasing disaccharides. *Mucin*, a glycoprotein, gives saliva its somewhat slippery character. Mucin secretions help food slide easily through the smaller openings of the digestive system as the food is swallowed.

The **tongue** is a large muscle that covers the floor of the mouth. *Taste buds*, clusters of sensory cells that react to the chemical composition of food, are found on the surface of the tongue. There are two distinct sets of muscle fibers within the tongue, giving it tremendous flexibility in movement. This allows the tongue to be used for manipulating food and swallowing, as well as for such nondigestive functions as talking, singing, and whistling.

During chewing, the tongue mixes food with saliva and gradually forms it into a ball called a **bolus,** which is swallowed via the combined action of the back of the tongue and the upper portion of the throat, the **pharynx.**

Swallowing

The pharynx (generally called the throat) serves as the passageway for both food and air. Separating the functions of breathing and swallowing is a critical matter, because the entry of just a small amount of food or water into the breathing passages can cause a blockage with serious consequences. (See Theory in Action: "Saving a Life" in Chapter 33.)

The channeling of food and air to the proper passageway is accomplished by the combined actions of several sets of muscles in the lower portion of the pharynx. When a bolus of food is swallowed (Figure 34.13), the soft palate is forced upward, temporarily sealing off the nasal passageway from the pharynx (that's why it is nearly impossible to swallow and breathe at the same time). Nerves in the pharynx sense contact with the bolus, and stimulate the **glottis,** the opening between the vocal cords, to close, helping to seal off the air passageway. The movement of the bolus itself down the throat pushes a small flap of tissue, the **epiglottis,** over the glottis to protect it further from the entry of food or water. As the bolus is swallowed, a circular band of muscle, the **upper esophageal sphincter,** relaxes to allow food to enter the esophagus, the tube that leads to the stomach.

Figure 34.13 Swallowing is a coordinated muscular reaction that traps food into a mass called a *bolus* and then forces it towards the back of the throat past the soft palate. As it passes the palate, a muscle relaxation opens the upper esophageal sphincter, allowing food to pass through and enter the esophagus. Once the bolus has passed, the sphincter closes again.

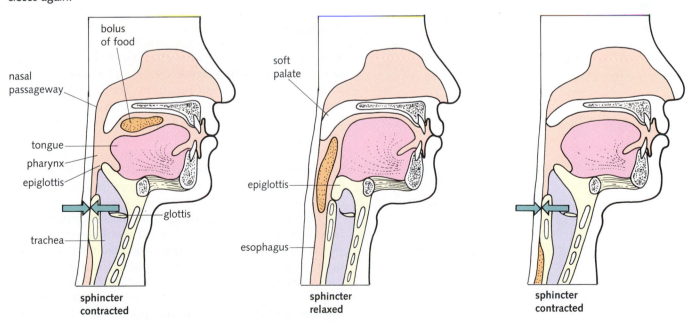

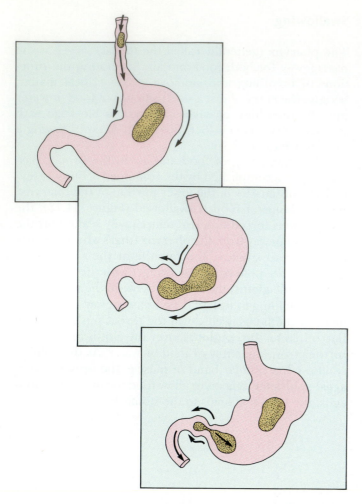

Figure 34.14 Peristaltic waves, caused by rhythmic contractions of the smooth muscles in the walls of the stomach, force material through the stomach as digestion proceeds. Similar contractions help to move food through the esophagus and intestines.

The Esophagus

Food moves through the **esophagus** into the stomach. Because we humans eat in an upright position, one might assume that gravity moves food down the esophagus into the stomach. However, that is not the case. Humans can swallow effectively while lying down, while standing on their heads, and even in the weightless environment of space. Grazing animals such as horses generally eat in positions that require their food to move through the esophagus *upward*, against the force of gravity.

Food is forced through the esophagus by a series of muscular contractions, known as **peristalsis,** in the wall of the tube (Figure 34.14). The circular muscles surrounding the esophagus contract just behind the bolus while the muscles just ahead of it relax. This produces a wave of muscular contractions that forces food into the stomach regardless of the direction of gravity. In humans a typical peristaltic wave takes about 8 sec to reach the stomach. Peristalsis then continues down the wall of the stomach, mixing the food with digestive fluids and moving it forward.

The wall of muscle surrounding the esophagus is particularly thick at the point where it enters the stomach, a region known as the **cardiac sphincter.** This sphincter maintains a seal between the stomach and the esophagus, although the seal is not perfect. Excess pressure on the abdomen or in the stomach can force fluid from the stomach up past the sphincter into the esophagus, and the acids in stomach fluid cause a painful sensation known as *heartburn* when they enter the esophagus. Eating too much food creates such pressure and is a major cause of heartburn.

The Stomach

The large muscular sac into which swallowed food empties is the **stomach** (Figure 34.15). The stomach performs several important functions: Its size enables it to take in a large amount of food at one time and send it, bit by bit, through the rest of the digestive system. This, in turn, makes it possible for us to eat a few large meals a day instead of having to nibble continuously. The stomach releases a potent combination of hydrochloric acid and digestive enzymes that work together to start the digestion of proteins.

The stomach lining, the *gastric mucosa,* contains millions of microscopic, pit-like structures known as *gastric glands* that release four major products. Cells near the upper regions of these glands produce mucus. Other cells (the *parietal cells*) at the base of the glands secrete hydrochloric acid, while cells called *chief cells* produce *pepsin,* an enzyme that cleaves proteins into polypeptide fragments.

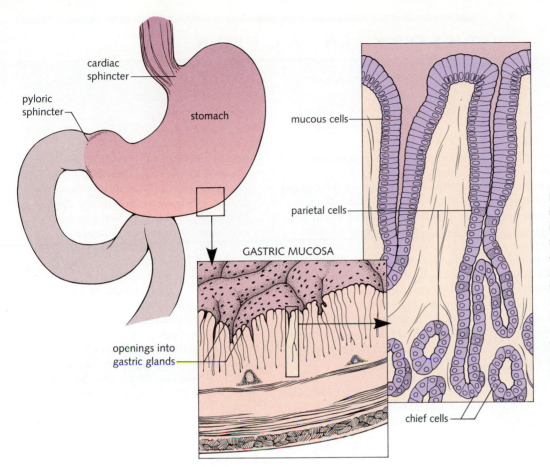

cardiac sphincter

pyloric sphincter

stomach

mucous cells

parietal cells

GASTRIC MUCOSA

openings into gastric glands

chief cells

Figure 34.15 This detailed drawing of the lining of the stomach illustrates the mucus cells, which secrete mucus to protect the stomach lining, and the gastric glands, groups of cells that secrete a variety of digestive enzymes. The parietal cells secrete hydrochloric acid, producing an acidic environment necessary for the chemical digestion of many foods. The chief cells release pepsinogen, which helps to break down proteins.

This enzyme is released in the form of *pepsinogen,* an inactive precursor that is converted into pepsin by the low pH produced by the hydrochloric acid. (Pepsin also works best in an acidic environment.) The lower portion of the stomach glands contains cells that secrete *gastrin,* a hormone that enters the circulation and then stimulates the cells lining the stomach to secrete still more digestive enzymes and HCl.

Gradually, the combined action of these secretions and the muscular contractions of the stomach wall, which are sometimes violent enough to be felt (and even heard), break the food down into smaller and smaller particles. As digestion in the stomach continues, the food becomes a nutrient-rich, milky slurry known as **chyme.** Chyme is forced toward the bottom of the stomach, where the pyloric sphincter controls the rate at which it passes from the stomach into the duodenum.

Ulcers Why don't the powerful enzymes and acids released into the stomach digest the stomach wall? The answer is that sometimes they do! In most cases, a thick layer of mucus protects the stomach lining against diges-tion by its own enzymes and acids. However, the stomach wall may become damaged, and tissue may be attacked by the digestive enzymes. As the tissue breaks down, cellular damage produces a wound in the wall of the stomach or intestine known as a **peptic ulcer.** The damage may become so severe that bleeding into the digestive system results, and extensive bleeding can become life-threatening. Foods that stimulate the production of digestive fluids, including the caffeine found in coffee and tea, aggravate the damage caused by a peptic ulcer and should be avoided to help the tissue heal.

The Duodenum

As the stomach fills with chyme, it gradually releases small amounts of material through the **pyloric sphincter** into the **duodenum,** the upper portion of the small intestine. The pressure of chyme against the sphincter causes it to open and close in a rhythmic fashion, timed to coincide with waves of peristaltic contractions across the stomach.

Alcohol: Metabolism and Illness

The liver is capable of detoxifying a wide range of drugs and poisons, and one of these is ethyl alcohol. Alcohol has been a part of human culture for thousands of years. Mesopotamian pottery dating to 4200 B.C. shows details of the fermentation process. The technique of distilling a beverage to increase its alcoholic strength has been known for nearly 2000 years. Alcoholic drinks are a part of popular culture throughout the world. Alcohol should be recognized as a powerful and addictive drug. At least 12 million Americans are active abusers of alcohol. Alcohol contributes directly to high rates of suicide, murder, assault, abuse, violence, traffic accidents, and mental illness. As if these difficulties were not enough,

alcohol has long-term destructive effects on the body.

Alcohol is absorbed rapidly across the lining of the stomach, which is one reason why drinking on an empty stomach has an immediate effect. When a drink is mixed with food, the alcohol concentration in the stomach is lower and absorption into the bloodstream is less rapid. Alcohol's most immediate effects are on the nervous system. It lowers brain activity, dulls reflexes, and interferes with sensory input. High doses of alcohol can be fatal.

Alcohol is broken down to acetaldehyde by enzymes in liver cells. Ultimately, acetaldehyde can be converted to a compound that enters the Krebs cycle (Chapter 25). In this limited

way, alcohol can supply energy like other foods. Acetaldehyde, however, is a toxic chemical that can damage mitochondria and upset the metabolism of liver cells. Long-term heavy drinking causes repeated damage to liver cells. The liver enlarges with swollen cells and fat globules. Patches of dead and dying cells appear throughout the organ, scar tissue builds up, and blood flow through the liver is impaired—a condition known as *cirrhosis*.

The effects of cirrhosis extend throughout the body. They include internal bleeding, abdominal swelling, and sexual disorders. There is no cure for cirrhosis, and the only way to arrest the damage is to stop drinking alcohol.

Gradually, over 3 to 6 hr, the entire contents of the stomach empties into the small intestine, in portions of about 50 mL (just under 2 teaspoons) at a time.

Most chemical work of the digestive process is done in the duodenum, and the breakdown of food molecules in this portion of the small intestine is the final preparation for absorption, which takes place in the remainder of the intestine.

Two major glands, the liver and the pancreas, produce secretions essential to this process (Figure 34.16). These secretions are added to the chyme through an opening at the *duodenal papilla*. Hormones triggered by the presence of food in the stomach help ensure that secretions from the liver and pancreas are released only when food passes into the duodenum.

Pancreatic secretions The **pancreas** is a double gland. As we saw in Chapter 31, the *endocrine* portion of the pancreas produces several hormones, two of which are glucagon and insulin. Each helps regulate the level of

sugar in the blood. The *exocrine* pancreas, which makes up 99 percent of the pancreas mass, produces a range of powerful digestive enzymes that break down proteins, fats, and carbohydrates.

Most of these enzymes are released from the cells of the pancreas in inactive forms. This prevents them from digesting each other or the tissues of the pancreas itself. They are activated by the environment of the intestine, and they rapidly begin to break down food molecules in the duodenum.

Other pancreatic cells produce *sodium bicarbonate,* which makes the secretions of the pancreas strongly alkaline (high pH). The release of bicarbonate neutralizes the strongly acidic chyme from the stomach. This pH change inactivates pepsin and activates enzymes from the pancreas and the duodenum.

The release of these materials from the pancreas is timed to coincide with the movement of food into the duodenum, enabling the pancreas to store its digestive enzymes until they are needed. How is this timing accomplished? The release of acidic chyme from the stomach to the duodenum stimulates the duodenum to release hormones. These hormones, in turn, cause the pancreas to dispense its mixture of bicarbonate and digestive enzymes. The neutralization of stomach acid that occurs when pancreatic fluid enters the duodenum inhibits further release of duodenal hormones, making the whole process self-regulating.

Liver secretions   Unlike the pancreas, the liver does not produce digestive enzymes. It does, however, produce a fluid that is absolutely vital to the successful digestion of fatty foods. This fluid is known as **bile.** Bile is a mixture of salts and lipids (including *cholesterol*) that aids in the digestion of fats. The liver also secretes *bilirubin,* a breakdown product of hemoglobin, into the bile to be eliminated from the body.

The best way to explain the function of bile is to compare it to a soap or detergent. Fats from foods are in the form of thick globules by the time they reach the small intestine. Like droplets of grease, these globules do not dissolve in water, so the fat-digesting *lipase* enzymes produced by the pancreas are unable to reach many of the fat molecules.

The bile salts and phospholipids in bile, however, contain both *hydrophilic* (water-soluble) and *hydrophobic* (water-insoluble) parts (see Chapter 19), just as detergents do. The hydrophobic portions of the bile compounds interact with fat molecules, while the hydrophilic portions interact with water. This enables bile compounds to break fat droplets into smaller and smaller pieces, a process known as *emulsification.*

Gradually, fats are emulsified into tiny lipid droplets known as *micelles* that are subject to direct attack by pancreatic lipase. Without bile to break fats into small droplets, as much as 90 percent of the fat in food would pass through the intestine undigested.

The secretions of bile from the liver accumulate in a sac known as the **gall bladder.** As a meal is digested, bile from the gall bladder passes through the same duct as pancreatic secretions to enter the duodenum. The gall bladder is not a vital organ, for the movement of bile into the digestive system is unimpaired by its absence. Many animals, in fact, have no gall bladder, and in humans it must be removed if it becomes blocked or infected.

Remarkably, the total content of bile salts in the body at any one time is only about 50 percent of the amount needed to emulsify the fats in a large meal. The digestive system copes with this undersupply of bile salts by rapidly *recycling* the bile salts it does have. As bile is used up to help emulsify fats, the bile salts are rapidly absorbed in the small intestine, where fat itself is also absorbed and returned to the blood. The liver retrieves the bile salts and cholesterol from the blood and re-secretes them as

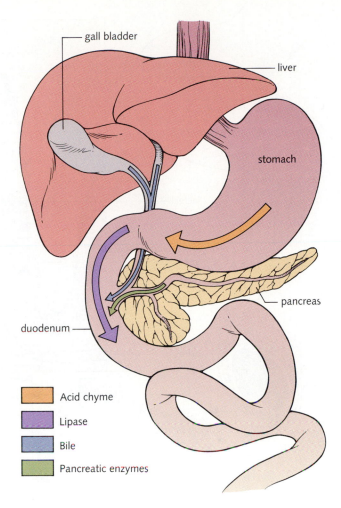

Figure 34.16   The duodenum is one of the most critical regions of the digestive system. Secretions from the liver and pancreas enter the digestive system in the duodenum, where they mix with acidic chyme from the stomach.

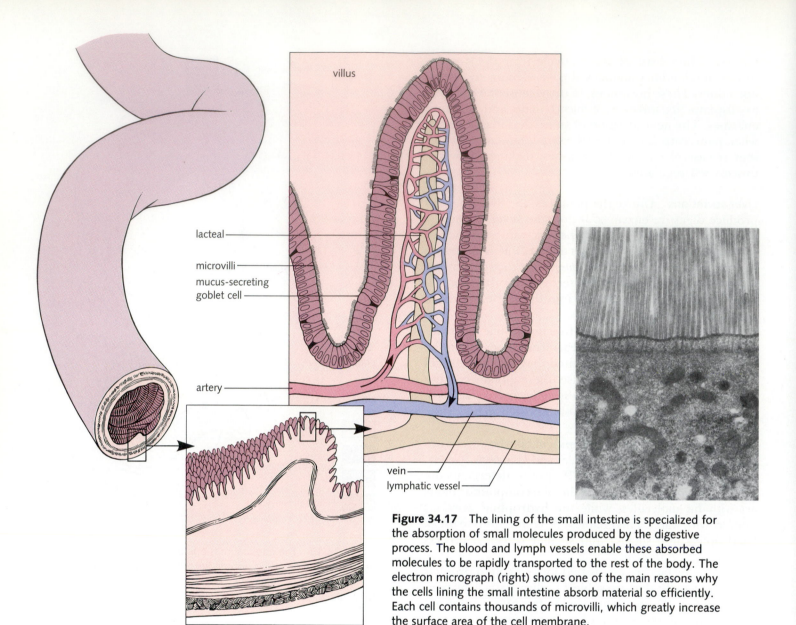

lacteal

microvilli

mucus-secreting
goblet cell

villus

artery

vein

lymphatic vessel

Figure 34.17 The lining of the small intestine is specialized for the absorption of small molecules produced by the digestive process. The blood and lymph vessels enable these absorbed molecules to be rapidly transported to the rest of the body. The electron micrograph (right) shows one of the main reasons why the cells lining the small intestine absorb material so efficiently. Each cell contains thousands of microvilli, which greatly increase the surface area of the cell membrane.

new bile, which is added to the digestive tract to emulsify the next batch of fatty food moving into the duodenum. This rapid recycling makes a limited resource go a long way.

The Small Intestine

By the time food passes through the duodenum into the rest of the small intestine, nearly all the important *chemical* work of digestion has been completed. What remains is the task of absorbing the nutrient molecules now available within the small intestine.

The rate at which material can be absorbed across a barrier (such as the wall of the small intestine) depends on the surface area of that barrier. All other things being equal, the larger the surface area, the greater the rate of absorption. For this reason, the inner surface of the small intestine is wrinkled and folded, and these folds are covered with finger-like projections called villi (Figure 34.17). Although these foldings produce a tremendous increase in surface area, the actual surface is even larger. Epithelial cells lining the intestine are covered with **microvilli**—tiny projections of the cell membrane that dramatically increase its surface area.

Nutrient molecules pass into these cells from the intestinal tract via a number of mechanisms. Free fatty acids, simple lipids, and a few sugars (such as fructose) diffuse across the epithelial cell membranes in response to a concentration gradient: Their high concentration within the intestine drives them by passive diffusion into the epithelial cells. Other molecules, including glucose, nucleic acids, amino acids, and even vitamins, are taken into the epithelial cells by active transport. As a result of osmotic pressure, a great deal of water follows the movement of nutrients into cells lining the small intestine.

Nutrient molecules don't remain long within the epithelial cell but are rapidly transported from the intestine to the bloodstream. Each villus of the small intestine has a fine network of capillaries that lies just below a single layer of epithelial cells. The close contact makes it possible for nutrient molecules to move into the bloodstream within a matter of minutes after the first food arrives.

Fats are transported by a different mechanism. Although most fat is taken into epithelial cells in the form of fatty acids, very few fatty acids leave those cells. Instead, epithelial cells convert fatty acids into triglycerides that form lipid droplets within the endoplasmic reticulum. These droplets, known as **chylomicrons,** are released from the inner surfaces of the epithelial cells.

Chylomicrons are prevented from entering the capillaries by a thick *basement membrane* of polysaccharide molecules. However, there is no basement membrane surrounding the *lacteals* (extensions of the lymphatic system), and chylomicrons pass into them rapidly. This allows the lymphatic system to absorb nearly all of the fat molecules taken up in digestion. After transport by the lymphatic system, fat molecules eventually enter the venous circulation.

Material is mixed in the small intestine by rhythmic contractions of the rings of smooth muscle that surround the organ. It is interesting to contrast these contractions, which are known as **segmentation movements,** to the peristaltic waves that move along the digestive tract, propelling its contents. Segmentation contractions squeeze the food and move it *back and forth*, which helps to ensure that all the contents of the intestine eventually come into contact with the absorptive wall.

The capillaries that absorb nutrients in the villi of the small intestine do not lead to the general circulation. Instead, they converge into larger and larger vessels that drain into the hepatic portal vein, leading to the liver (Figure 34.18).

Figure 34.18 Veins leaving the digestive system do not pass directly back into the general circulation. Instead, they collect into the hepatic portal vein, ensuring that newly absorbed food passes through the liver before it enters the general circulation. The *hepatic portal system*, as this circulation pattern is known, allows the liver to control the entry of food molecules into the bloodstream.

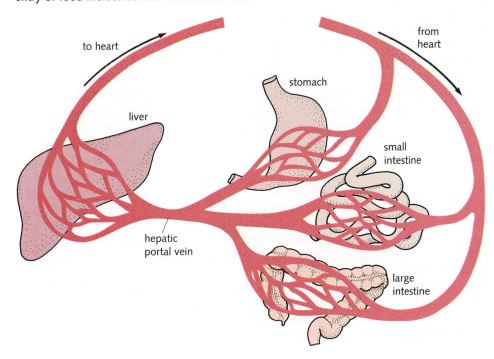

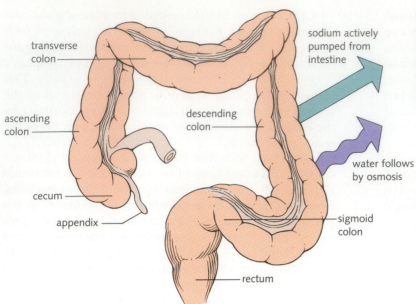

transverse colon

ascending colon

cecum

appendix

descending colon

sigmoid colon

rectum

sodium actively pumped from intestine

water follows by osmosis

Figure 34.19 The large intestine is the final absorptive organ of the digestive system. As material passes through it, sodium is actively pumped out of the digestive track into the circulatory system. Osmosis causes water to flow with the sodium, conserving water and producing a concentrated feces.

The Liver and the Hepatic Portal Circulation

Functionally, the liver is one of the most complex organs in the body. Its functions include the detoxification of drugs and alcohol, the formation of urea from nitrogen-containing wastes, the synthesis of several important blood plasma proteins, and, as we have seen in this chapter, the emulsification of fats.

The importance of the hepatic portal circulation that delivers blood to the liver cannot be overemphasized. If the circulatory system in the small intestine emptied directly into the general circulation, the levels of sugars, amino acids, and nucleic acids in the blood would rise dramatically every time food was consumed and would fall precipitously a short time thereafter. A quick snack or candy bar would cause, among other things, a rush of sugar into the circulation. This might not seem like much of a problem, but many of the cells of the body, including those of the nervous system, are extremely sensitive to the level of sugar in the blood.

The primary organ that regulates blood sugar is the liver. During the digestion of a large meal, the liver removes much of the sugar that comes to it via the hepatic portal vein. This ensures that the concentration of sugar in the blood does not rise rapidly. Then, when the meal is digested and the absorption of sugar from the small intestine has ceased, the liver gradually releases its store of carbohydrate to keep blood sugar levels from falling.

In this way, the liver acts as a sort of gatekeeper between the digestive system and the circulatory system, allowing only an appropriate amount of sugar and other nutrients to enter the general circulation and doling out its reserves of those compounds when the rest of the body needs them.

The Large Intestine

The large intestine, or **colon,** is about 1.3 m long and consists of three parts: The *ascending colon* lies along the right side of the body, the *transverse colon* crosses the top of the abdomen from right to left, and the *descending colon* goes down the left side of the body.

The small intestine dispenses its chyme into the colon in an area known as the *caecum* (Figure 34.19). A small finger-like pouch, the *vermiform appendix* (popularly known simply as the appendix), extends from the caecum. The vermiform appendix is found in primates and in some rodents. Generally, it is not present in herbivores. In many herbivores, the caecum is a large organ in which bacterial action breaks down cellulose, releasing usable carbohydrates and vitamins. In humans, however, the appendix is not an important organ and generally merits our attention only when it has become infected, a condition known as *appendicitis*, and needs to be surgically removed.

The 500 mL of chyme that enter the colon every day contain indigestible material that has neither been digested nor absorbed, large amounts of intestinal bacteria, and a substantial amount of water. Throughout the large intes-

Diarrhea: When the Large Intestine Fails

For most of us, the digestive system is something that works so well we seldom think about it. However, when something *does* go wrong with the digestive system, it becomes impossible not to think about it. Among these potential problems is the failure of the large intestine to remove adequate amounts of water from the chyme that enters it. The result is *diarrhea,* the elimination of a large volume of watery feces, which is unpleasant and annoying at best. At worst, diarrhea can be life-threatening.

Many forms of diarrhea result from infections that impair the ability of the large intestine to pump salts out of the intestinal tract. The inability to transport salt has two effects. Because there is no movement of salt across the intestinal wall, there is little osmotic pressure to force water to leave the tract. The result is a tremendous loss of fluid through the digestive system. A secondary problem is caused by the loss of salt itself. Sodium is actively secreted into the digestive system in the stomach, and the failure of the large intestine to reabsorb it causes a rapid loss of sodium throughout the body.

Mild food poisoning resulting from the presence of bacteria or their metabolic products in food is a common cause of diarrhea. A number of serious diseases, including *cholera,* also cause diarrhea. In heavily populated areas where sanitation is not efficient, large volumes of watery feces constitute a severe public health problem. Water supplies can become contaminated, causing the further spread of serious diseases. In most cases, diarrhea does not last long, but if it goes unchecked for more than a few days, diarrhea can cause death.

Chronic diarrhea leads to life-threatening dehydration and to a loss of salts so severe that they literally cannot be replaced. In many areas of the world, diarrhea is the leading cause of death among infants and young children.

---

tine, cells lining the tract actively transport sodium from the chyme into the blood. As sodium ions move in this direction, osmotic forces cause most of the remaining water to follow. Gradually, nearly all of the remaining water is removed, and what remains are dense *feces* concentrated in the last few centimeters of the tract. The descending portion of the colon enters the *rectum,* a muscular passageway wherein the feces may be stored for a time before they are eliminated by defecation.

Timing

In digestion, like politics, timing is everything. How does the digestive system coordinate the release of its secretions so that they are produced only when they are needed? Some responses of the digestive system to food are obvious, such as the way in which your mouth waters with saliva when the scent of your favorite food wafts by. The sight or scent of food can also trigger the release of hydrochloric acid and pepsinogen in the stomach, causing the stomach to "churn" even when it is empty.

As you recall, when food enters the stomach, receptors in the stomach wall react to the pressure by secreting the hormone *gastrin,* which causes the cells lining the stomach to secrete more acid (Table 34.4). The release of gastrin helps to ensure that just the right amount of acid is released. A large meal causes the secretion of more gastrin, which stimulates the release of more acid. Just as the stomach reacts to a volume of food, the duodenum reacts to the presence of chyme by releasing the hormones *secretin, cholecystokinin,* and *enterogastrone.* Secretin causes the pancreas to secrete bicarbonate (which neutralizes the acidic chyme) and cholecystokinin stimulates

Table 34.4 *Functions and Secretions of Digestive and Accessory Organs*

Organ	Function	Secretion
Digestive tube		
Mouth	Chops and grinds food	
Pharynx	Channeling of food and air to proper passageway	
Esophagus	Propels food to stomach	
Stomach	Stores and churns food Releases HCL and enzymes to initiate proteolytic breakdown	HCl, activates enzymes, breaks up food, kills germs Gastrin, stimulates HCl secretion Pepsin, cleaves protein Peptidase cleaves proteins
Small intestine	Completes digestion Absorbs nutrients	Sucrases cleave sugars Amylase cleaves starch and glycogen Lipase cleaves lipids Nuclease cleaves nucleic acids
Large intestine	Reabsorbs water and sodium ions Stores wastes	
Appendix	No known digestive function Contains cells of the immune system	
Rectum	Expels waste	
Anus	Opening for waste elimination	
Accessory Organs		
Tongue	Mixes food with saliva	
Salivary glands	Moisten food	Salivary amylase cleaves starch Mucin makes saliva slippery Bile aids in lipid digestion
Liver	Synthesis of blood plasma proteins Regulates blood sugar Formation of urea Secretes bile Detoxifies drugs and alcohol	
Gallbladder	Stores bile	
Pancreas	Adds digestive anzymes Regulates blood glucose levels Neutralizes stomach acid	Bicarbonate neutralizes stomach acid Trypsin and chymotrypsin cleave proteins Carboxypeptidase cleaves peptides Amylase cleaves starched glycogen Lipase cleaves lipids Nucleases cleave nucleic acids Glucagon and insulin regulate level of blood sugar

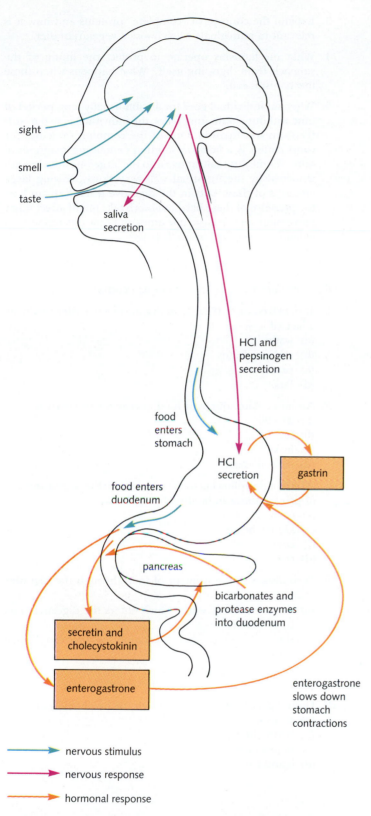

sight

smell

taste

saliva
secretion

HCl and
pepsinogen
secretion

food
enters
stomach

HCl
secretion

food enters
duodenum

gastrin

pancreas

bicarbonates and
protease enzymes
into duodenum

secretin and
cholecystokinin

enterogastrone

enterogastrone
slows down
stomach
contractions

→ nervous stimulus

→ nervous response

→ hormonal response

Figure 34.20 The digestive process is controlled by the senses, signals from the brain, and hormones secreted by the stomach and intestines.

the gall bladder to contract and release bile. Enterogastrone travels through the bloodstream to the stomach where it inhibits muscular activity in the stomach, ensuring that chyme is not forced too quickly into the small intestine. Working together, these hormones ensure that the organs of the digestive system release the proper sequences of secretions to break down even the most complex lunch into a series of sugars, amino acids, and lipids that can be absorbed in the small intestine. In addition, one of these hormones, cholecystokinin, is also a *neural transmitter substance* that acts on the hypothalamus to reduce appetite. This is an important signal that eating should stop (Figure 34.20).

SUMMARY

Animals are heterotrophs that eat a varied diet to obtain the nutrients they need to survive. On a cellular level, food molecules are needed to provide a source of cellular energy and the building blocks from which new molecules can be synthesized as needed for growth and repair. Food for energy can be derived from carbohydrate, fat, or protein. However, food for growth and repair requires a balanced diet that includes appropriate amounts of carbohydrates, fats, and proteins; small amounts of essential organic molecules known as vitamins; and a number of essential minerals.

Digestive systems are specialized for the intake, processing, and absorption of food. Simpler animals have simple digestive systems, beginning with those organisms in which digestion is a cellular process. More complex organisms have gastrovascular digestive systems, in which food enters and waste leaves by the same opening. Higher animals, including vertebrates, have digestive systems built around alimentary canals, which have two openings.

Food enters the human digestive system through the mouth, where it is chopped and ground by the teeth, and is then swallowed into the esophagus, where it is forced into the stomach by means of peristaltic waves of muscle contraction. The stomach releases digestive enzymes in an extremely acidic (low-pH) environment. As food enters the small intestine, it is joined by secretions from the liver and the pancreas, and these digestive juices complete the final chemical breakdown of material. Within the small intestine, epithelial cells transport nutrients into the circulatory system and the lymphatic system. The large intestine removes sodium ions and water from the contents of the gastrointestinal tract, producing dense feces containing very little water.

After studying this chapter, you should be able to:

- Describe the basic organs and tissues of the digestive system.
- Appreciate how food is broken down into small molecules that can be used for cellular activities.
- Explain the interactions between the digestive system and other systems of the body, including the nervous system and the circulatory system.
- Describe the basic requirements of the human diet, and explain how to establish a sound program of nutrition.
- Understand the medical dangers of high cholesterol levels in the bloodstream.

SELECTED TERMS

calorie *p. 647*
basal metabolic rate *p. 647*
undernourishment *p. 648*
malnourishment *p. 649*
vitamins *p. 651*
minerals *p. 651*
gastrovascular cavity *p. 661*
alimentary canal *p. 661*
anus *p. 661*
mouth *p. 661*

mucosa *p. 662*
palate *p. 663*
pharynx *p. 665*
esophagus *p. 666*
peristalsis *p. 666*
chyme *p. 667*
duodenum *p. 667*
pancreas *p. 668*
gall bladder *p. 669*
chylomicron *p. 671*

REVIEW

Discussion Questions

1. Outline the basic nutritional requirements for humans. How many general categories of nutrients are in the minimal list?

2. Drinking a large amount of liquid after a heavy meal can often cause heartburn. From what you know about the functioning of the stomach and the causes of heartburn, explain why.

3. Explain the concept of "complete" proteins and how it is relevant to individuals who choose a vegetarian diet.

4. What mechanisms operate to protect the lining of the stomach from digesting itself? What happens when these mechanisms fail?

5. When an individual goes for an abnormally long period of time without eliminating feces, the individual is said to suffer from constipation. One of the common remedies for constipation is a laxative solution containing magnesium salts. Magnesium salts move into the large intestine but are absorbed by the intestinal wall very slowly, leaving large amounts of these salts in the digestive tract. Explain how the presence of these hard-to-absorb salts might affect water movements in the large intestines and how this might contribute to the effect of the laxative.

Objective Questions   (Answers in Appendix)

6. If deprived of nutrients, an organism will suffer *first* from a lack of
   (a) water.
   (b) carbohydrates.
   (c) proteins.
   (d) fats.

7. An immediate source of food energy for the body is
   (a) protein.
   (b) fat.
   (c) lipid.
   (d) carbohydrate.

8. Which of the following food items contain the largest amount of protein that can be used by the body?
   (a) corn
   (b) kidney beans
   (c) eggs
   (d) rice

9. Peristalsis is the movement of food through the digestive tract by
   (a) the pressure of water which enters the digestive tract by diffusion.
   (b) the contraction and relaxation of the esophageal sphincter.
   (c) secretions of mucus.
   (d) wavelike contractions of the smooth muscle.

10. Gastric glands are responsible for the production of
    (a) trypsin.
    (b) hydrochloric acid.
    (c) pepsinogen.
    (d) b and c only

CHAPTER 35

Regulation of the Cellular Environment

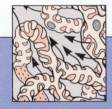

*"Water, water every where,
Nor any drop to drink"*
—Samuel Taylor Coleridge
THE RIME OF THE ANCIENT MARINER

The mariner in Coleridge's great poem found himself adrift at sea and lacking water to drink. The fact that seawater cannot support human life has been known since prehistoric times, and the mariner knew that he could easily die of thirst in the middle of the ocean. He might even have known that if he *did* drink seawater, he would die of dehydration far sooner than he would if he drank no water at all! Not surprisingly, he found the situation ironic.

With plenty of time on his hands, the mariner could well have wondered about the same questions that occur to us today when we think of other animals that live in and around the ocean. How can sea birds, mammals, and marine fishes thrive on the same saltwater that would cause us to die of thirst? For that matter, how do desert animals—from rattlesnakes to camels—survive in intensely hot places with little water or none at all?

In this chapter we will seek the answers to these and several other questions pertaining to water balance, the regulation of salt, and the control of body temperature. All of these are variations on the great theme of organismal physiology: the need to maintain constant internal conditions in the face of a changing external environment.

WATER, WASTES, AND TEMPERATURE

As you'll recall from several earlier chapters, the physiological systems of multicellular animals have been shaped by the requirements of life at the cellular level. Among those requirements are three whose functions are intertwined: the *excretion* of metabolic wastes, the *control* of salt and water balance within the body, and the *regulation* of body temperature.

It is a chemical fact of life that cellular metabolism produces molecules that must be removed from the body. As cells utilize nutrients, they produce many small molecules that are toxic to the organism and are released as *wastes*. One major waste product, carbon dioxide, is removed by the respiratory system.

But the handling of many other waste molecules is more complex. Furthermore, when animals ingest food they may also consume other substances, ranging from complex toxic compounds synthesized by plants to simple sodium salts. Many of these molecules and ions, if allowed to accumulate in the body, disrupt the delicate balance required to maintain the internal environment within narrow limits.

At the same time, animals need a sufficient supply of water in their bodies. The task of maintaining that supply is doubly complicated. For one thing, the elimination of many waste products requires the simultaneous elimination of at least some water. And water is also required by two other homeostatic functions: *respiration* and the *regulation of body temperature*.

We will begin our exploration of these physiological balancing acts performed by multicellular animals by examining a universal problem of all animals, the elimination of toxic nitrogenous wastes.

677

Toxic Ammonia

Several important metabolic pathways, such as the breakdown of amino acids, release nitrogen-containing compounds. When the amino group ($-NH_2$) is removed from an amino acid, a process known as *deamination,* ammonia (NH_3) is released. A water molecule consumed in the process provides the extra hydrogen that converts NH_2 to NH_3 (Figure 35.1).

For this reason, the complete metabolism of proteins releases ammonia—a potent poison—into the body. Few cells can survive high concentrations of ammonia, so it must be eliminated from living tissues as quickly as possible or converted to less toxic compounds. The elimination of metabolic wastes, such as ammonia, from the body is known as **excretion.**

Dealing with Ammonia: Three Strategies

Animals excrete nitrogenous wastes in one of three ways: as **ammonia,** as **uric acid,** or as **urea.**

Ammonia excretion　The simplest strategy, and the one used by most single-celled organisms, is to let ammonia diffuse out into the environment. In fact, because ammonia is quite soluble in water, many aquatic invertebrates simply allow ammonia to diffuse out of their tissues into the surrounding water.

Freshwater fishes, whose skin and scales present formidable barriers to ammonia diffusion, eliminate ammonia in two ways. First, they lose ammonia continually across their permeable gill membranes. Second, their kidneys collect ammonia from the bloodstream and release it regularly in thin, very dilute *urine.* Anyone who keeps pet fishes should know that water in a fish tank quickly becomes loaded with ammonia. In successful aquaria and in natural ecosystems, beneficial bacteria convert ammonia into far less toxic nitrate. Filling the tank with underwater plants, which absorb ammonia as a source of nitrogen, also helps to minimize ammonia poisoning.

Uric acid　Land animals, on the other hand, do not have the luxury of allowing ammonia to diffuse away. Furthermore, because most terrestrial animals must also conserve body water, they can't afford to flush ammonia away in a constant stream of dilute urine. But because ammonia is so toxic, it cannot be allowed to accumulate and become concentrated anywhere in their bodies.

One solution, used by reptiles, birds, and insects, is to remove ammonia from the circulation in the liver and convert it to **uric acid** (Figure 35.2). Uric acid is much less toxic than ammonia, and it is also much less soluble. This low solubility, though it might seem to be a disadvantage, actually offers an opportunity to conserve a great deal of water.

In birds and reptiles, for example, uric acid is removed from the circulation by the kidneys and passed through ducts to the **cloaca,** a cavity through which material from the excretory, reproductive, and digestive systems leaves the body (*cloaca* is the Latin word for "sewer"). The lining of the cloaca absorbs water from this fluid, causing the

Figure 35.1　When amino acids are metabolized, they are *deaminated*: the amino groups are removed, producing ammonia. Because ammonia is toxic at high concentrations, it is essential that it be removed or converted to a less dangerous form.

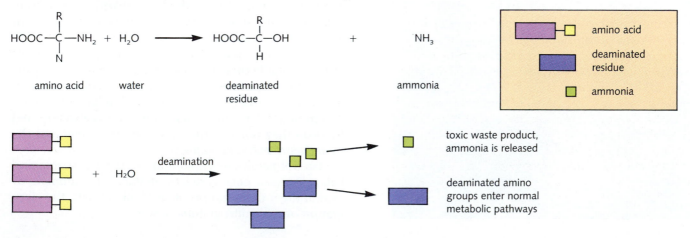

Terrestrial Animal	Nitrogen Waste Excreted
birds	
insects	
reptiles	uric acid
mammals	
some amphibians	urea

Figure 35.2 (left) Land animals usually convert ammonia to less toxic compounds, such as uric acid or urea. Urea is highly soluble and can be eliminated as urine. Uric acid is much less soluble, and is eliminated as a thick, sticky paste. (right) The large amounts of uric acid left by these sea birds on coastal rocks have produced *guano* deposits.

uric acid to precipitate in the form of white crystals. As more and more uric acid precipitates, it leaves behind more water than the cloaca can resorb. The nearly dry, milky-white paste called **guano** that birds finally excrete is composed primarily of uric acid. The insolubility of uric acid makes these familiar calling cards of birds almost impossible to wash away.

Urea In mammals and some amphibians, the liver removes ammonia from the bloodstream and produces **urea.** The liver returns urea, which is highly soluble in water, to the blood. Urea is then removed from the bloodstream by the kidneys and concentrated to form an excretory fluid known as **urine.** Although urine can contain quite a high concentration of urea, this method of excreting nitrogenous waste still requires significant amounts of water to keep urea in solution. The kidneys, in addition to removing nitrogenous wastes, also control the excretion of salts, water, and other metabolites, thus maintaining the requisite balance of water and solutes in the circulatory system.

EXCRETORY SYSTEMS

The simplest solutions to the problems of excretion and water balance are found in unicellular organisms that eliminate wastes by simple diffusion. Some multicellular aquatic organisms, particularly those (such as jellyfish) with thin, gelatinous bodies, can also eliminate their nitrogenous wastes by simple diffusion.

As an organism increases in size, however, the volume of its tissues (that produce ammonia) increases more quickly than its surface area (across which diffusion might take place). Thus even among aquatic animals, greater body size usually precludes the elimination of ammonia by simple diffusion.

For that reason, most larger organisms have specialized **excretory systems** that control water balance and handle the excretion of nitrogenous wastes. These systems may take a variety of forms. A freshwater organism without a rigid cell wall must deal with a persistent inflow of water that is due to osmosis. In pure distilled water, the influx of water can be so rapid that a cell may double

Figure 35.3 The contractile vacuole of *Paramecium* is an organelle that collects water from the cytoplasm (top) and expels it to the exterior (bottom).

its volume in only a few minutes. One cellular solution to the problem of excess water is the **contractile vacuole.** The contractile vacuole gets its name from its behavior. Under a microscope we can watch the transparent organelle grow larger and larger, then suddenly contract as it releases its contents from the cell. How does the contractile vacuole help to solve the problem of water balance?

Osmosis causes water to flow into the cell, diluting the concentration of dissolved material in the cytoplasm (Figure 35.3). The vacuole is surrounded by smaller vesicles that swell with fluid that is *isotonic* to the cytoplasm. Solutes are pumped out of these vesicles (a process that requires energy), producing a fluid that is nearly pure water. These vesicles then dump their contents into the contractile vacuole, causing the vacuole to swell as it fills

with water. When the vacuole is swollen with fluid accumulated in this way, a sudden contraction of the cytoplasm around it forces its contents outside the cell, releasing the accumulated fluid. Furthermore, although nitrogenous wastes from such organisms can diffuse directly through the cell membrane, some of the wastes may be eliminated with the water in the contractile vacuole.

Flatworms (platyhelminths) do not have circulatory systems, but they do have an excretory system. A network of *excretory canals* runs throughout the body, opening to the exterior through small pores known as *nephridiopores.* Water and nitrogenous wastes are pumped into the tubules throughout the body, and fluid is moved through the system by ciliated cells found at the ends of bulb-like passages in the system. The constant movement of cilia in these cells gives the impression of a flickering fire, and they are known as **flame cells** (Figure 35.4*a*). The flame cell system is one of the earliest structures that arose in the evolution of an excretory system.

Earthworms have excretory systems that take advantage of their closed circulatory systems by processing metabolic wastes collected by the circulating blood. In each segment of the worm's body, there is a pair of excretory organs called **nephridia** (Figure 35.4*b*). Each is composed of a tangle of capillaries that surrounds a duct of tissue leading to the outside of the body. Each *nephridium* collects metabolic wastes and excess water from the coelomic fluid and sends these to a bladder-like reservoir, which is periodically emptied to the exterior through a nephridiopore. By working in concert with the circulatory system, these isolated organs are able to excrete wastes from every segment of the earthworm's body.

Insects, which have open circulatory systems, remove nitrogenous wastes through a system of organs known as **Malpighian tubules.** These tubules, whose mode of operation resembles that of the flatworms' tubular system, thread throughout the posterior half of the body. The movement of fluid through the animal's open circulatory system helps to sweep waste materials toward the Malpighian system. Unlike the excretory canals of the flatworm, however, these tubules do not exit directly to the exterior. The Malpighian tubules empty into the digestive system just before the hindgut (Figure 35.4*c*). Water and salts are resorbed by the lining of the gut, leaving very dry feces that pass out through the anus.

Vertebrates eliminate nitrogenous wastes and maintain water balance by means of **kidneys,** paired organs located in the abdominal cavity. Kidneys have a rich blood supply, and their functions include the removal of urea or uric acid from the bloodstream. Vertebrate kidneys function in several different ways related to their environments. The need to conserve water has led to a variety of adaptations we will explore later in this chapter.

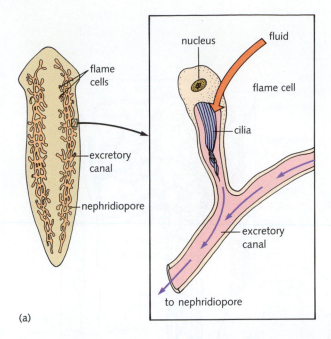

(a)

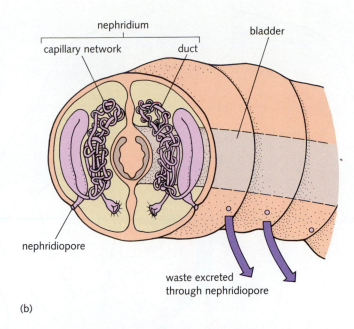

(b)

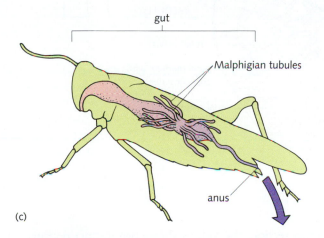

(c)

Figure 35.4 Excretory systems vary widely from one organism to another. **(a)** The flame cells of flatworms propel fluid through a network of excretory canals. **(b)** Earthworms contain pairs of excretory organs in each segment known as nephridia that remove wastes from the bloodstream and the coelomic fluid and pass them to the exterior. **(c)** Insects contain a network of Malphigian tubules that collects wastes and directs them into the digestive system.

THE HUMAN EXCRETORY SYSTEM

In the strictest sense, a complete inventory of the human excretory system includes the *lungs,* through which water vapor and carbon dioxide are eliminated during breathing, and the *skin,* which eliminates water and salt during sweating. The lungs and skin do not *regulate* water loss in response to water balance, however, so we will confine our attention here to the excretory and regulatory functions of the kidney.

The Kidney

Human kidneys are paired organs located near the posterior wall of the abdominal cavity (Figure 35.5). Each kidney is about the size of a fist and weighs roughly 170 grams. Each kidney is connected to the circulatory system by a large **renal artery** that brings blood into the organ and by a **renal vein** that returns blood to the circulatory system (*renal* is an adjective we apply to structures and functions associated with the kidney).

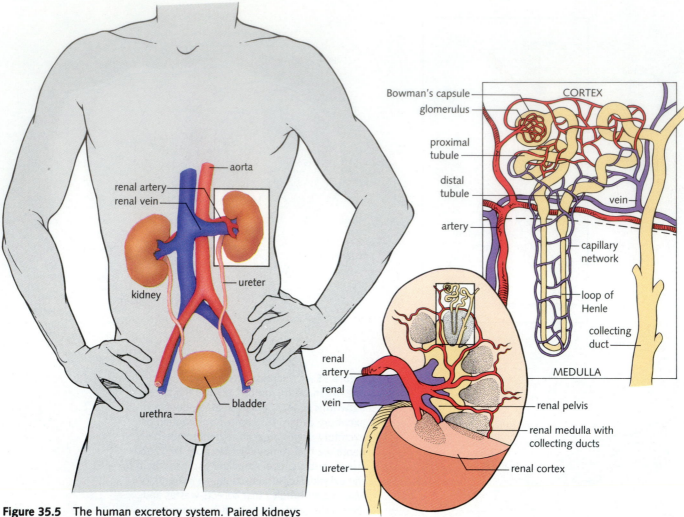

Figure 35.5 The human excretory system. Paired kidneys remove waste from the circulation and concentrate them in urine that is stored in the bladder prior to elimination. A cross-sectional view shows the internal organization of a kidney, as well as the structure of a nephron, one of the kidney's functional units. More than 1 million nephrons are present in a human kidney.

In cross section, the kidney shows two distinct regions: the *renal cortex*, the tissue near the outer edge of the organ, and the *renal medulla*, which lies inside the cortex and borders a cavity known as the *renal pelvis*.

The kidney removes water and waste products from the blood and forms urine. Urine passes through tiny **collecting ducts** that drain into the renal pelvis that empties into vessels called **ureters.** The ureters lead to the **bladder,** where urine is stored. Urine is eliminated from the body by passing through the **urethra** to the outside.

The Nephron

Each human kidney contains roughly 1 million functional units known as **nephrons.** Figure 35.5 illustrates the basic structure of a typical nephron and its relationship to the flow of blood through the kidney.

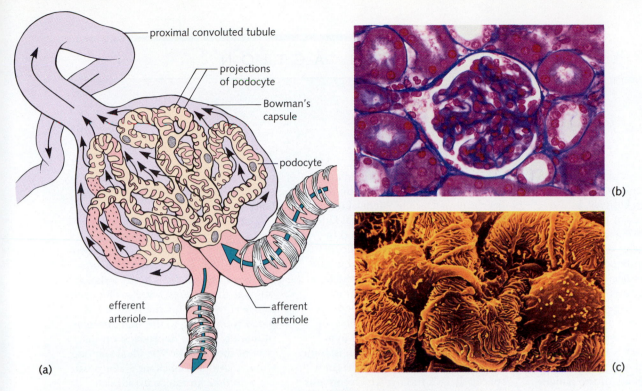

(a)

(b)

(c)

Figure 35.6 **(a)** The glomerulus is the site at which the primary filtrate is produced. Fluid passes from the bloodstream through a filtration barrier between the processes of podocytes, and enters the proximal convoluted tubule. **(b)** The meshwork of glomerular capillaries is enclosed within Bowman's capsule. **(c)** This scanning electron micrograph shows a web of podocytes covering the capillaries of a glomerulus.

The close association of capillaries to the collecting tubes within each nephron might lead you to conclude that the function of the nephron is simple: Blood passes through the capillaries, wastes and some water are removed, and urine is produced. That is indeed what happens, but four distinct steps are involved:

1. Formation of a primary filtrate
2. Resorption of the filtrate
3. Concentration of the urine
4. Tubular secretion

Formation of the filtrate Blood entering the nephron first passes through a roughly spherical network of thin capillaries known as a **glomerulus.** The glomerulus is surrounded by a cup-shaped structure known as **Bowman's capsule** (Figure 35.6a, b).

Here a fluid called the **primary filtrate** is produced through an essentially mechanical filtration process. The walls of glomerular capillaries are exceptionally "leaky." Because the arterial blood entering the glomerulus is under fairly high pressure, a substantial proportion of the blood plasma seeps through the walls of the glomer-

ulus like water through the walls of a leaky garden hose. This filtration process is far from haphazard, of course; cells called *podocytes* ("foot cells") cover the capillary walls with millions of tiny projections (Figure 35.6c). A filter-like extracellular layer extends between these projections, and the primary filtrate is formed by passage through this layer. This primary filtrate passes into Bowman's capsule, flows through the tubules of the nephron, and eventually is collected, along with fluid from thousands of other nephrons, to form urine.

Blood cells and platelets cannot pass through the tiny openings in the membrane. Neither can large protein or lipoprotein molecules. Smaller molecules (including salts, amino acids, and sugars) do pass through the membrane and become part of the filtrate.

The removal of the primary filtrate from the blood, however, is only part of the story. Blood leaving the glomerulus passes through two other places where dense capillary beds surround the tubule before it returns to the venous circulation. Why should the primary filtrate run through a series of tubes that are so closely associated with capillaries? The answer to this question is the key to understanding the function of the kidney.

The Artificial Kidney

The role the kidneys play in purifying the blood is essential to life. Fortunately we have two kidneys, so if one is affected by injury or disease, a functional kidney remains. The capacity of this organ is so great that one kidney can easily perform all of the filtering and concentrating functions the body requires. If both kidneys are lost or damaged, there are only two ways to keep an individual alive. The best solution is a kidney transplant. A kidney is taken from a healthy donor and carefully implanted in the body of the patient suffering from kidney failure. If the transplanted tissue is a close match to the tissues of the patient, the kidney may live and become a permanent part of the recipient's body.

In many cases, however, an appropriate donor is not available or surgery is not advisable. Medical technologists have developed artificial kidney machines that purify the blood by simulating the pathway that normal blood takes through the kidney. The accompanying figure shows the basic principles of the artificial kidney. Blood is removed from the body through a tube inserted in the arm and pumped through tubing in a chamber filled with a solution isotonic to blood. The blood moves through special dialysis tubing, which contains tiny pores. These openings allow salts and small molecules to pass through the walls of the tubing. When blood laden with nitrogenous wastes moves through the dialysis machine, these wastes diffuse away into the chamber, and purified blood is returned to the body.

Ingenious as it is, the artificial kidney is not a perfect solution to the problem of kidney failure. It is expensive and the process is time-consuming, occupying several hours a day as often as three times a week. The need to insert needles for the removal of blood is irritating to the skin and may ultimately lead to depression on the part of the patient. The ideal solution when a kidney transplant is not possible would be an internal artificial organ to take the place of the kidney. Work toward developing such an artificial kidney continues.

A kidney dialysis machine allows wastes to diffuse out of the bloodstream across an artificial membrane.

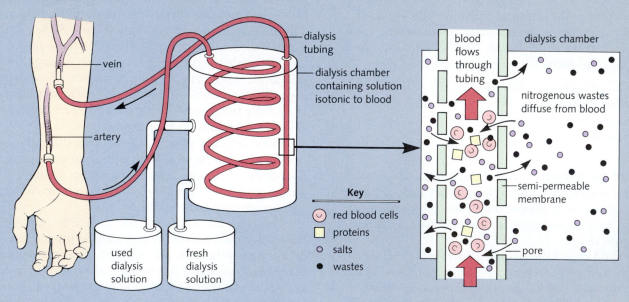

vein

artery

dialysis tubing

dialysis chamber containing solution isotonic to blood

used dialysis solution

fresh dialysis solution

Key

red blood cells
proteins
salts
wastes

blood flows through tubing

dialysis chamber

nitrogenous wastes diffuse from blood

semi-permeable membrane

pore

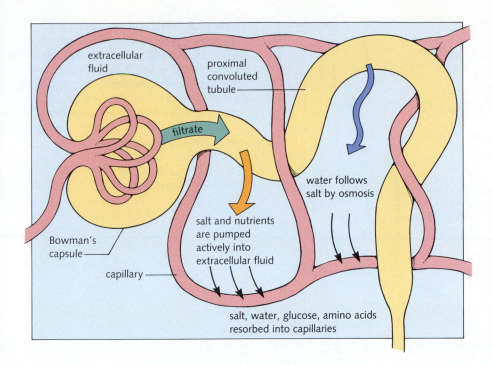

Figure 35.7 Most of the primary filtrate is resorbed in the proximal convoluted tubules. Salt is actively pumped into the extracellular fluid and resorbed into the capillaries. Osmosis causes water to follow the flow of salt, greatly reducing the volume of the filtrate.

Resorption   The rate at which primary filtrate is removed from the blood is very large: as much as 125 mL per minute, or 180 L per day. That's nearly 3 times the body weight of an average-sized adult. Fortunately, we don't urinate 180 L per day. Average daily urination is much closer to 1.5 L—less than 1 percent of the filtrate's original volume. This reduction is accomplished by a process of **resorption** in which 70 percent of the primary filtrate is returned to the blood.

Resorption begins in the first portion of the tubular system leading from the glomerulus, the **proximal convoluted tubule** (*proximal* means "close to" the glomerulus; *convoluted* means "twisted"). The walls of the tubule actively transport salt molecules and nutrients from the filtrate within to the extracellular fluid surrounding it. This energy-requiring transport removes most salts from the primary filtrate. When these salt molecules are pumped from the tubule, water follows by osmosis. The salt, nutrients, and water moving into the fluid are then *resorbed* by capillaries that follow the tubule pathway. Urea is not resorbed, however, so its relative concentration within the tubule increases as water is lost. This process alone returns to the blood nearly 65 percent of the water from the primary filtrate (Figure 35.7).

Concentration   The remaining fluid still has about the same total concentration of liquid and dissolved substances as the blood. If the tubule were now to lead directly to the bladder, the concentration of dissolved material in the urine would be the same as in blood plasma. If urine were always to be released at the same concentration as plasma, the kidney would not be able to get rid of excess water and could not conserve water when it was in short supply. Terrestrial animals, because they must conserve water, have a mechanism built into the structure of the nephron that enables them to produce a concentrated urine.

Concentrating the filtrate might seem like an easy job; all that is needed is a system to remove water from the filtrate. But water is not pumped out of the nephron directly. Instead, as we saw when we studied the operation of plant root hairs, cells lining the nephron move water from the filtrate by generating an osmotic gradient across the tubule wall. The kidney concentrates urine by actively transporting salt ions and allowing water to follow the salt by osmosis.

But that process alone is not powerful enough to concentrate urine sufficiently. In order for animals to survive in habitats where fresh water is in short supply, their excretory systems must be able to produce urine that is much more concentrated than their blood. To accomplish that feat, the mammalian kidney amplifies the abilities of its membranes to create an osmotic gradient by using a **countercurrent multiplier system.** In order to understand how this system works, you must understand the peculiar configuration of the tubule system.

Down the line from the proximal convoluted tubule, urine passes through a hairpin loop called the **loop of Henle.** Fluid enters the descending portion of this loop and descends from the cortex into the medulla. After

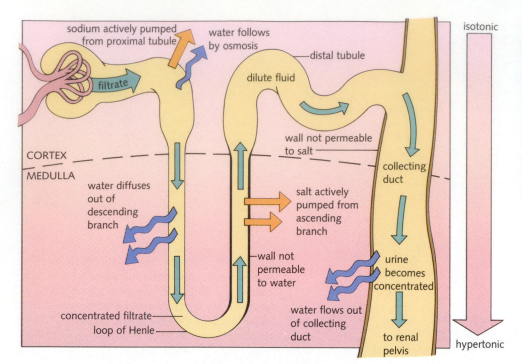

sodium actively pumped from proximal tubule

water follows by osmosis

filtrate

distal tubule

dilute fluid

isotonic

wall not permeable to salt

CORTEX

MEDULLA

collecting duct

water diffuses out of descending branch

salt actively pumped from ascending branch

wall not permeable to water

urine becomes concentrated

concentrated filtrate

loop of Henle

water flows out of collecting duct

to renal pelvis

hypertonic

Figure 35.8 Fluid and ion movements occur between the filtrate and its surroundings as the filtrate moves through the nephron. The fact that the wall of the collecting tubule is not permeable to salt allows the force of osmosis to draw water out of the filtrate, forming a more concentrated urine. This countercurrent flow pattern, in which fluids move through different segments of the tubule in opposite directions, is responsible for the kidney's great efficiency.

making a sharp turn, it then rises through the ascending portion of the loop, back into the cortex, and into the distal convoluted tubule.

The ascending and descending portions of the loop of Henle have different properties with respect to the movement of water and salt (Figure 35.8). The ascending portion actively pumps salt from the urine into the extracellular fluid around it. Because the wall of this portion of the loop is impermeable to water, water cannot follow the salt here. The net result of this process, therefore, is to increase the salt concentration outside the tubules.

The activity of the ascending portion of the loop then creates a zone of high salt concentration that includes the entire medullary region of the kidney. The flow of water through the loop, combined with continued pumping of salt across the ascending tubule wall, sets up a sharp concentration difference in the extracellular fluid. The saltiest portion is near the bottom of the loop of Henle, and the least salty portion is in the cortex. The strength of this gradient is increased by the geometry of the loop itself. This is a countercurrent multiplier system, in which a reverse flow process increases the effect of ion pumping by one part of the system.

Why should this be of any significance? Look at Figure 35.8 and you will see that both the descending portion of the loop and the collecting ducts pass through the medulla. As urine flows through the *descending* loop of Henle, the walls of which are permeable to water, water

leaves the tubule by osmosis into the surrounding region of high salt concentration. The urine thus decreases in volume and becomes more concentrated.

Because salts are pumped out of the *ascending* portion of the loop, the urine becomes less concentrated but stays at roughly the same volume. You might think that nothing has been accomplished: The fluid is no more concentrated than it was when it entered the system. But the "purpose" of the loop is not to concentrate the urine in the loop itself; it is to establish a concentration gradient in the extracellular fluid surrounding the tubule.

What function does that gradient serve? As dilute fluid leaves the loop, it enters the **distal convoluted tubule** and then flows into a **collecting duct.** The collecting duct flows back down through the extracellular gradient that is set up by the loop. The walls of the duct are not permeable to salt. They are, however, permeable to water. During this final trip through the medulla, therefore, the urine loses a great deal of water through the tubule walls and becomes more and more concentrated as it approaches the end of the tubule. By the time the fluid leaves the duct and passes toward the kidney pelvis, its concentration of dissolved material may be as much as four times that of blood plasma.

The remarkable ability of the loop of Henle to produce a water-concentrating salt gradient means that a terrestrial animal need devote only a small amount of water to eliminating nitrogenous waste. Nevertheless, a

Urine Testing

The ability of the kidney tubule to secrete complex molecules into the urine provides the body with a mechanism to eliminate drugs and poisons that cannot be detoxified by the liver. It also provides medical workers with an informative window on what's going on inside the body. When the body is unable to regulate blood sugar levels, for example, the sugar content of the blood may become so high that not all of the sugar in the primary filtrate can be resorbed. The presence of this excess sugar in the urine is one of the classic symptoms of *diabetes mellitus,* a serious metabolic disorder.

The presence of certain hormones in the urine is the first sign that a woman may be pregnant. Toward the end of a pregnancy, physicians monitor the condition of the placenta by testing for the presence of metabolic wastes from the fetus in the mother's urine. Sudden changes may indicate that the placenta has begun to degenerate and that immediate action is necessary to save the fetus. The presence of proteins in the urine may indicate kidney malfunction or myeloma, a form of cancer. The most controversial application of urine testing, however, is the search for traces of illegal drugs.

The kidney has the ability to eliminate virtually any drug from the body, making it possible to test for the presence of steroids, amphetamines, opiates, and cocaine. Traces of these drugs remain in the body for many days after their use, and modern detection methods can reveal them. Many companies have begun to make urine tests a condition of employment, and legal authorities have begun to demand such tests after motor vehicle accidents. There is no telling how urine testing will ultimately affect the framework of personal and privacy rights that support a free society, but it will certainly raise serious legal and personal issues in the years ahead.

Drug Type	Familiar Name	Approximate Life Span*
Amphetamines	Speed, Ecstasy, Crystal	2 days
Barbiturates	Phenobarbital, Butabarbital, Secobarbital, Pentobarbital	1 day—short-acting 2-3 weeks—long-acting
Benzodiazepines	Valium®, Librium®	3 days if dosage is swallowed
Cannabinoids	Marijuana	10 days, acute 21 days, chronic
Cocaine	"Coke"	2 to 4 days
Methadone	Methadone	3 days
Methaqualone	Quaalude®	12-14 days
Opiates	Heroin, Codeine, Morphine	2-4 days
Phencyclidine	PCP, Angel Dust	2-8 days
Propoxyphene	Darvon®	8 hours to 2 days

*Drug life span also depends on the type of drug, the amount taken, and the metabolic rate of the user.

certain amount of water must be used to dissolve urea and other waste products. In humans, this minimal volume is about 400 mL of water per day. This means that even under conditions where water intake is limited, this amount of water is lost from the body each day. This is one of the serious problems facing humans and most other mammals in dry environments.

Secretion Besides filtration and resorption, there is a third process at work in kidney tubules. **Secretion** is the active transport of specific molecules from capillaries into the tubules. Unlike filtration, secretion is very specific. Secretion pumps potassium (K^+) and Hydrogen (H^+) ions into the filtrate. Kidney tubules also secrete a number of foreign substances, which include chemicals,

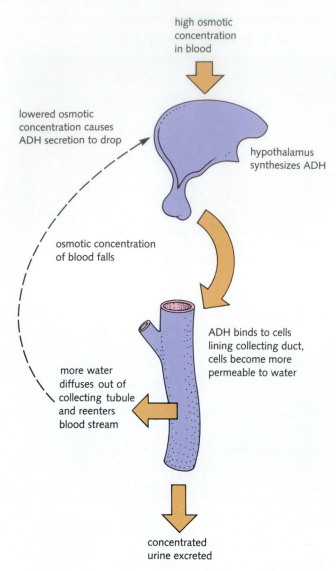

high osmotic
concentration
in blood

lowered osmotic
concentration causes
ADH secretion to drop

hypothalamus
synthesizes ADH

osmotic concentration
of blood falls

ADH binds to cells
lining collecting duct,
cells become more
permeable to water

more water
diffuses out of
collecting tubule
and reenters
blood stream

concentrated
urine excreted

Figure 35.9 The regulation of water balance by antidiuretic hormone, ADH.

ammonia, and drugs, into the filtrate. Secretion is the process that "clears" many drugs and poisons from the circulation, enabling them to be excreted in the urine.

Urination Let us now turn our attention briefly to the concentrated urine that leaves the collecting duct. This urine flows into the renal pelvis and is carried by the ureters into the bladder. The bladder is a flexible organ surrounded by smooth muscle. The bladder can accommodate as much as 800 mL of urine, a bit more than two cans of soda or beer. The bladder wall begins to stretch when the bladder accumulates about half this amount (400 mL). The extension of smooth muscle causes nerve cells in the wall of the organ to communicate a sensation to the central nervous system, and the individual becomes aware that the bladder is full. The bladder is emptied through the *urethra* when a sphincter of muscle at the boundary between the bladder and the urethra is relaxed.

Control of Water Balance

Conserving water when conditions demand it is only one role of the kidney. As we all know, when water intake is greater than normal, the kidney is able to produce large amounts of dilute urine. This prevents body fluids from becoming diluted when liquid intake is high. How can the same system that concentrates urine when water is in short supply produce a dilute urine when too much water is present?

ADH and the collecting ducts One key to the versatility of the kidney in maintaining water balance is the responsiveness of the collecting ducts to a hormone known as **antidiuretic hormone,** or **ADH.** A diuretic is an agent or drug that causes the urine to be more dilute and increases the rate at which water is removed from the body. An *anti*diuretic hormone, then, acts to concentrate the urine and conserve water.

The system works as follows. ADH, when present, acts on the wall of the collecting ducts to increase their permeability to water. The more permeable the walls of the duct, the more water leaves the tubules to reenter the bloodstream, and the more concentrated the resulting urine becomes. When ADH is not present, the wall of the tubule is much less permeable; more water remains in the tubule, and a dilute urine is produced. Thus the concentration of ADH determines how much water is extracted from the tubule and how much is retained in the body (Figure 35.9).

The concentration of ADH in the bloodstream is controlled by the *hypothalamus* and the *posterior pituitary,* two extremely important regions of the brain that we will discuss in more detail in Chapter 36. Cells in the hypo-thalamus monitor the osmotic concentration of the blood. When the concentration of material in the blood is too high (and thus the concentration of water too low), the hypothalamus synthesizes ADH and causes it to be released into the blood from the posterior pituitary. When ADH binds to the cells lining the collecting ducts within the kidney, it stimulates the resorption of water from the urine. When the concentration of dissolved material in the blood is too low (when there is excess water), the release of ADH is inhibited, preventing the retention of water. The system thus responds to changes in the blood's water balance, acting to conserve water when it is in short supply and to remove water when there is too much of it.

Thirst   Changes in blood water content affect the central nervous system as well as the kidneys. The sensation of *thirst* seems to be produced in the hypothalamus by the same factors that increase the release of ADH (Figure 35.10). In this way the hypothalamus regulates water balance, controlling the excretory system to reduce water loss and modifying behavior to increase water intake. Experiments with laboratory animals have shown that artificial stimulation of certain brain regions causes an insatiable thirst that leads the animals to drink continually.

Anyone who has ever stood near the restrooms at a baseball park on a hot day knows that there is a relation-ship between beer and urination. The casual drinker may think that this is due to the amount of liquid he has drunk, but this is only partly true. A heavy drinker feels "hung over" the next morning, and one of the symptoms of the hangover is a powerful thirst. The need to visit the rest-room and the hangover thirst are related. Alcohol inhib-its the release of ADH, and this increases the rate at which water is lost from the kidneys. Excessive drinking there-fore causes a great deal of water to be lost, producing an unusually large amount of urine and dehydrating the body. Consider the case of two fans sitting side-by-side at the ball park, one drinking beer and the other drinking an equal volume of soft drink. The non-alcohol drinker will not only see more of the game (fewer visits to the restroom) but will also feel better in the morning!

Aldosterone   Although ADH and drinking behavior both help regulate the concentration of water and salt in the blood, they do not directly control the total volume of blood. A second hormone system acts in concert with ADH to help regulate blood volume. When blood vol-ume is low, blood pressure falls, causing certain cells in the kidney to release an enzyme known as **renin.** Renin converts *angiotensinogen,* an inactive hormone, to its active form, *angiotensin.* Angiotensin has two effects. First, it causes smooth muscle surrounding blood vessels to con-

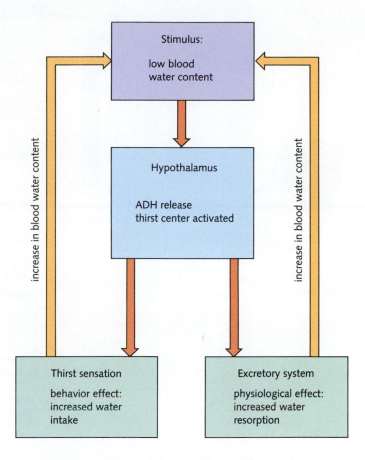

Figure 35.10   Feedback pathways affecting thirst and water loss as they regulate blood water content.

strict, increasing blood pressure. Second, it stimulates the **adrenal cortex** to release **aldosterone.** This hormone affects the distal convoluted regions of the tubule and increases the resorption of sodium and the secretion of potassium. Because this action increases the sodium resorption from the distal tubule, the salt concentration of the blood increases. This in turn increases the amount of water retained in the blood in the kidney.

High blood pressure, on the other hand, turns off this system, leading to the excretion of salt and a lowering of blood pressure. Figure 35.11 summarizes the effects of this feedback system.

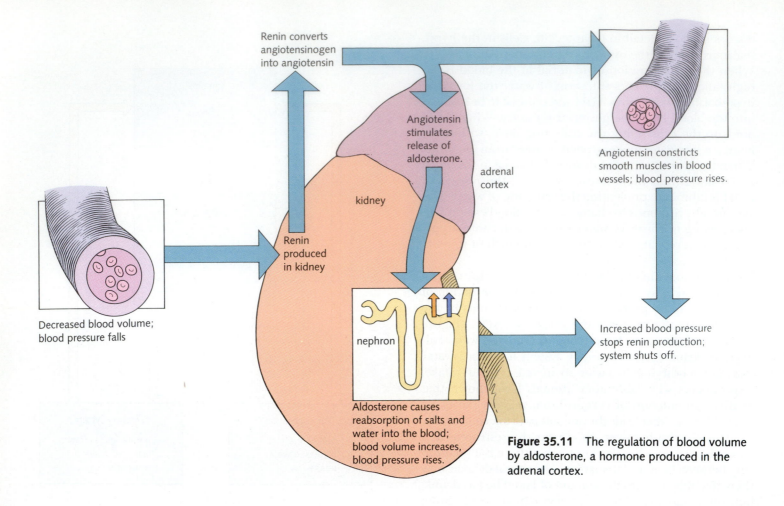

Renin converts angiotensinogen into angiotensin

Angiotensin stimulates release of aldosterone.

adrenal cortex

kidney

Renin produced in kidney

Angiotensin constricts smooth muscles in blood vessels; blood pressure rises.

Decreased blood volume; blood pressure falls

nephron

Increased blood pressure stops renin production; system shuts off.

Aldosterone causes reabsorption of salts and water into the blood; blood volume increases, blood pressure rises.

Figure 35.11 The regulation of blood volume by aldosterone, a hormone produced in the adrenal cortex.

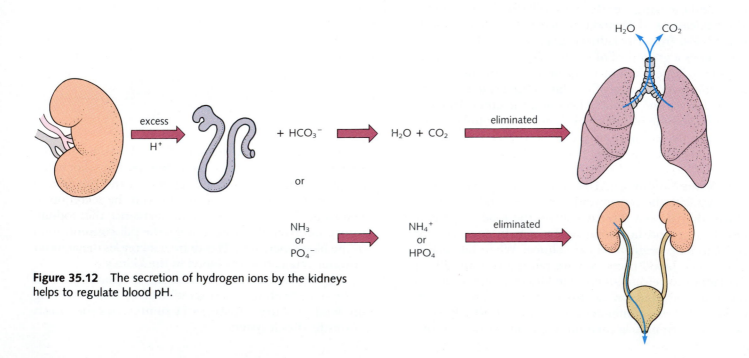

excess H^+

$+ HCO_3^-$ → $H_2O + CO_2$ → eliminated → H_2O CO_2

or

NH_3 or PO_4^- → NH_4^+ or HPO_4 → eliminated

Figure 35.12 The secretion of hydrogen ions by the kidneys helps to regulate blood pH.

Dealing with Nitrogen: Another Style of Life

The liver is capable of detoxifying a wide range of harmful substances. Animals excrete nitrogenous waste by releasing concentrated wastes in forms such as urine that are easily eliminated from the body. Do plants face the same problem? What happens to the metabolic wastes of a plant?

The fact that plants don't eliminate nitrogen in the same ways as animals is really a reflection of a profound difference in internal metabolism between plants and animals. Animals are able to synthesize only a few of the amino acids they require; the rest must be obtained through the diet. Therefore, most animals are unable to do anything useful with the ammonia (NH_3) molecules that are released when amino acids are used as food. Most plants, however, do not take up proteins as food. Plants are able to make most of their own amino acids, and the chemical pathways by which they do this *require* nitrogen to complete the synthesis of the amino acid. Therefore, plants have no need to eliminate nitrogen produced by metabolic pathways. Instead, they recycle much of the nitrogen that they produce, use it to synthesize their own amino acids, and never experience that great need common to so many animals at regular intervals: to empty the bladder.

Acid–base regulation The normal pH of arterial blood is 7.4. Venous blood has a slightly lower pH (about 7.35) because of its higher content of carbon dioxide, which reacts with water to produce carbonic acid and bicarbonate:

$$CO_2 + H_2O \rightleftharpoons H_2CO_3 \rightleftharpoons HCO_3^- + H^+$$
$$\text{carbonic} \quad \text{bicarbonate}$$
$$\text{acid}$$

The presence of these bicarbonate ions buffers the blood against a rapid decrease in pH when hydrogen ions (H^+) are produced by normal metabolic processes. If there were no mechanisms to eliminate H^+, however, the limited buffering capacity of the blood would soon be overwhelmed. Although the majority of pH regulation occurs in the lungs, the kidneys prevent the blood from becoming too acidic by secreting excess H^+ into kidney tubules. Some H^+ recombines with bicarbonate (HCO_3^-) in the tubule to produce water and CO_2, which is eliminated through the lungs. If blood pH is very low, then HCO_3^- may not be available, and the excess H^+ may be combined with ammonia or phosphate salts and excreted in the urine (Figure 35.12). The H^+-secreting ability of the tubule wall is so great that the pH of tubule fluid may drop as low as 4.5, signaling an H^+ concentration almost 1000 times higher than that of the blood.

WATER BALANCE IN OTHER ANIMALS

As we saw at the beginning of the chapter, humans have wondered for thousands of years how some animals can survive in salty environments. Our own kidneys enable us to survive in a variety of environments, conserving water when we have little to drink and excreting excess water when we drink more than we need. Animals in different environments display a variety of adaptations that enable them to maintain a proper balance of water and salt.

Aquatic Animals

Freshwater fishes live in an environment of nearly pure water. Therefore, they tend to gain water by osmosis. Although their bodies are coated with a thick mucus that tends to reduce the influx of water, water still enters through the gills. Without some means of counterbalancing the influx of water, the cellular salt concentration would be diluted to a dangerously low level. The kidneys of freshwater fishes remove large amounts of water from the blood and release a dilute, ammonia-containing urine. The kidneys resorb nearly all salts into the bloodstream, enabling the animal to conserve essential minerals (Figure 35.13).

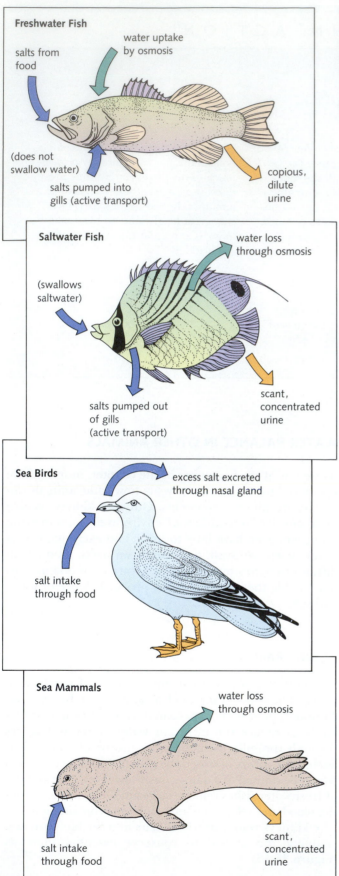

Freshwater Fish

salts from food

water uptake by osmosis

(does not swallow water)

salts pumped into gills (active transport)

copious, dilute urine

Saltwater Fish

water loss through osmosis

(swallows saltwater)

salts pumped out of gills (active transport)

scant, concentrated urine

Sea Birds

excess salt excreted through nasal gland

salt intake through food

Sea Mammals

water loss through osmosis

salt intake through food

scant, concentrated urine

Figure 35.13 Freshwater fishes must retain salts and release a dilute urine, while saltwater fishes must deal with an overabundance of salts. They produce a dense, concentrated urine and eliminate some salt by active transport from the gills. Sea birds use very little water in the elimination of nitrogen wastes, and remove excess salt from the bloodstream by means of salt glands. Sea mammals produce an extremely concentrated, salty urine, freeing them from the need to drink fresh water.

Saltwater fishes must deal with the opposite problem: **dehydration.** In spite of the fact that they are surrounded by water, the salt concentration of seawater is a bit higher than that of their body fluids. Therefore, these fishes tend to lose water to their environment by osmosis. They compensate for this in two ways. First, their kidneys produce a dense, concentrated urine that contains very little water. Second, specialized cells in their gills remove salt from the bloodstream by means of active transport and release it to the outside. By using their gills to excrete some salt, these fishes are able to maintain a proper water balance (Figure 35.13).

Sea birds, like other birds, produce a thick, uric-acid–laden guano to remove nitrogenous waste that contains very little water. This helps to minimize their water loss. However, the concentration of salt in seawater would still produce dehydration were it not for special glands that pump salt out of the bloodstream (Figure 35.13). These salt glands enable sea birds to drink seawater and extract the excess salt. The glands dump their secretions just above the bill, and crystals of dried salt are often seen in this region.

Sea mammals drink little or no seawater. They obtain small amounts of water from the food they eat, and they produce an extremely salty, concentrated urine that minimizes water loss (Figure 35.13). Why can't land mammals drink seawater? Our kidneys are unable to produce a urine concentrated enough to remove both the excess salt and metabolic wastes that our bodies generate.

Land Animals

Animals on land face another set of problems. Because they live in air, water gain or loss through osmosis is not a problem. However, land animals in dry environments lose large amounts of water during respiration and therefore must conserve water. They also face the serious problem of how to remove metabolic wastes in a dry environment. Some animals deal with these problems by presenting tough barriers against water loss, such as the insect's chitinous exoskeleton and the reptile's leathery skin. Others have evolved lifestyles that minimize water loss. Thin-skinned amphibians live in or near the water. The delicate mosquito, which loses water easily, searches for prey only in humid weather (Figure 35.14). A mosquito contains less than 3 microliters of water—about one-twentieth of a drop. If a mosquito were to take to the air in dry weather for more than a few minutes without finding a meal, its water loss from increased respiration would be fatal.

The most extreme adaptations of land animals for water conservation are seen in desert animals such as the kangaroo rat (Figure 35.15), which is found in the American southwest. As we saw in Chapter 33, the long nasal passageways of this animal help to minimize water loss during respiration. To escape the sweltering desert heat, this animal spends most of the day deep in its cool burrow, emerging only at night. Its kidneys hold the world record for efficiency, producing urine 25 times more concentrated than the kangaroo rat's blood. Not surprisingly, the nephrons of the rat have extremely long loops of Henle, providing a very long surface for water resorption. The large intestines of this desert dweller are equally efficient, extracting so much water that the feces it produces are nearly dry. With so many adaptations for water conservation, how much water does the kangaroo rat need to drink? None! The kangaroo rat obtains most of its water from a diet of dry seeds. You will remember from Chapter 25 that the oxidation of glucose produces carbon dioxide and water. Its metabolic water, produced by the oxidation of food carbohydrates, is enough to meet the kangaroo rat's needs. Efficiency personified.

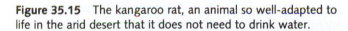

Figure 35.14 Water loss is a critical problem for small insects like the mosquito.

TEMPERATURE REGULATION

Regulation of fluid balance is an important component of internal homeostasis, but other factors are also controlled. One of these is the temperature of the body. Our review of basic chemistry emphasized the fact that chemical reactions, including enzyme-catalyzed ones, are affected by the temperatures at which they occur. The wide variety of local environments (and local temperatures) on the earth has made it important for many animals to control their internal body temperatures, a process called **thermoregulation,** whether they live in the Arctic or in the jungle.

Chemical reactions generally occur more quickly at higher temperatures, so the basic metabolic rate increases as the temperature increases. This phenomenon leaves an animal with two "choices." If it allows the external environment to influence its internal temperature, its metabolic rate will change as the external temperature changes. The alternative is to adjust heat-producing processes within the body so that a relatively constant

Figure 35.15 The kangaroo rat, an animal so well-adapted to life in the arid desert that it does not need to drink water.

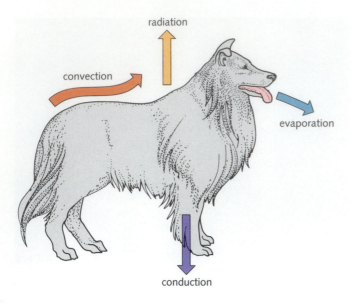

radiation

convection

evaporation

conduction

Figure 35.16 An animal may gain or lose heat in a number of ways. Rapid panting on a hot day helps this dog to lose heat by evaporation.

internal temperature is maintained while the external temperature varies.

Every organism has a source of heat available to it: its own metabolism. As we noted in Chapter 19, the metabolic pathways that are used to produce ATP from glucose capture only about 40 percent of the total available energy in the glucose molecule. The remaining 60 percent is lost as heat, and when ATP is used to do cellular work, much of the energy available in ATP is also released as heat. These facts of chemistry enable organisms to control the sources of heat production within their own bodies.

An organism may gain heat from or lose heat to its environment in a number of ways, as illustrated in Figure 35.16. (1) Heat gain from *radiation* occurs whenever an animal absorbs direct or indirect sunlight. Because all warm objects emit infrared radiation, radiation may also cool an organism. (2) *Conduction,* the direct transfer of heat between two bodies, occurs whenever an animal's temperature is different from its surroundings. Conduction may either heat or cool an animal, depending on the temperature of the environment. (3) Warm air is lighter than cool air, and this physical fact results in *convection,* in which rising warm air produces air currents. Convection currents increase the efficiency of cooling by bringing cooler air in contact with the surface of an organism. Convection may also occur in water. (4) Finally, *evaporation,* which involves water changing from a liquid to a gas phase, consumes a great deal of energy and is very effective in surface cooling.

Ectotherms

Most organisms are **ectotherms**—that is, their body temperatures are largely determined by their interactions with the environment in which they live. Virtually all invertebrates are ectothermic, as are all amphibians, most fishes, and most reptiles. Although such animals are commonly called "cold-blooded," this term is really a misnomer; the body temperatures of such animals may actually be quite high.

For small animals living in the water, ectothermy is almost unavoidable. The density of water is much greater than that of air, and as a result, much more energy is required to change the temperature of water a few degrees than is required to change air temperature by the same amount. In addition, the temperature of most bodies of water is relatively constant, so animals living in them can depend on a stable environmental temperature. This relative stability enables fishes and other organisms to adapt to the temperatures of their surroundings and be free of the need to regulate their own temperatures.

One consequence of this adaptation is that many ectothermic organisms can live only in narrow temperature ranges. Anyone who has neglected the temperature controls on a tank of tropical fish has found that even slight variations in water temperature can mean the difference between life and death.

Land ectotherms, including amphibians and reptiles, face a far more challenging environment. In many terrestrial habitats, daily variations in air temperature can be as large as 25°C. These variations affect ectotherms dramatically, for they must adjust their behavior in ways that keep their body temperatures within a suitable operating range.

Many active ectotherms can, in fact, control their body temperatures remarkably well by moving around; they seek shade or dip into ponds on hot days and bask in the sun on cool ones (Figure 35.17). Still, these animals are slow and sluggish during cool nights and must take great care to lower their body temperature on warm days. On cool, sunny days in the spring and fall, it is not unusual to find snakes curled up on rocks in the sunshine until their body temperature increases to the point where they are able to be more active.

Endotherms

Endotherms are animals that regulate their body temperature to a large extent by physiological means and maintain a more or less constant internal temperature regardless of environmental fluctuations. Endotherms generally maintain a body temperature greater than the temperature of their environment. Actual body temperatures vary from one endothermic species to the next,

Figure 35.17 Many ectotherms help to control their body temperature by their behavior, swimming on hot days and basking in the sun on cool ones.

but the normal human body temperature of 37°C is a typical value.

Maintaining a constant temperature gives endotherms a metabolic advantage over other organisms. Biochemical pathways and the enzymes that catalyze them often have their maximum efficiencies at temperatures near this point, and this enables endothermic organisms to remain active even in cold environments. Birds and mammals are endotherms.

Adaptations for temperature regulation Endotherms have evolved many adaptations that help them maintain a constant body temperature (Figure 35.18). Mammals that live in cold environments have thick coatings of insulating fur and body fat to minimize the rate at which heat is lost to the outside. The blubber of aquatic mammals like whales and dolphins is such an adaptation, enabling them to cope with the problems of living in a cool ocean environment where the heat capacity of water makes thermoregulation especially difficult.

Birds, for their part, have several layers of feathers that serve to insulate their relatively small bodies and prevent heat loss. Birds from particularly cold environments have an especially thick layer of down feathers; humans have exploited the insulating ability of down for many years.

Other organisms have adaptations to hot environments. Horses and many other animals cool themselves by sweating, others by panting. Some of these adaptations are behavioral. A bird lifts its wings to increase the surface area available for heat loss. Burrowing animals stay in the cool ground on hot days.

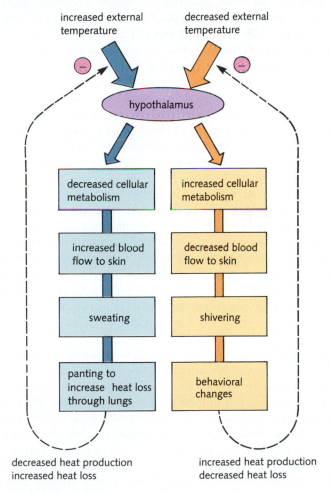

Figure 35.18 The hypothalamus coordinates the regulation of body temperature.

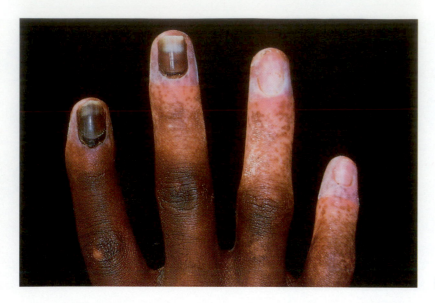

Figure 35.19 Frostbite

Thermoregulation in Humans

Humans, like other mammals, maintain a constant internal temperature regardless of local environmental temperature. The system of temperature regulation is so precise that a change in body temperature of even a few degrees is usually a signal of serious trouble.

Body temperature can be regulated by adjusting two key variables: the rate of heat production and the rate of heat loss or gain with respect to the environment. For the systems and organs of the body to work effectively, there must also be a temperature-sensitive center that coordinates and controls these heat-producing and heat-loss functions.

In humans, this center is the region of the brain known as the *hypothalamus*. The hypothalamus has two sources of temperature information: input from temperature-receptor neurons in the skin and input from receptors in the hypothalamus itself that monitor changes in blood temperature. These two sources of information alert the hypothalamus when a change in temperature occurs. This enables the body to respond quickly to the two types of temperature stress: cold and heat.

Response to cold The hypothalamus reacts to low temperatures by coordinating activities of both the nervous system and the endocrine system. Blood vessels near the skin constrict, limiting heat loss from the body. There is a gradual increase in skeletal muscle tone which may eventually lead to shivering, helping to produce heat. The body releases two hormones, epinephrine and thyroxine, that increase the body's metabolic rate: Cells throughout the body break down carbohydrates more quickly, fat reserves are mobilized, and sugar is released into the bloodstream from the liver. These changes pro-

duce heat in the tissues of the body, tending to elevate body temperature.

To minimize heat loss through the skin, certain smooth muscles surrounding blood vessels contract, restricting blood flow to the limbs. This is one of the reasons why your fingers and toes may seem extremely cold on a winter day. This response is not without risk. As reduced blood flow lowers the temperature of the extremities, they may approach the freezing point. In very cold weather, ice crystals may form in the tissues, causing a condition known as **frostbite** (Figure 35.19). At first, one experiences little pain in tissue numbed by frostbite, even though the growing ice crystals rupture cells and break blood vessels. (Freezing deadens nerve cells that would otherwise warn of tissue damage.) But later, as the tissue thaws, the pain may be intense and there is danger of permanent injury. One should avoid frostbite by leaving the cold immediately when tissue goes numb. Frostbitten tissue must be warmed slowly to minimize further injury as ice crystals melt.

The hypothalamus may also initiate **shivering,** an involuntary response that helps produce heat throughout the body. Some of the major sources of internal heat are the muscles of the body. Muscles make up as much as 40 percent of the total body mass, and they are major energy users. Shivering is a series of uncontrolled skeletal muscle contractions. These contractions do little mechanical work, because the body itself often remains stationary. Nearly all of the energy released by shivering contractions is converted into heat. A few minutes of vigorous shivering can increase the heat output of surface muscles by a factor of five or six! The cold response may also include "goose bumps," which result from a contraction of the adductor muscle in the root system of a hair. In fur-coated animals, the contraction of adductor muscles causes some of their fur to stand out straight and increases the amount of air trapped within an already thick fur coat, making it an even better insulator. In humans, the heat produced by the formation of goose bumps is not important. Our goose bumps may be only a reminder of the thicker hair that some of our evolutionary antecedents bore.

Response to heat High body temperature causes the hypothalamus to initiate a series of responses that slow down heat production and increase the rate at which heat is lost from the body. The same chemical responses that stimulate metabolic activities under cold stress are now reversed to check the breakdown of fats and carbohydrates. These changes may cause one to feel sluggish and tired on a very hot day.

In addition, circulation to the skin increases to maximize heat loss, and the increased blood flow may give the hands and faces of fair-skinned people a flushed, reddish appearance. The skin begins to produce *sweat,* a secretion of water and salts that cools the body by evaporation. Breathing, which results in heat loss through the lungs during gas exchange, increases to some extent in humans. In those mammals unable to sweat, such as dogs and cats, temperature loss through the lungs and via the tongue is increased by rapid *panting.*

In extreme cases, the body may lose the ability to maintain its own temperature. Body temperature may rise so high that cells in the skin are unable to function, producing a life-threatening condition known as **heat stroke.** A heat stroke patient ceases to sweat and has hot, dry skin. He or she may collapse or hallucinate. Immediate first aid to lower the body temperature is necessary to save a victim of heat stroke, and the most effective measure is often to place the victim in a tub of cool water.

Homeostasis and Thermoregulation

Thermoregulation is an example of a homeostatic mechanism that regulates a single variable of the internal environment—the temperature. Each aspect of temperature response, whether to cold or to heat, can be understood as part of a negative feedback loop.

In the case of cold, low temperatures initiate a series of responses that raise the body temperature. Once it has been elevated, further increases in temperature have a negative feedback effect on the hypothalamus.

A similar situation exists when the body responds to high temperatures. In each case, the body responds to the external temperature as though it maintained a specific temperature, or "set point," in its "biological thermostat" within the hypothalamus. Although biologists do not fully understand the nature of this thermostat, it is clear that one does exist, at least in a functional sense.

An interesting exception to the homeostatic tendency of the body to maintain a fixed temperature is **fever,** a long-term elevation in body temperature caused by illness. Fever can be understood as a response in which the body changes its thermostatic "set point" because of special circumstances. Normally, as body temperature begins to rise, the hypothalamus initiates a series of responses that slow cellular metabolism to reduce heat production. When it is faced with disease, however, the last thing the body needs is a reduction in cellular activity while the cells of the immune system are working to defeat an infection. The 2°C or 3°C elevation in body temperature associated with fever, therefore, is part of an integrated defense mechanism against disease. Allowing body temperature to rise for a few days while the immune system works to deal with the infection may make us feel miserable, but it is an effective way of combating disease.

SUMMARY

Organisms living in water must deal with the forces of osmosis. In fresh water they must retain salts and deal with the inward pressure of osmosis; in salt water they excrete salt and maintain an osmotic balance with the water around them. Animals living on land have evolved ways to conserve water and to eliminate waste products. The ammonia wastes produced by protein breakdown are especially difficult to deal with because the ammonia that is formed in these pathways is toxic. Ammonia may be converted into less toxic compounds and removed from the body by the excretory system. Excretory systems vary tremendously from one group of organisms to another. Flatworms contain flame cells that move wastes to the exterior; earthworms have nephridia in each segment; and insects have a system known as Malpighian tubules that adds metabolic wastes directly to the digestive system. In vertebrates, kidneys remove metabolic wastes and excess water from the circulation and produce urine, which is excreted.

The kidney is made up of hundreds of thousands of functional units known as nephrons. The filtration process involves the formation of a primary filtrate that contains much of the dissolved material from the bloodstream. This filtrate passes through the convoluted tubule system, in which much of the water and dissolved material is resorbed into the bloodstream. This process is assisted by the osmotic gradient created by the loop of Henle. The amount of water resorbed into the bloodstream is regulated by the amount of antidiuretic hormone (ADH), which is released from the posterior pituitary.

Temperature regulation in warm-blooded (endothermic) and cold-blooded (ectothermic) organisms is achieved by a number of mechanisms, including control of the overall metabolic rate. The internal temperature of an organism is determined by its ability to respond to rapid changes in its own internal activity and in the temperature of the external environment.

After studying this chapter, you should be able to:

■ Explain the role of the excretory system in maintaining water balance, and cite the differing demands placed on the system in various environments.

■ Describe the structure and physiology of the human excretory system and its relationship to other systems in the body.

■ Discuss the process of thermoregulation.

SELECTED TERMS

uric acid *p. 678*
urea *p. 679*
contractile vacuole *p. 680*
flame cell *p. 680*
nephridia *p. 680*
Malpighian tubules *p. 680*
kidneys *p. 680*
renal artery *p. 681*
renal vein *p. 681*
ureter *p. 682*
bladder *p. 682*
urethra *p. 682*
nephron *p. 682*
Bowman's capsule *p. 683*
glomerulus *p. 683*
convoluted tubule *p. 685*
loop of Henle *p. 685*
countercurrent system *p. 685*
ectothermic *p. 694*
endothermic *p. 694*

REVIEW

Discussion Questions

1. Describe the differences between the functions of the two capillary beds (glomerulus and net around tubules) in each nephron.

2. Some "crash" diets call for protein-rich food and the complete elimination of carbohydrates from the diet. People on these diets often complain of frequent urination and increased thirst. Why might they have these symptoms?

3. At what stage of the excretory process in the kidney are foreign substances removed from the blood?

4. If you did not drink any water for 24 hours, would you expect your level of ADH to increase or decrease? Why?

5. Animals with an extreme need to conserve water often have larger kidneys than related organisms that inhabit damp environments. The increased size of their kidneys corresponds to an increase in the distance between the glomerulus and the bottom of the loop of Henle. How does this structural change assist in water conservation?

6. Explain the key mechanisms of thermoregulation in humans, and indicate which mechanisms are subject to negative feedback loops.

Objective Questions   (Answers in Appendix)

7. Deamination refers to the
   (a) breakdown of ammonia.
   (b) failure of the kidneys to keep a water balance in the body.
   (c) filtering of amino acids through the liver.
   (d) removal of an $-NH_2$ group from an amino acid.

8. Terrestrial organisms must excrete nitrogenous wastes in the form of urea or uric acid because
   (a) ammonia cannot be excreted in crystalline form.
   (b) ammonia is toxic.
   (c) too much water would be needed for removal of ammonia.
   (d) All of the above are correct.

9. During the process of resorption
   (a) dissolved solutes are returned to the bloodstream.
   (b) specific molecules are transported from the bloodstream to the tubules.
   (c) urea is removed from the tubule.
   (d) water is produced as an end product.

10. During the process of filtration
    (a) most of the water and solutes that enter the tubules are transported to the bloodstream.
    (b) blood is moved from the glomerulus to the Bowman's capsule.
    (c) water is removed from the filtrate.
    (d) the walls of the descending loop of Henle become more permeable to water.

11. When blood volume is high, blood pressure rises. This results in
    (a) renin, angiotensin, and aldosterone levels all falling.
    (b) renin, angiotensin, and aldosterone levels all rising.
    (c) renin levels rising, and angiotensin levels falling.
    (d) renin and angiotensin levels falling, but aldosterone levels rising.

12. The kangaroo rat relies mostly on _____ in order to keep a healthy water balance.
    (a) food containing water
    (b) living close to a stream
    (c) oxidative processes
    (d) searching for prey only in humid weather

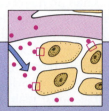

CHAPTER 36

Chemical Communication

Your brain requires a steady, uninterrupted supply of glucose to survive; if the flow of this vital sugar stops—even for a few moments—so does brain activity. Yet you can get along perfectly well by eating just three times a day, and most people can fast for several days with no lasting ill effects (Figure 36.1). How does your body maintain a steady supply of glucose to the brain over a 24-hour period when food intake is restricted to three brief meals a day?

An inquisitive toddler touches the outside of a barbecue grill containing red-hot coals. Almost instantaneously, before he is even aware of what has happened, he jerks his hand away. A second or two later, he feels pain and begins to cry, but long before that he has removed his fingers from danger, preventing more extensive damage to his body. How did that immediate withdrawal reaction occur before the child knew he was in pain?

These are just two examples of the need for coordination and interaction among the trillions of cells that make up your body. In order for you to survive, those cells must coordinate their activities in ways that enable them to serve not only their own needs as individual cells but also your needs as a multicellular organism.

Long-term, sustained responses to environmental conditions—such as the regulation of blood glucose level—require steady, dependable coordination of cellular and organ-level activities over periods ranging from a few minutes to days, weeks, months, and even years. *Short-term, rapid responses to events in the environment*—such as the hand's quick withdrawal from extreme heat—require swift reactions and precise coordination over periods ranging from milliseconds to seconds, minutes, or hours.

THE BODY'S VITAL MESSENGERS

Both the types of responses we have described require that cells or groups of cells somehow affect the actions of other cells, and that, in turn, requires **communication**. The bodies of most multicellular animals, from insects to humans, contain two major organ systems specializing in communication: the **nervous system** and the **endocrine system.**

The nervous system generally handles messages that must be delivered quickly, that generally (though not always) produce rapid responses, and that need not last for very long periods of time. These sorts of messages enable animals to catch prey (or to avoid being preyed

Figure 36.1 Students in China on a protest hunger strike. The brain requires a constant level of glucose in the bloodstream. Chemical messengers within the body make it possible to maintain this level even during fasting.

on) and to detect and respond in a timely fashion to events in the world around them. As we will see when we discuss the nervous system at length in Chapters 40 and 41, a typical nerve cell transmits messages at great speed over long distances by means of *electrochemical impulses*. These messages are carried quickly across the tiny gaps that separate adjacent nerve cells by chemical messengers known as *neurotransmitters*.

The endocrine system generally handles messages that can be delivered more slowly but that must have long-lasting effects. Such messages enable young animals to grow, mature animals to repair damage caused by disease and injury, and all of us to regulate body water content and blood nutrient levels. These messages are carried by chemical signals known as **hormones,** a term derived from a Greek verb that means "I arouse." Because hormones often cause dramatic changes in the cells they affect, and because they can influence the activity of millions of cells at once, their name is quite appropriate. Hormones travel through the bloodstream and can diffuse through body tissues to reach most cells of even large organisms.

The functions of these two communication systems overlap a great deal. In many situations, animals must make long-term changes in physiology and behavior in response to environmental conditions that can be detected only by the nervous system. Birds, for example, may need to prepare for migration as days lengthen or shorten. And when an animal is threatened, it may need to respond immediately, with reactions so swift that they must be mediated through the nervous system. At the same time, however, that animal may need to mobilize its resources for long periods of intense physical exertion to battle intruders or to flee over long distances.

In these circumstances, information is received by the nervous system, acted on by the nervous system, and simultaneously translated into the chemical language of the endocrine system. It should not surprise you, therefore, to learn that there are critical links between the two systems in several places, notably in a region of the brain known as the *hypothalamus*. Once you learn more about the nature and function of the body's chemical messengers, you will also discover that several compounds act both within the nervous system as neurotransmitters and in the body at large as hormones.

THE ENDOCRINE SYSTEM

In 1849, a researcher named A. A. Berthold removed the testes from several immature roosters and noted that the castrated birds failed to develop the comb and wattles normally found in mature males. Furthermore, they showed no interest in female birds. This by itself was no real surprise; it had long been a common practice among farmers to castrate some of their roosters to minimize fights among males scrambling to establish mating territories in the barnyard.

But Berthold took his experiment one step further. He transplanted testes back into the abdominal cavities of castrated birds and found that male plumage and male mating behavior developed in a matter of weeks. Berthold knew that the transplanted testes had no direct connection to any system in the recipient birds. So how had those free-floating gonads had any effect on the birds?

Berthold concluded that the testes were releasing a substance or group of substances into the bloodstream that produced male characteristics and male behavior in the birds. Those substances, of course, were male sex hormones. We now recognize Berthold as the first scientist to demonstrate definitively the activity of an *endocrine gland*.

The Principle of Chemical Signaling

The phenomenon Berthold observed is widespread, not only in the animal kingdom, but in both simple and complex members of other kingdoms as well. The cellular slime mold *Dictyostelium discoidium*, for example, offers a perfect example of primitive chemical communication in action (Figure 36.2). *Dictyostelium* spores, which are scattered on the surface of the soil, give rise to individual ameboid cells that forage independently of one another. When the food supply available to the ameboid cells begins to fail, however, the individual cells aggregate into clusters that take on the appearance of a thick slug (a *pseudoplasmodium*) that moves along the surface of the soil. After moving several centimeters, the pseudoplasmodium stops and the cells form a *fruiting body* that matures to release new spores and begin the cycle again.

This striking convergence of independent cells into a single, organized, multicellular structure requires the coordination of thousands of cells. How is it accomplished? By the release of chemical messages. As their nutritional situation worsens, individual cells send out pulses of a chemical known as **cAMP** (cyclic adenosine monophosphate), which diffuses through the moist soil surface and attracts other cells. As a few cells unite to form a cluster, they begin to release cAMP in coordinated pulses, and still more cells are drawn toward the center of the pulse. These small molecules carry the essential signal that enables the many cells of *Dictyostelium* to act as a single organism. Each cell can produce the chemical message, and each can respond to it. If *Dictyostelium* did not have this ability, it is difficult to see how it could regulate a life cycle involving a multicellular phase.

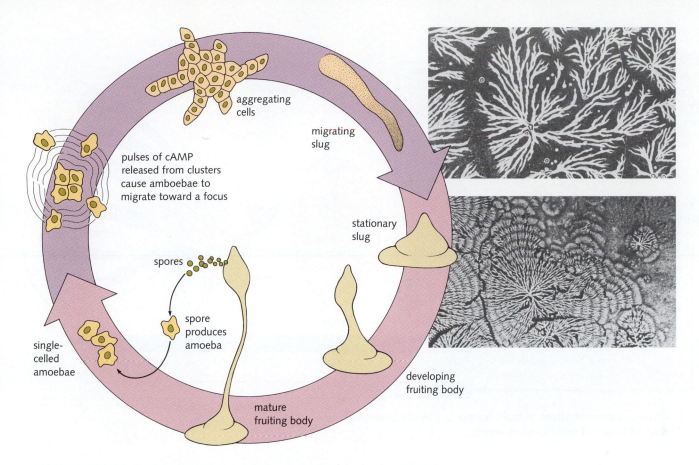

Figure 36.2 (left) The life cycle of *Dictyostelium discoidium,* the cellular slime mold. The aggregation of individual amoeba-like cells to form a migrating slug occurs in response to chemical signals. (right) The mass migration of thousands of *Dictyostelium discoidium* cells forms branch-like patterns leading to clusters where migrating slugs will form.

The simple chemical message used by *Dictyostelium* offers clues to the origin of complicated hormonal control systems. In this slime mold, the message "calls" independent cells together and primes them to act as a single unit. Endocrine systems in more complex organisms provide the communication links necessary to produce coordinated responses in widely separated cells.

Hormonal Control in Insects

Insects, with their complex cycles of growth and larval development, provide both interesting and important examples of hormones in action. As you recall from Chapter 15, insects are covered by a tough exoskeleton that serves several important functions. But this external skeleton also creates a serious problem: Insects cannot grow unless that exoskeleton is periodically discarded and replaced in a process called **molting** or **ecdysis.**

Molting is a complex process that involves coordination of many cells throughout the organism. During a molt, the animal's epidermis partially digests the existing exoskeleton. The insect then splits the weakened shell, pulls its body out, and grows a new, larger exoskeleton to replace the old one. Because the exoskeleton covers virtually every body surface of the organism (including part of its gut), and because molting is accomplished all at once, the process must be precisely orchestrated.

This coordination of metabolic functions among cells scattered across the body is precisely the sort of need that is well served by chemical signals. The specific signal that organizes the molting process is a hormone called **ecdysone,** produced and released by an organ called the *prothoracic gland,* which is located just behind the head of the larva. Early experiments showed that molting could be prevented if this gland were surgically removed. Conversely, extracts prepared from the gland induce molting when injected into other larvae.

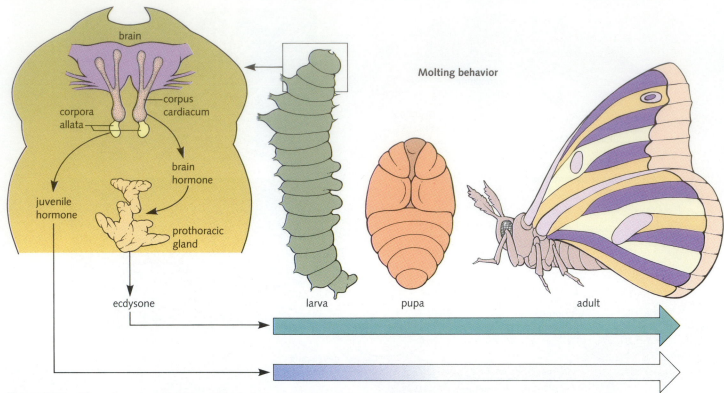

Figure 36.3 Changing concentrations of brain hormone, juvenile hormone, and ecdysone regulate molting and maturation in insects.

The production of ecdysone is controlled, in turn, by a substance known as **brain hormone,** produced and released by a small portion of the brain known as the *corpus cardiacum.* Thus, although ecdysone actually controls the molting process itself, the production of brain hormone determines the timing of the molt.

There is, however, an important aspect of insect development that cannot be explained solely by the system we have described so far. Many insects, such as butterflies, pass through several larval stages, each of which ends in a molt. During the first several molts, the larva simply grows larger. But at some point, the nature of the molt changes, and a *pupa* is formed instead of a larger larva. The development of adult tissues takes place within the pupa, and ultimately an adult insect emerges from it.

If all molts are triggered by ecdysone, what makes the final molt differ from those that precede it? The answer to this question is found in a different region of the brain: the paired structures known as the *corpora allata.* These small organs produce a substance known as **juvenile hormone (JH),** which controls the effect that ecdysone has on the larva. When JH is present during an ecdysone burst, the next stage of development is another larva.

During the final molt, however, JH is not released, and an adult develops (Figure 36.3).

Over time, biologists have learned to exploit this system for help in the control of insect pests. Because reproduction occurs only in the adult stage, any compound that mimics the action of juvenile hormone can prevent the final molt and keep an insect in its larval stage indefinitely. One strategy for controlling manure flies in stables today, for example, is to add JH-like chemicals to the manure. Fly larvae grow in the manure but never molt into mature flies; hence the reproductive cycle is broken and the fly population controlled.

COMPONENTS OF THE ENDOCRINE SYSTEM

The word **endocrine** comes from two Greek words: *endo,* which means "inside," and *krinein,* which means "to divide or separate." As the word implies, endocrine glands release their secretions *inside* the body—in most cases into the bloodstream. At one time endocrine glands were referred to as *ductless glands,* because they do not have ducts or passageways that lead outside the body or into the digestive tract. The term *endocrine* distinguishes these glands from **exocrine glands,** such as sweat glands or digestive glands, that release their products outside the body or into the digestive tract.

The basic structure of the endocrine system is different from that of other organ systems. The human endocrine system, for example, is made up of a series of apparently independent organs found in various places throughout the body. These scattered organs do not have a common developmental origin and do not respond to the same types of stimulation. They are small, specialized glands, each of which produces a particular product or group of products for release. The major human endocrine organs are shown in Figure 36.4. They are the **thyroid,** the **parathyroids,** the **endocrine pancreas** (islets of Langerhans), the **adrenals,** the **thymus,** the **pituitary,** and the **gonads: ovaries** in females, **testes** in males.

At first the isolation of endocrine organs from one another is puzzling: How can widely scattered organs form a single system? But if you keep in mind the fact that the endocrine system functions by means of chemical messages, the separation seems much less problematic. Even though the glands are scattered throughout the body, their ability to synthesize and release hormones permits them to act in concert, to respond to subtle changes in the body, and to maintain the stability of the body's internal environment.

TYPES OF HORMONES

Hormones secreted by endocrine glands are biochemical messages destined for distant targets throughout the body. Hormones' ability to travel through the blood enables them to reach virtually any tissue. It is important to keep in mind, however, that the term *hormones,* like the term *vitamins,* has a functional rather than a chemical definition. Hormones range from compounds as simple as ethylene (in plants) to simple peptides, lipids, and complex glycoproteins.

Ecdysone, for example, is a **steroid hormone.** As shown in Figure 36.5, its chemical structure is similar to that of other steroids, including the important human sex hormones *testosterone* and *estrogen.* Steroids are lipids, and steroid hormones have all the characteristics of lipid molecules, among them the ability to pass easily across the lipid bilayers that form cell membranes.

Many hormones are **proteins** or **polypeptides;** one of these is the brain hormone that regulates ecdysone production in insects. Polypeptide hormones may range in size from a few amino acids to several hundred, and many are complex glycoproteins with molecular weights

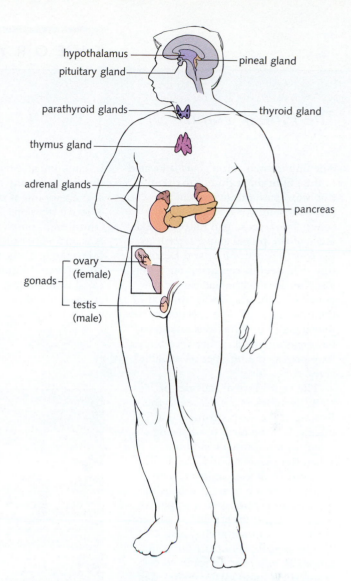

Figure 36.4 The organs of the human endocrine system.

The Pineal Gland

The tiny human *pineal gland* is located in the midbrain region of the skull (the word *pineal* derives from the gland's resemblance to a small pine cone). The Greeks had noted the existence of the pineal gland as early as the fourth century B.C., and Descartes (the seventeenth-century French philosopher and mathematician) suggested that the gland was "the seat of the rational soul." The function of the pineal gland in humans is unknown, but experiments on other vertebrates have provided a number of valuable clues.

The reproductive activities of many vertebrates are geared to the seasons of the year, ensuring that mating and the birth of offspring occur under optimal conditions. When the pineal gland is surgically removed from such animals, their ability to remain in synchrony with the seasons is often lost. The pineal seems to adjust an organism's physiology to environmental changes.

The pineal produces a hormone known as *melatonin*. In some vertebrates (fishes and salamanders), daily surges of melatonin adjust skin color by controlling the dispersal of pigment granules in skin cells, making the animals less readily visible as their environment changes from bright light to total darkness. Melatonin is also produced in mammals, including humans, but it does not regulate skin color in these organisms. Nonetheless, the onset of darkness brings about an increase in melatonin production that lasts until the first light of morning.

Some scientists have found evidence that they believe suggests that the nervous system can "tell" the pineal gland when darkness has begun even while the organism is asleep! Interestingly, although the human pineal is buried deep within the skull, in many reptiles it is located at the very top of the brain case. There it develops as a miniature "third eye," complete with transparent lens and a layer of rudimentary photoreceptors. Thus, at least in these ancient vertebrates, the pineal was capable of sensing the presence and absence of light directly.

Migrating animals, including these geese, must precisely sense the seasons of the year for their movements to be successful.

A number of effects have been suggested for melatonin, and there is some evidence linking its levels to an internal "clock" that regulates daily cycles of sleeping and waking and is responsible for the disorientation known as "jet lag" experienced by travelers who pass through several time zones in just a few hours. Other experiments have suggested that the deep depression many people feel during the darkest days of winter, especially near the polar regions where only five or six hours of daylight occur, results from more than a romantic longing for the rebirth of spring. The diminished hours of daylight may cause such high melatonin levels that normal cycles are disrupted.

The best therapy for such bouts of depression, not surprisingly, may be larger doses of light. A trip to an equatorial region, with its longer days, is one solution; another more reasonably priced alternative seems to be higher levels of artificial lighting during the hours normally spent indoors during dark winter evenings. This may be a solution that goes directly to the source of the problem—the pineal gland.

In mammals with seasonal mating patterns, the changes of season may influence melatonin in a way that makes the gonads shrink during the times of the year when mating is not appropriate. Does melatonin have similar effects in humans? There is no evidence for seasonal variation in human mating patterns. But it may be significant that some studies show that human melatonin levels decline as much as 75 percent during the transition from childhood to early puberty. A signal for sexual maturation? Only time and more research will tell.

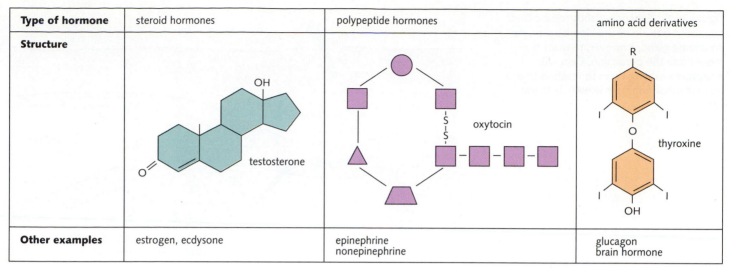

Type of hormone	steroid hormones	polypeptide hormones	amino acid derivatives
Structure	testosterone	oxytocin	thyroxine
Other examples	estrogen, ecdysone	epinephrine nonepinephrine	glucagon brain hormone

Figure 36.5 Chemical structures of several major classes of hormones.

exceeding 20,000. *Oxytocin,* which is released from the pituitary gland, is a representative polypeptide hormone (again, see Figure 36.5).

Another important class of hormones called **amino acid derivatives** consists of chemically modified versions of common amino acids. Among the best-known amino acid derivatives is *thyroxine,* which is synthesized by the thyroid gland from the amino acid tyrosine, as illustrated in Figure 36.5.

THE NATURE OF HORMONE ACTION

In studying hormones and their activities, researchers long ago realized that any theory of hormone action had to explain four intriguing phenomena.

First, hormones exert powerful effects in very small amounts. Even when we say that a hormone is present at "high" levels, the concentration of that hormone within the organism is still usually less than one part per million. Somehow, the effects of relatively few hormone molecules are amplified to cause many far-reaching changes.

Second, a single hormone can have different effects on several different processes in the same cell. For example, the hormone *adrenalin* stimulates the conversion of glycogen to glucose in a cell, while simultaneously slowing down the production of glycogen from glucose.

Third, a single hormone may affect only a single cell type or it may affect several cell types. And a hormone that affects several cell types may affect them in similar

ways or it may affect them differently. *Insulin,* for example, both promotes the conversion of glucose to glycogen in liver cells and stimulates the conversion of glucose to fats in adipose cells.

Fourth, different hormones can affect processes in the same cell differently. Often, one or more hormones act to speed up a particular reaction, whereas other hormones act to slow it down.

The Importance of Receptors

The ability of a particular cell to respond to a hormone depends on whether that cell has a **receptor,** a molecule that binds to the hormone and directs a cellular response to it. Cells containing hormone receptors are referred to as **target cells** for the hormone that affects them. If cells in different tissues carry receptors for a certain hormone, that hormone can affect all those cell types. Cells that lack such receptors, on the other hand, are not influenced by the hormone. This phenomenon explains how hormones can act throughout the body and yet influence only certain groups of cells. It is this ability that enables the endocrine system to regulate a range of body activities (Figure 36.6).

It should not surprise you to learn that hormones as different as simple amino acids and steroids exert their effects through different types of receptors and act in very different ways. Before we examine the effects of several specific hormones and study how they are produced and regulated, it will be helpful to learn how different classes of hormones influence cells.

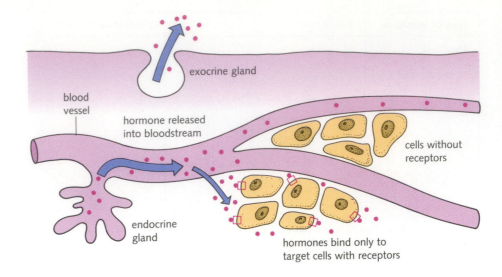

Figure 36.6 **Exocrine** glands release their secretions outside of the bloodstream. An **endocrine** gland, however, releases hormones into the circulation. Cells with **receptors** enabling them to respond to a particular hormone are known as **target cells.**

exocrine gland

blood vessel

hormone released into bloodstream

cells without receptors

endocrine gland

hormones bind only to target cells with receptors

Steroids

As lipids, steroid hormones pass through cellular membranes easily. Once they are inside, the ability of steroid hormones to influence a cell depends on the presence of receptor proteins that are usually found within the nucleus. If a cell contains the appropriate receptor, hormone and receptor bind to each other to form a **hormone–receptor complex,** as shown in Figure 36.7.

The shape of the complex enables it to bind very tightly to specific regions of chromatin, and that is where the hormone exerts its effect. Hormone–receptor complexes can either activate or deactivate specific genes or groups of genes within the nucleus. Note that neither the receptor protein nor the hormone alone possesses this regulatory ability; only the *combination* of hormone and receptor activates or deactivates the genes.

Through this type of direct effect on gene expression, steroid hormones can cause fundamental changes in the molecular biology of target cells. The binding of a single hormone–receptor complex to chromatin may increase by a factor as great as 1000 the rate at which certain mRNA molecules are produced. For example, an increase in mRNA levels occurs in the chicken oviduct following introduction of the steroid hormone estrogen. The effect is also highly specific—only some genes are activated.

Peptides and Their Derivatives

Many other classes of hormones, including polypeptides and amino acid derivatives, do not enter their target cells. Instead they bind to receptor molecules located on the cell surface. These receptors, which form part of the cell membrane, are very specific; in most cases, they allow one, and only one, hormone to bind.

The binding of a hormone molecule at the cell surface triggers a sequence of biochemical events that affects the target cell. Figure 36.8 shows an example of such a sequence of events. The binding of a hormone to its receptor causes it to form a complex with a so-called *"G protein,"* a membrane protein that binds *GTP* (guanosine triphosphate). Together, the G protein and the receptor cause a change in **adenylate cyclase,** an enzyme bound to the inner surface of the cell membrane.

When the hormone-binding site in the receptor is empty, adenylate cyclase is *inactive.* But the binding of the hormone to its receptor switches the enzyme to its active form, and it begins to catalyze a chemical reaction in which cAMP is formed from ATP:

$$ATP \rightarrow cAMP + P\sim P$$

By activating an enzyme such as adenylate cyclase, a single hormone molecule can stimulate the synthesis of several hundred cAMP molecules in a very short period of time. Why is this important? cAMP functions as a **second messenger,** a molecule that carries the signal from the primary messenger (the hormone) into the cell.

Second Messengers as Amplifiers

The power of this system is further amplified because the activation process does not end with the production of cAMP. Many proteins, including certain enzymes called **protein kinases,** are activated by cAMP. Thus each acti-

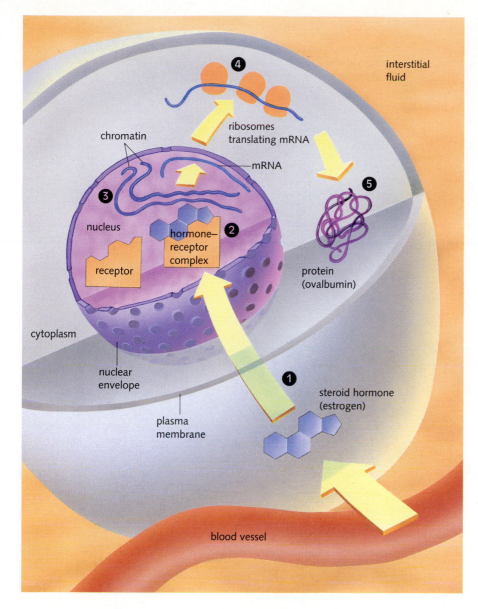

Figure 36.7 Steroid hormones enter their target cells (1) and bind to a protein receptor molecule. This forms a hormone receptor complex (2) that can bind to chromatin (3) and alter the pattern of gene expression (4). In chickens, the steroid hormone estrogen results in a 1000-fold increase in the amount of ovalbumin mRNA (5) produced by cells lining the oviduct.

vated protein kinase molecule can activate hundreds of other enzymes.

Ultimately, a single hormone can set in motion a "cascade" of events that amplifies by several orders of magnitude the message from a single hormone molecule (Figure 36.9). This amplification process is one of the reasons why small amounts of a hormone can cause drastic changes in body chemistry.

At the same time, hormones that participate as intermediates in the cascade can affect other hormones in different ways. Protein kinase, for example, simultaneously activates enzymes that catalyze the conversion of glycogen to glucose and *de*activates other enzymes that

catalyze the conversion of glucose to glycogen. Thus the activation of adenylate cyclase can have different effects on different reactions in the same cell.

cAMP is one of the best-understood second messengers, but it is not the only one. Many hormones cause the release of calcium (Ca^{2+}) into the cell after they bind. The rapid change in calcium concentration then affects a wide variety of cellular functions through a calcium-binding protein called **calmodulin.** Because calmodulin binds to—and therefore influences the properties of—several enzymes and cellular proteins, the activities of those cellular components are indirectly controlled by calcium level.

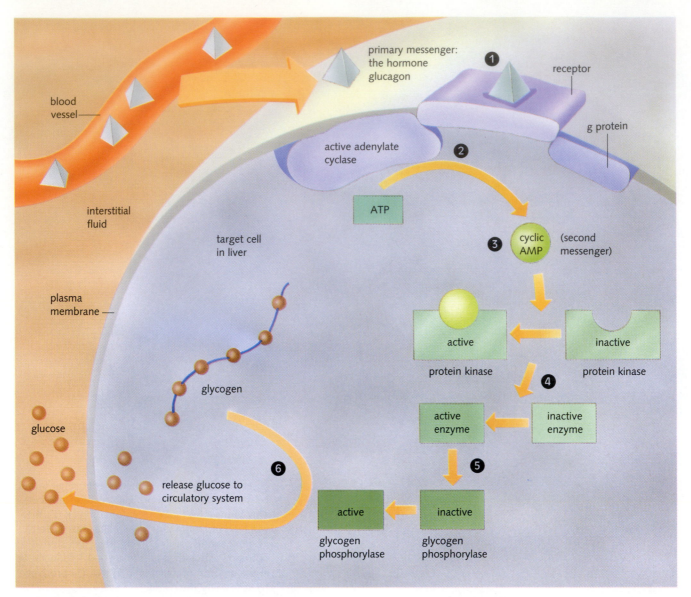

**primary messenger:
the hormone
glucagon**

receptor

blood
vessel

g protein

active adenylate
cyclase

interstitial
fluid

ATP

target cell
in liver

cyclic
AMP

(second
messenger)

plasma
membrane

active

inactive

protein kinase

protein kinase

glycogen

active
enzyme

inactive
enzyme

glucose

active

inactive

release glucose to
circulatory system

glycogen
phosphorylase

glycogen
phosphorylase

Figure 36.8 Peptide and protein hormones, such as glucagon, bind to receptors at the cell surface (1). Hormone binding causes the production of a **second messenger** (2), such as
cyclic AMP (cAMP) (3). cAMP activates protein kinases (4)
which may phosphorylate enzymes (5) that in turn control
the activity of other enzymes. In the case of glucagon, this
"cascade" of events results in the activation of a large number of enzyme molecules that break down the storage carbohydrate glycogen (6) and release glucose into the circulation.

The mechanism by which certain hormones affect
calcium levels has only recently come to be understood.
Hormone binding stimulates a membrane-bound enzyme
(called phospholipase *c*) to cleave **phosphatidyl inositol,**
a lipid found in the cell membrane, into two fragments
(Figure 36.10). One fragment, *inositol triphosphate*, causes
calcium to be released from storage in the endoplasmic
reticulum. The other fragment, *diacylglycerol*, remains in
the membrane and activates a protein kinase that then
activates a series of cellular enzymes by phosphorylation.
A number of hormones, including *antidiuretic hormone*
(ADH) and *serotonin* release calcium as a second messenger by means of phosphatidyl inositol.

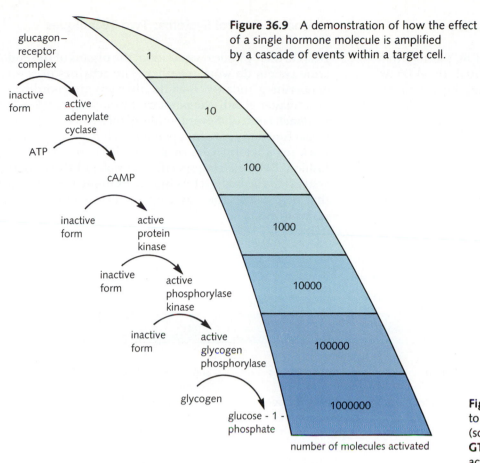

glucagon–receptor complex

inactive form → active adenylate cyclase

ATP

cAMP

inactive form → active protein kinase

inactive form → active phosphorylase kinase

inactive form → active glycogen phosphorylase

glycogen

glucose - 1 - phosphate

1
10
100
1000
10000
100000
1000000

number of molecules activated

Figure 36.9 A demonstration of how the effect of a single hormone molecule is amplified by a cascade of events within a target cell.

Figure 36.10 The binding of some hormones to their receptors (1) activates a "G-protein" (so named because it contains a binding site for **GTP,** guanosine triphosphate). The G-protein activates a membrane-bound phospholipase enzyme (2) that cleaves a membrane lipid, phosphatidylinositol, into two parts: inositol triphosphate (3) and diacylglycerol (4). Each of these compounds may then act as a second messenger (5).

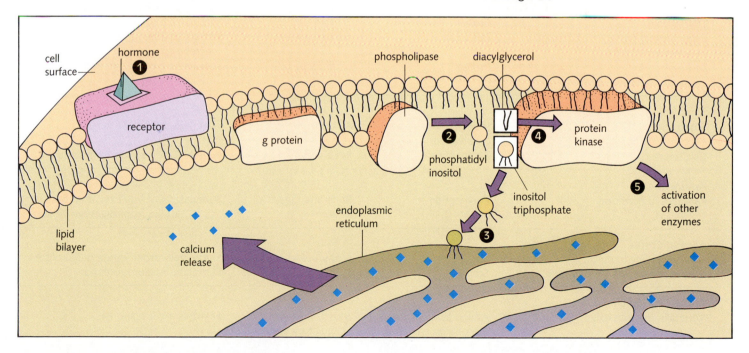

cell surface

hormone ①

receptor

g protein

phospholipase

phosphatidyl inositol

diacylglycerol

④

protein kinase

② ⑤ activation of other enzymes

inositol triphosphate

③

endoplasmic reticulum

lipid bilayer

calcium release

CONTROL OF THE ENDOCRINE SYSTEM

For many years scientists were puzzled by the fact that endocrine glands were so widely separated. In addition, many hormones seem to have overlapping functions, whereas other hormones seem to work in opposition to each other (Table 36.1). For example, the secretions of at least three glands directly affect the uptake and release of sugar: insulin and glucagon from the pancreas, glucocorticosteroids from the adrenal cortex, and adrenalin from the adrenal medulla. If each endocrine gland did, in fact, work independently, coordinated control of body chemistry would be next to impossible.

Hormonal Control Systems: Two Techniques

Despite their physical separation, the organs of the endocrine system do work together. The activities of several endocrine glands (such as the thyroid) are coordinated by a "master gland," the pituitary, through a process called **feedback control.** Several other glands (such as the pancreas) function more independently, but their hormones work together in pairs or groups in a complementary fashion. Before giving specific examples of these control systems in the body, let us briefly examine two analogies that help clarify the way each of these control mechanisms works.

Table 36.1 *The Endocrine Organs and the Hormones They Secrete*

Hormone	Type of Hormone	Target Tissue(s)	Effects
Posterior Pituitary			
Oxytocin	Polypeptide	Breasts, uterus	Stimulates milk release, uterine contractions
Antidiuretic hormone (vasopressin)	Polypeptide	Kidney, blood vessels	Regulates water balance, blood vessel constriction
Anterior Pituitary			
Melanin-stimulating hormone (MSH)	Polypeptide	Pigment cells in skin	Function in human unclear
Follicle-stimulating hormone (FSH)	Protein	Gonads	Stimulates gonads of both males and females; necessary for gamete cell development
Luteinizing hormone (LH)	Protein	Gonads	Stimulates gonads to produce sex hormones (estrogen and testosterone)
Prolactin	Protein	Breasts	Stimulates milk production
Thyrotropin	Protein	Thyroid	Stimulates thyroid activity
Corticotropin (adrenocorticotropic hormone: ACTH)	Polypeptide	Adrenal cortex	Stimulates adrenal cortex to release secretions
Somatotropin (growth hormone: GH)	Protein	All growing cells	Stimulates growth, increases metabolic rate
Thyroid			
Thyroxine	Iodinated amino acid derivative	All cells	Stimulates metabolic activity
Calcitonin	Polypeptide	Bone cells	Inhibits Ca^{2+} release from bone
Parathyroid			
Parathyroid hormone	Polypeptide	Bone and digestive tract	Stimulates Ca^{2+} release from bone and Ca^{2+} uptake from digestive system
Adrenal Cortex			
Corticosteroids (cortisol)	Steroid	Many tissues	Stimulates carbohydrate metabolism
Mineralocorticoids (aldosterone)	Steroid	Kidney and blood	Stimulates sodium retention
Adrenal Medulla			
Adrenalin, noradrenalin	Amino acid derivative (modified tyrosine)	Muscles, liver, circulatory system	Flight-or-fight response

Negative feedback loops Negative feedback control operates in much the same way as a thermostat in a home heating system. In such a system, when the temperature falls below a certain level, the thermostat turns on the heat. Then, when the system has raised the temperature above a certain level, the thermostat turns the heat off. Depending on the design of both the heating system and the thermostat, temperature may fluctuate a bit during the regulatory cycle, but it neither rises too high nor falls too low.

Later in this chapter you will see how a negative feedback system involving the pituitary gland controls blood concentrations of thyroid hormone. Then, in Chapter 37, you will see how a more complex version of the same system regulates the hormones of the mammalian reproductive system.

Complementary hormone action As Table 36.1 shows, two or more hormones often influence the same biochemical pathways in opposite ways. Such "opposing" actions regulate body chemistry in much the same way in which the brake and accelerator pedals in a car control its speed. Theoretically, a good driver can control the speed of a car on an open highway by using only the accelerator. (Especially if you drive a standard shift, you can, in fact, get quite proficient at this.) But in the more

Hormone	Type of Hormone	Target Tissue(s)	Effects
Pancreas			
Glucagon	Polypeptide	Liver and other cells	Stimulates glycogen breakdown, increases blood sugar
Insulin	Polypeptide	Liver and other cells	Stimulates glycogen synthesis, decreases blood sugar
Pineal			
Melatonin	Amino acid derivative	Nervous system, gonads	Regulates light/dark cycles, reproduction function in humans unclear
Ovaries			
Estrogens	Steroid	Cells throughout body	Female sexual characteristics, development of uterine lining
Progesterone	Steroid	Uterus	Growth of uterine lining
Testes			
Testosterone	Steroid	Cells throughout body	Male sexual characteristics, sperm development
Thymus			
Thymosin	Polypeptide	T-lymphocytes	Stimulates growth and development of T-lymphocytes
Digestive (GI) Tract			
Gastrin	Polypeptide	Stomach lining	Stimulates HCl production
Cholecystokinin	Polypeptide	Pancreas	Stimulates release of digestive enzymes and bicarbonate
Secretin	Polypeptide	Pancreas	Stimulates release of digestive enzymes and bicarbonate
Enterogastrone	Polypeptide	Stomach lining	Inhibits smooth-muscle contraction and HCl release
Kidney			
Erythropoietin	Glycoprotein	Bone marrow	Stimulates production of red blood cells
Heart			
Atrial Natriuretic Hormone	Polypeptide	Kidneys	Stimulates removal of sodium from blood

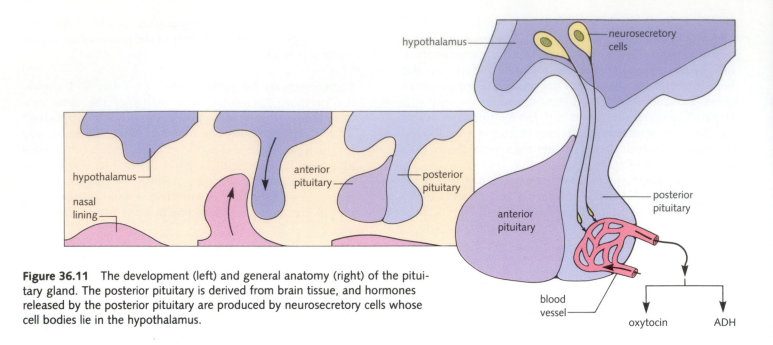

Figure 36.11 The development (left) and general anatomy (right) of the pituitary gland. The posterior pituitary is derived from brain tissue, and hormones released by the posterior pituitary are produced by neurosecretory cells whose cell bodies lie in the hypothalamus.

complex maneuvers of driving around town, even a good driver relying solely on the accelerator would quickly get into trouble; there are too many situations that require the opposing action of the brake to slow things down.

In much the same way, several endocrine regulatory functions rely not only on the negative feedback control of a single hormone, but also on the complementary effects of two opposing hormones. This sort of control system regulates levels of both calcium and glucose in the bloodstream.

The Pituitary Gland

The **pituitary gland** is a small structure about the size of a large bean located beneath a region of the brain known as the **hypothalamus.** It is found near the center of the skull, in one of the most protected locations in the body. Interestingly, the pituitary does not develop as a single structure during embryological development. Instead, its *anterior* and *posterior* portions are assembled from two tissues of completely different origin (Figure 36.11). As befits their separate origins, the two parts of the pituitary function as completely independent glands and work in distinctly different ways.

Each part of the pituitary, however, is closely tied to specific regions within the hypothalamus. Because of these connections, we now know that the pituitary does not function by itself as the supreme regulatory gland it was

once thought to be. Instead, the hypothalamus and pituitary together provide vital links between the nervous system and other endocrine glands.

The Posterior Pituitary and Its Hormones

The **posterior pituitary** grows down from the base of the brain as an extension of the hypothalamus and consists, in large part, of specialized **neurosecretory cells.** As their name implies, these cells act both as nerve cells and as hormone-producing cells. Their cell bodies—where hormones are produced—are located within the hypothalamus. From these cell bodies, long processes run several centimeters through the stalk that attaches the pituitary to the brain. Those cell processes end in a network of branches that are wrapped in a dense network of fine capillaries.

Because these cells are "rooted" in the brain, they are under direct control of the nervous system. Hormones produced in the cell bodies of these cells are transported down through the cell processes into their terminal branches. When the appropriate region of the hypothalamus is stimulated, the neurosecretory cells release these hormones into the capillary network, which carries them into the general body circulation.

The posterior pituitary produces two polypeptide hormones: **antidiuretic hormone (ADH)** and **oxytocin.** ADH affects the kidneys, where it stimulates water

The Heart Is an Endocrine Gland Too

Not very long ago, it was possible to summarize the workings of the endocrine system with a list of seven or eight glands and a dozen or so important hormones that they produced. In recent years, however, new evidence has forced us to alter this simple description of the endocrine system. Not only has the discovery of prostaglandins shown that tissues throughout the body are capable of producing hormones with powerful effects, but recent research suggests that many organs that nominally are part of other systems also have endocrine functions.

The heart, an organ that is often assigned the purely mechanical role of pumping blood for the circulatory system, is one of the latest organs shown to have an endocrine function. Its two atria filling with blood on every stroke, the heart is in an ideal position to "sense" the volume of blood in the circulatory system. When too much blood swells the atria, the blood volume should be reduced; when too little blood rushes in, the volume of blood should be increased. The endocrine function of the heart brings these changes about.

Tiny granules are found in certain cells within the atria. These granules contain a polypeptide hormone that is released when the walls of the atria are stretched, either artificially or by an unusually large volume of blood. This polypeptide hormone, called atrial natriuretic factor, or *ANF* (*natriuretic* means "sodium-eliminating"), acts on the kidneys to stimulate the removal of sodium from the blood. Water follows the flow of sodium, and the blood volume is decreased. The hormone also acts on the smooth muscle cells that line many blood vessels, helping to lower blood pressure, and it affects receptors in the hypothalamus, inhibiting the release of ADH (antidiuretic hormone). This remarkable system not only brings to light a new hormone produced by a "nonendocrine" organ but also suggests that many other endocrine mechanisms have yet to be discovered lurking in systems and organs.

(Based on M. Cantin and J. Genest, February 1986, The heart is an endocrine gland, *Scientific American*, 254:76 – 81.)

resorption. (A *diuretic* is anything that stimulates water loss through the kidneys, resulting in increased urine production. Because ADH does the opposite, preventing water loss, it is known as an *antidiuretic*.) As we saw in Chapter 35 (excretion), when blood becomes too dilute, the hypothalamus responds by not releasing ADH, allowing the kidneys to remove water from the blood. When the blood becomes too concentrated, the hypothalamus causes the release of ADH from the posterior pituitary, preventing further water loss. This self-regulating system forms a classic feedback loop that regulates the concentration of water in the blood with great precision.

Oxytocin, the other posterior pituitary hormone, has two important effects in women. It stimulates *labor*, the smooth-muscle contractions in the uterus that lead to childbirth. Physicians who feel that they must "induce" labor for one reason or another administer injections of oxytocin. Oxytocin is also important after childbirth, when its release causes milk to flow toward the nipples in the mammary glands. This "letdown reflex" is familiar to any woman who has nursed a baby.

The Anterior Pituitary and Its Hormones

The **anterior pituitary,** which forms from the lining of the mouth, grows up to meet the posterior pituitary as it descends from the base of the brain. Because of this difference in developmental origin, the anterior pituitary does not have the same sort of direct neural connection with the hypothalamus that the posterior portion has. Researchers tried to find such a direct nerve link for many years but found none. Yet there was abundant evidence of some functional link with the brain, because electrical stimulation of the hypothalamus caused the release of several hormones from the anterior pituitary.

Endocrinologists found a clue to the operation of this system in a pattern of blood circulation around the pituitary. A tiny arteriole circulates blood through a dense bed of capillaries that make intimate connections with a small section of the hypothalamus. These capillaries flow together to form a larger vessel, but then, instead of leaving the brain immediately, divide again to form a second network of capillaries that surround and penetrate the anterior pituitary. This is a most unusual pattern, and it suggested an unusual sort of connection between the hypothalamus and the pituitary.

Releasing hormones We now know that a group of neurosecretory cells in the hypothalamus produces small molecules (some are merely tripeptides) known as **releasing hormones.** Under appropriate stimulation from the brain, neurosecretory cells discharge hormones into the surrounding capillary network from which they are carried directly to the second capillary bed in the anterior pituitary. There these chemical messengers bind to receptors on anterior pituitary cells, where they regulate the release of other hormones (Figure 36.12).

Anterior pituitary hormones The anterior pituitary produces seven of the most important hormones in the body. Several of these are known as **tropic hormones** because their target organs are several other endocrine glands. A thorough understanding of these hormones requires knowledge of their target glands as well, so we will return later in this chapter, and again in Chapter 37, to discuss in detail their participation in the regulatory process.

Thyrotropic hormone (known as **TSH** because it is also called thyroid-stimulating hormone), is a glycoprotein hormone with target cells only in the thyroid gland itself. When TSH binds to receptors in the thyroid, it increases the release of thyroxine, the major thyroid hormone.

Figure 36.12 Hormones are released from the anterior pituitary in response to **releasing hormones** produced in the hypothalamus.

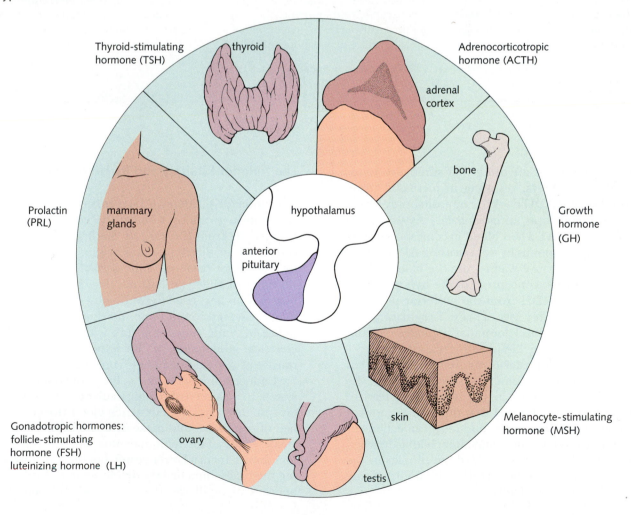

Follicle-stimulating hormone (FSH) and **luteinizing hormone (LH)** are called **gonadotropins** because their main targets are the gonads. These glycoprotein hormones are named for their effects in the female, although the same two hormones are present in males, where they perform slightly different functions.

Adrenocorticotropic hormone (ACTH) controls the activity of the adrenal cortex. This hormone, a peptide consisting of 39 amino acids, seems to be released in response to the level of circulating *corticosteroids* in the blood. The other part of the adrenal gland, the adrenal medulla, is not directly affected by any of the pituitary hormones. Instead, signals for the release of adrenalin from the medulla come directly from the nervous system.

Prolactin (PRL), as its name implies, affects the development of the breasts in females and the production of milk. When levels of prolactin drop too low, milk formation in the breasts stops. Males also produce prolactin, but it is not clear whether the hormone plays any role in the male.

Growth hormone (GH), or somatotropin, has target cells throughout the body. This small protein (191 amino acids) exerts powerful effects on cell growth. The hormone is *anabolic*—that is, it switches cellular metabolism in favor of reactions that *build up* larger molecules such as proteins and complex carbohydrates. The levels of growth hormone control the rate at which the body increases in size, and its effects on the growth of the skeletal system help determine the ultimate size to which an individual grows. Abnormally low levels of GH produce midgets; abnormally high levels of the hormone produce "pituitary giants" (Figure 36.13).

Melanocyte-stimulating hormone (MSH) is also produced by the anterior pituitary. In lower vertebrates this hormone causes an increase in skin pigmentation. Its function in humans is not clear.

MAJOR ENDOCRINE GLANDS AND THEIR HORMONES

The Thyroid

The thyroid gland produces several hormones, but the most important of these is **thyroxine** (Figure 36.14). Thyroxine is synthesized from the amino acid *tyrosine* in a reaction that requires *iodine*. This is the main reason why iodine is needed in the diet. Thyroid cells contain a powerful active-transport mechanism that can pick up trace amounts of iodine from the bloodstream and concentrate it in specialized storage proteins.

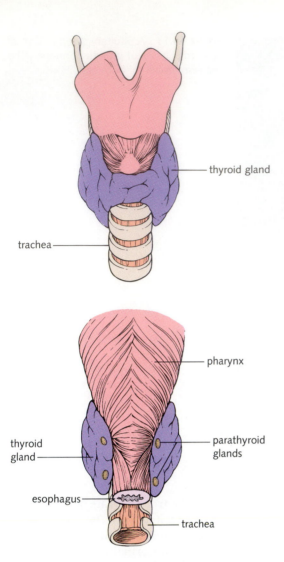

Figure 36.14 The thyroid and parathyroid glands.

Nearly every cell in the body is a target cell for thyroxine. This hormone helps to regulate metabolic rate—the rate at which cells use food and oxygen—and also the rate at which they grow. Increased levels of thyroxine stimulate greater metabolic activity; lower levels depress the metabolic rate.

Maintaining the right level of thyroxine is a critical job. When levels rise abnormally high, a syndrome known as **hyperthyroidism** results, with symptoms that include irritability, weight loss, high blood pressure and increased pulse rate, and bulging eyes (caused by an increase in pressure of the fluid within the eye). When too little thyroxine is produced, the result is **hypothyroidism:** lowered heart rate and blood pressure, sleepiness, loss of energy, and weight gain.

Hypothyroidism is often caused by a lack of iodine in the diet. As the thyroid tries to remedy this by increasing in size and attempting to extract even more iodine from the bloodstream, a noticeable swelling of the gland results, a condition known as *goiter*. The cure for this condition is to increase the amount of iodine in the diet, which causes the gland to return to normal size. Enlarged thyroids are not uncommon today in the inland areas of developing countries where seafood, which tends to be rich in iodine, is not available. Goiter was once common in parts of the American midwest until it became standard practice to add small amounts of iodine to table salt.

Underactivity of the thyroid gland early in life can produce a form of dwarfism. A deficiency in thyroxine lowers metabolic rates in the cartilage tissue that forms the ends of growing bones, and a permanent decrease in stature results. Thyroid dwarves also suffer from mental retardation caused by the low metabolic rate of the brain as it develops.

The thyroid produces another major homeostatic hormone known as *calcitonin,* which affects calcium levels in the blood. Calcitonin does not control calcium levels alone but rather works together with another calcium-regulating hormone produced by the parathyroid glands and discussed in more detail later in this chapter.

Feedback control of thyroxine levels The medical conditions caused by too much or too little thyroxine emphasize an important point that applies to other hormones as well: Regulating hormone levels is critical to normal growth, development, and body maintenance. We can now examine the control of thyroxine as an example of the way negative feedback makes possible this sort of precise regulation.

The feedback control system that regulates thyroxine concentration consists of three parts: the thyroid itself, the hypothalamus, and the anterior pituitary (Figure 36.15). As we have said, most cells in the body are target cells for thyroxine. In the vast majority of these cells, thyroxine stimulates an increase in metabolic activity. But thyroxine also *inhibits* cells in both the hypothalamus and the anterior pituitary. Thus when the blood thyroxine level drops, metabolic activity throughout the body slows a bit. The hypothalamus responds to this drop in activity and also to the lowered level of thyroxine by producing **thyroid releasing hormone (TRH).** This TRH is carried through the special capillary network we described earlier to the anterior pituitary, where it causes the release of thyroid-stimulating hormone (TSH). TSH, in turn, stimulates the thyroid to release thyroxine, raising the level of thyroxine in the bloodstream and increasing metabolic activity throughout the body.

But because thyroxine inhibits both the hypothalamus and the anterior pituitary, an increase in thyroxine concentration causes the hypothalamus to decrease its production of TRH, thus decreasing TSH release by the anterior pituitary and ultimately reducing the further release of thyroxine. This elegant self-regulating system ensures that the level of thyroxine is appropriate to maintain normal metabolic activity.

The Parathyroids

The **parathyroids** are four tiny glands embedded in the posterior surface of the thyroid. One might expect that these glands are related to the major function of the thyroids, and medical workers believed this for years, but this notion has proved to be mistaken. These glands produce a hormone known as **parathyroid hormone (PTH),** a polypeptide containing 83 amino acids.

PTH acts to increase blood calcium concentrations by stimulating the release of calcium from several sources. The major calcium reservoirs are the bones, where deposits of calcium phosphate make up the greatest portion of the tissue. Increased PTH levels cause calcium to be released from this reservoir into the bloodstream. PTH also converts vitamin D into its active form, which increases the rate at which calcium ions are absorbed from the intestines.

The effects of PTH are counterbalanced by **calcitonin,** a hormone produced by the thyroid, in an example of the second type of hormonal control mechanism discussed above. Calcitonin's effects are precisely the opposite of those of PTH. Calcitonin *reduces* blood calcium levels by preventing calcium removal from the bones. As the other cells of the body take up calcium, blood levels of the ion drop.

The release of both calcium-regulating hormones is controlled directly by the glands that produce them. Each gland responds directly and independently to calcium levels in the blood. The regulated release of these two complementary hormones maintains precise control over the level of calcium in the blood. Precise control is important, because several important cell types require steady calcium levels for their metabolic activities. Nerve cells are particularly sensitive to calcium levels, and muscle contraction and blood clotting require calcium ions. Diseases that result in the overproduction of PTH (or the underproduction of calcitonin) can severely weaken the skeleton by causing a chronic loss of calcium.

The Pancreas and Control of Blood Glucose

The **pancreas** is a large gland located in the abdominal cavity between the stomach and the duodenum (Figure 36.16). Most of the pancreas is an *exocrine gland* that produces digestive enzymes and releases them through ducts that empty directly into the digestive tract. Scattered

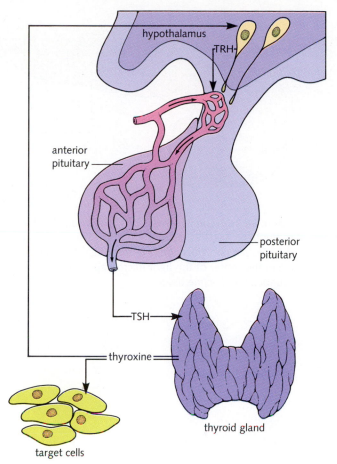

Figure 36.15 Thyroid activity is regulated by a feedback loop involving the hypothalamus and the anterior pituitary.

throughout the pancreas, however, are small clusters of *endocrine* cells known as **islets** of the pancreas. Islet tissue contains several different types of cells and produces a number of important hormones. The best known of these are two important hormones that control the level of glucose in the bloodstream, **insulin** and **glucagon.**

To appreciate fully the complex and complementary functions of these hormones, recall for a moment the design and activity of the digestive tract. Remember that after a large meal, blood leaving the intestines carries large amounts of glucose that fuel the metabolism of all the body's cells from neurons to skin cells. Between meals, on the other hand, the intestine adds little or no glucose to the blood passing through it, although the body's other cells keep utilizing that vital energy source. Thus if there were no "sugar buffering system" in the body, blood glucose levels would rise and fall cyclically (Figure 36.17), with the sort of catastrophic results seen in severe cases of diabetes.

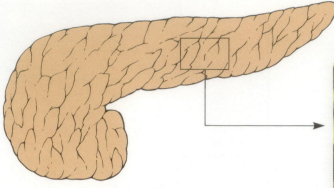

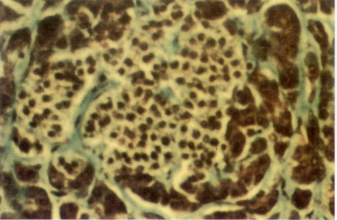

Figure 36.16 The pancreas (left). The endocrine tissues of the pancreas, located in the Islets of Langerhans, produce insulin and glucagon. An Islet of Langerhans is visible in the center of this light micrograph (right), surrounded by exocrine tissue.

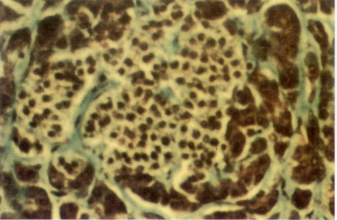

But there *are* tissues that can serve as "sugar buffers." After meals, the liver can remove glucose from the blood and convert it into *glycogen*, a polymer that may contain thousands of glucose molecules that can be stored for future use. The liver can also use excess glucose to produce lipids that are released into the bloodstream and stored by fat cells. Muscle cells use glucose to fuel their contractions, and a wide variety of other cells use glucose as an energy source. Between meals, both liver and adipose cells reverse their activity, converting glycogen and fat, respectively, back into glucose. As a result, glucose concentrations in the blood remain quite stable.

These complementary activities (and several related functions) are controlled by the complementary actions of insulin and glucagon. Not surprisingly, these hormones have several important targets—the liver, muscle tissue, and adipose cells.

Insulin, produced by islet cells known as *beta cells*, is released whenever a large amount of sugar is absorbed into the bloodstream. In short order, insulin stimulates the uptake of sugar by all three types of target cells. In both liver and muscle cells, insulin-stimulated cells allow sugar molecules to pass across their cell membranes, convert them into glycogen, and store them. In adipose cells, insulin increases the storage of fat. Insulin also stimulates the production of proteins and fats, and it simultaneously inhibits their metabolic breakdown, thus increasing growth in many cell types. Releasing insulin at just the right time and in just the right amount also prevents extra sugar from being lost through the kidneys. All these activities, in concert, ensure

Figure 36.17 Blood leaving the intestines after a meal is loaded with sugar from food absorption. Between meals, blood sugar tends to fall as cells throughout the body use the sugar for energy.

that the valuable chemical energy in sugar is either utilized or stored for future use.

Glucagon, a hormone produced by the *alpha cells* of the islets, is released when blood sugar levels begin to drop. This hormone causes liver cells to break down glycogen and release the resulting glucose into the bloodstream. Thus a proper balance of insulin and glucagon, released at the appropriate times, keeps blood sugar levels relatively constant and makes certain that sugar is effectively used and stored by the body.

Diabetes Unfortunately, we have direct evidence of the importance of this balance: Important medical complications occur when it fails. By age 65, nearly 3 Americans out of every 100 suffer from a metabolic disease known as **diabetes mellitus.** Although diabetes most often develops later in life, the tendency to develop the disease seems to be inherited, and the most severe form of the disorder, *juvenile onset diabetes,* appears at puberty.

Most forms of diabetes are caused by the failure of the pancreas to produce enough insulin on cue, although there are rare forms caused by the inability of the body to respond to the insulin molecule. Either of these conditions spells trouble. When a diabetic eats a meal, sugar is not taken from the blood into the liver and other tissues as quickly as it should be. The kidneys remove much of the excess sugar, and a great deal of water along with it. This produces one of the principal symptoms of diabetes: production of a large amount of sugar-containing urine. The Greeks noticed that bees would gather around the urine of diabetics; hence the name *diabetes mellitus* (from Greek words meaning "passing-through" and "honey"). Even today, a high sugar level in the urine is often the first symptom of diabetes noticed by physicians.

In addition to surges of blood sugar after eating, diabetics can experience dangerously low blood sugar levels between meals, because their tissues have not built up stores of sugar. Some cells and tissues suffer local episodes of malnutrition as a result, and damage to parts of the body can accumulate over many years. Muscle cells, for example, are nearly impermeable to glucose in the absence of insulin, so muscle wasting and weakness are typical symptoms of the disease.

Fortunately, most diabetics are able to make some insulin, so some cases of diabetes are relatively mild. In most cases, *adult onset diabetes* can be controlled by a careful diet that avoids sudden surges of sugar into the blood. In more severe cases, the only effective therapy is the administration of carefully controlled doses of insulin. This enables most diabetics to live quite normally, but it is not a perfect solution. It is impossible by this means to fine-tune the level of blood sugar the way a healthy pancreas does, and fluctuations of therapeutic insulin can cause damage to several tissues.

The Adrenal Glands

The **adrenal glands** are paired organs that sit atop the kidneys (the name *adrenal* means "next to the kidney"). Each adrenal can be thought of as two separate endocrine glands: The outer part of the gland, the **adrenal cortex,** is made up of tissue with a completely different function from that of the inner portion, the **adrenal medulla** (Figure 36.18).

The adrenal cortex The adrenal cortex makes up about three-quarters of the total mass of each adrenal gland. The cells of the cortex produce a wide variety of steroid hormones essential to normal body function. For this reason, removal of the adrenal cortex is fatal. These hormones, known as **corticosteroids,** are synthesized from cholesterol, and they fall into three broad classes: **glucocorticoids, mineralocorticoids,** and **sex hormones,** which include testosterone and estrogens.

Mineralocorticoids affect salt transport in the kidneys. In the absence of these hormones, the kidneys excrete dangerously high amounts of sodium.

Glucocorticoids regulate certain reactions to stress and infection, such as inflammation and wound healing, and also elevate blood sugar levels in ways similar (although not identical) to glucagon. **Cortisol,** the most important glucocorticoid present in humans, activates chemical

Figure 36.18 Location and internal structure of an adrenal gland.

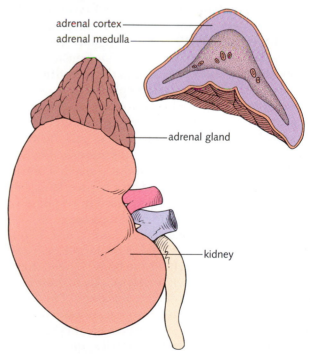

adrenal cortex

adrenal medulla

adrenal gland

kidney

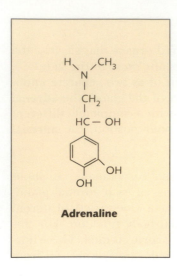

Figure 36.19 The chemical structure of adrenalin, which is synthesized from the amino acid tyrosine.

Adrenaline

pathways in adipose and muscle cells that produce glucose from fats and proteins. By complementing the effects of insulin in these tissues, cortisol makes more glucose available as a source of ready energy. Stress increases the release of cortisol and other glucocorticoids from the adrenals. During periods of difficulty, when we experience physical danger or psychological stress, the release of cortisol makes extra energy available. Cortisol also suppresses some functions of the immune system, including the inflammatory response. This may be one of the reasons why prolonged stress increases our susceptibility to disease.

The adrenal medulla The adrenal medulla does not produce steroid hormones. It produces a series of hormones by chemically modifying the amino acid tyrosine. The best known of these are the compounds **adrenalin** and **noradrenalin** (Figure 36.19). The word *adrenalin* comes from the same Latin roots as *adrenal* and means "near the kidney." (Greek also has its fans in the scientific community; hence the synonyms **epinephrine** and **norepinephrine.**) Historically, scientists working on the endocrine system have used the term *adrenalin,* whereas scientists working on the nervous system call the same compound epinephrine. Adrenalin is synthesized by the cells of the medulla and stored until it is needed. Small amounts of adrenalin are released on a regular basis, but a flood of the hormone can be produced in response to signals to the adrenals from the nervous system.

Why should the nervous system have a direct connection to the adrenal medulla? The target cells for adrenalin are found throughout the body. When adrenalin enters the bloodstream, it travels quickly to these cells and causes a number of immediate reactions. The heart rate increases, blood pressure rises, sugar reserves are released from the liver, blood is diverted from the intestine and directed toward skeletal muscles, the pupils of the eyes widen, and metabolic rates throughout the body increase. These changes prepare the body for vigorous physical activity. They are part of what is often called the *"flight-or-fight"* response (Figure 36.20).

The nervous system triggers the release of adrenalin in response to fear or anger, and the set of responses that adrenalin brings about prepares the body to act. The pallid skin and rapid pulse you may feel in stressful situations are a result of this response. Adrenalin release may be thought of as a survival mechanism. It is one of the reasons why the strength and determination of a trapped wild animal are to be feared and one of the reasons why humans are often capable of feats of great strength and endurance in life-or-death situations.

The Gonads

The **gonads,** ovaries in females and testes in males, produce large amounts of steroid hormones. The most common of these are the **estrogens** in females and **testosterone** in males. These sex hormones are responsible for regulating the reproductive process and producing the sexual characteristics that distinguish males from females. Because they are closely related to the reproductive process, it is best to discuss their functions in conjunction with the reproductive activities they control. We will therefore explore these hormones fully in Chapter 37.

The Thymus

The **thymus** is a gland located near the center of the chest. It reaches its maximum size about the time of puberty and becomes significantly smaller with age. The gland is generally considered part of the immune system and is important in the maturation of *lymphocytes,* specialized white blood cells that help to fight infections. It is included in a listing of the endocrine glands because it produces **thymosin,** a polypeptide hormone, which stimulates the development of certain lymphocytes.

Other Endocrine Glands

A number of organs that are usually *not* associated with the endocrine system produce hormones. The kidneys, for example, produce **erythropoietin,** a hormone that stimulates the development of erythrocytes (red blood cells) in the bone marrow. In Chapter 34 we saw how a series of hormones were produced by cells in the digestive system, helping to regulate the release of digestive enzymes. Recent studies have shown that the heart is also an endocrine organ (see Theory in Action: "The Heart Is an Endocrine Gland Too," p. 713).

Figure 36.20 Adrenalin release is part of the "fight-or-flight" response, which plays a role in the actions of predator and prey.

Prostaglandins

The definition of a hormone is a fairly general one, and it leaves open the possibility that other types of chemical messengers may be discovered as we learn more about individual organs and tissues. More than 50 years ago, a group of researchers discovered that a component of semen was able to produce contractions in the uterine muscles of the female. Gradually this substance was purified and shown to be a chemical derivative of fatty acids (Figure 36.21). Assuming that the material was contributed by the prostate gland (one of the glands that helps to produce the fluid portion of semen), they named the substance **prostaglandin.** This name has persisted, but it is a misnomer: The prostaglandins found in semen are actually synthesized in the *seminal vesicle*. Prostaglandins produced in the lining of the uterus during childbirth stimulate the smooth muscle contractions that occur during labor.

After the first group of prostaglandins in semen had been characterized, researchers began to look elsewhere for these hormones, and prostaglandins were discovered in great numbers. The list of tissues and organs that produce them has grown steadily, and it is now clear that prostaglandins can raise or lower blood pressure, cause the air passageways leading to the lungs to expand or contract, control several aspects of the immune response to foreign organisms and substances, and even regulate the level of pain sensed by the nervous system. Many of these effects are recent discoveries, and there is great hope that prostaglandins will be useful in treating a number of difficult medical conditions.

Though all of the data are not yet in, it appears that prostaglandins exert their effects on target cells by binding to a specific receptor molecule and producing a second messenger within the cell. Until prostaglandins were discovered, the mechanism by which aspirin (the most commonly used drug in many Western countries) works had been a mystery. There is good evidence that aspirin is a strong inhibitor of prostaglandin synthesis, and this action may help to dampen the threshold for pain sensation. It also explains the anti-inflammatory action of aspirin and the ability of aspirin to reduce the tendency of blood to clot. Both of these reactions involve prostaglandin hormones.

Figure 36.21 The structure of a representative prostaglandin: PGE2.

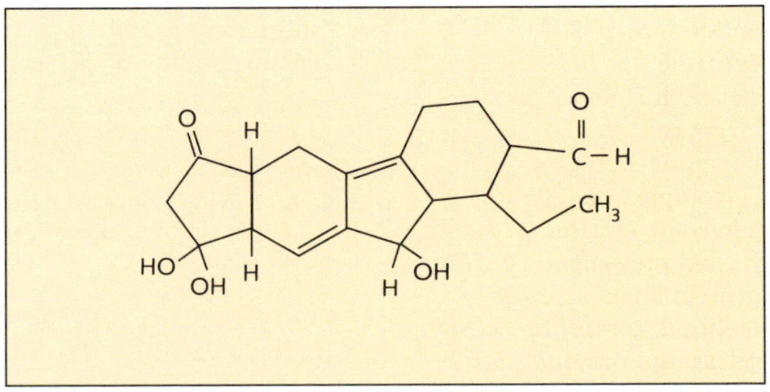

Cellular communication is mediated by two major systems, the nervous system and the endocrine system. Messages sent by the endocrine system are carried in the bloodstream by chemical messengers called hormones. Target cells that possess specific receptor molecules respond to hormone messages. The ability of cells to send and respond to chemical messages is found in many organisms, even unicellular forms such as the slime mold. A hormonal system that we analyzed as being representative coordinates the molting process in insects by means of a hormone known as ecdysone.

The major chemical classes of hormones include steroids, proteins, or polypeptides, and amino acid derivatives. Steroid hormones affect their target cells by binding to protein receptor molecules, forming a hormone–receptor complex within the cell nucleus. Steroid hormone–receptor complexes bind to specific sites in chromosomal DNA, where they activate specific genes and change the cellular pattern of RNA transcription. Other classes of hormones bind to receptors at the cell membrane and cause a system of proteins in the membrane to produce second messengers.

The major organs of the human endocrine system are the thyroid, parathyroids, pancreas, adrenals, gonads, thymus, and pituitary. The hormones produced by these glands affect a variety of cellular functions in tissues throughout the body, regulating blood sugar levels, metabolic activity, and growth processes. The pituitary, which consists of anterior and posterior portions, regulates several tissues by means of its hormone products, and it also controls the secretions of other glands via a series of trophic hormones. The release of trophic hormones is further controlled by releasing hormones produced in the hypothalamus. A final class of hormones are the prostaglandins, which are produced by cells throughout the body and have powerful local and long-range effects.

SELECTED TERMS

hormones *p. 700*
cAMP *p. 700*
ecdysone *p. 701*
juvenile hormone *p. 702*
endocrine *p. 702*
exocrine *p. 703*
steroid hormone *p. 703*
target cell *p. 705*
hormone–receptor
  complex *p. 706*
adenylate cyclase *p. 706*
second messenger *p. 706*
protein kinase *p. 706*
feedback control *p. 710*
releasing hormone *p. 714*

trophic hormones *p. 714*
insulin *p. 717*
glucagon *p. 717*
diabetes mellitus *p. 719*
adrenalin *p. 720*
prostaglandin *p. 721*

After studying this chapter, you should be able to:

■ Explain the concept of chemical messengers and describe the roles they play in integrating the activities of an organism.

■ Describe the major organs and hormones of the human endocrine system.

■ Give and explain specific examples of hormone action.

■ Examine the modes of hormone action at the cellular level and appreciate the diversity of hormone effects.

REVIEW

Discussion Questions

1. What makes a hormone a hormone? What requirements does a molecule have to meet to be classified as a hormone?

2. Why are hormones associated with homeostasis? In what respect do hormones contribute to homeostasis?

3. How is the method of action of steroid hormones different from that of other hormones?

4. Describe the basic types of second messenger systems. Why does a second messenger system enhance the effect of a single hormone molecule? Do steroid hormones have the equivalent of a second messenger?

5. In what respect does the link between the nervous system and the release of adrenalin by the adrenal medulla increase the effectiveness of a nervous response?

Objective Questions (Answers in Appendix)

6. Which organ or tissue is both an endocrine and an exocrine gland?
   (a) pineal gland
   (b) heart
   (c) pancreas
   (d) parathyroid gland

7. Pain resulting from the action of prostaglandins may be relieved by
   (a) aspirin.
   (b) oxytocin.
   (c) somatotropin.
   (d) thyroxine.

8. In animals, hormones are chemical messengers that are transported through _____ and have _____ effects.
   (a) the bloodstream / stimulating or suppressive
   (b) the ducts / stimulating
   (c) the lymphatic system / only stimulating
   (d) the nervous system / stimulating or suppressive

9. Neurosecretory cells in the hypothalamus produce
   (a) trophic hormones.
   (b) mineralocorticoids.
   (c) corticosteroids.
   (d) releasing hormones.

CHAPTER **37**

Reproduction

It is in this way that everything mortal is preserved, not by remaining the same, which is the prerogative of divinity, but by undergoing a process in which the losses caused by age are repaired by new acquisitions of a similar kind . . . ; it is in order to secure immortality that each individual is haunted by this eager desire and love.
—Diotima to Socrates in Plato's *Symposium*

An elaborately shaped orchid flower entices a male bee with colors and odors that mimic the female of his species. A peacock spends his entire life dragging an impossibly large tail behind him—all for the few brief moments when he succeeds, through its size and color, in attracting the attention of a peahen. And two humans—after more than a decade of adolescent role playing and earnest searching—collapse, exhausted, in each others' arms.

The drive to reproduce is a fundamental part of life on earth. Among bacteria and plants it occurs as matter-of-factly as growth; in humans it is often accompanied by an addictive mixture of angst and ecstasy. But in every case, the act of reproduction is an organism's link with both past and future and is the key to its species' survival.

FORMING A NEW GENERATION

Among many single-celled organisms, reproduction is as simple as cell division; a lone bacterium can divide with staggering speed in a culture flask, filling it with billions of cells in a few days. Organisms that undergo **asexual reproduction** give rise to offspring that are genetically identical to themselves, sometimes blurring the distinction between parent and offspring. Asexual reproduction is highly efficient, for every member of an asexual population, even isolated individuals can reproduce.

Among many multicellular creatures, however, **sexual reproduction** becomes so complex that its requirements govern many aspects of life from birth to death. At first glance, sexual reproduction seems not only more complicated than asexual reproduction but less efficient as well. Because two individuals are required for reproduction, but only one can actually produce eggs, sexual reproduction seems to cut the number of potential offspring per individual in half. Why, then, is this pattern of reproduction so widespread?

The classic view holds that sex has become common because the shuffling and recombination of genes during meiosis, combined with the addition of "new" genes from another individual, produces new—and potentially useful—combinations of genetic information in offspring. This genetic diversity, as we have discussed in earlier chapters, produces variability in a population which may be useful over evolutionary time as that population faces competition and environmental challenges. Although this is probably true, some biologists view it as a secondary benefit of—rather than the primary impetus for—the evolution of sex.

Although Diotima spoke of love, procreation, and immortality from a philosophical point of view, her

723

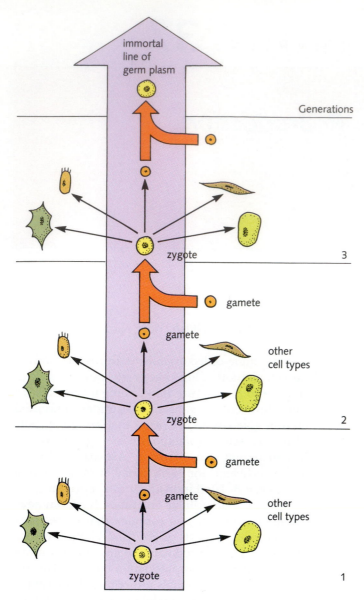

immortal
line of
germ plasm

Generations

zygote — 3

gamete

gamete

other
cell types

zygote — 2

gamete

gamete

other
cell types

zygote — 1

Figure 37.1 Of all the cells in an adult, only the reproductive cells have a chance to become part of a new generation. These germ line cells are part of an unbroken succession of reproductive events that links each organism with its ancestors.

in DNA replication and damage from environmental mutagens accumulate in somatic cell DNA throughout an organism's lifetime. Though some of these errors can be repaired during mitosis, others cannot, and both cell lines themselves and the organisms in which they exist age as damage mounts.

The cells of the reproductive system, however, are endowed with a potential that no other cells possess: to become part of the next generation. In that restricted sense, these cells have a shot at immortality. Recognizing this possibility, biologists sometimes refer to the cells of the reproductive system as being part of the **germ line**— a line of cells that are passed from one generation to another (Figure 37.1).

Furthermore, researchers have recently learned that the pairing and recombination of chromosomes during meiosis offers an opportunity for a fundamentally different sort of DNA repair than normally occurs during mitosis. For this reason, as germ cells are produced during meiosis, even serious, double-stranded damage to DNA can often be repaired. In effect, this powerful repair mechanism—together with the addition of new genetic material from another parent—does indeed repair the "losses caused by age" to DNA. In the process, it offers cells of the germ line the closest thing on earth to immortality.

The unique position of the reproductive system in the formation of the next generation leads to some interesting paradoxes. Though they are essential to the survival of the species, the reproductive organs form the only system in the body that is not essential to the survival of the individual. An injury to the reproductive system is usually not life-threatening, and it is even possible to remove the entire system without reducing an individual's chances for survival. Yet functioning reproductive systems do much more than produce gametes; their hormones (in addition to preparing the body physically for reproduction) influence the nervous system, producing patterns of behavior tending to ensure that every individual will do its best to pass its genetic material on to the next generation.

thoughts closely parallel a new hypothesis on the importance of sexual reproduction that is based on our increased understanding of molecular biology. We have known for some time that every body cell of a multicellular organism must eventually die, for no organism is immortal. Even when cells (other than cancer cells) are removed from animals and reared under perfect culture conditions, the cell line eventually becomes senile and dies. One reason for this unavoidable mortality is that errors

THE MALE REPRODUCTIVE SYSTEM

In humans and other mammals, the male reproductive system has the task of producing and delivering **sperm cells,** the male gametes. Sperm are produced in the **testes** and suspended in a fluid known as **semen,** which is produced by glands that line the reproductive tract. Semen is stored in the reproductive tract and released into the female reproductive system through the male copulatory organ, the **penis,** during sexual intercourse.

The Testes

Reproductive systems include *primary sexual organs*, which produce sperm and eggs, and *secondary sexual organs*, which are important to the reproductive process but do not actually produce gamete cells. In the male, the primary sexual organs are the testes (Figure 37.2). Although they develop within the abdomen, human testes descend at birth into a sac known as the **scrotum.** Sperm cannot be produced at normal body temperatures, and the external location of the testes lowers their temperature by 2–3°C, just enough to allow sperm to be produced.

The testes are composed of tightly coiled tubes known as **seminiferous** ("sperm-bearing") **tubules.** The combined length of human seminiferous tubules (nearly 250 meters) is lined with sperm-producing cells. Sperm development along the walls of these tubules is a continuous process. Cells called **spermatogonia,** found near the periphery of the tubules, divide by mitosis, providing a steady source of new cells that are available for sperm development. A few of these spermatogonia move away from the wall of the tubule and increase in size, becoming **primary spermatocytes.** Meiosis now begins, and each primary spermatocyte passes through two meiotic divisions. The two cells formed by the first meiotic division are known as **secondary spermatocytes,** and the four cells formed by the second meiotic division are known as **spermatids.**

At the completion of meiosis the haploid spermatid has only 23 chromosomes (one of each pair), whereas the diploid primary spermatocyte possessed 46. However, the spermatid is not yet ready for the reproductive process. It must first pass through a complex developmental process in which the appearance of the cell undergoes a complete change. Its nuclear chromatin is condensed into a compact *headpiece*, and a long, powerful *flagellum* develops from the centriole. The Golgi apparatus of the spermatid produces at the tip of the sperm a flattened vesicle known

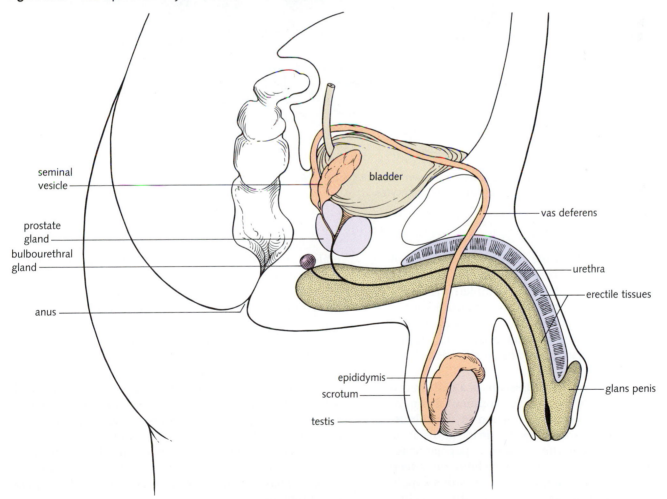

Figure 37.2 The reproductive system of the human male.

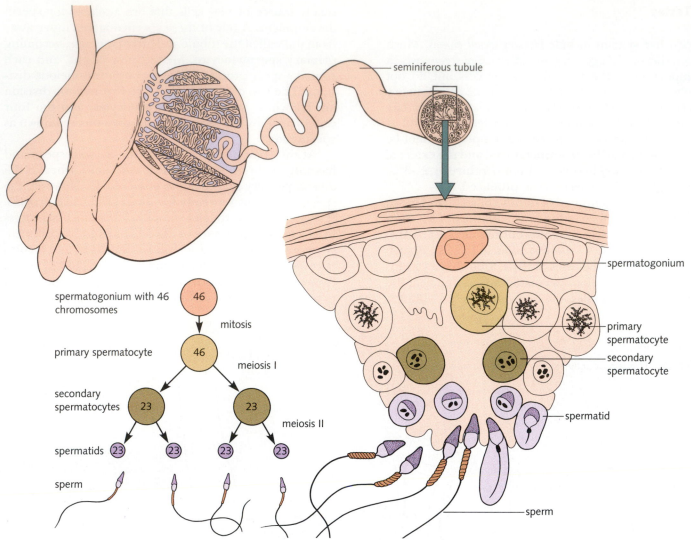

seminiferous tubule

spermatogonium with 46 chromosomes

46

mitosis

primary spermatocyte

46

meiosis I

secondary spermatocytes

23

23

meiosis II

spermatids

23

23

23

23

sperm

spermatogonium

primary spermatocyte

secondary spermatocyte

spermatid

sperm

Figure 37.3   Sperm are produced within the testes in the seminiferous tubules. Meiosis takes place as mature sperm cells are produced from spermatocytes.

as the **acrosome.** The acrosome contains a collection of enzymes that will help to break through the protective layers surrounding the egg cell. The complete process of development, from spermatogonium to mature sperm, takes approximately 72 days (Figure 37.3). Released when their development is complete, sperm travel through a system of collecting ducts that leads to the rest of the reproductive system.

The production of sperm in humans is a continuous process that occurs in "waves" of development throughout the testes. Within any single portion of the tubule, nearly all of the primary spermatocytes divide at the same time. In a nearby segment, the secondary spermatocytes may be dividing. These waves of development sweep through the tubules, ensuring that a continuous supply of mature sperm is available at all times. In a healthy

male, a single drop of semen contains between 5 and 10 million active sperm (Figure 37.4).

Many mammals do not reproduce year-round, as humans are capable of doing, but instead have seasonal reproductive cycles. In males of such species, sperm development throughout the tubules is coordinated in such a way that sperm develop together in preparation for the breeding season.

The testes contain two other cell types: **Sertoli cells** are found within the seminiferous tubules, where they aid in sperm development by providing nutrients. **Interstitial cells of Leydig** (*interstitial* means "in between") are found between the tubules and are not directly involved in sperm production, but they produce *testosterone*, the male sexual hormone that is necessary for sperm development.

The Male Reproductive Tract

The other organs and tissues of the male reproductive system are primarily concerned with the storage and delivery of sperm cells. They include the **epididymis,** through which sperm are collected from the testes and channeled to the **vas deferens,** which leads into the lower portion of the abdomen. The **seminal vesicle** secretes fluids into the vas deferens, and the **prostate** and **bulbourethral glands** secrete fluids directly into the **urethra,** the tube through which urine is passed from the body. These fluids produce semen, a suspension of salts, nutrients, and sperm.

As shown in Figure 37.2, just beneath the prostate gland the vas deferens merges with the urethra. During sexual intercourse, semen is released into the urethra and leaves through the tip of the penis.

Male Sex Hormones

The testes are directly affected by two hormones produced by the anterior pituitary: **LH** (luteinizing hormone) and **FSH** (follicle-stimulating hormone), both of which were first discovered (and named) in females. These two hormones are the major **gonadotropins** (gonad-controlling hormones) in males as well as females (Figure 37.5). **Gonadotropin-releasing hormone** (GnRH), produced by the hypothalamus, controls the release of FSH and LH from the pituitary. The target cells for LH are the interstitial cells of the testes. LH stimulates interstitial cells to produce the steroid hormone **testosterone.**

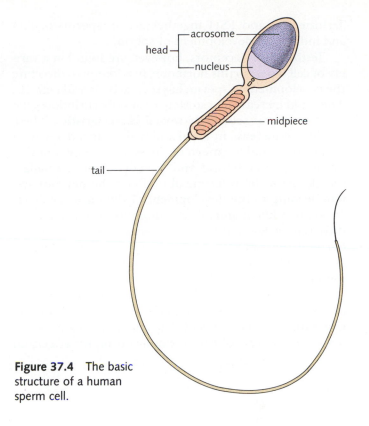

Figure 37.4 The basic structure of a human sperm cell.

Figure 37.5 At puberty, a boy's hypothalamus produces gonadotropin-releasing hormone (GnRH) (1) which causes the release of FSH (follicle-stimulating hormone) and LH (luteinizing hormone) from the anterior pituitary (2). These hormones act directly on cells in the testes. LH causes the interstitial cells to produce testosterone (3). Testosterone and FSH are both required for sperm development (4). Testosterone in the bloodstream (5) produces male secondary sexual characteristics and feeds back at the hypothalamus to inhibit the release of GnRH.

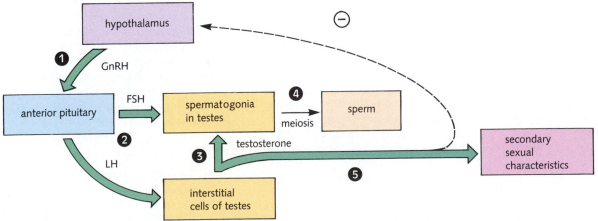

Testosterone and FSH together act on spermatogonia and induce the development of sperm.

Testosterone receptors, however, are found in a variety of cell types, so this hormone, in addition to directing the development of sperm, has several body-wide effects. The rapid increase in testosterone at puberty induces the development of **secondary sexual characteristics.** These include an increase in facial and body hair, a deepening of the voice, and a pattern of physical development that includes greater stature and a tendency to accumulate muscle mass. Testosterone also affects the nervous system, leading to the development of the male sex drive. In many other mammals the intensity of the male sex drive is directly—and exclusively—dependent on the level of circulating testosterone. In humans, however, the connection between hormones and sexual behavior is complicated by many psychological factors.

Because of testosterone's stimulatory effects on muscle development, young athletes are often tempted to take supplemental doses of this drug to increase muscle mass. But because of this hormone's complex effects on many body tissues—including the central nervous system—such steroid treatments pose serious hazards to both physical health and mental well-being.

Puberty in Males

In young boys, sexual organs are immature. Beginning during embryological development, the testes produce very low levels of testosterone—not enough to affect secondary sexual characteristics, but enough to direct the proper differentiation of male reproductive structures. Then, sometime between the ages of 11 and 14, a period known as **puberty,** the reproductive tissues develop and the sex organs become fully developed.

We do not understand for certain how puberty begins, but it is likely that during childhood the hypothalamus is extremely sensitive to feedback inhibition by testosterone. During early life, therefore, the hypothalamus does not secrete gonadotropin-releasing hormone (GnRH). Then, at the onset of puberty, the sensitivity of the hypo-

---

THEORY IN ACTION

For Want of a Receptor . . .

We often equate testosterone and estrogen with "maleness" and "femaleness," respectively, when thinking about the secondary sexual characteristics that these primary sex hormones help to produce. However, there is more to the process of sexual development than just the hormones themselves. Consider the case of a genetic abnormality known as *TF syndrome*. Individuals with this disorder are seemingly normal females. They have no apparent medical problems until puberty, when it becomes apparent that something is wrong. Although they seem to mature sexually as young women, they fail to menstruate. It is at this point that patients with TF syndrome often seek help from a physician.

If a karyotype is performed, the physician discovers that the young woman is actually a genetic male: 46XY. The gonads within the patient's abdomen are testes, not ovaries. A check of blood hormone levels shows that estrogen levels are low but that the blood contains a normal level of testosterone for a young man.

The initials TF stand for *testicular feminization*. Individuals with TF syndrome have spent their childhoods believing themselves to be females, and there is no reason for that conviction to change once the syndrome is diagnosed. Like about 10 percent of married women, they are unable to have children. They remain infertile, but they can lead perfectly normal lives as women.

TF syndrome results from a genetic defect in the receptor protein for testosterone. Testosterone, like all steroid hormones, binds to receptor proteins that are present in target cells. Although the testes of a TF patient produce large amounts of testosterone, potential target cells throughout the body lack the receptor. Therefore, they cannot respond to the hormone, and male secondary sexual characteristics do not develop. TF syndrome emphasizes the central role that hormones play in sexual development—and reminds us that hormones cannot act without their receptors.

---

thalamus to testosterone decreases, and GnRH is released. This releasing hormone stimulates the secretion of FSH and LH from the pituitary. LH in turn stimulates the interstitial cells of the testes to produce testosterone. At that point, spermatogonia begin to mature, secondary sexual characteristics develop, and the individual becomes sexually mature.

THE FEMALE REPRODUCTIVE SYSTEM

The female reproductive system also produces gametes, but the mammalian pattern of reproduction places special demands on the female system: Offspring develop within the body of the female and are born alive. If fertilization occurs, the zygote becomes implanted in a portion of the reproductive tract and develops into an embryo. The mammalian female reproductive system, therefore, must not only produce mature eggs but also prepare the lining of the *uterus* to receive and nurture the developing

embryo. This task requires a variety of complex reproductive organs and an equally complex system of hormonal controls to synchronize the system's activities.

The Ovaries

The primary female sex organs, the **ovaries,** develop from the same embryonic tissues as the testes do in males. The ovaries retain their abdominal location, although, like the testes, they are associated with a system of ducts that leads to the exterior (Figure 37.6). The basic internal organization of the ovaries is quite different from that of the testes, as shown in Figure 37.7, which reflects the fact that a woman produces fewer than 500 egg cells in her lifetime, contrasting sharply with the millions of sperm found in every drop of semen.

The female reproductive cells, or **ova,** develop from cells known as **oogonia,** the equivalents of spermatogonia in the male. Unlike spermatogonia, however, oogonia complete their process of mitosis before a woman

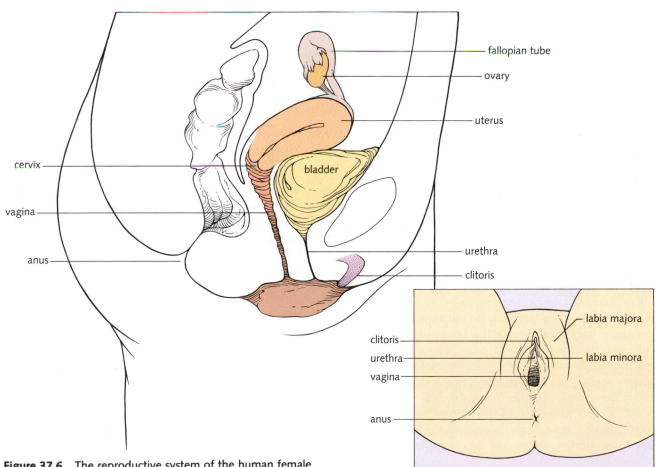

Figure 37.6 The reproductive system of the human female.

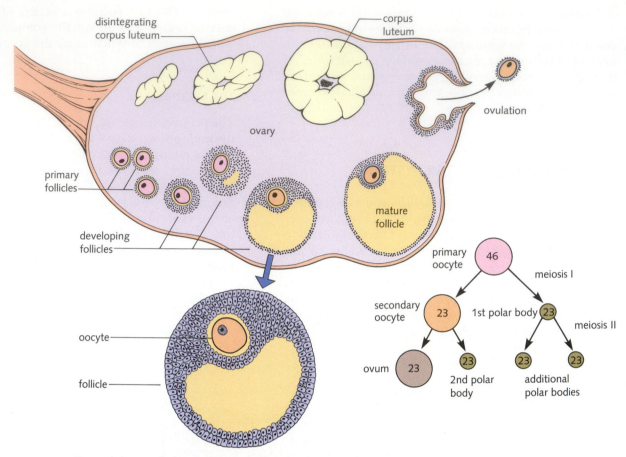

Figure 37.7 A diagrammatic view of the internal organization of the ovary. Ova develop within structures called follicles. The development of a secondary oocyte within its follicle is shown as a sequence of stages, beginning at the lower left and moving counterclockwise. After ovulation, the release of a secondary oocyte, the remaining follicular tissue forms a structure known as the corpus luteum.

is born. At birth the ovary contains about 400,000 **primary oocytes,** each of which is contained within a single layer of cells known as a *primary follicle,* ready to begin its first meiotic division. Prophase I of meiosis is not completed, however, until just before an ovum is released from an ovary.

Roughly once a month in a mature woman, a single follicle and its primary oocyte begin to enlarge. The cells of the follicle, known as *granulosa cells*, divide rapidly as the follicle enlarges, and they secrete the *zona pellucida,* a dense layer of material that surrounds the primary oocyte. The first meiotic division takes place within the follicle, producing a **secondary oocyte** and a much smaller cell with very little cytoplasm, the *first polar body.* The secondary oocyte enters prophase of the second meiotic division, and it is in this state that the oocyte is released from the ovary. **Ovulation,** the release of the oocyte, occurs

when a mature follicle ruptures at the surface of the ovary, allowing the oocyte to escape (Figure 37.8). Fluid movements sweep it into the **fallopian tube,** where cilia and smooth-muscle contractions slowly direct the oocyte toward the *uterus.* The second meiotic division, which produces the ovum and a secondary polar body, does not occur unless the secondary oocyte is fertilized by a sperm cell.

The Female Reproductive Tract

The fallopian tubes gradually deliver the ovum into the **uterus,** a thick, muscular organ where a fertilized ovum implants and begins embryonic development. The lining of the uterus, a tissue known as the **endometrium,** has a rich blood supply that provides the ideal environment

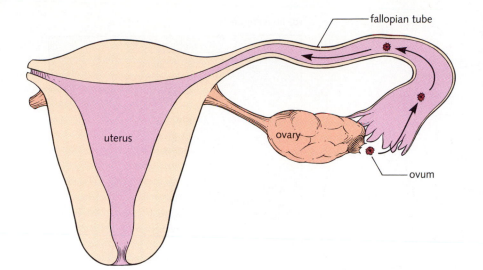

fallopian tube

uterus

ovary

ovum

Figure 37.8 Ova released from the ovaries travel down the fluid-filled fallopian tube toward the uterus.

for early development. The uterus opens into the **vagina,** which leads to the exterior of the body. The connection between the uterus and the vagina, a muscular opening known as the **cervix,** secretes mucus that may block the passageway during pregnancy, helping to isolate the developing embryo. The external opening of the vagina is covered by folds of skin known as the **labia majora** and **labia minora.** As we will see later, the physiology of male and female increases the likelihood that sexual intercourse will lead to successful fertilization.

Female Sex Hormones and the Reproductive Cycle

Human ova are released once every 28 days, so the female reproductive tract must prepare for a possible pregnancy at 28-day intervals. Because this preparation requires the coordination of many organs and tissues, a chemical control system is necessary to ensure that all the events involved occur in their proper sequence. Naturally enough, the endocrine system provides this control.

It is easiest to understand the female reproductive cycle when we begin as we did with males—at puberty. Puberty begins in girls sometime between the ages of 10 and 14 (on the average, about a year earlier than in boys). At that time, an increase in GnRH secretion by the hypothalamus initiates the release of FSH from the pituitary (Figure 37.9a). Under the influence of FSH, a few of the many thousands of immature primary follicles begin to grow. Usually, one follicle soon takes the lead and the others stop growing, allowing only a single ovum to develop to maturity at one time. (Sometimes two follicles develop and, if fertilized, yield nonidentical (fraternal) twins.)

At the same time, joint action of FSH and LH causes the granulosa cells of the follicles to secrete **estrogen,** the first of two principal female steroid sex hormones. Estrogen produces female secondary sexual characteristics, including the development of breast tissue and mammary glands, the appearance of hair under the arms and in the genital region, and a typically female pattern of bone and muscle growth in which the pelvis widens and deposits of body fat appear on the hips and thighs.

At the same time, estrogen stimulates extensive cell division in the lining of the uterus. The uterine endometrium develops an extensive blood supply and large numbers of glands. Meanwhile, the follicle continues to grow until, for some reason that is not completely understood in humans, a surge of luteinizing hormone is released from the anterior pituitary. This LH surge stimulates ovulation.

Then, as the ovum begins its descent through the fallopian tubes, the follicle that released it changes into a new structure called the **corpus luteum.** The corpus luteum secretes some estrogen, but it is primarily responsible for the secretion of **progesterone,** the second female sex hormone that causes the maturation of the blood vessels and glands in the lining of the uterus. For this reason, progesterone is essential in the final preparation of the uterus for implantation of the embryo.

The Conclusion of the Cycle

During the two days following its release into the reproductive tract, the ovum is available for fertilization by sperm cells deposited into the reproductive tract via sexual intercourse. What happens next depends on the presence or absence of viable sperm during this period.

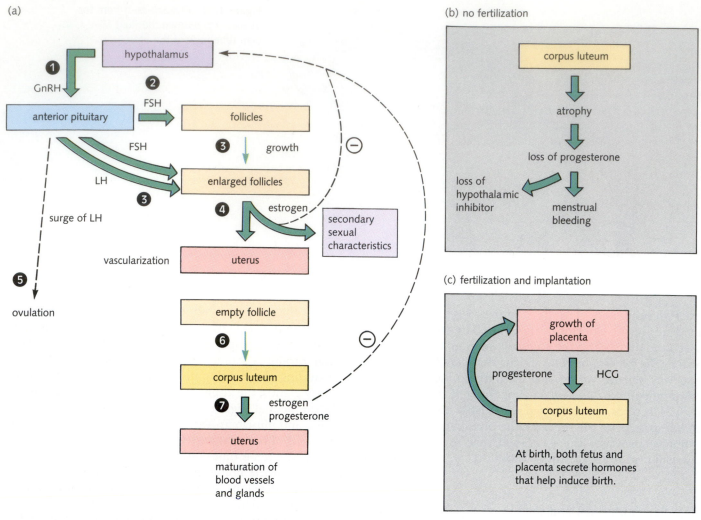

Figure 37.9 **(a)** A summary of hormonal regulation of the human menstrual cycle (steps 1-7). GnRH released from the hypothalamus stimulates the anterior pituitary to release FSH and LH. These hormones cause a follicle to enlarge and begin to mature. As the follicle grows, it releases estrogen into the bloodstream. Estrogen stimulates the uterus, causing the vascular lining of the uterus to grow. **(b)** If fertilization does not occur, progesterone levels fall as the corpus luteum breaks down. **(c)** If fertilization does occur, progesterone levels stay high enough to maintain the uterine lining.

■ If a viable sperm cell makes contact with the ovum, a fertilized zygote forms. If that zygote is successfully implanted in the uterine lining, pregnancy results.

■ If no sperm are present, the unfertilized ovum is soon flushed out of the uterus into the vagina, the uterine lining disintegrates, and a new menstrual period results.

Menstruation: when fertilization does not occur Following ovulation, the lifetime of the hormone-producing cells of the corpus luteum is limited. If no fertilization

and implantation occur within seven to ten days after ovulation, the corpus luteum begins to disintegrate, and its production of both estrogen and progesterone falls (Figure 37.9b). This fall in progesterone levels has two effects.

First, because the enlarged uterine lining is a hormone-sensitive tissue, the decrease in estrogen and progesterone levels causes much of the newly developed tissue to be resorbed. The rest of this tissue detaches from the uterine lining, and as its blood vessels rupture, menstrual bleeding begins. This is why the onset of a "period" usually signals that a pregnancy has not occurred.

Second, the continued drop in circulating levels of estrogen and progesterone frees the hypothalamus from the inhibition caused by these hormones. The hypothalamus then releases GnRH, which stimulates the pituitary to secrete FSH and LH, and the entire cycle begins again.

In the absence of fertilization, this cycling of the reproductive system continues at regular intervals until **menopause,** when follicle development and the release of ova cease. Like the time of puberty, the time of menopause is highly variable; it may occur anywhere between the ages of 43 and 55.

Pregnancy: when fertilization does occur

 If fertilization occurs, the zygote begins cell division and forms a **morula,** a ball of 30–60 cells by the time it reaches the uterus. Gradually, a cavity develops in the middle of this ball of cells, and the embryo is then known as a **blastocyst.**

The blastocyst implants in the lining of the uterus and grows rapidly just within the lining. The presence of the embryo causes changes to occur in the uterine cells that surround it. At this time, tissues from the embryo join with those of the mother to produce the organ called the **placenta** (Figure 37.9c). In addition to facilitating the exchange of nutrients, gases, and waste products between developing fetus and mother, the placenta itself serves as a vital, though temporary, endocrine gland.

The first major hormone secreted by the placenta, **human chorionic gonadotropin (HCG),** is essential to continuation of pregnancy. The action of HCG maintains the corpus luteum and its vital secretion of progesterone. High levels of progesterone, in turn, simultaneously preserve the uterine lining and inhibit the hypothalamus from releasing more GnRH during the early part of the pregnancy. Later on in pregnancy, the placenta itself begins to synthesize progesterone, ensuring that the uterine lining is maintained during pregnancy (Figure 37.10).

This maintenance of the uterine lining gives a woman her first hint that she may be pregnant: Her monthly

---

Other Reproductive Cycles

The human menstrual cycle and the pattern of behavior that goes with it are all but unique in the animal world. Very few mammals, for example, are receptive to mating throughout the year, as humans are. More typical is a pattern closely linked to the seasons in which mating is timed to bring offspring into the world at the most favorable time of the year. In many domesticated animals there is no longer any practical connection between their mating cycles and the seasons, yet they still are receptive to mating only at certain times.

Domestic dogs and cats, for example, mate only during the few days immediately before and after ovulation—a period known as *estrus,* or "heat." Dog breeders recognize this time by noting a swelling of the female's genitalia and the release of a bloody discharge from her vagina.

Some amateurs mistakenly take this discharge as the equivalent of a menstrual period and assume it means that the time of greatest fertility has passed. This is not the case. Instead, bleeding results from the rapid growth and enlargement of the uterine lining just prior to ovulation. In such animals, therefore, vaginal bleeding marks the beginning rather than the end of a period of high fertility.

North American deer have one of the most intricate of all sexual cycles. Male deer pass through a yearly cycle in which the level of testosterone in circulation rises and falls rhythmically. Testosterone stimulates the growth of *antlers,* a male secondary sexual characteristic important in the social ritual combat that precedes mating in many groups of deer. Throughout the summer the antlers, encased in a soft

layer of skin known as *velvet,* grow rapidly until the mating season begins in the autumn. As the season begins, the deer shed the velvet from their antlers, marking the point at which they are ready to mate with females. By the end of the mating season, another set of hormones causes the antlers to fall off and be lost.

Some of the most efficient reproductive cycles are found in cats and rabbits. In these animals, the activity in the nervous system that is caused by the act of sexual intercourse itself triggers the release of gonadotropic hormones from the pituitary. These hormones cause ovulation to occur within a few hours of intercourse, virtually ensuring that the newly deposited sperm will encounter ova. The expression "breeding like a rabbit" really does have a scientific basis!

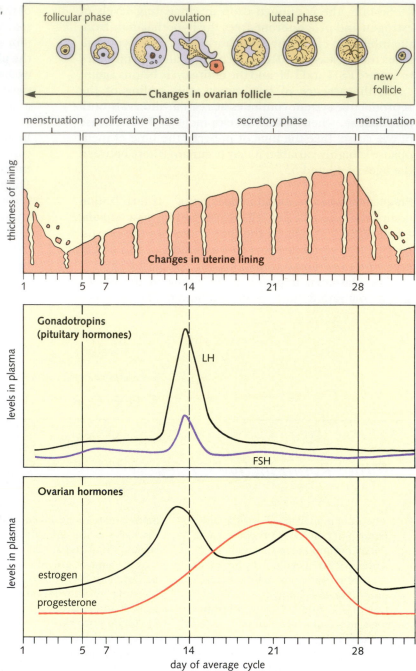

Figure 37.10 Changes in follicle structure, the uterine lining, and hormone levels associated with the menstrual cycle.

menstrual period does not occur. Home pregnancy tests generally contain chemicals that test for the appearance of HCG in the urine, often making it possible to detect a pregnancy in its early stages.

Timing of the menstrual cycle The average length of the human menstrual cycle is 28 days. The length of the cycle is determined by a number of factors, including the rate at which the corpus luteum disintegrates following ovu-

lation, the speed with which a new follicle develops, and the responsiveness of the hypothalamus to circulating estrogen levels in the blood stream. Because these individual factors are variable, the length of the menstrual cycle is variable too. Only about 30 percent of women have menstrual cycles that match the 28-day average; others vary from 21 to 48 days, and in many, the length of the menstrual cycle fluctuates from month to month (Figure 37.11).

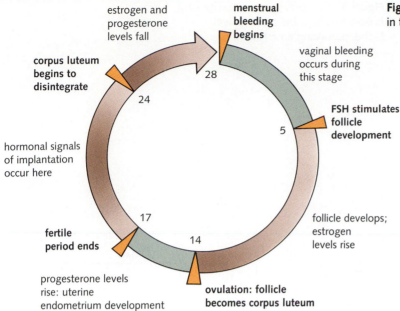

estrogen and
progesterone
levels fall

menstrual
bleeding
begins

Figure 37.11 A simplified view of events in the menstrual cycle.

corpus luteum begins to disintegrate

28

vaginal bleeding occurs during this stage

24

FSH stimulates follicle development

5

hormonal signals of implantation occur here

follicle develops; estrogen levels rise

17

14

fertile period ends

progesterone levels rise: uterine endometrium development becomes intense

ovulation: follicle becomes corpus luteum

Hormones, Behavior, and Reproduction in Other Mammals

In most mammals, there are closer and much more definite connections between sex hormone concentrations, sexual behavior, and the release of ova. As ova are prepared for release, larger amounts of the female sex hormones are released. These hormones prepare the reproductive tract for pregnancy and also cause the animal to become behaviorally receptive to mating, a condition known as "heat" or **estrus.** The **estrous cycle** is this sequence of endocrine, physical, and behavioral changes.

Higher primates are almost unique among mammals in that they do not exhibit an estrous cycle. Humans and other higher primates are physiologically and psychologically capable of mating at all times, although the release of ova and the preparation of the uterine lining occur at regular intervals.

SEXUAL INTERCOURSE

The reproductive system is more than a collection of organs and tissues: It is associated with behavior integral to the psychology of both males and females. Part of this is a strong attraction for members of the opposite sex. When the presence of the opposite sex induces sexual excitement, the nervous system causes a series of subconscious changes that facilitate sexual intercourse.

In females, smooth muscle around the opening of the vagina relaxes and the walls of the vagina secrete a fluid that lubricates the passageway and has the right combination of pH, salts, and nutrients to help sperm cells survive. As sexual excitement increases, hollow cavities in the **clitoris,** a small structure (about 2 centimeters in length) near the front of the vaginal opening, fill with blood, causing it to become erect.

In the male, sexual excitement causes blood vessels leading from the penis to partially close, restricting the flow of blood. Although its exit from the penis is restricted, blood continues to flow into the organ at its normal rate. Thus spongy tissue in the penis fills with blood and it becomes stiff and erect, which enables it to pass between the labia of the female into the vagina (Figure 37.12). During sexual intercourse, or **coitus,** the penis is inserted into the vagina, the feelings of pleasure in both male and female may rise and fall gradually, eventually leading toward a climax known as **orgasm** (Figure 37.13).

The tip of the penis is richly endowed with sensory receptors, making it one of the most sensitive areas of the body. As the penis is thrust back and forth within the vagina, mechanical stimulation brings the male to climax. At a point where sexual pleasure intensifies, smooth muscle surrounding the male reproductive tract contracts, expelling as much as a teaspoonful or more of semen into the vagina. The release of semen is known as **ejaculation.** The contraction of smooth muscles during the period of ejaculation is involuntary, so it cannot be controlled by the male.

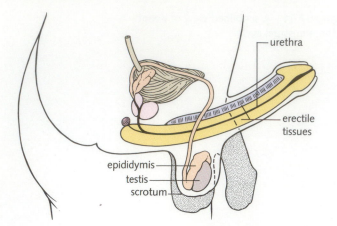

Figure 37.12 During sexual excitement, hollow tissues in the penis fill with blood, producing an erection.

urethra

erectile tissues

epididymis

testis

scrotum

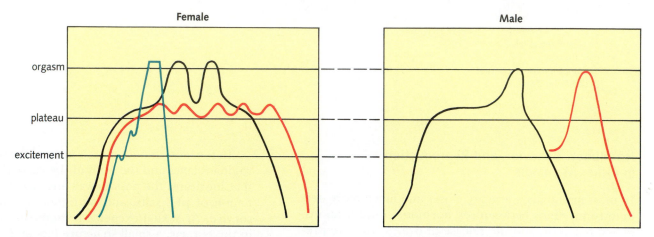

Female

Male

orgasm

plateau

excitement

Figure 37.13 Levels of sexual excitement in both male and female usually rise quickly and reach a plateau. Sexual excitement may then rise again, reaching a **climax** that may be repeated several times in the female. In the male, orgasm is generally followed by a refractory period before a second peak of sexual excitement is possible. The different tracings show several possible outcomes for sexual arousal.

Orgasm in the female also occurs at the height of sexual excitation, and it is accompanied by smooth muscle contractions in the vagina and uterus that are believed to assist the entry of sperm into the uterus. Female orgasm is not associated with a distinct physical event equivalent to male ejaculation, but like the male orgasm, it is associated with feelings of satisfaction and relaxation. The sensation of orgasm is often centered in the *clitoris*, and orgasm may begin with a series of powerful contractions of the uterine and vaginal walls, which spread a warm, pleasurable sensation throughout the pelvis. Note that orgasm in human females is not connected with the release of ova in any way and is not necessary for fertilization to occur. It is possible for the male to penetrate the reproductive tract, ejaculate sperm, and fertilize an ovum without inducing climax in the female.

CONTRACEPTION

The biological organization of the entire reproductive system facilitates mating and reproduction. Although there are only one or two days a month when it is possible for a woman to become pregnant, the fact that sperm cells may live within the reproductive tract for several days makes it possible for fertilization to occur a day or two after intercourse. This increases the likelihood of **conception**—the successful fertilization of the ovum.

There are many **contraceptive** (conception-preventing) techniques that work within the physiology of the reproductive system. The simplest and most reliable technique, of course, is **abstinence.** Pregnancy does not occur without the entry of sperm into the reproductive tract. **Withdrawal,** in which the penis is withdrawn just prior to male orgasm, is one of the least reliable contraceptive techniques, because several drops of semen (containing millions of sperm cells) are often released before orgasm without the male being aware of it.

A timing technique, sometimes known as the **rhythm method,** is reliable if a woman is able to determine the exact point at which she ovulates each month and avoids intercourse for several days before and after ovulation. This is often possible, because ovulation is accompanied by a slight change in body temperature. However, the irregularity of the menstrual cycle in many individuals makes this method uncertain at best.

Some of the most useful contraceptive methods involve a physical barrier, such as a **condom** or **diaphragm,** that prevents sperm and ovum from meeting. The reliability of these methods is increased when a **spermicide** is placed in the vagina to destroy any sperm that may escape through the barriers. The ability of condoms to block the passage of disease-causing organisms is an equally important benefit.

In the mid-1960s an alternative method of contraception became available for the first time—the so-called **birth control pill.** The artificial birth control pill contains a synthetic form of estrogen. Taking the pill on a daily basis keeps a steady high level of estrogen in the circulation. After 21 days, the woman either stops taking pills or takes "blanks" containing no hormone for 7 days. This causes estrogen levels to drop low enough to trigger a normal menstrual period, but not low enough to stimulate the release of FSH. Because FSH release does not occur, a follicle does not develop, and pregnancy is prevented because no ova are released. The estrogen-based birth control pill has a very high reliability; fewer than 1 percent of women who use the pill on a regular basis become pregnant in a given year.

Research is also under way on male birth control pills that block sperm development. **Gossypol,** a compound first extracted in China from cotton seeds, has shown promise, and it may eventually be marketed as the first birth control pill for men.

The most widely used method of birth control in the United States at the present time is **surgical sterilization.** In 1982, nearly 16.5 million American couples reported sterilization as their primary method of birth control (10 million reported using the pill). Whether in male or female, the goal of sterilization is to prevent the delivery of sperm or egg cells to the reproductive tract. In women this can be accomplished by closing the fallopian tubes that lead from the ovaries to the uterus, making it impossible for ova to reach any area accessible to sperm. However, **tubal ligation,** as this procedure is called, is a major surgical operation that is usually done under general anesthetic and therefore carries with it some risk.

The anatomy of the male makes a similar medical procedure much easier. A small incision is made in the scrotum near each testis, and a small portion of the vas deferens is removed. This procedure, known as a **vasectomy,** prevents sperm cells from leaving the testes and causes the male reproductive tract to produce semen that is normal in every respect except for the absence of sperm cells (Figure 37.14). One disadvantage of surgical sterilization is the fact that it cannot be easily reversed. Therefore, it is most appropriate for couples who have decided not to have any children or any more children.

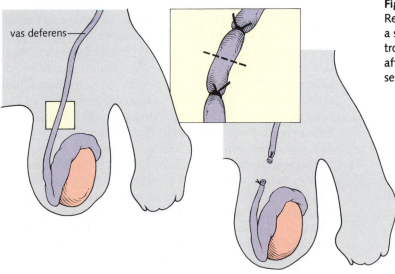

Figure 37.14 A surgical vasectomy. Removing a portion of the vas deferens is a safe and effective method of birth control for males. The operation does not affect sex drive or the ability to have sexual intercourse.

vas deferens

Table 37.1 *The Effectiveness of Widely Used Contraceptive Techniques*

Method	Rationale	Effectiveness	Problems
Withdrawal	Withdrawal of penis before ejaculation prevents sperm from entering vagina	75%	Premature ejaculation may limit effectiveness
Rhythm	Timing intercourse to avoid fertile periods	80%	Irregularity of cycles may limit effectiveness
Condom	Sheath that covers the erect penis traps semen	85%	Must be applied before intercourse; a small percentage of condoms leak
Diaphragm	Flexible membrane covers the cervix and prevents sperm from entering the uterus	80–90%	Must be applied before intercourse
The pill	Synthetic estrogen tablets prevent follicle development	99%	Must be taken regularly; does not provide a barrier against infection
Tubal ligation	Oviducts are surgically closed	100%	Does not protect against infection; sometimes irreversible
Vasectomy	Vas deferens is surgically closed	100%	Does not protect against infection; sometimes irreversible
IUD (intrauterine device)	Foreign object interferes with uterine implantation	90–95%	Must be inserted by physician; some risk of uterine puncture and infection
Douche	Cleansing of vagina with a special solution washes out sperm	70%	Ineffective if not done *immediately* after ejaculation; not a reliable method
Spermicidal foams and jellies	Suspension that kills sperm is placed inside vagina	80%	Must be inserted before intercourse; more effective if used with a barrier device
Sponge	Inserted dry, soaks up semen; diffusable spermicide in sponge kills sperm	*	Difficult to insert

*Assumed to be highly effective; no statistics available

Another contraceptive method that has shown much promise is the **intrauterine device,** or **IUD.** An IUD is a small metal or plastic structure that is implanted in the uterus by a physician. Because of its size and shape, it remains within the uterus indefinitely. The method by which an IUD prevents pregnancy is not fully understood, but it probably interferes with implantation by slightly disturbing the uterine lining. Although the IUD is reliable and does not require daily attention (as the pill and barrier methods do), some IUD designs have proved harmful to the women using them, producing uterine irritations, infections, and even sterility. These problems have led to a decrease in the use of IUDs, although a well-designed and properly inserted IUD is a safe and effective method of birth control.

Table 37.1 summarizes the rationale, relative effectiveness, and problems associated with each method of birth control.

Infertility

Pregnancy cannot occur unless both the male and female reproductive systems are functioning properly. Roughly 10 percent of the couples in the United States are **sterile**—unable to have children. Another 10 percent are **infertile**—able to have only one or two children when they would like to have more. One common cause of sterility or infertility is a low sperm content in semen. Some men produce defective sperm that are unable to fertilize ova, but many others produce too few sperm to be effective in sexual intercourse. Low sperm counts do not necessarily consign a couple to sterility. Many physicians employ a technique in which the sperm from several ejaculates are concentrated in the laboratory to produce semen with a near-normal sperm count. This concentrated semen can then be used to inseminate the female, and it is often successful in producing a pregnancy.

In other cases, physical obstructions in the vas deferens or the fallopian tubes prevent the delivery of reproductive cells, and these problems can be corrected by surgery. Occasionally, low progesterone levels make the uterine endometrium less receptive to implantation than it should be. This condition can be corrected by the administration of oral doses of progesterone. In many women, a lack of sensitivity to LH and FSH prevents the development of a mature follicle each month, making conception impossible. A number of synthetic "fertility drugs" that mimic the effects of these gonadotropins have been developed, and they often induce the ova to mature. Sometimes the drugs are *too* effective: One of their side effects is the possibility of multiple births resulting from the maturation of several follicles at once.

Endometriosis, a disorder produced by fragmentation of the uterine lining, has become an increasingly serious problem affecting fertility in recent years. Fragments of the endometrium (the lining of the uterus) break off and become lodged in other tissues such as the uterus or ovary. These scattered fragments are affected by hormones regulating the menstrual cycle in the same way as the rest of the uterine lining, producing periodic bleeding and disruption throughout the abdomen. The cause of endometriosis is not clear, but it is most common among women who have put off pregnancy until their thirties. Although extreme cases may require surgical removal of the ovaries, many cases of endometriosis can be successfully treated by administration of the hormones contained in oral contraceptives.

SEXUALLY TRANSMITTED DISEASES

The general pattern of mammalian reproduction includes internal fertilization. This pattern is effective and efficient, but it also creates a perfect opportunity for the spread of specialized parasites from one individual to another. The diseases caused by such organisms are known as **venereal diseases** (from Venus, the Greek goddess of love) or **sexually transmitted diseases (STDs).**

THEORY IN ACTION

In vitro *Fertilization*

Although human fertilization normally occurs within the female reproductive tract, fertilization is a cellular event, and there is no technical reason why it cannot occur in a laboratory dish. In 1978 this possibility became a reality when a baby girl named Louise Brown was born in England. Her parents had been unable to conceive because of a blockage in her mother's fallopian tubes. This blockage prevented ova from reaching the uterus and made her infertile. Physicians solved this problem by surgically removing an ovum from the surface of Mrs. Brown's ovaries. This ovum was then mixed with Mr. Brown's sperm, and fertilization took place, closely monitored under a microscope.

The fertilized cell began to divide and, a few hours later, was carefully removed from the dish and placed in Mrs. Brown's uterus. It successfully implanted in the lining of the uterus, and normal development took over. Nine months later Louise was born. The process that the Browns helped to pioneer is called *in vitro* fertilization (*in vitro* means "in glass"). To date, it has helped produce more than 100 children for couples who might never have conceived without it. The technical ease with which *in vitro* fertilization can be accomplished is stunning, and it is the product of our increasing knowledge of human reproduction.

As successful as it has been in helping childless couples, *in vitro* fertilization may soon raise new ethical problems. It is technically possible for a fertilized ovum to be implanted in the uterus of someone other than the donor of the ovum, so the possibility exists that individuals could hire others to bear their children. It is also possible to imagine *in vitro* screening for desirable characteristics, including gender, before an embryo is implanted. This would raise the prospect of "ordering" an embryo with the desired characteristics and discarding embryos with other characteristics. Like many advancing technologies, *in vitro* fertilization is already challenging us to develop an ethical framework for considering whether, and under what circumstances, we want to do some of the things science has made possible.

The Spread of STDs

Over the last 25 years, substantial changes in sexual behavior have taken place throughout the United States and other Western countries. The introduction of reliable contraceptive techniques has enabled people to engage in sexual intercourse without fear of unwanted pregnancy. However, contact with multiple sexual partners provides an ideal environment for the spread of STDs (Table 37.2).

Adding to these problems is the fact that none of the STDs produces a lasting immunity to reinfection. Although several STDs can be cured if recognized early, others cannot. Furthermore, several STDs—even those that can be cured if detected early—may cause sterility and other sorts of irreparable internal damage before their symptoms become apparent. For this reason, the only sensible method of dealing with STDs is to take every precaution necessary to *prevent* them.

Syphilis One of the most serious sexually transmitted diseases is **syphilis,** which is caused by a spirochete, a corkscrew-shaped bacterium known as *Treponema pallidum*. The spirochete is a delicate organism that doesn't survive well outside the body; it dies quickly when exposed to air or sunlight. Yet because the organism lives in the moist membranes of male and female reproductive tracts, sexual intercourse provides a perfect opportunity for the spirochete to be passed from one individual to another.

When a person is first infected with the spirochete, a hard sore known as a *chancre* forms at the point where the organism passes through the skin. The sore disappears after a few days or weeks, but this does not mean that the disease has passed. Instead, it has entered a *latent* phase in which it silently spreads throughout the body. It will appear in the reproductive tract, allowing it to be passed to other sexual partners. In addition, it will gradually infect the circulatory system and the nervous system and may eventually cause death. To make matters worse, the spirochete can be passed from a mother to her unborn child, so children may be infected with the disease before birth (congenital syphilis). In an attempt to guard against this, blood tests for syphilis are required in many states before a marriage license is granted.

Table 37.2 *The Most Common and the Most Serious Sexually Transmitted Diseases*

Disease	Organism	Mode of Transmission	Remarks
Syphilis	Syphilis spirochete (bacterium)	Sexual intercourse; infection occurs through penis or vagina	First sign is a small sore, or chancre; may cause brain damage, blindness, or death in final stages; 100,000 new cases in U.S. per year
Gonorrhea	*Neisseria gonorrhoeae* (bacterium)	Sexual intercourse; infection occurs in urinary tract	Produces a burning sensation during urination; may infect oviducts and cause sterility; 3 million new cases per year
Herpes	Herpes type II (virus)	Sexual intercourse	Painful inflammation of genital area; cannot be cured; 3 million new cases per year
Chlamydia	*Chlamydia trachomatis* (bacterium)	Sexual intercourse	Urinary inflammation; may cause sterility; difficult to detect; 3–10 million new cases per year
AIDS	Human immunodeficiency virus (HIV)	Sexual intercourse, intravenous drug use, transfusion with contaminated blood	Causes death by destruction of the immune response; 23,000 new cases in 1987; as many as 1 million infected with virus per year
Genital warts	Several forms of wart viruses	Sexual intercourse	Painful and unsightly; clear link with cervical cancer; 2 million new cases per year

Syphilis infections can generally be cured in their early stages by massive doses of such antibiotics as penicillin, but in its later stages, the disease can cause permanent physical damage that cannot be repaired by drugs.

Gonorrhea Another sexually transmitted disease caused by a bacterium is **gonorrhea.** This disease is produced by *Neisserea gonorrhoeae* bacteria, which live in the reproductive and urinary tracts, giving rise to one of the principal symptoms of the disease in males: a painful irritation associated with urinating. Blood and pus are often discharged in the urine as the infection persists. In some individuals, "clap" (as gonorrhea is sometimes known) does not cause any other serious problems. In most women, for example, the disease is virtually without symptoms until the infection produces *pelvic inflammatory disease*, which may result in sterility or death in some cases. The bacterium can infect a newborn child during its passage through the vagina at birth, and it often causes blindness. This is one of the reasons why it is now standard practice to medicate the eyes of all newborns.

Until recently, gonorrhea—once detected—responded readily to such common antibiotics as penicillin. Unfortunately, however, in several parts of the world where prostitution is widespread, active prostitutes have adopted the practice of regularly dosing themselves with antibiotics as "preventive medicine." You already know enough biology to guess what has happened: These circumstances have bred several strains of antibiotic-resistant gonorrhea organisms that are slowly spreading across the world. This dangerous development may make this disease more difficult to control in the future.

Chlamydia A serious infection known as **chlamydia** is caused by the bacterium *Chlamydia trachomatis* and can be passed on through sexual contact. The symptoms of chlamydia infection are subtle, and individuals—particularly females—often carry the disease for long periods without being aware of it. The bacterium infects the urinary tract and produces a painful sensation during urination. The results of long-term infection can be serious, and chronic infection is one of the common causes of sterility in women: The bacterium causes severe inflammation of the pelvic area and produces irreversible damage to the reproductive tract. Because of its long asymptomatic period and the ease with which it is transmitted, chlamydia is becoming distressingly common among sexually active individuals. Some studies suggest that as many as 20 percent of college-age women in some areas of the United States are infected with chlamydia.

Herpes Syphilis and gonorrhea are the "classic" sexually transmitted diseases, but several other disorders have caused increasing concern in recent years. **Herpes** is a family of viruses that cause a number of ailments in var-

ious locations in the body. *Herpes type I* is often responsible for cold sores that frequently appear on the lips. But Herpes virus, especially *Herpes type II,* can also infect the genital area. It causes painful, persistent infections that appear as small red blisters on the skin within a few days after infection and generally vanish within two to three weeks.

The virus itself, however, does not disappear. It remains latent, infecting the skin and nerve cells in the genital area. Herpes infections may then reappear without warning, brought on by stress, illness, exposure to sunlight, or even the excitement of sexual intercourse. Active virus particles are released from these painful sores, and by means of these particles the infection can be passed on to a sexual partner. A number of treatments, including the antiviral agent *acyclovir*, can reduce the severity of the blisters, but as yet nothing can cure the infections. Childbirth poses an additional hazard to women infected with herpes. The virus can be passed on to the newborn child as it passes through the vagina during birth, causing, in the most severe cases, brain damage or death.

Genital warts In recent years **genital warts,** which are caused by viral infections, have become an increasing medical problem. Although many warts can be treated with special drugs or surgical removal, they are a painful and unsightly problem. The viruses that are responsible for genital warts are spread through sexual contact, and researchers have now established a clear link between genital warts and several forms of cervical cancer.

AIDS In 1980 and 1981, a number of patients who were suffering from a series of unusual infections appeared at hospitals in New York and San Francisco. The physicians who treated them realized that the immune systems of these patients had all but stopped functioning, making them susceptible to pathogenic organisms that normally are not able to cause disease. Eventually this disorder came to be known as **AIDS (acquired immune deficiency syndrome).**

As physicians and public health officials followed the spread of AIDS, it soon became apparent that this disease, for which the long-term mortality rate is close to 100 percent, can be transmitted by sexual activity from members of either sex to members of either sex. AIDS will be discussed more fully in Chapter 39, but one aspect of the disease should be stressed here: AIDS, which is currently incurable, is *caused by a virus that can be transmitted by sexual contact.*

AIDS can also be spread by transfusions with infected blood and by the sharing of hypodermic needles with an infected person. The rapid increase in the number of AIDS patients since 1981, along with the fatal nature of the disease, has added a new imperative to the prevention of sexually transmitted diseases.

The Prevention of Sexually Transmitted Diseases

Researchers are actively working on treatments for each of these diseases, but prevention is always better than cure. Two strategies are effective in the control of sexually transmitted diseases. The first is a reduction in the number of one's sexual partners. The danger of contracting a disease transmitted by sexual contact increases in direct proportion to the number of sexual partners, and a population in which the number of sexual partners is limited provides the fewest opportunities for the spread of such diseases.

The second preventive approach is a change in the most commonly used types of contraception. The pill, surgical sterilization, and IUDs offer no barriers to the entry of disease-causing organisms, including the AIDS virus. The only method that does is the condom, and societies in which the use of condoms is highest (Japan, for example) have the lowest rates of sexually transmitted diseases. Although the condom is not 100 percent safe, it does provide a barrier against the entry of viruses and bacteria and helps prevent the spread of diseases associated with sexual contact. Public health officials agree that proper use of condoms will help to stop the spread of sexually transmitted diseases.

CURRENT CONTROVERSIES

Is Sex the Cornerstone?

From a purely biological point of view, one of the most remarkable aspects of human sexuality is the fact that men and women are receptive to mating throughout the year. What evolutionary pressures could have led to the loss of seasonal estrus in our species? Is it possible that the loss of estrus was a consequence of selective pressure brought about by the beneficial aspects of being able to mate at any time?

Some biologists have speculated that the loss of estrus is at least partially responsible for the basic social fabric of even the most primitive human communities. Human offspring are born in a condition in which they require a greater amount of care for a longer period of time than the offspring of any other organism. If humans mated only during a distinct breeding season, these biologists reason, there would be little incentive for males to remain with their mates once the season was over. The loss of estrus, however, enables sexuality to form a cohesive bond between males and females, increasing the likelihood that males will remain with their mates and assist in the process of forming a community in which the young can be raised to maturity.

This may not be a flattering way to think about the relations between the sexes: that men must be lured into social groups by the promise of sexual relations. It does, however, suggest a way to explain the evolution of the menstrual cycle to replace estrus. Females who were able to attract the presence of a male through their continuous sexual receptivity were more likely than other females to be successful in raising their own offspring. Similarly, males who were willing to participate in the formation of family units were more likely to pass their genes into the next generation, inasmuch as their presence increased the chances that their children would survive.

Although this theory is pure speculation, a few scientists have suggested that sexuality may have been the source of the rapid evolutionary success of the human species. By providing the bond that held individuals together in social groups, sexuality may have produced a situation wherein human culture was able to flourish. How remarkable to think that it may not have been our big brains or our dexterous hands that were the keys to our evolution, but rather our constant attraction to the opposite sex!

The reproductive system passes the genetic heritage of each species into the future by producing gamete cells from which offspring are formed. The primary sex organs of the male reproductive system are the testes. Sperm cells are produced within the seminiferous tubules of the testes and suspended in a fluid known as semen. Semen is delivered into the female reproductive tract by the remaining organs of the system. Sexual development in the male is under the control of pituitary gonadotropins, which are themselves regulated by releasing hormones derived from the hypothalamus. LH stimulates the production of testosterone, the male sex hormone, by the interstitial cells of the testes. Testosterone and FSH promote the maturation of sperm cells, and testosterone has the added effect of producing the male secondary sexual characteristics.

The female reproductive system produces ova in its primary sex organs, the ovaries. Under the influence of the endocrine system, a single ovum matures each month within a cluster of cells known as a follicle. Development of a follicle is stimulated by the pituitary hormones FSH and LH. The follicle cells release estrogens, female sex hormones, which are responsible for the production of secondary sexual characteristics. Ovulation, the release of the mature ovum, occurs when the follicle has completely developed, leaving behind the corpus luteum. The ovum travels down one of the fallopian tubes toward the uterus. If the ovum is not fertilized, disintegration of the corpus luteum causes the uterine lining to be sloughed off in a process known as menstruation. If fertilization occurs, the resultant zygote becomes implanted in the lining of the uterine wall. The implantation of a zygote prevents the corpus luteum from disintegrating and triggers a series of hormonal changes.

Sexual intercourse is a complex process involving a series of emotional and physiological changes that lead to feelings of intense pleasure in both male and female. Fertilization can be prevented by a series of contraceptive methods, including surgical, chemical, and barrier techniques. The act of sexual intercourse also creates an opportunity for disease-causing organisms to be passed from one partner to another.

STUDY FOCUS

After studying this chapter, you should be able to:

- Explain in detail the physiology of the male and female reproductive systems.
- Describe the interactions between the endocrine system and the reproductive system, and understand the extent to which the reproductive system depends on endocrine messages for its normal functioning.
- Discuss the most widely used methods of artificial contraception, citing the pros and cons of each.
- Explain the prevalence and mode of transmission of sexually transmitted diseases.

SELECTED TERMS

asexual reproduction p. 723
sexual reproduction p. 723
testes p. 724
semen p. 724
seminiferous tubules p. 725
gonadotropins p. 727
testosterone p. 727
ovaries p. 729
ovulation p. 730
fallopian tube p. 730
estrogen p. 731
corpus luteum p. 731
progesterone p. 731

menstruation p. 732
menopause p. 733
HCG (human chorionic gonadotropin) p. 733
estrus p. 735
orgasm p. 735
contraceptive p. 736
infertile p. 738

REVIEW

Discussion Questions

1. Describe the basic characteristics of the male and female reproductive systems.

2. List the primary male and female secondary sexual characteristics. Which hormones are responsible for these differences?

3. Some scientific writers have suggested that germ plasm is immortal. Although this cannot be strictly true, explain why the statement makes some sense.

4. Efforts to develop a male birth control pill have focused on drugs that would block the actions of FSH. Explain why it would be undesirable to block the actions of LH.

Objective Questions   (Answers in Appendix)

5. When blood testosterone levels fall below normal,
   (a) LH production in the pituitary ceases.
   (b) sperm production increases.
   (c) GnRH release ceases.
   (d) the pituitary releases LH.

6. What is the sequence of the developmental stages for sperm cells (starting with the earliest form)?
   (a) spermatogonia, sperm, spermatocytes, spermatids
   (b) sperm, spermatogonia, spermatocytes, spermatids
   (c) spermatogonia, spermatocytes, spermatids, sperm
   (d) spermatids, spermatocytes, spermatogonia, sperm

7. Ovulation occurs in response to increases in the blood in the levels of
   (a) progesterone.   (c) estrogen.
   (b) LH.   (d) testosterone.

8. Which birth control method is the least effective?
   (a) rhythm   (c) birth control pill
   (b) condoms   (d) IUD

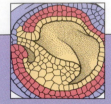

The most important event in your life is not birth, nor death, nor marriage. It is gastrulation.

—Lewis Wolpert

Every day of the year, in every corner of the earth, countless acts in an unseen drama play themselves out on the stage of life. Certain scenes in that drama unfold just beneath the quiet surfaces of ponds and streams. Other acts are set within the confines of eggshells. And still others transpire shrouded in the darkness of a mother's womb.

Yet despite their individual differences, all these dramas are but variations on a theme that unites all animal life: the formation of a new organism from a single, fertilized egg cell (Figure 38.1). This orderly process of **development**—which begins with the union of sperm and egg and continues throughout the life of the organism—is still the heart of one of biology's greatest mysteries: How does the complexity of a multicellular organism emerge from a process that begins with a single cell?

At the core of that mystery is a phenomenon that is at once both elementary and baffling. Every cell descended from a fertilized egg contains the same set of genes. Yet somehow, during development, each individual's genome directs cells to change in form and function in scores of different ways. The net result, if no errors occur, is a new individual composed of highly specialized cells, each of which performs a different task.

CHAPTER 38

Embryology and Development

THE PATTERNS OF DEVELOPMENT

Life begins with the fusion of sperm and egg to form a **zygote,** which then develops into an **embryo.** Embryos of different species follow a variety of developmental patterns. We will learn the most about our own development by studying the embryology of our closest relatives, other chordates.

The Egg: Protecting and Nurturing the Embryo

The different types of *ova,* or egg cells, that animals produce reflect a variety of evolutionary responses to two overriding requirements of development: the protection of the hapless embryo and the provision of adequate nutrients and energy to support intense cellular activity. In one way or another, we can understand all of the different types of chordate ova as adaptations in response to these demands.

The ancestors of terrestrial vertebrates were small aquatic organisms, most nearly represented today by such animals as the lancelet *Branchiostoma.* These simple creatures produce small eggs that remain in water as they develop. The eggs contain a modest amount of stored food material known as yolk. The tiny lancelets that hatch from these eggs develop quickly enough to escape the notice of most predators, and only a small amount of yolk is necessary to support development.

Amphibians such as frogs also reproduce in the water, but they pack much more yolk into their eggs and support a longer period of development. The egg first develops into an immature or *larval* stage known as a tadpole. Later, the tadpole goes through a rapid developmental period known as *metamorphosis* and then emerges as a mature frog.

Development in the lancelet: a small egg gives rise to a small organism that resembles the adult.

Development in the frog: medium-sized egg develops into a tadpole, tadpole metamorphoses into a frog.

Figure 38.1 These vertebrates employ a range of reproductive strategies. Vertebrate egg sizes range from the small to the very large. Egg types include those that may or may not be protected by an outer cover. The eggs of most mammals develop inside the mother's body.

Birds and reptiles produce eggs with very large amounts of yolk. During their long developmental period, they are protected with a tough outer shell.

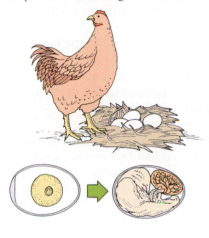

Mammals develop within the bodies of their mothers. This pattern enables mammals to produce small eggs which develop over a long period.

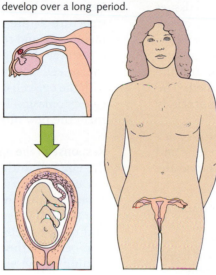

Reptiles, birds, and mammals have become independent of water for reproduction, but they have done so in ways that emphasize their evolutionary ties to our aquatic ancestors. Birds and reptiles produce large, fluid-filled eggs provided with enormous amounts of stored yolk and enclosed in tough covering for protection. As the embryo develops within, it produces special tissues that invade the mass of yolk and draw nutrients from it.

Most mammals exhibit a different adaptation: They produce small eggs with very little stored yolk, but embryos produced from these eggs develop and obtain nourishment while protected within fluid-filled membranes in the body of the mother. These mammals have evolved an organ known as the **placenta,** a connection that allows food and oxygen to be passed from mother to embryo.

Animals that lay eggs outside the body of the mother are said to be *oviparous,* a term derived from the Latin words for "egg" and "to give birth" (Figure 38.2). Those that nourish their young inside their bodies are *viviparous* ("to give birth" "alive"). The eggs of a few vertebrates,

Figure 38.2 Developing embryos are visible within a cluster of eggs of *Hyla ebraccata,* a tree frog found in the rainforest of Costa Rica.

including many common snakes, develop in a pouch within the mother's body but live off their yolk supplies and receive no further nourishment directly from the mother. These species are said to be *ovoviviparous.*

EARLY DEVELOPMENT

The single-celled zygote develops into a complex, multicellular organism through a process called **differentiation,** during which genetically identical cells develop along different pathways. Ultimately, the single fertilized egg produces the highly specialized cells that make up tissues as diverse as muscle, nerve, skin, and bone. But development is more than just cellular differentiation; embryonic tissues must grow within the limits of a set *pattern.* The organs must develop in certain positions, and the proportions of the limbs must be carefully scaled. The events in different parts of the embryo must be precisely timed.

Embryonic development can be divided into six distinct stages. We will review these stages one at a time and will follow the major structural changes that occur as the embryo develops.

gametogenesis	formation of gametes (egg and sperm)
fertilization	fusion of egg and sperm
cleavage	the first rounds of cell division
gastrulation	formation of the primary germ layers
neurulation	formation of the nervous system
organogenesis	formation of the major organ systems

Gametogenesis

Development begins with *gametogenesis,* the formation of sperm and egg cells. As we saw in Chapters 10 and 37, the genetic events of meiosis are nearly identical for male and female gametes. Other cellular events that occur during gametogenesis, however, are quite different, for sperm and egg cells are specialized for reproduction in very different ways.

While the sperm cell develops a motile flagellum, the cytoplasm of the larger egg cell is being loaded with extensive supplies of nutrients and raw materials to support the early growth of the embryo. In most vertebrates, yolk proteins are synthesized in the liver and carried to the ovaries through the bloodstream. *Follicle cells,* which surround the oocytes, absorb these proteins from the bloodstream and pass them to the developing egg. Furthermore, the active transcription of oocyte DNA fills the cytoplasm with messenger RNA molecules, some of which are stored and used to direct protein synthesis only after the egg has been fertilized.

The egg is surrounded by the *vitelline envelope,* known in mammals as the *zona pellucida.* This layer of glycoprotein serves not only as a protective covering but also as a species-specific recognition site to which sperm can bind.

Fertilization

When a sperm cell makes contact with the glycoprotein layers surrounding the egg, receptors on the membranes of both sperm and egg cell are activated, with virtually immediate and dramatic results.

Response of the sperm   Contact causes the *acrosome,* a structure at the tip of the sperm, to fuse with its cell membrane and to release enzymes that begin to dissolve a pathway through the layers surrounding the egg. The sperm cell penetrates through this opening and fuses with the egg cell membrane, releasing its nucleus into the egg cytoplasm. The haploid gamete nuclei then fuse to form the new diploid nucleus of the zygote.

Response of the egg   In some organisms, the reaction of the egg cell to the first sperm contact is spectacular. In sea urchins, fertilization activates thousands of *cortical granules* scattered around the periphery of the egg (Figure 38.3). These granules fuse with the cell membrane and release their contents to the exterior. A rapid reaction between the granule contents and the glycoprotein coating of the egg cell produces a *fertilization membrane* that surrounds the egg in a matter of a few seconds. This fertilization membrane completely blocks the entry of any other sperm, preventing *polyspermy,* or multiple fertilization.

Mammalian ova do not form fertilization membranes, but fertilization causes rapid changes in the zona

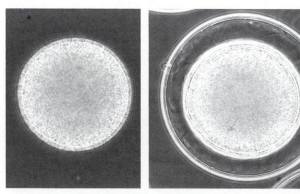

Figure 38.3   The cortex of a sea urchin egg (left) contains thousands of cortical granules. These granules are released when fertilization takes place, quickly forming a tough fertilization membrane (right) that prevents other sperm from entering the egg.

pellucida. These changes mask sperm-binding sites and ensure that only a single sperm cell is able to fertilize the ovum.

In some organisms, changes in the structure of egg cytoplasm occur at the point where the sperm cell enters the ovum. These changes have profound effects on further development. In frog eggs, for example, the penetration of the sperm causes a movement of the cytoplasm that changes the distribution of pigment granules in the egg cell (Figure 38.4). This produces a lightly pigmented region known as the gray crescent *opposite* the point where the sperm entered. The position of the gray crescent indicates the anterior–posterior axis of the embryo: the direction in which the head and tail of the tadpole will emerge from the egg.

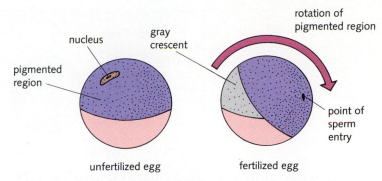

Figure 38.4 Sperm entry into the frog egg causes a movement of pigmented cytoplasm, producing a distinct region called the gray crescent *opposite* the point of sperm entry.

Cleavage in the Embryo: Cell Division Begins

After fertilization, the zygote goes through rapid mitotic divisions that partition the single cell into individual cells known as *blastomeres.* In species where the egg cell is large, these first divisions can be dramatic, and early embryologists chose the word **cleavage** ("splitting") to emphasize that fact. Differences in the distribution of yolk affect the way in which a zygote begins development (Figure 38.5).

In eggs with small amounts of yolk, including those of most mammals, sea urchins, and the lancelet, *complete cleavage* occurs, dividing the zygote into cells of roughly equal size. Complete cleavage also occurs in amphibians. However, a complication is introduced by the fact that amphibian eggs contain a higher concentration of yolk at one end of the cell, which is designated as the **vegetal pole.** The opposite, yolk-poor end is called the **animal pole.** At first, cleavage in amphibians produces cells of equal size. Before long, however, a more rapid rate of cell division at the animal pole of the egg produces cells much smaller than those formed at the vegetal pole.

By contrast, in eggs with large amounts of yolk—including those of birds, reptiles, and insects—cell division can occur only at the surface of the egg, a process known as *superficial cleavage.*

The blastula As cleavage proceeds, the number of cells in the zygote doubles with each round of mitosis. Gradually the dividing cells form a **morula,** a tight mass of cells named for its resemblance to the mulberry fruit (the Latin word *morula* means "mulberry") (Figure 38.6). In *Branchiostoma,* the cluster of cells gradually develops a hollow core. This stage of development is known as the **blastula** (the Greek word *blastos* means "bud"), and the cavity enclosed by the single layer of cells is called the *blastocoel.* A similar structure forms in all vertebrate embryos, but the different patterns of egg size and yolk

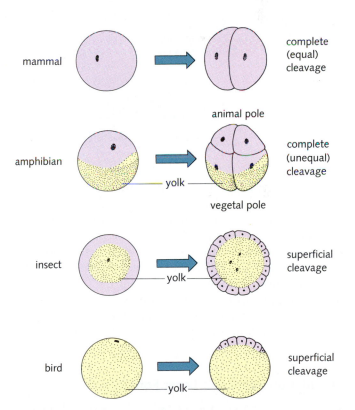

Figure 38.5 In small eggs (amphibians and placental mammals), complete cleavage follows fertilization. Insect eggs undergo superficial cleavage around the periphery, and superficial cleavage of the large eggs of birds and reptiles occurs in a small disk above the massive yolk.

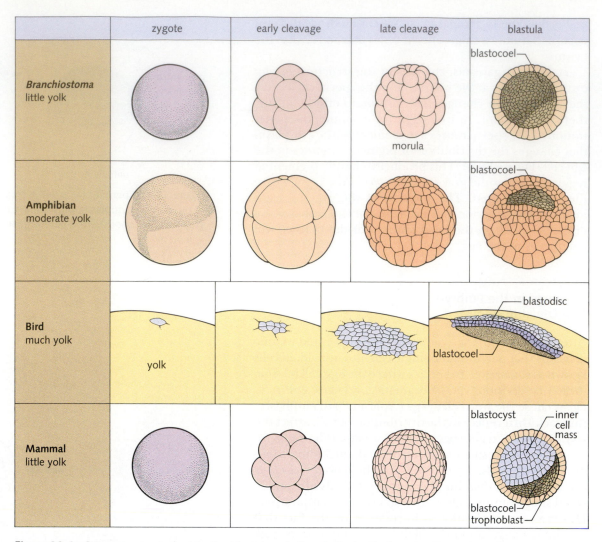

	zygote	early cleavage	late cleavage	blastula
Branchiostoma little yolk			morula	blastocoel
Amphibian moderate yolk				blastocoel
Bird much yolk	yolk			blastodisc / blastocoel
Mammal little yolk				blastocyst / inner cell mass / blastocoel / trophoblast

Figure 38.6 Structures equivalent to the blastula, a hollow ball of cells, form in all vertebrates.

distribution prevent a blastula of uniform structure from developing in all species.

In mammals, the equivalent of a blastula phase is the **blastocyst.** The blastocyst consists of two parts. The *inner cell mass* will give rise to the tissues of the embryo. The outer *trophoblast* will form part of the **placenta,** an organ that draws nourishment from the mother. In birds the early cleavages produce a **blastodisc,** a double layer of cells atop the yolk mass that gradually separates into two layers with a hollow *blastocoel* between them.

Gastrulation and Germ Layer Formation

Development in chordates produces three germ layers that ultimately form organs and tissues. The outermost layer, the **ectoderm** ("outer skin"), eventually forms the skin and the nervous system. The inner layer, the **endoderm** ("inner skin"), forms the lining of the digestive tube

and organs of the digestive system. The layer between the other two is the **mesoderm** ("middle skin"), source of muscle and bone tissue, the circulatory system, and other internal organs and tissues.

During the developmental stage known as **gastrulation,** the first steps are taken to produce these layers and to establish the basic body plan of the embryo. By the time gastrulation is complete, the cells that will produce the three germ layers can be clearly identified, and the embryo is known as a **gastrula.** Details of gastrulation differ among embryo types, but in each case they involve massive migrations of cells inward and on the surface (Figure 38.7). The simplest form of chordate gastrulation, exemplified by *Branchiostoma,* involves an inward movement of cells from the surface of the blastula. Like a soft rubber ball squeezed in at one side, the cells at one end of the blastula begin to fold in to form a new cavity, the **archenteron,** that is open to the exterior through the **blastopore.** The *archenteron* (from the Greek words for

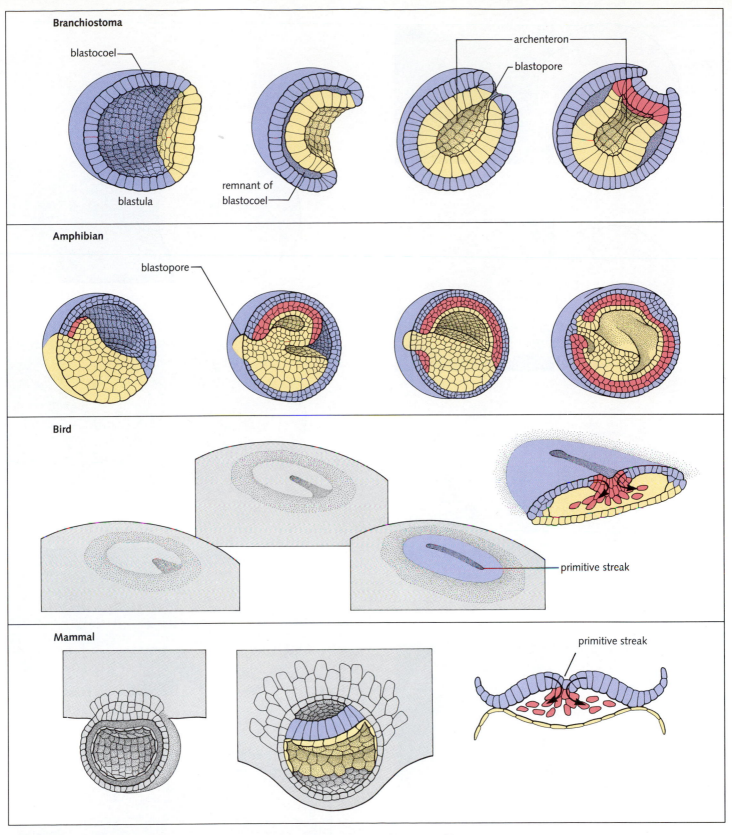

Figure 38.7 Gastrulation involves the inward migration of cells from the blastula. When gastrulation is complete, the three germ layers, ectoderm, mesoderm, and endoderm, have been formed. In lower chordates, mesoderm develops from early endoderm. In birds and mammals, mesoderm is formed by the inward migration of cells between ectoderm and endoderm.

endoderm

ectoderm

mesoderm

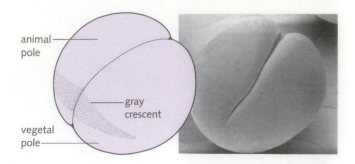

animal pole

gray crescent

vegetal pole

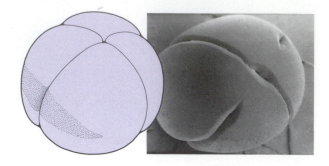

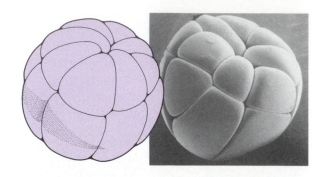

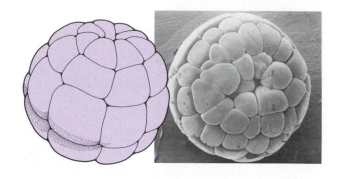

blastocoel

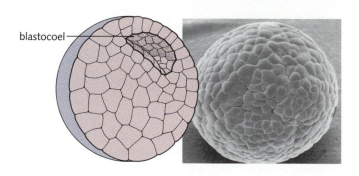

blastocoel

dorsal lip

archenteron

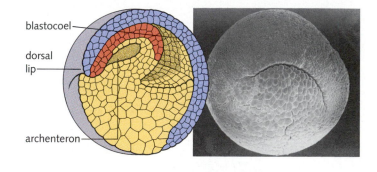

archenteron

yolk plug

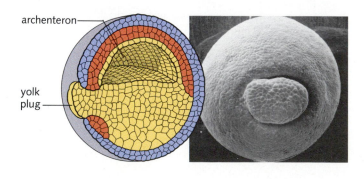

neural fold

notochord

gut cavity

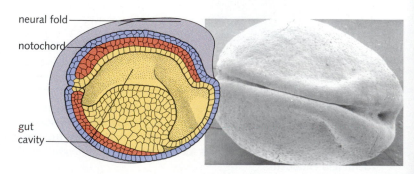

Figure 38.8 Blastulation and gastrulation in amphibians.

"primitive" and "intestine") will form the cavity of the digestive system of the embryo. The early gastrula consists of two distinct cell layers, the ectoderm and endoderm, surrounding the archenteron. Within a matter of hours the mesoderm forms between these two.

Gastrulation in amphibians also involves the inward movement of cells. A slit-like blastopore forms at the site of the gray crescent, and cells migrate inward to produce an endoderm-lined archenteron. Further cell movements produce mesodermal layers that give rise to the notochord and internal organs.

In birds and mammals gastrulation is different. Rather than moving inward toward a blastopore, certain embryonic cells migrate through a zone known as the **primitive streak,** which leads to the interior of the embryo. During gastrulation, movement of these cells toward and through the primitive streak is rapid, resulting in the immediate formation of mesoderm between the two existing cell layers. By the time gastrulation is complete, a small disk of three cell layers has been formed. Figure 38.8 shows, in greater detail, the blastula and gastrula stages of development in amphibians.

Neurulation

One of the most important steps in chordate development is **neurulation,** the development of the nervous system. During the formation of the three germ layers, a block of mesoderm begins to develop into the **notochord.** (All chordates, at some stage in their development, possess a notochord; in vertebrates, the notochord is eventually replaced by the spinal column.) As the noto-

chord begins to take shape, there is a noticeable change in the ectoderm just above it, in the region known as the *neural plate*. Along the length of the notochord, ectodermal cells thicken and change shape in a way that produces paired cell ridges along the length of the organism. Gradually, these ridges fuse to produce a **neural tube** from which the spinal cord is formed.

As neurulation proceeds, clusters of ectodermal cells break off from the folding neural plate and occupy the region between the neural tube and the epidermal ectoderm (Figure 38.9). These cells, *neural crest cells,* migrate throughout the body and are some of the most remarkable cells in the embryo. Some neural crest cells develop into sensory neurons, others form the *glial* cells that surround some nerve fibers, others become secretory cells in the adrenal medulla, and still others form the melanin-containing pigment cells that are responsible for skin coloration.

In humans, failure to complete neurulation successfully occurs in about 1 birth in 3000, causing a birth defect known as *spina bifida*. This incomplete closure of the neural tube may leave the nervous system open to the exterior. Spina bifida can be corrected by surgery to enclose the neural tube with skin and guard against infection.

Organogenesis

Before neurulation is finished, organ development, or **organogenesis,** begins. Organogenesis starts with the formation of **somites,** distinct blocks of mesodermal tissue on either side of the neural tube (Figure 38.10). These

Figure 38.9 Neurulation. Ectodermal tissue forms a hollow tube, which gives rise to the spinal cord and the brain. A few cells break off during neurulation to form neural crest cells, which migrate to various locations throughout the body.

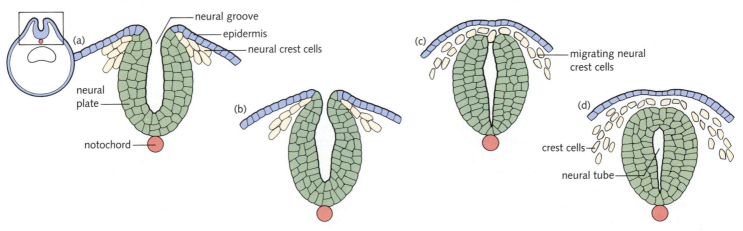

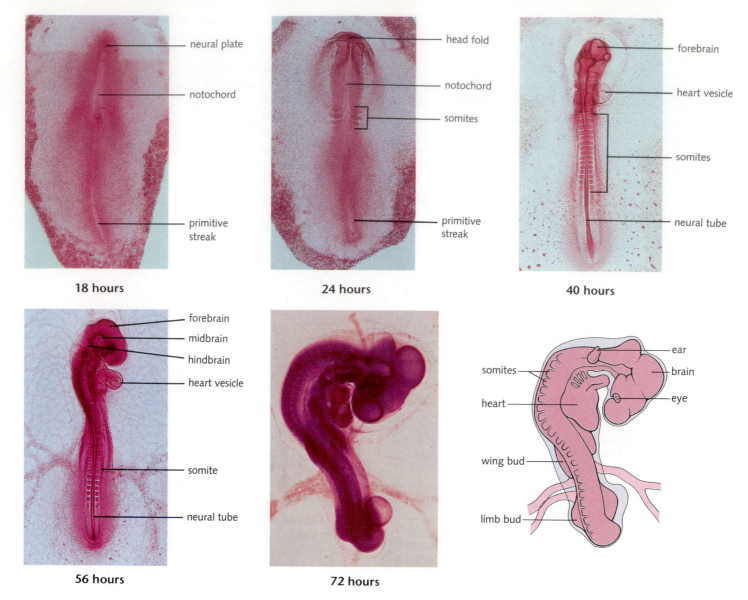

Figure 38.10 Photographs of embryonic development in the chick. Note the formation of the first somites, blocks of tissue from which bone and muscle will form. The diagram clearly shows the positions of major structures after 72 hours of development.

somites develop into bone, muscle, and certain internal organs.

At this point, organ development begins to involve cells and tissues from several layers. Epidermal ectoderm and nervous tissue, for example, both contribute to the formation of the eyes. Similarly, ectoderm and mesoderm form the *limb buds* from which front and back limbs develop, while endoderm and mesoderm form the pancreas.

As a summary, a comparison of the major developmental stages in the frog, chick, and human is shown in Figure 38.11.

Mechanisms of Development

As you have seen, many events during embryonic development are rapid and dramatic. In order for the systems of developing embryos to achieve their final form and function, hundreds or thousands of cells must perform specific operations in a highly coordinated fashion. Although there is much about this coordination that we do not understand, there are several developmental mechanisms and control systems that we can observe and describe, at least superficially.

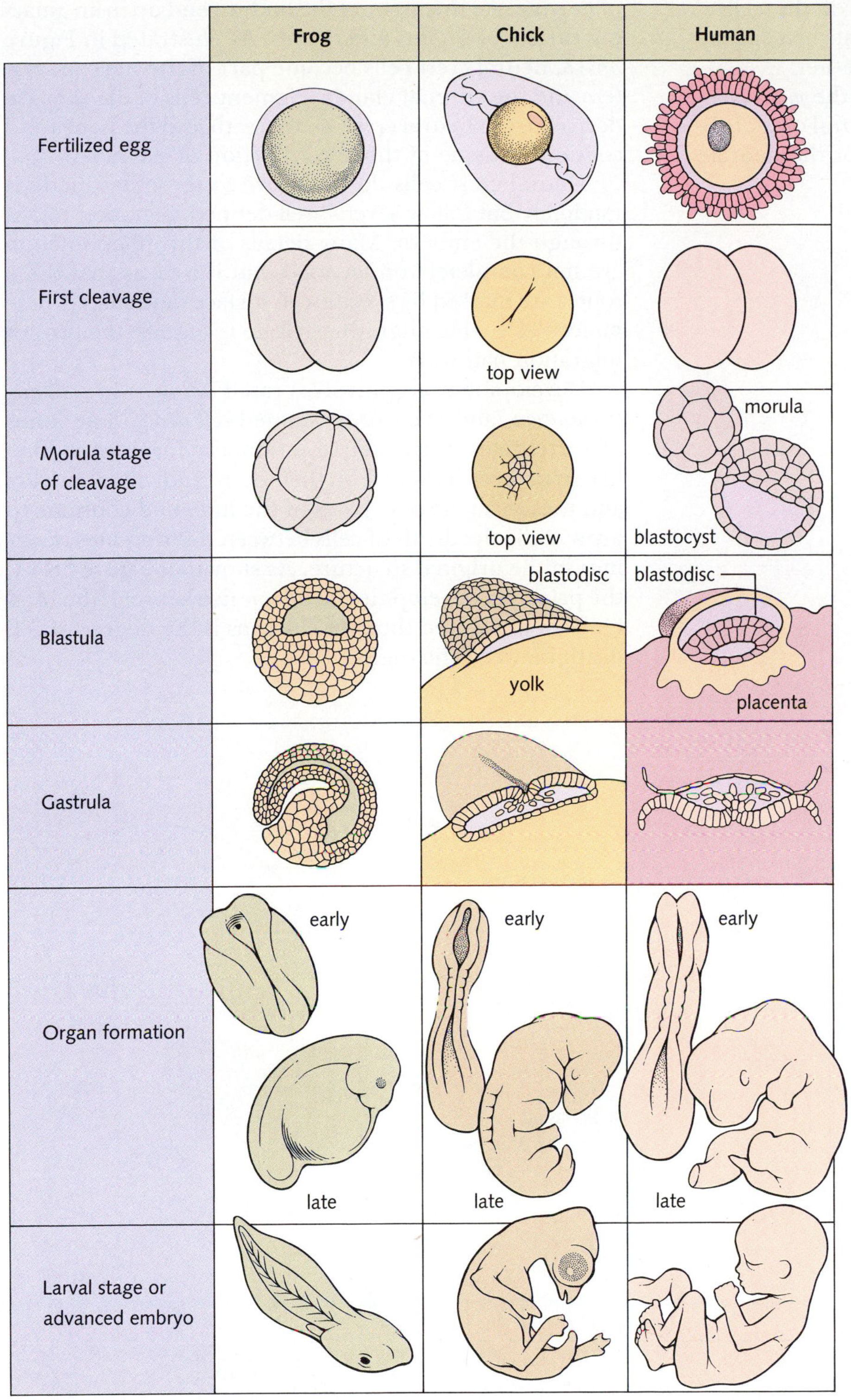

Figure 38.11 A summary of major developmental stages in the frog, chick, and human.

	Frog	Chick	Human
Fertilized egg			
First cleavage		top view	
Morula stage of cleavage		top view	morula / blastocyst
Blastula		blastodisc / yolk	blastodisc / placenta
Gastrula			
Organ formation	early / late	early / late	early / late
Larval stage or advanced embryo			

The relatively undifferentiated cells of early embryos undergo several processes during which they become organized into recognizable body systems. These processes, which occur at the cellular level, are often called cellular mechanisms of development.

Neurulation, for example, involves a series of coordinated and carefully timed *changes in cell shape*. Ectodermal cells at the ends of the neural plate first become elongated and then constricted at one end to produce the neural tube (Figure 38.12). These changes in shape involve microtubules that run the length of the cells; constriction of a ring of actin microfilaments acts like a pursestring to draw one end of the cell together.

Other developmental processes involve the *migration* of cells through the embryo. Cells of the neural crest, for example, which all originate at the edges of the neural plate, migrate throughout the body to end up in an amazing variety of organs and tissues. As illustrated in Figure 38.13, neural crest cells become part of the nervous system and the adrenal glands. Pigment cells in the skin are derived from neural crest, as are teeth and the bones and connective tissue of the lower portion of the head.

Neural crest cells do not move to these destinations randomly but follow several well-defined migration routes through the embryo. Many details of this phenomenon are not completely understood, but it is clear that these routes are marked by specific cell surface molecules. These molecules enable migrating cells to recognize the proper migration pathways.

Development also involves rapid *cell division*, *cell differentiation*, and even programmed *cell death*. The limbs of vertebrates, for example, originally form as paddle-like structures flattened on their ends. Individual fingers and toes form as some parts of the limb bud continue to grow while the death of cells between them creates openings in the original structure. As shown in Figure 38.14, the principal developmental difference between the hind limbs of ducks and those of chickens is the degree of cell death between the digits.

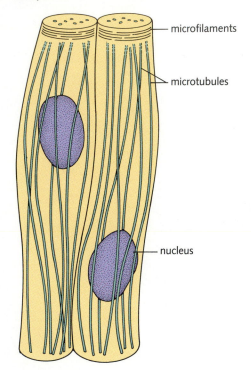

Figure 38.12 Individual changes in cell shape are responsible for neurulation. Cell elongation is the first in a series of changes that produce the neural tube.

microfilaments

microtubules

nucleus

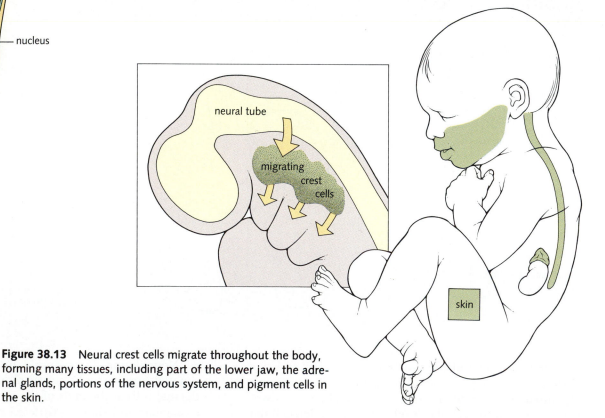

Figure 38.13 Neural crest cells migrate throughout the body, forming many tissues, including part of the lower jaw, the adrenal glands, portions of the nervous system, and pigment cells in the skin.

neural tube

migrating crest cells

skin

HUMAN EMBRYOLOGY

We can now follow this entire process in detail by tracing human development from start to finish.

Gametogenesis and Fertilization

As we saw in Chapter 37, human sperm and ova are highly specialized cells produced by meiosis and differentiation. Although both gametes are haploid, the differences between the formation of sperm and ova are very great, reflecting the differences in their roles in the reproductive process.

Sperm are specialized for service as actively moving carriers of genetic information. A sperm cell's sole mission is to locate the ovum, bind to its outer surface, and deliver the sperm's haploid nucleus. Sperm are therefore little more than "stripped down" chromosome carriers. Their genetic material is condensed and packaged into a streamlined head, which is attached to an energy-generating midpiece and a long flagellum. Each sperm carries an acrosome whose enzymes help it penetrate the zona pellucida when it contacts the ovum.

The tasks of the ovum, on the other hand, are more complex. In addition to providing its own haploid nucleus, the ovum must provide almost all of the cytoplasm for the zygote, including ribosomes, transfer RNAs, and preformed messenger RNAs. The ovum also contains a store of food material in the form of yolk, which will provide energy for the zygote and support its development until it is able to draw nourishment from the mother. Thus, although the human ovum is quite small compared to the eggs of amphibians or birds, it is many thousands of times larger than sperm. The process of fertilization, as described earlier in this chapter, generally occurs in the fallopian tubes. The fusion of a sperm cell with an egg causes the nucleus of the ovum to complete meiosis, and the haploid male and female *pronuclei* join to form the diploid zygote nucleus.

The First Month of Development

Human development follows a pattern that is typical of mammals. The first cell division typically occurs within 36 hours of fertilization (Figure 38.15). Other cell divisions follow as the zygote moves down the fallopian tube, and within 4 days the *morula,* a ball of about 75 cells, has formed. Gradually the morula develops an internal cavity; when this occurs, the embryo has reached the *blastocyst* stage, and cell division slows down. In a sense, the embryo waits for the next critical step, **implantation,** which

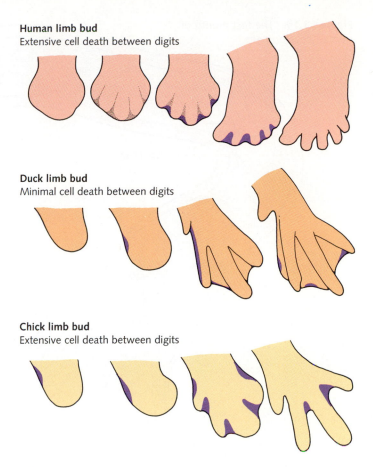

Human limb bud
Extensive cell death between digits

Duck limb bud
Minimal cell death between digits

Chick limb bud
Extensive cell death between digits

Figure 38.14 As limb buds develop, a genetically-determined pattern of cell death helps to sculpt the individual digits.

involves the binding of the blastocyst to the *endometrium,* the lining of the uterus.

Cell surfaces in the embryo carry glycoprotein molecules that bind to molecules on cells lining the endometrium. This enables the blastocyst to attach tightly to the endometrium. Once the embryo has made this attachment, cell division resumes as rapid growth begins again.

As with other mammals, the blastocyst contains the inner cell mass that eventually grows into the new individual—in this case, a human being. The inner cell mass is surrounded by a series of tissues that make direct contact with the uterus and already help to channel nourishment to the rest of the embryo. The outer tissues, initially known as the trophoblast, penetrate the tissues of the mother. As they develop, they form a distinct layer known as the **chorion.**

Figure 38.15 The first month of human development.

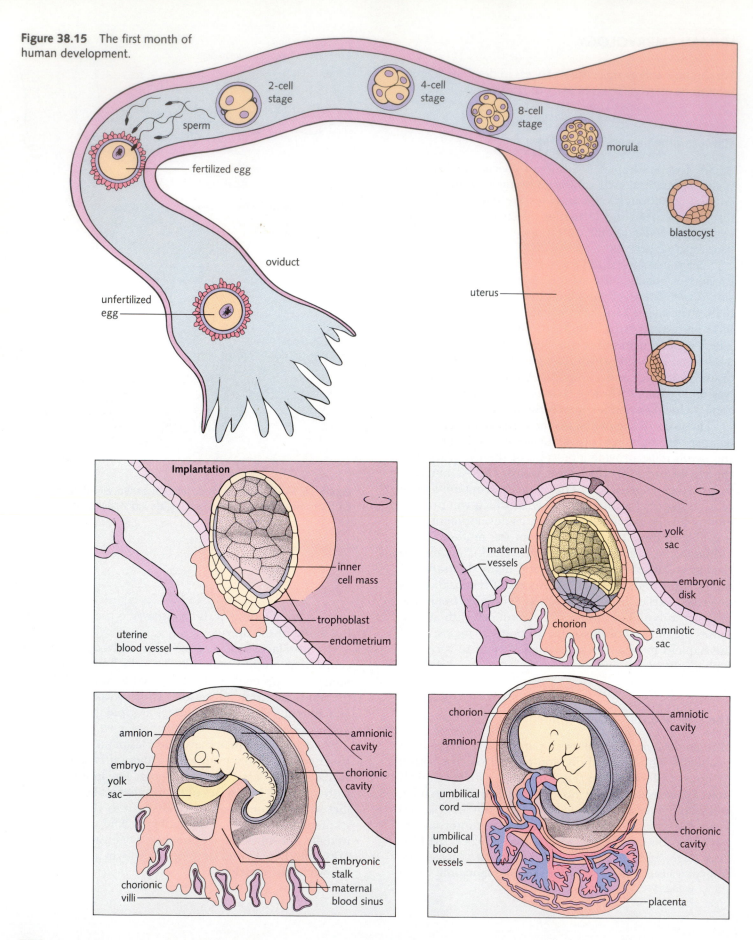

After implantation, the embryo literally invades the tissues of the endometrium, and within 2 weeks after fertilization it is totally embedded in the uterine wall. In addition to making direct contact with the mother's cells, the trophoblast has another important function. By releasing **human chorionic gonadotropin (HCG),** it prolongs the lifetime of the corpus luteum. The corpus luteum, in turn, prevents menstruation by continuing the secretion of estrogen and progesterone.

As Figure 38.15 shows, the implanted embryo goes through a number of dramatic changes in tissue organization. First the inner cell mass produces an embryonic disk with two cell layers. Next gastrulation occurs, resulting in the formation of three germ layers. As you will note, the embryo is already associated with two cell-lined *extraembryonic* ("outside the embryo") sacs: the **yolk sac** and the **amniotic sac.** In birds, as you recall, the yolk sac contains yolk that nourishes the embryo during development. Embryos of placental mammals, though they have very little yolk, form an empty yolk sac from which embryonic blood cells are formed, retracing their evolution from organisms that produced yolk-bearing eggs.

Shortly after gastrulation, the amniotic sac expands so that it surrounds the entire embryo. The yolk sac and another extraembryonic sac, the allantois (which, in birds and reptiles, stores metabolic wastes), are squeezed together in a thin stalk of tissue that passes through the growing amniotic sac to connect the embryo to the mother. This is the beginning of the umbilical cord.

Gradually, blood vessels from the developing embryo grow through the umbilical cord, and a new organ, the placenta, is formed from both maternal and embryonic tissue. Figure 38.16 shows how the circulation of blood from mother and embryo is organized within the placenta. Tiny *villi* containing embryonic blood vessels extend into open *sinuses* of maternal blood. Although the two circulations come into very close contact, they normally do not mix. Food, oxygen, carbon dioxide, metabolic wastes, drugs, antibodies, and viruses cross the placental barrier, but whole cells generally do not.

The production of two parallel but separate circulatory systems is an important feature in the development of placental mammals. It prevents many infections from spreading through the circulatory system and enables both mother and developing embryo to be prepared for the day when the connection between them will be broken. In essence, the placenta serves the developing embryo as its principal organ of excretion, nourishment, and respiration.

In the fourth week of development, two prominent ridges of ectoderm begin to form running the length of the embryo's dorsal side. These ridges roll up to form a

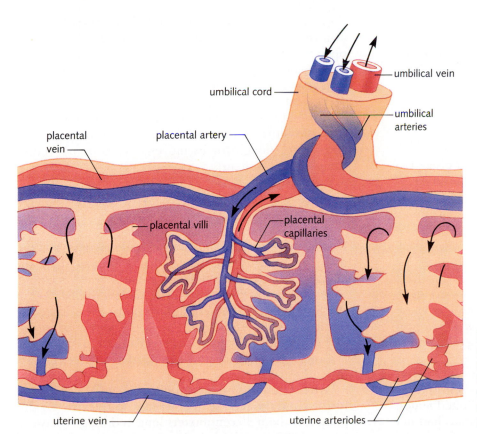

Figure 38.16 A schematic view of the placenta. Fetal and maternal blood do not mix within the placenta. Nonetheless, they do come into very close contact, allowing oxygen, carbon dioxide, food, and waste products to pass between mother and child.

Figure 38.17 Neurulation in the human embryo. After four weeks of development, the neural tube is completely enclosed.

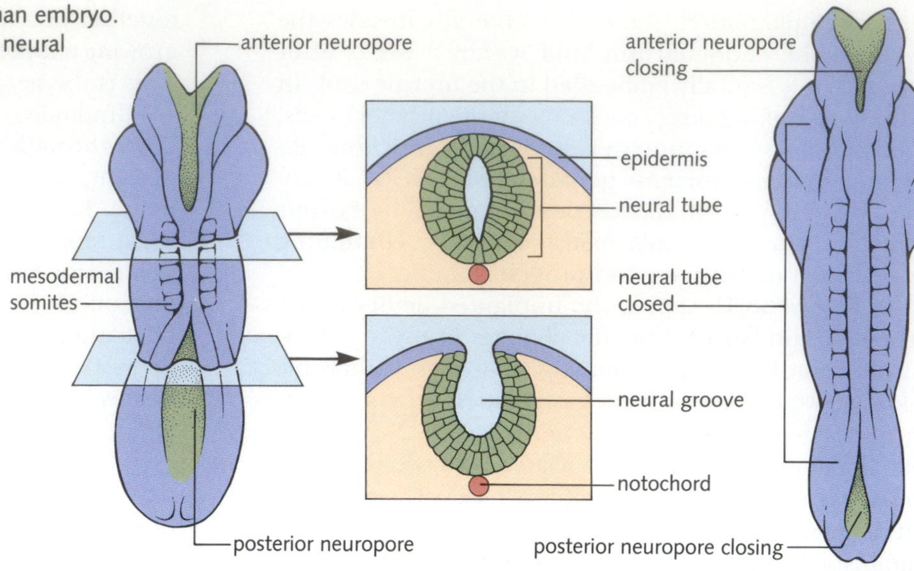

anterior neuropore

anterior neuropore closing

mesodermal somites

epidermis

neural tube

neural tube closed

neural groove

notochord

posterior neuropore

posterior neuropore closing

deep, hollow cleft, and neurulation begins (Figure 38.17). This process is similar to that in other vertebrates, and by the end of the fourth week the neural tube has formed and become enclosed within the embryo.

By this time all the major cell layers of the embryo have been formed, and most of its major organs are either present or well on their way to formation. Small buds have formed where the limbs will emerge, and the embryo is about half a centimeter in length. All this has happened in a very short time; the mother's menstrual period has been delayed by only about 10 days, and she may have begun to wonder whether she is pregnant.

The First Trimester

For reasons of convenience, physicians often break the process of human development into three **trimesters,** 3-month periods of development. During the first trimester the embryo takes on the distinct appearance of a human child, both internally and externally (Figure 38.18).

Because all body systems are established during the first trimester, this is a particularly sensitive time for development, and a number of external factors can affect the embryo. *Rubella* (German measles) infections can cause the heart to develop improperly, injure the eyes, or cause deafness. Drugs, such as the tranquilizer *thalidomide,* can prevent normal limb development, resulting in crippling malformations.

Thalidomide was introduced in the early 1960s. It proved safe in all tests on animals and was released for sale in Europe. Its sale was blocked in the United States, however, because its safety in pregnant animals had not been established. This proved to be a fortunate precau-

tion. The drug prevents normal limb formation, and women who took the drug and who were carrying fourth- and fifth-week embryos (the time when the limb buds form) gave birth to babies with severe limb malformations. Small flipper-like structures were produced in place of hands or feet, and thousands of children were born crippled for life.

By the end of 8 weeks, the embryo is officially known as a **fetus.** Its skeleton and muscle systems have developed to the point where it begins to make a few halting movements, and the sexual organs have begun to develop (it is now possible to tell a boy from a girl). By the end of 12 weeks the fetus is nearly 10 centimeters long. For centuries, the end of the first trimester was known as the time the fetus "quickened" in the womb (became large and active enough for its movements to be noticed by the pregnant woman).

The Second Trimester

During this phase, existing organs and systems grow and mature. The fetus is covered with a layer of soft hair (the lanugo, a term derived from a Latin word meaning "down"). By the end of the fifth month, the beating of the fetal heart is loud enough to be heard outside the mother with a stethoscope. The fetus's lungs are enlarging, its digestive system is almost ready to function, and it shows reflexes in response to sudden changes in light or loud sounds.

After 6 months of development, all body systems are formed and the fetus is nearly ready to be born. By this point it is about 35 centimeters long and weighs between 500 and 800 grams. A baby born at the end of the second

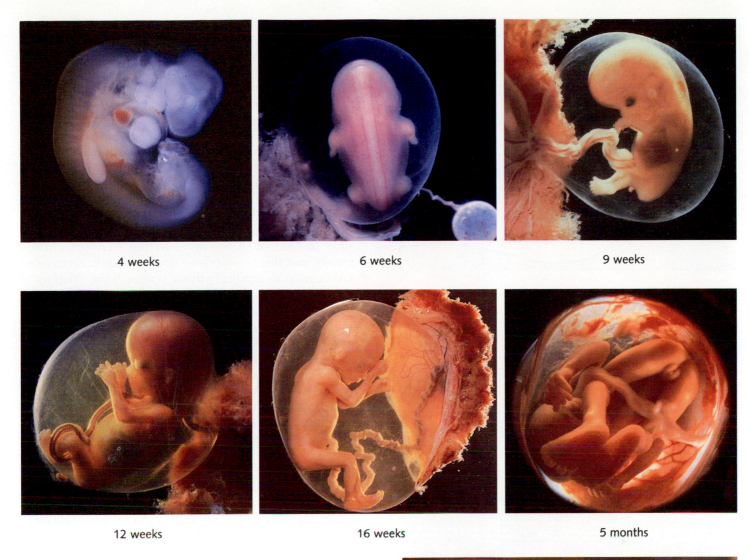

4 weeks

6 weeks

9 weeks

12 weeks

16 weeks

5 months

trimester has a chance of surviving. The greatest problems faced by prematurely born infants at this stage are temperature regulation (because they are so small) and breathing (because their lungs are not quite ready to take over gas exchange for the whole body). Such infants suffer from chronic respiratory distress.

Despite these difficulties, the care of premature infants is improving almost yearly, and intensive-care wards at major hospitals are now able to save a large fraction of babies born near the end of the second trimester.

The Third Trimester and Birth

The third trimester sees the most dramatic weight gains of the developmental period. A mother is well aware that her baby has gone from 600 grams to an average of 3000 grams during those last 3 months. During this period, all the major organ systems develop to the point where

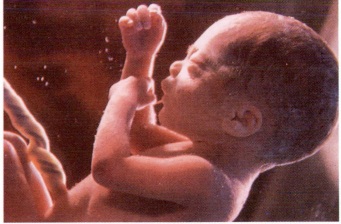

9 months

Figure 38.18 Photographs made of human development within the uterus at different periods.

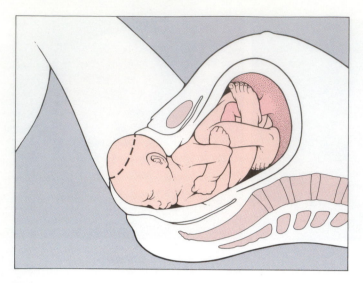

Figure 38.19 The position of a baby as it passes from the uterus through the birth canal during delivery.

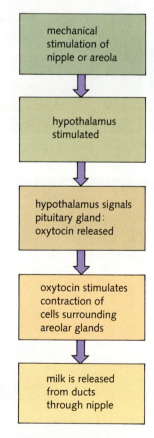

mechanical stimulation of nipple or areola

↓

hypothalamus stimulated

↓

hypothalamus signals pituitary gland: oxytocin released

↓

oxytocin stimulates contraction of cells surrounding areolar glands

↓

milk is released from ducts through nipple

Figure 38.20 Successful nursing requires the involvement of the nervous system and the endocrine system to release the milk produced by exocrine glands in the breast.

they are prepared to function independently of the mother. By the end of the eighth month, development has progressed so far that an early birth presents little danger. The major change physicians watch for at this stage is any sign that the placenta might be degenerating and ceasing to function.

Finally, approximately 266 days after it has begun, the process of embryonic development is complete, though it is not clear exactly how the body "knows" that it is time to send the fetus along to a life of its own.

Slowly at first, the large smooth muscles surrounding the uterus begin a series of rhythmic contractions known as **labor.** These contractions are triggered by **oxytocin,** a hormone released from the posterior pituitary. (It is possible to "induce" labor by injecting oxytocin intravenously, and this is sometimes done for medical reasons.) Other muscles gradually dilate (enlarge) the opening of the cervix, and little by little it becomes large enough (10 centimeters) for the fetus to pass through it.

The contractions of labor continue, becoming stronger and closer together. When the cervix has dilated to a full 10 centimeters and the contractions occur at intervals of 2–3 minutes, birth is ready to begin. The stronger contractions burst the amniotic sac, and the fluid within it rushes through the cervix and out of the body through the vagina. Gradually the fetus is forced through the action of contractions, usually head-first, out of the uterus and through the cervix into the vagina, or birth canal (Figure 38.19).

The baby is said to have *crowned* when its head is visible in the cervix. At this point, more contractions generally take from 2 to 90 minutes to force the baby through the cervix and out through the birth canal, still attached to its mother via the umbilical cord. The cord is just long enough to allow the infant to pass completely through the birth canal while still receiving nourishment and oxygen from the mother.

As the newborn makes contact with the outside world, it may cough or cry to loosen some of the fluid that has filled its lungs, and then it takes its first independent breath. Blood vessels leading to the placenta begin to close, and the placenta itself detaches from the uterine wall and is released, traveling through the birth canal as the **afterbirth.**

In larger hospitals, the placenta and part of the umbilical cord are recovered—the skin that covers them is useful in treating burn patients. (The cells of the placenta are specialized, in a sense, to invade the tissues of another human without triggering the immune reaction. Hence they do not trigger such a reaction so readily as other tissues that might be used to cover burns.)

By nursing her child soon after it is born, a mother helps bring the birthing process to conclusion (Figure

38.20). Hormonal feedback caused by nursing releases more oxytocin, which now restricts blood flow to the uterus. This helps slow down any bleeding that might have been induced when the placenta pulled out of the uterine wall.

In addition, the first secretions from the breast after birth contain not milk but **colostrum,** a fluid rich in nutrients containing a special class of antibodies that help protect the infant from infection for many weeks.

Nursing a newborn has many advantages for both mother and child. Human milk is the perfect food for a baby, and it is always sterile, is always the right temperature, and doesn't require a special trip to the store. In addition, there is little chance that it will provoke the allergic reaction to feeding that sometimes occurs when cow's milk is substituted.

Maturation: Development Continues

Human development does not end with birth, for the cells in many body tissues continue to proliferate and differentiate for weeks, months, or even years. The immune system is only partially mature at birth; it is not fully functional for several weeks into infant life. The nervous system continues to develop as it interacts with the environment, a fact that emphasizes the importance of providing infants with stimulating and interesting surroundings. Clearly, the reproductive organs are not fully functional until puberty. And the growth of the muscular and skeletal systems continues for 16 to 18 years after birth until the full adult height is reached.

It might be tempting to say that human development is complete at that point (or shortly thereafter). And it is true that beyond a certain point, most cells in most tissues stop dividing and differentiate into their final specialized form. From that time on, many cells of the brain and major nerves, for example, normally divide only when damaged. But in several body tissues, development never stops. In many tissues—including those of the liver, the skin, and the lining of the digestive system—continuous production of new cells is required to keep up with normal wear and tear. Bone marrow, too, continues to produce several different classes of blood cells throughout life.

You will see shortly that the processes of cellular reproduction, growth, and differentiation are regulated by several sorts of intercellular messengers. Normally, the actions of these messengers maintain body-wide homeostasis in mature individuals. If any cells escape this regulatory control system for some reason, one or another form of cancer develops.

Aging

For a number of years after physical maturity is reached, body systems continue to operate at peak efficiency. But after several decades, the flexibility of the skeletal system decreases, muscle strength ebbs, the skin becomes wrinkled and more transparent, and the reproductive system becomes less active. In females **menopause,** which usually occurs before age 50, marks a complete halt in the ability to produce ova and bear children. Researchers often mark age 60 as the beginning of **senescence,** or old age.

Despite intense investigation into the cellular basis of aging, this process is still poorly understood. Explanatory theories include the accumulation of cellular waste products over time, the accretion of errors in DNA replication, and a preprogrammed genetic cycle of physical decline.

Whatever the ultimate causes may be, old age does not have to be a period of inactivity. Average life expectancy in developed countries is approaching 75 years, and it is clear that people are capable of enjoying life and making important contributions to society for each and every one of those years. With attention to good health, including a sensible diet and regular exercise, older people can make the last phase of the human life cycle very rewarding (Figure 38.21).

Figure 38.21 Four generations in the same family. Good diet, exercise, and attention to medical care help to ensure that every stage of human life can be full and rewarding.

Fertilized Egg	Nuclear Division	Nuclear Migration	Blastoderm Formation		Gastrula	Early Organ Formation	Larva

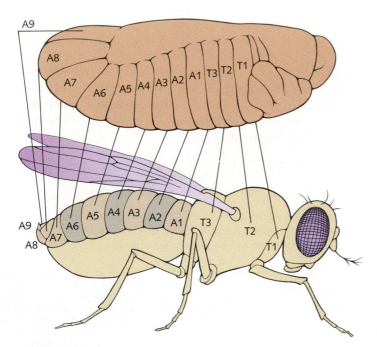

Figure 38.22 (top) Stages in *Drosophila* development. (bottom) There is a direct correspondence between segments in the embryo and those in the adult fly.

COORDINATION OF DEVELOPMENT

Developmental processes as complex as those we have been discussing could not occur if individual cells—or even groups of cells—operated independently. Clearly, the behavior of embryonic cells and the fates of their descendants must be controlled and coordinated with great precision.

Yet each cell in an embryo contains exactly the same genes as all other cells around it. Each cell thus contains all the information it needs to develop into any of the tissue types found in the adult. This statement raises questions that are at once simple and profound. How do various cells "know" what sort of tissue to become? And if all cells are genetically identical, how does the process of controlled differentiation begin?

If you think about this problem for a moment, you will realize that the fate of an embryonic cell depends on the ways in which some of its genes are turned on while others are turned off. For that reason, research on the cutting edge of developmental biology today focuses on the patterns and mechanisms of gene expression during development. Before we delve into that fascinating research, however, we need to consider briefly the organisms chosen for this work and the strategy used by many researchers to investigate the intricate pathways of development.

The complexity of vertebrate embryos has led researchers to search for simpler organisms in which to apply the tools of genetics and molecular biology. One such organism is the fruit fly, *Drosophila melanogaster,* for which a rich library of genetic information has been accumulated over the years

Development in *Drosophila*

Development in *Drosophila* differs from the vertebrate pattern. After several rounds of nuclear division, the nuclei of the zygote move toward the periphery of the embryo. Cell membranes then separate the individual nuclei, forming a monolayer of cells known as the **blastoderm** (Figure 38.22).

Next, during gastrulation, the three germ layers are produced and the embryo becomes fragmented longitudinally into a series of distinct *segments* from which the adult body is derived. A normal embryo contains 15 segments: 3 head segments, 3 thorax segments, and 9 abdomen segments.

This embryo hatches from the egg into a *larva* that retains these 15 segments and, after a period of intense feeding, growth, and molting, passes through a *pupal* stage and becomes an adult fly by the process of *metamorphosis*.

The limbs of the adult fly are not derived directly from the segments of the larva but arise from a series of small pouches in the larval epithelium that are known as **imaginal disks.** Disks on either side of the body produce the eyes and antennae, the three pairs of legs, and the wings. Experiments in which cells were transplanted from one part of the embryo to another have shown that in the blastoderm, all cells are already committed to their particular segments—their developmental fates have already been determined. Once this discovery was appreciated, investigators began to search for genes that controlled the pattern of segmentation and cell determination in *Drosophila*.

Caenorhabditis: The World's Best-Known Embryo?

There is another organism, little known to the public, that plays an important role in developmental research. In the 1960s, British scientist Sidney Brenner suggested that the tiny roundworm *Caenorhabditis elegans* might be an ideal species in which to study development.

This organism, less than a millimeter in length, affords several experimental advantages. It can be grown easily on culture plates in the lab. Its generation time is less than three days, so several generations can be studied within a month. And its size and transparency make it possible to observe under a laboratory microscope all the embryonic events that characterize this species.

J. E. Sulston and his associates have painstakingly followed every single cell in the *C. elegans* embryo and established a complete *lineage map* for the organism, a roadmap for the ancestry of each and every one of the (exactly!) 959 cells in the adult (Figure 38.23). This unique achievement—*C. elegans* is the only organism with such a map—has made it possible to isolate genes that affect nothing more than the development of a single cell. It is clear that research on this organism will increase our understanding of genetic control of cell development.

Developmental Anomalies as Research Tools

As you may recall from our discussion of cholesterol in Chapter 34, it is often easier for researchers to "get a handle" on a physiologically complex system if they can

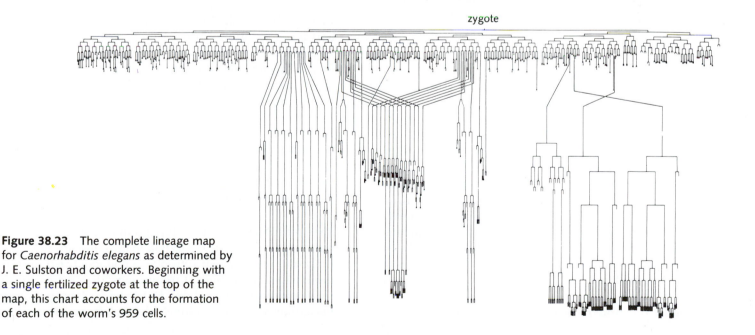

zygote

Figure 38.23 The complete lineage map for *Caenorhabditis elegans* as determined by J. E. Sulston and coworkers. Beginning with a single fertilized zygote at the top of the map, this chart accounts for the formation of each of the worm's 959 cells.

Figure 38.24 Siamese twin girls who share a single heart muscle.

Master Control Genes

Thus far we have considered genes largely as isolated "bits" of information; one codes for blue eyes, another codes for brown hair, others control blood type, and a few working in tandem color human skin anywhere from alabaster to ebony. Yet the body is formed, not from independent parts, but from tissues whose growth and differentiation are clearly coordinated in some way. How is that coordination accomplished?

As far back as 1915, researchers hypothesized the existence of some sort of "master control" genes. Such genes would somehow regulate the activity of hundreds or even thousands of other genes, turning them on and off in sequence to produce the patterns we know as a fully formed embryo.

The first evidence for such genes appeared in *Drosophila,* where certain bizarre mutations produce differences far more dramatic and far-reaching than changes in eye color or bristle number. One such gene is a mutation that results in *fushi tarazu,* which means "too few segments" in Japanese. Flies with the *fushi tarazu* mutation have half the usual number of segments, and they die before they are able to hatch from the egg.

Homeotic mutants   Another series of mutant genes, known as *homeotic mutations,* transforms the body part produced on one segment into another body part that is normally found on a different segment. One such mutation produces legs on the head where antennae should be; another substitutes antennae in place of wings, and still another turns mouthparts into feet (Figure 38.25).

Dozens of homeotic genes have been found, and most of these have been mapped to two clusters in the genome. One cluster affects the head and anterior thorax segments, whereas the other cluster controls the abdomen and posterior thorax. Researchers have even been able to pick up a homeotic mutant gene, insert it into the germ line of normal flies, and turn it on at will, essentially redesigning the fly in the process.

We are slowly learning how these genes exert their effects; at least some of them code for proteins that are found within cell nuclei, suggesting that their function is to bind to DNA and regulate the expression of other genes. These homeotic genes may bind to DNA sites throughout the genome, activating (or deactivating) whole groups of genes at once.

The homeobox sequence   Intriguingly, a surprising biochemical similarity seems to link many—though not all—homeotic genes. When researchers bound labeled mRNA molecules to *Drosophila* chromosomes, they discovered a short piece of DNA, about 180 bases in length, whose nucleotide sequence is almost identical from one gene to

compare its proper functioning to a case in which some part of the system is missing or malfunctioning. For that reason, one important research strategy is to examine cases in which normal developmental process go awry. In such situations, either naturally occurring mutations or experimental manipulations produce developmental anomalies: Siamese twins, animals with two heads, or individuals with malformed or duplicated limbs (Figure 38.24). You will see as our discussion proceeds how important these unfortunate individuals have been in expanding our understanding of development.

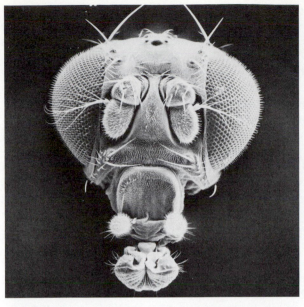
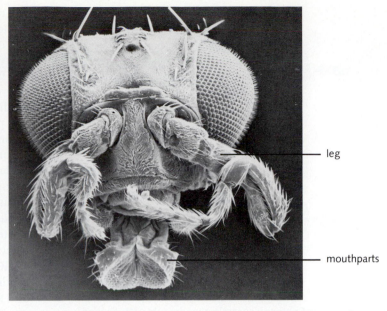
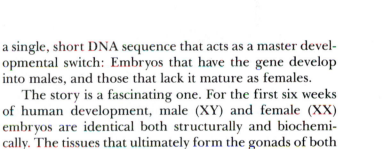
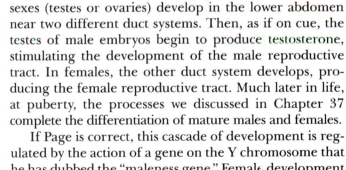

leg

mouthparts

Figure 38.25 Compare the normal fly (left) with the mutated fly (right). The fly shown in the scanning electron micrograph on the right possesses a homeotic mutation that produces legs where its antennae should be.

the next. Because of its presence in these homeotic mutants, Swiss scientist Walter Gehring coined the name *homeobox* to refer to this piece of DNA.

Are homeoboxes in fruit fly genes special cases that mark master control genes only in flies? Definitely not. Sequences closely related to the *Drosophila* homeobox have now been found in yeast, sea urchins, and frogs, and in 1984 they were found in humans as well. It is now suspected that genes containing the homeobox DNA sequence control the timing of gene expression in a wide variety of animals.

The universality of this sequence is suggestive; the homeobox may be a universal marker for at least one class of master control genes. This remarkable molecular similarity offers the tantalizing suggestion that at least some aspects of genetic control are shared among organisms that took separate evolutionary paths more than 600 million years ago.

Sex determination in humans Although several homeotic mutants of *Drosophila* survive to develop and emerge as adults, the same is not true of mammalian embryos. Either such mutants do not exist in mammalian systems because of different developmental patterns, or all of them are fatal in early embryonic life. Recently, however, researcher David Page at M.I.T. discovered a system that may prove to be of major importance for studying regulatory genes in mammals. Page believes he has located

a single, short DNA sequence that acts as a master developmental switch: Embryos that have the gene develop into males, and those that lack it mature as females.

The story is a fascinating one. For the first six weeks of human development, male (XY) and female (XX) embryos are identical both structurally and biochemically. The tissues that ultimately form the gonads of both sexes (testes or ovaries) develop in the lower abdomen near two different duct systems. Then, as if on cue, the testes of male embryos begin to produce testosterone, stimulating the development of the male reproductive tract. In females, the other duct system develops, producing the female reproductive tract. Much later in life, at puberty, the processes we discussed in Chapter 37 complete the differentiation of mature males and females.

If Page is correct, this cascade of development is regulated by the action of a gene on the Y chromosome that he has dubbed the "maleness gene." Female development occurs when this gene is absent, and male development takes place when it is present. Page used molecular techniques to isolate a small portion of the chromosome that seemed to contain the "maleness" factor. Analysis of the DNA sequence in this region showed that it coded for a polypeptide capable of binding to DNA. Even more remarkable is the fact that a similar DNA sequence seems to reside on the X chromosome. Is this the "femaleness" factor? Only more research on these intriguing candidates for master control genes in humans will tell.

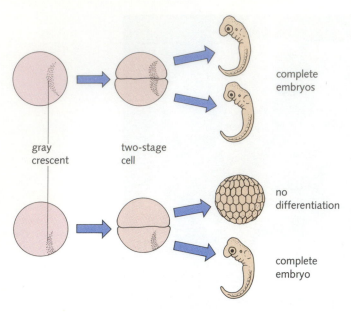

Figure 38.26 Hans Spemann's classic experiments with divided embryos showed that material in the gray crescent was necessary for complete development.

The Influence of Egg Cytoplasm

As fascinating and powerful as master genes may be, however, their presence alone cannot explain the formation of patterns in early embryos. All cellular descendants of the fertilized ovum, after all, contain the same genes. Thus even master control genes themselves must be differentially activated in different embryonic cells in order to direct those cells along various developmental paths. The question thus becomes "What controls the controllers?" It turns out that many key influences on the earliest stages of development are contained within the highly structured cytoplasm of the egg cell.

At the turn of the century, Hans Spemann showed that one region of the frog egg, the gray crescent, was absolutely essential to development (Figure 38.26). By carefully dividing a two-cell embryo, he showed that when each half of the embryo contained part of the gray crescent, two normal tadpoles developed. However, any cell that contained none of the gray crescent failed to develop and produced only an undifferentiated ball of cells. Even simpler manipulations of certain eggs have equally dramatic results; merely centrifuging frog or trout eggs for 4 minutes within half an hour of fertilization produces "Siamese twins."

These classic experiments suggest that some sort of key developmental influence resides in the egg cytoplasm. More recent studies using molecular techniques have shown that the egg cytoplasm of many organisms contains a presynthesized store of mRNA molecules that

are not translated until development begins. These molecules direct the synthesis of a host of proteins that are not coded by the embryonic nuclei but nonetheless direct many aspects of early development.

This situation has interesting consequences. Embryonic cells usually do not start to synthesize large amounts of their own mRNA until gastrulation. For that reason, any preformed messages that are present in the egg cytoplasm at fertilization can exert powerful influences on embryonic cellular activities. Furthermore, the differential distribution of mRNA during early cell division can induce those cells to develop along different paths. This influence is not restricted to mRNA, of course; differential distribution of any other substance that affects either protein synthesis or gene expression could have the same effect.

Recent research has shown that there is, in fact, a gradient of various compounds in early *Drosophila* embryos and that these compounds affect—either directly or indirectly—the expression of homeotic genes in different embryonic segments.

Embryonic Induction

No cell is an island; in any multicellular environment, the way a cell behaves very much depends on its surroundings. Thus, although preformed material in the egg cytoplasm is important in the earliest stages of development, it is not long before interactions among the cells of the embryo itself become critical in determining developmental changes. Several classes of experiments have demonstrated that various embryonic tissues exert powerful effects on one another.

During neurulation, how do the ectodermal cells "know" that they will become part of the nervous system? An experiment suggests one possibility. If cells derived from the gray crescent are removed from the dorsal lip of a frog blastula and transplanted inside a second blastula, they give rise to a block of mesodermal tissue that produces a second notochord. When neurulation occurs, two nerve tubes form—one above the original notochord and one above the transplanted tissue (Figure 38.27). What has happened? The ectoderm above the transplanted mesodermal tissue would normally have produced skin. But somehow the presence of notochord mesoderm at that point in development *induces* nearby ectoderm to develop into nerve tissue.

This process of **embryonic induction** is not an isolated phenomenon. It is a general pattern of development, and it is repeated in the formation of many different structures. Another well-known example is the *induction* of the lens of the eye by the **optic cup,** shown in Figure 38.28. A lobe of the developing brain called the *optic vesicle* grows close to the ectoderm at the anterior

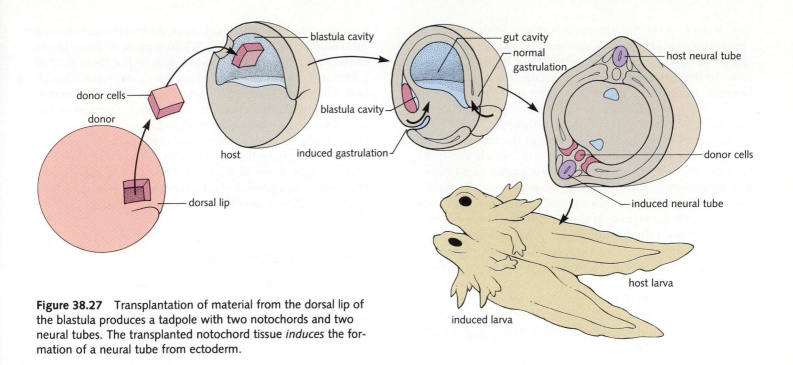

Figure 38.27 Transplantation of material from the dorsal lip of the blastula produces a tadpole with two notochords and two neural tubes. The transplanted notochord tissue *induces* the formation of a neural tube from ectoderm.

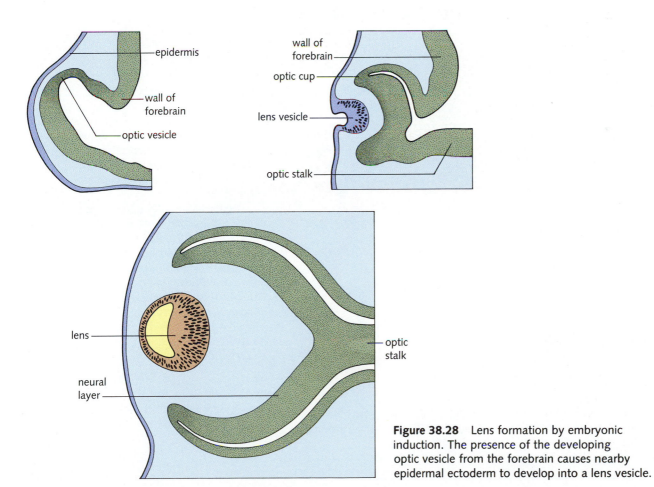

Figure 38.28 Lens formation by embryonic induction. The presence of the developing optic vesicle from the forebrain causes nearby epidermal ectoderm to develop into a lens vesicle.

end of the embryo. Gradually, this vesicle folds in to form the optic cup. This optic cup then seems able to induce *any* piece of ectoderm, *anywhere* on the body, to form a lens. In one experiment, an optic cup transplanted just beneath the skin in the belly region of a frog induced belly ectoderm to form a lens and a complete eye structure on the ventral surface of the tadpole!

In some tissues, induction occurs as the result of more complex signals, including hormones, which are quite specific and affect only certain target cells and tissues. But induction can also occur in response to very subtle cues, such as a change in local pH, a change in salt concentration, or the release of simple carbohydrates or other simple compounds. Although the most complex of these interactions have not been explored, there are a few cases in which we have begun to understand the complexities of induction.

One compound that seems to play an important role in inducing changes in embryonic tissue is *retinoic acid,* a derivative of vitamin A. One series of experiments dem-onstrated the effects of this compound on the developing limb bud of chick embryos. When a piece of filter paper soaked in retinoic acid is placed next to the front part of a limb bud, it causes the bud to produce a duplicate (though "mirror-image") set of digits (Figure 38.29). Further experiments showed that a gradient of retinoic acid occurs naturally in the limb bud and that this gradient normally controls the differentiation of different parts of the wing.

Intriguingly, retinoic acid also affects the development of limb buds in mammals. As we will see shortly, this simple compound even has inducing effects on several other classes of cells in mature individuals.

Further Control During Maturation

As we mentioned earlier, when an animal reaches maturity, rapid growth and differentiation give way to homeostasis in most tissues. Throughout the body, cells communicate with each other constantly, both through cell

---

THEORY IN ACTION

Hen's Teeth!

Embryonic induction is not limited to tissues of the same species. Hans Spemann and Oscar Schotté had shown this in a classic experiment in 1932 when they switched the oral ectoderm of a frog and salamander embryo. The mesoderm of each embryo induced the ectoderm to produce a mouth, but the type of mouth reflected the origin of the ectoderm. The salamander larva had a frog mouth, and the tadpole had salamander teeth. We might explain this by saying that the mesoderm "told" the ectoderm to make a mouth, and the ectoderm "obeyed" by making the only kind of mouth it could.

Much more recently, Kollar and Fisher used the same technique to play a revealing trick. They transplanted epithelium from the jaw-forming region of a chick embryo

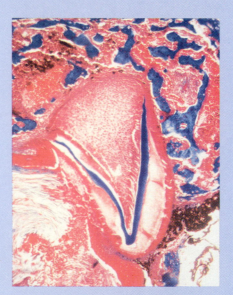

A tooth produced by chicken ectoderm transplanted next to the jaw mesoderm of a mouse.

onto the oral mesoderm of mouse embryos. What happened? The chicken tissue, responding to the instructions of the mesoderm, produced teeth. Hen's teeth!

Remarkably, although chicken cells have not formed teeth for millions of years, they still retain the genetic information to do so. All that was necessary was an induction signal from mesoderm to "make teeth." The fallout from this experiment is as much literary as scientific. The phrase "scarce as hen's teeth," will probably survive this assault, but it will never carry quite the same impact.

surface molecules and through more mobile chemical messengers such as peptide hormones and retinoic acid. Somehow, this collective interaction not only determines how many of each cell type there should be but also controls how much of which products are made by which cells. A number of studies in various systems have shown that this sort of communication is mediated by both positive and negative controls: Some messengers stimulate cells to grow or to produce certain products, and others induce them to stop dividing or halt synthesis.

As we mentioned earlier, however, there are several tissues in which growth and differentiation continue throughout adult life. Not unexpectedly, these tissues are controlled by the same general sorts of intracellular messengers that regulate embryonic development.

The various types of circulating blood cells, for example, all have limited life spans. For that reason, they must be produced throughout life. But the various kinds of blood cells are not always needed in the same relative numbers. Injury may create a need for more red blood cells, for example, or infection may increase the demand for certain types of lymphocytes.

Where do new and replacement blood cells come from and how is their production controlled? It turns out that all classes of blood cells are produced by a single marvelously versatile class of **stem cells** (Figure 38.30). These stem cells—which may either circulate in the bloodstream or settle in the bone marrow, spleen, or liver—

are potentially immortal; they continue to divide and reproduce for the life of the organism.

Through a complex process involving several stages, stem cells are induced by a variety of circulating growth factors to produce one or another type of blood cell. When more red blood cells are required, for example, the kidney releases *erythropoietin*, and under its influence, stem cells produce more erythrocytes. The production of the various classes of white blood cells is controlled by a series of similar growth factors, several of which have been identified.

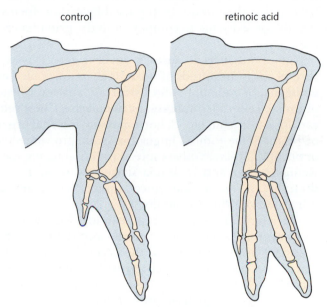

control   retinoic acid

Figure 38.29 The application of retinoic acid to the front edge of this developing limb bud has produced two "mirror image" sets of digits.

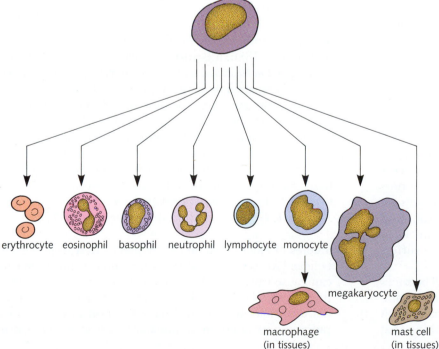

stem cell

erythrocyte   eosinophil   basophil   neutrophil   lymphocyte   monocyte

megakaryocyte

macrophage (in tissues)

mast cell (in tissues)

Figure 38.30 The wide range of blood cells is produced from a single class of pluripotent ("many powers") stem cells.

CANCER: WHEN CELLS REBEL

Cancer is one of humanity's oldest enemies, and it is still one of the most formidable. Signs of cancer have been found in 5000-year-old Egyptian mummies, and it claimed the lives of half a million people in the United States last year. Nearly 1 person in 3 will develop some form of cancer during his or her lifetime, and present statistics indicate that 50 percent of those will die from the disease within 10 years after diagnosis.

The origin of the use of the Latin word *cancer,* which means "crab," to describe the disease is obscure. Some scholars have suggested that the name is derived from the crab-like shape of some cancerous tissue, and others believe that the intense pain caused by certain forms of the disease led some to compare it to the grip of a crab's claw.

Cancer has myriad causes; it can be triggered by exposure to any of hundreds of noxious chemicals called **carcinogens,** by radiation, and by certain viruses. Clusters of cancer cells called **tumors,** or **neoplasms** ("new cells") can appear in skin, blood, bone marrow, internal organs, or brain. Some tumors linger for years, growing slowly or not at all, whereas others spread with frightening speed. Retinal cancer and leukemia strike mostly young children; breast cancer, skin cancer, lung cancer, and many others attack primarily adults.

Because of this variety of causes and effects, it seemed for years that "cancer" was actually a collection of more than 100 different diseases, each caused by a different carcinogen and each associated with a different set of symptoms. But over the last few decades, the techniques of molecular biology have proved that beneath its diversity, *cancer is a disease of the genes.* There is now good evidence that at the root of all cancers lie **oncogenes**— stretches of DNA capable of transforming normal body cells into tumor cells.

Failure to Control Cell Growth

Cancer arises when the genetic material in a single cell is damaged in a manner that frees the cell from normal constraints on growth and reproduction. Certain cancers appear when cells fail to respond to the signals that normally prevent mitosis. Other cancers arise when cells act as though they were constantly stimulated to divide. And still other cancers are caused by cells that fail to heed body messengers that normally induce them to differentiate and stop reproducing.

Once transformed in this manner, a cancer cell becomes like a parasite reproducing out of control, spawning millions of similar cells that differ from the normal parent tissue by as little as a single gene. This is one thing that makes cancer so difficult to treat. There is no foreign organism that can be made the target of special drugs or medicine; the disease-causing cells are the cells of the victim, growing in the absence of normal controls.

Oncologists—physicians and scientists who study and treat tumors—classify them into two groups differing in behavior and in medical consequences. **Benign tumors** (from the Latin word for "kind") are those in which the growth of cells is restricted to the area of the tumor itself. Despite their name, benign tumors can cause serious problems, particularly when they develop in delicate areas such as the brain, where tumor growth can deprive normal cells of nutrients and put pressure on blood vessels and nerves. But the cells of benign tumors tend to stay together in one mass, making it possible to remove them neatly and completely through surgery.

The cells of **malignant** ("evil") **tumors,** on the other hand, tend to break off from the original mass, enabling them to **metastasize,** or spread through the body. This property makes malignant tumors especially dangerous. If such a tumor is not removed before it undergoes metastasis, dozens of tumors may spring up throughout the body, blocking the movement of materials, disrupting the functions of vital organs, and ultimately killing their host.

Cancer Traced to Altered Genes

The nature of the damaged genes that cause cancer, and the mechanisms by which their altered gene products free cells from growth restraints, are being brought to light by four separate lines of inquiry that have converged only in the last 20 years.

Studies of cancer that seemed to be inherited Over a century ago, in 1886, a British physician reported that survivors of *retinoblastoma,* a particularly virulent form of eye cancer, bore an unusually high percentage of children who were also struck by the disease. Then in 1910, it was noted that retinoblastoma survivors also exhibited a much higher than normal incidence of an otherwise rare form of bone marrow cancer later in life. These findings alerted researchers that some forms of cancer have a genetic basis, although many decades passed before techniques to explore this lead were developed. (See Theory in Action, A Specific Gene for a Heritable Cancer, page 774).

Studies of environmental carcinogens Researchers have believed for decades that radiation and certain chemicals can cause cancer, and it has recently become clear that they do so by causing changes in DNA. Radiation, from the ultraviolet light in ordinary sunlight to the gamma radiation released in a nuclear explosion, can cause

Metastasis: How does a Cancer Cell Invade?

If cancer cells were not capable of metastasis, spreading throughout the body, cancer would not be such a serious medical problem. Localized tumors can be removed by surgery, but invasive ones spread to scores of places before they can be detected. Their ability to spread through the body is remarkable. A metastasizing cell must break its connections to other tumor cells, penetrate into the bloodstream, evade the immune system, and find a suitable place to grow. What gives malignant cancer cells this deadly ability? Researchers have tried to find out by analyzing individual cells derived from malignant tumors. One of the first findings was that some cells from a tumor are much more invasive than others.

These most invasive cells are armed with a formidable bag of tricks. In the bloodstream they attract macrophage cells and induce them to release an enzyme that helps the tumor cells break through blood vessel walls. Researchers have discovered that even if the malignant cells are detected by the immune system, the most invasive among them have cell surface molecules that enable them to avoid destruction by turning off the immune response. Finally, the most dangerous cancer cells produce a substance known as *tumor angiogenesis factor* (TAF). TAF affects nearby blood vessels and causes them to grow in the direction of a new tumor, providing a blood rich in nutrients to support further growth. The discovery of these tricks could not be called good news in the battle against the disease, but they have led to experimental strategies that may help to control metastasis by directly attacking the most invasive, and therefore the most dangerous, cells.

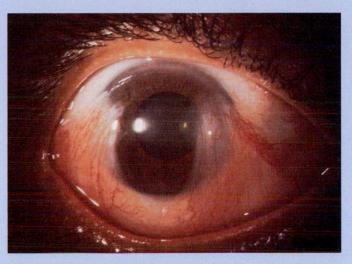

The rich growth of blood vessels associated with this tumor makes its active growth possible.

molecular changes in DNA. Similar effects are noted with carcinogenic chemicals. Hundreds of chemicals have been shown to cause mutations, and nearly all of these are also able to produce cancers under the right conditions.

Studies of cancer-causing viruses In 1910, a scientist named Peyton Rous discovered that a virus could produce tumors in chickens and that the virus could be isolated and used to spread the disease to other chickens (Figure 38.31). This virus, called the *Rous sarcoma virus* (RSV) belongs to a large family of oncogenic viruses, known as **retroviruses,** that contain RNA as their genetic material instead of DNA. After infecting a cell, retroviruses produce a *DNA* copy of their genome, which is inserted into the host cell's DNA. Most retroviruses produce diseases of other sorts (Chapter 39), but RSV and more than 40 other members of its family have been shown to cause cancer in animals.

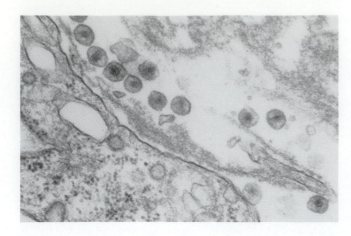

Figure 38.31 (left) These retrovirus particles are capable of producing cancer in the tissues they infect. (below) RSV and related retroviruses contain 4 key genes. One of these, the src gene, is essential for causing cancer.

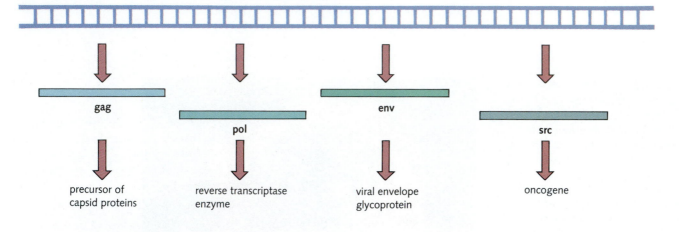

gag — precursor of capsid proteins

pol — reverse transcriptase enzyme

env — viral envelope glycoprotein

src — oncogene

Although the significance of this work by Rous was not immediately recognized (he received the Nobel Prize decades later), genetic analysis of RSV and other oncogenic viruses during the 1970s showed that most contain a critical gene that causes cancer when inserted into the host genome. Appropriately, these DNA sequences were named *oncogenes*. The oncogene from RSV, discovered by Raymond Erickson and Marc Collett at the University of Colorado, is known as *src* (pronounced "sarc" (Figure 38.31)).

Remarkably, *src* and several of the 30 other viral oncogenes that have since been discovered are nearly identical to genes found not only in human cells but also in the cells of many other species ranging from yeasts to chimpanzees. Clearly, to have retained a common form through hundreds of millions of years of independent evolution, these genes must have very basic and important functions.

Studies of gene transfer using tumor cells Using a procedure similar to that used by Oswald Avery and his colleagues (Chapter 21), several investigators transferred pieces of DNA from animal and human tumor cells into healthy cells growing in cell culture. Some of the healthy cells were transformed into cancer cells that caused tumors when injected into healthy mice. The "instructions" that transformed healthy cells into cancer cells must have been carried from one cell to another by pieces of DNA.

Theories of Oncogene Action

What could oncogenes be doing within a normal cell, and how is it that they cause cancer? No universal pattern characterizes oncogenes, and we are not yet sure of the function of each and every one. Nonetheless, some general patterns have become clear. First, it is possible for a *point mutation*—a substitution involving only a single base among hundreds in each gene—to turn a normally functioning gene into an oncogene.

Second, many oncogenes have been shown to code for proteins that play important roles in regulating cell functions. Some, including *src*, code for protein kinase enzymes, the molecular targets of many powerful peptide hormones (Chapter 36). Thus oncogene products may regulate the activities of scores of other enzymes. Some oncogenes code for products that regulate the timing of the cell cycle, and others seem to code either for growth factors that stimulate cell division or for the membrane proteins that serve as receptors for those growth factors.

This means that certain oncogenes could create cancer by causing cells (1) to produce too much of a growth-stimulating substance, (2) to act as though a growth stimulator were present even when it isn't, or (3) to *ignore* the message from a *growth-inhibiting* or *differentiation-inducing* substance. Preliminary work shows that all of these mechanisms probably exist in one cancer or another.

Treatment for Cancer

Despite intensive medical effort, drugs and other cancer treatments have had only limited success. For solid tumors, including breast cancer and prostate cancer (two leading causes of death among women and men, respectively), the best hope lies in early diagnosis. If these tumors can be removed by surgery before they have metastasized, the chances for a complete cure are excellent.

Other approaches represent attempts to take advantage of the fact that cancer cells pass through rounds of cell division much more often than most normal cells. Therefore, if a poison can be targeted to destroy rapidly dividing cells, it may be able to kill most or all of the cells in a tumor while killing relatively few normal cells. This approach is known as **chemotherapy.** Medical researchers have developed a host of drugs that act on such cells, and in a few cases the results have been spectacular. Several forms of childhood leukemia, a form of cancer that attacks white blood cells, now have cure rates approaching 80 percent.

In other cases the results have been less encouraging, and many of the drugs have serious side effects. Because they tend to poison *all* rapidly dividing cells, many healthy cells in important organs are affected by the chemotherapy, and the patient may become quite ill as a result. However, there have been dramatic improvements in the design and testing of such drugs.

Many cancers can also be treated via **radiation therapy:** Tumor regions are irradiated with high-intensity radiation to help kill the rapidly dividing cancer cells. This treatment can be effective because rapidly growing cells, which must replicate their DNA more frequently than other cells, have less time to repair radiation-induced damage. Therefore, they tend to accumulate serious mutations and eventually to die when the number of genetic defects induced by the radiation becomes too great. Radiation therapy is used only after a cancer has been located that is a direct threat to life.

Currently, several experimental cancer treatments focus on trying to get the body's own immune system to fight malignant cells. Because the immune system recognizes and destroys any proteins not normally found in the body (Chapter 39), and because many tumor cells display unusual proteins on their surfaces, it should theoretically be possible to "beef up" the body's own defenses to attack cancer cells. In 20–30 percent of the cases in which the most promising of these treatments was tried, it was spectacularly successful, but in most cases its effects were either negligible or insufficient. **Immune therapy** seems most successful with skin cancers called **melanomas,** but other types of cancer are highly resistant to immune therapy. Researchers are working hard to find out why this is true.

Ultimately, researchers studying the genetic basis of cancer hope to devise a new generation of cancer treatments based not on the brute-force methods of radiation and chemotherapy but on drugs specifically targeted at the mechanisms by which oncogenes release cells from normal growth controls. Unfortunately, because much of this research is still in its infancy, no one anticipates the discovery of such a cure in the immediate future.

There are some promising leads, however. One tantalizing series of experiments at Boston's Children's Hospital has recently shown that the application of retinoic acid can cause a certain type of cancer cell to differentiate and stop growing, thus curing the cancer. Intriguingly, at an early stage in the differentiation process of those cancer cells, researchers detected the expression of genes containing a homeobox.

Prevention and Control of Cancer

The fact that viruses have been linked to certain types of cancer should not be taken to mean that cancer can be acquired through contact with infected individuals. Only in the case of a single type of virally caused cancer of the cervix has the cancer been shown to be communicable, and in that case sexual intercourse is the only known means of transmission. There is no evidence that prolonged personal contact with any other class of cancer patient increases the risk of contracting the disease. Close contact with a cancer patient, in fact, is one of the most beneficial things that friends and relatives of a victim can do to aid in that patient's recovery.

There is strong evidence that nearly all forms of human cancer are caused by exposure to specific agents in the environment. As an example, let's look at two serious cancers of the digestive system. Stomach cancer is common in Japan but rare in the United States (where it occurs at about one-fifth its rate of incidence in Japan). Intestinal cancer, on the other hand, is much more common in the United States than it is in Japan (where it occurs at about one-quarter the U.S. rate). A scientist might examine this information and suggest that the dif-

A Specific Gene for a Heritable Cancer

Although the oncogene hypothesis gives us a basis to say that cancers are caused by mutations in a few cellular genes, there are relatively few places where workers have been able to pinpoint the exact nature of the change that leads to the transformation into a cancer cell. But those few important exceptions may teach us a great deal about the genetic changes responsible for cancer.

Retinoblastoma is an inherited form of cancer. It is caused by a single recessive allele and therefore is expressed only in individuals with two copies of the allele. In retinoblastoma, a tumor originates within the retina of the eye and grows rapidly, often damaging the eye and causing blindness. Earlier genetic studies had concluded that the retinoblastoma allele was *dominant* for the *tendency* to develop the tumor, and this finding provided some insights into how the disease begins. Most often, children inherit the gene defect from only one parent. Then, when a mutation dam-

ages the other normal gene in one of the hundreds of thousands of retina cells, a tumor develops.

Nearly 20 years ago, Jorge Yunis at the University of Minnesota found a patient with the tumor who also had a

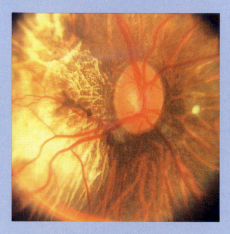

A retinoblastoma tumor is visible as the pale oval in this photograph of a human retina taken through the front of the eye.

deletion on chromosome 13. Soon other patients were found to have similar deletions (though not all of them did), and a link between the tumor and the loss of material on chromosome 13 seemed to be clear. Little by little the affected region on that chromosome was narrowed down until, in 1986, two research groups in Boston were able to isolate the entire retinoblastoma gene. Now that this gene has been isolated, the really interesting work will begin. Already experiments have shown that abnormalities in the retinoblastoma gene are associated with other kinds of tumors. Gradually, researchers should be able to develop a test for the gene deletion that causes retinoblastoma, enabling them to look for something more revealing than a deletion on chromosome 13. But the most interesting result of this development is the chance to analyze the product of the gene. It will be fascinating to examine a protein whose *absence* allows a cancer to develop.

ferences in cancer rates could be due either to environmental factors or to differences in genetic makeup between Americans and Japanese.

It is easy to choose between these two possibilities. Hundreds of thousands of Americans are of Japanese ancestry, and public health statistics for these Japanese–Americans show an interesting trend. The first generation of immigrants have cancer rates very close to those displayed by the homeland Japanese population. The second generation, those born in the United States, have rates midway between the homeland Japanese and general U.S. rates, and third- and fourth-generation Japanese–Americans have rates for these two types of cancer that match those of the general U.S. population.

It is clear that genetics does not play an important role in these cancers. General environmental agents, such as air and water, are probably not to blame either; this is suggested by the fact that first-generation immigrants exhibit a cancer rate matching that of their homeland. Studies of the lifestyles of this population group suggest that the most important factor in altering the rates for these two types of cancers is diet: the change from traditional Japanese food to typical American fare. Clearly, each form of diet seems to have its drawback—and the best solution would be to identify the best (and worst) aspects of each diet.

Armed with these kinds of data, health researchers have tried to identify dangerous chemicals in the diet or

Table 38.1 *The Most Potent Chemical Carcinogens*

Carcinogen	Exposure	Principal Cancers
Aflatoxin B1	Dietary	Liver
4-Aminobiphenyl	Occupational	Bladder
Arsenic	Occupational	Skin, lung
Asbestos	Occupational	Lung
Benzene	Occupational	Leukemia
Benzidine	Occupational	Bladder
Bis(chloromethyl) ether	Occupational	Lung
Chlornaphazine	Medical	Bladder
Cigarette smoke	—	Lung, bladder, pancreas
Diethylstilbesterol (DES)	Medical	Vagina
Estrogens	Medical	Corpus uteri
High-energy radiation	Medical, occupational, wartime	Various
Mustard gas	Occupational	Bladder
2-Naphthylamine	Occupational	Bladder
Ultraviolet light	Sunlight	Skin
Vinyl chloride	Occupational	Liver, brain (?)

SOURCE: Mathew S. Meselson, 1979, *Chemicals and Cancer*, Boulder, Colo.: Colorado Associated University Press.

in the immediate environment that might be responsible for cancer. The goal of this research is to identify chemicals we encounter in our everyday lives that may be responsible for the many cases of cancer that seem to have no direct cause (Table 38.1). The best way to combat this dread disease is to eliminate its causes, and for many years health researchers have been trying to do just that. Interestingly, many of the most potent carcinogens do not cause cancer until changed by one of the body's enzymes into a carcinogenic form.

The Clearest Cause of Cancer

Some carcinogens in the environment are present in such low concentrations that their effects can scarcely be noted. But one carcinogenic agent has an effect so strong that it cannot be ignored: cigarette smoking. Lung cancers, nearly 90 percent of which are caused by smoking, are now the leading cause of death from cancer among men and women in the United States. A decision to avoid smoking virtually ensures that an individual will not suffer from this disease (Figure 38.32).

Given such evidence, it is difficult to understand why the habit of smoking, with all of its deadly consequences, persists at all. The best advice that the authors of this book can give you is to avoid smoking entirely and not add your name to the list of lung cancer statistics.

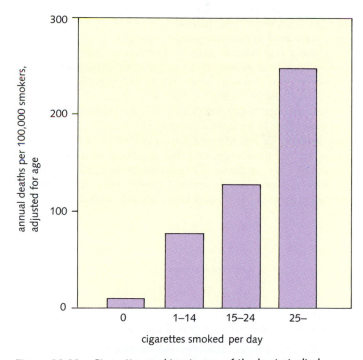

Figure 38.32 Cigarette smoking is one of the best-studied cases of a direct link between a behavior and cancer. This graph shows the rising incidence of lung cancer deaths with increased smoking.

SUMMARY

In vertebrates, all patterns of development follow a basic plan by which gastrulation results in the formation of the three primary germ layers: ectoderm, mesoderm, and endoderm. The details of this process differ dramatically, influenced by the size of the ovum and the mechanism by which it is protected and nurtured during development. The event that initiates development is fertilization, the fusion of sperm and egg. Cleavage occurs, and the zygote rapidly becomes a mass of cells known first as a morula and then as a blastula. Gastrulation is a process of cell migration in which the inward movement of the blastula surface establishes the internal germ layers of the embryo. Neurulation is the formation of a primitive nerve tube from ectoderm, an event that is completed after the major cell layers of the embryo have been formed.

Human development proceeds via a pattern typical for placental mammals: After the blastocyst implants in the uterine wall, an embryonic disk forms that is surrounded by outer cell layers that gradually invade the surrounding uterine tissue. As gastrulation and neurulation proceed, the extraembryonic membranes and placenta are formed, protecting the embryo and enabling it to draw nourishment from the mother.

Increasingly, biologists have turned to molecular models for development, including those derived from work with *Drosophila*, in which the techniques of classical genetics and molecular biology have been used to produce large numbers of developmental mutants. Analyses of these mutants have shown that certain DNA sequences, including one known as the homeobox, are found in many developmental genes.

Cancer is a cellular disease that results from the loss of control over cell growth in a multicellular organism. Tumors that develop in normal tissue may be either benign or malignant, depending on whether they tend to invade neighboring tissues and organs, a process known as metastasis. Cancers seem to be the result of heritable changes (mutations) in cellular DNA that lead to the loss of control over growth. Studies on oncogenic viruses have suggested that a certain set of genes, the oncogenes, are particularly important in the regulation of cell growth, and much cancer research now focuses on understanding what these genes do and how they are controlled.

STUDY FOCUS

After studying this chapter, you should be able to:

- Explain the process of embryonic development in representative vertebrates.
- Compare different types of embryos.
- Examine in detail the process of human embryonic development.
- Pose some of the fundamental questions of developmental biology, including the issue of gene activation in early development.

SELECTED TERMS

placenta *p. 745*	organogenesis *p. 751*
differentiation *p. 746*	somites *p. 751*
gametogenesis *p. 746*	chorion *p. 755*
cleavage *p. 747*	amniotic sac *p. 757*
blastula *p. 747*	carcinogen *p. 770*
blastocyst *p. 748*	oncogene *p. 770*
blastodisc *p. 748*	benign tumor *p. 770*
gastrula *p. 748*	malignant tumor *p. 770*
archenteron *p. 748*	metastasize *p. 770*
neurulation *p. 751*	retrovirus *p. 771*

REVIEW

Discussion Questions

1. Describe the basic vertebrate body plan and distinguish among the three primary germ layers. Describe how the layers are formed in amphibian, bird, and mammalian embryos.

2. What are the primary means by which the endocrine system is signaled that a pregnancy has begun? Why do these signals prevent ovulation until the pregnancy is completed?

3. Which aspects of the trophoblast and the placenta are of particular interest to researchers who want to learn how to prolong the survival of skin grafts and other tissue transplants?

4. The primary signal for male sexual differentiation is testosterone. In the absence of testosterone, the female sexual pattern develops. Why is the production of estrogen by the embryonic ovary initially of little significance in sexual development?

Objective Questions (Answers in Appendix)

5. The three primary germ layers (ectoderm, mesoderm, and endoderm) are formed during
   (a) gastrulation. (c) cleavage.
   (b) neurulation. (d) embryonic induction.

6. The yolk sac in the human embryo
   (a) contains a yolk.
   (b) is not present.
   (c) produces the mesoderm.
   (d) gradually disappears as the fetus matures.

7. The lining of most of the digestive system originates from
   (a) mesoderm. (c) endoderm.
   (b) ectoderm. (d) chorion.

8. In vertebrates, the notochord is replaced by
   (a) the digestive tube.
   (b) the spinal column.
   (c) somites.
   (d) limb buds.

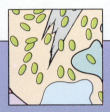

CHAPTER **39**

Immunity and Disease

In different parts of the world, four children suffer: one with measles, one with the flu, one with sleeping sickness, and one with AIDS. The prognosis for these children—and the millions of other people around the world whom their cases represent—are dramatically different. The first will recover quickly and never contract measles again. The second will recover also but will undoubtedly battle influenza ("the flu") several times throughout life. The third will fight a long and difficult battle, perhaps winning, perhaps not. And the fourth, at least so far, is doomed to an early death (Figure 39.1).

Why does the body have the ability to fight off some diseases? Why do some diseases, such as measles, never recur? How does AIDS so completely overwhelm the body that a young life is brought to an early end?

We know today that each of the children described above is fighting a war with a **pathogenic** organism: one of the many bacteria, viruses, and protozoans that can infect our bodies and cause disease. We also know that our body battles these invaders through the **immune system,** an assortment of cells and tissues coordinated through a complex network of molecular messages and interacting genes (*immunis* means "exempt," *immunity* means "freedom from infection"). From birth, the immune system learns to distinguish "self" (cells and proteins native to the body) from "nonself" (organisms and compounds from outside the body) and mobilizes to attack any alien intruders. The immune system also "remembers" each infection; once having been encountered, any invader is dealt with more rapidly and more efficiently the next time around.

DISEASE THROUGH THE AGES

For most of human history, as you learned in Chapter 2, diseases were mysterious scourges for which either poison or supernatural causes were often blamed. By the sixteenth century, physicians began to understand conditions that *spread* disease, but they had no clue to what *caused* illness.

The first useful theories, for example, attributed illness to *miasma,* or "bad air." Miasma could be generated by natural and unnatural disorders, including earthquakes, unburied corpses, too much sexual activity, and the wrath of God. Once miasma existed, the theory went, it transmitted disease from the sick to the healthy; it hovered around infected persons, houses, and towns and could be carried in clothes, bedding, and baggage.

Though incorrect, this concept inspired sensible responses to epidemics. Contact with sick people was avoided, infected houses were fumigated, and contaminated bedding was burned. But the microbial causes of disease remained unknown for centuries, and medical breakthroughs usually followed leaps of intuition rather than methodical investigation.

The First Vaccination: A Shot in the Dark

Up through the late eighteenth century, a virulent disease called **smallpox** had terrorized Europe, leaving millions dead in its wake. Those who survived were left with disfiguring scars, but they also gained a mysterious protection against further smallpox infection. They became *immune*—resistant to infection. A similar but far milder disease called **cowpox** occurred among milkmaids and dairy hands. Interestingly, contracting cowpox seemed to confer the same sort of immunity to smallpox.

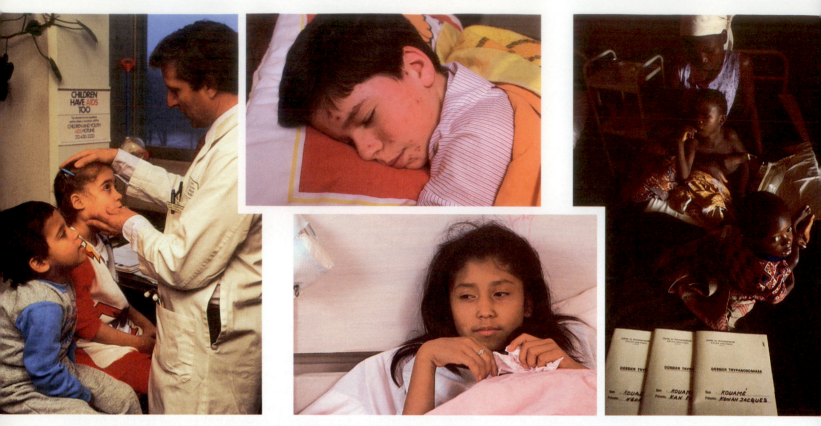

Figure 39.1 These children are suffering from four different infectious diseases: (from left to right) AIDS, measles, flu, and sleeping sickness.

In 1796, the English physician Edward Jenner tested a revolutionary idea: He infected a patient with cowpox to see whether he could produce immunity to smallpox. Jenner took fluid and pus from a cowpox lesion on a milkmaid and mixed them into a shallow cut he had made in the arm of an 8-year-old boy, Jamie Phipps. Several weeks later, he performed the same procedure, but this time he used material from a smallpox patient. Fortunately, this unorthodox procedure worked. Jamie did not come down with smallpox. He had become *immune* to the disease.

Societal response to Jenner's work was mixed. Though many hailed the procedure, others, including physicians who feared its mysterious actions and doubted its effectiveness, lobbied against it (Figure 39.2). (After all, not only was the immune system unknown, but viruses hadn't been discovered yet!) In fact, the modern name for this procedure, **vaccination** (*vacca* means "cow"), was originally a derogatory term that literally meant "encowment."

Jenner prevailed, however, and by the beginning of the nineteenth century, thousands throughout Europe had been vaccinated against smallpox. Today we understand why vaccination works; we know that cowpox and smallpox are caused by similar viruses and that the immune system "remembers" infection in a way that helps it head off subsequent attacks before the virus causes much damage.

The Agents of Disease

The connection between microorganisms and disease gradually became clearer as both scientists and the public began to associate particular sources of contagion (such as polluted water) with specific diseases (such as cholera). In the late nineteenth century, pioneering microbiologists, including Louis Pasteur and Robert Koch, paved the way for a useful scientific understanding of several diseases. Bacteria were shown to cause cholera, anthrax, and tuberculosis. Bacteria spread by fleas from rodents to humans were revealed to be the cause of bubonic plague. As researchers continued to discover and describe pathogens and their carriers, public health measures curtailed the spread of epidemics.

The early decades of the twentieth century saw the identification of viruses as the causes of such diseases as smallpox, polio, and the flu. Then, from the 1940s on, researchers discovered antibiotics and developed several effective vaccines. Smallpox, polio, measles, and diphtheria were all but eradicated, and infections from injuries and surgery became both less common and less life-threatening.

The last few decades, however, have suggested that medicine is not omnipotent. In the tropics, malaria, sleeping sickness, and yellow fever continue to resist control. In industrialized countries, pathogenic bacteria have developed resistance to penicillin and to other commonly used antibiotics with worrisome speed. Table 39.1 summarizes the loss of human life associated with several epidemics of the past and present.

Figure 39.2 A cartoon ridiculing Jenner's cowpox vaccine for smallpox.

Table 39.1 *Losses of Human Life to Epidemics*

Epidemics of the Past	Effects
Bubonic plague	One outbreak of the plague (1347–1350) killed between 17 million and 28 million people, one-third to one-half of Europe's population.
Influenza ("the flu")	Killed 22 million people worldwide in 1917–1918.
Smallpox	Killed 400,000 Europeans in the nineteenth century alone.
Polio	Infected 400,000 Americans, of whom 22,000 died between 1943 and 1956.

Epidemics of the Present	Effects
Measles	Effectively controlled in industrialized countries, but kills at least 1.5 million people annually in the Third World.
Tuberculosis	Kills 500,000 people annually worldwide.
AIDS	Over 100,000 diagnosed and 60,000 deaths in the United States alone.
Diarrhea (largely cholera)	Kills 10 million people annually in Third World countries.
Malaria	Kills 1.2 million people annually in Third World countries.

PATHOGENIC ORGANISMS

What Makes a Pathogen?

The world around us teems with microscopic life. Scores of invisible organisms enter the body with every breath, and millions cover the surface of the skin, ready to enter even the tiniest scrape or cut. Most of these present no problems; they interact with our tissues in ways that are either harmless or beneficial. But a relatively small number of microorganisms—those called **pathogens**—cause disease. They harm the body by causing direct physical damage to tissues, by releasing toxic chemicals, or by taking over cells' genetic machinery to use it for their own ends.

Because disease involves specific interactions between host and pathogen, and because various hosts differ a great deal in physiology, many microorganisms that cause serious disease in some species grow harmlessly in others. Many potential pathogens, too, are normally kept out of the body (or prevented from surviving if they get in) by the normal functioning of the immune system. Such pathogens can cause problems only when the body's natural defenses are compromised.

Researchers are now studying pathogens at the molecular level. They seek to understand how pathogens establish infections, which genes regulate the pathogen life cycle, what molecular mechanisms cause the symptoms of disease, and how these molecular mechanisms might be exploited by new drugs.

The Diversity of Pathogens

A wide variety of organisms can act as pathogens, including viruses, bacteria, protozoa, fungi, and worms. Table 39.2 lists a few important pathogens and the diseases they cause.

 Viruses, as we noted in Chapter 22, are tiny, infectious particles composed of a core of nucleic acid (either DNA or RNA) surrounded by a protein coat. Viruses enter living cells and grow within them, usurping control of the cells' biochemical machinery to such an extent that they cause infected cells to die.

Viruses infect host cells by binding to the cell membrane and then injecting genetic material into the cell. DNA viruses may express their genetic material directly within an infected cell. RNA viruses called **retroviruses,** on the other hand, cross the cell membrane and then use an enzyme called **reverse transcriptase** to create a DNA copy of their RNA core. After that point, both DNA viruses and RNA viruses may do one of two things:

1. *The virus may immediately produce an active infection.* In this case, viral DNA replicates and viral genes direct the host cell to synthesize new viruses. Some viruses do this slowly, gradually budding off new virus particles from the cell surface while the host cell survives. Other viruses overwhelm infected cells by forcing them to create hundreds of new viruses rapidly. The host cell bursts, releasing viruses that attack neighboring cells. Rhinoviruses, flu viruses, and measles viruses follow this course of action.

2. *The virus may produce a latent infection.* In such cases, instead of replicating, viral DNA (or the DNA copy of retroviral RNA) inserts itself into host cell chromosomes and "disappears" for periods ranging from a few weeks to many years. Psychological stress seems to activate some latent viral infections, such as herpes. Infection with other pathogens seems to trigger the activity of other viruses, such as the AIDS virus. But we do not know precisely what molecular events control latency and activation. Once acquired, latent viral infections stay with a host for life; there is no known way to remove viral DNA without killing the host cells.

Free virus particles may be passed from one host to another, spreading the infection. Some virus particles, such as those that cause colds and flu, can survive outside the body and may be spread by casual contact and by coughs and sneezes. Other viruses, including those that cause herpes and AIDS, cannot survive outside the body and must be spread through direct contact or through the exchange of body fluids such as blood, mucous secretions, or semen. We will examine the action of the AIDS virus (so far as it is known) at the end of this chapter.

In the minds of many people, all **bacteria** are pathogenic. This popular conception of bacteria is misguided, for only a few members of this diverse group cause humans any harm. Those few do, however, cause some of our worst diseases (Table 39.2). Pathogenic bacteria may attack and destroy cells directly, they may grow so quickly that they deprive host cells of necessary nutrients, or they may release poisonous waste products or **toxins** that injure host tissues.

Protozoa cause a number of serious diseases. The protozoan genus *Trypanosoma,* for example, includes several flagellates that live in their hosts' blood and release poisons that damage host cells seriously enough to cause death. One species, carried by the tsetse fly, causes **African sleeping sickness** in both humans and cattle. This disease is virulent enough to make it almost impossible to raise cattle in some parts of Africa. Not only have these parasites resisted our efforts to develop a vaccine to protect either livestock or humans, but they are also nearly invulnerable to the body's normal defenses. We will explain why this is so when we discuss the way the immune system works.

Another serious disease, **malaria,** is produced by the protozoan *Plasmodium,* which, as we noted earlier, spends

Table 39.2 *Pathogens*

Pathogen	Examples	Diseases	Notes
Viruses			
DNA viruses	Rhinovirus	Common cold	More than 50 types known
	Flu virus	Influenza	New strains arise every few years
	Measles virus	Measles	Preventible by vaccine
RNA viruses	Herpes simplex II	Genital herpes	Spread by sexual contact
	Rous sarcoma virus	Cancer (in chickens)	First oncogenic virus discovered
	HIV	AIDS	Spread by blood or sexual contact; fatal
Bacteria	*Mycobacterium tuberculosis*	Tuberculosis	Treatable with drugs; a limited vaccine available
	Mycobacterium leprose	Leprosy	Causes long-term infections
	Streptococcus pneumoniae	Bacterial pneumonia	Before antibiotics, nearly 1/3 of patients died; now, 95% recover
	Corynebacterium diphtheriae	Diphtheria	Serious childhood disease in 19th century; now preventible by vaccine
	Yersinia pestis	Bubonic plague	Still present in Western U.S.; carried by prairie dogs
	Treponema pallidum	Syphilis	Spread by sexual contact
	Vibrio cholerae	Cholera	A leading cause of infant death in Third World
Protozoa	*Trypanosoma*	African sleeping sickness	Spread by tsetse fly
	Plasmodium	Malaria	Prevalent in tropics; a serious health problem
Flatworms (Platyhelminthes)	*Schistosoma mansoni*	Schistosomiasis	Disease spread by coinfection in snails; a very serious health problem in rice-growing areas
	Taenia saginata	Beef tapeworm	Spread by contaminated meat

part of its complex life cycle in the *Anopheles* mosquito. Infections are spread from human to human when an infected mosquito injects saliva containing the malaria organism as it begins to bite. *Plasmodium* reproduces within red blood cells, causes them to burst, and produces severe fever and chills.

Parasite is the term commonly used to describe multicellular pathogens. Some parasites, such as tapeworms, live within the digestive systems of their hosts. There they absorb food and release eggs into fecal material. Others, including the schistosomes, or flukes, live within host tissues and have complicated life cycles involving two or more hosts, which may include humans, snails, and freshwater fishes (Chapter 15). The widespread use of human feces for fertilizer in Asia distributes the eggs of this species widely in such agricultural areas as flooded rice paddies.

THE IMMUNE SYSTEM: DEFENSE AGAINST INFECTION

Nonspecific Defenses

The body's first defense against pathogens is simple: It tries to keep them out. Generally, the first shield consists of **nonspecific defense mechanisms,** which are physical, chemical, or biological barriers that act against a wide variety of pathogens.

The skin The largest and most obvious barrier against infection, the **skin,** is a classic example of a nonspecific defense (Figure 39.3). Pathogens can usually enter the body only where skin is broken or in places where the skin is exceptionally thin and delicate, such as the mucous membranes of the nose, mouth, or genitals. Nowhere is

Figure 39.3 Skin is the primary barrier against infection. An outer layer of epidermis lies just above the dermis, which contains sensory nerves, glands, and hair follicles. Subcutaneous tissue (literally meaning "under the skin") lies below the dermis.

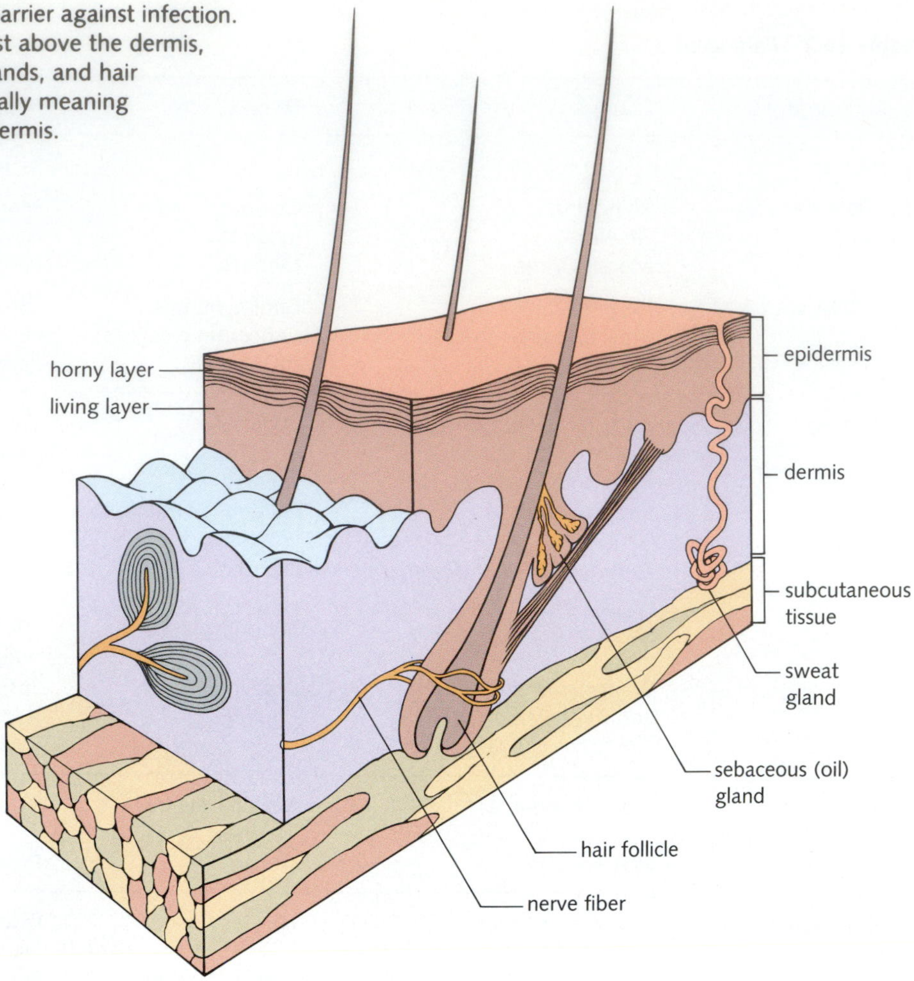

horny layer
living layer
epidermis
dermis
subcutaneous tissue
sweat gland
sebaceous (oil) gland
hair follicle
nerve fiber

skin's protective value more apparent than in serious burn victims, who face major risks of infection. The best sanitary precautions of modern hospitals cannot match the protective value of healthy skin, and medical workers try to grow or graft skin onto the burned areas as quickly as possible.

In addition to providing a physical barrier, the skin is an active and integral part of the immune system. Skin cells respond to infection by creating scales, rashes, scabs, scars, and blisters. **Sebaceous glands** produce an oily material known as **sebum,** which contains several compounds that inhibit the growth of bacteria and fungi. The pH of skin, influenced by sweat and oil glands, is quite acidic, ranging from 3 to 5. This acidic pH prevents the growth of many potential pathogens. Skin cells also produce a variety of potent chemical messenger molecules that influence other cells in the immune system.

Enzymatic defenses Although the skin is much thinner and more delicate at the mouth, the anus, and the genitals, the body has other nonspecific defenses at these openings. Respiratory passageways are coated with mucus that traps inhaled particles (including microorganisms) and enables ciliated cells to push them toward the alimentary canal and the digestive system, where they are destroyed by digestive enzymes. Most bacteria that enter the mouth or urogenital openings are killed by an enzyme known as **lysozyme,** which breaks down their cell walls. Lysozyme is also produced by the tear ducts, helping to protect the surface of the eye.

Inflammation When skin is cut or torn, bacteria enter the wound almost immediately. These pathogens grow and spread rapidly at first, releasing toxins that kill surrounding body cells. But this invasion is quickly challenged by a powerful second line of nonspecific defense, which is known as the **inflammatory response** because in severe cases the skin turns a flaming color (Figure 39.4). **Mast cells** near the skin surface release a substance called **histamine** that causes blood vessels near the wound to expand. Fluid leaks into the wounded area from these vessels, and the wound begins to swell. Histamine and other compounds released from the wound site attract and direct the actions of several types of white blood cells.

Phagocytes—cells that can engulf and destroy foreign cells—swarm over the infected area, engulfing and digesting bacteria. **Monocytes,** small roundish cells found throughout the circulation, are attracted to the wound, where they change into larger cells called **macrophages** that also engulf bacteria.

The toxins produced by the bacteria kill many defending cells, and as hours go by the debris of battle—fluid filled with dead bacteria and white blood cells—builds up. The increased cellular activity in the wound area may raise its temperature several degrees. It is reddened, hot, and tender to the touch. Bit by bit, however, the infection is brought under control and the wound begins to heal (Figure 39.4).

Specific Defenses

The nonspecific defenses at the surface of the body are backed by an immune system capable of powerful actions against *specific* invaders. The specific defenses of the immune system are triggered by *antigens,* foreign substances which may be part of the surface coat of viruses, bacteria, or other pathogens. Any substance, including lipids, carbohydrates, or proteins, that can stimulate the specific defenses of the immune system is considered an antigen. The specific defense mechanisms of the immune system are targeted against individual pathogens at two levels.

First, the immune system produces a series of soluble proteins known as **antibodies** that circulate in the bloodstream and bind to sites on foreign molecules. These

Figure 39.4 Cellular events associated with a simple splinter wound. Even a small wound brings about a dramatic response that we recognize as inflammation.

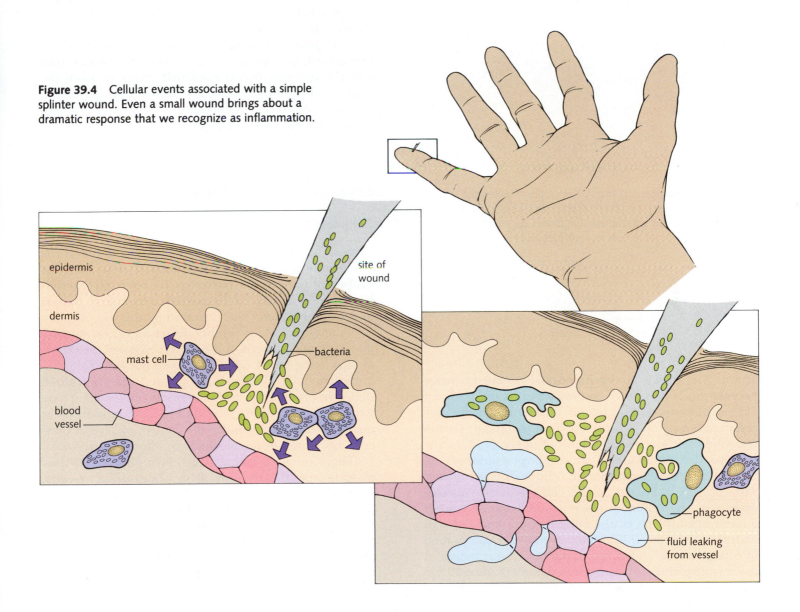

epidermis

dermis

mast cell

blood vessel

site of wound

bacteria

phagocyte

fluid leaking from vessel

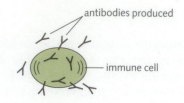

Humoral immune response

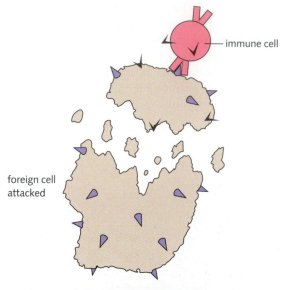

antibodies produced

immune cell

immune cell

foreign cell attacked

Cell-mediated immune response

Figure 39.5 The humoral immune response involves the production of antibodies, protein molecules that help to fight infection. In contrast, the cell-mediated immune response involves a direct attack by immune cells against foreign cells.

actions are known as the **humoral immune response** because antibodies move in the *humors* of the body, an old word for blood and lymph. The humoral immune response is especially effective against bacteria or viruses that find their way into the fluids of the body.

Second, the immune system organizes a series of cells that act directly against foreign cells or infected tissues. This process is known as the **cell-mediated immune response.** The cell-mediated immune response can act directly against cells of the body that are infected with bacteria or viruses. It is also involved in the recognition of foreign tissue and is responsible for the rejection of tissue transplants from unrelated donors (Figure 39.5).

The humoral and cell-mediated response of the immune system against invaders is a process that serves three important functions:

1. *Recognition.* The immune system recognizes an invader as foreign. Despite the enormous diversity of pathogens and foreign tissues, the immune system can mount a response against virtually any organism. The ability to recognize such a wide variety of invaders implies that, in effect, the immune system maintains some sort of "catalogue" of which macromolecules belong in the body, and somehow uses that catalogue to determine which do not.

2. *Reaction.* After identifying a foreign antigen, the immune system reacts by preparing humoral and cell-mediated reactions to the challenge. Very often, the reaction involves the stimulation of specific cells that attack the invader. These cells may go through many rounds of mitosis, creating a large group of cells that produce antibodies and act directly against the foreign tissue.

3. *Disposal.* The activated immune system destroys the foreign cells or tissues. This destruction may occur from a direct and specific cell-mediated attack or from the binding of antibodies and soluble proteins to foreign antigens. When the actions are effective, the foreign invader is destroyed and disposed of.

COMPONENTS OF THE IMMUNE SYSTEM

Cells and Organs

Because the immune system must be able to patrol the entire body to guard against infection, its cells cannot be confined to any single region of the body. Like undercover police, the cells of the immune system must be able to travel throughout the organism so that no infection goes unnoticed. It should therefore come as no surprise that the principal cells of the immune system are classes of circulating blood cells called **lymphocytes.** Lymphocytes, like other blood cells, originate from generalized stem cells produced in bone marrow (Chapter 38). These stem cells grow, divide, and differentiate into different classes of blood cells under the control of several powerful hormones, some of which we will discuss shortly.

There are two principal types of lymphocytes: **B-lymphocytes** or **B-cells** and **T-lymphocytes** or **T-cells.** Although both B-cells and T-cells originate in bone marrow, T-cells mature after they have passed through the *thymus gland*, which is where the T (for "thymus-derived") comes from. These two cell types are nearly identical in appearance but differ in function. B-cells are responsible for the production of antibodies, the humoral immune response. T-cells are primarily responsible for the cell-mediated immune response.

A third type of cell, the **macrophage** ("big eater") engulfs and digests pathogens ranging from viruses to bacteria. Macrophages then display small pieces of foreign proteins in specialized regions of their cell membranes where the antigens are easily recognized by lymphocytes.

Lymphocytes are found in the lymphatic system and the lymph nodes, as well as in the other organs of the immune system, especially the spleen and thymus (Figure 39.6). It makes sense that the primary cells of the immune system should be associated with the lymphatic system. Remember that plasma leaks slowly from capillaries into body tissues, drains into the lymphatic system, and passes through a number of lymph nodes (which are especially rich in lymphocytes) before being returned to the circulatory system. This flow pattern ensures that a few bacteria or virus particles from any infection are swept along with the flow and make direct contact with lymphocytes. Once that contact is made, the body-wide immune response begins.

Antibodies: The Primary Molecules of the Immune System

The key to the body's ability to respond to pathogens is a set of special proteins, called **antibodies**, that are synthesized by B-cells. Antibodies may remain bound to lymphocyte cell membranes or they may be released into the blood or other body secretions, such as mucus and saliva.

Each of these extraordinarily diverse antibody molecules is fashioned specifically to bind to one particular foreign molecule or **antigen**. Almost any large molecule can serve as an antigen, or "*anti*body-*gen*erator." Molecules on the surfaces of viruses, bacteria, and parasites can serve as antigens, as can isolated macromolecules ranging from proteins and glycoproteins to complex carbohydrates.

There are five major classes of antibodies, or **immunoglobulins,** as these proteins are formally known (Table 39.3). Each immunoglobulin (Ig) class is abbreviated by a letter. Antibodies found in body secretions (including sweat, saliva, and milk) belong to the IgA class; the largest group of antibodies in the bloodstream of an adult is the IgG class.

Each type of antibody interacts with different molecules and cells of the immune system, but all antibodies share certain characteristics. Each has a **variable region** that recognizes a specific antigen, and each has a **constant region** that controls how the molecule interacts with other parts of the immune system. Functionally, variable

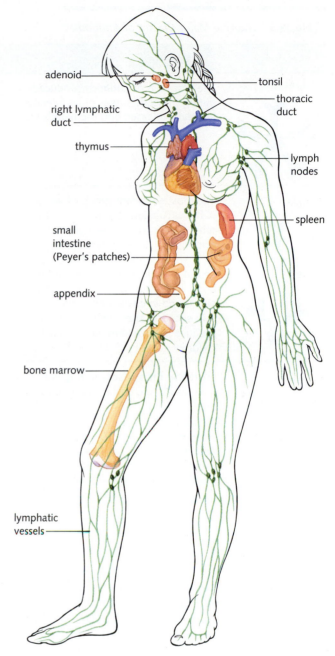

Figure 39.6 Lymphocytes, the cells of the immune system, are found throughout the body. Lymphocytes are present in large numbers in the lymphatic system, the bone marrow, thymus, spleen, tonsils, adenoids, blood, and in the digestive system.

Table 39.3   *The Five Major Classes of Antibodies*

Antibody	Characteristics
IgM	The first class of antibody produced during the early stages of an infection; generally released into the bloodstream
IgG	The most abundant group of antibodies in the circulation; can penetrate to other tissues as well; is produced in response to serious infections
IgA	Found in secretions throughout the body, including sweat, tears, and saliva; also found in colostrum, a mother's first milk secreted after childbirth; helps produce passive immunity in nursing children
IgD	Found on the surfaces of B-lymphocytes; may play a role in the immune response itself
IgE	Associated with allergic reactions and asthma; found on the surfaces of histamine-carrying mast cells

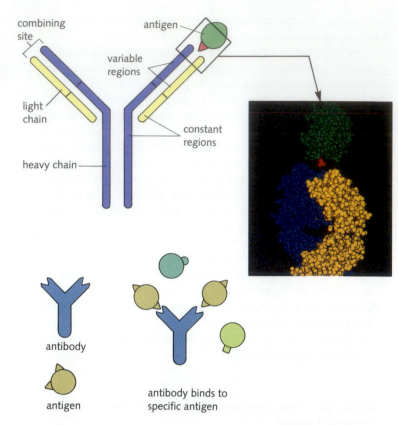

Figure 39.7   Antibodies, like this IgG, are composed of heavy and light chains. The variable region of the molecule contains an antigen-binding, or combining, site. (right) A computer-generated model shows the close fit between antigen and the combining site of the antibody. The critical portion of the antigen recognized by the antibody is shown in red.

regions act like "glue" that attaches the antibody to the antigen. The constant regions at the other end of the antibody molecule act like a "signal flag" to direct the responses of other components of the immune system (Figure 39.7).

An IgG antibody, for example, is a Y-shaped molecule composed of two identical **light chains** and two identical **heavy chains** linked together. Both types of chains form part of each variable region and the constant region. An IgG contains two **combining sites** formed from the variable regions. These combining sites bind to a specific antigen. The binding is remarkably specific: An antibody can distinguish between two molecules that differ by only a few atoms.

Ultimately, each mature B-cell produces only a single type of antibody targeted against a specific antigen. The fascinating process by which B-cell response is finely tuned in this manner begins with an encounter between the antigen and B-cells.

THE IMMUNE SYSTEM IN ACTION

Soon after a pathogen enters the body, normal processes of circulation and lymph movement cause a chance encounter between B-cells and one or more of the antigens associated with that pathogen. In some cases, B-cells encounter intact pathogens with antigens on their surfaces. In other cases, antigens are "presented" to lymphocytes by macrophages that have already engaged the invader.

What happens next depends on the B-cells involved in the encounter. Each immature B-cell carries on its cell surface a single type of antibody. One cell, for example, might carry copies of an antibody that binds to a coat protein of the measles virus, whereas another cell might produce an antibody that binds to carbohydrates on a cell membrane of a trypanosome. Although this might seem like a fairly hit-or-miss strategy, the population of B-cells is large and diverse. This diversity ensures that the immune

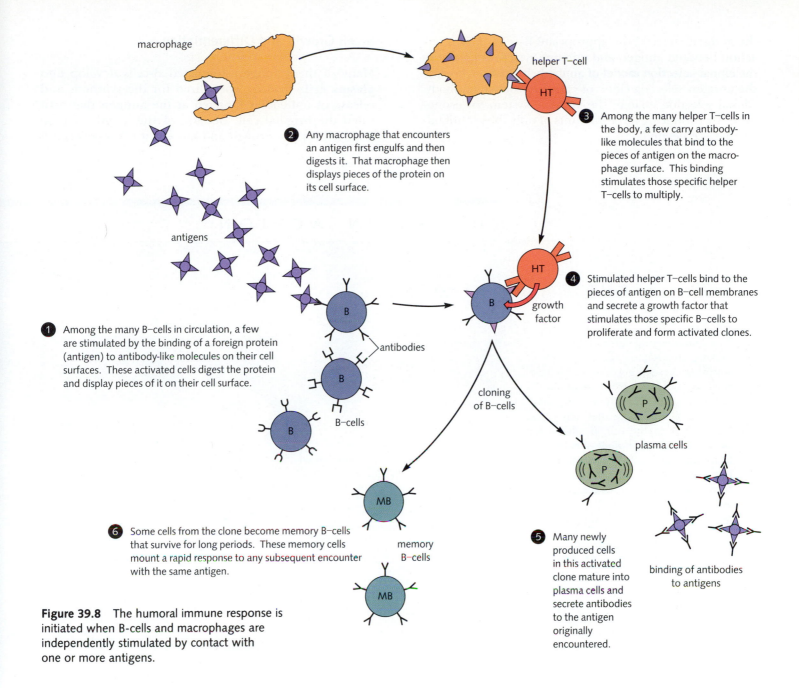

macrophage

2 Any macrophage that encounters an antigen first engulfs and then digests it. That macrophage then displays pieces of the protein on its cell surface.

helper T-cell

3 Among the many helper T–cells in the body, a few carry antibody-like molecules that bind to the pieces of antigen on the macrophage surface. This binding stimulates those specific helper T–cells to multiply.

antigens

4 Stimulated helper T–cells bind to the pieces of antigen on B–cell membranes and secrete a growth factor that stimulates those specific B–cells to proliferate and form activated clones.

growth factor

1 Among the many B–cells in circulation, a few are stimulated by the binding of a foreign protein (antigen) to antibody-like molecules on their cell surfaces. These activated cells digest the protein and display pieces of it on their cell surface.

antibodies

B-cells

cloning of B–cells

plasma cells

6 Some cells from the clone become memory B–cells that survive for long periods. These memory cells mount a rapid response to any subsequent encounter with the same antigen.

memory B–cells

5 Many newly produced cells in this activated clone mature into plasma cells and secrete antibodies to the antigen originally encountered.

binding of antibodies to antigens

Figure 39.8 The humoral immune response is initiated when B-cells and macrophages are independently stimulated by contact with one or more antigens.

system contains cells capable of responding to virtually any antigen the body may encounter.

Antigen Stimulation

If a B-cell happens, strictly by chance, to encounter an antigen to which its particular antibody is matched, the antibody binds to that antigen. This binding process activates the B-cell and initiates a chain of events that sets the immune response in motion.

The B-cell immediately digests the antigen (in this case, a protein) into smaller peptides, which it places on the outer surface of its cell membrane. If one of those peptides is recognized (again by chance) by a receptor molecule on a **helper T-cell,** that T-cell binds to the B-cell. The helper T-cell releases molecules known as **interleukins,** which stimulate the B-cell to proliferate into a clone of cells, each of which produces antibodies against the antigen (Figure 39.8).

Because the antigen activates the very B-cells that are prepared to make antibodies against it, we might say that the antigen "selects" the appropriate B-cells. This interaction between antigen and immune system is known as the **clonal selection model** of antibody formation, because the antigen selects a clone of cells to produce antibody. Clonal selection enables the immune system to "economize" by producing large amounts of only those antibodies that are needed.

B-cell Growth and Differentiation

Many of these antigen-stimulated B-cells develop into **plasma cells** that are specialized for the synthesis and release of antibodies targeted at the antigen that activated the original cell. Because of the length of time required for cell growth and mitosis, it takes between 5

THEORY IN ACTION

Millions of Antibodies from a Few Genes

The immune system is capable of making an antibody against almost anything. Research workers who have purified molecules from biological systems can be virtually certain that if they inject those molecules into a rabbit or a mouse, the animal will quickly begin to make antibodies in response to the injection. The *breadth* of the immune response (its ability to manufacture an almost limitless variety of antibodies) gives rise to an interesting question: Because antibodies are very large proteins, and because the amino acid sequences of proteins are coded for in DNA, shouldn't the genomes of mammalian cells be dominated by millions of different genes that code for different antibodies? Putting it another way, how does an organism with an immune system manage to find room for anything else in its DNA?

Until very recently this question baffled immunologists and geneticists alike. But in 1976, a scientist named Susumu Tonegawa made an interesting discovery. Using a DNA copy of the mRNA that coded for the variable and constant regions of an antibody from an IgG-producing cell, he discovered that the corresponding DNA in an embryonic cell of the same species was in several pieces. Why would DNA that was in one piece in the adult IgG-producing cell be in pieces in an embryonic cell? The only simple

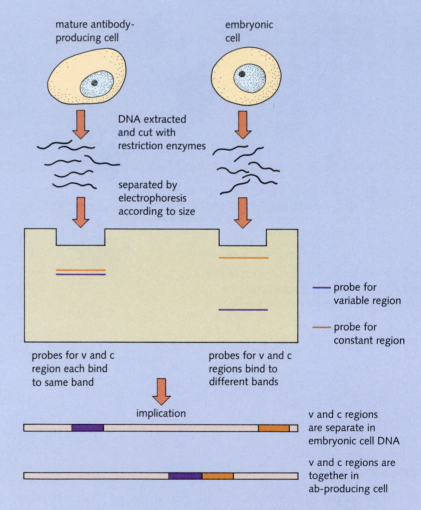

mature antibody-producing cell

embryonic cell

DNA extracted and cut with restriction enzymes

separated by electrophoresis according to size

— probe for variable region

— probe for constant region

probes for v and c region each bind to same band

probes for v and c regions bind to different bands

implication

v and c regions are separate in embryonic cell DNA

v and c regions are together in ab-producing cell

Susumu Tonegawa used DNA hybridization and gel electrophoresis to discover that antibody genes move during the course of development.

and 10 days for the first significant amounts of antibody to appear after an initial infection.

Other antigen-stimulated B-cells proliferate into clones of **memory cells.** Memory cells are stored in the spleen, where they are ready to respond rapidly should the body ever encounter that same antigen again. The existence of these memory cells makes a second response of the immune system to the same antigen swifter and much more powerful than the first (Figure 39.9). This is the basis of *permanent immunity* to certain diseases.

Somatic mutation   Immunologists had thought that the antibodies produced by a plasma cell clone were exactly the same as those produced by the original antigen-

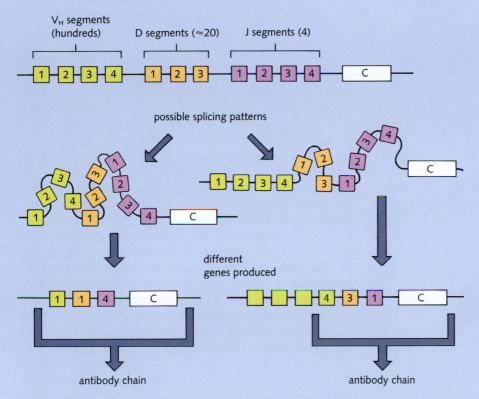

Complete antibody chain genes are assembled from four different segments. Cutting and splicing brings together one copy each of the V, D, J, and C segments to form the complete gene. Because the exact pattern of splicing is a chance event, it happens in a slightly different way in each cell, ensuring that the immune system as a whole will be able to produce thousands of completely different antibody chains.

conclusion was that some sort of rearrangement of the DNA must have taken place as the cells developed.

Tonegawa and other scientists quickly explored these clues, and we now have a clear picture of how diversity is built into the immune system. Each developing cell contains several gene regions that code for the constant regions of antibody chains, several hundred that code for the variable regions of the chains, and several more that code for D and J regions connecting the two in the finished antibody molecule. As shown in the accompanying figure, a series of cutting and splicing events during development produces a final immunoglobulin gene that contains an apparently random collection of C, D, J, and V segments. The genes for antibodies are assembled almost as though by shuffling a deck of genetic possibilities from the segments present in each cell.

Because these events occur almost at random, the final immunoglobulin gene is different in every cell that goes through the process of cutting and splicing. Because millions of cells are involved in the immune system, this process guarantees that hundreds of millions of different possibilities (in the form of antibody genes) are built up and that the system is ready to form antibodies against any antigen.

Figure 39.9 The presence of memory cells makes the second response to an antigen much more powerful than the first.

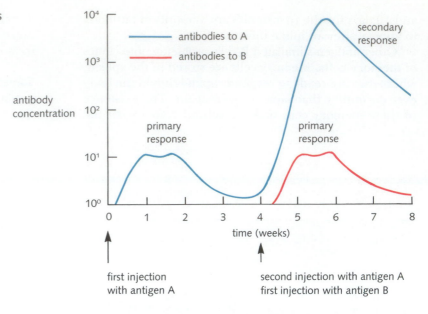

first injection with antigen A

second injection with antigen A
first injection with antigen B

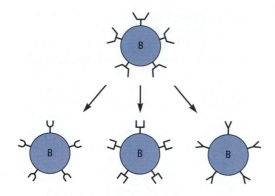

mutation produces variation in binding sites

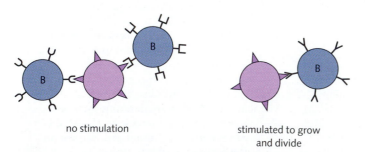

no stimulation

stimulated to grow and divide

Figure 39.10 Small mutations in the antibody-producing genes produce a B-cell population that is constantly selected (and improved) for its ability to bind antigen. Those B-cells whose antibodies fit the antigen best are stimulated to grow and divide.

stimulated B-cell. Recently, however, they have discovered that as a stimulated B-cell develops, it actually gives rise to a variety of clones which produce antibodies that fit the antigen better and better as time goes on!

How and why does this occur? The best explanation seems to be that stimulation causes an extremely high rate of mutation in the DNA regions that code for the most variable portions of the antibody combining site. Because of these rapid mutations, the daughter cells of a single B-cell produce a range of slightly different antibodies. Some of these antibodies fit the antigen better than the original antibody, and some of them fit it more poorly. Those B-cells producing antibodies that bind the antigen most tightly are stimulated to divide more rapidly than those that bind less tightly (Figure 39.10).

This process of mutation and antigen stimulation allows the immune response to be fine-tuned to each and every antigen, making it more versatile and more effective. The fine-tuning continues throughout the proliferative phase of clonal selection and ends when mature plasma cells are formed.

Effects of the Humoral Response

The antibodies produced in response to a foreign antigen defend the individual against a foreign invader in several ways. Because antibodies have two combining sites, they can *crosslink* antigen-carrying viruses or bacteria into a network, as shown in Figure 39.11a. Crosslinked bacteria or virus particles cannot infect cells, and they are easy targets for **phagocytes** ("eating cells") that engulf and destroy them.

To make this response even more effective, the constant region of the antibody molecule—the area we earlier referred to as a "signal flag"—carries "instructions" to other molecules and cells in the body on how to handle the crosslinked package. If the antigen is a free-floating protein, the particular antibody bound to it may single it out for excretion by the kidney. If the antigen is part of a virus or bacterium, the constant region attracts phagocytes that engulf and destroy the invader (Figure 39.11b).

Finally, certain antibodies activate a group of blood proteins known as **complement proteins.** When these circulating proteins encounter an antibody bound to the surface of a cell, they attach to the cell membrane and create a pore, causing the cell to burst (Figure 39.11c). Complement proteins thus make possible the destruction of cells too large to be eaten by phagocytes.

T-Cells and the Cell-Mediated Immune Response

As important as antibody formation is, it is only part of the immune response. T-lymphocytes mount a direct cell-to-cell campaign, or **cell-mediated immune response,** against foreign organisms. Like B-cells, T-cells have antibody-like molecules on their cell surfaces. Unlike B-cells, however, T-cells do not secrete these molecules.

There are three kinds of T-cells, each of which performs a different function in cell-mediated responses: **helper T-cells, cytotoxic** or **"killer" T-cells,** and **suppressor T-cells.**

T-cell actions   When *killer T-cells* are stimulated by contact with an antigen on the surface of a foreign cell, they attack the antigen-bearing cell directly and destroy it by disrupting the cell membrane (Figure 39.12). Other T-cells, known as *suppressor T-cells,* help to suppress, or moderate, the immune response by slowing down the rate of cell division and limiting the production of antibodies once the infection is under control. Perhaps the most important of all classes of T-cells, however, are the *helper T-cells* we have already encountered. In addition to secreting interleukins that stimulate B-cells, helper T-cells secrete several other substances that affect a wide range of responses that occur throughout the immune system (Figure 39.12).

The cell-mediated response is extremely important in viral infections, because although circulating antibodies can trap free virus particles, they are unable to attack viruses that have entered cells. Active viruses, however, cause their host cells to manufacture viral proteins, which the host cells often "display" on their surfaces. Killer T-cells can detect those viral proteins on the surfaces of

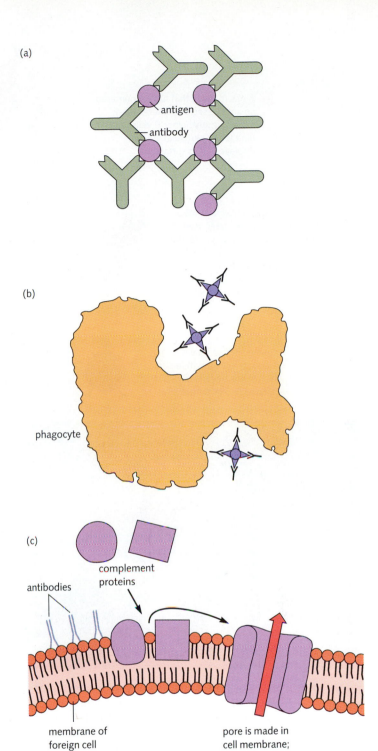

Figure 39.11 Antibodies can (**a**) crosslink and immobilize antigens, attract phagocytes (**b**) that seek out and destroy foreign cells, and start the binding of complement (**c**) to cells that carry antigen markers. Complement proteins bind to such cells and produce destructive pores in their cell membranes.

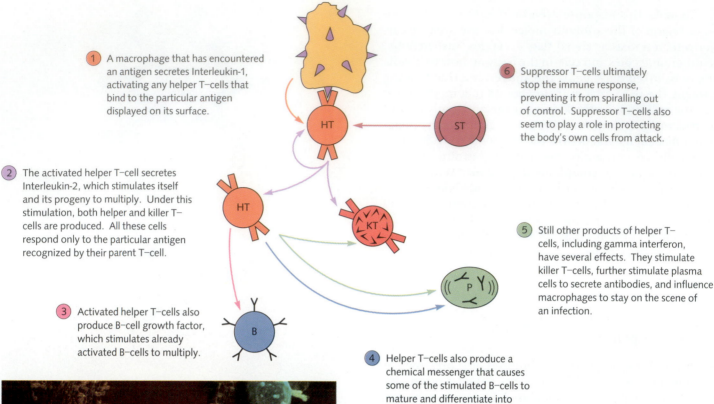

1. A macrophage that has encountered an antigen secretes Interleukin-1, activating any helper T–cells that bind to the particular antigen displayed on its surface.

2. The activated helper T–cell secretes Interleukin-2, which stimulates itself and its progeny to multiply. Under this stimulation, both helper and killer T–cells are produced. All these cells respond only to the particular antigen recognized by their parent T–cell.

3. Activated helper T–cells also produce B–cell growth factor, which stimulates already activated B–cells to multiply.

4. Helper T–cells also produce a chemical messenger that causes some of the stimulated B–cells to mature and differentiate into plasma cells that produce antibodies.

5. Still other products of helper T–cells, including gamma interferon, have several effects. They stimulate killer T–cells, further stimulate plasma cells to secrete antibodies, and influence macrophages to stay on the scene of an infection.

6. Suppressor T–cells ultimately stop the immune response, preventing it from spiralling out of control. Suppressor T–cells also seem to play a role in protecting the body's own cells from attack.

Figure 39.12 (top) The entire immune response is coordinated by protein messengers that influence the growth, differentiation, and activities of several classes of cells. Because T-cells secrete several of these messengers, they are absolutely essential to the body's defense against disease. (bottom) Killer T-cells attack a cancer cell in this artificially colored scanning electron micrograph.

infected cells and kill them before the virus can replicate and infect other cells. Both the humoral and cell-mediated responses to viral antigen are summarized in Figure 39.13.

Self and Non-Self

The effectiveness of the cell-mediated response in attacking foreign cells raises the question of how the immune system avoids attacking the cells of the body itself. In short, how does the system distinguish between self and non-self? The answer is found in a series of marker proteins at the cell surface that enable T-cells to identify cells that "belong" to the body and cells that do not. The genes that code for these marker proteins are found in a region of the genome known as the **major histocompatibility complex (MHC complex).** The proteins themselves are sometimes known as **MHC markers.**

Because of genetic recombination, no two individuals (except identical twins) have exactly the same set of MHC markers. Foreign MHC markers activate T-cells that bring on the cell-mediated immune response (Figure 39.14). The greater the differences between the MHC markers of the foreign cells and those of the body, the stronger and more effective the response.

Encounter:

Macrophages encounter virus particles, ingest them, digest them, and display pieces of viral proteins (antigens) on their surfaces.
These exposed antigen fragments selectively bind to antibody-like molecules on the surfaces of certain helper T–cells. T–cells thus "selected" are activated by a secretion of the macrophage. Meanwhile, certain B–cells encounter and selectively bind to antigen fragments. These molecular fragments may be either floating around in the plasma or attached to the surface of macrophages.

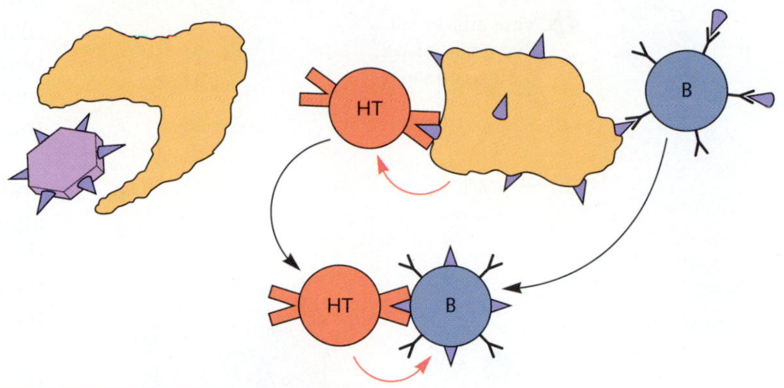

Mobilization:

Activated helper T–cells proliferate, forming clones of both activated helper T– and killer T–cells. Helper T–cells secrete chemical messengers that stimulate the multiplication of those B–cells and killer T–cells that have also been activated by contact with antigens from the invading virus.
As B–cells proliferate, further secretions from helper T–cells induce them to mature and secrete antibodies to the activating antigen.

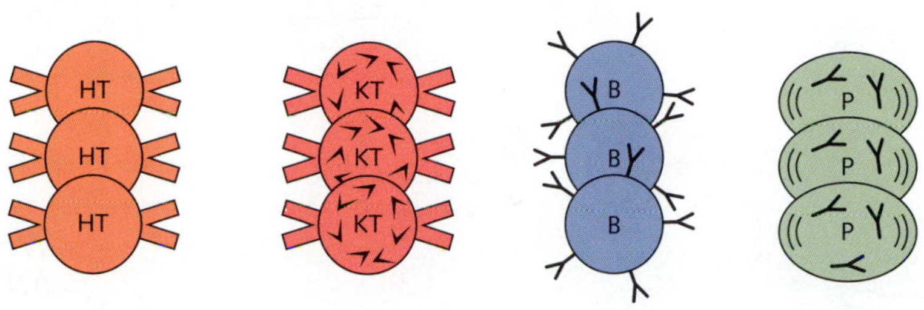

Attack:

Killer T–cells multiply and destroy those body cells infected with active virus. This sacrificial maneuver prevents further viral replication. Mature plasma cells secrete free antibodies that bind directly to free virus particles, preventing those viruses from attacking other cells. Antibodies bound to the surfaces of virus-infected cells also direct the action of other chemical reactions that destroy those cells.

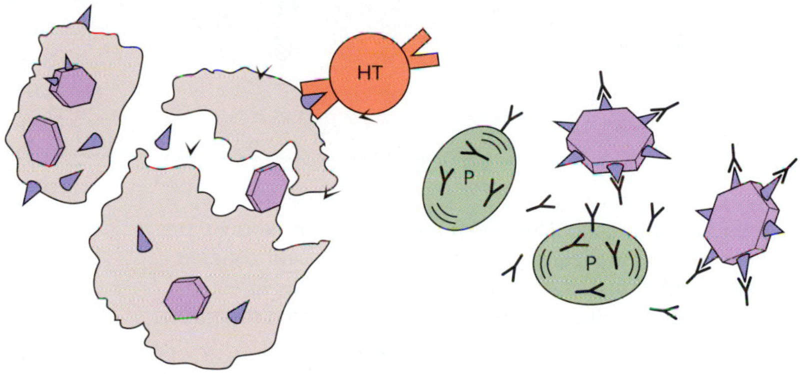

Cease-fire:

Suppressor T–cells halt the actions of both plasma cells and other T–cells, preventing their self-activating chain reactions from getting out of control.
Some B–cells and T–cells from activating clones differentiate into memory cells. These memory cells can respond rapidly to future infection for nearly the entire life span of the individual.

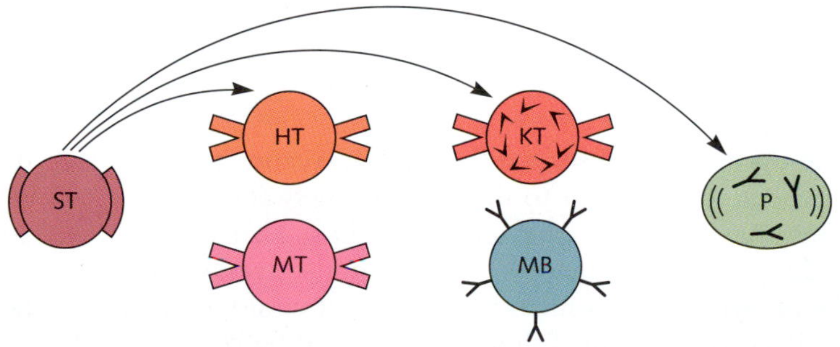

Figure 39.13 Humoral and cell-mediated immune responses combine to combat an invading virus. For this figure, we have arbitrarily divided the complex series of events into four stages and labelled them as encounter, mobilization, attack, and cease-fire. Note also that several chemical messengers and processes have been omitted here for simplicity.

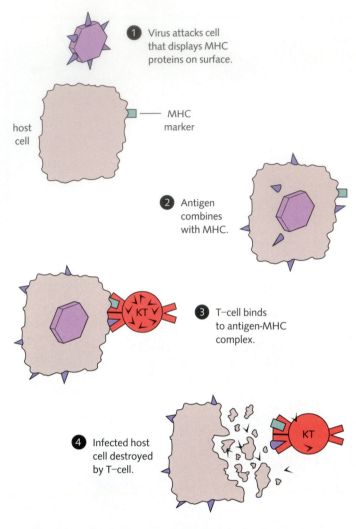

① Virus attacks cell that displays MHC proteins on surface.

host cell

MHC marker

② Antigen combines with MHC.

③ T–cell binds to antigen-MHC complex.

④ Infected host cell destroyed by T-cell.

Figure 39.14 Interactions between marker proteins known as the major histocompatibility complex (MHC) and viral antigens allow the immune system to attack and identify virus-infected cells. MHC markers also help the immune system to distinguish foreign cells from those that are part of the body.

The **thymus gland** plays a fascinating role in "teaching" the immune system to discriminate between those proteins that belong in the body and those that don't. According to current theory, the enormous diversity of T-cells present at birth includes many that could attack the body's own proteins if they survived. But before any damage can occur, the thymus gland singles out those T-cells and eliminates them.

How does the thymus accomplish this feat? This remarkable gland contains specialized cells that absorb a representative sample of the body's proteins, digest them, and display their component peptides to the T-cell population. Any developing T-cell carrying receptors that bind to one of these "self" peptides is killed in the thymus. By some time after birth, therefore, all the T-cells that might attack our own tissues are eliminated.

Transplant rejection The cell-mediated immune response against foreign MHC markers is responsible for the rejection of tissues and organs transplanted from one individual to another. When the transplanted cells contain MHC markers that can be recognized by a population of T-cells, these lymphocytes mount a cell-mediated response that destroys the transplanted tissue. Physicians have tried two strategies to make transplants work: finding a "compatible" donor (usually a family member whose MHC markers are very similar to those of the recipient) and suppressing the function of the immune system with drugs such as *cyclosporin*.

WHEN THE IMMUNE SYSTEM WORKS AND WHEN IT FAILS: CASE STUDIES

Although the immune system is always activated in the same way by foreign organisms, the results of that activation may vary greatly. In some cases the system is able to deal with a pathogen effectively, but in other cases it cannot do so. Let us return to the first three children in the opening scene of this chapter—one suffering from measles, one from the flu, and one from sleeping sickness—and consider why the system works in some cases and fails in others.

Recovering from Disease with Permanent Immunity

The immune system reacts to specific antigens on the surfaces of pathogens, including the bacteria that cause tuberculosis and diphtheria and the viruses that cause smallpox and measles. When the body first encounters these pathogens, it produces a powerful immune response that attacks them with antibodies and produces a large clone of memory cells. If a pathogen carrying those same antigens enters the body again, these memory cells produce antibody so quickly that the viruses are destroyed before an infection can occur (Figure 39.15a). Luckily for us, many pathogens (including the virus that causes measles) change very little, even over long periods of time. For that reason, the measles virus your body encounters today will be nearly identical to one you may first have encountered as a child, and the memory cells "on file" can prevent reinfection.

Monoclonal Antibodies

When an antigen stimulates the immune response, antibodies are made against specific sites on the antigen molecule. These sites can be very small—often just three or four amino acids in the case of a protein antigen—and they are distributed along the exposed surfaces of the molecule. Because an individual protein may have hundreds of such sites along its surface, hundreds of different B-cells may be stimulated to produce antibodies when the immune system reacts to that protein. This leads to a *polyclonal* immune response, because many clones of B-cells develop in response to the antigen, each clone producing a different antibody in response to a different site.

However, because the production of an antibody is a cellular event, it is possible to select a *single clone* of B-cells that will produce *monoclonal* antibodies against a single site on the antigen—a perfect supply of exquisitely specific antibodies.

That was the thinking of Cesar Milstein of Great Britain as he followed suggestions about the immune response first made by George Kohler in Switzerland. Milstein and Kohler knew that two problems would arise in growing single clones of antibody-producing cells: selecting the individual single B-cells or plasma cells that might react to an antigen from the millions around them that did not, and coaxing those cells to grow and divide in culture. Milstein developed a technique to do both, and it is now widely used in research work.

The antigen is injected into mice, and after a few days, millions of lymphocyte cells are removed from the mouse. Somewhere among these are a few cells that have been stimulated by the antigen. In order to ensure that

inject mouse
with antigen "A"

The preparation of monoclonal antibodies by the fusion of mouse lymphocytes with melanoma cells.

lymphocytes

myeloma cells

mix cells—some fuse
transfer to specific medium

Hybridoma cells

unfused cells die;
fused (hybridoma)
cells grow

hybridoma cells
cultured in
separate wells
for analysis
of antibodies

positive clones used to produce large
quantities of monoclonal antibodies

the antibody-producing cells will grow and divide efficiently, the investigator mixes the lymphocytes with cells of a mouse *myeloma*, a cell line produced from a tumor of the antibody-producing plasma cell. The mixture of cells is treated under conditions wherein cell *fusion* occurs, producing hybrid cells between the mouse B-cells and the myeloma cells.

These "*hybridoma*" cells are now immortal and will grow almost without limit in culture. The investigator grows individual clones of the hybridomas and tests them, one at a time, for reactivity with the original antigen. The clones that test positive are

set aside and grown under conditions in which they produce large amounts of antibody.

These antibodies are said to be *monoclonals*, because they are produced by a single clone of cells derived from the reaction to one specific site on the antigen. Monoclonal antibodies are much more specific than polyclonals, and they can be used to detect the presence of individual antigens on proteins, virus particles, and tumor cells. Monoclonal antibodies, because of their extreme specificity, are already used in an experimental test for the AIDS virus and in targeting drugs to cancer cells.

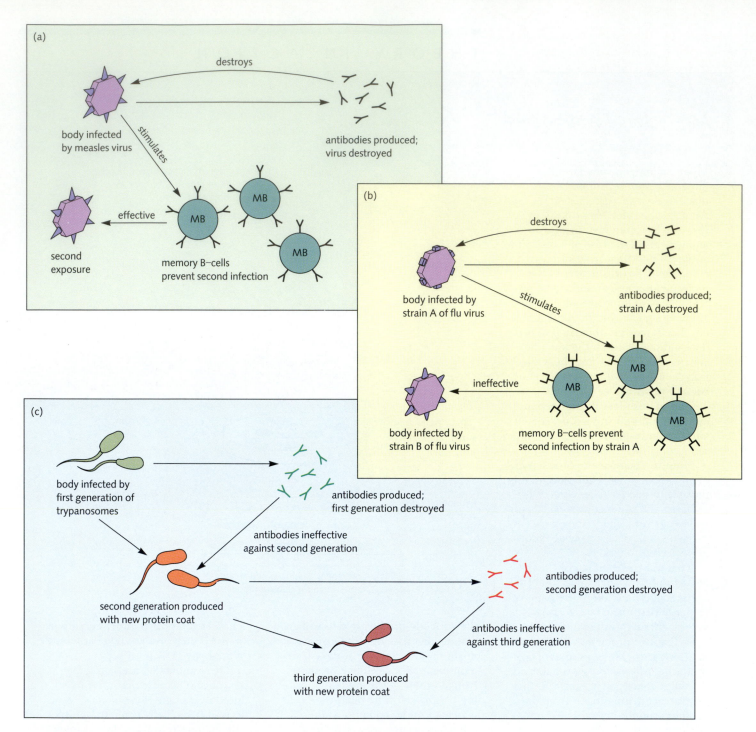

Figure 39.15 The memory cells of the immune response are effective in preventing a second infection by measles virus (**a**). However, the coat proteins of flu virus undergo frequent mutations, so that memory cells may not be effective against a second infection by a different strain of the flu virus (**b**). Trypanosomes undergo rapid rounds of mutation as they grow in the body, ensuring that every time the immune response rises to destroy the parasite, a few cells with different coat proteins evade the response and continue the infection (**c**).

This, of course, is also the reason why vaccines work to protect us against certain diseases. Louis Pasteur realized this and applied his insight to the control of several diseases for which no mild form (such as Jenner's cowpox) existed. Pasteur discovered that many pathogens could be grown in a manner that limited their ability to produce disease but did *not* affect their ability to stimulate the immune response. This strategy is the basis for today's vaccines against diseases such as measles, whooping cough, and polio. It is also possible to produce some vaccines by killing the pathogen before injection.

Furthermore, it is now possible to use the techniques of molecular biology to identify a single, specific protein from the pathogen that triggers a suitable immune response. That protein is next either cloned for use as a vaccine by itself or inserted into the genome of a harmless microorganism that is then cultured and injected into the body. This fascinating technique holds great promise for controlling some of the world's most persistent diseases.

Recovering from Disease with Temporary Immunity

The second child in our opening discussion will mount a strong immune reaction against the flu and will probably recover completely within a few days. He will produce a clone of memory cells against the flu virus and, for the time being, will be protected from reinfection by that particular strain of the virus.

But the viruses that cause the flu and the common cold, unlike those that cause measles, mutate rapidly in ways that change the nature of the proteins in their outer coat. As a result, every few months (in the case of colds) or every few years (in the case of flu) new strains arise that carry completely different coat proteins. Naturally, memory cells specialized to respond to the "old" viral strains are useless against these new strains (Figure 39.15b). That's why our patient (and all the rest of us) cannot acquire permanent immunity to the flu or the common cold. That is also why it has been impossible to develop a lifetime flu or cold vaccine.

A Pathogen That Defeats the Immune System

Our third patient, suffering from sleeping sickness, is in much greater difficulty. Trypanosomes, which live in the bloodstream, escape the immune response by having more than 100 sets of genes coding for the proteins that coat their outer surfaces. The immune system produces a powerful antibody response against the pathogen only a few days after infection begins. But just as the immune system gears up to attack the trypanosomes that invaded the body, a new generation of pathogens uses a different set of genes to make a different set of surface proteins. As far as the immune system is concerned, it is as though the invaders had disappeared; the body must start from scratch to mount a response against the new antigens. And by the time that new response is geared up, the trypanosome switches coats again, always staying one jump ahead of the body's defenses (Figure 39.15c).

That's why this patient will struggle through fever after fever, suffering brain inflammation and damage to the nervous system. In most cases, the immune system loses the battle and the patient dies.

DISORDERS OF THE IMMUNE SYSTEM

Allergies

Although the immune system evolved to protect us, when it overreacts to a foreign molecule in the environment, serious problems called **allergies** can result. We still do not know why some people suffer from allergies and others do not, but we do understand in principle how the allergic response proceeds.

The offending molecule can be part of a pollen grain, an animal hair, the toxin from a bee sting, or even house dust. The immune system is always capable of stimulating an inflammatory response when antigens are discovered, and these reactions are caused by a class of immunoglobulin known as IgE. When IgE binds to its antigen, it activates a special type of immune cell known as a **mast cell** (Figure 39.16). Activated mast cells release *histamine granules* that initiate the inflammatory response we described above. The runny nose, sneezing, and itchy eyes of hay fever are all the result of histamine release by mast cells activated by IgE reactions to pollen antigens. Although there is no effective cure for such symptoms, many individuals are able to get some relief by taking *antihistamines,* drugs that counteract the effects of histamines. (Note that cold medications containing antihistamines do *not* fight infection; they simply minimize the inflammatory response in nose and sinuses to help you breathe while the viral infection runs its course.)

The smooth-muscle reactions caused by histamines are particularly dangerous. If a strong allergic reaction develops in the breathing passageways, they contract and begin to fill with mucus. This serious condition, known as **asthma,** may make breathing almost impossible. Asthma attacks are usually triggered by exposure to a particular substance to which the sufferer is allergic, and the best

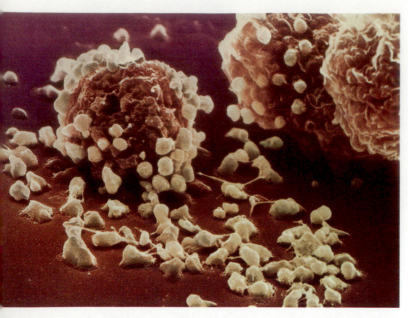

Figure 39.16 Mast cells releasing histamine granules.

medical advice is to avoid exposure to that substance. Immediate relief from a severe asthmatic attack is provided by inhaling a mist of *epinephrine* (usually from a pocket-sized inhaler), which causes smooth muscles in the passageways to relax and permits normal breathing.

Autoimmune Diseases

As we have said, a properly functioning immune system must distinguish self from non-self. When this ability to recognize the body's own molecules is lost, the immune system's power to destroy may be turned against the body's own tissues to cause an **autoimmune disease.**

One such disorder is **myasthenia gravis,** in which victims gradually grow weaker and weaker, lose control of their voluntary muscles, and may become seriously crippled. This disease is caused by an immune system response against molecules that form part of the neuromuscular junction (Chapter 42). In **multiple sclerosis,** the immune system attacks the protective myelin sheaths around nerve cells (Chapter 40), whereas in **rheumatoid arthritis,** a misguided immune response destroys cells in the body's joints (Chapter 42). One current theory holds that these autoimmune diseases are caused by renegade macrophages that pick up and display self molecules instead of non-self molecules, eventually triggering a destructive T-cell response.

Sometimes infection can trigger autoimmune diseases, as exemplified by **scarlet fever,** a disease that results when toxin-producing strains of *Streptococcus* infect the throat. Although the infection is not serious in itself, the reaction of the immune system to the bacterium can cause permanent damage. Apparently, some of the surface molecules of certain strains of the bacterium are similar to cell surface proteins found in the heart muscle. Therefore, in some cases of scarlet fever, a complication known as **rheumatic fever** results from the inadvertent attacks of the immune system on the heart muscle.

AIDS

And I looked and beheld a pale horse: and his name that sat on him was Death. —Revelation 6:8

The fourth child described in the opening of this chapter suffers from a disease that first appeared in tropical Africa. That a new disease should appear is not in itself surprising; the jungles of equatorial Africa spawn diseases new to science with alarming regularity. But this disease was caused by a virus close to perfection in terms of pathogen evolution. Not only had this pathogen developed ways to avoid every single defense mechanism of the immune system, but it also attacked that system directly, lived inside it, and destroyed it. This was the virus that came to be called **HIV**—the *human immunodeficiency virus.*

HIV spread through central Africa, where it went unrecognized for years, both because its victims died of other, already recognized diseases and because sophisticated medical care was usually wanting. It spread slowly to Europe and to Haiti, and then to the United States, Latin America, and Asia. There the first isolated cases—a few in Europe and one or two in the United States in the mid-1970s—were puzzles that left physicians wringing their hands helplessly and scratching their heads in amazement. Their patients were dying of infections from microorganisms they should have fought off easily, and no one could imagine why.

Finally, the virus found its way into two Western populations whose behaviors ensured its explosive spread: the gay community, where sexual freedom was a way of life, and the legions of intravenous drug users whose ritual sharing of needles spread infected blood with deadly speed and efficiency. At about the same time, the disease began to spread among heterosexually active individuals throughout central Africa.

These recognized modes of HIV transmission—through sexual activity and via infected blood—seriously interfered with world society's ability to react to and ultimately to control this new disease. Many medical researchers and science journalists note that AIDS has brought out both the best and the worst in everyone connected with the epidemic. On the one hand, we have learned more about this insidious virus in a shorter period of time than we have about nearly any other disease in history. On the other hand, fear, prejudice, and denial—which we discussed in a societal context in Chapter 2—created major obstacles to effective action to protect the health of those at risk of infection. Because AIDS remains incurable, and because it presents a threat to the health of many Americans, we will try to correct some of the misconceptions about AIDS by reviewing what is now known about its history and mode of action.

The History of AIDS in the United States

In the early 1980s, physicians in California and New York noticed an increasing number of patients with unusual infections and a form of skin cancer so rare that most cancer specialists had never even seen a case before. The infections, including a type of pneumonia attributed to *Pneumocystis carinii* (a protozoan), were caused by organisms that are common in the environment and cause serious disease only in extremely rare cases. The purple skin tumors were identified as *Kaposi's sarcoma*, a cancer normally seen only in elderly patients with impaired immune systems or in patients whose immune systems had been suppressed with drugs such as cyclosporin.

Thoughtful physicians soon recognized that these were **opportunistic infections**: infections that were successful only because the immune system of the patient was impaired for some other reason. The first six recognized cases of the disease were described in Los Angeles in 1981, and the name **AIDS** (for **acquired immune deficiency syndrome**) was suggested for the disorder.

Because the disease was first reported among gay men, physicians—rather than looking immediately for an infectious agent—assumed that something about these men's lifestyles led to the breakdown of the immune response. But a few epidemiologists conducted studies among the growing number of gay and IV-drug-using AIDS patients. They suggested by the fall of 1981 that AIDS was a contagious disease that might be spread through blood (by the sharing of contaminated needles) and sexual intercourse. By 1982, the first cases of AIDS were reported among infants born to infected mothers and among hemophiliacs and surgical patients who had received transfusions of blood or blood products.

Then, in 1983, Luc Montagnier and his associates at the Institut Louis Pasteur in France announced that a virus had been identified as the cause of AIDS. This discovery was quickly confirmed by a team led by Robert Gallo at the National Institutes of Health in the United

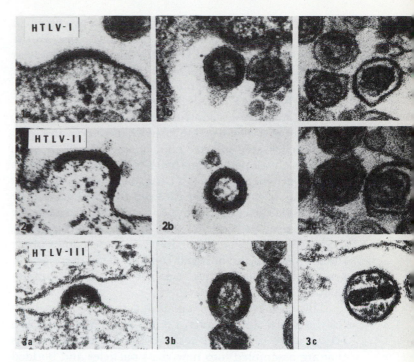

Figure 39.17 The three known types of AIDS virus particles budding from cells grown in laboratory culture. HTLV III has been renamed HIV.

States (Figure 39.17). Federal funds were released to support development of a test to detect AIDS in donated blood, and all blood donated in the United States has been screened for antibodies to the virus since 1986. Research on the disease is now progressing in laboratories all over the world, and we have learned a great deal about the biology of AIDS in a relatively brief period of time.

The Biology of HIV Infection

HIV is a retrovirus that infects populations of lymphocytes and certain epithelial cells throughout the body. Many researchers have called HIV the most insidiously "clever" parasite ever to evolve. Not only does it avoid the defenses of the immune system, but it also subverts the communication network that coordinates the activities of several classes of lymphocytes and "co-opts" that network in a manner that destroys the system. The effects of active HIV infection serve as a grisly lesson in the importance of a functional immune system in ordinary day-to-day life.

We now know that HIV is composed of a glycoprotein envelope surrounding a dense, cylindrical core. In the core are two molecules of viral RNA that contain at least eight genes (Figure 39.18). Some of these genes code for the envelope proteins, some code for the core proteins, one directs the synthesis of the enzyme reverse transcriptase, and three genes control the activity of the virus.

At first it was believed that the virus selectively—and nearly exclusively—infected populations of T-cells, because one of its envelope proteins binds to a receptor molecule on the T-cell membrane. It is clear that this sort of attachment does occur and that it causes the viral core to be taken inside the cell. There the virus synthesizes reverse transcriptase, the enzyme that makes numerous DNA copies of its RNA genome. Some of those copies insert themselves into the host cell DNA and stay there *permanently,* while other copies remain free in the cytoplasm (Figure 39.19).

This viral DNA may remain inactive in the host cell for widely varying periods of time. When stimulated into action, it directs the synthesis of viral RNA and proteins that are assembled into new virus particles and budded off into the bloodstream. Active HIV infection can kill not only the cell it has infected, but many other T-cells as well. Because they carry the critical viral protein on their cell membrane, infected cells bind to other, healthy T-cells, killing them *en masse.*

It is now clear that HIV can also infect another important class of immune cells, macrophages. This aspect of HIV infection is particularly insidious, because active viruses seem to be able to "hide" inside macrophages for weeks, months, or possibly even longer without triggering other responses by the immune system. In fact, the virus seems to be able to grow and reproduce indefinitely inside a macrophage without killing the host cell. Furthermore, by hiding inside monocytes that cross the blood–brain barrier, the HIV infection can also enter the central nervous system, where it can cause serious brain damage.

Part of the difficulty in tracing and predicting the spread of HIV arises from its long, highly variable latent period, which appears to range from a few weeks to as many as 10–15 *years*. During that time an individual may have no warning that the disease is about to develop. When the virus is first activated, the signs of immune suppression may be subtle or invisible; they include night sweats, swollen lymph nodes, a fungal infection of the throat called "thrush," and severely reduced numbers of helper T-cells in the blood.

A diagnosis of full-blown AIDS is made when the patient develops either Kaposi's sarcoma or *Pneumocystis* infection. The average time from infection to full-blown AIDS seems at this writing to be as long as 8 years, which means that some infected persons do not develop the disease for 16 years (or longer) after infection. Long-term studies of HIV-positive individuals in San Francisco seem to indicate that between 95 and 99 percent of infected individuals ultimately develop the disease. At present, there is no medical test that can tell for certain when or how soon AIDS will develop in an infected individual.

Figure 39.18 (left) A schematic model of the HIV retrovirus, showing its membrane-like outer coat and internal core structure. (right) The HIV genome.

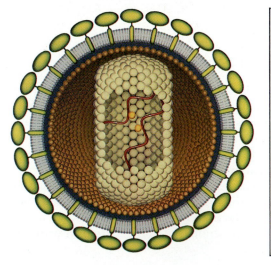

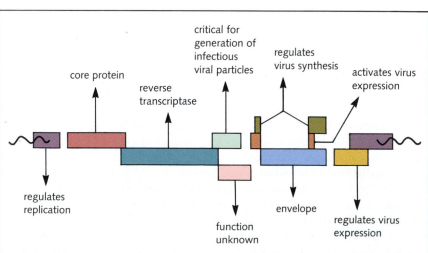

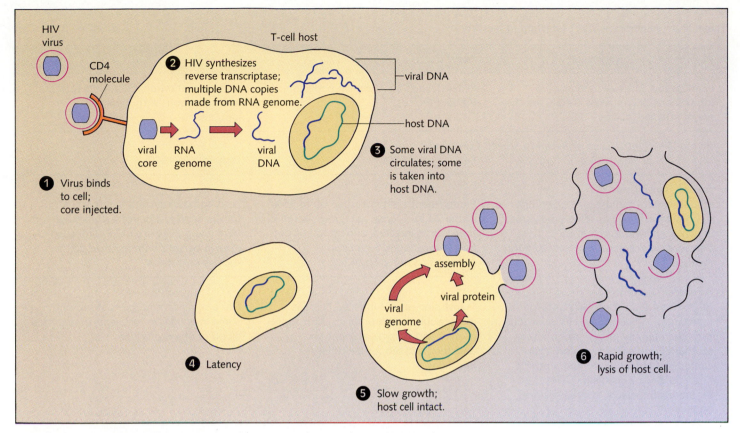

Figure 39.19 Infectious cycle of the HIV retrovirus. The RNA genome of the virus is copied into DNA by reverse transcriptase, and becomes part of the host cell DNA. The viral genome may remain latent within the cell for long periods of time. It may also direct the synthesis of new virus particles, or may grow rapidly and destroy its host cell.

The Spread of HIV Infection

Although a cure for AIDS has not yet been developed, we are certain how the disease is spread: through blood, through sexual intercourse, and from infected mother to unborn child. It is important to realize that AIDS is not exclusively a "gay and bisexual" disease. An unknown number of heterosexual transfusion recipients, hemophiliacs, intravenous drug users, and the sexual partners of all these individuals are currently carrying the virus though they show no signs of infection. These asymptomatic carriers are probably concentrated in major metropolitan areas, but the mobility of American society makes it likely that significant numbers of them have traveled widely across the nation. Because there is so much confusion and inaccurate information about the transmission of AIDS, we will examine the biologically relevant facts in detail.

How AIDS Is *Not* Transmitted

Although AIDS is one of the deadliest diseases known, it is *not* easy to catch. Numerous studies conducted over several years have shown that AIDS is *not* transmitted through casual contact. There is *no* evidence that HIV can spread via food or water or through coughing, sneezing, dry kissing, hugging, or sharing clothing, bedding, or eating utensils. Everyday contact with AIDS victims, including shaking hands and engaging in close conversation, poses no risk of infection. There is, therefore, neither purpose in nor need of quarantining or isolating either HIV-positive individuals or AIDS patients.

There is no evidence that AIDS can be transmitted by mosquitoes. Recall from our earlier discussion that mosquito-borne diseases are extremely host-specific; the mosquito that transmits malaria cannot carry yellow fever, and vice versa. Furthermore, the bite of the mosquito

Table 39.4 *Patterns of HIV-1 Infection in the World*

Pattern 1	Pattern 2	Pattern 3
Homosexual/bisexual men and intravenous drug abusers (IVDA) are the major affected groups	Heterosexuals are the main population group affected	More recent introduction with spread among persons with multiple sex partners
Period When Introduced or Began to Spread Extensively Mid-1970s or early 1980s	Early to late 1970s	Early to mid-1980s
Sexual Transmission Predominantly homosexual. Over 50% of homosexual men in some urban areas infected. Limited heterosexual transmission occurring, but expected to increase	Predominantly heterosexual. Up to 25% of the 20-to-40-year age group in some urban areas infected, and up to 90% of female prostitutes. Homosexual transmission not a major factor	Both homosexual and heterosexual transmission just being documented. Very low prevalence of HIV infection even in persons with multiple partners, such as prostitutes
Parenteral Transmission Intravenous drug abuse accounts for the next largest proportion of HIV infections, even the majority of HIV infections in southern Europe. Transmission from contaminated blood or blood products not a continuing problem, but tens of thousands of persons infected before 1985	Transfusion of HIV-infected blood is major public health problem. Non-sterile needles and syringes account for undetermined proportion of HIV infections	Not a significant problem at present. Some infections in recipients of imported blood or blood products
Perinatal Transmission Documented primarily among female IVDA, sex partners of IVDA, and women from HIV-1 endemic areas	Significant problem in those areas where 5 to 15% of women are HIV-1 antibody-positive	Currently not a problem
Distribution Western Europe, North America, some areas in South America, Australia, New Zealand	Africa, Caribbean, some areas in South America	Asia, the Pacific Region (minus Australia and New Zealand), the Middle East, Eastern Europe, some rural areas of South America

SOURCE: P. Piot, F. A. Plummer, F. S. Mhalu, J. Lamboray, J. Chin, and J. M. Mann, February 1988, AIDS: an international perspective, *Science,* 239:576.

transmits malaria only because the disease-causing organism enters the insect's salivary glands. There is no evidence that HIV either lives in the mosquito's system or enters its salivary secretions. Because the insect does not actually inject blood from a previous meal into new victims, there is no way for a mosquito bite to transmit this particular disease.

How AIDS Can Be Transmitted

AIDS is transmitted through specific high-risk behaviors. During the early stages of the epidemic, both medical

authorities and the media often referred to "high-risk groups," by which they meant gay and bisexual men and intravenous drug users. It is now clear that the disease is not restricted to any specific group of people. Rather, the disease can be contracted by anyone who engages in specific high-risk *behaviors.* Those behaviors are clearly identifiable: sexual intercourse of any kind and the sharing of hypodermic syringes (Table 39.4).

Sexual transmission   The initial prevalence of AIDS in the United States among non-drug-using gay men, many of whom practice anal intercourse, misled many people into thinking that AIDS is not transmitted through vagi-

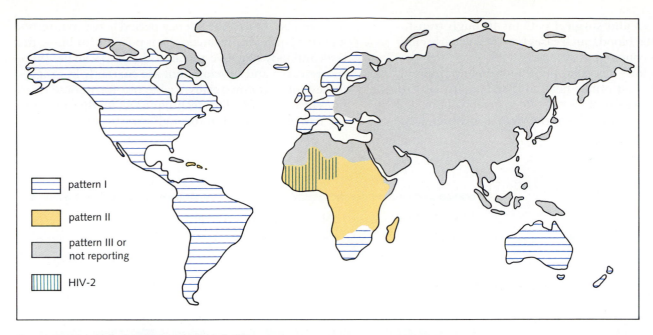

Figure 39.20 Global patterns of HIV infection.

pattern I

pattern II

pattern III or
not reporting

HIV-2

nal intercourse. Across most of Africa, however, heterosexual intercourse is the predominant mode of transmission and the ratio of infected females to infected males is close to 1:1 (Figure 39.20). Some individuals insist that differences in sexual practices between Africa and the United States make it unlikely that heterosexual transmission has occurred or will occur here. Most responsible physicians and researchers, however, now agree that this is not the case. Estimates of the number of Americans already infected with HIV range from 500,000 to nearly 2 million. Estimates of the global HIV-infected population also vary widely, in part because many countries have not reported the disease to the World Health Organization until recently for fear of being stigmatized.

It does appear that anal sex may be more effective than vaginal sex in transmitting the disease. It also appears that the presence of other, untreated sexually transmitted diseases in either partner increases the risk of transmission during vaginal intercourse. The statistical risk of transmission in any single sexual act is very small—somewhere between 1 in 100 and 1 in 1000. But a substantial number of non-drug-using sexual partners of HIV-infected individuals have become infected, some of them through a single sexual encounter. And the virus has clearly been transmitted both from male to female and from female to male.

Luckily, it does not appear that the virus is spreading explosively through the population at large, as some researchers feared it would; the number of infected individuals outside of those participating in high-risk behaviors is still small. But because of the virus's long latency period, because of uncertainty about the rate of transmission through heterosexual intercourse, and because of an almost total lack of knowledge about the sexual activities of the American public, prudent epidemiologists advise caution.

Sharing of syringes   The AIDS virus is clearly transmitted in the infected blood spread from person to person through the use of shared needles among intravenous drug users. In 1988, estimates of the percentage of intravenous drug users who are infected with the AIDS virus ranged from 50 to 75 percent in various parts of the country. Because AIDS is spread through sexual activity, the sexual partners of intravenous drug users are directly at risk and—if sexually active themselves—may serve as conduits for the virus into non-drug-using populations.

Transmission from mother to fetus   The AIDS virus easily crosses the placenta from the mother's blood supply to that of the fetus. The percentage of babies born to infected mothers, many of whom are non-drug-using sexual partners of intravenous drug users, is rising annually. By early 1988, nearly 2 percent of the babies born in New York State tested positive for HIV.

Transfusion-borne AIDS   Before 1985, there were numerous cases in which the HIV infection was transmitted to hemophiliacs and surgical patients through transfusions of infected blood. Such cases have been nearly eliminated by two concurrent strategies:

1. The nation's blood supply is now screened for antibodies to HIV, and public health officials assure us that the nation's blood supply is safe. Because not all infected persons produce antibodies to the virus at all times, however, and because of the latency period between infection and first antibody production, it is possible for blood that tests antibody-negative to carry the AIDS virus. Intensive research and development are currently in progress to develop a commercially usable test for minute quantities of viral RNA. This technique will further minimize the risk of transmission through transfusion.

THEORY IN ACTION

Protecting Yourself Against AIDS

The very best protection against contracting AIDS from sexual activity is not to have sex at all. The next best strategy is to form a monogamous relationship with another uninfected individual and to have no sexual activity outside of that relationship.

Those individuals to whom these are unacceptable courses of action should be aware that the risks of unprotected sexual intercourse are well established and that the magnitude of the risks increases with the number of an individual's sexual partners. Most contraceptive methods, including the pill and IUDs, do nothing to prevent the spread of the AIDS virus or of other sexually transmitted diseases.

Only *condoms*, which prevent contact with semen, seem to be effective in preventing the transmission of HIV during intercourse, and they are effective only if used during all sexual contact and with great care. The use of condoms by sexually active heterosexuals and homosexuals (and their limiting the number of their sexual contacts) may be the only way to slow the spread of the disease until a cure or treatment is found.

It is true that the statistical chance of getting infected with AIDS through a single act of unprotected intercourse with a member of the

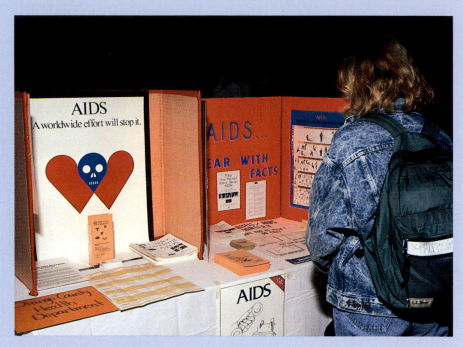

Although AIDS cannot be cured, it can be prevented. AIDS education is a critical weapon in the fight against the disease.

general population may be as low as one in a million. (This is because of AID's relatively low infectivity and because the rate of infection among the population at large seems to be quite small.) But remember the implications of the virus's long latency period. As some AIDS experts say, if you have sex with an individual, you are sharing in that individual's entire sexual history for the last 15 years. Remember too that, by most estimates, AIDS is ultimately a fatal disease.

2. The blood industry is doing everything it can to discourage potentially infected individuals from donating blood in the first place.

Note that because all licensed health care providers use sterile equipment when handling blood and blood products, it is virtually impossible to contract AIDS when giving blood at a recognized and licensed blood-donating facility.

The Outlook for Development of Vaccines and Cures

Unfortunately, the genes that code for HIV's protein envelope exhibit very high rates of mutation. This means that new strains of the virus are continually emerging, a fact that makes vaccine development extremely difficult.

The only useful tool we have at present is a test for antibodies *against* the AIDS virus. The presence of antibodies against the virus does not mean that the individual with such antibodies is immune to the virus. Rather, it implies that she or he has been exposed to the virus. Therefore, a negative result in a test for such antibodies is a reasonably safe indicator that a person has not been infected with the virus; as we have noted, the meaning of a positive test is less clear.

As of this writing, no *cure* for AIDS has been found. In other words, we know of no treatment that eliminates the virus from infected individuals. There are, however, several drugs that appear to hinder viral activity, to slow the progression of the disease, and to boost the number of healthy T-cells in infected individuals. A few of these drugs have been approved for general distribution to people with AIDS, and several others are being subjected to the sorts of controlled trials discussed in Chapter 2.

The first drug to be generally recognized for its value in AIDS treatment was AZT, a modified thymidine molecule. AZT is effective because it acts as a "stop signal" in the synthesis of viral DNA by reverse transcriptase. Because AZT is not incorporated by most body cells in their DNA synthesis, it is tolerated by many patients. However, AZT does interfere with DNA synthesis in the bone marrow of some patients, causing serious side effects. Another drug, called peptide T, is reported to interfere with HIV infection by occupying the T-cell receptor sites to which the virus normally binds.

At least a dozen other drugs are being tested at this writing, both alone and in combination with one another. Some of these drugs are aimed at HIV itself; others are directed at the organisms that cause life-threatening opportunistic infections. One such drug, pentamidine, is administered in aerosol form to prevent and treat *Pneumocystis* pneumonia.

SUMMARY

Disease is a fact of life for all organisms. One of the challenges that humans and other large organisms face in their day-to-day existence is coping with the diversity of smaller organisms that can cause disease. These pathogenic organisms include viruses, bacteria, eukaryotic cells, and multicellular organisms. They cause diseases in a number of ways, including the direct destruction of living cells and the release of toxins—poisons that interfere with normal functions. The primary barrier against infection from these organisms is the skin, the single largest organ in the human body. The skin forms a tough and effective barrier, and defenses that exist just beneath the skin come into play if the skin is broken.

The immune system comes into play when the skin is breached. Cells from the immune system play a key role in the inflammatory response, which deals with such invasions. The immune response, mediated by T- and B-lymphocytes, is a cellular mechanism for recognizing and dealing with foreign molecules. The humoral immune response, which is triggered by the introduction of foreign antigens, results in the formation of soluble antibody proteins. The stimulation of B-cells seems to occur by clonal selection of the cells that react with a new antigen. A cell-mediated immune response also is mounted in response to antigens, and its actions include a direct cell-to-cell process in which activated T-cells help to destroy large cells or even tissue transplants from another individual. The immune response can be manipulated by vaccination, a process in which weakened or altered antigen from a pathogen is injected into a healthy individual to stimulate, safely, the immune response against that pathogen. Vaccinations have helped conquer many of the world's most dread diseases, including polio and smallpox. The immune response also can cause problems for otherwise healthy individuals, including the production of allergies and autoimmune diseases, in which some aspects of the immune response seem to be directed against an individual's own cells. The importance of the immune response is illustrated tragically by the rise of AIDS, a viral disease that results in the destruction of some aspects of the immune system to the point where death from other infections is all but certain to follow. AIDS is transmitted by blood (especially via the sharing of hypodermic needles) and during sexual intercourse.

STUDY FOCUS

After studying this chapter, you should be able to:

- Review disease through the ages and the development of ways to combat and prevent disease.
- Explain the essential features of the immune response and the components and functions of the immune system.
- Describe various disorders of the immune system, paying particular attention to AIDS.

pathogen *p. 780*
parasite *p. 781*
histamine *p. 782*
monocyte *p. 783*
antibodies *p. 783*
B-lymphocyte *p. 784*
T-lymphocyte *p. 784*
antigens *p. 785*
macrophage *p. 785*
helper T-cell *p. 787*
plasma cell *p. 788*
phagocytes *p. 790*
complement proteins *p. 791*
cytotoxic T-cell *p. 791*
suppressor T-cell *p. 791*
thymus gland *p. 794*
allergies *p. 797*
autoimmune disease *p. 798*
HIV *p. 798*
AIDS *p. 799*

R E V I E W

Discussion Questions

1. Explain why public health scholars argue that society's understanding of and reactions to disease are at least as important as medical practices in controlling infectious disease. How has public reaction to the AIDS epidemic helped or hindered our efforts to contain and cure this disease?

2. Explain how, without having a genome of infinite size, it is possible for the immune system to manufacture antibodies against a seemingly endless parade of pathogens it has never seen.

3. Explain how the actions of the immune system might act as agents of natural selection in affecting the evolution of bacteria, viruses, and multicellular parasites over time. What are some of the strategies that parasites have evolved to evade the immune system?

4. Many people have trouble understanding that, even though AIDS is a communicable disease that is fatal, it is not easily transmitted from one person to another. Can you explain?

Objective Questions (Answers in Appendix)

5. Which of the following immunological defenses in vertebrates is highly specific?
   (a) skin
   (b) inflammatory response
   (c) humoral immune response
   (d) mucus membranes

6. Which of these symptoms is *not* a manifestation of the inflammatory response?
   (a) phagocytosis by white blood cells
   (b) increased blood flow at location of skin cut
   (c) production of specific antibodies
   (d) release of histamines

7. What are the granulocytes?
   (a) particles released by mast cells in response to pathogens
   (b) circulating white blood cells that engulf invading bacteria
   (c) antibodies in body secretions
   (d) blood proteins that attach to a cell membrane and puncture it, causing it to burst

8. The incubation period of the AIDS virus ranges from
   (a) 1–6 months.
   (b) 24 hours–3 weeks.
   (c) several months to decades.
   (d) no one knows.

9. Certain viruses are called retroviruses because
   (a) their genetic code is carried on their DNA.
   (b) they can make DNA from an RNA template.
   (c) they cause T-cells to move in reverse.
   (d) they cause B-cells to attack T-cells.

CHAPTER 40

Nervous Control

The mind of man is capable of anything—because everything is in it, all the past as well as the future.

—Joseph Conrad
HEART OF DARKNESS

It's a desperate scene that plays itself out in endless variations every day across the nation: An addict buys a fleeting glimpse of paradise in a vial of crack cocaine. The first high, users say, is like nothing they have experienced before—euphoria, an overwhelming sense of self-confidence, and a combination of physical, emotional, and sexual pleasures that overshadows all delights they have ever known.

But this ticket to an ephemeral Eden, they soon discover, is a true devil's bargain. The effect of crack on the mind is so great that even a single dose can prove irrevocably addicting. You may have seen crack's effects on the psyche explained on television or in magazine articles. One user compared the cocaine high to "being Adam and having God blow life into your nostrils." But when the brief high wears off, reality returns. The "superman" or "wonderwoman" dissolves once again into the same, tired body. Depression sets in. The real pleasures of everyday life—friends, parents, children, music, learning, food, even sex—pale in comparison to the drug-induced high. With frightening speed, crack undermines rational thought; for the addict, money to get high becomes more precious than life itself. Lying, cheating, stealing (from relatives, if necessary) and even committing murder seem preferable to being deprived of the drug.

THE MIRACLE OF MIND

Let us take a moment to frame this question in biological terms. What is the human mind that a simple compound extracted from a plant can wreak such havoc upon it? What is pleasure? What is pain? Why do we respond to events around us as we do? And what, for that matter, are the thoughts that are running through your mind as you read this?

Any search for the human mind leads to the brain, an organ awe-inspiring in its complexity. We know fairly well what the brain is composed of; it contains more than 100 billion nerve cells, each of which makes as many as 1000 contacts with other nerve cells and receives input from 10,000 more. We are familiar with the list of things the brain does; within its physical substance, thoughts materialize, emotions fly, moods dance, creativity springs, and dozens of automatic systems monitor and control the body's life-sustaining processes. We have traced the physical "wiring patterns" of nerve cells into and through the brain in exquisite detail; accumulated information on the anatomy and interconnections of brain areas fills many volumes far larger than this one.

Yet when asked how the miracle of mind emerges from this mass of neurons, most neurobiologists simply shrug; the human nervous system is a whole that is far greater than the sum of its individual parts. We have yet to understand fully the workings of our neural connections or to decipher the messages those networks create and carry.

Of course, neurobiologists don't give up after admitting that they don't have all the answers. Though the workings of the intact brain are still a mystery to us, those aspects of the nervous system that have been illuminated serve as fascinating and useful objects of study. We may not know how nerve impulses form thoughts, but we do know enough about brain chemistry to understand how crack and other drugs affect normal neural function. We do not yet know how you recognize a photograph of your mother, but we are learning enough about the brain's visual centers to guide us in designing computers that can see.

The human nervous system, of course, is among the most complex in the animal kingdom. In addition to the experimental difficulties this complexity presents, ethical considerations restrict the sorts of experiments we can perform on one another's gray matter. Luckily, however, our shared evolutionary history with less complex animals unites our nervous system with theirs through bonds of common chemistry and physiology. For that reason, researchers have learned a great deal by studying the tiny brains of sea snails and the giant nerve cells of squid. It is largely through work on these less complicated systems that neurobiologists have accumulated the knowledge and techniques that enable us to study our own nervous systems today.

WHY NERVOUS SYSTEMS?

You have seen how the body's chemical communication system uses hormones to mediate long-term processes such as digestion, metabolism, growth, and reproduction—processes that occur over seconds, months, or even years. But even the simplest have more immediate needs that hormonal control systems alone cannot meet. Animals must constantly gather and analyze information about their environment and respond immediately and appropriately if they are to maintain homeostasis and survive in changing, and often hostile, environments.

Even single-celled organisms respond to a variety of stimuli. The undifferentiated cytoplasm of *Amoeba* somehow mediates that protist's negative responses to strong light, and specialized organelles in many other single-celled organisms detect such stimuli as light, noxious chemicals, and dissolved compounds emitted by food. Locomotory organelles such as flagella or cilia respond to these stimuli by propelling the organism either toward a stimulus or away from it.

But this type of simple response is inadequate for large, multicellular animals that must be able to pass information from cell to cell within their bodies. The complex behaviors of higher animals also require more detailed sensory information and more coordinated movement than simple organelles can provide. So just as higher organisms evolved specialized tissues and organs for respiration and digestion, they evolved multicellular nervous systems.

Components of Nervous Systems

The principal components of the nervous system proper are the many classes of nerve cells or **neurons,** several of which we will examine in detail in the next three chapters. A fully functioning nervous system, however, also depends on many other, nonneural cell types. Nerve cells throughout the body, for example, are intimately associated with specialized **glial cells,** or **neuroglia** (from the Greek word for "glue"). Neuroglia were once thought to form only passive, supporting structures in the nervous system, but we know now that they are intimately involved with the metabolic processes of neurons they surround.

The responses and behaviors that neural activity triggers also depend on cells that themselves are not neurons. A tiger's decision to lunge at prey, for example, is carried out by muscle cells that produce movement. Other neurons exert their effects through tissues of the endocrine, respiratory, circulatory, and excretory systems. Collectively, nonneural cells whose actions are under neural control are known as *effectors,* and the organs they form, such as muscles and glands, are called *effector organs.*

Types of neurons Neurons can be grouped into several classes according to their location and function in the chain of information processing and response (see Figure 40.1 below).

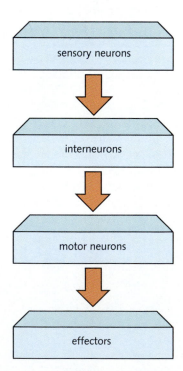

Figure 40.1 Functional components of the nervous system.

- **Sensory neurons** respond to light, heat, pressure, or chemicals in the environment and translate that information into the electrochemical language of the nervous system.

- **Interneurons** collect and relay information gathered from sensory neurons. Certain interneurons found in the spinal cord simply connect sensory neurons with the next link in a neural chain. Other interneurons in the eye and brain process and compare information from a variety of receptors, evaluate the incoming information, and determine the appropriate response(s).

- **Motor neurons** carry the "directions" issued by interneurons to the muscles that generate appropriate movements of the limbs, heart, lungs, gut, or other parts of the body.

Two other functional terms are often used to describe nerve cells: **afferent neurons** (*affere* means "to carry to") carry information toward the brain. **Efferent neurons** (*effere* means "to carry out") conduct information from the brain or spinal cord to effector organs. Motor neurons are efferent neurons, as are neurons that control nonmuscular effectors such as glands.

Neurons can also be classified into three broad groups—unipolar, bipolar, and multipolar—according to the arrangement of their main cellular parts: the **cell body,** the finely branched **dendrites** (**dendron** means "tree"), and the long, slender **axon** (from the Greek *axis*) (Figure 40.2). Usually, one or more dendrites pick up information and carry it toward the cell body, while the axon conducts information away from the cell body. Often, the far end of the axon divides into one or more branches that distribute the information either to other neurons or to effector cells.

Axons rarely run singly through the bodies of higher vertebrates; they are usually gathered together with accompanying glial cells into bundles called **nerves.** Some nerves contain only a few neurons, others contain thousands. Although most individual axons carry information in only one direction, a single nerve can contain both afferent and efferent neurons.

Neural circuits   No neuron functions in isolation. Neurons connect and communicate with each other and with effector cells through specialized structures called **synapses** (*synapsis* means "to clasp"). You can think of synapses as "connectors" that transport information from one neuron to the next. Synapses tie neurons together into *neuronal circuits,* collections of cells that work together to process information and to orchestrate such complex functions as coordinated movement, thought, and creativity (Figure 40.3).

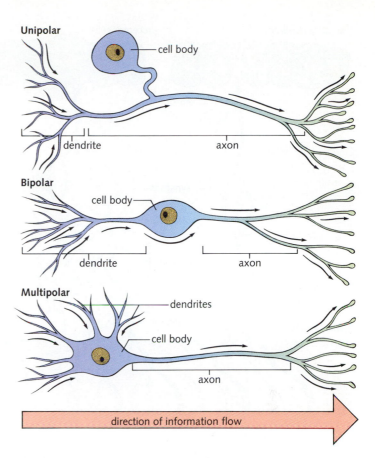

Figure 40.2   A unipolar neuron, bipolar neuron, and multipolar neuron classified according to structure.

One of the simplest neural circuits, the **reflex arc,** is responsible for actions such as the knee-jerk reflex in humans. This reflex, initiated by a quick, sharp tap below the kneecap, requires no conscious activity and occurs faster than you could intentionally move your leg (Figure 40.4). The sudden stretching of the upper thigh muscle is detected by sensory neurons called *stretch receptors.* The responses of those stretch receptors are relayed through the spinal cord to motor neurons that cause the stretched muscle to contract. Simple, unconscious reflexes such as this one are essential in nearly all coordinated movement.

The far more intricate neuronal circuits in the brain have thus far defied our efforts to understand their function as clearly as we understand the knee-jerk reflex. But scientists studying the smaller nervous systems of insects and snails, and the more accessible neuronal networks of mammalian eyes are uncovering tantalizing clues to the actions of ever-more-complex circuits, as you will see in this chapter and the next. Before examining what we know about complex neural circuits, however, we must first describe how individual neurons work.

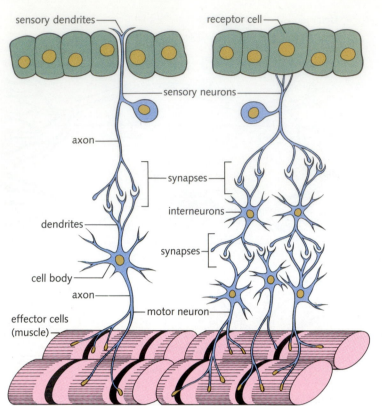

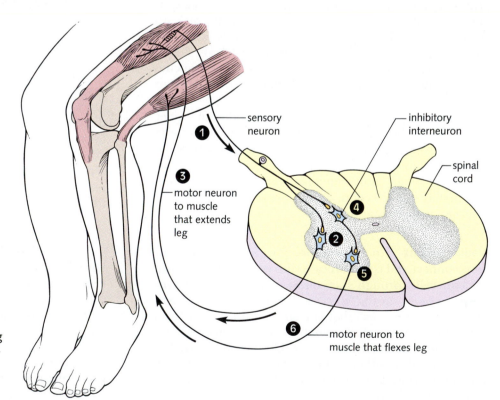

Figure 40.3 Simple neural circuits. (left) In this circuit, a sensory neuron connects directly to a motor neuron, which directly stimulates muscle cells. Circuits this simple are common in invertebrates, less common in higher vertebrates. (right) In most higher animals, specialized sensory cells gather information, one or more interneurons process that information, and several motor neurons allow for more precise control of effector organs.

Figure 40.4 The knee-jerk reflex is controlled by one of the simplest neural circuits in humans. A sudden stretch of the upper thigh muscle stimulates a stretch receptor that, in turn, stimulates a motor neuron in the spinal cord. That motor neuron stimulates the thigh muscle to contract, extending the leg. At the same time, the stretch receptor causes an inhibitory neuron to prevent firing in the motor neuron leading to muscles that flex the leg.

HOW NEURONS WORK

Neurons transmit information by using specialized proteins in their plasma membranes to create and manipulate electric currents. But nerve cells do *not* transmit messages by sending electric currents down their axons the way telephones send electrical pulses through wires. To understand the way neurons work, we must understand a few basic facts about electrical phenomena in living tissue.

The Difference Between Potential and Current

Electric current is defined as any flow of charged particles. Normally we think of current as a flow of negatively charged electrons through a conducting material such as a wire. But in living systems, electric current can be generated by the movement of either positive ions (such as Na^+, K^+, and Ca^{2+}) or negative ions (such as Cl^-), all of which are common in both cytoplasm and extracellular fluid.

Electrical potential is a force that causes charged particles to move and is measured in volts (V). Standard flashlight batteries, for example, have an electrical potential of 1.5 V between their poles. Batteries develop that potential because they are put together in such a way that electrons are drawn away from one pole and attracted to the other.

We can create a similar situation by placing an impermeable membrane between two solutions of ions in water. If each solution contains equal numbers of positive and negative ions, no potential exists between them (Figure 40.5a). If, however, we place more positive ions on one side of the membrane, an electrical potential exists across the membrane (Figure 40.5b). Note that electrical potentials exist in these situations in the absence of any current flow. When charged ions *do* flow in response to an electrical potential, they move in a way that reduces that potential. That's why batteries run down.

MEMBRANE PUMPS, LEAKS, AND ELECTRICAL POTENTIALS

Nerve cells develop electrical potentials because specialized proteins in their plasma membranes use energy from ATP to pump sodium ions out of the cell as they pump potassium ions into the cell. The action of this protein, which is called the **sodium–potassium pump,** raises the concentration of sodium ions outside the cell to ten times the concentration of sodium ions inside. At the same time, the pump moves potassium ions into the cell.

But many of the potassium ions pumped into the cell do not remain inside because the nerve cell membrane is not absolutely impermeable. Instead, the membrane is slightly "leaky" to sodium ions and far more leaky to potassium ions. Because the membrane leaks potassium,

Figure 40.5 **(a)** A membrane separating two solutions containing equal numbers of positively and negatively charged ions. No electrical potential exists here. **(b)** This membrane separates solutions with different relative concentrations of positive and negative ions. An electrical potential exists across this membrane.

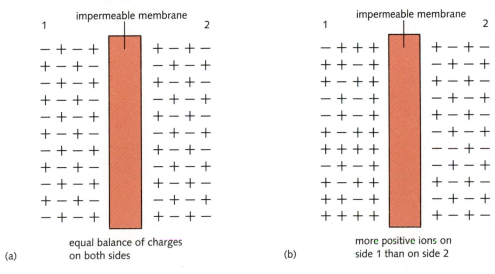

and because potassium is more concentrated inside the cell than outside, those potassium ions tend to diffuse out. In the equilibrium that results, there are both more positive ions and more sodium ions outside the cell than inside (Figure 40.6). For this reason, we say that the interior of the cell is negative with respect to its immediate surroundings. (Under normal circumstances, the membrane is less permeable to negatively charged chloride ions and almost totally impermeable to large, negatively charged organic molecules found inside the cell. These ions therefore cannot cross the membrane to balance out the charges.)

This electrical potential across the nerve cell membrane is called the **resting potential,** because it is characteristic of nerve cells at rest. By placing one very fine electrode inside a nerve cell and another electrode alongside the cell, we can measure that resting potential, which turns out to be about -70 millivolts (mV). (That's 7×10^{-2} or 0.070 volts, compared with the battery's 1.5 volts.) The minus sign indicates that the inside of the cell is negatively charged with respect to the outside of the cell.

If the interior of the nerve cell becomes *more negative* than it normally is at equilibrium—in other words, if the resting potential changes from -70 mV to, for example, -100 mV—we say that the cell has been **hyperpolarized** (*hyper* means "more"). The entry of negatively charged chloride ions into the nerve cell, for example, would hyperpolarize it.

If the interior of the cell becomes *less negative* than it normally is at equilibrium—in other words, if the resting potential changes from -70 mV to, for example, -40 mV—we say that the cell has been **depolarized** (*de* means "less"). Allowing positively charged sodium ions to enter a cell would depolarize it.

Ion Channels and Excitability

The existence of a resting potential is not unique to nerve cells. What distinguishes nerve (and muscle) cells from other cells is their ability to create and transmit disturbances in that resting potential that carry information. This ability depends on two classes of pores in the cell membrane that are called **gated ion channels** because they open and close like gates in a fence. (Note that these pores exist in addition to, and operate independently of, the sodium–potassium pump.)

Electrically controlled channels are pores that open and close when the electrical potential across the cell membrane changes. At a normal resting potential of -70 mV, for example, the channels that control the entry of sodium into the cell are closed, and virtually no sodium can flow passively across the membrane. If the cell becomes sufficiently depolarized, however, these electrically gated channels open and allow sodium to pass through readily (Figure 40.7a).

Small depolarizations do not affect these channels, but if the neuron becomes less negative than it normally is at equilibrium to a level called the **threshold,** many sodium channels spring open at once. Electrically con-

trolled channels are important in the generation of the nerve impulse, as we will see shortly.

Chemically controlled channels are pores that open only when activated by compounds called **neurotransmitters,** chemical messengers that transmit information across synapses (Figure 40.7b). Neurotransmitters and chemically gated channels interact in much the same way as hormones and their receptors. The only difference is that hormones generally travel through the bloodstream before reaching their receptors, whereas neurotransmitters and their receptors are separated only by the small distance across a synapse. As we noted in Chapter 36 and we will see again shortly, several compounds, such as adrenalin (also called epinephrine), act both as circulating hormones and as neurotransmitters.

Figure 40.7 **(a)** Electrically gated sodium channels are composed of proteins that span the cell membrane and can switch back and forth between different shapes. At normal resting potential, the shape of these proteins keeps the channel closed. When the interior of the cell depolarizes sufficiently, the channel proteins shift into a different shape, opening the channel. **(b)** Chemically gated channels are normally closed. When bound to their specific neurotransmitter molecules, however, they change conformation and open.

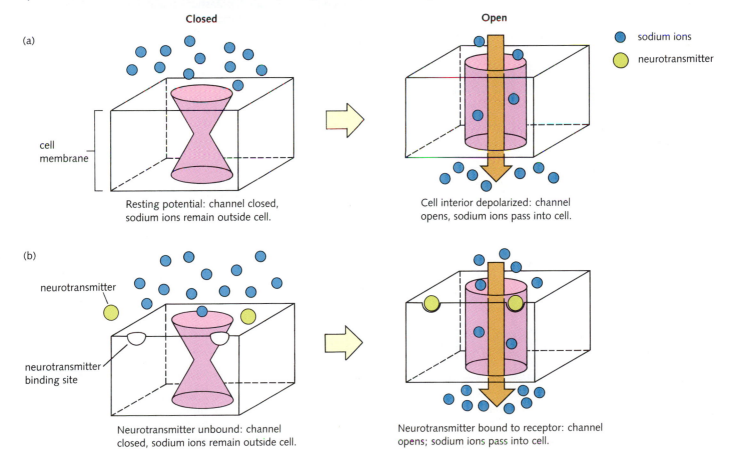

Closed   Open

(a)

cell membrane

Resting potential: channel closed, sodium ions remain outside cell.

Cell interior depolarized: channel opens, sodium ions pass into cell.

sodium ions

neurotransmitter

(b)

neurotransmitter

neurotransmitter binding site

Neurotransmitter unbound: channel closed, sodium ions remain outside cell.

Neurotransmitter bound to receptor: channel opens; sodium ions pass into cell.

THE ACTION POTENTIAL: ELECTRICALLY GATED CHANNELS IN ACTION

Neurons with long axons carry information by means of **action potentials**—brief changes in the membrane's electrical potential that sweep along the axon. Although action potentials are also called *nerve impulses*, they are not pulses of electricity that travel through the axon; they are disturbances in the resting potential that move along the membrane like ripples passing along the surface of a quiet stream. Once an impulse is initiated, it doesn't need to be "pushed" in any way. It is **self-propagating**, which means it travels down the axon with no further input of energy from the cell.

Depolarization and Threshold

The action potential starts when part of a nerve cell is depolarized to threshold. As soon as threshold is reached, large numbers of electrically controlled sodium channels spring open, and positively charged sodium ions rush into the cell. Sodium ions move for three reasons: They are attracted by the negative charges inside the cell, they are repelled by the positive ions outside the cell, and they diffuse from an area of high sodium concentration to an area of low sodium concentration.

Remember that when current flows because of an electrical potential, the potential decreases. As sodium ions rush into one area of a cell, therefore, the resting potential in that region drops to zero and then actually

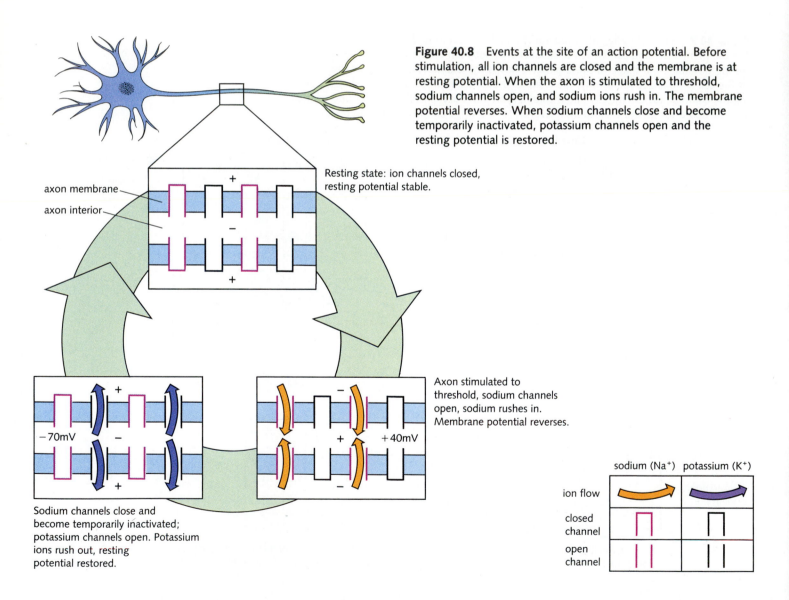

Figure 40.8 Events at the site of an action potential. Before stimulation, all ion channels are closed and the membrane is at resting potential. When the axon is stimulated to threshold, sodium channels open, and sodium ions rush in. The membrane potential reverses. When sodium channels close and become temporarily inactivated, potassium channels open and the resting potential is restored.

axon membrane
axon interior

Resting state: ion channels closed, resting potential stable.

−70mV

Axon stimulated to threshold, sodium channels open, sodium rushes in. Membrane potential reverses.

+40mV

Sodium channels close and become temporarily inactivated; potassium channels open. Potassium ions rush out, resting potential restored.

	sodium (Na⁺)	potassium (K⁺)
ion flow		
closed channel		
open channel		

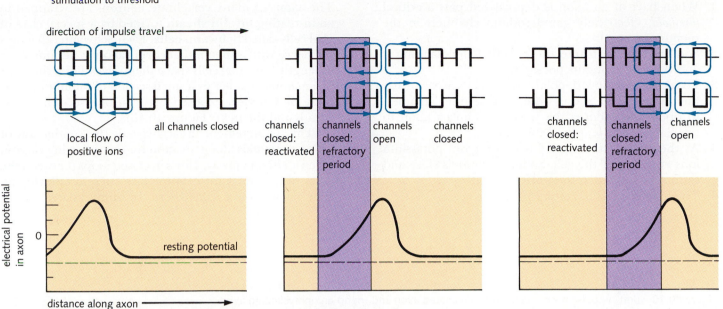

(a) Schematic of ion flow in axon at instant of stimulation to threshold

(b) Ion flow 1 millisecond after stimulation

(c) Ion flow 2 milliseconds after stimulation

Figure 40.9 Propagation of the action potential. (**a**) On stimulation to threshold, inward movement of positive ions creates an electric current that depolarizes nearby parts of the membrane. (**b**) The newly activated channels open, allowing sodium to enter farther down the axon. At the same time, the first channels to open close and become inactivated. (**c**) As channels open and close in sequence along the axon, they create a traveling wave of electrical disturbance.

reverses briefly, changing from -70 mV to $+40$ mV in about 2 milliseconds (Figure 40.8). Then, almost as quickly, the membrane potential reverses again and returns to -70 mV, as the sodium channels close and other channels open, allowing positive potassium ions to rush out.

After the sodium channels have opened and closed, they cannot be activated for a brief but finite period of time called the **refractory period.** This sequence of depolarization and repolarization constitutes the action potential.

The Traveling Wave: An All-or-Nothing Event

The action potential travels in one direction along the axon because of the way sodium channels work. As sodium ions rush in, these moving charged particles create a flow of current that depolarizes nearby parts of the axon (Figure 40.9a). That depolarization is strong enough to open the electrically controlled sodium channels in the adja-

cent section of the axon (Figure 40.9b). Because sodium channels *behind* the disturbance cannot reopen for a few milliseconds, the wave does not spread in two directions but travels "one way" down the axon (Figure 40.9b, c).

Because all the electrically gated sodium channels of a neuron have the same threshold for opening, either all the channels in a region open in response to a depolarization or none of them do. As a result, action potentials are **all-or-nothing events;** either they happen or they don't. Additionally, all action potentials generated in a particular neuron are exactly the same size. If the stimulus that produces an action potential is larger than threshold or lasts for a long time, it doesn't generate larger action potentials but instead gives rise to a series of many impulses, one after another.

Compared to the total number of ions near the membrane, the actual number of ions involved in an action potential is small. This fact and the constant activity of the sodium–potassium pump "reset" the resting potential so quickly that the axon is ready to fire again as soon as the brief refractory period is past.

Summary of the Action Potential

1. When part of an axon is depolarized past a critical *threshold,* electrically gated sodium channels in the membrane open.
2. Sodium ions rush into the cell, first depolarizing it and then reversing its potential.
3. Local current flow depolarizes adjacent areas of the membrane. This causes additional sodium channels to open and induces the wave of depolarization to travel down the length of the axon.
4. As the peak of the wave passes a region, potassium ions rush out of the cell. Sodium channels close and cannot be reopened for a brief period.
5. The area returns to resting potential.

The Role of Myelin

The axons of many vertebrate neurons are wrapped in an insulating **myelin sheath** formed by a special class of glial cells called **Schwann cells.** Each Schwann cell wraps the nerve with a sheath that covers about 1 millimeter of its length. The glial cells space themselves along the axon in a manner that leaves a small section of membrane, the **node of Ranvier,** exposed between one section of the sheath and the next (Figure 40.10).

This arrangement greatly speeds the conduction of action potentials along an axon for the following reason. Electric current always flows in closed loops (between the poles of a battery, for example), so current entering an axon must leave it somewhere. In unmyelinated axons,

Figure 40.10   (left) Impulse transmission in a myelinated axon and (right) an unmyelinated axon.

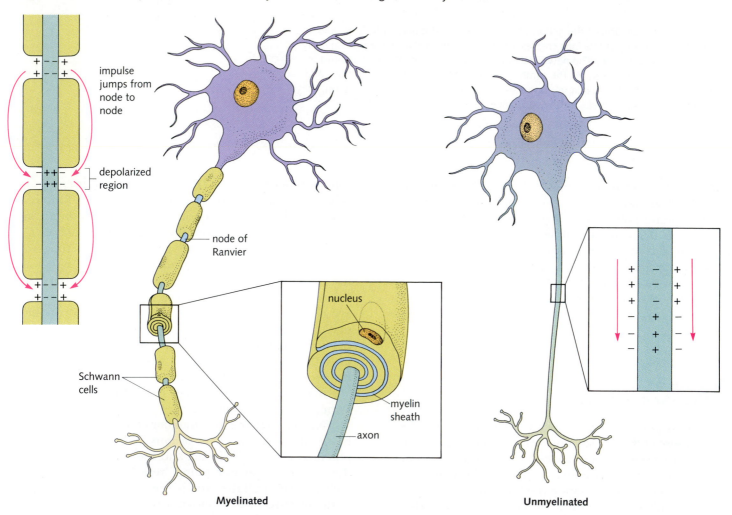

impulse jumps from node to node

depolarized region

node of Ranvier

Schwann cells

nucleus

myelin sheath

axon

Myelinated

Unmyelinated

the current entering the neuron at the site of the action potential flows through very tight local loops. Thus the action potential moves down unmyelinated axons in rapid but very small steps that result in conduction speeds of between 0.5 and 2 meters per second.

In myelinated axons, on the other hand, the regions wrapped by the myelin sheath have very high resistance to current flow. Current entering the axon at one node of Ranvier can exit only at the next node, where it activates the electrically gated channels. The action potential thus "jumps" quickly from one node to the next—a process called **saltatory conduction** after the Latin word for "jump"—rushing down the axon. The increase in speed is substantial; myelinated axons can conduct impulses at speeds up to 120 meters per second!

SYNAPSES: SITES OF NEURAL CONTROL AND INTEGRATION

As we have seen, action potentials have definite advantages for carrying information along axons. But a series of action potentials in an axon is neither more nor less than a collection of identical signals. Because those impulses do not change in size, shape, or speed, regardless of recent activity in the nerve, they cannot be responsible for the sorts of changes in neural activity that make learning and memory possible. And action potentials cannot interact with other action potentials to process sensory information. Where, then, do these critical events occur?

As we mentioned earlier, individual neurons are connected into neural circuits by **synapses.** It is at these synapses that virtually all the essential work of comparing, integrating, and processing neural information occurs. Changes in the response of synapses to incoming action potentials under different conditions allow us to compare sensory stimuli, evaluate the meaning of those stimuli, and determine the appropriate course of action. And it is becoming clear that changes over time in the structure and function of certain types of synapses underlie the processes of learning and memory. For that reason, we will examine synapses closely.

Electrical Synapses

The simplest way for neurons to connect to one another is through direct electrical connections called **electrical synapses.** These connections, which are also called **gap junctions,** allow the free flow of ions between cells. Cells connected by electrical synapses are tied together into a single, continuous electrical unit.

Transmission of impulses across electrical synapses is essentially instantaneous, a feature that makes these connections valuable in certain situations. Electrical synapses linking the giant axons of lobsters, for example, enable those animals to respond with all possible speed to escape predators. (The *chemical* synapses you will learn about, by contrast, introduce a delay of several milliseconds between neurons.) Electrical synapses do have limitations, however. They have relatively little ability to process information so they are thought to play only minor roles in learning and other flexible types of behavior. It should not surprise you that in the nervous systems of animals that engage in complex, modifiable types of behavior, a different sort of neural connection is involved.

Chemical Synapses

Chemical synapses, of which there are several types, are both structurally and functionally more complex than electrical synapses. Action potentials do not simply pass across chemical synapses as they do across electrical synapses. Instead, when an action potential reaches a chemical synapse, it is converted into a chemical signal: a pulse of neurotransmitter molecules that diffuse across the synapse. On the far side of the synapse, that chemical signal is converted back into an electrical signal. Once you understand how this process works, you will have a glimpse into the mechanism that allows virtually all of the information processing that occurs in the nervous system.

A schematic representation of a simple chemical synapse is shown in Figure 40.11. The neuron through which the action potential arrives, the **presynaptic neuron,** is separated from the **postsynaptic neuron,** the cell on the other side of the synapse, by a gap called the **synaptic cleft.** Inside the presynaptic neuron, *synaptic vesicles* store neurotransmitter molecules that will be released into the synapse when the presynaptic neuron is stimulated. In the membrane of the postsynaptic neuron, a host of chemically gated ion channels act as receptors for neurotransmitter molecules.

The sequence of events that occurs when an action potential reaches this type of synapse is as follows:

1. When an action potential arrives at a synapse, it triggers the opening of electrically gated calcium channels in the presynaptic cell near the synapse.

2. Calcium ions outside the cells diffuse inward, drawn to the cell's negative interior.

3. The calcium ions inside the presynaptic cell cause synaptic vesicles to fuse with the cell membrane and release neurotransmitter into the synaptic cleft.

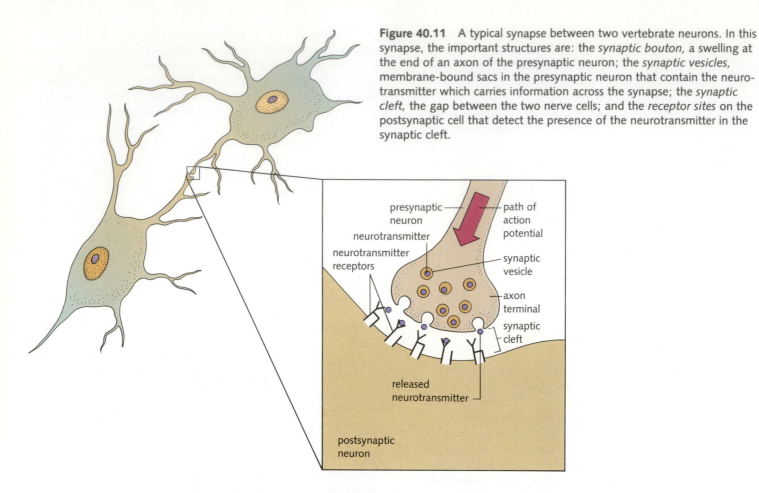

Figure 40.11 A typical synapse between two vertebrate neurons. In this synapse, the important structures are: the *synaptic bouton,* a swelling at the end of an axon of the presynaptic neuron; the *synaptic vesicles,* membrane-bound sacs in the presynaptic neuron that contain the neurotransmitter which carries information across the synapse; the *synaptic cleft,* the gap between the two nerve cells; and the *receptor sites* on the postsynaptic cell that detect the presence of the neurotransmitter in the synaptic cleft.

presynaptic neuron

path of action potential

neurotransmitter

neurotransmitter receptors

synaptic vesicle

axon terminal

synaptic cleft

released neurotransmitter

postsynaptic neuron

4. Neurotransmitter molecules diffuse across the synaptic cleft and bind with receptor sites on the postsynaptic membrane.

5. The opening of various chemically gated channels allows either positive or negative ions into the postsynaptic cell, causing either a local depolarization or hyperpolarization.

6. Transmitter remaining in the synapse is eliminated, either by chemical reactions that inactivate it or by transport mechanisms in the presynaptic membrane and glial cells that remove it. This clears the synapse for the arrival of the next impulse.

What happens to the postsynaptic membrane depends both on the specific transmitter(s) released at a synapse and on the character of the chemically controlled channels it binds to. This is because different transmitters can have different effects on the postsynaptic cell. Although neurobiologists once believed that a single neuron could manufacture and release only a single transmitter, we now know of several neurons that use at least two. For the sake of clarity, we will look first at a relatively simple chemical synapse that uses a single transmitter.

A Simple Synapse: The Neuromuscular Junction

The best-understood of all synapses is the **neuromuscular junction,** the connection between motor neurons and muscle cells (Figure 40.12a). (As you will see in Chapter 42, muscle cell function depends on membrane action potentials that are virtually identical to nerve impulses.) Neuromuscular synapses are large and simple, and they rely on a single neurotransmitter, a compound called **acetylcholine (ACh).** Stimulation of a neuromuscular junction by the neurotransmitter acetylcholine nearly always excites the muscle cell.

When an action potential reaches a neuromuscular junction, the presynaptic cell releases acetylcholine into the synaptic cleft. The transmitter diffuses across the cleft, binds to its receptors, and then opens channels for ions in the muscle cell membrane. Because the neuromuscular junctions release a substantial amount of acetylcholine, enough channels are usually opened by a single action potential to depolarize the muscle cell to threshold. At that point, the muscle cell membrane generates an action potential that causes that cell to contract (Figure 40.12b).

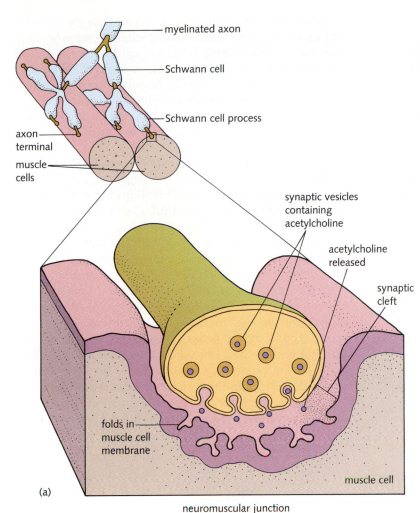

myelinated axon

Schwann cell

Schwann cell process

axon terminal

muscle cells

synaptic vesicles containing acetylcholine

acetylcholine released

synaptic cleft

folds in muscle cell membrane

muscle cell

(a)

neuromuscular junction

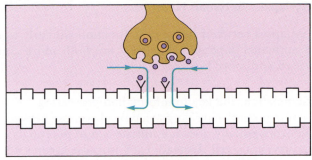

Stimulated nerve ending releases acetylcholine; acetylcholine-controlled channels are opened.

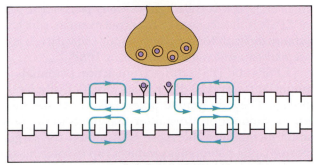

Channels opened chemically allow ion flow; this depolarizes membrane; electronically-controlled channels nearby are opened.

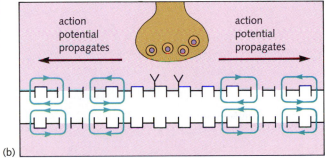

action potential propagates

action potential propagates

(b)

Neurotransmitter released into the synapse cannot continue to affect the postsynaptic cell for long, because the muscle cell should be stimulated only when its motor neuron is actively firing. For that reason, transmitter must be removed from the synapse or inactivated after it has performed its function. In synapses using acetylcholine as a transmitter, this is accomplished by a compound called *acetylcholinesterase*, an enzyme that destroys virtually all free molecules of ACh within a few hundred microseconds. The vital role played by acetylcholinesterase is dramatically demonstrated by the action of several nerve gases such as *sarin* and insecticides such as *parathion* that interfere with its activity. This interference results in the accumulation of acetylcholine in neuromuscular junctions, wildly unregulated postsynaptic activity, and loss of muscular coordination. Exposure to sufficient doses of acetylcholinesterase inhibitors can also cause death from respiratory arrest by preventing relaxation of the diaphragm and other muscles essential to breathing.

Figure 40.12 **(a)** The structure of a neuromuscular junction between a motor neuron and a major body muscle under conscious control. **(b)** Schematic summary of events at a neuromuscular junction. When an action potential arrives, the release of acetylcholine nearly always initiates an action potential in the muscle.

Synapses Between Neurons

Synapses between nerve cells function similarly to neuromuscular junctions, but they are both more complicated and more flexible.

Summation: the key to information processing A single action potential arriving at a typical synapse between two neurons usually does not initiate action potentials in the postsynaptic cell. Instead, each impulse creates a small, temporary, local change in potential called a **postsynaptic potential.** If that change is a depolarization, it brings the postsynaptic cell closer to threshold (and hence closer to firing), so it is called an *excitatory postsynaptic potential* or **EPSP.** If the change is a hyperpolarization, which makes it more difficult for the postsynaptic cell to fire, it is called an *inhibitory postsynaptic potential* or **IPSP.**

EPSPs and IPSPs differ from action potentials in two important respects. First, they are not all-or-nothing events. They are **graded potentials** whose size depends on the amount of transmitter that binds to the postsynaptic membrane; the more transmitter, the larger the postsynaptic potential it produces (up to some maximum value). Second, IPSPs and EPSPs do not travel; they decrease in intensity with increasing distance from the synapse, and they disappear with time (Figure 40.13).

Graded potentials are important because their effects add up in a process called **summation.** Remember that most nerve cells receive synaptic input from thousands of other neurons (Figure 40.14). A single EPSP at any one of these synapses usually doesn't trigger an action potential, but many EPSPs arriving simultaneously can cause the postsynaptic cell to fire. Similarly, rapid, repeated stimulation of a few synapses can also add up to cause firing.

When a neuron receives simultaneous EPSPs and IPSPs, it responds according to the algebraic sum of those positive and negative inputs. This ability to combine excitatory and inhibitory inputs underlies all processing of information in the nervous system.

You might wonder at this point just how IPSPs and EPSPs allow integration and control. As an example of a phenomenon with which you are already familiar, recall from Chapter 32 that the heart's natural "pacemaker," the S-A node, can be induced either to speed up or to slow down the heartbeat. This adjustment occurs as a result of interactions between excitatory and inhibitory transmitters released onto the excitable cells of the S-A node. Excitation of the S-A node causes the heart rate to increase, and its inhibition causes the heart to slow down.

Similarly, the ability of chemical synapses to compare inputs from different presynaptic neurons enables sen-

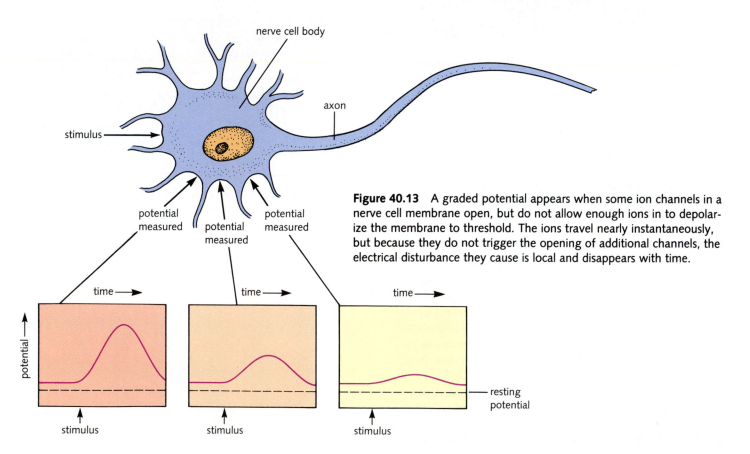

Figure 40.13 A graded potential appears when some ion channels in a nerve cell membrane open, but do not allow enough ions in to depolarize the membrane to threshold. The ions travel nearly instantaneously, but because they do not trigger the opening of additional channels, the electrical disturbance they cause is local and disappears with time.

sory processing centers in the brain to weigh and interpret input from many sensory receptors. As we will see during the discussion of the visual system in the next chapter, this sort of comparison allows the eye and brain to judge distance, to perceive movement, and to see in color. Similar interactions appear to underlie the processing of auditory information and information gleaned from other senses. Far more complicated versions of similar processes underlie thought and other higher mental functions.

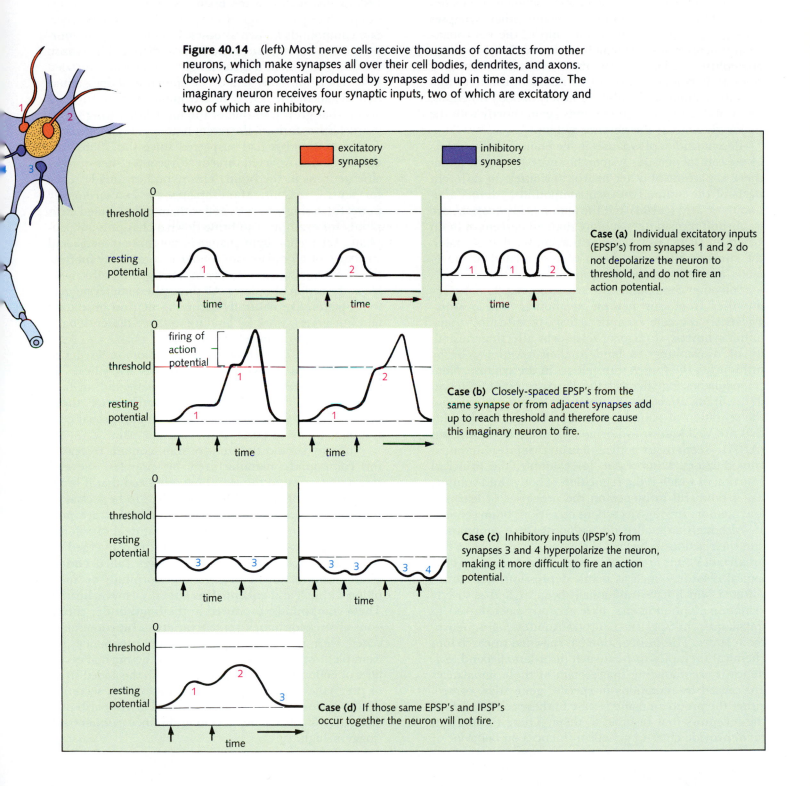

Figure 40.14 (left) Most nerve cells receive thousands of contacts from other neurons, which make synapses all over their cell bodies, dendrites, and axons. (below) Graded potential produced by synapses add up in time and space. The imaginary neuron receives four synaptic inputs, two of which are excitatory and two of which are inhibitory.

excitatory synapses

inhibitory synapses

Case (a) Individual excitatory inputs (EPSP's) from synapses 1 and 2 do not depolarize the neuron to threshold, and do not fire an action potential.

Case (b) Closely-spaced EPSP's from the same synapse or from adjacent synapses add up to reach threshold and therefore cause this imaginary neuron to fire.

Case (c) Inhibitory inputs (IPSP's) from synapses 3 and 4 hyperpolarize the neuron, making it more difficult to fire an action potential.

Case (d) If those same EPSP's and IPSP's occur together the neuron will not fire.

A Diversity of Transmitters

For many years, researchers thought that the entire nervous system employed only three neurotransmitters: acetylcholine, epinephrine (adrenaline), and norepinephrine (noradrenaline). This assumption was natural enough, because these three transmitters act at many synapses throughout the nervous system. Acetylcholine, for example, is found in the brain, in many other synapses throughout the nervous system, and in the neuromuscular junctions of skeletal muscles. Epinephrine and norepinephrine—which we have already encountered in their roles as hormones—also act as neurotransmitters that can either stimulate or inhibit a variety of body functions. (We will discuss these compounds again shortly with the major divisions of the nervous system.)

Recent studies of synapses in the brain and other parts of the central nervous system, however, have shown that there are a host of other neurotransmitters in different parts of the brain. Two such compounds, gamma aminobutyric acid (*GABA*) and the amino acid *glutamate,* are important transmitters at more than 50 percent of brain synapses. *Dopamine* is another transmitter that is vital to the function of several brain centers related to movement, arousal, and the emotions.

Still other compounds affect synapses but do not carry impulses from one neuron to another. These *neuromodulators* regulate synaptic activity by affecting transmitter synthesis, storage, or release, by affecting the ability of transmitters to bind to their receptors, or by influencing the fate of transmitters in the synapse. Neuromodulators, together with the diverse transmitters found in the brain, affect mood, attention, learning, sleep, and the sensation of pain. Interestingly, certain compounds that are well known as hormones—such as vasopressin (ADH)—seem to act both as neurotransmitters and neuromodulators. Vasopressin, for example, the principal function of which is the regulation of body fluid volume, has a powerful influence on the processes of learning and memory through its actions on various brain centers.

Although a detailed discussion of these compounds and their actions is beyond the scope of this book, a few important points are worth noting. First, several psychological disorders such as manic depression and schizophrenia, which have traditionally been treated as strictly "emotional" phenomena, have recently been linked to imbalances in specific neurotransmitters and neuromodulators. This information has spawned research into chemical therapies to treat such disorders. Second, evidence is accumulating that certain of these imbalances are caused by mutations in specific genes that control either the production and release of these compounds or the receptors that respond to them. These findings, in turn, provide clues to genetic influences on behavior of the sort that we will discuss in Chapter 43. Finally, it is by aiding, replacing, or interfering with neurotransmitters and neuromodulators that mind-altering drugs—from hallucinogens to stimulants and antidepressants—exert their influence (see Theory in Action, Drugs, Synapses, and the Brain).

Endorphins: pain and the brain   Although many drugs can produce psychological and physical dependence, certain compounds known as **opiates,** such as opium, morphine, and heroin, are particularly addictive. The reason for this phenomenon is startlingly simple, but it puzzled researchers for many years. The epidemic of drug abuse that began in the 1970s fueled many studies on drug action that grew into one of the most fascinating scientific detective stories in the recent history of brain research.

Neuroscientists had suspected since the 1950s that opiates somehow acted directly at receptor sites on specific neurons in the brain. This hypothesis made sense for several reasons. Many opiates are extraordinarily powerful; a single dart coated with an opiate called etorphine, for example, can bring down a charging bull elephant. Yet even a slight change in the three-dimensional structure of an opiate can render it completely ineffective. And finally, there are several drugs that effectively block the action of opiates. All these phenomena suggest that opiates, like naturally occurring neurotransmitters and hormones, work by binding to three-dimensionally specific receptors on cell membranes.

Then, in 1972, Candice Pert and Solomon Snyder of Johns Hopkins University proved this hypothesis by showing that radioactively labelled opiates did, in fact, bind to specific neural receptors. Furthermore, those binding sites are located in specific brain areas that control perceptions of pleasure and chronic pain.

But why should the human brain contain receptors for compounds manufactured by poppies? Several researchers around the world hypothesized that if brain receptors for opiates existed, there had to be a class of yet-undiscovered, opiate-like compounds that are normally found in the brain. After a great deal of intense (and competitive) work, three laboratories converged on the trail of opiate-like compounds they called **endorphins** (a contraction of "endogenous morphine" or "morphine within") and **enkephalins** (literally, "in the head"). These compounds, produced in the brain under conditions of stress or injury, serve a variety of functions associated with relief from pain. During pregnancy, for example, women produce eight times the normal quantities of endorphins, perhaps to cope with the added stress of pregnancy, labor, and delivery. Endorphin secretion during intense and prolonged physical activity also produces the blissful state that long-distance runners call "joggers' high."

Drugs, Synapses, and the Brain

The use of mind-altering or *psychoactive* drugs is part of nearly every human culture, for better and for worse. South American jungle tribes induce healing trances by drinking or smoking plant extracts that cause hallucinations. Artists and writers ranging from Samuel Taylor Coleridge to Arthur Conan Doyle and Timothy Leary have credited drugs with enhancing sensitivity and spurring creativity. And legions of unfortunate addicts testify to the extraordinary power of such drugs as opium and heroin to cause not only short-term changes in mood but also long-term alterations in personality and physiology.

Psychoactive drugs alter brain function either by overstimulating synapses (causing rapid, uncontrolled firing of postsynaptic neurons) or by turning synapses off (blocking the normal flow of information through the brain). Each of these actions can be accomplished in several ways.

Synapses can be overstimulated by drugs that mimic the actions of transmitter molecules, cause transmitters to be released in the absence of normal stimulation, or interfere with inactivation of transmitters in the synapse. Amphetamines (speed), for example, act as stimulants by causing rapid release of the transmitter noradrenaline at brain synapses. Nicotine, a major active component in cigarette smoke, acts as a stimulant because it mimics the effects of acetylcholine at other synapses.

Synapses can also be "turned off" by substances that block receptor sites on the postsynaptic membrane, block the synthesis or storage of transmitter in the presynaptic cell, or prevent the normal release of transmitters. Valium, often prescribed as an anti-

Drugs that act at synapses are often quite specific in their effects. This schematic, for example, shows the action of an antidepressant that blocks the action of the transmitter norepinephrine but allows another transmitter, serotonin, to act unhindered.

depressant or antianxiety drug, enhances the effects of GABA, an inhibitory transmitter at many brain synapses. Several commonly used tranquilizers, such as chlorpromazine, attach to postsynaptic receptors for both acetylcholine and noradrenaline and block the activity of those transmitters.

Although many psychoactive drugs (including nicotine and alcohol) can produce psychological and physical dependence, certain compounds (such as opium and crack cocaine) are particularly addictive. This is because these drugs' short-term chemical effects that flood the brain with "feel-good" signals also cause long-term changes in synaptic function. Crack, for example, causes a sudden increase in the release of dopamine onto neurons in brain areas that control emotions, the sensation of pain, and the general sensation of "pleasure." This stimulation, however, causes a drop in the amount of dopamine stored in presynaptic neurons for later release. After a few doses of crack, stimuli that would normally cause pleasure no longer do so, and the absence of dopamine in these brain areas between "hits" leads rapidly to depression. Soon, the only way to feel "normal"—not to mention happy—is to rely continually on the drug. This vicious dependency has made crack a serious social problem.

Finally, all the pieces of the puzzle had fallen into place. Morphine, opium, and heroin bind to the brain's receptors for natural pain-relieving compounds, flooding the brain with "pleasure" signals and inhibiting response to pain. But that flood of stimulation takes its toll. In some cases, repeated stimulation with opiates causes either a decrease in the production of naturally occurring transmitters or a decline in the number of receptor sites. Consequently, when addicts stop taking drugs, their brains are deprived of both natural *and* injected "feel-good" signals. The result is the agonizing pain and mental agony of withdrawal, which lasts until the brain recovers its normal function.

Changes at the Synapse: Learning and Memory

> You have to begin to lose your memory, if only in bits and pieces, to realize that memory is what makes our lives. Life without memory is no life at all. . . . Our memory is our coherence, our reason, our feeling, even our action. Without it, we are nothing. . . .
>
> —Luis Buñuel

How do we learn? How do we remember past events? These are among the most fundamental and exciting questions in neurobiology and psychology. Neurobiologists have long assumed that memory resulted from changes in the number, structure, and function of brain synapses, but they had no way to study the changes they imagined. Today, tissue culture, molecular genetics, and recombinant-DNA technology are ushering in a new era of molecular neurobiology. The proteins that form certain gated channels and the genes, mRNAs and tRNAs, that control them have been isolated and cloned. This allows neurobiologists to examine changes in synaptic function at a level it was impossible to achieve even five years ago.

Functionally, there are at least two different kinds of memory. **Short-term memory** enables us to remember information for a few minutes or hours; we use this facility when we memorize phone numbers for immediate use or "cram" for an examination. Some (but not all) information stored in short-term memory ultimately enters **long-term memory,** where it remains accessible for days, months, or years. How might changes in synaptic structure and function make these processes possible?

Experience and synaptic function   When similar thoughts or actions are repeated, certain groups of synapses are stimulated repeatedly. We now know that this kind of continued stimulation can affect the release of transmitter from the presynaptic cell, the expression of genes that

control transmitter production, and even the genes that control the production of gated channels. Thus synapses that are repeatedly stimulated may end up making or releasing more (or less) transmitter than similar synapses that are not stimulated. In the same way, repeated stimulation may either increase or decrease the number of functional chemically gated channels in the postsynaptic neuron, thereby increasing or decreasing that neuron's responsivity to neurotransmitters. Additionally, certain neurotransmitters and neuromodulators, through their actions on various cell receptors, act as second messengers and affect long-term metabolic processes in cells just as hormones do (see Chapter 34).

Experience and change in neural structures   Experience and the brain activity it generates can also affect the expression of genes that control nerve growth and survival. In many cases, both in developing brains and in mature brains, neurons that are stimulated grow and expand their synaptic connections, while those that are not stimulated stay the same, regress slightly, or wither away and die. Those changes in nerve growth within our brain structure clearly play some role in creating the physical records that store our memories.

FROM NEURON TO BRAIN: THE EVOLUTION OF NERVOUS SYSTEMS

Nervous systems have evolved steadily during the history of multicellular animals (Figure 40.15). The simplest multicellular animals, such as *Hydra*, have diffuse **nerve nets** with no defined control center. In many of these animals, the system can conduct action potentials in either direction.

As bilateral symmetry became more pronounced, animals developed a tendency to move in one direction. The front end of the animals became increasingly important, and both sensory cells and neural relay centers became concentrated there. At the same time, various neurons (such as sensory receptor cells) became more specialized, different parts of the nervous system became specialized for different functions, and the number of interneurons in neural circuits increased rapidly.

The accumulation of sensory neurons and interneurons in the head region, a process called **cephalization,** led to collections of nerve cell bodies called *ganglia,* from which one or two *nerve cords* ran down the length of the body. In most invertebrates, cephalization progressed to a certain point and then stopped, leaving several secondary segmental ganglia distributed along the animals' nerve

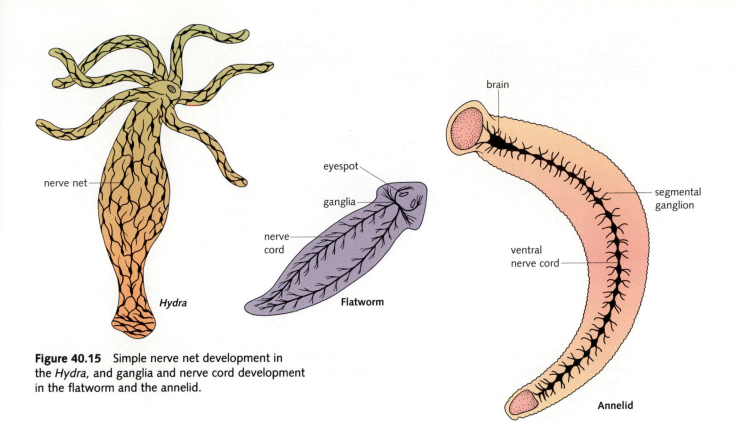

nerve net

eyespot

ganglia

nerve
cord

Hydra

Flatworm

brain

segmental
ganglion

ventral
nerve cord

Annelid

Figure 40.15   Simple nerve net development in
the *Hydra*, and ganglia and nerve cord development
in the flatworm and the annelid.

cords. In some animals, such as flatworms, these ganglia
have retained so much control over essential body functions that decapitated animals can survive long enough
to grow a new head. In others, such as arthropods, ganglia along the nerve cord coordinate sensory input and
muscular activity.

In the line leading to vertebrates, however, cephalization continued, resulting ultimately in the evolution of
the vertebrate brain, a three-part swelling at the anterior
end of a dorsal nerve cord. In the simplest vertebrates,
such as fishes, those three original regions are easy to
see: the **hindbrain** (*rhombencephalon*), the **midbrain** (*mesencephalon*), and the **forebrain** (*prosencephalon*). Even in
higher primates, whose brains are much larger, those three
regions are visible during early development (Figure
40.16).

In fishes and other lower vertebrates, the forebrain
is concerned primarily with detecting and integrating
information from the chemical senses of smell and taste.
In higher vertebrates, including humans, the forebrain
not only integrates incoming sensory information about
the environment but also initiates voluntary movement
and performs the mysterious processes we call "thinking."

THE HUMAN NERVOUS SYSTEM

The human nervous system is customarily divided into
two main parts according to location and function (Figure 40.17). The **central nervous system (CNS),** housed
almost entirely within the bony protective structures of
the skull and vertebral column, consists of the brain and
spinal cord. The **peripheral nervous system (PNS),** distributed throughout the body, consists of afferent and
efferent fibers and sense organs.

The peripheral and central nervous systems interact
constantly. The CNS integrates information about internal and external environments that is provided to it by
the sense organs and afferent neurons of the PNS. Once
the CNS has "decided" on a course of action, it operates
by distributing "instructions" to effector organs ranging
from glands to blood vessels to fingers and toes through
the efferent neurons of the PNS.

Both divisions of the nervous system are therefore
essential for either *voluntary* actions (such as putting on
a sweater before you go out into the cold) or *involuntary*
activities (such as shivering when you don't dress warmly
enough).

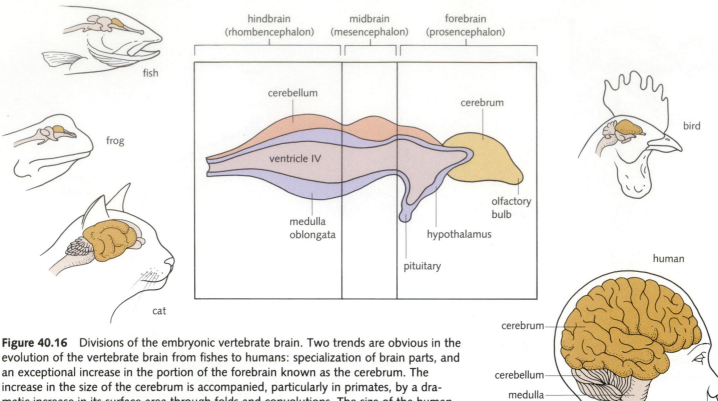

Figure 40.16 Divisions of the embryonic vertebrate brain. Two trends are obvious in the evolution of the vertebrate brain from fishes to humans: specialization of brain parts, and an exceptional increase in the portion of the forebrain known as the cerebrum. The increase in the size of the cerebrum is accompanied, particularly in primates, by a dramatic increase in its surface area through folds and convolutions. The size of the human cerebrum, however, is not simply the result of larger animals growing larger brains; there is a clear increase not only in the absolute size of the brain, but in the ratio of brain size to body weight. Living mammals and birds have brains about 15 times larger than those of most similarly sized lower vertebrates.

The Peripheral Nervous System

The peripheral nervous system contains both afferent and efferent neurons. The afferent portion includes *somatic sensory neurons* in all major sense organs (except the eyes) as well as *visceral sensory neurons,* which are isolated sensory cells that detect blood pressure, the contents of the digestive system, and the relative positions of various body parts. The efferent portion of the peripheral nervous system is divided functionally into the voluntary or **somatic nervous system** and the involuntary or **autonomic nervous system,** depending on whether the responses they mediate are under voluntary or involuntary control.

It is important to remember that the PNS and CNS interact constantly and that many body functions are under both voluntary and involuntary control. Breathing, for example, is normally controlled involuntarily. But during activities ranging from speaking to swimming, we can exert considerable voluntary control over the same muscles. Even heartbeat, which Western physicians consider an entirely involuntary function, can be controlled consciously by those trained in Eastern techniques of yoga and meditation.

The somatic nervous system The somatic nervous system includes most sensory neurons from the skin and skeletal muscles and most motor neurons leading to skeletal muscle. This division of the nervous system is responsible for both voluntary movement and reflex arcs of the sort described earlier.

The autonomic nervous system The autonomic nervous system is subdivided into the **sympathetic division** and the **parasympathetic division.** Sympathetic and parasympathetic neurons make synapses with glands and with smooth and cardiac muscles, where they release different neurotransmitters that often (but not always) have complementary (opposite) effects (Figure 40.18). Specific effector organs controlled by autonomic neurons usually mediate long-term physiological processes such as secretion of saliva, narrowing or widening of blood vessels, and secretion and muscular activity in the digestive tract. The control functions of the autonomic nervous system thus overlap with those of many hormones (see Chapter 36).

The sympathetic division prepares the body for intense physical exertion, the same "flight-or-fight" response mediated by adrenaline. In fact, when activated by the CNS through fear, anger, or the perception of danger, the sympathetic division simultaneously stimulates the release of adrenaline (epinephrine) and reinforces that hormone's actions by increasing heart rate, widening capillaries in major muscle groups, mobilizing glucose from the liver for quick energy, and slowing down digestive activity. At the same time, the sympathetic system releases noradrenaline onto its effector organs.

The parasympathetic division, on the other hand, helps the body adjust to the relaxed state one enters after a large meal; it slows down heart rate, decreases blood flow to muscles and skin, and increases secretory and circulatory activity in the digestive tract. Parasympathetic neurons release acetylcholine onto their effector organs.

The Central Nervous System

The brain and spinal cord share several physical characteristics. Both are protected by bony coverings and cushioned by three layers of connective tissue called **meninges.** Both brain and spinal neurons have cells organized in such a way as to produce regions called **gray matter** and **white matter.** The glistening white matter—located in the interior of the cerebral cortex and on the exterior of the spinal cord—is composed predominantly of long axons whose color is due to their myelin sheaths. The grey matter—found on the outer layer of the forebrain and in the center of the spinal cord—is composed predominantly of cell bodies (Figure 40.19).

Both brain and spinal cord have internal cavities: the four *ventricles* of the brain and the *central canal* of the spinal cord. These cavities are lined with ciliated cells and filled with **cerebrospinal fluid,** which circulates through the CNS carrying oxygen, glucose, white blood cells, and hormones. Cerebrospinal fluid picks up these substances from blood capillaries, but not in the same way as other tissues do. Brain capillaries are not as "leaky" as normal

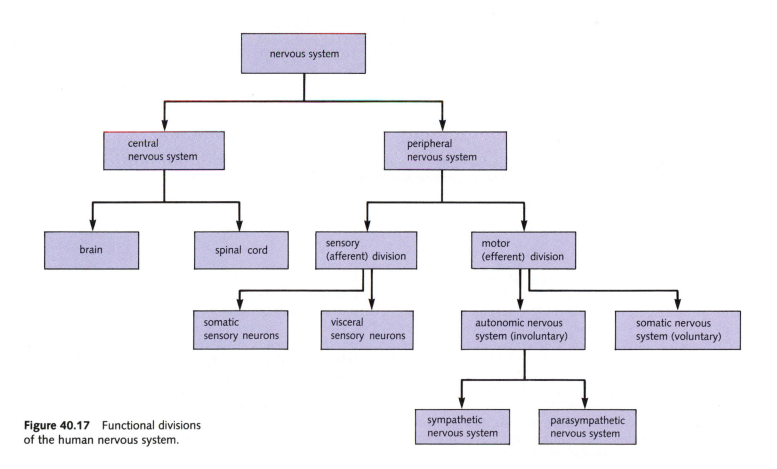

Figure 40.17 Functional divisions of the human nervous system.

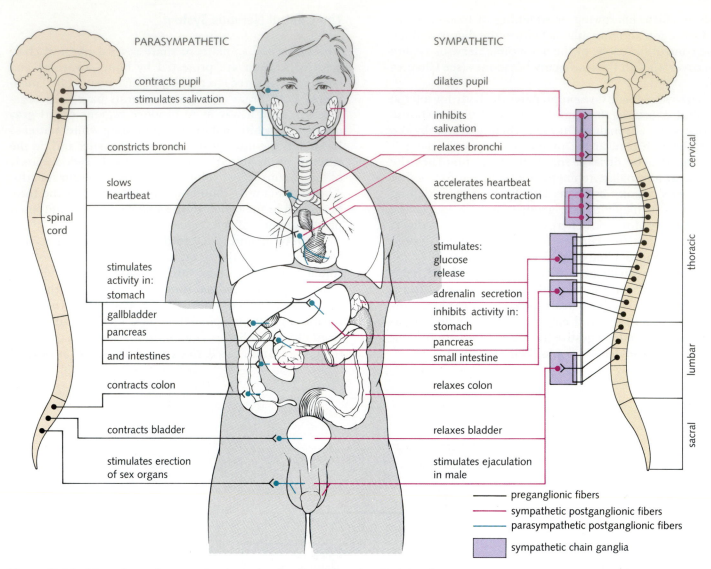

PARASYMPATHETIC

contracts pupil
stimulates salivation
constricts bronchi
slows heartbeat

spinal cord

stimulates activity in:
stomach
gallbladder
pancreas
and intestines
contracts colon
contracts bladder
stimulates erection of sex organs

SYMPATHETIC

dilates pupil
inhibits salivation
relaxes bronchi
accelerates heartbeat strengthens contraction
stimulates:
glucose release
adrenalin secretion
inhibits activity in:
stomach
pancreas
small intestine
relaxes colon
relaxes bladder
stimulates ejaculation in male

cervical
thoracic
lumbar
sacral

— preganglionic fibers
— sympathetic postganglionic fibers
— parasympathetic postganglionic fibers
◼ sympathetic chain ganglia

Figure 40.18 The autonomic nervous system, showing the different paths taken by sympathetic and parasympathetic neurons and their complementary effects.

capillaries are. They form a highly selective membrane called the **blood–brain barrier.** That barrier does not pass many large proteins, including certain hormones and some drugs, but it does allow the passage of other drugs, such as nicotine, alcohol, and cocaine.

The Brain

The forebrain: cerebrum and thalamus As the embryonic human brain develops, the leading edge of the forebrain enlarges to form the **cerebrum,** the largest part of the

brain and the part that has changed most through evolution. The surface of the cerebrum, the **cerebral cortex,** is grey matter because it contains the cell bodies of the cerebral cells. Interestingly, gray matter in primitive brains is internal, as it is in the spinal cord. It seems likely that relocating the cell bodies on the outside was important because of the phenomenal increase in the number of cells in the cerebrum: Many cells could be added to the outside of the brain without disrupting the axons that led to and from them.

The human cortex is folded so extensively to increase its surface area (and hence the number of cells it can

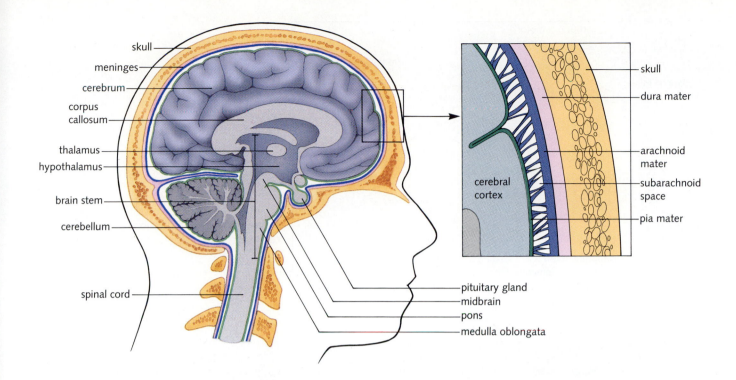

Figure 40.19 The gross anatomy of the brain and spinal cord.

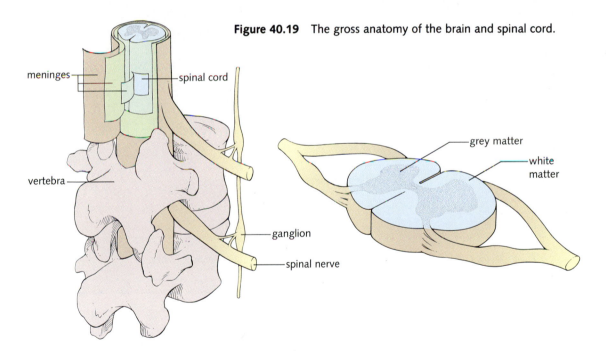

hold) that if it were laid out flat, it would cover an entire square meter. Different parts of the cerebral cortex are specialized for different functions (Figure 40.20). Sensations from specific body areas are "felt" by well-defined regions, and commands to muscles that move those body parts originate from other cortical areas.

The cerebrum is divided into two halves, or **hemispheres,** connected by a narrow band of axons called the **corpus callosum.** Oddly enough, the sensations and motor areas of the two sides of the body are reversed in the cerebrum: the left side of the body is represented on the right side of the brain, and vice versa. Certain broad

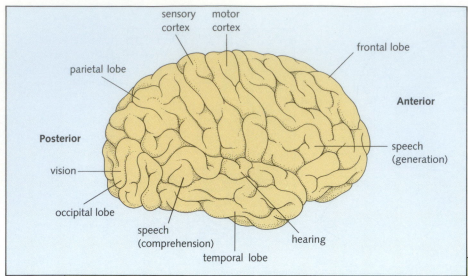

Figure 40.20 Specializations of the cerebral cortex.

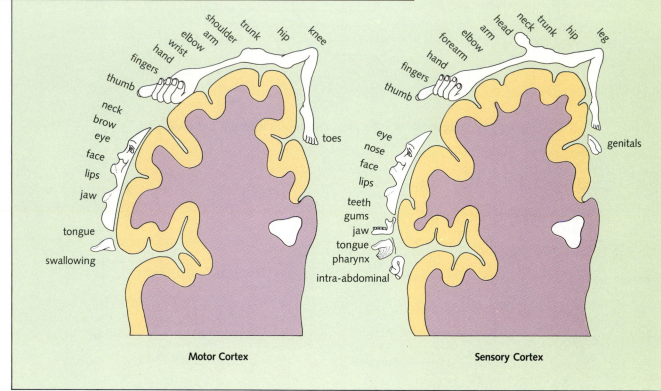

Motor Cortex

Sensory Cortex

classes of higher cerebral functions seem to be localized in different hemispheres as well. The right hemisphere seems to govern creativity and artistic ability, while the left hemisphere plays the more dominant role in analytical, verbal, and mathematical ability.

A part of the forebrain just behind the cerebrum grows into a smaller structure called the **thalamus,** an important "relay center." Here, incoming messages from both olfactory and visual receptors and from many other parts of the nervous system are relayed to the cortex.

During the development of the thalamus, two buds from the thalamus grow outward to form the neural portions of the eyes. Another, smaller thalamic area grows upward to form the *pineal body,* a structure often called the "third eye" that functions in regulating day/night cycles of sleep and wakefulness. Still another part of the thalamus grows down to produce the **hypothalamus** and **posterior pituitary**—the vital links between the nervous system and the hormonal system that we considered in Chapter 34.

The midbrain The midbrain in humans does not enlarge proportionately as much as the cerebrum; it functions predominantly to relay incoming information from eyes and ears to the cortex and to connect the cerebrum with other parts of the brain. This region contains parts of two small but indispensable functional units, the **limbic system** and the **reticular formation** (Figure 40.19a, b).

The limbic system controls the emotions. Stimulation of various parts of the limbic system can evoke anger, hostility, or pleasure. The reticular formation monitors sensory input to the brain and determines which incoming stimuli the cortex attends to or ignores on a moment-by-moment basis. Thus it is the reticular formation that "wakes up" the rest of the brain when it detects certain noises or odors. Electrical stimulation of sensory areas in the cortex alone does not wake a sleeping individual, but stimulation of the reticular formation generates waves of impulses throughout the cortex, stimulating it to wakefulness. Other parts of the reticular formation suppress those "wake-up" centers and make sleep possible.

The hindbrain The hindbrain develops into three distinct regions: the **cerebellum,** the **pons,** and the **medulla oblongata.** Each of these regions acts as a neural "switchboard," controlling the flow of nerve impulses between the brain and the rest of the body. The medulla, for example, controls a number of functions, including breathing, blood pressure, heart rate, and coughing.

The cerebellum coordinates body movements by performing several sophisticated integrating functions. The cerebellum regulates general equilibrium (balance) by interpreting information from the organs of balance within the ear. It also controls muscle tone, both at rest and during action; this is essential to control muscles' resistance to stretch and their ability to stabilize joints. The cerebellum also coordinates the actions of opposing muscle groups by determining when to start and stop contraction and how much force to apply at each moment. To accomplish all these functions, the cerebellum processes input from skin, eyes, ears, stretch receptors in skeletal muscle, and the cerebrum.

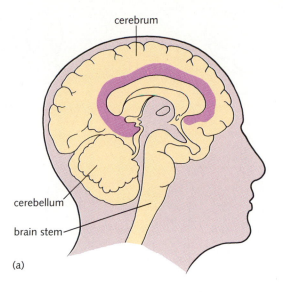

(a)

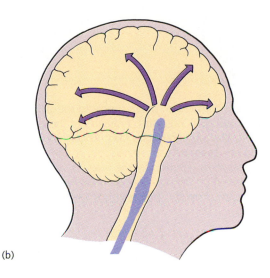

(b)

Figure 40.21 **(a)** The limbic system. **(b)** The reticular formation.

The Spinal Cord

Perhaps nowhere in the body is the heritage of our segmented invertebrate and lower vertebrate ancestors more evident than in the structure of the spinal cord. Here, at regular intervals, efferent neurons leave the CNS and afferent neurons enter along what are known as **spinal nerves.** Some afferents and efferents synapse directly with one another, such as those of the knee-jerk reflex that we discussed earlier. Others connect through interneurons that travel along the spinal cord to and from the brain. The pattern in which spinal nerves connect with receptors in the skin reflects the fate of tissue derived from embryonic segments during development (Figure 40.22).

It is interesting that the brain tends to confuse certain kinds of sensory information from nerves originating from the same body segment. This is particularly true if certain neurons from a segment are regularly stimulated and others are not. The heart and the region of skin indicated in blue in Figure 40.22, for example, are served

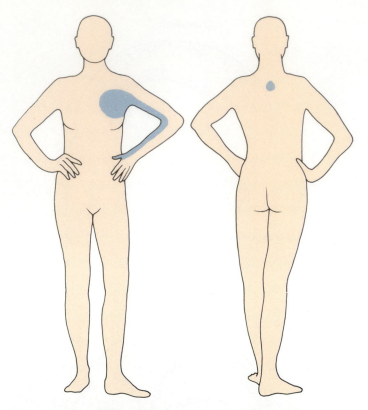

Figure 40.22 Patterns of nerve supply to the skin. Neurons leave the spinal cord in groups; branches of these spinal nerves supply both skin and internal organs. Because healthy internal organs do not generate pain signals, discomfort in those organs is often mistakenly reported by the brain as originating in the skin area served by the same nerve as the organ in question. The colored regions show the areas confused with pain from the heart.

by nerves that enter the spinal cord together. During a heart attack—a condition we rarely encounter—the brain often interprets heart pain as pain in the left armpit and the inner surface of the left arm, areas that are stimulated more often. This is known as *referred pain*.

SO ELEGANT AN ENIGMA: THE BIOLOGY OF THE MIND

Neurologists and psychologists have long been challenged by the difficulty of determining how the brain functions. For many years, researchers have been able to pick up traces of the brain's electrical activity though electrodes applied to the scalp. Typically, these patterns have been displayed in the form of an **electroencephalogram,** or **EEG.** Because the brain exhibits different patterns of activity during different states of wakefulness and sleep, EEGs have been useful in categorizing the levels of sleep and in investigating certain kinds of sleep disorders (Figure 40.23).

Today, however, medical science has far more powerful ways to see what is going on inside the brain. **CAT** scans (**c**omputer-**a**ssisted **t**omography) take finely focused X rays of the brain from many different angles, analyze the amount of radiation absorbed by different parts of the brain, and convert that information into a series of "maps" that are like visual cross sections of the brain (Figure 40.24).

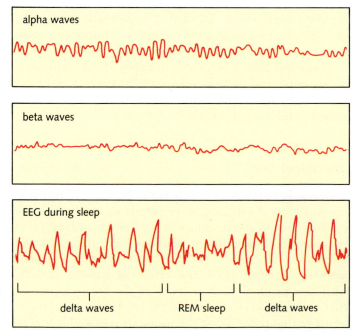

Figure 40.23 Electroencephalograms (EEG) can be used to monitor brain activity during states of wakefulness and sleep.

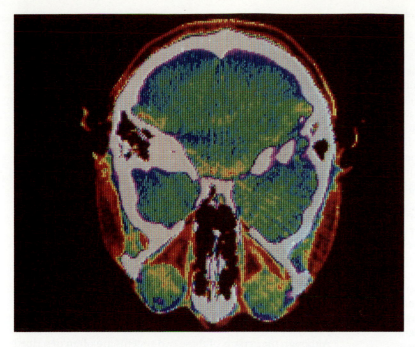

Figure 40.24 CAT scan of a normal human brain.

PET scans (**p**ositron **e**mission **t**omography) go a step further than EEGs and CAT scans by monitoring and offering physicians a record of the physiological activity of the brain. By administering doses of short-lived radioactive isotopes that are metabolized by the brain, physicians can create maps of metabolic activity in the brain during different activities (Figure 40.25).

Finally, the recently developed **BEAM** scans (**b**rain **e**lectrical **a**ctivity **m**apping) use the sort of sophisticated data processing techniques used in CAT and PET scans to translate the jagged lines of EEG recordings into a color map of brain function. With these and other tools at their disposal, both basic and clinical researchers hope to continue unraveling the mysteries of the intact brain.

Figure 40.25 (left) PET scan patient and (right) PET scan.

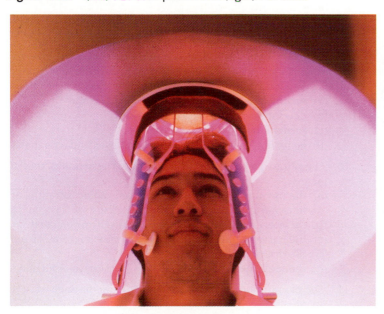

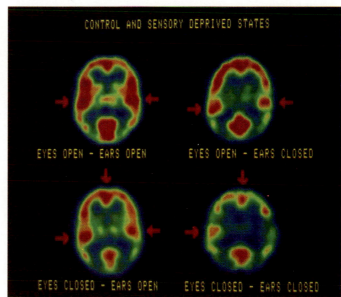

SUMMARY

The nervous system receives information through sensory neurons, carries and processes information through interneurons, and directs responses by stimulating muscles and glands called effectors. Neurons carry information as action potentials, waves of electrical disturbance that travel along axons. Action potentials are made possible by ion pumps and ion channels in the nerve cell membrane. Neurons communicate at synapses, where electrical activity is converted into pulses of chemical neurotransmitters. Neurotransmitters may either excite or inhibit the next neuron in the circuit.

The simplest nervous systems are diffuse nerve nets with few specialized cells. Among higher invertebrates and vertebrates, however, axons are gathered into afferent and efferent nerves and central nerve cords. Collections of cell bodies and synapses called ganglia serve as information processing centers. As higher vertebrates evolved, the largest of these—the cerebral ganglion—grew into a highly developed brain. The human nervous system is divided functionally and structurally into *central* and *peripheral nervous systems*. Functionally, the peripheral nervous system is divided into the voluntary or somatic nervous system and the involuntary or autonomic nervous system. The autonomic nervous system is further divided into sympathetic and parasympathetic divisions.

STUDY FOCUS

After studying this chapter, you should be able to:

■ Describe the principal components of the nervous system and how these components are interconnected.
■ Understand the basic facts about electrical phenomena in living tissue.
■ Explain how the nervous system transmits information by describing the mechanisms by which action potentials are generated and chemical synapses receive and transmit stimuli.

SELECTED TERMS

neuron *p. 808*
dendrite *p. 809*
axon *p. 809*
nerve *p. 809*
synapse *p. 809*
reflex arc *p. 809*
sodium–potassium
   pump *p. 811*
neurotransmitter *p. 813*
action potential *p. 814*
Schwann cell *p. 816*

saltatory conduction *p. 817*
gap junction *p. 817*
neuromuscular junction *p. 818*
acetylcholine *p. 818*
summation *p. 820*
endorphins *p. 822*
cephalization *p. 824*
hypothalamus *p. 830*
limbic system *p. 831*
reticular formation *831*

REVIEW

Discussion Questions

1. Explain the series of events that result in the generation and propagation of an action potential.

2. It is tempting to compare action potentials traveling down axons to pulses of electric current traveling through insulated wires. Why is this not an accurate comparison?

3. Neurons are rarely connected to one another by direct electrical connections. Why? How are chemical synapses important to the functioning of nervous systems?

4. Explain the important differences between graded potentials and action potentials. What are the advantages and disadvantages of each?

5. How do the central and peripheral nervous systems differ from one another? How are they similar?

6. Compare and contrast the activities of the sympathetic and parasympathetic divisions of the autonomic nervous system. Although the actions of these two systems are often described as "antagonistic," it is more accurate in some ways to describe them as complementary. Why?

Objective Questions (Answers in Appendix)

7. When a neuron is not transmitting an impulse, a(n) _____ is present across its cell membrane.
   (a) impermeable barrier to K^+ ions
   (b) actional potential
   (c) resting potential
   (d) depolarized current

8. When a stimulus comes to a neuron cell membrane, the membrane becomes _____permeable to _____ ions.
   (a) more/sodium   (c) less/potassium
   (b) less/sodium   (d) more/potassium

9. An evolutionary trend in the nervous system of invertebrates that is seen in a more advanced form in vertebrates is
   (a) accumulation of ganglia in the head region.
   (b) development of a ventral nerve cord.
   (c) formation of a diffused nerve net.
   (d) increasing control of body functions by secondary ganglia.

10. The phrase "all or none" as it relates to the action potential means
   (a) the membrane generates a complete action potential or does not generate any action potential.
   (b) the movement of nerve impulses is self-propagated.
   (c) the nerve impulse doesn't lessen or end as it moves from the origin of the stimulus.
   (d) the movement of the nerve is by saltatory conduction.

11. The blood-brain barrier does not allow the passage of
   (a) alcohol.   (c) cocaine.
   (b) nicotine.   (d) all hormones.

The Sensory System

If the doors of perception were cleansed, every thing would appear to man as it is, infinite.
For man has closed himself up 'til he sees all things through narrow chinks of his cavern.

—William Blake

Make believe for a moment that you are in your favorite corner of the natural world—a tropical beach, a Colorado mountain ridge, a lush New England forest. Now immerse yourself in that setting as completely as possible. Imagine the warmth of the sun on your skin, the feather touch of the breeze on your face, the tang of salt air, the bracing scent of pine needles, the subtle odors of moss and wet soil. Conjure up the images of wildflowers blazing in an alpine meadow, butterfly wings flashing in the sun, crickets chirping. Savor these sensations. Yes, this is your favorite spot, just as you remember it.

If you think the way most of us do, the collection of sensory images you've just conjured up *is* that place for you. Your perceptions *are* your world, for perceptions are all of the world any of us can ever know. Each of our senses gathers information about part of our immediate environment, and together they inform us of changes that occur around us.

Sensory systems, human and nonhuman, are of incalculable importance in daily life. In this chapter we will explore the form and function of human senses and the senses of other animals. In the process we will learn to view the senses as evolutionary adaptations to the advantages and disadvantages of various sensory modalities in different environments.

THE WORLD, THE SENSES, AND REALITY

The world we create from our perceptions—the environment that is so real to us—is an insular and idiosyncratic "reality" based on only a fraction of the phenomena present in the environment. The total reality of that spot you've chosen is far more complex and varied than human senses can fathom; there are sounds we cannot hear, objects and colors we cannot see, smells and tastes of which we are completely unaware. And altogether beyond our sensory capabilities are events we can describe and detect with instruments but cannot experience directly. The earth's magnetic field envelops us. The ground beneath us carries countless faint vibrations. And underwater, both inanimate objects and living organisms emit minute electric currents that surround them with faint electromagnetic auras as personal as fingerprints.

All these phenomena are potential sources of environmental information to animals with appropriately specialized sensory receptors. In some cases animal senses are more acute than, but basically similar to, our own. Dogs track odors of which we are unaware. Bees can see ultraviolet light. Bats and dolphins "see" by emitting sounds far above our range of hearing. But other animals have senses that are completely different from anything we possess. Certain fishes communicate by generating electric currents, for example, and many birds navigate in part by detecting the earth's magnetic field.

Thus, if we accept the metaphor of our senses as windows on the world, then we must also say that we perceive the world only "through a glass, darkly." We can devise microscopes and telescopes, extending that glass's range with lenses. But the glass remains stubbornly in place, coloring our view of the world so intimately that we can conceive of different world views only abstractly.

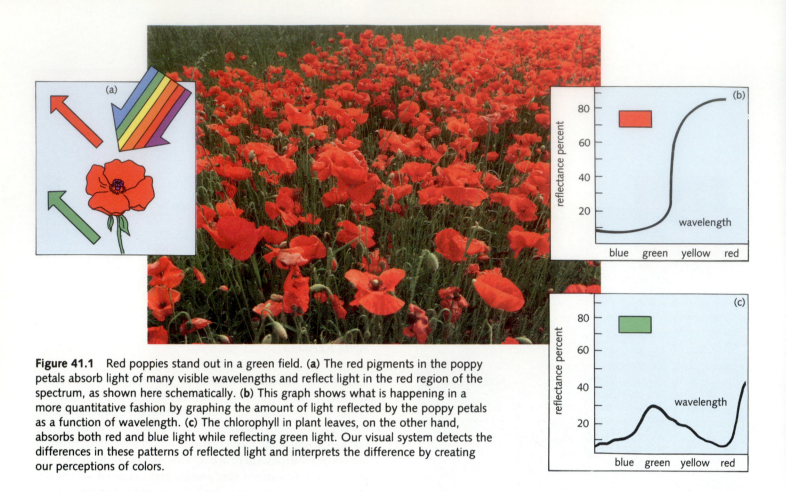

Figure 41.1 Red poppies stand out in a green field. (**a**) The red pigments in the poppy petals absorb light of many visible wavelengths and reflect light in the red region of the spectrum, as shown here schematically. (**b**) This graph shows what is happening in a more quantitative fashion by graphing the amount of light reflected by the poppy petals as a function of wavelength. (**c**) The chlorophyll in plant leaves, on the other hand, absorbs both red and blue light while reflecting green light. Our visual system detects the differences in these patterns of reflected light and interprets the difference by creating our perceptions of colors.

THE NATURE OF SENSORY STIMULI

Information about the environment is carried by **sensory stimuli**—environmental factors to which human or animal sensory receptors can respond. Many sensory stimuli are in the form of energy and contain information encoded in variations in the form, frequency, or intensity of that energy. We see vibrantly colored flowers in a field, for example, because their petals reflect different wavelengths of light energy from the grass around them (Figure 41.1). We hear the annoying buzz of mosquitoes because the vibrations of their wings create sound waves of particular intensity and frequency.

Other sensory stimuli are composed of atoms or molecules. Because many flowers release fragrant compounds that travel through the air, you can distinguish between a rose and a dandelion with your eyes closed. Similarly, information about the differences between salt water and fresh water, and between sugar solution and milk, is carried by atoms and molecules as well as by the liquids' visual appearance.

ESSENTIALS OF SENSORY FUNCTION

For animals to respond appropriately to environmental stimuli, their sensory systems must perform two interrelated tasks. First they must convert, or *transduce,* sensory stimuli into the electrochemical "language" of the nervous system. Then they must process the resulting neural information in a manner that enables the central nervous system to identify and evaluate important stimuli. As you will see, part of this essential sensory processing often takes place in the sensory organs themselves, but a great deal occurs in higher brain centers.

Sensory Transduction

Transducing stimuli into neural activity is the task of *receptor cells,* specialized neurons that detect stimuli of one type or another. Each type of receptor cell is typically specialized to detect only one particular kind of stimulus; photoreceptors respond to light, whereas taste receptors respond to chemicals. Upon receipt of an appropriate stimulus, the receptor cell generates a **receptor potential,** a graded potential like those discussed in Chapter 40. In most types of receptors, the magnitude of this receptor potential varies with the intensity of the stimulus.

For sensory information to be acted on, it must be transmitted to the central nervous system for processing. But, as you recall from the last chapter, graded potentials (including receptor potentials) cannot carry information very far from their point of origin. The sensory system handles this problem in two different ways.

Some receptor cells, such as the stretch receptors in muscle, convert receptor potentials directly into action potentials. When such receptors are stimulated, they fire impulses that travel down long axons wired into the central nervous system. Note that because all action potentials produced by a particular cell are identical, receptors cannot communicate the intensity of the stimulus by changing the *size* of those impulses. Instead, they encode the intensity of stimulation in the *rate* at which they fire.

Stretch receptors called **muscle spindles,** for example, are seldom totally inactive, for even in a muscle at rest they are slightly stretched. Resting spindles produce action potentials at a slow, steady rate called their **basal firing rate** or **rate of spontaneous activity.** When the muscle around them stretches, they fire more often, and when the muscle around them contracts, their basal firing rate decreases (Figure 41.2).

This pattern of increases and decreases in basal firing rate is common throughout the nervous system. The pattern enables individual neurons to signal either an increase or a decrease in stimulation from a relatively stable resting state.

Other receptor cells, such as hair cells (which respond to motion and vibration), produce only graded potentials when stimulated. The cell bodies of these receptors (which usually lack axons) synapse directly onto one or more interneurons that produce action potentials in response to those graded potentials. In the hair cells of the ear, only a single interneuron is involved (Figure 41.3). In the eye, there may be as many as five interneurons between photoreceptors and higher processing centers. These peripheral interneurons play a very important role in sensory processing, as we will soon see.

Figure 41.2 Stimulation and firing rate in stretch receptor cells of vertebrate skeletal muscle. Each vertical line in the graphs below the diagram indicates a single action potential. The horizontal axis indicates time. In a muscle at rest, muscle spindles are slightly stretched, and fire impulses slowly. In a muscle stretched beyond its resting length, the receptors increase their basal firing rate. In a muscle that has just contracted, the receptors decrease their firing rate. Note: muscle spindles in higher vertebrates are more complex.

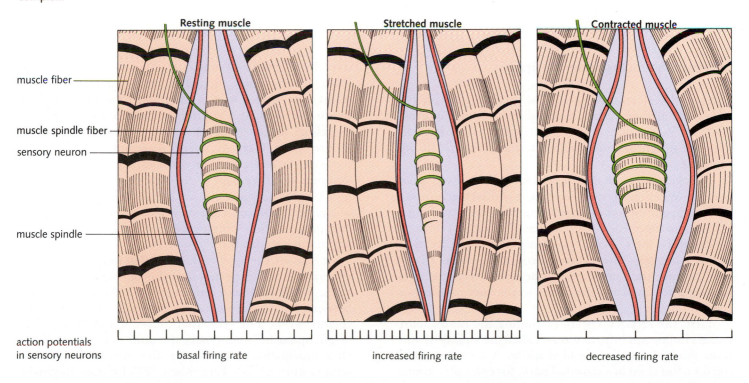

muscle fiber

muscle spindle fiber

sensory neuron

muscle spindle

action potentials in sensory neurons

basal firing rate   increased firing rate   decreased firing rate

Resting muscle   Stretched muscle   Contracted muscle

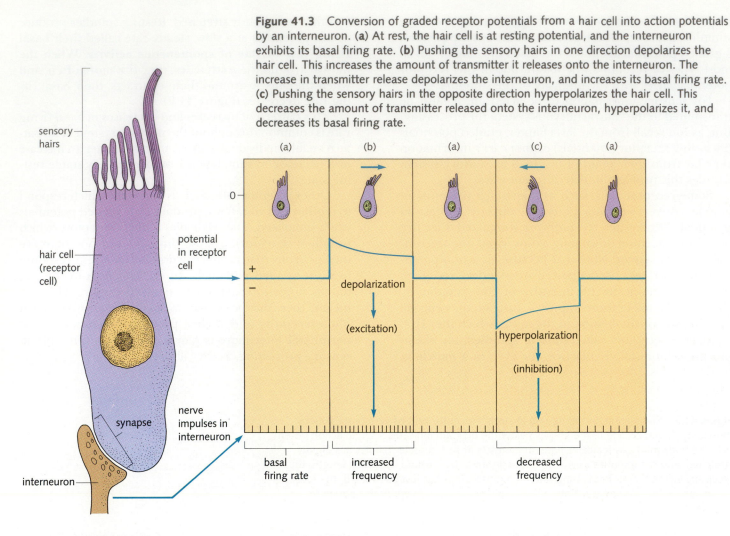

Figure 41.3 Conversion of graded receptor potentials from a hair cell into action potentials by an interneuron. (**a**) At rest, the hair cell is at resting potential, and the interneuron exhibits its basal firing rate. (**b**) Pushing the sensory hairs in one direction depolarizes the hair cell. This increases the amount of transmitter it releases onto the interneuron. The increase in transmitter release depolarizes the interneuron, and increases its basal firing rate. (**c**) Pushing the sensory hairs in the opposite direction hyperpolarizes the hair cell. This decreases the amount of transmitter released onto the interneuron, hyperpolarizes it, and decreases its basal firing rate.

Information Processing in the Brain

It is said that Bertrand Russell, the great English philosopher and mathematician, once visited his dentist in great distress. "Where does it hurt?" the dentist asked. Russell, ever the insightful philosopher, replied crisply, "In my mind, of course."

Biologically, as well as philosophically, Russell was quite correct. Whether sensory receptors are located in our teeth or in our eyes, they do nothing more than transduce stimuli into neural activity. The actual process of sensory perception—the response to a toothache or the creation of visual images—always takes place in the brain. This is true because the "raw data" gathered by receptor cells must be integrated and processed by interneurons in two ways at once.

First, the sensory system (including the brain) must extract important information—called the **signal**—from irrelevant information—called **noise.** A parent searching for a lost child in a crowded park, for example, strains to pick out that child's voice (the signal) from among all others in the noisy throng. Second, animals are constantly bombarded with far more information than their brains can possibly pay attention to at once. The CNS, therefore, must selectively attend to certain stimuli and ignore others. Harried parents searching for a child may focus so much attention on the auditory search that they do not notice they have badly stubbed their toe until the little tyke is safely in hand. We can also generate that kind of concentration consciously; when intent on reading, for example, we often screen out such stimuli as a radio playing in the background.

It is also at the level of the CNS that moods, emotions, experience, and expectations color our ultimate perceptions, which differ not only between individuals but also between specific moments within each individual's life (Figure 41.4). The odor that causes a child to respond "Ugh! Spinach!" might make the same individual salivate with anticipation several years later. A sunny room at a temperature of 70° Fahrenheit (21° Celsius) might be perceived as uncomfortably warm by an eskimo in winter, as pleasant by a Bostonian in May, and as chilly by a Bahamian in July. What is the "truth" about these stimuli? Is the odor of spinach pleasant or repugnant? How warm is a 70° room? The only answer, thanks to the vagaries of perception, is "It depends." Information processing is bewilderingly complex, and though we are beginning to understand some of the neural mechanisms that make perception possible, others remain a mystery.

THE DIVERSITY OF SENSORY SYSTEMS

Senses and Environments

Animal senses have evolved over time as adaptations to the nature of sensory stimuli in different environments, for each environment has physical characteristics that affect both energy and molecules that carry sensory information. Our environment, for example, consists of the media that surround us: the air around our bodies and the earth beneath our feet. Other animals may be immersed in the clear water of a mountain stream, the turbid water of a swamp, or the salty water of the sea.

Each of these media allows only certain environmental stimuli to pass. Air transmits visible light very well, for example, and conducts sound waves relatively efficiently. But air passes little or no electrical energy and carries only a limited assortment of small molecules detectable via the sense of smell. Water, on the other hand, carries sound both faster and farther than air, and it dissolves and carries a much wider range of chemicals, including sugars, amino acids, and large proteins. Water—especially sea water—is also an excellent conductor of electricity, but it absorbs (and hence fails to transmit) many wavelengths of visible light. Not surprisingly, animal sense organs have evolved in ways that relate to the environment in which they must function.

Similarities Among the Senses

Cataloguing all the evolutionary adaptations among the senses and explaining the mechanisms by which the information they detect is processed would fill several books. But behind differences in stimuli and in the nature of receptor cells, many underlying similarities unite all the senses. For each sense, there is a fascinating story of environmental information, the evolutionary adaptation of receptor cells to detect that information, and neural processing in the CNS to make the use of that information possible. In this chapter, we have selected the visual system to serve as a detailed example of these major principles, though we will briefly examine other human and animal senses.

Figure 41.4 Look for a moment at this scene on an island shore, devoid of human life. Or is it? Look again. Now look still again, focusing on the space between the gnarled trees on the left. Isn't that Napoleon standing in a characteristic pose? Notice how what you see in this case is very much determined by what you are looking for. Notice also that you probably would not see the Napoleonic silhouette unless you had been exposed to similar images in the past; this is just one example of the way current perceptions are affected by expectations and experience.

VISION: A MODEL OF SENSORY FUNCTION

Flowers, leaves, fruits, and the iridescent wings of tropical butterflies all differ in the way they absorb and reflect the electromagnetic energy of the sun. From honeybees searching for nectar-bearing flowers to fishes engaged in courtship dances, most animals depend heavily on patterns of reflected light to provide information about their environments and about each other. The elegance of design that enables vertebrate eyes to serve these purposes so impressed Darwin that he discussed them in a chapter of *On the Origin of Species* that he called "Difficulties of the theory: organs of extreme perfection and complication."

To understand why eyes are so complicated, and to appreciate the functions of visual systems, we will examine the characteristics of visual stimuli in nature and will explore the evolutionary responses of animal visual systems to the demands of vision in different environments.

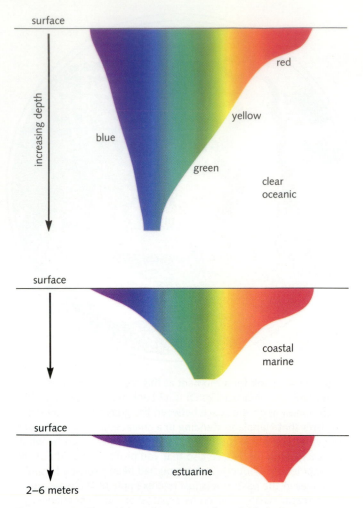

Figure 41.5 The color and intensity of available light varies greatly among different aquatic habitats. Seawater absorbs long-wave and extreme shortwave light, and thus acts as a turquoise color filter placed over sunlight. Water along temperate sea-coasts and in freshwater lakes carries dissolved and suspended materials that absorb light of short wavelengths; the light that persists is thus colored greenish-yellow. The minimal light available in many rivers, swamps, and marshes is deep reddish-brown.

Because the visual system is among the best-understood of the sensory systems, we will also study briefly the first stages of visual information processing. This examination will serve as a specific example of the way chemical synapses and neural wiring patterns work together to process information in the nervous system. It is important to remember that although we do not have the time or space to examine each of the other senses in this much detail, all senses perform tasks analogous to these.

Light and the Visual Environment

Visible light is our name for the portion of the electromagnetic spectrum that functions as an adequate stimulus for vision in humans. Describing the physical nature of light is difficult, for it has properties of both waves and particles. Physicists recognize both sets of properties and describe light simultaneously in terms of waves and in terms of packets of energy called **photons.** Our visible spectrum ranges from wavelengths of about 400 nanometers (nm), which we perceive as violet, to about 700 nm, which we perceive as deep red.

With the exception of the few organisms that produce their own light, natural objects are visible only by virtue of the sunlight they reflect. "White" objects reflect roughly equal amounts of all visible wavelengths, whereas "colored" objects reflect more light of certain wavelengths than of others (as we saw in Figure 41.1).

But the light that illuminates objects in nature varies enormously at different times and in different places. Between midday in June and a moonless night, for example, the intensity of natural light varies by a factor of 10^{15}. And between dawn and midday, from summer to winter, and in different habitats, the color of available light changes enormously (Figure 41.5).

Because of these changes in natural illumination, the physical stimulus the eye receives from an object at different times and in different places also changes greatly. Visual systems are therefore confronted with the formidable task of providing constant, reliable information under highly variable conditions; a ripe strawberry should look red regardless of when or where we see it. The eye and brain accomplish this task with such accuracy that we rarely even notice the necessary adjustments in progress.

Structures and Functions of the Eye

Light-sensitive organs have evolved independently in a wide variety of animals, and they vary from the simple "eyespots" of protozoa to the intricate visual sense organs and processing centers of higher vertebrates. One of the simplest such structures, the paraflagellar body of *Euglena,* is shown in Figure 41.6a. The most highly developed eyes among invertebrates are found in the arthropods and cephalopod molluscs (Figure 41.6b, c). The **compound eyes** of arthropods are very different from ours, for they consist of dozens to thousands of identical units called **ommatidia** (Figure 41.6d). Each ommatidium is a complete miniature eye with its own lens-like structure and *retinula,* or "little retina." Each retinula, in turn, contains eight or more light-sensitive **photoreceptor cells.**

It is difficult to imagine how the thousands of separate images produced by ommatidia are combined by the insect brain into a coherent picture of the world. Behavioral studies have shown that although insect eyes cannot provide so fine-grained an image of the world as our eyes can, they can detect movements far more rapid than those our visual systems can perceive. Many insects also have well-developed color vision.

Cephalopod eyes, on the other hand, though they evolved completely independently from those of vertebrates, show remarkable convergence in structure with vertebrate eyes (Figure 41.6e, f). Both cephalopod and human eyes have a clear outer covering, or **cornea,** and a **lens** that aids in focusing the image on a **retina** lined

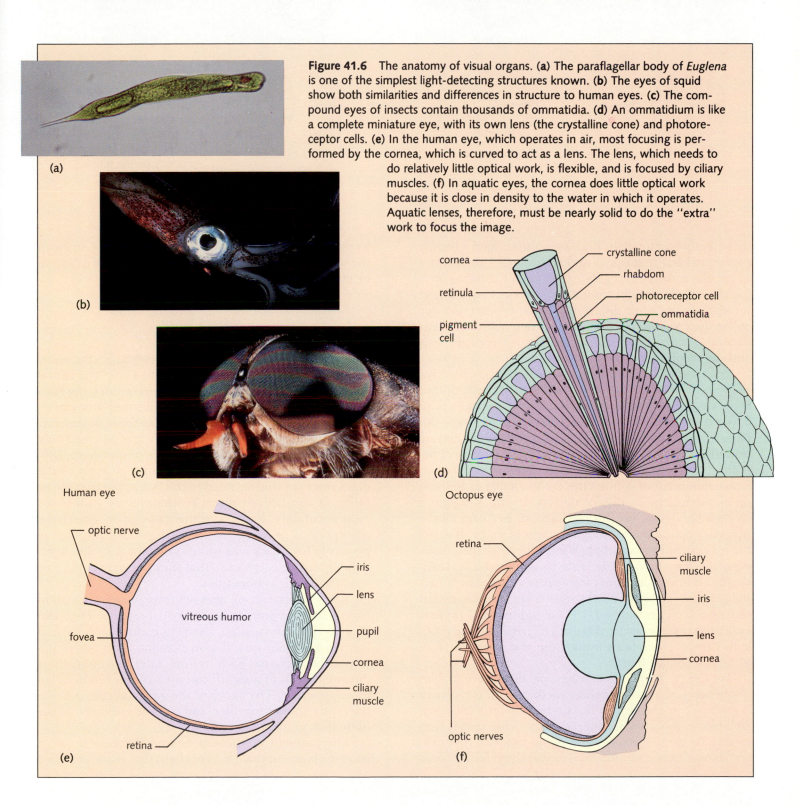

Figure 41.6 The anatomy of visual organs. (**a**) The paraflagellar body of *Euglena* is one of the simplest light-detecting structures known. (**b**) The eyes of squid show both similarities and differences in structure to human eyes. (**c**) The compound eyes of insects contain thousands of ommatidia. (**d**) An ommatidium is like a complete miniature eye, with its own lens (the crystalline cone) and photoreceptor cells. (**e**) In the human eye, which operates in air, most focusing is performed by the cornea, which is curved to act as a lens. The lens, which needs to do relatively little optical work, is flexible, and is focused by ciliary muscles. (**f**) In aquatic eyes, the cornea does little optical work because it is close in density to the water in which it operates. Aquatic lenses, therefore, must be nearly solid to do the "extra" work to focus the image.

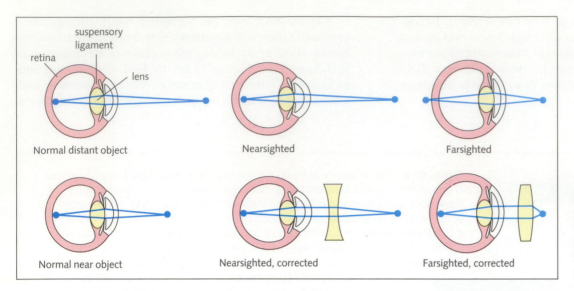

Figure 41.7 Myopia (near-sightedness) and hyperopia (far-sightedness) result when an improperly shaped lens fails to focus images properly on the retina.

The image labels read: suspensory ligament, retina, lens, Normal distant object, Nearsighted, Farsighted, Normal near object, Nearsighted, corrected, Farsighted, corrected.

with photoreceptor cells. Both eyes also contain a heavily pigmented **iris** that opens and closes under the control of circular muscles to regulate the amount of light entering the eye. The major optical structures of aquatic and terrestrial eyes work differently because water is more dense than air and because the focusing ability of a lens depends on the difference in density between the lens and the medium in which it operates.

For eyes (such as our own) that operate in air, the lens-shaped cornea actually does a good deal of the "work" in focusing light on the retina. Because the cornea is so important in this respect, irregularities in its shape can cause the visual distortions called **astigmatism.** Sitting behind the cornea, our lens is responsible mostly for "fine-tuning" the focus of the visual image. This process, known as **accommodation,** adjusts the eye to enable us to focus on either nearby or distant objects. In this situation a moderately dense, yet flexible lens can perform all the necessary optical work through changes in shape. If either the lens or the eye as a whole is improperly shaped, however, the lens may not be able to accommodate sufficiently to bring all images into proper focus. In "near-sightedness" (**myopia**), nearby objects are seen clearly but distant objects are blurred. In "far-sightedness" (**hyperopia**), the reverse occurs (Figure 41.7). All of these problems can be corrected through the use of proper eyeglasses or contact lenses.

For eyes that operate in water, the difference in density between that much denser medium and the cornea is minimal, so the cornea does relatively little focusing. For this reason a highly curved, very dense lens is necessary. Aquatic lenses are so dense, in fact, that they cannot change shape as ours do. Instead these lenses focus by moving back and forth in front of the retina, in much the same way as we might focus the glass lens of a camera or microscope.

The Retina

The vertebrate retina is composed of two layers, the *neural retina* and the *pigment epithelium* (Figure 41.8). The neural retina consists of two classes of photoreceptors, the **rods** and **cones;** several classes of glial cells; and four classes of interneurons, the *horizontal cells, bipolar cells, amacrine cells,* and *ganglion cells.* Cones, which function well only in fairly bright light, are responsible for the sharp, full-color visual images we depend on through most of the day. Cones are concentrated in the **fovea,** the part of the retina we instinctively direct at whatever we wish to see most accurately. Rods, sensitive enough to detect the absorption of a single photon of light, are responsible primarily for "black and white" vision in very dim light, although there is evidence that they participate in color vision at low light intensities.

The four classes of interneurons relay information from photoreceptors to higher visual centers. They also process that information, extracting clues to color and form in a manner we will soon examine. This processing is possible because each retinal interneuron gathers information from a significant area of the retina called its *receptive field.* The axons of the ganglion cells traverse the retina, gather together, and exit the eye through the **optic disk** to form the **optic nerve.**

The pigment epithelium behind the neural retina contains pigment that prevents light that is not absorbed

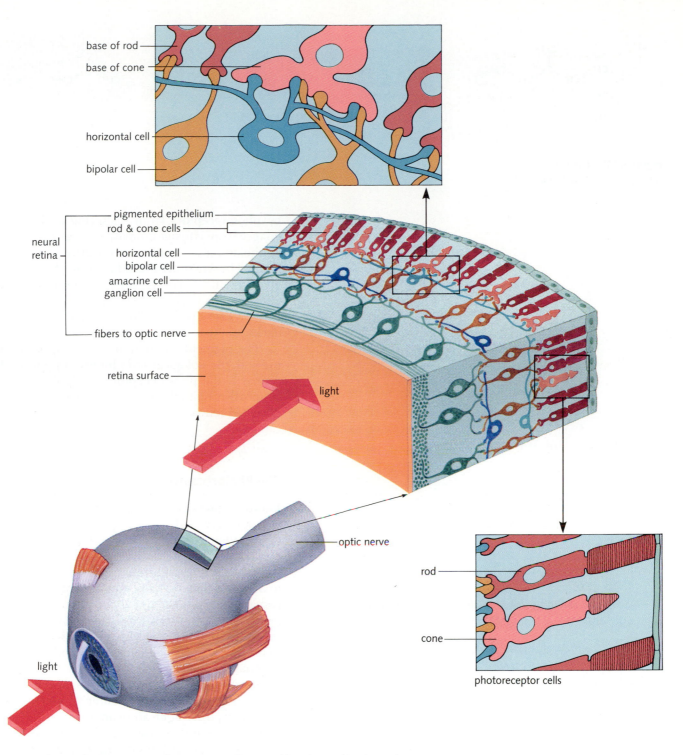

Figure 41.8 Vertebrate photoreceptor cells synapse with several bipolar and horizontal cells. Both bipolar and horizontal cells receive input from many photoreceptors. Several bipolar cells, in turn, synapse with each ganglion cell. These three types communicate using only graded potentials. The axons of ganglion cells (which do produce action potentials) lead out of the eye through the optic nerve. The dendrites of a retinal bipolar cell gather information from an area called the cell's receptive field.

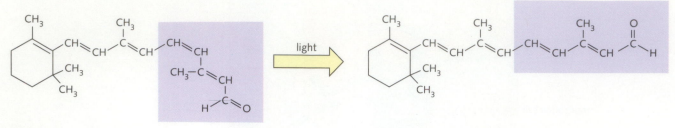

Figure 41.9 The effects of light absorption on visual pigments.

by photoreceptors from scattering and degrading the image. In nocturnal animals such as cats, however, the epithelium also contains a reflective layer called the *tapetum*. The tapetum reflects light onto the photoreceptors, giving them a "second chance" to absorb the limited light available. This tapetum creates the "eye shine" you see when a cat's eyes are illuminated by a spotlight in the dark.

The pigment epithelium is also important in providing both physiological and physical support to the neural retina. If the photoreceptor layer pulls away from this supportive epithelium, creating a condition known as **detached retina,** retinal damage and impairment of vision can result.

The Photoreceptors

Photoreceptors owe their light sensitivity to a class of compounds called **visual pigments,** which are composed of a protein group or **opsin** combined with a derivative of vitamin A_1 or A_2 called **retinal.** When visual pigment molecules absorb light, they change shape (Figure 41.9) in a way that alters the resting potentials of the cells that contain them. After the shape changes, the opsin and retinal separate—a process called *bleaching* that temporarily inactivates the visual pigment. Reactivation requires that opsin and retinal be rejoined and returned to their original shape.

Visual pigments are concentrated in the region of the receptor cells known as the outer segment. In vertebrate photoreceptors, light absorption by visual pigments in these outer segments causes a graded *hyperpolarization* of the receptor, but it does not initiate an action potential.

Most animals have a single class of rod photoreceptors, all of which contain the same visual pigment. Diurnal (day-active) animals may possess as many as four classes of cone cells (humans have three), each of which contains a different visual pigment. All visual pigments absorb light of most visible wavelengths to some degree, but each particular visual pigment is maximally sensitive to light of certain wavelengths (Figure 41.10). Extensive neural processing of signals from different classes of cone cells

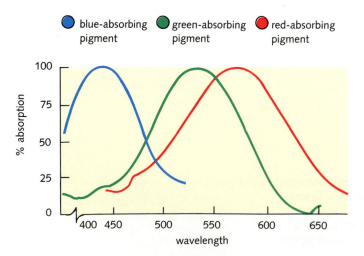

Figure 41.10 The absorption spectra of visual pigments from human rods and cones. Each pigment has a characteristic absorption spectrum with a well-defined peak, referred to as its wavelength of maximum sensitivity.

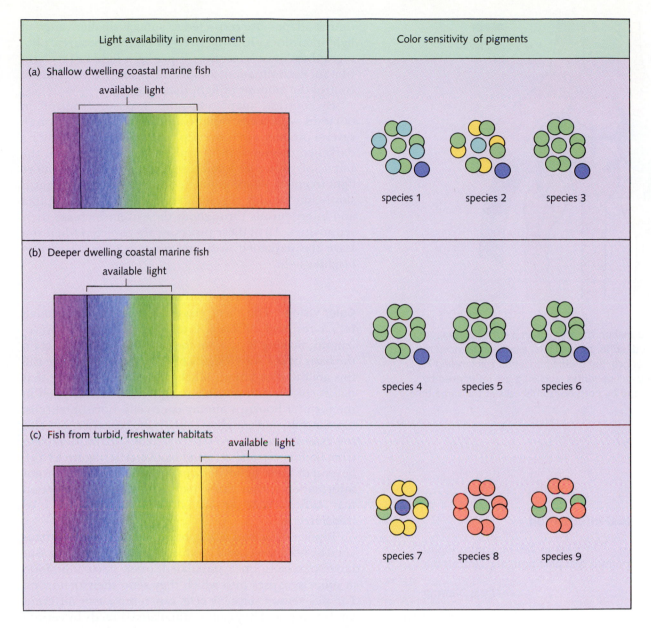

Light availability in environment	Color sensitivity of pigments

(a) Shallow dwelling coastal marine fish

available light

species 1   species 2   species 3

(b) Deeper dwelling coastal marine fish

available light

species 4   species 5   species 6

(c) Fish from turbid, freshwater habitats

available light

species 7   species 8   species 9

Figure 41.11 Visual pigments and environments. Aquatic animals living where light is dim have evolved visual pigments whose sensitivities match the color of available light. The colored dots for each species represent the colors to which their visual pigments are maximally sensitive.

enables the visual system to tell objects apart on the basis of both brightness and color.

Although several mammals, such as cats and other nocturnal species, don't see color well, most other animals (including insects, fishes, turtles, birds, and monkeys) have well-developed color vision. This is undoubtedly because color vision provides a great deal of important visual information about the environment. So useful is color vision, in fact, that it seems to have evolved inde-

pendently in different groups of animals at least four times. Some animals, such as bees, can see wavelengths invisible to humans because their visual pigments absorb ultraviolet light. Other animals, such as turtles and birds, have better color vision than we do; they can see more subtle differences in colors because they have more classes of cone receptors. Many species, particularly aquatic ones, have evolved visual pigments attuned to the colors of light present in their environments (Figure 41.11).

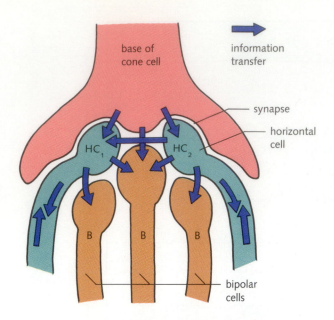

Figure 41.12 Information transfer within a cone cell synapse. The horizontal cell dendrite marked HC₁ can be influenced not only by the receptor cell, but by the horizontal cell labelled HC₂ as well. The bipolar cell dendrites labelled B are influenced by the photoreceptor and by both horizontal cells at the same time.

In figure labels: base of cone cell, information transfer, synapse, horizontal cell, HC$_1$, HC$_2$, B, B, B, bipolar cells.

Processing of Visual Information

> But to determine . . . by what modes or actions [light] produceth in our minds that phantasm of colours is not so easie.
>
> —Isaac Newton

The absorption of light by photoreceptors is the first step in a long series of events that ultimately results in vision. The complexity of information processing involved in these events is mirrored in the intricacy of neural networks in the retina and higher visual centers. Here we will investigate two phenomena: light adaptation and color vision. Like the other material discussed in this section, each of these visual phenomena has analogs in other sensory mechanisms.

Adaptation One of the most remarkable properties of the visual system is its capacity to operate over an enormous range of light intensities. Faced with widely varying conditions of light and dark, the eye adjusts its sensitivity through a process known as **sensory adaptation** (not to be confused with evolutionary adaptation). By comparison, photographic emulsions used in camera films—which

cannot adapt—are restricted to a much smaller range of light intensities. To avoid over- or underexposures, photographers must select the proper high-light or low-light film for each situation, and set their cameras carefully to control the amount of light that enters.

By contrast, adaptation—the "exposure control" in our eyes—proceeds rapidly and automatically through several processes that overlap in time. When you enter a darkened theater from a sunlit street, for example, the pupils of your eyes immediately expand to allow more light to enter. Visual pigment bleached in strong light is slowly but steadily regenerated. Finally, photoreceptors and interneurons throughout the system increase their sensitivities. All of these processes are reversed when you leave the darkened area and emerge once again into brighter light.

Color vision The three classes of cones in the human retina respond differently to light of different wavelengths, making it possible for us to tell objects apart on the basis of the wavelengths of light they reflect. We call this ability color vision. Much of the neural wiring that makes color vision possible is right in the retina itself in the form of hookups among cone cells, horizontal cells, bipolar cells, and ganglion cells. At the bases of cone cells, for example, are large synapses that contain dendrites from both horizontal cells and bipolar cells (Figure 41.12). Several of these dendrites can both receive and transmit information within this synapse. And that information may either excite or inhibit the cell to which it is transferred.

Bipolar cells and horizontal cells are usually stimulated by photoreceptors to which they attach directly. But horizontal cells, once stimulated, often cause inhibition at other synapses into which they send their dendrites. For this reason, bipolar cells can receive two different kinds of input: They can be stimulated directly by certain photoreceptors, and they can be inhibited by horizontal cells stimulated by other photoreceptors (Figure 41.13a).

This kind of wiring enables certain bipolar and ganglion cells to function as "spot detectors." Light falling on photoreceptors in the center of such a ganglion cell's receptive field stimulates bipolar cells. Those bipolar cells in turn stimulate ganglion cells, which respond by increasing the rate at which they fire action potentials (Figure 41.13b). If a light falls on photoreceptors outside that central area, on the other hand, those photoreceptors stimulate horizontal cells that *inhibit* central bipolar cells and ultimately *decrease* the rate at which the ganglion cell fires (Figure 41.13c).

These responses make it easier for the visual system to detect a spot of light against a background. And if several of these "spot detectors" are hooked together in

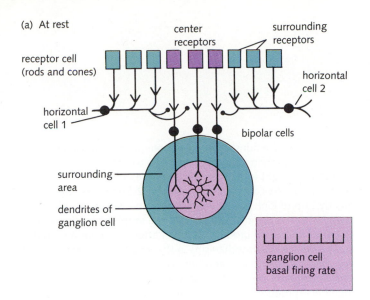

(a) At rest

center receptors

surrounding receptors

receptor cell (rods and cones)

horizontal cell 2

horizontal cell 1

bipolar cells

surrounding area

dendrites of ganglion cell

ganglion cell basal firing rate

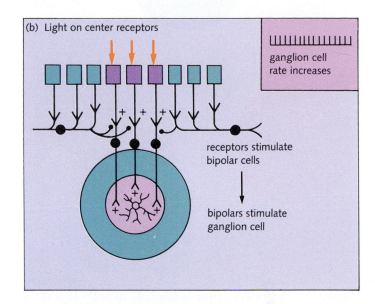

(b) Light on center receptors

ganglion cell rate increases

receptors stimulate bipolar cells

bipolars stimulate ganglion cell

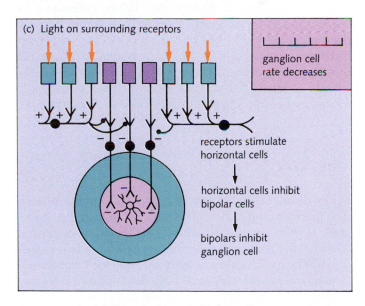

(c) Light on surrounding receptors

ganglion cell rate decreases

receptors stimulate horizontal cells

horizontal cells inhibit bipolar cells

bipolars inhibit ganglion cell

groups, they function as "edge detectors" that help the visual system detect the edges of large objects.

This sort of information processing is called *opponent processing*, because inputs from different cells are hooked up in opposition to one another. A type of opponent processing is also involved in color vision, as the demonstration shown in Figure 41.14 illustrates. This color-opponent interaction is the first step in a long line of processes that ultimately results in the perception of a full-color image. Step by step, in ways that we still do not completely understand, all the neurally coded data about color, shape, and movement are "assembled" as the information passes through several brain centers to the visual cortex. Precisely where and how visual perception occurs is still unknown.

Note that any sort of defect in this chain—from the visual pigments that first absorb light to the layers of neural processing that occur in higher neural centers—can result in visual deficiencies. The most common deficiencies in color vision (often labeled colorblindness) result when an individual is missing one or more of the cone visual pigments.

Lessons from vision applied to the other senses We have just outlined the basic functions of the visual system and have seen a few examples of how structures that serve these functions have evolved to meet different needs in different environments. All of these points should be viewed, not simply as information about the visual system in particular, but as principles demonstrated in all sensory systems.

In all species, sense organs and sensory receptor cells have adapted in ways that enable them to function in particular environments and for particular purposes.

Figure 41.13 (a) Neural wiring and information processing in the retina. Let us assume that we are monitoring the response of the single ganglion in the center of the figure as shown. Note that in these schematic illustrations, we show functional (rather than anatomical) connections between cells. Excitatory connections are shown by "+", and inhibitory connections are shown by "−". (b) Light on the "center" receptors stimulates bipolar cells, which cause our central ganglion cell to increase its basal firing rate. (c) Light on surrounding receptors stimulates horizontal cells 1 and 2, which inhibit those bipolar cells, lowering the ganglion cell's firing rate.

Figure 41.14 Demonstration of opponent color processing. Place this illustration next to a piece of plain white paper under a strong, white light (not a cool-white fluorescent). Stare intently at the "X" in the center of the figure for 2-3 minutes, and then quickly turn to look at the white paper. You will see an after-image of colors complementary to those in the figure. Looking at a strong red light produces a green after-image because red- and green-sensitive mechanisms in the visual system are wired in opponent fashion to one another. Similarly, looking at a strong yellow stimulus—which stimulates red- and green-sensitive cells together—produces a blue after-image because those cells are hooked up in opposition to blue-sensitive cells.

Similarly, opponent processing of neural signals is important because it allows higher brain centers to compare input from different sources. It is found not only in the visual system but also in many other senses and throughout the central nervous system. As just one example, opponent processing enables the auditory system to compare the sounds heard by each ear, thereby enabling the hearer to judge the location of a sound source.

AUDITION

After vision, hearing is probably the most important sense for humans and other primates. For numerous insects and several mammals, sound is the primary sensory channel for finding food, avoiding being made into food by others, and communicating as well (see Chapter 43). Sound is a superb carrier of long-distance messages and, unlike light, can travel through soil and cloudy water and can move around tree trunks and foliage with little loss of energy.

What Is Sound?

Sound is a mechanical disturbance created in a gas, liquid, or solid by a vibrating or moving object. Physicists describe two components of sound waves, far-field sound and near-field sound.

Far-field sound consists of pressure waves and is so named because it can travel over long distances. When a loudspeaker produces sounds, for example, its cone vibrates, alternately pushing and pulling the air molecules next to it (Figure 41.15). This repeated alternation of high and low pressure is transmitted to nearby air molecules, spreads outward from the sound source, and forms the stimuli we detect with our ears. What we perceive as the pitch of the sound depends on the frequency of its vibrations; the higher the frequency, the higher the pitch we perceive.

Sound frequency is expressed in units of cycles per second, or **hertz** (Hz). The human ear can detect sounds ranging in frequency from about 16 Hz to about 20,000 Hz. Dogs, by comparison, can hear frequencies up to 40,000 Hz. (That's why dog whistles work; they produce sound within dogs' hearing range but above ours.) Other species' hearing abilities extend even farther; moths and bats, for example, hear frequencies of up to 100,000 Hz. We call sound above our hearing range **ultrasound.**

Near-field sound consists of the actual back-and-forth movement of molecules in the sound path. Near-field effects are rarely important to terrestrial animals because, as its name implies, near-field sound falls off rapidly with distance in air. Although we have shown the movement of air molecules in Figure 41.15, such movements are usually undetectable more than a few inches from the speaker. Just about the only place a human can experience near-field effects is in a disco; if you walk up to a large, floor-mounted loudspeaker playing at very high volume, roll up your sleeve, and put your arm in front of the speaker cone, you will feel the hairs on your arm move back and forth during loud base passages. Aquatic animals often experience near-field effects, however, because even near-field sound carries over long distances in water.

Hearing in Air: The Human Ear

The human ear uses both physical and neural elements to transduce pressure waves in air into sequences of neural impulses (Figure 41.16). The **outer ear,** which is shaped like a funnel, channels pressure waves into the **ear canal,** amplifying them as it does so. At the end of the ear canal, the pressure waves set the **tympanic membrane** (or "eardrum") in motion. The vibrations of the eardrum are transmitted through the **middle ear,** a linked system of three bones (the **malleus** or "hammer," the **incus** or "anvil," and the **stapes** or "stirrup") to the **oval window.** Vibrations of the oval window in turn create pressure waves in the fluid-filled **cochlea** of the **inner ear.**

Within the long, coiled cochlea, vibrations of different frequencies are spread out along the **basilar membrane,** which separates the inner ear's two fluid-filled compartments. On the basilar membrane sits the **organ of Corti,** a structure that contains the **hair cells.** These hair cells are sensory receptors connected to the neurons that leave the ear to form the auditory (cochlear) nerve. As the

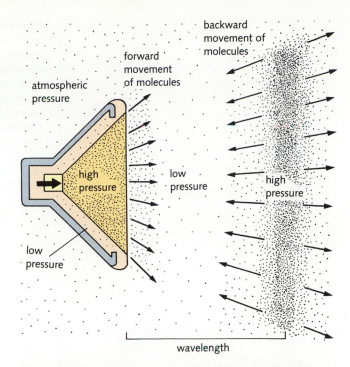

Figure 41.15 The generation of sound by a loudspeaker. Movement of air molecules is shown by arrrows, and areas of high and low pressure are shown by the amount of space between dots.

Figure 41.16 The anatomy of the human ear.

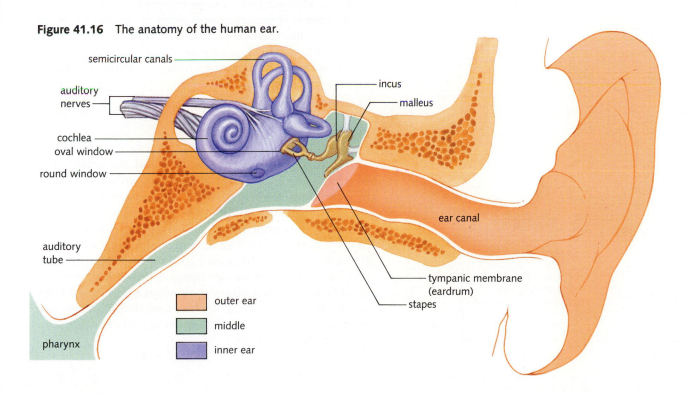

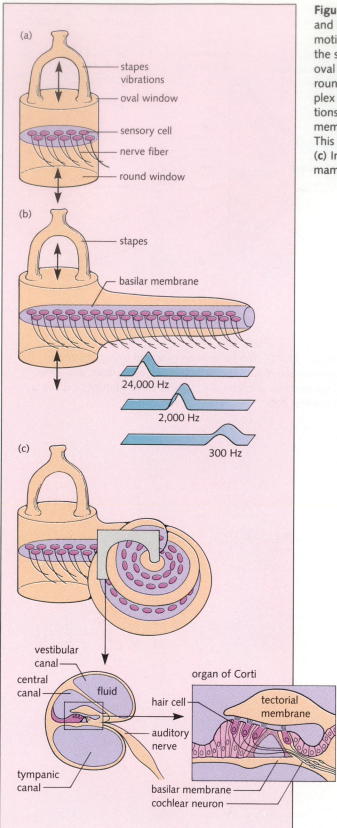

Figure 41.17 These schematics show how the cochlea is constructed and how vibrations of the oval window set the basilar membrane in motion. (**a**) This simplified version of the inner ear, which resembles the simple ears of certain reptiles, shows the relationships between the oval window, basilar membrane, and round window. The flexible round window allows the container to vibrate. (**b**) In this more complex system, the basilar membrane has stretched out. Because vibrations of different frequency are localized along different parts of this membrane, more accurate discrimination among sounds is possible. This development is characteristic of advanced reptiles and birds. (**c**) Inner ears with coiled cochleas, such as this one, are unique to mammals.

basilar membrane vibrates, the hair cells brush against the **tectorial membrane,** bending their sensory hairs as shown in Figure 41.17. The method by which hair cells transduce vibrations into neural impulses is illustrated in Figure 41.3.

Sound in Water: The Lateral Line Sense

In water, near-field sounds are common, although pressure ripples from water currents make it hard to draw a distinction between touch and hearing. Confounding the distinction between ears and other organs are the **lateral lines** of fishes and amphibians—organized collections of hair cells clustered in open-ended canals just beneath the scales or skin surface. These receptors enable aquatic vertebrates to detect minute movements of water relative to their bodies (Figure 41.18). This ability helps schooling fishes maintain their relative positions in their schools and enables many predatory fishes to detect the movements of nearby prey.

Echolocation

Two groups of mammals, the bats and the whales and dolphins, have evolved remarkably sophisticated systems for **echolocation**—a method of using echoes of their own cries to locate and identify prey and to navigate around obstacles. Bats emit pulses of ultrasound, whereas dolphins emit clicks that contain both audible and ultrasonic frequencies. So sophisticated is this echolocation system that bats and dolphins can navigate through mazes and identify inanimate objects even when blindfolded (Figure 41.19). In nature, both bats and dolphins also use their sensitive auditory systems to identify sounds generated by moving prey.

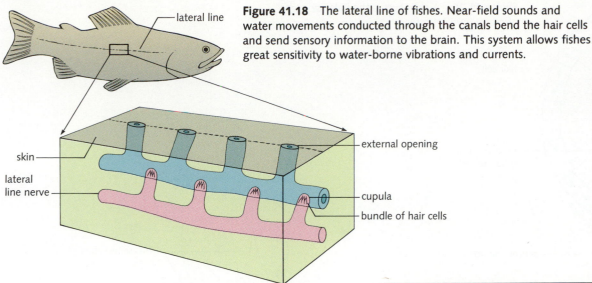

Figure 41.18 The lateral line of fishes. Near-field sounds and water movements conducted through the canals bend the hair cells and send sensory information to the brain. This system allows fishes great sensitivity to water-borne vibrations and currents.

BALANCE AND ACCELERATION

Nearly all organisms, aquatic and terrestrial, need to know which end is up or, more precisely, which end is down; they need to keep track of their body's position relative to the pull of gravity. In all organisms, this function is served by yet another receptor organ using hair cells as the sensory receptors. In many invertebrates, spherical organs called **statocysts** are lined with hair cells and equipped with either sand grains or other small particles. The particles settle to the lowest part of the chamber, where their pressure stimulates the hair cells (Figure 41.20).

The human sense of balance or **equilibrium** is mediated by hair cells located in two fluid-filled chambers within the inner ear, the **utricle** and the **saccule** (Figure 41.21). The cilia of these hair cells are covered with a layer of jelly-like material studded with particles of calcium carbonate. Because the jelly-like mass is denser than the surrounding fluid, it presses down on the hair cells beneath it, exerting pressure that changes whenever the head changes position. Those changes in pressure deform the hair cells' cilia, causing changes in neural activity that are reported to the brain.

The **semicircular canals,** a set of three mutually perpendicular, fluid-filled tubes, enable the brain to monitor precisely any sudden movements of the head. At the base of each canal is a tuft of hair cells; their cilia are embedded in a stiff yet pliable gelatinous mass called the cupula. When the head is moved suddenly, the fluid in the canals

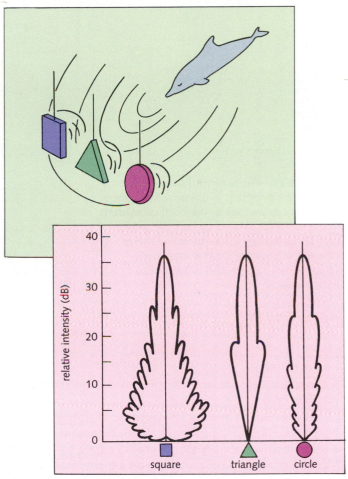

Figure 41.19 Echo patterns from three test objects during discrimination trials with an experimentally blindfolded bottlenose dolphin. Detecting the differences in sound reflection patterns from these objects, the dolphin correctly selected the circular target in 90% of the trials.

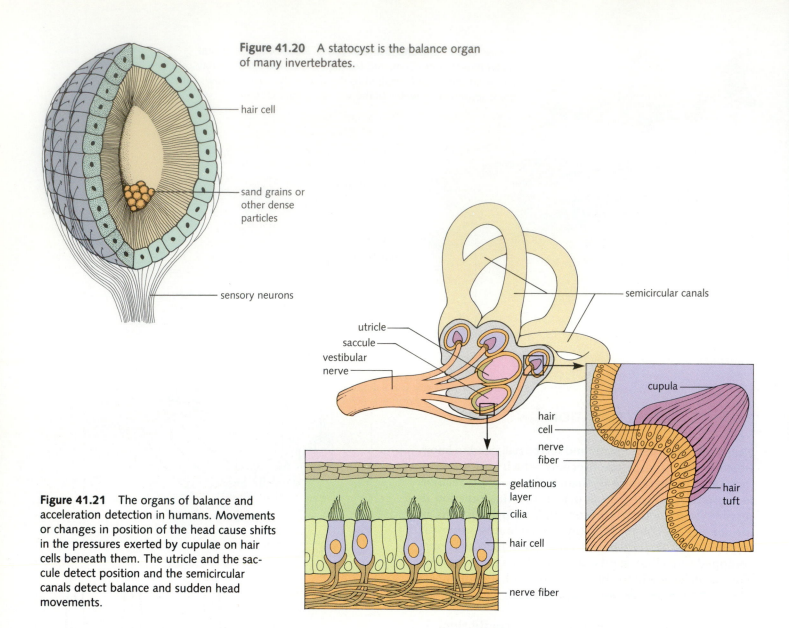

Figure 41.20 A statocyst is the balance organ of many invertebrates.

hair cell

sand grains or other dense particles

sensory neurons

semicircular canals

utricle
saccule
vestibular nerve

cupula

hair cell
nerve fiber

hair tuft

gelatinous layer
cilia
hair cell

nerve fiber

Figure 41.21 The organs of balance and acceleration detection in humans. Movements or changes in position of the head cause shifts in the pressures exerted by cupulae on hair cells beneath them. The utricle and the saccule detect position and the semicircular canals detect balance and sudden head movements.

tends to lag behind, bending the cupula and stimulating the hair cells.

In addition to serving the purpose of preserving balance and equilibrium in the brain, information from the semicircular canals provides sensory input to the *vestibulo-ocular reflex,* a response involving the muscles that position the eyes. The vestibulo-ocular reflex automatically adjusts the position of the eyes when the head is moved suddenly.

These sensory systems constantly gather information about movement and acceleration, whether we want them to or not. All of us have experienced times when we would have preferred to ignore that sensory input—times when the provocative movement of a car, boat, or plane has

led to the intense discomfort of motion sickness. Researchers do not all agree on the specific processes that generate motion sickness, but most implicate the receipt of conflicting or "mismatched" cues about motion from different senses. When you are in the cabin of a tossing ship, for example, your senses of balance and acceleration insist that you are being tossed up and down, but your eyes report that things around you are stable. At other times, the mismatch may be produced by motion that causes the utricle and saccule to generate information that conflicts with reports from the semicircular canals. Large amounts of alcohol can alter the responses of the semicircular canals and can make a moderately bumpy ride in an automobile a great deal more unpleasant.

THE CHEMICAL SENSES

More than a century ago, French naturalist Jean Henri Fabré found his den invaded by male Peacock moths. That morning, he had placed a newly emerged female moth into a wire-gauze cage, and as night fell he discovered "Coming from every direction and apprised I know not how . . . forty lovers eager to pay their respects to the marriageable bride born this morning. . . ." And more than a century before that, Napoleon sent a brief note to Josephine from the battle front. "Ne te lave pas. Je reviens," he wrote; "Don't wash. Coming home."

Both of these behaviors reflect animals' ability to detect and identify a wide variety of chemical substances ranging from individual inorganic ions—such as sodium and chloride—to complex organic compounds—including amino acids, hormones, and other proteins. Chemical senses are vital to animals in many ways; they make it possible to detect and identify potential mates by their odors, to locate prey, and/or to detect predators.

The senses that make these discriminations possible are usually divided into three categories: **olfaction** (smell), **gustation** (taste), and a variety of specialized receptors grouped together as the **common chemical sense.** In order for any of these receptors to detect an atom or compound, it must interact in solution with the receptor cell membrane. For this reason, all chemical sense receptors require a moist environment, which is usually provided by a mixture of liquid and mucus secreted by supporting cells and glands.

All these senses, though based on the actions of different receptors, are linked by neural processing in the CNS in such a way that the sensations they produce interact and overlap. Though served by distinctly different receptors from taste, for example, olfactory stimuli play an important role in determining the flavors of foods we eat. You must have experienced, for example, the major apparent loss of taste that accompanies a head cold. In this situation, your taste buds are functioning perfectly well, but your olfactory receptors are blocked by nasal congestion. Furthermore, the body's internal common chemical sense receptors can influence what odors and tastes animals find attractive. Olfactory centers in the brains of rats are excited by the odor of food when the rats are hungry. Yet if those animals are injected with sugars that satisfy the body's caloric needs, the olfactory centers of their brains show far less activity even when normally attractive food odors are present.

Olfaction

The sense of smell is the animal kingdom's long-distance chemical sense. Olfactory receptors are stimulated by **odors**—minute concentrations of chemicals carried to the receptors through air or water. The sensitivity of some olfactory receptors is extraordinary. Fabré's male moths responded to a sex attractant released by the females from as far away as 11 kilometers. To locate females from that distance, males had to respond to concentrations as low as one molecule of attractant in 10^{15} molecules of air. That's roughly equivalent to being able to taste a single grain of sugar dissolved in an 8-ounce glass of water!

The moths' sex attractant is an example of a **pheromone,** an important class of compounds used in animal communication. As you will learn in Chapter 43, animals ranging from arthropods to primates use pheromones and other odors in urine and glandular secretions to locate and identify mates and relatives and to mark home territories. Among mammals, specific body odors are often vital to bonding between mothers and newborns. Many human mothers, for example, can identify blankets and items of clothing belonging to their infants by odor alone. Within a few weeks of birth, nursing infants can differentiate between their mothers' breasts and those of other lactating women. And although most Americans today feel differently from Napoleon about the sexual attractiveness of natural body odor, our use of perfumes and deodorants testifies to the importance of odor in communication. Curiously, many popular scents include musk, a powerful sex attractant distilled from the body secretions of other mammals.

Olfactory receptors in invertebrates may be located in a variety of places; moths generally carry them on their antennae (Figure 41.22). In vertebrates they are found exclusively in the moist epithelium within the nasal passages (Figure 41.23). Olfactory receptors are stimulated when odor molecules interact with receptor proteins in ways that alter their membrane potentials. Olfactory cells convert those receptor potentials directly into action potentials, which travel through the olfactory nerve to the olfactory bulbs of the central nervous system.

Gustation

Our sense of taste is based on four types of taste receptors that are sensitive to sweet, bitter, salty, and sour compounds. But much as the visual system combines the responses of three types of cones to produce innumerable colors, the chemical sense somehow combines these primary sensations with olfactory clues to register a nearly endless variety of flavors. Several hypotheses attempt to explain the evolution of the four basic tastes, and there are clear advantages to detecting certain components in edible objects. Sweet foods, for example, contain sugar, an easy source of available energy. Bitter tastes are often associated with secondary compounds in plants that may be toxic. And salt is essential to balancing the levels of body ions.

43

Figure 41.22 Olfactory receptors. Each of the hairs on this Australian moth's antennae carries many still finer hairs. Each of those contains the animal's exquisitely sensitive olfactory receptors.

Human taste receptors are grouped into **taste buds** that line grooves in the surface of the tongue (Figure 41.24). Although young humans are equipped with as many as 10,000 taste buds, that number declines to as few as 3000 with age. (This phenomenon may account for the increased appetite of older individuals for salt and certain spices.) Other animals may carry their taste receptors in surprising locations. In arthropods they may be on mouthparts, antennae, or walking legs; in fishes they may be scattered over the entire body surface.

Ever since the Romans discovered the first artificial sweetener nearly 2000 years ago, the food industry has invested a great deal of effort in determining how various molecules stimulate taste receptors. We now know that many compounds interact with sensory cells by fitting into specific receptor sites on the receptor cell membrane. Food chemists or "flavorists," as they are sometimes called, have become adept at synthesizing compounds that "fool" taste buds by mimicking the shapes of natural food components. Such compounds as saccharin and aspartame, for example, are used to stimulate sugar receptors without providing the calories of sugar.

The Common Chemical Sense

Other chemical receptors scattered throughout the body are collectively called the common chemical sense. Some of these produce generalized irritation responses when stimulated by noxious chemicals. Chlorine gas and cigarette smoke, for example, stimulate mucus production and choking responses in the respiratory tract. Other chemical receptors monitor the body's internal condition and report directly to higher neural centers. Chemical receptors within the carotid artery, for instance, monitor oxygen and carbon dioxide concentrations in the blood (Chapter 32), and others monitor blood glucose levels (Chapter 34).

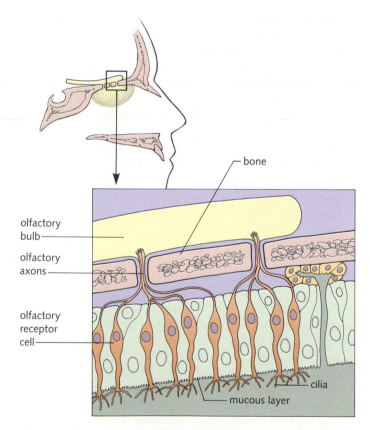

Figure 41.23 The human nasal epithelium carries our olfactory receptors. The cilia of these cells, which must be kept moist, are protected by a layer of mucus.

THE GENERAL SENSES

The general senses, another group of miscellaneous receptors, are often broken into two groups: exteroceptors and interoceptors. **Exteroceptors** monitor external conditions impinging on the body and are responsible for the sensations of pain, warmth, cold, and light touch (Figure 41.25). These receptors initiate appropriate reflexes when stimulated. The stimulation of cold receptors, for example, produces body responses to conserve heat. Pain is reported by a variety of free nerve endings in the skin that respond to mechanical, chemical, or thermal stimuli. Certain injuries remain painful for some time

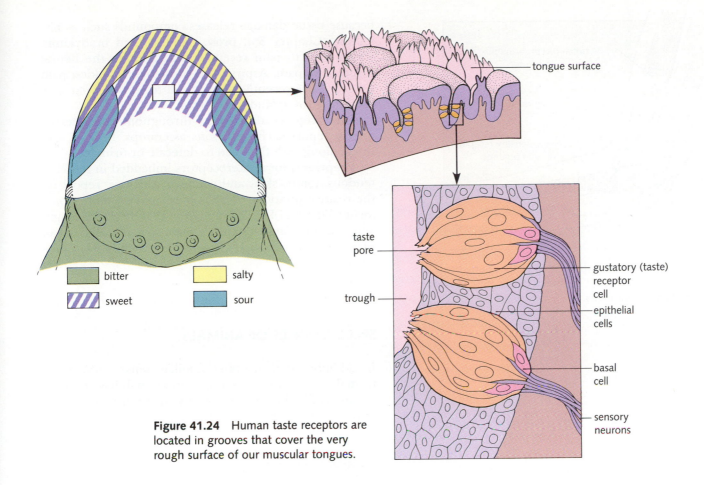

Figure 41.24 Human taste receptors are located in grooves that cover the very rough surface of our muscular tongues.

bitter

salty

sweet

sour

tongue surface

taste pore

trough

gustatory (taste) receptor cell

epithelial cells

basal cell

sensory neurons

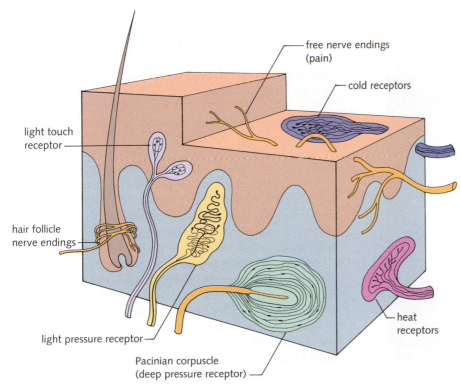

free nerve endings (pain)

cold receptors

light touch receptor

hair follicle nerve endings

light pressure receptor

Pacinian corpuscle (deep pressure receptor)

heat receptors

Figure 41.25 Our skin contains a host of receptors that provide information about our immediate environment. Pacinian corpuscles respond to pressure and touch. The end bulbs of Krause are believed to respond to cold, while the Ruffini corpuscles probably respond to heat. Free nerve endings mediate responses to pain. Hair-follicle nerve endings respond to displacements of body hairs by touch or air movement.

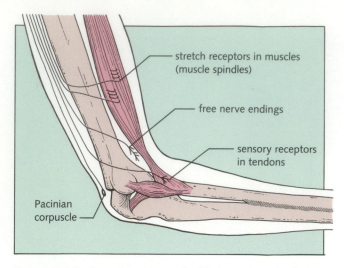

Figure 41.26 Proprioceptors, imbedded in various body tissue, continuously monitor the amount of stretch in our muscles and the relative positions of our joints.

stretch receptors in muscles (muscle spindles)

free nerve endings

sensory receptors in tendons

Pacinian corpuscle

because tissue damage releases compounds such as histamine (Chapter 36), prostaglandins, and bradykinin, which bind to pain receptors and amplify the neural response to pain. Aspirin, one of humanity's oldest pain relievers, has recently been found to work by partially blocking prostaglandin secretion.

Interoceptors distributed throughout the internal organs report such sensations as cramps, hunger, and thirst, along with the need to defecate or urinate.

Proprioceptors, interoceptors imbedded in muscles, tendons, joints, skin, and connective tissues, report on the relative positions of body parts and on their movements (Figure 41.26). *Stretch receptors* that detect the length and tension of muscles and *Pacinian corpuscles* that detect pressure deep in body tissues are well-known proprioceptors.

SPECIAL SENSES OF ANIMALS

In addition to developing familiar senses differently from the way our species has, many animals have evolved senses totally different from any in the human repertoire.

Infrared Detection

A number of snakes—pit vipers, for example—have heat-sensitive pits located on either side of their snouts (Figure 41.27). The *infrared receptors* in these pits assemble, and report to the snake's brain, a thermal map of the environment much like the visual map produced by the visual system. Snakes use this sense to track warm-blooded prey, such as small rodents, in the dark.

Electroreception

Sharks and several other fishes can detect minute electric currents in the water around them. The electroreceptive cells that make this sense possible, the *ampullae of Lorenzini,* are evolutionarily related to hair cells and are concentrated in a series of canals and pits around the animal's snout.

Electrodetection is a very useful ability for salt-water predators, because the essential life processes of most animals generate small electric currents. The exchange of ions across the gills of fishes as they breathe, for instance, generates electric currents, as do the movements of their respiratory muscles. By homing in on these currents, sharks are able to detect even prey that are completely

Figure 41.27 Pit organs, infra-red receptors of snakes, can detect the body heat of their small-mammal prey.

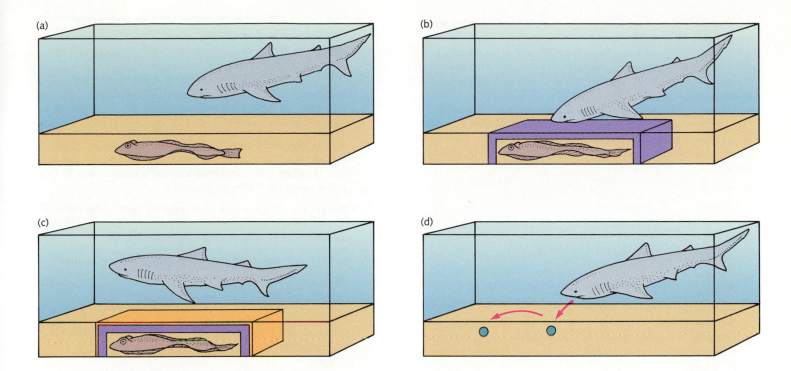

Figure 41.28 (a) A shark easily finds live fish buried out of sight by homing in on the tiny electrical impulses produced by the prey's breathing movements. (b) The shark detects just as easily a fish covered with agar—which blocks scent cues but not electric currents. (c) An agar chamber, which blocks scent cues, covered with an electrically insulating film successfully hides the fish. (d) A shark dives for electrodes that simulate the electric field of a living fish in the absence of any scent cues.

buried in the sand (Figure 41.28). In addition, sharks apparently use these receptors to help them navigate across large stretches of open ocean.

The exquisitely sensitive electrodetecting sense of deep-water sharks has made them a serious hazard to long-distance undersea cables. Those cables need to have amplifiers at regular intervals along their length. Unfortunately, these amplifiers generate small electric currents in the seawater around them that attract sharks and encourage them to attack the cables!

Two other groups of fishes from turbid habitats in Africa and South America generate their own electric fields by using specially modified muscle tissue. Such fishes can hunt, navigate around obstacles, and communicate with one another by detecting changes in these electric fields.

SUMMARY

By responding to natural phenomena, the senses of humans and other animals gather information about the environment. Though the specifics of sensory processing vary among senses, the principles of sensory function are always similar. Specialized neurons called receptor cells transduce such natural phenomena as light energy and sound waves into graded receptor potentials. Receptor potentials are converted—either by receptor cells or by interneurons—into action potentials that are processed by the CNS, where perception occurs. Animals often have abilities very different from ours because of the design of their sensory systems; bees can see ultraviolet light, for example, and dogs can detect ultrasound. Other animals have senses for which there are no human equivalents, such as sharks' ability to detect electric current.

Receptor cells of the human eye, which are stimulated when they absorb photons, fall into two classes: cones, of which there are three types, and rods. Each type of cone is maximally sensitive to light in one part of the visible spectrum. By comparing the signals from these types of cones, our visual system is able to distinguish among objects on the basis of color.

Hearing, the senses of balance and acceleration, and the lateral line sense of fishes and amphibians are all mediated by sense organs based on receptors known as hair cells. Hair cells produce generator potentials when their projecting cilia are deflected.

STUDY FOCUS

After studying this chapter, you should be able to:

- Explain the roles and functions of sensory systems, human and nonhuman.
- Describe how information is processed by using the visual system as an example, and identify the similarities that unite all senses.
- Explain how the physical characteristics of different environments influence the transmission of sensory information.

SELECTED TERMS

receptor potential *p. 837*
basal firing rate *p. 837*
compound eyes *p. 840*
retina *p. 841*
accommodation *p. 842*
rods *p. 842*
cones *p. 842*
visual pigments *p. 844*
sensory adaptation *p. 846*
hertz (Hz) *p. 848*
lateral line *p. 850*
echolocation *p. 850*

statocyst *p. 851*
equilibrium *p. 851*
olfaction *p. 853*
gustation *p. 853*
pheromone *p. 853*
taste bud *p. 854*
exteroceptor *p. 854*
interoceptor *p. 856*

REVIEW

Discussion Questions

1. How do the differences in structure between the human eye and a fish eye reflect the different operational demands of vision in air and water?

2. Explain why the stimuli available to the chemical senses of taste and smell are different for terrestrial and aquatic animals. What sorts of compounds could a lobster "smell" that we can only taste? Why?

3. Choose any single sense and explain how differences in receptor cell and/or sense organ structures among different animals result in different sensory capabilities.

4. Describe one animal sense that humans lack altogether.

Objective Questions   (Answers in Appendix)

5. In the compound eye of arthropods, there are dozens to thousands of complete miniature eyes known as
   (a) foveas.
   (b) rods.
   (c) ommatidia.
   (d) cones.

6. In humans, chemical sense receptors of airborne substances are primarily found in the
   (a) semicircular canals.
   (b) olfactory epithelium.
   (c) nerve endings in the skin.
   (d) oval window.

7. Mechanical vibrations of the air are converted into neural impulses by sensory receptors in the ear called
   (a) hair cells.
   (b) horizontal cells.
   (c) olfactory receptors.
   (d) ear canals.

8. When our ears respond to different _____ of sound waves, we are perceiving the _____ of the sound.
   (a) lengths / frequency
   (b) frequencies / pitch
   (c) pitches / frequency
   (d) lengths / pitch

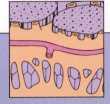

CHAPTER 42

The Musculo-Skeletal System

To move things is all mankind can do, and for such the sole executant is muscle, whether in whispering a syllable or in felling a forest.

—Sir Charles Sherrington

An eagle soars in graceful arcs on an updraft, controlling its glide with subtle adjustments of its wings. A cheetah chases a gazelle, expertly avoiding obstacles and adjusting to irregularities in the terrain beneath his feet as he races along at speeds up to 95 km (60 miles) an hour. A ballerina pirouettes on point, her body exquisitely balanced and positioned. And an athlete, all his muscles toned and working in harmony, sprints with cat-like grace toward a high jump. All these feats—flying, running on four legs, and balancing on a few toes— are accomplishments of the vertebrate *musculo-skeletal system,* the bones and muscles that together enable animals to move efficiently.

THE WONDER OF CONTROLLED MOVEMENT

Muscle tissue shortens when stimulated and by itself can generate the force necessary to pump blood or to move food through the intestines. But to permit movement through the environment, muscular force alone is of limited use. Jellyfish, for example, contract vigorously, but they are hardly high-speed swimmers. An evolutionary step in the right direction was taken by annelid worms, which—though they lack rigid body parts—use hydraulic principles to operate **hydrostatic skeletons** that enable them to crawl and burrow efficiently (Figure 42.1).

But for animals to run, they must have rigid body parts to push against the ground. To fly or swim, they must apply force against air or water. Although muscle can supply the *force* to those body parts, only a skeleton can supply the necessary *support.* One highly successful support system is the external skeleton, or exoskeleton, of arthropods (Figure 42.2; see also Chapter 15). The alternative solution in vertebrates is the **endoskeleton,** a system of internal bones and joints (Figure 42.2). Both of these systems of bones, joints, and muscles are called musculo-skeletal systems, and both strategies have advantages and disadvantages (Table 42.1).

If rigid skeletons were all fashioned from one piece, of course, they would be useless; both endoskeletons and exoskeletons are composed of individual hard parts connected by **joints** that allow them to move relative to one another. Muscles are attached to bones by strong **tendons.** Many joints are actually held together both by ligaments attached to bones on either side of the joint and by muscles and tendons that stretch across the joints and act to stabilize them. As you can see in Figure 42.2, bones and joints act as a system of levers and hinges that translate muscle contraction into body movement.

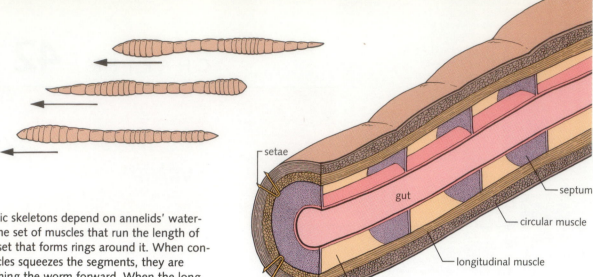

Figure 42.1 Hydrostatic skeletons depend on annelids' watertight body segments, one set of muscles that run the length of the body, and another set that forms rings around it. When contraction of circular muscles squeezes the segments, they are forced to elongate, pushing the worm forward. When the long muscles shorten the segments, they become broader.

setae

gut

septum

circular muscle

longitudinal muscle

coelom

Table 42.1 *Relative Advantages of Exoskeletons and Endoskeletons*

Advantages	Disadvantages
Exoskeleton	
Provides good support for small animals	Too heavy to support large animals
Readily adaptable into wings, flippers, or claws	Must be shed during growth
Allows wide variety of movements	
Protects soft tissues from damage and desiccation	
If broken, can be replaced at next molt	
Endoskeleton	
Protects brain and some internal organs	Exposes soft tissues to mechanical damage and desiccation
Allows for steady growth	If broken, cannot be replaced and must be repaired
Provides good support per unit weight; can carry large animals	

Because muscle tissue can generate force only by *shortening,* it can *relax,* but it cannot forcibly extend itself. In order to move body parts back and forth, therefore, muscles in both insects and vertebrates are arranged in **antagonistic pairs** that pull in opposite directions across skeletal joints. Many complex joints, such as the knee and shoulder joints, are actually operated by **antagonistic groups** of muscles, rather than by single pairs. Around the human elbow joint, for example, the primary (but not the only) muscles involved are the biceps and triceps groups. The biceps, which bends the elbow, acts as a *flexor* muscle. The triceps, which straightens the arm, acts as an *extensor.*

STRUCTURE AND FUNCTION IN MUSCLE TISSUE

As we saw in Chapter 31, there are three basic types of muscle tissues in vertebrates. All three contract in essentially the same way, but under a microscope they can be distinguished from one another by the organization of their cellular parts. (These three muscle types are illustrated in Figure 31.7.)

Skeletal Muscle

Skeletal muscle, primarily responsible for voluntary movement, is also called **striated muscle** because of the alternating light and dark bands, or striations, that are produced by the organization of molecules within its cells. Vertebrate skeletal muscles are controlled by motor neurons through the specialized synapses called neuromuscular junctions that we discussed in Chapter 40. Because these synapses are strictly excitatory, stimulation of motor neurons always leads to muscle contraction. (There is no way, in other words, that a neural command can induce skeletal muscle to relax.)

Skeletal muscle is divided into two subtypes. **Red muscle,** also called *slow-twitch* muscle, can work for long periods without fatigue. These muscles, which obtain their energy primarily through aerobic respiration, contain many mitochondria. The dark coloration of red muscle results from biochemical and structural adaptations to its need for oxygen; these tissues contain a substantial quantity of dark, oxygen-storing myoglobin (Chapter 33) and are richly supplied with blood by dense networks of capillaries. Red muscle is found throughout the human body and forms the "dark meat" in poultry drumsticks and in portions of fish such as tuna. (Chickens and turkeys are primarily ground-dwelling birds, and they normally run more often and for longer periods than they fly. For this reason the legs of these birds contain mostly red muscle. Similarly, red muscles in tuna enable those fish to swim continuously.)

White muscle, also called *fast-twitch* muscle, contracts more rapidly and generates more force than red muscle, but it obtains its energy primarily from anaerobic glycolysis and hence fatigues rapidly. The breast muscles of chickens and turkeys, which operate the seldom-used wings of these ground-dwelling birds, are mostly white muscle.

Smooth Muscle

Smooth muscle, found primarily in and around internal organs, arteries, and the digestive tract, lacks striations and tends to form sheets, rather than the bundles that skeletal muscle forms. Smooth muscle contracts much more slowly than either type of skeletal muscle, but it can maintain contractions for a long time. Both sympathetic and parasympathetic neurons supply smooth muscle, and because these two classes of neurons release different transmitters at their neuromuscular junctions, they have complementary effects. If sympathetic stimulation causes a particular smooth muscle to contract—as it does for most muscles of the circulatory system, for example—parasympathetic stimulation causes it to relax.

Cardiac Muscle

Cardiac muscle, discussed in detail in Chapter 32, has characteristics of both smooth and skeletal muscle. The tight electrical connections among cardiac muscle cells allow excitatory impulses to spread across the heart.

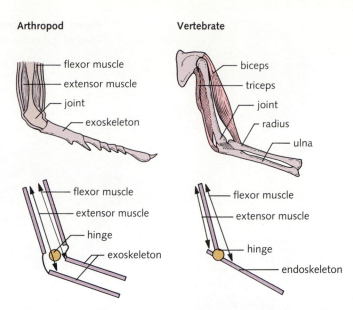

Figure 42.2 Arthropods wear their skeletons on the outside; muscles operate within its protective covering. Vertebrate bones are internal, so they are surrounded by the muscles that move them. The biceps and triceps muscles of the upper arm are antagonists; contraction of the biceps flexes the arm, whereas contraction of the triceps extends it.

The Structure of Skeletal Muscle

We can progressively dissect a vertebrate skeletal muscle into smaller and smaller units to understand how it operates (Figure 42.3). Each whole muscle is formed from numerous parallel bundles surrounded by connective tissue sheaths that attach the muscle to tendons at both ends. Each of those units, in turn, is composed of many still smaller parallel units called **muscle fibers.** Each muscle fiber is functionally a single cell, although numerous nuclei show that each was formed from the fusion of many embryonic cells. (Some of these fused-cell fibers are extremely long; single muscle cells in the long muscles of the leg may be half a meter in length.) Thus the terms "muscle cell" and "muscle fiber" are used interchangeably.

Inside the muscle cell membrane (called the **sarcolemma**) is a still smaller nested set of units called **myofibrils,** each of which is made up of two types of **myofilaments.** *Thin filaments* are twisted strands of three types of proteins: **actin, troponin,** and **tropomyosin.** *Thick filaments* are made from a protein called **myosin.** Many myosin molecules, each of which looks somewhat like a match stick with its head bent sideways, line up in a staggered parallel array to form each thick filament.

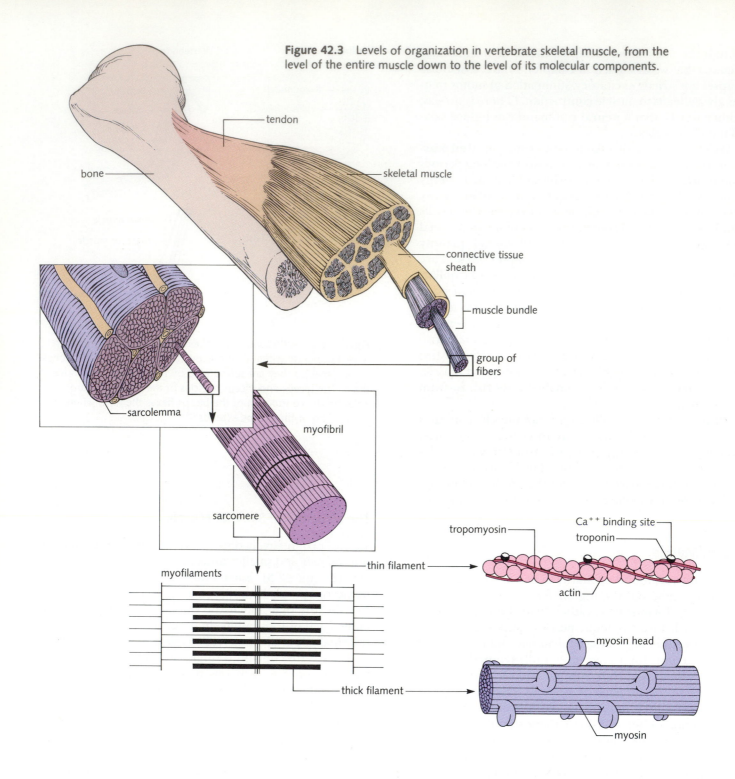

Figure 42.3 Levels of organization in vertebrate skeletal muscle, from the level of the entire muscle down to the level of its molecular components.

Thick and thin filaments are organized into units called **sarcomeres** ("muscle parts"), whose overlapping banded structures create the striations visible in micrographs (Figure 42.4). Sarcomeres are bounded on each end by fibrous structures called *Z lines* to which numerous thin filaments are attached. The Z line and the light area immediately on either side of it consist exclusively of actin filaments and form the *I band*. In the middle of the sarcomere, the myosin filaments that form the *A band* are linked to one another at the *M line*. The center of the A band, which contains only myosin filaments, is called the H band. Thick and thin filaments overlap extensively, the area of overlap becoming the darkest, most dense region of the sarcomere.

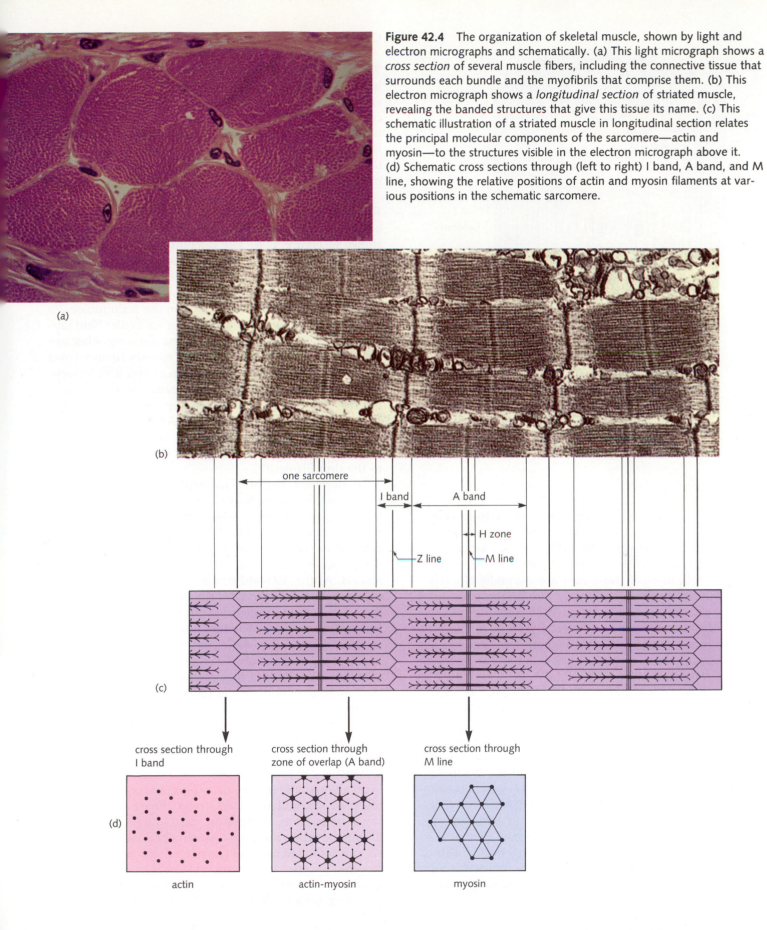

Figure 42.4 The organization of skeletal muscle, shown by light and electron micrographs and schematically. (a) This light micrograph shows a *cross section* of several muscle fibers, including the connective tissue that surrounds each bundle and the myofibrils that comprise them. (b) This electron micrograph shows a *longitudinal section* of striated muscle, revealing the banded structures that give this tissue its name. (c) This schematic illustration of a striated muscle in longitudinal section relates the principal molecular components of the sarcomere—actin and myosin—to the structures visible in the electron micrograph above it. (d) Schematic cross sections through (left to right) I band, A band, and M line, showing the relative positions of actin and myosin filaments at various positions in the schematic sarcomere.

(a)

(b)

one sarcomere

I band

A band

H zone

Z line

M line

(c)

cross section through
I band

cross section through
zone of overlap (A band)

cross section through
M line

(d)

actin

actin-myosin

myosin

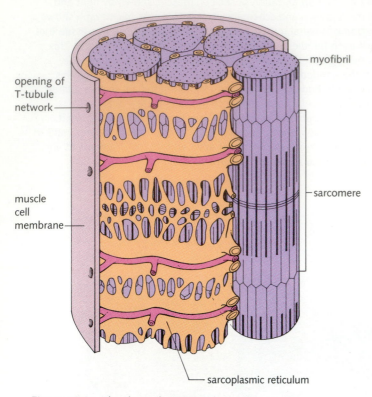

opening of
T-tubule
network

muscle
cell
membrane

myofibril

sarcomere

sarcoplasmic reticulum

Figure 42.5 The three-dimensional structure
of a muscle cell.

When the sarcomere is at rest, the heads of the myosin molecules project toward the thin actin filaments, but most do not touch them. Although there are *active sites* on the thin filaments to which the myosin heads would readily attach, most of those sites are normally blocked by troponin and tropomyosin molecules. In this position, the myosin heads are bound to ADP and phosphate derived from previously hydrolyzed ATP. The energy released in that process has "set" the myosin head like a loaded spring now poised to release that stored energy during contraction.

To understand how muscle contraction works, we must back up a bit and place the sarcomere in context within the muscle cell. As shown in Figure 42.5, the sarcomeres run down the length of the myofibrils within the muscle cell. All myofibrils are surrounded by two networks of tubules, the **sarcoplasmic reticulum** and the **T-tubule** network. The sarcoplasmic reticulum is a closed network of tubes and storage sacs within the cell that collects and stores calcium ions when the muscle is at rest. The T tubules are extensions of the cell membrane that penetrate into the cell and run between the reservoirs of the sarcoplasmic reticulum. The cell membrane itself is an excitable membrane that generates and conducts action potentials just as nerve cells do.

Contraction in Individual Muscle Cells

The stimulus Muscle contraction begins when a nerve impulse reaches the neuromuscular junction. There, enough acetylcholine is released to depolarize the muscle cell membrane and fire off an action potential. That action potential races along the cell membrane and dives down into the cell along the T-tubule network, where it causes the sarcoplasmic reticulum to become "leaky" to calcium ions, which flood out into the myofibrils.

The force generators: sliding filaments The process that occurs next, according to the **sliding filament model,** is shown in Figure 42.6. The calcium ions released by the sarcoplasmic reticulum bind in large numbers to the troponin–tropomyosin complex on the thin filaments. This binding changes the shape of those molecules sufficiently to uncover the active sites on the actin molecules themselves. The heads of the myosin molecules bind rapidly to the closest open active sites, forming what are called **cross bridges** between each myosin filament and the actin filaments around it. As soon as they have formed, these cross bridges release their stored energy by bending toward the center of the sarcomere and pulling the actin molecules along with them. As each myosin head completes this motion, it releases the ADP and phosphate to which it had been attached but remains attached to the thin filament.

If there are any ATP molecules in the vicinity (and there usually are in healthy muscle), the head quickly binds one, hydrolyzes it, releases the active site, and resets to its original "spring-loaded" position. As long as enough calcium ions remain bound to the troponin–tropomyosin complex, this process repeats, as rapidly as 5 times each second. As the 350-odd heads on each myosin filament tug in this fashion, the actin filaments from both sides of the sarcomere are pulled toward the center, shortening the sarcomere. As all the sarcomeres in the entire fiber contract at once, the whole fiber shortens.

Contraction continues for as long as the muscle is stimulated, or until it fatigues (see below). When no further impulses stimulate the cell, the sarcoplasmic reticulum resorbs calcium ions by active transport, the troponin and tropomyosin settle back over the active sites on actin, and the sarcomere rests (lengthens) once again.

Contractions of whole muscles Each muscle cell responds in an all-or-nothing fashion to an action potential. When the fiber is stimulated, it contracts. There is no large or small contraction, because the muscle action potential is always the same. But entire muscles don't work that way; we can hammer nails and stroke a baby with the same muscles because we can control the force our muscles generate. The nervous system exercises that control by adjusting both the *rate* at which individual muscle cells

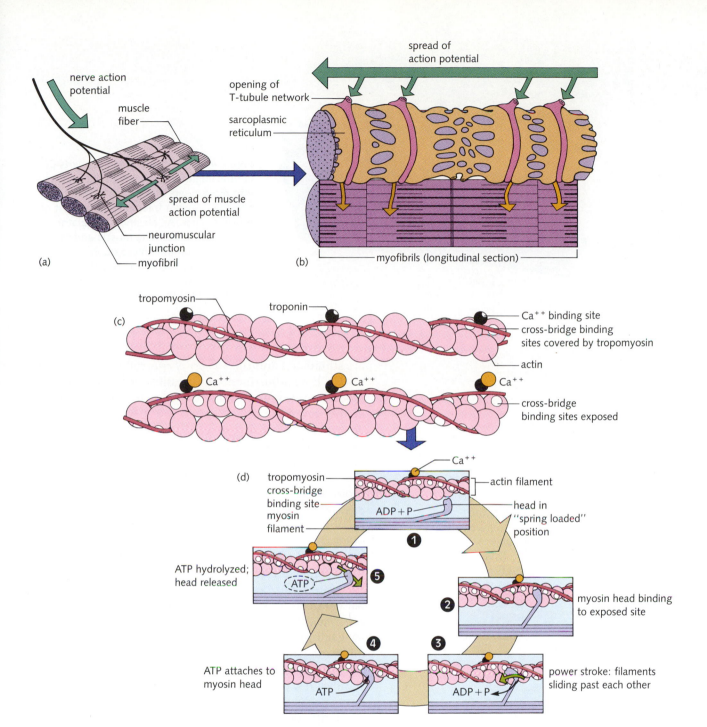

Figure 42.6 Events in muscle contraction. (**a**) An action potential arrives at the neuromuscular junction. (**b**) The muscle cell action potential enters the interior of the muscle fiber via the T-tubule network. (**c**) Ca$^+$ ions attach to the troponin-tropomyosin complex, exposing active sites on the actin filaments. (**d**) The molecular events during muscle contraction according to the sliding filament model. (1) Calcium ions flooding into the sarcomere expose the myosin binding sites on the actin filaments. (2) Myosin heads attach to those binding sites and change shape, (3) pulling the ends of the sarcomere closer together. (4) ATP combines with the myosin heads and (5) causes them to release and reset.

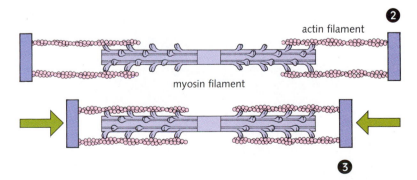

Figure 42.7 A physiograph converts the contraction of an isolated, intact muscle into movements of a magnetically driven pen that leaves a trace on a long chart of moving paper. An electrical stimulator is hooked up to both the muscle and to a second pen on the physiograph. Whenever the stimulator excites the muscle, this second pen leaves a trace on the drum at the exact moment the stimulus is applied.

are stimulated and the *number of cells* called upon to contract. Those adjustments are possible because of the way motor neurons are connected to muscle cells and because of the way those cells respond to repeated stimulation. To explain these phenomena, we use records of intact, isolated muscle activity produced by a device called a **physiograph** (Figure 42.7).

Motor units Each motor neuron branches as it reaches its target muscle and forms synapses on numerous muscle fibers (cells). A single motor neuron and the collection of muscle fibers it serves form a unified, functioning element called a **motor unit** (Figure 42.8). All the cells in a motor unit contract simultaneously whenever that unit is stimulated, but because action potentials do not spread between cells in skeletal muscle, each motor unit can be controlled individually.

Each motor nerve contains many motor neurons, so different patterns of activity among those neurons can stimulate the target muscle to contract to varying degrees. If only a few motor neurons fire, for example, only those muscle cells connected to them are stimulated, and the resulting contraction represents only a fraction of the muscle's total potential strength. As more neurons fire, more fibers are called into action, and the overall contraction of the muscle increases in intensity.

We can see this by conducting an experiment that directly stimulates a muscle with electrical pulses, as shown in Figure 42.9. If the stimulus is small enough, nothing happens, but we can slowly increase the strength of the stimulus until we activate a few muscle fibers near the stimulating electrodes. When those few fibers fire, they contract and generate the first small contraction, or **muscle twitch,** shown in Figure 42.9. This simulates what

Figure 42.8 Muscle innervation and motor units. Each motor neuron, together with the several muscle fibers it innervates, can function independently, and is called a motor unit.

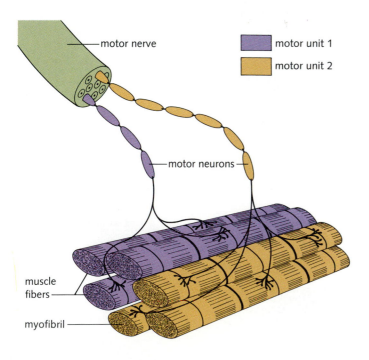

motor nerve

motor unit 1

motor unit 2

motor neurons

muscle fibers

myofibril

happens when only a few motor units are stimulated. If we increase the stimulus intensity still further, we fire more fibers, and the muscle contracts more strongly. The stronger we make the stimulus, the more fibers fire, and the stronger the contraction, up to a certain maximum.

Muscles that perform very different tasks in the body differ both in size and in the "wiring" of their motor units. In some of the body's smallest muscles—such as those that control the precision movements of the eye—each motor unit contains only a few muscle cells. This allows for extremely accurate control of muscle activity. In other, much larger and more powerful muscles—such as those of the thigh—each motor unit consists of many muscle cells. Though this wiring pattern results in coarser control, it is an efficient way to serve the hundreds of thousands of fibers in large muscles.

Single twitch, summation, and tetanus

If we speed the movement of the paper chart on our apparatus, we can see clearly the components of a single twitch (Figure 42.10). Each twitch has a *latency period*, a *contraction time*, and a *relaxation time*. Fast-twitch muscle and slow-twitch muscle respond differently to stimulation (Figure 42.10).

So far in our demonstrations, we have always waited until the end of the muscle's relaxation time before delivering the next stimulus. Under these conditions, identical stimuli produce identical twitches (Figure 42.11a). If we deliver identical stimuli rapidly enough to overlap the relaxation period, however, the muscle responds by contracting more strongly, as shown in Figure 42.11b. This increase in response to closely spaced stimuli is called **summation.** By stimulating the muscle more and more rapidly (represented by more closely spaced stimuli in the figure), we can increase the amount of summation until all the responses fuse into the steady, powerful contraction called **tetanus** (Figure 42.11c).

Thus the central nervous system can adjust the strength and extent of muscle contraction both by controlling the *number of motor units* firing at any given time and by controlling *the rate at which each motor unit is stimulated.* Even in muscles that are not actively involved in exercise, a certain proportion of motor units are always stimulated. This basal level of stimulation produces a resting tension called **muscle tone** that helps maintain body posture and muscular readiness for action.

Muscle Physiology: The Power Behind the Force

Powering muscle contractions over extended periods requires more energy than can be stored effectively in muscle in the form of ATP. Rather, muscles' ATP stores are replenished from several sources during and after

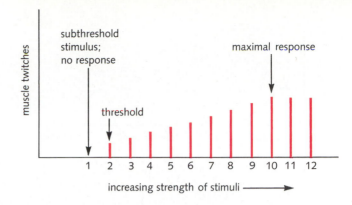

Figure 42.9 These traces show the response of an isolated, intact muscle to stimuli of progressively higher intensity. Note that the muscle twitch reaches a certain maximum, beyond which increasing stimulus intensity has no further effect.

Figure 42.10 These traces of single twitches in fast (**a**) and slow (**b**) muscle show the differences in their response speed. The latency period is the time between stimulation and the beginning of the twitch. The relaxation time is the time it takes the muscle to return to its resting length after contraction.

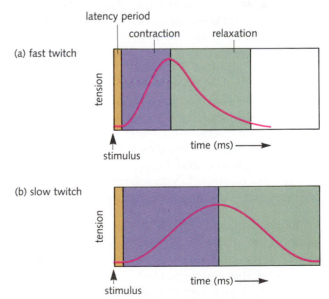

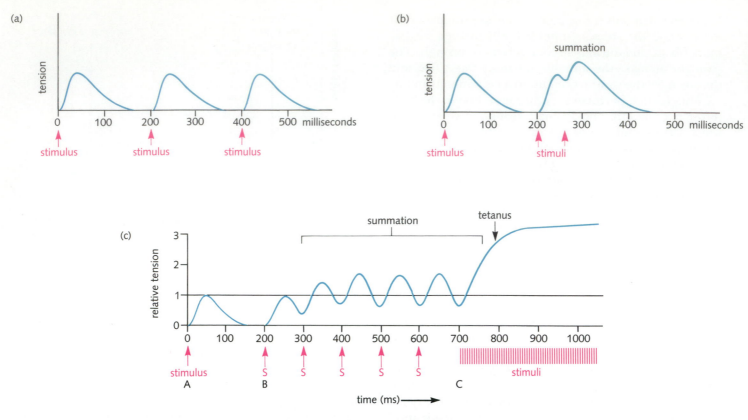

Figure 42.11 Summation in muscle preparation. **(a)** Identical, widely spaced stimuli produce identical twitches. **(b)** When the same stimuli are spaced so that the second occurs during the relaxation period of the first twitch, the second twitch generates more force. **(c)** When many stimuli are closely spaced, the resulting twitches fuse and increase in intensity, producing a response called a tetanus.

contraction. Most of the quickly available energy in vertebrate muscle is stored in **creatine phosphate,** a compound that, like ATP, contains a high-energy phosphate bond. Both the energy and the phosphate can be rapidly transferred to ADP during muscle contraction, maintaining the concentration of ATP during brief periods of muscle activity.

If contractions continue for very long, however, the supply of creatine phosphate runs low, and the muscle tissue must obtain energy from other sources. Red muscle, as we noted earlier, relies mostly on oxidative phosphorylation of glucose and fatty acids. This is a very energy-efficient pathway, as you will recall from Chapter 25, but it requires substantial amounts of oxygen and is relatively slow. (It takes 2 to 3 minutes for a muscle to "gear up" to produce maximal power aerobically. ATP production from anaerobic glycolysis, on the other hand, can reach maximum power in less than 5 seconds.) Accordingly, if the muscle is worked beyond a certain point, it switches pathways and forms ATP without the use of oxygen through glycolysis.

Fatigue But muscle power has its limits. If we stimulate our muscle to tetanus for a long time, at some point the muscle response begins to fall off (Figure 42.12). This failure to sustain contraction is called **muscle fatigue.** (Not to be confused with mental fatigue, this term refers strictly to impaired muscle performance.) The full story of why muscles in intact animals fatigue is complex and is still not completely understood, despite decades of intensive research.

It was once believed that muscles fatigued when they ran out of ATP, but it is not so simple. When muscle *really* runs out of ATP, the myosin cross bridges cannot disconnect from actin, and the muscle is bound in a rigid state called **rigor.** The only time this normally happens is after death, when all the body's muscles stiffen in **rigor mortis.**

Most muscles, when overworked, fail to contract long before ATP is completely exhausted. This failure is at least partially due to a buildup of lactic acid, the end product of glycolysis. Lactic acid buildup in muscle tissue lowers pH (increases the H^+ concentration) within the muscle tissue, producing the short-term muscle pain familiar to anyone who exercises strenuously. There is also evidence that the increase in H^+ ions inhibits two enzymes in the glycolytic pathway and interferes with the activation of thin filaments by calcium. At the same time, repeated or constant contractions may cause a buildup of potassium ions around the muscle cells, altering the membrane potential and decreasing the sensitivity of the neuromuscular junction.

This process explains these symptoms during and just after intense exercise, but it cannot account for fatigue experienced after brief workouts. Nor can it explain the weakness, stiffness, and muscle aches experienced long after unusually heavy or prolonged activity. It appears that muscular strength and endurance are related, not only to structures and biochemical events within muscles themselves, but also to the state of the blood supply to individual muscles, to the overall cardiovascular fitness of the individual, and to both long-term and short-term intake of food in various forms. We will return to these issues when we discuss endurance and the effects of exercise.

THE SKELETAL SYSTEM: LEVERS AND HINGES

The skeletal systems of vertebrates are composed of bones and joints that form moveable and adaptable backbones, jaws, limbs, and even the grasping tails of some primates. In physical terms, bones act as levers that apply the force generated by skeletal muscles to other bones and to the environment. Our leg bones, for example, apply force to the ground, enabling us to stand, walk, or run, whereas the bones in birds' forelimbs apply the power of their breast muscles to the air, enabling them to fly. So closely linked are skeletal elements, the muscles that power them, and the functions they perform that paleontologists can often reconstruct both the muscles of extinct animals and their habits by studying the areas of fossil bones to which muscles and tendons were once attached.

In vertebrates, the skeleton is divided into axial and appendicular portions.

Axial Skeleton

The **axial skeleton** (Figure 42.13) is composed of the central supporting elements, including the skull, the vertebral column (backbone), the ribs, and the sternum (breast bone). Most of the bones that form the skull are joined by immobile joints called *sutures,* although a few, such as the bones of the lower jaw, can move. The skull rests on the upper two bones of the vertebral column, the *atlas* and the *axis,* which allow the head to nod up and down

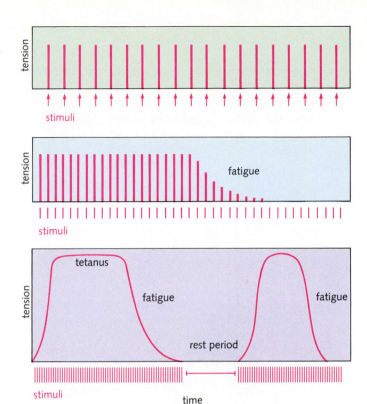

Figure 42.12 When muscles are stimulated to tetanus for prolonged periods, response falls off as muscle fatigue sets in.

and to swivel from side to side. The bones of the vertebral column are firmly but flexibly connected to each other by ligaments and cushioned by pads called intervertebral disks.

Appendicular Skeleton

The **appendicular skeleton** consists of the arm and leg bones, together with the **pectoral** and **pelvic** girdles that attach them to the axial skeleton. The pectoral girdle, composed of the paired **clavicles** ("collar bones") and **scapulae** (shoulder blades), are held together and to the head of the **humerus** (upper arm bone) by tendons and ligaments to form the shoulder joint. The clavicles are firmly attached to the breastbone in front; the scapulae are hung in back by ligaments, tendons, and muscles.

The bones of the forearm, the **radius** and **ulna,** are arranged such that the radius swivels at the elbow and the ulna swivels at the wrist. This allows us to turn our arms and position our hands. The wrist bones and hand

Exercise and Skeletal Muscles

Adult muscle cells cannot divide, but they can change in length and diameter under the influence of exercise. In some way not yet understood, repeated stimulation of muscles under heavy load close to tetanus alters—possibly through tearing—the fine structure of myofibrils. This process, which is thought to play a role in delayed muscle soreness, stimulates regenerating muscle fibers to increase in diameter, producing the muscle growth (hypertrophy) seen in athletes. Disuse of muscles, on the other hand, leads to a decrease in fiber diameter (atrophy).

Increased endurance under training involves both physical and chemical changes within muscle fibers, changes in the gross structure of muscles, and changes in the cardiovascular system. Endurance training not only improves the energy reserves of muscles but also minimizes fatigue by stimulating growth in the muscles' local blood supply, by improving the muscles' ability to extract oxygen from blood, and by increasing both heart volume and lung efficiency.

It is well known that weight lifters and sprinters rely on fast-twitch fibers for quick power, and marathon runners depend on slow-twitch fibers for endurance. Fiber proportions vary, however, even among different classes of runners. Olympic-class marathoners' muscles are 80–90 percent slow-twitch fibers; sprinters' muscles can be up to 70 percent fast-twitch.

Marathoners reap two benefits from their predominantly slow-twitch muscles. On the one hand, they gain the greater long-term muscle power of predominantly slow-twitch fibers. But they may also enjoy greater freedom from muscle pain, because it is the anaerobic, fast-twitch fibers that overload with lactic acid.

Genetic factors strongly influence the relative numbers of fast- and slow-twitch muscle fibers an individual develops early in life. It is not clear, however, whether exercises chosen to prepare for different sports can convert one existing fiber type into another or whether world-class athletes stumble (or are coached) into the events best suited to their genetically determined muscle composition.

In an effort to clarify this situation, researchers in Stockholm recently documented changes in the muscles of an extraordinary 46-year-old man who ran more than 3500 kilometers in only 7 weeks. With the athlete's permission, researchers removed and examined small samples of muscle tissue before and after his feat. These samples confirmed that his muscles were composed mostly of the slow-twitch fibers at the beginning

of the run. But during the 2-month run, his muscles changed dramatically; fiber shapes were altered, and he developed even higher percentages of slow-twitch muscle.

These changes apparently did not occur because of the sort of fiber damage seen in weight lifting, but because of insufficient blood supply during the run. However, the researchers could still not say for certain whether pre-existing fast-twitch fibers had converted to slow-twitch, or whether those fast-twitch fibers had simply died from lack of oxygen, to be replaced by regenerating slow-twitch fibers.

Many athletes prepare for marathons and other endurance events by spending the two days preceding the event alternating exercise training with the intake of massive meals of high carbohydrate foods. This strategy is known as *carbohydrate loading.* Although there are no clear-cut, quantitative data on the actual benefits athletes derive from this procedure, it is based on chemical events understood at the molecular level.

Recall that the energy for prolonged muscular work is largely provided by glycogen stored in muscle tissue. On the average, human muscles store about 1.5 grams of glycogen per 100 grams of muscle. It has been shown, however, that muscle glycogen stores can be increased to as much as 4 or 5 grams per 100 grams of muscle through dietary control. Stored glycogen is increased the most by eating only protein and fat for several days and then gorging on carbohydrates for two days prior to the athletic event. Because creatine phosphate levels are not affected by this procedure, however, it probably does not improve performance in short-term activities such as sprinting.

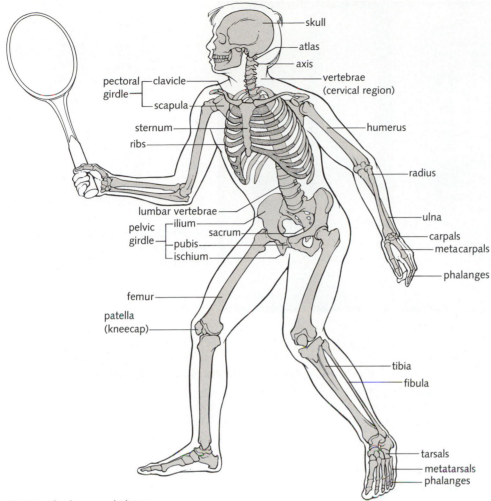

Figure 42.13 The human skeleton.

bones (the **carpals** and **metacarpals**) and finger bones (**phalanges**) form the versatile human hand.

The pelvic girdle (hip bone) is created by the fusion of paired bones, the **ilium,** the **ischium,** and the **pubis.** This assembly, in turn, is tied firmly to the **sacrum** at the base of the vertebral column by a network of ligaments. The largest bone in the body, the **femur** (thigh bone), is connected to the **tibia** (shin bone) at the knee joint, one of the most complex and important joints. The much smaller **fibula** runs alongside the tibia. The ankle and foot bones (**tarsals** and **metatarsals**) are in turn attached to the small **phalanges** of the toes.

Cartilage

Cartilage is a dense, fibrous connective tissue (Chapter 31) that absorbs shocks and provides support for body parts that don't carry much weight. The skeleton of most vertebrates is predominantly cartilage at birth, but bone gradually replaces the cartilage as the animal grows (Figure 42.14). During growth, cartilage becomes restricted to areas called *epiphysial plates* at either end of the bones. In adults, epiphysial plates disappear, and cartilage remains only in the nose, ears, larynx, trachea, where it serves for support, and around and inside joints, where it cushions the impact of movement and aids in lubricating bone surfaces as they slide past one another.

Bone

Bone is a complex tissue in which living cells and long, twisted collagen fibers are supported by crystals of *hydroxyapatite,* a mineral formed from calcium, phosphate, and water. Like steel bars imbedded in reinforced concrete, this conglomerate of flexible and rigid elements gives bone a remarkable combination of strength,

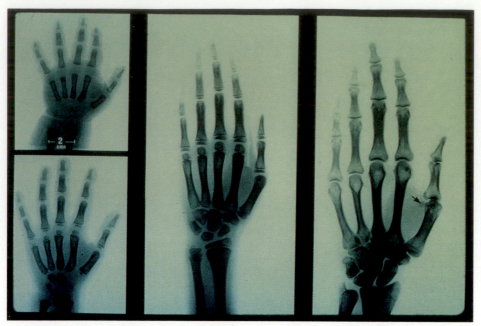

Figure 42.14 The first X-ray in this series shows just how much of the skeleton of a two-year-old boy is still cartilage. Note how the lighter cartilage is replaced by bone at each stage from 2 years, to 3 years, 14 years, and 60 years.

rigidity, and resistance to impact that exceeds the characteristics of any of its components alone. Compact bone can take nearly as much stress as cast iron, though bone weighs only one-third as much. (The breaking stress of bone is 15.5 metric tons per square inch, compared with 18 metric tons per square inch for iron.) This strength is necessary because during many normal activities, bones and joints are subjected to much more stress than you might expect. Sprinting, for example, subjects leg bones to forces equal to nearly five times the runner's weight.

Most bones contain several distinctly different types of bone tissue, as shown in a longitudinal section of a young, long bone from the arm or leg (Figure 42.15a). The outer bone shaft is composed of dense **compact bone,** and most of the interior is made up of less dense **spongy bone.** At either end, where the bone forms part of a moveable joint, a network of **trabeculae** is arranged like the supporting elements in a bridge. Trabeculae transmit stress applied to the bone ends down onto the compact bone along the shaft (Figure 42.15b). In the center of long bones is a soft tissue called **bone marrow,** the source of the cells that ultimately give rise to both red blood cells and white blood cells of all types (Chapter 32).

The structure of compact bone Mature compact bone consists of many roughly cylindrical units called **Haversian systems.** At the center of each is a tube called the *Haversian canal,* which contains a bundle of blood vessels and nerves and is surrounded by concentric rings of bone tissue. Within those rings, spaces or *lacunae* house living bone cells (Figure 42.15c).

Bone growth and remodeling Although we tend to think of bones as inert (like fingernails and hair), bone is a dynamic, living tissue that grows, remodels itself, and repairs itself. Active skeletal growth begins during embryonic life and continues through roughly age 25. This growth is under the control of three different factors: the action of growth hormone during early life, the combination of stresses that exercise and gravity apply to the skeleton on a daily basis, and the levels of circulating calcium in the bloodstream, which are determined by diet and general body physiology.

During embryological development, nearly the entire skeleton of most vertebrates is present in a sort of "scale model" made out of cartilage. Each part of that model grows, but as new cartilage is added, the older cartilage is degraded and replaced by bone (Figure 42.16).

The rigid bone matrix is first laid down by bone cells called **osteoblasts** near the bone surface. Some of these cells become imbedded in the bone and mature into **osteocytes** that continue to live within the bone matrix. These cells together control the formation of compact and spongy bone that replaces the original cartilage model, beginning in the center of the bone shaft and continuing toward the ends.

Near each end of the bone, the cartilage is organized into the epiphyseal plates, which separate the growing bone shaft from the very end of the bone. In the epiphyseal plates, new cartilage is continually produced and steadily replaced by new bone tissue. At the same time, the addition of bone by cells around the outside of the bone enables it to grow in girth. As bone growth nears

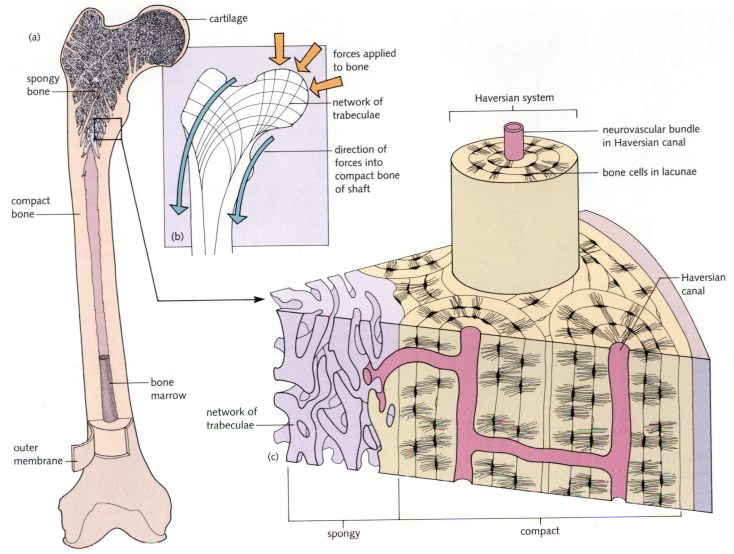

Figure 42.15 Anatomy of a human long bone. (**a**) In this cross section of long bone, notice the relative positions of compact bone, spongy bone, and marrow. (**b**) The trabeculae that comprise the spongy bone at the top of the femur channel the combination of vertical and sideways forces on the hip joint into mostly vertical force onto the shaft. This allows the compact bone to absorb forces that would snap it if applied laterally. (**c**) This schematic of bone fine structure shows the location of living bone cells and blood vessels within the haversian systems of compact bone.

completion, cartilage production in the epiphyseal plates slows down and finally stops, and the cartilage is completely replaced by bone, except within the joints.

Bone growth, however, is not a one-way process. Cells called **osteoclasts,** derived from macrophage cells in the bloodstream, can literally tunnel through bone by dissolving hydroxyapatite and releasing calcium and phosphate into the blood. Osteoblasts, osteocytes, and osteoclasts establish a dynamic balance of bone formation and destruction.

Day-to-day activity and the pull of gravity produce physical stresses essential to maintaining normal bone structure, although the mechanisms that control this response are not understood. Bone, particularly in young children, can change markedly in response to changes in forces applied to them, so any unusual, long-term changes in stress on growing bones can dramatically change their final shape. (That's why you were told to sit up straight when you were a child; chronically bad posture can result in permanent deformation of the vertebral column.) When

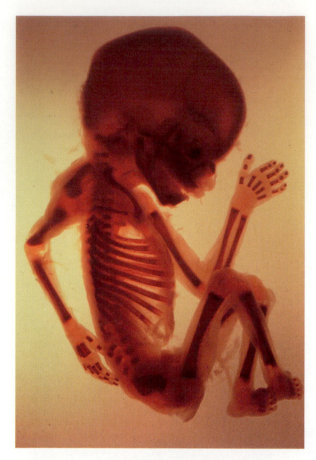

Figure 42.16 Early centers of bone growth in the cartilage of a human fetus about three months old.

stress on bones is removed altogether, as it is during weightlessness in space travel, osteoclasts begin to dissolve bone minerals. Even in young astronauts at peak physical condition, prolonged weightlessness triggers bone resorption, though the balance is rapidly restored upon the individual's return to earth. The ability of bone cells to remodel the matrix around them also allows broken bones to heal.

Additionally, as we noted in Chapter 36, bone serves as a major body reservoir for calcium. Should calcium levels fall, parathyroid hormone stimulates osteoclasts to dissolve bone; when calcium levels rise, calcium is redeposited by osteoblasts via the action of the hormone calcitonin.

Changes in the balance between bone production and destruction generally cause bone mass to decline slowly after age 20 to 30. In some elderly individuals, particularly women, mineral loss can so weaken bones that they break under minimal stress. This syndrome, called **osteoporosis,** has several physiological causes, but it is also related to decreased activity level. In numerous clinics around the country, regular, supervised, moderate exercise in aging men and women has been shown to slow bone loss significantly, probably by applying beneficial stress to the skeleton.

Joint Structure and Function

Bones meet at **joints,** whose structure determines the nature and extent of possible movement. Joints are just as important to movement as muscle and bone are; if you have ever injured one of your joints, you know that we usually take their remarkable structures and functions for granted. There are three major classes of joints: *fibrous* or "fixed" joints, *cartilaginous* or "slightly moveable" joints, and *synovial* or "freely moveable" joints.

The most important type of fibrous joint is the *suture,* an immobile joint found only among the bones of the skull. At birth, while the skull is still partly cartilage, the sutures between bones are slightly flexible. (This property enables the fetus's head to flex as it squeezes through the birth canal.) During the 18 months following birth, however, the skull solidifies and the edges of the bones become as irregular as pieces of a jigsaw puzzle, interlocking with one another, ensuring stability. After bone growth stops, the membrane lining the suture is replaced by bone.

Cartilaginous joints are common throughout the axial skeleton, and they range from the tight, scarcely moveable joints between ribs and sternum to the more flexible joints between the vertebrae in the spinal column. The ribs, though firmly attached to the vertebrae and sternum by ligaments and cartilage, can move slightly during breathing, and they can be pried apart by surgeons who must gain access to the chest cavity. The cartilaginous joints of the spinal column, designed for support and resilience rather than extensive movement, can be damaged if bent too far under pressure. Lifting heavy weights with the back bent, for example, can damage the cartilaginous intervertebral disks, causing a painful condition known as a herniated disk.

Synovial joints are remarkable both for their strength and for the freedom and efficiency of movement they allow. Synovial joints consist of a combination of tendons and ligaments that form a fibrous *joint capsule,* which helps hold the bones together. Inside the capsule, bone surfaces are lined with resilient cartilage and "oiled" by an extraordinarily effective natural lubricant called *synovial fluid.*

Synovial joints vary enormously in structure. The shoulder joint, whose capsule is reinforced by tendons of the many muscles that run across it, allows exceptional range and freedom of movement. The knee joint, one of the most complicated and critical joints in the body, not only allows a great range of motion but also absorbs powerful shocks during such exercises as jumping and running (Figure 42.17).

Flattened sacs called *bursae,* also filled with synovial fluid, often occur to ease friction in places where muscle, tendons, or ligaments rub against each other or against bones.

Joints are subject to several kinds of injuries and diseases. **Bursitis,** an inflammation of the bursa, and **arthritis,** a more general set of problems, are common, both in young athletes who push themselves too far in training and in elderly individuals who undertake too little exercise. In both of these cases, moderate exercise can often eliminate discomfort while maintaining joint flexibility. More serious types of arthritis result from genetic defects, acquired disease (such as untreated gonorrhea or Lyme disease), or malfunctions of the immune system.

MUSCLES AND BONES TOGETHER: A DYNAMIC SYSTEM

Neural control of movement is a complex phenomenon that is only partially understood. As you have already learned, conscious movement is initiated in the motor areas of the cerebral cortex, whereas coordination involves the cerebellum and relay areas in the brainstem (Figure 42.18). Thus, although what we call *voluntary* movement is initiated consciously, the carrying out of that movement is accomplished only through a great deal of unconscious, or involuntary, central nervous system activity.

Once movements are started, feedback from stretch receptors in muscle, and from other receptors in tendons, joints, skin, and inner ears, provides constantly changing information about body position, acceleration, and balance. Some of this information we are conscious of; other parts of it are monitored subconsciously. When we learn any task involving movement, we are modifying neural pathways in a manner similar to the more abstract processes of learning. Whether the task involved is walking, writing, driving a car, or skiing, what were once conscious activities become more and more automatic—and hence more efficient.

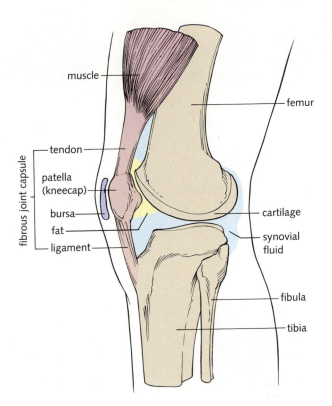

Figure 42.17 The knee joint, which absorbs forces several times body weight during running, is reinforced both internally and externally by ligaments and tendons, and is cushioned by a fluid-filled sac and pads of resilient fat.

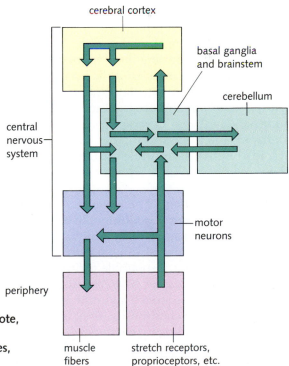

Figure 42.18 Neural pathways involved in body movements. Note, even in this highly simplified schematic, that coordinated movement requires a constant exchange of information among muscles, cortex, cerebellum, and other relay centers within the brain.

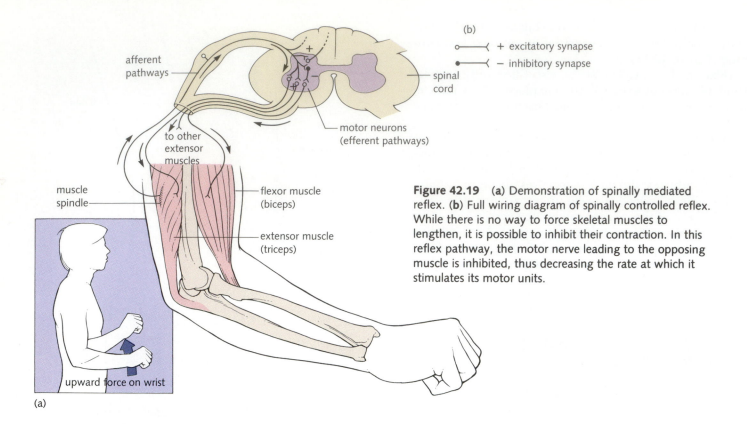

afferent
pathways

o━┤ + excitatory synapse

●━┥ – inhibitory synapse

spinal
cord

to other
extensor
muscles

motor neurons
(efferent pathways)

muscle
spindle

flexor muscle
(biceps)

extensor muscle
(triceps)

upward force on wrist

(a)

Figure 42.19 (a) Demonstration of spinally mediated reflex. (b) Full wiring diagram of spinally controlled reflex. While there is no way to force skeletal muscles to lengthen, it is possible to inhibit their contraction. In this reflex pathway, the motor nerve leading to the opposing muscle is inhibited, thus decreasing the rate at which it stimulates its motor units.

Coordination of Muscular Activity

The task of coordinating muscular activity is made somewhat easier by the fact that many complex motor patterns are "hard-wired" into the spinal cord and operate without any input from the brain. Try the following demonstration: Have a friend hold his or her arm at a right angle, with all the muscles of the upper arm tensed rigidly in place (Figure 42.19a). Note that both biceps and triceps stand out and are rigid to the touch. Now place your hand at the point shown by the arrow in the figure and apply force upwards without warning. Note that your friend's biceps relaxes while the triceps contracts to hold the arm in place.

Reciprocal inhibition The demonstration above is an example of **reciprocal inhibition,** in which stretching an extensor muscle (the triceps) inhibits the flexor muscle (the biceps). Reciprocal inhibition is controlled by the same sort of circuit responsible for the knee-jerk reflex we examined in Chapter 40. Recall that the sensory neurons that detect the stretch and stimulate motor neurons leading to the stretched muscle simultaneously inhibit the motor neuron leading to that muscle's antagonist (Figure 42.19b).

This sort of circuit, which works at most of the body's joints, helps keep antagonistic muscles and muscle groups working with each other smoothly. When someone hands you an unexpectedly heavy object, for example, your arms respond to the extra load before your brain realizes what has happened. Similarly, if you trip or stub your toe on an unseen object while walking, reflex leg movements help you keep your balance.

Walking: A Complex "Simple" Activity

Watching a baby trying to toddle across a room suggests just how complicated the "simple" act of walking really is. Most of us don't remember, of course, but it took us a good year of regular practice to walk unaided on two limbs instead of four. This is because walking is controlled by scores of muscles organized into several complementary groups that stretch from our toes all the way up to the center of our bodies. The contractions of all these elements must follow one another like clockwork. Those movements eventually become second nature; we can walk—and even run—while listening to music, talking, and thinking of other things.

876 PART 7 *Animal Systems*

The musculo-skeletal system that enables us to move efficiently is composed of several important components. Muscles generate contractile power by hydrolysing ATP. Bones provide rigid structural supports that apply the force generated by muscles to the environment, allowing the animal to run, swim, or fly. Joints between bones permit the skeleton to move. Tendons and ligaments control the application of muscular force within the body and, together with the shapes of joint surfaces, determine the range of possible movements.

Muscles, like nerves, are composed of excitable cells that conduct action potentials. According to the sliding filament theory of muscle contraction, action potentials in the muscle cell membrane enter the cell interior through a network of T tubules. There they stimulate the release of calcium from storage sites in the sarcoplasmic reticulum. Through a series of events, this calcium causes actin and myosin filaments to slide past one another, shortening the muscle. Both the structure and the physiology of muscles can be altered through exercise.

Bone, though largely composed of crystalline calcium salts, is an active and dynamic tissue. Living bone cells can lay down or resorb bone throughout most of an individual's life, reshaping the skeleton in response to applied stress. Bone also serves as an important reservoir of calcium for the body.

STUDY FOCUS

After studying this chapter, you should be able to:

■ Explain the structure and function of muscle and describe the process of skeletal muscle contraction.
■ Describe the major components of the skeletal system and explain how bone is a dynamic living tissue.
■ Explain the interrelationship between muscles and bones.

SELECTED TERMS

endoskeleton *p. 859*
sarcolemma *p. 861*
actin *p. 861*
troponin *p. 861*
tropomyosin *p. 861*
myosin *p. 861*
sarcomere *p. 862*
sliding filament
    model *p. 864*
motor unit *p. 866*
summation *p. 867*

tetanus *p. 867*
creatine phosphate *p. 868*
rigor mortis *p. 868*
axial skeleton *p. 869*
appendicular
    skeleton *p. 869*
osteoblasts *p. 872*
osteoclasts *p. 873*
osteoporosis *p. 874*
arthritis *p. 875*
reciprocal inhibition *p. 876*

Discussion Questions

1. Why are both muscles and some sort of skeletal system required for effective movement? Can you imagine how a "spineless" person—if one existed—could move effectively on land?

2. Compare and contrast the advantages and disadvantages of endoskeletons and exoskeletons. Why do you think there will never be any spiders 10 meters tall in terrestrial environments? Why can arthropods grow so much larger in aquatic habitats than on land?

3. Explain, in order, the events that are involved in muscle contraction, according to the sliding filament theory.

4. What is the phenomenon called fatigue as shown in laboratory muscle preparations? How is it different from the fatigue you feel hours after heavy exercise? What do we know about exercise-induced fatigue?

Objective Questions   (Answers in Appendix)

5. The cells of skeletal muscle
   (a) contain actin and myosin.
   (b) are generally under involuntary control.
   (c) are arranged in sheets.
   (d) are found in internal organs.

6. Each muscle fiber is
   (a) a nucleus.
   (b) a sarcomere.
   (c) a cell.
   (d) a sarcolemma.

7. As an action potential moves along a skeletal muscle cell,
   (a) phosphate ions are released by the T-tubules.
   (b) energy is stored in the creatine phosphate.
   (c) calcium ions are stored in the T-tubule network.
   (d) calcium ions are released by the sarcoplasmic reticulum.

8. The sliding filament theory refers to the movement of _____ by each other.
   (a) sarcolemmas
   (b) ATP molecules
   (c) calcium ion binding sites
   (d) actin and myosin

9. Which muscle region almost disappears when a muscle myofibril contracts?
   (a) A band
   (b) I band
   (c) H band
   (d) Z line

CHAPTER 43

Animal Behavior

On the arid grasslands of Ethiopia, a male baboon guards a small group of females who have associated with him for several months. The offspring he sired with those females play with one another or cling to their mothers' backs. Suddenly a foreign male appears, battles the resident male, and displaces him. The newcomer then methodically kills the nursing infants in the troop, over their mothers' vocal but ineffectual protests.

This is but one striking example of **animal behavior,** the response of animals to stimuli in their environments. As we have seen in earlier chapters, behavior is essential to animals in finding food, in selecting habitats, in mediating predator–prey and symbiotic interactions, and in creating the reproductive isolation necessary for speciation. In this chapter we will discuss the events that underlie behavior and examine them from an evolutionary perspective.

THE BEHAVIORAL SCIENCES

Behavior ranges from simple responses of protozoans to light and chemicals to complex interactions within chimpanzee and human societies. These diverse phenomena attract researchers ranging from biochemists to anthropologists—each of whom brings to the study of behavior different techniques and preconceptions.

This chapter examines behavior primarily from the perspective of **ethology** (*ethos* means "manner or behavior"). Ethologists concentrate on the normal behavioral repertoire of animals, either in their natural environment or in laboratory settings that simulate natural conditions as closely as possible. Modern ethology was founded by Konrad Lorenz, Niko Tinbergen, and Karl von Frisch, who shared a Nobel prize in 1973. But ethology's roots date back to Darwin, who observed that individual *variations* in behavior made behavioral evolution possible.

Terminology and Approach in Behavioral Studies

Behavioral scientists are often criticized for describing animal behavior in human terms. In most cases, these criticisms result either from a misunderstanding of the manner in which ethologists use such terms or from a lack of appreciation of the complexities of animal behavior.

As an example of the first case, there are species of ants that invade the nests of related species and carry the helpless pupae back to their own nests. When those pupae emerge, they become workers in the colony of the invading species. Although the technical name for this behavior is "dulosis," it is usually referred to as "slave making."

Use of such language does *not* ascribe human motives to the ant "captors." Neither does it justify slavery in

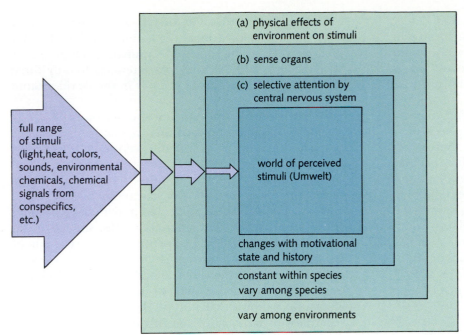

full range
of stimuli
(light, heat, colors,
sounds, environmental
chemicals, chemical
signals from
conspecifics,
etc.)

(a) physical effects of
environment on stimuli

(b) sense organs

(c) selective attention by
central nervous system

world of perceived
stimuli (Umwelt)

changes with motivational
state and history

constant within species
vary among species

vary among environments

Figure 43.1 This schematic diagram illustrates the physical and physiological "filters" that allow some environmental stimuli to enter an animal's awareness while blocking the perception of others. (**a**) Certain potential stimuli pass through the environment in which the animal lives, while others cannot. Light cannot pass through soil, for example, and electricity does not travel through air. (**b**) The sense organs of each species detect certain stimuli and fail to detect others. Human eyes, for example, cannot detect ultraviolet light, but the eyes of honeybees can. (**c**) The central nervous system sifts through the reports of sense organs, paying attention to some stimuli and ignoring others, depending on the individual's motivational state.

humans by implying that slavery is a "natural" phenomenon. Such language is employed *strictly* as a convenient shorthand. To remind you of that point in this chapter, we will place such words in quotation marks. Similarly, when discussing the evolution of behavior, we might say "it is advantageous for the female to act this way" or "the best strategy for the female is to behave in this manner." This language does *not* mean that any individual female "knows" her behavior is advantageous; it is a convenient way of saying "females that behave in this manner will leave more offspring, so, over time, this trait is favored by selection."

Most behavioral researchers scrupulously avoid ascribing human emotions to animals. This is easy when dealing with lower organisms, for no one would suggest that a *Paramecium* swims toward a light because it "knows" it will find food there. But animals such as chimpanzees often present problems. When observing a female chimpanzee whose infant has been killed, for example, some field researchers find the simplest explanation to be that the female "loved" the infant and "mourns" its death.

ELEMENTS OF BEHAVIOR

Sensory Worlds

As we noted in Chapter 41, animals are bombarded with information about their environment. Species differ in their ability to receive that information, because each has its own sense organs and information-processing abilities

that enable it to attend to certain stimuli while ignoring others.

German ethologists, aware of the importance of this phenomenon, coined the term *Umwelt* (*Um* means "around"; *Welt* means "world"; hence *Umwelt* means "the world around") to describe the sensory world perceived by an organism at any time. An animal's Umwelt may vary not only with its location but also with its age and *motivational state* or "mood" (whether, for example, it is hungry) (Figure 43.1).

Researchers must consider carefully the Umwelt of their subjects when designing and interpreting experiments. Early in the study of honeybee behavior, for example, one researcher wondered whether bees had color vision. To test this hypothesis, he designed a wooden box with two circular holes. He left one hole open and blocked the other hole with a transparent piece of glass. When he placed bees in the dark interior of the box, they immediately tried to escape by flying toward the light. Naturally, they could get out through the open hole but not through the glass.

To test whether or not the bees could discriminate between two different colors, the experimenter shone light of one color through the open hole and light of a different color through the hole blocked with glass. He hypothesized that if the bees could discriminate between the two colors, they would associate one color with freedom and learn to fly toward light of that color to make good their escape. The bees, however, never learned which color signaled the way out of the box. The investigator concluded that bees are colorblind.

Karl von Frisch, on the other hand, reasoned that bees *must* have color vision. Why else would bee-pollinated flowers be so brightly colored, and how else could bees single out from a distance individual types of flowers in fields containing many species? So he designed an experiment to test honeybee color vision in as natural a situation as possible (Figure 43.2). Von Frisch trained bees to feed on sugar water in a glass dish placed on a colored square. He then cleaned the square thoroughly (to remove any odor cues that might be left over) and placed it among other squares of different colors and shades of grey. Identical dishes on *all* squares held only distilled water, but the bees flew unerringly toward the color on which they had been fed earlier. Thus, when operating in "food-gathering mode" in a setting similar to that in which they normally use color vision, bees can distinguish colors. When trying to escape from a dark box, however, they simply ignore color information.

Simple Behaviors: Programmed Responses

In certain species, an individual's Umwelt is restricted to a simple sensory input at any point in time, and its responses are totally preprogrammed and inflexible. An extreme example occurs in the females of one species of tick, which require a meal of mammalian blood to lay their eggs. From the time these animals hatch until they lay their eggs and die, they pay attention to only three sensory stimuli of all those present in the world around them.

When a female tick hatches, she responds positively to light by climbing upward on grass and shrubs. Once perched, she ignores all stimuli except one: the odor of butyric acid, a compound found in mammalian sweat. Time is apparently not relevant to the tick at this stage of her life cycle, for she may wait, immobile, up to 20 years for the scent indicating that a mammal is passing beneath her. When that stimulus comes, she lets go of the leaf and drops—if she is fortunate—onto the skin of the mammal below. She then responds blindly to heat (a sign of blood near the skin surface) and upon finding a warm, hairless spot, she burrows in. At this stage, experiments have shown that she will drink any warm liquid, regardless of taste, until she fills to capacity. She then drops to the ground, lays her eggs, and dies.

Each stimulus in the tick's sensory world (light, butyric acid, and heat) serves as what ethologists call a **sign stimulus** or **releaser;** each triggers a specific pattern of behavior. Although we usually associate the word *sign*

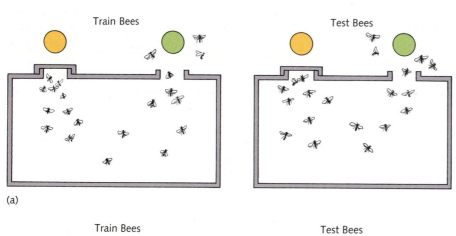

(a)

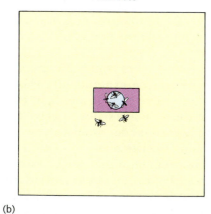

(b)

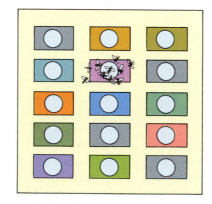

Figure 43.2 The importance of context in animal behavior experiments, shown by tests of color vision in bees. **(a)** One early experimenter placed bees in a dark box and attempted to teach them that following light of one color would lead to freedom, but flying towards light of another color would not. The bees failed to learn this lesson. **(b)** Karl von Frisch trained bees to find sugar water in a glass dish placed on a background of a particular color. He then attempted to "hide" an identical dish and colored square amidst a host of identical dishes placed on differently-colored backgrounds. The bees, however, unerringly selected the color to which they had been trained.

with a visual sign, sign stimuli can occur in any sensory channel.

A neurally preprogrammed, invariant response to a sign stimulus is called a **fixed-action pattern.** Some fixed-action patterns are as simple as the responses of the female tick, while others may be quite involved. One of the best known among higher vertebrates is the egg-retrieving behavior of the greylag goose, first reported by Lorenz and Tinbergen. When a goose incubating her nest notices an egg nearby, she rises, stretches out her neck, rolls the egg neatly into the nest, and settles down upon it. Even if the egg is removed after she has begun to stretch out her neck, the animal completes the sequence, gingerly retrieving and incubating an egg that is no longer there! Once begun, this preprogrammed behavior pattern continues to completion automatically, even in the absence of further sensory cues.

Furthermore, this behavior can be triggered by many objects (including beer cans and baseballs) that resemble eggs in certain ways. The goose's nervous system apparently evolved to respond with retrieval behavior to convex objects with rounded edges. Because geese in nature never encounter beer cans or baseballs, there was no selective pressure to distinguish and exclude them as stimuli, and in the birds' Umwelt for this behavior, all these objects are identical. Interestingly, a separate behavioral program, which controls removal of empty eggshells and foreign objects from the nest, focuses on a different set of sensory cues. In that program, the carefully retrieved beer can is recognized as a "non-egg" and ejected from the nest.

Figure 43.3 Presented with a breast, a newborn human infant will instinctively begin suckling.

Complex Behaviors: Inherited and Acquired

Some simple behaviors (such as those of the ticks described above) are controlled by genetically preprogrammed wiring in the nervous system and are therefore called **innate behaviors** or **instincts.** The existence of instinct enables animals to perform certain acts important to their survival without ever having had the opportunity to learn them.

For many years, behavioral scientists believed a sharp distinction existed between such instinctive behaviors and acquired behaviors, which they classified as **learning.** To many, a behavior had to be *either* instinctive *or* learned. Ethologists, viewing behavior from an evolutionary perspective, emphasized the importance of instinctive behaviors to survival in nature. Behavioral psychologists, interested in mechanisms of learning under controlled conditions, believed that all complex behaviors are acquired by a nervous system that starts off at birth as a clean slate. To some, instinct seemed nearly irrelevant.

This philosophical conflict between the importance of genes (or "nature") and the significance of environ-

mental factors and learning during development (or "nurture") raged fiercely for many years. We now know that both nature and nurture are important in shaping many complex behaviors. Genetic instructions guide the growth of the neurons, effectors, and synapses that make behavior possible, but individual experience can change both the structure and the function of cells and synapses. The behaviors that mature animals display, like many other phenotypic characteristics, depend on both genes and environment. To understand this still-controversial area, we must look closely at both instinctive behaviors and learned responses.

Instinctive behaviors Many instinctive behaviors are relatively rigid and stereotyped in form. Even these responses, such as the fixed-action patterns of ticks and the suckling response of newborn human infants (Figure 43.3), may involve the action of numerous muscles over long periods of time.

Other instincts are far more complex. Both bird nests and spider webs, for example, can be surprisingly elaborate in design and construction, yet the animals begin building these structures the first time with no prior experience (Figure 43.4). Somehow, local sensory cues guide the animals in combining many short, stereotyped actions into long behavioral routines that create a nest architecture well suited to a precise location. Over the course of construction, however, the animals do learn, and their performance becomes more skilled as time goes on.

Learning The ability to modify behavior through experience, a process known as **learning,** is critical to the success of animals from honeybees to humans. Learning specialists often divide learning into several related phenomena.

Habituation (also called extinction), a simple form of learning, occurs when an animal decreases or stops its response to "insignificant" stimuli. A sea anemone presented with a small disk of paper on one of its tentacles, for example, bends that tentacle to place the disk in its mouth. The anemone finds the paper neither rewarding (it is indigestible) nor noxious (it contains no poisons or irritants). After several presentations, the anemone "learns" to ignore the paper as irrelevant.

Classical conditioning (also called associative learning) occurs when animals associate one sensory stimulus with the occurrence of another stimulus. The most famous example of classical conditioning was noted by the Russian physiologist Ivan Pavlov, who studied digestion in dogs. Pavlov first determined that dogs salivate when presented with meat or meat extract. He then discovered that if he repeatedly rang a bell just before he presented food or food odors, the dogs learned to salivate when they heard the bell alone. The dogs thus associated the sound of the bell with the presentation of food.

Operant conditioning (also called trial-and-error learning) occurs when an animal learns to perform a particular operation to receive a reward (**positive reinforcement**) or to avoid some painful experience such as an electric shock (**negative reinforcement**). The classic examples are the rats used in learning experiments performed by psychologist B. F. Skinner. Skinner trained rats to press a lever in their boxes, in response to sound or light stimuli, either to receive a food pellet or to avoid an electric shock. Pigeons can be similarly trained to peck at various objects or colored lights in order to receive food.

Insight learning occurs when an animal applies prior experience in a completely new situation without trial and error. Neither positive nor negative reinforcement is involved in the original learning process; rather, the animal "thinks" about what to do and then acts. Insight learning is most pronounced in our own species and has been documented in other primates, but it is rare in the rest of the animal kingdom. Chimpanzees and orangutans, for example, spontaneously use objects as tools in novel situations to obtain food or to play tricks on one another (Figure 43.5). It is not unusual for chimpanzees to pile up objects to reach a food item hanging out of reach above their heads, even though they have never encountered that situation before.

Figure 43.4 The construction of a weaver-bird nest involves a series of individual knots and weaving patterns determined by the shape of the specific nesting site. This complex instinctive behavior can be performed with no prior experience, yet becomes more accomplished with practice.

Instinctively Guided Learning

It is now clear that instinct and learning, rather than operating independently, often work together, for learning is often shaped or guided by instinct. The fact that important components of complex behaviors can be inherited allows those behaviors to evolve over time. The flexibility of acquired behaviors, on the other hand, enables individual organisms to adjust to specific conditions they encounter.

Imprinting is learning in which an animal "customizes" an inherited behavior on the basis of environmental information it encounters, usually during a brief period early in life. This information is often learned without the benefit of any immediate reward or punishment. Imprinting was first described by Lorenz in experiments with newly hatched geese that begin to waddle behind their mothers a day or so after hatching. This act of following an adult indelibly imprints the image of the parent on the young birds' minds. Thereafter, they follow wherever their parents go.

Goslings will, in fact, trundle after any object that presents sign stimuli even remotely resembling the movement and sound of a goose. Lorenz discovered, for example, that by mimicking goose parenting calls, he could imprint on himself goslings that had been hatched away from their mothers. The result (in addition to numerous important ethological papers) was a series of endearing photographs of the ethologist followed by his "brood" (Figure 43.6).

It has since been determined that many birds are most sensitive to imprinting within a fairly narrow period of time after hatching. Birds not imprinted by the end of that period may never imprint.

Though imprinting usually occurs early in life, the stored information is often used again much later. While goslings imprint on adults as objects to follow, for example, they are also learning to identify members of their species as future mates. When Lorenz's human-imprinted birds matured, they repeatedly attempted to breed with him. Though the birds had continuously associated with others of their species, when the time came to choose mates, their brains called up the imprinted image of Lorenz!

Migration and homing Imprinted information is also used by animals that migrate over long distances. Salmon, for example, hatch in freshwater streams but migrate hundreds of miles out to sea where they feed and grow for a year or more. When the fish mature, they return to breed in the stream of their birth. Among the processes that facilitate this remarkable feat is an imprint of their home stream odor that they carry throughout their lives. By searching out this olfactory stimulus, each fish locates

Figure 43.5 A chimpanzee uses a plant stem as a tool to probe into a termite nest for food.

Figure 43.6 Geese, imprinted on Konrad Lorenz at birth, follow him across a field just as they would normally follow their mother.

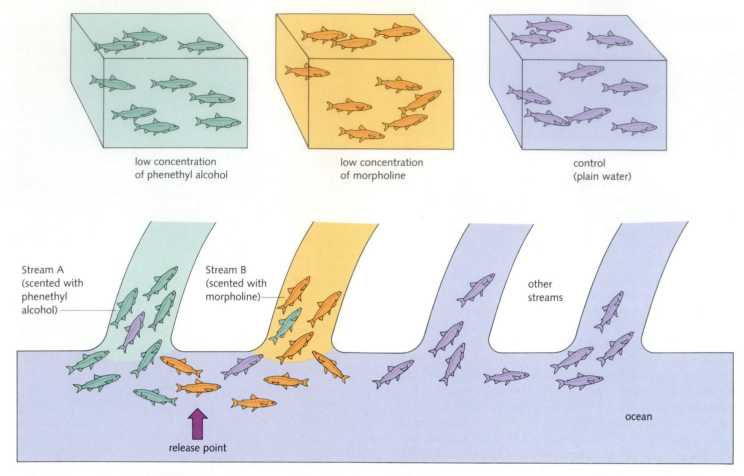

Figure 43.7 Experiments performed by Arthur Hasler on salmon in Lake Michigan. Young fish were raised in holding tanks where 1/3 were exposed to very low concentrations of the chemical morpholine, 1/3 were exposed to phenethyl alcohol, and controls were not exposed to any artificial odor. Fish were released into the sea from a spot on the shore between two streams. At spawning season, one of these streams was "flavored" with morpholine, the other with the alcohol. The vast majority of fish "homed" to the streams flavored with the compound on which they had imprinted.

the precise stream in which it was born. An experiment demonstrating the role of olfaction in salmon homing is shown in Figure 43.7.

Genetic Influences on Song Learning in Birds

Woods and forests are alive with bird song in early spring, as male birds of many species court females and stake out their nesting and feeding territories. The songs of each species are distinct and identifiable; white-crowned sparrows, chaffinches, red-winged blackbirds, and scores of other birds sing species-specific songs. But those songs vary subtly from place to place and from individual to individual within each species. Because many birds use their songs to identify themselves not only as members of their species but also as members of local populations and as individuals, each song must somehow encode that information.

But how can birds create individualized songs without losing the critical patterns that identify their species? Sophisticated experiments over the years have shown that in many birds there is an intricate interaction between innate mental processes and individual experience in song learning.

Normally, young birds hear the songs of their parents and other members of their species as hatchlings. Later, as the males mature and begin to sing, they produce what is called *subsong*—a rambling, babbling song that crystallizes over a period of weeks into a fully formed adult song. This adult song, though it is specific to the species, often displays both regional variations, or *dialects,* and individual variations from male to male.

Hearing adult song is important to song learning in many species; white-crowned sparrows reared in isolation sing when they mature, but they produce only stilted, simplified versions of their species' song (Figure 43.8). If, however, experimenters play tapes of recorded adult songs to young birds during a critical learning period, the birds learn from those tapes.

But if songs are learned, how do birds in the wild learn only their own species' songs and not those of other birds they also hear while young? The answer seems to be that the birds' brains contain some sort of "song-learning program" that accepts input only from its own species's song. Given a choice, birds favor songs of their own species as models. Exposed only to the "wrong" song, they fail to learn at all.

Behavioral Genetics

Innate influences on learning and complex behaviors that appear in the absence of experience provide solid evidence of a strong genetic component to behavior. But the only way to *prove* that behaviors are genetically influenced is to show that behaviors can be inherited in the same ways as other phenotypic characteristics, such as flower color in Mendel's experiments with peas. Ideally, one could observe the behavioral phenotype of hybrids and cross them back to parental types to see how the traits assort. Ultimately, these kinds of experiments could show not only *which* behaviors are shaped by genes but also *how* that influence is produced.

Genes and mating behavior   The activities by which individuals mate and rear their young are of utmost importance to the survival of species and are therefore good candidates for genetic influence. We have already discussed one important example of such behavior: species-specific song, which is used to attract and identify mates.

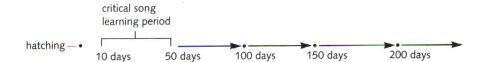

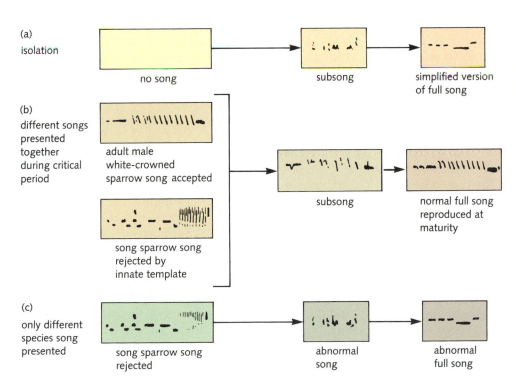

Figure 43.8   **(a)** Exposed to no song at all, birds produce subsong, but develop only a rudimentary version of their species' normal song. **(b)** Exposed to tapes of both their own species' song and that of the related song sparrow, they produce more complex subsong and a fully developed song characteristic of their own species. **(c)** Exposed only to the other species' song, they fail to learn.

(a) Species A
(parent)

(b) Species B
(parent)

(c) A♀ × B♂

F₁ hybrids

(d) A♂ × A♀

Figure 43.9 The dynamics of cricket mating (top) involves complex mating songs. (**a**) Males of species A begin each mating song with a five-pulse chirp and follow with roughly nine, two-pulse chirps. (**b**) Males of species B begin their mating songs with a six-pulse chirp and follow with one or two chirps of about eleven pulses each. (**c,d**) Hybrids are intermediate in number of pulses/chirp and total chirp number.

In an effort to study the genetics behind mating songs in crickets, one group of researchers performed laboratory crosses of two cricket species that produce different calls. The resulting hybrids produced calls intermediate between those of their parents: strong evidence of genetic control (Figure 43.9). Hoping that only a few genes were involved in controlling song production, the researchers made back-crosses to the parent species. Those F_2 hybrids, however, were intermediate between the F_1 hybrids and the parents. Further experiments strongly indicated that, unlike the trait of color in peas, this behavior is subject to the control of several genes.

Genes and learning The complexity of neural function and the number of interacting genes involved in even the simplest innate behaviors make it difficult to conceive of how individual genes, or even small groups of genes, can have powerful effects on complex behaviors. Genes, after all, code for structural proteins and enzymes that affect behavior, not for the behaviors themselves. Some experiments in behavioral genetics, however, remind us that certain gene products can indeed affect learning and memory.

Wild-type fruit flies, for example, can be trained to avoid unpleasant stimuli, but two single-locus (single-allele) mutants, called *dunce* and *amnesiac*, cannot learn so efficiently. Homozygous dunce mutants simply do not learn. Homozygous amnesiac mutants, although they appear to *learn* normally, *forget* much more quickly than wild-type flies. Molecular studies have shown that the dunce allele codes for an enzyme in the cyclic AMP pathway, which—as you may recall from the discussion of second messengers in Chapter 40—is involved in the sorts of changes in synaptic activity that accompany learning.

Extrapolating these results to mammalian behavior, however, is both difficult and dangerous, as shown by a series of experiments involving maze-running ability in rats. The experimenter ran many rats through a particular type of maze and recorded the number of errors the animals made in learning the correct route. He then selected the animals with the very best scores and those with the very worst scores, inbred them within each group, and continued this artificial selection for several generations. He had no problem demonstrating that one group of progeny did far better in that maze than the other.

This researcher called his groups "bright" and "dull," implying that he had bred lines of smart and stupid rats. But when other researchers used the same rats in a different kind of maze, the scores of the two groups were indistinguishable. Though the original experiments had clearly manipulated genetic influences on behavior, they had manipulated only some influence on ability to migrate through a single sort of maze, not "intelligence" overall.

SOCIAL BEHAVIOR

Animals that reproduce sexually cannot live alone. They are forced to link destinies for at least part of their lives with one or more of their fellows to form social units ranging from mated pairs to highly complex societies. Social life has significant costs, but as the existence of animal social systems proves, benefits often outweigh those costs. Mated pairs of animals, for example, can often gather food, defend territories, protect one another and their young from predators, and raise offspring far more successfully than a single animal working alone (Figure 43.10). And the most complex animal societies, such as those of ants, bees, termites, and other types of *social insects,* can accomplish tasks that far exceed the abilities of any single individual.

Communication

The coordination of activity between individuals requires **communication,** the passing from one animal to another of information that influences subsequent behaviors of the recipient of the information. In *social communication* within a species, information exchange is often two-way and facilitates cooperation and coordination between breeding males and females or parents and offspring, for example, or among members of a larger society. Communication may also occur between animals of different species; in such cases, the purpose of communication is not always to transmit accurate information (Figure 43.11).

Intraspecific communication Communication within species often revolves around a variety of **social releasers,** sign stimuli that trigger socially related behaviors in mates or members of the signaler's social group. As is the case with all behaviors, responses to social releasers depend heavily on the individual's motivational state at the time the message is received. Social releasers can take the form of signals presented to any of the senses; virtually all sensory modalities—from vision to electroreception—are used in the intraspecific communication systems of one species or another. In general, however, the most important social signals are presented to the senses of vision, hearing, and olfaction. In this general discussion, we will use examples taken from the world of chemical communication, but the concepts discussed apply to all types of social signals.

Olfactory communication is extremely important in many species and is often mediated by **pheromones,** chemicals released by one animal that influence the behavior of other individuals of the same species. Some animals produce only one pheromone, but others—particularly social insects such as ants, bees, and termites—are practically walking pheromone factories (Figure 43.12).

Releaser pheromones trigger immediate behavioral responses in the individuals that receive them. Sexually mature female moths of several species, for example, release a *sex pheromone* that attracts males from miles away.

Figure 43.10 Elephants care for one another and protect their young as shown in this defensive formation of adults and calves.

Figure 43.11 Deceit in animal communication. (top) A leaf katydid conceals itself from predatory birds. (bottom) Visual camouflage is used adroitly by a scorpionfish to blend in with its surroundings.

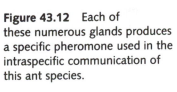

Figure 43.12 Each of these numerous glands produces a specific pheromone used in the intraspecific communication of this ant species.

Figure 43.13 Queen bees are constantly surrounded by workers who feed her, groom her, and provision the eggs she lays within the honeycomb. The queen, marked here by a researcher for purposes of observation, constantly excretes pheromones that are spread among the workers through mutual grooming.

Ants, on the other hand, use many releaser pheromones to coordinate the activities of their colony. A worker ant that finds food outside the nest, for example, returns home dragging her abdomen and leaving a trace of *trail pheromone* behind her. Her nestmates follow that trail from the nest back to the food. Ants and bees also have *alarm pheromones* that, released in times of distress, signal threats to the colony.

Primer pheromones act like externally applied hormones by initiating long-term physiological and behavioral changes in the animal that detects them. Among honeybees, for example, the queen bee releases a primer pheromone called *queen substance* to control her workers (Figure 43.13). Although the entire population of the hive is female for most of the year, only the queen can lay eggs because queen substance suppresses the worker bees' ovaries. Primer pheromones are common in mammals as well. Female mice raised in complete isolation from males, for example, often fail to enter estrus. But when the odor of male mouse urine is introduced into their cages, a pheromone it contains causes the female to begin reproductive cycling immediately.

Courtship behavior In many species, courtship involves an intricate exchange of successive signals called a **stimulus–response chain,** in which each behavior by one partner serves as a stimulus that releases the next behav-

ior in the other partner. Often, stimuli received by the animals' nervous systems interact through the hypo-thalamo-pituitary axis (Chapter 37) to synchronize mating with hormonal cycles that control reproductive organs (Figure 43.14).

Male–male competition and female choice In the half-light of dawn on an American prairie, scores of male sage grouse fight fiercely with one another on a communal mating ground, jockeying for the preferred position in the center of the arena. Physical competition ensures that the coveted center spot will be held by the largest, strongest, and most experienced male. This raucous display, though amusing to humans, is irresistible to female grouse, who wander through the melee eyeing the competitors. The females bypass the younger, weaker males at the periphery and head straight for the center, where one after another mates with the top-ranking male (Figure 43.15).

This display is a prime example of two widespread behaviors that together create a phenomenon known as **sexual selection:** *male–male competition* for females and *female choice* in mating. The situation arises because reproductively, females of most species represent a "limiting resource" in the ecological sense.

Because sperm take so little energy to produce, males can produce millions of them easily. Females, on the other hand, expend far more energy to produce far fewer eggs. This difference in *parental investment* in offspring can have profound evolutionary consequences.

In situations where females by themselves can successfully raise offspring, males can maximize their reproductive success (and Darwinian fitness) by mating as frequently as possible. Females' reproductive success, on the other hand, is limited by the number of eggs they can produce and the number of offspring they can raise. For that reason, the evolutionarily most advantageous strategy for females is to select the "best" male available.

The evolutionary consequence of this basic situation (which has many variations depending on the characteristics of individual species) is selective pressure on males to advertise their sexual readiness, to compete with one another for mates, and to mate with any female who comes along. As a corollary of this competition, males often carry either sexual ornaments to attract the attention of females or weapons to battle competitors (Figure 43.16).

Females, on the other hand, tend to be "choosy" about selecting their mates—hence the concept of *female choice.* Some, such as grouse females, allow competition among males in an arena to pinpoint the most powerful, aggressive individuals. Others, such as peacock females, select their mates on the basis of individual displays. Still others, such as the females of many grazing animals, do not choose the male himself, but rather the territory he has "won"

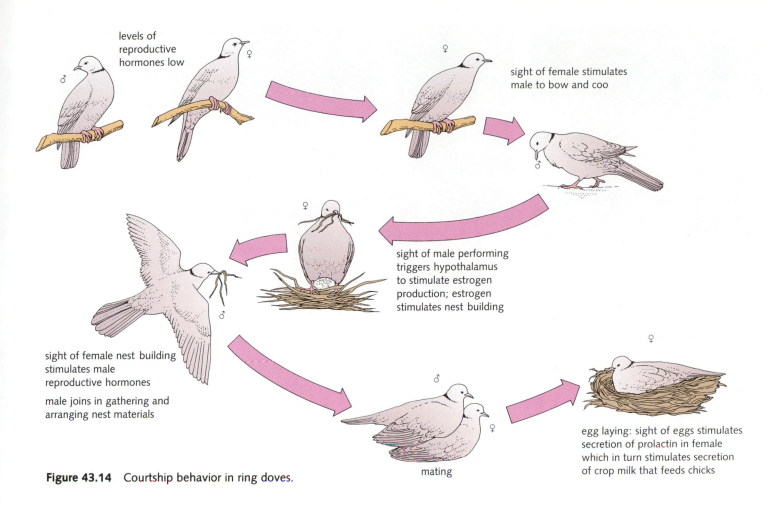

levels of reproductive hormones low

sight of female stimulates male to bow and coo

sight of male performing triggers hypothalamus to stimulate estrogen production; estrogen stimulates nest building

sight of female nest building stimulates male reproductive hormones

male joins in gathering and arranging nest materials

mating

egg laying: sight of eggs stimulates secretion of prolactin in female which in turn stimulates secretion of crop milk that feeds chicks

Figure 43.14 Courtship behavior in ring doves.

Figure 43.15 Bird mating rituals can be both graceful, as shown by these courting Japanese Red-Crowned cranes (left), and fierce, as displayed on this sage grouse lekking ground (right). These males compete with each other for the attention of females.

Figure 43.16 The results of sexual selection. (left) Male peacocks carry large, gaudy tails, used only in mating behavior. (right) Red deer carry antlers used in fierce combat over feeding territory.

in competition with other males. Wandering about to feed, females spend the most time in the most fertile and hospitable territory and thus mate most frequently with the males guarding those territories.

Exceptions to this "promiscuous" mating behavior occur when neither male nor female alone can provide sufficient care to ensure the survival of offspring. Sexual selection still operates in such cases; males advertise and court, while females make the ultimate mating decision. But because no young survive unless both parents are involved in raising them, behavioral evolution has channeled these species into more permanent relationships. Among many bird species, for example, hatchlings must be fed prodigious amounts of food that only *both* parents can provide. In such cases, males and females mate—often for life—to form **monogamous** pairs.

Among many mammals, on the other hand, the young must be fed milk, which only females can provide, but they often require at least the part-time protection of males. In such cases several females often choose to form long-term associations with individual males, a situation called **polygyny.**

It is important to realize that these relationships between care of offspring and mating systems are powerful but not absolute; mating systems ranging from monogamy and polygamy to strictly casual reproductive encounters are found in many vertebrate groups from fishes all the way through primates. Some birds, such as red-winged blackbirds, are polygynous. By the same token, monogamy—though rare among mammals—is by no

means unheard of. Lest the exceptions obscure the rule, however, it is worth noting that nearly 93 percent of bird species that have been studied are monogamous. In contrast, only 10 percent of mammalian species are monogamous.

Language

Ever since Darwin, most biologists have assumed that animal communication is little more than a collection of simple social releasers and fixed-action patterns that express simple mental states such as fear or sexual arousal. Philosophers and linguists alike built a wall between human-style communication and that of other species, asserting that *Homo sapiens* alone uses abstract symbols (words) to represent objects in communicating to its fellows.

Although there are important differences between human language and all other forms of animal communication (no other species comes close to ours in grammatical complexity and vocabulary), a host of experiments over the years has given ethologists great insight into and respect for the complexity and subtlety of animal communication.

The dance language of the bees   One of the most famous and best-studied examples of animal communication is the "dance language" of the bees—a combination of tactile, olfactory, and visual communication elements that enables scout bees to direct their nestmates to food sources they have discovered. This information is encoded in a manner so complex and so unusual that von Frisch published his description of the behavior only after 20 years of experiments.

To decode bees' language, von Frisch first trained various scout bees to feed on dilute sugar–water solution at different locations around the hive. (When quality of food is low, scouts do not recruit other nestmates to it.) He then marked these bees with identification tags and enriched the food source sufficiently that the scouts began to recruit their nestmates.

Von Frisch determined that the recruitment process is simple when food is close to the hive. Returning scouts perform a maneuver called the *round dance*, vibrating and circling on the honeycomb within the hive (Figure 43.17). This action, combined with the scent of food carried on the scouts' bodies, directs nestmates to fly out and search around for food nearby.

But for distant food sources (and single scouts search areas over 25 square miles), encouraging nestmates to search around randomly would be very inefficient. When scout bees locate rich food sources farther than about 80 meters from their hive, therefore, they return and perform a different dance, the *waggle dance*, on the honey-comb (Figure 43.17). This waggle dance conveys to the scout's nestmates both the approximate distance and the direction from the hive to the food source.

Distance from the hive is coded in the number of waggles in the dance and in the total length of time each dance circuit takes. The longer the dance, the longer the distance to the food source.

Direction from the hive to the food source is coded by the angle between the waggling part of the dance and an imaginary vertical line down the honeycomb. Apparently the bee measures the angle between the sun and the food source as she leaves the hive and flies toward the food. She then duplicates that angle on the honeycomb with her dance. As the sun moves across the sky, the dancing bees change the angle they depict on the hive, thus keeping their information accurate.

Language in primates   Our closest relatives, the higher primates, use a wide variety of gestures, body postures, facial expressions, and vocalizations to communicate within and among their social groups. In the past few years, a series of elegant experiments with apes and monkeys in the wild has revealed great sophistication in these animals' use of vocalization.

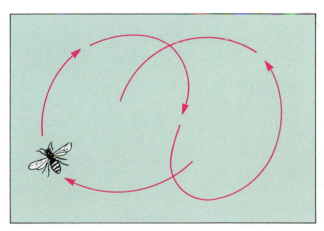

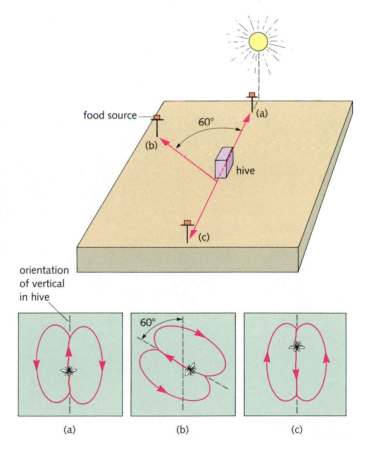

Figure 43.17   The dance language of honeybees. (left) The round dance. (right) The waggle dance contains specific information on the location of the food source. In this imaginary depiction of von Frisch's experiments, different groups of scouts have been trained to feed at locations A, B, and C. The relationship between the food source and the sun for each location is translated into waggle dances on the honeycomb.

Peter Marler of Rockefeller University knew that vervet monkeys (like many other animals) emit alarm calls when they spot predators, but he decided to determine whether the calls carried any more specific information (Figure 43.18). To many people's surprise, Marler and his co-workers discovered different calls for each of the predators that threaten vervets: snakes, eagles, and leopards. Upon hearing the "snake alarm," the monkeys jump on their hind legs and retreat, searching in the grass where the snake might hide. Upon hearing the eagle alarm call, on the other hand, they look skyward, trying to spot the bird as they move toward cover.

Other researchers have studied the alarm calls of rhesus monkeys, which live in large troops. As is common in primate societies, rhesus troops develop a pecking order called a **dominance hierarchy** through aggressive interactions. That hierarchy is continually tested and reshaped by regular squabbles in which numerous individuals participate.

Whenever fights break out among troop members, the individual being attacked emits an alarm call that was long thought to be either a simple emotional outlet or a behavior intended to appease the attacker. In fact, a substantial vocabulary of alarm calls carries a great deal of specific information; the calls inform the members of the troop not only which animal is being attacked but also by whom it is being attacked. Other animals react to that information and, on the basis of their relationships to the combatants, decide whether or not to join the fray.

The Evolution of Social Behavior

Most behaviors we have discussed can be easily explained in evolutionary terms, for most of them directly enhance organisms' Darwinian fitness—their genetic contribution to the next generation. Certain behaviors, however, seem to fly in the face of evolutionary logic. Sounding a predator alarm, for example, benefits the social group but places the *caller* at greater risk. Alarm calls, sharing of food, and joining fights among other group members are all examples of **altruistic behaviors:** actions that are either apparently self-destructive or place an individual at risk to benefit others.

The subfield of behavior called **sociobiology,** which concerns itself with the special problems of social behavior, explains such behaviors in evolutionary terms through two related phenomena. The concept of **inclusive fitness** states that an individual can increase the number of its genes that survive, not only by reproducing itself, but also by helping *relatives* (who share some of those genes) to survive and reproduce. The theory of **kin selection** holds that an altruistic behavior can be adaptive if the extent to which it enhances an animal's inclusive fitness is greater than the loss of personal fitness the animal incurs.

Two siblings (brothers and/or sisters), for example, share on average half of their genes, whereas first cousins share one-eighth of their genome. The logic of inclusive fitness and kin selection argues that it is evolutionarily

Figure 43.18 Upon seeing the proverbial snake in the grass, vervet monkeys produce a call that clearly means "snake," and not just "danger" or "run." The call issued here has warned nearby individuals to stand up and search the ground for the python.

advantageous for an organism to sacrifice its life for a sibling if that sacrifice doubles the sibling's success in raising offspring. A similar sacrifice for a cousin would be advantageous for the altruist only if it allowed that cousin to rear eight times as many offspring.

There is no way to prove that kin selection is universal, but a host of observations on social behavior suggest that it exists in many species. The most extreme and widespread self-sacrificing behaviors involve parents and their offspring. When caring for their young in a nest, for example, many birds (such as nighthawks and wood ducks) decoy predators away from the nest by feigning near-fatal injury and hopping along the ground as though crippled with a broken wing. Once the predator has been led astray by the performance, the bird flies off and returns circuitously to the nest (Figure 43.19).

Altruism involving more distant relatives is common only among animals (such as birds and mammals) whose complex behaviors and social systems enable siblings, parents, half-siblings, and cousins to keep track of one another. We now know, for example, that members of many bird flocks, lion prides, elephant herds, and monkey troops are composed of close relatives—usually parents and one or more generations of their offspring (Figure 43.20). We also know that many animals have an uncanny ability to recognize relatives among many individuals that appear identical to the human eye.

There is a dark side to relatedness and behavior, however, for it also explains such behaviors as the killing of infants described at the beginning of this chapter. There is no evolutionary advantage in protecting the offspring of unrelated males. In such cases, a new male that gains control of a female group eliminates the offspring of his predecessor, inducing the females to enter estrus and bear *his* young. Among some primates, analysis of the situation at takeover time is remarkably sophisticated; males seem to "know" that infants born within a few weeks of their arrival do not belong to them—and they kill those infants as well.

Insect Societies

The most extreme cases of altruistic behavior occur among the social insects. For most of the year, colonies of ants, bees, and wasps are composed entirely of female workers, all of whom are daughters of a single queen and who are therefore referred to as "sisters." (The queen produces male offspring only seasonally, and these are not tolerated within the colony for long.) Workers devote their lives not to reproducing themselves but to helping their mother raise yet more sisters.

This unusual turn of events, according to sociobiological theory, has been brought about by the peculiar

Figure 43.19 A killdeer distracts a potential predator by feigning a broken wing.

Figure 43.20 Lions, like elephants, are matriarchal. Lion prides, which may contain between 4 and 37 individuals, center on a core of several related adult females and their cubs as shown here. Females cooperate and hunt in groups to bring down large prey. Males (after watching the kill) push in rudely to eat their fill first. Males do, however, assist in defending the group against foreign males and other intruders.

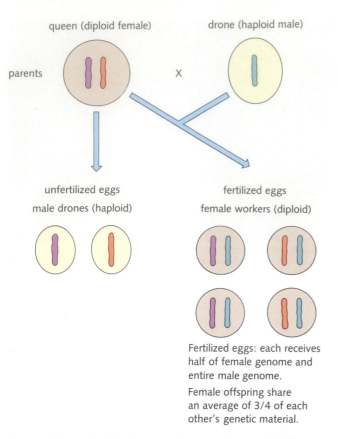

parents

queen (diploid female)   drone (haploid male)

X

unfertilized eggs   fertilized eggs
male drones (haploid)   female workers (diploid)

Fertilized eggs: each receives
half of female genome and
entire male genome.

Female offspring share
an average of 3/4 of each
other's genetic material.

Figure 43.21 Fertilization, sex determination, and relatedness among social insects. In these insects, males (produced from unfertilized eggs) are haploid; females (from fertilized eggs) are diploid. Daughters of the same queen share 3/4 of their genes with one another.

genetics of social insects and the effects of that genetics on inclusive fitness and kin selection. For in most hymenoptera, all fertilized eggs develop into diploid females, while unfertilized eggs develop into haploid males—a condition known as **haplodiploidy.**

Because workers in a colony share half the genes they received from their mother but share *all* the genes they received from their haploid father, the average number of genes shared among sisters is 3/4 (Figure 43.21). Haplodiploid reproduction thus creates a situation wherein a worker gains more in inclusive fitness by "helping" her mother raise sisters than by raising her own offspring. The results in terms of behavioral and social evolution are little short of extraordinary.

The insect colony   If you looked into an ant or bee colony, you would see thousands of insects scurrying around, buzzing, clicking, and rubbing antennae. These apparently random meetings transmit information that connects individual insects, creating an entirely new level of biological organization—the colony. And that colony is more intelligent and adaptable than individual insects could ever be. For just as cells in multicellular organisms differentiate to serve varied functions, individuals in social insect colonies develop into strikingly divergent **castes** whose body forms and behaviors serve different purposes within the colony (Figure 43.22).

Individual workers isolated from their sisters seem to be capable of little more than random activities. Join those

Figure 43.22   (left) These bloated members of a honey ant nest spend their lives as "living honey casks" that store liquid food for distribution to their sisters during lean times. (right) The termite caste system includes soldiers, nymphs, and workers.

workers into a colony, however, and complex, highly ordered group behaviors emerge. Tropical leaf-cutter ants, for example, don't eat the leaves they strip from plants. Instead they take them back to their nest, chew them into a pulp, and use them as food to grow a colony of fungus on which they feed (Figure 43.23). That fungus garden is tended, watered, "weeded," and harvested as meticulously as any human farmer could do it. In addition to growing fungus gardens, several species of termites build complex, highly ordered, three-dimensional nests with built-in "air conditioning" ducts whose convection currents keep the colony from overheating (Figure 43.24).

Some researchers view the insect colony not merely as a social unit but also as a sort of "super-organism" because of its extreme degree of specialization and integration. Both ant and bee colonies, for example, appear to be able to "learn" and respond to environmental changes in ways that individual workers cannot. There are, in fact, striking parallels between the way individual insects join to form a colony and the way individual nerve cells connect to form the brains of higher animals.

Figure 43.23 These tropical leaf-cutter ants are carrying leaves back to their nest where they will be used as food for a fungus garden whose produce feeds the ant colony.

Figure 43.24 Schematic drawing of an African termite nest. Remarkably, the entire nest is constructed in a manner that continually refreshes the air inside. Warm air from the nest's central core rises and enters a network of channels close to the outer nest walls. There it cools and sinks, picking up oxygen and losing carbon dioxide by diffusion as it completes the circuit.

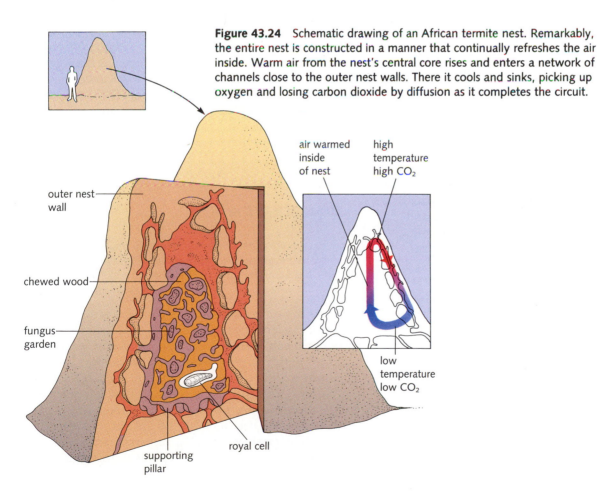

outer nest wall

chewed wood

fungus garden

supporting pillar

royal cell

air warmed inside of nest

high temperature high CO_2

low temperature low CO_2

Figure 43.25 Research on primate societies, as primatologist Sarah Hrdy demonstrates here, requires countless hours of patient observation (often over several years) and great skill at recognizing differences among individual animals. The langurs under observation here in India exhibit a number of surprising behaviors, including infanticide.

Primate Societies

Some of the most fascinating studies in animal behavior—and among the most difficult to interpret—involve members of our own order, the primates (Figure 43.25). Many of the most important observations on primates in the wild have been conducted by three primatologists sometimes nicknamed the "trimates": Jane Goodall, who works with chimpanzees in Tanzania; Birute Galdikas, who observes orangutans in Borneo; and Dianne Fossey, who studied and protected the mountain gorillas of Rwanda until her death at the hands of poachers.

These and other primate researchers have shown that many higher primates have personalities as varied and individual as those of humans. Like humans, many apes develop distinctive styles of interpersonal relations and varied strategies for survival within their troops. Some fight their way to top positions in the dominance hierarchy; others gain status and protection by making "friends" with dominant individuals or by forming coalitions within the troop. The animals cooperate in hunting, squabble among themselves, appear to exhibit such emotions as "jealousy," "affection," and "grief," and—in the case of chimpanzee troops—make war on one another (Figure 43.26).

It is often difficult not to see primate behavior in human terms. Pygmy chimps, for example, seem to "play games." In the San Diego Zoo, they often amuse themselves by climbing down a chain that hangs into the moat surrounding their enclosure. Periodically, when the dominant male in the group is alone in the moat, an adolescent runs over to the chain and pulls it up, trapping the older male at the bottom. Several younger males then gather at the top and make play-faces (the chimpanzee equivalent of laughing) at the dominant below. This suggests to some that the pranksters can place themselves in the position of the older male sufficiently to realize that they're playing a trick on him.

HUMAN BEHAVIOR

What a piece of work is man! how infinite in faculty! in form and moving how express and admirable! in action how like an angel! in apprehension how like a god!
—Shakespeare, *Hamlet*

Homo sapiens is a behaviorally complex primate whose behavior concerns us a great deal. We approach the study of our own behavior from the perspectives of several disciplines: anthropology, sociology, psychology, and—of particular interest to us here—ethology and sociobiology.

We have seen that the spectrum of animal behaviors ranges from rigidly preprogrammed fixed-action patterns to flexible combinations of instinct and learning. From a biological perspective, researchers ask whether humans stand at some locatable point on this continuum

Figure 43.26 (left) Two young snow monkeys at play. Many young animals, particularly mammals, engage in play for varying lengths of time. (right) Grooming, seen here among yellow baboons, is a common behavior among many primates. In addition to removing bothersome parasites, grooming helps cement positive relationships among troop members.

or have leapt off the scale altogether. As individuals, how much of what we think, feel, and do today is programmed within our DNA?

Sociobiologists emphasize that our genes have shaped most of our evolutionary history, and they argue that despite our behavioral flexibility, it is unlikely that we have outgrown completely the influences of genotype on behavioral phenotype. Other researchers counter forcibly that the relatively recent innovations of language and culture have introduced unparalleled behavioral flexibility and variety—a "nurture" that overwhelms the influences of any genetically predetermined "human nature."

These fascinating and provocative positions have been debated fiercely for centuries. Unfortunately, the history of studies into genetic influences on the human mind includes a great deal of bad science, distorted by the sorts of prejudice we discussed in Chapters 1 and 2. For it is particularly difficult for anyone to investigate links between genes and brain activity in humans without falling prey to their own preconceptions.

Central to this issue is the question of how single genes or groups of genes could possibly influence behavior patterns as complex as those found in humans. It is one thing to find a single gene that affects learning in a fruit fly or genes that affect mating calls in crickets. It is entirely

another to postulate (and extremely difficult—some would say impossible—to *prove*) that genes—which can code directly only for structural or regulatory proteins and not for behavior—control such difficult-to-define human traits as "intelligence," "aggressiveness," and "dominance."

It is now clear that identifiable single mutations in the human genome can powerfully influence complex human behaviors. A decades-long study of the Amish, for example, has yielded remarkably strong evidence for genetic predisposition to manic depression in that tightly inbred, socially cohesive group. Other studies are investigating links between single genes (or closely linked genes) and behaviors as diverse as alcoholism, childhood depression, schizophrenia, and Alzheimer's disease.

Yet the question remains: What do normal genes affecting the brain have to do with the essence of being human? Molecular biology is now providing the tools to permit a new generation of experiments correlating, for example, the genetically influenced production of neurotransmitter molecules and receptors with brain activity and behavior. The answers these experiments ultimately provide to questions probing the mysteries of the human mind may tell us more about who we are than some of us want to know. What we as a society choose to do with the information will tell us even more.

The activities of the nervous and muscular systems in response to environmental stimuli are ultimately reflected in animal behaviors vital to survival and reproduction. One approach to the study of behavior is ethology, which emphasizes the importance of examining animals in a natural context.

Behaviors that are governed primarily by genetically programmed neural wiring and require little or no experience to develop are called innate or instinctive. Behaviors that depend on changes in the nervous system as a result of experience are called learned. Among invertebrates and lower vertebrates, strong genetic influences on many behaviors can be demonstrated through breeding experiments. In higher animals, most behaviors result from complex interactions between nature and nurture. The fact that many behaviors are influenced by genes means that behavior can respond to natural selection by evolving over time. The influence of genes on human behavior and the extent to which such behaviors as language and emotion are uniquely human are subjects of heated debate.

Social behavior, coordinated by intraspecific communication through visual, chemical, auditory, tactile, and even electric signals, includes courtship, mating, parental care of young, and more intricate activities of complex animal societies. The sophisticated organization of social insect colonies enables those societies to build elaborate nests and to exchange information.

STUDY FOCUS

After studying this chapter, you should be able to:

- Discuss behavior as a critical part of evolutionary adaptation.
- Explain "innate" and "learned" behavior.
- Describe the evolution of social behavior.
- Explore the biological roots of human behavior.

SELECTED TERMS

sign stimulus/releaser *p. 880*
fixed-action pattern *p. 881*
innate behavior *p. 881*
instincts *p. 881*
learning *p. 881*
habituation *p. 882*
classical conditioning *p. 882*
operant conditioning *p. 882*
insight learning *p. 882*
imprinting *p. 883*
communication *p. 887*

social releasers *p. 887*
pheromones *p. 887*
stimulus–response
 chain *p. 888*
sexual selection *p. 888*
dominance
 hierarchy *p. 892*
altruistic behavior *p. 892*
inclusive fitness *p. 892*
kin selection *p. 892*
castes *p. 894*

REVIEW

Discussion Questions

1. What is an animal's Umwelt? How and why can Umwelt vary from species to species, from individual to individual, and from time to time?

2. What is "instinctive" behavior? Are any behaviors in any animals *totally* instinctive? If so, how can such behaviors appear in the absence of experience?

3. Describe the four basic kinds of learning and give examples of each.

4. Explain how many complex behaviors are shaped both by "nature" and by "nurture."

5. Explain the phenomenon of female choice in shaping courtship and mating behavior.

6. What is the biological definition of altruistic behavior? How is altruism in nonhuman animals explained in terms of the principles of inclusive fitness?

Objective Questions   (Answers in Appendix)

7. Any stimulus that activates a specific motor response is called a(n)
   (a) ritualized cue.   (c) releaser.
   (b) imprinting cue.   (d) pheromone.

8. The different sexes of song sparrows look alike. A pair mates only once in a season, and both help in nest building, food gathering, and defense. These behaviors are examples of maximizing reproductive success
   (a) through monogamy.
   (b) through polygamy.
   (c) by not competing with other mates for females.
   (d) by selecting the best female available.

9. Using the concept of inclusive fitness, an organism is "fit" if
   (a) its genes are passed to the offspring through its own efforts or that of its relatives.
   (b) it produces many healthy sex cells.
   (c) it survives beyond its statistical life expectancy.
   (d) it is nurtured for a long time by its parents.

10. Applying prior experience to a new situation is called
    (a) habituation.
    (b) classical conditioning.
    (c) operant conditioning.
    (d) insight learning.

11. Chemicals that are released by one animal and that influence the behavior of other individuals of the same species are known as
    (a) imprints.   (c) castes.
    (b) pheromones.   (d) insights.

A P P E N D I X

Answers   **A–2**

Genetics Problem Set: Mitosis, Meiosis, and Mendelian Genetics   **A–4**

Solutions to Genetics Problem Set   **A–5**

The Metric System   **A–7**

Classification Scheme   **A–8**

Readings   **A–10**

Glossary   **A–13**

Illustration Credits   **A–30**

Index   **A–35**

ANSWERS

Chapter 1

4. c; 5. d; 6. c; 7. b

Chapter 2

6. b; 7. a; 8. a; 9. c; 10. a

Chapter 3

7. a; 8. a; 9. b; 10. b; 11. d

Chapter 4

7. c; 8. a; 9. a; 10. a; 11. a

Chapter 5

8. b; 9. a; 10. c; 11. d; 12. c

Chapter 6

7. d; 8. a; 9. a; 10. b; 11. a; 12. a

Chapter 7

7. c; 8. b; 9. c; 10. a; 11. d

Chapter 8

11. a; 12. d; 13. b; 14. d; 15. a

Chapter 9

5. a; 6. a; 7. d; 8. b; 9. b

Chapter 10

6. d; 7. a; 8. b; 9. c; 10. c

Chapter 10 In-Text Questions

Page 186   The ideal bird, of course, is the white one—the double-recessive one.

Page 203   Parakeets are not epistatic. In fact, the two loci are codominant because each is expressed.

Chapter 11

6. b; 7. a; 8. a; 9. c; 10. a

Chapter 12

6. d; 7. a; 8. a; 9. d; 10. c

Hardy–Weinberg Problems

Problem 1

Remember that $p^2 + 2pq + q^2 = 1$.

We assume, because of the nature of the disease as explained in the problem, that the population is in Hardy–Weinberg equilibrium. (Nothing in the data presented suggests any effects of the gene on the reproductive success of either homozygous or heterozygous individuals.)

Here the homozygous recessives in the population number 1 in 5000, so, expressed as a percentage of the total population, those homozygotes represent

$$q^2 = 1/5000 = 0.0002.$$

Therefore,

$$q = 0.014142.$$

Then, because $p + q = 1$,

$$p = 0.98585.$$

The number of heterozygotes in the population is

$$2pq = 0.027884$$

so the total number of heterozygotes in a population of 5000 is 0.027884(5,000) or roughly 139 individuals.

To show that this result is not trivial, imagine a similar sample population that differs only in the number of $A'A'$ homozygotes as follows:

| 1627 | 1469 | 138 | 20 |
| total | AA | AA' | A'A' |

In this case, the observed percentages of the total population for each genotype are

| 0.90289 | 0.08481 | 0.01229 |
| AA | AA' | A'A' |

so

$$p^2 = 0.90289 \text{ and } p = 0.9502$$
$$q^2 = 0.01229 \text{ and } q = 0.1109.$$

Thus the expected percentage of the heterozygotes in the population is

$$2pq = 0.2175$$

or more than twice the observed percentage of heterozygotes. If this were indeed the case, what hypotheses could you devise to explain this difference between expected distribution and observed distribution? What experiments could you design to test your hypotheses?

Problem 2

	Individuals collected	% of total population
Total population	1612	100
AA	1469	0.91129
AA'	138	0.08561
A'A'	5	0.003102

There are two ways to determine whether this population is close to H–W equilibrium.

Most simply, for the population sampled,

$$AA = p^2 = 0.91129, \quad \text{so } p = 0.9546$$
$$A'A' = q^2 = 0.00310, \quad \text{so } q = 0.0556$$

Then $AA' = 2pq = 0.1061$, the calculated percentage of heterozygotes, whereas the observed percentage was 0.08561.

This result indicates that the population conforms reasonably well to theoretical expectations, which, however, predict a slightly higher percentage of heterozygotes at equilibrium.

Chapter 13

6. b; 7. a; 8. b; 9. a; 10. c

Chapter 14

7. a; 8. b; 9. b; 10. b; 11. a

Chapter 15

6. d; 7. a; 8. a; 9. c; 10. c

Chapter 16

6. a; 7. b; 8. a; 9. d; 10. b

Chapter 17

5. a; 6. c; 7. d; 8. c; 9. b

Chapter 18

6. a; 7. d; 8. c; 9. b; 10. c

Chapter 19

6. a; 7. d; 8. b; 9. a; 10. b; 11. c

Chapter 20

5. b; 6. d; 7. b; 8. a; 9. d; 10.b;
11. a; 12. b

Chapter 21

8. a; 9. a; 10. b; 11. c; 12. d

Chapter 22

7. b; 8. c; 9. b; 10. d

Chapter 23

7. b; 8. b; 9. a; 10. a; 11. c

Chapter 24

4. b; 5. b; 6. c; 7. d

Chapter 25

5. c; 6. d; 7. b; 8. a; 9. a; 10. d; 11. b

Chapter 26

4. b; 5. a; 6. c; 7. c; 8. c

Chapter 27

7. b; 8. a; 9. c; 10. d; 11. d

Chapter 28

5. a; 6. d; 7. b; 8. d

Chapter 29

5. c; 6. a; 7. d; 8. a; 9. b

Chapter 30

5. c; 6. c; 7. b; 8. a; 9. a; 10. c

Chapter 31

7. d; 8. b; 9. c; 10. b; 11. a

Chapter 32

6. a; 7. c; 8. a; 9. b; 10. a

Chapter 33

6. b; 7. a; 8. d; 9. c; 10. a

Chapter 34

6. a; 7. d; 8. c; 9. d; 10. d

Chapter 35

7. d; 8. d; 9. a; 10. b; 11. a; 12. c

Chapter 36

6. c; 7. a; 8. a; 9. d

Chapter 37

5. d; 6. c; 7. b; 8. a

Chapter 38

5. a; 6. d; 7. c; 8. b

Chapter 39

5. a; 6. c; 7. b; 8. c; 9. b

Chapter 40

7. c; 8. a; 9. a; 10. c; 11. d

Chapter 41

5. c; 6. b; 7. a; 8. b

Chapter 42

5. a; 6. c; 7. d; 8. d; 9. c

Chapter 43

7. c; 8. a; 9. a; 10. d; 11. b

GENETICS PROBLEM SET
Mitosis, Meiosis, and Mendelian Genetics

Problem Set

1. Distinguish between two sister chromatids and two homologous chromosomes. Distinguish between a diploid nucleus and a haploid nucleus.

2. If we let X be the minimum amount of genetic material that carries all the information of a species, then in an organism with a diploid chromosome number of 2, how much DNA (X, $2X$, $4X$, etc.) is found in each of the following? an egg nucleus, a sister chromatid, a daughter nucleus following mitosis, a homologue following mitosis, a nucleus at the onset of mitotic coiling

3. A karyotype of a mitotic nucleus from a female cat shows 76 sister chromatids. What are the diploid and haploid chromosome numbers of the cat? How many homologous chromosome pairs are present?

4. Consider an organism with a haploid chromosome number of 7. How many sister chromatids are present in its mitotic metaphase nucleus? In its meiotic metaphase I nucleus? In its meiotic metaphase II nucleus?

5. Consider an organism with a haploid chromosome number of 3. Write out the eight possible combinations of maternal and paternal homologues in its gametes (for example, 1M2M3P). For each case, diagram the alignments of the nonhomologues with respect to the metaphase I plate.

6. Cite six human traits that you would classify as wild-type traits (that is, highly invariant) and six traits that commonly appear in a number of variant forms in the human population.

7. A pair of alleles is always observed to segregate at the first meiotic division. How might this observation be explained?

8. A cross was made between two *Drosophila* that exhibited the wild phenotype. Their progeny were found to have the same phenotype. A sample was taken of the progeny, each of which was crossed with a fly that had purple eye color. Half of the crosses gave only wild-type flies, and the other half gave 50 percent wild-type and 50 percent purple-eyed progeny. What were the genotypes of the original pair of wild-type flies?

9. Thalassemia is a type of human anemia that is rather common in Mediterranean populations but relatively rare in other peoples. The disease occurs in two forms, minor and major; the latter is much more severe. Persons with thalassemia major are homozygous for an aberrant recessive gene; mildly affected persons (with thalassemia minor) are heterozygous; persons normal in this regard are homozygous for the normal allele.
   (a) A man with thalassemia minor marries a normal woman. With respect to thalassemia, what types of children, and in what proportions, can they expect? (Let t = the allele for thalassemia minor and T = its normal allele.)
   (b) Both father and mother in a particular family have thalassemia minor. What is the chance that their baby will be severely affected? Mildly affected? Normal? Diagram the possible unions of germ cells in this family.
   (c) An infant has thalassemia major. According to the information given, what results would you expect if you checked the infant's parents for anemia?
   (d) Thalassemia major is almost always fatal in childhood. How does this fact modify your answer to part (c)?

10. In humans the most frequent type of albinism (itself quite rare) is inherited as a simple recessive characteristic. Standard symbols are C = normal pigmentation and c = albino. Assume that the genes for thalassemia and albinism assort independently.
   (a) A husband and wife, both normally pigmented and neither afflicted with severe anemia, have an albino child who dies in infancy of thalassemia major. What are the probable genotypes of the parents?
   (b) If these people have another child, what are its chances of being phenotypically normal with respect to pigmentation? Of having entirely normal (that is, nonthalassemic) blood? Of being phenotypically normal in both regards? Of being homozygous for the normal alleles of both genes?

11. A cross was made between two wild-type *Drosophila*. Their progeny were 187 males with raspberry eye color, 194 wild-type males, and 400 wild-type females. Is *ras* a sex-linked gene? Explain. What are the parental genotypes? What are the genotypes of the F_1 wild-type females, and what is their ratio?

12. A cross was made between a female heterozygous for the recessive genes *ct* (cut wings) and *se* (sepia eye color) and a sepia male. Among their female progeny, half the phenotypes were wild-type and half sepia. Among their

male progeny, the phenotypes were 1/4 wild-type, 1/4 cut, 1/4 sepia, and 1/4 cut and sepia. Is either of these genes sex-linked? What are the genotypes of the parents and their offspring?

13. If a woman with normal color vision has a colorblind father, what is the probability that her sons will be colorblind if she marries a man with normal color vision? What genotypes are possible among her male and female offspring? What is the probability of her having a colorblind child if she marries a colorblind man, and does this probability differ depending on whether the child is a male or a female?

14. Hemophilia is a sex-linked recessive trait. A son born to phenotypically normal parents has Klinefelter syndrome and hemophilia. In which meiotic division of which parent is nondisjunction of the X chromosome most likely to have occurred? Explain and state your assumptions.

15. Neither Tsar Nicholas II, his wife Empress Alexandra, nor their daughter Princess Anastasia had the disease hemophilia. However, their son, the Tsarevich Alexis, did have the disease. Can one automatically assume that Anastasia was a carrier? Why or why not?

16. In *Drosophila,* the gene for red eye is dominant to its white allele, and the gene for long wing is dominant to its vestigial allele. The $+/w$ locus is on the X chromosome; the $+/vg$ locus is not on a sex chromosome. Two red-eyed, long-winged flies are bred together and produce offspring in the following ratios:

Females 3/4 red long, 1/4 vestigial
Males 3/8 red long, 3/8 white long, 1/8 red vestigial, 1/8 white vestigial

What are the genotypes of the parents?

Solutions to Genetics Problem Set

1. Sister chromatids are the identical copies of a single chromosome produced by DNA replication. Homologous chromosomes are two chromosomes that contain genes determining the same traits. The genes on two

homologous chromosomes need not be identical! A diploid nucleus has a complete set of all chromosomes, including *both* copies of each pair of homologous chromosomes. A haploid nucleus has only *one* representative of each pair of homologous chromosomes.

2. It depends on what you mean by "minimum amount"! We'll take it to be the haploid amount, OK?

Egg nucleus	X
Sister chromatid	X
Daughter nucleus	$2X$
Homologue after mitosis	$2X$
Nucleus before mitosis	$4X$

3. Diploid number, 38. Haploid number, 19. There are 19 pairs of homologous chromosomes present at metaphase.

4. The haploid number is 7, so the diploid number is 14.

Metaphase of	
mitosis	28 chromatids
Meiosis I	28
Meiosis II	14

5. Here are the possible combinations, written as though they were aligned along a single metaphase plate running right to left:
MMM MMP MPM MPP PPP
PPM PMP PMM

6. This one you can work out for yourselves. Invariant traits (generally) include such traits as number of eyes, fingers, or arms. Highly variable traits in our species include skin color, height, and body structure.

7. Both alleles must be both located on the same pair of homologous chromosomes. Because they separate after the first meiotic division, the alleles would always be observed to segregate as well.

8. If we use P for purple eyes and W for wild-type eye color, we can answer this pretty easily. The half of the offspring that gave *only* wild-type flies when crossed with the purple ones must have been WW. The other half must have been WP (and the purple flies PP).

9. (a) Normal children (Tt), 50%; mild anemia (Tt), 50%.
   (b) (Tt) × (Tt) = 25% severe (tt), 50%

mild (Tt), 25% normal (Tt).
   (c) The parents must *both* be either Tt or TT.
   (d) They must *both* be Tt.

10. (a) The fact that the child was both albino and tt for thalassemia means that the child's genotype must have been $cctt$. Therefore, each parent must have at least one copy of each recessive gene. However, both parents were phenotypically normal. This means that they must both have been heterozygous for each trait: $CcTt$.
    (b) Normal pigmentation: 3 chances in 4 (75% chance)
    Normal blood: 1 chance in 4 (25% chance)
    Phenotype normal in both: 9 chances in 16
    Homozygous normal for both alleles: 1 in 16

11. You bet *ras* is a classic sex-linked gene! The fact that no females are *ras* and 1/2 of the males are *ras* is a classic sex-linked pattern. It suggests that 1/2 of the males have the *ras* gene on their X chromosomes. The phenotypes of the parents must have been XrX and XY. The wild-type females of the F_1 generation are 50% XX and 50% XrX (r represents the *ras* gene.)

12. This is tough. Let's look at each gene individually to get the answer. First, sepia. 1/2 of the offspring are sepia without regard to sex. Therefore, sepia must not be sex-linked. The parents must have been $se/+$ (female) and se/se (male). No problem so far.

Now, cut wings. Note that none of the females have cut wings. However, 1/2 of the males do have cut wings. Cut wings must be sex-linked. Here are the genotypes of parents and offspring:

PARENTS
Female $XctX$
        $se/+$
Male X
        se/se

OFFSPRING

Female	XX	$XXct$	XX	$XXct$
	se/se	se/se	$se/+$	$se/+$
Male	Xct	Xct	X	X
	se/se	$se/+$	se/se	$se/+$

13. The "colorblind gene" is on the *X* chromosome. Because a woman gets one of her *X* chromosomes from her father, and he has only one *X* chromosome, this woman must have the colorblind gene on one of her *X* chromosomes.

Father with
normal vision: 50% of sons
                    color-blind

Possible
genotypes: Male *XY* and
                    XcbY
                    Female *aXX* and
                    XXcb

Colorblind
father: 50% of both sons
                    and daughters
                    colorblind

14. This one takes a little close reasoning. A person with Klinefelter syndrome has 47 chromosomes and the genotype 44*XXY*. This poor fellow also has hemophilia. He couldn't have gotten his hemophilia genes from his father, however (or Dad would have had hemophilia too). Mom must have been a carrier for the disease (*XXh*), but how did he get two *Xh* chromosomes? The homologous *X* and *Xh* chromosomes from Mom separated at the first meiotic division. Then at the second meiotic division, the two identical *Xh* chromosomes must have failed to separate when the haploid gametes were formed. Those two *Xh* chromosomes gave this poor boy both Klinefelter syndrome and hemophilia.

15. No. Because Alexis got his only *X* chromosome from his mother, it must have had the hemophilia gene on it, meaning that Alexandra must have been a carrier. However, Anastasia got one *X* from Nicholas that didn't have the gene, and she got the other *X* from her mother. Because Alexandra was *XXh*, she had a 50–50 chance of giving the good *X* to her daughter. (The events of October 1918 made all of this somewhat academic. Putting it briefly, this family left no offspring.)

16. This is a nice exercise in genetic analysis. We know that one gene is sex-linked and one is not. Let's look at the vestigial gene first. 1/4 of the offspring were vestigial. This number is part of the classic 3:1 ratio for crosses of heterozygotes. Thus the original flies, both with long wings, must both have been +/*vg* at the wing locus. So far so good. Now, we know that red eyes (+) are dominant over white eyes (*w*). The male in our cross must therefore have been *X, Y* (he had red eyes, remember). Similarly, the female must have been *X, Xw*.

PARENTAL GENOTYPES

Sex	*X, Y*	*X, Xw*
Somatic	+/*vg*	+/*vg*

THE METRIC SYSTEM

History

The original basis for the metric system was the diameter of the earth. The original length of the meter, the basic metric unit, was chosen such that the distance from one pole to the equator would be approximately 10,000,000 meters.

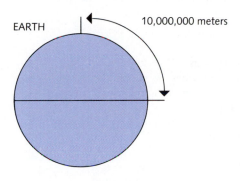

EARTH   10,000,000 meters

Commonly Used Metric Units

Metric	Prefixes	Mass	Volume	Distance
mega	10^6	—	—	—
kilo	10^3	kilogram	—	kilometer
hecto	10^2	—	—	—
deka	10	—	—	—
—	1	gram	liter	meter
deci	10^{-1}	—	—	—
centi	10^{-2}	—	—	centimeter
milli	10^{-3}	milligram	milligram	millimeter
micro	10^{-4}	—	—	—
nano	10^{-9}	—	—	nanometer
pico	10^{-12}	—	—	—

Relationship Between Volume, Distance and Mass

One *liter* is the volume enclosed by a cube 10 *centimeters* on a side.

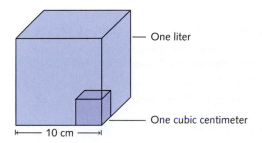

One liter

One cubic centimeter

10 cm

One liter (1000 ml) of water has a mass of one kilogram. One *cubic centimeter* of water, with a volume of 1 *milliliter* (1 ml) under standard conditions, has a mass of 1 gram.

Square Units

Square meter = 1 meter on each side
Hectare = 100 meters on each side
Square kilometer = 1000 meters on each side

Temperature

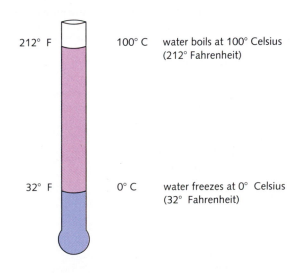

212° F   100° C   water boils at 100° Celsius (212° Fahrenheit)

32° F   0° C   water freezes at 0° Celsius (32° Fahrenheit)

KINGDOM MONERA: Prokaryotes
(under constant revision; assigned many more divisions by some researchers concentrating on biochemical pathways and genetics)

Division Archaebacteria: "Ancient" bacteria
Division Eubacteria: "True" bacteria
Division Cyanobacteria: "Blue-Greens"

KINGDOM PROTISTA: Predominantly unicellular and colonial eukaryotes (A "catch-all" category for predominantly single-celled eukaryotes; extensively revised or even eliminated by some researchers)

PHYLUM Mastigophora: Flagellates
PHYLUM Sarcodina: Amoebae
PHYLUM Sporozoa: Sporozoans
PHYLUM Euglenophyta: Euglenas
PHYLUM Chrysophyta: Diatoms, golden-brown algae
PHYLUM Pyrrophyta: Dinoflagellates

KINGDOM FUNGI: Predominantly multicellular, heterotrophic eukaryotes with cell walls containing chitin

Division Zygomycota: Common molds
Division Ascomycota: Sac fungi and yeasts
Division Basidiomycota: Mushrooms
Division Oomycota: Water molds
Division Deuteromycota: Imperfect fungi

KINGDOM PLANTAE: Predominantly multicellular, photosynthetic, autotrophic eukaryotes; cell walls containing cellulose

Division Chlorophyta: Green algae
Division Rhodophyta: Red algae
Division Phaeophyta: Brown algae
Division Bryophyta: Mosses, liverworts, hornworts
Division Lycophyta: Club mosses
Division Sphenophyta: Horsetails
Division Pterophyta: Ferns
Division Cycadophyta: Cycads
Division Ginkgophyta: Ginkgoes

Division Coniferophyta: Conifers
Division Anthophyta: Flowering plants
   Class Monocotyledones: Monocots (lillies, grasses, etc.)
   Class Dicotyledones: Dicots (roses, deciduous trees, etc.)

KINGDOM ANIMALIA: Multicellular, heterotrophic eukaryotes

PHYLUM Porifera: Sponges
PHYLUM Cnidaria
   Class Hydrozoa: Hydras
   Class Scyphozoa: Jellyfishes
   Class Anthozoa: Sea anemones and corals
PHYLUM Platyhelminthes
   Class Turbellaria: Flatworms
   Class Trematoda: Flukes
   Class Cestoda: Tapeworms
PHYLUM Nematoda: Roundworms
PHYLUM Mollusca: Snails, clams, and squid
   Class Polyplacophora: Chitons
   Class Gastropoda: Snails and slugs
   Class Bivalvia: Clams, oysters, and scallops
   Class Cephalopoda: Octopi, squid, and nautilus
PHYLUM Annelida
   Class Polychaeta: Feather duster worms, etc.
   Class Oligochaeta: Earthworms
   Class Hirudinea: Leeches
PHYLUM Arthropoda
   Subphylum Trilobita: Trilobites (extinct)
   Subphylum Chelicerata: Scorpions, spiders, and mites
   Subphylum Crustacea: Shrimps, crabs, and lobsters
   Subphylum Uniramia: Centipedes, millipedes, and insects
      Class Chilopoda: Centipedes
      Class Diplopoda: Millipedes
      Class Insecta: Insects
PHYLUM Echinodermata
   Class Asteroidea: Starfish
   Class Echinoidea: Sea urchins
   Class Crinoidea: Sea lillies
   Class Holothuroidea: Sea cucumbers
   Class Ophiuroidea: Brittle stars

PHYLUM Chordata
(major groups with at least 1 living member)
Subphylum Urochordata: Tunicates
Subphylum Cephalochordata: Lancelets
Subphylum Vertebrata:
Class Agnatha*: Jawless fishes
(lampreys, hagfishes)
Class Chondrichthyes: Cartilaginous fishes
(sharks, skates, and rays)
Class Osteichthyes: Bony fishes
Subclass Actinopterygii: Ray-finned fishes
(contains most modern fishes)
Subclass Sarcopterygii: Fleshy-finned fishes
Order Dipnoi: Lungfishes
Order Crossopterygii: Coelacanths (and ancestors
of land vertebrates)

Class Amphibia: Amphibians
Order Anura: Frogs and toads
Order Urodela: Salamanders and newts
Order Apoda: Limbless amphibians
Class Reptilia: Reptiles
Subclass Anapsida: Turtles
Subclass Lepidosauria: Snakes, lizards
Subclass Archosauria: Dinosaurs, crocodilians
Class Aves: Birds
Class Mammalia: Mammals
Subclass Protheria: Egg-laying mammals
Subclass Theria
Infraclass Metatheria: Pouched mammals
Infraclass Eutheria: Placental mammals (contains many
orders, including our own, the Order Primates)

*"Agnatha," though now rarely used by ichthyologists as a formal class,
is still a convenient grouping under which to list these fishes.

READINGS

Part One

Burnham, John C. *How Superstition Won and Science Lost*. New Brunswick, NJ: Rutgers University Press, 1987. A well-documented and perceptive explanation of at least some reasons for the decline in science education in this country.

Goodfield, June. *Reflections on Science and the Media*. Washington, DC: AAAS, 1981. A slim but thought-provoking volume essential for anyone interested in the way mass media affect public perceptions of science.

Gould, Stephen Jay. *The Mismeasure of Man*. New York: Norton, 1981. A masterfully written, enlightening, and often infuriating exposé of abuses of science in society over the last century.

Levins, Richard, and Richard Lewontin. *The Dialectical Biologist*. Cambridge, MA: Harvard University Press, 1985. Not easy going, but a fascinating sociological and psychological perspective on both historical and contemporary interactions between science and society.

Lovelock, J. E. *Gaia: A New Look at Life on Earth*. New York: Oxford University Press, 1979. This book is a gem—the product of an unfettered, inventive mind. Though its central premise is accepted by relatively few biologists, the questions it raises have inspired many fascinating and productive research projects.

Morrison, P., and P. Morrison. *Powers of Ten*. New York: W. H. Freeman, Co., 1982 (distributed by Scientific American Books). A magnificently produced journey from the universe to the nucleus of an atom.

Part Two

Brown, Lester, et al. *State of the World*. New York: Norton, annually since 1984. Published each year, these volumes offer a unique combination of ecological and economic analysis of progress toward a sustainable global society.

Kaufman, Les, and Kenneth Mallory. *The Last Exchange*. Cambridge, MA: M.I.T. Press, 1986. A brilliant introduction and six chapters by different authors draw on the researchers' personal experiences to address issues involving global pollution and habitat destruction.

Teal, John, and Mildred Teal. *Life and Death of a Salt Marsh*. New York: Ballantine, 1969. A timeless classic that combines American history with estuarine ecology.

Worster, Donald. *Nature's Economy*. Garden City, NY: Doubleday, 1979. A fascinating and engaging history of ecological and economic thought from the eighteenth century to the present.

Part Three

Eldredge, Niles. *Life Pulse*. New York: Facts on File, 1987. A sprightly, well-documented presentation of modern evolutionary theory and the process of science in evolutionary biology.

Gould, Stephen Jay. *Ever Since Darwin* (1977), *The Panda's Thumb* (1980), and *Hen's Teeth and Horse's Toes* (1983). New York: Norton. If you never read any other essays on natural history, don't miss these collections of Gould's entertaining and scholarly essays first published in his column in *Natural History* magazine.

Stanley, Steven. *The New Evolutionary Timetable*. New York: Basic Books, 1981. A brilliant summary of the important aspects of punctuated equilibrium theory and the way it fits into current evolutionary thought.

Part Four

Brush, S. G. "Nettie M. Stevens and the discovery of sex determination by chromosomes." *ISIS* 69 (1978): 162–172. A superb sketch of the life and scientific achievements of the discoverer of sex chromosomes.

Crow, J. F. "Genes that violate Mendel's rules." *Scientific American* 240 (February 1979): 134–146. In principle, all genes are sorted according to Mendel's rules as they pass through meiosis. However, there are a few genes that "cheat" by affecting meiosis itself. This article explores how these genes act.

Fuchs, F. "Genetic amniocentesis." *Scientific American* 242 (June 1980): 47–53. A basic description of the uses of and possibilities inherent in prenatal genetic screening.

John, B. "Myths and mechanisms of meiosis," *Chromosoma* 54 (1976): 295–325. Some aspects of meiosis can be confusing. This clearly written article explores meiosis in a way that focuses on the essential aspects of the process and its central relationship to Mendelian genetics.

Lewin, R. "Cultural diversity tied to genetic differences." *Science* 212 (1981): 908–910. A short but provocative article reviewing *Genes, Mind, and Culture,* by E. O. Wilson and C. Lumsden, which explores the possibility that there is a genetic basis for culture.

Mange, A. P., and E. J. Mange. *Genetics: Human Aspects*. Philadelphia: W. B. Saunders, 1980. A comprehensive study of the subject.

Mazia, D. "The cell cycle." *Scientific American* 230 (January 1974): 53–64. This article contains an excellent description of the cell cycle by one of the pioneers of cell biology.

Patterson, D. "The causes of Down syndrome." *Scientific American* 257 (August 1987): 52–61. Only recently have the actual causes of Down syndrome become clear. This well-written article explores the mechanisms that produce this important human disorder.

Pickett-Heaps, J. D. Tippit, and K. R. Porter. "Rethinking mitosis," *Cell* 29 (1982): 729–744. This article describes the mitotic structure and offers some intelligent speculation on the evolution of mitosis and meiosis. Difficult in parts, but thought-provoking.

Sloboda, R. D. "The role of microtubules in cell structure and cell division." *American Scientist* 68 (1980): 290–298. Microtubules are the major structural components of the mitotic spindle. This article explores some theories of how they behave during cell division.

Stern, C., and E. Sherwood, eds. *The Origin of Genetics.* San Francisco: Freeman, 1966. A classic text on the development of genetics as a science. It includes an English translation of Mendel's original papers and also contains papers by T. H. Morgan and other pioneers of genetics.

Witkop C. "Albinism." *Natural History* 84 (1975): 48–59. A complete description of the many forms of albinism. This article makes interesting connections between the biochemical, genetic, and physiological aspects of albinism.

Part Five

Anderson, W. F., and E. G. Diacumakos. "Genetic engineering in mammalian cells." *Scientific American* 245 (July 1981): 106–121. A description of the prospects for genetic manipulation of mammals.

Doolittle, R. F. "Proteins." *Scientific American* 253 (October 1985): 88–99. An excellent discussion of the fundamentals of protein organizations.

Felsenfeld, G. "DNA." *Scientific American* 253 (October 1985): 58–67. A cogent article focusing on the interactions of DNA with proteins.

Frieden, E. "The chemical elements of life." *Scientific American* 227 (July 1972): 52–60. A description of the key elements found in living systems.

Gunning, B. E. S., and M. W. Steer. *Ultrastructure and the Biology of Plant Cells,* London: Edward Arnold Publishers, 1975. A superbly illustrated account of the main structural features of plant cells.

Hinkle, P. C., and R. E. McCarthy. "How cells make ATP." *Scientific American* 238 (March 1978): 104–123. A description of ATP synthesis in both mitochondria and chloroplasts according to the chemiosmotic theory.

Karplus, M., and J. A. McCammon. "The dynamics of proteins." *Scientific American* 254 (April 1986): 42–51. Proteins are not static structures. This article describes the importance of changes in shape for protein function.

Lawn, R. M., and G. A. Vehar. "The molecular genetics of hemophilia." *Scientific American* 254 (March 1986): 48–54. An excellent account of the molecular basis of a major human genetic disorder.

Miller, K. R. "The photosynthetic membrane." *Scientific American* 241 (October 1979): 102–113. This article describes the basic structural organization of the photosynthetic membrane in higher plants.

———. "The photosynthetic membrane in prokaryotic and eukaryotic cells." *Endeavor* 9 (1985): 175–182. This article addresses certain aspects of photosynthetic membrane organization in prokaryotic and eukaryotic cells.

Ourisson, G., P. Albrecht, and M. Rohmer. "The microbial origin of fossil fuels." *Scientific American* 251 (August 1984): 44–51. An outline of the chemistry involved in the conversion of living matter into fossil fuels.

Portugal, F. H., and J. S. Cohen. *The Century of DNA: A History of the Discovery of the Structure and the Function of the Genetic Substance.* Cambridge: M.I.T. Press, 1977. A comprehensive historical treatment of DNA.

Sharon, N. "Carbohydrates." *Scientific American* 243 (November 1980): 90–116. A straightforward description of the structure of carbohydrates and the biological roles they play.

Sibley, C. G., and J. E. Ahlquist, "Reconstructing bird phylogeny by comparing DNAs." *Scientific American* 254 (February 1986): 82–93. The application of molecular phylogeny to an important problem in evolution.

Singer, S. J., and G. Nicolson. "The fluid mosaic model of the structure of cell membranes." *Science* 250 (1972): 78–94. A classic paper describing the key elements of the fluid mosaic model of membrane organization. This closely reasoned summary recounts the experimental evidence that led to our current understanding of cell membrane organization.

Watson, J. D. *The Double Helix.* New York: Mentor Books, 1968. A lively and intensely personal account of the development of the double helix model by one of those who developed it. No one interested in the nature of scientific discovery should overlook this book.

Watson, J. D., and F. H. C. Crick. "Molecular structure of nucleic acids: A structure of deoxyribose nucleic acid." *Nature* 171 (1953): 737. This is the article that announces its authors' discovery of the double-helical structure of DNA.

White, R., and C. T. Caskey. "The human as an experimental system in molecular genetics." *Science* 240 (1988): 1483–1488.

Wilson, A. C. "The molecular basis of evolution." *Scientific American* 253 (October 1985): 164–173. A description of the molecular principles that govern evolutionary change.

The following two official reports may also be of interest:

Mapping and Sequencing the Human Genome. Washington, DC: National Academy Press, 1988. An authoritative and exhaustive report on the practical, scientific, and ethical questions surrounding a concerted effort to determine the DNA sequence of the human genome.

The New Human Genetics. Bethesda, MD: National Institute of General Medical Sciences, 1984. An informative booklet that summarizes the role molecular biology can play in combatting human disease.

Part Six

Barrett, S. C. H. "Mimicry in plants." *Scientific American* 257 (September 1987): 76–83. A remarkable description of the ways in which plants mimic other plants and animals, fooling many potential predators, including humans.

Chilton, M. D. "A vector for introducing new genes into plants." *Scientific American* 248 (June 1983): 51–59. An excellent explanation of the use of transforming bacteria in plant genetic engineering.

Core, H., W. Cote, and A. Day. *Wood Structure and Identification.* Syracuse, NY: Syracuse University Press, 1979. An expert can identify a tree merely from the color and texture of its wood. This book describes the individual differences between woods from various trees, and it contains some remarkable scanning electron micrographs of wood.

Goldberg, R. B. "Plants: Novel developmental processes." *Science* 240 (1988): 1460–1468. A forward-looking article on the unique aspects of plant growth and development that may furnish important experimental systems in the years ahead.

Gould, J. L. "How bees remember flower shapes." *Science* 227 (1984): 1492–1494.

Heslop-Harrison, Y. "Carnivorous plants." *Scientific American* 238 (February 1978): 104–115. A fascinating look at plants that trap animals for food.

Mandoli, D. F., and W. R. Briggs. "Fiber optics in plants." *Scientific American* 251 (August 1984): 90–98. Even plant tissues that are beneath the soil can sense the direction of sunlight. This article describes the surprising physical principles behind this response.

Niklas, K. J. "Aerodynamics of wind pollination." *Scientific American* 257 (July 1987): 90–95.

Rosenthal, G. A. "The chemical defenses of higher plants." *Scientific American* 254 (January 1986): 94–99. Plants are not so defenseless as they may seem! This article describes the formidable chemical weapons various plants marshal against their predators.

Sisler, E. C., and S. F. Yang. "Ethylene, the gaseous plant hormone." *Bioscience* 34 (April 1984): 234–238. A complete description of the mode of action of one of the most important plant hormones, and the only one which acts as a gas.

Part Seven

Brown, M. S., and J. L. Goldstein. "How LDL receptors influence cholesterol and atherosclerosis." *Scientific American* (November 1984): 58–67.

Buisseret, P. D. "Allergy." *Scientific American* (August 1982): 86–95.

Downer, John. *Super Sense*. New York: Henry Holt, 1988. An intriguing, well-illustrated, and broad-ranging survey of the sensory capabilities of animals ranging from humans to bats and electric fishes.

Edelson, R. L., and J. M. Fink. "The immunologic function of skin." *Scientific American* (June 1985): 46–53.

Premack, David. *Gavagai! or The Future History of the Animal Language Controversy.* Cambridge, MA: M.I.T. Press, 1986. An intensely personal and stimulating view of the similarities and differences between human language and the communication systems of a variety of animals.

Robinson, T. F., S. M. Factor, and E. H. Sonnenblick. "The Heart as a suction pump." *Scientific American* (June 1986): 84–91.

Sacks, Oliver. *The Man Who Mistook His Wife for a Hat*. New York: Harper & Row, 1987. A delightfully written and fascinating collection of stories about the relationship between neurological disorders of the brain and bizarre psychological syndromes among the patients of a most perceptive physician.

Schnapf, J. L., and Baylor, D. A. "How photoreceptor cells respond to light." *Scientific American* (April 1987): 40–47.

Snyder, S. H. "The molecular basis of communication between cells." *Scientific American* (October 1985): 132–141.

Wassarman, P. M. "The biology and chemistry of fertilization." *Science* 235 (1987): 553–560.

"What science knows about AIDS." *Scientific American* (October 1988). A complete issue (10 articles) devoted to the AIDS epidemic.

Young, J. D., and Cohn, Z. A. "How killer cells kill." *Scientific American* (January 1988): 38–44.

Zapol, W. M. "Diving adaptations of the Weddell seal." *Scientific American* (June 1987): 100–107.

ABO gene The hereditary unit containing the A, B, and O alleles that determine the configuration of antigens on the surface of red blood cells.

abscission layer (*ab-SIH-zhun*) A corky cell barrier that forms when hormonal action at the base of a leaf stalk shuts off the flow of water and nutrients to the leaf; this layer is the "break point" when the leaf falls off.

accommodation Adjustments in the lens of the eye that focus images on the retina.

acetylcholine (*a-SEE-till-KOH-leen*) A neurotransmitter released and hydrolyzed at certain nerve endings, notably at neuromuscular junctions.

acid rain Rain that, because of acid-forming industrial pollutants introduced into the atmosphere, has a pH lower than normal. Acid rain is currently causing serious environmental damage in the northern United States and Europe.

acquired immune deficiency syndrome (AIDS) A disorder of the immune system, caused by a retrovirus, that renders victims vulnerable to a variety of infections. At the time of this writing, AIDS is incurable and fatal.

acrosome (*A-krow-soam*) The vesicle at the forward end of a sperm containing enzymes that help break through protective layers surrounding the egg cell.

actin With myosin, one of the two major proteins making up the thin filaments of muscle; functions in muscle contraction.

action potential An impulse in a nerve cell; a sudden, brief change in the electric charge across the plasma membrane that is propagated along the membrane.

activation energy The energy necessary to initiate a chemical reaction.

active site The place on an enzyme where substrate molecules bind during catalysis.

active transport A process in which energy is used in the movement of ions and molecules across a cell membrane against a concentration gradient.

adaptation (*a-dap-TAY-shun*) A structure, behavior, or other feature of an organism that enables it to survive or reproduce more successfully than it otherwise would under the existing environmental conditions.

adaptive radiation The evolution, from a common ancestor, of a number of species specialized for survival in diverse environments.

adenosine triphosphate (ATP) (*a-DEN-a-seen try-FOSS-fate*) A major energy-carrying molecule in cells, this nucleotide can release the energy stored in its phosphate bonds; the energy is used in many metabolic activities.

adenylate cyclase (*a-DEN-a-late SIGH-klase*) An enzyme that, when activated, transforms ATP into cyclic AMP.

adhering junction An attachment between adjacent animal cells that strongly binds them together but permits materials to move in the spaces between the attachments; common in tissues subject to mechanical stress, such as skin. Also called **desmosome** (*DEZ-mo-sowm*).

adipose tissue (*AD-ih-pos*) A type of connective tissue that serves as a storage reservoir for fat; found throughout the body.

adrenal cortex (*a-DREE-nul KOR-tex*) The outer part of the adrenal gland; produces a wide variety of steroid hormones essential to normal body function, including glucocorticoids, mineralocorticoids, and cortical sex hormones.

adrenalin (*a-DRE-na-lin*) A hormone produced by the adrenal medulla and released into the bloodstream when danger threatens; also called *epinephrine*.

adrenal medulla (*a-DREE-nul me-DULE-uh*) The inner part of the adrenal gland; produces a variety of hormones, including adrenalin.

aerobe (*AIR-obe*) An organism that requires oxygen to carry on respiration.

aerobic (*air-OH-bik*) Requiring free oxygen.

aggregate fruit A fruit formed from multiple unfused ovaries within a single flower, as in a raspberry, strawberry, or pea.

agnathans (*ag-NAY-thuns*) A group of jawless fishes, fossils of which are the earliest evidence of vertebrate life forms.

albinism (*al-BYE-nizm*) A recessive gene disorder in which the affected individual cannot synthesize the pigment melanin, which is responsible for most human skin and hair color.

alimentary canal (*al-im-EN-ta-ree*) The passageway that extends from the mouth through a variety of specialized digestive organs to the anus.

allantois (*a-LAN-twahz*) An extraembryonic sac that contributes to the formation of the umbilical cord and placenta; develops from the primitive gut.

alleles (*a-LEELZ*) Two or more forms of a single gene.

allergy A disorder in which the immune system overreacts to a foreign molecule in the environment, initiating an inflammatory response.

allopatric speciation (*al-lo-PAT-rick spee-shee-AY-shun*) The development of new species resulting from the physical separation of two or more populations of an organism such that gene flow ceases between these populations.

allosteric site (*AL-oh-STEER-ik*) A location on an enzyme that is not an active site but that affects the shape of the active site when a regulatory molecule binds to it.

alpha helix A coiled chemical configuration of the secondary structure of proteins in which hydrogen bonds link every fourth peptide bond.

alternation of generations A pattern in the life cycle of plants in which a diploid sporophyte stage alternates with a haploid gametophyte stage.

altruistic behavior (*AL-true-ISS-tik*) Behavior that benefits others of the same species but is destructive or potentially destructive to the individual that performs it.

alveoli (singular: **alveolus**) (*al-VEE-a-lie*) The small, thin-walled sacs in the lungs where gas exchange takes place.

amino acid (*a-MEE-no*) A molecule containing an amino group (-NH2), a carboxyl group (-COOH), and an "R-group" that varies from one amino acid to another. Amino acids are the building blocks of proteins.

amniotic egg (*AM-nee-AWE-tik*) A watertight egg produced by internal fertilization and wrapped in three protective membranes.

amniotic sac (*AM-nee-AWE-tik*) An extraembryonic sac filled with fluid that protects the embryo from desiccation and physical shock.

amoeba (plural: *amoebae* or *amoebas*) (*a-MEE-bah*) Simple animal-like protists that generally live in water or soil habitats or occur as parasites. Includes radiolarians and foraminiferans, important members of plankton communities.

amphibians A class of four-legged ectothermic vertebrates, many of which exhibit an aquatic larval phase and a semiterrestrial or terrestrial adult phase.

anabolic pathway (*AN-a-BOL-ic*) A chemical pathway in which smaller, simpler molecules

anabolic pathway (*continued*)
are assembled into larger, more complex molecules.

anaerobe (*AN-ur-obe*) An organism capable of living in an environment in which there is no oxygen.

analogous structures Organic structures of similar function (and sometimes similar appearance) that result from convergent evolution rather than common ancestry.

anaphase (*AN-a-faze*) The third stage of mitosis, during which the centromeres break and the chromosomes move to the poles of the nuclear spindle.

angiosperms (*AN-jee-oh-sperms*) The flowering plants; a group of vascular plants that produce seeds enclosed in ovaries.

anion (*AN-eye-un*) A negatively charged ion.

annelid (*AN-el-id*) A segmented worm. Members of this class are characterized by setae, body segments sealed off by septa, and a hydrostatic "skeleton." The earthworm is an annelid.

anthropoids (*AN-thra-poyd*) The more advanced branch of the order of primates; includes the New World and Old World monkeys and the great apes.

antibody (*AN-tee-bod-ee*) A Y-shaped protein capable of binding to an antigen and thereby tagging it for destruction by the body's defense systems.

anticodon (*AN-tee-CODE-on*) A string of three nucleotide bases on a tRNA molecule that is complementary to a particular mRNA codon.

antigen (*AN-tih-jun*) A molecule recognized as foreign by the body's immune system.

anus (*AY-nus*) The opening at the end of the alimentary canal through which waste material is expelled.

aorta (*ay-OR-tuh*) The thick artery that carries oxygenated blood from the left side of the heart to all parts of the body except the lungs.

apical dominance (*AY-pih-kul*) The tendency for plant growth to be most vigorous at the tip, or apex, of branches.

apical meristem (*AY-pih-kul MARE-ih-stem*) The undifferentiated embryonic tissue in root tips or stem buds where cells are produced, causing increases in length.

appendicular skeleton (*AP-pen-DIK-you-lar*) The arm and leg bones and the pectoral and pelvic girdles that attach them to the axial skeleton.

aquaculture (*AH-kwuh-kul-chur*) The cultivation of aquatic organisms to harvest for human use.

aquifer (*AH-kwi-fer*) An underground rock stratum that holds water.

archaebacteria (*ARK-ee-bak-TEER-ee-a*) Three groups of bacteria—the methanogens, thermoacidophiles, and halophiles—that resemble the first prokaryotes to evolve on earth.

archenteron (*ark-EN-ter-on*) The cavity formed during gastrulation that ultimately forms the embryo's digestive system.

arthritis Inflammation of the tissues of a moveable skeletal joint.

arthropods (*AR-throw-podz*) A group of organisms characterized by jointed legs, a segmented body, and a hard exoskeleton that must be shed to permit new growth; the phylum includes the insects, crustaceans, spiders, millipedes, and centipedes.

artificial selection Specialized breeding by humans of plants or animals possessing valued characteristics.

ascomycetes (*ASS-koh-MY-seets*) The sac fungi, a diverse group of unicellular or multicellular fungi that produce spores inside a sac, or ascus; includes the yeasts and most of the fungi that live symbiotically in lichens.

asexual reproduction Reproduction mediated not by the union of gametes but by budding, binary fission, fragmentation, or other nonsexual means.

atherosclerosis (*ATH-er-row-skle-ROE-sis*) The buildup of fatty tissue on the inner surface of arteries, which can slow down or stop blood flow and result in strokes or heart attacks.

atomic number The number of protons in an atomic nucleus.

atomic orbital The volumes around an atom's nucleus wherein an electron is most likely to be found.

atrioventricular (AV) node (*AY-tree-oh-ven-TRIK-yoo-lar*) The area of the heart where the fibers originate that conduct the impulse to contract from the atria to the ventricles.

atrioventricular valves (*AY-tree-oh-ven-TRIK-yoo-lar*) The heart valves that permit blood to flow from the atria to the ventricles but not in the other direction.

atrium (plural: *atria* or *atriums*) (*AY-tree-um*) A chamber of the heart that receives blood from a vein and pumps it into a ventricle. The human heart has two atria.

australopithecines (*AWE-strah-low-PITH-a-seens*) Members of a genus of early bipedal hominids thought by many to have been the common ancestor of all other hominids.

autoimmune disease (*AWE-toe-i-MUNE*) A disorder in which the immune system malfunctions and turns its destructive power against the body's own tissues.

autosomal chromosomes (*AWE-toe-SOHM-ul*) Generally, chromosomes that are not sexlinked; in humans, the 22 pairs of homologous chromosomes.

autotroph (*AWE-toe-trowfe*) An organism, such as a plant, capable of producing its own food from inorganic materials and an environmental energy source.

auxin (*AWEK-sin*) Any of several plant hormones that can produce a variety of effects in different types of cells and in different concentrations, including promoting cell growth and, in some circumstances, inhibiting it.

axial skeleton (*AXE-ee-ul*) The central supporting skeleton, including the skull, backbone, ribs, and breastbone.

axon (*AXE-on*) The long, slender process that conducts impulses away from a cell body.

Barr body The condensed, inactive *X* chromosome found within the nuclei of female cells during embryonic development.

basal firing rate The slow, steady pace at which a resting muscle spindle produces action potentials.

basal metabolic rate The minimal rate of energy use that is required to sustain vital functions in an organism at complete rest.

basidiomycetes (*bass-ID-ee-oh-MY-seets*) Fungi that produce spores in reproductive structures known as basidia; this group includes the mushrooms, puffballs, and shelf fungi.

Batesian mimicry (*BAIT-see-un*) A form of mimicry in which a harmless species resembles a poisonous or otherwise protected species; the harmless species thereby gains protection.

behavioral isolation A prezygotic isolating mechanism that functions when two species do not interbreed because of behavioral incompatibility.

benign tumor (*bih-NINE*) A type of tumor that does not metastasize and can therefore often be removed surgically.

beta carotene (*BAY-ta CARE-a-teen*) A reddish-orange accessory photosynthetic pigment that absorbs light in regions of the spectrum where chlorophyll absorbs less.

beta particle (*BAY-ta*) A fast-moving electron such as those released during radioactive decay.

beta sheet (*BAY-ta*) A chemical configuration of the secondary structure of proteins in which hydrogen bonds link regions of the polypeptide chain that run in opposite directions.

bilateral symmetry (*bye-LAT-er-al*) Symmetrical organization in which, when an organism is divided down a central longitudinal plane, the two halves are mirror images of one another. Worms and many higher organisms exhibit bilateral symmetry.

binary fission (*BYE-na-ree*) A process of prokaryotic cell division. DNA is replicated while attached to the cell membrane; the membrane between the DNA molecules grows, separating the DNA molecules; ultimately, the cell splits, each daughter cell receiving a copy of the DNA.

binomial nomenclature (*bye-NOME-ee-al NO-men-clay-chur*) The naming system in which each organism is designated by two Latin names, one for genus and one for species.

biogeochemical view The thesis that the physical and chemical characteristics of the earth and/or its atmosphere are governed by both physical and biological processes.

biogeography (*BYE-oh-jee-OG-ra-fee*) The study of the geographic distribution of species.

biological diversity The variety of organisms in an area, encompassing diversity in such realms as genetics, species, and ecosystems, among others.

biological magnification The process by which substances such as toxic pollutants come to be found in increasing concentrations in the tissues of organisms at higher trophic levels.

biome (*BYE-ome*) A major ecological community inhabited by certain characteristic types of plants and animals. The biomes of the world include tundra, taiga, temperate forest, desert, grassland, and tropical rain forest.

biosphere (*BYE-oh-sfeer*) Collectively, the regions of the earth that support life, including the lower atmosphere, the earth's surface, and water environments.

biotic environment (*bye-AH-tik*) The living aspects of an organism's environment.

bivalve (*BYE-valve*) A clam, oyster, or scallop; a mollusc that has two shells, or *valves*; most bivalves are sedentary filter-feeders.

bladder In an excretory system, the membranous sac in which urine is stored before being excreted.

blastocyst (*BLASS-toe-sist*) In mammals, the equivalent of the blastula phase of embryonic development, in which an inner cell mass produces embryo tissues and an outer trophoblast produces the tissues that draw nourishment from the mother.

blastodisc (*BLASS-toe-disk*) In birds, the equivalent of the blastula phase of embryonic development, in which a double layer of cells on top of the yolk mass gradually separates into two layers with a hollow blastocoel between them.

blastula (*BLASS-chew-lah*) The hollow ball of cells that results from cleavage in early embryonic development.

B-lymphocyte (*B-LIM-foe-site*) A type of white blood cell that produces antibodies.

bone A type of strong, rigid connective tissue made up of a matrix of collagen that has gradually been mineralized by calcium crystals. The skeleton consists of bones, which provide structural support and protect internal organs.

Bowman's capsule In the kidney, the cup-shaped structure that much of the blood plasma enters after being filtered by the glomerulus.

bronchioles (*BRONG-kee-oles*) The progressively finer tubes into which the bronchi split; bronchioles bear alveoli at their tips.

bryophytes (*BRI-oh-fights*) The mosses, liverworts, and hornworts; a large, ancient group of small, nonvascular plants commonly found in moist terrestrial habitats.

buffer A substance that tends to stabilize pH by maintaining the relative concentrations of hydrogen and hydroxyl ions in solution. Buffers occur in living cells.

calorie (*KAL-a-ree*) The amount of heat energy needed to raise the temperature of one gram of water one degree Celsius.

Calvin cycle A stage of the dark reactions of photosynthesis in which two three-carbon molecules (PGA) are produced by adding a carbon dioxide molecule to a five-carbon sugar; the reaction is catalyzed by the enzyme RuBP carboxylase. Energy for the reaction is provided by NADPH and ATP produced by the light reactions.

cambium (*KAM-bee-um*) The lateral meristem, the cells in vascular plants that are responsible for lateral growth. Vascular cambium produces secondary xylem and phloem; cork cambium produces cork that fills spaces where phloem tissue has been split by growth.

cancer An invasive, uncontrolled growth of certain cells of an organism at the expense, and to the detriment, of other cells and the organism.

cannibalism The eating of members of one's own species.

capillary action (*KA-pill-air-ee*) The movement of a liquid along a tube caused by adhesion of the liquid's molecules to the tube walls and those molecules' cohesion to each other. Plants rely on capillary action to bring liquids to their upper branches.

carbohydrate (*KAR-bow-HIGH-drate*) An organic molecule in which carbon, hydrogen, and oxygen are present in a ratio of about 1:2:1. Sugars, starches, and cellulose are carbohydrates.

carbon cycle The cycle that carbon follows through the biosphere by such processes as photosynthesis, cellular respiration, and the decomposition and combustion of inorganic compounds.

carbon monoxide (*KAR-bun mun-OX-ide*) A colorless, odorless gas (CO) produced when fuel is burned under conditions of limited oxygen availability; can supplant oxygen at the binding sites of hemoglobin and cause rapid death.

carcinogen (*kar-SIN-oh-jen*) A chemical capable of causing cancer.

carnivore (*KAR-nih-vore*) An animal that subsists on other animals.

carpel (*KAR-pel*) The female portion of a flower, which contains the ovary, style, and stigma.

carrying capacity The maximum population of a species that a given environment can sustain for an extended period of time.

cartilage (*KAR-tih-lidg*) A type of tough, resilient connective tissue composed of collagen and other fibers; in humans, cartilage occurs at bone joints and in flexible structures such as the ears and nose.

Casparian strip (*kass-PAR-ee-un*) A thin, waxy strip contained in endodermal cell walls that prevents water from moving between cells.

caste Among social insects, a class of individuals of similar body form and behavior that performs a particular type of activity within the colony.

catabolic pathway (*CAT-a-BALL-ik*) A chemical pathway in which larger, more complex molecules are broken down into simpler, smaller molecules.

catalyst (*CAT-a-list*) A chemical that changes—especially that increases—the rate of a reaction without being consumed in the process. In effect, catalysts lower the activation energy of the reaction.

catastrophism (*KAT-a-STROFF-ism*) The view that accounted for the fossil record by contending that, after creation, life was destroyed in a series of catastrophes and repopulated by survivors or created anew.

cation (*CAT-eye-un*) A positively charged ion.

cell division The splitting of a cell to form two cells.

cell plate The double membrane that forms between the two halves of a dividing plant cell and produces the new cell wall.

cell theory The theory that all living organisms are composed of individual, living, self-reproducing structures known as cells.

cellular respiration Collectively, the pathways of glycolysis, the Krebs cycle, and electron transport, which produce ATP through the consumption of oxygen in reaction with an organic fuel.

centrioles (*SEN-tree-oles*) Barrel-shaped structures near the cell nucleus that serve as foci of the poles of mitotic spindles in animal cells.

centromere (*SEN-troe-mere*) The specialized region of a chromosome to which two sister chromatids are joined during mitosis.

cephalization (*SEFF-a-lies-ZAY-shun*) The evolutionary concentration of sensory and nerve tissue at one end of an organism, forming a head (or head end).

cephalopod (*SEFF-a-low-pod*) An octopus, squid, or nautilus, which are predatory marine molluscs. Octopi and squid are shell-less, have good vision and advanced nervous systems, and are good swimmers.

cerebral cortex (*su-REE-brul KOR-tex*) The convoluted outer layer of gray tissue on the surface of the cerebrum that is responsible for most of the higher functioning of the nervous system.

C4 pathway A pathway of carbon fixation in which carbon dioxide is first incorporated into oxaloacetate (which contains four carbon atoms) before the Calvin cycle is initiated.

chelicerate (*ke-LISS-er-ate*) A member of a group of arthropods characterized by a two-part body consisting of a cephalothorax and abdomen and by the presence of chelicera and pedipalps. This group includes the ticks, mites, scorpions, spiders, and horseshoe crabs.

chemical energy A form of potential energy stored in molecular structure, such as the energy in gasoline.

chemiosmosis (*KEM-ee-oz-MOE-sis*) The production of ATP from ADP through the operation of an electrochemical gradient across a cell membrane. The gradient results from

chemiosmosis (*continued*)
the buildup of protons pumped through the membrane by the electron transport chain.

chimera (*ki-MEER-a*) In genetic engineering, a combination of a DNA fragment and a plasmid; generally, an organism that has tissues from two or more genetically distinct parents.

chiton (*KYE-tin*) A member of a class of marine herbivorous molluscs with shells divided into eight connected plates.

chlorophyll (*KLOR-oh-fill*) Any of several green pigment molecules that absorb light during the process of photosynthesis.

chloroplast (*KLOR-oh-plast*) A eukaryotic organelle that contains chlorophyll and other substances associated with photosynthesis.

chlorosis (*klor-OH-sis*) Yellowing of older leaves resulting from a magnesium deficiency that prevents the plant from replacing damaged chlorophyll quickly enough to maintain the normal green color.

cholesterol (*kol-ESS-tur-awl*) An essential steroid component of animal cell membranes and a precursor of other steroids.

Chondrichthyes (*kon-DRIK-thee-eez*) The class of fishes with flexible skeletons made of cartilage rather than bone; includes the sharks and rays.

chordates (*KOR-dates*) A phylum of animals that have a hollow dorsal nerve cord and, at least in embryonic stages, a notocord, a series of pharyngeal gill slits, and a tail that continues past the end of the digestive tract.

chorion (*KOR-ee-on*) The outermost layer of tissue surrounding the embryos of mammals, birds, and reptiles; in mammals it is a major component of the placenta.

chromatids (*KROW-ma-tids*) The two strands of a duplicated chromosome that are connected to the centromere before and during nuclear division.

chromatin (*KROW-ma-tin*) The complex of protein and DNA in the eukaryotic nucleus.

chromosome (*KROW-ma-sowm*) The structure in the nucleus of a eukaryotic cell, consisting of protein and DNA, that contains part or all of an organism's genetic inheritance.

chrysophytes (*KRICE-oh-FIGHTS*) Photosynthetic protists, including single-celled algae such as diatoms, golden-brown algae, and yellow-green algae. Many are flagellated; diatoms have silica shells.

chylomicrons (*KI-low-MY-krons*) Lipid droplets in the endoplasmic reticulum, formed from fatty acids that have been resynthesized into triglycerides. Chylomicrons are absorbed by the lacteals of the lymphatic system and eventually enter the general circulation.

chyme (*KIME*) The product of the digestive processes of the stomach, a nutrient-rich milky slurry that is sent to the small intestine.

ciliates (*SILL-ee-ates*) Protists characterized by the presence of many rapidly beating hairs, or cilia; may be either free-living or sedentary.

cilium (plural: **cilia**) (*SILL-ee-um*) A whip-like structure less than 20 micrometers long and used in cell locomotion.

circulating tissue The blood and lymph, including both cellular and fluid components.

classical conditioning The association of an initially neutral stimulus with a stimulus that elicits a particular response such that the formerly neutral stimulus alone also comes to elicit the response.

cleavage (*KLEE-vej*) The rapid division of cells of the early embryo from the stage of the fertilized ovum to the blastula or blastocyst stage.

cleavage furrow (*KLEE-vej*) The progressive constriction in the cytoplasm between two nuclei that forms during the process of cytokinesis in animal cells.

climate The weather conditions, including temperature, rainfall, wind, hours of sunlight, and other factors, that *on average* or *typically* prevail in a particular area; compare *weather*.

climax community The ultimate stage in a succession sequence; during this stage, the ecological community remains stable as long as environmental conditions do not change.

clone (*KLOHN*) A group of genetically identical cells descended from a single ancestor cell.

clotting factor One of the proteins needed for blood clotting; absent in hemophiliacs.

codominance (*koh-DOM-ih-nance*) In genetics the situation in which two alleles of a gene are both expressed in the heterozygote.

codon (*KOH-dawn*) A nucleotide triplet of three bases that codes for insertion of an amino acid or for chain termination at a particular location during protein synthesis.

coelacanth (*SEEL-a-kanth*) A primitive fleshy-finned fish long thought to be extinct but rediscovered in 1938.

coenzyme (*KOH-en-zime*) A cofactor that is an organic molecule. Many vitamins that humans require in their diets are coenzymes that they cannot synthesize themselves.

coenzyme A (*KOH-en-zime*) A molecule of adenine, ribose, pantothenic acid, and sulfur that permits two carbon atoms at a time to enter the Krebs cycle by passing the acetyl group to a molecule of oxaloacetic acid.

coevolution (*KOH-ehv-eh-LEW-shun*) The evolution of two species in close association such that they reciprocally influence one another's adaptations.

cofactor (*KOH-fak-tore*) A molecule or ion required by some enzymes to function properly.

columnar epithelium (*call-UM-nar ep-ih-THEE-lee-um*) A tissue composed of large, roughly columnar cells that function in secretion and absorption; found in such locations as the stomach and intestinal linings.

commensalism (*kum-MEN-sul-izm*) A type of symbiosis in which one organism benefits and the other is neither harmed nor benefited.

common descent Descent of diverse types of plants or animals from a shared ancestor

communication The passage of information from one animal to another such that the subsequent behavior of the recipient is affected.

competition Rivalry between organisms that require the same or similar resources over access to these resources when they are in limited supply.

competitive exclusion The principle that when two species live in the same environment at the same time and both require one or more of the same limiting resources, one species always drives the other to extinction locally.

competitive inhibitor A molecule that blocks the active site of an enzyme and prevents it from being occupied by a substrate.

complement proteins A group of circulating blood proteins that, when they encounter an antibody bound to a cell surface, attach to the cell membrane and puncture it, causing the cell to burst. Complement proteins destroy cells too large to be eaten by phagocytes.

compound A substance that is composed of two or more elements in fixed proportions and the characteristics of which generally differ from those of its constituent elements.

compound eye The type of eye found among arthropods, consisting of many individual "eyes" called ommatidia, each with a lens-like structure and a "retinula," or small retina, containing photoreceptor cells.

conception (*kon-SEP-shun*) Fertilization of an ovum.

condensation reaction A chemical reaction common in the formation of biological polymers in which monomers join through the removal of water molecules. One monomer loses a hydrogen ion and the other loses a hydroxyl group; the monomers form a new bond, and the hydrogen and oxygen can join to form a water molecule.

cones One of the two classes of photoreceptors in the neural retina. Cones function well only in relatively bright light and are responsible for the sharp, full-color visual images we receive in daylight.

conifers (*KON-ih-ferz*) A group of gymnosperms, the reproductive structures of which are contained in male and female cones; the conifers include the pines.

connective tissue The tissue that makes up the basic support structures of the body, including the bones, ligaments, and tendons.

contact inhibition The phenomenon commonly observed in cells growing in tissue culture that is marked by cessation of cell division when the cells come in contact with each other.

continental shelf A relatively shallow, downward-sloping, shelf-like extension of the continental shoreline into the sea, terminating abruptly at a steep, downward declination into the oceanic abyss.

contraceptive (*KON-trah-SEP-tiv*) A device or technique designed to prevent conception.

contractile vacuole (*kon-TRAK-tile VAK-you-ole*) An organelle found in some single-celled organisms that accumulates the cell's excess water and contracts, pumping it out of the cell.

control In an experiment, the subject or group to which the experimental variable is not applied.

controlled experiment A scientific procedure in which two parallel trials are performed, their conditions differing in only one introduced factor. Variations in results can then be reliably attributed to the introduced factor.

convergent evolution The process by which unrelated organisms exposed to similar environments and selection pressures come to resemble one another.

convoluted tubule (*KON-va-LOO-tid TUBE-yule*) The kidney tubule into which the primary filtrate empties; functions in reabsorption of salt and water molecules into the blood.

coral reef A large limestone structure constructed by the secretions of small animals in nutrient-poor tropical seas; home to a diverse community of plant and animal life.

coronary thrombus (*KOR-a-NAIR-ee THRAHM-bus*) A small blood clot that becomes lodged in one of the coronary arteries, blocking blood flow to the heart.

corpus luteum (*KOR-pus LOO-tee-um*) A structure that develops from a ruptured ovarian follicle after that follicle produces an egg; secretes estrogen and progesterone.

countercurrent flow The flow of adjacent fluids in opposite directions. In gills, countercurrent flow maximizes rates of gas exchange; in kidneys, a countercurrent system functions to concentrate waste products in urine.

coupling factor An enzyme that connects electron transport to the synthesis of ATP during photophosphorylation.

covalent bond (*koh-VAY-lent*) A chemical bond formed when two atoms share a pair of electrons.

creatine phosphate (*KREE-a-teen*) The compound in which most of the quickly available energy in vertebrate muscles is stored; contains a high-energy phosphate bond.

cristae (*KRISS-tee*) Intrusive convolutions of the inner membrane of mitochondria.

Cro-Magnon (*krow-MAG-nun*) An early form of *Homo sapiens* that appeared in Africa about 100,000 years ago and migrated throughout Europe and Asia.

C3 pathway In photosynthesis, a pathway of carbon fixation in which the first stable compound produced has three carbons; the compound is created through the Calvin cycle.

cuboidal epithelium (*kyew-BOY-dull ep-ith-EEL-ee-um*) A tissue composed of roughly cubical cells that generally function in secretion;

found in such locations as the kidneys and a variety of glands.

cyanobacteria (*sy-AN-oh-bak-TEER-ee-uh*) Prokaryotic bacteria that carry on photosynthesis; some are capable of nitrogen fixation.

cyclic adenosine monophosphate (CAMP) (*SICK-lick ah-DEN-a-seen mon-oh-FOSS-fate*) A molecule that functions as a chemical messenger in slime molds; in vertebrate endocrine systems, CAMP functions as an intracellular second messenger.

cyclic electron transport A pattern of electron transport in which excited electrons move out of a photosynthetic reaction center, pass through a series of carrier molecules, and are then returned to reduce the reaction center from which they originated. Involves photosystem I and produces ATP.

cystic fibrosis A heritable, progressive, chronic disease affecting primarily the lungs and pancreas and leading to death, usually in early childhood.

cytokinesis (*SY-toe-kie-NEE-sis*) The process by which the cytoplasm of a cell divides following mitosis.

cytokinins (*SY-toe-KYE-nins*) Plant hormones that work with auxin and complement its effects, stimulating cell division and regulating plants' shape and growth.

cytoskeleton (*SY-toe-SKEL-e-ton*) A skeleton-like structure in the cell, composed of microtubules, microfilaments, and intermediate filaments, that helps support the cell and aids in locomotion.

cytotoxic T-cell (*SY-toe-TOK-zik*) A type of T-cell that directly attacks an antigen-bearing foreign cell and destroys it by attacking the cell membrane.

dark reactions The second stage of photosynthesis, in which the ATP and NADPH produced in the first stage are used to make sugars and other compounds; these reactions occur in the stroma of the chloroplasts.

day-neutral plant A plant the flowering of which is not affected by the duration of daylight.

DDT (*dichlorodiphenyltrichloroethane*) An insecticide that persists in the environment, accumulates in fatty tissues, and can have harmful environmental consequences; once widely used in the United States but now banned.

deductive reasoning Reasoning in which a conclusion by necessity follows from a given premise. (In the scientific method, the hypothesis derived by induction represents the premise used in deduction, and the conclusion that follows may represent a prediction).

deforestation The destruction of forests; some experts contend that deforestation to produce land for agriculture and cattle ranching will destroy all the world's tropical rain forests by the year 2000.

deletion An alteration in chromosome structure in which part of a chromosome is missing.

demography (*de-MAWE-gra-fee*) Study of the characteristics of a population, such as size, age distribution, fertility, mortality, and survivorship.

dendrite (*DEN-drite*) One of the many finely branched processes extending from the cell body of a neuron. Generally, one set of dendrites picks up information from the environment and carries it to the cell body; another set receives information from axons and distributes it to other neurons or to effector cells.

deoxyribonucleic acid (DNA) (*DEE-oxx-ee-RYE-bow-new-CLAY-ik*) A double-stranded helical nucleic acid molecule containing genetic information capable of replicating itself and thereby passing on genetic instructions from one generation of cells to the next.

dermal tissue (*DER-mul*) The outer layer of cells that protects a plant.

desert A biome characterized by very low annual rainfall and by plants and animals specially adapted to living in dry conditions.

desertification The conversion of forest, grassland, or other ecological communities to desert, often through habitat destruction or poor farming methods.

deuterostome (*DEW-ter-oh-stome*) A member of one of two evolutionary lines distinguished by differences in embryonic development. In deuterostomes, cell division produces a radial arrangement, the ultimate function of each cell being determined relatively late in development. The coelom is formed by buds that branch and separate from the embryonic gut. The embryonic gut opening becomes the anus, and a new opening forms the mouth.

diabetes mellitus (*die-a-BEE-tis MEL-ih-tus*) A disorder caused by failure of the pancreas to produce enough insulin, which prevents tissues from absorbing sugar from the blood at the proper rate.

diaphragm (*DI-a-fram*) The muscular partition that separates the thoracic and abdominal cavities; functions in expanding and contracting the thoracic cavity to carry out respiration.

diastole (*die-AS-ta-lee*) The state of the heart when the ventricles relax.

differentiation The process through which cells produce specialized structures or come to perform specialized functions over the course of development

diffusion The process through which a substance moves passively from areas where it is in higher concentration to areas of lower concentration.

dinoflagellates (*DINE-oh-FLAJ-a-lates*) "Armored" flagellates, protists with bodies encased in tough cellulose plates. Most are capable of photosynthesis; they are an important component of phytoplankton.

diploid (*DIP-loid*) Having the full complement of two sets of chromosomes, one set from each parent.

directional selection A form of natural selection that favors individuals at one end of a range of phenotypic expression and thereby shifts the phenotype of the population as a whole.

disruptive selection A form of natural selection that reduces the numbers of individuals that exhibit the moderate range of a trait and favors expression at the extremes of the range.

disulfide bond (*die-SUL-fide*) A covalent bond between two cysteines that under certain conditions links two regions of a polypeptide chain. The two cysteine molecules may be remote from one another in the primary sequence of the molecule.

dominance hierarchy A social "pecking order" among members of a group.

dominance, law of The principle that when two or more forms of the same gene exist, one expresses itself rather than the other(s).

dominant gene A gene that expresses itself in the phenotype when paired with either a dominant or a recessive gene.

Down syndrome A genetic disorder in humans caused by the presence of three copies of chromosome 21; results in mental retardation, reduced resistance to disease, and greatly lowered life expectancy.

double-blind study An experimental procedure for ensuring the objectivity of an observation by concealing from the observer, until the conclusion of the experiment, the identity of both the experimental sample and the control sample.

duodenum (*doo-oh-DEE-num*) The upper portion of the small intestine that is connected to the stomach.

ecdysone (*EK-die-sone*) A hormone that is produced by the arthropod prothoracic gland and induces molting.

echinoderms (*e-KINE-oh-derms*) A phylum of marine invertebrates characterized by radial symmetry, often spiny skin, and a water-vascular system used in feeding, locomotion, and respiration. Examples include starfish and sea urchins.

echolocation (*EK-oh-low-KAY-shun*) A system through which bats, marine mammals, and certain other animals use echoes of their own cries to navigate and to locate prey.

ecological isolation (*eek-oh-LOJ-ih-KUL*) A prezygotic isolating mechanism that functions when two species live in close proximity but are adapted to radically different environments.

ecological pyramid (*eek-oh-LOJ-ih-KUL*) The pyramid-shaped representation of successive trophic levels, each of which is smaller than the one below because only about 10 percent of the energy stored in one level is ultimately contributed to the biomass of organisms at the next higher level.

ectoderm The outermost tissue or "germ" layer of triploblastic embryos.

ectotherm (*EK-toe-therm*) An animal that depends on external sources of energy to regulate its body heat.

effector (*ee-FEK-tore*) A molecule that regulates the activity of an enzyme, either increasing or decreasing the rate of an enzyme-catalyzed reaction.

egg cell The gamete, produced by the female gametophyte, with which a sperm nuclei fuses to form a diploid zygote

electron transport chain A sequence of enzymes on the inner mitochondrial membrane that is critical to oxidation-reduction reactions as electrons are carried from one enzyme to the next. The process releases energy used in ATP formation and in other reactions.

electrophoresis (*eh-LEK-troh-fer-EES-iss*) The migration of weakly charged molecules or colloidal particles through a fluid or gelatinous medium under the influence of an electric field imposed on that medium.

embolus (*EM-buh-luss*) A detached blood clot that drifts through the circulatory system until it lodges in an artery too small for it to pass.

embryo sac The seven-celled female gametophyte of an angiosperm, in which the embryo develops.

emigration The departure of individuals from a population or geographic area.

endemic species (*en-DEM-ik*) A species that is native to a particular location and occurs nowhere else.

endocrine (*ENN-doh-krinn*) Pertaining to a ductless gland that secretes into the blood or tissue fluids, or to the secretion (a hormone) of such a gland.

endoderm The innermost tissue layer of triploblastic embryos.

endorphins (*en-DOR-fins*) Polypepides secreted by the pituitary and brain that mimic opiates in alleviating pain.

endoskeleton (*EN-doe-skell-e-ton*) A system of internal bones and joints, such as that of vertebrates, that functions in support and locomotion.

endosperm (*EN-doe-sperm*) The triploid nutritive tissue developed by flowering plants that nourishes the embryo during early growth.

endotherm (*EN-doe-therm*) An animal capable of sustaining a constant body temperature by generating heat in its tissues.

entropy (*EN-troe-pee*) A measure of the degree of disorganization of a system. As the energy in a system is dispersed, the entropy increases.

environment The external biological and physical conditions that influence an organism.

enzyme (*EN-zime*) A substance that serves as a biological catalyst, allowing chemical reactions to take place at normal cell temperatures at rates that would otherwise be far too slow to support life.

epiglottis (*EP-ih-GLOT-iss*) A small flap that closes over the entrance of the trachea and prevents food from entering it during swallowing.

epithelial tissues (*EP-ih-THEE-lee-al*) The tissues that form surfaces on and within the body. These surfaces include the skin; the linings of the digestive system, blood vessels, and body cavities; and many secretory glands. Such tissues are composed of sheets of tightly packed cells.

equilibrium 1. The stage in a chemical reaction at which the rates of the forward and reverse reactions are equal, such that the net amount of products or reactants is constant. 2. The sense of balance. In humans, the sense of equilibrium is mediated by hair cells in chambers in the inner ear. A jelly-like material presses down certain hair cells whenever the head changes position, causing a neural message to be sent to the brain.

erythrocytes (*ur-RITH-row-sites*) The red blood cells, which are produced by cells in the bone marrow and lack nuclei.

esophagus (*ee-SOFF-a-gus*) The muscular tube that transports food from the pharynx to the stomach.

estrogen (*ESS-troe-jen*) A female steroid sexual hormone that is produced in a follicle in the ovary and helps prepare the uterus lining for pregnancy. Estrogen is also active in the development of secondary female sexual characteristics.

estrus (*ESS-truss*) In nonprimate mammals, the period when a female is sexually receptive

estuary (*ESS-chew-air-ee*) A highly fertile coastal area where freshwater runoff from land mixes with seawater.

ethology (*ee-THOL-oh-gee*) The study of animal behavior in its natural environment.

ethylene (*ETH-a-leen*) A gaseous plant hormone that influences fruit formation and ripening. Its production is triggered by auxin.

eubacteria (*YOU-bak-TEER-ee-a*) A diverse group of prokaryotic bacteria, including many of importance to humans. Most of the bacteria that exist today are eubacteria.

euglenids (*you-GLEE-nids*) Flagellated photosynthetic protists, often common in fresh water. Members of the genus *Euglena* can function as autotrophs in sunlight but may exist as heterotrophs in darkness and under certain other conditions.

eukaryotic cell (*YOU-car-ee-AWE-tik*) A cell containing a true nucleus and other organelles that are individually bounded by membranes.

eutherians (*you-THEER-ee-uns*) The "true" mammals, each female of which produces a placenta for nourishing a developing fetus.

eutrophic (*you-TROFE-ik*) Containing many nutrients. Because of heavy phytoplankton growth, eutrophic lakes are usually not clear.

evolution The process by which populations of organisms change over time through such mechanisms as mutation, natural selection, and inheritance.

evolutionary adaptation A genetically controlled characteristic that increases an organism's evolutionary fitness.

evolutionary fitness The measure of an organism's capacity to pass on its genes to the next generation.

evolution through inheritance of acquired characteristics Lamarck's theory that changes in organisms over time resulted from the passing on to offspring of characteristics that individuals developed during their lives.

exocrine (*EX-o-krinn*) Pertaining to a gland that secretes onto a free surface either directly or via a duct, or to the secretion (a hormone) of such a gland.

exon (*EX-on*) The parts of a eukaryotic gene code that are expressed.

experimental variable In an experiment, the factor introduced into the experimental group (but not into the control group) to assess the factor's influence.

exponential growth Population growth in which the growth rate increases as the size of the population increases. For example, a pair of rabbits might have 6 offspring, which if unrestrained by environmental factors might have 36 offspring, which could have 216 offspring, and so on. Also known as *logarithmic growth*.

exteroceptor (*ex-TARE-oh-sep-tor*) A receptor that monitors conditions impinging on the organism from the exterior, such as taste, temperature, light, touch, and certain kinds of pain.

extracellular matrix A layer of fibers and a ground substance surrounding and produced by the cells of connective tissues; may be fluid or solid, loose or dense; responsible for many of the properties of the various connective tissues.

facilitated diffusion A form of diffusion in which molecules cross membranes by passing through special pores or by using transport molecules that facilitate their passage. Like simple diffusion, facilitated diffusion is driven by a concentration gradient.

fallopian tubes (*fah-LO-pee-un*) In the female reproductive system, the tubes that convey the mature ova from the ovaries to the uterus.

fatty acid A monomer consisting of a long hydrocarbon chain with a carboxyl group at one end. Fatty acids are a building block of most lipids.

feedback control A self-regulatory mechanism by which the effect of a given activity or function influences that very activity or function. In negative feedback, the activity is inhibited or depressed. In positive feedback, the activity is enhanced.

fermentation A pathway of carbohydrate metabolism that produces relatively small amounts of ATP from glucose without an electron transport chain. A substance such as ethanol or lactic acid is also produced.

fertility rate The rate at which organisms in a given population produce offspring.

fertilizers Essential nutrients applied by humans to improve the condition of soil for growing crops.

fibroblast (*FYE-bro-blast*) A type of cell, common in loose connective tissue, that produces the fiber-like proteins of the extracellular matrix.

fitness The physical and behavioral characteristics of an organism that enable it to survive and reproduce in a particular environment.

fixed action pattern A neurally preprogrammed, unvarying response to a sign stimulus.

flagellates (*FLAJ-e-lates*) Simple animal-like protists, many of which are parasites, such as members of the genus *Trypanosoma* that cause African sleeping sickness.

flagellum (plural: *flagella*) (*fla-JEL-um*) A whip-like structure between 20 and 100 micrometers long; used in cell locomotion.

flame cell A component of the excretory system of flatworms that pumps out of the body the water and nitrogen wastes collected in tubules that extend throughout the organism.

flower The reproductive organ of an angiosperm, typically made up of sepals, petals, stamens, and carpels.

fluid mosaic model The most widely accepted theory of cell membrane structure, which posits that the membrane consists of a fluid lipid bilayer in which individual protein molecules drift.

food chain A sequence of consumption among organisms; for example, a plant may be eaten by a beetle, which is eaten by a bird, which is eaten by a cat.

food web The complex set of interrelationships among members of different trophic levels in an ecosystem.

founder effect The influences on the genetic differentiation of an isolated population that result from the fortuitous genetic configuration of its founders.

frameshift mutation A mutation caused by the insertion or deletion of a number of nucleotides that is not a multiple of 3. Frameshift mutations disrupt the reading of every condon "downstream" from the point of the alteration.

fruit The ripened ovary of a gymnosperm, which serves as a protective covering for seeds and, in many plants, aids in seed dispersal.

fungi (*FUN-jie*) The kingdom of nonphotosynthetic unicellular or multicellular plants. The fungi exhibit a broad range of adaptations, functioning as decomposers, symbionts, predators, and parasites. Includes the yeasts, molds, and mushrooms

gall bladder A sac that accumulates secretions of bile from the liver and passes them on to the duodenum.

gamete incompatibility (*GAM-eet*) A prezygotic isolating mechanism that functions when two species cannot interbreed because their sperm and eggs cannot successfully unite in fertilization.

gametes (*GAM-eets*) The haploid germ cells (sperm and egg) that unite during fertilization to form a zygote.

gametogenesis (*gam-ee-toh-JENN-eh-siss*) The process by which gametes, or sex cells, are produced.

gametophyte (*gam-EET-oh-fight*) The haploid, gamete-producing stage of the plant life cycle.

gap junction A small, pore-like protein channel that connects the cytoplasm of one cell to that of another and permits certain compounds to pass between them. A gap junction functioning as an electrical synapse permits the free flow of ions between cells and thereby avoids the transmission delay of several milliseconds that occurs in chemical synapses.

gastropod (*GAS-troe-pod*) A snail or slug. Members of this class have a large, fleshy foot. Many have a spiraled shell, and most are herbivorous and aquatic. Some carnivorous and many land-dwelling forms also exist.

gastrovascular cavity (*GAS-troe-VAS-kyou-lar*) A sac-like digestive cavity with only one opening to the exterior, which is used both to ingest food and to expel undigested material; found in simple organisms such as hydras and planarians.

gastrula (*GAS-true-la*) A stage in embryonic development in which cells migrate and differentiate to produce distinct layers, generally ectoderm, endoderm, and mesoderm.

gene A unit of hereditary information located on a chromosome.

gene pool The totality of genes available for reproduction in a given population at a given time.

genetic drift Changes in a population's gene pool that result from chance alone.

genetic map A representation of the physical locations of genes on a chromosome.

genotype (*JEEN-oh-type*) An organism's hereditary makeup.

genus (*JEEN-us*) The taxonomic group just one step more inclusive than the species level, designated by the first of the two parts of an organism's Latin name.

geochemical view The thesis that the physical and chemical characteristics of the earth and/or its atmosphere are governed by strictly physical processes.

geographic isolation A prezygotic isolating mechanism that functions when two populations of a species live in different locations

geographic isolation (*continued*)
separated by a physical barrier, such as a mountain range or body of water.

geotropism (*JEE-oh-TROE-pizm*) A plant's response to gravity. Roots are positively geotropic, growing in the direction from which the force of gravity is exerted; stems are negatively geotropic, growing in the direction opposite to the force of gravity.

germination The reactivation of growth in a plant embryo that leads to the plant's sprouting from its seed.

gibberellin (*JIB-er-ELL-in*) A plant hormone thought to regulate the elongation of the internode regions of plant stems; also functions in seed germination.

gills Respiratory organs specialized for gas exchange in water; they collect oxygen and release carbon dioxide.

glomerulus (*glow-MARE-you-luss*) The net of thin capillaries through which blood plasma is filtered into the Bowman's capsule in the kidney.

glucagon (*GLEW-ka-gone*) A polypeptide hormone that is secreted by the pancreas and causes cells in the liver to release sugar into the bloodstream.

glucose (*GLEW-kose*) A six-carbon sugar that is a major source of nutrients for cells. Starch, cellulose, and glycogen are all polymers of glucose.

glycerol (*GLISS-er-all*) A molecule with three carbons, to each of which a hydroxyl group is covalently bonded. Glycerol is a component of many triglycerides and has a variety of commercial uses.

glycolysis (*gly-KOL-lih-sis*) The initial chemical pathway involved in the breakdown of glucose; common to all unicellular and multicellular plants and animals, glycolysis results in a relatively small net gain of ATP for the cell and produces pyruvate.

Golgi apparatus (*GOAL-jee*) A stack of flattened vesicles associated with the rough endoplasmic reticulum that function in modifying and storing secretions within the cell.

gonadotropins (*go-NAD-oh-TROE-pins*) Hormones, produced by the anterior pituitary, that control male and female sex organs. The major gonadotopins are luteinizing hormone (LH) and follicle-stimulating hormone (FSH).

gradualism The view that evolution proceeds through the gradual accumulation of small changes within species.

grassland A biome found in temperate and tropical areas with low to moderate rainfall. Grasses are the dominant plants. Also called *savannah*.

greenhouse effect Global warming as a consequence of elevated concentrations of carbon dioxide, water vapor, and certain other gases in the atmosphere that trap the heat of solar radiation.

greenhouse effect Earth's atmosphere's functioning like the glass of a greenhouse to retain heat. Solar energy in the form of visible and ultraviolet light is absorbed by plants and other materials and reradiated as infrared energy, which is trapped by certain gases in the atmosphere and thus prevented from escaping into space.

green revolution A dramatic increase in world agricultural output that began in the 1950s.

ground tissue Plant tissue that provides structural support; also contains photosynthetic cells.

ground water Water that exists beneath the earth's surface in porous rock formations.

gustation (*gus-TAY-shun*) The sense of taste.

gymnosperms (*JIM-no-sperms*) A group of vascular plants that produce "naked" seeds not enclosed in ovaries; includes the conifers.

habitat The location within an environment in which an organism actually lives or is commonly found.

habituation (*ha-BIT-you-AY-shun*) A simple form of learning in which an organism learns to ignore a stimulus that is of no importance to it.

half-life The time it takes for half of a radioactive substance to decay into a lighter substance.

halophiles (*HAL-oh-files*) Archaebacteria that inhabit extremely salty environments.

haploid (*HAP-loyd*) Having only one set of unpaired chromosomes.

Hardy–Weinberg law The principle that sexual reproduction does not change the relative frequencies of genes in a population.

helper T-cell The kind of T-cell that secretes the interleukin that stimulates B-cells to differentiate and produce antibodies or to secrete several other substances that influence immune system response.

heme (*HEEM*) In the hemoglobin molecule, the ring-like, iron-containing component that binds with oxygen.

hemoglobin (*HEE-moh-glow-bin*) The reddish protein in red blood cells that functions in carrying oxygen.

hemophilia (*HEE-moh-FILL-ee-yah*) Disease characterized by failure of blood to clot. Hemophilia is caused by a defective gene on the male's *X* chromosome and hence is transmitted from the mother.

herbicide (*ER-biss-ide*) A substance used to kill plants.

herbivore (*ER-biv-ore*) An animal that subsists solely on plants.

heritable (*HER-it-a-bull*) Capable of being passed on genetically by an organism to its offspring.

hertz (Hz) (*HURTS*) A unit of frequency equal to 1 cycle per second.

heterocysts (*HET-er-oh-sists*) Specialized cells, found in some cyanobacteria, that carry out nitrogen fixation.

heterotroph (*HET-er-oh-trofe*) An organism incapable of producing its own food from inorganic materials; such organisms depend, directly or indirectly, on primary producers to meet their food requirements.

heterozygous (*HET-er-oh-ZYE-gus*) Having two alleles for a single trait.

histamine (*HISS-ta-meen*) A substance released in response to injury by most cells near the skin surface; causes blood vessels near the wound to expand and attracts and directs the actions of white blood cells.

histone (*HISS-tone*) A member of a group of simple proteins that contain large numbers of positively charged amino acids, which bind to DNA strands making up chromatin.

homeostasis (*HOME-ee-oh-STAY-sis*) The maintenance of stable internal conditions by an organism despite variations in the external environment.

hominoids (*HOM-ih-noids*) An advanced group of anthropoids that includes modern and extinct ancestral humans and the living apes.

homologous chromosomes (*home-ALL-a-gus*) Chromosomes that possess genes for the same characteristics at the same loci. Sexually reproducing organisms generally inherit one such chromosome from the male parent and one from the female parent.

homologous structures (*home-ALL-a-gus*) Structures found in different species that have similar form or configuration because of the species' common ancestry. An example is the human arm and the bird wing.

homozygous (*HOME-oh-ZYE-gus*) Having identical genes for a particular trait.

hormone A substance that is produced in one part of an organism and affects the physiology of another part.

hormone–receptor complex A unit consisting of a hormone bound to a receptor in a target cell. Such a complex can activate or deactivate specific genes or groups of genes in the cell's nucleus.

human chorionic gonadotropin (HCG) (*kor-ee-ON-ik go-NAD-oh-TROPE-in*) A hormone, produced by a developing embryo and the surrounding tissue, that prepares the body to deal with pregnancy and prevents the hypothalamus from inducing development of new ova.

human immunodeficiency virus (HIV) (*im-MYUNE-oh-dee-FISH-en-see*) A retrovirus that causes AIDS by impairing the body's immune system.

humus (*HYUME-us*) Decomposing organic materials, which serve as an important source of nutrients and water-retaining capacity in soil.

Huntington's disease A rare human genetic disorder, caused by a dominant allele on the fourth chromosome, that leads to degeneration of the nervous system and death.

hybrid infertility A postzygotic isolating mechanism that functions when two species can reproduce, but their offspring are sterile.

hybrid inviability (*in-VYE-a-BILL-a-tee*) A postzygotic isolating mechanism that functions when two species can interbreed and produce a fertilized egg, but the embryo or newborn organism cannot survive to reproduce.

hydrogen bond A weak chemical bond in which an electronegative atom is attracted to a hydrogen atom that is already involved in a polar covalent bond.

hydrological cycle (*HIGH-droe-LOJ-ih-kull*) The cycle that water follows through an ecosystem via evaporation, condensation, precipitation, and runoff.

hydrolysis (*high-DRAHL-ih-sis*) A chemical reaction in the breakdown of polymers in which bonds between molecules are broken through the consumption of water molecules. Hydrogen becomes bonded to one monomer, and a hydroxyl group joins the adjacent monomer.

hydrophilic molecules (*HIGH-droe-FILL-ik*) Polar molecules that interact strongly with water and dissolve freely.

hydrophobic molecules (*HIGH-droe-FOE-bik*) Nonpolar molecules that do not interact with water and tend to cluster together; an example is oil in water.

hydroponic culture (*HIGH-droe-PAWN-ik*) A technique for growing plants in liquids; used to assess a plant's needs for nutrients and also to grow certain commercial crops.

hyphae (singular: **hypha**) (*HIGH-fee*) The thin filaments that compose the mycelium of a mushroom.

hypothalamus (*HIGH-poe-THAL-a-muss*) The part of the brain that lies below the thalamus and functions in the regulation of temperature and various other autonomic activities.

hypothesis (*high-POTH-ih-sis*) An assertion that can be tested through experimentation.

hypothyroidism (*HIGH-poe-THIGH-royd-ism*) A condition resulting from insufficient thyroxine levels; often caused by lack of iodine in the diet. Symptoms include lowered heart rate and blood pressure, sleepiness, energy loss, and weight gain.

ideal types In Greek philosophy, the idea that actual objects are imperfect manifestations of the mental images we create of their perfect forms.

imbibition (*IM-bie-BIH-shun*) A seed's absorption of water, which causes it to swell and initiates germination.

immigration Movement of organisms born elsewhere into a population of their own kind.

imprinting A type of learning in which an animal "customizes" an inherited behavior on the basis of the environmental information it encounters, often during a brief, critical period early in life.

inclusive fitness The idea promulgated by sociobiologists that an individual can increase the number of its genes that survive not only by reproducing itself but also by helping its relatives (and thus their genes) survive and reproduce.

incomplete dominance The condition in which neither gene for a trait is dominant and the offspring's trait blends characteristics of the trait found in both parents.

independent assortment, law of The principle that each pair of alleles segregates independently during formation of gametes.

inductive reasoning A process of reasoning whereby a general principle or hypothesis is derived from a series of specific observations.

infertile Incapable of producing functional gametes.

inhibitor A molecule that blocks the active site of an enzyme and prevents it from being occupied by a substrate.

innate behavior A behavior that is genetically preprogrammed such that, given the proper stimulus, an animal can perform it without having learned it or having been exposed previously to the stimulus that elicits it. Also called *instinct*.

insectivorous (*IN-sek-TIV-er-us*) Insect-eating. Certain plants trap and digest insects as a source of nutrients.

insects An enormously diverse class of arthropods; all have six legs and a body divided into three major sections, and many have two pairs of wings.

insight learning Learning that occurs when an animal applies previous experience to a completely new situation without trial and error.

instinct Behavior that is genetically preprogrammed such that, given the proper stimulus, an animal can perform it without having learned it or having been exposed previously to the stimulus that elicits it. Also called *innate behavior*.

insulin (*IN-su-lin*) A hormone, produced by beta cells in the islets of Langerhans, that stimulates cells to remove excess sugar from the blood and store it.

interspecific competition Competition among members of two or more species.

intraspecific competition Competition among members of a single species.

intron (*IN-tron*) A noncoding sequence of nucleotides in a eukaryotic gene.

inversion An alteration in chromosome structure in which part of a chromosome is turned upside down with respect to the rest of the chromosome.

invertebrate An animal that lacks a backbone.

ionic bond (*eye-ON-ik*) A chemical bond formed between oppositely charged ions.

isotopes (*ICE-oh-topes*) Forms of the atoms of an element that have differing numbers of neutrons.

juvenile hormone A hormone, produced by the corpora allata of arthropods, that prevents the development of adult characteristics in larvae.

karyotyping (*CARE-ee-oh-type-ing*) A technique for examining human chromosomes that entails adding colchicine to cultured white blood cells; the cells become locked in metaphase and the chromosomes can be analyzed in photomicrographs. The technique is often used to diagnose genetic disorders.

kelp A type of giant cold-water alga that forms large coastal "forests" that support a distinctive food web.

keystone predator The top carnivore in a food chain; such predators often have a powerful influence on the structure of the communities in which they live.

kidneys In vertebrates, the two organs located in the abdominal cavity that remove nitrogenous waste, salts, and excess water from the blood and form urine.

kinetic energy (*ki-NET-ik*) Energy of motion.

kin selection theory The theory that altruistic behavior can be adaptive if its benefit to an animal's inclusive fitness is greater than the loss to the animal's personal fitness that it entails.

Klinefelter syndrome A genetic nondisjunction abnormality in which the individual has two *X* chromosomes and one *Y* chromosome (47*XXY*); afflicted persons are male, but they are sterile, often unusually tall, and frequently mentally retarded.

Krebs cycle A cyclic chemical pathway, occurring in mitochondria, in which acetyl coenzyme A from pyruvic acid is used in the production of two molecules of carbon dioxide and four pairs of hydrogen atoms.

***K*-selected species** Species that maximize their carrying capacity. Such species exhibit late maturity, infrequent mating, and the production of relatively few offspring, to the rearing of which they may devote a good deal of care and energy. Also known as *equilibrium species*.

***lac* operon** (*LACK OP-er-on*) A set of protein-coding genes, regulated as a cluster, that affects the metabolism of lactose in *E. coli*.

larynx (*LARE-inks*) The upper part of the windpipe, which contains the vocal cords.

lateral line A sensory organ of fishes and amphibians that permits them to detect minute movements in water.

learning The modification of behavior as a result of past experience.

leukocytes (*LEW-ko-sites*) The white blood cells, which function in the immune system.

lichen (*LYE-kun*) An organism formed by a symbiotic partnership between a fungus and an alga or cyanobacterium.

light microscope An instrument that makes use of the magnifying properties of lenses to produce an enlarged image of minute objects.

light reactions The first stage of photosynthesis, in which the energy of sunlight is trapped and transformed into a chemical

light reactions (*continued*)
form, producing ATP and reducing NADP⁺ to NADPH.

limbic system (*LIM-bik*) The section of the midbrain responsible for the control of emotions.

limiting factors, law of The principle that the growth of an organism is limited when any factor essential to its growth is lacking, regardless of the quantity available of other factors.

lipids (*LIP-ids*) A diverse group of waxy or oily, generally hydrophobic substances that are soluble in organic solvents; most are hydrocarbons.

lipid bilayer (*LIP-id*) The structural foundation of a biological membrane, consisting of lipids arranged in two layers with their hydrophilic groups facing outward at the two surfaces of the bilayer and their hydrophobic groups gathered in the center of the bilayer.

logarithmic growth (*LOG-a-rith-mik*) Population growth in which the rate of growth increases as the size of the population increases. For example, a pair of rabbits might have 6 offspring, which if unrestrained by environmental factors might have 36 offspring, which could have 216 offspring, and so on. Also known as *exponential growth*.

logistic growth (*low-JIST-ik*) A pattern of population growth. Growth is initially slow because the number of reproducing individuals is low; the growth rate increases as the population grows, and it subsequently decreases as environmental limitations begin to have an effect. Growth ultimately levels off as the environment's carrying capacity is reached.

long-day plant A plant that flowers only when daylight lasts longer than a certain minimum. Such plants flower in late spring and early summer.

loop of Henle A section of the nephron (kidney) tubule, consisting of an ascending and a descending branch, that functions in the concentration of dissolved waste products in urine.

lungs The organs that carry out respiration in terrestrial vertebrates and certain other organisms. Gas exchange is conducted as the blood in the lungs flows past the thousands of tiny air sacs into which the lungs have ramified.

lycophytes (*LIKE-oh-fights*) The club mosses; an ancient group of seedless vascular plants dominant during the Carboniferous period. About 1000 small species survive today.

lymph (*LIMF*) The fluid in the vessels of the lymphatic system, which returns accumulated interstitial fluids to circulation, absorbs fat from the digestive tract, and helps move white blood cells to aid in the immune response.

Lyonization (*LYE-on-eye-ZAY-shun*) The model proposed by Mary Lyon suggesting that in a female cell, one of the two *X* chromosomes is inactive during embryonic development.

lysosome (*LYE-so-soam*) A small, membrane-bounded organelle filled with enzymes that function in digestion, in the destruction of damaged organelles, and in the killing of cells in locations where space for development is needed.

macromolecule (*MAK-roe-MOL-a-kyool*) A giant molecule such as occurs in living cells, formed by the aggregation of smaller molecules. Examples include nucleic acids, proteins, carbohydrates, and lipids.

macronutrients (*MAK-roe-NEW-tree-ents*) The nine elements that must be present in relatively large amounts for a plant to thrive: oxygen, carbon, hydrogen, nitrogen, potassium, calcium, magnesium, phosphorus, and sulfur.

macrophage (*MAK-roe-fayj*) A phagocyte that engulfs and digests pathogens, thereby beginning a process that helps T-cells recognize antigens and that galvanizes the immune system into action.

malignant (*ma-LIGG-nant*) In pathology, a term descriptive of the life-threatening, uncontrolled growth and invasive character of certain tumors.

malnourishment The condition of receiving an inadequate supply of one or more essential nutrients.

Malpighian tubules (*mal-PIG-ee-un TOOB-yules*) In insects, the system of ducts that removes nitrogen wastes, which are emptied into the digestive system and excreted through the anus.

mammals A class of endothermic vertebrates the members of which possess body hair and produce milk for their offspring.

mandibles (*MAN-dih-bulls*) Jaw-like structures found on the second or third body segment of crustaceans and uniramians.

mangrove (*MANG-grove*) A type of salt-tolerant tree or shrub found growing in shallow waters along tropical and subtropical coasts. Mangrove forests host distinctive communities of marine and terrestrial life.

marsupials (*mar-SOO-pee-als*) An order of mammals among whom newborn offspring are reared in a maternal pouch.

mass extinction event An episode in the history of life when large numbers of species became extinct.

mechanical isolation A prezygotic isolating mechanism that functions when two species cannot interbreed because their reproductive organs are structurally or functionally incompatible.

meiosis (*my-OH-sis*) The two-stage process by which the number of chromosomes in a cell nucleus is halved during the formation of gamete cells.

menopause (*MEN-oh-pawz*) The end of the menstrual cycle, which signifies that a woman's reproductive potential has ended.

menstruation (*MEN-strew-AY-shun*) The process by which the lining of the uterus is broken down and discharged if pregnancy has not occurred.

mesoderm (*MEZ-o-derm*) The middle tissue or cellular layer of triploblastic embryos.

metabolism (*meh-TAB-oh-lizm*) The sum of all the chemical reactions associated with life processes.

metaphase (*MET-a-faze*) The second stage of mitosis, during which the chromosomes come to be aligned along the mitotic spindle equator.

metastasize (*me-TAS-ta-size*) To spread from one site to another throughout the body, as malignant tumors may.

metatherians (*meh-ta-THEER-ee-uns*) The order of mammals also known as *marsupials*. Its members, such as kangaroos, rear newborn offspring in a maternal pouch.

methanogens (*meth-AN-a-jens*) A type of archaebacterium that lives in anaerobic mud and produces methane as a metabolic by-product.

metric system The decimal system of length and mass used by scientists. The meter (about 39 inches) is the standard unit of length, and the kilogram (about 2.2 pounds) is the standard unit of mass.

microclimate (*MIKE-roe-kly-met*) The climate of a microhabitat, including such characteristics as temperature, humidity, and wind speed.

microenvironment (*MIKE-roe-en-VYE-run-ment*) A small or minute environment. Ecologists study microenvironments because all of the conditions relevant to a particular organism may exist in a very small area.

microfilaments (*MIKE-roe-FILL-a-ments*) Fibers in the cytoskeleton, composed of the protein actin, that function in stabilizing cell shape, in cell movement, and in cell growth.

micronutrients (*MIKE-roe-NEW-tree-ents*) Elements essential to plant growth that need be present only in minute amounts to meet the plant's requirements: boron, chlorine, copper, iron, manganese, molybdenum, and zinc.

microtubules (*MIKE-roe-TOOB-yules*) Tube-like structural components of the cytoskeleton, composed of the protein tubulin, that are involved in providing support for the cell surface in mitosis and in constructing the cell's motile structures.

mimicry (*MIM-ik-ree*) The resemblance of one species to another; evolves through natural selection when such a resemblance has survival value for the mimic.

mineral nutrients In biology, the various inorganic nutrients necessary for an organism's health. Humans require at least 18 mineral nutrients.

minichromosome (*MIN-ee-KROME-oh-sowm*) A structure used in the transformation of yeast cells; similar to the plasmids used for the transformation of prokaryotic cells.

mitogen (*MY-toh-jen*) A substance that stimulates the initiation of cell division.

mitosis (*my-TOE-sis*) A process in eukaryotic cell division in which chromosomes in the cell nucleus are duplicated such that each daughter cell receives a complete set.

mitotic spindle (*my-TOT-ik*) A structure composed of microtubules that extends from one pole of a cell to another during mitosis and functions in chromosome distribution.

molecule (*MOLL-ih-kule*) A bonded aggregation of atoms of one or more elements. A molecule is the smallest unit of a compound that displays the characteristics of that compound.

mollusc (*MOLL-usk*) One of a diverse group of more than 100,000 soft-bodied animals, including slugs, snails, oysters, mussels, octopi, and squid. Molluscs are characterized by a visceral mass containing the internal organs; a muscular foot; a mantle, that may secrete a shell; and, in many cases, a head.

monoculture (*MON-oh-kull-chur*) A form of agriculture in which a genetically uniform variety of a single species of crop is grown over large areas.

monocyte (*MAHN-oh-site*) Small, roundish cells in the circulation that, when attracted to a wound, change into larger granulocytes that engulf bacteria.

mortality In a life table, the percentage of a population that can be expected to die at a certain age.

mosaic evolution (*moe-ZAY-ik*) The evolution of different sets of traits at different rates as organisms exploit new ecological opportunities.

motor unit A functional unit consisting of a single motor neuron and the collection of muscle fibers that it serves.

mouth The opening at the beginning of the alimentary canal through which an organism ingests food.

mucosa (*mew-KOE-sa*) The innermost layer of the gastrointestinal tract, consisting of a layer of epithelial cells that releases mucus, digestive enzymes, and other substances into the tract; a thin layer of connective tissue; and a thin layer of muscle.

Müllerian mimicry (*mew-LAIR-ee-un*) A form of mimicry in which two or more species that share a similar defense mechanism (such as unpalatability) resemble one another.

multiple fruits Fruits formed from several separate flowers clustered tightly together, such as a pineapple or fig.

muscle tissue Tissue composed of cells specialized to contract and cause internal and external movement. Muscles also produce heat and so are important in homeostasis. There are three types: smooth, skeletal, and cardiac.

mutagen (*MEW-ta-jen*) An agent capable of causing a mutation, such as radiation and certain chemicals.

mutation (*mew-TAY-shun*) A heritable alteration in an organism's gene structure.

mutualism (*MEW-chew-al-izm*) A form of symbiosis in which both organisms benefit.

mycelium (*my-SEEL-ee-um*) The tangled, branched structure of hyphae that makes up the main body of many fungi.

mycorrhizae (*MY-ko-RYE-za*) Fungi that live in a mutualistic relationship with the roots of many plants, providing the roots with moisture and mineral nutrients. The plant, in turn, provides the fungi with organic nutrients.

myocardial infarction (*my-oh-KARD-ee-ul in FARK-shun*) A heart attack, which occurs when a vessel that supplies blood to the heart is blocked and part of the heart muscle dies.

myoglobin (*MY-oh-GLOW-bin*) A molecule resembling hemoglobin that binds oxygen in muscle cells and releases it as needed during vigorous activity.

myosin (*MY-a-sin*) The protein that makes up the thick filaments of the myofibrils; myosin functions with actin to produce contractions.

natural selection The process through which organisms that exhibit the variations that make it best suited to thrive in the environment survive and produce more offspring than other organisms of the same species.

Neanderthals (*nee-AN-der-thawls*) An extinct subspecies of *Homo sapiens* that originated in Africa about 500,000 years ago.

negative feedback system A mechanism that maintains homeostasis in which a shift in a physiological variable ultimately produces conditions that inhibit the process that led to the shift.

nephridia (*nef-RID-ee-a*) A pair of excretory organs, found in each segment of an earthworm's body, that collect metabolic wastes and excess water and send them to a bladder for excretion.

nephrons (*NEF-rons*) The structures in the kidney that remove excess water and nitrogen wastes from the blood and produce urine. Each human kidney possesses about 1 million nephrons.

nerve A bundle of neurons in a sheath of connective tissue; nerves function in communication between the central and peripheral nervous systems.

nervous tissue Specialized tissue that senses stimuli and rapidly transmits information between parts of the body.

neuromuscular junction (*NURE-oh-MUSS-kyou-ler*) The junction between motor neurons and muscle cells.

neuron (*NURE-on*) A nerve cell, the principal component of the nervous system, consisting of a cell body, dendrites, and a long, slender axon.

neurotransmitter (*NURE-oh-TRANZ-mit-er*) A chemical messenger that diffuses from a presynaptic neuron across the synaptic cleft to the membrane of the postsynaptic cell, which it binds to and stimulates.

neurulation (*noor-yu-LAY-shun*) The formation of a neural tube in chordate animals.

neutral mutations Mutations that do not change the characteristics of the protein that the genes specify.

niche (*NISH*) The range of physical and biological conditions within which an organism can exist, and the manner in which the organism makes use of these conditions.

nitrogen cycle The cycle that nitrogen follows through organisms and the environment by such processes as nitrification, ammonification, nitrogen fixation, and denitrification.

noncompetitive inhibitor A negative effector; a molecule that decreases the rate of an enzyme-catalyzed reaction.

noncyclic electron flow (*non-SIH-klick*) An electron flow route that occurs during the light reactions of photosynthesis; involves both photosystem I and photosystem II and produces oxygen, ATP, and NADPH.

nondisjunction (*NON-dis-JUNK-shun*) Failure of the two sex chromosomes to separate properly during the first meiotic division; causes a variety of problems, including Turner and Klinefelter syndromes.

nonrenewable energy Energy derived from limited sources such as oil, coal, and natural gas.

notochord (*NO-toe-kord*) A flexible supporting structure that runs the length of the body between the gut and the nerve cord.

nuclear envelope (*NEW-klee-er*) The double membranes that surround the eukaryotic cell nucleus.

nuclear pores (*NEW-klee-er*) Openings in the nuclear envelope that permit materials to move in and out of the nucleus without passing directly through a membrane.

nucleic acid (*new-CLAY-ik*) A polymer composed of nucleotide monomers; the two main classes, ribonucleic acid and deoxyribonucleic acid, play an important role in the transmission of hereditary information.

nucleolus (*new-KLEE-oh-lus*) The section of the nucleus in which ribosomes are assembled from ribosomal RNA and the appropriate proteins.

nucleosome (*NEW-klee-oh-soam*) A tightly organized, bead-like structure containing a double coil of DNA wound around eight protein molecules; thought to function in packing DNA and perhaps in controlling the expression of genes.

nucleus (*NEW-klee-us*) In eukaryotic cells, the organelle that contains the DNA and controls various cell processes.

olfaction (*ole-FAK-shun*) The sense of smell.

oligotrophic (*ol-ih-go-TROE-fik*) Containing few nutrients. Oligotrophic lakes are clear and have little phytoplankton growth.

oncogene (*ONG-koe-jeen*) A gene capable of transforming normal body cells into cancerous cells.

operant conditioning (*OP-er-ent*) Conditioning in which the subject learns to perform a particular operation in order to receive a reward or to avoid a painful experience.

organ A structure made up of more than one tissue and specialized for performing a particular function or group of related functions; examples include the brain, liver, skin, pancreas, nerves, and lungs.

organelles (*or-gan-ELZ*) Specialized structures in the cytoplasm of eukaryotic cells that perform various functions.

organism A living thing.

organogenesis (*or-GAN-oh-JEN-ih-sis*) The origin and development of organs in an embryo.

organ system A group of organs that are physically or functionally related; the mouth, stomach, and large intestine, for example, form part of the digestive system.

orgasm (*OR-gazm*) The culmination of sexual excitement, accompanied by ejaculation in the male and by pleasurable sensations in both sexes.

osmosis (*oz-MOE-sis*) The movement of water molecules across a semipermeable membrane in response to a concentration gradient.

Osteichthyes (*OS-tee-IK-thee-eez*) The bony fishes; the class made up of fishes with skeletons of bone rather than cartilage; contains about 30,000 species.

osteoblast (*OS-tee-oh-blast*) A cell that secretes bone tissues.

osteoclasts (*OSS-tee-oh-CLASTS*) Bone-dissolving cells that play a vital role in the modeling of developing bone and as an adjunct to calcium metabolism.

osteoporosis (*OS-tee-oh-pore-OH-sis*) A condition found in older persons in which mineral loss results in weakened bones that break under minimal stress.

ostracoderms (*os-TRAK-a-DERMS*) An extinct type of jawless fish. The bodies of ostracoderms were encased in an armored covering of bony plates.

ovary (*OH-var-ee*) In animals, the primary female sexual organs that produce the female gametes and sex hormones; in plants, the part of the pistil that contains the ovules.

overexploitation (*OH-ver-EX-ploy-TAY-shun*) The harvesting of stock (such as fish) at a rate exceeding the stock's ability to replenish its numbers.

ovulation (*oh-vyou-LAY-shun*) The release of a mature ovum from an ovarian follicle.

oxidation (*OX-ih-DAY-shun*) Any chemical reaction in which electrons are removed from an atom or a compound.

oxygen-evolving apparatus In plants, a set of enzymes in the photosynthetic membranes that are capable of removing electrons from water; these electrons are used to replace those lost by the chlorophyll in photosystem II.

oxytocin (*OX-ee-TOE-sin*) A hormone released from the anterior pituitary that stimulates uterine contractions during labor.

ozone (*OH-zone*) O^3, a gas each molecule of which consists of three atoms of oxygen. The ozone layer in the earth's atmosphere, which protects the planet's surface from potentially dangerous ultraviolet radiation, is currently threatened by compounds, such as some found in aerosol sprays, released into the atmosphere.

pacemaker The sinoatrial node, the region of the right atrium that initiates the contraction of the heart.

palate (*PAL-et*) The roof of the mouth. Near the front of the oral cavity, the palate is hard and bony; at the rear, it is soft and fleshy. The soft palate functions in separating breathing and swallowing.

palindrome (*PAL-in-drome*) A sequence of nucleotides in DNA that reads the same in one direction as in the other.

pancreas (*PAN-kree-us*) A "double" gland that produces hormones that regulate blood sugar and produces digestive enzymes that break down proteins, fats, and carbohydrates.

parapatric speciation (*PARE-a-PAT-rik*) The development of new species that occurs when a small population on the fringe of a large population diverges despite a modest degree of gene flow between the two populations.

parasite (*PARE-a-site*) An organism that lives and feeds on or in a host organism for at least part of its life cycle; a parasite may or may not kill its host.

parathyroid hormone (PTH) (*PARE-a-THIGH-royd*) A hormone, secreted by the parathyroid, that promotes release of calcium into the bloodstream.

partial pressure The pressure that can be directly attributed to one gas in a mixture of gases.

pedigree (*PED-ih-gree*) Lineage, ancestry, or genealogical chart or table.

pericycle (*PARE-ih-sigh-kul*) A core of cells within the endodermis of a root. During rapid growth, the pericycle is the source of new tissue as the root sends out lateral branches.

peristalsis (*PARE-ih-STALL-sis*) The series of muscular contractions that move food through the esophagus.

petals The often brightly colored segments of flowers that in many plants function in attracting pollinators.

pH A measure of the degree of acidity or alkalinity of a solution. On the pH scale, 7 is neutral, an acid solution has a pH of less than 7, and a basic solution has a pH of greater than 7.

phagocyte (*FAG-oh-site*) A blood cell that is capable of engulfing and digesting a foreign cell or particle in the bloodstream.

pharyngeal gill slits (*fa-RINJ-ee-al*) Openings or pouches in the lining of the upper digestive tract of chordates.

pharynx (*FARE-inks*) The throat, a muscular passage that leads from the nasal cavities to the esophagus and trachea; used in moving both food and air.

phenotype (*FEE-no-type*) The observable form of an organism; phenotype reflects both genetic inheritance and environmental factors.

pheromone (*FER-o-mone*) A hormone-like, volatile chemical substance that is secreted by one individual and elicits a physiological or behavioral response from another individual of the same species.

phloem (*FLOW-em*) The vascular tissue that circulates nutritive sap throughout a plant.

phospholipid (*FOSS-foe-LI-pid*) A polar lipid consisting of a glycerol molecule covalently bonded to two fatty acids and a phosphate group to which another molecule may be attached. Phospholipids are an important component of cell membranes.

photolysis (*foe-TAHL-a-sis*) The splitting of water molecules, fueled by radiant energy, during the first stage of noncyclic photophosphorylation. Photolysis yields electrons for reducing photosystem II, yields oxygen, and releases a pair of protons into the thylakoid sac.

photomorphogenesis (*FOE-toe-MOR-foe-JEN-ih-sis*) The process through which exposure to light influences the shape and growth pattern of a plant.

photoperiodism (*FOE-toe-PEER-ee-ud-izm*) A plant's physiological response to variations in duration of daylight.

photorespiration A metabolic pathway used by plants in bright, arid circumstances when oxygen concentrations in cells exceed carbon dioxide concentrations. The process reduces the efficiency of photosynthesis, releases carbon dioxide, consumes oxygen, and does not produce ATP.

photosynthesis (*foe-toe-SIN-the-sis*) The process by which plants use the energy of sunlight to construct organic molecules, primarily from water and carbon dioxide.

photosynthetic electron transport (*FOE-toe-sin-THET-ik*) The process by which electrons ejected from a photosystem pass through a sequence of molecules in a series of oxidation–reduction reactions, releasing energy that is used to attach an inorganic phosphate to ADP, thereby forming ATP.

photosystem I The photosynthetic reaction center in the thylakoid membrane that absorbs light with a maximum wavelength of 700 nm.

photosystem II The photosynthetic reaction center in the thylakoid membrane that absorbs light with a maximum wavelength of 680 nm.

phototropism (*FOE-toe-TROE-piz-em*) A plant's differential growth in response to light.

physical environment The nonliving components of an organism's surroundings.

phytochrome (*FY-toe-krome*) A plant pigment molecule that is activated and deactivated by certain wavelengths of light and functions in seed germination, flowering, leaf expansion, and other processes of plant growth.

pigment A molecule that absorbs most wavelengths of light and reflects light of particular wavelengths, which gives it a distinctive color. Chlorophyll and hemoglobin are pigments.

placenta (*pla-SEN-ta*) An internal organ possessed by female mammals that nourishes a developing fetus.

plasma (*PLAZ-ma*) The fluid in which blood cells are suspended. Plasma consists of water in which plasma proteins, gases, salts, sugars, and other substances are dissolved.

plasma cell A descendant of an activated B-cell that synthesizes and releases antibodies targeted at the antigen that originally activated the cell.

plasmid (*PLAZ-mid*) A type of "minichromosome" found in some bacteria that replicates independently of the bacteria's single DNA molecule.

platelets (*PLATE-lets*) Small fragments of cells with sticky surfaces that function in blood clotting; also called *thrombocytes*.

plate tectonics (*tek-TAWN-iks*) The drifting of large slabs that make up the earth's crust, producing a variety of geological effects.

pleiotropy (*PLEE-oh-troe-pee*) The capacity of a single gene to have several effects on an organism's phenotype.

pleural sac (*PLUR-al*) The two sacs, made up of epithelial membrane, that contain the lungs and reduce friction against the walls of the thoracic cavity.

point mutation A mutation caused by a change in one or two bases on a chromosome.

polar molecule A molecule in which different parts have different electrical charges.

pollen The immature male gametophytes produced by the anther of a flower.

pollen grain The immature male gametophyte produced by angiosperms and gymnosperms.

pollen sacs The chambers in the anther of a flower that contain the microspore mother cells.

polychlorinated biphenyls (PCBs) (*POL-ee-KLOR-in-ate-ed buy-FEN-uls*) Toxic compounds used in electronics parts manufacture that have persistent and damaging environmental consequences.

polymerization (*PAWL-ee-mer-eyes-AY-shun*) The formation of large molecules (polymers) by the bonding of many small molecules (monomers). The four major classes of biological macromolecules produced through polymerization are carbohydrates, proteins, lipids, and nucleic acids.

polymorphism (*PAWL-ee-MORF-izm*) The condition of two or more distinctly different manifestations of form.

polypeptide (*PAWL-ee-PEP-tide*) A chain of amino acids linked by peptide bonds between the nitrogen and carbon atoms of adjacent amino acids.

polysaccharide (*PAWL-ee-SAK-ah-ride*) A group of three or more bonded monosaccharides.

polytene chromosome (*PAWL-ee-teen*) A thickened, banded chromosome resulting from repeated chromosome replication without cell division or separation of sister chromatids. The copies remain packed together in parallel rows, creating distinctive bands easily seen with a microscope.

population A group of individuals of a single species that interact and interbreed in a particular area.

positive feedback system In biology, a mechanism that operates such that a shift in a physiological variable produces conditions that intensify the process that led to the shift.

posterior pituitary (*pih-TOO-ih-tair-ee*) The posterior lobe of the pituitary gland, which produces two neurohormones, ADH and oxytocin. ADH increases water retention in the kidneys; oxytocin stimulates uterine contractions during labor.

postzygotic isolating mechanism (*post-zye-GOT-ik*) A mechanism that promotes reproductive isolation by preventing hybrid zygotes from maturing prenatally or from surviving after birth.

potential energy Stored energy capable of being released under the proper conditions.

power of 10 The number 10 multiplied by itself a certain number of times; indicated by the use of an exponent, such as 10^{-1} (0.1) or 10^3 (1000).

prezygotic isolating mechanism A mechanism that promotes reproductive isolation by preventing eggs and sperm from uniting to form a zygote.

primary consumer An organism that eats plants; an herbivore.

primary growth Plant growth that occurs in the apical meristems.

primary producer An organism that can produce its own food from inorganic materials and an environmental energy source; an autotroph.

primary root The major "trunk" root that a plant sends into the soil.

primates An order of mammals characterized by highly dexterous digits, sophisticated binocular vision, a complex central nervous system, few offspring per pregnancy, and social systems.

progesterone (*pro-JES-te-rone*) A female sexual hormone, produced by the corpus luteum, that helps prepare the uterus for pregnancy.

prokaryotic cell (*pro-KAR-ree-OT-ik*) A type of cell that lacks membrane-bounded organelles and a distinct nucleus.

prophase (*PRO-faze*) The first stage of mitosis, during which the chromosomes gather near the center of the cell.

prosimians (*pro-SIH-mee-uns*) The more primitive branch of the order of primates; includes the tarsiers and lemurs.

prostaglandins (*PRAHS-tuh-GLANN-dinz*) A somewhat heterogeneous group of modified fatty acids secreted by a variety of tissues and having a variety of hormonal-like effects.

protein (*PRO-teen*) A diverse class of macromolecules formed by the polymerization of amino acids. Proteins perform a variety of important functions in living cells.

protein kinase (*KYE-nase*) A protein that is an inactive enzyme until cAMP binds with it, at which time the kinase is activated. The activated kinase catalyzes chemical reactions that activate other enzymes.

protostome (*PRO-toe-stome*) A member of one of two evolutionary lines distinguished by differences in embryonic development. In protostomes, cell division produces a spiral arrangement of cells, the ultimate function of each cell being determined early in development. The coelom is formed by the splitting of solid masses of mesoderm tissue, and the embryonic gut opening becomes the animal's mouth.

prototherians (*PRO-toe-THEER-ee-yuns*) The egg-laying mammals: the platypus and the echidna.

pterophytes (*TAIR-oh-fights*) The ferns; an ancient group of vascular plants that have compound leaves (fronds) and reproduce with spores. About 12,000 species exist today.

punctuated equilibrium The proposition that evolution proceeds through long periods during which species change very little and brief periods during which change is relatively rapid.

purine (*PURE-een*) Nitrogenous bases that occur in nucleotides and have two carbon–nitrogen rings in their structure; examples include adenine and guanine.

pyrimidine (*pih-RIM-ih-deen*) Nitrogenous bases that occur in nucleotides and have one carbon–nitrogen ring in their structure; examples include thymine and cytosine.

pyrrophytes (*PYE-roe-fights*) Dinoflagellates, or "armored" flagellates; protists with bodies encased in tough cellulose plates. Most are capable of photosynthesis, and they are an important component of phytoplankton.

radial symmetry (*RAY-dee-ul*) Symmetrical organization around a central axis, like the spokes of a bicycle wheel; sponges, starfish, and sea anemones are radially symmetrical.

radioisotope (*RAY-dee-oh-EYE-soh-tohp*) The unstable form of an element that emits radioactive radiation; a radioactive isotope-receptor. A molecule to which a hormone can bind, permitting the cell to respond to the hormone's "message."

receptor potential A brief change in voltage across the plasma membrane of a receptor

receptor potential (*continued*)
cell. Receptor potentials are graded potentials; that is, the size of the potential varies with the intensity of the stimulus.

recessive gene A gene that is expressed in the phenotype when paired with another recessive gene but not when paired with a dominant gene.

reciprocal inhibition The characteristic of antagonistic muscle pairs whereby stretching an extensor muscle inhibits the flexor muscle; mediated by the spinal cord.

recombinant DNA (*re-KOM-bin-ent*) DNA created through the recombination of genes from different sources.

reduction A chemical reaction in which electrons are added to an atom or compound.

reflex arc A sequence of several neurons that regulates a simple body movement without requiring conscious thought. A reflex arc is responsible for the knee-jerk reflex in humans.

releasing hormones Hormones that bind to receptors on cell membranes in the anterior pituitary, where they regulate the release of other hormones.

renal artery (*REE-nul*) The artery that brings blood to a kidney, where wastes and water are removed from the blood.

renal vein (*REE-nul*) The vein that returns blood to the circulatory system after wastes have been removed by the kidneys.

renewable energy Energy derived from replenishable sources such as sunlight, wind, and plants.

replication (*REP-li-KAY-shun*) The process by which DNA produces an exact duplicate of itself.

reproductive isolation The absence of interbreeding between members of different populations of the same species.

reptiles A class of ectothermic, egg-laying, usually terrestrial vertebrates that have a scaly skin and breathe by means of lungs.

resolution limit The ultimate limit of minuteness at which a microscope can distinguish two or more structures a certain distance apart.

restriction enzyme (*EN-zime*) An enzyme, produced by certain bacteria, that limits the types of DNA sequences that can survive in the bacteria and thereby helps protect the bacteria from invading viruses.

reticular formation (*re-TIK-you-lar*) The net-like structure of nerve cells in the midbrain that determines which stimuli the cortex attends to and which it ignores.

retina (*RET-in-uh*) The light-receptive membrane at the back of the eye. The vertebrate retina consists of two layers, the neural retina and the pigment epithelium.

retrovirus (*RE-troe-vye-rus*) An RNA virus that constructs a DNA molecule from its RNA. AIDS and many types of tumors are caused by retroviruses.

reverse transcriptase (*tran-SKRIP-tase*) An enzyme encoded by certain RNA viruses that can synthesize DNA from an RNA template.

RFLP Acronym for restriction fragment length polymorphism. A change in the length of a restriction fragment in DNA that is due to inactivation of an adjacent restriction site by mutation.

Rh factor A blood antigen the gene for which has two alleles, positive and negative. Rh-factor incompatibility between mother and fetus can cause the mother's immune system to attack the fetus's blood if the proper medical treatment is not instituted.

ribonucleic acid (RNA) (*RYE-boe-new-CLAY-ik*) A single-stranded nucleic acid, containing ribose as its pentose sugar, that functions in protein synthesis.

ribosome (*RYE-boe-soam*) An organelle consisting of two subunits, each composed of proteins and RNA, that is the site of protein synthesis in the cell.

rigor mortis (*RIG-or MOR-tiss*) A condition of muscle rigidity that occurs after death when muscles run out of ATP. When this occurs, myosin crossbridges cannot disconnect from actin.

rods One of the two classes of photoreceptors in the neural retina; responsible for "black and white" vision in very dim light.

root hairs Hair-like projections from the surface of a root that greatly increase the surface area in direct contact with the soil.

rough endoplasmic reticulum (*en-doe-PLAZ-mik ri-TIK-you-lum*) The portions of the endoplasmic reticulum that are associated with ribosomes.

***r*-selected species** Species that maximize their intrinsic rate of reproduction. Such species reproduce frequently and at a young age, produce many offspring, devote little care to their offspring, and are often found in unpredictable environments. Also called *opportunistic species*.

salinization (*SAL-in-eye-ZAY-shun*) The buildup of salts in irrigated soil in arid regions; salinization can result in dramatic reductions in crop yields.

saltatory conduction (*SAWL-ta-tor-ee*) Conduction down a myelated axon in which the action potential "jumps" from one node of Ranvier to the next.

salt marsh A saltwater shoreline ecosystem, characterized in temperate areas by the growth of marsh grass, in which the soil is alternately inundated and drained by the tides.

sarcolemma (*SAR-koh-LEM-a*) The plasma membrane that surrounds a muscle fiber.

sarcomere (*SAR-koh-MERE*) The muscle unit responsible for contraction; it is composed of thick and thin filaments and bounded on both ends by Z lines.

scanning electron microscope A microscope that produces an image by scanning an electron beam across an object's surface and dislodging electrons, which are measured by a detector and the image displayed on a video screen.

Schwann cell (*SHWAHN*) A type of glial cell that forms an insulating myelin sheath around the axons of many vertebrate neurons; functions in increasing the speed at which action potentials are conducted along the axon.

scientific method. The systematic approach to investigation characteristic of scientific research. It typically entails making observations (or recognizing a question), developing a hypothesis, testing the hypothesis through observation and experiment, and drawing and reporting conclusions.

scientific notation The system used by scientists to denote numbers efficiently. The system is based on powers of 10; 4,800,000 is expressed as 4.8×10^6, for example.

secondary consumer A carnivore that eats herbivores (primary consumers).

secondary growth Lateral plant growth, produced by the cambium, that results in the thickening of plant stems or roots.

secondary roots The smaller roots that branch off a primary root.

second messenger A molecule such as cAMP that conveys a signal from a primary messenger (a hormone) across a cell membrane into the cell interior.

secretory vesicle (*sih-KREE-tore-ee VES-ik-al*) A vesicle involved in releasing proteins from the Golgi bodies to the exterior of the cell.

seed A plant reproductive structure consisting of an embryo and a food supply encased in a protective coat.

segregation, law of The principle that even though an adult organism has two genes for each trait, the two genes separate when reproductive cells are formed.

semen (*SEE-men*) The fluid in which sperm are suspended. Semen is produced by glands in the reproductive tract and expelled through the penis during sexual intercourse.

seminiferous tubules (*SEM-in-IF-er-us TOOB-yules*) Tightly coiled tubes in the testes that produce sperm cells in their cell walls.

senescence (*se-NESS-ense*) Generally, a state of agedness. Leaf senescence is the process by which plant leaves begin to cease metabolic activity and drop from branches in the fall.

sensory adaptation The process by which the eye adjusts its sensitivity to varying degrees of light and darkness. It involves pupil expansion and contraction, bleaching and reactivation of visual pigment, and adjustments in the sensitivity of photoreceptors and interneurons.

sepals (*SEE-puls*) Leaf-like structures that enclose the flower bud during early development and remain green throughout the life of the flower.

sex chromosomes Generally, chromosomes that differ in the two sexes; in humans, the *XX* chromosomes in females and the *XY* chromosomes in males.

sexual reproduction The process by which two parents generate offspring through fusion of gametes (eggs and sperm).

sexual selection A natural selection process in which traits that give an animal an advantage in the competition for mates tend to be passed on to offspring; often manifested in male–male competition and female choice in mating.

short-day plant A plant that flowers only when daylight is briefer than a certain maximum. Such plants flower in late summer and early fall.

sickle-cell anemia An inherited disorder of the hemoglobin in red blood cells that can cause fatal internal bleeding and blood clots in the lungs in persons who receive the sickle-cell gene from both parents.

sieve tubes The elongated, interconnected cells in phloem that convey phloem sap.

sign stimulus An external stimulus that elicits a particular pattern of behavior; a releaser.

simple dominance The condition in which one gene for a trait is dominant and the other is recessive.

simple fruit A fruit produced from a single carpel or from several fused carpels, such as a tomato or grape.

sinoatrial (SA) node (*SIGH-no-AY-tree-ul*) The region in the right atrium that initiates the contraction of the heart.

sliding filament model A model of muscle activity suggesting that muscles contract (shorten) when thin actin filaments slide over thick myosin filaments, pulling the Z lines together.

smooth endoplasmic reticulum (*EN-doe-PLAZ-mik re-TIK-you-lum*) The portions of the endoplasmic reticulum that do not have ribosomes attached.

social releasers Sign signals that trigger socially related behaviors in mates or members of the signaler's social group.

sodium–potassium pump A protein in nerve cell plasma membranes that uses energy from ATP to pump sodium ions out of and cell and potassium ions into the cell.

soil The medium, composed of pulverized rock and organic substances, in which plants grow.

somites (*SOE-mites*) Distinct blocks of mesodermal tissue in the embryo that develop into bone, muscle, and internal organs.

specialization The process through which undifferentiated cells differentiate to perform a variety of functions.

speciation (*spee-shee-AY-shun*) The process by which populations of organisms diversify genetically and form new species.

species (*SPEE-shees*) A natural population or group of interbreeding natural populations that produces fertile offspring and is reproductively isolated from other groups.

spermatogonia (*sper-MAT-oh-GOE-nee-a*) Cells that are found near the periphery of the seminiferous tubules and divide by mitosis to produce the primary spermatocytes.

sphenophytes (*SFEEN-oh-fights*) The horsetails; an ancient group of seedless vascular plants. The sphenophytes were a diverse group during the Carboniferous; only about 20 species survive today.

spiracle (*SPEER-ih-kul*) An opening in the exoskeleton of an insect through which air enters and gases are exchanged.

sporophyte (*SPORE-oh-fight*) The diploid, spore-producing stage of the plant life cycle.

sporozoa (*SPORE-oh-ZOE-a*) Parasitic protists, many of which have complex life cycles involving two hosts.

squamous epithelium (*SKWAY-muss EP-ih-THEEL-ee-um*) A tissue made up of flat, scale-like cells that functions in diffusion and filtration; found in such locations as the air sacs of lungs and the linings of blood vessels.

stabilizing selection A form of natural selection that reduces the prevalence of extreme expressions of a trait and favors a more moderate manifestation of the trait.

stamen (*STAY-men*) The male portions of a flower, which produce pollen in the anthers.

statocyst (*STAT-oh-sist*) The spherical organ found in many invertebrates that is used to convey information about the body's orientation relative to the pull of gravity. The interior of the statocyst is typically lined with hair cells, which are stimulated by small particles that gravitate to the lowest part of the statocyst's interior.

steroid (*STEER-oid*) Member of a class of lipids characterized by a nucleus of four-carbon ring structures with various side groups and including a number of hormones, such as sex hormones and hormones of the adrenal cortex.

stimulus–response chain A sequence of signals, characteristic of the courtship of many animals, in which one partner's behavior serves as a stimulus that releases the next behavior in the other partner, and so on.

stomata (singular: **stoma**) (*sto-MAH-tah*) Openings in leaf epidermis that allow evaporation of water and intake of carbon dioxide. Two guard cells around the opening regulate the stoma's closing and opening.

stroma (*STROH-mah*) The semifluid matrix within the chloroplast in which the dark reactions of photosynthesis take place.

stromatolite (*stroh-MAT-oh-lite*) Fossilized domes consisting of layered sediment that were produced by early prokaryotic organisms; the fossils of these organisms are the earliest known evidence of life on earth.

subatomic particle (*SUB-a-TOM-ik*) The particles of which atoms are composed: electrons, protons, and neutrons.

substrate (*SUB-strait*) A reactant in an enzyme-catalyzed reaction.

substrate phosphorylation (*SUB-strait foss-FOUR-a-LAY-shun*) The formation of ATP by the direct transfer of a phosphate group from a substrate molecule to ADP.

succession The gradual process of species replacement in response to changes in the environment.

summation A characteristic of the transmission of action potentials between neurons whereby the membrane potential of the postsynaptic cell is affected by the total activity of all excitatory and inhibitory potentials of presynaptic neurons.

suppressor T-cell A type of T-cell that moderates the immune response by slowing down the rate of cell division and limiting the production of antibodies once an infection has been brought under control.

surface tension The resistance to rupture of the surface of a liquid caused by the mutual attraction of molecules. Water has high surface tension.

survivorship In a life table, the percentage of a population that can expect to live to a given age.

symbiosis (*sim-bee-OH-sis*) A protracted association of two organisms in an intimate relationship.

sympatric speciation (*sim-PAT-rik spee-shee-AY-shun*) The development of a new species in the midst of its parent populations as a result of genetic or other barriers.

synapse (*SIN-naps*) The point of junction between the axon of one neuron and the cell body or dentrite of another.

synapsis (*sin-AP-sis*) The pairing of homologous chromosomes during the first stage of meiosis.

systematics The classification of organisms on the basis of degree of relationship and evolutionary history.

systole (*SIS-ta-lee*) The state of the heart when the ventricles contract and push blood into the aorta and major arteries.

taiga (*TYE-guh*) A biome of predominantly evergreen forest characterized by harsh winters, warmer and longer summers than tundra areas experience, and a diverse, complex array of plants and animals.

target cell A cell that contains receptor sites for a particular hormone molecule.

taste bud The site of the taste receptor cells in grooves in the surface of the tongue. The actual sensation of taste occurs when a particular molecular shape binds to a receptor molecule.

taxa (singular: **taxon**) (*TAX-a*) The categories used in naming and classifying organisms, including phylum, class, order, family, genus, and species.

taxonomy (*tax-ON-a-mee*) The science of naming and classifying organisms.

Tay–Sachs disease (*TAY-SAKS*) A relatively rare, heritable, debilitating, lethal disease leading progressively to blindness, emaciation, and early death.

teleology (*TEEL-ee-AWL-a-jee*) The study of the ultimate purpose or goal of natural processes. Lamarck's theory of evolution was teleological in that he thought evolution was purposeful, motivated by organisms' striving for "perfection" as they attempted to rise through the "chain of being."

telophase (*TEE-loe-faze*) The final stage of mitosis, the stage during which the chromosomes organize into two cell nuclei.

temperate forest A biome found in the middle latitudes, characterized by predominantly broad-leaved forests, relatively rich soil, ample rainfall, and a great diversity of plant and animal life.

template (*TEM-plit*) A pattern for making accurate reproductions. In DNA replication, each strand of the double helix serves as a template for synthesis of a new strand.

temporal isolation A prezygotic isolating mechanism that functions when two species cannot interbreed because they breed at different times.

ten, rule of The assumption that roughly 10 percent of the energy produced by organisms at one trophic level is ultimately stored by organisms at the next higher level.

tertiary consumer (*TUR-shee-AIR-ee*) A carnivore that eats other carnivores.

test cross In genetics, the mating of individuals of uncertain genotype with individuals that are homozygous recessive for the genes in question. The phenotypes of the offspring reveal what the uncertain genotype must have been.

testes (singular: **testis**) (*TES-teeze*) The primary male sexual organs, which produce the male gametes (sperm) and sex hormones.

testosterone (*tes-TOSS-ter-own*) The male sexual hormone, which acts on spermatogonia to induce sperm development and, at puberty, induces the development of secondary sexual characteristics.

tetanus (*TET-nus*) The powerful, sustained muscle contraction that results from a steady, rapid succession of action potentials caused by constant stimulation.

theory A set of systematically organized ideas that explains a process or phenomenon.

therapsids (*thir-AP-sidz*) A group of terrestrial animals intermediate between reptiles and mammals; flourished during the Permian.

thermal stratification In bodies of water, the condition in which surface layers are separated from deeper layers by sharp differences in temperature (thermoclines,) creating distinctly different environments.

thermoacidophiles (*THUR-moe-ass-ID-oh-files*) Archaebacteria that inhabit hot, acid environments such as sulfur springs and volcanic vents.

thermocline (*THUR-moe-kline*) An area of rapid temperature change separating surface and deeper water masses.

thermodynamics, first law of (*THUR-moe-dye-NAM-iks*) The principle of conservation of energy, which states that the total amount of energy available in the universe does not change.

thermodynamics, second law of (*THUR-moe-dye-NAM-iks*) The principle that, in the universe as a whole and in any isolated system not at equilibrium, the energy becomes more dispersed and less organized such that, once it has been used to do work, it is less available to do additional work.

thermoregulation (*THUR-moe-REG-you-LAY-shun*) The process by which an organism controls its body temperature.

thigmotropism (*THIG-moe-TROE-pizm*) A plant's growth or movement in response to contact with a solid object.

thoracic cavity (*thoe-RASS-ik*) The sealed cavity in the chest where the lungs are located.

thymine dimer (*THIGH-mean DYE-mer*) A molecule of two covalently bonded thymidine molecules that cannot be copied properly during DNA replication, resulting in a pair of adjacent base substitutions.

thymus gland (*THIGH-mus*) A gland that, according to current theory, functions in the development of the immunological system by eliminating T-cells that would otherwise attack the host's tissues.

tight junction An attachment between adjacent cells that seals off the space between the cells and thereby prevents materials from leaking between them; important in the lining of fluid-filled cavities.

T-lymphocyte (*LIMF-oh-sight*) A type of lymphocyte capable of mounting a cell-to-cell attack on a foreign organism. There are three kinds of T-cells (helper, killer, and suppressor) that differentiate and mature only after passing through the thymus gland.

trachea (*TRAY-key-a*) The windpipe, a tube leading from the larynx to the two bronchi that enter the lungs.

tracheids (*TRAY-key-idz*) Specialized elongated cells in vascular plants that function in the conduction of fluids and in providing structural support.

tracheoles (*TRAY-key-ohlz*) In arthropods, minute branches of the trachea, which make direct contact with many cells and function in gas exchange.

transcription The synthesis of RNA as the complement of one strand of a double-stranded DNA molecule.

transformation The process by which a cell acquires genetic material from an external source.

transgenic organism (*tranz-JEEN-ik*) In genetics, an organism that incorporates functional DNA derived from another organism, usually one of a different species.

translation The conversion of the information from an RNA molecule into a particular series of amino acids in a protein.

translocation An alteration in chromosome structure in which part of a chromosome is broken off and attached to another chromosome.

transmission electron microscope A microscope that produces an image by using magnetic lenses to project an electron beam through a specimen and ultimately onto a screen of photosensitive film.

transpiration (*TRANS-purr-AY-shun*) The loss of water from a plant as a result of evaporation.

transposable elements (*tranz-POHZ-uh-bul*) Relatively short segments of DNA that are replicated at one part of a chromosome and inserted in another part.

trichloroethylene (TCE) A toxic, colorless liquid and a suspected carcinogen discovered in the groundwater water supplies of several states.

triglyceride (*try-GLISS-er-ide*) A lipid consisting of a single glycerol molecule covalently bonded to three fatty acids.

trilobite (*TRY-loe-bite*) One of an extinct group of marine arthropods that were common in early Cambrian seas.

trisomy-21 (*TRY-so-mee*) See *Down syndrome*.

trophic hormones (*TROH-fik*) Hormones that stimulate the secretion of other hormones.

trophic level (*TROE-fik*) A stage in a food chain. Primary producers represent one trophic level, herbivores a second, carnivores who eat herbivores a third, and so on.

tropical rain forest A biome found in tropical areas and characterized by substantial rainfall, lush growth, and an extraordinary diversity of plant and animal life.

tropism (*TROE-pizm*) A plant growth response to an environmental factor such as gravity, light, a chemical concentration, or physical contact with an object. The response typically entails differential rates of cell elongation, causing the plant organ as it grows to move toward or away from the source of the stimulus.

tropomyosin (*TROE-poe-MY-oh-sin*) One of the three proteins that compose the myofilaments of the myofibrils.

troponin (*TROE-poe-nin*) One of the three proteins that compose the myofilaments of the myofibrils.

tumor Any abnormal mass of cellular tissue, usually benign, that may or may not be circumscribed and may or may not become malignant.

tundra (*TON-drah*) A biome found in polar regions and at high altitudes and characterized by permanently frozen subsoil, a brief summer, and low-growing plants.

Turner syndrome A genetic nondisjunction abnormality in which the individual has only one *X* chromosome. Afflicted persons are

Turner syndrome (*continued*)
female, but they are infertile, do not reach sexual maturity, and have a variety of other abnormal characteristics.

undernourishment The condition of having less than the amount or kind of food required for normal health or development.

uniformitarianism (*YOUNE-if-for-mih-TARE-ee-un-izm*) The theory proposed by Lyell that natural laws remain constant and operate gradually and that, as a result, past events can be explained in terms of processes occurring today.

urea (*yur-EE-a*) In mammals and some amphibians, the soluble nitrogenous waste released by the liver into the bloodstream, from which it is removed by the kidneys and excreted in urine.

ureters (*yur-IH-ters*) The tubes that transport urine from the kidneys to the bladder.

urethra (*yur-EE-thra*) The tube (channel) that transports urine from the bladder out of the body.

uric acid (*YUR-ik*) In birds, insects, and reptiles, the insoluble nitrogenous waste into which ammonia is processed for excretion.

vaccination (*VAK-sin-AY-shun*) Inoculation with a weakened or dead form of a pathogen, which causes the body to develop antibodies for the pathogen and thereby produces immunity to it.

variations Differences among individuals in a species. Before Darwin, variations were regarded as imperfections in a world of fixed species types, but Darwin realized that variation was a natural characteristic of all organisms.

vascular bundles Parallel bundles of xylem and phloem tissue running lengthwise in the stems of tracheophytes.

vascular cylinder (*VAS-kew-ler*) The cylinder of water-transporting endodermis in the central region of a root or stem.

vascular tissue (*VAS-kew-ler*) Tissue specialized for carrying fluids.

vector pollination The transfer of pollen from the anthers to the stigmata of flowers by another organism, such as a bee.

vegetative reproduction (*VEJ-ih-TAY-tiv*) Asexual reproduction in plants through such means as adventitious roots, runners, cuttings, and rhizomes.

ventricle (*VEN-tre-kul*) A muscular chamber in the heart that pumps blood out of the heart. The human heart has two ventricles.

vertical zonation The pattern of horizontal bands in which plants and animals are distributed along shorelines between the upper and lower reaches of the tide.

vestigial structure (*veh-STIJ-ee-ul*) A degenerate or rudimentary form of an organic structure that functioned more fully in ancestors that occurred in earlier stages of the organism's evolution.

virus (*VYE-russ*) A disease-causing agent, consisting of a core of nucleic acids surrounded by a coat of protein, that cannot carry on its own metabolism but can direct the metabolic processes of host cells to its own use and can reproduce only within the cells of a host.

visual pigments Photosensitive compounds concentrated in the outer segment of the receptor cells of the eye, the molecules of which change shape when they absorb light.

vitamins Various complex organic compounds needed in small amounts in the diet, most of which function as cofactors in catalyzing chemical reactions.

weather The atmospheric conditions (rainfall, cloudiness, humidity, sunlight, wind speed, and other factors) that exist in an area at a given time; compare *climate*.

xylem (*ZYE-lem*) The vascular tissue that conducts water and minerals up from a plant's roots.

zygomycetes (*ZYE-go-MY-seets*) Members of a group of terrestrial or aquatic saprophytic or parasitic fungi. An example is *Rhizopus*, the common black bread mold.

zygote (*ZYE-goat*) The diploid cell formed by the fusion of gametes.

The following abbreviations are used for sources that are referred to frequently: Animals Animals–AA; Bruce Coleman, Inc.–BC; Biological Photo Service–BPS; Earth Scenes–ES; Grant Heilman–GH; Peter Arnold, Inc.–PA; Phototake–PHT; Photo-Nats–PN; Photo Researchers–PR; Tom Stack & Associates, Inc.–TSA; Visuals Unlimited–VU.

Part One Earth Satellite Corporation

Chapter 1

Fig. 1.1 (left) Paul Shoul / Impact Visuals; *Fig. 1.1 (right)* Donna Binder / Impact Visuals; *Fig. 1.2* and *detail p. 2* NASA; *Fig. 1.3 (left to right)* Dennis Kunkel / PHT, David M. Phillips / VU, Arthur Siegelman / VU; *Fig. 1.4* Patrice Rossi; *Fig. 1.5* Richard Feldman; *Fig. 1.6* P. Rossi; *Box p. 7* Dick Canby / DRK; *Fig. 1.7* Dr. Alvin O. Bellak Philadelphia / The Philadelphia Museum of Art; *Fig. 1.8* Stephen Jay Gould, *The Mismeasure of Man*, W. W. Norton & Co. New York, 1981, *p. 93*; *Fig. 1.9* The Bettmann Archive; *Fig. 1.10*, *Fig. 1.11*, and *Fig. 1.15*, P. Rossi; *Fig. 1.16* Jeff Schultz.

Chapter 2

Fig. 2.1 and *p. 19 detail* Roger Ressmeyer / Starlight; *Fig. 2.2* Metropolitan Museum of Art; *Fig. 2.3* Wellcome Institute Library, London; *Fig. 2.4* Lynn Johnson / Black Star; *Fig. 2.5* Michel de la Sabliere / Hillstrom Stock Photo, Inc.; *Fig. 2.6* David J. Cross; *Fig. 2.7* Doug Wechsler / ES; *Fig. 2.8* Time; *Fig. 2.9* Patrice Rossi; *Fig. 2.10* Michael Woods / Linden Artists Ltd.; *Fig. 2.11 (clockwise)* Brian Parker / TSA, Walt Anderson / TSA, David J. Cross.

Part Two Steve C. Wilson / Entheos

Chapter 3

Fig. 3.1 Boyd Norton; *Fig. 3.2* William H. Amos / BC; *Fig. 3.4* Patrice Rossi; *Fig. 3.6 (top)* Bruce Davidson / AA; *Fig. 3.6 (middle)* Jeff Foott / TSA; *Fig. 3.6 (bottom)* Zig Leszczynski / AA; *Fig. 3.7* P. Rossi; *Fig. 3.8* Kevin Schafer / TSA; *Fig. 3.9 (top to bottom)* Walter E. Harvey / PR, Inc., Stephen Dalton / PR, Inc., Stephen J. Krasemann / PR, Inc.; *Fig. 3.10* Ed Reschke / PA, Inc.; *Fig. 3.11* David Overcash / BC; *Fig. 3.12* Dave Woodward /

TSA; *Fig. 3.13 (left)* Jeff Rotman; *Fig. 3.13 (right)* P. Rossi; *Fig 3.14* Marlene Den-Houter; *Fig. 3.15* and *Fig. 3.16* P. Rossi; *Fig. 3.17* and *Fig. 3.18* Michael Woods / Linden Artists Ltd.; *Fig 3.19* P. Rossi; *Box p. 49* Sanderson Associates; *Fig. 3.20* and *p. 34 detail* Michael Woods / Linden Artists Ltd.; *Fig. 3.21* Michael Woods / Linden Artists Ltd.; *Fig. 3.22* Lyrl Ahern.

Chapter 4

Fig. 4.1 (top) R. Ingo Reipl / ES; *Fig. 4.1 (bottom)* Jeff Rotman; *Fig. 4.2* and *detail p. 54* Andrew Robinson / Virgil Pomfret Agency; *Fig. 4.3* through *Fig. 4.6* Patrice Rossi; *Fig. 4.7* Patti Murray / ES; *Fig. 4.8* Jeff Rotman; *Fig. 4.9* P. Rossi; *Box p. 62–63* P. Rossi; *Fig. 4.10* P. Rossi; *Fig. 4.11* Sanderson Associates; *Fig. 4.12* and *Fig. 4.13* P. Rossi; *Fig. 4.14* Sanderson Associates; *Fig. 4.15* P. Rossi; *Fig. 4.16 (top)* Gene Ahrens / BC; *Fig. 4.16 (bottom left)* Breck P. Kent / ES; *Fig. 4.16 (bottom right)* Arthur Gloor / ES; *Fig. 4.16 (center)* Sanderson Associates; *Fig. 4.17* Sanderson Associates; *Box p. 69* Richard Legekis; *Fig. 4.18 (top)* P. Rossi; *Fig. 4.18 (bottom)* Sanderson Associates; *Fig. 4.19 (left)* P. Rossi; *Fig. 4.19 (right)* Zig Leszczynski / ES.

Chapter 5

Fig. 5.1 Ray Pfortner / PA, Inc.; *Fig. 5.2* through *5.9* Patrice Rossi; *Fig. 5.10* Jack D. Swneson / TSA; *Fig. 5.11* and *Fig. 5.12* P. Rossi; *Box p. 84* and *detail p. 74* P. Rossi; *Fig. 5.13* Anne Wertheim; *Fig. 5.14* P. Rossi; *Box p. 87* P. Rossi; *Fig. 5.15 (top)* Jeff Rotman; *Fig. 5.15 (bottom)* E. R. Degginger / ColorPic, Inc.; *Fig. 5.16* Zig Leszczynski / AA; *Fig. 5.17* P. Rossi; *Fig. 5.18 (both)* Division of Entomology / CSIRO.

Chapter 6

Fig. 6.1 Sanderson Associates; *Fig. 6.2* and *Fig. 6.3* Patrice Rossi; *Fig. 6.4* Stephen J. Krasemann / PA; *Fig. 6.5* Peter Arnold / PA; *Fig. 6.6 (left)* Richard Thom / TSA; *Fig. 6.6 (right)* Scott Blackman / TSA; *Fig. 6.7* Jeff Foott / TSA; *Fig. 6.8* Jocelyn Burt / BC; *Fig. 6.9* and *detail p. 93* Stephen J. Krasemann / PA; *Fig. 6.10* UPI / Bettmann Newsphotos; *Fig. 6.11 (left)* Lynette Cook; *Fig. 6.11 (right)* Mickey Gibson / TSA; *Fig. 6.12* and *Fig. 6.13* P. Rossi; *Fig. 6.14* Alan D. Briere / TSA; *Fig. 6.15* M. Timothy O'Keefe / TSA; *Fig. 6.16*

and *Fig. 6.17* Marlene DenHouter; *Box p. 106* Loren McIntyre; *Fig. 6.18* P. Rossi; *Fig. 6.19* Jeff Rotman; *Fig. 6.20* Jeff Foott / BC; *Fig. 6.21 (both)* Jeff Rotman.

Chapter 7

Fig. 7.1 (both) Photri; *Fig. 7.2* Marlene DenHouter; *Fig. 7.3* © 1975 Aileen and W. Eugene Smith / Black Star; *Fig. 7.4* Sanderson Associates; *Fig. 7.5* Patrice Rossi; *Fig. 7.6* Michael Stewart / Project Lighthawk; *Fig. 7.7 (both)* K. H. Mills, from D. W. Schindler et al., *Science*, vol 228, pp. 1395–1401, 21 June 1985 (First Light, Toronto); *Fig. 7.8* P. Rossi; *Fig. 7.9* NASA; *Fig. 7.10 (both)* NASA; *Fig. 7.11* Frans Lanting; *Fig. 7.12* GH; *Fig. 7.13* P. Rossi; *Fig. 7.14* Tom Sobolik / Black Star; *Box p. 126* and *detail p. 112* P. Rossi; *p. 127 (top)* M. DenHouter; *p. 127 (bottom)* BC; *Fig. 7.15 (both)* T. G. Laman; *Fig. 7.16* through *7.19* P. Rossi; *Fig. 7.20* J.Donoso / Sygma; *Fig. 7.21* Comstock; *Box p. 134* Kjell B. Sandved; *Fig. 7.22* Comstock.

Part Three Howard Sochurek / Woodfin Camp & Associates, Inc.

Chapter 8

Fig. 8.1 Royal College of Surgeons / Charles Darwin House; *Fig. 8.2* Patrice Rossi; *Fig. 8.3 (both)* James R. Hill, 3rd / New York State Museum; *Fig. 8.4 (left)* P. Rossi; *Fig. 8.4 (right)* The British Museum of Natural History; *Box p. 145* National Portrait Gallery, London; *Fig. 8.5* Sanderson Associates; *Fig. 8.7 (top)* Hans and Judy Beste / AA; *Fig. 8.7 (bottom)* Barrie E. Watts / OSF / AA; *Fig. 8.8 (left)* L. L. Rue, 3rd / BC; *Fig. 8.8 (right)* W.H. Hodge / PA; *Fig. 8.9* The Royal Geographical Society / Bridgeman Art Library; *Fig. 8.10 (left top to bottom)* David M. Cavagnaro; *Fig. 8.11* and *detail p. 140* Enid Kitschnig / Scientific American; *Fig. 8.12 (top)* Marlene DenHouter; *Fig. 8.12 (bottom)* The Illustrated London News; *Fig. 8.13* Andrew Robinson / Virgil Pomfret Agency; *Fig. 8.14* The Bettmann Archive / BBC Hulton; *Fig. 8.16* M. DenHouter; *Fig. 8.17* Stephen Dalton / AA; *Fig. 8.18* and *Fig. 8.19* M. DenHouter; *Fig. 8.20 (left)* Patti Murray / ES; *Fig. 8.20 (right)* John M. Trager / VU; *Fig. 8.21 (left)* P. Rossi; *Fig. 8.21 (right)* Jack W. Dykinga / BC; *Fig. 8.22* P. Rossi; *Fig. 8.23 (top)* © Charles Marden Fitch; *Fig. 8.23 (bottom)* Catherine Pringle / BPS; *Fig. 8.24* Laurence Gilbert; *Fig. 8.25 (clockwise)* E. R. Degginger / ColorPic, Inc., L. West / BC, Breck P. Kent / AA; *Fig. 8.26 (both)* Lincoln Brower / Univ. of Florida, Gainesville; *Fig. 8.27* P. Rossi.

Chapter 9

Fig. 9.1 (left) Arleen Frasca; *Fig. 9.1 (right)* Ken Miller; *Fig. 9.2 (clockwise)* David M. Phillips / VU, L. L. Sims / VU, © Paul W. Johnson 1987 / BPS, Peter Parks / OSF / AA; *Fig. 9.3 (top)* © CNRI / PHT; *Fig. 9.3 (bottom)* Richard H. Gross / Biological Photography; *Fig. 9.4* K.G. Murti / VU; *Fig. 9.5* Patrice Rossi; *Fig. 9.6* Horvath & Cuthbertson; *Fig. 9.7 (left)* Hans Ris, Univ. of Wisconsin; *Fig. 9.7 (right)* H&C; *Fig. 9.8* H&C; *Fig. 9.9* Keith Porter; *Fig. 9.10* P. Rossi; *Fig. 9.11* Andrew S. Bajer, Univ. of Oregon; *Fig. 9.12 (left)* H&C; *Fig. 9.12 (right)* Conly L. Rieder; *Fig. 9.13 (left)* Andrew S. Bajer, Univ. of Oregon; *Fig. 9.13 (right)* H&C; *Fig. 9.14* Andrew S. Bajer, Univ. of Oregon; *Fig. 9.15 (left)* Andrew S. Bajer, Univ. of Oregon; *Fig. 9.15 (right)* Jeremy Pickett-Heaps; *Fig. 9.16* and *Fig. 9.17 (all)* Andrew S. Bajer, Univ. of Oregon; *Fig. 9.18* David Phillips / VU; *Box p. 176* Peter Hepler; *Fig. 9.19 (left)* P. Rossi; *Fig. 9.19 (right)* Peter Hepler; *Fig. 9.20* P. Rossi.

Chapter 10

Fig. 10.1 Hans Reinhard / BC; *Fig. 10.2* Moravske Muzeum, Brne Czechoslovakia; *Fig. 10.3* through *Fig. 10.13* Patrice Rossi; *Box p. 192* Dr. Tokuyasu, Univ. of California, San Diego; *Fig. 10.14* through *Fig. 10.17* P. Rossi; *Fig. 10.18* Marsha Goldberg; *Fig. 10.19* M. L. Pardue, M.I.T.; *Fig. 10.20* and *Fig. 10.21* P. Rossi; *Fig. 10.22 (left)* P. Rossi; *Fig. 10.22 (right)* Cabisco / VU; *Box p. 200* Carnegie Institute of Washington; *Fig. 10.24 (photos)* John Elseley; *Fig. 10.25 (photos)* J. D. Cunningham / VU; *Fig. 10.25 (top right)* Richard Kolar / AA; *Fig. 10.25 (bottom left and right)* Hans Reinhard / BC; *Fig. 10.26 (photos top to bottom)* Henry Ausloos / AA, Robert Pearcy / AA, Jean-Claude Carton / BC, Ralph A. Reinhold / AA; *Box p. 204* P. Rossi; *Fig. 10.27* and *Fig. 10.28* P. Rossi; *Fig. 10.29* Photri / A. Novak.

Chapter 11

Fig. 11.1 NASA; *Fig. 11.2* H&C; *Fig. 11.3* Dr. Anne Richardson; *Fig. 11.4 (top)* Universitetsbiblioteket, Oslo; *Fig. 11.5 (top)* Richard Dranitzke / PR; *Fig. 11.5 (bottom)* H&C; *Fig. 11.6* Ken Miller; *Fig. 11.7* Lester V. Bergman; *Fig. 11.8* H&C; *Fig. 11.9* David Falconer / BC; *Fig. 11.11* Ken Miller; *Fig. 11.12* Patricia Barry Levy / Profiles West; *Fig. 11.13* Patrice Rossi; *Fig. 11.15* and *Fig. 11.16* H&C; *Fig. 11.18* Sally Myers; *Fig. 11.19* and *Fig. 11.20* H&C; *Fig. 11.21 (left)* Michael and Elvan Habicht / ES; *Fig. 11.21 (right)* Dr. Ram S. Verma, Interfaith Medical

Center, Div. of Cytogenetics / PHT; *Box p. 224* Mickey Senkarik; *Fig. 11.22* P. Rossi; *Fig. 11.23* Ken Miller; *Fig. 11.24* and *Fig. 11.25* P. Rossi; *Fig. 11.26 (both)* The Children's Hospital of Philadelphia.

Chapter 12

Fig. 12.1 H&C; *Fig. 12.2* Patrice Rossi; *Fig. 12.3* F. Blakeslee, *Journal of Heredity*, 1914; *Fig. 12.4* P. Rossi; *Fig. 12.5* H&C; *Fig. 12.6 (clockwise)* M.A.Chappell / AA, David Cavagnaro, Phyllis Greenberg / Comstock; *Fig. 12.7* through *Fig. 12.19* P. Rossi; *Fig. 12.20 (both)* Breck P. Kent / AA.

Chapter 13

Fig. 13.1 © 1987 Laura Riley / BC; *Fig. 13.2 (top)* Frank T. Awbrey / VU; *Fig. 13.2 (bottom)* Dan Suzio; *Fig. 13.3* through *Fig. 13.5* Patrice Rossi; *Fig. 13.6* H&C; *Box p. 253* Robert Noonan; *Fig. 13.7* through *Fig. 13.12* P. Rossi.

Part Four Andrew McClenaghan / Science Photo Library / PR

Chapter 14

Fig. 14.1 Giraudon / Art Resource; *Fig. 14.2 (left)* Tom Branch / PR; *Fig. 14.2 (center)* Nell Bolen / PR; *Fig. 14.2 (right)* Bruce Cushing / VU; *Fig. 14.3* and *Fig. 14.4* Patrice Rossi; *Fig. 14.5* Steven Brook Studios; *Fig. 14.6 (top)* Richard Weiss / PA; *Fig. 14.6 (bottom)* William Ferguson; *Box p. 272* P. Rossi; *p. 273* WHOI / D. Foster / VU; *Fig. 14.7 (left)* P.W. Johnson & J. McN. Sieburth, Univ. of Rhode Island / BPS; *Fig. 14.7 (right)* Sinclair Stammers / Science Photo Library / PR; *Fig. 14.8* Dr. Paul Strother, Boston Univ.; *Fig. 14.9 (clockwise)* Dr. G. Kite, J. Robert Waaland, Univ. of Washington / BPS, Linda Thomashow, Washington State Univ. / BPS; *Fig. 14.10* H&C; *Fig. 14.11* P. Rossi; *Fig. 14.12 and detail p. 264* Arleen Frasca; *Fig. 14.13* H&C; *Fig. 14.14* David Phillips / VU; *Fig. 14.15* Mickey Senkarik; *Fig. 14.16* BPS; *Fig. 14.17* S. Sharnoff / VU; *Fig. 14.18* David Phillips / VU; *Fig. 14.19* William Patterson / VU; *Fig. 14.20 (top)* C. Rosebush Visions Corp. / PHT; *Fig. 14.20 (bottom left)* K. G. Murti / VU; *Fig. 14.20 (center)* Science / VU; *Fig. 14.20 (right)* G. Musil / VU.

Chapter 15

Box p. 287 (photo) BPS; *Box p. 287 (art)* Arleen Frasca; *Box p. 288 (photo)* Field Museum of Natural History; *Box p. 288 (art)* A. Frasca; *Fig. 15.1* Charles Seaborn; *Box p. 290 (photo)* Charles Seaborn; *Box p. 290 (art)* Mickey Senkarik; *Box p. 291 (art)* M. Senkarik; *Fig. 15.2* Charles Seaborn; *Box p. 293 (art)* M. Senkarik; *Box p. 293 (right photo)* Runk / Schoenberger / GH; *Box p. 293 (bottom photo)* Denise Tackett / TSA; *Table 15.1* A. Frasca; *Box p. 294 (left)* M. Senkarik; *Box p. 294 (right)*

H&C; *Box p. 296 (art)* M. Senkarik; *Box p. 296 (photo)* Charles Seaborn; *Fig. 15.3 (left)* Charles Seaborn; *Fig. 15.3 (bottom)* Gregory Dimijian / PR; *Fig. 15.3 (right)* John Cunningham / VU; *Box p. 298 (art)* M. Senkarik; *Box p. 298 (left photo)* Runk / Schoenberger / GH; *Box p. 298 (right photo)* Stan Elems / VU; *Box p. 300 (art)* M. Senkarik; *Box p. 300 (left photo)* C. Allan Morgan / PA; *Box p. 300 (center photo)* Peter Bryant / BPS; *Box p. 300 (right photo)* William Ferguson; *Fig. 15.4 (top)* A. Frasca; *Fig. 15.4 (bottom)* T. A. Wiewandt / DRK; *Fig. 15.5 and detail p. 286* Ed Robinson / TSA; *Box p. 302 (art)* M. Senkarik; *Box p. 302 (left photo)* Brian Parker / TSA; *Box p. 302 (center & right photo)* Charles Seaborn; *Box p. 303* Michael Fogden / OSF / AA; *Box p. 304* Field Museum of Natural History; *Box p. 305 (art)* M. Senkarik; *Box p. 305 (photo)* Brian Parker / TSA; *Fig. 15.6 (top)* D. P. Wilson / Eric & David Hosking / PR; *Fig. 15.6 (center)* Robert Evans / PA; *Fig. 15.6 (bottom)* William Ferguson.

Chapter 16

Box p. 309 (photo) Field Museum of Natural History; *Box p. 309 (art)* Arleen Frasca; *Box p. 311 (art)* Mickey Senkarik; *Box p. 311 (photos)* William Ferguson; *Fig. 16.1* A. Frasca; *Fig. 16.2* William Ferguson; *Box p. 313 (art)* M. Senkarik; *Box p. 313 (photos)* R. Arndt / VU; *Fig. 16.3* Marlene DenHouter; *Fig. 16.4 (left)* D. Wilder / TSA; *Fig. 16.4 (right)* G. R. Roberts; *Fig. 16.5 (top)* William Ferguson; *Fig. 16.5 (bottom)* R. L. Ferguson / BPS; *Fig. 16.6* M. Senkarik; *Box p. 317 (art)* M. Senkarik; *Box p. 317 (left photo)* Charlie Ott / PR; *Box p. 317 (center)* Stephen Krasemann / DRK; *Box p. 317 (right)* Ray Coleman / PR; *Box p. 319* M. Senkarik; *Fig. 16.7* A. Frasca; *Box p. 321 (art)* M. Senkarik; *Box p. 321 (photo)* Charles Seaborn / Odyssey Productions; *Box p. 322 (art)* M. Senkarik; *Box p. 322 (photo)* Patrice / VU; *Fig. 16.8* A. Frasca; *Fig. 16.9* H&C.

Chapter 17

Box p. 326 (art) Arleen Frasca; *Box p. 326 (photo)* Field Museum of Natural History; *Box p. 327 (art)* Mickey Senkarik; *Box p. 327 (top photo)* Carl Roessler / BC; *Box p. 327 (bottom photo)* M. P. Kahl / BC; *Fig. 17.1* Patrice Rossi; *Fig. 17.2* Patrice / VU; *Box p. 329 (art)* M. Senkarik; *Box p. 329 (photo)* Charles Seaborn / Odyssey Productions; *Fig. 17.3* Peter Scoones / Planet Earth Pictures; *Fig. 17.4* Hans Reinhard / BC; *Fig. 17.5 (left)* Smithsonian Institution / Div. of Palentology; *Fig. 17.5 (right)* Marlene DenHouter; *Fig. 17.6 and detail p. 325* John MacGregor / PA; *Box p. 332 (left art)* A. Frasca; *Box p. 332 (right art)* M. Senkarik; *Box p. 332 (left photo)* John Burnley / PR; *Box p. 322 (right photo)* Jane Burton / BC; *Box p. 333* Field Museum of Natural History; *Fig. 17.7 (left)* M. Den-

Houter; *Fig. 17.7 (right)* M. Senkarik; *Box p. 335 (left)* G. R. Roberts; *Box p. 335 (right)* Tom McHugh / PR; *Box p. 336 (art)* M. Senkarik; *Box p. 336 (top photo)* Michael Ederegger / DRK; *Box p. 336 (bottom)* C. P. Hickman / VU; *Fig. 17.8* M. DenHouter; *Box p. 339 (photo)* Douglas Henderson; *Fig. 17.9 (art)* P. Rossi; *Fig. 17.9 (top left)* William Ferguson; *Fig. 17.9 (right)* Douglas Henderson; *Fig. 17.9 (bottom left)* G. R. Roberts; *Fig. 17.10* Mark Hallett, Ranger Rick's Dinosaur Book, © 1984 National Wildlife Federation; *Fig. 17.11 (top)* S. S. Krasemann / PA; *Fig. 17.11 (bottom)* John Shaw / TSA; *Box p. 343 (art)* M. Senkarik; *Box p. 343 (left photo)* Townsend Dickinson / PR; *Box p. 343 (right)* J. Anderson / Taurus; *Box p. 344* M. DenHouter.

Chapter 18
Box p. 348 and detail p. 347 NMNH / PR; *Fig. 18.1 and Fig. 18.2* H&C; *Fig. 18.3* Hans and Judy Beste / AA; *Fig. 18.4* Kavid C. Fitts / AA; *Fig. 18.5* Milton Tierney, Jr. / VU; *Fig. 18.6* Patrice Rossi; *Fig. 18.7* R. Van Nostrand / PR; *Fig. 18.8 (photos)* David Brill, © National Geographic Society; *Fig. 18.8 (art)* Marlene DenHouter; *Fig. 18.9* David Brill / © National Geographic Society; *Fig. 18.10* M. DenHouter; *Fig. 18.11* H&C; *Fig. 18.12* David Brill, © National Geographic Society; *Fig. 18.13* David Brill, National Museum of Kenya; *Fig. 18.15* Denver Museum of Natural History.

Part Five Louis R. and Arlene Wolberg Institute, Inc.

Chapter 19
Fig. 19.1 Biblioteque Royal Albert, Brussells; *Fig. 19.2 and Fig. 19.3* H&C; *Fig. 19.4* SIU School of Medicine / PA; *Fig. 19.5* H&C; *Fig. 19.6 (left)* H&C; *Fig. 19.6 (right)* Bruce Iverson; *Fig. 19.7 and Fig. 19.8* H&C; *Fig. 19.9 (left)* NASA / Color-Pic Inc.; *Fig. 19.9 (right)* H&C; *Fig. 19.10 through Fig. 19.12 and detail p. 362* H&C; *Fig. 19.13 (left)* Patrice Rossi; *Fig. 19.13 (right)* Herman Eisenbeiss / PR.

Chapter 20
Fig. 20.1 H&C; *Fig. 20.5 (art)* H&C; *Fig. 20.5 (photos top to bottom)* Dr. Jeremy Burgess / Science Photo Library / PR, R. Rodewald / U. of Virginia / BPS, Clyde H. Smith / PA, Stephen J. Krasemann / PA; *Fig. 20.8 through Fig. 20.10* H&C; *Fig. 20.11* J. R. Eyerman, Life Magazine © 1958 Time Inc.; *Fig. 20.12, 20.13, and detail p. 377* H&C; *Fig. 20.14* Lyrl Ahern; *Fig. 20.15* H&C; *Fig. 20.16* Patrice Rossi; *Fig. 20.17* Marlene DenHouter; *Fig. 20.18* H&C; *Fig. 20.19* M. Goldberg; *Fig. 20.20 (left)* H&C; *Fig. 20.20 (right)* M. DenHouter.

Chapter 21
Fig. 21.1 The Image Bank; *Fig. 21.2 and Fig. 21.2* H&C; *Fig. 21.3* The Bettmann Archive; *Fig. 21.4* H&C; *Fig. 21.5* © S. Nielsen / DRK; *Fig. 21.6 through Fig. 21.15* H&C.

Chapter 22
Fig. 22.1 and Fig. 22.2 H&C; *Fig. 22.3 (art)* H&C; *Fig. 22.3 (photo)* Laboratory of Molecular Biology, Medical Research Council, Cambridge, England; *Fig. 22.5* Dr. James Watson, Cold Spring Harbour Laboratory; *Fig. 22.6* Laboratory of Molecular Biology, Medical Research Council, Cambridge, England; *Fig. 22.7 (left)* Patrice Rossi; *Fig. 22.7 (right)* NIH / Science Source / PR; *Box p. 414* Alberts, 2nd edition; *Molecular Biology of the Cell,* Garland Publishing Co.; *Fig. 22.8 through Fig. 22.11* H&C; *Fig. 22.12 (top)* Dr. James Lake, UCLA, reprinted with permission from James A. Lake, *Journal of Molecular Biology,* 105 (1976): 131-50; *Fig. 22.12 (bottom)* P. Rossi; *Fig. 22.13 and Fig. 22.14* P. Rossi; *Box p. 423* Oscar Miller; *Fig. 22.15 (left)* Jonathan King / MIT; *Fig. 22.15 (right)* Dr. Timothy S. Baker / Purdue Univ.; *Fig. 22.16 through 22.18* H&C; *Fig. 22.19* From H. Delius and A. Worcel, *Journal of Molecular Biology,* 82 (1974): 108. Reprinted with permission.; *Fig. 22.20 (top)* Ada L. Olins and Donald E. Olins; *Fig. 22.20 (bottom)* H&C; *Fig. 22.21 and Fig. 22.22* H&C; *Fig 22.23* Alberts, *Molecular Biology of the Cell,* 2nd edition: Garland Publishing Co.; *Fig. 22.24* Illustrious, Inc.; *Fig. 22.25 through Fig. 22.27* Lyrl Ahern.

Chapter 23
Fig. 23.1 Lucien Neumeyer; *Fig. 23.2 and Fig. 23.3* H&C; *Fig. 23.4* Nina Fedorof / Carnegie Institute; *Fig. 23.5* H&C; *Box. p. 440 (left)* George Wilder / VU; *Box p. 440 (right)* David M. Phillips / VU; *Box p. 440 (map)* H&C; *Fig. 21.4 through Fig. 23.10* H&C; *Fig. 23.11* Lyrl Ahern; *Fig. 23.11 (photo)* Susan Dibartolomeis; *Box p. 448 Discover; Fig. 23.12 and Fig. 23.13* H&C; *Fig. 23.14 (top)* "Genetically Engineering Plants for Crop Improvement" Gasser, C. S. and Robert T. Fraley, Science, Vol 244, pp. 1281-1288, Fig. I, 16 June 1989. © 1989 by AAAS; *Fig. 23.14 (bottom)* Robert Farley / *Science; Fig. 23.15* Thomas Hovland / GH; *Fig. 23.16 and Fig. 23.17* L. Ahern.

Chapter 24
Fig. 24.1 and Fig. 24.3 Patrice Rossi; *Fig. 24.4* Fritz Goro; *Fig. 24.5* P. Rossi; *Fig. 24.6* K. Miller; *Fig. 24.7 (art)* Charles Boyter; *Fig. 24.7 (photos)* Ken Miller; *Box p. 464* P. Rossi; *Fig. 24.8* Marlene DenHouter; *Fig. 24.9 (art)* M. DenHouter; *Fig. 24.9 (photo)* Daniel Branton and Leigh Engstrom, Harvard Univ., The Biological Laboratories; *Fig. 24.10* M. DenHouter; *Fig. 24.11 through*

Fig. 24.16 P. Rossi; *Fig. 24.17 (all)* © Dennis Kunkel / PHT; *Fig. 24.18* P. Rossi; *Fig. 24.19 (left)* Ken Miller; *Fig. 24.19 (right)* M. DenHouter; *Fig. 24.20 (top left)* Helen Padykula / Shirwin Pockwins; *Fig. 24.20 (top right)* Nigel Unwin, Cambridge, England; *Fig. 24.20 (bottom left)* M. DenHouter; *Fig. 24.21 (left)* Dr. Joachim Frank; *Fig. 24.21 (right)* George E. Palade, Yale Univ.; *Fig. 24.22* Daniel S. Friend, M.D., Univ. of California, SF; *Fig. 24.23 and detail p. 457* Mickey Senkarik; *Fig. 24.24* BPS; *Fig. 24.25 (top)* James Jamieson / Dept. of Cell Biology / Yale Univ.; *Fig. 24.25 (bottom)* Sanders / BPS; *Box p. 477* A. M. Siegelman / VU; *Fig. 24.26 (top)* P. M. Novikoff; *Fig. 24.26 (bottom)* M. DenHouter; *Fig. 24.27 (top left)* M. Schliwa / VU; *Fig. 24.27 (top right)* P. Rossi; *Fig. 24.27 (bottom left and center)* Lewis Tilney; *Fig. 24.27 (bottom right)* K.G. Murti / VU; *Fig. 24.28 (top left)* Michael Abbey / PR; *Fig. 24.28 (top right)* Biophoto Associates / Science Source / PR; *Fig. 24.28 (bottom left)* Zosia Rybkowski; *Fig. 24.28 (bottom right)* P. Rossi; *Fig. 24.29 and Fig. 24.30* David M. Phillips / VU; *Fig. 24.31 (left)* Robert Craig; *Fig. 24.31 (right)* P. Rossi; *Fig. 24.32* Lan BoChen; *Fig. 24.33 (top, center)* Ken Miller; *Fig. 24.33 (bottom)* Lloyd Matsumoto, Rhode Island College.

Chapter 25
Fig. 25.1 Patrice Rossi; *Fig. 25.3* P. Rossi; *Fig. 25.4* Lyrl Ahern; *Fig. 25.5* P. Rossi; *Fig. 25.6* L. Ahern; *Fig. 25.7* P. Rossi; *Fig. 25.8* L. Ahern; *Fig. 25.9 through Fig. 25.11* P. Rossi; *Fig. 25.12* L. Ahern; *Fig. 25.13 through Fig. 25.15* P. Rossi; *Fig. 25.16 (top)* P. Rossi; *Fig. 25.16 (bottom)* Kenneth Miller; *Fig. 25.17 (left)* L. Ahern; *Fig. 25.17 (right)* R. Btatnasar / VU; *Fig. 25.18* P. Rossi; *Fig. 25.19 and Fig. 25.20* L. Ahern; *Fig. 25.21* P. Rossi; *Fig. 25.22 (left)* P. Rossi; *Fig. 25.22 (right)* Peter Menzel; *Fig. 25.23 (top)* David Madison / Duomo; *Fig. 25.23 (bottom)* P. Rossi.

Chapter 26
Fig. 26.1 Lyrl Ahern; *Fig. 26.2 (left)* Marlene DenHouter; *Fig. 26.2 (right)* Ken Miller; *Fig. 26.3 (left)* Peter A. Simon / PHT; *Fig. 26.3 (right)* Illustrious, Inc.; *Fig. 26.4 (art)* Patrice Rossi; *Fig. 26.4 (photo)* Peter Goro; *Fig. 26.7* Larry Ulrich / DRK; *Fig. 26.8 through Fig. 26.10* M. DenHouter; *Fig. 26.11 and Fig. 26.12* P. Rossi; *Fig. 26.13 (top)* W. H. Hodge / PA; *Fig. 26.13 (bottom)* Liz Ball / PN; *Fig. 26.14* P. Rossi.

Part Six David Cavagnaro

Chapter 27
Fig. 27.1 Gregory Dimijian / PR; *Fig. 27.2* Patrice Rossi; *Fig. 27.3 (top left)* Wayne Lynch / DRK; *Fig. 27.3 (top right)* William Ferguson;

Fig. 27.3 (bottom) Zig Leszczynski / ES; Fig. 27.4 Dr. Ursula Goodenough, Washington Univ.; Fig. 27.5 (top) Peter Parks / Oxford Scientific Films; Fig. 27.5 (bottom) D.P. Wilson / Eric David Hosking / PR; Fig. 27.6 (both) William Ferguson; Fig. 27.7 William Ferguson; Box p. 526 Hans Reinhard / BC; Fig. 27.8 and Fig. 28.9 P. Rossi; Fig. 27.10 (left) William Ferguson; Fig. 27.10 (right) Lynette Cook; Fig. 27.10 (bottom right) P. Dayanandar / PR; Fig. 27.11 and Fig. 27.12 P. Rossi; Fig. 27.13 (left) R. Moore / VU; Fig. 27.13 (right) P. Rossi; Fig. 27.14 P. Rossi; Box p. 533 Raymond Mendez / AA; Fig. 27.15 P. Rossi; Fig. 27.16 L. Cook; Fig. 27.17 (top) P. Rossi; Fig. 27.17 (bottom) Richard Weiss / PA; Fig. 27.18 (bottom Left) David Stone / PN; Fig. 27.18 (right) Marcia Smith; Fig. 27.19 (top) Bruce Iverson; Fig. 27.19 (bottom) M. Fogden / BC; Fig. 27.20 (art) M. Smith; Fig. 27.20 (photos) Dr. Jeremy Burgess / Science Photo Library / PR.

Chapter 28
Fig. 28.1 Townsend Dickinson / PR; Fig. 28.2 (left) William Ferguson; Fig. 28.2 (right) Runk / Schoenberger / GH; Fig. 28.3 and Fig. 28.4 Patrice Rossi; Fig. 28.5 Marcia Smith; Fig. 28.6 Dr. Jeremy Burgess / Science Photo Library / PR; Fig. 28.7 Lynette Cook; Fig. 28.8 (top) Tom Bean / DRK; Fig. 28.8 (bottom) P. Rossi; Fig. 28.9 (top left) J. Barden / PN; Fig. 28.9 (center) Valorie Hodgson / PN; Fig. 28.9 (bottom right) L.L.T.Rhodes / ES; Fig. 28.9 (art) P. Rossi; Box p. 552 Dennis Kunkel / BPS; Fig. 28.10 (left) William Ferguson; Fig. 28.10 (center) James Carmichael / BC; Fig. 28.10 (right) Gay Bumgarner / PN; Fig. 28.11 (both) N.A.S. / M.W.F. Tweedie / PR; Box p. 555 Phil Gates / BPS; Fig. 28.12 Wayne van Kinen / DRK; Fig. 28.13 Merlin Tuttle / PR; Fig. 28.14 Leonard Lee Rue III / BC; Fig. 28.15 (left) Gay Bumgarner / PN; Fig. 28.15 (right) G. R. Roberts; Box p. 559 Ken Miller.

Chapter 29
Fig. 29.1 Lynette Cook; Fig. 29.2 Barry Runk / GH; Fig. 29.3 Patrice Rossi; Fig. 29.4 Holt Studios / ES; Fig. 29.5 and Fig. 29.6 P. Rossi; Fig. 29.7 (both) E.R. Degginger / ColorPic, Inc.; Box p. 569 W. B. Rudman / Australian National Museum; Fig. 29.8 (left) Dana Richter / VU; Fig. 29.8 (right) Runk / Schoenberger / GH; Fig. 29.9 (top) Hugh Spencer / PR; Fig. 29.9 (bottom) Science Photo Library / PR; Fig. 29.10 (left) Bruce Coleman; Fig. 29.10 (center) William Ferguson; Fig. 29.10 (right) Denny Barret / PN.

Chapter 30
Fig. 30.1 Patrice Rossi; Fig. 30.2 E. R. Degginger / ColorPic, Inc.; Fig. 30.3 Runk / Schoenberger / GH; Fig. 30.4 William Ferguson; Fig. 30.5 through Fig. 30.8 P. Rossi; Fig. 30.9 Runk / Schoenberger / GH; Fig. 30.10 P. Rossi; Fig. 30.11 Alfred Owczarzak / Taurus Photos; Fig. 30.12 Wide World Photos; Fig. 30.13 E. R. Degginger / ColorPic, Inc.; Fig. 30.14 Steve Raye / Taurus Photos; Fig. 30.15 Breck P. Kent / ES; Fig. 30.16 and Fig. 30.17 P. Rossi; Fig. 30.18 (both) William Ferguson; Fig. 30.19 P. Rossi; Fig. 30.20 D. Cavagnaro / DRK; Fig. 30.21 Dr. Jeremy Burgess / Science Photo Library / PR; Fig. 30.22 (top) C. Raymond / PR; Fig. 30.22 (bottom) P. Rossi; Fig. 30.23 David Ow / Science.

Part Seven Polaroid Corporation / Photomicrograph by Dr. Dennis Kunkel / Univ. of Washington.

Chapter 31
Fig. 31.1 Biblioteque de Geneve; Fig. 31.2 (art) Mickey Senkarik; Fig. 31.2 (top left) Robert Knauft / BioMedia / PR; Fig. 31.2 (top right) Ed Reschke / PA; Fig. 31.2 (center left) Manfred Kage / PA; Fig. 31.2 (center right) Ed Reschke / PA; Fig. 31.2 (bottom left) Ed Reschke / PA; Fig. 31.2 (bottom right) John Cunningham / VU; Fig. 31.3 (top) Arleen Frasca; Fig. 31.3 (photos) L. A. Stacehelin and B. E. Hull / Univ. of Boulder, CO; Fig. 31.4 (left) A. Frasca; Fig. 31.4 (right) Manfred Kage / PA; Fig. 31.5 Ed Reschke / PA; Fig. 31.6 (top) M. Senkarik; Fig. 31.6 (bottom) Ed Reschke / PA; Fig. 31.7 (art) M. Senkarik; Fig. 31.7 (photo) Manfred Kage / PA; Fig. 31.8 (photo) Ed Reschke / PA; Fig. 31.8 (art) M. Senkarik; Fig. 31.9 (photos: top) M. I. Walker / PR; Fig. 31.9 (middle) Dwight Kuhn; Fig. 31.9 (bottom) Eric V. Grave / PR; Fig. 31.9 (art) M. Senkarik Fig. 31.10 Patrice Rossi; Table 31.1 and Fig. 31.11; M. Senkarik; Fig. 31.12 through Fig. 31.14 P. Rossi; Fig. 31.15 David Madison; Fig. 31.16 P. Rossi; Fig. 31.17 A. Frasca.

Chapter 32
Fig. 32.1 Arleen Frasca; Fig. 32.2 Mickey Senkarik; Fig. 32.3 A. Frasca; Fig. 32.4 and Fig. 32.5 M. Senkarik; Fig. 32.6 through Fig. 32.8 Charles Boyter; Fig. 32.9 Dr. Don W. Fawcett; Fig. 32.10 A. Frasca; Fig. 32.11 through Fig. 32.13 M. Senkarik; Fig. 32.14 Patrice Rossi; Fig. 32.15 Sloop-Ober / VU; Box p. 621 (art) P. Rossi; Box p. 621 (photo) Mary Ann Fittipaldi; Fig. 32.16 M. Senkarik; Fig. 32.17 P. Rossi; Box p. 624 Hank Morgan / PR; Fig. 32.18 (left) Martin M. Rotker / Taurus Photos; Fig. 32.18 (right) Bill Longcore / Science Source / PR; Fig. 32.19 (all) Alfred Owczarzak / Taurus Photos; Fig. 32.20 (top) P. Rossi; Fig. 32.20 (bottom) David Phillips / VU.

Chapter 33
Fig. 33.1 and Fig. 33.2 Patrice Rossi; Fig.33.3 (left) Marlene DenHouter; Fig. 33.3 (right) R. Calentine / VU; Fig. 33.4 and Fig. 33.5 M. DenHouter; Fig. 33.6 Mickey Senkarik; Fig. 33.7 and Fig. 33.8 P. Rossi; Fig. 33.9 M. DenHouter; Fig. 33.10 (art) M. Senkarik; Fig. 33.10 (photo left) Cabisco / VU; Fig. 33.10 (right) R. Dirksen / VU; Fig. 33.11 and Fig. 33.12 P. Rossi; Box p. 639 Bettmann Newsphotos; Fig. 33.13 through Fig. 33.16 P. Rossi; Box p. 644 P. Rossi; Fig. 33.17 (both) Oscar C. Williams.

Chapter 34
Fig. 34.1 and Fig. 34.2 Patrice Rossi; Fig. 34.3 Wide World Photos; Box p. 652 and Fig. 34.4 P. Rossi; Fig. 34.5 Douglas Kirkland / Contact Press Images; Fig. 34.6 (clockwise) Steven C. Kaufman / PA, Hans Pfletschinger / PA, Fred Bavendam / PA, John Cancalori / PA; Fig. 34.7, P. Rossi; Fig. 34.8 (left) P. Rossi; Fig. 34.8 (right) Michael Abbey / PA; Fig. 34.9 P. Rossi; Fig. 34.10 and Fig. 34.11 Mickey Senkarik; Fig. 34.12 (left) David M. Stone / PN; Fig. 34.12 (right) Tom McHugh / PR; Fig. 34.13 and Fig. 34.14 P. Rossi; Fig. 34.15, detail p. 647, and Fig. 34.16 M. Senkarik; Fig. 34.17 (art) M. Senkarik; Fig. 34.17 (photo) Dr. Barbara Hull / Wayne State Univ.; Fig. 34.18 P. Rossi; Fig. 34.19 M. Senkarik; Fig. 34.20 P. Rossi.

Chapter 35
Fig. 35.1 Patrice Rossi; Fig. 35.2 (left) P. Rossi; Fig. 35.2 (right) M. P. Kahl / PR; Fig. 35.3 (both) Cabisco / VU; Fig. 35.4 P. Rossi; Fig. 35.5 Mickey Senkarik; Fig. 35.6 (left and detail p. 677) M. Senkarik; Fig. 35.6 (top right) Biophoto Associates / PR; Fig. 35.6 (bottom right) David M. Phillips / VU; Box p. 684 M. Senkarik; Fig. 35.7 and Fig. 35.8 M. Senkarik; Fig. 35.9 and Fig. 35.10 P. Rossi; Fig. 35.11 M. Senkarik; Fig. 35.12 and Fig. 35.13 P. Rossi; Fig. 35.14 Stephen Dalton / PR; Fig. 35.15 Alford W. Cooper / PR; Fig. 35.16 P. Rossi; Fig. 35.17 (left) M. Warren Williams / Taurus; Fig. 35.17 (center) Joe McDonald / VU; Fig. 35.17 (right) Leonard Lee Rue / PR; Fig. 35.18 P. Rossi; Fig. 35.19 Custom Medical Stock.

Chapter 36
Fig. 36.1 Richard Vogel / Gamma Liaison; Fig. 36.2 (left) Mickey Senkarik; Fig. 36.2 (right both) Dr. Peter Newell / Oxford Univ.; Fig. 36.3 and Fig. 36.4 M. Senkarik; Box p. 704 Greg Crisci / PN; Fig. 36.6 and detail p. 699 Patrice Rossi; Fig. 36.7 through Fig. 36.10 P. Rossi; Fig. 36.11 and Fig. 36.12 M. Senkarik; Fig. 36.13 (left) © Keith Hamshere / Shooting Star; Fig. 36.13 (right) Carol Publishing Group; Fig. 36.14 and Fig. 36.15 M. Senkarik; Fig. 36.16 (left) M. Senkarik; Fig. 36.16 (right) Arthur M. Siegelman; Fig. 36.17 P. Rossi; Fig. 36.18 M. Senkarik; Fig. 36.20 Mitch Reardon / PR.

Chapter 37

Fig. 37.1 Patrice Rossi; *Fig. 37.2 through Fig. 37.4 and detail p. 723* Mickey Senkarik; *Fig. 37.5* P. Rossi; *Fig. 37.6 through Fig. 37.8* M. Senkarik; *Fig. 37.9 through Fig. 37.11* P. Rossi; *Fig. 37.12* M. Senkarik; *Fig. 37.13* P. Rossi; *Fig. 37.14* M. Senkarik.

Chapter 38

Fig. 38.1 Patrice Rossi; *Fig. 38.2* Michael Fogden, Oxford Scientific Films / AA; *Fig. 38.3 (both)* Carol Vater and Robert Jackson / Dartmouth; *Fig. 38.4 and Fig. 38.5* P. Rossi; *Fig. 38.6 and Fig. 38.7* Mickey Senkarik; *Fig. 38.8 (photos)* Dr. Richard G. Kessel and Dr. C. Y. Shin; *Fig. 38.8 (art)* M. Senkarik; *Fig. 38.9* M. Senkarik; *Fig. 38.10 (photos)* Carolina Biological Supply Company; *Fig. 38.10 (art)* M. Senkarik; *Fig. 38.11 through Fig. 38.17* M. Senkarik; *Fig. 38.18 (top and middle rows)* Lennart Nilsson, *Bonna Fakta from A Child Is Born*, Dell Publishing Company; *Fig. 38.18 (bottom)* Lennart Nilsson, Nestle / PR; *Fig. 38.19* M. Senkarik; *Fig. 38.20* P. Rossi; *Fig. 38.21* Stanley Schoenberger / GH; *Fig. 38.22* M. Senkarik; *Fig. 38.24* © Lori Grinker / Contact; *Fig. 38.25 (both)* F. R. Turner / BPS; *Fig. 38.26* P. Rossi; *Fig. 38.27 and Fig. 38.28* M. Senkarik; *Box p. 768* Dr. Edward Kollar / Univ. of Connecticut Health Center; *Fig. 38.29 and Fig. 38.30* M. Senkarik; *Box p. 771* Camera M.D. Studios; *Fig. 38.31 (top)* K. G. Murti / Boston Univ.; *Fig. 38.31 (bottom)* P. Rossi; *Fig. 38.32* P. Rossi; *Box p. 774* © 88 J. Gilman/Custom Medical Stock.

Chapter 39

Fig. 39.1 (left) Alan Reiniger / Contact Press Images; *Fig. 39.1 (top)* Mary Ann Fittipaldi; *Fig. 39.1 (bottom)* © Mary Ann Fittipaldi; *Fig. 39.1 (right)* Georg Gerster / Comstock; *Fig. 39.2* Beinecke Library, Yale Univ.; *Fig. 39.3, Fig. 39.4, and detail p. 777* Mickey Senkarik; *Fig. 39.5* Illustrious, Inc.; *Fig. 39.6* M. Senkarik; *Fig. 39.7 (photo)* Dr. Amit and Dr. Poljak, Pasteur Institute Paris; *Fig. 39.7 (art)* Patrice Rossi; *Fig. 39.8* Illustrious, Inc.; *Box pp. 788-789* P. Rossi; *Fig. 39.9* P. Rossi; *Fig. 39.10 and Fig. 39.11* Illustrious, Inc.; *Fig. 39.12 (top)* Illustrious, Inc.; *Fig. 39.12 (bottom)* © Boehringer Ingelheim International, GMBH, Lennart Nilsson, photographer; *Fig. 39.13 and Fig. 39.14* Illustrious, Inc.; *Box p. 795* P. Rossi; *Fig. 39.15* Illustrious, Inc.; *Fig. 39.16* © Boehringer Ingelheim International, GMBH, Lennart Nilsson, photographer; *Fig. 39.17 (all)* Robert Gallo; *Fig. 39.18* George Kelvin, *Scientific American; Fig. 38.18 (right)* Lyrl Ahern; *Fig. 39.19 and Fig. 39.20* P. Rossi; *Box p. 804* © Billy E. Barnes.

Chapter 40

Fig. 40.1 Horvath & Cuthbertson; *Fig. 40.2 through Fig. 40.4* Mickey Senkarik; *Fig. 40.5 through Fig. 40.9* H&C; *Fig. 40.10 through Fig. 40.12* Patrice Rossi; *Fig. 40.13 and Fig. 40.14* H&C; *Box p. 823* Ciba–Geigy Corporation; *Fig. 40.15 and Fig. 40.16* P. Rossi; *Fig. 40.17* M. Goldberg; *Fig. 40.18 and Fig. 40.19* M. Senkarik; *Fig. 40.20 and Fig. 40.21* P. Rossi; *Fig. 40.23* H&C; *Fig. 40.24* Ohio–Nuclear Corporation, Science Photo Library / PR; *Fig. 40.25 (left)* © Roger Ressmeyer 1988; *Fig. 40.25 (right)* UCLA Medical School.

Chapter 41

Fig. 41.1 (photo) and p. 835 detail A. J. Deane / BC; *Fig. 41.1 (art)* H&C; *Fig. 41.2* Mickey Senkarik; *Fig. 41.3* Patrice Rossi; *Fig. 41.4* Reproduced from Barlow, H.B. and Mollon, J.D. (1982) *The Senses*, Cambridge Univ. Press; *Fig. 41.5* P. Rossi; *Fig. 41.6 (top)* Richard H. Gross / Biological Photography; *Fig. 41.6 (middle)* Lynn Funkhouser / PA; *Fig. 41.6 (bottom)* © 1979 Darwin Dale / PR; *Fig. 41.6 (art)* Marlene DenHouter; *Fig. 41.7* Illustrious, Inc.; *Fig. 41.8* M. Senkarik; *Fig. 41.10* H&C; *Fig. 41.12* M. DenHouter; *Fig. 41.13* H&C; *Fig. 41.15* H&C; *Fig. 41.16 through Fig. 41.18* M. Senkarik; *Fig. 41.19* H&C; *Fig. 41.20* M. DenHouter; *Fig. 41.21* M. Senkarik; *Fig. 41.22* Frithfoto / BC, *Fig. 41.23 through Fig. 41.25* M. Senkarik; *Fig. 41.26* M. DenHouter; *Fig. 41.27* J. Cancalosi / PA; *Fig. 41.28* P.Rossi.

Chapter 42

Fig. 42.1 and Fig. 41.2 Marlene DenHouter; *Fig. 41.3* Mickey Senkarik; *Fig. 42.4 (top)* Ed Reschke / PA; *Fig. 42.4 (art)* Patrice Rossi; *Fig. 42.5 and Fig. 42.6* M. Senkarik; *Fig. 42.7* Narco Bio-Systems; *Fig. 42.8* M. Senkarik; *Fig. 42.9 through Fig. 42.12* P. Rossi; *Box p. 870* R. Leatherwood / FPG; *Fig. 42.13* M. Senkarik; *Fig. 42.14* Biophoto Associates / PR; *Fig. 42.15* M. Senkarik; *Fig. 42.16* © 1986 Scott Camazine / PR; *Fig. 42.17* M. Senkarik; *Fig. 42.18* P. Rossi; *Fig. 42.19* M. Senkarik.

Chapter 43

Fig. 43.1 and Fig. 43.2 Patrice Rossi; *Fig. 43.3* David Austin / Stock Boston; *Fig. 43.4* P. Rossi; *Fig. 43.5* Hugo van Lawick / Nature Photographers; *Fig. 43.6* Nina Leen © 1964 Time, Inc.; *Fig. 43.7* P. Rossi; *Fig. 43.8 (left)* P. Rossi; *Fig. 43.8 (right)* Jeff March / PN; *Fig. 43.9 (top)* K. G. Preston–Mafham / AA; *Fig. 43.9 (bottom)* P. Rossi; *Fig. 43.10* Stephen G. Mora / PN; *Fig. 43.11 (top)* Michael Fogden / AA; *Fig. 43.11 (bottom)* Carl Roessler / BC; *Fig. 43.12* P. Rossi; *Fig. 43.13* OFS / BC; *Fig. 43.14* P. Rossi; *Fig. 43.15 (left)* © 1986 Steven C. Kaufman / PA; *Fig. 43.15 (right)* Charles G. Summer Jr. / TSA; *Fig. 43.16 (left)* Dwight Kuhn; *Fig. 43.16 (right)* Y. Arthus–Bertrand / PA; *Fig. 43.17* P. Rossi; *Fig. 43.18* Wrangham / AnthroPhoto; *Fig. 43.19* Edward Hodgson / PN; *Fig. 43.20* Priscilla Connell / PN; *Fig. 43.21* P. Rossi; *Fig. 43.22 (left)* BC; *Fig. 43.22 (right)* Donald Specker / AA; *Fig. 43.23* David M. Dennis / TSA; *Fig. 43.24* P. Rossi; *Fig. 43.25* Anthrophoto; *Fig. 43.26 (left)* David C. Fritts / AA; *Fig. 43.26 (right)* Duncan Anderson & Rachel Wilder / AA.

Note: Italicized page numbers indicate terms defined in the text, and boldfaced page numbers indicate material in tables, figures, and illustrations.

A band, 862
Abalone, 109
Abdomen, **319**
Abdominal thrust, 644
Abdominopelvic cavity, 602
Abscisic acid, **580**, *581*, 585
Abscission layer, *586*
Absorption of light, *504*
Abstinence, as contraceptive method, 736
Acceleration, 851–52
Accessory pigment, *42*, 304, 305, 306, 585
Accommodation, *842*
Acetabularia, 166, **167**
Acetaldehyde, 668
Acetyl Coenzyme A, 491
Acetylcholine (ACh), *818*, 822
Acetylcholinesterase, 819
Achondroplasia, *213*
Achromatopsia, *221*
Acid, **372**
Acid phosphatase, gene for, 227
Acid rain, *118–19*
Acoelomate, 292
Acquired immune deficiency syndrome. *See* AIDS
Acritarch, 274
Acrosome, *726*, 746
Actin, 481, **482**, *861*, **863**
Actinomycetes, *278*
Action potential, *814*, 815–16
graded potential converted to, **838**
propagation of, **815**
reaching synapse, 817–18
Activation energy, *398*
Active site, *400*
Active transport, 468, **469**, 532
in plants, 565
Acyclovir, 741
Adaptation, *151*, 230
Adaptive radiation, *154*, 286, 323
Adenine, *410*, 416

Adenosine diphosphate. *See* ADP
Adenosine monophosphate. *See* AMP
Adenosine triphosphate. *See* ATP
Adenovirus, **284**
Adenylate cyclase, *706*
ADH, 604–5, 708, **710**, *712–13*, 822
function of, 688–89
Adhering junction, *596*
Adhesion, 530
Adipose tissue, *596*
ADP, *397*, 488, 489, 496
in muscle, 864, 867
Adrenal cortex, *689*
hormones of, **710**, 719–20
Adrenal gland, *703*
Adrenal medulla, hormones of, **710**, 720
Adrenalin, 705, **710**, *720*, **721**, 813, 822
Adrenocorticotropic hormone (ACTH), **710**
Adventitious root, 543
Aerobe, *60*
Aerobic pathway, 493
Afferent neuron, *809*
Africa
deserts in, 33
grain production in, **129**
irrigation in, 129
savannah of, **43**
soil degradation in, 93
African sleeping sickness, 280, *780*, **781**
African violet, reproduction of, 543, **544**
Afterbirth, *760*
Age distribution, *78–79*
Agent orange, 579
Aggregate fruit, *550*, **551**
Aging, 761
Agnathan, *322*, 325
Agriculture, 125–30

prospects for, 129
sustainable, 129–30
Agrobacterium, 452, 586, **587**
AIDS, **3**, **740**, *741*, 777, **779**, **781**
global epidemiology of, **803**
HIV and, 426, 798–99
misconceptions about, 801–2
potential cures of, 805
protection against, 804
society and, 22–24
spread of, 799
transmission of, 802–5
treatment of, *15–17*
Air pollution, 117–18
acid rain and, 118–19
fossil fuels and, 118
ozone layer and, 119–20
satellite evidence of, **120**
smog, 118
types of, **119**
Air sac, 634, **635**
Ajuga remota, 576
Akiapolaau, **156**
Albinism, *211*
Albumins, *623*
Alchemy, **362**
Alcohol, 823
formula for, 374
liver and, 668
Alcoholic fermentation, 499–500
Aldehyde, **375**
Aldosterone, *689–90*, **710**
Aleurone layer, 563
Alexandra, Tsarina, 220
Alexis, Prince, 220
Algae, 34, **35**, 282, 303–4, 520, *522*
bloom of, 569
part of lichen, 315
in symbiosis, 569
Alice, Princess, 220
Alimentary canal, *661*
Allantois, 757
Alleles, *183*, *184*

multiple, *202*, 213–15
Allergy, *797*
Allochemical, *576*
Allopatric speciation, *249–50*, 253
case study of, 254–55
All-or-nothing event, *815*
Allosaurus, 341
Allosteric site, *405*
Alpha cell, 719
Alpha helix, *385*
Alpha linkage, **37**
Alternation of generations, *305*, 544–45
Altitude, temperature and, **96**
Altruistic behavior, *892–93*
Alveoli, 634, **636**, 638
Amacrine cell, 842
Amanita, 316, **317**
Amazon
delta of, 104
fish in, *106*
rubber harvest in, 24
Amazon basin, rainfall in, 68
Amber, 533
Amchitka Island, 109
Amino acid, *381–83*
derivatives of, *705*, 706
essential, *649*
Amino group, *374–75*, 383
Ammonia, 50
as bodily waste, *678–79*
plant use of, 691
Ammonification, 51
Amniocentesis, 224
Amniotic egg, 335
Amniotic sac, 757
Amoeba, **167**, 280, 808
digestion in, 659
Amoebocyst, 289
AMP, *397*, 488
cyclic, 700
learning and, 886
Amphetamine, 823
Amphibian, 331–33
excretion by, 679

Amphibian (*continued*)
  heart of, 612
  temperature regulation by, 694
Ampullae of Lorenzini, 856
Amylase, 381, 665
Anabolic reaction, 485
Anabolic steroid, 715
Anaerobe, *60*
Anaerobic fermentation, 271
Anal pore, 659
Analogous structures, *155*
Anaphase, *174–75*
Anatomy, 591
Anchovy catch, El Niño and, **123**, 130
Andrews, Tommie Lee, 448
Angiosperm, *342*, **343**, *344–45*, *527*
  alternation of generations in, 545
  fruit of, 550, **551**
  reproduction in, **548**
Angiotensin, 689
Angiotensinogen, 689
Angraecum orchid, 159
Anhidriotic dysplasia, 225
Animal behavior. *See* Behavior
Animal pole, 747
Animalia kingdom, **268**
Animals, seed dispersal by, 558
Anion, *367*
Annelid worm, 298–99, **825**, 859, **860**
  circulator system of, 609
Anomalocaris, 299
Anopheles mosquito, 781
Anorexia nervosa, 658
Ant, behavior of, **894**, **895**
Antagonistic groups, *860*
Antarctica
  climate changes in, 68
  continental drift in, 70
Anterior pituitary, **710**, *713–15*, **732**
  thyroxine and, 716–17
Anther, *546*
Antheridium, **311**, **313**
Anthias, 110
Anthocyanin, 585
Anthophyta, Division, 342–45
Anthropoid, *351*
Antibody, *215*, *783*, *785*
  classes of, **786**
  function of, **791**
  monoclonal, 795
Anticodon, 418
Antidiuretic hormone (ADH), 604–5, 708, **710**, *710*, *712–13*
  function of, 688–89
Antigen, *783*, *785*
  stimulating immune response, 787–88
Antihistamine, 797
Antipodals, 547

Antlers, 733
Anus, 661, 674
Aorta, *612*, *614*
Aortic arch, 609
Aortic valve, 614
Apatosaurus, 341
Apical dominance, 578, **579**
Apical meristem, 534
Apomixis, 559
Appendicitis, 672
Apple, propagation from cutting, 543
  seed dispersal of, *558*
Appropriate technology, *123*
Aquaculture, 130–31
Aquatic ecosystem, *102–9*
Aquifer, *114*, 46
ARC, **16**, 17
Archaebacteria, Division, **269**, 277–78
Archaeopteryx, 341, 344
Archegonium, **311**, **313**
Archenteron, 748, 751
Argentina, animals and fossils of, 146
Aristotle, 140, 263, 264, 267
Arteriole, *610*
Artery, *610*
  coronary, *614*
  pulmonary, *614*
Arthritis, *875*
  rheumatoid, *798*
Arthropod, 299–301, 318–21
  circulatory system of, 609
  skeleton of, **861**
Artificial selection, 150
Ascaris, **294**
Ascocarp, **317**
Ascomycetes, *316*
Asexual reproduction, *723*
  plant, 542–44, 559
Aspartame, 854
Aspirin, 721, 856
Associative learning, 882
Asteroids, affecting climate, 68
Asthma, 644, 797
Astigmatism, *842*
Atherosclerosis, 619, *620*, 652
Atlas, 869
Atmosphere, nitrogen in, 51
Atom, *363*
Atomic mass, *365*
Atomic number, *363*, 365
Atomic orbital, *363*
Atomic weight, *365*
ATP, *397–98*, 532, 647, 811
  in active transport, 468, **469**
  in cellular movement, 480
  in cellular respiration, 483
  cyanide blocking production of, 639
  glucose and, 487–99 passim, 503
  and heat production, 694
  in muscle, 864, 867–68

synthesis in plants, 507, 509–11, 515
  synthesis of, 490
ATPase, 468
Atrial fibrillation, 621
Atrial natriuretic hormone (ANH), **710**, 713
Atrioventricular (AV) node, *615*
Atrioventricular (AV) valve, 613–14, 616
Atrium, *611*
  human, 613
Atrophy, 870
Audition. *See* Hearing
Australia
  animals native to, 146
  El Niño and, 69
  rabbit population in, 74, 86, 88
Australopithecine, 353–56
Autoimmune disease, 798
Autolysis, *478*
Autonomic nervous system, 826–27, **828**
Autosomal chromosome, *208*
Autosomal dominant inheritance, 212–13
Autosomal gene, *210*
Autosomal recessive inheritance, 211–12
Autosomal trisomy, 223–24
Autotroph, *34*
Autumn, 585–86
Auxin, *576–79*, 580–81, 585
Avery, Oswald, 409–10, 449, 772
Axillary bud, *535*
Axis, 869
Axon, *809*
  myelinated, **816**, 817
  unmyelinated, 816–17
Azidothymidine (AZT), *15–17*, 805

B cell, *784*, 786
  somatic mutation of, 789–90
Backbone, 869
Bacteria, 778–779
  divisions of, 276–79
  in human body, 279
  intestinal, 664
  in nature, 279
  in nitrogen cycle, 50, 51
  nitrogen-fixing, 127
  pathogenic, *780*, **781**
  photosynthetic, *507*, 510
  useful, 279
  virulent vs. nonvirulent, 409
Bacteriophage, 424–25, **426**
Bakker, Robert, 338
Balance, 851–52
Balanus, 85
Bali
  agriculture in, 126

settlement of, 112–13
Banana, **551**
Bark, tree, *536*
Barley, sprouting of, 580
Barley grass, **342**
Barnacles, competition among, 85
Barr body, 225
Basal body, *481*
Basal firing rate, *837*
Basal metabolic rate, *647–48*, 656
Base, **372**
Base pairing, *413*
Base substitution, *437*
Basement membrane, 671
Basidia, **317**
Basidiomycetes, *316*
Basilar membrane, *849*, **850**
Bass, physiology of, *62*
Bat, pollination by, 556
Batesian mimicry, *160*, **161**
Bateson, William, 192
Bdelovibrio, 275
HMS *Beagle*, 144
BEAM scan, *833*
Bean, 550, **551**
Beef, grain equivalent of, 130
Bee
  behavior of, 879–80
  dance language of, 890–91
  pheromones affecting, 888
Behavior, 878
  altruistic, *892–93*
  elements of, 879–86
  genetics and, 885–86
  human, 896–97
  inherited vs. acquired, 881–84
  programmed, 880–81
  social, 886–96
  studies of, 878–79
Behavioral isolation, *248*
Bellflower, 555
Bends, the, 643
Benign tumor, *770*
Benthic organisms, **107**
Beriberi, 651
Bernal, J., 270
Berner, Robert, 533
Berry, Hugh, 87
Berthold, A. A., 700
Beta cell, 718
Beta linkage, **37**
Beta sheet, *386*
Bhopal disaster, 639
Bicarbonate ion, 372–73, 642, 665, 691
Bicuspids, *663*
Bidi-bio, seed dispersal of, **558**
Bilateral symmetry, *292*, 600, 824
Bilayer, 463–65
Bilirubin, 669
Binary fission, 275

Binet, Alfred, 217
Binocular vision, 351
Binomial nomenclature, *264–65*
Biochemistry
  evolution and, 160, 162
  link with genetics, 406
Biodiversity, 122–23
Biogeochemical cycle, *46*
Biogeochemistry, view of global warming, *28–29*
Biological clock, 575
Biological diversity, *93*
Biological magnification, **114**, 115
Biology, importance of, *2–3*
Biomass, power generated from, 132
Biome, *94*, **95**
  types of terrestrial, 96–102
Biosphere, *4, 33*
Biotin, **654**
Bipolar cell, 842, 846, **847**
Biramous appendate, **319**
Bird
  beak size in, **232**, 233–37
  courtship and mating of, **889**
  excretion by, 678
  genetic influence on song, 884–85
  heart of, 612
  learning by, 882, 883
  migratory, 99
  pollination by, 556
  reproduction of, 745
  respiration of, 634
  temperature regulation by, 695
  water balance in, 692
Birth, human, 760–61
Birth control pill, 737, **738**
Birth rate, 75
Biston betularia moth, 244–45
Bivalency, 189
Bivalvia, Class, *297*
Black-eyed Susan, **437**
Bladder, *682*, 688
Blade, **313**, *538*
Blake, William, 835
Blastocoel, *747*
Blastocyst, *733, 748*, 755
Blastoderm, *763*
Blastodisc, *748*
Blastomere, 747
Blastopore, *748*
Blastula, *747–48*
Blastulation, **750**
Bleaching, 844
Blendell, James, 213
Blight, 316
Blood, 609
  cells in, 623–26
  clotting of, **626**, 627
  composition of, 623–27

flow of. *See* Circulatory system
flow through heart, 613–14
gas exchange with tissues, *640–41*
glucose in, 699
pH of, **690**, *691*
plasma characteristics, 623
sugar in, 623
as tissue, *598, 609*
volume of, 619
Blood cell, **769**
Blood groups, 214
Blood pressure, *618–19*
  prostaglandins and, 721
Blood vessel, 609, 610–11
Blood-brain barrier, *828*
Blue dragon, 569
Blue-bell, genera of, **265**
Body cavity, *292*, 602
Body fluid, transport of, *309*
Body temperature
  adaptations for regulation of, 695
  homeostasis in regulation of, 697
  in ectotherm, 694
  in endotherm, 694–95
  in human, 696–97
  regulation of, 677, 693–97
Bog, 103
Bolivia, conservation in, 135
Bolus, *665*
Bond
  chemical, *367*
  ionic vs. covalent, 367–69
Bone, *597, 871–72*
  growth of, 872–74
Bone marrow, *597, 872*
Book lung, **319**
Borthwick, H., 582
Boveri, Theodor, 191
Bowman's capsule, *683*
Brackish-water ecosystem, *103–4*
Bradykinin, 856
Brain
  embryonic, **826**
  evolution of, 824–25
  information processing in, 838–39
  makeup of, 807
  nourishment of, 699
  physiology of, 827–31
Brain electrical activity mapping, *833*
Brain hormone, 702
Bran, 563
Branchiostoma, 320–21, 744
Brazil
  economic problems of, 133, 135
  environmental problems in, 30
  soil degradation in, 93

Breastbone, 869
Breathing, *636*
Brenner, Sidney, 763
Brian, C. K., 356
Brightfield optics, 477
Broca, Paul, 9
Broccoli, evolution of, **150**
Bronchi, *636*
Bronchioles, *636*
Bronchitis, 643
Brontosaurus, 341
Brown, Louise, 739
Brown, Michael, 652
Brownian movement, 467
Brussels sprouts, evolution of, **150**
Bryophyte, 310, **311**, **524**, *525*
Bubonic plague, **779**, **781**
Büchner, Eduard, 399
Büchner, Hans, 399
Bud, of stem, *535*
Buffer, **372**
Bugleweed, 576
Bulbourethral gland, *727*
Bulbs of Krause, **855**
Bulimia, *658*
Bundle of His, 615
Bundle sheath, 515
  in C4 plant, 540
Buñuel, Luis, 824
Burma, forests of, 34
Bursa, 875
Bursitis, 875i
Butterfly, pollination by, 556

Cabbage, evolution of, **150**
Cactus, **99**, *156*
Caecum, 672
Caenorhabditis, 763
Cairns-Smith, G., 270
Calcitonin, **710**, 716, 717
Calcium
  in bone, 874
  as nutrient, 651, **655**
  regulation of, 716, 717
California
  El Niño and, 69
  irrigation in, 129
Calmodulin, 707
Calorie, *397, 647*
Calvin, Melvin, 512
Calvin cycle, 512–14, *515*
Cambium, *535, 536, 537*
Cambrian period, 288–89
  advent of modern phyla, 289
  ecological diversity in, 289
cAMP, *700*, 706–8
Campanula, 555
Cancer
  altered genes and, 770–73, 774
  causes of, 770, 775
  cell division and, 178
  in chickens, **781**

diet and, 774–75
  history of, 770
  lung, 643–44
  prevention and control of, 773–75
  smoking and, 775
  treatment of, 526, 773
Candida albicans, 318
Cannibalism, 89
Capillary, *609, 610*
  in kidney, 683
  material exchange in, *617–18*
Capillary action, 373, *530*
Capillary bed, 617
Carbohydrate, *34, 36, 36–37, 377, 379–81, 649*
Carbohydrate loading, 870
Carbon
  importance of, 374
  in Krebs cycle, 493
Carbon cycle, *47–48*
Carbon dioxide
  climatic effects of, *27–30*, 118
  in photosynthesis, 503, 511–14, 538, 540
  transport of, *642*
Carbon monoxide, 642–43
Carbon-14 dating, 158
Carbonic acid, 372
Carbonic anhydrase, 642
Carboniferous period, 333
Carbonyl group, *374, 375*
Carboxyl group, *374*, 383
Carboxypeptidase, *400*
Carcinogen, 770, **775**
  environmental, 770–71
Cardiac cycle, *616*
Cardiac muscle, *599*, **615**, *861*
Cardiac sphincter, *666*
Carnivore, types of, **39**
Carotene, *507, 508*, 585
Carotenoids, 305
Carpals, *871*
Carpel, 546
Carpellate flower, 553
Carrying capacity, 76–78
Carson, Rachel, 25
Cartilage, *597, 871*, **872**, 873, **874**
Cartilaginous joint, 874
Casparian strip, *528, 532*
Caste, **894**
Cat
  mosaicism in, 225
  reproductive cycle in, 733
CAT scan, *832*, **833**
Catabolic reaction, 485
Catalyst, *399*
Catastrophism, *143*
Catfish, in aquaculture, 130
Catharanthus, 526
Cation, *367*
Cech, Thomas, 404
Cell, 4, **5**
  auxin and, 577–78

Cell (*continued*)
blood, 623–26
cytoskeleton of, 478–82
differentiation of, 593
discovery of, 459
diversity of, 165–68
division of, **169**, 170, 177
endoplasmic reticulum of, 474–76
fusion of, 795
genetics and, 186, 188
Golgi apparatus of, **475**, 476
growth of, *168–69*
inflammation and, **783**
as interdependent unit, 592
intermediate filaments in, 482
life cycle of, 169–70
lysosome of, *478*
mediating immune response, *784*, *791–92*, **793**
microfilaments in, 481, **482**
microtubules in, 479, 481
nucleus of, 472–73
organelles of, 472–78, 483
organization of, 462
osmosis and, 471, **472**
plant, **521**
reproduction of, 168
respiration of, 497–99
ribosome of, 473–74, 475
secretory vesicles of, 476–77
structure of, 165–66
types of, 166–68
types of animal, 593–99
vacuole in, *478*
wall of, *471*
See also specific types of cell
Cell cycle, *170–71*
Cell death, 754
Cell differentiation, 754
Cell division, 754
Cell junction, 594–95
Cell membrane, 5, 462
Cell migration, 754
Cell plate, *175*, **177**
Cell theory, 459
Cell wall, **166**
Cellulose, *37*, 381
Cenozoic era, 347–49
Centipede, **300**, 320
Central canal, spinal cord, 827
Central cell, 547
Central nervous system, 825, 826, 827, 831
Centriole, *172*, **173**, 174, *481*
Centromere, *172*, **173**
Cephalanthera, 555
Cephalization, **292**, *824*
Cephalochordata, Subphylum, 320
Cephalopod, *297*
eyes of, 841–842
Cephalothorax, **300**, **319**
Cerata, 569
Cercaria, **293**
Cerebellum, *831*

Cerebral cortex, *828*, **830**
Cerebral ganglion, *319*
Cerebrospinal fluid, *827*
Cerebrum, *828–30*
Cereus cactus, 575
Cervix, *731*
changes in childbirth, 760
CFCs, 119–20
C4 pathway, *514*
C4 photosynthesis, *514–15*, **515**
Channel, ion, *812–13*
Chaos, 280
Chargaff, Erwin, 410
Chargaff's rule, *410*, 412
Chase, Martha, 424–25
Chaterjee, Sankar, 341
Chelicerae, *319*
Chelicerata, Subphylum, 299, 301, 318
Cheliped, **319**
Chemical bond, *35*
Chemical communication. *See* Communication
Chemical element, *364*
Chemical energy, *395*
Chemical formula, 366
Chemical group, 374–75
Chemical mutagen, *438*
Chemical property, *364*
Chemical reaction
catalyst for, 399
energy for, 398
reversibility of, 401
Chemically controlled channel, *813*
Chemiosmosis, 494–97
in chloroplast, 510
Chemistry
inorganic, 373
organic, 373–74
Chemosynthesis, bacterial, 272
Chemotherapy, *773*
Chernobyl disaster, 132
Cherry, 550, **551**
Chewa religion, 8
Chicken, embryo of, **752**, **753**
Chief cell, 666
Childbirth, 604
Chilopoda, Class, 320
Chimera, *449*, **450**
Chimpanzee
learning by, 882, **883**
play behavior of, 896
China, aquaculture in, 130
Chitin, 299, 381
Chiton, *297*
Chlamydia, **278**, **740**, *741*
Chlamydium trichomonas, 279
Chlamydomonas, 304, 305, 522–24
Chlorine, as nutrient, 651, **655**
Chlorophyll, 305, 504, 521
chemistry of, **506**
phytochrome and, 585
and sunlight, 504

synthesis of, 565
types of, *504*
Chlorophyta, 303, 304–6, 543
Chloroplast, **166**, 275, 483, *504*
chemiosmosis in, 510
compared with mitochondria, 515–16
photosynthesis in, **507**, **509**
Chlorosis, 565
Cholecystokinin, 673, 675, **710**
Cholera, 279, **779**, **781**
Cholesterol, *392*, 669
and heart attack, 620, 652–53
Choline, **654**
Chondrichthyes, Class, 325–26, **327**
Chondrocyte, **597**
Chondrus crispus, 306
Chordata, Phylum, 320–23
Chorioallantois, 631
Chorion, 755
Chorionic villi biopsy, 224
Chromatid, *172*, **173**, *174*, 188
Chromatin, *427*, **428**, 725
Chromosomal inheritance, 222–25
Chromosomal mutation, *438*
Chromosome, *170*, 174, 188–91
autosomal, *208*
deletion, inversion, and translocation in, 226–27
eukaryotic, 427–28, 431
giant, 196
mapping of, 225–27
prokaryotic, 426–27, 431
sex, 197, *208*, **218**
in spermatid, 725
Chromosome puff, 196
Chrysanthemum, 159
Chrysophyta, phylum, 282
Chthamalus, 85
Chylomicron, *671*
Chyme, 667, 672
Cilia, **480**, *481*, 851, **854**
Ciliata, Phylum, 281
Circadian rhythm, *540*
Circulating tissue, *598*
Circulation
double, *612*
pulmonary, *612*, 614, *616*
single, *611*
systemic, *612*, 614, *616–17*
Circulatory system, 288, 598, *608*
control of, 617
diseases of, 619–22
human, **601**, 616–17
open vs. closed, 609
vertebrate, 609–16
Cirrhosis, 668
Citicorp, 135
Citrate, **491**, 492
Citric acid cycle, 491
Cladists, 266, **267**
Clam, 297

Clark, Barney, 624
Classical conditioning, *882*
Clavicle, *869*
Cleavage, 746, *747–48*
Cleavage furrow, *175*
Climate, *63*
astronomically caused changes in, *68*
biologically caused changes in, *70–72*
effect of carbon dioxide on, 118
effect of vegetation on, *67–68*
geologically caused changes in, *68–70*
Climax, 735, 736
Climax community, *71*, *72*
Clitellum, **298**
Clitoris, 735, 736
Cloaca, *678–79*
Clonal selection model, 788
Clone, 543
Cloning, 450–51
Clonorchis fluke, **293**
Closed circulatory system, *609*
Clotting factor, 219
Cloud, Preston, 10
Clovis people, 358
Club moss, 312, **333**, 525
Cnidaria, Phylum, 289, **291**, 292, 303
CNS, 825, 826, 827, 831
Coacervate droplet, 270
Coal, U.S. reserves of, 132
Cochlea, *849*, **850**
Cocklebur, 584, **585**
Coconut, seed dispersal of, 557
Coding, proteins used in, 408
Codominance, *201–2*, *214*
Codon, *417*
Coelacanth, 259, *330*
Coelomate, 292
Coenzyme, *405*, 651
Coenzyme A, 491
Coenzyme Q, **494**
Coevolution, *159*
gene-culture, 356
Cofactor, *405*, 651
Cohesion, *373*, 530
Coitus, 735
Cold
common, **781**
response to, 696–97
Coleoptera, Order, **334**
Coleoptile, 577
Coleridge, Samuel Taylor, 677, 823
Collagen, *596*
Collar cell, 289
Collecting duct, *682*, *686*
Collett, Marc, 772
Collins, J., 140
Colon, 672–73
Color vision, **836**, 846–47, **848**

Colorblindness, *220–21*, 437, 847
Colostrum, *761*
Columnar epithelium, *594*
Combining site, *786*
Commensalism, **88**, 89, *160*
Common descent, 152
Communication, 887
  courtship, 888–89
  deceit, 887
  intraspecific, 887–88
  language, 890–92
  organ systems for, 699
Compact bone, *872*
Companion cell, 531
Competition, 323
  intraspecific and interspecific, *82*
  laboratory study of, **83**
  male-male, 888
  in nature, *83, 85*, 87
  and niche, *82–83*
Competitive exclusion, *82*
Competitive inhibitor, *403*
Complement protein, *791*
Complete cleavage, 747
Compound, chemical, *366*
Compound eye, *840*, **841**
Computer-assisted tomography, *832*, **833**
Concentration gradient, 467
Conception, 736
Condensation reaction, 379–81
Conditioning, classical and operant, *882*
Condom, 737, **738**, 742
Conduction, heat, 694
Cone
  of eye, *842*
  pine, 337
Cone cell, 846
Cone cell synapse, **846**
Conformation, *405*
Congo River delta, 104
Conifer, **336**, 337
Connective tissue, 596
Connell, Joseph, *85*, 108
Conrad, Joseph, 807
Conservation International, 135
Conservation of energy, *395*
Constant region, *785*
Contact inhibition, *177*, **178**
Continental drift, 68, 70, **326**, **349**
Continental shelf, *121*
Continuous variation, *206*
Contraception, 736–38
Contractile vacuole, 471, 680
Contraction time, muscle, 867
Contrast, image, 477
Control, *11*
Controlled experiment, *11*
Convection, 694
Convergent evolution, *155*, 348
Copernicus, Nikolaus, 9

Coral, **291**, 292, 569
Coral reef, **109**, 110
Cork cambium, *536*
Corn, 514
  cross-fertilization in, 257
  flower of, **553**
  phototropism of, 575
Corn blight, **122**
Cornea, *841*, 842
Coronary artery, *614*
Coronary thrombus, *620*
Corpora allata, 702
Corpus callosum, *829*
Corpus cardiacum, 702
Corpus luteum, *731–32*, 733, 757
Correns, Carl, 188
Cortex
  adrenal, *689*, **710**, 719–20
  cerebral, *828*, **830**
  stem, *535*
Cortical granules, 746
Corticosteroids, **710**, 715, *719*
Corticotropin, **710**
Cortisol, **710**, *719–20*
Corynebacterium diphtheriae, **781**
Costa Rica, forest in, **521**
Cotyledon, 527
Countercurrent exchange system, *685–86*
Countercurrent flow, *632*
Coupling factor, 510
Courtship, 888, **889**
Covalent bond, 368–69
Cowpox, 777
Crab, 318, 320
Crabgrass, 514
Crack cocaine, 823
Crane, mating behavior of, **889**
Cranial cavity, 602
Creatine phosphate, *868*
Cretaceous period, 338, 339, 342
  extinctions in, 345
Crick, Francis, 412–13, 414, 416
Cricket, mating behavior of, 886
Cristae, **495**
Crocodile, **340**
Cro-Magnon man, 357–58
Crop genetics, 122–23
Cross bridge, *864*
Crossing, 181, 257
Crossing over, *194*
Crow, James F., 192
Cruciferae, Family, 576
Crustacea, Subphylum, 318, 320
C3 pathway, *512*, 514
Cuboidal epithelium, *594*
Cupula, 851
Curare, 526
Current, oceanic, 65, **66**
Cuspids, *663*
Cuticle, *309, 525*
  of leaf, *538*

Cuvier, Georges, 141, 154
Cyanide, 639
Cyanobacteria, 51, *273–74*, 278
  part of lichen, 315
Cycad, **335**, 337
Cyclic AMP, *700*, 706–8
Cyclic electron flow, 510
Cyclosporin, 794
Cypress, 342
Cystic fibrosis, *212*, *454–55*
Cytochrome f, 510
Cytochromes, **494**
Cytokinesis, *175*, **176**, 177
Cytokinin, *580–81*
Cytopharynx, 650
Cytoplasm, *168*
Cytosine, *410*, 416
Cytoskeleton, *478–82*
Cytotoxic T-cell, 791

Daddy-longlegs, *300*
Dandelion, 559
Darkfield optics, 477
Darwin, Charles, 10, 12, 139, 140, *140–43*, 351, 436, 576–78, 839, 890
  biological observations of, 146–48
  early publications, 147–48
  evolution theory, 149–58 passim
  geological observations of, 144
  on natural selection, 152–53
  paleontological observations of, 146, 157–58
  study of finches, 233
  view of speciation, *257, 259*
Darwin, Erasmus, 141, 144
Darwin, Francis, *576–78*
Darwin tubercle, *212*, **213**
Day-neutral plant, 584
DDT, 114
  resistance to, 231, 245
Deamination, 678
Death rate, 75
Decomposer, *40*
Deductive reasoning, *11*
Deer
  reproductive cycle in, 733
  sexual selection among, 890
Defense mechanism
  nonspecific, 781–83
  specific, 783–84
Deforestation, *120*, **128**
Dehydration, 649, 692
Deletion, chromosome, *227*
Demographic transition, 84
Demography, *78–81*
Dendrite, *809*
Denitrifying bacteria, 50
Density-dependent population factor, 81

Density-independent population regulation, *81, 89*
Deoxyribonucleic acid. *See* DNA
Depolarization, *812*, 814–15, 820
Dermal tissue, *527*, **528**
Dermis, **782**
Descartes, René, 704
Desert, *99–100*
Desertification, *120*, **121**
Desiccation, 308
Desmosome, **595**, *596*
Detached retina, *844*
Detritus, 41, 105
Deuteromycetes, *318*
Deuterostome, 294, **295**
Development, *744*
  anomalies in, 763–64
  coordination of, 762–69
  early, 746–52
  human. *See* Human development; Human embryology
  mechanisms of, 752–54
  patterns of, 744–46
Devonian period, 325, 330, 333
DeVries, William, 624
Dewi Danu, 126
Diabetes, 687, *719*
Diacylglycerol, 708, **709**
Dialysis machine, 684
Diaphragm, *636*
  contraceptive, 737, **738**
Diastole, *618*
Diastolic pressure, *618*
Diatom, 107, 282
Dicotyledon, *342, 527*, **538**, 549
  seed of, 561, **562**
Dictyostelium discoidium, 700–701
Dieldrin, 245
Diet, cancer and, 774–75
Differentiation, cell, *593*, *746*
Diffraction, 459
Diffusion, *467*, 678, 679
  across membrane, 468–69
  facilitated, 468, **469**
  laws of, 467
  simple, 468, **469**
Diffusion equilibrium, 467
Digestion
  cellular, 659
  extracellular, 661
  intracellular, 661
  physiology of, 666–72, 674
  timing of, 673
Digestive enzyme, *37*
Digestive system, 600
  hormones of, **711**
  human, 601, 662–75
  simple, 661
Digitalis, 526
Dikaryotic hyphae, **317**
Dinoflagellate, 107, *283*
Dinosaur
  body temperature of, 338

Dinosaur (*continued*)
  extinction of, 345
  types of, 341
Dioecious plant, 553
Diotima, 723
Dioxin, 579
Diphtheria, **781**
Diploid cell, *188*
Diploidy, *305*
Diplopoda, Class, 320
Diptera, Order, **334**
Directional selection, 232, *233*, 236, 238
Disaccharide, *379*
Discontinuous variation, *206*
Disease
  agents of, 778–779
  agricultural, *122*
  AIDS. *See* AIDS
  appendicitis, 672
  autoimmune, *798*
  cancer, 178, 526, 770–75
  of circulatory system, 619–22
  cirrhosis, 668
  diabetes, 687, 719
  eating disorders, 658
  epidemic, **779**
  fever to combat, *697*
  fungal, 318
  genetic, *210–13*, *219–25*, 437, 453–55, 627, 728
  history of, 777–779
  of lymphatic system, 622
  nematode-caused, 294
  nutritional, 650–656 passim
  of reproductive system, 739–42
  respiratory, 642–45
  sexually transmitted, 739–42
  society and, 20–22
  tetanus, *867*
  of thyroid, 716
  ulcers, 667
  viral, **284**
Disjunction abnormality, 222
Dispersal strategy, 310
Disposable economy, 135
Disruptive selection, 232, *233*, 238, **239**
Distal convoluted tubule, *686*
Disulfide bond, *386*, **387**
Diversity, importance of, *93–94*
DNA, 5, *169*
  antibody production and, 788–89
  base composition of, 410
  cancer-causing, 770, 772, 780
  chimeric, **450**
  in chloroplast, 483, 516
  cloning of, 450–51
  as coding molecule, 416
  cytokinin as breakdown product of, 580
  in egg, 746

of eukaryote, 170, *171*, 427–28
as "fingerprint," *448*
in genetic engineering, 445–52
homeobox, 764–65
intervening sequence in, *433*
in mitochondria, 483, 516
models of, **414**
mutagens in, 724
mutations in, 436–37
operator sequence of, *429*, **430**
origin of, 270
palindromic sequence in, *446*
and plant genetics, 586–88
in plasmid, 426–27
of prokaryote, 170, 426–27
promoter sequence of, *429*, **430**
reading of, 446–449
recombinant, *449*, **450**
repair of, 724
replication of, 415
and sex determination, 765
in sperm, 755
structure of, *392*, 410–14
in transformation, 409–10
in virus, 284, 423–25, 780
X-ray of, 410–12
DNA polymerase, *415*
Dog
  learning by, 882
  reproductive cycle in, 733
  temperature regulation by, **694**, 697
Dogwood, **342**
Dolphin, temperature regulation by, 695
Dominance, *184*, 186, 201
  codominance, *201–2*
  incomplete, *201*
Dominance hierarchy, 892
Dominant trait, *183*
Donne, John, 165
Dopamine, 822
Dormancy, 561–63
Dorsal fin, 325
Dorsal hollow nerve cord, 320
Double circulation, *612*
Double covalent bond, *368*, **369**
Double fertilization, 547
Double helix, 412–14
Double-blind study, *14–15*
Douche, as contraceptive method, **738**
Dove, courtship behavior of, **889**
Down syndrome, *223*, 224, **225**
Doyle, Arthur Conan, 268, 823
Drip irrigation, 129
Drosophila
  chromosomes of, 196, *197*
  development of, **762**, *763*

directional vs. disruptive selection in, 238, 257
experiment with, 204
gene expression in, **431**
learning by, 886
meiosis in, 189, **190**
mutations in, 764, **765**
sex chromosomes of, 198–200
Drug
  psychoactive, 823
  urine testing for, 687
Duchenne muscular dystrophy, *221*
Duck-billed platypus, 349, **350**
Dugesia, **293**, **294**
Dulosis, 878
Duodenum, *667–68*, **669**
Duplication, 168
Dust bowl, **101**
Dutch elm disease, 316
Dwarfism, 242, 716
Dynein, 480
Dysentery, 280

Ear, anatomy of, 849–50
Ear canal, *849*
Earthworm
  excretion by, 680
  respiration of, 630–31
Easter Island, settlement of, 112–13
Ecdysis, 299, *701*
Ecdysone, *701–2*, 703, 705
Echidna, 349
Echinoderm, 301–2
  circulatory system of, 609
Echinosphaerium, **479**
Echolocation, *850*, **851**
Eco RI, 445
Ecological isolation, 249, 250–51, 253
Ecological pyramid, *45–46*
Ecology, 25–27, *33*
  civilization and, 112–13
  functional units of, *74–75*
  human. *See* Human ecology
Economics, 25
  affected by ecology, 133, 135
  affecting ecology, 133
The Economist, 26
Ecosystem
  aquatic, *102–9*
  energy flow in, *44–46*
  interactions involving, *110*
  resiliency of, *93–94*
  terrestrial, *94–96*
Ecosystem diversity, *93*
Ectoderm, *593*, 748, **749**
Ectotherm, *61*, 338, *694*
  case study of, 62
Ecuador, El Niño and, 69
Edema, 622, 651
Ediacaran fauna, 287

EEG, *832*
Effector, *403*, 808
Effector organ, 808
Efferent neuron, *809*
Egg, 744–46
  cytoplasm of, 766
  response to sperm, 746
Egg cell, 744
Egypt
  agriculture in, **129**
  aquaculture in, 130
Einstein, Albert, 257
Ejaculation, 735
El Niño, 69, 110
  anchovy catch and, **123**, 130
Eldredge, Niles, 257, 259
Electric current, *811*
Electric potential, *811*, 812
Electrically controlled channel, *813*
Electrocardiogram (EKG), 621
Electrode, 35
Electroencephalogram, *832*
Electrolysis, *36*
Electromagnetic radiation, *504*
Electron, *363*
  flow in thylakoid membrane, 510
Electron microscope, 460
  scanning, 460–61
  transmission, 460, 461
Electron transport, 493–97
Electron transport chain, 494, **497**
Electrophoresis, 446
Electroreception, 856–57
Electrostatic forces, *367*
Elephantiasis, 622
Embryo, *744*
  chick, **752**, **753**
Embryo sac, *547*
Embryology, 154
  human. *See* Human embryology
Embryonic induction, *766*, **767**, 768
Emigration, 88
Emphysema, 643, 645
Emulsification, 669
Encysted metacercaria, **293**
Endemic species, 253
Endergonic reaction, *398*
Endocardium, 613
Endocrine system, 699
  chemical signaling in, 700–701
  components of, 702–3
  control of, 710–15
  discovery of, 700
  functions of, 700
  hormones in, 703, 705–9
  human, **601**, 703
  and reproductive cycle, 727–28, 731
Endocytosis, **476**, 477–78

Endoderm, *593, 748,* **749**
Endometriosis, *739*
Endometrium, *730,* 755, 757
Endoplasmic reticulum, 474–76
Endorphins, *822*
Endoskeleton, *859,* **860, 861**
Endosperm mother cell, 547, 549
Endothelium, 610
Endotherm, *61,* 338, 348, *694–95*
Energy, *394*
  activation, *398*
  autotrophs and, 34
  chemical, *395*
  chemical balance of, 397–98
  chemical generation of, 647
  conservation of, *395*
  flow of, *44–46*
  from food, 647–48
  kinetic, *394*
  life and, *41–42*
  nonrenewable, *131–32*
  nuclear, *132*
  potential, *395,* **396**
  renewable, *132–33*
  storage of, **34–36**
  thermodynamics, 395–97
Energy crisis, 135
Engels, Friedrich, 152
England, rabbit population in, 74
Enkephalins, *822*
ENSO, 69
Entamoeba histolytica, 280
Enteric bacteria, **278**
Enterocoelous organism, **295**
Enterogastrone, 675, **710**
Entropy, *396*
Environment, *55*
  biologically controlled, **54**
  biotic, *55*
  effects on gene, 206
  human control of, 91
  internal, *592,* 602–3
  long-term changes in, *68–72*
  perception of, *879–80*
  physical, *55*
  physically controlled, **54**
  plant responses to, 574–75
  predictable, 90, 91
  problems in, 124–31
  remedies for problems of, 126, 127, 129–31
  unpredictable, 90–91
Environmental policy, *135–36*
Environmental variable, *54–55*
  impact of, *61–67*
  response to, **64**
Environmentalism, 25–27
Enzyme, *399–400*
  capabilities of, 405–6
  characteristics of, 400–403
  conformation of, *405*
  as defense mechanism, 782

digestive, *37*
feedback and, *405*
hydrolytic, 586
inhibited, 403
and mass-action principle, 403
nonprotein, 404
regulation of, 403
restriction, *445,* 446
Enzyme induction, *429*
Enzyme-substrate complex, *399*
Epicardium, 613
Epidemic, **779**
Epidermal tissue, 538
Epidermis, **291, 782**
  of root, *528*
Epididymis, 727
Epiglottis, *636,* 665
Epilimnion, **102**
Epinephrine, *720,* 798, 813, *822*
Epiphyseal plate, *871, 872–73*
Epiphyte, 101
Epistasis, *203*
Epithelial tissue, 593–96
EPSP, *820*
Equilibrium, 76, **90,** *402–3, 851*
  diffusion, *467*
ER, 474–76
Erickson, Raymond, 772
Erythrocyte, *623–24,* **625**
  kidney and growth of, 720
Erythropoietin, **710,** *720,* 769
Escherichia coli, 166, 168, 169, 276
  electron micrograph of, **427**
  genes of, 428–29
Esophageal sphincter, *665*
Esophagus, *666,* 674
Essential nutrient, *46*
Estrogen, 703, **710,** 719, 720, 731
Estrus, 733, 734, 742
Estuary, *103–4,* **105**
  habitat destruction, *120–21*
Ethanol, 374, 499
Ethiopia
  desertification in, **121**
  undernourishment in, 648
Ethology, *878*
Ethylene, *580, 581–82,* 585
Eubacteria, Subkingdom, **269,** **278,** 279
Euchromatin, 473
Euglena, **167,** 282
  photoreceptors of, **841**
Euglenophyta, Phylum, 282
Eukaryote
  adaptive radiation of, 286
  chromosome in, *427–28,* 431
  gene regulation in, 430–34
  mRNA in, 431
  origin of, 275–76
  ribosome in, 431
  rise of, 274–75
  single-celled, 276, 279–83
Eukaryotic cell, 166, 168

Euphrates valley, salinization of, 129
Europe, grain production in, **129**
Eurypterids, 301
Eutherian mammal, *350*
Eutrophic lake, *103*
Evaporation, 694
Evolution, *140, 230*
  genetic engineering, 445–52
  hominid, 353–58
  human-caused, 244–45
  intron's role in, 443
  molecular change and, 444
  mosaic, *341*
  mutations and, 442–44
  and natural selection, 230–33
  primate, 351–52
  tracing of, 443–44
  vertebrate, 323, **328**
Evolutionary adaptation, *230*
Evolutionary fitness, *230*
Evolutionary theory
  biochemistry and, 160, 162
  biological significance of, 153
  coevolution and, 159
  current debate on, 257–60
  evidence in favor of, 153–62
  evolution of, 162–63
  historical background of, 141–43
  natural selection and, 149–52, 230–32
  paleontology and, 158
  predecessors of, 143–44
  and social sciences, 153
Excitatory postsynaptic potential, *820*
Excretory canal, 680
Excretory system, 679–80
  human, **601,** 681–91
Excurrent pore, **290**
Exergonic reaction, *398*
Exocrine glands, *703*
Exocytosis, **475, 476,** *477*
Exon, *434*
Exoskeleton, 299, 692, 701, 859, **860, 861**
Experiment, controlled, *11*
Experimental set, 11
Experimental variable, *11*
Exponential growth, *75–76*
Exteroceptor, *854, 856*
Extracellular fluid, 602
Extracellular matrix, *596*
Extraembryonic sac, 757

Fabré, Jean Henri, 853
Facilitated diffusion, 468, **469**
FAD, 493
FADH$_2$, 493
Fallopian tube, *730,* 755
Far-field sound, 848

Fast-twitch muscle, *861,* **867,** 870
Fat, *596,* 650–51
Fatigue, *868,* **869**
Fatty acid, *389–90*
  essential, 650
Feces, 673
Feedback, 603–4
Feedback control, 710–11
Feedback inhibition, *405*
Feedback pathway, *405*
Female choice, 888
Female reproductive system, 729–35
Femur, *871*
Fermentation, *499*
  by-products of, 500
  types of, 499–500
Fern, 312, **333,** 525
Fertility, *79–80*
Fertilization, 545, 733–34, 746, 747
  double, 547
  human, 755
  in vitro, 739
  plant, 547
Fertilization membrane, 746
Fertilizer, 126, 127, 567–68
Fetus, 758
Fever, 697
Fibrin, 627
Fibrinogen, *623,* 627
Fibroblast, 432, **462,** *596*
Fibrous joint, 874
Fibula, *871*
Fig, **551**
Fight-or-flight response, 720, **721**
Filial generations, 182
Filter feeder, 41
Filtrate, kidney, *683,* 685
Finch
  allopatric speciation of, 254–56
  beak size in, 233–34, **237,** 255, **256**
Fingerprints, genetic basis of, 215–16
Fins, types of, 325, **329**
Fir, **336**
Fire ant, taxonomy of, **266**
Firefly, 586
First law of thermodynamics, *395*
Fish
  bony, 326–30
  cartilaginous, 325–26
  excretion by, 678, 692
  fleshy-finned, 327, 331
  heart of, 611
  jawed, 322–23
  jawless, 322
  ray-finned, 327, **329**
  respiration of, 631–33

Fish (continued)
    temperature regulation by, 694
    water balance in, 691–92
Fisher, 768
Fisheries, 130
Fission, binary, 275
Fitness, 147, 230
Fixed-action pattern, 881
Flagella, 480, 481, 725
Flagellate, 280
Flame cell, 680, 681
Flatworm, 292, 293, 680, 681, 781, 825
Flavine adenine dinucleotide, 493
Flavine mononucleotide, 494
Flavoprotein, 494
Fleming, Alexander, 10
Florigen, 585
Flower, 542
    incomplete, 553
    specialized, 553
Fluid mosaic model, 466
Fluke, 292, 293
FMN, 494
Folic acid, 654
Follicle, 730
Follicle cell, 746
Follicle-stimulating hormone (FSH), 710, 715, 727, 729, 732, 733
Food
    caloric value of, 647–48
    feeding techniques, 658–59
    nutrients in, 648–49
    production of, 125–30
Food chain, 42, 44
Food web, 42, 43, 44
Ford, E. B., 245
Ford, Henry, 152
Forebrain, 825, 828–30
Fossey, Dianne, 896
Fossil, trace, 287
Fossil fuels, 131–32
    causing air pollution, 118
    and global warming, 30
Fossil water, 46
Founder effect, 242
Fovea, 842
Foxglove, 526
Fragmentation, 543
Frameshift mutation, 438
Franklin, Rosalind, 411, 412, 414
Freeze-etching, 464–65
Freshwater ecosystem, 103
Freud, Sigmund, 153
Frisch, Karl von, 878, 880, 891
Frog
    embryo of, 745, 753
    life cycle of, 332
Frontal plane, 600
Frost, bacteria genetically engineered to retard, 452

Frostbite, 696
Fructose, 379
Fructose-1,6-diphosphate, 489
Fructose-6-phosphate, 489
Fruit, 345, 542
    types of, 550, 551
Fruit fly
    chromosomes of, 196, 197
    development of, 762, 763
    directional vs. disruptive selection in, 238, 257
    experiment with, 204
    gene expression in, 431
    learning by, 886
    meiosis in, 189, 190
    mutations in, 764, 765
    sex chromosomes of, 198–200
Fruiting body, 700
Fucoxanthin, 306
Fuller, R. Buckminster, 5
Fungi, 268, 314
    groups of, 316–18
    mycorrhizae, 569–70
    types of, 314–15
Fungi imperfecta, 318
Futuyama, Douglas, 152

G protein, 706, 709
Gaia hypothesis, 6, 7
Galactose, 379
Galactosemia, 212
Galapagos Islands, 147–48, 233, 236, 254–55, 272
Galapagos woodpecker finch, 156
Galdikas, Birute, 896
Gall bladder, 669, 674
Gallo, Robert, 799
Gamete, 183, 188, 544
Gamete incompatibility, 248
Gametogenesis, 746
    human, 755
Gametophyte, 305, 314, 330, 544–45
Gamma aminobutyric acid (GABA), 822, 823
Ganges River delta, 104
Ganglion, 293, 824
Ganglion cell, 842, 846, 847
Ganglioside, 212
Gap junction, 595, 614, 817
Gar, 327
Gas exchange. See Respiration
Gastric gland, 666
Gastric mucosa, 666
Gastrin, 667, 673, 710
Gastrointestinal tract, 662
Gastropod, 297
Gastrovascular cavity, 291, 292, 293, 661
Gastrula, 748
Gastrulation, 746, 748–51
    human, 757

Gated ion channel, 812
Gatun Lake, 134
Gause, G. F., 82, 83
Gauss, Karl F., 9
Gehring, Walter, 765
Gene, 183
    activating or repressing other genes, 432
    antibody-producing, 788–89
    autosomal, 210
    electron micrograph of, 423
    environmental effects on, 206
    expression in prokaryote, 428–30
    influence on learning, 886
    influence on mating, 885–86
    "jumping," 438
    mapping, 225–26
    master, 764–65
    nature of, 408–10
    oncogene, 770, 772–73, 774
    regulation in eukaryote, 430–34
    "reporter," 587
    sd, 192
    sex-linked, 210, 218–21
Gene linkage, 192–95
Gene pool, 229
Gene-culture coevolution, 356
Generative nucleus, 547
Genetic abnormality, 210
Genetic code, 416–17, 418
Genetic disease, 210, 437, 453–55
    defective cholesterol receptor, 653
    hemophilia, 627
Genetic diversity, 93
Genetic drift, 241, 254
Genetic engineering
    basis of, 445–47, 449
    combining plasmid and DNA, 449
    NIH guidelines, 452
    plant, 586–88
    safety of, 452
    transformation, 449–50
Genetic map, 195–96
Genetic recombination, 195
Genetic variation
    observable, 230–31, 235–36
    studies of, 243–45
Genetics
    behavioral, 885–86
    birth of, 181–82
    cell and, 186, 188
    dominance, 183–184, 201–2
    gene interaction, 202–3
    genetic map, 195–96
    genetic variation, 201–3, 205–6
    genotype vs. phenotype, 184–85
    human. See Human genetics
    independent assortment, 185

    link with biochemistry, 406
    meiosis and, 186–95, 191
    Mendelian principles, 184
    particulate inheritance, 182–83
    polygenic system, 203, 205–6
    problem solving with, 186, 204
    sex inheritance, 197–200
Genital warts, 740, 741
Genome, structure of, 434
Genotype, 184
Genus, 265
Geochemistry, view of global warming, 28
George's Bank
    algae on, 34
    photic zone of, 107–08
Geotropism, 574, 578, 579
Germ layer, 593, 748
Germ line, 724
Gibberella fungus, 580
Gibberellin, 580
Gibbon, 352
Gilbert, Walter, 446
Gill, 288, 296, 631–33
    in water balance, 692
Gill slits, 320
Ginkgo, 335, 337
Glial cell, 751, 808
Global air movement, 64–67
Global 2000 report, 113–14
Global warming, 27–30
    biogeochemical view, 28–29
    fossil fuel use and, 29–30
    geochemical view, 28
    human factors in, 29–30
Globulins, 623
Glomerulus, 683
Glottis, 665
Glucagon, 708, 710, 717, 719
Glucose, 34, 35, 378, 379, 486
    ATP and, 487–99 passim, 503
    in blood, 623, 699
    breakdown of, 488–91, 495
    energy potential of, 486
    oxidation of, 487, 498
    regulation by pancreas, 717–19
    structure of, 485
Glucose-6-phosphate, 489
Glutamine, 822
Glycera, 298
Glycerol, 390
Glycine, 360, 383
Glycogen, 381, 623, 718, 870
Glycolipid, 466
Glycolysis, 488–91, 495
Glycoprotein, 755
Glyptodon, 146
Goiter, 716
Goldstein, Joseph, 652
Golgi apparatus, 166, 475, 476
Gonad, 703, 720
Gonadotropin, 715, 727

artificial, 739
Gonadotropin releasing
   hormone (GnRH), 727,
   728–29, 731–733
Gondwanaland, **304**, 308, 339,
   348–50
Gonorrhea, 279, **740**, *741*
Goodall, Jane, 896
Goose
   behavior of, 881
   learning by, 883
   migration of, 704
Gorter, 463–64
Gossypol, 737
Gould, Stephen Jay, 143, 257,
   259
Grabowski, Paula, 404
Graded potential, *820*, **838**
Gradualism, *259*
Gram, Hans, 279
Gram molecular weight, 367
Gram stain, 279
Grana, *504*
Grant, Peter, 234, 236
Grant, Rosemary, 234, 236
Granulosa cell, 730
Grape, 550, **551**
   grafting of roots, 543
Grass, reproduction of, 543
Grasshopper, **610**
Grassland, *100*
Gravity, plant response to, *574*
Grazer, 38
Great Chain of Being, 140, 141,
   144
Green Revolution, *125*, *127*
Greenhouse effect, *27*, **28**
Greenland
   climate changes in, 68
   continental drift in, 70
Grendel, 463–64
Grey matter, *827*
Griffith, Frederick, 408–10, 449
Ground tissue, *527*, **528**
Groundwater, *114*
   contamination of, 117
Growth
   exponential, *75–76*
   logistic, *76*
Growth factor, *177*
Growth hormone (GH), **710**,
   715
Guanidine triphosphate (GTP),
   420, 493
Guanine, *410*, 416
Guano, *679*, 692
Guanosine triphosphate, 493
Guard cell, 538
Guatemala, soil depletion in,
   128
Guilding, Rev. Lansdown, 303
Gull, allopatric speciation of,
   251
Gustation, 853–54
Gut, *661*

Gymnosperm, *337*, *344*, *527*
   pollination of, 553–54
Gypsy moth, 74, **75**
Gyrodinium, 275

H band, 862
Habitat, *555*
Habitat destruction
   deforestation, 120
   estuarine, 120–21
   marine, 120–21
   species loss in, 121–23
Habituation, *882*
Haemanthus flower, **173**
Hagfish, 322
Hair cell, *849*
Half-life, *366*
Hallucigenia, **288**
Halophile, *278*
Halophyte, *534*
Hamburg, solid waste
   production in, 136
Hanford, Mass., **20**
Haploid cell, *188*
Haplodiploidy, *201*, *894*
Haploidy, *305*
Hardin, G., 82
Hardy, George H., 238
Hardy-Weinberg law, 238–41
Hare, as prey, 86
Hasler, Arthur, 884
Hatch, 514
Haversian systems, *872*
Hawaii
   carbon dioxide levels in, **48**
   water crisis on, *49*
Hawaiian honeycreeper, **252**
HCG, 733, 734, *757*
Headpiece, 725
Hearing, 848
Heart, *609*
   artificial, 624
   blood flow through, 613–14
   four-chambered, *612–13*
   hormones of, **711**, 713, 720
   human, 613–16
   muscle of, *599*, **615**
   S-A node, 820
   structure of, 611–12
   three-chambered, *612*
   two-chambered, *611*
Heart attack, 608, 620
   risk factors in, 620, 652–53
Heartbeat, control of, 614–15
Heat, response to, 697
Heat capacity, 373
Heat stroke, *697*
Heavy chain, *786*
Heimlich maneuver, 644
Helper T-cell, *787*, *791*
Heme, 388, **389**, *639*
Hemiptera, Order, **334**
Hemispheres, brain, *829*
Hemocoel, *609*

Hemoglobin, 388, **389**, **442**,
   624, *639*, 651
   CO poisoning and, 642–43
   deoxygenated, 642
   oxygen transport by, 639–41
Hemophilia, *219–20*, 437, *627*
Hendricks, 582
Henking, H., 200
Henslow, J. S., 144
Hepatic portal circulation, *672*
Hepler, Peter, 176
Herbicide, 579, 588
Herbivore, 37, 344
   diversity of, **38**
   types of, 38–39
Heritable trait, *149*, 182
Heroin, 824
Herpes virus, **284**, **740**, *741*,
   781
Herrick, James, 440
Hershey, Alfred, 424–25
Hertz (Hz), *848*
Heterochromatin, 473
Heterocyst, *273–74*, 278
Heteropolymer, **378**
Heterotroph, 36–37, 486
   ecological strategies of, 37–41
   nutrition of, 647
   saprophyte as, 40
Heterozygote advantage, 242
Heterozygous organism, *184*
High-density lipoprotein (HDL),
   623, 652–53
Hindbrain, *825*, 831
Hindenberg airship, *396*
Hindu religion, 8
Histamine, *782*, 797, **798**, 856
Histone, *427*
Hitler, Adolf, 153
HIV, **740**, **781**, *798*
   biology of, 799–800
   infection cycle of, **801**
   spread of, 798
   transmission of, 798–99, 801
   types of, **799**
Hodgkin's Disease, **122**
Holbrook, Mass., 19–20
Holdfast, 306
Holdfast cell, 523
Hollyhock, 584
Holt, S. B., 215
Homarus, 109
Homeobox sequence, 764–65
Homeostasis, 7, 61, *603*
   feedback in, 603–4
   sugar balance, 606
   water balance, 604–5
Homeotic mutation, 764
Homing, 883–84
Hominid
   radiation of, 353
   evolution of, 353–58
Hominoid, *352*
Homo, 354, 356–58
   See also Human

Homologous chromosomes, *188*
Homologous structures, *154*,
   155
Homopolymer, **378**
Homozygous organisms, *184*
Honey ant, **894**
Hooke, Robert, 459
Horizontal cell, 842, 846
Hormone, 392, 576, **700**
   action of, 705–9
   complementary action of,
      711–12
   control of endocrine system
      by, 710–15
   plant, 575–82
   receptors for, *705*
   releasing, *714*
   trophic, *714*
   types of, 703, 705
Hormone-receptor complex,
   706
Hornwort, 310, **311**
Horse
   evolution of, **258**
   temperature regulation by,
      695
Horseshoe crab, 301
Horsetail, 312, **330**, **333**
Howell, Steven, 587
Human
   eyes of, **841**
   heart of, 613–16
   as herbivore, 39
   as omnivore, 40
   taxonomy of, **266**
Human behavior, 896–97
Human body
   plan of, 600–601
   temperature of, 695–97
   See also names of specific
      organs and systems
Human chorionic gonadotropin
   (HCG), 733, 734, *757*
Human development
   aging, 761
   embryology, 755–61
   maturation, 761
   sex determination, 765
Human ecology
   civilization and, 112
   economics and, 133–35
   energy production, 131–33
   food production, 125–30
   Island Earth concept, 112–13
   population growth, 124–25
   water pollution, 114–17
   worldwide, 113–14
Human embryology, **753**
   early, **756**
   fertilization, 755
   gametogenesis, 755
   intrauterine, **759**
Human evolution, 351–58
   brain size, 356

Human genetics
  abnormalities and disorders, *210*, 211–13, 219–21
  chromosome mapping, 225–27
  genes, 210–18
  genetic counseling, 212
  karyotyping, *209*
  modern, 453–55
  multiple alleles, 213–15
  neutral mutations, 442–43
  polygenic traits, 215–18
  prenatal, 224
Human growth hormone, 451
Human immunodeficiency virus. *See* AIDS; HIV
Humerus, *869*
Hummingbird, pollination by, 556
Humoral immune response, *784*, 790–91, **793**
Humus, *95*
Huntington, George, 213
Huntington's disease, *213*, 453–54; test for, 455
Hutton, James, 143
Huxley, T. H., 1
Hybrid cell, *226*
Hybrid infertility, *249*
Hybrid inviability, *248*
Hybridoma, 795
Hydra, 292, 608, **609**, 824, **825**; digestion in, **660**, *661*
Hydraulics, 591
Hydroelectric power, 132
Hydrogen bond, *371*
Hydrogen ion, **372**
Hydrological cycle, *46–47*
Hydrolysis, 381
Hydrolytic enzymes, 586
Hydrophilic molecule, *371*
Hydrophobic molecule, *371*
Hydroponic culture, *564*
Hydrostatic skeleton, 299, *859*, **860**
Hydroxyapatite, 871
Hydroxyl ion, **372**, *374*
Hyla, **745**
Hymenoptera, Order, **334**, 344
Hyperopia, *842*
Hyperpolarization, *812*, 820, 844
Hypertension, *619*
Hyperthyroidism, *716*
Hypertonic solution, *470*, **471**, 618
Hypertrophy, 870
Hyphae, *314*, **317**
Hypolimnion, **102**
Hypophosphatemia, *221*
Hypothalamus, 604, 688–89, 700, 712, *727*, 732–34 passim, 830

in temperature regulation, **695**, 696–97
  thyroxine and, 716–17
Hypothesis, 8, *10*
Hypothyroidism, *716*
Hypotonic solution, *470*, **471**
Hypoxia, *641*
Hypsilophodont, **340**

I band, 862
Ice fish, 633
Ichthyosaurus, 341
Ichthyostega, **331**
Ideal type, 140
Identical twins, 215
IgA, 785, **786**
IgD, **786**
IgE, **786**
IgG, 785, **786**
IgM, **786**
Iguana, **340**
Ilium, *871*
Imaginal disk, *763*
Imbibition, *563*
Immigration, *88*
Immune system, 598, *625*, 777, 782–84
  actions of, 786–94
  cell-mediated response, *784*, *791–92*, **793**
  components of, 784–86
  disorders of, 797–98
  function and failure of, 794, 797
  gene in, 788–89
  human, **601**
  humoral response, *784*, 790–91, **793**
  prostaglandins and, 721
  specific defenses, 783–84
Immune therapy, *773*
Immunoglobulin, *785*
  gene locus, *214*
Implantation, *755*
Imprinting, 883
Inca religion, 8
Incisors, *663*
Inclusive fitness, *892*
Income, by nation, *125*
Incomplete dominance, *201*
Incomplete flower, 553
Incus, *849*
Independent assortment, *185*, 186, 189
Indeterminacy, 573, **574**
Indolacetic acid, 577
Indonesia, El Niño and, 69
Inducer, *429*
Induction, embryonic, *766*, **767**, 768
Inductive reasoning, *10–11*
Industrial melanism, 244–45
Infertility, *738–39*

Inflammatory response, *782*, **783**
Infiltration, *47*
Influenza, **779**, **781**
Infrared receptor, 856
Inhalation, **637**
Inheritance, 180–81
Inhibitory neuron, **810**
Inhibitory postsynaptic potential, *820*
Initiation complex, 420
Initiation factor, 419
Innate behavior, *881*
Inner cell mass, 748
Inner ear, *849*
Inorganic chemistry, 373
Inositol triphosphate, **70**, 708
Insect, 320
  colonial behavior, 894–95
  excretion by, 680
  eyes of, **841**
  features of, 334
  hormonal control in, 701–2
  plant defenses against, 576
  pollination by, 554, 556
  respiration of, 633–34
  society, 893–94
  types of, 334–35
  water balance in, 692
Insectivorous plant, 571
Insight learning, 882
Instinct, *881*, 881–82
Insulin, 381, 606, 705, **710**, *717*
Integral membrane protein, 466
Integument, 546, 549
  See also Skin
Intelligence, factors affecting, 217
Intelligence quotient (IQ), 217
Interleukin, *787*
Intermediate filament, 482
Internal transport, 608–9
Interneuron, *809*, **838**
  of retina, 842
Internode, *535*
Interoceptor, *856*
Interphase, *170*, 172
Interspecific competition, *82*
Interstitial cells of Leydig, *726*
Interstitial fluid, *617*
Intervening sequence, *433*, **434**
Intestinal flora, 664
Intestine, *661*
  large, 672–73, 674, 693
  small, 670–71, 674
Intolerance, zone of, 63
Intraspecific communication, 887
Intraspecific competition, *82*
Intrauterine device, 738
Intrinsic rate of growth, *75*
Intron, *433*, **434**, 443
Inversion, chromosome, *227*
Invertebrate, *289*
  Annelida, 298–99

Arthropoda, 299–301
Cnidaria, 289, **291**, 292
Echinodermata, 301–2
  higher, 294–95
Mollusca, 296–97
Nematoda, 294
Platyhelminthes, 292–93
Porifera, 289, **290**
  worm-like, 292
Involuntary action, 825
Iodine
  as nutrient, 651, **655**
  thyroid use of, *715–16*
Ion, *367*
Ion channel, *812*
Ionic bond, 367–68, 369
Ionizing radiation, *439*
IPSP, *820*
Ireland, potato famine in, 318
Iris, *841*
Irish moss, 306
Iron, as nutrient, 651, **655**
Irrigation, 129
Ischium, *871*
Islets of Langerhans, *703*, 717
Isolation, types of, 247–248, 249
Isoleucine, 405
Isomers, 379
Isoptera, Order, **334**
Isotonic solution, *470*, **471**
Isotopes, 365–66
IUD, 738
Iwanowski, Dimitri, 283, 423

J-shaped growth curve, *75*
Jacob, Francois, 429
Japan, cancer incidence in, 773–74
Jarvik, William, 624
Jarvik-7 artificial heart, 624
Java man, 356
Jellyfish, 289, 291–92, 859
  excretion by, 669
Jenner, Edward, 778, **779**
Jensen, Arthur, 217
Jogger's high, 822
Johansen, Donald, 354
Joint, *859*, 874–77
Josephine (Empress), 853
"Jumping" gene, 438
Jung, Karl, 153
Jurassic period, 338, 339, 341, 344
Juvenile hormone (JH), *702*
Jyeshtha, smallpox goddess, 21

K-selected species, 91
Kahn, Herman, 114
Kalahari Desert, 99
Kangaroo, 350
Kangaroo rat, 635, **636**
  water balance in, 693

Kaposi's sarcoma, 799, 800
Karyotype, *209*
Kelp, 108–9
Keratin, 381, 386
Ketone, **375**
Kettlewell, H. B. D., 245
Keystone predators, *108*
Kidney
  artificial, 684
  excretion through, 678, 679, 680
  fish, 691–92
  function of, 683, 685–88
  hormones of, **711**, 720
  human, *681–82*, 683, 685–91
  hydrogen ion secretion by, **690**
  tubules of, *685–687*
Killdeer, behavior of, **893**
Killer T-cell, 791–92
Kilocalorie, *647*
Kin selection, *892*
Kinetic energy, *394*
Kinetochore, *172*, **173**
King, Samuel, 242
Kingdom, taxonomic, 266, *267–68*, **269**
Klinefelter syndrome, *223*
Knee-jerk reflex, 809, **810**
Knee joint, 874, **875**
Koch, Robert, 778
Kohler, George, 795
Kollar, 768
Krebs, Hans, 491
Krebs cycle, **488**, 491–93, 515
Kubo, Isao, 576
Kurosawa, Eiichi, 580
Kwashiorkor, 650

Labia majora and minora, *731*
Labor, human birth, *760*
Labrador retriever, genetics of, 202–3
lac operon, *429–30*
lac repressor, *429*
Lactate fermentation, 500, **501**
Lacteal, 671
Lactic acid, 868, 870
Lactose, 379, *428*
Lake, 103
  as climate moderator, *66–67*
Lamarck, Jean Baptiste de, 144, 230
Lamellae, *631*
Lamprey, 322
Lancelet, *320–21*
Landis, Gary, 533
Landsteiner, Carl, 214
Language, 890
  bee, 890–91
  primate, 891–92
Lanugo, 758
Large intestine, 599
Large subunit, 420

Larva, 701–2, 744, 763
Larynx, *636*
Latency period, muscle, 867
Lateral bud, *535*, 578
Lateral line, *850*, **851**
Lateral meristem, *535*
Laurasia, 339, 348
Law of tolerance, *61, 63*
Leaf, *527*
  gas exchange in, 538–40
  plumular, 563
  structure of, 538
  tissue of, 540
Leaf eater, 38
Leaf mesophyll, *540*
Leaf vein, 540
Leaflet, 538
Leakey, Richard, 356
Learning, *881, 882*
  instinct-guided, 883
Leary, Timothy, 823
Leeuwenhoek, Anton van, 458–59
Lens, 457–58, *841*
  microscope, 459
Leopold, Prince, 220
Lepidoptera, Order, **334**
Leprosy, **781**
Lesch-Nyhan disease, *221*
Leukocyte, *624–26*
Lewis structure, **368**
Liana, 101
Lichen, *315*
Life
  attitudes towards, 561
  chemical nature of, 362–63
  early animal, 318–20
  early forms of, 271–74
  early multicellular, 286–87
  early vertebrate, 320–23
  elements important to, **365**
  energy and, 394–95
  eukaryote dominance, 274–76
  origins of, 268, 270, **277**
  origins on land, 308–10
Life table, *79*
Light
  in aquatic ecosystem, *102–3*
  nature of, 840
  plant response to, *574*
  See also Sunlight
Light chain, *786*
Light microscope, 477
  optics of, 459
  resolution limit of, 460
Limb bud, 752, **755**
Limbic system, *831*
Limulus, 301
Linné, Carl von (Linnaeus), 25, 141, 264, 266
Lion, social behavior of, **893**
Lipase, 669
Lipid, 362, *377*

formation of, 388–90
  neutral, *390*
  polar, *390–91*
Lipid bilayer, **391**, 463–65
Lipoprotein, *623*
Liposome, **391**
Liquid feeder, 40
Liver, 606, 668, 669, 674
  in ammonia excretion, 678, 679, 691
  hepatic portal circulation, *672*
Liverwort, 310, **311**, **524**, *525*
Loam, *567*
Lobster, 318, **319**, 320
Locus, *202*
Logarithmic growth, 75–76
Logistic growth, 76
Long-day plant, *584*
Long-term memory, *824*
Loop of Henle, *685–86*, 693
Loose connective tissue, *596*
Lorenz, Konrad, 878, 881, 883
Lovelock, James, 7
Low-density lipoprotein (LDL), 623, 652–53
Luciferase, 586–88
Lucy (Australopithecine), 354
Lung, *634, 635*
  damage to, 653–55
  as excretory organ, 681
  prostaglandins and function of, 721
Lungfish, 327
Lupinus, 561
Luteinizing hormone (LH), **710**, *715*, 727, 729, 732, 733
Lycopod, 312, **330**, **333**
Lyell, Charles, 143, 144, 145, 157
Lylisothiocyanate, 576
Lymph, *622*
Lymph vessel, *622*
Lymphatic system, 622, 626
Lymphocyte, *625*, 784, 785
  thymus and growth of, 720
Lynx, as predator, 86
Lyon, Mary, 225
Lyonization, *225*
Lysosome, 478, 659
Lysozyme, 782

M line, 862
Macroevolution, 247
Macromolecule, *366*
Macronutrient, *564*, **566**
Macrophage, **4**, *783, 785*
Madagascar, **122**
Madagascar aye-aye, **156**
Madagascar periwinkle, 526
Madreporite, **302**
Major histocompatibility complex, 792, **794**
Malaria, 281, **779**, *780–81*
  sickle cell and, 242, **243**, 441

Malate, 515
Malawi, Lake, allopatric speciation at, 253
Male reproductive system, 724–29
Male-male competition, 888
Malignant tumor, 770
Malleus, *849*
Mallotus villosus, 633
Malnourishment, 648
Malpighian tubule, *680*
Malthus, Thomas, 81, 145, 148, 149
Malthusian Doctrine, 81
Maltose, 379
Mammal
  early, 341–42
  egg-laying, 349
  in equilibrium population, 99
  eutherian (placental), *350*
  excretion by, 679, 692–93
  heart of, 612
  pouched, 349–50
  radiations of, **349**
  reproduction of, 745–46
  water balance in, 692–93
Manaus, Brazil, clearcutting in, **30**
Mandible, 320
Mangrove, *104, 105*
Mantle, *296*
Maple, **585**
Margulis, Lynn, 7, 276, 516
Marigold, 526
Marine habitat destruction, *120–21*
Marler, Peter, 892
Marrow, *597*
Marsh, 103, *105*
Mars, temperature of, 27
Marsupial, 349–50
Marx, Karl, 152
Mass extinction, *260*
Mass-action principle, 403
Massachusetts, gypsy moths in, 74
Mast cell, *782, 797*, **798**
Mastigophora, Phylum, 280
Materialism, 152
Mating, 888, 888–90, **889**, 890
  genes and, 885–86
Maturation, 761
  control during, 768–69
Mauna Loa, carbon dioxide levels on, **48**
Maxam, Alan, 446
McClintock, Barbara, 438, *439*
McClung, Clarence, 200
Mealworm, chromosomes of, 200
Measles, **779**, **781**
Mechanical isolation, *247*
Medulla oblongata, *831*
Medusa, **291**
Megagametophyte, 546, 547

Megakaryocyte, *627*
Megaspore, 545
Megaspore mother cell, 546
Meiosis, 188, 191, **222**, 544, 545
  divisions of, 189–91
  segregation disorder and, 192
Mekong River delta, 104
Melanin, *211*, 216
Melanism, 244–45
Melanocyte, 216
Melanocyte-stimulating
  hormone (MSH), **710**, *715*
Melanoma, *773*
Melatonin, *704*, **710**
Membrane, **166**
  active transport, 468–69
  components of, **466**
  diffusion across, 467–69
  diversity of, 466
  functions of, 466–67
  osmosis across, 470–71
  photosynthetic, 507, **509**
  protein of, 464–66
  structure of, 463–66
Memory, *824*
Memory cell, *789*, *794*, **796**
Mendel, Gregor, 139, 163,
  181–88 passim, 197, 198,
  204, 206, 436, 885
  See also Genetics
Mendeleev, Dmitri, 364
Mendez, Francisco (Chico), 24
Meninges, *827*
Menopause, *733*, *761*
Menstrual cycle, 732, 734
Merkel's disks, **855**
Mesencephalon, 825
Mesoderm, 292, *593*, *748*, **749**
Mesophyll, 514, 540
Mesozoic era, 338–39
Messenger, chemical. *See*
  Communication
Messenger RNA (mRNA), *416*,
  419, 421, 431, 474
Metabolism, 485–86
Metacarpals, *871*
Metamorphosis, 701–2, 744,
  763
Metaphase, *174*
Metastasis, *770*, 771
Metatarsals, 871
Metatherian, 349–50
Methanogen, *277*, **278**
Methanol, 374
Mexico, demography of, **80**, 84
MHC complex, *792*, **794**
Miasma, 777
Micelle, **465**, 669
Michel, Hartmut, 507
Microclimate, *55–57*
Microenvironment, *55*
Microfilament, 481, **482**
Microgametophyte, 547
Micronutrient, *565*, **566**
Micropyle, 546, 549

Microscope
  electron, 460–61
  elements of, 457–58
  invention of, 458–59
  light, *459–60*, 477
  Nomarski interference, 477
  phase-contrast, 477
  polarizing, 477
  resolution of, 460, 461, 477
  staining for, 477
  visibility in, 477
Microspore, 545, 547
Microspore mother cell, 547
Microtubule, *172*, 479, 481
Microvilli, 670
Midbrain, *825*, 831
Middle ear, *849*
Midwest, water crisis in, *49*
Migration, 704, 883–84
  cell, 754
Mildew, 316
Milkweed, seed dispersal of, **557**
Miller, Carlos, 580
Miller, Stanley L., 270
Millipede, 320
Milstein, Cesar, 795
Mimicry, *160*
Mind, nature of, 807
Mineral, nutrient, 651, 654
Mineralocorticoids, **710**, *719*
Mini-chromosome system, *451*
Mitchell, Peter, 494–96
Mite, 318
Mitochondria, 483, **495**, **496**
  compared with chloroplast,
  515–16
Mitochondrion, **166**, 275
Mitogen, *177*
Mitosis, *170*, 176, **222**, 544
  stages of, 171–75
Mitotic spindle, *172*, **173**
Model, 11
Molars, *663*
Mold, 316
Mole (gram molecular weight),
  367
Molecular biology, *416*
Molecular clock, 352
Molecular systematics, 267
Molecular weight, 366
Molecule, *366*
Molière, 647
Mollusc, 296–97
  circulatory system of, 609
  in Ordovician, 303
Molting, 299, *701*
Monarch butterfly, 160, **161**
Monera, Kingdom, **268**, 276–79
Money magazine, 26
Monkey, 351
  play behavior of, **897**
Monoclonal antibody, 795
Monocotyledon, *342*, *527*, **538**,
  549
  seed of, 561, **562**

Monoculture, *127*
Monocyte, *783*
Monod, Jacques, 429
Monoecious plant, 553
Monogamous pair, *890*
Monomer, *377*
Monomoy National Wildlife
  Refuge, *104*
Monosaccharide, *34*, 36–37,
  379
Monozygotic twins, *215*
Montagnier, Luc, 799
Monteverde cloud forest, **521**
Mood, 879
Morgan, T. H., 195, 198, 200
Morphine, 824
Morphogenesis, 582
Mortality, *79*
Morula, *733*, *747*, 755
Mosaic evolution, *341*
Mosaicism, 225, **226**
Mosquito
  evolution of insecticide-
    resistant, 245
  water balance in, 693
Moss, 310, **311**, **524**, *525*
Moth, pollination by, 556
Motion sickness, 852
Motivational state, 879
Motor neuron, *809*
Motor unit, *866*
Mountain, affecting wind and
  rain, 66
Mouse
  pheromones affecting, 888
  reproductive cycle in, 733
  species variation of, 255
Mouth, 599, 661, *663*, 674
Movement, controlled, 859–60
mRNA, *416*, 419, 421, 431, 474
Mucin, 665
Mucosa, *662*
Mud salamander, 331
Mulcahy, David, 552
Mulcahy, Gabriella, 552
Mule, infertility of, *249*
Muller, Hermann J., 361, 436
Müllerian mimicry, *160*, **161**
Multicellular life, origin of,
  286–87
Multigene family, 434
Multiple alleles, *202*, 213–15
Multiple fruit, *550*, **551**
Multiple sclerosis, 798
Muscle, 580, 593, *598–99*
  cardiac, 861
  cell structure of, **864**
  contraction of, 864–67
  exercise and, 870
  fast- vs. slow-twitch, *861*, **867**,
    870
  fatigue in, *868*, **869**
  fibers of, *861*
  innervation of, **866**
  intercostal, 636

  physiology of, 867–69
  red vs. white, *861*
  skeletal, *860–61*, 861–64
  smooth, *861*
Muscle spindle, 837
Muscle tone, 867
Muscle twitch, *866*
Muscular dystrophy, *221*
Muscular system, 600, **601**
Muscularis externa, *663*
Musculo-skeletal system
  coordination of, 875–76
  movement and, 859–60
  muscle tissue, 860–69
  muscular system, 600, **601**
  skeletal system, 869, 871–75
Mushroom, 316, *317*
Mustard family, 576
Mutagen, *436*
  chemical, *438*
  physical, *438*
Mutation, *436*
  chromosomal, *438*
  DNA and, 436–37
  in *Drosophila*, 764
  effects of, 437–38
  and evolution, 442–44
  frameshift, *438*
  homeotic, 764
  in intron, 443
  neutral, 442–43
  point, *437*, 453, 772
  somatic, 789–90
Mutualism, 89, *160*, 570
Myasthenia gravis, 798
Mycelium, *314*, **317**
Mycobacterium leprose, **781**
Mycobacterium tuberculosis, **781**
Mycoplasma, 165, **167**
Mycorrhizae, *315*, 569–70
Myelin sheath, *816*
Myeloma, 795
Myocardial infarction, *620*
Myocardium, 613
Myofibril, *861*
Myofilament, *861*
Myoglobin, *641*, 651
Myopia, *842*
Myosin, *861*, **863**
Mytilis, 108

NAD$^+$, 489–99 passim
NADH, 490–99 passim
NADPH, 507–14 passim
Napoleon, **839**, 853
Nariokotome boy, *357*
Nasal epithelium, **854**
National Geographic, 26
Natural selection, 151
  Darwin and, 152–53
  evolution and, 230–33
  experimental studies of, 236,
    238

and phenotype, 232–33
variation and, 149–50, 230–38
Nautilus, 297
Neanderthal, 357
Near-field sound, 848
Neblinaria, 253
Neblinichthyes, 253
Negative effector, *403*
Negative feedback, 603–4, *711*
Negative geotropism, *574*
Negative reinforcement, *882*
Negev Desert, 99
Neisseria gonorrhoeae, 270, **740**, 741
Nekton, **107**
Nematocyst, **291**, 292
Nematode, 294
Neoplasm, *770*
Nephridia, *680*
Nephridiopore, 680
Nephron, *682–83*
Nerve, *809*
  of muscle, *866*
  optic, *842*
  spinal, *831*
Nerve cord, **293**, **319**, 824
Nerve endings, **855**
Nerve net, 289, **291**, 824
Nervous system, 288, *699–700*
  components of, 808
  evolution of, 824–25
  function of, 808
  human, **601**
  plant, 573
Nervous tissue, *598*
Neural circuit, *809*, **810**
Neural crest cell, 751, **754**
Neural plate, 751
Neural retina, 842
Neural tube, *751*
Neuroglion, 808
Neuromodulator, 822
Neuromuscular junction, *818–19*
Neuron, *598*, *808*
  mechanism of, *811*
  presynaptic and postsynaptic, *817*
  synapse and, 820–21
  types of, 808–9
Neurosecretory cell, 712
Neurotransmitter, 700, *813*
  types of, 822–24
Neurulation, 746, *751*, 753–54, 766
  human, **758**
Neutral lipid, *390*
Neutral mutation, 442–43
Neutron, *363*
New Guinea striped opossum, **156**
New Scientist, 23
New World monkey, 351
New York

pollution of harbor, 62
sewage treatment in, 116–17
solid waste production in, 136
New Zealand, rabbit population in, 74
Newton, Isaac, 153, 257
Niacin, **654**
Niche, 57, **59**
  biological aspects of, *57–58*
  competition and, *82–83*
  fundamental, 58
  physical aspects of, *57*
  realized, 58
Nicholas II, 220
Nicotinamide adenine dinucleotide, 489–99 passim
Nicotinamide adenine dinucleotide phosphate, 507–14 passim
Nicotine, 823
Nilsson, L. Anders, 555
Nitrate, 50
Nitrification, *50*
Nitrite, 50
Nitrobacter bacterium, 50
Nitrogen, as bodily waste, *678–79*, 691
Nitrogen cycle, *48*, 50–51
Nitrogen narcosis, 643
Nitrogen-fixing bacteria, 51, 127, 570–71
Nitrosomonas bacterium, 50
Node, *535*
Node of Ranvier, *816*
Noise, *838*
Nomarski interference microscope, 477
Noncompetitive inhibitor, *403*
Noncyclic electron flow, 510
Nondisjunction, *222*, **223**
Nonpolar molecule, *371*
Nonteleological evolution, *151*
Nonvirulent bacteria, *409*
Noradrenalin, **710**, *720*, 822
Norepinephrine, 720, 822, 823
Notochord, *320*, *751*
Nucellus, 546
Nuclear envelope, 176, *472*
Nuclear pore, *472*
Nuclear power, **3**, *132*
Nucleic acid, 377, *392*
Nucleolus, *473*
Nucleosome, *427*, *428*
Nucleotide, 392
Nucleus, 5, *165*, **166**, **173**, *363*, *472–73*
Nudibranch, 569
Nutrient
  as environmental variable, *60–61*
  essential, *46*
  limitation of, *52*
Nutrient cycle, 46
  interlocking, **52**

Nutrition
  malnourishment, 648
  overnutrition, 657–58
  proper, 648–56
  undernourishment, 648
Nylon, **378**

Obesity, 657–58
Objective of microscope, 459
Observation, 10
Ocean
  climate modification by, *66–67*
  destruction of marine habitats, 120
  energy and life cycle in, *41–42*
  life in, 107–8
  nitrogen cycle in, **51**
  plankton populations in, 78
  provinces of life in, **107**
Oceanic current, *65*, **66**
Octopus, 297
Odonata, Order, **334**
O'Donnell family, 19–20
Odor, *853*
Ogalalla Aquifer, **49**
Old World monkey, 351
Olfaction, *853*, **854**
Oligotrophic lake, *103*
Olins, Ada, 427
Olins, Don, 427
Ommatidia, 840, **841**
Omnivore, 40
Oncogene, 770, 772–73, 774
  action of, 772–73
Oncologist, *770*
Onychophora, Phylum, 299, 303
Oocyte, *730*
Oogonia, 729
Oomycetes, *316*
Oparin, A. I., 270
Open circulatory system, *609*, **610**
Operant conditioning, *882*
Operator sequence, *429*, **430**
Operon, *429*
Ophrys, 555
Opiate, *822*
Opium, 823, 824
Opponent processing, 847–48
Opportunism, **90**
Opportunistic infection, *799*
Opsin, *844*
Optic cup, 766, 768
Optic disk, *842*
Optic nerve, *842*
Optic vesicle, 766, 768
Opuntia englemanii, 522
Orangutan, 352, **353**
Orchid, 555
Ordovician period, 289, 303–4
Organ, *599*
Organ of Corti, *849*

Organ system, *599*, 600
Organelle, 5, *165*, 462
  energy-producing, 483
  synthetic, 473
  types of, 472–78
Organic chemistry, 373–74
Organism, 4
Organogenesis, 746, *751–52*
Orgasm, 735, 736
Origin of Species, 140, 839
Origin of the Continents and Oceans, 68
Orthomyxovirus, **284**
Orthoptera, Order, **334**
Osculum, **290**
Osmosis, 470–71, 618, 679–80
  effect on cell, 471, **472**
Osmotic pressure, 471, **472**
Osteichthyes, Class, 326–27, 330
Osteoblast, *872*
Osteoclast, *873*
Osteocyte, *597*, *872*
Osteoporosis, *874*
Ostia, 609
Ostracoderm, *322*
Ostrom, John, 338, 341, 344
Otter, 109
Outer ear, *849*
Ova, 729–30, 744, 755
Oval window, *849*
Ovalbumin, 433
Ovary, 546, 549, *703*
  anatomy of, 729–30
  hormones of, **711**, *720*
Overexploitation, *130*
Oviparous organism, 745
Ovoviviparous organism, 745
Ovulation, *730*
Ovule, *546*
Oxaloacetate, 493, 514, **515**
Oxidation, *486–87*
Oxidative phosphorylation, 494
Oxygen
  atmospheric, 273, 286, 533
  in Cambrian atmosphere, 288
  as environmental variable, *60*
  eukaryotes and, 274
  evolution of in plants, *508*
  hemoglobin and, 639–41
  in Krebs cycle, 493–94
  in Precambrian atmosphere, 274
  respiration of. *See* Respiration
  toxicity of, 273
  turnover of, **102**
Oxygen debt, 500
Oxytocin, 705, **710**, *712–13*, *760*
Oyster, 297
Ozone layer, 26, 119–20

Pacemaker, *615*
Pacinian corpuscles, **855**, 856
Page, David, 765

Pain
  modulation of, 822
  prostaglandins and, 721
  referred, 832
Paired fins, 325–26
Palate, *663*
Paleontology, 158
Paleozoic Era, 288–89, 303–4, 308–9, 325–26
Palindromic sequence, *446*
Palisade cell, 540
Pampa, *100*
Panama, rain forest in, 134
Pancreas, *668–69*, 674, *703*
  hormones of, **711**, 717–19
Pangaea, 325, 331, 333, 335, 337, 339
Panting, 697
Pantothenic acid, **654**
Papavirus, **284**
Papilio polyxenes, 576
Papillary muscle, 613
Parakeet, **180**
Paramecium, 281, **472**
  contractile vacuole of, **680**
  digestion in, 659, **660**
  growth of, **78**, 82, **83**
Paramyxovirus, **284**
Parapatric speciation, *256*
Parasite, *781*
Parasitism, 86, *160*
Parasympathetic division, *826*, 827
Parathion insecticide, 819
Parathyroid, *703*
  hormones of, **710**, 717
Parental generation, 182
Parental investment, 888
Parietal cell, 666
Partial pressure, *640*
Particulate inheritance, *182–83*
Parvovirus, **284**
Passiflora (passion flower) vine, 159, 160, **553**
Pasteur, Louis, 778, 797
Pathogenic organism, 777, 780–781
Pauling, Linus, 385
Pavlov, Ivan, 882
PCBs, 115
Pea, **551**
  genetic experiments with, 181–85
Peach, 550, **551**
Peacock, sexual selection in, **890**
Pear, propagation from cutting, 543
Pectoral girdle, *869*
Pedigree, *210*
Pedipalp, *301*
Peking man, 356
Pelagic organisms, **107**
Pelvic girdle, *869*, 871
Pelvic inflammatory disease, 741

Penicillin, *403*
Penicillium mold, 263, 318
Penis, *724*
Pepsin, 666
Pepsinogen, 667, 673
Peptic ulcer, *667*
Peptide bond, *383*
Peptide T, 805
Perennial plant, 535
Pericardium, 613
Pericycle, *528*, 534
Peripatus, 303
Peripheral membrane protein, 466
Peripheral nervous system, 825, 826–27
Peristalsis, *666*
Permafrost, *96*
Permanent immunity, 789, 794, 797
Permian period, 335, 337, 341
Pert, Candice, 822
Peru
  anchovy catch in, **123**, 130
  El Niño and, 69
  photic zone off coast of, 108
Pesticides, 126, 128
PET scan, *833*
Petal, *546*
Petiole, 538
pH scale, 372
Phaeophyta, 303, 306
Phagocyte, *625*, *783*, *790*
Phagocytosis, **476**, **478**
Phalanges, *871*
Pharyngeal gill slits, *320*
Pharynx, *293*, *636*, *665*, 674
Phase transition, 373
Phase-contrast microscope, 477
Phenetics, 266, **267**
Phenol, 536
Phenotype, *184*
Phenylketonuria, 454
Pheromone, *853*, *887–88*
Philippines, agriculture in, 127
Phipps, Jamie, 778
Phloem, *309*, *528*, 529, *531–32*, 535, 536, 540
Phloem sap, 531, 533
Phosphate, **375**
Phosphatidyl inositol, *708*
Phosphoglycerate (PGA), *512–13*
Phosphoglycolate, 514
Phospholipase *c*, 708
Phospholipid, *390–91*
Phosphorus, as nutrient, 651, **655**
Phosphorylation, *398*, 490
  oxidative, 494
  in plants, 510
Photic zone, 42, 51, 103
  concentration of life in, **107**
Photolysis, *508*
Photomorphogenesis, *582–83*

Photon, *504*, *840*
Photoperiodism, **583**, *584*
Photophosphorylation, *510–11*
Photoreceptor, *840*, **843**, *844–45*
  and color vision, 846
  types of, 842, **843**
Photorespiration, *514*
Photosynthesis, *34*, *503*
  by algae, 304
  bacterial, 271
  carbon dioxide in, 503, 511–14, 538, 540
  C4, *514–15*
  chloroplast in, 483
  experiments on, 512
  light-dependent reactions, 505–11
  light-independent reactions, 511–15
  membrane for, 507, **509**
  origins of, 273
  prokaryote, 273
  reaction center, 507–8
  compared with respiration, 503
  undersea, 41–42
Photosynthetic electron transport, *508*
  noncyclic flow, 510
  oxygen evolution, 508, 510
Photosynthetic membrane, **166**
Photosystem I, 508–10
Photosystem II, 508–10
Phototropism, *574*, 576–78
Photovoltaic cell, 35–36
Phragmoplast, **177**
Phycocyanin, 306
Phycoerythrin, 306
Physical mutagen, *438*
Physical properties, *364*
Physical support, 309–10
Physics, laws of, 406
Physiograph, *866*
Physiology, 591, *603*
Phytochrome
  chlorophyll and, 585
  discovery of, *582–83*
Phytophora, 318
Phytoplankton, *41*, 107–8
Picornavirus, **284**
Pierce, Franklin, 136
Pigeon, learning by, 882
Pigment, *504*
  accessory, 585
Pigment epithelium, 842
Pine, **336**
Pineal body, 830
Pineal gland, 704
  hormones of, **711**
Pineapple, 550, **551**
Pisaster, 108
Pit, in xylem, 529–30
Pit organ, 856
Pitcher plant, 571

Pith, *535*
Pituitary, 599, 604–05, *703*
  anatomy of, 712
  anterior, **710**, *713–15*, **732**
  posterior, 688–89, **710**, *712–13*, 830
  thyroxine and, 716–17
PKU, 454
Placenta, *350*, 641, *733*, *745*, *748*, 757
Placoderm, *322*, 325
Plague, 20–22
Planaria, digestion in, 661
Plant
  asexual reproduction in, 542–44, 559
  autumn changes, 585–86
  cell culture of, 586
  cells of, **521**
  chemical composition of, 564
  chemical defenses of, 576
  day-neutral, 584
  defined, 520–22
  early, 310
  early vascular, 312
  embryonic development, 547–50
  flowering, 342–45, 542–58
  fossil, **543**
  fruit of, 542, 549, 550–51
  genetic engineering of, 586–88
  germination of, 563
  growth of, 563, 573–75
  hormones of, 575–82
  hydroponic culture of, *564*
  insectivorous, 571
  land, 524–25
  long-day, *584*
  medical uses of, 526
  mimicry by, 555
  need for sunlight, 520–21
  need for water, 522
  nitrogen and, 691
  nutrition of, 564–71
  pollination and fertilization of, 547
  regulation of growth in, 582–86
  reproduction of, 314, 522
  reproductive organs of, 546
  respiration of, 631
  seed, 342
  sexual reproduction in, 544
  short-day, *584*
  spores and gametes of, 546–47
  structure of, 527, **528**
  vascular, 525, 527
Plant hardiness, **64**
Plant-herbivore coevolution, 159
Plant-pollinator coevolution, 159
Plantae, Kingdom, **268**, 520
Plantlet, 586

Planula, **291**
Plasma, *618, 623*
Plasmid, *426, 449*, 586
Plasmodesmata, 531
Plasmodium, 281, 780, **781**
Plastocyanin, 510
Plastoquinone, 510
Plate tectonics, *68*
Platelet, *626–27*
Platinum, 399
Plato, 140, 723
Platyhelminth, 292–93, **781**, **825**
  excretion by, 680, **681**
Platypus, 349, **350**
Pleiosaur, 341
Pleiotropy, *242*
Pleural sac, *636*
Plumular leaf, 563
Pneumocystis carinii, 799, 800
Pneumonia, bacterial, **781**
PNS, 825, 826
Podocyte, 683
Poinsettia, 584
Point mutation, *437, 453, 772*
Polar body, 730
Polar lipid, *390–91*
Polar nuclei, 547
Polarity, *369, 371*
Polarizing microscope, 477
Polio, **779**
Pollen, *330, 546, 547*, **549**, 552
Pollen tube, 547, 552
Pollination, 547
  self-, 556
  techniques of, 553–57
Pollution
  air, 117–120
  water, 114–17
Polyacetylene, 526
Polyclonal immune response, 795
Polydactyly, *213*
Polygenic system, *203*, 205–06, 215–18
Polygyny, *890*
Polymer, *377*
  condensation reaction in formation of, 379, 381
Polymerization, *377*, **378**
Polymorphism, *243*
Polynucleotide, *410*
Polyp, **291**
Polypeptide, *383*, 384
  as hormone, *703*, 706, **708**
Polyplacophora, Class, *297*
Polypodiacea, **525**
Polysaccharide, *381*
Polyspermy, 746
Polytene chromosome, *196*
Polyunsaturated fatty acid, *390*
Pond, 103
Pongidae, Family, 352
Pons, *831*
Population, *74–75*

age distribution of, 78–79
dynamics of, 74–78, 80–81
growth of, 81, 84, 124–25
regulation of, 81–89
zero growth, 77
Porifera, Phylum, 289, **290**
Porphyridium, 168
Porphyrin ring, *504–05*
Positive effector, *403*
Positive feedback, *604*
Positive geotropism, *574*
Positive reinforcement, *882*
Positron emission tomography, *833*
Posterior pituitary, **710**, *712–13*, 830
Postsynaptic neuron, *817*
Postsynaptic potential, *820*
Postzygotic isolating mechanisms, *248–49*
Potassium, as nutrient, 651, **655**
Potassium pump, 540
Potassium-40 dating, 158
Potato blight, 318
Potential energy, *395*, **396**
Potential. *See* Electrical potential
Poxvirus, **284**
Prairie, *100*, **101**
Precambrian era, *286–87*
Precipitation, *66*, **67**, **95**
Predation, *85–86*, 87
Pregnancy
  fetal respiration, *641*
  urine test for, 687
Prehensile tail, 351
Pressure-flow hypothesis, *532*
Presynaptic neuron, *817*
Prezygotic isolating mechanisms, *247–48*
Pribilof Islands, reindeer in, **78**
Primary consumer, *37*
Primary growth, in plants, *535*, **536**
Primary producer, *34*
Primary sewage treatment, 116
Primary structure of protein, *384*
Primary succession, *71*
Primate
  language of, 891–92
  society of, 896
Primer pheromone, *888*
Primitive streak, *751*
Primula, 584
Principles, 10
Product, *399*
Progesterone, **710**, *731*, 733
  infertility and, 739
Prokaryote
  chromosome in, 426–27, 431
  gene expression in, 428–30
  mRNA in, 431
  ribosome in, 431
Prokaryotic cell, 166, 168
Prolactin (PRL), **710**, *715*

Promoter sequence, *429*, **430**
Pronucleus, 755
Prophase, *172*, **173**, **174**
Proprioceptor, *856*
Prosencephalon, 825
Prosimian, *351*
Prostaglandin, *650, 721*, 856
Prostate, 727
Prosthetic group, 388
Protease, 673
Protein, *377, 649*
  complete vs. incomplete, *650*
  deficiency in, *649–50*
  function of, *381, 383*
  in genetic coding, 408
  histone, *427*
  as hormone, *703*, **708**
  membrane, *464–66*
  nonhistone, *427, 428*
  functioning as pore, *468*
  regulatory, **433**
  secretory, *475*
  structure of, *384–88*
  synthesis of, *383*, 416, 419–22
Protein kinase, 706–7
Proteinoid microsphere, 270
Proterozoic Era, *286–87*
Prothoracic gland, 701
Prothrombin, *623*, 627
Protista, Kingdom, **268**, *279–83*
Protoavis, 341, 344
Proton, *363*
Protonema, 310, **311**
Protonephridia, **293**
Protoplast, *586*
Protostome, 294, **295**
Prototherian, 349, **350**
Protozoa, **4**, *780*, **781**
Proximal convoluted tubule, *685*
Pseudocoelomate, 292
Pseudocopulation, 555
Pseudomonas, **278**
  genetically engineered, 452
Pseudoplasmodium, 700
Pseudostratified epithelium, *594*
Pteraeolidia nudibranch, 569
Pteranodon, 341
Pterophyte, 312
Puberty, **727**, *728–29*, 731
Pubis, *871*
Puerto Rico, phosphorescent bay of, 283
Pulmonary artery, 614
Pulmonary circulation, *612*, 614, *616*
Pulmonary valve, 614
Pulse, *618*
Punctuated equilibrium, *259–60*
Punnett, R. C., 192
Punnett square, *183*, **187**
Pupa, 702, 763
Purine, *412*
Pyloric sphincter, 667

Promoter sequence, *429*, **430**
Pronucleus, 755
Prophase, *172*, **173**, **174**
Proprioceptor, *856*
Prosencephalon, 825
Prosimian, *351*
Pyramid
  ecological, *45–46*
  of numbers, 45
Pyrethrum, 159
Pyrimidine, *412*
Pyrrophyta, Phylum, 283
Pyruvate, *490–92*

Quantitative data, 11
Quantum, *504*
Quaternary structure of protein, *386*, 388
Queen substance, 888
Quercus tree, 249, 250, 251

Rabbit
  genetics of, 202
  reproductive cycle in, 733
Radial symmetry, *289*
Radiation, heat, 694
Radiation therapy, *773*
Radioisotope, *158*
Radius, *869*
Radula, **296**
Ragweed, 584
Rain forest, 101–2
Rain shadow, 66
Rainfall, *66*, **67**
  characteristic of biome, **95**
Raspberry, 550, **551**
Rat
  evolution of warfarin-resistant, 245
  learning by, *882*
Rate of spontaneous activity, *837*
Rauwolfia, 526
Ray, 326, **327**
Reactant, *399*
Reaction, chemical, *367*
Reasoning, inductive and deductive, *10–11*
Receptive field, 842
Receptor, *576*
  hormone, *705*
Receptor cell, 836
Receptor potential, *837*
Receptor site, postsynaptic, **818**
Recessive trait, *183*
Reciprocal inhibition, *876*
Recombinant-DNA technology, *449*, 452
Recommended Daily Allowance (RDA), 651
Rectum, 661, 673, 674
Recycling, 135
Red cell, *623–24*, **625**
Red muscle, *861*
Red Sea, **54**
Red tide, 283
Redox reaction, *486, 505*
Reduction, *486–87*
Reduction division, 188–89

Redwood, 342
Reevaluation, 12
Referred pain, 832
Reflection of light, *504*
Reflex arc, *809*
Refractory period, *815*
Reinforcement, *882*
Relaxation time, muscle, 867
Releaser, *880*
Releaser pheromone, *887*
Releasing hormone, *714*, 716–17
Religion, science and, *6–8*
Renal artery and vein, *681*
Renal cortex, medulla, and sinus, 682
Renewable energy, *132–33*
Renin, *689*
Rennin, 451
Reovirus, **284**
Replica plating, **231**
Replication, *408*
"Reporter" gene, 587
Reproduction, asexual vs. sexual, *723*
Reproductive isolation, 247–49, 254
  development of, 249–57
Reproductive polyp, **291**
Reproductive strategy, 310
Reproductive system, **601**
  contraception and, 736–38
  diseases of, 739–42
  female, 729–35
  infertility in, 738–39
  intercourse and, 735–36
  male, 724–29
  in mammals, 733
  primary and secondary sexual organs, 725
Reptile, 335
  adaptive radiation of, 339, **340**
  excretion by, 678
  heart of, 612
  reproduction of, 745
  temperature regulation by, 694
  water balance in, 692
Research, developmental anomalies in, 763–64
Reserpine, 526
Resolution limit, 460
Resorption, *685*
Resource utilization curve, 58, **83**
The Resourceful Earth, 114
Respiration, *498–99*
  air movement in, 638
  cellular, 629–30
  compared with photosynthesis, 503
  control of, 642
  gas composition in, 638

gas exchange in, 638–39, 640–41
  water in, 677
Respiratory surface, 630
Respiratory system, *629*
  damage to, 642–45
  human, **601**, 636–42
  types of, 630–35
Resting potential, *812*
Restriction enzyme, *445*, 446
Restriction fragment length polymorphism (RFLP), *453*
  and gene defect detection, 454
  and genetic disease, 453–54
Reticular formation, *831*
Retinal, *844*
Retina, *841*, *842*, *844*
  detached, *844*
  information processing in, **847**
Retinitus pigmetosa, *221*
Retinoblastoma, 770, 774
Retinoic acid, 768, **769**
Retrovirus, **284**, *426*, *771*, **772**, *780*
  See also HIV
Reverse transcriptase, *426*, *780*
Reversible reaction, 401
Rh factor, *214–15*
Rhesus monkey, 214
Rheumatic fever, *798*
Rheumatoid arthritis, 798
Rhinovirus, **781**
Rhizobium bacteria, 570–71
Rhizoctonia fungus, 570
Rhizoid, *525*
Rhizome, **313**
Rhizopus, 316
Rhodophyta, 303, 306
Rhombencephalon, 825
Rhynia, **312**
Rhyniophyta, 312
Rhythm, as contraceptive method, 737, **738**
Ribonucleic acid. *See* RNA
Ribose, 379, **417**
Ribosomal RNA, 419
Ribosome, *418–21*, 431, 473–74, 475
Ribs, 869
Ribulose bisphosphate (RuBP), *512–13*
Rigor, *868*
Rigor mortis, *868*
RNA
  of AIDS virus, 800
  in egg, 746, 755, 766
  function of, 416
  origin of, 270
  in protozoan, 404
  of retrovirus, 771, 780
  structure of, 392, **417**
  of virus, 284, 780
RNA polymerase, 416

Rods, 842
Rome, solid waste production in, 136
Root, *525*, *527*
  adventitious, 543
  auxin and growth of, 579–80
  growth pattern of, 534
  primary vs. secondary, *528*, **529**, 534
  structure of, 528–29
  water movement into, 532, 534
Root cap, 534
Root hairs, *528*
Rosy periwinkle, and Hodgkin's disease, **122**
Rough endoplasmic reticulum, *474*
Roundworm, 294
Rous, Peyton, 771–72
Rous sarcoma virus, 771, **772**, **781**
Roux, Wilhelm, 591
rRNA, 419
RSV, 771, **772**, **781**
Rubber tapping, 24
Rubella, 758
Rudman, William, 569
Ruffini corpuscles, **855**
Rule of 10, 45
Russell, Bertrand, 838
Russula mushroom, **317**

S-shaped growth curve, 76
Sac fungi, 316
Saccharin, 854
Saccule, *851*, 852
Sachs, Bernard, 211
Sacrum, *871*
Sage, Richard, 255
Sage grouse, mating behavior of, **889**
Sagittal plane, *600*
Sahara Desert, growth of, **121**
Salinity, *59–60*
Salinization, *129*
Saliva, 663
Salivary gland, *663*, 674
Salmon, migration of, 883–84
Salt, balance in body, 677
Salt gland, 692
Salt marsh, *104*
Saltatory conduction, *817*
Sanger, Frederick, 446, **447**
Saprophyte, *40*
Saprophytic fungi, *314*
Sarcodina, Phylum, 280
Sarcolemma, *861*
Sarcomere, *862*, **863**
Sarcoplasmic reticulum, 864
Sarin insecticide, 819
Saturated fatty acid, *390*
Savannah, *100*
Scallop, 297

Scanning electron microscope, 460–61
Scapula, *869*
Scarlet fever, *798*
Schizocoelous organism, **295**
Schleiden, Matthias, 459
Schotté, Oscar, 768
Schrödinger, Erwin, 406
Schwann, Theodore, 459
Schwann cell, *816*
Science
  concerns of, 8–9
  impact on society, 19–20
  nature of, 6–10
  religion and, 6–8
  society and, 9–10
Scientific method, 9, *10–12*
  case studies of, 12–17
Scorpion, **300**, 318
Scrotum, *725*
Sea
  climate modification by, *66–67*
  destruction of marine habitats, 120
  energy and life cycle in, *41–42*
  life in, 107–8
  nitrogen cycle in, **51**
  plankton populations in, 78
  provinces of life in, **107**
Sea anemone, 289, 291–92
  learning by, 882
Sea cucumber, 301, **302**
Sea lettuce, 305
Sea lily, 301
Sea perch, **321**
Sea scorpion, 301
Sea urchin, 109, **302**, **746**
Sealth (Seattle), Chief, 136
Sebaceous glands, *782*
Sebum, *782*
Second law of thermodynamics, *395*
Second messenger, 706–8
Secondary consumer, 39
Secondary growth, in plants, *535*, **536**
Secondary phloem, *536*
Secondary sewage treatment, 116
Secondary structure of protein, *385*, **388**
Secondary succession, *71*
Secondary xylem, *536*
Secretin, 673, **710**
Secretion, 687–88
Secretory gland, 593
Secretory protein, 475
Secretory vesicle, **475**, 476–77
Seed, **330**
  dormancy of, 561–63
  fruit, and berry picker, 38–39
  germination of, 563
  imbibition, 563

monocot vs. dicot, 561
Seed coat, 549
Seed dispersal, 550
  techniques of, *557–58*
Segmentation movements, *671*
Segregation disorder (sd), 192
Segregation of alleles, *183, 184,*
  186, 189
Selective permeability, 466–67
Self-incompatibility, 556–57
Self-propagating impulse, *814*
Semen, *724*
  sperm count of, 726, 738
Semicircular canal, *851*
Semilunar valve, 614
Seminal vesicle, 721, 727
Seminiferous tubules, *725,* **726**
Senescence, *761*
  leaf, *586*
Sensory neuron, *809*
Sensory system
  animal senses, 856–57
  balance and acceleration,
    851–52
  behavior and, 879–80
  common chemical sense, 853,
    854
  diversity of, 839
  environment and, 839
  exteroceptor, *854,* 856
  interoceptor, *856*
  proprioceptor, *856*
  stimuli to, *836*
  taste and smell, 853–54
  transduction, 836–37
  unity of, 839
  vision, 839–47
  as "window on the world," 835
Sepal, *546*
Separation, 168
Septum, 299
Serosa, *663*
Serotonin, 708, 823
Sertoli cell, *726*
Seta, 299
Sewage, 115–17
Sex, determination of, 197–201,
  765
Sex chromosome, *208*
  genetic map of, **218**
  inheritance of, 197–98
Sex hormone. *See* Estrogen;
  Testosterone
Sex pheromone, 887
Sex-linked disorders, *219–21*
Sex-linked gene, *210,* 218–21
Sex-linked inheritance, *218*
Sex-linked trait, 198–200
Sexual intercourse, 735–36
Sexual reproduction, 274–75,
  723
  in plant, 544
Sexual selection, *888*
Sexually transmitted diseases
  (STDs), 739–42

prevention of, 742
Shakespeare, William, 896
Shark, 326, **327**
Sherrington, Charles, 859
Shistosomiasis, **781**
Shivering, 604, 696
Shockley, William, 217
Short-day plant, *584*
Short-term memory, *824*
Shrimp, 318
Siamese twins, **764,** 766
Sickle-cell anemia, *212,* 224,
  242, **243,** 437, 440–41
Sierra primrose, 584
Sieve plate, **302**
Sieve tube, *531*
Sign stimulus, *880*
Signal, *838*
Silurian period, 308
Simon, Julia, 114
Simple diffusion, 468, **469**
Simple epithelium, *594*
Simple fruit, *550,* 551
Simple sugar, 377, 379
Sinai Desert, **55,** 99
Single circulation, *611*
Sinigrin, 576
Sinoatrial (SA) node, *615*
Sinus, **296,** 757
Sinus venosus, 611
Siphonoptera, Order, **334**
Size, genetic basis of, 215
Skate, 326
Skeletal muscle, *599, 860–61,*
  861–64
  structure of, 861
Skeleton
  appendicular, *869,* 871
  axial, *869*
  human, 600, **601**
  hydrostatic, *859,* **860**
Skin, 593, 599, *781–82*
  color of, 216, 218
  as excretory organ, 681
  nerve supply to, **832**
  receptors of, **855**
  respiration through, 630
Skinner, B. F., *882*
Skoog, Folke, 580
Skull, 869
Slack, 514
Slavery, 878–79
Sleeping sickness, 280, *780,* **781**
Sliding filament model, *864*
Slime mold, 281–82
  chemical signaling in,
    700–701
Slow-twitch muscle, *861,* **867,**
  870
Slug, 297
Small-ribosomal subunit, 419
Smallpox, 777, **779**
Smell, *853,* **854**
Smith, Adam, 25
Smog, *118*

Smoking
  and cancer, 775
  and heart attack, 620
  and lung damage, 643–45
Smooth endoplasmic reticulum,
  474
Smooth muscle, *598–99, 861*
Snail, 297
  evolution of, 158
Snakeroot, 526
Snapdragon, dominance in, **201**
Snyder, Solomon, 822
Social behavior, 886
  communication, 887–90
  evolution of, 892–93
  in insect society, 893–95
  language, 890–92
  in primate society, 895
Social Darwinism, 152–53
Social insects, 886
Social releaser, *887*
Society
  AIDS and, 22–24
  disease and, 20–22
  effects on science of, 20
Sociobiology, 892
Sodium, as nutrient, 651,
  654–56
Sodium-potassium pump,
  811–12
Soft palate, *663*
Soil, *94–96,* 565
  acidity of, **57**
  characteristic of biome, **95**
  fertilization of, 567–69
  nutrients in, 565–67
Soil depletion, 128
Solar energy, 42, 132
  See also Sunlight
Solar system, **6**
Solubility, *371*
Somatic cell, 188
  hybridization of, *226,* **227**
Somatic mutation, 789–90
Somatic nervous system, 826
Somatotropin, **710,** *715*
Somite, *751–52*
Sonoran Desert, 99
Sori, **313**
Sound
  defined, *848*
  generation of, 849
South America, El Niño and, 69
Southern Oscillation, 69
Soybean, 584
Spaceship Earth, 5
Spadefoot toad, **332**
Spartina, 104
Spawning grounds, *105*
Speciation, 247, *247–49*
  allopatric, *249–50,* 253–55
  mechanisms of, 249–57
  parapatric, *256*
  sympatric, *256–57*
Species, 229, 265

diversity of, *93*
  endemic, 253
Spectrin, 466
Spemann, Hans, 766, 768
Spencer, Herbert, 144, 152
Sperm, **4,** *724,* 744, 888
  human, 755
  response to egg, 746–47
  structure of, **727**
Sperm nucleus, 547
Spermatid, *725*
Spermatocyte, *725*
Spermatogonia, *725*
Spermicide, 737, **738**
Sphagnum moss, 103
Sphenophyte, 312
Sphincter, 661
Sphygmomanometer, 618
Spicule, **290**
Spider, 318, **319**
Spina bifida, 751
Spinal cord, 827, 831–32
Spinal nerve, *831*
Spinneret, **319**
Spiny anteater, 349
Spiracle, *633*
Spirochete, **278**
Sponge
  animal, 289, **290**
  as contraceptive method, **738**
Spongin, **290**
Spongy bone, *872*
Spongy mesophyll, 540
Sporangia, **282**
Spore, 544
Spore mother cell, 545
Sporophyte, *305, 314, 544*
Sporozoa, Phylum, 281
Squamous epithelium, *594*
Squid, 297
  eyes of, 841–842
src, 772
Stabilizing selection, 232, *233,*
  238
Staining, for microscope, 477
Stamen, *546*
Staminate flower, 553
Stanley, Steven, 259
Stapes, *849*
Starch, *37,* 381
Starfish, 301, **302**
Start signal, 419
Statocyst, *851*
Steady state, 76
Stegosaurus, 341
Stem, *527*
  growth pattern of, 535–37
  structure of, 535
Stem cell, 769
Sterility, *738*
Sterilization, as contraceptive
  method, 737
Sternum, 869
Steroid, *392, 703, 706,* **707**
Stevens, Nettie, 198, 200

Sticky ends, *449*
Stigma, *546*
Stimulus-response chain, *888*
Stipe, **313**, **317**
Stoma, *309*
  of leaf, *538*, **539**
Stomach, 599, 661, *666–67*, 674
Stop codon, *417*
Stratified epithelium, *594*
Strawberry, 550, **551**
  reproduction of, 543, **544**
Streptococcus, 798
  pneumoniae, **781**
Stress, zone of, 63
Stretch receptor, 809, 856
Striated muscle, *860*
Striation, 599
Stroke, *622*
Stroma, *507*, *511*
Stromatolite, *271*, 289
Structural isomers, 379
Sturgeon, 327
Sturtevant, Alfred, 195
Style, *546*
Subatomic particle, *363*
Subcutaneous tissue, **782**
Submarine hot springs, 272–73
Submucosa, *662*
Subsong, 885
Substrate, *399*
Substrate phosphorylation, 490
Succession, *70–71*
Sucrose, 379
Sudan, undernourishment in, 648
Sugar, *34*, 36–37, **37**
  metabolism of, 672
  simple, 377, 379
  See also Glucose
Sugar balance, 606
Sulfur, as nutrient, 651, **655**
Summation
  in muscle response, *867*, **868**
  of potentials, *820*
Sundew, **571**
Sunflower, taxonomy of, **266**
Sunlight
  and climate change, 68
  chlorophyll and, 504–5
  energy from, **397**
  as environmental variable, *57*, 58
  heating Earth's surface, **65**
  photosynthesis and, 503–11, 520–21
  refraction of, **505**
  temperature and, 64
Supercontinents, 70
Superficial cleavage, 747
Suppressor T-cell, 791
Surface tension, 373
Surinam toad, **332**
Survivorship, *79*
Suspensor, 549
Sustainability, *123*

Sutton, Walter, 191
Suture, 869, 874
Swallowing, *665*
Swallowtail butterfly, 576
Swamp, 103
Sweat, 604, 697
Sweden, demography of, **80**, 84
Swim bladder, **329**, 634
Symbiosis, 89, 160
  and plant nutrition, 569–71
Symbiotic fungi, *314*
Symmetry, 600
Sympathetic division, *826*, 827
Sympatric speciation, *256–57*
Synapse, *809*, *817*
  chemical, *817–18*
  drug effect on, 823
  electrical, *817*
  example of, 818–19
  interneuronal, *820–21*
  and learning and memory, 824
  types of, *817–18*
Synapsis, *189*
Synaptic bouton, **818**
Synaptic cleft, *817*, **818**
Synaptic vesicle, **818**
Synergid, 547
Synovial joint, 874–75
Synthetic organelle, 473
Syphilis, 279, 740–41, **781**
Systematics, 266–67
Systemic circulation, *612*, 614, *616–17*
Systole, *618*
Systolic pressure, *618*, 619

T-cell, *784*
  helper, *787*
  in immune response, 791–92
  types of, 791
T-tubule, 864
Tadpole, 744
Taenia, **781**
TAF, 771
Taiga, *97–98*
Tapeworm, 292, **293**, **781**
Taraxacum officinale, 559
Target cell, 576, 578, *705*, **709**
Tarsals, *871*
Tasmania, sheep in, **78**
Taste, *853–54*
Taste bud, 665, *854*, **855**
Taxa, *266*
Taxonomy, 264–66
Tay, Warren, 211
Tay-Sachs disease, *211–12*, 224
TCE, 117
Technology, appropriate, *123*
Tectorial membrane, *850*
Teeth, *663*, **664**
Telephone, *175*
Temin, Howard, 425
Temperate forest, *98–99*

Temperature
  altitude and, **96**
  of aquatic ecosystem, *102–3*
  body, 677, 693–97
  characteristic of biome, **95**, **96**
  as environmental variable, *57*, *59*
  greenhouse effect and, *27–30*
  sunlight and, 64
Template, **415**, *416*
Temporal isolation, *248*
Temporary immunity, 797
Ten percent rule, 45
Tendon, *859*
Tenebrio worm, 200
Terminal bud, *535*, 586
Termination factor, 421
Termite, nest of, **895**
Tertiary consumer, *39*
Tertiary sewage treatment, 116
Tertiary structure of protein, *386*, **387**, **388**
Test cross, *184*, **185**, **186**, **187**
Testes, *703*, *724*, *725–26*
  hormones of, **711**, *720*
Testosterone, 703, **710**, 719, 720, *726*, *727–28*
  receptor for, 728
Tetanus, *867*, **869**
Tetrad, *189*
Tetrahymena, 404
Tetrapod, 330, 331
TF (testicular feminization syndrome), 728
Thailand, aquaculture in, 131
Thalamus, *830*
Thalidomide, 758
Theory, *12*
Theraspid, 337–38
Theria, Subclass, 349
Thermal stratification, *102–3*
Thermoacidophile, *278*
Thermocline, *102–3*
Thermodynamics, 395–97
Thermoregulation, *61*, *693*, 697
Thiamine, 651
Thick filament, 861
Thigmotropism, *574*, **575**
Thin filament, 861
Third World
  economic concerns of, 135
  energy needs of, 131, 133
  food production in, 125, 127
Thirst, 689
Thoracic cavity, 602, *636*
Thoreau, Henry David, 25
Three Mile Island reactor, 132
Threonine, 405
Threshold, *813*, *814–15*
Thrombin, *627*
Thrombocyte, *627*
Thromboplastin, *627*
Thrush, 318

Thylakoid, *504*, 507, 510
Thymidine kinase (TK), *226*
Thymine, *410*
  in dimer, *439*
Thymosin, **710**, 720
Thymus, *703*, *784*, *794*
  hormones of, **711**, 720
Thyroid, **366**, *703*
  diseases of, 716
  hormones of, **710**, 715–17
  stimulation of, 714
Thyroid releasing hormone (TRH), 716–17
Thyrotropic hormone (TSH), **710**, 714
Thyroxine, 651, 705, **710**, *715–17*
Ti plasmid, 586–87
Tibia, *871*
Tick, behavior of, 880
Tight junction, *595*
Tigris valley, salinization of, 129
Time magazine, 26
Timothy, seed dispersal of, 558
Timothy grass, **549**
Tinbergen, Niko, 878, 881
Tissue, *593*
Tissue culture, 544
Toad, **332**
Tobacco mosaic virus, 423
Togavirus, **284**
Tomato, 550
Tonegawa, Susumu, 788
Tongue, *665*, 674
Toxic waste, 19–20
Toxin, *780*, 783
Trabeculae, *872*
Trace fossil, *287*
Trachea, 609
  human, *636*
  respiration through, *633–34*
Tracheid, *525*, 529
Tracheole, *634*
Tracheophyte, 525, 527
Trade wind, 65
Transcription, *416*
Transcription factor, 432–33
Transduction, 836–37
Transfer RNA (tRNA), *417–18*, **419**, 420–421
Transformation, *409*, 449
Transfusion, 213
  AIDS and, 803–5
Transgenic organism, *451*
Translation, *419*
Translocation, *227*, 531
Transmission electron microscope, 460, 461
Transmitter. *See* Neurotransmitter
Transparency, 373
Transpeptidase, *403*
Transpiration, **47**, *531*
Transplant rejection, 794
Transport molecule, 468

Transposable element, *438*
Transverse plane, *600*
Tree
  growth of, 536, *537*
  injury to, 536–37
  rings of, *536*, **537**
Treponema pallidum, 279, 740, **781**
Trial-and-error learning, 882
Triassic period, 338, 339, 341
Tricarboxylic acid cycle, 491
Triceratops, 341
Trichinosis, 294
Triglyceride, 390
Trilobite, *299*
  extinction of, 303
Trimester
  first, 758
  human gestation, *758*
  second, 758–59
  third, 759–61
Triops cancriformis, 259
Triple covalent bond, *368*, **369**
Triploid endosperm cell, *547*
Trisomy 21, *223*
Tristan da Cunha, 221
Triticum, 584
tRNA, *417–18*, 420–21
Trophic hormone, *714*
Trophic level, *45–46*
Trophoblast, 748, 755
Tropical rain forest, 101–2, 127, **128**
Tropomyosin, *861*
Troponin, *861*
Truffle, 316
Trypanosoma, 280, 780, **781, 796,** 797
Tschermak, Erich von, 188
Tubal ligation, *737*, **738**
Tube feet, 301
Tube nucleus, 547
Tuberculosis, **779, 781**
Tubulins, 479
Tumbleweed, seed dispersal of, 557
Tumor, *178, 770*
Tumor angiogenesis factor, 771
Tundra, *96–97*
Tunicate, *320*
Turgor, 540
Turner syndrome, 222, 224
Turnover, oxygen, **102**
Twins, 215
2,4-D, 579
Tympanic membrane, *849*
Tyrannosaurus Rex, 341
Tyrosine, 715

Ubiquinone, **494**
Ulcers, 667
Ulna, *869*
Ultrasound, **139,** *848*
Ulva, 304, **305, 306,** 522–24

Umbilical cord, 757
Umwelt, 879
Undernourishment, 648
Uniformitarianism, *143*
Uniramia, Subphylum, 318, 320
United States
  agricultural surpluses in, 127
  cancer incidence in, 773–74
  climate changes in, 68
  demography of, **80**
  energy crisis of 1970s, 135
  environmental regulation in, 135–36
  irrigation in, 129
  pollution in, 33
  soil depletion in, 128
Unsaturated fatty acid, *390*
Upwelling, 51
Uracil, *416*, **417**
Urea, 50, *679*, 685
Ureter, *682*
Urethra, *682*, 688, 727
Uric acid, *678*
Urine, 678, *679*, 688, 691–92
  drug testing of, 687
Urochordata, Subphylum, 320
Ussher, James, 8
Uterus, *730*, 732, 755, 757
  contraction of, 604
Utricle, *851, 852*

Vaccination, *778*
Vacuole, **166,** *478*
Vagina, *730*
Valdez tanker, 26
Valiela, Ivan, 11–12
Valium, 823
Valve, *611*
  aortic, 614
  atrioventricular (AV), 613–14, 616
  pulmonary, 614
  semilunar, 614
Variable, *11*
Variable region, *785*
Variation, *149–50*
Vas deferens, 727
  removal of, 737, **738**
Vascular bundle, *535*, 540
Vascular cambium, *536*
Vascular cylinder, *532*
Vascular tissue, *309*, 525, 527, **528**
Vasectomy, 737, **738**
Vasopressin, **710,** 822
Vector pollination, 554
Vegetal pole, 747
Vegetation, effect on climate, *67–68*
Vegetative reproduction, *542,* 543
Vein, *610–11*
Venereal disease, 739

Venezuela, Neblina plateau of, 253
Ventilation, *636*
Ventral fin, 325
Ventricle, *611*
  brain, 827
  human heart, 613
Ventricular fibrillation, 621
Venule, *610*
Venus, temperature of, 27
Venus flytrap, 571
Vermiform appendix, 672, 674
Vertebral column, 869
Vertebrate, 321–22
  early, 320–23
  reproductive strategies, **745**
Vertical zonation, *108*
Vesalius, Andreas, **593**
Vesicle, 176
Vessel, in xylem, 530
Vestibulo-ocular reflex, 852
Vestigial structure, *156*
Vibrio cholerae, **781**
Viceroy butterfly, 160, **161**
Victoria, Queen, 220
Villi, 757
Vincristine, 526
Virulent bacteria, *409*
Virus, *283, 423–24*, 779
  bacterophage, 424–25, **426**
  causing cancer, 771–72
  diversity of, 284
  mechanism of, 424–26, 780
  retrovirus, *426*
  types of, **781**
Visibility, in microscope, 477
Vision
  binocular, 351
  color, **836,** 846–47
  information processing, 846–47
  light and, 840
Visual pigment, 844–45
Vitalism, 361
Vitamin, *405, 651,* **654**
Vitelline envelope, 746
Viviparous organism, 745
Vocal cords, *636*
Voluntary action, 825
Volvox, 304, 522

Waldsterben, 118
Walker, Alan, 354, 356
Wall Street Journal, 26
Wallace, Alfred Russel, 144, 145
Warfarin, 245
Waste
  ammonia, 678–79
  as environmental variable, *60*
  excretion of, 679–80
  as population factor, 89
Waste feeder, 40
Water
  as environmental variable, *59*

conservation in respiratory system, 635–36
  cycling of, *46–47*
  importance of, 369–71
  as nutrient, *649*
  plant need for, 521–22
  respiration in, 631–33
  seed dispersal by, 557–58
  shortages of, *49*
  as solvent, 372
Water balance, 604–5, 677
  control in aquatic animal, 691–92
  control in human, 688–91
  control in land animal, 692–93
Water hyacinth, seed dispersal of, 557
Water lily, seed dispersal of, 557
Water pollution
  industrial, 114
  sewage, 115–17
  surface, 114
Water-vascular system, 301, **302**
Watson, James D., 412–13, 414, 416
Weather, *63–64*
Wedgewood, Josiah, 141
Wegener, Alfred, 68
Weinberg, Wilhelm, 238
Went, Frits, 577–78
Whale, temperature regulation by, 695
What is Life?, 406
Wheat, 584
White cell, *624–26*
White matter, *827*
White muscle, *861*
Whitman, Walt, 519
Whittaker, R. H., 268
Wilkins, Maurice, 411, 414
Wilson, Edward O., 356
Wind, *65–66*
  pollination by, 554
  power generated from, 132
  seed dispersal by, 557
Withdrawal, as contraceptive method, 736–37, **738**
Wolpert, Lewis, 744
Wood, *536*
Woodcock, Christopher, 427
Woodpecker, evolution of, **151**
World Bank, conservation fostered by, 135

X-ray diffraction, 410–12
Xanthophyll, 306
Xenopus frog, 432, **433**
Xerophyte, 99
Xylem, *309,* 528–31, 535, 536, 540
Xylem sap, 531
  cytokinin in, 581

Yeast, 316, 499, 500
Yersinia pestis, **781**
Yolk, 744
Yolk sac, *757*
Yunis, Jorge, 774

Z lines, 862
Zaug, Arthur, 404
Zea mays, genetic map of, **196**
Zero population growth, 77, **80**
Zona pellucida, 730, 746
Zonation, 108
Zone of elongation, 534
Zone of intolerance, 63
Zone of maturation, 534
Zone of stress, 63
Zooplankton, *41,* 107–8
Zooxanthellae, 109
Zygomycetes, *316*
Zygospore, 316
Zygote, *188,* 732, *744*
   plant, 545, 547
Zymogen, **476**